CONTENTS

Blood, coagulants and anticoagulants, 1
Blood fractionation, 25
Boron, elemental, 62
Boron compounds, 67
Brake linings and clutch facings, 202
Brighteners, fluorescent, 213
Bromine, 226
Bromine compounds, 243
BTX processing, 264
Burner technology, 278
Butadiene, 313
Butyl alcohols, 338
Butylenes, 346
Butyraldehyde, 376
Cadmium and cadmium alloys, 387

Cadmium compounds, 397
Calcium and calcium alloys, 412
Calcium compounds, 421
Calorimetry, 449
Carbamic acid, 475
Carbides, 476
Carbohydrates, 535
Carbon, 556
Carbonated beverages, 710
Carbon dioxide, 725
Carbon disulfide, 742
Carbonic and chloroformic esters, 758
Carbon monoxide, 772
Carbonyls, 794
Carboxylic acids, 814
Cardiovascular agents, 872

EDITORIAL STAFF FOR VOLUME 4

Executive Editor: **Martin Grayson**
Associate Editor: **David Eckroth**
Production Supervisor: **Michalina Bickford**
Editors: **Galen J. Bushey**   **Loretta Campbell**   **Anna Klingsberg**
   Lorraine van Nes

CONTRIBUTORS TO VOLUME 4

Edward Abrams, *Chemetron Corp., Chicago, Illinois,* Carbonic and chloroformic esters
George T. Armstrong, *National Bureau of Standards, Washington, D.C.,* Calorimetry
W. Robert Ballou, *C & I Girdler Incorporated, Louisville, Kentucky,* Carbon dioxide
Charles M. Bartish, *Air Products and Chemicals, Inc., Allentown, Pennsylvania,* Carbon monoxide
Friederich Benesovsky, *Metallwerke Plansee A. G., Reutle/Tyrol, Austria,* Survey Industrial heavy-metal carbides both under Carbides
N. L. Bottone, *Union Carbide Corp., Cleveland, Ohio,* Carbon and artificial graphite under Carbon
Joseph G. Bower, *U.S. Borax & Chemical Corp., Sanford Station, California,* Boron, elemental
J. R. Brotherton, *U.S. Borax Research Corp., Anaheim, California,* Boric acid esters under Boron compounds
R. M. Bushong, *Union Carbide Corp., Cleveland, Ohio,* Carbon and artificial graphite under Carbon

viii   CONTRIBUTORS TO VOLUME 4

S. C. Carapella, Jr., *ASARCO Incorporated, South Plainfield, New Jersey,* Cadmium and cadmium alloys
Ared Cezairliyan, *National Bureau of Standards, Washington, D.C.,* Calorimetry
Eli M. Dannenberg, *Cabot Corporation, Billerica, Massachussetts,* Carbon black under Carbon
Daniel J. Doonan, *U.S. Borax Research Corp., Anaheim, California,* Boron oxides, boric acid, and borates under Boron compounds
Gerald M. Drissel, *Air Products and Chemicals, Inc., Allentown, Pennsylvania,* Carbon monoxide
Gary B. Dunks, *Union Carbide Corp., Tarrytown, New York,* Boron hydrides, heteroboranes, and their metalloderivatives (commercial aspects) under Boron compounds
Leon Goldman, *Lederle Laboratories, American Cyanamid Co., Pearl River, New York,* Cardiovascular agents
William F. Hauschildt, *Amoco Chemicals Corp., Naperville, Illinois,* Butylenes
Heinz Hefti, *Ciba-Geigy Ltd., Basel, Switzerland,* Brighteners, fluorescent
Paul L. Henkels, *United States Gypsum Co., Des Plaines, Illinois,* Calcium sulfate under Calcium compounds
Melvern C. Hoff, *Amoco Chemicals Corporation, Naperville, Illinois,* Butylenes
M. L. Hollander, *ASARCO Incorporated, South Plainfield, New Jersey,* Cadmium and cadmium alloys
E. D. Howell, *The Carborundum Co., Niagara Falls, New York,* Silicon carbides under Carbides
Judith K. Hruschka, *Ohio Northern University, Ada, Ohio,* Blood, coagulants and anticoagulants
Un K. Im, *Amoco Chemicals Corporation, Naperville, Illinois,* Butylenes
M. P. Ingham, *Shell International Chemical Co., Ltd., London, England,* Trialkylacetic acids under Carboxylic acids
M. G. Jacko, *Bendix Corporation, Southfield, Michigan,* Brake linings and clutch facings
Robert W. Johnson, Jr., *Union Camp Corporation, Savannah, Georgia,* Branched-chain acids; Fatty acid from tall oil both under Carboxylic acids
Martha B. Jones, *Royal Crown Cola Company, Columbus, Georgia,* Carbonated beverages
B. R. Joyce, *Union Carbide Corp., Cleveland, Ohio,* Carbon and artificial graphite under Carbon
Richard Kieffer, *Technical University, Vienna, Vienna, Austria,* Survey; Industrial heavy-metal carbides, Cemented carbides all under Carbides
Isidor Kirschenbaum, *Exxon Research and Engineering Co., Linden, New Jersey,* Butadiene
T. M. Korzekwa, *The Carborundum Co., Niagara Falls, New York,* Silicon carbides under Carbides
Charles J. Kunesh, *Pfizer, Inc., Easton, Pennsylvania,* Calcium and calcium alloys
S. M. Kuntz, *The Carborundum Co., Niagara Falls, New York,* Silicon carbides under Carbides
Richard H. Lepley, *Pfizer, Inc., Easton, Pennsylvania,* Calcium carbonate under Calcium compounds
J. C. Long, *Union Carbide Corp., Cleveland, Ohio,* Carbon and artificial graphite under Carbon
Loren D. Lower, *U.S. Borax Research Corp., Boron, California,* Boron oxides, boric acid, and borates under Boron compounds

CONTRIBUTORS TO VOLUME 4

J. T. Meers, *Union Carbide Corp., Cleveland, Ohio,* Carbon and artificial graphite under Carbon

H. C. Miller, *Super-Cut, Inc., Chicago, Illinois,* Diamond, natural under Carbon

George E. Moore, *Consultant, Schenectady, New York,* Burner technology

J. W. Parker, *Shell International Chemical Co., Ltd., London, England,* Trialkylacetic acids under Carboxylic acids

P. D. Parker, *AMAX Base Metal Research and Development, Inc., Carteret, New Jersey,* Cadmium compounds

Edward G. Perkins, *University of Illinois, Urbana, Illinois,* Analysis and standards under Carboxylic acids

E. L. Piper, *Union Carbide Corporation, Cleveland, Ohio,* Carbon and artificial graphite under Carbon

Ralph H. Potts, *Armak Company, McCook, Illinois,* Manufacture under Carboxylic acids

Everett H. Pryde, *United States Department of Agriculture, Peoria, Illinois,* Introduction; Economic aspects both under Carboxylic acids

Imre Puskas, *Amoco Chemical Corporation, Naperville, Illinois,* Butylenes

Derek L. Ransley, *Chevron Research Company, Richmond, California,* BTX processing

R. L. Reddy, *Union Carbide Corp., Cleveland, Ohio,* Carbon and artificial graphite under Carbon

Charles E. Reineke, *Dow Chemical U.S.A., Midland, Michigan,* Bromine

S. K. Rhee, *Bendix Corporation, Southfield, Michigan,* Brake linings and clutch facings

Carl L. Rollinson, *University of Maryland, College Park, Maryland,* Survey under Calcium compounds

Ralph W. Rudolph, *University of Michigan, Ann Arbor, Michigan,* Boron hydrides, heteroboranes, and their metalloderivatives under Boron compounds

R. Russell, *Union Carbide Corp., Cleveland, Ohio,* Carbon and artificial graphite under Carbon

P. M. Scherer, *Union Carbide Corp., Cleveland, Ohio,* Carbon and artificial graphite under Carbon

Sherwood B. Seeley, *Consultant, Marco Island, Florida,* Natural graphite under Carbon

Walker L. Shearer, *Dow Chemical U.S.A., Midland, Michigan,* Calcium chloride under Calcium compounds

Paul Dwight Sherman, Jr., *Union Carbide Corp., South Charleston, West Virginia,* Butyl alcohols; Butyraldehyde

Noel B. Shine, *Shawinigan Products Dept. Gulf Oil Chemicals, Houston, Texas,* Calcium carbide under Carbides

H. D. Smith, Jr., *Virginia Polytechnic Institute and State University, Blacksburg, Virginia,* Organic boron nitrogen compounds under Boron compounds

Richard A. Smoak, *The Carborundum Company, Niagara Falls, New York,* Silicon carbide under Carbides

R. W. Soffel, *Union Carbide Corp., Cleveland, Ohio,* Carbon and artificial graphite under Carbon

V. A. Stenger, *Dow Chemical U.S.A., Midland, Michigan,* Bromine compounds

Martin H. Stryker, *The New York Blood Center, New York, New York,* Blood fractionation

David M. Stuart, *Ohio Northern University, Ada, Ohio,* Blood, coagulants and anticoagulants

Robert W. Timmerman, *FMC Corporation, Princeton, New Jersey,* Carbon disulfide

CONTRIBUTORS TO VOLUME 4

R. J. Turner, *Shell International Chemical Co., Ltd., London, England,* Trialkylacetic acids under Carboxylic acids
H. F. Volk, *Union Carbide Corp., Cleveland, Ohio,* Carbon and artificial graphite under Carbon
Frank S. Wagner, *Strem Chemicals, Inc., Newburyport, Massachussetts,* Carbonyls
Alan A. Waldman, *The New York Blood Center, New York, New York,* Blood fractionation
Robert J. Wenk, *United States Gypsum Co., Des Plaines, Illinois,* Calcium sulfate under Calcium compounds
R. H. Wentorf, Jr., *General Electric Company, Schenectady, New York,* Refractory boron compounds under Boron compounds; Diamond, synthetic under Carbon
Roy L. Whistler, *Purdue University, West Lafayette, Indiana,* Carbohydrates
L. L. Winter, *Union Carbide Corp., Cleveland, Ohio,* Carbon and artificial graphite under carbon
J. H. Woode, *Shell International Chemical Co., Ltd., London, England,* Trialkylacetic acids under Carboxylic acids
Reinhard Zweidler, *Ciba-Geigy Ltd., Basel, Switzerland,* Brighteners, fluorescent
John R. Zysk, *Purdue University, West Lafayette, Indiana,* Carbohydrates

NOTE ON CHEMICAL ABSTRACTS SERVICE REGISTRY NUMBERS AND NOMENCLATURE

Chemical Abstracts Service (CAS) Registry Numbers are unique numerical identifiers assigned to substances recorded in the CAS Registry System. They appear in brackets in the *Chemical Abstracts* (CA) substance and formula indexes following the names of compounds. A single compound may have many synonyms in the chemical literature. A simple compound like phenethylamine can be named β-phenylethylamine or, as in *Chemical Abstracts,* benzeneethanamine. The usefulness of the Encyclopedia depends on accessibility through the most common correct name of a substance. Because of this diversity in nomenclature careful attention has been given the problem in order to assist the reader as much as possible, especially in locating the systematic CA index name by means of the Registry Number. For this purpose, the reader may refer to the CAS Registry Handbook-Number Section which lists in numerical order the Registry Number with the Chemical Abstracts index name and the molecular formula; eg, **458-88-8,** Piperidine, 2-propyl-, (*S*)-, $C_8H_{17}N$; in the Encyclopedia this compound would be found under its common name, coniine [*458-88-8*]. The Registry Number is a valuable link for the reader in retrieving additional published information on substances and also as a point of access for such on-line data bases as Chemline, Medline, and Toxline.

In all cases, the CAS Registry Numbers have been given for title compounds in articles and for all compounds in the index. All specific substances indexed in *Chemical Abstracts* since 1965 are included in the CAS Registry System as are a large number of substances derived from a variety of reference works. The CAS Registry System identifies a substance on the basis of an unambiguous computer-language description of its molecular structure including stereochemical detail. The Registry Number is a machine-checkable number (like a Social Security number) assigned in sequential order to each substance as it enters the registry system. The value of the number lies in the fact that it is a concise and unique means of substance identification, which is independent of, and therefore bridges, many systems of chemical nomenclature. For polymers, one Registry Number is used for the entire family; eg, polyoxyethylene (20)sorbitan monolaurate has the same number as all of its polyoxyethylene homologues.

Registry numbers for each substance will be provided in the third edition index (eg, Alkaloids will show the Registry Number of all alkaloids (title compounds) in a table in the article as well, but the intermediates will have their Registry Numbers shown only in the index). Articles such as Absorption, Adsorptive separation, Air conditioning, Air pollution, Air pollution control methods will have no Registry Numbers in the text.

Cross-references have been inserted in the index for many common names and for some systematic names. Trademark names appear in the index. Names that are incorrect, misleading or ambiguous are avoided. Formulas are given very frequently in the text to help in identifying compounds. The spelling and form used, even for industrial names, follow American chemical usage, but not always the usage of *Chemical Abstracts* (eg, *coniine* is used instead of *(S)-2-propylpiperidine, aniline* instead of *benzenamine,* and *acrylic acid* instead of *2-propenoic acid*).

There are variations in representation of rings in different disciplines. The dye industry does not designate aromaticity or double bonds in rings. All double bonds and aromaticity will be shown in the Encyclopedia as a matter of course. For example, tetralin has an aromatic ring and a saturated ring and its structure will appear in the

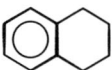

Encyclopedia with its common name, Registry Number enclosed in brackets, and parenthetical CA index name, ie, tetralin, [*119-64-2*] (1,2,3,4-tetrahydronaphthalene). With names and structural formulas, and especially with CAS Registry Numbers, the aim is to help the reader have a concise means of substance identification.

CONVERSION FACTORS, ABBREVIATIONS, AND UNIT SYMBOLS

SI Units (Adopted 1960)

A new system of measurement, the International System of Units (abbreviated SI), is being implemented throughout the world. This system is a modernized version of the MKSA (meter, kilogram, second, ampere) system, and its details are published and controlled by an international treaty organization (The International Bureau of Weights and Measures) (1).

SI units are divided into three classes:

BASE UNITS

length	meter[†] (m)
mass[‡]	kilogram (kg)
time	second (s)
electric current	ampere (A)
thermodynamic temperature[§]	kelvin (K)
amount of substance	mole (mol)
luminous intensity	candela (cd)

[†] The spellings "metre" and "litre" are preferred by ASTM; however "-er" will be used in the Encyclopedia.
[‡] "Weight" is the commonly used term for "mass".
[§] Wide use is made of "Celsius temperature" (t) defined by

$$t = T - T_0$$

where T is the thermodynamic temperature, expressed in kelvins, and $T_0 = 273.15$ K by definition. A temperature interval may be expressed in degrees Celsius as well as in kelvins.

FACTORS, ABBREVIATIONS, AND SYMBOLS

SUPPLEMENTARY UNITS

plane angle	radian (rad)
solid angle	steradian (sr)

DERIVED UNITS AND OTHER ACCEPTABLE UNITS

These units are formed by combining base units, supplementary units, and other derived units (2–4). Those derived units having special names and symbols are marked with an asterisk in the list below:

Quantity	Unit	Symbol	Acceptable equivalent
*absorbed dose	gray	Gy	J/kg
acceleration	meter per second squared	m/s^2	
*activity (of ionizing radiation source)	becquerel	Bq	1/s
area	square kilometer	km^2	
	square hectometer	hm^2	ha (hectare)
	square meter	m^2	
*capacitance	farad	F	C/V
concentration (of amount of substance)	mole per cubic meter	mol/m^3	
*conductance	siemens	S	A/V
current density	ampere per square meter	A/m^2	
density, mass density	kilogram per cubic meter	kg/m^3	g/L; mg/cm^3
dipole moment (quantity)	coulomb meter	C·m	
*electric charge, quantity of electricity	coulomb	C	A·s
electric charge density	coulomb per cubic meter	C/m^3	
electric field strength	volt per meter	V/m	
electric flux density	coulomb per square meter	C/m^2	
*electric potential, potential difference, electromotive force	volt	V	W/A
*electric resistance	ohm	Ω	V/A
*energy, work, quantity of heat	megajoule	MJ	
	kilojoule	kJ	
	joule	J	N·m
	electron volt[†]	eV[†]	
	kilowatt-hour[†]	kW·h[†]	

[†] This non-SI unit is recognized by the CIPM as having to be retained because of practical importance or use in specialized fields (1).

Quantity	Unit	Symbol	Acceptable equivalent
energy density	joule per cubic meter	J/m^3	
*force	kilonewton	kN	
	newton	N	kg·m/s^2
*frequency	megahertz	MHz	
	hertz	Hz	1/s
heat capacity, entropy	joule per kelvin	J/K	
heat capacity (specific), specific entropy	joule per kilogram kelvin	J/(kg·K)	
heat transfer coefficient	watt per square meter kelvin	W/(m^2·K)	
*illuminance	lux	lx	lm/m^2
*inductance	henry	H	Wb/A
linear density	kilogram per meter	kg/m	
luminance	candela per square meter	cd/m^2	
*luminous flux	lumen	lm	cd·sr
magnetic field strength	ampere per meter	A/m	
*magnetic flux	weber	Wb	V·s
*magnetic flux density	tesla	T	Wb/m^2
molar energy	joule per mole	J/mol	
molar entropy, molar heat capacity	joule per mole kelvin	J/(mol·K)	
moment of force, torque	newton meter	N·m	
momentum	kilogram meter per second	kg·m/s	
permeability	henry per meter	H/m	
permittivity	farad per meter	F/m	
*power, heat flow rate, radiant flux	kilowatt	kW	
	watt	W	J/s
power density, heat flux density, irradiance	watt per square meter	W/m^2	
*pressure, stress	megapascal	MPa	
	kilopascal	kPa	
	pascal	Pa	N/m^2
sound level	decibel	dB	
specific energy	joule per kilgram	J/kg	
specific volume	cubic meter per kilogram	m^3/kg	
surface tension	newton per meter	N/m	
thermal conductivity	watt per meter kelvin	W/(m·K)	
velocity	meter per second	m/s	
	kilometer per hour	km/h	
viscosity, dynamic	pascal second	Pa·s	
	millipascal second	mPa·s	
viscosity, kinematic	square meter per second	m^2/s	

Quantity	Unit	Symbol	Acceptable equivalent
	square millimeter per second	mm^2/s	
volume	cubic meter	m^3	
	cubic decimeter	dm^3	L(liter) (5)
	cubic centimeter	cm^3	mL
wave number	1 per meter	m^{-1}	
	1 per centimeter	cm^{-1}	

In addition, there are 16 prefixes used to indicate order of magnitude, as follows:

Multiplication factor	Prefix	Symbol	Note
10^{18}	exa	E	
10^{15}	peta	P	
10^{12}	tera	T	
10^9	giga	G	
10^6	mega	M	
10^3	kilo	k	
10^2	hecto	h[a]	
10	deka	da[a]	
10^{-1}	deci	d[a]	
10^{-2}	centi	c[a]	
10^{-3}	milli	m	
10^{-6}	micro	μ	
10^{-9}	nano	n	
10^{-12}	pico	p	
10^{-15}	femto	f	
10^{-18}	atto	a	

[a] Although hecto, deka, deci, and centi are SI prefixes, their use should be avoided except for SI unit-multiples for area and volume and nontechnical use of centimeter, as for body and clothing measurement.

For a complete description of SI and its use the reader is referred to ASTM E 380 (4) and the article Units and Conversion Factors which will appear in a later volume of the *Encyclopedia*.

A representative list of conversion factors from non-SI to SI units is presented herewith. Factors are given to four significant figures. Exact relationships are followed by a dagger. A more complete list is given in ASTM E 380-76(4) and ANSI Z210.1-1976 (6).

Conversion Factors to SI Units

To convert from	To	Multiply by
acre	square meter (m^2)	4.047 × 10^3
angstrom	meter (m)	1.0 × 10^{-10}†
are	square meter (m^2)	1.0 × 10^2†
astronomical unit	meter (m)	1.496 × 10^{11}
atmosphere	pascal (Pa)	1.013 × 10^5
bar	pascal (Pa)	1.0 × 10^5†
barrel (42 U.S. liquid gallons)	cubic meter (m^3)	0.1590
Bohr magneton μ_β	J/T	9.274 × 10^{-24}
Btu (International Table)	joule (J)	1.055 × 10^3

† Exact.

FACTORS, ABBREVIATIONS, AND SYMBOLS

To convert from	To	Multiply by
Btu (mean)	joule (J)	1.056×10^3
Btu (thermochemical)	joule (J)	1.054×10^3
bushel	cubic meter (m^3)	3.524×10^{-2}
calorie (International Table)	joule (J)	4.187
calorie (mean)	joule (J)	4.190
calorie (thermochemical)	joule (J)	4.184†
centipoise	pascal second (Pa·s)	1.0×10^{-3}†
centistoke	square millimeter per second (mm^2/s)	1.0†
cfm (cubic foot per minute)	cubic meter per second (m^3/s)	4.72×10^{-4}
cubic inch	cubic meter (m^3)	1.639×10^{-5}
cubic foot	cubic meter (m^3)	2.832×10^{-2}
cubic yard	cubic meter (m^3)	0.7646
curie	becquerel (Bq)	3.70×10^{10}†
debye	coulomb·meter (C·m)	3.336×10^{-30}
degree (angle)	radian (rad)	1.745×10^{-2}
denier (international)	kilogram per meter (kg/m)	1.111×10^{-7}
	tex‡	0.1111
dram (apothecaries')	kilogram (kg)	3.888×10^{-3}
dram (avoirdupois)	kilogram (kg)	1.772×10^{-3}
dram (U.S. fluid)	cubic meter (m^3)	3.697×10^{-6}
dyne	newton (N)	1.0×10^{-5}†
dyne/cm	newton per meter (N/m)	1.00×10^{-3}†
electron volt	joule (J)	1.602×10^{-19}
erg	joule (J)	1.0×10^{-7}†
fathom	meter (m)	1.829
fluid ounce (U.S.)	cubic meter (m^3)	2.957×10^{-5}
foot	meter (m)	0.3048†
footcandle	lux (lx)	10.76
furlong	meter (m)	2.012×10^{-2}
gal	meter per second squared (m/s^2)	1.0×10^{-2}†
gallon (U.S. dry)	cubic meter (m^3)	4.405×10^{-3}
gallon (U.S. liquid)	cubic meter (m^3)	3.785×10^{-3}
gauss	tesla (T)	1.0×10^{-4}
gilbert	ampere (A)	0.7958
gill (U.S.)	cubic meter (m^3)	1.183×10^{-4}
grad	radian	1.571×10^{-2}
grain	kilogram (kg)	6.480×10^{-5}
gram force per denier	newton per tex (N/tex)	8.826×10^{-2}
hectare	square meter (m^2)	1.0×10^4†
horsepower (550 ft·lbf/s)	watt (W)	7.457×10^2
horsepower (boiler)	watt (W)	9.810×10^3
horsepower (electric)	watt (W)	7.46×10^2†
hundredweight (long)	kilogram (kg)	50.80
hundredweight (short)	kilogram (kg)	45.36
inch	meter (m)	2.54×10^{-2}†
inch of mercury (32°F)	pascal (Pa)	3.386×10^3

† Exact.
‡ See footnote on p. xiv.

FACTORS, ABBREVIATIONS, AND SYMBOLS

To convert from	To	Multiply by
inch of water (39.2°F)	pascal (Pa)	2.491×10^2
kilogram force	newton (N)	9.807
kilowatt hour	megajoule (MJ)	3.6†
kip	newton (N)	4.48×10^3
knot (international)	meter per second (m/s)	0.5144
lambert	candela per square meter (cd/m^2)	3.183×10^3
league (British nautical)	meter (m)	5.559×10^3
league (statute)	meter (m)	4.828×10^3
light year	meter (m)	9.461×10^{15}
liter (for fluids only)	cubic meter (m^3)	1.0×10^{-3}†
maxwell	weber (Wb)	1.0×10^{-8}†
micron	meter (m)	1.0×10^{-6}†
mil	meter (m)	2.54×10^{-5}†
mile (U.S. nautical)	meter (m)	1.852×10^{3}†
mile (statute)	meter (m)	1.609×10^3
mile per hour	meter per second (m/s)	0.4470
millibar	pascal (Pa)	1.0×10^2
millimeter of mercury (0°C)	pascal (Pa)	1.333×10^{2}†
minute (angular)	radian	2.909×10^{-4}
myriagram	kilogram (kg)	10
myriameter	kilometer (km)	10
oersted	ampere per meter (A/m)	79.58
ounce (avoirdupois)	kilogram (kg)	2.835×10^{-2}
ounce (troy)	kilogram (kg)	3.110×10^{-2}
ounce (U.S. fluid)	cubic meter (m^3)	2.957×10^{-5}
ounce-force	newton (N)	0.2780
peck (U.S.)	cubic meter (m^3)	8.810×10^{-3}
pennyweight	kilogram (kg)	1.555×10^{-3}
pint (U.S. dry)	cubic meter (m^3)	5.506×10^{-4}
pint (U.S. liquid)	cubic meter (m^3)	4.732×10^{-4}
poise (absolute viscosity)	pascal second (Pa·s)	0.10†
pound (avoirdupois)	kilogram (kg)	0.4536
pound (troy)	kilogram (kg)	0.3732
poundal	newton (N)	0.1383
pound-force	newton (N)	4.448
pound per square inch (psi)	pascal (Pa)	6.895×10^3
quart (U.S. dry)	cubic meter (m^3)	1.101×10^{-3}
quart (U.S. liquid)	cubic meter (m^3)	9.464×10^{-4}
quintal	kilogram (kg)	1.0×10^{2}†
rad	gray (Gy)	1.0×10^{-2}†
rod	meter (m)	5.029
roentgen	coulomb per kilogram (C/kg)	2.58×10^{-4}
second (angle)	radian (rad)	4.848×10^{-6}
section	square meter (m^2)	2.590×10^6
slug	kilogram (kg)	14.59

† Exact.

To convert from	To	Multiply by
spherical candle power	lumen (lm)	12.57
square inch	square meter (m^2)	6.452×10^{-4}
square foot	square meter (m^2)	9.290×10^{-2}
square mile	square meter (m^2)	2.590×10^6
square yard	square meter (m^2)	0.8361
stere	cubic meter (m^3)	1.0[†]
stokes (kinematic viscosity)	square meter per second (m^2/s)	1.0×10^{-4}[†]
tex	kilogram per meter (kg/m)	1.0×10^{-6}[†]
ton (long, 2240 pounds)	kilogram (kg)	1.016×10^3
ton (metric)	kilogram (kg)	1.0×10^{3}[†]
ton (short, 2000 pounds)	kilogram (kg)	9.072×10^2
torr	pascal (Pa)	1.333×10^2
unit pole	weber (Wb)	1.257×10^{-7}
yard	meter (m)	0.9144[†]

[†] Exact.

Abbreviations and Unit Symbols

Following is a list of commonly used abbreviations and unit symbols appropriate for use in the *Encyclopedia*. In general they agree with those listed in *American National Standard Abbreviations for Use on Drawings and in Text* (ANSI Y1.1) (6) and *American National Standard Letter Symbols for Units in Science and Technology* (ANSI Y10) (6). Also included is a list of acronyms for a number of private and government organizations as well as common industrial solvents, polymers, and other chemicals.

Rules for Writing Unit Symbols (4):

1. Unit symbols should be printed in upright letters (roman) regardless of the type style used in the surrounding text.

2. Unit symbols are unaltered in the plural.

3. Unit symbols are not followed by a period except when used as the end of a sentence.

4. Letter unit symbols are generally written in lower-case (eg, cd for candela) unless the unit name has been derived from a proper name, in which case the first letter of the symbol is capitalized (W,Pa). Prefix and unit symbols retain their prescribed form regardless of the surrounding typography.

5. In the complete expression for a quantity, a space should be left between the numerical value and the unit symbol. For example, write 2.37 lm, *not* 2.37lm, and 35 mm, *not* 35mm. When the quantity is used in an adjectival sense, a hyphen is often used, for example, 35-mm film. *Exception:* No space is left between the numerical value and the symbols for degree, minute, and second of plane angle, and degree Celsius.

6. No space is used between the prefix and unit symbols (eg, kg).

7. Symbols, not abbreviations, should be used for units. For example, use "A," not "amp," for ampere.

8. When multiplying unit symbols, use a raised dot:

N·m for newton meter

In the case of W·h, the dot may be omitted, thus:

$$Wh$$

An exception to this practice is made for computer printouts, automatic typewriter work, etc, where the raised dot is not possible, and a dot on the line may be used.

9. When dividing unit symbols use one of the following forms:

$$m/s \text{ or } m \cdot s^{-1} \text{ or } \frac{m}{s}$$

In no case should more than one slash be used in the same expression unless parentheses are inserted to avoid ambiguity. For example, write:

$$J/(mol \cdot K) \text{ or } J \cdot mol^{-1} \cdot K^{-1} \text{ or } (J/mol)/K$$

but *not*

$$J/mol/K$$

10. Do not mix symbols and unit names in the same expression. Write:

$$\text{joules per kilogram } or \text{ J/kg } or \text{ J} \cdot \text{kg}^{-1}$$

but *not*

$$\text{joules/kilogram } nor \text{ joules/kg } nor \text{ joules} \cdot \text{kg}^{-1}$$

ABBREVIATIONS AND UNITS

A	ampere	amt	amount
A	anion (eg, H*A*)	amu	atomic mass unit
a	atto (prefix for 10^{-18})	ANSI	American National Standards Institute
AATCC	American Association of Textile Chemists and Colorists	AO	atomic orbital
		APHA	American Public Health Association
ABS	acrylonitrile–butadiene–styrene	API	American Petroleum Institute
abs	absolute	aq	aqueous
ac	alternating current, *n*.	Ar	aryl
a-c	alternating current, *adj*.	ar-	aromatic
ac-	alicyclic	as-	asymmetric(al)
ACGIH	American Conference of Governmental Industrial Hygienists	ASHRAE	American Society of Heating, Refrigerating, and Air Conditioning Engineers
ACS	American Chemical Society		
AGA	American Gas Association		
Ah	ampere hour	ASM	American Society for Metals
AIChE	American Institute of Chemical Engineers	ASME	American Society of Mechanical Engineers
AIP	American Institute of Physics	ASTM	American Society for Testing and Materials
alc	alcohol(ic)		
Alk	alkyl	at no.	atomic number
alk	alkaline (not alkali)	at wt	atomic weight

av(g)	average	dp	dew point; degree of polymerization
bbl	barrel		
bcc	body-centered cubic	dstl(d)	distill(ed)
Bé	Baumé	dta	differential thermal analysis
bid	twice daily	(*E*)-	entgegen; opposed
BOD	biochemical (biological) oxygen demand	ϵ	dielectric constant (unitless number)
bp	boiling point	*e*	electron
Bq	becquerel	ECU	electrochemical unit
C	coulomb	ed.	edited, edition, editor
°C	degree Celsius	ED	effective dose
C-	denoting attachment to carbon	emf	electromotive force
		emu	electromagnetic unit
c	centi (prefix for 10^{-2})	eng	engineering
ca	circa (approximately)	EPA	Environmental Protection Agency
cd	candela; current density; circular dichroism	epr	electron paramagnetic resonance
cgs	centimeter–gram–second		
CI	Color Index	eq.	equation
cis-	isomer in which substituted groups are on same side of double bond between C atoms	esp	especially
		esr	electron-spin resonance
		est(d)	estimate(d)
		estn	estimation
cl	carload	esu	electrostatic unit
cm	centimeter	exp	experiment, experimental
cmil	circular mil	ext(d)	extract(ed)
cmpd	compound	F	farad (capacitance)
COA	coenzyme A	f	femto (prefix for 10^{-15})
COD	chemical oxygen demand	FAO	Food and Agriculture Organization (United Nations)
coml	commercial(ly)		
cp	chemically pure		
CPSC	Consumer Product Safety Commission	fcc	face-centered cubic
		FDA	Food and Drug Administration
D-	denoting configurational relationship	FEA	Federal Energy Administration
d	differential operator	fob	free on board
d-	dextro-, dextrorotatory	FPC	Federal Power Commission
da	deka (prefix for 10^1)	fp	freezing point
dB	decibel	frz	freezing
dc	direct current, *n.*	G	giga (prefix for 10^9)
d-c	direct current, *adj.*	g	gram
dec	decompose	(g)	gas, only as in $H_2O(g)$
detd	determined	*g*	gravitational acceleration
detn	determination	*gem*-	geminal
dia	diameter	glc	gas-liquid chromatography
dil	dilute	g-mol wt;	
dl-; DL-	racemic	gmw	gram-molecular weight
DMF	dimethylformamide	grd	ground
DOE	Department of Energy	Gy	gray
DOT	Department of Transportation		

FACTORS, ABBREVIATIONS, AND SYMBOLS

H	henry	log	logarithm (common)
h	hour; hecto (prefix for 10^2)	LPG	liquefied petroleum gas
ha	hectare	lx	lux
HB	Brinell hardness number	M	mega (prefix for 10^6); metal (as in MA)
Hb	hemoglobin		
HK	Knoop hardness number	M	molar
HRC	Rockwell hardness (C scale)	m	meter; milli (prefix for 10^{-3})
HV	Vickers hardness number	m	molal
hyd	hydrated, hydrous	m-	meta
hyg	hygroscopic	max	maximum
Hz	hertz	MCA	Manufacturing Chemists' Association
^{i}(eg, Pr^i)	iso (eg, isopropyl)		
i-	inactive (eg, i-methionine)	MEK	methyl ethyl ketone
IACS	International Annealed Copper Standard	meq	milliequivalent
		mfd	manufactured
ibp	initial boiling point	mfg	manufacturing
ICC	Interstate Commerce Commission	mfr	manufacturer
		MIBC	methylisobutyl carbinol
ICT	International Critical Table	MIBK	methyl isobutyl ketone
ID	inside diameter; infective dose	min	minute; minimum
		mL	milliliter
IPS	iron pipe size	MLD	minimum lethal dose
IPT	Institute of Petroleum Technologists	MO	molecular orbital
		mo	month
ir	infrared	mol	mole
ISO	International Organization for Standardization	mol wt	molecular weight
		mom	momentum
IUPAC	International Union of Pure and Applied Chemistry	mp	melting point
		MR	molar refraction
IV	iodine value	ms	mass spectrum
J	joule	mxt	mixture
K	kelvin	μ	micro (prefix for 10^{-6})
k	kilo (prefix for 10^3)	N	newton (force)
kg	kilogram	N	normal (concentration)
L	denoting configurational relationship	N-	denoting attachment to nitrogen
L	liter (for fluids only) (5)	n (as n_D^{20}	index of refraction (for 20°C and sodium light)
l-	levo-, levorotatory		
(l)	liquid, only as in NH$_3$(l)		
LC$_{50}$	conc lethal to 50% of the animals tested	^{n} (as Bu^n),	
		n-	normal (straight-chain structure)
LCAO	linear combination of atomic orbitals		
lcl	less than carload lots	n	nano (prefix for 10^{-9})
LD$_{50}$	dose lethal to 50% of the animals tested	na	not available
		NAS	National Academy of Sciences
liq	liquid		
lm	lumen	NASA	National Aeronautics and Space Administration
ln	logarithm (natural)		
LNG	liquefied natural gas	nat	natural

NBS	National Bureau of Standards	PVC	poly(vinyl chloride)
neg	negative	pwd	powder
NF	*National Formulary*	qv	quod vide (which see)
NIH	National Institutes of Health	R	univalent hydrocarbon radical
NIOSH	National Institute of Occupational Safety and Health	(R)-	rectus (clockwise configuration)
		rad	radian; radius
nmr	nuclear magnetic resonance	rds	rate determining step
NND	New and Nonofficial Drugs (AMA)	ref.	reference
		rf	radio frequency, n.
no.	number	r-f	radio frequency, adj.
NOI-(BN)	not otherwise indexed (by name)	rh	relative humidity
		RI	Ring Index
NOS	not otherwise specified	RT	room temperature
nqr	nuclear quadrople resonance	^{s} (eg, Bu^s);	
NRC	Nuclear Regulatory Commission; National Research Council	sec-	secondary (eg, secondary butyl)
NRI	New Ring Index	S	siemens
NSF	National Science Foundation	(S)-	sinister (counterclockwise configuration)
NTSB	National Transportation Safety Board	S-	denoting attachment to sulfur
O-	denoting attachment to oxygen	s-	symmetric(al)
o-	ortho	s	second
OD	outside diameter	(s)	solid, only as in $H_2O(s)$
OPEC	Organization of Petroleum Exporting Countries	SAE	Society of Automotive Engineers
OSHA	Occupational Safety and Health Administration	SAN	styrene–acrylonitrile
		sat(d)	saturate(d)
owf	on weight of fiber	satn	saturation
Ω	ohm	SCF	self-consistent field
P	peta (prefix for 10^{15})	Sch	Schultz number
p	pico (prefix for 10^{-12})	SFs	Saybolt Furol seconds
p-	para	SI	Le Système International d'Unités (International System of Units)
p.	page		
Pa	pascal (pressure)		
pd	potential difference	sl sol	slightly soluble
pH	negative logarithm of the effective hydrogen ion concentration	sol	soluble
		soln	solution
		soly	solubility
pmr	proton magnetic resonance	sp	specific; species
pos	positive	sp gr	specific gravity
pp.	pages	sr	steradian
ppb	parts per billion	std	standard
ppm	parts per million	STP	standard temperature and pressure (0°C and 101.3 kPa)
ppt(d)	precipitate(d)		
pptn	precipitation		
Pr (no.)	foreign prototype (number)	SUs	Saybolt Universal seconds
pt	point; part	syn	synthetic

xxiv FACTORS, ABBREVIATIONS, AND SYMBOLS

^{t} (eg, But), t-, tert-	tertiary (eg, tertiary butyl)	Twad	Twaddell
		UL	Underwriters' Laboratory
		USP	*United States Pharmacopeia*
T	tera (prefix for 10^{12}); tesla (magnetic flux density)	uv	ultraviolet
		V	volt (emf)
t	metric ton (tonne) temperature	var	variable
		vic-	vicinal
TAPPI	Technical Association of the Pulp and Paper Industry	vol	volume (not volatile)
		vs	versus
tex	tex (linear density)	v sol	very soluble
THF	tetrahydrofuran	W	watt
tlc	thin layer chromatography	Wb	Weber
TLV	threshold limit value	Wh	watt hour
trans-	isomer in which substituted groups are on opposite sides of double bond between C atoms	WHO	World Health Organization (United Nations)
		wk	week
		yr	year
		(Z)-	zusammen; together

Non-SI (Unacceptable and Obsolete) Units   *Use*

Å	angstrom	nm
at	atmosphere, technical	Pa
atm	atmosphere, standard	Pa
b	barn	cm^2
bar†	bar	Pa
bhp	brake horsepower	W
Btu	British thermal unit	J
bu	bushel	m^3; L
cal	calorie	J
cfm	cubic foot per minute	m^3/s
Ci	curie	Bq
cSt	centistokes	mm^2/s
c/s	cycle per second	Hz
cu	cubic	exponential form
D	debye	C·m
den	denier	tex
dr	dram	kg
dyn	dyne	N
erg	erg	J
eu	entropy unit	J/K
°F	degree Fahrenheit	°C; K
fc	footcandle	lx
fl	footlambert	lx
fl oz	fluid ounce	m^3; L
ft	foot	m
ft·lbf	foot pound-force	J
gf den	gram-force per denier	N/tex
G	gauss	T
Gal	gal	m/s^2
gal	gallon	m^3; L

† Do not use bar (10^5Pa) or millibar (10^2Pa) because they are not SI units, and are accepted internationally only for a limited time in special fields because of existing usage.

Non-SI. (*Unacceptable and Obsolete*) Units		Use
Gb	gilbert	A
gr	grain	kg
hp	horsepower	W
ihp	indicated horsepower	W
in.	inch	m
in. Hg	inch of mercury	Pa
in. H$_2$O	inch of water	Pa
in.·lbf	inch pound-force	J
kcal	kilogram-calorie	J
kgf	kilogram-force	N
kilo	for kilogram	kg
L	lambert	lx
lb	pound	kg
lbf	pound-force	N
mho	mho	S
mi	mile	m
MM	million	M
mm Hg	millimeter of mercury	Pa
mμ	millimicron	nm
mph	miles per hour	km/h
μ	micron	μm
Oe	oersted	A/m
oz	ounce	kg
ozf	ounce-force	N
η	poise	Pa·s
P	poise	Pa·s
ph	phot	lx
psi	pounds-force per square inch	Pa
psia	pounds-force per square inch absolute	Pa
psig	pounds-force per square inch gage	Pa
qt	quart	m^3; L
°R	degree Rankine	K
rd	rad	Gy
sb	stilb	lx
SCF	standard cubic foot	m^3
sq	square	exponential form
thm	therm	J
yd	yard	m

BIBLIOGRAPHY

1. The International Bureau of Weights and Measures, BIPM, (Parc de Saint-Cloud, France) is described on page 22 of Ref. 4. This bureau operates under the exclusive supervision of the International Committee of Weights and Measures (CIPM).
2. *Metric Editorial Guide (ANMC-75-1)*, American National Metric Council, 1625 Massachusetts Ave. N.W., Washington, D.C. 20036, 1975.
3. *SI Units and Recommendations for the Use of Their Multiples and of Certain Other Units (ISO 1000-1973)*, American National Standards Institute, 1430 Broadway, New York, N. Y. 10018, 1973.
4. Based on *ASTM E 380-76 (Standard for Metric Practice)*, American Society for Testing and Materials, 1916 Race Street, Philadelphia, Pa. 19103, 1976.
5. *Fed. Regist.*, Dec. 10, 1976 (41 FR 36414).
6. For ANSI address, see Ref. 3.

R. P. LUKENS
American Society for Testing and Materials

B *continued*

BLOOD, COAGULANTS AND ANTICOAGULANTS

Hemostasis or stoppage of blood flow can be shown as the disturbance of a delicately poised system of two processes—coagulation and fibrinolysis. Under normal circumstances blood remains fluid, but if vascular damage occurs or if certain abnormal physiological states develop, steady states in one or both of the processes are disturbed and hemostasis results. Figure 1 is a simplified model of this system. This article reviews coagulants, anticoagulants, and fibrinolytics used for the management of coagulation and thrombotic disorders.

Coagulants

Biochemically, blood coagulation is the result of polymerization of a polypeptide monomer (modified fibrinogen) into a cross-linked mesh of fibrils, an insoluble gel called fibrin. Eleven plasma proteins participate in the process; a deficiency of any of nine of them may be associated with a bleeding disorder. This number includes those proteins that have been historically known to participate in coagulation; they are termed *factors*. Recently other proteins, eg, kallikrein and high molecular weight kininogen, have been recognized as playing roles in coagulation.

The theory accounting for the coagulation of blood has evolved from a two-step mechanism to the present model; this depicts coagulation as a multistep cascade of activations of protein factors that culminate in fibrin formation (1) as illustrated in Figure 2. For a brief historical review of the development of blood coagulation theory see reference 2. Two pathways of activation—an intrinsic pathway involving only blood factors and an extrinsic pathway that requires the participation of a tissue lipoprotein (tissue factor)—can be demonstrated. These pathways converge at the level of activation of factor X. The operation of both seems to be necessary for effective hemostasis; deficiencies of factors in either result in a hemorrhagic state.

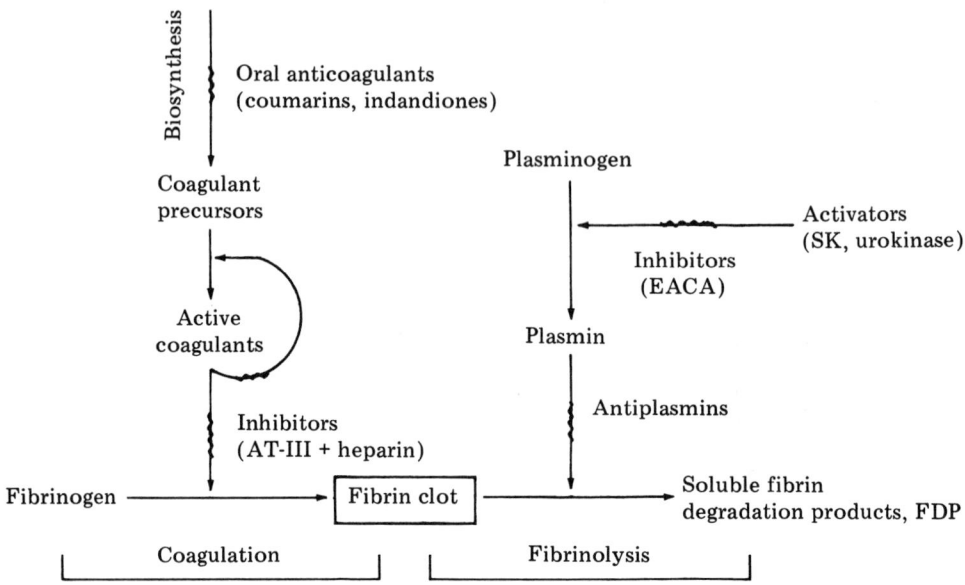

Figure 1. Relation of coagulation, anticoagulation, and fibrinolysis to hemostasis. The development of hemostasis may be viewed as dependent on the relative rates of the reactions leading toward and away from the fibrin clot. (SK = streptokinase; AT-III = antithrombin III; and EACA = ϵ-amino caproic acid)-

Many of the reactions share a requirement for phospholipids and Ca^{2+} for rapid reaction rates. Tissue factor (see below) most likely provides the lipid for the extrinsic mechanism; platelets supply phospholipid (platelet factor 3) for the intrinsic pathway. The phospholipid is thought to play an important role in organizing and localizing the participants in the reactions (3) and perhaps in preventing the inactivation of active forms of the factors by naturally occurring inhibitors in plasma (4). Many of the factors involved in phospholipid-requiring reactions (IX, X, VII, and prothrombin) contain γ-carboxyglutamyl residues which are necessary for the factor's attachment to the negatively charged phospholipid surface through calcium ion bridges. The rate of conversion of prothrombin is increased five hundred times by the presence of phospholipid (3).

The complexity of the coagulation process and the initial difficulty in separating the various components led to a confusing redundancy of names in the literature. For simplification, Roman numerals have been assigned to each participating substance (5). It is common practice to refer to factors V through XIII by Roman numerals, and the subscript (a) is used to indicate the activated form. Factors I–IV are generally referred to by common names, ie, fibrinogen (fibrin), prothrombin (thrombin), tissue factor, and calcium. Table 1 shows a list of synonyms of factors which have been used. A more detailed system of nomenclature that systematizes the activation products and fragments has recently been proposed (6).

The Coagulation Factors. All but one of the coagulation factors (III, tissue factor) circulate in the blood plasma. Table 2 illustrates certain properties of the various factors. It should be noted that most of these factors exist as zymogens (factor precursors) and must be modified by hydrolytic cleavage to achieve full activity. Cleavage

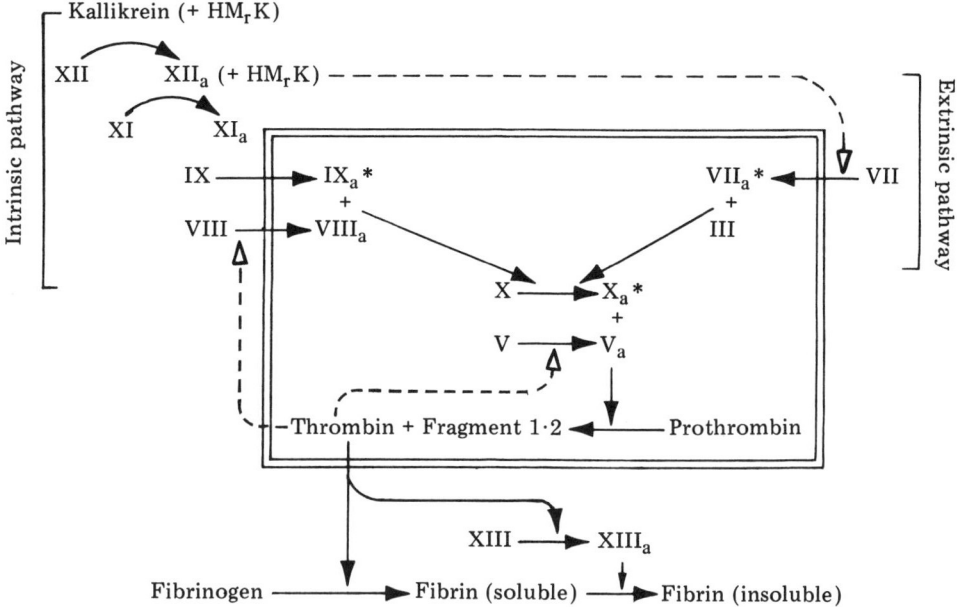

Figure 2. Cascade mechanism of blood coagulation. Box encloses reactions requiring phospholipid and Ca^{2+}; * enzymatic component when more than one factor is involved in an activation; dotted arrows show feedback mechanisms; and HM_rK = high molecular weight kininogen.

is accomplished by the factor immediately preceding in the coagulation cascade; consequently, most of the activated factors are endopeptidases (XIIa, XIa, Xa, IXa, VIIa, IIa). These are remarkably similar in structure; nearly all are two-chain (factor XI is four-chain) serine proteases with substrate specificities similar to trypsin. They have significant sequence homology with trypsin (see Table 3), which suggests common evolutionary origin. The bond specificities of the proteases are also trypsin-like (arginyl-X, where X is often valine or isoleucine), but substrate specificity is much narrower than that of the pancreatic proteases. Cleavage of the zymogen results in activation, probably because it generates an additional —NH_2 terminus. As with trypsin, this new —NH_2 may interact with the aspartate residue close to the active serine to uncover the active site.

Because of the proteolytic nature of many substances and the low concentrations of the factors in plasma, much difficulty was encountered in isolation of the intact zymogens. The problem of degradation has been overcome primarily by the use of large amounts of proteolytic inhibitors such as benzamidine, DFP (diisopropyl fluorophosphate), or polybrene during processing. A property of factors II, VII, IX, and X, useful in their purification, is adsorption on $BaSO_4$. All of the factors have now been adequately isolated and purified in zymogen and activated form. Techniques are typical of those for protein purification and are detailed in reference 31.

Factors in the intrinsic pathway are typically assayed by the partial thromboplastin time test (PTT), with the use of plasma deficient in the factor to be assayed as substrate. Extrinsic pathway substances are assayed by the Quick prothrombin time test (PT) (31). Plasmas with deficiencies of factors II, V, VII, VIII, IX, X, XI, or XII are commercially available in the lyophilized form for these assays (Dade

4 BLOOD, COAGULANTS AND ANTICOAGULANTS

Table 1. Nomenclature of Coagulation Factors

Factor number	CAS Registry No.	Common names
I	[9001-32-5]	fibrinogen
II	[9001-26-7]	prothrombin; activated form: thrombin
III	[9002-05-5]	tissue factor, tissue thrombokinase, extrinsic thromboplastin
IV	[14127-61-8]	calcium ion (Ca^{2+})
V	[9001-24-5]	proaccelerin, labile factor, plasma accelerator globulin, plasma Ac-globulin, accelerator factor, plasma prothrombin convertin factor, thrombogene; activated form: accelerin
VII	[9001-25-6]	proconvertin, stable factor, prothrombinogen, autoprothrombin I; activated form: serum prothrombin, conversion accelerator (SPCA), convertin
VIII	[9001-27-8]	antihemophilic factor (AHF), antihemophilic globulin (AHG), hemophilic factor A, platelet cofactor 1, thromboplastinogen, thrombocytolysin
IX	[9001-28-9]	Christmas factor, antihemophilic factor B, autoprothrombin II, plasma thromboplastin component (PTC), platelet cofactor 2, thromboplastinogen B; activated form: prephase accelerator (PPA)
X	[9001-29-0]	Stuart factor, Prower factor, Stuart-Prower factor, autoprothrombin III; activated form: plasma thromboplastin, autoprothrombin C, thrombokinase
XI	[9013-55-2]	plasma thromboplastin antecedent (PTA), antihemophilic factor C, activated form: third prothromboplastic factor
XII	[9001-30-3]	Hageman factor, surface factor, contact factor, clot-promoting factor
XIII	[9013-56-3]	fibrin stabilizing factor (FSF), fibrin stabilizing enzyme, Laki-Lorand factor, fibrinase, plasma transglutaminase, fibrinoligase
XIV[a]	[42617-41-4]	Protein C[a]; activated form: autoprothrombin II-A

[a] Factor XIV usage has been recently proposed; it is not in common usage

Coagulation Factor Deficient Substrate Plasma). Rabbit brain phospholipids are used in the PTT as a substitute for platelet factor 3 activity and are available in lyophilized form (Platelet Factor Reagent, BBL; Platelin, General Diagnostics). Other preparations also contain a particulate activator for determination of activated PTT times; these include Platelin Plus Activator, General Diagnostics; Dade Activated Cephaloplastin; and Fibrolet Activated Platelet Factor Reagent.

For comprehensive reviews in the area of blood coagulation factors see references 1–2 and 32–34.

Factor XII. Factor XII from both human (35) and bovine (30) sources has been partially characterized. Factor XII is difficult to maintain as the zymogen but can be stored in purified form for at least 9 mo at 4°C in a solution of 6 mM acetate, pH 5.0, 0.5 mM EDTA, and 0.16 M NaCl (19).

Factor XII is activated *in vitro* both by contact with various solid materials bearing a negative charge (eg, collagen, glass, kaolin), and by proteases such as kallikrein [9001-01-8], trypsin [9002-07-7], plasmin [9001-90-5], and factor XIa [37203-62-6]. Maximum rates of activation are achieved *in vitro* if two plasma proteins, prekallikrein [9055-02-1] and high molecular weight kininogen [12244-26-7] (HM$_r$K), plus an activating surface as supplied by kaolin [1332-58-7] are present (36–37). Evidence indicates the following mechanism: factor XII, after binding to kaolin and combining in stoichiometric ratios with HM$_r$K, is readily susceptible to cleavage by the serine protease, kallikrein. Kallikrein, a γ-globulin, is reciprocally generated by cleavage of plasma prekallikrein by factor XIIa. Figure 3 summarizes these interactions. The same type of mechanism probably occurs *in vivo* with collagen in vascular subendothelial tissue providing the activating surface.

Fluid-phase activation of human factor XII by proteases produces five overlapping fragments with the smallest (ie, 28,000 mol wt) containing the active site (35). The kaolin binding sites exist on the other section(s) of the zymogen molecule. In contrast to fluid-phase activation, surface-bound human factor XII is activated by a single enzymic cleavage by kallikrein with the production of only two fragments: 28,000 and 48,000 mol wt (19). Purified bovine factor XIIa also consists of two chains, 28,000 and 46,000 mol wt, connected by disulfide bond(s) with the active site located on the 28,000 mol wt fragment. See Table 3 for the amino acid sequences.

Factor XII is not uniquely involved in coagulation; it hydrolyzes several physiological proteins. Its activity is accelerated by HM_rK as a cofactor. In coagulation, it is responsible for the activation of factor XI in the intrinsic pathway, and recently has been shown capable of activating factor VII (10). Other actions include initiating fibrinolytic activity (see p. 18) and kinin generation as shown in Figure 3.

Factor XI. An effective isolation procedure for the human zymogen (16) and its mode of activation (36) have recently been reported. Trypsin, as well as factor XIIa, activates XI *in vitro*.

Factor IX. The third step in the intrinsic pathway is the activation of factor IX by a two-step process in which a 9000 mol wt activation fragment is released (38). In addition to the properties listed in Table 2, purified factor IX can be stored for months at $-20°C$ in 50% glycerol (13).

Factor VIII. Factor VIII has been traditionally known as the antihemophilic factor (AHF) because a lack of activity by this protein is the defect occurring in classical hemophilia (hemophilia A). A review by Graham and Barrow deals with the physiology of factor VIII (39). This factor has received much attention, but, in spite of intense efforts to definitively characterize the molecular species with factor VIII activity, some questions have not been resolved.

Factor VIII activity, along with platelet aggregative activity, has been shown to accompany a 1.1×10^6 mol wt glycoprotein. This high mol wt species has been dissociated by reducing agents such as 2-mercaptoethanol into subunits of 200,000 mol wt (40–41). The subunit preparations retain both activities (42–43). Recently, however, factor VIII and platelet aggregating activities have been separated by high salt concentrations; one preparation consisted of a 100,000 mol wt species (44–45). Others have been unable to duplicate this separation with a highly purified sample of VIII (42).

Factor VIII activity is greatly enhanced by addition of the proteolytic enzymes thrombin and Xa (6), as well as plasmin and trypsin. No mol wt change is observed during activation of the 200,000 mol wt subunits (40–42). Although continued action of the enzymes eventually leads to a loss of coagulant activity, the activation of VIII by thrombin *in vivo* may be important in hemostasis and may be a communicating link between the intrinsic and extrinsic pathways, ie, a small amount of thrombin produced via the extrinsic pathway may potentiate VIII, resulting in rapid escalation of coagulation via the intrinsic pathway (34).

Most evidence indicates that factor VIIIa is a catalytic cofactor in the activation of factor X by factor IXa [*37316-87-3*]. Factor VIII has never been shown to have enzymatic activity (1). In the presence of calcium ions and phospholipid, IXa can activate X in the absence of VIII, however, the presence of factor VIII accelerates the reaction by a magnitude of 1000 (1).

Table 2. Properties of Protein Coagulation Factors[a]

Factor	Plasma conc, mg/100 mL	Zymogen	Activated form Total	Activated form A chain[b]	Activated form B chain[b]	Inhibitors of activated form	Stability[c]	Deficiency state[d]	Refs.[e]
			A. Serine Proteases						
II	10	bovine: 72,000 single chain 10–14% CH$_2$O human: 68,700 single chain 8% CH$_2$O	α: 38,000	6,000	32,000	DFP, hirudin AT-III	±	prothrombin deficiency Coumarin anticoagulant therapy liver disease	3, 8
VII	0.1	bovine: 45,500 single chain 9.1% CH$_2$O	45,500 9.1% CH$_2$O	18,000	27,000	DFP AT-III + heparin	+	factor VII deficiency Coumarin anticoagulant therapy liver disease	8–12
IX	0.5	bovine: 55,400 single chain 26% CH$_2$O human: 70,000 single chain	46,000 15% CH$_2$O	16,000	27,300	SBTI AT-III[f]	+	Christmas disease hemophilia B Coumarin anticoagulant therapy liver disease	8, 13–14
X	1	bovine: 55,000 two chains A: 38,000, 10% CH$_2$O B: 17,000, 0% CH$_2$O	16% CH$_2$O X$_a$α: 45,300 2.1% CH$_2$O X$_a$β: 42,600 0% CH$_2$O	30,000 27,000	17,000	DFP SBTI AT-III[f]	+	factor X deficiency Coumarin anticoagulant therapy liver disease	3, 8, 14–15
XI	0.4	bovine: 130,000, 11% CH$_2$O 2 subunits of 55,000 mol wt disulfide linked human: 160,000, 12% CH$_2$O 2 subunits of 83,000 mol wt disulfide linked	83,000 for each subunit	50,000	33,000	AT-III[f]	±	PTA deficiency	16–18
XII	2.9	bovine: 74,000 single chain 15% CH$_2$O human: 76,000 single chain	74,000 76,000	46,000 48,000	28,000 28,000	DFP LBTI AT-III[f] C1-INH		Hageman defect	4, 19–20

		Molecular weight and structure	B. Cofactors	C. Other	
III		apoprotein: bovine: 56,000, 15–30% CH_2O; human: 52,000, 6.3% CH_2O; dog: 80,000			21–24
V	0.5–1.0	bovine: 290,000–400,000; 2 chains: 125,000 and 70,000, 10–20% CH_2O; on cleavage by thrombin, 125,000 chain loses 80,000 mol wt fragment	–	factor V deficiency	25
VIII	1.0	bovine and human: 1.1×10^6 mol wt consisting of similar or identical 200,000 mol wt subunits disulfide linked; 9% CH_2O (bovine); 6% CH_2O (human)	–	hemophilia A; von Willebrand's disease	8

		Zymogen		Activated form	
I	170–400	340,000, $\alpha_2\beta_2\gamma_2$ disulfide linked α: 63,000; β: 56,000; γ: 47,000 4–5% CH_2O	+	mol wt 3% less than fibrinogen; peptides cleaved from α and β; $\alpha_2\beta'_2\gamma_2$ polymer cross-linked.	
XIII	trace	bovine and human plasma: 300,000, $\alpha_2\beta_2$; α: 75,000; β: 88,000 5% CH_2O in β chain (human) human platelets: 160,000, α_2		human α'_2: 140,000; α': 71,000	26

[a] Cited in ref. 1 unless otherwise specified. Abbreviations: CH_2O = carbohydrate; AT-III = antithrombin III; DFP = diisopropylfluorophosphate; SBTI = soy bean trypsin inhibitor; LBTI = lima bean trypsin inhibitor; and C-1 INH = C-1 esterase inhibitor.
[b] A and B refer to sequence of chains from NH_2 terminal of protein. Except for human factor XII, the chains in the activated form are linked by disulfide bonds. The active site serine is present on the B chain in all cases.
[c] Stability *in vitro* in plasma. See also ref. 27.
[d] Ref. 7.
[e] The reference column lists references pertaining to each particular column.
[f] Accelerated by heparin.

Table 3. Sequence Homologies for Serine Proteases Involved in Coagulation and Fibrinolysis[a,b]

Amino-terminal sequence of the zymogens[c]

		ref.
prothrombin	Ala → Asn → Lys → Gly → Phe → Leu → Glu → Glu → Gla → - - - → Val → Arg → Lys → Gly → Asn → Leu	1
factor IX	Tyr → Asn → Ser → Gly → Lys → Leu → Glu → Glu → Phe → Val → Arg → - - - → Gly → Asn → Leu	1
factor X	Ala → Asn → Ser → - - - → Phe → Leu → Glu → Glu → - - - → Val → Lys → Gln → Asn → Leu	1
factor VII	Ala → Asn → - - - → Gly → Phe → Leu → Gla → Gla → Leu → Leu → Pro → - - - → Gly	10

Amino-terminal sequence of the B chains of active enzymes

thrombin	Ile → Val → Glu → Gly → Gln → Asp → - - - → Ala → Glu → Val → Gly → Leu → Ser → Pro → Trp → Gln	28
factor IX$_a$	Val → Val → Gly → Gly → Glu → Asp → - - - → Ala → Glu → Arg → Gly → Glu → Phe → Pro → Trp → Gln	28
factor X$_a$	Ile → Val → Gly → Gly → Arg → Asp → Cys → Ala → Glu → - - - → Gly → Glu → Cys → Pro → Trp → Gln	28
factor XII$_a$	Val → Val → Gly → Gly → Leu → Val → - - - → Ala → Leu → Pro → Gly → Ala → ? → Pro → Tyr → Ile	28
plasmin	Val → Val → Gly → Gly → Cys → Val → - - - → Ala → His → Pro → His → Ser → Trp → Pro → Tyr → Gln	29
factor VII$_a$	Ile → Val → Gly → Gly	10

Sequence of the active site[d]

                    180   185   190   195

		ref.
thrombin	Phe → Cys → Ala → Gly → Tyr → Lys → Pro → Gly → Glu → Gly → Lys → Arg → Gly → Asp → Ala → Cys → Glu → Gly → Asp → Ser → Gly → Gly → Pro → Phe	10
factor IX$_a$	Phe → Cys → Ala → Gly → Tyr → His → - - - → Glu → Gly → Gly → Lys → - - - → Asp → Ser → Cys → Gln → Gly → Asp → Ser → Gly → Gly → Pro → His	10
factor X$_a$	Phe → Cys → Ala → Gly → Tyr → Asp → Thr → Gln → Pro → Glu → - - - → Asp → Ala → Cys → Gln → Gly → Asp → Ser → Gly → Gly → Pro → His	10
factor XI	Val → Cys → Ala → Gly → Tyr → Arg → - - - → Glu → Gly → Gly → Lys → - - - → Asp → Ala → Cys → Lys → Gly → Asp → Ser → Gly → Gly → Pro → ?	10
factor XII	Leu → Cys → Ala → Gly → Phe → Leu → - - - → Glu → Gly → Gly → Thr → - - - → Asp → Ala → Cys → Gln → Gly → Asp → Ser → Gly → Gly → Pro → Leu	10
plasmin[e]	Leu → (Gly → Ala) → His → Leu → Ala → Cys → Asn → (Gly → Thr) → - - - → - - - → Ser → Cys → Gln → Gly → Asp → Ser → Gly → Gly → Pro → Leu	29
factor VII	Phe → Cys → Ala → Gly → Tyr → Thr → - - - → Asp → Gly → Thr → Lys → - - - → Asp → Ala → Cys → Lys → Gly → Asp → Ser → Gly → Gly → Pro → His	10

[a] All are bovine species, except plasmin (human).
[b] Gla: γ-carboxyl glutamic acid residues. Dashes (- - -) are used to bring sequences into alignment for greater homology.
[c] Sequences have been determined for plasminogen and bovine factor XII, and they are not homologous with the other serine proteases. The amino terminal sequence of Factor XII is homologous, however, with the active site of several naturally occurring protease inhibitors (30).
[d] Numbering corresponds to chymotrypsin with active serine at 195.
[e] The position of residues in parentheses was not definitely established. The second set has been reversed from that originally reported.

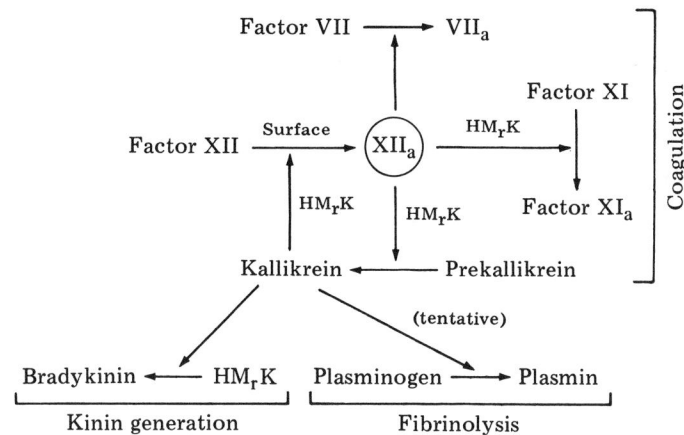

Figure 3. Factor XII, activation and physiological activities.

Factor X. Two forms of bovine factor X, X_1, and X_2, have been separated (46); these two forms display different phospholipid binding characteristics (47) but have other identical properties and may differ only in carbohydrate content (48). This factor is soluble in water (pH 4–8) and is stable for several days at pH 6.1–9.0, at room temperature. The purified factor is stable for several months at −20°C in a 50% glycerol solution with DFP present (15).

The activation of factor X is due to cleavage of an arginyl–isoleucine bond, with the release of an 11,000 mol wt fragment from the amino terminal end of the heavy chain. An additional 4000 mol wt polypeptide may be cleaved from the carboxyl-terminal end with no loss in activity. An alternative route to the second activated form exists (49). The conversion of X to Xa is the first reaction common to the intrinsic and extrinsic pathways (Fig. 2).

Factor Xa, with cofactor V, is responsible *in vivo* for the conversion of prothrombin to thrombin. In addition, factors VII and X act as substrates for Xa without need of V. Factor VII is cleaved at two specific sites and X is attacked at three sites (50).

Factor VII. Factor VII is unstable below pH 5 and above pH 9; at neutral pH, it remains stable for several hours at 0–37°C, if in the presence of protein at concentrations greater than 20 µg/mL (8).

Because serum VII contains ten times the activity of plasma VII it has been assumed that factor VII, like other factors, is activated during coagulation. Evidence now exists to support this assumption. An activation of VII has been shown to occur by proteolytic cleavage. Factor VIIa is further degraded to an inactive form by thrombin (10) and factor Xa (9) but cleavage by factor XIIa is limited to activation (10). Factor VII, with tissue factor, is responsible for the activation of factor X by the extrinsic pathway.

Tissue Factor (Factor III). Tissue factor was originally called tissue thromboplastin and is the one component of blood coagulation that is not normally present in blood. It is a lipoprotein which occurs in all tissues, although most abundantly in the placenta, brain, and lung. It has been shown by immuno- and enzyme-histochemical techniques to be concentrated in the intima of large and small vessels and to be localized at plasma membranes of parenchymal cells, including vascular endothelial cells where it would be immediately available for initiating coagulation if vascular injury occurs (51).

The chemical and physiological description of tissue factor is one of the most incomplete of the coagulation factors. Its description as a lipoprotein arises from a crude preparation that is 30–45% phospholipid. The protein component has been isolated from human, dog, and bovine brain tissues, and shows no procoagulant activity, although it must be responsible for the specificity of factor III.

Tissue factor, as thromboplastin, is used chiefly in conjunction with other materials to test blood coagulation time. One combination (Thrombotest, BBL) is used to determine the combined activities of factors VII, IX, X, and prothrombin. The mixture is composed of tissue factor from bovine brain, crude cephalin extracted with ether from human brain or soybean extract, bovine plasma which has factors VII, IX, X, and prothombin removed by adsorption, and an optimum concentration of calcium (0.025 M CaCl$_2$). The lyophilized mixture is indefinitely stable at room temperature. Tissue factor from rabbit brain in lyophilized or liquid form, with an optimum concentration of Ca^{2+}, is also available for the prothrombin time test (Ortho Brain Thromboplastin, Ortho Diagnostics; Fibroplastin, BBL; Dade Thromboplastin Reagent; Simplastin, General Diagnostics). A variation of Simplastin, Simplastin A, contains additional quantities of factors V and fibrinogen for testing plasmas over 4 h old, and for the prothrombin–proconvertin test.

Factor V. Factor V, an α-globulin, is most stable at pH 6.5–7.0 and unstable below pH 4.0 or above pH 10.5. It is very labile; 50% of its activity disappears from human plasma at 23°C in 2 h (25). It has been difficult to study this protein since its lability is even greater during purification. It is known to be a large glycoprotein that seems to consist of two chains. Thrombin increases the activity of V by releasing an 80,000 mol wt fragment from the heavy chain. Enhanced activity also results from the cleavage of the heavy chain by a purified protease from Russell viper venom (52).

Factor Va acts as an accelerating cofactor in the conversion of prothrombin to thrombin. Factor Va, but not V, binds to prothrombin fragment 2 in the presence of Ca^{2+}; unlike other coagulant factors that have roles mediated by Ca^{2+}, factor V contains no γ-carboxyl glutamate residues, and also unique is its interaction with phospholipid as no exogenous Ca^{2+} is required. This binding may occur at the carbohydrate side chains of factor Va and/or with endogenous calcium that has been detected in the molecule (25).

Prothrombin (Factor II). Prothrombin, thrombin, and its activation fragments have been extensively studied (see Tables 2 and 3).

The conversion of prothrombin to thrombin occurs on a phospholipid surface with the participation of factors Xa, Va, and calcium, and may occur by either of two pathways (Fig. 4). This is possibly the most complicated of the various activations and a central event in control of coagulation since thrombin, the product of the reaction, has been implicated in the activation and inactivation of other factors. The polypeptide fragments that are released may also play physiological roles. One of the interesting aspects in prothrombin activation is the removal of the portion of prothrombin which contains γ-carboxylated glutamate residues. These negatively charged residues are responsible for binding prothrombin to the lipoprotein activation complex. The thrombin formed by cleavage contains none of the γ-carboxyl glutamate residues; therefore, it cannot bind directly to the phospholipid but may be freed into the plasma where it converts fibrinogen to fibrin monomers. However, thrombin also binds to

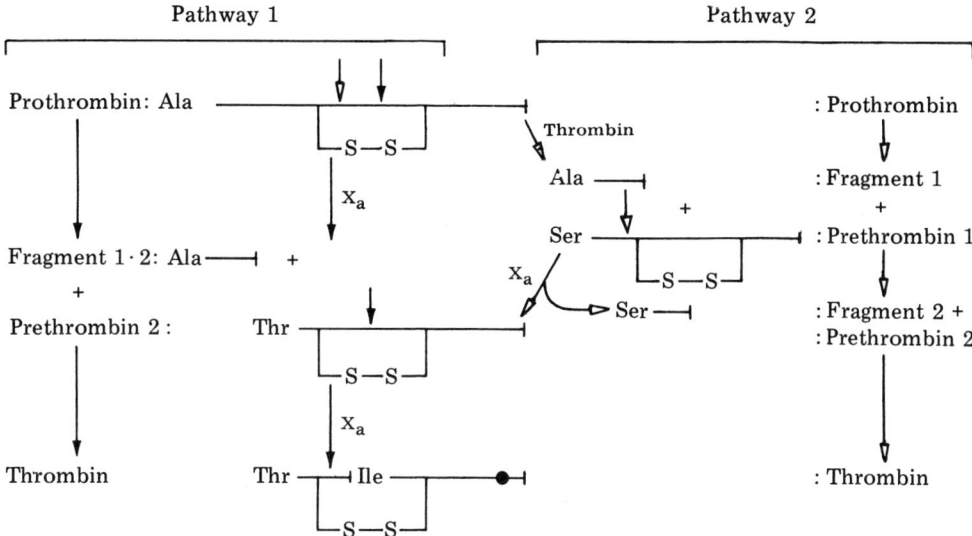

Figure 4. Formation of thrombin and activation fragments; heavy dots indicate location of serine in exposed active sites.

prothrombin fragment 2, which in turn binds to Va, part of the phospholipid activation complex.

In addition to converting fibrinogen to fibrin, the thrombin that is released may play other important roles in coagulation. It has been shown to potentiate factors VIII, V, VII, and X, although continued attack on factors VIII, VII, and V results in inactivity. Thrombin is also responsible for activating factor XIII, the fibrin-stabilizing factor, and enhances the aggregation of platelets and triggers their release reaction.

Thrombin may also exert some negative effects on coagulation. It modifies prothrombin, decreasing its affinity for factor V.

Prothrombin fragment 1·2 which is released on conversion of prothrombin to prethrombin 2 has been shown to accelerate the conversion of prothrombin to thrombin. It contains the γ-carboxyl glutamate residues originally present on prothrombin and remains bound to the phospholipid surface which organizes the elements in activation of prothrombin. For further information on prothrombin and thrombin see refs. 1, 3, and 52–53.

Thrombin (Topical Thrombin, Parke Davis) is not used intravascularly as a clotting agent, but is used externally as a powder or in solution to prevent oozing of blood from capillaries and small venules. In addition, it is useful for the control of bleeding in parenchymatous tissue; it is available as a diagnostic reagent for fibrinogen determination (Dade Data-FI Thrombin Reagent) and quantitative testing of fibrinolysis (Buffered Thrombin in Date Data-FI Euglobulin Lysis Reagent set).

Fibrinogen (Factor I). Fibrinogen consists of three pairs of polypeptide chains (α_2-β_2-γ_2) covalently linked by disulfide bonds with a total mol wt of 340,000. It is the only coagulation protein of appreciable concentration in plasma, 200–400 mg/100 mL. Fibrinogen is converted to fibrin through proteolysis by thrombin. An activation peptide, fibrinopeptide A (human) [25442-31-5] is cleaved from the amino-terminus of each α chain; fibrinopeptide B (human) [36204-23-6] from the amino-terminus of each β chain. The resulting monomer spontaneously polymerizes to a fibrin gel which is soluble in 6 M urea and dilute acids and bases. Further stabilization of the fibrin polymer to an insoluble, mechanically strong form, requires cross-linking by factor XIII as described below. Fibrinogen and its conversion to fibrin have been extensively reviewed (54).

In addition to its therapeutic use, fibrinogen reagents are important in laboratory analysis of coagulation defects (Fibrinogen, General Diagnostics, in lyophilized form from bovine plasma).

Factor XIII. Factor XIII is a plasma glycoprotein which has also been isolated from human platelets and placenta. The plasma form is a tetramer of two nonidentical subunits α and β (α_2–β_2). The platelet and placental form are dimers of a subunit which appears identical to plasma α (55).

Factor XIII is converted to XIIIa [9067-75-8] by thrombin. XIIIa cross-links the γ chains of fibrin by transglutaminase activity, forming ϵ- (γ-glutamyl) lysine cross-links. The α chains of fibrin also may be cross-linked. Most of the stability of the clot seems to be due to γ–γ cross-links although α–α links have been reported to make the polymer more resistant to degradation by plasmin (55). This transglutaminase activity can be destroyed by alkylation of a cysteine residue on one of the α' subunits.

Coagulation and Vitamin K. It has been known for many years that vitamin K exerts a procoagulant effect *in vivo*; a deficiency of vitamin K results in a bleeding tendency. It has been shown that this vitamin plays a role in the biosynthesis of factors VII, X, IX, and prothrombin (the vitamin K-dependent coagulation factors), factors involved in reactions that occur in lipoprotein complexes. Their binding to the lipids is dependent on the presence of γ-carboxyl glutamate residues and vitamin K is believed to be a required cofactor in the post-translational carboxylation of glutamate residues of these factors. For more information on the procoagulant effect of vitamin K, see ref. 56 (see also Vitamins).

Coagulant Factor Replacement Therapy. The traditional treatment of hemophilia and other bleeding disorders due to the lack of activity of coagulation factors has been plasma transfusion. A more recent development is the use of plasma concentrates; because 90% of congenital hemorrhagic states are the cause of deficiencies in activity of factors VIII (classical hemophilia, von Willebrand's disease) and IX (Christmas disease), concentrates containing these factors have received the most attention (7,57–59).

AHF concentrates can be made locally at blood banks by cryoprecipitation (see Blood fractionation). Fresh plasma is rapidly frozen and then thawed slowly (18–24 h period) at 4–8°C. About 3% of the protein does not redissolve; this cryoprecipitate contains about 50% of the fresh plasma VIII activity and can be isolated by centrifugation. AHF concentrates are available commercially (Profilate, Abbott; Factorate, Armour; Koate, Cutter; Humafac, Parke Davis; and Hemofil, Hyland) as lyophilized powder in single dosage form, to be reconstituted with a small amount of diluent for injection or infusion. They typically contain 250–500 AHF units per dose with one AHF unit being the activity present in 1 mL of plasma pooled from at least 10 donors and tested within 3 h of collection. A concentrate is also prepared by the American Red Cross. Some of the products have been further purified by glycine precipitation or polyethylene glycol (PEG) subfractionation or both (60).

Table 4 contains a list of concentrates available for treatment of factor IX deficiency. These concentrates are referred to as prothrombin complex preparations as

Table 4. Factor IX Concentrates Available[a]

Name of preparation	Place of manufacture	Starting material	Absorbent	Concentration of IX, U/mL[b]
PPSB	Paris and Edinburgh	EDTA plasma	$Ca_3(PO_4)_2$	25
Oxford Type C	Oxford	residue from factor VII, γ globulin, and albumin fractionation	$Ca_3(PO_4)_2$	5–10
Oxford Type DE	Oxford	cryosupernatant	DEAE–cellulose	25–68
Edinburgh DE factor IX	Edinburgh	cryosupernatant	DEAE–cellulose	30
Konyne (Cutter)	California	Cohn fraction I supernatant	DEAE–cellulose	25
Proplex (Hyland)	California	Cohn fraction III	$Ca_3(PO_4)_2$	8–12
Prothrombal	Leiden	cryosupernatant	$Al(OH)_3$ gel	
Supernine	Edinburgh	DE·F·IX	DEAE–cellulose	100–150

[a] Refs. 7 and 58.
[b] U = unit.

they contain not only factor IX, but factors X, VII, and prothrombin. They may be used in factor X and VII deficiencies also; however, in the case of VII, administration of fresh-frozen plasma is the treatment of choice (7). Caution must be exercised in any use of prothrombin complexes as some preparations cause diffuse intravascular thrombosis. Prothrombin complexes have been traditionally isolated by adsorption onto alkaline earth precipitates, but a variety of methods are used to prepare the factor IX concentrates.

Fibrinogen concentrates are used in treatment of the congenital hemorrhagic states, afibrinogenemia, dysfibrinogenemia, and in acquired hypofibrinogenemia. The concentrate may be prepared from Cohn fraction I which consists of approximately 60% clottable protein or it may be prepared by alcohol precipitation and polyethylene glycol subfractionation of the supernate from cryoprecipitation. Crude fibrinogen is difficult to prepare and maintain because of its rapid conversion to fibrin. Commercial preparations are available as Fibrinogen, (Hyland and Merck, Sharp & Dohme) and Parenogen (Cutter).

Miscellaneous Hemostatics. Carbazochrome salicylate [13051-01-9] (1), the sodium salicylate complex of the oxidation product of epinephrine–adrenochrome (Adrenosem Salicylate, Beecham Labs), is administered systemically by the intramuscular or oral route to reduce oozing and bleeding after some surgical procedures. Its effect is not on the coagulation process; rather, it reduces capillary permeability, thus controlling capillary bleeding.

(1) carbazochrome salicylate

Other products are used locally to control oozing. These include oxidized cellulose (Oxycel, Parke Davis and Surgicel, Johnson & Johnson) in gauze strips, cloth pads and pledgets, and a specially treated gelatin product available in sponge form or powder (Gelfoam, Upjohn), or as a film (Gelfilm, Upjohn), for use in brain or lung surgery as dural or pleural implants or for use in some procedures in eye surgery. Gelfoam may also be administered orally for control of gastroduodenal hemorrhage. Microfibrillar collagen, a powder derived from bovine corium collagen (Avitene, Avicon), is a newer product that is used as an adjunct to ligation during surgery. It attracts platelets, promoting platelet adhesion and release reaction.

Thrombin and ϵ-aminocaproic acid, also regarded as hemostatic agents, are discussed on pp. 10, 18, and 22, respectively.

Anticoagulants

In Vitro **Anticoagulants.** Substances that remove Ca^{2+} from plasma have long been used to prevent the coagulation of blood samples. Ethylenediamine tetraacetic acid (EDTA) [60-00-4] (Sequestrene) complexes with Ca^{2+} and is effective as an anticoagulant at 1 mg of the disodium salt per mL blood. Oxalate and fluoride ions form insoluble salts with Ca^{2+}; salts containing these ions that are used as anticoagulants include dilithium oxalate [553-91-3], $Li_2C_2O_4$, 1 mg/mL blood; disodium oxalate [62-76-0], $Na_2C_2O_4$, 2 mg/mL blood; dipotassium oxalate monohydrate [6487-48-5], $K_2C_2O_4 \cdot H_2O$, 2 mg/mL blood; and sodium fluoride [7681-49-4], NaF, 2 mg/mL blood. Sodium polyanetholsulfonate [52993-95-0] (Liquoid, Hoffmann-La Roche Inc.) has recently become available in the United States and is used at concentrations of 1–2.5 mg/mL blood.

Heparin, 0.2 mg/mL blood, is also used as an anticoagulant *in vitro*.

Plasma Anticoagulants. Several plasma proteins have been shown to inhibit the activity of thrombin and other serine proteases. Among these are Antithrombin III (AT-III, antithrombin–heparin cofactor), α_2-macroglobulin, α_1-antitrypsin, and C-1 esterase inhibitor. A major portion of the antithrombin activity of plasma (70–90%) has been shown by immunoprecipitation to be due to AT-III.

AT-III also inhibits (by a stoichiometric ratio of 1:1) other serine proteases of the coagulation pathways: XIa, IXa (but not IX), Xa (but not X), and XII (14,61). Factor VII is also inhibited by AT-III but only in the presence of heparin. Heparin is not required for inhibition of the other factors; however, its presence dramatically accelerates their rates of inhibition. By chemical modification of AT-III, Rosenberg (61) has shown that the serine proteases combine with AT-III at arginine residues. The serine proteases typically cleave peptide bonds involving arginine. It was also shown that the polyanionic heparin molecule probably combines with the cationic lysine residues of AT-III. This complex is responsible for the anticoagulant properties of heparin.

Therapeutic Anticoagulants. *Heparin.* There has been intense research activity with heparin in the last decade (62–63). Heparin (mol wt 6,000–25,000) is a glycosaminoglucuronan (acid mucopolysaccharide) that occurs in most tissues; human liver, lung, and mast cells are especially rich sources. Its presence in normal circulating blood remains in doubt (64). Chemically, it is not homogenous, but is a family of linear polymers which differ in chain length and mol wt, but have similar disaccharide units. The disaccharide units are composed of D-glucuronic or L-iduronic acids in 1,4-glycosidic linkage to glucosamine; each unit contains two sulfate esters and one N-sulfate group.

Heparin is obtained commercially from a variety of animal tissues, particularly bovine lung tissue and intestinal mucosa of pigs. It is isolated by various processes (65); most include initial tissue autolysis, precipitation by acidification of alkaline extracts, and enzymatic protein hydrolysis. In the United States, the commercial product must have a potency of at least 120 USP units per mg; potencies are often as high as 150 USP units and may range up to 170 units per mg. USP units are defined by the ability of a reference standard of heparin obtained from porcine intestinal mucosa to prevent clotting of sheep plasma. An international reference standard from bovine lung tissue is maintained by the World Health Organization; its potency is 130 IU/mg. Table 5 lists the dosage forms and clinical properties of heparin.

Physiologically heparin displays several types of activity related to hemostasis. It interferes with the coagulation mechanism, enhances fibrinolysis, and, depending on the circumstances, can potentiate or inhibit platelet aggregation and release reaction (64). Heparin interferes with coagulation by dramatically increasing the rate of activity of the serine protease inhibitor, AT-III, also called antithrombin–heparin cofactor. Attachment to AT-III occurs at lysine residues with a stoichiometry of 1:1 (61). It also has an antilipemic effect mediated by lipoprotein lipase.

Heparin has been used for several decades as an anticoagulant; it has been particularly useful in immediate treatment of deep vein thrombosis and pulmonary and systemic embolism, and recently has been shown efficacious in low doses for prophylaxis of post operative thrombo-embolism. Low dosage regimens may also be useful for control of coronary heart disease and for prophylactic control of deep vein thrombosis after myocardial infarction (66).

Coumarins. These synthetic anticoagulants were developed after the isolation of dicumarol [66-76-2] (3,3'-methylenebis[4-hydroxy-2H-1-benzopyran-2-one]), from spoiled sweet clover in 1939. A tabulation of agents that are commercially available is shown in Table 5; structures are shown in Figure 5. The parent compound for synthesis of these substances is 4-hydroxycoumarin which can be easily synthesized from methyl salicylate by acetylation and cyclization (67). The derivatives are prepared as illustrated in Figure 5.

The coumarin anticoagulants, in contrast with heparin, are active *in vivo* but not *in vitro*. They produce their effect by a competitive inhibition of vitamin K which leads to a reduction in the activity, in blood, of factors VII, IX, X, and prothrombin through interference with the synthesis of these agents in the liver. Since their anticoagulant effect is not immediate, it is often desirable to initiate therapy with heparin and maintain the effect with a long-acting coumarin derivative. A recent review of anticoagulant therapy is ref. 68.

Indandiones. These agents (see Tables 5 and 6) are similar to the coumarins in structure and mechanism of action; however, they take longer to produce effects and have a longer duration of action. They were first introduced into therapy in 1953 and were used extensively in Great Britain. It has been recommended not to use them clinically because of a high percentage of adverse and even fatal effects (68).

Anticoagulant Antagonists. Heparin, due to its strongly acidic character, is readily precipitated by strongly basic agents; thus, protamine sulfate [9009-65-8] is used widely as a heparin antagonist. Protamine [9012-00-4] is a substance derived from the ripe sperm of salmon and other species of fish. It is a low mol wt protein (8000) with 58 amino acid units of which 40 are arginine. Protamine has no effect on the coagulation time of normal individuals but a 40–50-mg dose will neutralize a 50-mg dose of heparin

Table 5. Clinical Properties of Some Anticoagulants

Generic and trade name	CAS Registry No.	Dosage forms available	Dosage[a]				Latent period, h	Maximum effect, h	Duration of effect, d
				1st d	2nd d	maintenance (daily)			
heparin Hepathrom, Fellows Heplock, Elkins-Sinn Heprinar, Armour Lipo-hepin, Riker Liquaemin, Organon Panheprin, Abbott	[9005-49-6]	100, 250, 1,000, 2,500, 3,000, 4,000, 5,000, 6,000, 7,500, 10,000, 15,000, 20,000, 40,000 U/mL in vials, ampules, and disposable syringes in 1–10 mL lots	subcutaneous: 8,000–10,000 U every 8 h or 15,000–20,000 U every 12 h intermittent iv: 10,000 U, then 5,000–10,000 every 4–6 h continuous iv: 20,000–40,000 U/d				immediate	immediate	1–3
bishydroxycoumarin Dicumarol, various	[66-76-2]	25, 50, 100 mg capsules or tablets		200–300		50–200	24	60–96	5
ethyl biscoumacetate Tromexan, Geigy	[548-00-5]	150 and 300 mg tablets		900–1500		150–1000	8–12	18–30	2–3
cyclocoumarol	[518-20-7]	50 mg tablets		100–200	25–50	10–50	18–24	60–96	5
acenocoumarin Cumopyran, Abbott	[152-72-7]	4 mg tablets		16–28	8–16	2–10	8–12	24–36	3–4
phenprocoumon Sintrom, Geigy	[435-97-2]	3 mg tablets		24		0.75–6.0	12–24	30–36	5–6
warfarin sodium Liquamar, Organon Coumadin, Endo Panwarfin, Abbott Athrombin, Purdue Fred	[81-81-2]	2, 2.5, 5, 7.5, 10, and 25 mg tablets		40–60		2–10	8–12	24–36	6
phenindione Hedulin, Merrell National Danilone, Schieffelin	[83-12-5]	50 mg tablets		300	200	100	2–8	48–60	4–5
diphenadione Dipaxin, Upjohn	[82-66-6]	5 mg tablets		20–30	10–15	3–5	12–24	48–72	15–20
anisidione	[117-37-3]	50 mg tablets		300	200	100 3rd d, 25–250 thereafter	24–24	48–72	1–3

[a] These are average ranges; the dose for each patient should be determined individually and the effects carefully monitored. Note that iv = intravenous; U = unit (IU = international unit).

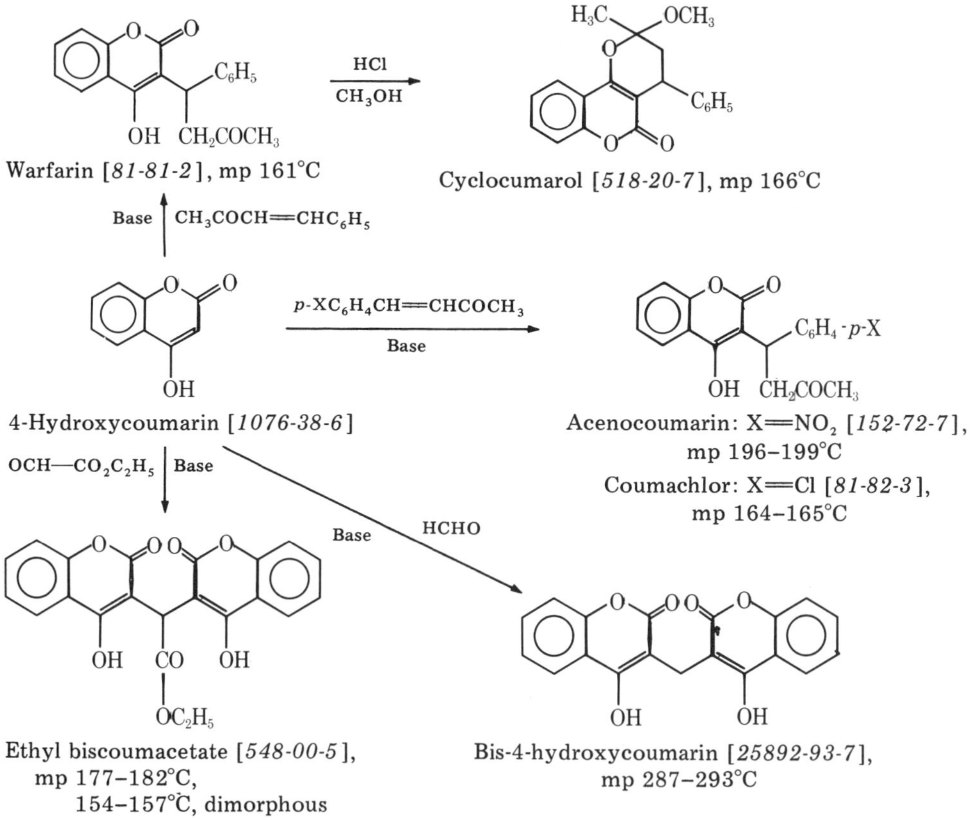

Figure 5. Structures and synthesis of coumarin derivatives (67).

Table 6. Structures and Melting Points of Indandione Anticoagulants

Compound	Mol wt	mp, °C	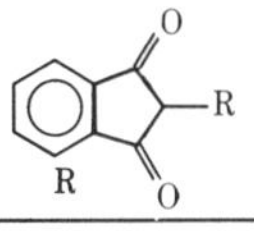 R
phenindione [83-12-5] (2-phenyl-1H-indene-1,3(2H)-dione)	222.23	149–151	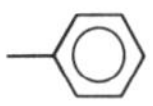 —C₆H₅
diphenadione [82-66-6] (2-diphenyl-acetyl-1H-indene-1,3(2H)-dione)	340.36	146–147	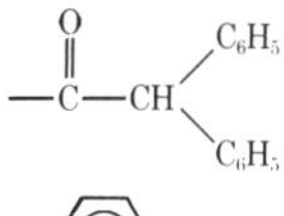 —C(O)—CH(C₆H₅)₂
anisindione [117-37-3] (2-(4-methoxy-phenyl)-1H-indene-1,3(2H)-dione)	252.26	156–157	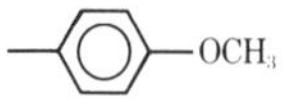 —C₆H₄—OCH₃

within 5 min. Protamine, especially in high doses, has a tendency to lower the blood pressure in some individuals.

The antidote for overdosage of the coumarin and indandione drugs is vitamin K. The naturally occurring vitamin K [84-80-0] (2) (K₁, phylloquinone) is much more effective than the water soluble forms (menadione [58-27-5] (3)) in counteracting the effects of the oral anticoagulants. Vitamin K is also administered when hypoprothrombinemia develops as a result of vitamin K deficiency; a deficiency is usually due to the administration of drugs that suppress the growth of intestinal microorganisms, an important source of vitamin K.

(2) vitamin K₁

(3) menadione

Fibrinolysis

Like coagulation, fibrinolysis involves a number of components, many of which are proteases. Some components play roles in both systems (eg, factor XII). Figure 1 outlines the fibrinolytic system.

The culminating event of the fibrinolytic system is the dissolution of a clot by the degradation of fibrin to water soluble fibrin degradation products (FDP). This hydrolysis is accomplished by a serine protease, plasmin, which is generated from its zymogen, plasminogen, by substances termed plasminogen activators, most of which are also proteases. Plasmin can use a wide range of plasma proteins as substrates (including many coagulation factors), thus it is important for it to be generated only in the vicinity of fibrin formation and not circulate in significant quantities in blood. Accordingly, plasminogen has a high affinity for fibrin, and native plasminogen activators have been shown to preferentially activate plasminogen bound to fibrin (gel phase). Antiplasmins circulating in plasma police the blood by combining with any free plasmin and forming inactive complexes.

Activators responsible for the generation of plasmin from plasminogen by proteolytic cleavage are present in many body tissues and fluids. Factor XII plays a role in initiating fibrinolysis and recent studies show that this activity may be mediated by kallikrein (Fig. 3) (69).

Plasminogen (Profibrinolysin). Plasminogen [9001-91-6], Pg, is a monomeric, single-chain protein of 80,000–90,000 mol wt. It exists in two distinct biophysical forms, a soluble-phase form in plasma and extracellular fluids, and a gel-phase form in thrombi and other fibrinous deposits (70–71). The purified protein has been shown to assume three molecular conformations which are interconvertible by use of acid or ε-aminocaproic acid (72). Two molecular forms (Glu–Pg and Lys–Pg) have been isolated from human plasma and serum; each of these two forms has been resolved into multiple isoelectric species. Lys–Pg has a mol wt about 2% less than that of Glu–Pg and is believed to be a partially degraded form of Glu–Pg. Other mammalian plasminogens also exhibit multiple isoelectric forms and vary with respect to NH₂-terminal residues: rabbit (Glu), bovine and cat (Asp), and dog (blocked X) (29).

Plasminogen can be isolated from plasma, serum, or plasma III or III$_{2,3}$ fractions by affinity chromatography with L-lysine-substituted Sepharose (see Chromatography, affinity). It is eluted by ε-aminocaproic acid and further purified by ammonium sulfate precipitation (29). Plasminogen is assayed by plasmin assay methods after conversion to plasmin.

Plasmin (Fibrinolysin). Plasmin [9001-90-5], Pm, is a serine protease with typical trypsinlike specificity for lysyl and arginyl bonds. Many plasma proteins as well as a variety of other proteins act as substrates (73), however, it is doubtful whether they are significant substrates under normal circumstances *in vivo*. Nevertheless, because of this wide range of specificity, the term plasmin is often used instead of the older term fibrinolysin. Human plasmin has 3–5 major molecular isoelectric forms; other mammalian species also have multiple enzyme forms. The primary structures of various plasminogens and plasmins are at least partially homologous with other serine proteases (Table 3) (29).

Human plasmin (Lys–Pm) consists of two chains linked by a single disulfide bond. The A chain is a single isoelectric form with 48,000 mol wt. The B chain is smaller (25,700 mol wt) and has 3 isoelectric forms; it also contains the active site serine (29).

The mechanism of activation of plasmin from plasminogen by a typical proteolytic activator is shown in Figure 6. Under most circumstances *in vitro* (and probably *in vivo*) Lys–Pm is formed during activation, but if trasylol [9039-77-4], a plasmin inhibitor, is present Glu–Pm is the enzyme produced (74).

Plasmin is commonly assayed by determining either its esterolytic or proteolytic activities. Substrates used include casein, fibrin (fibrin clot assay or plate assay), and synthetic esters (29,75).

Plasminogen Activators. Plasminogen activators are known to be present in nearly all body tissues (heart and lung have particularly large amounts) as well as body fluids including tears, breast milk, urine, and plasma. It is not known if the activators are chemically identical although urinary, tissue culture, blood vessel, and human heart activators are immunologically identical (76). Plasma and urinary activators, however, have been shown to be immunologically different (77).

The urinary activator, urokinase [9039-53-6], has been best characterized; it is probably synthesized in kidney cells (78) and is isolated from urine. Urokinase seems to be a serine protease which has a physiological specificity limited to a single bond (arginyl–valine) in one substrate, plasminogen (Fig. 6). The activator has a high gel phase–fluid phase plasminogen activation ratio, as would be predicted for a native activator from the current theory of *in vivo* fibrinolysis. Two forms, 31,300 mol wt and 54,700 mol wt, have been reported (79). The enzyme is fairly stable, except in salt

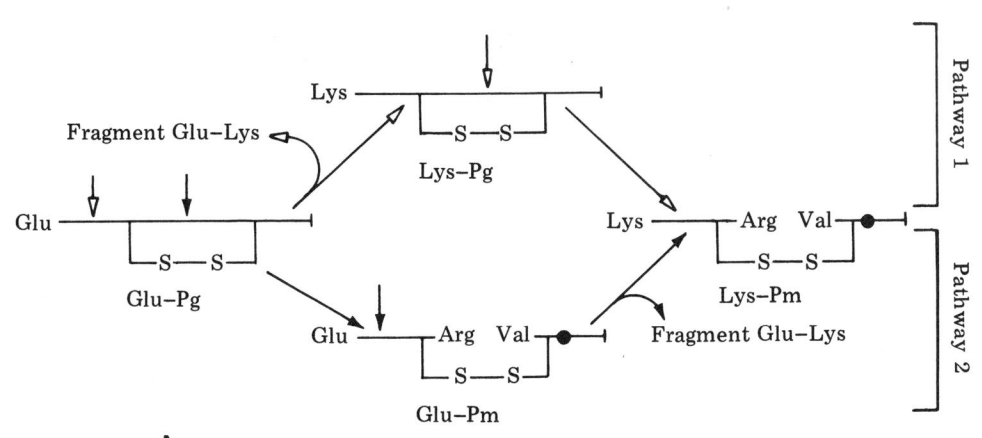

Figure 6. Activation of plasminogen to plasmin by proteolytic activators such as urokinase. Pg = plasminogen; Pm = plasmin; heavy dots indicate location of serine in exposed active sites.

concentrations less than 0.03 M NaCl. Barlow has recently reviewed this substance (80).

Streptokinase [9002-01-1], SK, is an exocellular protein product of hemolytic streptococci. Its role in the generation of plasmin is quite unlike that of the native activators previously discussed. Streptokinase alone has never exhibited enzymatic activity, but by combining in stoichiometric ratios with plasminogen it uncovers plasminogen's active site. The SK·Pg complex is a potent plasminogen activator and can act inter- or intramolecularly. It acts rapidly intramolecularly, ie, it is short-lived, generating SK·Pm, also an excellent plasminogen activator. Streptokinase is also cleaved while in the complex.

Streptokinase shows different activation effects on plasminogen derived from different biological species. Human plasminogen is most sensitive to activation; cat, dog, and rabbit plasminogen follow. It is assayed by its activation of plasminogen and the subsequent plasmin activity generated. One unit of SK is defined as the amount required to lyse a standard clot in 10 min. It is a single-chain, monomeric protein of 45,000–50,000 mol wt. It is unstable above pH 9.0 however, if purified, it is stable indefinitely when lyophilized or frozen in aqueous solution (29). Other physical and chemical properties are described in reference 81.

Streptokinase has been used extensively in the experimental investigation of the plasminogen–plasmin system. It is used clinically in combination with streptodornase (Varidase, Lederle) orally or intramuscularly as an antiinflammatory agent where such conditions exist in close proximity with contusions and bruises; and it has some usefulness as a thrombolytic agent.

Thrombolytic Agents. The formation of intravascular clots (thrombi) is a life-threatening situation that occurs in many abnormal physiological states. Various substances have been used in an attempt to hydrolyze existing clots and to prevent formation of thrombi. These agents supplement the role of therapeutic anticoagulants, although they are aimed not at preventing coagulation, but at enhancing fibrinolysis and thrombolysis. They have not yet achieved the status of the anticoagulants in management of thrombotic disorders.

In view of the current concept of *in vivo* fibrinolysis, the ideal thrombolytic agent would be one which activates thrombi-bound (gel-phase) plasminogen. It would, therefore, be classified as a plasminogen activator and would have a high gel-phase–soluble-phase plasminogen activation ratio. Alternatively, the agent could act in a less direct manner and stimulate the release of plasminogen activator *in vivo*.

Before the present concept of thrombolysis was established, other approaches to thrombolytic therapy were tested. Among these was the use of proteases which hydrolyze fibrin *in vitro,* ie, trypsin and chymotrypsin. These are still used locally for wound debridement and have been claimed as being useful systemically in reducing inflammation due to accidental trauma. Another approach was the use of plasmin and plasmin–plasminogen activator mixtures. Although these drugs were of some success, their efficacy was shown to be due to the presence of plasminogen activators, either as ingredient or contaminant. A serious disadvantage was that plasmin introduced into the circulation would hydrolyze fibrinogen and other coagulation factors, resulting in a hemorrhagic state. A commercial preparation of plasmin (+ plasminogen activator) is still available as Thrombolysin (Merck, Sharp & Dohme) but its use has been largely supplanted by purified plasminogen activators.

Two plasminogen activators, streptokinase and urokinase, have undergone extensive clinical trials and are commercially available as thrombolytic agents.

Streptokinase has a relatively low gel-phase–soluble-phase plasminogen level. Thus the desired thrombolytic levels cannot be achieved unless almost all the circulating plasminogen has been activated. Consequently, it has been reported that: adequate therapy produces blood coagulation defects to a mild to moderate and an occasional severe degree; it is a relatively inflexible therapeutic agent; fibrin formed during treatment or shortly thereafter is plasminogen free and unavailable for thrombolysis; and antibodies in the patient may delay effective thrombolysis for several hours (82). Streptokinase is available as Kabikinase (Kabi, Sweden) and Streptase (Behringwerke, Germany) in some countries; however, it is not approved for use in the United States.

Urokinase has a high gel-phase–soluble-phase plasminogen ratio. In spite of this fact, the product has not been introduced commercially until now because of several factors: preparations showing negligible toxicity in animals are difficult to produce; it has not been possible until recently to produce a product with a high specific activity—3500 CTA urokinase units/mg protein; preparations with a lack of viral activity have been difficult to prepare; and it has been hard to eliminate known thromboplastic substances (82). When preparations of the drug meet these requirements, urokinase appears to be an ideal thrombolytic agent with none of the disadvantages of streptokinase previously mentioned. However, care should be exercised in following the prescribed dosage regimen to prevent transient blood hypercoagulability upon initial infusion. Urokinase, purified from urine, has just recently become available commercially as Win-Kinase (Winthrop) and Abbokinase (Abbott).

Many pharmacological agents have been observed to increase fibrinolytic activity *in vivo,* presumably through stimulation of the release of endogenous plasma plasminogen activator(s). Epinephrine (qv), nicotinic acid, and vasopressin were some of the earlier substances discovered to have this action. Subsequently, other drugs that stimulate the release of epinephrine (eg, insulin, oral hypoglycemics, corticosteroids, and anabolic steroids) have also demonstrated an ability to enhance fibrinolytic activity. Unfortunately, in most cases the effect is transient and resistance develops after a few weeks of treatment. However, one of the anabolic steroids, Stanazol, has shown promise as a useful therapeutic agent (83).

Fletcher and Alkjaersig (82) and Prentice and Davidson (84) have recently reviewed thrombolytic agents.

Inhibitors of Fibrinolysis. There are two types of inhibitors, those that inhibit plasminogen activation and those that inhibit plasmin directly. Table 7 lists the naturally occurring and synthetic inhibitors and their proposed mechanisms of action. The first four listed are present in plasma. In addition, an α_2-globulin has been tentatively identified as the primary *in vivo* plasma regulator of fibrinolysis (85). Antiplasmin activity is also present in platelets but to a much lesser degree. This activity can be induced by collagen or thrombin (86). Antiplasmin activity can also be extracted from mammalian spleen, kidney, liver, bladder, skin, muscle, and blood vessel tissue (87). Placental tissue has been shown to be a rich source of two different kinds of urokinase inhibitors (88–89). Bennett and Ogston (90) have reviewed this subject.

Of the inhibitors named above, only the synthetic amino acids have had any wide clinical use. In a sense, they may be viewed as coagulants. The rationale for their use is that they enhance formation of clots by preventing lysis of incipient fibrin deposits. Table 8 lists the types of agents commercially available. They may be indicated in

22   BLOOD, COAGULANTS AND ANTICOAGULANTS

Table 7. Inactivators of Fibrinolysis

Name	Mol wt of protein	Mechanism of inhibition
α_1-antitrypsin	55,000	noncompetitive inactivation of plasmin; retarded in the presence of enzyme substrate; antiplasmin
α_2-macroglobulin	820,000	binds active enzyme; esterolytic activity less inhibited than proteolytic
C_1'-inactivator	γ-globulin neuraminoglycoprotein	inhibits plasmin proteolytic activity by competitive substrate inhibition; antiplasmin
antithrombin-III	65,000	inhibits serine proteases
ϵ-aminocaproic acid (EACA)		inhibits plasminogen activation by urokinase at concentration of $10^{-2}M$ and higher; non-competitive inhibition of plasmin in concentration of $5 \times 10^{-5}M$ and higher
tranexamic acid (AMCA)		
p-aminomethylbenzoic acid (PAMBA)		
platelet inhibitors		inactivates urokinase
aprotinin (Trasylol)	6,200	competitive inhibitor of urokinase and streptokinase; noncompetitive inhibitor of plasmin

Table 8. Therapeutic Fibrinolytic Inhibitors

Name and trade name	CAS Registry No.	Dose	Chemical structure
ϵ-aminocaproic acid, 6-aminohexanoic acid, Amicar, Lederle	[60-32-3]	4–6 g loading dose followed by 1 g/h iv or oral	$H_2N-CH_2-(CH_2)_4-COOH$
tranexamic acid, trans-4-(aminomethyl)-cyclohexanecarboxylic acid, Amstat, Lederle	[1197-18-8]	10 mg/kg 3 times daily iv or 10–20 mg/kg orally	$H_2N-CH_2-\bigcirc-COOH$
aprotinin, trypsin inhibitor, trasylol, Trasylol, Delbay	[9039-77-4]	initial dose of 100,000 KIU[a] iv followed by 100,000 KIU iv/h	protein mol wt 6200

[a] Kallikrein international units.

pathological conditions where hemostasis does not occur, ie, postprostatectomy bleeding, menorrhagia, ulcerative colitis, hereditary angio-edema, and dental extraction from patients with hemophilia (90). In addition, because of possible inhibitory effects on the complement system, they may be useful in immunologically mediated phenomena such as organ transplants, asthma (90), and blood transfusion induced hemolytic states (91).

BIBLIOGRAPHY

"Coagulants and Anticoagulants" in *ECT* 2nd ed., Vol. 5, pp. 586–605, by David M. Stuart, Neisler Laboratories, Inc.

1. E. W. Davie and K. Fujikawa, *Ann. Rev. Biochem.* **44,** 700 (1975).
2. J. W. Suttie and C. M. Jackson, *Physiol. Rev.* **57,** 1 (1977).

3. C. M. Jackson, in V. V. Kakkar and D. P. Thomas, eds., *Heparin, Chemistry and Clinical Uses*, Academic Press, London, 1976.
4. R. D. Rosenberg in ref. 3, p. 101.
5. "The Nomenclature of Blood Clotting Factors," *J. Am. Med. Assoc.* **180,** 733 (1962).
6. C. M. Jackson, *Thromb. Haemosts.* **38,** 567 (1977).
7. C. R. Rizza, *Clin. Haematol.* **5,** 113 (1976).
8. R. Radcliffe and Y. Nemerson in L. Lorand ed., *Methods in Enzymology*, Vol. XLV, Academic Press, Inc., London, 1976, p. 49.
9. R. Radcliffe and Y. Nemerson, *J. Biol. Chem.* **251,** 4797 (1976).
10. W. Kisiel, K. Fujikawa, and E. W. Davie, *Biochemistry* **16,** 4189 (1977).
11. H. D. Godal, *Thromb. Res.* **5,** 773 (1974).
12. Y. Nemerson, *Biochemistry* **5,** 601 (1966).
13. K. Fujikawa and E. W. Davie in ref. 8, p. 74.
14. B. Osterud and co-workers, *Thromb. Haemosts.* **35,** 295 (1976).
15. K. Fujikawa and E. W. Davie in ref. 8, p. 89.
16. B. N. Bouma and J. H. Griffin, *J. Biol. Chem.* **252,** 6432 (1977).
17. T. Koide, H. Kato, and E. W. Davie in ref. 8, p. 65.
18. P. S. Damus, M. Hicks, and R. D. Rosenberg, *Nature (London)* **246,** 355 (1973).
19. J. H. Griffin and C. G. Cochrane, in ref. 8, p. 56.
20. T. Koide, H. Kato, and E. W. Davie in ref. 10, p. 2279.
21. D. T. H. Liu and L. E. McCoy, *Thromb. Res.* **7,** 199 (1975).
22. H. Gonmori and Y. Takeda, *Am. J. Physiol.* **229,** 618 (1975).
23. F. A. Pitlick and Y. Nemerson in ref. 8, p. 37.
24. E. Bjorklid and co-workers, *Biochem. Biophys. Res. Commun.* **55,** 969 (1973).
25. R. W. Colman and R. M. Weinberg in ref. 8, p. 107.
26. C. G. Curtis and L. Lorand in ref. 8, p. 177.
27. S. Coccheri in H. F. Conn, ed., *Current Therapy 1977*, W. B. Saunders Co., Philadelphia, Pa., 1977.
28. K. Fujikawa, K. Kurachi, and E. W. Davie in ref. 10, p. 4182.
29. K. C. Robbins and L. Summaria in ref. 8, p. 259.
30. K. Fujikawa, K. A. Walsh, and E. W. Davie in ref. 10, p. 2270.
31. L. Lorand in ref. 8, Sec. II and III.
32. B. Bennett and A. S. Douglas, *Clin. Haematol.* **2,** 3 (Feb. 1973).
33. E. W. Davie and co-workers in D. W. Ribbon and K. Brew, eds., *Proteolysis and Physiological Regulation*, Academic Press, Inc., New York, 1976, p. 189.
34. Y. Nemerson and F. A. Pitlick, *Prog. Hemostasis Thromb.* **1,** 1 (1972).
35. S. D. Revak and C. D. Cochrane, *J. Clin. Invest.* **57,** 852 (1976).
36. J. H. Griffin and C. G. Cochrane, *Proc. Nat. Acad. Sci. U.S.A.* **73,** 2554 (1976).
37. H. L. Meier and co-workers, *J. Clin. Invest.* **60,** 18 (1977).
38. K. Fujikawa and co-workers, *Biochemistry* **13,** 4508 (1974).
39. E. M. Barrow and J. B. Graham, *Physiol. Rev.* **54,** 23 (1974).
40. M. E. Legaz and co-workers, *J. Biol. Chem.* **248,** 3946 (1973).
41. G. A. Shapiro and co-workers, *J. Clin. Invest.* **52,** 2198 (1973).
42. M. E. Legaz and co-workers, *Ann. N.Y. Acad. Sci.* **240,** 43 (1975).
43. c. D. Forbes and C. R. M. Prentice, *Nature (London) New Biol.* **241,** 149 (1973).
44. H. A. Cooper, T. R. Griggs, and R. H. Wagner, *Proc. Nat. Acad. Sci. U.S.A.* **70,** 2326 (1973).
45. T. R. Griggs and co-workers, in ref. 44, p. 2814.
46. C. M. Jackson and D. J. Hanahan, *Biochemistry* **7,** 4506 (1968).
47. P. V. Subbaiah and co-workers, *Biochem. Biophys. Acta* **444,** 131 (1976).
48. C. M. Jackson, *Biochemistry* **11,** 4873 (1972).
49. K. Fujikawa and co-workers in ref. 38, p. 5290.
50. J. Jesty and Y. Nemerson in ref. 8, p. 95.
51. S. M. Zeldis and co-workers, *Science* **175,** 766 (1972).
52. K. G. Mann in ref. 8, p. 123.
53. R. L. Lundblad, H. S. Kingdon, and K. G. Mann in ref. 8, p. 156.
54. R. F. Doolittle, *Adv. Protein Chem.* **27,** 1 (1973).
55. M. L. Schwartz and co-workers, *J. Biol. Chem.* **248,** 1395 (1973).
56. R. A. O'Reilly, *Ann. Rev. Med.* **27,** 245 (1976).
57. J. G. Watt in ref. 7, p. 95.

58. J. Wallace in ref. 32, p. 129.
59. C. D. Forbes and J. F. Davidson in ref. 58, p. 101.
60. Brit. Pat. 1,178,958 (Jan. 28, 1970), (to Baxter Laboratories, Inc.).
61. R. Rosenberg, *N. Engl. J. Med.* **292,** 146 (1975).
62. V. V. Kakkar and D. P. Thomas, eds., *Heparin, Chemistry and Clinical Uses,* Academic Press, London, 1976.
63. R. Bradshaw and S. Wessler, eds., *Heparin, Structure, Function and Clinical Implications,* Plenum Press, New York, 1975.
64. P. N. Walsh in ref. 62, p. 125.
65. H. G. Hind, *Man. Chem.* **34,** 510 (1963).
66. C. Warlow in ref. 62, p. 293.
67. S. Divald and M. M. Joullie in A. Burger, ed., *Medicinal Chemistry,* 3rd ed., Part II, John Wiley & Sons, Inc., New York, 1970, p. 1092.
68. M. Nussbaum and C. B. Moschos, *Med. Clin. N. Am.* **60,** 855 (1976).
69. R. Mandle, Jr., and A. P. Kaplan, *J. Biol. Chem.* **252,** 6097 (1977).
70. N. Alkjaersig, A. Fletcher, and S. Sherry, *J. Clin. Invest.* **38,** 1086 (1959).
71. N. Alkjaersig, *J. Clin. Invest.* **38,** 1096 (1959).
72. N. Alkjaersig, *Biochem. J.* **93,** 171 (1964).
73. S. Sherry, A. P. Fletcher, and N. Alkjaersig, *Physiol. Rev.* **39,** 343 (1959).
74. L. Summaria and co-workers, *J. Biol. Chem.* **252,** 3945 (1977).
75. F. J. Castellino and J. M. Sodetz in ref. 8, p. 273.
76. M. B. Bernik and co-workers, *J. Clin. Lab. Med.* **84** (1974).
77. C. Kucinski, A. Fletcher, and S. Sherry, *J. Clin. Invest.* **47,** 1238 (1968).
78. M. Bernik and H. C. Kwaan, *J. Clin. Lab. Med.* **70,** 650 (1967).
79. W. R. White, G. H. Barlow, and M. M. Mozer in ref. 12, p. 2160.
80. G. H. Barlow in ref. 8, p. 239.
81. F. J. Castellino and co-workers in ref. 8, p. 244.
82. A. P. Fletcher and N. Alkjaersig in W. S. Root and N. I. Berlin, eds., *Physiological Pharmacology,* Vol. 5, Academic Press, Inc., London, 1974, p. 132.
83. J. F. Davidson and co-workers, *Br. J. Haematol.* **22,** 543 (1972).
84. C. R. M. Prentice and J. F. Davidson in ref. 32, p. 159.
85. N. Aoki and co-workers, *J. Clin. Invest.* **60,** 361 (1977).
86. J. H. Joist, S. Niewiarowski, and J. F. Mustard in E. F. Maniman, G. F. Anderson, and M. I. Barkhart, eds., *Thrombolytic Therapy,* Schattauer, New York, 1972, pp. 113–119.
87. M. B. Bernik and H. C. Kwaan, *Am. J. Physiol.* **221,** 916 (1971).
88. T. Kawano, K. Morimoto, and Y. Uemura, *Nature (London)* **217,** 253 (1968).
89. M. Uszynski and V. Abildgaard, *Thromb. Diath. Haemorrh.* **25,** 580 (1971).
90. B. Bennett and D. Ogston, in ref. 32, p. 135.
91. F. B. Taylor and H. Fudenburg, *Immunology* **7,** 319 (1964).

DAVID M. STUART
JUDITH K. HRUSCHKA
Ohio Northern University

BLOOD FRACTIONATION

Blood is a solution of salts and proteins in which are suspended a variety of specialized cell types. The properties and composition of human blood are shown in Table 1. Recent advances in understanding of the function of blood cell types and plasma proteins and the deficiencies involved in a variety of blood disorders, combined with improvements in techniques for fractionation and storage of the major cellular and protein components of human blood, have resulted in increased utilization of specific subfractions of human blood, rather than whole blood, for therapeutic purposes. The most commonly prepared components and their usage are presented in Table 2.

Obtaining Source Material

In the United States, all facilities that draw blood or prepare blood components are licensed by or registered with the federal government, and are regulated by the Bureau of Biologics of the FDA. Strict guidelines for donor selection, as well as for collection and handling of blood and blood components, are contained in the Code of Federal Regulations (CFR) (5), which is published annually. Detailed operations manuals are available (6–7) and can be useful in establishing the quality control protocols and records required by the bureau. The regulations require that the donor be a healthy adult. Special emphasis is placed on determining if the individual might have, or might recently have been exposed to, diseases transmissible by blood transfusion, especially viral hepatitis, syphilis, and malaria. It should be noted that the hepatitis tests presently available, which measure hepatitis B surface antigen, do not detect all hepatitis B carriers, and are not applicable to non-B hepatitis.

Depending upon the method used, either whole blood or only specific components of blood are collected during the donation process. The standard volume of whole human blood taken during a single donation is 450–500 mL, and is referred to as one unit of whole blood. In general, a donor cannot give more than one unit every eight weeks. If only the plasma portion of blood is to be taken during the donation by a plasmapheresis procedure, larger volumes of material may be taken with greater frequency. The regulations, which set the upper limits, are that no more than 1000 mL of blood can be removed for processing at any one plasmapheresis procedure or within a 48-h period (for individuals above 79.4 kg, the limit is 1200 mL), and that no more than 2000 mL of blood can be removed for processing within a 7-d period (for individuals above 79.4 kg, the limit is 2400 mL). The regulations also require that no more than 500 mL (600 mL for persons over 79.4 kg) of whole blood can be outside the body at any given time during the procedure. Because of the larger volumes and increased frequency of donation allowed, additional regulations for plasmapheresis are that the donor have a total serum protein of no less than 6.0 g per 100 mL of serum, and that the immunoglobulin composition of the donor's plasma or serum be checked at frequent intervals. When plateletpheresis or leukapheresis techniques are used, to selectively obtain platelets or leukocytes, many blood unit volumes may be processed at one donation with no danger to the donor (8–9), as long as all other components are returned and the procedure used has been approved by the Bureau of Biologics. The collection of multiple components, eg, both plasma and platelets, is limited by the regulations concerning the most constrained component, in the above case, plasma.

Table 1. Properties and Composition of Human Blood[a]

Property	Men	Women
total blood volume, mL	body weight × 77	body weight × 67
total red cell volume, mL	body weight × 36	body weight × 26
total plasma volume, mL	body weight × 41	body weight × 41
hematocrit value[b]	46.2 (43.2–49.2)	40.6 (35.8–45.4)
hemoglobin (A) g/100 mL blood	16.0 (12.0–20.0)	14.0 (10.0–18.0)
pH[c]	7.35 (7.27–7.43)	7.37 (7.31–7.43)
specific gravity of whole blood	1.055–1.064	1.050–1.056
specific gravity of plasma	1.025–1.029	1.025–1.029

Cellular components	Counts	%
erythrocytes (red cells)		
men	5.4 (4.6–6.2) × 10^6 per mm^3	
women	4.8 (4.2–5.4) × 10^6 per mm^3	
thrombocytes (platelets)	260 (145–375) × 10^3 per mm^3	
leukocytes (white cells)		
neutrophils	4.4 (1.8–7.7) × 10^3 per mm^3	59
lymphocytes	2.5 (1.0–4.8) × 10^3 per mm^3	34
monocytes	0.3 (0–0.8) × 10^3 per mm^3	4
eosinophils	0.2 (0–0.45) × 10^3 per mm^3	2.1
basophils	0.045 (0–0.20) × 10^3 per mm^3	0.5

Plasma protein fractions	Plasma, g/100 mL	%
albumins	4.04	60
globulins	2.34	35
α-globulins		
$α_1$-globulins	0.31	4.6
$α_2$-globulins	0.48	7.1
β-globulins	0.81	12.1
γ-globulins	0.74	11.0
fibrinogen [9001-32-5]	0.34	5.1
Total	6.72	100

Other components	Whole blood[d]		Plasma or serum[d]	
carbohydrates				
glucosamine		(60–82)	67	(61–78)
glucose	85.5	(75.5–95.5)	91	(75–107)
glucuronic acid	6.7	(4.1–9.3)	1.71	(1.13–2.29)
glycogen	5.5	(1.2–16.2)	6.8	(4.8–8.9)
electrolytes				
calcium	9.7		5.0	(4.5–5.5) meq/L
CO_2–HCO_3		(19.9–25.1) mmol/L		(24.2–30.5) mmol/L
chloride	84	(77–88) meq/L	103	(99–111) meq/L
iron		(38–45)		(50–170) μg/100 mL
magnesium	4.24		1.85	(1.5–2.5) meq/L
phosphate	2.95	(2.4–3.5)	3.9	(3.1–4.9)
potassium	45.5	meq/L	4.4	(3.5–5.5) meq/L
sodium	85.4	(79.3–91) meq/L	140	(133–144) meq/L
sulfate		(3.84–5.06)	3.38	(2.95–3.75)
glutathione	36.8	(36.6–37.0)		
lipids				
total lipids	652			(521–730)
total fatty acids	364	(283–442)		(294–341)
total cholesterol		(140–215)		(230–272)

Table 1. (continued)

Other components	Whole blood[d]			Plasma or serum[d]
nitrogen				
total N	3430	(3000–4100)		(1200–1430)
nonprotein N	33.8	(27–47)		(23–37)
urea N	11.0	(6–16)	12.5	(7–18)

[a] Mean value regardless of sex if only one value given (1–3). Body weight is in kg. Ranges are in parentheses.
[b] Volume (mL) packed red blood cells/100 mL whole blood.
[c] Venous blood of men and women 15–60 years old.
[d] Values are in mg/100 mL unless otherwise noted.

The Code of Federal Regulations (CFR) requires that both the container and the anticoagulants used for collection be sterile and pyrogen-free, and that the anticoagulant be sterilized in the container. Only certain anticoagulants are recognized for use with human blood or plasma. Whole blood may be collected into anticoagulant citrate dextrose (ACD), anticoagulant citrate phosphate dextrose (CPD) or anticoagulant heparin. Recently, it has been proposed that citrate phosphate dextrose adenine (CPD-A) solution be added to the list of approved anticoagulants for blood collection (10). At present, ACD and CPD, which allow whole blood to be processed into plasma for further products, are the most commonly used anticoagulants for whole blood collection in the United States. CPD-A has been used in Europe. When only plasma is to be collected, approved anticoagulants are ACD, CPD, or anticoagulant sodium citrate, but not anticoagulant heparin or CPD-A. Anticoagulant sodium citrate is the anticoagulant most commonly used for plasmapheresis. In the United States, blood is collected almost exclusively in plastic containers, which are easy to manipulate and can be prepared and sterilized with several satellite containers attached.

Methods of Blood Fractionation

Current procedures for large (at least one unit) scale fractionation of human blood into plasma and cellular components are based upon one of two principles, sedimentation or adhesion. Sedimentation techniques take advantage of differences in density between plasma and suspended cells and of differences in size and density among the cell types. Methods utilizing centrifugation to enhance sedimentation are the most common. Such methods can separate whole blood into 4 fractions—plasma, platelet concentrate, white cell (leukocyte) concentrate, and leukocyte-poor packed red cells. Adhesion techniques take advantage of variations in the ability of different cell types to bind to solid supports. Methods based on these techniques have been developed for removing leukocytes from packed red cells (11) and for selectively isolating granulocytes from whole blood (12–13). The following components must be isolated within 4 h of collection of the source blood: platelets, leukocytes, and plasma which is to be used for the preparation of labile coagulation factors.

Table 2. Blood Components: Use and Storage[a]

Component	Contents	Indications for use	Shelf life	Storage temperature
red cells	red cells, some plasma, some WBC, and platelets or their degradation products	increase patient red cell mass	21 d closed system	1–6°C
leukocyte-poor red cells	red cells, some plasma, and few WBC	prevent febrile reaction from leukoagglutinins	21 d closed system; 24 h open system	1–6°C; 1–6°C
frozen red cells	red cells, no plasma, minimal WBC and platelets	increase red cell mass, prevent tissue antigen sensitization, prevent febrile or anaphylactic IgA[b] reaction, provide rare bloods	3 yr frozen; 24 h thawed	frozen, −65°C or −196°C; thawed, 1–6°C
leukocyte concentrate	WBC, few platelets, and some RBC	agranulocytosis	24 h	1–6°C
platelet concentrate	platelets, few WBC, and some plasma	bleeding due to thrombocytopenia	72 h	20–24°C or 1–6°C
single donor plasma	plasma with no labile coagulation factors	blood volume expansion	5 yr	frozen, −18°C
fresh frozen plasma	plasma with all coagulation factors and no platelets	treatment of coagulation disorders	1 yr frozen; 2 h thawed	frozen, −18°C; thawed (not above 37°C)
cryoprecipitate	coagulation factors I and VIII	hemophilia A and von Willebrand's disease, fibrinogen deficiency	1 yr	frozen, −18°C
antihemophilic factor	factor VIII	hemophilia A	up to 2 yr	lyophilized, 1–6°C
prothrombin complex	factors II, VII, IX, and X	hemophilia B (Christmas disease)	1 yr	lyophilized, 1–6°C
albumin	albumin	blood volume expansion, replacement of protein	3 yr	room temperature (not above 37°C)
plasma protein fraction	albumin, α- and β-globulin	blood volume expansion	3 yr	room temperature (not above 30°C)
fibrinogen	fibrinogen	hypofibrinogenemia	5 yr	lyophilized, 1–6°C
immune serum globulin	γ-globulin	disease prophylaxis or attenuation, agammaglobulinemia	3 yr	1–6°C
Rh$_0$(D) immune globulin	γ-globulin from sensitized donors	prevention of Rh$_0$(D) sensitization	6 mo	1–6°C

[a] Ref. 4.
[b] IgA is one of the γ-globulins.

Isolation of Platelets and Plasma. Platelets are the smallest and least dense of the cell types of blood (Table 3) and are the last cell type to sediment. The following discussion describes the techniques involved in recovering both platelets and plasma from a single unit of whole blood, and presents methods for adapting these techniques to selective isolation on a single unit or larger scale.

A unit of blood which is to be fractionated into plasma, platelets, and packed cells is collected into a plastic container to which are attached, in a y-formation, two satellite bags. The procedure involves two centrifugation steps: the first yields platelet rich plasma, the second separates platelet-rich plasma (PRP) into platelet concentrate and platelet-poor plasma (PPP). The centrifugation conditions for maximum platelet recovery are determined experimentally by exposing a series of whole blood units to centrifugal fields of different force for varying lengths of time at 20–24°C, and by then determining platelet recovery in the supernatant, which is PRP. Such a series of centrifugation experiments gives a pattern of recovery similar to that in Figure 1 (16). The recovery curve exhibits a peak which represents the appropriate balance between the conditions necessary to clear red and white blood cells from the plasma (increasing recovery towards the peak) and to sediment platelets into the packed red cell mass (decreasing recovery after the peak).

Following centrifugation, the PRP is removed. In the closed bag system this is done by breaking the seal between the main bag and one of the two satellite bags and by exerting gentle, uniform upward pressure so that the packed cell volume is not disturbed during expression of the PRP. A commercially available, spring-loaded V-shaped expressor (Fenwal Laboratories) can be used for reproducible recovery of the PRP. After expression, the tubing leading from the main bag is clamped, and the paired satellite bags, one of which contains the PRP, are detached. This can be done sterilely by heat-sealing the tubing. Commercial heat-sealers are available. The recovery of platelets and plasma that can be obtained in the PRP from one unit of whole human blood, under a variety of optimal centrifugation conditions, has been recently established by Kahn and co-workers (17).

Platelet concentrate (PC) and platelet-poor plasma (PPP) are then prepared from PRP by centrifugation. The optimal conditions for this procedure are determined experimentally by varying the time and speed of centrifugation of the paired satellite bags, and by measuring the extent of removal of platelets from the supernatant plasma. The platelets form a packed pellet, and platelet removal approaches 100%. By plotting the data obtained from the centrifugation experiments, one can obtain the minimum time of centrifugation necessary for maximum removal of platelets at a given speed. One can then determine the most efficient procedure by comparing the platelet yield

Table 3. Specific Gravity and Diameter of Blood Elements [a]

Component	Specific gravity	Diameter, μm
erythrocytes	1.093–1.096	7.2–7.9
granulocytes	1.087–1.092	10–14
monocytes	ca 1.075–1.080	15–22
lymphocytes	1.070	ca 10 (up to 20)
platelets	1.040	1–2
plasma	1.025–1.029	

[a] Values for specific gravity from ref. 14, and values for diameter from ref. 15.

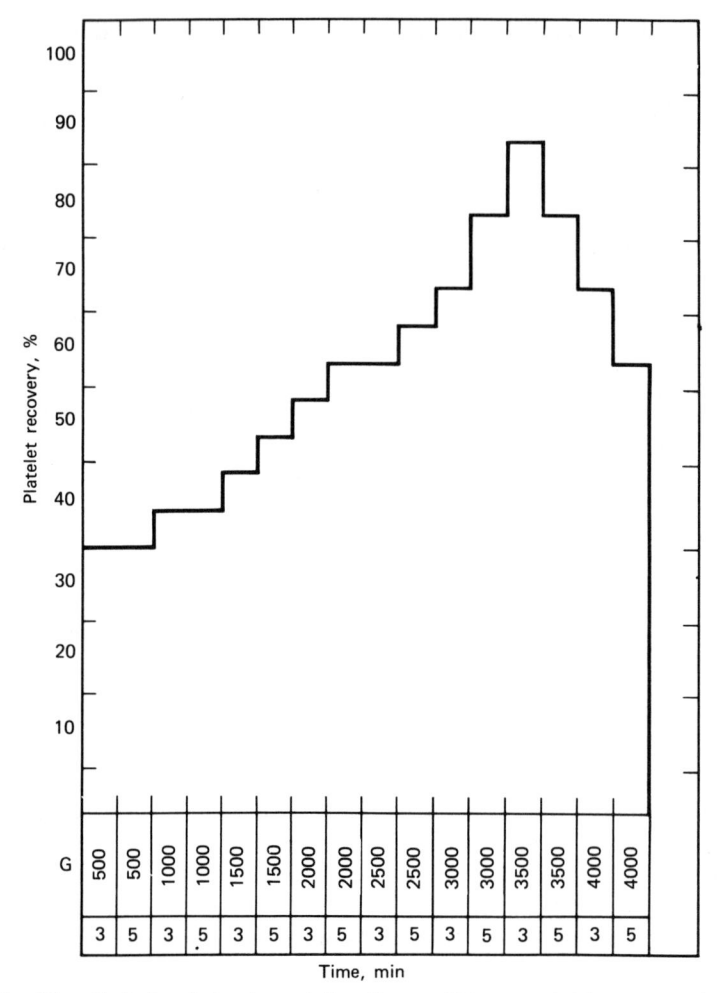

Figure 1. The effect of variation in centrifugation conditions on platelet recovery in platelet-rich plasma (16).

to the total time of centrifugation. The results of such a study, reported by Kahn and co-workers (17), show platelet yields are about 95%. Following centrifugation, the PPP is physically separated from the platelet pellet. This is done by breaking the seal between the attached satellite bags, and by expressing the PPP into the empty bag, taking care not to disturb the platelet pellet. Depending upon the conditions of platelet concentrate (PC) storage to be used, either 20–30 mL or 30–50 mL of PPP is left with the platelets. The bags are then sealed and detached. Federal regulations require that at least 4 units of PC be tested every month (18). At least 75% of the units must have total platelet counts above 5.5×10^{10} per unit and a pH above 6.0 after 72 h.

The above procedure can be modified to produce only plasma from a single unit of whole blood (plasmapheresis) by performing a single centrifugation under conditions similar to those used for the preparation of PC from PRP. The sedimented cells are reinfused to the donor. Two units of whole blood may be taken from a donor for processing into plasma during a 48-h period as long as only one unit of red cells is outside the donor at any given time. Following reinfusion, therefore, another unit of whole blood may be taken and similarly fractionated.

Plateletpheresis, ie, the processing of whole blood for the preparation of platelets only, is performed rather infrequently, unlike plasmapheresis which is performed routinely at a number of centers in order to obtain plasma as a source material for fractionation. Plateletpheresis is used to collect large amounts (equivalent to several units) of platelets from a single donor for transfusion to a thrombocytopenic patient who has become refractory to the platelets in concentrates obtained from random blood donations. As this refractoriness is frequently due to antibodies formed by the patient to allogeneic platelets or human leukocyte antigens (HLA), platelets obtained from an immunologically compatible donor are still effective (19). Plateletpheresis can be done either by repetitive performance of the basic technique described above or through the use of commercially available machines. The machines most commonly used are the Haemonetics Models 10 and 30 (Haemonetics Corp), which operate on a principle that has been termed discontinuous flow centrifugation. In brief, ACD-anticoagulated blood is allowed to flow into a specially designed centrifuge bowl (Latham bowl) (21), which is spinning at a fixed speed throughout the filling process. The separation of components is, therefore, continuous. Six to eight units can be processed in 2 h. An overall recovery of approximately 65% with respect to the platelets available has been reported for such a Haemonetics system (21). The Latham bowl, as well as the tubing necessary for collection and reinfusion, are disposable, and are used only once. It has been suggested (22) that an National Cancer Institute (NCI)-IBM type continuous flow centrifuge (23), exemplified by the Aminco Celltrifuge (American Instrument Co.), can also be used for plateletpheresis. The continuous flow centrifuge was originally developed for white cell isolation (see Centrifugal separation).

Isolation of Leukocyte Concentrates and Leukocyte-Poor Packed Red Blood Cells. Visual inspection of the packed cell mass remaining after centrifugation of a unit of whole human blood and removal of the PRP, as described above, will reveal the presence of a white buffy coat consisting of white blood cells (WBC, leukocytes) above the packed red blood cells (RBC). Red blood cells tend to sediment faster than leukocytes, owing to their higher density (Table 3) and ability to form rouleaux (24). However, only a partial separation of red blood cells from white blood cells or of white cell types from each other can be obtained by centrifugation. The following discussion describes the results obtained from a standard, single unit, closed-bag system, and presents the methods available for further purification and separation of the red and white cells.

Single units of whole blood which are processed by centrifugation into leukocyte concentrates and leukocyte-poor red cells must be collected into plastic bags having at least two satellite bags, as the supernatant platelet-rich plasma (PRP) will be expressed into one of the satellite bags, and the leukocyte concentrate into the other. However, the large demand for platelets makes it preferable to use plastic collection bags having three satellite bags (two of which are attached in a y-formation), so that the PRP can then be further fractionated into platelet concentrate (PC) and platelet-poor plasma (PPP), as described in the previous section. In order to optimize platelet and plasma recovery from the supernatant phase and minimize platelet contamination of the packed cell mass, the centrifugation of the whole blood is carried out under the conditions normally used to prepare PRP. Following the centrifugation of the unit, the PRP is expressed into a satellite bag and the tubing leading to that bag is clamped. The leukocyte concentrate is then obtained by opening the connection between the main bag and the remaining satellite bag and by expressing, with gentle upward pressure, the top 35–50 mL volume of the packed cell mass. This includes the buffy coat volume as well as the neighboring red blood cells. The residual packed red cells represent the leukocyte-poor fraction.

The dependence of the composition of leukocyte concentrates on the volume expressed has been examined (25). A picture of the total white blood cells (WBC) and of the granulocytes (polymorphonuclear leukocytes or PMNs), lymphocytes, and monocytes which are obtained as the volume of a leukocyte concentrate is increased is shown in Figure 2. A comparison of the recovery of total WBC and of each WBC type in the top 50 mL and in the residual packed cell volume is presented in Table 4. Over 95% of the mononuclear cell types (lymphocytes and monocytes) are obtained in the top 50 mL (ca 16% of the packed cell volume), whereas only 33% of the PMNs are obtained. There is significant contamination of the leukocyte concentrates with RBC. As the volume is increased from 10 to 50 mL, the hematocrit increases linearly from 18 to 48%. This loss of red cells represents a real limit to the volume that can be taken from the packed cells, since a blood bank distributes the packed red cells for use in treatment of anemia.

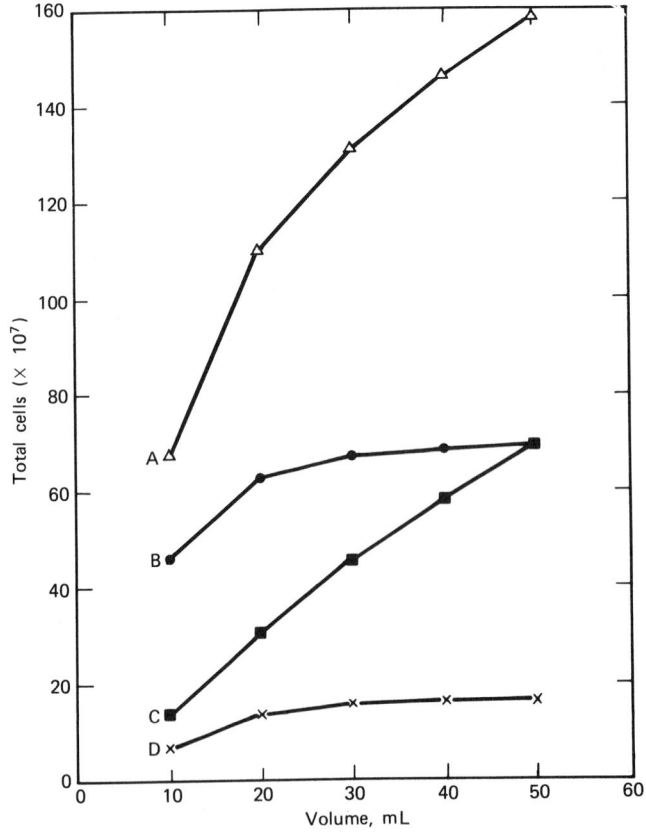

Figure 2. Contents of leukocyte concentrates of differing volumes (25). A, total white cells; B, lymphocytes; C, granulocytes; D, monocytes.

Table 4. Distribution of White Cells in Packed Cells[a]

Cell type	Top 50 mL, %	Residual (av = 267 mL), %
total white cells	51.6	48.4
granulocytes	33.1	66.9
lymphocytes	95.4	4.6
monocytes	97.1	2.9

[a] Ref. 25.

The efficient recovery of mononuclear leukocytes in the single unit leukocyte concentrates makes the concentrates excellent starting material for the preparation of interferon (26) or transfer factor (27) (products of possible clinical significance but as yet not purified nor tested on a large scale). However, the rather low recovery of granulocytes in such concentrates makes them a poor source of granulocytes for transfusion into neutropenic patients, which is their most common use (see Table 3). In addition, effective granulocyte therapy requires large numbers of granulocytes per transfusion (20–24 units of leukocyte concentrates per day for an adult) which increases the risk of alloimmunization of the recipient when random donor sources are used. The development, since the early sixties, of machines that allow isolation of granulocytes from a single donor in numbers equivalent to the granulocyte content of 12 or more units of whole blood, has resulted in the availability of more satisfactory preparations of granulocytes for transfusion, and random donor single unit leukocyte concentrates are now used only when no such single donor source is available. The machines used for leukapheresis, ie, for the processing of blood for the isolation of leukocytes, are based on the principles of continuous flow centrifugation, discontinuous flow centrifugation, or reversible leukoadherence.

The IBM cell separator and the Celltrifuge, both based upon the original NCI-IBM continuous flow centrifuge (23), allow for continuous separation of cellular components. In brief, under standard protocol, blood from a donor who has been treated with heparin anticoagulant is allowed to flow into a specially constructed bowl. The bowl is spinning throughout the flow, and as the blood enters along the axis of the bowl and is forced out and around an inner core, the blood components separate (28–29). ACD-anticoagulant is also added to the blood as it enters the bowl. Automatic monitoring systems are available to locate the plasma-packed cell interface (30). Plasma, white blood cells and red blood cells are continuously removed through separate ports. The red blood cells and plasma are reinfused to the donor; only the white blood cells are removed. If no other additions are made to the blood during centrifugation, the white blood cells are essentially those in the buffy coat (mainly lymphocytes); the bulk of the granulocytes sediment into the red cell mass. Increasing the efficiency of granulocyte recovery by removing the upper layer of the packed red cell mass would expose the donor to the risk of anemia. By adding red cell rouleaux-enhancing agents such as hydroxyethyl starch (31–32), dextran (33), or modified gelatin (34–35) to the whole blood as it enters the bowl, the sedimentation rate of the red blood cells is increased, and the granulocytes can be readily separated. Pretreatment of the donor with steroids to raise the donor's circulating granulocyte count was also found an effective means of increasing granulocyte collection by continuous flow centrifugation (31–32). Granulocyte yields similar to those obtained in the continuous flow systems can be obtained in the discontinuous flow Haemonetics system when the donors are pretreated with steroids and when red cell rouleaux-enhancement agents are added to the whole blood (36). Both centrifugation systems yield leukocyte concentrates which contain, in addition to granulocytes, significant amounts of lymphocytes.

An alternative system based upon the selective and reversible adherence of granulocytes to nylon fibers or other solid supports (11,37–38) has been developed by Djerassi and co-workers (12–13). This system isolates granulocytes in high yield and is considerably cheaper than either of the centrifugation systems (39).

In brief, the donor's blood, anticoagulated with heparin, is allowed to flow through a nylon fiber filter. Granulocytes and monocytes bind to the fibers, while plasma and the other blood cell types pass through the filter and are returned to the donor. The flow is continuous, and the binding of the granulocytes is highly efficient ($\geq 70\%$) (40). The binding of granulocytes to nylon fibers is dependent upon divalent cations (41). After washing the filters with saline solution to remove any trapped red cells, the granulocytes are recovered by passing ACD-containing plasma over the filters. The ACD chelates divalent cations and release the bound cells. Complete elution of the granulocytes requires vigorous tapping of the filters, and early studies on granulocytes obtained with this technique found defects in both *in vivo* and *in vitro* granulocyte functions (41–42). Harris and co-workers (43) have recently reported a modified system which allows collection of granulocytes under relatively mild elution conditions, and which appears to yield granulocytes that retain complete *in vitro* function. The ability of the recovered granulocytes to function *in vivo* after transfusion has yet to be determined.

Overall, the future of this promising procedure has been compromised by findings that components of the complement system are activated during the process of filtration leukapheresis (44–45).

In addition to attempts by many laboratories to improve the leukocyte yields obtained during leukapheresis procedures, the main emphasis of present leukocyte fractionation research is the development of techniques for the selective isolation of specific leukocyte cell types. The most popular method for the preliminary fractionation of leukocyte cell types is that of Bøyum (46), which uses Ficoll-Hypaque discontinuous density gradients to separate mononuclear cell types from red cells and granulocytes. Further separation of the monocytes and lymphocytes can be obtained by allowing monocytes to adhere to solid supports under conditions where lymphocytes remain in suspension (47–48). Another common technique is to allow monocytes to phagocytize iron particles, which increases the monocyte density sufficiently for their separation from lymphocytes by a repetition of the Bøyum procedure (46). Newer methods, based on continuous density gradients, not only separate the monocytes and the lymphocytes but also permit recovery of both cell types in suspension (49–51), reducing the risk of alteration of the monocyte surface. A system for counterflow centrifugation introduced by Beckman Instruments has been reported to yield relatively pure populations of monocytes and lymphocytes (52–53) and should be adaptable to large-scale preparations. Procedures for fractionation of the lymphocytes into subclasses (B-cell, T-cell, etc) take advantage of differences in the cell surface antigens and receptors among the various subclasses (54–55), and have been reviewed in detail (56). It can be expected that preparation and utilization of specific lymphocyte subclasses will expand with increasing understanding of the immune system and its disorders (57). The buffy coat from a unit of centrifuged whole blood serves as an excellent starting material for these procedures.

Buffy-coat-poor packed red cells still have approximately 50% of the original leukocyte content of the blood unit (Table 4). The residual leukocytes are over 90% granulocytes and can be removed by passing the packed cells over nylon fiber columns (58). As the adherence of granulocytes to nylon fibers is dependent on the presence of calcium ion (59), anticoagulant heparin must be present. Halterman and co-workers (60) have shown that blood can be collected in ACD and packed red cells can be prepared and stored for up to 10 d before heparin and calcium are added for the nylon filtration procedure. As this technique cannot be performed in a closed system, the sterility of the red cell fraction must be considered compromised, and the cells should be used within 24 h. An alternative procedure for decreasing the leukocyte content of packed red cells is inverted centrifugation, which can be performed in a closed system. The method involves the resuspension by inversion of the packed cell volume following removal of the platelet-rich plasma, and recentrifugation of the unit at relatively high centrifugal forces with the port-side down (61). During this inverted centrifugation, the leukocytes are concentrated to the top, leaving a volume of red cells almost free of white cells directly above the ports. While the inverted position is maintained, red cells are then drained or expressed into the attached satellite bag. The disadvantage of this procedure is that it entails a loss of 20–30% of the red cells, which are left with the leukocytes in the original collection bag. Greater than 80% removal of leukocytes has been reported for this procedure (61–62). As discussed concerning leukapheresis procedures, certain macromolecules, such as hydroxyethyl starch, dextran, and modified fluid gelatin, have the ability to enhance the erythrocyte sedimentation rate, and can be used for the separation of leukocytes and red cells. Several laboratories have described methods (63–65) using erythrocyte sedimentation rate enhancers which result in a packed red cell volume containing less than 10% of the starting white cells in the unit, at a loss of less than 10% of the starting red cells. These methods, as described, require compromising the sterility of the system, and the product must be used within 24 h of preparation.

A similar time limit exists for red cell concentrates washed by either manual or automated techniques, as washing cannot be performed in a closed system. The saline washing procedures, a series of repetitive suspensions and recentrifugations of the red cells, can be used to reduce the leukocyte content of the packed red cell products if one removes the buffy coat region after each centrifugation step (66–67). The reported extent of leukocyte removal from fresh red cell concentrates is not greater than that obtained by the simpler procedures described above. However, when red cells are frozen and thawed before washing, the red cell products are almost leukocyte-free (68–69). It has been suggested that the improved removal of leukocytes from previously frozen red cells is due to damage and destruction of the leukocytes during the glycerolization and the freeze-thaw processes (70–71). Even these red cell preparations, which have over 98% of the leukocytes removed, have been found to have viable white cells or white cell debris in amounts perhaps sufficient to cause sensitization to human leukocyte antigens (HLA) (72–73). Overall, although none of the presently available methods will yield red cell preparations free of all other blood constituents, washed, previously frozen red cells come very close to this ideal, being almost leukocyte-free and having a plasma protein content less than 0.2% of the original concentrate (69). However prepared, leukocyte-poor red cells can be considered a derivative of packed red cells and, as such, are subject to the same regulations as packed red blood cells.

Packed Red Blood Cells. Packed red cells are defined as red cells remaining after separating plasma from human blood (74). This component is used for all routine transfusions requiring replacement of red cell mass, and has effectively replaced the use of whole blood for this purpose. Packed red cells can be prepared either by centrifugation or by sedimentation anytime within the storage period for the parent whole blood (21 d for blood collected in ACD or CPD, 35 d for blood collected in CPD-A). Any centrifugation conditions which clearly separate plasma and red cells may be used (74), as long as sufficient plasma is left with the packed cells to ensure optimal cell preservation. Packed red cells even with hematocrits of 90% or greater are still quite viable, more than satisfying the accepted criterion that greater than 70% of the transfused cells must still circulate 24 h after transfusion (75–77).

Storage of Blood Components

The conditions and maximum length of storage for the major blood components and derivatives are presented in Table 2 and, except in the case of leukocyte concentrates, are mandated by the CFR (5). Whole blood collected into ACD or CPD has a storage life of 21 d at 1–6°C, and whole blood collected into anticoagulant heparin has a storage life of 48 h at 1–6°C. When CPD-A can be used as an approved anticoagulant, whole blood collected in CPD-A will have a storage life of 35 d at 1–6°C. Whole blood should be refrigerated within 4 h of collection, in order to preserve red cell function. Plasma from outdated, ACD- or CPD-anticoagulated blood can be used for the preparation of all plasma protein fractions except the labile coagulation factors, eg, factor VIII. Packed red cells may be prepared by sedimentation or by centrifugation of ACD-, CPD-, or CPD-A-anticoagulated whole blood at any time during the storage period of the whole blood.

The 21-d storage life of red cells can be markedly extended by freezing the cells after adding a cryoprotective agent. There are three accepted techniques, all of which use glycerol as the protective agent (78–80). In the Huggins (78) and Meryman (79) procedures, the storage temperature is $-85°C$, and in the Rowe (80) procedure, it is -160 to $-196°C$ (in liquid-N_2). The red cells should be frozen within 5–6 d of source blood collection. Storage life is 3 yr in the frozen state. Once thawed and washed to remove cryoprotective agent, the red cells have a storage life of 24 h at 1–6°C. Cryopreservation of other blood cell types is still in the research stage (81). Leukocyte concentrates should be prepared within 4 h of source blood collection. Recent evidence indicates that granulocyte function is maintained for at least 24 h of storage at 1–6°C (82–83), and that leukocyte concentrates can be used as a source of granulocytes for transfusion during that period. Platelet concentrates (PC) should also be prepared within 4 h of source blood collection. Whole blood collected in anticoagulant heparin or CPD-A may not be processed into platelet concentrates. There are two accepted sets of storage conditions: platelets in 30–50 mL plasma, stored at 20–24°C with continuous agitation, or platelets in 20–30 mL plasma, stored at 1–6°C with agitation optional (84). Platelet concentrates have a storage life of 72 h under either set of conditions. Single donor plasma for blood volume expansion can be prepared within 5 d of source collection. When frozen at $\leq -18°C$, the storage life of this component is 5 yr. Fresh-frozen plasma (FFP), which is to be used as a source of the labile coagulation factors, factors V and VIII, must be prepared within 4 h of the collection of source blood and must be frozen within 2 h of separation of the source plasma. In the frozen state (at $\leq -18°C$), the storage life of this component is 1 yr. Once thawed, it must be used or further processed within 2 h. Cryoprecipitate, which is derived from FFP, should either be used or refrozen immediately. In the frozen state (at $\leq -18°C$), its storage life is 1 yr. Once thawed, cryoprecipitate should be used or processed further within 6 h. More purified plasma protein fractions have extended storage lives, up to 3 yr in many cases (see Table 2).

Plasma Fractionation

Two methods have been used to classify the plasma proteins. The first method is based upon solubility in salt solutions. On this basis there are three categories: (*1*) euglobulins, those proteins insoluble in water at their isoelectric point and precipitated by 33% saturated (1.4*M*) ammonium sulfate; (*2*) pseudoglobulins, those proteins soluble in water at their isoelectric point and precipitated at ammonium sulfate concentrations between 33% and 48% saturation (2*M*); and (*3*) albumins, those proteins soluble in water at their isoelectric point and requiring ammonium sulfate concentrations greater than 50% saturation (2.06*M*) for precipitation. The second classification method is based upon electrophoretic mobility. The Schlieren pattern obtained from free boundary electrophoresis of normal human plasma has 6 peaks. The pattern obtained with normal human serum has 5 peaks. Each peak represents a family of proteins having similar electrophoretic mobility. By precipitating all proteins insoluble in 50% saturated ammonium sulfate, the fastest moving peak was identified as albumin. The peak present in plasma but absent in serum was identified as fibrinogen. The remaining peaks were identified as globulins and named α_1-, α_2-, β-, and γ-globulins in order of decreasing mobility. Each of these classifications represents a heterogeneous mixture of proteins grouped on the basis of solubility or electrophoretic mobility.

The plasma proteins serve a wide variety of functions in the human organism. Their roles in the maintenance of blood volume and other physical characteristics of blood, such as viscosity, are extremely important because blood, in order to perform any of its numerous functions, must be a rapidly circulating medium. If the volume of plasma falls, the pumping action of the heart is strained and there is increased resistance to flow owing to the increased concentration of the red cells relative to the plasma. The blood volume depends on the balance between the hydrostatic pressure of the blood in the capillaries, which tends to expel liquid from the blood into the tissues, and the osmotic pressure (owing to the plasma proteins), which tends to draw liquid back into the blood. The major contribution to the osmotic pressure of plasma is from albumin because of its concentration and properties. Albumin comprises more than 50% of the plasma proteins by weight. It has a relatively low molecular weight and a high net negative charge at physiological pH (see Table 1). Albumin solutions have relatively low viscosity because of the spherical shape of the molecule. Many of the plasma proteins are enzymes and have a role in the various regulatory and metabolic pathways of the body. Plasma proteins act as carriers for many vital nonprotein molecules which are transported in the blood such as hormones, vitamins, cholesterol, bile pigments, phospholipids and fatty acids. There are specific metal-carrying proteins. A large group of the plasma proteins is concerned with the immune response. The immune serum globulins (γ-globulins) include antibodies directed against many disease causative agents. The isoagglutinins and the complement components are other proteins involved in immune reactions. The coagulation factors are a group of plasma proteins which participate in the control of blood loss after vascular damage. These proteins, many of which are proteolytic enzymes or their precursors, act in conjunction with platelets in the process of hemostasis. The biochemistry of the coagulation factors has recently been reviewed (85–87). Table 5 is a compilation of plasma protein data. There are currently more than 60 plasma proteins that have been isolated, highly purified, and well characterized. The 10 proteins present in the highest concentrations constitute 84% of the total protein content of plasma.

Plasma is fractionated to produce therapeutic materials containing one or more of the plasma proteins in concentrated and purified form in order to achieve optimal clinical usefulness. The major therapeutic fractions currently produced are (see Table 2): albumin, in several degrees of purity; immune serum globulin (ISG), both normal and specific; antihemophilic factor (AHF) (factor VIII); prothrombin complex (PTC) (factors II, VII, IX, and X); and fibrinogen (factor I). The major impetus for the development of fractionation methods was the need to provide large amounts of a blood volume expander for the treatment of battlefield injuries during World War II. A product was desired that would provide the required oncotic action, not require refrigeration, and be free from the transmission of disease. Human albumin was found to be the most acceptable therapeutic fraction. A fractionation method was developed during the 1940s by a group headed by Cohn at the Harvard Medical School (89). The procedures were scaled up at the Harvard pilot plant and made available to commercial laboratories under contract to the U.S. Navy to provide blood derivatives for the Armed Forces. The methods developed during this period, with some modifications, are still the most popular methods for the preparation of albumin and ISG.

Methods for the Production of Plasma Fractions

Cold Ethanol Methods for Albumin Production. The monumental paper of Cohn and co-workers (89) describes methods for the separation and purification of the protein and lipoprotein components of human plasma (Fig. 3). Method 6 for the preparation of albumin is described in detail. In each of these methods there was an initial separation of the protein components of plasma into a small number of fractions in which the major components are separated, and then into a large number of subfractions into which these components are further concentrated and purified. The methods involved lowering the solubility of proteins by reducing the dielectric constant of the solution by the addition of ethanol. Thus separations could be carried out in the range of low ionic strengths at which the interactions of proteins with electrolytes differ from each other markedly.

Cohn and co-workers (89) used certain solubility ranges as guidelines for the determination of fractionation conditions. The protein to be separated must have a high solubility when most other components of the system have low solubilities, or the converse. Solubilities of 0.01–0.1 g/L serve to separate a component as a precipitate. Solubilities of 10 g or more per liter suffice if other components of the system are to be precipitated. A five variable system (pH 4.4–7.4, ionic strength 0.001–0.16, ethanol concentration 0–40%, protein concentration 0.2–66 g/L, and temperature 0 to $-10°C$) was used for plasma fractionation.

Figure 3 diagrammatically represents the variations in pH and ethanol concentration to obtain each of the major fractions in methods 1, 2, 5, and 6. Figure 4 is a flow chart for the production of albumin by method 6. The plasma used in the development

Table 5. Molecular Parameters of Purified Human Plasma Proteins[a]

Protein	$s^0_{20, w}(S)$[b]	Mol wt	pI[c]	Electrophoretic mobility	Amount in normal serum (plasma), mg/100 mL
prealbumin (thyroxine-binding)	3.9	54,980	4.7	7.6	10–40
retinol-binding protein	2.3	21,000		α_1	3–6
albumin	4.6	66,000	4.7	5.92	3500–4500
α-globulins					
α_1-acid glycoprotein	3.5	40,000	2.7	5.7	55–140
α_1-antitrypsin	3.41	54,000	4.8	5.42	200–400
α_1-fetoglobulin	4.5	64,000		α_1	1 μg/100 mL
9.5S α_1-glycoprotein	9.5	308,000		α_1	3–8
Gc-globulin	3.7	51,000	5.0	α_2	40–70
ceruloplasmin	7.1	151,000	4.4	4.6	15–60
3.8S histidine-rich α_2-glycoprotein	3.8	58,500		α_2	5–15
α_2-macroglobulin	19.6	725,000	5.4	4.2	150–420
4S α_2-β_1-glycoprotein	4.1	59,600	4.0	3.9	trace
α_1B-glycoprotein	3.8	50,000		α_1	15–30
α_1T-glycoprotein	3.3	60,000		α_1	5–12
α_1-antichymotrypsin		68,000		α_1	30–60
Zn-α_2-glycoprotein	3.2	41,000	3.8	4.2	2–15
α_2HS-glycoprotein	3.3	49,000	4.2	4.2	40–85
8S α_3-glycoprotein		220,000		α_3	3–5
serum cholinesterase	12	348,000	3.0	3.1	0.5–1.5
thyroxine-binding globulin	3.9	58,000		α_1	1–2
inter-α-trypsin inhibitor	6.4	160,000		α_2	40–70
transcortin		55,700			7
haptoglobin					
type 1-1	4.4	100,000	4.1	4.5	100–220
type 2-1	4.3, 6.5	200,000	4.1		160–300
type 2-2	7.5	400,000			120–260

Table 5 (continued)

Protein	$s_{20,w}^0(S)^b$	Mol wt	pI^c	Electro-phoretic mobility	Amount in normal serum (plasma), mg/100 mL
β-globulins					
hemopexin	4.8	57,000		3.1	50–100
transferrin	5.3	76,500	5.5	3.1	200–320
pregnancy-specific	4.6	120,000		β_1	variable
β_1-glycoprotein					
β_2-microglobulin	1.6	11,818		β	trace
β_2-glycoprotein I	2.9	40,000		1.6	15–30
β_2-glycoprotein II (C3-proactivator)		63,000			12–30
β_2-glycoprotein III		35,000			5–15
C-reactive protein		118,000		β	<1
Low molecular weight proteins					
lysozyme		14,000			0.5–1.5
basic protein B1	1.4	11,000		+	<1
basic protein B2	1.3	8,800	10.1	+1.3	<1
0.6S γ_2-globulin	0.6	5,100			
2S γ_2-globulin	1.9	14,000	5.9	1.5	0.1
3S γ_1-globulin (carbonic anhydrase B)	2.9	25,000	5.6	2.2	variable
complement components					
C1q-component	11.1	400,000		γ_2	10–20
C1r-component	7.5	180,000		β	
C1s-component (C1 esterase)	4.5	86,000		α_2	1–2
C2-component	4.5	117,000		β_1	2–3
C3-component	9.5	180,000		β_2	160
C4-component	10.0	206,000		β_1	65
C5-component	8.7	180,000		β_1	8
C6-component	5.5	95,000		β_2	7

C7-component	6.0		110,000	β_2	6
C8-component	8.0		163,000	γ_1	8
C9-component	4.5		79,000	α	23
other complement factors					
C1 esterase inhibitor	3.7	2.1	104,000	α_2	15–35
properdin	5.2		184,000	γ_2	2–3
coagulation proteins					
antithrombin III	4.1		65,000	α_2	17–30
prothrombin			72,000	α_2	(5–10)
antihemophilic factor (factor VIII)	24		1,100,000		
plasminogen	4.2	5.6	87,000	3.7	15–35
fibrin-stabilizing factor (factor XIII)			320,000		(1–4)
fibrinogen	7.9	5.5	340,000	2.1	200–450
immunoglobulins					
immunoglobulin G	6.6	5.8–7.3	160,000	1.1	800–1800
immunoglobulin A	7, 10, 13, 15		160,000 and polymers		90–450
immunoglobulin M	19, 26		950,000 and polymers	2.1	60–250
immunoglobulin D	7.86		170,000		<15
immunoglobulin E			190,000		<0.06
κ Bence Jones protein	3.4	4.6–6.7	23,000; 46,000	1.0–4.7	trace
λ Bence Jones protein	3.4	4.6–6.7	23,000; 46,000	1.0–4.7	trace

[a] Ref. 88.
[b] Sedimentation coefficient.
[c] Isoelectric point.

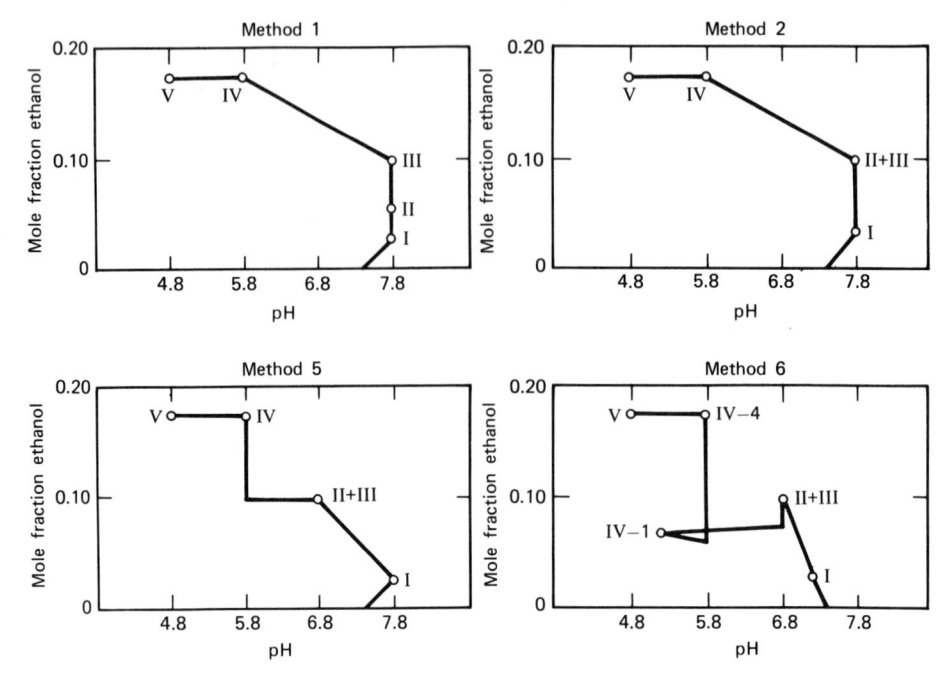

Figure 3. Ethanol concentration and pH for the separation of plasma fractions by the methods of Cohn and co-workers (89).

of these methods was obtained from blood collected into sodium citrate anticoagulant. Acetate and carbonate buffer systems were used to adjust pH and ionic strength. Precipitation was carried out at the lowest convenient ethanol concentration and temperature, and at the optimum pH and ionic strength for each separation. The protein components of some of the fractions obtained in the cold ethanol precipitation methods are shown in Table 6.

The recovery of albumin from plasma was not increased in method 6 as compared to method 5, but remained constant at approximately 80%. In the original procedures ethanol was always added as a 53.3% ethanol–water mixture. This was done to cope more easily with the heat of solution generated by the addition of ethanol and to minimize denaturation by high local concentrations of ethanol during addition. The 53.3% ethanol–water mixture brought the protein concentration of the mixture to the desired value without further addition of water.

Many manufacturers interested primarily in the large-scale preparation of albumin from plasma have adopted simplified versions of methods 5 and 6. In these methods the number of fractions is reduced and the total volume of the system is smaller (90). Most of the more recent procedures include the addition of precooled 95% ethanol to reduce the volume of some of the suspensions.

As used in Sweden, such a revised method gives a better yield of albumin from both normal and outdated plasma than method 6 (91). The fractionation scheme used in general in the United States is shown in Figure 5 (92). In this method, as in the method used at The New York Blood Center, fractions I and II + III are precipitated in one step.

Although the fractional precipitation methods described above were found to

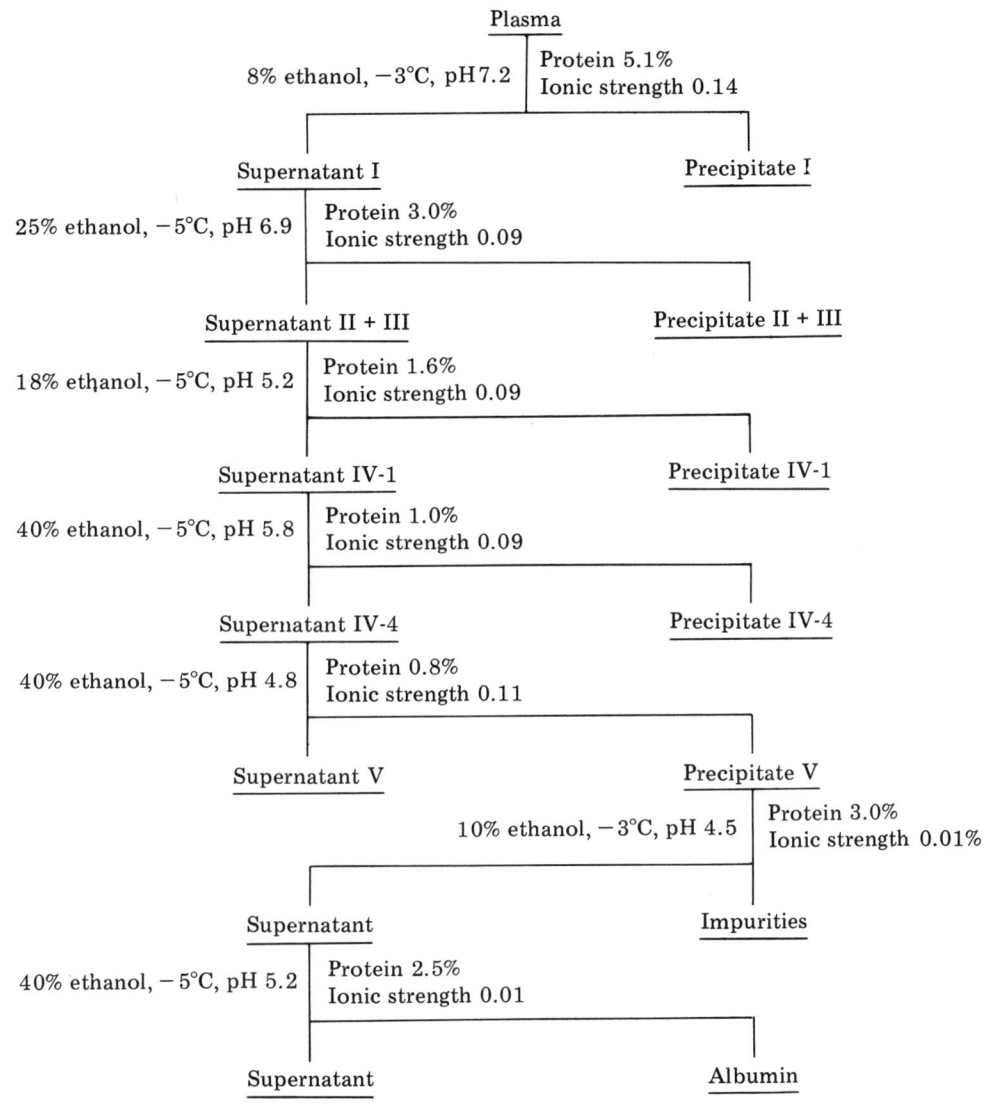

```
                                    Plasma
                                    ┬
                         8% ethanol, −3°C, pH 7.2 │ Protein 5.1%
                                                  │ Ionic strength 0.14
                         ┌──────────────────────────────────────┐
                  Supernatant I                           Precipitate I

25% ethanol, −5°C, pH 6.9 │ Protein 3.0%
                          │ Ionic strength 0.09
                ┌──────────────────────────────────────┐
         Supernatant II + III                   Precipitate II + III

18% ethanol, −5°C, pH 5.2 │ Protein 1.6%
                          │ Ionic strength 0.09
                ┌──────────────────────────────────────┐
           Supernatant IV-1                        Precipitate IV-1

40% ethanol, −5°C, pH 5.8 │ Protein 1.0%
                          │ Ionic strength 0.09
                ┌──────────────────────────────────────┐
           Supernatant IV-4                        Precipitate IV-4

40% ethanol, −5°C, pH 4.8 │ Protein 0.8%
                          │ Ionic strength 0.11
                ┌──────────────────────────────────────┐
            Supernatant V                           Precipitate V
                                         10% ethanol, −3°C, pH 4.5 │ Protein 3.0%
                                                                   │ Ionic strength 0.01%
                                    ┌──────────────────────────────┐
                              Supernatant                     Impurities

40% ethanol, −5°C, pH 5.2 │ Protein 2.5%
                          │ Ionic strength 0.01
                ┌──────────────────────────────────────┐
            Supernatant                               Albumin
```

Figure 4. Method 6 of Cohn and co-workers (89).

be adequate for the purpose of producing large amounts of therapeutic concentrates of plasma proteins, Cohn and co-workers (93) attempted to devise a method for the production of plasma proteins as little altered from their native state as possible. The new procedure, called method 10, took advantage of the increased stability of proteins in the solid state. All of the proteins are rapidly precipitated by a combination of the effects of ethanol and zinc ion. Separations from the solid state are made by fractional extraction. Specific metal–protein interactions favor the separation of undenatured proteins by reducing the extremes of pH and ethanol concentration. The lowest pH used is 5.5 and the highest ethanol concentration is 19%. Method 10 in its entirety has never been put into large-scale use. It has been used as an analytical method for small amounts of plasma (90,94).

Table 6. Major Components of Cold Ethanol Fractions

Fraction	Proteins
I	fibrinogen; cold insoluble globulin; factor VIII; properdin
II and III	IgG; IgM; IgA; fibrinogen; β-lipoprotein; prothrombin; plasminogen; plasmin inhibitor; factor V; factor VII; factor IX; factor X; thrombin; antithrombin; isoagglutinins; ceruloplasmin; complement $C'1$, $C'3$
IV-1	α_1-lipoprotein; ceruloplasmin; plasmin inhibitor; factor IX; peptidase; α- and β-globulins
IV-4	transferrin; thyroxine binding globulin; serum esterase; α_1-lipoprotein; albumin; alkaline phosphatase
V	albumin; α-globulin
VI	α_1-acid glycoprotein; albumin

Some of the conditions of method 10 have been used by the Swiss Red Cross for a large-scale production method (95–96), which has several important advantages over method 6 (89). The use of 95% ethanol reduces the amount of ethanol required and reduces the volume of solutions to be processed. The time required for the Swiss method is approximately one third that for method 6 and the recovery of albumin is approximately 90%. All of these factors make the Swiss method economically advantageous.

For many therapeutic indications a preparation called plasma protein fraction (PPF) is used interchangeably with albumin. PPF is albumin in a slightly less pure form than the albumin produced by the methods described above. The contaminants are α- and β-globulins and salts. The first large-scale method for the production of PPF was that of Hink and co-workers (97) which is shown schematically in Figure 6. In the fractionation scheme outlined in Figure 5, PPF can be produced by eliminating precipitation IV_4 and precipitating fractions IV_4 and V in a single step. If this is done, a filtration of supernatant phase IV_1 is required. PPF is more economical to produce than albumin and can be recovered in higher yield. Mealey and co-workers (98) have reported on a modification of method 6 which gives a heat-stable PPF in good yield. This method is based on increasing the ethanol concentration and decreasing the pH for the separation of fraction II + III. The optimal conditions were found to be: ethanol concentration 30–35%, pH 6.3–6.8, and −5°C. PPF in the form of fraction $IV_1 + IV_4 + V$ was precipitated by 40% ethanol at pH 4.85, and −5 to −6°C. Acetic acid (10N) and 95% ethanol were used to make adjustments in pH levels and ethanol concentrations, respectively.

Cold Ethanol Methods for Immune Serum Globulin Production. The procedure developed by the Harvard group, method 9 of Oncley and co-workers (99), has become the classic method for the production of immune serum globulin (ISG). Method 9 is diagrammatically represented in Figure 7. The starting material for the preparation of ISG by method 9 is fraction II + III of method 6. The goal of method 9 was to separate as many constituents of fraction II + III as possible into useful and stable concentrates. However, it is only the ISG of fraction II that has found large-scale clinical use.

A method was developed for the subfractionation of fraction II + III in which fraction II was obtained by extraction at high dilution as a single fraction rather than being precipitated as fraction II-3 and fraction II-1,2 as shown in Figure 7 (100). The

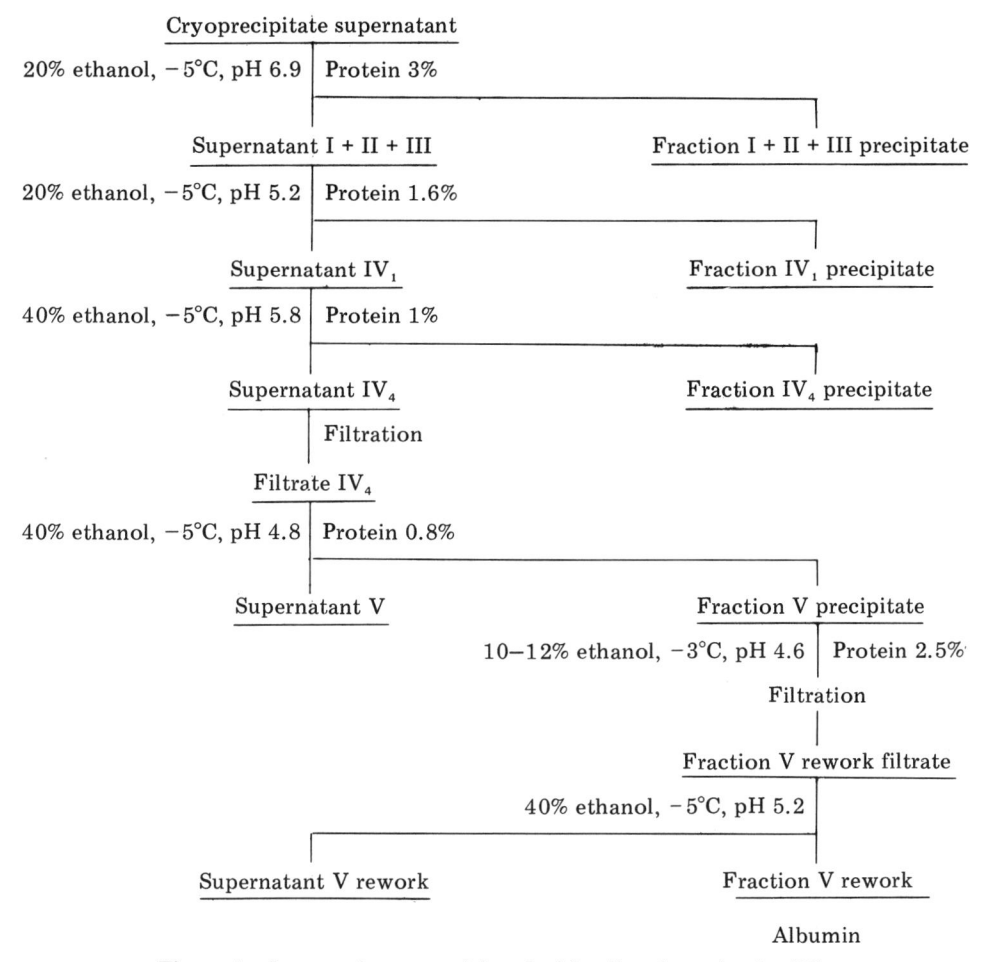

```
                    Cryoprecipitate supernatant
20% ethanol, −5°C, pH 6.9 │ Protein 3%
                          │
        ┌─────────────────┴─────────────────────────┐
Supernatant I + II + III                    Fraction I + II + III precipitate
20% ethanol, −5°C, pH 5.2 │ Protein 1.6%
                          │
        ┌─────────────────┴─────────────────────────┐
   Supernatant IV₁                           Fraction IV₁ precipitate
40% ethanol, −5°C, pH 5.8 │ Protein 1%
                          │
        ┌─────────────────┴─────────────────────────┐
   Supernatant IV₄                           Fraction IV₄ precipitate
                          │ Filtration
                   Filtrate IV₄
40% ethanol, −5°C, pH 4.8 │ Protein 0.8%
                          │
        ┌─────────────────┴─────────────────────────┐
   Supernatant V                             Fraction V precipitate
                                 10–12% ethanol, −3°C, pH 4.6 │ Protein 2.5%
                                                              │ Filtration
                                                Fraction V rework filtrate
                                 40% ethanol, −5°C, pH 5.2 │
                          ┌───────────────────────┴─────────┐
              Supernatant V rework                   Fraction V rework
                                                          Albumin
```

Figure 5. Large-scale commercial method for albumin production (92).

conditions for this extraction, which gives a higher yield of ISG than the method 9 of Figure 7, are 17% ethanol, pH 5.1, ionic strength 0.01, and −6°C. In large-scale use, method 9 has been modified to give a total fraction II by precipitating supernatant III at 25% ethanol, pH 7.4, ionic strength 0.05, −5°C, and a protein concentration 0.8%. This modification gives a recovery of ISG (ca 80%) comparable to that of the method shown in Figure 7.

The Swiss Red Cross (95–96) prepares ISG from precipitate A of their modified method 10 in a procedure which is a modification of the method of Figure 7. The recovery of ISG from plasma devoid of fraction I is approximately 80%. If precipitate A is prepared from whole plasma the recovery of ISG is nearly 90%.

Figure 8 is a diagrammatic representation of the fractionation scheme in general use in the United States for the production of ISG from fraction I + II + III (92). All ISG for therapeutic use is prepared from large pools of plasma from many donors so that the final product will contain a broad spectrum of antibodies.

46 BLOOD FRACTIONATION

```
                              Plasma
                              Protein 5.1 to 5.6%
  8% ethanol, −2 to −2.5°C, pH 7.2
                              Ionic strength 0.14

                    ┌───────────────────────────────┐
                Effluent I                      Fraction I
                    │
                    │         Protein 4.3%
  21% ethanol, −6°C, pH 6.8   Ionic strength 0.12

                    ┌───────────────────────────────┐
              Effluent II + III                Fraction II + III
                    │
                    │         Protein 3%
  19% ethanol, −6°C, pH 5.2   Ionic strength 0.11

                    ┌───────────────────────────────┐
                Effluent IV-1                   Fraction IV-1
                    │
                    │         Protein 2.5%
  30% ethanol, −7°C, pH 4.65  Ionic strength 0.094

                    ┌───────────────────────────────┐
                 Effluent                         PPF
                 Discarded
```

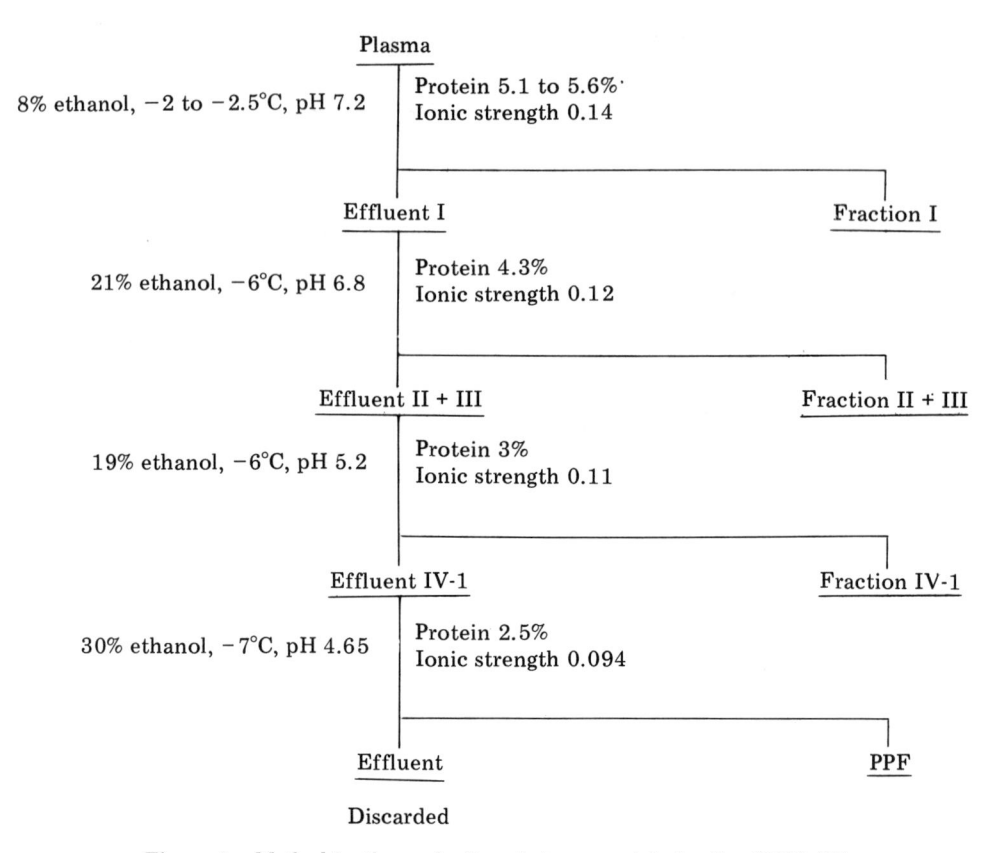

Figure 6. Method for the production of plasma protein fraction (PPF) (97).

Alternative Methods for the Production of Albumin and ISG. The search for alternatives to the cold ethanol plasma fractionation methods was spurred by the desire to decrease costs and to increase the recovery of undenatured therapeutic fractions. An economical method for the preparation of albumin, developed shortly after those discussed above, involved heat denaturation of the nonalbumin components of plasma (101–103). In this method, plasma or serum is heated to 70°C in the presence of caprylate ions, under which conditions the globulins and fibrinogen become denatured. The caprylate serves to stabilize the albumin against thermal denaturation. By manipulation of pH, all the denatured proteins are precipitated and removed, leaving albumin in solution. Heat denaturation as described above was not adopted as a large-scale plasma fractionation method because it results in the production of only one product: albumin. A modified method for the production of albumin by the heat denaturation of the nonalbumin components has been adopted. This method allows for the separation of the coagulation factors and ISG, if they are desired, whereas the isolation of albumin can begin at any step (104–105). The albumin produced is further concentrated by polyethylene glycol precipitation or ultrafiltration.

Several methods have been used for the preparation of heat stable plasma fractions rich in albumin, to be used as plasma volume expanders. Method 12 of the Harvard group (106) utilized zinc complexes for fractionation. A fraction obtained by desalting plasma with ion-exchange resins and thus precipitating euglobulins, has

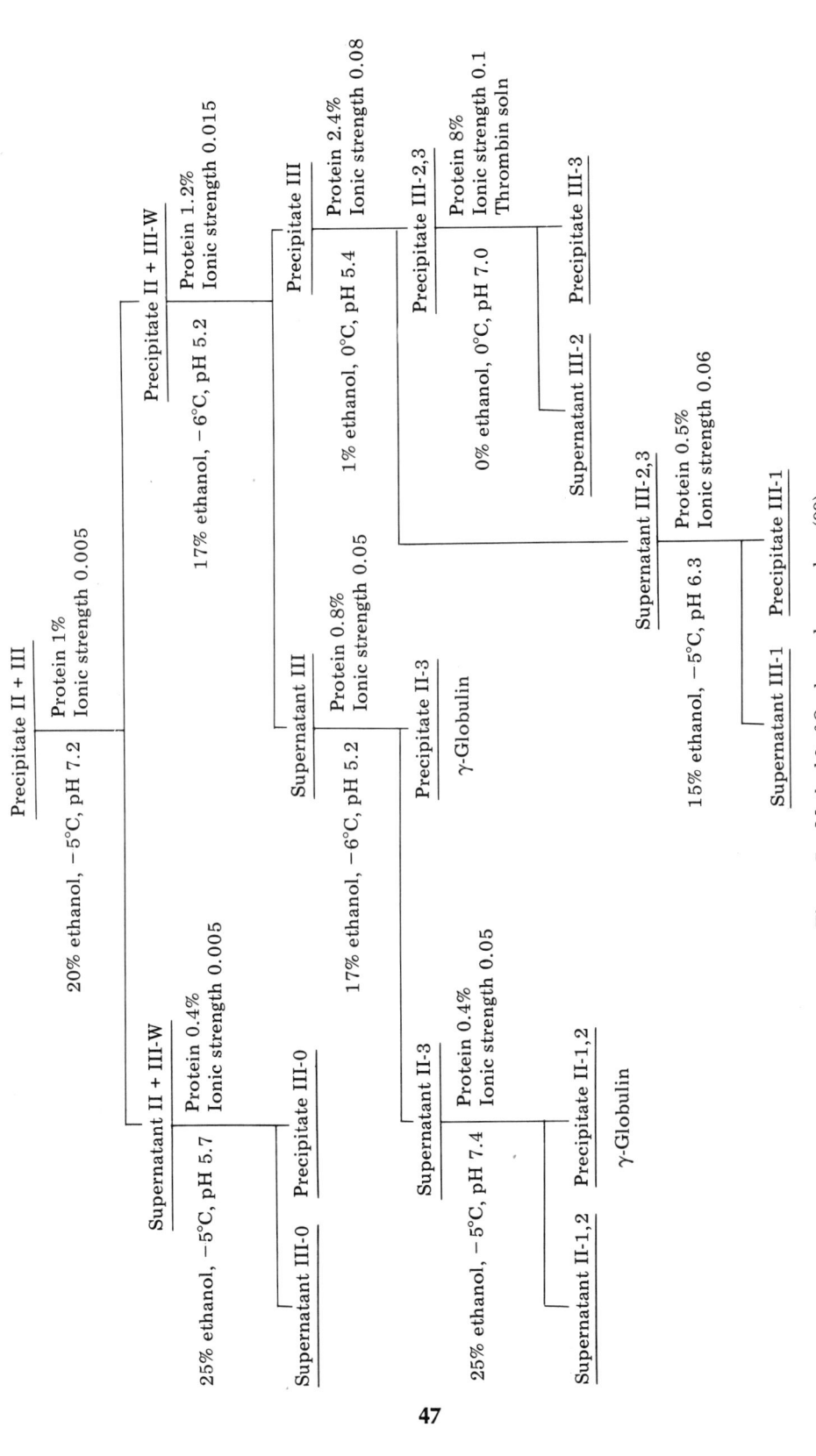

Figure 7. Method 9 of Oncley and co-workers (99).

48   BLOOD FRACTIONATION

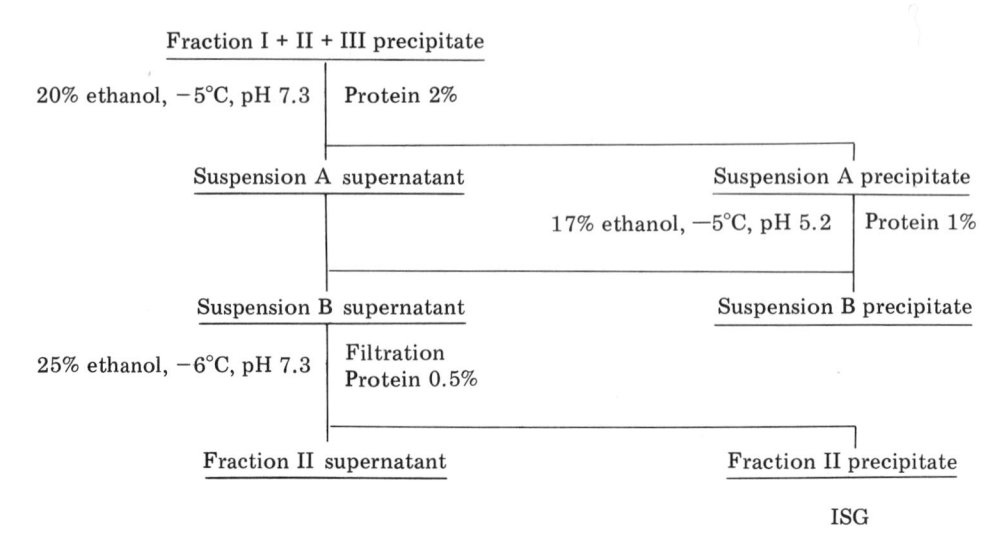

```
                    Fraction I + II + III precipitate
20% ethanol, −5°C, pH 7.3  │  Protein 2%
                           │
        ┌──────────────────┴────────────────────┐
  Suspension A supernatant              Suspension A precipitate
        │                    17% ethanol, −5°C, pH 5.2  │  Protein 1%
        │
        ├──────────────────────────────────────┐
  Suspension B supernatant              Suspension B precipitate
                            Filtration
25% ethanol, −6°C, pH 7.3  │ Protein 0.5%
        │
        ├──────────────────────────────────────┐
  Fraction II supernatant              Fraction II precipitate
                                                ISG
```

Figure 8.   Large-scale commercial method for immune serum globulin (ISG) production (92).

been described (107). A fractionation scheme using polyphosphate as a precipitant has also been used (108). None of these methods yield a fraction with a sufficiently high albumin content to meet regulations of the FDA for albumin or plasma protein fraction (PPF).

Polyethylene glycol (PEG) has become a very popular protein precipitant. It acts by concentrating the protein component in the inter-PEG spaces by a displacement mechanism. The mechanism of action of PEG as a protein precipitant and its use in plasma fractionation have been studied (109–112). Although PEG can be used in a complete fractionation scheme to obtain albumin, ISG, fibrinogen, and antihemophilic factor (AHF), it has been most effectively used to prepare high purity AHF and non-aggregated ISG. The most commonly used grades of PEG have been those with molecular weights of 6000 and 4000. However, recent evidence (113) has indicated that lower molecular weights of PEG may give greater specificity in fractionation.

Plasma fractionation schemes using precipitants other than ethanol or PEG for the isolation of albumin and ISG have been developed and used successfully, primarily in Europe. Many of these techniques have been reviewed (114–115). Ethyl ether has been used as a precipitant in England (116). Rivanol and ammonium sulfate have been used in conjunction at Behringwerke in Germany (117). In France, placental blood is fractionated with the use of chloroform, trichloroacetic acid, and ethanol as precipitants (118). Recently, Pluronic polyols (119) (BASF Wyandotte Corp) and solid-phase maleic anhydride polyelectrolytes (120) have been used successfully on an experimental scale.

Adsorption chromatography has been used for the purification of ISG (91). A large-scale method for the production of albumin utilizing PEG, adsorption chromatography, and gel chromatography has recently been developed (121). Continuous preparative electrophoresis (122–123), polarization chromatography (124), isotachophoresis and isoelectric focusing (125–126) are all promising techniques for the large-scale purification of plasma proteins.

Methods for Fibrinogen Production. With most precipitants, fibrinogen is generally the first plasma protein to precipitate. It is found in the cold ethanol fraction I (89). Glycine has been used to prepare purified fibrinogen from fraction I (127). Amino acids such as glycine (128) and β-alanine (129) have been used to prepare fibrinogen directly from plasma. Precipitation with PEG has been shown to be an effective method for the preparation of fibrinogen (112).

Methods for Antihemophilic Factor Production. Antihemophilic factor (AHF, factor VIII) is a protein present in trace amounts in normal plasma. It is defined by its activity, which is deficient in patients with hemophilia A. One unit of factor VIII activity is defined as the amount in one mL of pooled, fresh, normal, citrate-treated plasma.

It was recognized early in the development of cold ethanol fractionation of plasma that fraction I was rich in AHF (89). Fraction I was used in the treatment of hemophilia A (130) but its potency was so low that it never became a valuable therapeutic material in the United States. In England, the ether fraction I of Kekwick and Mackay (115) was used to produce an AHF concentrate (131–132) which was later supplanted by fraction I-0 of Blombäck (133). Fraction I-0 is prepared from cold ethanol fraction I by glycine–citrate–ethanol extraction (133–135). A therapeutic concentrate prepared by this method is currently produced in Sweden.

In the United States, transfusion with fresh or fresh-frozen plasma was the predominant mode of treatment for hemophilia A until the introduction of the single donor cryoprecipitate. The technique of preparing an AHF concentrate from a single unit of blood is based on the observation that when frozen plasma is completely thawed, the cold-insoluble residue contains a considerable fraction of the parent plasma AHF (136). From this observation Pool and Shannon (137) developed a method for the production of an AHF concentrate in a closed-bag system from a single unit of blood. The method consists of rapid separation of the plasma from its formed elements in a closed-plastic-bag system, freezing of the plasma, thawing at 0–4°C, and centrifugation and drainage of the supernatant plasma, leaving the AHF-rich precipitate in a sterile plastic bag. There have been several recent studies concerned with producing single donor cryoprecipitates (138–143). It should be possible to produce single donor cryoprecipitates having an average potency of 100 AHF units per unit of blood. This is demonstrated by the results shown by The Greater New York Blood Program (Fig. 9). The method for the production of single donor cryoprecipitates is simple enough for use in hospital blood banks. The practice at many European fractionation centers is to produce small-pool (2–8 donor) cryoprecipitates. These are prepared using aseptic techniques and lyophilized.

AHF is a very large molecule, with a molecular weight estimated at $(1-2) \times 10^6$ (145–146). Cryoprecipitation results from formation of regions of high concentration of solutes such as proteins and salts in the intercrystalline spaces of the ice lattice, during either freezing or thawing or both. High molecular weight solutes (AHF, fibrinogen) are precipitated by the action of high concentrations of salts and lower molecular weight proteins at low temperature (112,147).

Cryoprecipitation is the first step in most, if not all, methods in use today for the large-scale production of AHF concentrates. Fresh frozen plasma is pooled, thawed at 2°C, and the precipitate collected in continuous flow centrifuges. The cryoprecipitate thus obtained is then processed by a variety of methods. All of these methods have as their last steps sterile filtration and lyophilization. The simplest methods

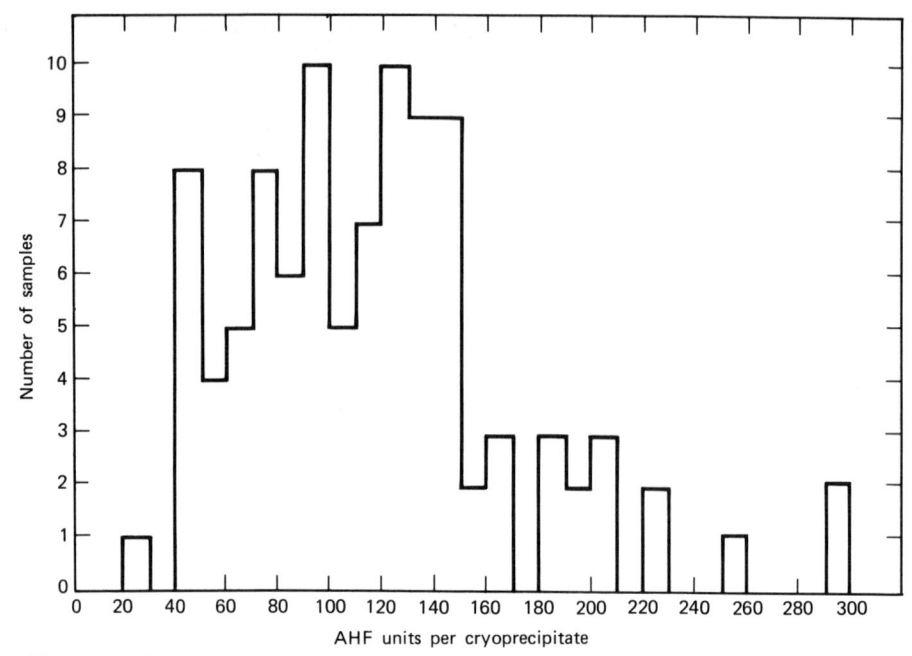

Figure 9. Distribution of antihemophilic factor (AHF) activity in single donor cryoprecipitates prepared by The Greater New York Blood Program in 1976 (144). Mean AHF units per cryoprecipitate = 116.9; s (standard deviation) = 53.1; n (number of single donor cryoprecipitates) = 100; mean plasma volume = 258.0 mL; mean recovery = 45.3%.

consist of an extraction of the cryoprecipitate with buffer, sometimes followed by an adsorption with aluminum hydroxide to remove prothrombin and other contaminants (148–150). The addition of ethanol (3%) to the frozen plasma to aid the precipitation of AHF (148–149) has been omitted in most subsequent methods.

High purity concentrates of AHF are produced from cryoprecipitate extracts by fractional precipitation with PEG (149), PEG and glycine (151–152) and ethanol (153–154). Most AHF concentrates are produced by commercial fractionation firms which do not release information concerning yields. However, a good approximation of the recovery of AHF from plasma is 200–400 units per liter for intermediate purity concentrates and 100–250 units per liter for high purity concentrates.

Methods for Prothrombin Complex Production. Coagulation factors II, VII, IX, and X are usually isolated together as a fraction called prothrombin complex. In contrast to the labile AHF (factor VIII), which is prepared from fresh-frozen plasma, prothrombin complex is more stable and can be prepared from plasma separated from its formed elements several days after collection. The usual starting materials for the preparation of prothrombin complex are the cryoprecipitate supernatant phase or residues of large-scale cold ethanol fractionation.

The first clinically effective prothrombin complex concentrate (PPSB), was prepared from blood collected over an ion-exchange resin (155) and was later prepared from blood collected into an anticoagulant containing ethylenediaminetetraacetic acid (EDTA) (156). The prothrombin complex is adsorbed from the plasma with tricalcium phosphate, eluted with citrate, and purified by ethanol precipitation of lipoproteins and other impurities. The final steps, as with all prothrombin complex preparations are sterile filtration and lyophilization. The problem with the PPSB method is that it limits the use of other components of the blood because of the presence of EDTA.

Precipitate B of Kistler and Nitschmann (96) is used to prepare prothrombin complex in England (157). The precipitate is suspended in saline and adsorbed with calcium phosphate. The prothrombin complex is then eluted with citrate and stabilized with heparin. Finally, lipoproteins are removed by ether fractionation. The addition of heparin is employed in many preparations to prevent the conversion of prothrombin to thrombin, the injection of which would be extremely dangerous.

Two prothrombin complex preparations are produced commercially in the United States. One preparation starts from cold ethanol fraction III which is suspended in saline and treated with calcium phosphate. The prothrombin complex is then eluted with citrate and freed of lipoproteins by cold ethanol fractionation (158). Precipitation with PEG is used to purify and concentrate the coagulation factors (159).

The second commercial preparation performed on a large-scale starts from cold ethanol fraction I supernatant (160). The prothrombin complex is adsorbed onto DEAE-Sephadex and eluted in several fractions. A similar procedure using the cryoprecipitate supernatant phase as starting material has been described in detail (161). This method is represented diagrammatically in Figure 10. Other methods of preparing prothrombin complex from cryoprecipitate supernatant or fraction I supernatant have utilized adsorption onto DEAE-cellulose (161–162). The active factors are then eluted in several fractions and can be further purified and concentrated by precipitation with PEG (105).

```
                        Cryoprecipitate supernatant
         DEAE-Sephadex  |
           (1.5 gm/L)   |  Stirring at 22°C
                        |
                        ├──────────────────────────────┐
                        DEAE-Sephadex              Supernatant
Wash three times with 0.2M NaCl–0.01M
    trisodium citrate, pH 7.0
                        |
                        ├──────────────────────────────┐
                        DEAE-Sephadex                 Wash
    Elute with 2M NaCl–0.01M
    trisodium citrate, pH 7.0
                        |
                        ├──────────────────────────────┐
                        DEAE-Sephadex                Eluate
           Dialyze against 0.01M trisodium citrate, pH 7.0
                                                       |
                                                Prothrombin complex
```

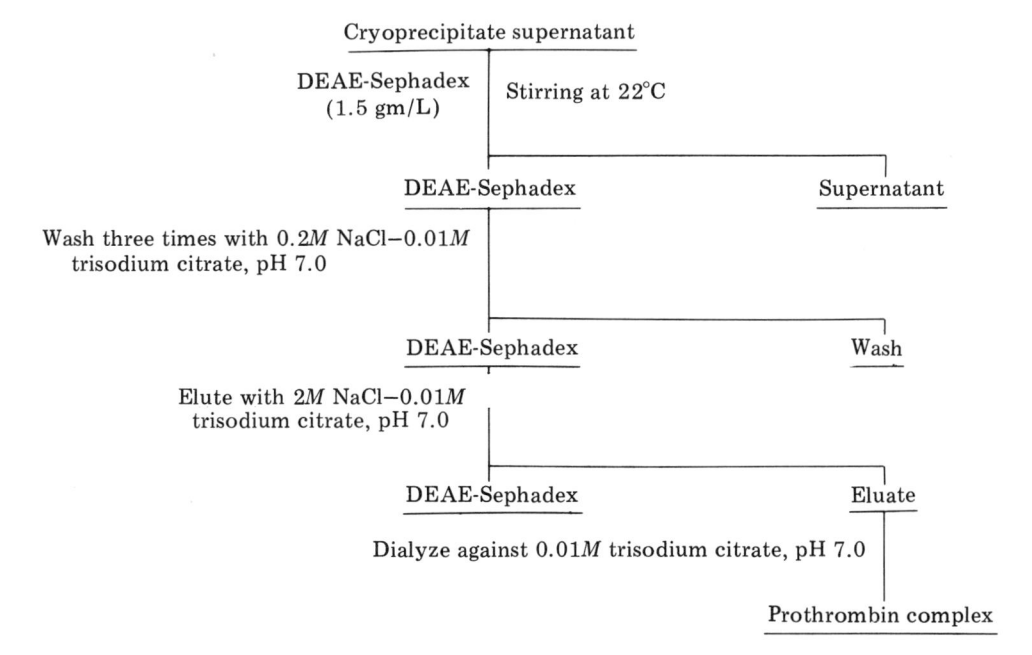

Figure 10. Preparation of prothrombin complex by adsorption onto DEAE-Sephadex (161).

Procedures, Equipment, and Reagents for Plasma Fractionation

Plasma fractionation is carried out in large tanks with jackets to allow circulation of fluid for precise control of temperature. The tanks are generally made of stainless steel and are constructed so that they may be cleaned in place or easily disassembled for thorough cleaning. Some tanks may be equipped with temperature control coils inside the vessels. The size of the tank is determined by the volume of the fractionation step requiring the highest dilution.

Most fractionation procedures in general use, such as the cold ethanol methods and the AHF methods requiring cryoprecipitation, call for low temperatures. For this reason most of the fractionation steps are carried out in refrigerated rooms at −5 to 4°C, depending on the procedure. Coolant is circulated in the tank jackets and coils as well as in coils inside the centrifuges used to harvest precipitates. Most procedures include lyophilization, which requires refrigeration, for drying and the removal of ethanol. Plasma for fractionation is generally stored at −20°C or below although plasma not intended for AHF production may be stored at 1–6°C. All of these requirements for cooling make refrigeration one of the most important and expensive factors in plasma fractionation (see Refrigeration).

Low températures serve to protect proteins against denaturation. However, the choice of equipment for fractionation is also very important in this regard. Stirrers are generally of the propeller type and should be chosen to provide efficient mixing without foaming. Pumps, generally of the centrifugal or peristaltic type, must be chosen to minimize shear which causes denaturation (163). The technique of reagent addition is critical in avoiding denaturation and in determining the physical characteristics of the precipitates formed. Reagents must be added slowly, with efficient mixing, to avoid local excesses of precipitant concentration, temperature, pH, and ionic strength. This is generally accomplished by adding the reagents as liquids through a device with multiple narrow-bore openings directly beneath the propeller blades of the stirrer. Alcohol is chilled before addition and is added slowly enough for the heat of solution to be dissipated by the refrigeration of the processing tank and the cold room.

Liquid–solid separations are critical for the efficiency of a plasma fractionation procedure. Filtration and centrifugation are the two methods generally used (164–165). Centrifugation is generally used to harvest large quantities of precipitated proteins whereas filtration is generally used for removal of small amounts of solids and the clarification of solutions. Continuous flow tubular bowl centrifuges are widely used in plasma fractionation. A typical centrifuge operates at 15,000 rpm and has a bowl with a capacity of 6–7 kg of precipitate. The precipitate contains 15–30 wt % solids, depending on the nature of the fraction. A Funda filter has been evaluated as a substitute for the continuous flow centrifuge in plasma fractionation (166). This filtration device has a mechanism for solids discharge whereas the centrifuges must be stopped, dismantled, and the precipitate removed manually whenever the bowl is filled. It was found that the Funda filter was effective and efficient for the separation of intermediate fractions, but less effective for the separation of purified albumin and ISG.

Filtration is used for the clarification of several of the supernatants obtained during cold ethanol fractionation. Asbestos filters were used for this purpose until the FDA placed severe restrictions on the use of fiber-releasing filters (167). The adsorptive characteristics of asbestos filters were particularly useful in plasma fractionation for the removal of lipids and denatured proteins. Asbestos filters have been successfully replaced by cellulose filters containing inorganic filter aids treated to provide an electrokinetic profile similar to asbestos (92). These filters are also effective as prefilters in the preparation of sterile therapeutic fractions. The sterilizing filters are membrane filters with an absolute retention of all particles larger than 0.22 μm (see Filtration).

Lyophilization is a standard procedure in plasma fractionation for removal of organic solvents and for drying. Fractions such as AHF, prothrombin complex, and fibrinogen are sterile-filtered, dispensed into vials, and lyophilized. They are reconstituted with sterile water immediately prior to use. Albumin, PPF, and ISG are lyophilized in bulk. Solutions can then be made to precise protein concentrations. Stabilizers are added and the solutions are sterile-filtered and dispensed into vials. Sodium acetyltryptophanate and sodium caprylate are the stabilizers used for albumin and PPF, and glycine for ISG. Albumin and PPF are pasteurized at 60°C for 10 h. ISG solutions generally contain a mercurial preservative as a bacteriostatic agent.

Several alternatives to lyophilization have been devised for the removal of ethanol from plasma protein fractions. Vacuum distillation (168) and molecular sieve chromatography (169–170) have been used successfully. Molecular sieve chromatography followed by ultrafiltration seems to be the most promising alternative to lyophilization for the preparation of albumin solutions (see Adsorptive separation).

The methods described above are all batch procedures, ie, the pH, ionic strength, temperature, protein concentration, and precipitant concentration are adjusted to the desired values and there is a period of equilibration before phase separation. Adjustments in pH are usually made by extrapolation of a sample titration performed under standard conditions. An electrode has recently been described which allows for direct measurement of pH under the conditions of cold ethanol fractionation (172). Various control systems are available for monitoring temperatures and dispensing reagents but the fractionation system is basically controlled by the operator. There has been considerable recent work on methods for converting the cold ethanol batch process to a semicontinuous flowing stream. Watt (173–175) has developed a continuous small volume mixing system which has been found to be amenable to computer control. A continuous flow system with a solids-ejecting centrifuge is under development (176). A continuous flow precipitating system with automated controls and rapid mixing devices has been shown to be capable of producing PPF equal in purity and yield to that obtained by the comparable batch method (177). It is hoped that continuous flow systems with automated controls will allow for more reproducible, efficient, and sanitary operation of plasma fractionation.

In the United States, the production of blood products is regulated by the Bureau of Biologics of the FDA. There are good manufacturing practices (GMP) regulations, general biologic product regulations, and additional regulations pertaining to individual products. Producers of blood products operate under establishment licenses granted by the FDA and must have specific product licenses for each fraction they produce. Samples of products, as well as production and quality control protocols, are examined by the FDA prior to release of the products for distribution. Extensive documentation of each step in production and quality control is required. The function of the regulatory agency and the quality control section of the producer is to assure the safety, potency, and efficacy of the products.

Reagents used in plasma fractionation must be completely removed from the final products or must be demonstrated to be safe. One of the problems in plasma fractionation is contamination of fractions by pyrogens, ie, substances which elicit a febrile response in the recipient. The pyrogens are generally bacterial endotoxins. Pyrogen testing is accomplished by injecting the material intravenously into rabbits and then monitoring their temperatures. Recent work has shown that an *in vitro* test utilizing limulus lysates may be an effective means of screening for bacterial endotoxins (178). Water for processing and washing as well as all chemical reagents are regularly tested for pyrogenic contamination. Water is distilled and then stored at high temperatures to inhibit bacterial growth. Chemical reagents are generally USP grade.

Quality control and quality assurance are essential functions in a blood fractionation operation. Sampling and testing are performed at each step of the fractionation process to provide information for further processing and to assure the quality of the final product. Extensive testing of final products is performed to assure their safety, potency, and efficacy. The quality control of final products includes tests for sterility, pyrogenicity, toxicity, protein composition, chemical composition, stability, identity, and potency. The potency tests include coagulation factor assays for AHF and prothrombin complex and antibody assays for ISG.

Benefits and Risks of Plasma Fractionation

Human blood is a precious and complex commodity. The indiscriminate transfusion of whole blood is no longer justified since most patients can be treated more effectively and safely with one of the blood components or derivatives. The use of therapeutically active plasma fractions eliminates the danger of hypervolemia and minimizes the side effects resulting from contaminating proteins. Adequate replacement therapy for patients with coagulation disorders is only possible through the use of coagulation factor concentrates. Antibody titers high enough for prophylaxis or therapy can be achieved only through the use of ISG concentrates.

The major hazard in producing fractions from large pools of plasma is the transmission of hepatitis. This is a danger both for the recipient of the fractions and for the workers in fractionation plants. It has been shown that fractionation workers, particularly those engaged in the preparation of plasma pools, are at high risk of developing hepatitis B (179–180).

Barker and Hoofnagle (181) have reviewed the transmission of hepatitis B by plasma derivatives from large pools. They have classified products as either high risk or low risk with respect to the transmission of hepatitis B. The high risk products are fibrinogen, AHF, and prothrombin complex. The low risk products are ISG, PPF, and albumin. The lack of infectivity of PPF and albumin is attributable to heating the final products at 60°C for 10 h. There have been no reports of overt hepatitis cases following administration of ISG prepared by cold ethanol methods. This fact has caused manufacturers and the FDA to be resistant to the adoption of new techniques for the purification of ISG.

It is now required in the United States that all donors of blood or plasma be tested for the presence of hepatitis B surface antigen by radioimmunoassay or reversed passive hemagglutination. This screening reduces but does not prevent the transmission of hepatitis B virus. A major problem is the transmission of non-B hepatitis, for which there is no screening test. Recent evidence indicates that non-A, non-B hepatitis also invokes a viral agent (182). The etiology and prevention of post transfusion hepatitis has been reviewed by Prince (183).

PPF has occasionally been associated with hypotensive reactions in recipients

(184). This had led to a contraindication for the use of PPF in patients on cardiopulmonary bypass and a recommendation that it not be infused at a rate greater than 10 mL/min. The cause of the hypotensive reactions seems to be the presence of components of the kinin system (185–186) and efforts are in progress to effect a removal of these agents from PPF.

Another hazard of plasma fractionation is the partial denaturation of some fractions such as ISG, caused by the fractionation methods (187–189). These denatured proteins may have toxic effects or may be immunogenic in the recipients (190). These undesirable side effects are more than compensated for by the increased stability and potency of concentrated fractions.

Therapeutic Uses of Plasma Fractions

Albumin and Plasma Protein Fraction (PPF). Albumin and PPF as 5% solutions are used primarily as blood volume expanders. Concentrated albumin, as a 25% solution, is also used as a plasma volume expander, because of its oncotic effect, in patients who are not dehydrated. Concentrated albumin is used to treat traumatic or hemorrhagic shock, extensive burns, pancreatitis, severe hypoalbuminemia, and other conditions in which there is a continued loss of protein.

Immune Serum Globulin (ISG). ISG is used for the treatment of congenital and acquired agammaglobulinemias. It has also been found useful for the prevention of poliomyelitis and for the prevention and modification of measles and hepatitis A. ISG is used as a general prophylactic against various infectious diseases. Its usefulness in the prevention of hepatitis B is questionable (183).

ISG is injected via the intramuscular route. This limits the amount that can be injected without causing severe discomfort to the patient. Attempts at intravenous injection resulted in frequent vasomotor reactions attributed to the anticomplement activity of aggregated IgG (188–190). Several methods have been examined in attempts to produce an intravenous injectable ISG (188–192). The most promising of these methods utilizes adsorption chromatography and affinity chromatography (qv) (193). As yet no preparation of intravenous injectable ISG is licensed for use in the United States.

Many specific ISG preparations have been produced from the plasma of donors with known high antibody titers. These are either naturally-occurring or the result of specific immunization. These specific ISG preparations include anti-$Rh_0(D)$ (194–195), antivaccinia, antirubella, antitetanus (196), antihepatitis B (197), and antipseudomonas (198).

Fibrinogen. The use of fibrinogen to treat hypofibrinogenemia is in question because it does not treat the underlying cause which is usually disseminated intravascular coagulation. Fibrinogen has been used to treat acute hemorrhage associated with fibrinolysis or defibrination and which has not responded to other treatment (whole blood, fibrinolytic inhibitors, or heparin). The use of fibrinogen prepared from fraction I has come into disfavor because of the high risk of hepatitis. Because of the doubtful efficacy and the hazards of the product the FDA has recently revoked all product licenses for the manufacture of fibrinogen (199). In the rare instances when fibrinogen is indicated, single donor cryoprecipitates are used because of the much lower hepatitis risk. Fibrinogen from pooled plasma is still produced in Europe.

Antihemophilic Factor (AHF). AHF is used for the treatment of hemophilia A (deficiency of factor VIII procoagulant activity). Single donor cryoprecipitate, intermediate purity AHF, and high purity AHF may be used. Single donor cryoprecipitate and intermediate purity AHF are sufficient for prophylactic therapy and for the treatment of most bleeding episodes. High purity AHF is required when large doses are necessary, eg, when surgery is required or in case of severe accidents, to prevent the infusion of large volumes of fluid and large amounts of fibrinogen. The lyophilized products, and especially the high purity preparations, are preferred by patients and physicians because of their convenience. Although AHF prepared from large pools has a high risk of being contaminated with hepatitis virus its benefits outweigh the risks. The multiply-transfused patients who receive these preparations usually develop an immunity to hepatitis without suffering an overt case of the disease. For the treatment of von Willebrand's disease single donor cryoprecipitates have been shown to be more effective than lyophilized concentrates (200–201).

Prothrombin Complex (PTC). Prothrombin complex is indicated primarily for the treatment of hemophilia B (factor IX deficiency, Christmas disease). In these patients, as with hemophilia A patients, the benefits of treatment with a concentrated therapeutic material outweigh the risk of hepatitis. Prothrombin complex is also indicated in the much rarer congenital deficiencies of factors II, VII, or X. Prothrombin complex may be used in the treatment of coagulation factor deficiencies induced by overdoses of coumarin-type anticoagulants. Its use in these cases must be weighed against the risks of hepatitis and of thromboembolic disease resulting from the presence of activated factors. Cryoprecipitate supernatant phase or fresh-frozen plasma may be preferred in acquired prothrombin complex deficiencies. The prothrombin complex may be deficient in severe liver disease, but again the single-donor products may be the preferred treatment because of their lower risks.

Economic Aspects

The blood fractionation industry has grown from two sources: the hospital blood bank and the pharmaceutical industry. At present in the United States, most single-donor blood products are produced and distributed by regional blood centers. These are nonprofit organizations which cooperate very closely with the hospitals that they serve. The cost of the products to the hospitals is the actual processing cost.

In the United States, plasma fractionation is controlled to a great extent by several large firms, most of which are divisions of pharmaceutical corporations; there are also some nonprofit plasma fractionators. A general summary of the cost of plasma derivatives and the change in cost in recent years is given in Table 7.

Table 7. Representative Prices of Plasma Derivatives

Product	1974 Price, $	1978 Price, $
normal serum albumin (NSA), 25% solution, 100 mL	66.90	108.75
plasma protein fraction (PPF), 5% solution, 500 mL	64.95	101.90
immune serum globulin (ISG), 16.5% solution, 10 mL	9.00	9.00
antihemophilic factor (AHF), 260 units	35.00	40.00
prothrombin complex (PTC), 400–500 units factor IX	49.00	56.85

In Europe, producers of single-donor blood products are either nonprofit or government-run organizations. Many of these organizations engage in plasma fractionation as well. There are also several large commercial plasma fractionators in Europe.

Abbreviations

ACD	anticoagulant citrate dextrose
AHF	antihemophilic factor
CFR	Code of Federal Regulations
CPD	anticoagulant citrate phosphate dextrose
CPD-A	citrate phosphate dextrose adenine
EDTA	ethylenediaminetetraacetic acid
HLA	human leukocyte antigens
ISG	immune serum globulin
NSA	normal serum albumin
PC	platelet concentrate
PEG	polyethylene glycol
PMNs	polymorphonuclear leukocytes, or granulocytes
PPF	plasma protein fraction
PPP	platelet-poor plasma
PPSB	prothrombin complex concentrate
PRP	platelet-rich plasma
PTC	prothrombin complex
RBC	red blood cells
WBC	white blood cells, leukocytes

BIBLIOGRAPHY

"Blood Fractionation" in *ECT* 1st ed., Vol. 2, pp. 556–584, by L. E. Strong, Kalamazoo College; "Blood Fractionation" in *ECT* 2nd ed., Vol. 3, pp. 576–602, by Laurence E. Strong, CBA Project, Earlham College.

1. B. A. Myhre and co-eds., *Blood Component Therapy: A Physicians's Handbook*, American Association of Blood Banks, Washington, D.C., 1975, p. 5.
2. K. Diem, ed., *Documenta Geigy Scientific Tables, 6th Edition*, Geigy Pharmaceuticals, Ardsley, N.Y., 1962, pp. 546–581.
3. P. L. Altman and D. S. Dittmen, eds., *Biology Data Book*, 2nd ed., Vol. 3, Federation of the American Societies for Experimental Biology, Bethesda, Md., 1974, pp. 1751–1753, 1805–1807, 1819–1820, 1830–1831, 1854–1857.
4. Ref. 1, pp. 26–27.
5. *Code of Federal Regulations 21 Parts 600–1299*, Office of the Federal Register, General Services Administration, U.S. Government Printing Office, Washington, D.C., 1977.
6. W. V. Miller and co-eds., *Technical Methods and Procedures*, 6th ed., American Association of Blood Banks, Washington, D.C., 1974.
7. B. A. Myhre, *Quality Control in Blood Banking*, John Wiley & Sons, Inc., New York, 1974.
8. R. F. Reiss and A. J. Katz, *Transfusion* 16, 312 (1976).
9. D. H. Buchholz and co-workers in J. M. Goldman and R. M. Lowenthal, eds., *Leucocytes: Separation, Collection and Transfusion*, Academic Press, Inc., New York, 1975, p. 177.
10. *Fed. Reg.* 43, 2890 (1978).
11. T. J. Greenwalt, M. S. Gajewski, and J. L. McKenna, *Transfusion* 2, 221 (1962).
12. I. Djerassi and co-workers, *J. Med. (Basel)* 1, 358 (1970).
13. I. Djerassi and co-workers in ref. 9, p. 123.
14. D. W. Huestis, J. R. Bove, and S. Busch, *Practical Blood Transfusion*, 2nd ed., Little, Brown and Co., Boston, Mass., 1976, p. 307.
15. W. J. Williams and co-workers, *Hematology*, McGraw-Hill, Inc., New York, 1972, pp. 16, 18–19.
16. Ref. 7, p. 148.

17. R. A. Kahn, I. Cossette, and L. I. Friedman, *Transfusion,* **16,** 162 (1976).
18. Ref. 5, p. 103.
19. M. Wiekowicz, *Transfusion* **16,** 193 (1976).
20. A. Latham, Jr., and G. F. Kingsley in Ref. 9, p. 203.
21. J. L. Tullis and co-workers, *Transfusion* **11,** 368 (1971).
22. N. M. Abelson, *Topics in Blood Banking,* Lea & Febiger, Philadelphia, Pa., 1974, p. 107.
23. E. J. Freireich in ref. 9, p. xxvii.
24. J. H. Cutts, *Cell Separation: Methods in Hematology,* Academic Press, Inc., New York, 1970, pp. 36–37.
25. A. A. Waldman, in press.
26. D. C. Burke, *Sci. Am.* **236,** 42 (1977).
27. H. S. Lawrence, *Harvey Lect.* **68,** 239 (1973).
28. C. J. Remenyik and co-workers in ref. 9, p. 3.
29. V. R. Kruger, K. B. McCredie, and E. J. Freireich in ref. 9, p. 14.
30. C. D. West and D. D. Willis in ref. 9, p. 30.
31. K. B. McCredie and co-workers, *Transfusion* **14,** 357 (1974).
32. J. M. Mishler and co-workers, *Transfusion* **14,** 352 (1974).
33. R. M. Lowenthal and co-workers in ref. 9, p. 499.
34. M. Benbunan and co-workers in ref. 9, p. 81.
35. L. Debusscher, R. Badjou, and R. Stryckmans in ref. 9, p. 349.
36. D. W. Huestis and co-workers, *Transfusion* **15,** 559 (1975).
37. J. E. Garvin, *J. Exp. Med.* **114,** 51 (1961).
38. Y. Rabinowitz, *Blood* **23,** 811 (1964).
39. W. Hawley and E. Blanda in *A Seminar on Component Preparation and Use, 11th Annual Meeting Ohio Association of Blood Banks, Columbus, Ohio, 1977,* pp. 45–50.
40. I. Djerassi and co-workers, *Transfusion* **12,** 75 (1972).
41. G. P. Herzig, R. K. Root, and R. G. Graw, Jr., *Blood* **39,** 554 (1972).
42. R. G. Graw, Jr., and co-workers, *Clin. Res.* **19,** 491 (1971).
43. M. B. Harris and co-workers, *Blood* **44,** 707 (1974).
44. J. Fehr in T. J. Greenwalt and G. A. Jamieson, eds., *The Granulocyte: Function and Clinical Utilization,* Alan Liss, New York, 1977, p. 243.
45. J. Nusbacher and co-workers, *Blood* **51,** 359 (1978).
46. A. Bøyum, *Scand. J. Immunol. 5, Suppl.* **5,** 9 (1976).
47. L. P. Einstein, E. E. Schneeberger, and H. R. Colten, *J. Exp. Med.* **143,** 114 (1976).
48. D. A. Horwitz and M. A. Garrett, *J. Immunol.* **118,** 1712 (1977).
49. T. G. Pretlow, II, and D. E. Luberoff, *Immunology* **24,** 85 (1973).
50. H. Loos and co-workers, *Blood* **48,** 731 (1976).
51. W. D. Johnson, Jr., B. Mei, and Z. A. Cohn, *J. Exp. Med.* **146,** 1613 (1977).
52. R. J. Sanderson and co-workers, *J. Immunol.* **118,** 1409 (1977).
53. A. M. Fogelman and co-workers, *Biochem. Biophys. Res. Commun.* **76,** 167 (1977).
54. G. J. V. Nossal in B. R. Brinkley and K. Porter, eds., *International Cell Biology, 1976–1977,* Rockefeller University Press, New York, 1977, p. 103.
55. D. H. Katz in ref. 54, p. 112.
56. J. B. Natvig, P. Perlmann, and H. Wigzell, eds., *Lymphocytes: Isolation, Fractionation and Characterization,* University Park Press, Baltimore, Md., 1976.
57. R. A. Good in C. F. Högman, K. Lindall-Kiessling, and H. Wigzell, eds., *Blood Leukocytes: Function and Use in Therapy,* Almquist and Wiksell Intl., Stockholm, Sweden, 1977, p. 78.
58. T. J. Greenwalt, M. S. Gajewski, and J. L. McKenna, *Transfusion* **2,** 221 (1962).
59. G. P. Herzig, R. K. Root, and R. G. Graw, Jr., *Blood* **39,** 554 (1972).
60. R. H. Halterman and co-workers, *Transfusion* **13,** 50 (1973).
61. W. V. Miller, M. J. Wilson, and H. J. Kalb, *Transfusion* **13,** 189 (1978).
62. S. Kevy in *Transfusion Therapy—A Technical Workshop,* American Assoc. of Blood Banks, Washington, D.C., 1976, p. 72.
63. I. Dorner and co-workers, *Transfusion* **15,** 439 (1975).
64. H. F. Polesky and co-workers, *Transfusion* **13,** 383 (1973).
65. F. J. Tenczar, *Transfusion* **13,** 183 (1973).
66. C. Jones and co-workers, *Transfusion* **8,** 323 (1968).
67. M. J. O'C. Wooten, *Transfusion* **16,** 464 (1976).
68. J. P. Crowley and co-workers, *Transfusion* **17,** 1 (1977).

69. *Technical Manual of the American Assoc. of Blood Banks*, 7th ed. American Assoc. of Blood Banks, Washington, D.C. 1977, p. 51.
70. H. T. Meryman and M. Hornblower, *Transfusion* 13, 388 (1973).
71. J. P. Crowley and C. R. Valeri, *Transfusion* 14, 188 (1974).
72. M. Telischi, E. Krmpotic, and G. Moss, *Transfusion* 15, 481 (1975).
73. M. Polesky in J. A. Griep, ed., *Clinical Uses of Frozen—Thawed Red Blood Cells*, Alan R. Liss, New York, 1976, p. 141.
74. Ref. 5, pp. 100–101.
75. I. O. Szymanski and C. R. Valeri, *N. Engl. J. Med.* 280, 281 (1969).
76. A. W. Boone, N. B. Whittemore, and J. L. Hutchinson, *Can. Med. Assoc. J.* 104, 788 (1971).
77. J. Umlas, *Transfusion* 15, 111 (1975).
78. C. E. Huggins, *Monogr. Surg. Sci.* 3, 133 (1966).
79. H. T. Meryman and M. Hornblower, *Transfusion* 12, 145 (1972).
80. A. W. Rowe, E. Eyster, and A. Kellner, *Cryobiology* 5, 119 (1968).
81. C. R. Valeri, *Blood Banking and the Use of Frozen Blood Products*, CRC Press, Inc., Cleveland, Ohio, 1976.
82. J. McCullough and co-workers, *Lancet* 2, 1333 (1969).
83. J. McCullough, S. J. Carter, and P. G. Quie, *Blood* 43, 207 (1974).
84. Ref. 5, p. 102.
85. E. W. Davie and K. Fujikawa, *Ann. Rev. Biochem.* 44, 799 (1975).
86. E. W. Davie and D. J. Hanahan in F. W. Putnam, ed., *The Plasma Proteins*, 2nd ed., Vol. 3, Academic Press, Inc., New York, 1977, p. 422.
87. M. P. Esnouf, *Br. Med. Bull.* 33, 213 (1977).
88. F. W. Putnam, *The Plasma Proteins*, 2nd ed., Vol. 1, Academic Press, Inc., New York, 1975, pp. 60–62.
89. E. J. Cohn and co-workers, *J. Am. Chem. Soc.* 68, 459 (1946).
90. R. B. Pennell in F. W. Putnam, ed., *The Plasma Proteins*, Vol. 1, Academic Press, Inc., New York, 1960, p. 9.
91. H. Björling, *Vox Sang* 23, 18 (1972).
92. S. Holst, *paper presented at Parenteral Drug Association Convention, San Francisco, Calif., 1976*.
93. E. J. Cohn and co-workers, *J. Am. Chem. Soc.* 72, 465 (1950).
94. S. Keller and R. J. Block in P. Alexander and R. J. Block, eds., *A Laboratory Manual of Analytical Methods of Protein Chemistry*, Vol. 1, Pergamon Press, Inc., Elmsford, N.Y., 1960, p. 1.
95. H. Nitschmann, P. Kistler, and W. Lergier, *Helv. Chem. Acta* 37, 867 (1954).
96. P. Kistler and H. Nitschmann, *Vox Sang.* 7, 414 (1962).
97. J. H. Hink, Jr., and co-workers, *Vox Sang.* 2, 174 (1957).
98. E. H. Mealey and co-workers, *Vox Sang.* 7, 406 (1962).
99. J. L. Oncley and co-workers, *J. Am. Chem. Soc.* 71, 541 (1949).
100. H. F. Deutsch and co-workers, *J. Biol. Chem.* 164, 109 (1946).
101. H. Hoch and A. Chanutin, *Arch. Biochem. Biophys.* 50, 271 (1954).
102. U.S. Pat. 2,705,230 (Mar. 29, 1955), A. F. Reid.
103. U.S. Pat. 2,765,299 (Oct. 2, 1956), J. D. Porsche, J. B. Lesh, and M. D. Grossnickle.
104. W. Schneider and co-workers, *Blut* 30, 121 (1975).
105. W. Schneider, D. Wolter, and L. J. McCarty, *Vox Sang.* 31, 141 (1976).
106. D. M. Surgenor and co-workers, *Vox Sang.* 5, 272 (1960).
107. H. Nitschmann and co-workers, *Vox Sang.* 1, 183 (1956).
108. H. Nitschmann, E. Rickli, and P. Kistler, *Vox. Sang.* 5, 232 (1960).
109. A. Polson and co-workers, *Biochim. Biophys. Acta* 82, 463 (1964).
110. P. R. Foster, P. Dunnill, and M. D. Lilly, *Biochim. Biophys. Acta* 317, 505 (1973).
111. I. R. M. Juckes, *Biochim. Biophys. Acta* 229, 535 (1971).
112. A. Polson and C. Ruiz-Bravo, *Vox. Sang.* 23, 107 (1972).
113. W. Hönig and M. R. Kula, *Anal. Biochem.* 72, 502 (1976).
114. K. Heide, H. Haupt, and H. G. Schwick in F. W. Putnam, ed., *The Plasma Proteins*, 2nd ed., Vol. 3, Academic Press, Inc., New York, 1977, p. 545.
115. M. Steinbuch, *Vox Sang.* 23, 92 (1972).
116. R. A. Kekwick and M. E. MacKay, *Medical Research Council Special Report No. 286*, H.M.S.O., London, 1954.

117. H. G. Schwick, J. Fischer, and H. Geiger in E. Merler, ed., *Immunoglobulins*, National Academy of Sciences, Washington, D.C., 1970, p. 116.
118. J. Liautaud and co-workers, *13th Int. Cong. IABS, Budapest, 1973, Part A: Purification of Proteins. Develop. Biol. Standard. 27*, Karger, Basel, 1974, p. 107.
119. L. A. Garcia and G. A. Ordonez, *Transfusion* **16**, 32 (1976).
120. A. J. Johnson and M. Semar, *Abstracts XIV Congress Int. Soc. Blood Trans.*, Helsinki, Finland, 1975.
121. J. M. Curling and co-workers, *Vox. Sang.* **33**, 97 (1977).
122. A. R. Thomson, P. Mattock, and G. F. Aitchison, *paper presented at The International Workshop on Technology for Protein Separation and Improvement of Blood Plasma Fractionation*, Reston, Virginia, 1977.
123. M. Bier in ref. 122.
124. E. N. Lightfoot in ref. 122.
125. A. Chrambach and co-workers in ref. 122.
126. M. Bier in ref. 122.
127. B. Blombäck and M. Blombäck, *Ark. Kemi* **10**, 415 (1956).
128. L. A. Kazal and co-workers, *Proc. Soc. Exp. Biol. Med.* **113**, 989 (1963).
129. W. Straughn, III, and R. H. Wagner, *Thromb. Diath. Haemorrh.* **16**, 198 (1966).
130. G. R. Minot and co-workers, *J. Clin. Invest.* **24**, 704 (1945).
131. R. A. Kekwick and P. Wolf, *Lancet. i*, 647 (1957).
132. R. A. Kekwick and P. L. Walton, *Br. J. Haematol.* **11**, 537 (1965).
133. M. Blombäck, *Ark. Kemi* **12**, 387 (1958).
134. J. E. Jorpes, B. Blombäck, and S. Magnusson, *Acta Med. Scand. Suppl.* **379**, 7 (1962).
135. M. Blombäck and B. Blombäck, *Thromb. Diath. Haemorrh. Suppl.* **35**, 21 (1969).
136. J. G. Pool and J. Robinson, *Br. J. Haematol.* **5**, 24 (1959).
137. J. G. Pool and A. E. Shannon, *N. Engl. J. Med.* **273**, 1443 (1965).
138. J. N. Shanberge and co-workers, *Transfusion* **12**, 251 (1972).
139. E. R. Burka and co-workers, *Transfusion* **15**, 307 (1975).
140. E. R. Burka, T. Puffer, and J. Martinez, *Transfusion* **15**, 323 (1975).
141. C. K. Kasper and co-workers, *Transfusion* **15**, 312 (1975).
142. R. F. Reiss and A. J. Katz, *Transfusion* **16**, 229 (1976).
143. S. J. Slichter and co-workers, *Transfusion* **16**, 616 (1976).
144. M. H. Stryker, in press.
145. O. D. Ratnoff, L. Kass, and P. D. Lang, *J. Clin. Invest.* **48**, 957 (1969).
146. M. E. Legaz and co-workers, *J. Biol. Chem.* **248**, 3946 (1973).
147. W. G. Owen and R. H. Wagner, *Thromb. Res.* **1**, 71 (1972).
148. A. J. Johnson, M. H. Karpatkin, and J. Newman, *Thromb. Diath. Haemorrh. Suppl.* **35**, 49 (1969).
149. J. Newman and co-workers, *Br. J. Haematol.* **21**, 1 (1971).
150. H. L. James and M. Wickerhauser, *Vox Sang.* **23**, 402 (1972).
151. K. M. Brinkhous and co-workers, *Blood* **30**, 855 (1967).
152. K. M. Brinkhous and co-workers, *J. Am. Med. Assoc.* **205**, 613 (1968).
153. J. G. Pool, E. J. Hershgold, and A. R. Pappenhagen, *Nature (London)*, **203**, 312 (1964).
154. E. J. Hershgold, J. G. Pool, and A. R. Pappenhagen, *J. Lab. Clin. Med.* **67**, 23 (1966).
155. P. Didisheim and co-workers, *J. Lab. Clin. Med.* **53**, 322 (1959).
156. J. P. Soulier and co-workers, *Thromb. Diath. Haemorrh. Suppl.* **35**, 61 (1969).
157. E. Bidwell and co-workers, *Br. J. Haematol.* **13**, 568 (1967).
158. G. S. Gilchrist and co-workers, *N. Engl. J. Med.* **280**, 291 (1969).
159. U.S. Pat. 3,560,475 (Feb. 2, 1971), L. Fekete and E. Shanbrom.
160. M. S. Hoag and co-workers, *N. Engl. J. Med.* **280**, 581 (1969).
161. J. Heystek, H. G. J. Brummelhuis, and H. W. Krijnen, *Vox Sang.* **25**, 113 (1973).
162. S. M. Middleton, I. H. Bennett, and J. K. Smith, *Vox Sang* **24**, 441 (1973).
163. S. E. Charm and B. L. Wong, *Biotechnol. Bioeng.* **12**, 1103 (1970).
164. H. F. Porter, J. E. Flood, and F. W. Rennie, *Chem. Eng.* **78**, 39 (1971).
165. C. M. Ambler, *Chem. Eng.* **78**, 55 (1971).
166. H. Friedli and co-workers, *Vox Sang.* **31**, 289 (1976).
167. *Code of Federal Regulations*, Title 21, Section 211.40(j), U.S. Government Printing Office, Washington, D.C., 1976.
168. J. K. Smith and co-workers, *Vox Sang.* **22**, 120 (1972).
169. H. Friedli and P. Kistler, *Chimia* **26**, 25 (1972).

170. A. J. Dickson and J. K. Smith, *Vox Sang.* **28,** 90 (1975).
171. H. Friedli and co-workers, *Vox Sang.* **31,** 283 (1976).
172. H. Friedli and co-workers, *Vox Sang.* **31,** 277 (1976).
173. J. G. Watt, *Vox Sang.* **18,** 42 (1970).
174. J. G. Watt, *Vox Sang.* **23,** 126 (1972).
175. J. G. Watt, *Clin. Haematol.* **5,** 95 (1976).
176. R. B. Pennell and co-workers, *Inclusive Report June 15, 1973–Mar. 25, 1976, New Methods of Plasma Fractionation, NIH Contract No. N01-HL-3-2968* Center for Blood Research, Boston, Mass.
177. J. Vandersande and co-workers, in press.
178. J. Cooper, J. Levin, and H. Wagner, *J. Lab. Clin. Med.* **78,** 138 (1971).
179. J. S. Taylor, E. Schmunes, and W. A. Holmes, *J. Am. Med. Assoc.* **230,** 850 (1974).
180. S. R. Cohen and co-workers, *J. Occup. Med.* **18,** 685 (1976).
181. L. F. Barker and J. H. Hoofnagle, *13th Int. Cong. IABS, Budapest 1973, Part A: Purification of Proteins. Develop. Biol. Standard 27, Karger, Basel, 1974,* p. 178.
182. A. M. Prince, *paper presented at the Second University of California at San Francisco Symposium on Viral Hepatitis, San Francisco, Calif., 1978.*
183. A. M. Prince in E. Ikkala and A. Nykänen, eds., *Transfusion and Immunology,* Vammala, Helsinki, Finland, 1975, p. 81.
184. J. H. Bland, M. B. Laver, and E. Lowenstein, *J. Am. Med. Assoc.* **224,** 1721 (1973).
185. S. Morichi and K. Izaka, *Transfusion* **16,** 178 (1976).
186. *Proceedings of the Workshop on Measurement of Potentially Hypotensive Agents,* Bureau of Biologics, FDA, Bethesda, Md., 1977.
187. R. H. Painter in Ref. 116, p. 174.
188. S. Barandum, P. Kistler, and F. Jeunet, *Vox Sang.* **7,** 157 (1962).
189. P. Schiff, S. K. Sutherland, and W. R. Lane, *Aust. Paediatr. J.* **4,** 121 (1968).
190. J. T. Sgouris, *Vox Sang.* **13,** 71 (1967).
191. J. T. Sgouris and M. J. Matz, *Vox Sang.* **13,** 59 (1967).
192. M. Hainski, J. H. Payne, and G. A. Ordonez, *Vox Sang.* **20,** 469 (1971).
193. A. F. S. A. Habeeb and R. D. Francis, *Vox Sang.* **32,** 143 (1977).
194. R. Finn and co-workers, *Br. Med. J.* **1,** 1486 (1961).
195. W. Pollack and co-workers, *Transfusion* **8,** 151 (1968).
196. H. W. Krijnen and co-workers, in Ref. 116, p. 322.
197. A. M. Prince and co-workers, *N. Engl. J. Med.* **285,** 933 (1971).
198. H. G. Schwick, *Vox Sang.* **23,** 82 (1972).
199. *Fed. Reg.* **43,** 131 (1978).
200. P. M. Blatt and co-workers, *J. Am. Med. Assoc.* **236,** 2770 (1976).
201. D. Green and E. V. Potter, *Am. J. Med.* **60,** 357 (1976).

MARTIN H. STRYKER
ALAN A. WALDMAN
The New York Blood Center

BLUE PRINTING. See Printing processes.

BORDEAUX MIXTURE. See Fungicides.

BORON, ELEMENTAL

Boron [7440-42-8], the fifth element in the periodic table, is composed of two stable isotopes with mass numbers of 10 and 11. Although widespread in nature, it has been estimated to constitute only 0.001% of the earth's crust, usually occurring as alkali or alkaline earth borates, or as boric acid (see Boron compounds, boron oxides).

Boron was discovered as an element almost concurrently by the English chemist Davy and by the French chemists Gay-Lussac and Thenard in June of 1808. They isolated a dark, combustible material, probably no more than 50% pure; almost 100 years passed before the element was obtained in 80% purity. In 1909, Weintraub was finally able to obtain high purity elemental boron (>99%) by passing repeated alternating-current arcs through a mixture of boron trichloride and hydrogen. Since that time, numerous other methods have been developed (see below); however, research continues for methods of obtaining commercial quantities of the pure element, particularly as a filamentary reinforcement for advanced composites.

Structure

Boron exists in a number of allotropic forms, including an amorphous form and six crystalline polymorphs of which the α- and β-rhombohedral forms are well established from single-crystal studies; their approximate preparation temperatures and observed densities are shown on p. 63. The α-rhombohedral structure degrades above 1200°C, and at 1500°C recrystallization to the β-rhombohedral form occurs. The amorphous form converts to β-rhombohedral above approximately 1000°C, and any type of pure boron ultimately transforms to the β-rhombohedral form when heated above the melting point and recrystallized. It has been suggested (1) in fact, that achievement of thermodynamic equilibrium in a system of pure boron at ordinary pressures would give uniformly, at all temperatures between 0 K and the mp, a single polymorph, the β-rhombohedral form. Approximately a dozen other polymorphs have been prepared, but it appears that they are monotropic forms. Many of these are nucleated with the inclusion of impurities and continue the impurity-induced structure although the available supply of impurity-atoms vanishes, leaving a pure boron monotropic polymorph which represents the limiting case of a nonstoichiometric boride. Others, such as the α-rhombohedral structure with the unique cross-linking three center (delta) bonds, exist simply because there is, at low temperatures (below 1100°C), no practicable mechanism to form the energically superior but more complex β-rhombohedral equilibrium structure. It has been suggested (2) that the α-tetragonal structure of boron, previously considered one of the best characterized crystallographic forms, was in all probability a network of boron atoms stabilized by small amounts of impurity atoms. This conclusion has been supported by recent evidence (3) based on studies of $B_{50}C_2$ and $B_{50}N_2$, both of which have the α-tetragonal structure.

Discrepancies in the crystallographic data on β-rhombohedral boron previously published can be explained in terms of the affinity of boron for many impurities (4).

Properties

Physical properties of boron are listed in Table 1. Additional physical properties are summarized in references 5 and 6. The major thermodynamic constants (heat content, heat capacity, Gibbs free energy, entropy, etc) have been determined over a wide range of temperatures in work related to rocket propulsion programs (3, 7–9). Thermal conductivity has also been studied over a broad range of temperatures ranging from 3 to 1000 K (10–11). Many electrical properties relating to semiconductor applications have been studied under varying conditions of temperature, pressure frequency, voltage, and impurity content. Among these are photoconductivity, thermoelectric power, Hall constant, and paramagnetic susceptibility (see references 5 and 6). The electrical resistivity of boron, about 10^6 at room temperature, varies from 10^{13} to 10 in the range -200 to $400°C$ (12–14).

Most measurements of the optical properties of elemental boron have been limited to the near and far ir regions for the β-rhombohedral and amorphous forms (13, 15–18). The α-rhombohedral modification has been reported to transmit yellow to red visible light, but precise data on corresponding optical parameters are lacking. The color of the opaque forms ranges from light yellow through brown to dull black for amorphous

Table 1. Physical Properties of Boron

Property	Value
atomic weight	10.811 ± 0.003
mp, °C	2190 ± 20
bp, °C	3660
coefficient of thermal expansion per °C	
from 25 to 1050°C	5×10^6 to 7×10^6
hardness	
Knoop, HK	2110–2580
Mohs, modified scale[a]	11
Vickers, HV	5000
density	
liquid[b]	2.08
α-rhombohedral crystals	2.46
filamentary boron	
tensile strength, MPa (psi)	3450–4830 (500,000–700,000)
Young's modulus, MPa (psi)	3040–3330 (440;000–480,000)
structural modifications[c]	
preparation temp, °C	
amorphous	800
α-rhombohedral	800–1100
α-tetragonal	1100–1300
β-rhombohedral	1300
density, g/cm³	
amorphous	2.3
α-rhombohedral	2.46
α-tetragonal	2.31
β-rhombohedral	2.35

[a] Diamond = 15.
[b] Just above melting point.
[c] The crystalline forms not listed here are designated tetragonal II (β-tetragonal), tetragonal III, and hexagonal.

boron, and is a consistent metallic gray-black for crystalline β-rhombohedral material. Large pieces of fused boron have a lustrous, black metallic appearance. Raman scattering work has also been reported (19–20).

Thermal neutron capture cross-section values of the two stable boron isotopes differ dramatically, averaging 3.84×10^{-25} m^2 for boron-10 and about 4×10^{-32} m^2 for boron-11 (21–23).

The chemical properties of elemental boron appear to depend on its morphology and particle size. In general, crystalline boron is relatively unreactive, whereas micron-sized, amorphous boron reacts readily, and sometimes violently, with a variety of chemical agents, as illustrated in Table 2. Boron forms borides with metals and metal oxides, boron carbide, B_4C, with carbon, boron nitride, BN, with nitrogen, and boron trihalides such as BF_3, BCl_3, and BBr_3 with halogens (see Boron compounds). Boron reacts with water at elevated temperatures to give boric acid and other products and reacts with its oxide, B_2O_3, above 1000°C to give boron monoxide or other boron suboxides. With mineral acids, depending on their concentration and temperature, reactions can be slow to explosive, with boric acid as the major product.

Owing to the use of boron in propulsion systems and pyrotechnical devices, its reactions with oxygen and oxidizing agents have been studied extensively (24–25) (see Explosives and propellants; Pyrotechnics). Several surface-controlled oxidation mechanisms have been proposed. The ignition temperature in air is 580°C. Reactions of boron with a wide variety of specific chemical compounds are listed in reference 6.

Preparation

Boron may be prepared from its compounds by chemical reduction with reactive elements, nonaqueous electrolytic reduction, or thermal decomposition. Summaries of preparative methods may be found in references 4, 5, 26 and 27.

Reduction of boron compounds, including borates, boron oxides and halides, fluoroborates, and borohydrides, has been carried out with H, Li, Be, Na, Mg, Al, Si, P, K, Ca, Fe, Zn, and Hg. The most common method for producing large amounts of elemental boron is the exothermic reduction of boron trioxide with magnesium, giving an amorphous product known as Moissan's boron. Reduction with hydrogen at high temperatures, especially hot-filament reaction with boron halides, is the conventional procedure for obtaining high-purity boron (99% and better).

Boron is also prepared by electrolysis of (1) fused melts of boron trioxide in potassium halides or oxides; (2) KBF_4–KCl melts; and (3) KBF_4–KF–KCl or KBF_4–

Table 2. Chemical Reactions of Boron

Reagent	Crystalline	Powdered, amorphous
hydrogen		reaction at 840°C
oxygen	stable at 750°C	pyrophoric at 700°C
chlorine	begins to react at 550°C	complete reaction at 400°C
water	inert at 100°C	slow reaction at 100°C
nitric acid, dil	inert	slow reaction
nitric acid, hot conc	reacts slowly	reacts vigorously
sodium hydroxide, fused	inert at 500°C	
sodium carbonate, fused	reacts completely at 850°C	reacts completely at 850°C

KCl–NaCl melts with a boron carbide anode. These methods, especially those developed after 1950 (28–30), give products with purity of 87–99.8%, whereas direct electrolysis of alkali or alkaline earth borates gives less pure products. However, electrolysis has not as of yet, been an important commercial method for the production of elemental boron.

Direct thermal decomposition of boron compounds to high purity boron is limited to halides and hydrides. Boron tribromide or triiodide, and boron hydrides (from diborane to decaborane) have been decomposed on a wide variety of substrates ranging from glass to tungsten at temperatures from 800–1500°C (5–6).

Purification. Ultrapure boron is purified by zone-refining (qv) or other thermal techniques, ie, impurities are removed by progressive recrystallization or volatilization at high temperatures. Limitations are imposed on these methods as much by difficulties in analyzing for impurities as by the unusually high reactivity of boron at high temperatures. Carbon, oxygen, hydrogen, and nitrogen are the most difficult impurities to detect as well as to remove. Nearly all are present in elemental boron, and few, if any, suitable crucible materials can be found that are free of these elements. However, impurities in ultrapure boron usually are below the 0.5% range.

Applications

In the early 1970s, the first large scale use of boron was studied in research and development facilities of the major aerospace corporations, at numerous United States Government laboratories, and at several academic institutions. A review of these studies is given in reference 6. Boron in the form of small-diameter filaments had been found to provide superior strength characteristics and was evaluated as a reinforcing material for composites in light-weight, high-stiffness applications in several models of commercial and military aircraft (see Composite materials). Exhaustive studies were made on mass production of the filaments by various methods and the cost was reduced to approximately $400/kg. Matrix materials for the composites varied from epoxy resins and polyamides to aluminum, titanium, and other metals. The filaments were often pretreated with coatings of various substances such as silicon carbide or boron nitride to maximize bonding between the filaments and the various matrix materials. Recently, because of their lower cost, graphite filaments are replacing boron filaments in many proposed uses, and it appears that boron filaments will be limited to highly specialized applications (see Ablative materials; Carbon and artificial graphite).

Electronic uses for boron as a major component have been quite limited. Several devices have been described, such as a neutron detector based on thermistors separately fabricated from the two boron isotopes, switches based on the current-voltage characteristics of β-rhombohedral boron, and resistors fabricated from boron–molybdenum disilicides. No commercial devices have as yet employed elemental boron except as a doping agent.

Applications in nuclear technology utilize thin films of boron for neutron counters, and dispersions of powdered boron in poly(vinyl chloride) or polyethylene castings which are effective for shielding against thermal neutrons (see Nuclear reactors). Metallurgical applications of the pure element are rare, but in some cases metal surfaces have been covered with amorphous boron and subsequently heated above 700°C, eg, to deoxidize copper or improve the grain structure of tungsten. Boron has been studied as a catalyst for olefin polymerization and dehydration of alcohols. As an abrasive it imparts superior durability to cut-off wheels. Finally, it has been found effective as the major component of an ultra-high-pressure gasketing composition.

BIBLIOGRAPHY

"Elemental Boron" under "Boron and Boron Alloys" in *ECT* 1st ed., Vol. 2, pp. 584–588, by Walter Crafts, Union Carbide and Carbon Research Laboratories, Inc.; "Elemental Boron" under "Boron and Boron Alloys" in *ECT* 2nd ed., Vol. 3, pp. 602–605, by J. G. Bower, U.S. Borax Research Corporation.

1. J. L. Hoard and R. E. Hughes, "Elemental Boron and Compounds of High Boron Content" in E. L. Muetterties, ed., *The Chemistry of Boron and Its Compounds,* John Wiley & Sons, Inc., New York, 1967.
2. E. Amberger and co-workers, *Proceedings of the Third International Symposium on Boron, Warsaw, Poland,* 1968, pp. 131–141.
3. R. P. Burns, A. J. Jason, and M. G. Inghram, *J. Chem. Phys.* **46,** 394 (1967).
4. T. Lundström, *Proceedings of t he Fourth International Symposium on Boron, Tbililsi,* 1972, pp. 44–51.
5. J. G. Bower in R. J. Brotherton and H. Steinberg, eds., *Progress in Boron Chemistry,* Vol. 2, Oxford, 1970, pp. 231–271.
6. "Elemental Boron" in *Mellor's Comprehensive Treatise, Boron Supplement Part C,* to be published by Longman, London.
7. *JANAF Thermochemical Tables,* The Dow Chemical Co., Midland, Mich., 1965 (Suppl.)
8. R. C. Paule and J. L. Margrave, *J. Phys. Chem.* **67,** 1368 (1963).
9. H. W. Kolsky, R. M. Gilmer, and P. W. Gilles, *J. Chem. Phys.* **27**(2), 494 (1957).
10. G. A. Slack, D. W. Oliver, and F. H. Horn, *Phys. Rev.* (B) 4(6), 1714 (1971).
11. A. V. Petrov and co-workers, *Fiz. Tverd. Tela* 11(4), 907 (1969).
12. T. N. Anderson, O. H. Berzijian, and H. Eyring, *J. Electrochem. Soc.,* **114,** 8 (1967).
13. W. Dietz and H. Herriman in *Boron,* Vol. 2, *Preparation, Properties and Applications,* New York, 1965, p. 107.
14. D. Geist, in ref. 13, p. 203.
15. I. R. King, F. E. Wawner, G. R. Taylor and C. P. Talley, in ref. 13, p. 45.
16. J. Jaumann and J. Schnell, *Z. Naturforsch.* **20A,** 1639 (1965).
17. H. Werheit, A. Hausen and H. Binnenbruck, *Phys. Stat. Solidi* 42(2), 733 (1970).
18. H. Binnenbruck and H. Werheit, *J. Less-Common Met.* **47,** 91 (1976).
19. W. Richter, A. Hansen, and H. Binnenbruck, *Phys. Status Solidi Boron* **60,** 461 (1973).
20. W. Richter, W. Weber, and K. Ploog, *J. Less-Common Met.* **47,** 85 (1976).
21. A. Prosdocimi and A. J. Deruytter, *J. Nucl. Energy (A and B)* 17(2), 83 (1963).
22. S. A. R. Wynchank, A. E. Cox, and C. H. Collie, *Nucl. Instr. Methods* 39(2), 350 (1966).
23. J. W. Meadows and J. F. Whalen, *Nucl. Sci. Eng.* **40,** 12 (1970).
24. E. Buchner and I. Husman, *Angew. Chem. Int. Ed.* 10(6), 421 (1971).
25. A. Macek and J. M. Semple, *U.S. Govt. Rept. SQUID-TR-ARC-13-PU,* 1970.
26. A. E. Newkirk and J. L. Hoard, *Adv. Chem. Ser.* **32,** 27 (1961).
27. A. E. Newkirk in R. M. Adams, ed., *Boron, Metallo Boron Compounds and Boranes,* New York, 1964, pp. 233–288.
28. N. P. Nies, *J. Electrochem. Soc.* **107**(10), 817 (1960).
29. K. Akashi and I. Egami, *Seisan Kenkyu* **15**(11), 427 (1965).
30. D. R. Stern, *J. Electrochem. Soc.* **107,** 441 (1960).

JOSEPH G. BOWER
U.S. Borax & Chemical Corp.

BORON COMPOUNDS

Boron oxides, boric acid, and borates, 67
Boric acid esters, 111
Refractory boron compounds, 123
Boron halides, 129
Boron hydrides, heteroboranes, and their metallo derivatives, 135
Boron hydrides, heteroboranes, and their metallo derivatives (commercial aspects), 183
Organic boron-nitrogen compounds, 188

BORON OXIDES, BORIC ACID, AND BORATES

This section includes the oxides and oxyacids of boron as well as a variety of hydrated and anhydrous metal borates. An alphabetical list of compounds referred to in the text and their Chemical Abstracts Service Registry Numbers are shown on page 107.

Nomenclature

The confusing and often ambiguous systems of nomenclature encountered in the literature of inorganic borates have been previously described (1); this situation has not been improved in recent years. The accumulation of detailed structural data for many of the crystalline compounds has led to derivation of more complex names and formulas in an effort to convey more precise information, eg, Chemical Abstracts has adopted a classification system based on a series of (usually hypothetical) boric acids.

A single example will suffice to illustrate the various terminologies. A few years ago, the compound with empirical formula $Zn_2B_6O_{11}.7H_2O$ would have been called dizinc hexaborate heptahydrate. Applying the resolved oxide system proposed by the IUPAC, the substance becomes $2ZnO.3B_2O_3.7H_2O$, known as zinc (2:3) borate heptahydrate. This latter system has gained wide acceptance in recent years, and will be followed throughout this article. However, a recent structural determination indicates a more precise formulation, $Zn(B_3O_3(OH)_5).H_2O$, ie, zinc triborate monohydrate, which is listed in Chemical Abstracts as boric acid, $H_7B_3O_8$, zinc salt [12429-73-1] (2). Since many authors continue to use the older formulations, a second listing has been devised by Chemical Abstracts for the same compound, ie, boric acid, $H_4B_6O_{11}$, zinc salt (1:2) heptahydrate [12280-01-2].

BORATE MINERALS

The principal borate minerals and their compositions are listed in Table 1. A much more complete listing is given by Gmelin (3).

At present, borax (tincal), colemanite, probertite, ulexite, and szaibelyite are the only borate minerals of commercial importance. Borax and colemanite are the most important. Present borate production comes mostly from five countries: the United States, Turkey, the U.S.S.R., Argentina, and China. Major deposit areas in these countries are shown in Table 2. Total Russian, Argentine, and Chinese reserves may be as high as 60, 15, and 30 million metric tons of B_2O_3, respectively (4).

BORON COMPOUNDS (OXIDES, ACID, BORATES)

Table 1. Borate Minerals

Mineral	CAS Registry Number	Composition	Wt %, B_2O_3
sassolite	[10043-35-3]	$B(OH)_3$	56.4
borax (tincal)	[1303-96-4]	$Na_2O.2B_2O_3.10H_2O$	36.5
tincalconite	[12045-88-4]	$Na_2O.2B_2O_3.5H_2O$	47.8
kernite	[12045-87-3]	$Na_2O.2B_2O_3.4H_2O$	51.0
inyoite	[12260-25-2]	$2CaO.3B_2O_3.13H_2O$	37.6
meyerhofferite	[57572-66-4]	$2CaO.3B_2O_3.7H_2O$	46.7
colemanite	[12291-65-5]	$2CaO.3B_2O_3.5H_2O$	50.8
priceite	[61583-61-7]	$4CaO.5B_2O_3.7H_2O$	49.8
ulexite	[1319-33-1]	$Na_2O.2CaO.5B_2O_3.16H_2O$	43.0
probertite	[12229-14-0]	$Na_2O.2CaO.5B_2O_3.10H_2O$	49.6
hydroboracite	[12046-12-7]	$CaO.MgO.3B_2O_3.6H_2O$	50.5
inderite	[12260-26-3]	$2MgO.3B_2O_3.15H_2O$	37.3
szaibelyite	[12447-04-0] and [36564-04-2]	$2MgO.B_2O_3.H_2O$	41.4

Table 2. Distribution of Borate Minerals—Major Producing Countries

Country	Area	Major minerals	Reserves, million metric tons B_2O_3	Reference
U.S.	Boron, Calif.	tincal, kernite	41–50	4
	Searles Lake, Calif.	brine	15	4
	Death Valley, Calif.	colemanite, ulexite, probertite	several	10
Turkey	Bigadic	colemanite, priceite, ulexite		
	Emet	colemanite	23	11
	Kirka	tincal, colemanite, ulexite	122	11
U.S.S.R.	Inder	szaibelyite		
Argentina	Tincalayu	tincal, kernite, ulexite		
China	Iksaydam			

Reports have recently been made concerning the minerals of the Searles Lake area (5), the Boron-Kramer area (6), and the Death Valley area (7) in the United States. The mineralogy of the Kirka (8) and Emet (9) districts of Turkey have been described. A number of general reviews on borate manufacture and economics give mineralogical data (10–11). A large review on boron geochemistry has recently been completed (12). The application of crystallographic data to borate geology has been described (13), and the history of borate mining and production in the United States has been reviewed (14–15).

BORON OXIDES

Boric oxide, B_2O_3, is the only commercially important oxide; however, one higher oxide and several suboxides have been reported.

Boron Monoxide and Dioxide. High temperature vapor phases of BO, B_2O_2, and BO_2 have been the subjects of a number of spectroscopic and mass spectrometric studies aimed at developing theories of bonding, electronic structures, and thermochemical data (1,16). Values for the principal thermodynamic functions have been calculated and compiled for these gases (17).

Vibrational emission spectra indicate that the B_2O_2 molecule has a linear O=B–B=O structure. Values of 782 and 502 kJ/mol (187 and 120 kcal/mol) were calculated for the respective B=O and B–B bond energies (18).

Two noncrystalline solid forms of BO have been prepared (1,16). Several polymeric $(BO)_n$ or $(B_2O_2)_n$ structures have been proposed for these materials. Although conclusive structural evidence is unavailable, the presence of B–B bonds appears likely. The low temperature form is a white, water-soluble powder produced by vacuum-dehydration of $B_2(OH)_4$. This product is irreversibly converted to an insoluble, light-brown modification on heating above 500°C. The latter material was also prepared by high-temperature reduction of B_2O_3 by boron, carbon, or boron carbides (19). Both BO polymorphs are strong reducing agents that decompose slowly in water to yield hydrogen gas and boric acid.

Lower Oxides. A number of hard, refractory suboxides have been prepared either as by-products of elemental boron production (1), or by the reaction of boron with boric acid at high temperatures and pressures (20). It appears that the various oxides represented as B_6O, B_7O, $B_{12}O_2$, and $B_{13}O_2$ may all be the same material in varying degrees of purity. A representative crystalline substance was determined to be rhombohedral boron suboxide ($B_{12}O_2$), usually mixed with traces of boron or B_2O_3 (20). A study has been made of the mechanical properties of this material which exhibits a hardness comparable to that of boron carbide (21). At temperatures above 1000°C, $B_{12}O_2$ gradually decomposes to B (s) and B_2O_2 (g).

Boric Oxide. Boric oxide, B_2O_3, (also known as diboron trioxide, boric anhydride, or anhydrous boric acid) is normally encountered in the vitreous state. This colorless, glassy solid is usually prepared by dehydration of boric acid at elevated temperatures. It is quite hygroscopic at room temperature, and the commercially available material contains ca 1% moisture as a surface layer of boric acid. The reaction with water:

$$B_2O_3 \text{ (glass)} + 3\ H_2O \rightarrow 2\ B(OH)_3$$

is exothermic, $\Delta H° = -75.94$ kJ/mol (-18.15 kcal/mol) B_2O_3 (22).

Boric oxide is an excellent Lewis acid. It coordinates even weak bases to form four-coordinate borate species. Reaction with sulfuric acid produces $H[B(HSO_4)_4]$ (23). At high temperatures (>1000°C) boric oxide dissolves most metal oxides and is thus very corrosive to metals in the presence of oxygen.

Molten boric oxide reacts readily with water vapor to form HBO_2 gas. For the reaction:

$$\tfrac{1}{2}\ B_2O_3 \text{ (glass)} + \tfrac{1}{2}\ H_2O\ (g) \rightarrow HBO_2\ (g)$$

a value of $\Delta H_{298} = -199.2 \pm 8.4$ kJ/mol (-47.61 ± 2.0 kcal/mol) has been calculated. This reaction is of considerable economic importance to glass manufacturers as B_2O_3 losses during glass processing are greatly increased by the presence of water. For this reason anhydrous borates or boric oxide are often preferred over hydrated salts (eg, borax) or boric acid. The presence of MgO has been found to reduce volatilization of B_2O_3 from glass charges (24).

The physical properties of vitreous boric oxide (Table 3) are somewhat dependent

Table 3. Physical Properties of Vitreous Boric Oxide

Property	Value	Reference
vapor pressure[a], 1331–1808 K	$\log P_{kPa} = 5.849 - \frac{16960}{T}$	32
heat of vaporization[b], ΔH, kJ/mol, 1500 K	390.4	33
298 K	431.4	33
boiling point, extrapolated	2316°C	22
viscosity, $\log \eta$, mPa.s (= cP)		
350°C	10.60	34
700°C	4.96	34
1000°C	4.00	34
density, g/cm^3, 0°C	1.8766	
18–25°C, well-annealed	1.844	
18–25°C, quenched	1.81	
500°C	1.648	34
1000°C	1.528	34
index of refraction, 14.4°C	1.463	
heat capacity (specific)[b], J/(kg·K)		
298 K	62.969	22
500 K	87.027	22
700 K	132.63	22
1000 K	131.38	22
heat of formation, ΔH_f, kJ, 298.15 K[b]		
for 2 B(c) + 3/2 O$_2$(g) = B$_2$O$_3$ (glass)	-1252.2 ± 1.7	22

[a] To convert kPa to torr, multiply by 7.5.
[b] To convert J to cal, divide by 4.184.

on its moisture content and thermal history. Much of the older physical data has been revised following development of more reliable techniques for sample preparation (25–27). Boric oxide glass softens on heating to about 325°C and becomes just pourable at 500°C. A nearly anhydrous material can be obtained by prolonged heating at 1200°C (25).

The historical debate over the structures of vitreous and molten boric oxide is now largely resolved (28). The solid glass is a branched network of planar boroxol (—BO—)$_3$ rings (1). The three oxygens outside the ring form bridges to neighboring rings or to planar BO$_3$ groups (29–31).

(1)

The branched network breaks down as the glass melts, and spectroscopic features due to the boroxol group (eg, the strong Raman line at 808 cm^{-1}) disappear as the liquid is heated to 800°C. It has been proposed (27) that above 800°C the liquid consists of discrete, but strongly associated, small molecules, conceivably the same monomeric B$_2$O$_3$ units observed in the vapor state (35).

Two crystalline forms of boric oxide have been prepared, and the structures of both materials have been determined by x-ray diffraction (23). The phase relationships between the liquid and crystalline forms have also been developed (35a). The more common crystal phase, B_2O_3-I (d = 2.46, mp = 455–475°C), is more stable than the vitreous phase. For the transformation B_2O_3-I → B_2O_3 (glass), $\Delta H° = +18.24$ kJ/mol (4.36 kcal/mol) (22). However, B_2O_3 glass does not crystallize readily in the absence of seed crystals or increased pressure. A second crystalline phase, B_2O_3-II (d = 3.11), can be obtained at high temperatures and pressures. The crystal lattice of B_2O_3-II consists of a highly compact network of BO_4 tetrahedra where the four apical oxygens are shared by either two or three boron atoms. The acidic character associated with trigonal BO_3 groups is thus masked in B_2O_3-II. Although this material is thermodynamically unstable under ordinary conditions, it reacts very slowly with Lewis bases such as water and fluoride ion.

Two industrial grades of vitreous boric oxide are currently produced in the United States. A high purity grade (99% B_2O_3) is produced by fusing refined, granular boric acid in a glass furnace fired by oil or gas. The molten glass is solidified in a continuous ribbon as the melt flows over chill-rolls. The solid product is crushed, screened, and packed in sacks or drums with moisture-proof liners. The price of this product remained nearly constant throughout the decade from 1960–1970, but increased manufacturing costs have brought about a threefold increase during the past seven years. The carload (>36 metric tons) price in July 1977 was $1230–1360 per metric ton (fob plant) depending on mesh size and packaging (36). A less costly process has been developed that eliminates the need for refining the boric acid intermediate (37). Mixtures of borax and sulfuric acid are fed to a fusion furnace where molten B_2O_3 and Na_2SO_4 separate into layers at temperatures above 750°C. The decanted boric oxide product, typically containing 96–97% B_2O_3, is of sufficient purity for many applications. It is marketed at a price about half that of the high purity boric oxide (36).

The major uses of boric oxide relate to its behavior as a flux, an acid catalyst, or a chemical intermediate. The fluxing action of B_2O_3 is important in preparing many types of glass, glazes, frits, ceramic coatings, and porcelain enamels. The largest single consumer by far is the textile fiberglass industry where boric oxide is used in producing the low-sodium continuous filaments (E-glass) required for glass-belted tires and fiberglass reinforced plastics (4) (see Glass).

Boric oxide is used as a catalyst in many organic reactions. It also serves as an intermediate in the production of boron halides, -esters, -carbide, -nitride, and metallic borides.

BORIC ACID

The name boric acid is usually associated with orthoboric acid, which is the only commercially important compound. Three forms of metaboric acid also exist.

Forms of Boric Acid

Orthoboric acid, H_3BO_3, crystallizes from aqueous solutions as white, waxy plates (triclinic; sp gr $_4^{14}$ = 1.5172). Its normal mp is 170.9°C, however, when heated slowly it loses water to form metaboric acid, HBO_2, which may exist in one of three crystal modifications. Orthorhombic HBO_2-III (d = 1.784 g/cm³, mp = 176°C) forms first around 130°C, and gradually changes to monoclinic HBO_2-II (d = 2.045 g/cm³, mp

= 200.9°C). Water vapor pressures associated with these decompositions are as follows:

Temp, °C	Vapor pressure of H_2O over H_3BO_3 and HBO_2-III, kPa (mm Hg)	Vapor pressure of H_2O over H_3BO_3 and HBO_2-II, kPa (mm Hg)
25	0.048 (0.36)	0.16 (1.2)
100	8.4 (63)	16 (121)
130	39.9 (299)	62.5 (469)
150	102 (768)	143 (1074)

At temperatures above 150°C, dehydration continues to yield viscous liquid phases beyond the metaboric acid composition (38). The most stable form of metaboric acid, cubic HBO_2-I (d = 2.49 g/cm^3, mp = 236°C) crystallizes slowly when mixtures of H_3BO_3 and HBO_2-III are melted in an evacuated, sealed ampoule and held at 180°C for several weeks (39).

The relationships between condensed phases in the B_2O_3–H_2O system are shown in Figure 1 (38). There is no evidence for stable phases other than those shown. B_2O_3 melts and glasses containing less than 50 mol % H_2O have mechanical and spectroscopic properties consistent with mixtures of HBO_2 and vitreous B_2O_3.

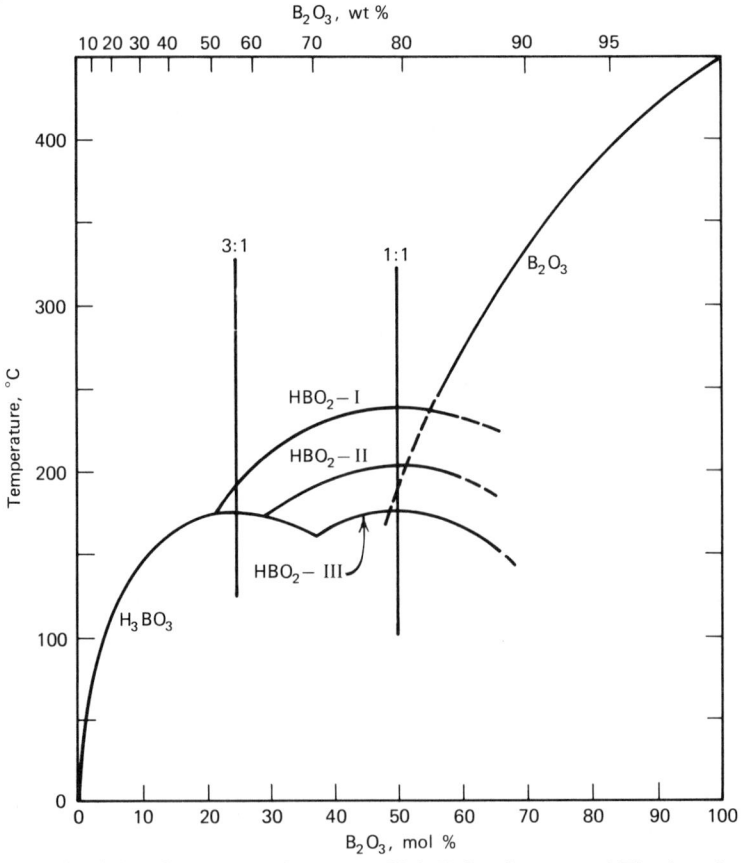

Figure 1. Solubility diagram for the system H_2O–B_2O_3. Courtesy of The American Journal of Science.

Vapor phases in the B_2O_3–H_2O system include water vapor and H_3BO_3 (g) at temperatures below 160°C. Appreciable losses of boric acid occur when aqueous solutions are concentrated by boiling (40). At high temperatures (600–1000°C), HBO_2 (g) is the principal boron species formed by equilibration of water vapor and molten B_2O_3 (41).

The crystal structure of orthoboric acid consists of planar sheets made up of hydrogen-bonded, triangular $B(OH)_3$ molecules. The stacking pattern of the molecular layers is completely disordered, indicative of relatively weak van der Waals forces between the planes. This accounts for the ease with which the crystals are cleaved into thin flakes (42). The structures of all three forms of metaboric acid are also known. The basic structural unit of HBO_2-III is the trimeric ring (2).

(2)

This trimer may persist to some extent in the vapor phase, but infrared spectra indicate that monomeric O=B–OH species predominate in gaseous metaboric acid (43).

Chemical Properties

The standard heats of formation of crystalline orthoboric and metaboric acids are (in kJ/mol): $\Delta H_f^\circ = -1094.3$ (H_3BO_3); -804.04 (HBO_2-I); -794.25 (HBO_2-II); and -788.77 (HBO_2-III) (44). Values for the principal thermodynamic functions of H_3BO_3 are given in Table 4 (22).

The solubility of boric acid in water (Table 5) increases rapidly with temperature. The heat of solution is somewhat concentration dependent. For solutions with molalities in the range 0.03–0.9 m, the molar heats of solution fit the empirical relation (45):

$$\Delta H = [22062 - 222\, m + 979\, e^{-1230m}] \text{ J/mol}$$

The presence of inorganic salts may enhance or depress the solubility of boric acid in water. It is increased by potassium chloride as well as potassium or sodium sulfate, but decreased by lithium and sodium chlorides. Basic anions and other

Table 4. Thermodynamic Properties of Crystalline Boric Acid (H_3BO_3)[a,b]

Temp, K	C_p°, J/(kg·K)	S°, J/K	$H^\circ - H^\circ{}_{298}$, J/mol
0	0	0	−13393
100	35.92	28.98	−11636
200	58.74	61.13	−6866
298	81.34	88.74	0
400	100.21	115.39	9284

[a] To convert J to cal, divide by 4.184.
[b] Ref. 41.

BORON COMPOUNDS (OXIDES, ACID, BORATES)

Table 5. Solubility of Boric Acid in Water and in Various Organic Solvents

Solvent	Temp, °C	% H_3BO_3 by wt of soln	g H_3BO_3/L
water	0	2.52	
	20	4.72	
	40	8.08	
	60	12.97	
	80	19.10	
	100	27.53	
glycerol, 98.5%	20	19.9	
86.5%	20	12.1	
ethylene glycol	25	18.5	
diethylene glycol	25	13.6	
ethyl acetate	25	1.5	
acetone	25	0.6	
acetic acid, 100%	30	6.3	
methanol	25		173.9
ethanol	25		94.4
n-propanol	25		59.4
2-methylbutanol	25		35.3
dioxane	25		ca 14.6
pyridine	25		ca 70
aniline	20	0.15	

nucleophiles (notably borates and fluoride) greatly increase boric acid solubility by forming polyions (41).

Boric acid is quite soluble in many organic solvents (Table 5). Some of these solvents (eg, pyridine, dioxane, diols) are known to form complexes with boric acid.

Dilute aqueous solutions of boric acid contain predominantly monomeric, undissociated $B(OH)_3$ molecules. The acidic properties of boric acid relate to its behavior as a base acceptor rather than a proton donor. For the reaction:

$$B(OH)_3 + H_2O \rightleftharpoons B(OH)_4^- + H^+$$

an equilibrium constant of 5.80×10^{-10} has been reported. However, calculated pH values based on this constant deviate considerably from measured values as the boric acid concentration is increased, as shown in Table 6. The increased acidity has been attributed to secondary equilibria involving condensation reactions between $B(OH)_3$ and $B(OH)_4^-$ [15390-83-7] to produce polyborates. A trimeric species $B_3O_3(OH)_4^-$ [17927-69-4] appears to be the most important of these complex ions (47).

Table 6. Comparison of Observed and Calculated pH Values for Boric Acid[a]

Concentration, M	pH observed	pH calculated
0.0603	5.23	5.23
0.0904	5.14	5.14
0.1205	5.01	5.08
0.211	4.71	4.96
0.422	4.22	4.80
0.512	4.06	4.76
0.753	3.69	4.54

[a] Ref. 46.

The apparent acid strength of boric acid is increased by strong electrolytes that modify the structure and activity of the solvent water, and also by reagents that form complexes with $B(OH)_4^-$ and/or polyborate anions. More than one mechanism may be operative when salts of metal ions are involved. In the presence of excess calcium chloride the strength of boric acid becomes comparable to that of carboxylic acids, and such solutions may be titrated with strong base to a sharp phenolphthalein endpoint. Normally titrations of boric acid are carried out following addition of mannitol or sorbitol which form stable chelate complexes with $B(OH)_4^-$ in a manner typical of polyhydroxy compounds. Equilibria of the type:

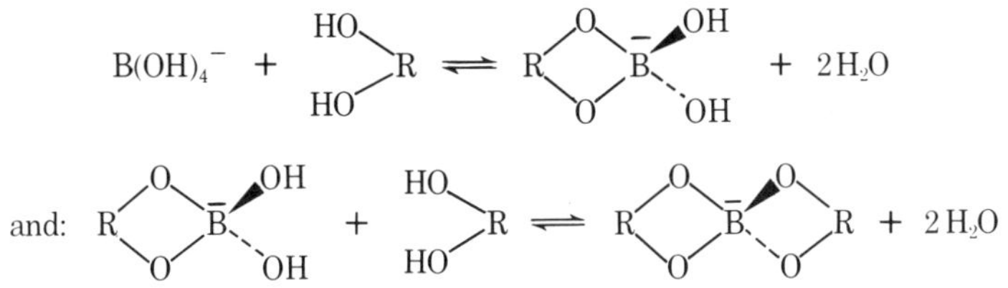

$$B(OH)_4^- + \begin{array}{c}HO\\HO\end{array}\!\!R \rightleftharpoons R\!\!\begin{array}{c}O\\O\end{array}\!\!\overset{-}{B}\!\!\begin{array}{c}OH\\OH\end{array} + 2H_2O$$

and:
$$R\!\!\begin{array}{c}O\\O\end{array}\!\!\overset{-}{B}\!\!\begin{array}{c}OH\\OH\end{array} + \begin{array}{c}HO\\HO\end{array}\!\!R \rightleftharpoons R\!\!\begin{array}{c}O\\O\end{array}\!\!\overset{-}{B}\!\!\begin{array}{c}O\\O\end{array}\!\!R + 2H_2O$$

have been exploited in other applications besides analytical determinations of boric acid (48). Ion-exchange resins containing polyols have been developed that are highly specific for removing borates from solution (49). A number of aliphatic and aromatic diols have been patented as extractants for borates and boric acid (50).

Boric acid reacts with fluoride ion to form a series of fluoroborates by displacement of OH^- by F^-. Stepwise formation of $BF(OH)_3^-$ [*32554-53-3*], $BF_2(OH)_2^-$ [*32554-52-2*], and $BF_3(OH)^-$ [*18953-00-9*] proceeds rapidly in acidic solutions, but BF_4^- [*14874-70-5*] forms slowly (51). A fluorosubstituted polyborate, $B_3O_3F_6^{3-}$ [*59753-06-9*], has also been identified (47).

Alcohols react with boric acid with elimination of water to form borate esters, $B(OR)_3$ (see Boric acid esters).

A wide variety of borate salts and complexes have been prepared by the reaction of boric acid with inorganic bases, amines, and heavy metal cations or oxyanions (41–42). Fusion with metal oxides yields anhydrous borates or borate glasses.

Manufacture

The majority of boric acid is produced by the reaction of inorganic borates with sulfuric acid in an aqueous medium. Sodium borates are the principal raw material in the United States. European manufacturers have generally used partially refined calcium borates (mainly colemanite from Turkey). This pattern may shift somewhat with the development of Turkish sodium borate reserves (52).

When granulated borax or borax-containing liquors are treated with sulfuric acid, the following reaction ensues:

$$Na_2B_4O_7 \cdot xH_2O + H_2SO_4 \rightarrow 4\,H_3BO_3 + Na_2SO_4 + (x-5)\,H_2O$$

The boric acid product is obtained by cooling the solution to the proper temperature. The by-product sodium sulfate may be recovered as the decahydrate by further cooling the mother liquor.

When boric acid is made from colemanite, the ore is ground to a fine powder and stirred vigorously with diluted mother liquor and sulfuric acid at about 90°C. The by-product calcium sulfate is removed by settling and filtration, and the boric acid is crystallized by cooling the filtrate.

A unique liquid–liquid extraction process for manufacturing boric acid has been operated at Searles Lake, Trona, Calif., since 1962. A number of chelating aromatic diols (eg, 3-chloro-2-hydroxy-5-(1,1,3,3-tetramethylbutyl)benzyl alcohol) have been developed and patented for use in this process (50). The currently preferred compound has not been disclosed. Weak brines containing sodium and potassium borates come in contact with a kerosene solution that carries the organic extractant. The organic phase becomes loaded with alkali metal salts of the anionic diol–borate complex. The spent brine is discarded and the loaded extractant is stripped with dilute sulfuric acid to yield an aqueous phase containing boric acid as well as sodium and potassium sulfates. Residual organics are removed by passing the aqueous stripping solution through a column of activated carbon. Crystalline boric acid and a mixture of sulfates are obtained sequentially when the acid solution is concentrated in a pair of evaporator-crystallizers (53) (see Chemicals from brine).

Boric acid crystals are separated from aqueous slurries by centrifugation and dried in rotary driers heated indirectly by warm air. To avoid overdrying, the product temperature should not exceed 50°C. Powdered and impalpable boric acid are produced by milling the crystalline material.

The principal impurities in technical grade boric acid are the by-product sulfates (<0.1%) and various minor metallic impurities present in the borate ores. A boric acid titer is not an effective measure of purity since overdrying may result in partial conversion to metaboric acid and lead to H_3BO_3 assays above 100%. High purity boric acid is prepared by recrystallization of technical-grade material.

Three grades of granular and powdered boric acid are manufactured in the United States. In July 1977 carload (ca 91 t) prices per metric ton of granular boric acid were: technical grade, $332; NF grade, $410; special quality grade, $688; and all prices fob plant for material packed in 45.4-kg multiwall sacks (36).

Uses

Boric acid has a surprising variety of applications in both industrial and consumer products (4,52). It serves as a source of B_2O_3 in many fused products, including textile fiber glass, optical and sealing glasses, heat-resistant borosilicate glass, ceramic glazes, and porcelain enamels. It also serves as a component of fluxes for welding and brazing (see Solders; Welding).

A number of boron chemicals are prepared directly from boric acid. These include synthetic, inorganic borate salts, boron phosphate, fluoborates, borate esters, and metal alloys such as ferroboron.

Boric acid catalyzes the air oxidation of hydrocarbons and increases the yield of alcohols by forming esters that prevent further oxidation of hydroxyl groups to ketones and carboxylic acids (see Hydrocarbon oxidation).

The bacteriostatic and fungicidal properties of boric acid have led to its use as a preservative in natural products such as lumber, rubber latex emulsions, leather, and starch products. It is also used in washing citrus fruits to inhibit mold, and in mildew-resistant latex paints.

NF-grade boric acid serves as a mild, nonirritating antiseptic in mouthwashes, hair rinse, talcum powder, eyewashes, and protective ointments (see Disinfectants). Although relatively nontoxic to mammals (54), boric acid is quite poisonous to insects, and has been used to control cockroaches and to protect wood against insect damage (see Insect control technology).

Inorganic boron compounds are generally good fire retardants (55). Boric acid, alone or in mixtures with sodium borates, is particularly effective in reducing the flammability of cellulosic materials. Applications include treatment of wood products, cellulose insulation, and cotton batting used in mattresses (see Flame retardants).

As boron compounds are good absorbers of thermal neutrons (owing to isotope ^{10}B), the nuclear industry has developed many applications. High purity boric acid is added to the cooling water used in high-pressure water reactors.

SOLUTIONS OF BORIC ACID AND BORATES

Polyborates and pH Behavior

As stated previously, boric acid is essentially monomeric in dilute aqueous solutions, but polymeric species may form at concentrations above 0.1 M. The conjugate base of boric acid in the water system is the tetrahydroxyborate anion, $B(OH)_4^-$. This species is also the principal anion in solutions of alkali metal (1:1) borates such as $Na_2O \cdot B_2O_3 \cdot 4H_2O$ (56). Mixtures of $B(OH)_3$ and $B(OH)_4^-$ would appear to form classical buffer systems where the solution pH is governed primarily by the acid/salt ratio, ie, $[H^+] = K_a [B(OH)_3/B(OH)_4^-]$. This relationship is nearly correct for solutions of sodium or potassium (1:2) borates (eg, borax), where the ratio $B(OH)_3/B(OH)_4^- = 1$, and the pH remains near 9.0 over a wide range of concentrations. However, for solutions that have pH values much greater or less than 9.0, the pH changes greatly on dilution as shown in Figure 2 (57).

This anomalous pH behavior is due to the presence of polyborates which dissociate

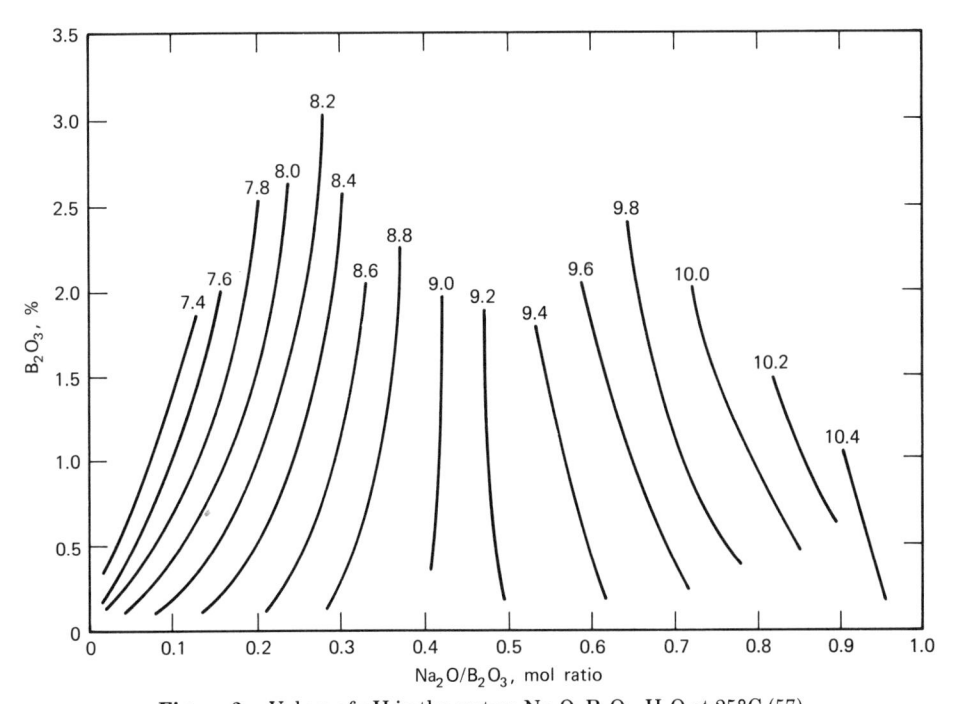

Figure 2. Values of pH in the system $Na_2O–B_2O_3–H_2O$ at 25°C (57).

into $B(OH)_3$ and $B(OH)_4^-$ as the solutions are diluted. It is clear that the principal polymeric species below pH 9.0 contain more $B(OH)_3$ than $B(OH)_4^-$ since the $B(OH)_3/B(OH)_4^-$ ratio is increased (pH decreases) on dilution. The reverse is true above pH 9.0 where the basic solutions become still more basic as excess $B(OH)_4^-$ is formed on dilution. There is a certain mol ratio (Na_2O/B_2O_3 = 0.41 at pH 9.00, or K_2O/B_2O_3 = 0.405 at pH 8.91) where the pH is independent of concentration. In such solutions the $B(OH)_3/B(OH)_4^-$ ratio must be the same as that which exists in the polyborate ions present under those conditions. This ratio and the pH associated with it have been termed the isohydric point of borate solutions (58).

The presence of metal salts, particularly those containing alkaline earth cations and/or halides, cause some shifts in the polyborate equilibria. This may be due to direct interaction with the boron–oxygen species, or to changes in the activity of the solvent water (59).

Solubility Trends

Formation of polyborates greatly enhances the mutual solubilities of boric acid and alkali borates. Solubility isotherms in the system $Na_2O–B_2O_3–H_2O$ are shown in Figure 3. When borax, $Na_2B_4O_7 \cdot 10H_2O$, is added to a saturated boric acid solution,

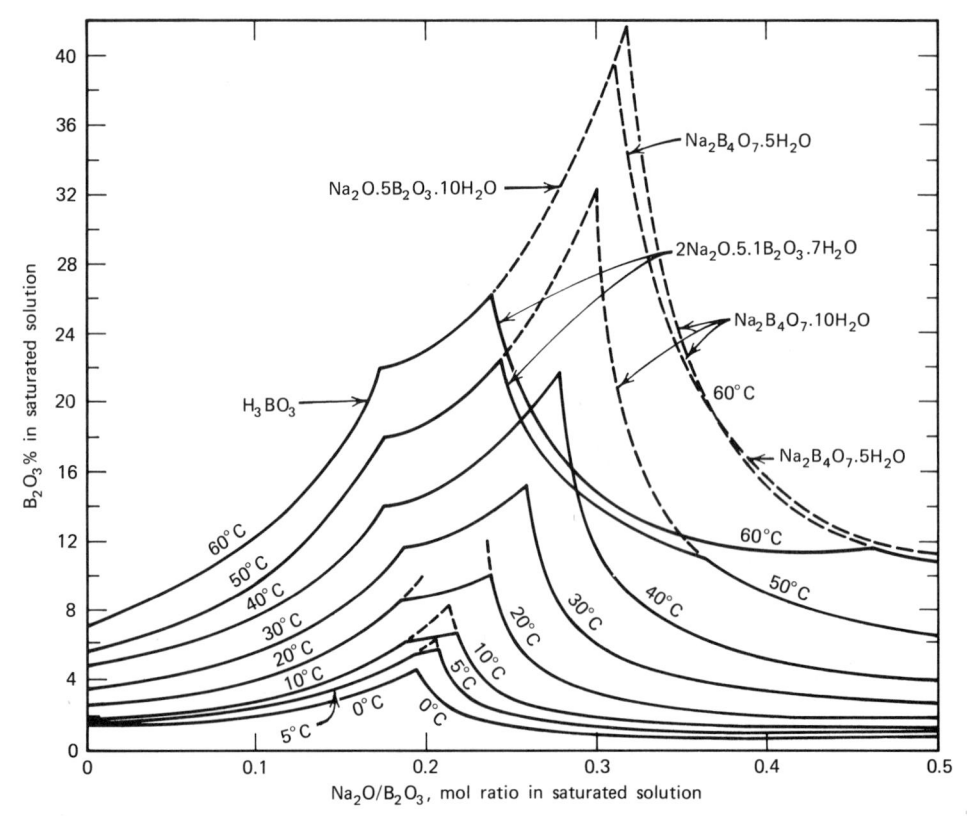

Figure 3. Solubility isotherms for the system $Na_2B_4O_7–B_2O_3–H_2O$ at 0–60°C. The compound $2 Na_2O \cdot 5.1B_2O_3 \cdot 7H_2O$ (Suhr's borate) usually does not appear since it crystallizes very slowly in the absence of seed.

or boric acid to a saturated borax solution, the B_2O_3 wt % in the solution greatly increases. Polymerization decreases the concentrations of $B(OH)_3$ and $B(OH)_4^-$ in equilibrium with the solid phases, thus permitting more borax or boric acid to dissolve. The evolution of heat that accompanies mixing of boric acid and borate solutions indicates that the formation of polyborates is exothermic.

Sodium borate solutions near the Na_2O/B_2O_3 ratio of maximum solubility can be spray dried to form an amorphous product with the approximate composition $Na_2O\cdot4B_2O_3\cdot4H_2O$ (60). This material will dissolve rapidly in water without any decrease in temperature to form supersaturated solutions. Such solutions have found application in treating cellulosic materials to impart fire-retardant and decay-resistant properties (see Cellulose; Flame retardants).

The Nature of Polyborate Species

From a series of very accurate pH studies, Ingri calculated a series of equilibrium constants involving the species $B(OH)_3$, $B(OH)_4^-$, and the polyions $B_3O_3(OH)_5^{2-}$ [12344-78-4], $B_3O_3(OH)_4^-$ [12344-77-3], $B_5O_6(OH)_4^-$ [12343-58-7], and $B_4O_5(OH)_4^{2-}$ [12344-83-1]. The relative populations of these species as functions of pH are shown in Figure 4 (61). It is clear that species containing 3, 4, and 5 borons are significant at intermediate pH values. The ratio between the total anionic charge and the number of borons per ion increases with increasing pH.

The polyions postulated by Ingri all have known structural analogues in crystalline borate salts; a recent investigation of the Raman spectra of borate solutions confirmed the presence of three of these species. The triborate, $B_3O_3(OH)_4^-$, tetraborate,

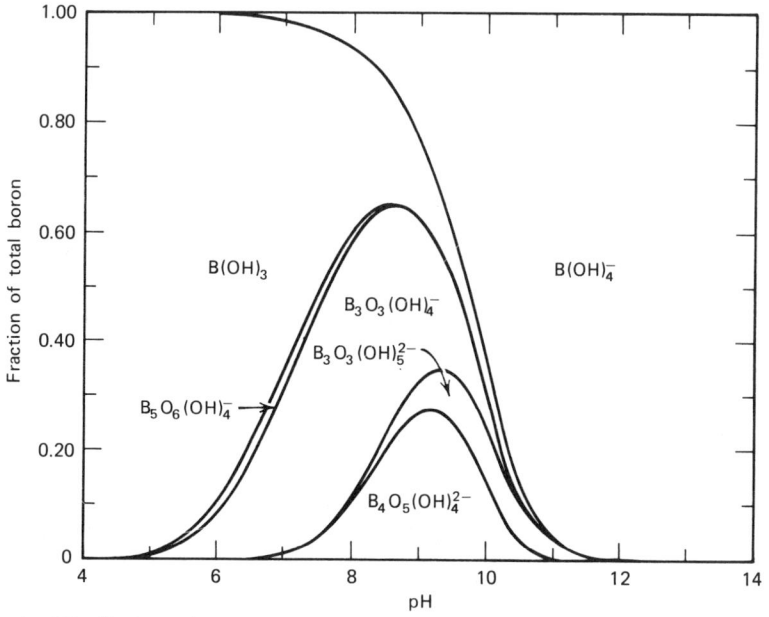

Figure 4. Distribution of boron in various ions (61). Total B_2O_3 concentration is 13.93 g/L. At a given pH, the fraction of the total boron in a given ion is represented by the portion of a vertical line falling within the corresponding range.

[$B_4O_5(OH)_4^{2-}$], and pentaborate, $B_5O_6(OH)_4^-$, ions were all identified in the predicted pH regions. The skeletal structures shown below were assigned, based on coincidences between the solution spectra and those of solid borates for which definitive structural data are available (47).

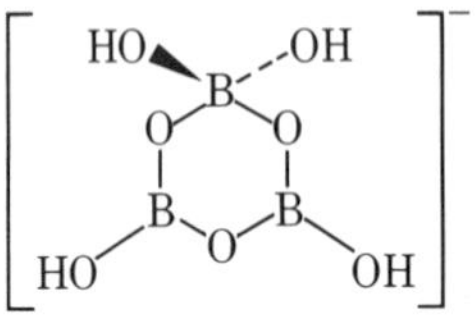

$B_3O_3(OH)_4^-$ 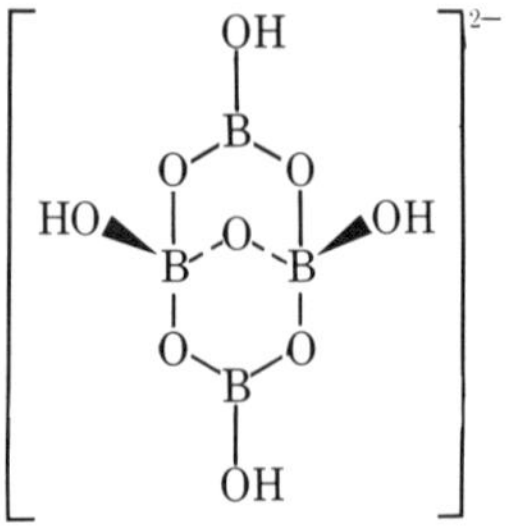 $B_4O_5(OH)_4^{2-}$ 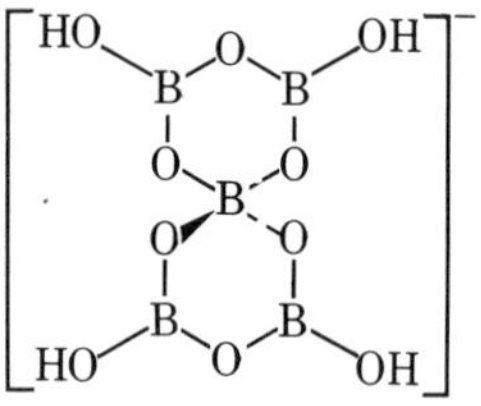 $B_5O_6(OH)_4^-$

A rapid equilibrium exists among the various polyborate species in aqueous solutions. The nmr spectra of boric acid–borate mixtures indicate a single time-averaged environment for the ^{11}B nucleus over a wide range of pH values. The measured chemical shift corresponds to a weighted mean of $B(OH)_3$ and $B(OH)_4^-$ (62).

SODIUM BORATES

Properties

Disodium Tetraborate Decahydrate (Borax Decahydrate). Disodium tetraborate decahydrate, $Na_2O \cdot 2B_2O_3 \cdot 10H_2O$, (formula wt, 381.43; monoclinic; sp gr, 1.71; specific heat, 1.611 kJ/(kg·K) [0.385 kcal/(g·°C)] at 25–50°C (63); heat of formation, −6.2643 MJ/mol (−1497.2 kcal/mol) (64)) exists in nature as the mineral borax. Its crystal habit may be changed by adding various substances and by altering other conditions (65).

The solubility–temperature curves for the Na_2O–B_2O_3–H_2O system are given in Figure 5 (also see Table 7). The solubility curves of the penta- and decahydrates intersect at 60.6–60.8°C, indicating that the decahydrate, when added to a saturated solution above this temperature, will dissolve with crystallization of the pentahydrate, and the reverse will occur below this temperature. This transition temperature may be lowered in solutions of inorganic salts. Heats of solution for borax have been determined (63,67) and the manufacturer quotes a value of 283 kJ/kg (122 Btu/lb) (36).

The vapor pressures of saturated borax solutions at various temperatures are as follows (67–68):

kPa (mm Hg)	Temp, °C
17.25 (129.4)	57.94
17.33 (130.0)	57.99
17.51 (131.4)	58.23
17.74 (133.1)	58.56
17.94 (134.7)	58.82
18.05 (135.4)	58.91
18.42 (138.2)	59.42

Table 7. Solubilities of Alkali Metal and Ammonium Borates at Various Temperatures

Compound	CAS Registry Number	0	10	20	25	30	40	50	60	70	80	90	100
Li$_2$O·5B$_2$O$_3$·10H$_2$O[a]	[37190-10-6]							20.88	24.34	27.98	31.79	36.2	41.2
Li$_2$O·2B$_2$O$_3$·4H$_2$O	[39291-91-3]	2.2–2.5	2.55	2.81	2.90	3.01	3.26	3.50	3.76	4.08	4.35	4.75	5.17
Li$_2$O·B$_2$O$_3$·16H$_2$O[b]	[41851-38-1]	0.88	1.42	2.51	3.34	4.63	9.40						
Li$_2$O·B$_2$O$_3$·4H$_2$O	[15293-74-0]						7.40	7.84	8.43	9.43	{10.58 9.75	11.8 9.7	13.4[h] 9.70
Na$_2$O·5B$_2$O$_3$·10H$_2$O	[12046-75-2]	6.28	10.55	12.20	13.75	17.40	21.80	26/9		32.25	37.84	43.80	50.30
Na$_2$O·2B$_2$O$_3$·10H$_2$O	[1303-96-4]	1.18	1.76	2.58	3.13	3.85	6.00	9.55	15.90				
Na$_2$O·2B$_2$O$_3$·5H$_2$O[c]	[12045-88-4]								16.40	19.49	23.38	28.37	34.63
Na$_2$O·2B$_2$O$_3$·4H$_2$O[d]	[12045-87-3]								14.82	17.12	19.88	23.31	28.22
Na$_2$O·B$_2$O$_3$·8H$_2$O[e]	[10555-76-7]	14.5	17.0	20.0	21.7	23.6	27.9	34.1					
Na$_2$O·B$_2$O$_3$·4H$_2$O	[16800-11-6]								38.3	40.7	43.7	47.4	52.4
K$_2$O·5B$_2$O$_3$·8H$_2$O	[12229-13-9]	1.56	2.11	2.82	3.28	3.80	5.12	6.88	9.05	11.7	14.7	18.3	22.3
K$_2$O·2B$_2$O$_3$·4H$_2$O	[12045-78-2]		9.02	12.1	13.6	15.6	19.4	24.0	28.4	33.3	38.2	43.2	48.4
K$_2$O·B$_2$O$_3$·2.5H$_2$O	[27516-44-5]		42.3	43.0	43.3	44.0	45.0	46.1	47.2	48.2	49.3	50.3	
Rb$_2$O·5B$_2$O$_3$·8H$_2$O	[37190-12-8]	1.58	2.0	2.67	3.10	3.58	4.82	6.52	8.69	11.4	14.3	18.1	23.75[i]
Cs$_2$O·5B$_2$O$_3$·8H$_2$O	[12229-10-6]	1.6	1.85	2.5	2.97	3.52	4.8	6.4	8.31	10.5	13.8	18.0	23.45[j]
Cs$_2$O·2B$_2$O$_3$·5H$_2$O[f]	[12228-83-0]												
Cs$_2$O·B$_2$O$_3$·7H$_2$O	[66634-85-3]			36.8[g]									
(NH$_4$)$_2$O·2B$_2$O$_3$·4H$_2$O	[10135-84-9]	3.75	5.26	7.63	9.00	10.8	15.8	21.2	27.2	34.4	43.1	52.7	
(NH$_4$)$_2$O·5B$_2$O$_3$·8H$_2$O	[12229-12-8]	4.00	5.38	7.07	8.03	9.10	11.4	14.4	18.2	22.4	26.4	30.3	

[a] Incongruent solubility below 37.5 or 40.5°C.
[b] Transition point to 4-hydrate, 36.9 or 40°C.
[c] Transition point to 10-hydrate, 60.7°C, 16.6% Na$_2$B$_4$O$_7$.
[d] Transition point to 10-hydrate, 58.2°C, 14.55% Na$_2$B$_4$O$_7$.
[e] Transition point to 4-hydrate, 53.6°C, 36.9% Na$_2$B$_2$O$_4$.
[f] Incongruent solubility.
[g] At 18°C.
[h] At 101.2°C.
[i] At 102°C.
[j] At 101.65°C.

Values for the specific heat of aqueous borax solutions as a function of weight % decahydrate are as follows (67):

wt %	kJ/(kg·K), (cal/(g·°C)
1.9	4.13 (0.987)
4.7	4.08 (0.975)
7.2	4.04 (0.965)
9.5	3.99 (0.956)
19.0	3.84 (0.918)
22.8	3.78 (0.903)
26.6	3.71 (0.887)
30.4	3.65 (0.872)
38.0	3.52 (0.842)
45.6	3.57 (0.854)
55.1	3.68 (0.880)

The pH of borax solutions is nearly independent of concentration (ie, 0.1 wt %, pH = 9.26; 0.5, 9.23; 1.0, 9.24; 2.0, 9.24; and 5.0, 9.32). Solubilities of borax in organic solvents are given below (36):

Solvent	Temp, °C	Wt % $Na_2O \cdot 2B_2O_3 \cdot 10H_2O$
glycerol, 98.5%	20	52.6
glycerol, 86.5%	20	47.1
ethylene glycol	25	41.6
diethylene glycol	25	18.6
methanol	25	19.9
acetone	25	0.60
ethyl acetate	25	0.14

If borax has been previously warmed to 50°C, it decomposes reversibly into the pentahydrate and H_2O vapor. The equilibrium vapor pressure for this transition at various temperatures is as follows (68–69): 15°C, 0.933 kPa (7.0 mm Hg); 19.8, 1.33 (10.0); 25, 1.87 (14.0); 59, 17.7 (133.0). If the decahydrate has not been warmed above 50°C, it develops a vapor pressure of only 0.213 kPa (1.6 mm Hg) at 20°C. In this case, when placed over P_2O_5, it does not form the crystalline pentahydrate but decomposes gradually to form an amorphous product of about 2.4 molecules H_2O content.

Heats of dehydration per mole of H_2O vapor are (68); deca- → pentahydrate, 54.149 kJ (12.942 kcal); and deca- → tetrahydrate, 54.074 kJ (12.924 kcal). Borax stored over a saturated sucrose–sodium chloride solution maintains exactly 10 mols of H_2O and can thus be used as an analytical standard. Commercial borax tends to lose water of crystallization if stored at high temperature or in dry air.

A single-crystal x-ray diffraction study has shown that the borate ion present in borax has formula $[B_4O_5(OH)_4]^{2-}$ and the structure shown above (70).

The same borate ion exists in the pentahydrate, explaining the ready interconversion of the penta- and decahydrates (71).

Borax reacts with acids to produce borates with an Na_2O/B_2O_3 mole ratio less than 0.5 and with base to produce borates with a ratio greater than 0.5.

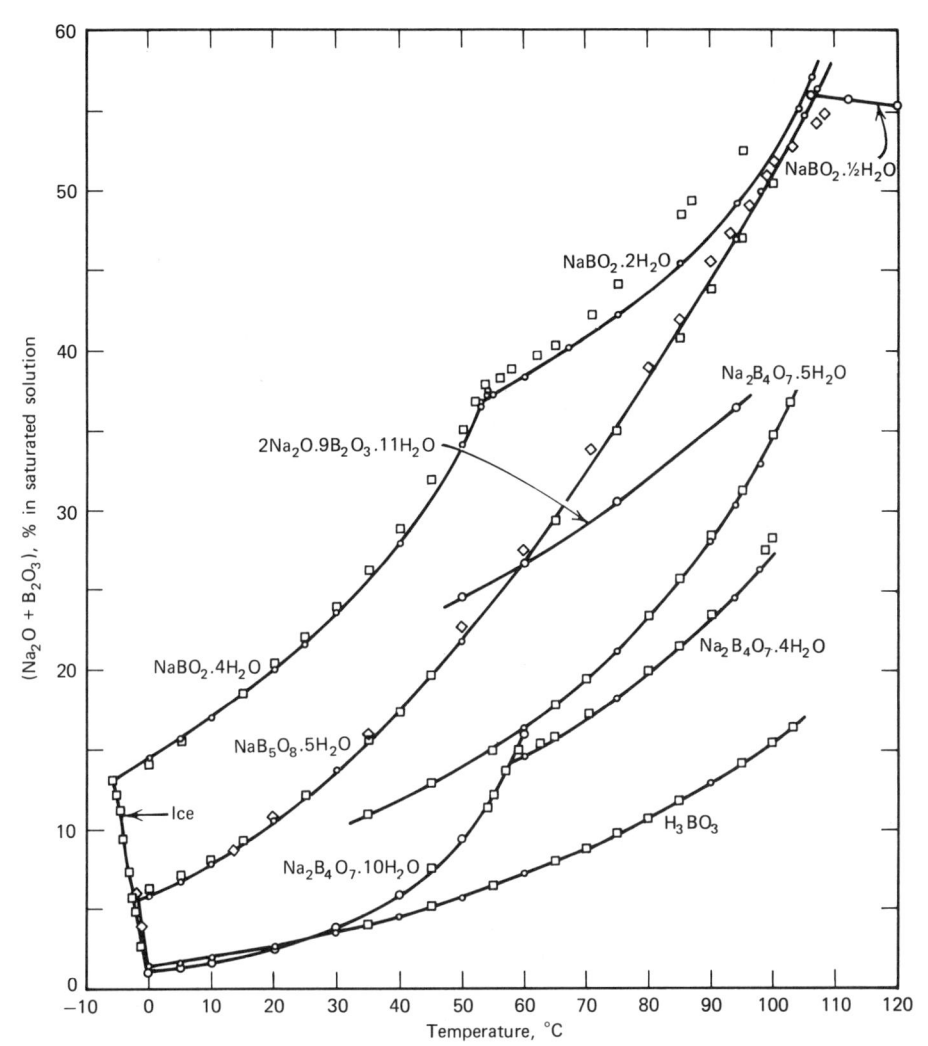

Figure 5. Solubility–temperature curves for boric acid, borax, sodium pentaborate, and sodium metaborate (66). Courtesy of The American Chemical Society.

Disodium Tetraborate Pentahydrate (Borax Pentahydrate). Disodium tetraborate pentahydrate, $Na_2O \cdot 2B_2O_3 \cdot 5H_2O$ (formula wt, 291.35; trigonal; sp gr, 1.88; specific heat, 1.32 kJ/(kg·K) [0.316 kcal/(g·°C)] (63); heat of formation, −4.7844 MJ/mol (−1143.5 kcal/mol) (64)) is found in nature as a fine-grained deposit formed by dehydration of borax.

Solubility data in H_2O are given in Figure 5 and in Table 7. Heats of solution in H_2O have been determined (63,67). The pentahydrate in contact with its aqueous solution above 39°C is metastable with respect to the tetrahydrate (kernite). Solubilities of the pentahydrate expressed as wt % $Na_2O \cdot 2B_2O_3 \cdot 5H_2O$, in organic solvents at 25°C, are as follows (57): methanol, 16.9%; ethylene glycol, 31.1; diethylene glycol, 10.0; and propylene glycol, 21.8.

Commercial pentahydrate usually contains about 4.75 mole of H_2O. The reason that it is not exactly five is not well understood at this time. At 88°C and 0.26 kPa (2 mm Hg) the pentahydrate is reversibly converted to an amorphous dihydrate, which can also be obtained by boiling with xylene (67,69). The heat of dehydration for the pentahydrate to tetrahydrate has been calculated to be 53.697 kJ (12.834 kcal) per mole of H_2O (68).

A single-crystal x-ray structure determination has shown that the borate ion in the pentahydrate and in borax are identical (71).

Disodium Tetraborate Tetrahydrate. Disodium tetraborate tetrahydrate, $Na_2O \cdot 2B_2O_3 \cdot 4H_2O$ (formula wt, 273.34, monoclinic; sp gr, 1.91; specific heat, ca 1.2 kJ/(kg·K) [0.287 kcal/(g·°C)] (57), heat of formation, -4.4890 MJ/mol (-1072.9 kcal/mol) (64)) exists in nature as the mineral kernite. The crystals have two perfect cleavages and when ground, form elongated splinters.

The H_2O solubility of kernite is demonstrated in Figure 5 and in Table 7. Kernite is the stable phase in contact with its solutions from 58.2 to ca 95°C (66). Its rate of crystallization is, however, much slower than that of the pentahydrate. Large kernite crystals can be grown by seeding saturated borax solutions.

At relative humidities above 70%, kernite absorbs H_2O to form borax. Kernite loses H_2O slowly over P_2O_5 in vacuum, forming a crystalline dihydrate, metakernite, which reverts to kernite at 60% relative humidity (68).

The structure of kernite consists of parallel infinite chains of the $[B_4O_6(OH)_2]_n^{2n-}$ ion (72). The polymeric nature of the anion explains the slow rates of dissolution and crystallization observed for kernite.

Disodium Tetraborate (Anhydrous Borax). Disodium tetraborate, $Na_2O \cdot 2B_2O_3$, (formula wt 201.27; sp gr (glass), 2.367; sp gr (α crystalline form), 2.27; heat of formation (glass), -3.2566 MJ/mol (-778.4 kcal/mol), heat of formation (α crystalline form), -3.2767 MJ/mol (783.2 kcal/mol) (22)) exists in several crystalline forms as well as a glassy form (67). The α crystalline form is obtained by dehydrating borax hydrates and is the stable form above 600–700°C (67). A large amount of heat capacity data has been reported (22,73–75). Anhydrous borax glass dissolves in H_2O more slowly than the hydrated forms. Heats of solution have been measured (63), and the manufacturer lists a value of -213.8 kJ/kg (-92 Btu/lb) (36). The solubilities of finely divided crystalline disodium tetraborate, expressed as wt % $Na_2O \cdot 2B_2O_3$, in methanol and in ethylene glycol are 16.7 and 30%, respectively (57).

Crystalline anhydrous borax takes up H_2O from moist air. It becomes anhydrous near 700°C, and melts at 742.5°C. The heat of hydration to borax has been calculated as 161 kJ/mol (38.5 kcal/mol) of $Na_2O \cdot 2B_2O_3$ (67,76). The heat of fusion has been reported as 81.2 kJ/mol (19.4 kcal/mol) (22).

A single-crystal x-ray diffraction study has shown that the borate anion in anhydrous borax is polymeric in nature, and is formed via oxygen bridging of triborate and dipentaborate groups (77). The chemistry of anhydrous borax has been reviewed (67,78).

Disodium Octaborate Tetrahydrate. The composition of a commercially available sodium borate hydrate, Polybor (60), corresponds quite closely to that of a hypothetical compound, disodium octaborate tetrahydrate ($Na_2O \cdot 4B_2O_3 \cdot 4H_2O$). This product dissolves rapidly in H_2O without the temperature decrease which occurs when the crystalline borates dissolve. The solubility of the product is compared with that of borax in the following table (36):

Temp, °C	Wt % of product	Wt % B$_2$O$_3$ in saturated solns of: product	borax
0	2.4	1.6	0.7
10	4.5	3.0	1.1
20	9.5	6.3	1.7
30	21.9	14.5	2.6
40	27.8	18.4	4.1
50	32.0	21.2	6.5
60	35.0	23.2	11.1
75	39.3	26.0	14.7
94	45.3	30.0	21.0

The pHs of aqueous solutions of the product as a function of its wt % are as follows (36): 1%, pH = 8.5; 2, 8.4; 5, 8.0; 10, 7.6; and 15, 7.3.

Sodium Pentaborate Pentahydrate. Sodium pentaborate pentahydrate, (NaB$_5$O$_8$·5H$_2$O) or Na$_2$O.5B$_2$O$_3$.10H$_2$O, (formula wt, 295.17; monoclinic; sp gr, 1.71) exists in nature as the mineral sborgite. Heat capacity measurements have been made at 15–345 K (79).

Sodium pentaborate can easily be prepared from a solution with an Na$_2$O/B$_2$O$_3$ ratio of 0.2. Its H$_2$O solubility (Fig. 5, Table 7) exceeds that of borax and boric acid. When a saturated pentaborate solution is agitated for some time at temperatures near boiling, the compound 2Na$_2$O.9B$_2$O$_3$.11H$_2$O will crystallize if seed is present. In the absence of seed crystals, however, the stable phase above 106°C shifts to pentaborates of lower hydration (67). The pHs of aqueous solutions (20°C) of the pentaborate as a function of its wt % are (36): 0.5%, pH = 8.5; 1.0, 8.4; 2.0, 8.2; 3.0, 8.0; 5.0, 7.7; 10.0, 7.1; and 14.0, 6.8.

Crystalline sodium pentaborate pentahydrate is stable in the atmosphere. When heated in vacuum, it is stable to 75°C; however, above 75°C, 4 of its 5 H$_2$O molecules are lost (67).

A single-crystal x-ray diffraction study has shown the pentaborate to contain the [B$_5$O$_6$(OH)$_4$]$^-$ ion, analogous to that found in the corresponding potassium compound (80).

Sodium Metaborate Tetrahydrate. Sodium metaborate tetrahydrate, NaBO$_2$.4H$_2$O or Na$_2$O.B$_2$O$_3$.8H$_2$O (formula wt, 137.88; triclinic; sp gr, 1.74) is easily formed by cooling a solution containing borax and an amount of sodium hydroxide just in excess of the theoretical value. It is the stable phase in contact with its saturated solution between 11.5 and 53.6°C. At temperatures above 53.6°C, the dihydrate, NaBO$_2$.2H$_2$O, becomes the stable phase. The water solubility of sodium metaborate is given in Figure 5 and in Table 7.

Heat capacity data for metaborate solutions have been reported (81). The pHs of aqueous solutions (20°C) as a function of wt % NaBO$_2$.4H$_2$O are as follows (36): 0.1%, pH = 10.52; 0.5, 10.84; 1.0, 11.00; 2.0, 11.18; 4.0, 11.38; 6.0, 11.52; 8.0, 11.64; 10.0, 11.76; 15.0, 11.86; and 18.0, 12.00. The solubility of sodium metaborate tetrahydrate in methanol at 40°C is 26.4% (57).

The relative humidity over a saturated solution of the tetrahydrate at 14–24°C is 90 ± 1%, and the humidity over mixtures of the tetra- and dihydrates is 39% at 19.3°C, 42% at 22°C, 43% at 24.8°C, and 45% at 27.0°C (82). The heat of hydration for the di- to tetrahydrate conversion has been calculated as 52.51 kJ (12.55 kcal) per mole of H$_2$O (82).

Sodium metaborate absorbs CO_2 from the atmosphere, forming borax and sodium carbonate. Crystals of the tetrahydrate melt in their H_2O of crystallization at about 54°C. The solid state structure of the tetrahydrate consists of discrete tetrahedral $B(OH)_4^-$ groups (83).

Sodium Metaborate Dihydrate. Sodium metaborate dihydrate, $NaBO_2 \cdot 2H_2O$ or $Na_2O \cdot B_2O_3 \cdot 4H_2O$ (formula wt, 101.84; triclinic; sp gr, 1.91) can be prepared by heating a slurry of the tetrahydrate above 54°C, by crystallizing hot metaborate solutions, or by dehydrating the tetrahydrate in vacuum. Large crystals can be grown by heating the solid in its mother liquor for several days. The dihydrate is the stable phase in contact with its saturated solution between 54 and 105°C. At higher temperatures a hemihydrate, $NaBO_2 \cdot 0.5H_2O$, is formed (67).

The H_2O solubility for the dihydrate is shown in Figure 5 and in Table 7. The pHs of H_2O solutions at 20°C as a function of its wt % in solution are (36): 0.1%, pH = 10.62; 0.5, 10.90; 1.0, 11.12; 2.0, 11.28; 4.0, 11.48; 6.0, 11.64; 8.0, 11.78; 10.0, 11.82; 15.0, 12.00; and 18.0, 12.20. The solubility of the dihydrate in ethanol is 0.3 wt % at boiling, and in methanol it is 17.8% at 22°C, 19.5% at 40°C, and 24.6% at 60°C (57).

The dihydrate loses water slowly at room temperature. Its heat of dehydration to $NaBO_2 \cdot 0.5H_2O$ has been calculated as 58.1 kJ/mol of H_2O (13.9 kcal/mol) (82). It reacts with atmospheric CO_2 to produce sodium carbonate and borax. The melting point is 90–95°C, compared to 54°C for the tetrahydrate.

Some crystallographic work has been done (84).

Test Methods

The alkali metal and ammonium borates are analyzed for M_2O and B_2O_3 content by dissolving the compound in H_2O, titrating the M_2O content with dilute HCl and determining the B_2O_3 content by complexation with mannitol followed by titration with dilute NaOH (85). The B_2O_3 content for calcium and other borates of low water solubility is determined by extraction into acid solution followed by mannitol complexation and titration with dilute base. The commercial hydrates are often overdried, leading to apparent assays over 100%.

The various impurities are determined by standard techniques. For example, Cl^- and SO_4^{2-} can be measured turbidimetrically. The o-phenanthroline method is often used for the determination of Fe.

Methods for analysis of industrial borate chemicals have been reviewed (86).

Manufacture

Borax Decahydrate and Pentahydrate. Borax decahydrate and pentahydrate are produced from sodium borate ores, from dry lake brines, from colemanite, and from magnesium borate ores.

Production from sodium borate ores takes place in the United States, Argentina, and Turkey. All United States production based on sodium borate ores is from the U.S. Borax & Chemical Corporation at Boron and Wilmington, Calif. Argentine production is carried out at Tincalayu, primarily by Boroquimica Samicaf. Turkish mining of tincal takes place at Kirka. This operation is under the control of the Turkish government and its representative, Etibank.

At its production facility in Boron, Calif., U.S. Borax operates an open pit borax mine and a refinery. This facility represents the largest single source of borate chemicals in the world. In the mine, overburden is blasted and hauled to storage areas in 91-metric ton electric trucks. The ore is then drilled, blasted, and trucked to an impact mill where it is crushed to <20.3 cm. The crushed ore, consisting primarily of tincal and clay, is transported up a belt conveyor to a surface stockpile. It is then crushed to <3.8 cm and sent by belt to a series of storage bins. Ore from the bins is blended to a constant B_2O_3 content and fed to a dissolving plant where it is mixed with hot recycle liquor. Liquor leaving the dissolvers is passed over vibrating screens that remove rocks and clay particles larger than 0.25 mm (60 mesh). The liquor and fine insolubles are then fed to a primary thickener for settling. There are four thickener stages operating in a countercurrent fashion so that the underflow from each thickener is washed by a progressively weaker borax liquor. Water is added to the fourth thickener to wash the underflow mud. Strong liquor from the primary thickener is pumped to Struthers Wells' continuous vacuum crystallizers to separate borax pentahydrate. Crystallizer capacity is also maintained for production of the decahydrate. The crystallizer slurries are dewatered on continuous centrifuges and the products are dried in Wyssmont or rotary driers. Fluid bed drying will be started in 1978. Combined plant capacity for all products expressed as B_2O_3 is 635,000 t/yr (4).

No coproducts are produced in the U.S. Borax plant, since the ore is processed only for its B_2O_3 content. Liquid effluent is pumped into sealed ponds. A dust abatement program has reduced airborne emissions from the plant by 90%. Sodium borate ores are mined in the United States under provisions of the Mineral Leasing Act of 1917 and 1920.

The source of Argentine production is an open pit tincal mine at 4000 m above sea level. A modern plant has recently been constructed at the mine site. Total production of sodium borates is approximately 78,000 t/yr (4).

Turkish production of sodium borates originates from an open pit tincal mine at Kirka. Present operations (10) include a plant at Bandirma designed for 54,000 t of sodium borates and 25,000 t of boric acid per year and a tincal washing plant with a capacity of 400,000 t/yr of upgraded product (35% B_2O_3). A plant capable of producing 180,000 t/yr of crude borax pentahydrate, 50,000 t of crude anhydrous borax, and 10,000 t of refined anhydrous borax is now, or will soon be, completed. The plant at Bandirma is a batch-wise operation. The Turkish tincal reportedly contains trace amounts of arsenic (87).

Kerr-McGee Corporation produces borax pentahydrate and decahydrate from Searles Lake brines at both Trona and West End, Calif. The 88 km^2 dry lake consists of two brine layers, the analyses of which are given in Table 8. Two distinct procedures are used for the processing of upper and lower lake brines.

Borax is produced at Trona from upper lake brines by an evaporative procedure involving the crystallization of potash and several other salts prior to borax separation. A portion of the Na–K–SO$_4$–CO$_3$–Cl–H$_2$O system, saturated with NaCl at 100°C, is shown in Figure 6. This system includes the major nonborate species of interest during the evaporation procedure. Raw lake brine is represented by point B and plant mother liquor (ML-2) by point M. A mixture of these two solutions in the proportions of about three parts of brine to one part of mother liquor is represented by point F. This mixture is fed to triple-effect evaporators.

During evaporation sodium chloride and burkeite (Na$_2$CO$_3$–2Na$_2$SO$_4$) are crystallized and removed continuously while the composition of the liquor travels along the line FA until saturation with Na$_2$CO$_3$.H$_2$O is reached at point A. The path of

Table 8. Typical Brine Analyses from Searles Lake

Constituent	Upper structure brine, wt %	Lower structure brine, wt %
KCl	4.90	3.50
Na_2CO_3	4.75	6.50
$NaHCO_3$	0.15	
$Na_2B_4O_7$	1.58	1.55
$Na_2B_4O_4$		0.75
Na_2SO_4	6.75	6.00
Na_2S	0.12	0.30
Na_3AsO_4	0.05	0.05
Na_3PO_4	0.14	0.10
NaCl	16.10	15.50
H_2O, by difference	65.46	65.72
WO_3	0.008	0.005
Br	0.085	0.071
I	0.003	0.002
F	0.002	0.001
Li_2O	0.018	0.009

crystallization then follows the line AC, approaching saturation with KCl at point C. Thus, only sodium salts, such as the carbonate, sulfate, and chloride, have been removed from solution during evaporation, while all of the potassium and borate components of the brine have been concentrated in solution in the hot liquor.

The hot, concentrated liquor is separated from any remaining suspended solids, and cooled rapidly in continuous vacuum crystallizers to crystallize crude potassium chloride. During this cooling, the solution becomes supersaturated with borax. Care is taken to avoid the crystallization of borax with the potash. This is done by quick processing, by avoiding the presence of seed, and by control of concentrations and temperatures. The potassium chloride crystals are removed by filtration. A mixture of crude borax decahydrate and sodium bicarbonate from the carbonation process is added to the mother liquor (ML-1) from the potash plant (Fig. 7). The bicarbonate prevents accumulation of sodium metaborate in the liquors. Sodium metaborate in solution is formed to a small extent by loss of carbon dioxide during the original evaporation (see Chemicals from brine).

The mother liquor is held for a short period in a large tank for deaeration to decrease foaming before going to the crude pentahydrate crystallizers. The incoming liquor, containing some borax pentahydrate crystals formed from the carbonation process borax, is mixed with a thick bed of pentahydrate seed crystals to produce borax pentahydrate. The overflow at the settling-cone lip normally contains 6–8% settled solids (small pentahydrate crystals) by volume. It goes next to an 18.3 m Dorr thickener where most of these crystals are settled.

Part of this thickener underflow is fed back to the crystallizers as seed, and the rest goes to the filter feed tank. The thickener overflow is mixed with filtrate from the crude borax filters to form ML-2. This is mixed with raw brine and passes through the evaporators again. The sludge stream from the bottom of the crystallizers is classified in cyclone separators in which the >0.25 mm (+60 mesh) crystals are thickened to 70–80% settled solids by volume and then flows to the filter feed tank. Liquor and <0.25 mm (−60 mesh) crystals return to the crystallizers. From the filter feed tank the crude borax pentahydrate slurry goes to two horizontal 4-m flat-bed filters. The

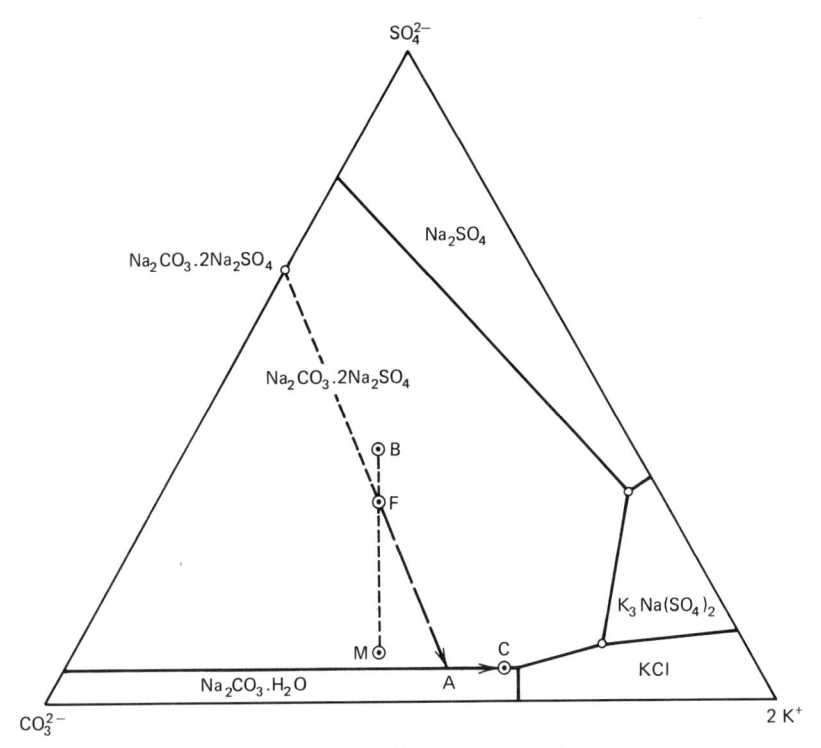

Figure 6. Portion of the system Na^+–K^+–SO_4^{2-}–CO_3^{2-}–Cl^-–H_2O saturated with NaCl at 100°C (88). B = raw lake brine, M = plant mother liquor, and F = mixture of B and M.

crude borax pentahydrate is contaminated with entrained mother liquor. It is repulped and dissolved in refined borax mother liquor (ML-3) with direct injection of steam to give a concentrated borax liquor (CBL) at 93.3°C. This is polished in pressure filters and sent to vacuum crystallizers, where either refined borax decahydrate or pentahydrate is produced, depending on the temperature. The crystal slurry flows into a centrifugal feed-settling cone for thickening before centrifuging. Centrifuge filtrate and washing and settling cone overflow (ML-3) are stored and pumped to the crude borax pentahydrate repulper as needed. Moist borax from the centrifuges goes either to rotary and shelf driers or directly to the boric acid and anhydrous borax plants. The borax goes through two rotary dryers in series. The discharge from no. 1 drops into the feed chute of no. 2. Hot air enters the discharge end of no. 1 and the feed end of no. 2. Dry borax discharges from all dryers at about 50°C.

A carbonation process is used by Kerr-McGee at both Trona and West End to derive borate values from lower lake brines. At Trona, flue gases, after being scrubbed, cooled, and compressed, enter the bottom of carbonating towers. Fine bubbles are maintained by means of rotating screens. Raw lower structure brine enters at the top of the towers. The absorbed carbon dioxide converts sodium carbonate to sodium bicarbonate, which crystallizes because it is only slightly soluble in the brine. At the same time, the pH of the brine is lowered, converting metaborate to tetraborate which is less soluble than metaborate. The slurry from the primary towers is sent to secondary towers where it is contacted by carbon dioxide returned from the bicarbonate calciners. The bicarbonate is removed by filtration, is calcined, and is recrystallized as sodium

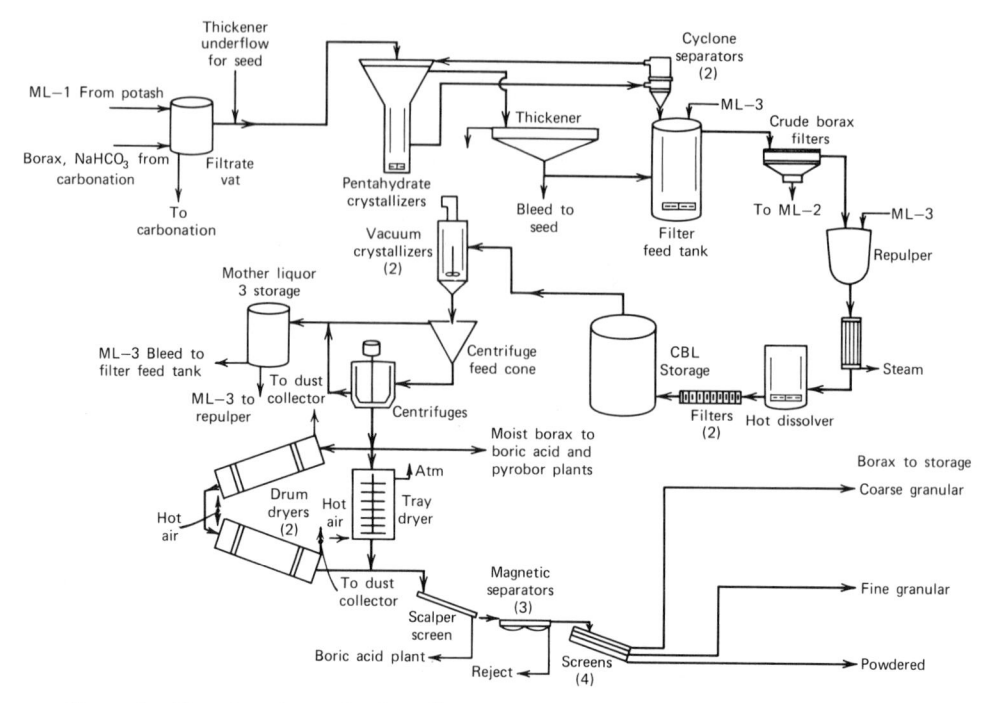

Figure 7. Flow sheet for production of borax at Trona, California (89). Courtesy of The American Chemical Society.

carbonate monohydrate. The filtrate, rich in borax, is neutralized with uncarbonated lake brine to minimize borax solubility, and cooled by refrigeration to crystallize borax. The borax and part of the sodium bicarbonate from the carbonation process go into the evaporation process as previously described. In the West End process the liquors are further refrigerated to crystallize sodium sulfate and the filtrate is returned to the lake. In the West End process the source of carbon dioxide is a lime kiln rather than boiler flue gas. Both plants use some liquid carbon dioxide for make-up.

Chinese production of borates is centered in the Iksaydam dry lake region.

Production of borax from the reaction of colemanite with sodium carbonate is carried out in Spain, Italy, and Poland. Turkish production from colemanite has been discontinued in favor of direct production from tincal ore. Sodium borates are produced in the U.S.S.R. from szaibelyite, $Mg(BO_2)OH$.

Anhydrous Borax. Anhydrous borax is produced from its hydrated forms by fusion. Calcining is usually an intermediate step in the process. The primary producers are U.S. Borax and Kerr-McGee. Yearly fusion capacities for the two companies are reported to be 227,000 and 69,000 metric tons B_2O_3, respectively. There is a Turkish plant designed for the production of 50,000 t/yr of crude anhydrous borax and 10,000 t/yr of refined anhydrous borax from tincal ore (10). Small quantities of anhydrous borax have been produced in Argentina.

In the Kerr-McGee process, borax decahydrate is first partially dehydrated in calciners 2.4 m in diameter and 21.3 m long (89). Hot exhaust gases from the fusion furnaces at 700–815°C are drawn into the calciners by direct induction fans. Flow is concurrent with the borax. Some dusting occurs during calcining; the dust-laden exhaust gases pass through a cyclone separator and then into wet scrubbers. Water in the wet scrubber recirculates and is continuously bled off and used as process water in the borax plant. Dust collected by the cyclone separators is added to the calciner discharge which then goes to the fusion furnaces.

Each of the fusion furnaces has a cylindrical firebox, fired by a natural gas burner at the top. There is a gap of about 20 cm between the outer bottom edge of the firebox and the top inner edge of the furnace bottom, which is bowl-shaped and water-jacketed as seen in Figure 8. Directly above this gap is the furnace feed ring. Three feed hoppers continuously circle the furnace on the feed ring and distribute calcined borax into the furnace bed so that it completely fills the gap between the firebox and the bottom.

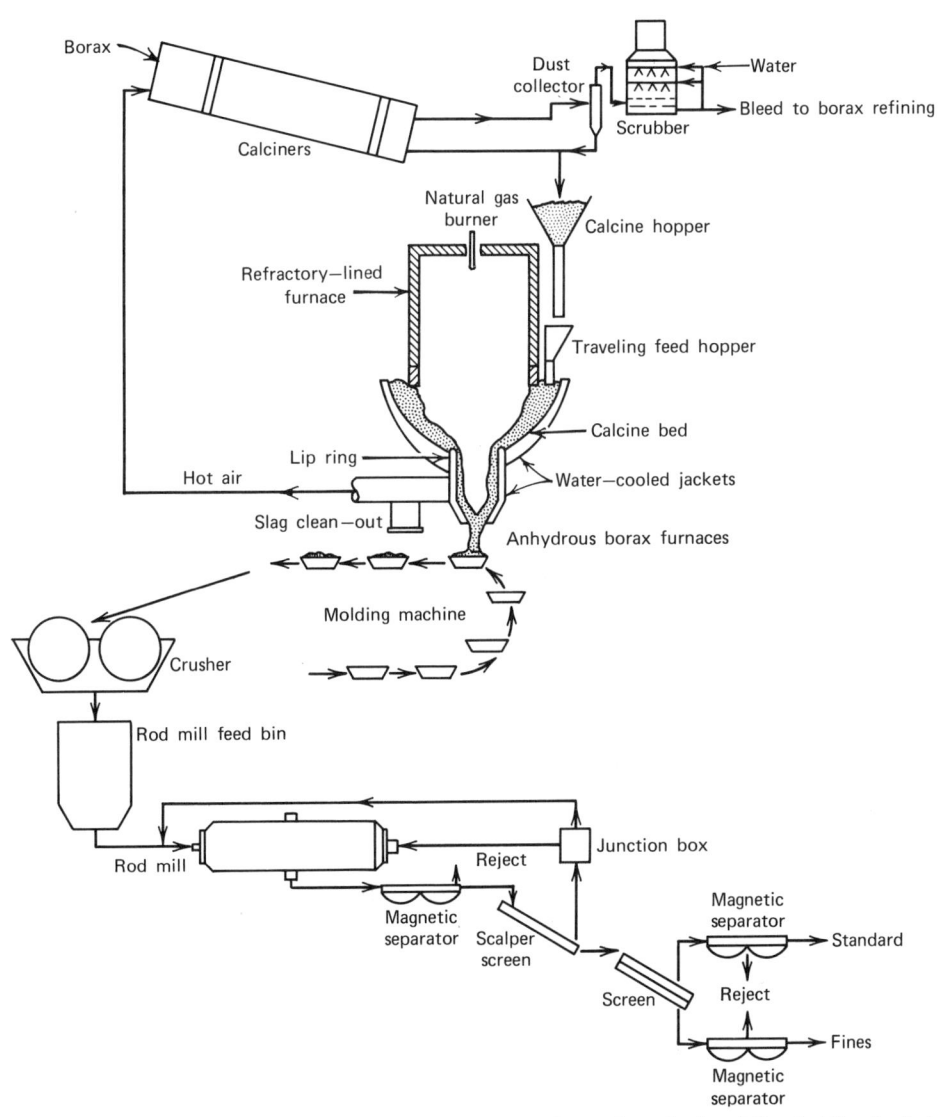

Figure 8. Flow sheet for the production of anhydrous borax (standard and fines) at Searles Lake, Trona, California (89). Courtesy of The American Chemical Society.

Each traveling feed hopper is refilled as it passes beneath the calcined borax storage bin.

At the bottom center of the borax bed is a water-jacketed lip ring. Molten borax flows down the calcine bed, over the lip ring, and into the furnace nozzle. The design of the furnace is such that the molten borax flows off the bed surface and out of the furnace as soon as it is liquid enough to do so. Thus the molten borax, which attacks refractories, actually comes in contact with a bed of calcined borax rather than the refractory furnace shell. The temperature may be from 1200–1440°C at the top of the firebox and 980°C in the fusion zone. Exhaust gases leave the furnace in a 1.2 m diameter flue leading from the nozzle and are conducted through flues to the calciners.

Commercial anhydrous borax is largely amorphous. The crystalline material grinds easily but the amorphous material is much more abrasive. To produce crystalline anhydrous borax, the molten borax is placed in molds which are actually the buckets of a continuous bucket conveyor. The hot borax remains in these molds long enough to crystallize; the ingots are then crushed.

To produce amorphous anhydrous borax, the molten borax is run between two large water-cooled rolls, forming sheets about 1.6 mm thick, which are then crushed and screened to the desired particle size. Since it is cooled rapidly by the rolls, it remains largely amorphous, although it may contain some crystalline anhydrous borax.

In general, the production of fused materials is much more energy intensive than that of hydrated products, and this difference is reflected in their prices.

Other Sodium Borates. Polybor (60), a proprietary product of U.S. Borax, has the approximate composition of disodium octaborate tetrahydrate, $Na_2O.4B_2O_3.4H_2O$. The material is produced by spray-drying mixtures of borax and boric acid.

Routine production of sodium pentaborate pentahydrate has been terminated. The compound is easily prepared by crystallizing a borax–boric acid solution with an Na_2O/B_2O_3 ratio of 0.2.

Sodium metaborate tetrahydrate can be prepared by cooling a solution containing borax and an amount of sodium hydroxide just in excess of the theoretical amount. The dihydrate is prepared by a patented process (90) by U.S. Borax & Chemical Corporation. This process involves the preparation of a mixture 46–52% in $Na_2O.B_2O_3$ by mixing appropriate quantities of a hydrated form of $Na_2O.2B_2O_3$ with aqueous NaOH. The mixture is then heated to about 90°C to dissolve all solids and slowly cooled to 60–75°C. Crystals of the dihydrate are then harvested and dried.

Product Specifications

Specifications for the maximum allowable impurity levels for the major borate products are given in Table 9. Where maximum levels are not set, the typical values are given in parentheses. Typical levels of the various impurities generally fall well below the maximum specification. Both borax decahydrate and pentahydrate are overdried in manufacture to give assays of over 100%.

USP borax decahydrate is manufactured to conform to standards of the USP, Revision XVIII, for sodium borate. Maximum allowable impurity levels for technical anhydrous borax are: SO_4, 0.41%; SiO_2, 0.21%; Al_2O_3, 0.14%; CaO, 0.03%; MgO, 0.15%; and Fe_2O_3, 0.02%.

Table 9. Maximum Impurity Specifications[a]

Chemical	Grade	Cl⁻	SO$_4^{2-}$	PO$_4^{3-}$	Fe$_2$O$_3$	Na⁺	Ca^{2+}	Heavy metals as Pb	H$_2$O insolubles	Basis
Na$_2$O·2B$_2$O$_3$·10H$_2$O	T[b]	0.07	0.06		0.003				0.02	%
	SQ	0.4	1.0	10	2.8		50	10	10	ppm
Na$_2$O·2B$_2$O$_3$·5H$_2$O	T	0.05	0.08		0.004					%
NaBO$_2$·4H$_2$O	T	0.1	0.1		0.003		(0.002[c])	(0.0005)	(0.002)	%
NaBO$_2$·2H$_2$O	T	0.1	0.1		0.007		(0.003)	(0.0005)	(0.002)	%
K$_2$O·2B$_2$O$_3$·4H$_2$O	T	0.05	0.05		0.0014	0.10	(0.002)	(0.0005)	(0.002)	%
KB$_5$O$_8$·4H$_2$O	T	0.05	0.05		0.003	0.10		(0.0005)		%
(NH$_4$)$_2$O·2B$_2$O$_3$·4H$_2$O	T	0.05	0.05		0.0014	(0.0026)				%
NH$_4$B$_5$O$_8$·4H$_2$O	T	0.05	0.05		0.0014			(<0.0001)		%
	SQ	0.4	1		5			2	10	ppm

[a] Ref. 36.
[b] T = technical and SQ = special quality.
[c] Parentheses denote typical values.

Many of the borate chemicals are sold in a variety of particle size distributions, and average size analyses for the various cuts are available from the manufacturer.

Shipping

Table 10 summarizes data on prices and shipping methods for the major borate products. The prices quoted for the sodium tetraborates are February 1977 rates. The values for the other chemicals represent carload rates in July 1977. Within the United States most shipments are by rail. Additional charges are made for split cars, truck shipments, palletized shipments, etc. Prices are fob Southern California. In general, the borates are stable solids and require no special handling techniques with the possible exception of dust collection. The alkali metal and ammonium borates are also available in research quantities (4).

Economic Aspects

Decahydrate, Pentahydrate, and Anhydrous Borax and Bulk Calcium Borates. The bulk borate products, borax decahydrate and pentahydrate, anhydrous borax, boric acid and oxide, and upgraded colemanite and ulexite, account in both tonnage and monetary terms for over 99% of sales of the boron primary products industry (52). Economic considerations for all these products are highly interrelated, and most production and trade statistics available do not clearly distinguish the various products.

Table 10. Prices Per Metric Ton for Carload Quantities (>36 t) of Borate Chemicals in 1977[a] in U.S. Dollars

Chemical	Grade	Form	Bulk	45.4 kg Paper sacks	Large drums
$Na_2O \cdot 2B_2O_3 \cdot 10H_2O$	T[b]	G	107	134	
	T	P		155	
	T	F		201	
	USP	G		292	364
	USP	P		338	410
	USP	I		475	547
	SQ	G		682	754
	SQ	P			988
$Na_2O \cdot 2B_2O_3 \cdot 5H_2O$	T	G	129	157	
	T	P		180	
$Na_2O \cdot B_2O_3$	T		293	322	
	T	glass		692	
$Na_2O \cdot 4B_2O_3 \cdot 4H_2O$ [c]	T			500[d]	
$NaBO_2 \cdot 4H_2O$	T			312	384
$NaBO_2 \cdot 2H_2O$	T			444	
$K_2O \cdot 2B_2O_3 \cdot 4H_2O$	T	G		919	991
	T	P		1023	1095
$KB_5O_8 \cdot 4H_2O$	T	G		860	931
	T	P		965	1036
$(NH_4)_2O \cdot 2B_2O_3 \cdot 4H_2O$	T	G		863	935
	T	P		1002	1074
$NH_4B_5O_8 \cdot 4H_2O$	T	G		746	
	T	P		870	941

[a] Ref. 36.
[b] T = technical, G = granular, F = fine powdered, I = impalpable, P = powdered, and SQ = special quality.
[c] Polybor, ref. 60.
[d] 22.7-kg sacks.

The total tonnages of borate chemicals produced by the world's five major producing countries are given in Table 11. The five nations listed produce nearly all of the world's borate supply. The industry has historically been dominated by the United States, but Turkey has recently become an important producer.

In 1975, approximately 8% of the United States B_2O_3 production was as borax

Table 11. World Borate Production, Thousands of Metric Tons[a]

	1967	1970	1973	1974	1975
United States	809	944	1190		
	(429)[b]	(510)	(602)	(562)	(547)
Turkey	266[c]	303	635		
U.S.S.R.	138[d]	141	181		
Argentina	16[d]	32	54		
China	27[d]	31	32		

[a] Refs. 4 and 11.
[b] Parentheses denote B_2O_3 value.
[c] Turkish production is ca 40% B_2O_3.
[d] Estimated.

decahydrate, 35% as the pentahydrate (crude and refined), 32% as anhydrous borax, and 6% as calcium-containing borates (4). Most of the remainder is accounted for by boric acid. The United States imported about 25,000 metric tons of Turkish colemanite in 1975.

Turkey produced 490,000 t of borates in 1974, most of which was colemanite. Sodium borate production in 1975 was about 45,000 t (4). Development of the large tincal deposit at Kirka will significantly increase Turkey's production. Although the Turkish borate industry has potential for large-scale expansion, the United States industry is unlikely to expand by over 30% (10).

Table 12 summarizes the export data for the United States borate industry.

Table 12. Production and Exports of United States Borates, Thousands of Metric Tons[a]

	1967	1970	1973	1974	1975
total production, gross	809	944	1190		
total production, B_2O_3	429	510	602	562	547
total exports, gross	431	503	594	599	590
total exports, B_2O_3	236	276	306	329	313
boric acid and refined sodium borates, gross exported	169	211	191	297	223
refined sodium borates, gross exported	148	164	153	194	192

[a] Refs. 4 and 11.

The export of crude ore concentrates accounts for the large difference between total exports and exports of refined sodium borates and boric acid. Most United States exports are shipped to Europe, via Rotterdam, the Netherlands. In 1974 the consumers of exported United States borates were (in decreasing order) (87): F.R. Germany, France, United Kingdom, Japan, Belgium, Spain, Italy, and the Netherlands.

Turkey exports about 85% of its production. In 1973 the percentages of total exports sent to various countries were (4): Italy, 21%; France, 16%; Japan, 10%; Switzerland, 10%; F.R. Germany, 10%; United Kingdom, 8%; and United States, 4%. Small quantities of borates are exported from the U.S.S.R. Most of Argentina's production is sold in South America.

The Turkish borate industry has an advantage over the United States industry in terms of cost for international distribution of products (52). Freight charges on Turkish colemanite to the United States' Atlantic Coast, Mediterranean ports, and Northern Europe were about $21.50 per metric ton, 1975; $11/t, 1977; and $15.40/t, 1977, respectively (4). Current shipping charges from the western United States to Rotterdam are approximately $72/t.

The cost of rail shipping within the United States has nearly doubled in the last ten years, and transportation charges represent a large percentage of total product price. For example, the freight charge from Boron, California to Charleston, South Carolina is $92.40/t for 27.2-t loads. Per ton charges for 86.2-t loads are about half this value (4).

The United States demand for borate products between 1967 and 1974 is given in Table 13.

A large increase in demand in the United States has been related to the use of borates in energy-conserving products (ie, insulation). Wang has described some of the other factors affecting demand (10). The United States, Western Europe, and

Table 13. United States Demand For Borate Products, Thousands of Metric Tons as B_2O_3 [a]

Year	1967	1970	1973	1974
demand	247	258	333	307

[a] Ref. 10.

Japan accounted for about 30, 42, and 7% of total world demand in 1974, respectively (4). World demand to the year 2000 has been projected (10).

Changes in per-ton-prices, since 1950, for bulk quantities of technical-grade products from United States producers are given in Table 14.

In 1975, colemanite was priced at about $110/t, which is approximately twice the 1970 level.

Other Sodium Borates. Prices for disodium octaborate tetrahydrate, sodium metaborate tetrahydrate, and sodium metaborate dihydrate in 1967 were $158 per metric ton, $128/t, and $183/t, respectively. Prices in 1977 were $500/t, $312/t, and $444/t, respectively (36).

Health and Safety

Toxicological studies have been reported for borax and boric acid. There are reports that death may result from the ingestion of 15–30 g of borax or 2–5 g of boric acid. However, others report that high doses have been administered during neutron capture therapy for brain tumors without severe toxic effects. Poisonings by boric acid have been reported following its use over large areas of burned or denuded skin (54). The handling of borax or boric acid is generally not considered dangerous.

Reviews on borate toxicology have appeared (91–92). Little data are available on borates other than borax and boric acid (91).

Uses

General. The primary uses of the various borate ores and their derivative products are given in Figure 9. A very complete list has recently been published (4).

In the United States over 40% of the total B_2O_3 consumption is for glass (qv) manufacture. Approximately 20% is used in fiberglass insulation where the borate is added at a level of 5–7 wt % B_2O_3 to increase fiber durability. Approximately 10% is used in textile fiberglass where the B_2O_3 is added at a level of 8–9 wt % to aid in weathering resistance. Other glass applications use 10%, eg, the production of borosilicate glasses, where the B_2O_3 imparts a low coefficient of thermal expansion when added at a level of 12–15 wt %.

Table 14. Prices Per Metric Ton for Technical-Grade Products, fob Works[a] in U.S. Dollars

Product	1950	1960	1970	1976	1977
$Na_2O.2B_2O_3.10H_2O$	34	47	55	100	107
$Na_2O.2B_2O_3.5H_2O$		64	74	121	129
$Na_2O.2B_2O_3$	71	91	103	273	293

[a] Refs. 4 and 36.

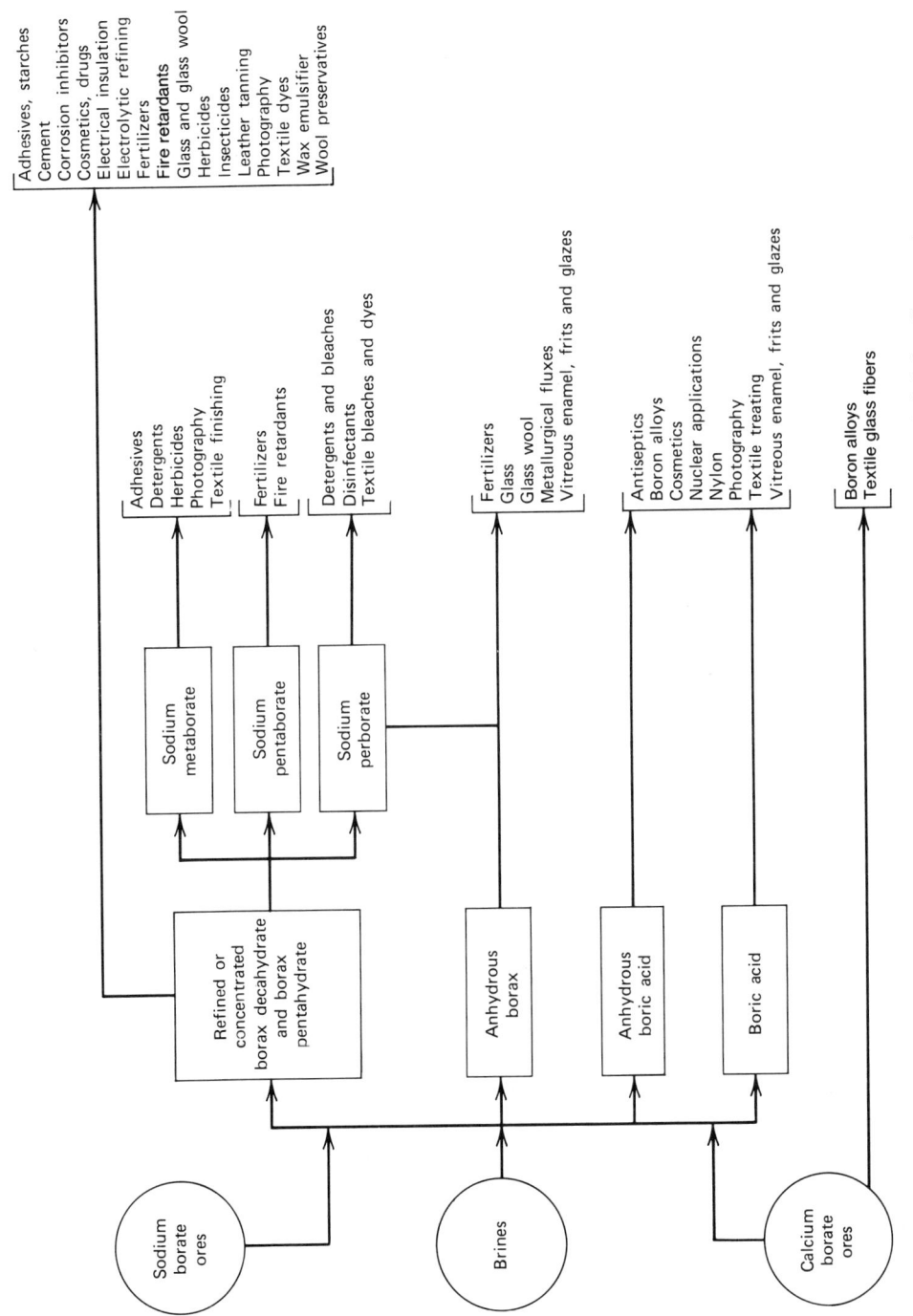

Figure 9. Main uses of boron compounds (52). Courtesy of British Sulphur Corporation.

Borates are used as fluxing agents for porcelain enamels and ceramic glazes. This market accounts for about 10% of the total usage for the United States.

Approximately 15% of total consumption goes into soap and cleaning compositions. Borax is sold both in pure form and as a major constituent in many household laundry aids and sweeteners. In these applications the germicidal and water conditioning properties of borax are put to use. Sodium perborate is the active ingredient in a number of dry, nonchlorine bleaches (see Bleaching agents; Surfactants).

Approximately 5% of the United States' consumption of B_2O_3 is in agriculture. Boron is a necessary trace nutrient for plants, and is added in small quantities to a number of fertilizers. Borates are also used in crop sprays for fast relief of boron deficiency.

Borates, when applied at relatively high concentration, act as nonselective herbicides. In the United States this use accounts for about 2–3% of total consumption (see Herbicides).

The use of boric acid as a catalyst in the air oxidation of hydrocarbons accounts for about 2% of United States consumption. Small quantities of borates are used in the manufacture of alloys and refractories. Molten borates readily dissolve other metal oxides; their use as flux in metallurgy is an important application. Other important, small volume applications for borates are in fire retardants for both plastics and cellulosic materials, in hydrocarbon fuels for fungus control, and in automotive antifreeze for corrosion control. Borates are used as neutron absorbers in nuclear reactors. The relative amounts of each major borate product used in various applications have been outlined (4).

The Western European pattern of borate consumption is different from that in the United States (10). Over 25% of consumption is as perborate bleach, 25% for enamels and ceramics, 10% for borosilicate glass, less than 10% for insulating fiberglass, and 5% for textile fiberglass.

Disodium Tetraborate Decahydrate, $Na_2O \cdot 2B_2O_3 \cdot 10H_2O$. In the United States, nearly all the refined borax is used for household cleaning products. Small amounts are used as fertilizers and herbicides. USP-grade borax is used in cosmetic and toilet goods, where purity is demanded. Special quality-grade borax is used in electrolytic capacitors, in nuclear applications, and as a laboratory chemical.

Disodium Tetraborate Pentahydrate, $Na_2O \cdot 2B_2O_3 \cdot 5H_2O$. Nearly 30% of the refined pentahydrate consumed in the United States is used in fertilizers. Significant amounts are also used in insulation fiberglass, glass, and herbicides. Smaller amounts are used in antifreeze, in ceramic glazes, and in cleaning agents. Most of the pentahydrate produced in the United States is exported as a crude concentrate. A large-scale application of this concentrate is in the preparation of perborate bleaches. Some of the pentahydrate is used in the production of boric acid.

Disodium Tetraborate, $Na_2O \cdot 2B_2O_3$. In the United States, anhydrous borax finds most application in the glass industry, for fiberglass insulation, borosilicate glass, and enamels. It is also used as a fire retardant, an antifreeze additive, and as an algicide in industrial water. Some anhydrous borax is used in fertilizers.

Disodium Octaborate Tetrahydrate, $Na_2O \cdot 4B_2O_3 \cdot 4H_2O$. Commercially available products, having the approximate composition of a hypothetical disodium octaborate tetrahydrate, have found application in wood preservatives, fertilizer sprays, herbicides, and fire retardants. In many applications the large water solubility of these products is an asset.

Sodium Pentaborate Pentahydrate, NaB$_5$O$_8$.5H$_2$O. Sodium pentaborate has found limited application in agricultural sprays and fire retardants.

Sodium Metaborate Tetrahydrate and Dihydrate, NaBO$_2$.4H$_2$O and NaBO$_2$.2H$_2$O. The sodium metaborates are components in textile finishing, sizing and scouring compositions, adhesives, and detergents. They are also used in many photographic applications. In agriculture they are used in both herbicides and fertilizer sprays. The dihydrate is less affected by heat.

ALKALI METAL AND AMMONIUM BORATES

Properties

Dipotassium Tetraborate Tetrahydrate. Dipotassium tetraborate tetrahydrate, K$_2$O.2B$_2$O$_3$.4H$_2$O, (formula wt, 305.51; orthorhombic; sp gr, 1.92) is much more soluble than borax in H$_2$O. Its solubility at various temperatures is given in Table 7.

Phase relationships in the system K$_2$O–B$_2$O$_3$–H$_2$O have been described and a portion of the phase diagram is given in Figure 10. The pHs of tetrahydrate solutions at 25°C as a function of its wt % in H$_2$O are as follows (36): 0.1%, pH = 9.18; 0.5, 9.14; 1.0, 9.15; 2.0, 9.20; and 5.0, 9.30.

The tetrahydrate can be dried at 65°C without loss of H$_2$O molecules of crystallization. It begins to dehydrate between 85 and 111°C, depending on the partial pressure of H$_2$O vapor in the atmosphere. This conversion is reversible, with a heat of dehydration of 86.6 kJ/mol (20.7 kcal/mol) of H$_2$O (95).

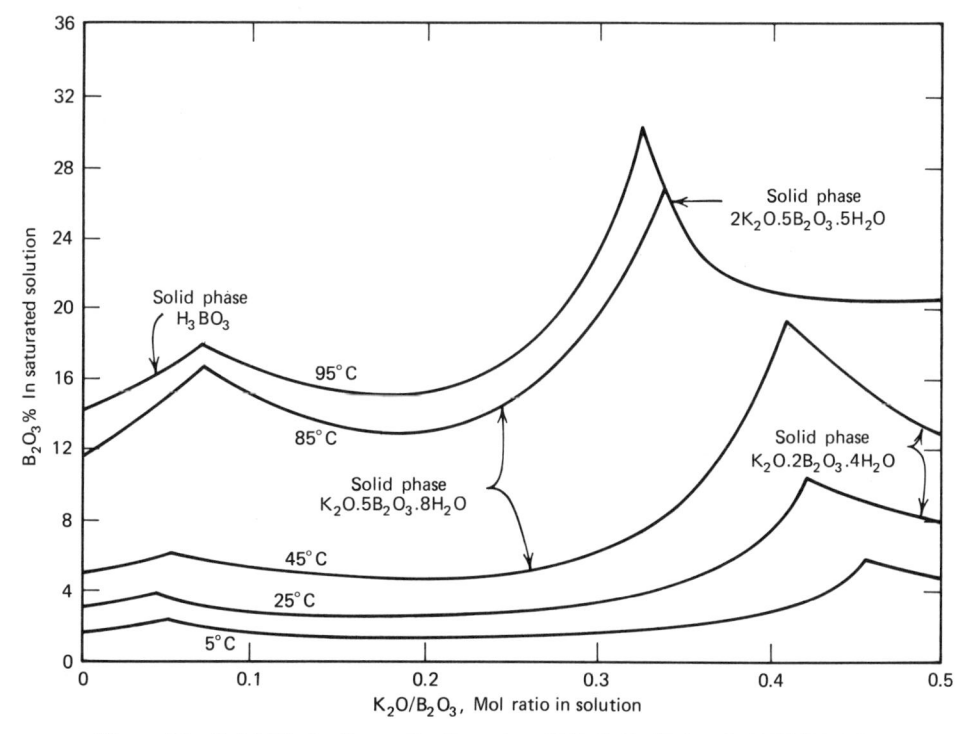

Figure 10. Solubility isotherms for the system K$_2$O–B$_2$O$_3$–H$_2$O at 5–95°C (93–94).

Single-crystal x-ray studies have shown that the borate ion in the potassium compound is identical to that found in borax (96).

Potassium Pentaborate Tetrahydrate. Potassium pentaborate tetrahydrate, $KB_5O_8 \cdot 4H_2O$ or $K_2O \cdot 5B_2O_3 \cdot 8H_2O$, (formula wt, 293.21; orthorhombic; sp gr, 1.74; heat capacity, 329.0 J/(mol·K) [78.6 cal/(mol·K)] at 296.6 K) is much less soluble than sodium pentaborate. Its solubility in H_2O at various temperatures is given in Table 7. Heat capacity measurements on the solid have been made over a broad temperature range (79). The pHs of tetrahydrate solutions at 25°C as a function of its wt % in H_2O are as follows (36): 0.29%, pH = 8.47; 0.58, 8.38; 1.17, 8.36; 2.93, 8.00; and 5.86, 7.60.

The tetrahydrate is stable under normal conditions of storage. Its heat of dehydration has been calculated as 110.8 kJ/mol (26.5 kcal/mol) between 106.5 and 134°C (95).

The solid state structure consists of $[B_5O_6(OH)_4]^-$ ions, which is analogous to that found in sodium pentaborate (97).

Diammonium Tetraborate Tetrahydrate. Diammonium tetraborate tetrahydrate, $(NH_4)_2O \cdot 2B_2O_3 \cdot 4H_2O$ (formula wt, 263.38; tetragonal; sp gr, 1.58) is readily soluble in H_2O (Table 7). The pHs of solutions of diammonium tetraborate tetrahydrate are nearly independent of concentration, ranging from 8.83 at 0.1 wt % to 8.87 at 10% (36). The compound is quite unstable, and exhibits an appreciable vapor pressure of ammonia. Phase relationships have been outlined (98).

Ammonium Pentaborate Tetrahydrate. Ammonium pentaborate tetrahydrate, $NH_4B_5O_8 \cdot 4H_2O$ or $(NH_4)_2O \cdot 5B_2O_3 \cdot 8H_2O$ (formula wt, 272.2; sp gr, 1.58; heat capacity, 359.4 J/(mol·K) [85.9 cal/(mol·K)] at 301.2 K) exists in two crystalline forms, orthorhombic (α) and monoclinic (β). Its heat capacity has been measured over a broad temperature range (79). Solubility data are given in Table 7. The pH of solutions of the tetrahydrate at 25°C as a function of its wt % in H_2O are as follows (36): 0.1%, pH = 8.48; 0.5, 8.44; 1.0, 8.35; 2.0, 8.16; 5.0, 7.74; and 10.0, 7.32.

Ammonium pentaborate tetrahydrate is very stable in respect to ammonia loss. Upon heating to 50°C, it loses 75% of its H_2O content, but less than 1% of the ammonia. At 200°C, under reduced pressure, the H_2O content drops to 1.15 mol, whereas only 2% of the ammonia is lost (57).

The pentaborate is shown by x-ray data to contain the $[B_5O_6(OH)_4]^-$ ion, analogous to that found in the sodium and potassium compounds.

Manufacture

Potassium tetraborate tetrahydrate may be prepared from an aqueous solution of KOH and boric acid with a B_2O_3/K_2O ratio of about 2.0, or by separation from a KCl–borax solution (99).

Potassium pentaborate is prepared in a manner analogous to that used for the tetraborate, but the strong liquor has a B_2O_3/K_2O ratio near 5.

Ammonium tetraborate tetrahydrate is prepared by crystallization from an aqueous solution of boric acid and ammonia with a $B_2O_3/(NH_4)_2O$ ratio of 1.8–2.1.

Ammonium pentaborate is similarly produced from an aqueous solution of boric acid and ammonia with a $B_2O_3/(NH_4)_2O$ ratio of 5. A process for the production of ammonium pentaborate by precipitation from an aqueous ammonium chloride–borax mixture has been patented (100).

Economic Aspects

The potassium and ammonium borates are low volume products with production figures of hundreds of tons per year for the tetra- and pentaborates. Prices during 1967 and 1977 are compared in Table 15.

Uses

Dipotassium tetraborate tetrahydrate is used to replace borax in applications where an alkali metal borate is needed but sodium salts cannot be used, or where a more soluble form is required. The potassium compound is used as a solvent for casein, as a constituent in welding fluxes, and a component in diazo-type developer solutions. Potassium pentaborate tetrahydrate is used in fluxes for welding and brazing of stainless steels and nonferrous metals. Diammonium tetraborate tetrahydrate is used when a highly soluble borate is desired, but alkali metals cannot be tolerated. It is used mostly as a neutralizing agent in the manufacture of urea–formaldehyde resins and as an ingredient in flameproofing formulations. Ammonium pentaborate tetrahydrate is used as a component of electrolytes for electrolytic capacitors, as an ingredient in flameproofing formulations, and in paper coatings.

CALCIUM-CONTAINING BORATES

Properties

Dicalcium Hexaborate Pentahydrate. Dicalcium hexaborate pentahydrate, $2CaO.3B_2O_3.5H_2O$ (formula wt, 411.16; monoclinic; sp gr, 2.42; heat of formation, -3.469 kJ/mol (-0.83 kcal/mol) (101), exists in nature as the mineral colemanite. Its solubility in H_2O is about 0.1% at 25°C. Heats of solution have been determined in HCl (101). Colemanite is formed on heating saturated solutions of inyoite, $2CaO.3B_2O_3.13H_2O$, or other higher hydrates.

The crystal structure of colemanite has been shown to contain $[B_3O_4(OH)_3]n^{2n-}$ chain polyanions. The structural relationships between colemanite and the other minerals of the series $2CaO.3B_2O_3.nH_2O (n = 1, 5, 7, 9, 13)$, and structural changes accompanying the ferroelectric transition of colemanite have been outlined (102).

Sodium Calcium Pentaborate Octahydrate. Sodium calcium pentaborate octahydrate, $NaCaB_5O_9.8H_2O$ or $Na_2O.2CaO.5B_2O_3.16H_2O$ (formula wt, 405.30; triclinic; sp gr, 1.95), exists in nature as the mineral ulexite. The compound can be prepared by seeding a solution of 110 g $CaB_2O_4.6H_2O$, 40 g H_3BO_3, 100 g borax, 450 g $CaCl_2$, and

Table 15. Prices Per Metric Ton for Potassium and Ammonium Borates—Carload Quantities (>36 t) in 45.4 kg Bags[a] in U.S. Dollars

Chemical	1967	1977
$K_2O.2B_2O_3.4H_2O$	412	919
$KB_5O_8.4H_2O$	282	860
$(NH_4)_2O.2B_2O_3.4H_2O$	377	863
$NH_4B_5O_8.4H_2O$	270	746

[a] Ref. 36.

2.5 L H$_2$O (103). Ulexite is slowly converted to NaCaB$_5$O$_9$.5H$_2$O, probertite, when seed is added to a moistened sample at 80–100°C. When crystals of ulexite are heated, 4 moles of H$_2$O are lost at 80–100°C, 8.5 more until 175°C, and the remaining 3.5 on heating to 450°C (104). Some crystallographic data have been recorded (105).

Sodium Calcium Pentaborate Pentahydrate. Sodium calcium pentaborate pentahydrate, NaCaB$_5$O$_9$.5H$_2$O or Na$_2$O.2CaO.5B$_2$O$_3$.10H$_2$O, (formula wt 351.25; monoclinic; sp gr 2.14) exists in nature as the mineral probertite. Probertite can be prepared by heating a mixture of 2 parts ulexite and 1 part borax to about 60°C (103). Some crystallographic data have been published (106).

Manufacture

The alkaline earth metal borates of primary commercial importance are colemanite and ulexite. Both of these borates are sold as impure ore concentrates. United States production is in Death Valley, Calif. Late in 1976, the American Borate Corporation purchased rights to the Death Valley colemanite and ulexite reserves which had been held by Tenneco Inc.; the major world producer is Turkey.

In the Death Valley operation (11), colemanite ore is blasted in a pit approximately 53 m deep. The ore is then hauled to the surface in trucks, dumped over a grizzly to remove large lumps, crushed, and screened. The screened ore is loaded from stockpiles into 18.1 t trucks and hauled to Lathrop Wells, Nevada, for further processing. Here the ore is sized to >0.25 mm (+60 mesh), <2.5 cm, and fed into a calciner which causes the colemanite to decrepitate and separate from the attached shale. The coarse shale is separated from the fine product, which now contains about 48% B$_2$O$_3$. Most of the product is <0.044 mm (−325 mesh). The nominal capacity of the Lathrop Wells facility is 63,500 t/yr of refined product.

Ulexite is also mined from open pits in Death Valley and trucked to Dunn, California for upgrading. The salable product is 26–28% B$_2$O$_3$.

Mining in certain areas of the Death Valley National Monument is forbidden due to environmental concerns. Colemanite mining on federal lands in the United States is governed by the Mining Laws of 1892.

Major colemanite and ulexite mining areas in Turkey are the Bigadic, Emet, and Mustafakemalpasa regions (4).

In the Bigadic region, several private concerns produce a total of about 91,000 metric tons per year of upgraded colemanite and 36,000 t/yr of upgraded ulexite. This production is glass grade material. The mining is both open-pit and underground.

At Hisarcik, in the Emet District, Etibank operates an open-pit mine and a colemanite washing plant. The mining operation has been somewhat converted from manual to modern techniques. The ore consists of colemanite nodules, closely packed with shale. The presence of high concentrations of arsenic sulfides has been indicated. Plant capacity is about 300,000 t/yr of upgraded product, 43% in B$_2$O$_3$. At Espey, Etibank operates an underground mine, producing about 36,000 t/yr of colemanite.

In the Mustafakemalpasa region, Mortas/Bortas produces about 54,000 t/yr of colemanite from an underground mine.

In the Turkish facilities the shale and colemanite are usually separated by either a washing or weathering procedure; a good deal of manual labor is often involved.

Specifications and Shipping

The colemanite which is to be used in the production of glass fibers must conform to the purchasers' specifications on Fe and As.

Colemanite is available in bags and bulk.

Uses

Colemanite, 2CaO.3B$_2$O$_3$.5H$_2$O. Colemanite is used in the production of boric acid and borax, as well as in several direct applications. It is a highly desirable material for the manufacture of the "E" glass used in textile glass fibers and plastic reinforcement (where sodium cannot be tolerated). High As or Fe levels in the ore concentrate prevent its use in this application. Colemanite has seen limited application as a slagging material in steel manufacture. It is also used in some fire retardants and as a precursor to some boron alloys.

Ulexite, NaCaB$_5$O$_9$.8H$_2$O, and Probertite, NaCaB$_5$O$_9$.5H$_2$O. Ulexite and probertite have found application in the production of insulation fiberglass and borosilicate glass as well as in the manufacture of other borates.

BORATE MELTS AND GLASSES

When alkali metal hydroxides or carbonates are fused with borax or boric acid (0–2 mol M$_2$O/mol B$_2$O$_3$), or when hydrated alkali metal borates are heated, much puffing occurs unless the heating is done very slowly. The melts become clear liquids when nearly anhydrous; if these liquids are high in boric oxide content, they become viscous on cooling and form glasses.

Most of the interest in alkali metal borate glasses has centered on reports which indicated the existence of maxima and minima in some of the physical properties (ie, viscosity, thermal expansion coefficient, etc) of the glasses with increasing metal oxide content. This phenomenon has been called the boron oxide anomaly (28). Modern theory on borate glass structure, however, indicates that the changes in the properties with alkali content are not anomalous, but are the result of well-defined structural changes in the glass at the moecular level. Krogh-Moe (107) theorized that four different borate structural groups are present in alkali borate glasses below 34 mol % M$_2$O. These groups are shown in Figure 11.

The triborate and pentaborate groups always occur in pairs (which are then referred to as tetraborate groups). According to Krogh-Moe, pure B$_2$O$_3$ consists of boroxol groups. On addition of M$_2$O, up to 20 mol %, tetraborate groups are formed. Between 20 and 34 mol % M$_2$O, diborate groups form at the expense of the tetraborate groups. Infrared (107) and laser Raman (31) data on the borate glasses and analogous crystalline anhydrous borates support Krogh-Moe's approach. Changes in the physical properties of the glass with M$_2$O content represent compromises between the effect of adding more metal ions to the system and the effect of making the borate structural entities more rigid by converting trigonally coordinated borons to tetrahedrally coordinated borons.

A number of reviews have appeared covering the various aspects of borate glasses. Nies (67) has reviewed the structure, physical properties, thermochemistry, reactions, phase equilibria, and electrical properties of alkali borate melts and glasses. Griscom (28) has recently reviewed the application of x-ray diffraction, nmr, Raman scattering,

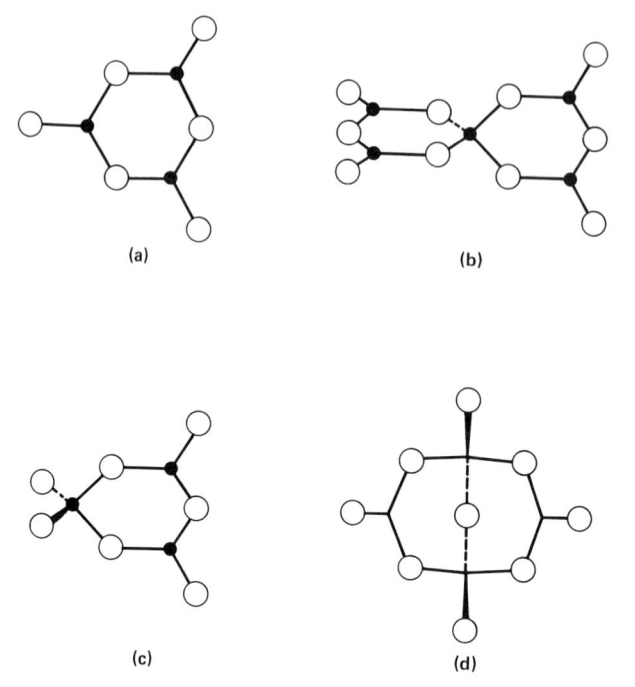

Figure 11. The borate glass structural groups (**a** = boroxol; **b** = pentaborate; **c** = triborate; **d** = diborate; ● = boron; and ○ = oxygen) (107).

ir spectroscopy, and esr to structural analysis. Phase equilibrium diagrams for a large number of anhydrous borate systems are included in a compilation by Levin (108). Thermochemical data on the anhydrous alkali metal borates have been compiled (22).

The borate glass compounds of commercial importance are B_2O_3 and $Na_2O \cdot 2B_2O_3$.

OTHER METAL BORATES

General Features

Borate salts or complexes of virtually every metal have been prepared. For most metals, a series of hydrated and anhydrous compounds may be obtained by varying the starting materials and/or reaction conditions. The individual compounds described in this section include only materials that have achieved some commercial importance.

In general, hydrated borates of heavy metals are prepared by mixing aqueous solutions or suspensions of the metal oxides, sulfates, or halides with boric acid or alkali metal borates (eg, borax). The precipitates formed from basic solutions are often sparingly-soluble, amorphous solids with variable compositions. Crystalline products are generally obtained from slightly acidic solutions.

Anhydrous metal borates may be prepared by heating the hydrated salts to 300–500°C, or by direct fusion of the metal oxide with boric acid or B_2O_3. Many binary and tertiary anhydrous systems containing B_2O_3 form vitreous phases over certain ranges of composition (108).

Barium Metaborate

At least four hydrates of barium metaborate, $BaO \cdot B_2O_3 \cdot xH_2O$ are known. The tetra- and pentahydrates both contain the $B(OH)_4^-$ anion, and are properly formulated as $Ba[B(OH)_4]_2 \cdot xH_2O$, where $x = 0$ or 1. These compounds crystallize when solutions of barium chloride and sodium metaborate are combined at room temperature. The higher hydrate is favored when excess sodium metaborate is used. Saturated aqueous solutions contain 12.5 g/L of $BaO \cdot B_2O_3 \cdot 4H_2O$ at 25°C. Both forms dehydrate at temperatures above 140°C (109). When the precipitation reaction is carried out at 90–95°C, a lower hydrate, $BaO \cdot B_2O_3 \cdot 2H_2O$, is formed. A monohydrate has also been described. Barium metaborate may also be prepared from barium sulfide formed by prior reduction of barium sulfate. The presence of sulfide impurities in the product may render it unsuitable for some applications (57).

Barium metaborate is used as an additive to impart fire retardant and mildew resistant properties to latex paints, plastics, textiles, and paper products (4). A recent patent describes the use of barium metaborate as a preservative in protein-based glues (110) (see Glue).

Cobalt and Copper Borates

Amorphous metaborates of copper and cobalt are precipitated when borax is added to aqueous solutions of the metal(II) sulfates or chlorides.

Cobalt metaborate, $CoO \cdot B_2O_3 \cdot xH_2O$, $x = 2$ or 3, is slowly converted to a hexaborate, $CoO \cdot 3B_2O_3 \cdot 10H_2O$, by stirring it with an aqueous solution of boric acid. A material described as technical grade cobalt tetraborate, $CoB_4O_7 \cdot 4H_2O$, is marketed as an acid catalyst by the Shepherd Chemical Company, Cincinnati, Ohio. This material may actually be a mixture of metaborates and hexaborates. The 1976 selling price was $9.70/kg for 230 kg lots (4).

Hydrated copper metaborate, $CuO \cdot B_2O_3 \cdot 2H_2O$, has been used as a fungicide for treatment of lumber and other cellulose materials (111). The anhydrous salt, $CuO \cdot B_2O_3$, is used as an oil pigment (4).

Manganese Tetraborate

Crystalline manganese tetraborate, $MnO \cdot 2B_2O_3 \cdot 9H_2O$, has been shown to have the structure $MnB_4O_5(OH)_4 \cdot 7H_2O$ (112). This compound is prepared via an intermediate hexaborate, $2MnO \cdot 3B_2O_3 \cdot xH_2O$, which forms when borax is added to an aqueous solution of manganese(II) sulfate. The tetraborate crystallizes when a stoichiometric quantity of boric acid is added to a stirred suspension of the hexaborate (113).

Two grades of manganese borates are marketed as $MnB_4O_7 \cdot xH_2O$ by the General Metallic Oxides Company, Jersey City, N.J. Prices quoted in 1976 were $3.70/kg for six 91-kg drums of pure powder, containing 18% manganese and $1.79/kg for 102-kg drums of technical powder, containing 6% manganese. Both materials are used as printing ink driers. The annual domestic demand is estimated at 2700–3600 kg (4).

Zinc Borates

A series of hydrated zinc borates have been developed over the past fifty years for use as fire-retardant additives in coatings and polymers (55,114). Worldwide consumption of these zinc salts is currently several thousand metric tons per year. A substantial portion of this total is used in vinyl plastics where zinc borates are added alone or in combination with other fire retardants such as antimony oxide.

The most commonly encountered zinc borate is $2ZnO.3B_2O_3.7H_2O$ which is formed when borax is added to solutions of soluble zinc salts. An x-ray structure determination has indicated that this compound is actually a triborate, zinc triborate monohydrate, $Zn[B_3O_3(OH)_5].H_2O$ (2). A compound with this composition is marketed as ZB-237 by the Humphrey Chemical Corporation, Aberdeen Proving Ground, Maryland. The same firm produces two other zinc borates, ZB-112 and ZB-325, prepared by heating dry blends of zinc oxide and boric acid. Apparently, these products that have the approximate compositions $ZnO.B_2O_3.2H_2O$ and $3ZnO.2B_2O_3.5H_2O$ contain $2ZnO.3B_2O_3.7H_2O$ as well as some amorphous product and unreacted starting materials. All of these zinc borates are dehydrated by heating from about 130–250°C (55). Prices for these products in 1976 ranged from \$1.28/kg for nine metric ton lots of ZB-237 to \$1.74/kg for 23 kg of ZB-325; prices are fob plant for material packed in 23-kg multiwall paper bags (115).

A different crystalline hydrate, $2ZnO.3B_2O_3.3.5H_2O$, is produced when the reaction between zinc oxide and boric acid is carried out at temperatures of 90–100°C (116). This patented product has also been crystallized from solutions containing borax, zinc chloride, and sodium hydroxide (117). It is marketed by the U.S. Borax & Chemical Corporation under the trademark Firebrake ZB. This compound has the unusual property of retaining its water of hydration at temperatures below 250°C. This thermal stability makes it attractive as a fire retardant additive for plastics that require high processing temperatures. The 1977 selling price for Firebrake ZB ranged from \$1.61/kg for 4.5 t lots of 113-kg drums to \$1.83/kg for a 45-kg drum (36) (see Flame retardants).

Since all the zinc borates are sparingly soluble in water and organic solvents, they are generally used as finely divided (2–10 μm) solids. A technique for impregnating porous materials such as wood products and textiles has been patented (118). In this process the zinc borates are prepared in aqueous ammonia where they form stable solutions containing complex ions. As material treated with these solutions is dried, the ammonia evaporates to leave an evenly distributed residue of insoluble zinc borates.

BORON PHOSPHATE

Boron phosphate, BPO_4, is a white, infusible solid that vaporizes slowly above 1450°C, without apparent decomposition. It is normally prepared by dehydrating mixtures of boric acid and phosphoric acid at temperatures up to 1200°C:

$$H_3BO_3 + H_3PO_4 \rightarrow BPO_4 + 3 H_2O$$

Complete dehydration requires temperatures above 1000°C.

The structure of boron phosphate prepared under normal atmospheric conditions consists of tetragonal bipyrimids analogous to the high cristobalite form of silica. Both the boron and phosphorus are tetrahedrally coordinated by oxygen. A quartz-like form of boron phosphate can be prepared by heating the common form to 500°C at 5.07 GPa (50,000 atm) (119).

The tri-, tetra-, penta-, and hexahydrates of boron phosphate have been reported. All of these decompose rapidly in water to give solutions of the parent acids. Anhydrous boron phosphate hydrolyzes in a similar fashion, although the reaction proceeds quite slowly for material that has been ignited at high temperatures.

As the principal application of boron phosphate has been as a heterogeneous acid catalyst, a high degree of chemical purity is often not important or even desirable. A variety of preparative methods have been devised in order to produce catalysts with optimum particle size, surface activity, and ease of regeneration. Methods employed to obtain modified catalytic properties include: incomplete hydration or partial rehydration; exposing the catalyst to inorganic acids; doping with alumina, silica, or transition metal oxides; and employing nonstoichiometric formulations containing slight excesses of either boron or phosphorus (120).

Boron phosphate has been frequently cited as a dehydration catalyst. Typical transformations include conversion of aliphatic alcohols to olefins, and the formation of piperylene from 2-methyltetrahydrofuran. The hydration and isomerization of olefins have also been reported. Other organic reactions catalyzed by boron phosphate include polymerization of aldehydes, preparation of nitriles from carboxylic acids and ammonia, and nitration of aromatic hydrocarbons (120).

Although boron phosphate is derived from two of the three most common glass-forming oxides, it exhibits little tendency to form a glass itself. Boron phosphate is a primary phase over a considerable portion of the B_2O_3–SiO_2–P_2O_5 system (121). The use of boron phosphate as a flux in silica-based porcelain and ceramics has been described, but no large-scale commercial applications have developed (120).

Table 16 is an alphabetical list of boron compounds referred to in the text:

Table 16. Oxides and Borates Referred to in Text

Compound	CAS Registry No.	Mol formula
ammonium pentaborate tetrahydrate	[12229-12-8]	$NH_4B_5O_8.4H_2O$
barium metaborate hydrate	[13701-59-2]	$BaO.B_2O_3.xH_2O$
boron dioxide	[13840-88-5]	BO_2
boron monoxide	[12505-77-0]	BO
boron oxide (6:1)	[11056-99-8]	B_6O
boron oxide (7:1)	[12447-73-3]	B_7O
boron oxide (13:2)	[56940-67-1]	$B_{13}O_2$
boron phosphate	[13308-51-5]	BPO_4
boron suboxide, boron oxide (12:2)	[54723-68-1]	$B_{12}O_2$
cobalt hexaborate decahydrate	[12447-18-6]	$CoO.3B_2O_3.10H_2O$
copper metaborate	[10290-09-2]	$CuO.B_2O_3$
copper metaborate dihydrate	[18851-84-8]	$CuO.B_2O_3.2H_2O$
diammonium tetraborate tetrahydrate	[10135-84-9]	$(NH_4)_2O.2B_2O_3.4H_2O$
diboron dioxide, boron oxide (2:2)	[13766-28-4]	B_2O_2
diboron trioxide, boron oxide (2:3)	[1303-86-2]	B_2O_3
dicalcium hexaborate pentahydrate	[12291-65-5]	$2CaO.3B_2O_3.5H_2O$
dipotassium tetraborate tetrahydrate	[12045-78-2]	$K_2O.2B_2O_3.4H_2O$
disodium octaborate tetrahydrate	[12280-03-4]	$Na_2O.4B_2O_3.4H_2O$
disodium tetraborate	[1330-43-4]	$Na_2O.2B_2O_3$
disodium tetraborate decahydrate or borax	[1303-96-4]	$Na_2O.2B_2O_3.10H_2O$
disodium tetraborate pentahydrate	[12045-88-4]	$Na_2O.2B_2O_3.5H_2O$
disodium tetraborate tetrahydrate	[12045-87-3]	$Na_2O.2B_2O_3.4H_2O$
dizinc hexaborate heptahydrate	[12280-01-2]	$Zn_2B_6O_{11}.7H_2O$
manganese tetraborate heptahydrate	[55200-92-5]	$Mn(B_4O_5(OH)_4).7H_2O$

Table 16. (continued)

Compound	CAS Registry No.	Mol formula
metaboric acid	[13460-50-9]	HBO$_2$
orthoboric acid	[10043-35-3]	H$_3$BO$_3$
potassium pentaborate tetrahydrate	[12229-13-9]	KB$_5$O$_8$.4H$_2$O
sodium calcium pentaborate octahydrate	[1319-33-1]	NaCaB$_5$O$_9$.8H$_2$O
sodium calcium pentaborate pentahydrate	[12229-14-0]	NaCaB$_5$O$_9$.5H$_2$O
sodium metaborate dihydrate	[16800-11-6]	NaBO$_2$.2H$_2$O
sodium metaborate tetrahydrate	[10555-76-7]	NaBO$_2$.4H$_2$O
sodium pentaborate pentahydrate	[12046-75-2]	NaB$_5$O$_8$.5H$_2$O
tetrazinc dodecaborate heptahydrate	[12513-27-8]	2ZnO.3B$_2$O$_3$.3.5H$_2$O
zinc diborate dihydrate	[27043-84-1]	ZnO.B$_2$O$_3$.2H$_2$O
zinc tetraborate pentahydrate	[12536-65-1]	ZnO.2B$_2$O$_3$.5H$_2$O
zinc triborate monohydrate	[12429-73-1]	Zn(B$_3$O$_3$(OH)$_5$).H$_2$O

BIBLIOGRAPHY

"Boron Oxides, Boric Acids, and Borates" under "Boron Compounds" in *ECT* 1st ed., Vol. 2, pp. 600–622, by M. H. Pickard, Pacific Coast Borax Co.; "Boron Compounds, Boron Oxides, Boric Acid, and Borates" in *ECT* 2nd ed., Vol. 3, pp. 608–652, by Nelson P. Nies, U.S. Borax Research Corporation.

1. N. P. Nies and G. W. Campbell, "Inorganic Boron–Oxygen Chemistry" in R. M. Adams, ed., *Boron, Metallo-Boron Compounds, and Boranes*, Interscience Publishers, a division of John Wiley & Sons, Inc., New York, 1964, pp. 192–194.
2. J. Ozols, I. Tetere, and A. Ievins, *Latv. PSR Zinat. Akad. Vestis Kim. Ser.* (1), 3 (1973).
3. G. Heller in K. Niedenzu and K. C. Buschbeck, eds., *Gmelin Handbuch der Anorganischen Chemie, Band 28, Teil 7*, Springer-Verlag, Berlin, 1975, pp. 2–4.
4. A. Ferguson and D. Treskon, "Boron Minerals and Chemicals" in *Chemical Economics Handbook*, Stanford Research Institute, Menlo Park, Calif., 1977, p. 717.1000A.
5. H. E. Pemberton, *Miner. Rec.* **6**, 74 (1975).
6. J. H. Puffer, *Miner. Rec.* **6**, 84 (1975).
7. J. F. McAllister, *Map Sheet 14*, California Div. of Mines and Geology, (1970).
8. K. Inan, A. C. Dunham, and J. Esson, *Inst. Min. Metall. Trans.* **B114**, (1973).
9. H. Helvaci and R. J. Firman, *Inst. Min. Metall. Trans.* **B142**, (1976).
10. K. P. Wang, "Boron" in *Minerals—Facts and Problems, U.S. Bureau of Mines Bulletin 667*, 1975, p. 173.
11. R. B. Kistler and W. C. Smith, *Industrial Minerals*, AIME 4th ed., 1975, p. 473.
12. V. Morgan, "Boron Geochemistry" in *Mellor's Comprehensive Inorganic Chemistry*, in press.
13. C. L. Christ, *J. Geol. Ed.*, **20**, 235 (1972).
14. W. E. VerPlanck, *Calif. J. Mines Geol.* **52**, 273 (1956).
15. W. A. Gale, "History and Technology of the Borax Industry" in ref. 1, p. 1.
16. Ref. 3, pp. 5–7.
17. D. D. Wagman and co-workers, *Nat. Bur. Stand. U.S. Tech. Note 270–2*, 26 (1966).
18. D. White and co-workers, *J. Chem. Phys.* **32**, 481 (1960).
19. F. A. Kanda and co-workers, *J. Am. Chem. Soc.* **83**, 1509 (1956).
20. D. R. Petrak, R. Ruh, and B. F. Goosey, *Proceedings of the 5th Materials Research Symposium on Solid State Chemistry, Nat. Bur. Stand. U.S. Special Pub. No. 364*, 605 (1972).
21. D. R. Petrak, R. Ruh, and G. R. Atkins, *Bull. Am. Ceram. Soc.* **53**, 569 (1974).
22. *JANAF Thermochemical Tables*, 2nd Ed., Nat. Stand. Ref. Data Ser., Nat. Bur. Stand. (U.S.), Washington, D.C., (1971).
23. Ref. 3, pp. 7–15.
24. G. S. Bogdanova, S. L. Antonova, and V. I. Kislyak, *Steklo. Keram.* **13** (1975).
25. J. Boow, *Phys. Chem. Glasses* **8**, 45 (1967).
26. K. H. Stern, *J. Res. Nat. Bur. Stand.* **A69**, 281 (1965).
27. L. L. Sperry and J. D. MacKenzie, *Phys. Chem. Glasses* **9**, 91 (1968).
28. D. L. Griscom, *Preconference Manuscript on Borate Glass Structure*, Alfred Conference on Boron in Glass and Glass Ceramics, Alfred, N.Y., 1977.

29. R. L. Mozzi and B. E. Warren, *J. Appl. Cryst.* **3**, 251 (1970).
30. G. E. Jellison, Jr. and P. J. Bray, *Solid State Commun.* **19**, 517 (1976).
31. W. L. Konijnendijk and J. M. Stevels, *J. Non-Cryst. Solids* **18**, 307 (1975).
32. J. R. Soulen, P. Sthapitanonda, and J. R. Margrave, *J. Phys. Chem.* **59**, 132 (1955).
33. D. L. Hildebrand, W. F. Hall, and N. D. Potter, *J. Chem. Phys.* **39**, 296 (1963).
34. A. Napolitano, P. B. Macedo, and E. G. Hawkins, *J. Am. Ceram. Soc.* **48**, 613 (1965).
35. P. L. Hanst, V. H. Early, and W. Klemperer, *J. Chem. Phys.* **42**, 1097 (1965); (a) J. D. MacKenzie and W. F. Claussen, *J. Am. Ceram. Soc.* **44**, 79 (1961).
36. *Industrial Products Catalog and Price Schedules,* U.S. Borax and Chemical Corp., Los Angeles, Calif., 1977.
37. U.S. Pat. 3,468,627 (Sept. 23, 1969), L. L. Fusby (to U.S. Borax & Chemical Corporation).
38. F. C. Kracek, G. W. Morey, and H. E. Merwin, *Am. J. Sci.* **35-A**, 143 (1938).
39. M. V. Kilday and E. J. Prosen, *J. Am. Chem. Soc.* **82**, 5508 (1960).
40. C. Feldman, *Anal. Chem.* **33**, 1916 (1961).
41. R. W. Sprague, "Properties and Reactions of Boric Acid" in ref. 12.
42. Ref. 1, pp. 67–69.
43. D. White and co-workers, *J. Chem. Phys.* **32**, 488 (1960).
44. Ref. 17, pp. 27–28.
45. J. Smisko and L. S. Mason, *J. Am. Chem. Soc.* **72**, 3679 (1950).
46. J. O. Edwards, *J. Am. Chem. Soc.* **75**, 6151 (1953).
47. L. Maya, *Inorg. Chem.* **15**, 2179 (1976).
48. B. R. Sanderson, "Coordination Compounds of Boric Acid" in ref. 12.
49. R. Kunin and F. Preuss, *Ind. Eng. Chem. Prod. Res. Dev.* **3**, 304 (1964); U.S. Pat. 3,887,460 (June 3, 1975), C. J. Ward, C. A. Morgan, and R. P. Allen (to U.S. Borax & Chemical Corporation).
50. U.S. Pat. 2,969,275 (Jan. 24, 1961), D. E. Garrett (to Am. Potash and Chemical Corporation); Ger. Pat. 1,164,997 (Mar. 12, 1964), D. E. Garrett and co-workers (to Am. Potash and Chemical Corporation); U.S. Pat. 3,479,294 (Nov. 18, 1969), F. J. Weck (to Am. Potash and Chemical Corporation); U.S. Pat. 3,424,563 (Jan. 28, 1969), R. R. Grinstead (to The Dow Chemical Co.); and U.S. Pat. 3,493,349, (Feb. 3, 1970), C. A. Schiappa and co-workers (to The Dow Chemical Co.).
51. R. E. Mesmer, K. M. Palen, and C. F. Bates, *Inorg. Chem.* **12**, 89 (1973).
52. K. A. L. G. Watt, *World Minerals and Metals, No. 12,* British Sulfur Corp. Ltd., 1973, pp. 5–12.
53. C. R. Havighorst, *Chem. Eng.* **70**(23), 228 (1963).
54. R. J. Weir and R. S. Fisher, *Toxicol. Appl. Pharmacol.* **23**, 351 (1972).
55. W. G. Woods, "Boron Compounds as Flame Retardants in Polymers" in V. M. Bhatnager, ed., *Advances in Fire Retardants, Part 2: Technomic,* 1973, pp. 120–153; J. P. Neumeyer, P. A. Koenig, and N. B. Knoepfler, *U.S. Agricultural Research Service, South Reg., ARS-S-64,* **70** (1975).
56. J. O. Edwards, G. C. Morrison, and J. W. Schultz, *J. Am. Chem. Soc.* **77**, 266 (1955).
57. Unpublished Data, U.S. Borax Research Corp.
58. Ref. 1, pp. 85–89.
59. S. Shishido, *Bull. Chem. Soc. Jpn.* **25**, 199 (1952).
60. *Technical Data Sheets,* U.S. Borax & Chemical Corp., 1977; U.S. Pat. 2,998,310 (Aug. 29, 1961), P. J. O'Brien and G. A. Connell (to U.S. Borax & Chemical Corp.).
61. N. Ingri, *Svensk. Kem. Tidskr.* **75**(4), 199 (1963).
62. M. J. How, G. R. Kennedy, and E. F. Mooney, *J. Chem. Soc. Chem. Commun.,* 267 (1969).
63. S. Scholle and M. Szmigielska, *Chem. Prumysl* **15**, 530 (1965).
64. U.S. National Bureau of Standards, *Circular 500,* 481 (1952).
65. D. E. Garrett and G. P. Rosenbaum, *Ind. Eng. Chem.* **50**, 1681 (1958).
66. N. P. Nies and R. W. Hulbert, *J. Chem. Eng. Data* **12**, 303 (1967).
67. N. P. Nies, "Alkali Metal Borates, Physical and Chemical Properties" in ref. 12.
68. H. Menzel and H. Schulz, *Z. Anorg. Chem.* **245**, 157 (1940).
69. H. Menzel and co-workers, *Z. Anorg. Chem.* **224**, 1 (1935).
70. N. Morimoto, *Mineral J. (Sapporo)* **2**, 1 (1956).
71. C. Giacovazzo, S. Menchetti, and F. Scordari, *Am. Mineral.* **58**, 523 (1973).
72. R. F. Giese, *Science* **154**, 1453 (1966).
73. E. F. Westrum and G. Grenier, *J. Am. Chem. Soc.* **79**, 1799 (1957).
74. E. F. Westrum, *Thermodynamic Transport Properties Gases, Liquids, Solids,* Papers Symposium, Lafayette, Ind., 1959, p. 275.
75. C. R. Fuget and J. F. Masi, *Thermodynamic Properties for Selected Compounds,* U.S. Atomic Energy Communication, CCC-1024-TR-263, 1957.
76. A. Predvoditelev, *Z. Phys.* **51**, 136 (1928).

77. J. Krogh-Moe, *Acta. Cryst.* **B30,** 578 (1974).
78. Ref. 1, p. 176.
79. G. T. Furukawa, M. L. Reilly, and J. H. Piccirelli, *J. Res. Nat. Bur. Stand.* **68A,** 381 (1964).
80. S. Merlino and F. Sartori, *Acta. Cryst.* **B28,** 3559 (1972).
81. S. N. Sidorova, L. V. Puchkov, and M. K. Federov, *Zh. Prikl. Khim.* **48,** 253 (1975).
82. H. Menzel and H. Schulz, *Z. Anorg. Chem.* **251,** 167 (1943).
83. S. Block and A. Perloff, *Acta. Cryst.* **16,** 1233 (1963).
84. Ref. 3, p. 131.
85. I. M. Kolthoff and E. B. Sandell, *Textbook of Quantitative Inorganic Analysis,* MacMillan, New York, 1952, p. 534.
86. F. D. Snell and C. L. Hilton, eds., *Encyclopedia of Industrial Chemical Analysis,* Wiley-Interscience, New York, 1968, p. 368.
87. K. P. Wang, "Boron" in *1974 Minerals Yearbook,* U.S. Bureau of Mines, 1976, p. 229.
88. W. A. Gale, *Ind. Eng. Chem.* **30,** 867 (1938).
89. G. H. Bixler and D. L. Sawyer, *Ind. Eng. Chem.* **49,** 322 (1957).
90. U.S. Pat. 3,300,278 (Jan. 24, 1967), N. P. Nies and P. F. Jacobs (to U.S. Borax and Chemical Corp.).
91. R. W. Sprague, "Toxicity of Boron Compounds" in ref. 12.
92. G. J. Levinskas, "Toxicology of Boron Compounds" in ref. 1, p. 693.
93. G. Carpeni, *Bull. Soc. Chim. Fr.,* 1327, (1955).
94. G. Carpeni, J. Haladjian, and M. Pilard, *Bull. Soc. Chim. Fr.,* 1634 (1960).
95. J. Haladjian and G. Carpeni, *Bull. Soc. Chim. Fr.,* 1629 (1960).
96. M. Marezio, H. Plettinger, and W. Zachariasen, *Acta. Cryst.* **16,** 975 (1963).
97. W. H. Zachariasen and H. A. Plettinger, *Acta Cryst.* **16,** 376 (1963).
98. Ref. 3, p. 149.
99. U.S. Pat. 2,776,186 (Jan. 1, 1957), F. H. May (to American Potash and Chemical Corp.).
100. U.S. Pat. 2,867,502 (Jan. 6, 1959), H. Stange (to Olin Mathieson Chemical Corp.).
101. V. M. Gurevich and V. A. Sokolov, *Geokhimiya,* **3,** 455 (1976).
102. J. R. Clark, D. E. Appleman, and C. L. Christ, *J. Inorg. Nucl. Chem.* **26,** 73 (1964).
103. C. Palache, H. Berman, and C. Frondell, *Dana's System of Mineralogy,* John Wiley & Sons, Inc., New York, 1957, p. 347.
104. Ref. 1, p. 129.
105. J. R. Clark and C. L. Christ, *Am. Min.* **44,** 712 (1959).
106. M. Kurbanov, I. M. Rumanova, and N. V. Belov, *Dokl. Akad. Nauk SSSR* **152,** 1100 (1963).
107. J. Krogh-Moe, *Phys. Chem. Glasses* **6,** 46 (1965).
108. E. M. Levin, H. F. McMurdie, and F. P. Hall, *Phase Diagrams for Ceramacists,* Part 1, 1956; Part II, 1959; Supplement I, 1964; and Supplement II, 1969; The American Ceramic Society, Columbus, Ohio.
109. Ref. 1, p. 133.
110. Ger. Pat. 2,336,052 (Feb. 7, 1974), R. W. Long.
111. A. Kalnins and co-workers, *Latv. PSR Zinat. Akad. Vestis* (8), 64 (1969); U.S. Pat. 3,431,059 (Mar. 4, 1969), C. J. Connor and G. S. Danna.
112. I. Berzina, J. Ozols, and A. Ievins, *Latv. PSR Zinat. Akad. Vestis* (6), 648 (1974).
113. Ref. 1, p. 138.
114. W. G. Woods and J. G. Bower, *Mod. Plast.* **47,** 140 (1970).
115. *Price List 267,* Humphrey Chemical Corp., Aberdeen Proving Grounds, Md., Jan. 12, 1976.
116. U.S. Pat. Re 27, 424 (1972), N. P. Nies and R. W. Hulbert (to U.S. Borax and Chemical Corp.); U.S. Pat. 3,718,615 (1973), W. G. Woods, J. G. Whiten, and N. P. Nies.
117. U.S. Pat. 3,649,172 (1972), N. P. Nies and R. W. Hulbert (to U.S. Borax & Chemical Corp.).
118. U.S. Pat. 3,524,761 (Aug. 18, 1970), S. B. Humphrey (to Humphrey Chemical Corp.).
119. Ref. 1, pp. 184–186.
120. B. P. Long, "B-O Compounds of Groups V and VI" in ref. 12.
121. W. J. Englert and F. A. Hummel, *J. Soc. Glass Technol.* **39,** 121T (1955).

Daniel J. Doonan
Loren D. Lower
U.S. Borax Research Corp.

BORIC ACID ESTERS

In the first reported synthesis of organic boron compounds Ebelman and Bouquet (1) reported in 1846 the preparation of the methyl, ethyl, and amyl esters of boric acid from the reactions of boron trichloride with the appropriate alcohols. Since that time the chemistry of boric esters has expanded significantly, and the related chemical literature is voluminous. Although the patent literature in this field has also grown rapidly, industrial applications of boric acid esters have remained at a relatively low level.

The term boric acid esters refers to compounds with the general formula $B(OR)_3$. Much of the chemistry of these compounds is related to the electrophilic nature of boron. A series of related compounds can be formed when electron deficient boron accepts a fourth nucleophilic substituent leading to tetrahedral boron structures such as $NaB(OR)_4$.

Nomenclature in the boric acid ester series is confusing at best. The IUPAC committee on boron chemistry has suggested the use of trialkoxy- and triaryloxyboranes (2) for compounds widely referred to in current literature as boric acid esters, trialkyl (or aryl) borates, trialkyl (or aryl) orthoborates, alkyl (or aryl) borates, alkyl (or aryl) orthoborates, and in the older literature as boron alkoxides and aryloxides. The lower molecular weight esters such as methyl, ethyl, phenyl, etc, are most commonly referred to as methyl borate, ethyl borate, phenyl borate, etc.

Although trialkoxy- and triaryloxyboranes represent by far the most common and thoroughly investigated class of boric acid esters, a variety of closely related cyclic esters of polyhydric alcohols and phenols such as (1–3) have been prepared. Examples of the anhydride type (1) and the biborate (2) have been produced and marketed in relatively large quantities as biocides, epoxy curing agents, and gasoline additives.

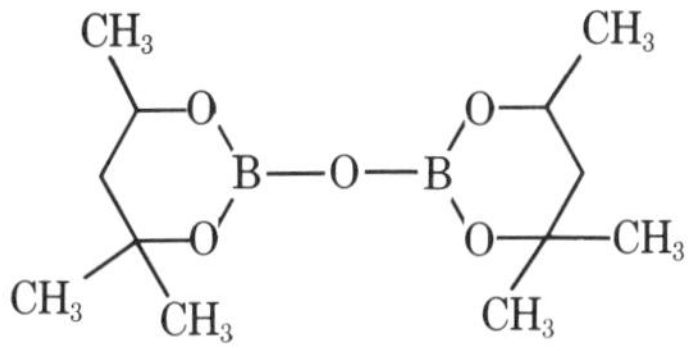

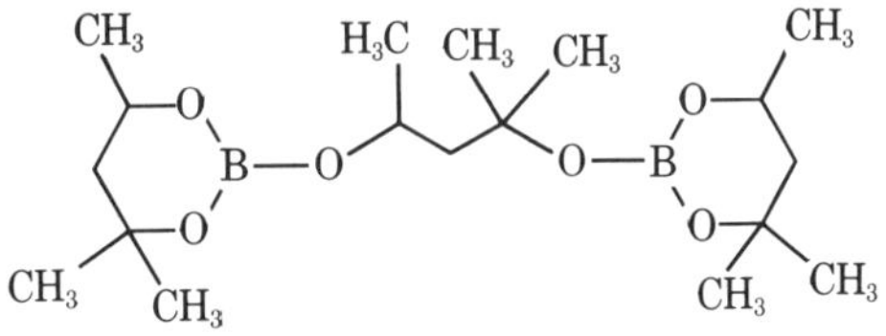

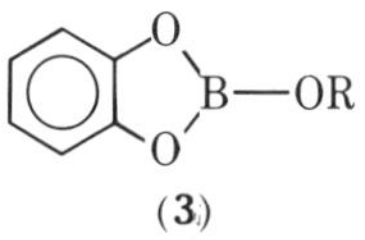

General classes related to boric acid esters but which have not found significant commercial application include partial esters (4) and boroxines or boroxins (5). Acyloxyborates such as $(CH_3COO)_3B$ have also been reported as well as a number of alkoxyboranes with substituents other than the alkoxy group on the boron atom such as $ROBH_2$, $(RO)_2BSR'$, $R_2NB(OR')_2$, and $(RO)_2BX$. Tetrahedral borates, $NaB(OR)_4$, peroxyborates such as $(ROO)_3B$, and borates with B—O—M bonds, eg, $(R_3SnO)_3B$ and $(R_3SiO)_3B$ are also in the general class of boric acid esters.

$(RO)_2BOH$

(4)

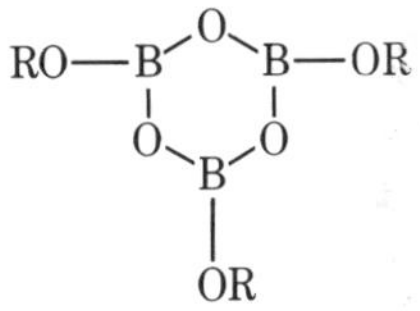

The simple boric acid esters or trialkoxy(aryloxy)boranes are far more frequently noted in the chemical and patent literature than these related classes. The most comprehensive reference available for additional specific details of the chemistry of boric acid esters of all types is Steinberg (3) which covers the literature exhaustively up to 1962 and remains the authoritative source in this field.

Properties

Physical Properties. Trialkoxy(aryloxy)boranes, which represent the most numerous and closely studied of the boric acid ester classes, range from colorless low boiling liquids such as trimethoxyborane (commonly referred to as methyl borate) to high melting solids. Examples of the most common trialkoxy(arloxy) boranes used in industrial applications as well as some commercial glycol borates, boroxines and others are included in Table 1. In some cases the reported ranges of physical properties make the choice of a single value difficult. For a more comprehensive tabulation and for additional physical properties see (3). Christopher and others have also compiled tables of densities (4), viscosities (4), dipole moments (5), surface tensions (6), vapor pressure and thermodynamic functions of a number of common borate esters (7–9).

Trialkoxy(aryloxy)boranes are typically monomeric, soluble in most organic solvents, and dissolve in water with accompanying hydrolysis to boric acid and the corresponding alcohol or phenol. This hydrolysis occurs rapidly except in cases where the boron atom is protected by bulky alkyl or aryl substituent groups. The boron atom in trialkoxy(aryloxy)boranes is in a trigonal coplanar state with sp^2 bond hybridization as shown, with a vacant p orbital along the three-fold symmetry axis perpendicular to the BO_3 plane. This vacant orbital is readily available for acceptance of nucleophiles such as water which have unshared electrons. Structural studies of trialkoxy and triaryloxyboranes have confirmed this structure with the angle between C—O bonds at 120° except in those cases where the angles are slightly distorted by bulky substituent groups. The susceptibility of the boron atom to attack by nucleophiles is similar to that of the carbonyl carbon in carboxylic acid esters and leads to the analogy with boric acid ester nomenclature.

Reactions. Trialkoxyboranes from straight-chain alcohols and triaryloxyboranes are stable to relatively high temperatures; eg, methyl borate is reported to be stable to 470°C (10). Attempts to use this potentially useful property in borate ester-based high temperature lubricants and heat transfer media have met with limited success because of the reactivity of these compounds to water and oxygen.

Trialkoxyboranes from branched-chain alcohols are much less stable, and boranes from tertiary alcohols decompose at relatively low temperatures; many are unstable at 100°C (11). Decomposition of branched-chain esters leads to mixtures of olefins, alcohols, and other derivatives.

The chemical property of borate esters which restricts their general utility more than any other is their ready hydrolysis in the presence of water. Atmospheric moisture

Table 1. Properties of Some Boric Acid Ester Derivatives

Compound	Name	CAS Registry Number	mp, °C	bp, °C (at 101.3 kPa[a])	d_4^t	n_D^t
B(OCH$_3$)$_3$	methyl borate	[121-43-7]	−29	68.0–68.5	0.920[20]	1.3548[25]
B(OCH$_3$)$_3$·CH$_3$OH	methyl borate azeotrope			54.3	0.8804[25]	1.3472[25]
B(OCH$_2$CH$_3$)$_3$	ethyl borate	[150-46-9]	−84.8	117–119	0.859[26]	1.3723[25]
B(OCH$_2$CH$_3$)$_3$·7.75CH$_3$CH$_2$OH	ethyl borate azeotrope			76.6		
B(OCH$_2$CH$_2$CH$_3$)$_3$	n-propyl borate	[688-71-1]		176–179	0.356[24]	1.3993[25]
B[OCH(CH$_3$)$_2$]$_3$	isopropyl borate	[17862-14-5]		139–140	0.815[23]	1.3750[25]
B(OCH$_2$CH$_2$CH$_2$CH$_3$)$_3$	n-butyl borate	[688-74-4]		227	0.856[25]	1.4077[25]
B(OC$_6$H$_5$)$_3$	phenyl borate	[1095-03-0]	71–81[b]	360–370		
B(OC$_6$H$_4$CH$_3$)$_3$	tricresyl borate (m and p) (isomers)	[26248-41-9]		224–230 (at 2.3)		
(1)	2,2′-oxybis(4,4,6-trimethyl-1,3,2-dioxaborinane)	[14697-50-8]		185–200 (at 0.27)	1.013[24]	1.4308[25]
(2)	2,2′-(1,1,3-trimethyltrimethylenedioxy)bis-(4,4,6-trimethyl-1,3,2-dioxaborinane)	[100-89-0]		114–115 (at 0.27)	0.932[21]	1.4381[25]
				274–276		
				170–172 (at 2.7)		
(8)	2,2′-(1-methyltrimethylenedioxy)-bis(4-methyl-1,3,2-dioxaborinane)	[2665-13-6]		207–213 (at 2.3)	1.071[25]	1.4464[17]
(5)	trimethoxyboroxine, R = CH$_3$	[102-24-9]		dec	1.2286[25]	
(5)	isopropoxyboroxine, R = Pr^i	[10298-87-0]	52–54	235–239 dec		
(9)	catechol borane	[274-07-7]	12	77 (at 13.3)		1.5070[20]
B(OCH$_2$CH$_2$)$_3$N	triethanolamine borate	[283-56-7]	235–239			

[a] To convert kPa to mm Hg, multiply by 7.5.
[b] Phenyl borate has been reported to melt at temperatures fom 38 to 146°C. Most values are in the range of 80–90°C.

is sufficient to hydrolyze most esters rapidly to boric acid and the respective alcohol or phenol:

$$(RO)_3B + 3 H_2O \rightarrow 3 ROH + H_3BO_3 \qquad (1)$$

Partially hydrolyzed derivatives such as $(RO)_2BOH$ are not isolated except in special cases involving large substituent groups. Comparative hydrolysis rates have been measured for a number of borate esters (3). Rates are significantly affected by the steric requirements of the alcohol or phenol, and most hydrolytically stable derivatives have this characteristic. Some particularly stable examples are tri-t-amyl borate [22238-22-8], tri-2-cyclohexylcyclohexyl borate [5440-19-7] and the general class of hindered phenolic borates such as the derivatives of 2,6-di-t-butyl phenol. This latter type is of particular interest because compounds such as (6) and (7) have been widely investigated as antioxidants (qv) and stabilizers (see Heat stabilizers). These and other examples are the subject of a series of composition patents (12) which describe their preparation and properties. Other borate esters exhibiting good hydrolytic stability are compounds such as triethanolamine borate and tri-isopropanolamine borate [101-00-8] in which the boron atom through its vacant orbital is coordinated with an internal nitrogen atom containing a free electron pair.

Boric acid esters react with amines to form complexes or, under specific conditions, tri(amino)boranes by alcohol displacement (13–14). The latter reaction (eq. 2) does not proceed readily except when the amines have boiling points much higher than the displaced alcohol and result in aminoboranes that are not severely sterically hindered.

$$(RO)_3B + 3 R'_2NH \rightarrow 3 (R'_2N)_3B + 3 ROH \qquad (2)$$

This reaction is also facilitated when the products formed are stabilized cyclic borazines or related compounds (15–17).

Formation of complexes between borate esters (Lewis acids) and amines or other bases has been reported with trialkoxyboranes (particularly in the presence of alcohol) and triaryloxyboranes. However, there is more recent evidence (18) showing that (with the possible exception of the addition of methyl borate to ammonia and methylamines) the complexes reported with lower alkyl borates and amines in the presence of catalytic amounts of the alcohol are tetraalkoxylborate salts, $R'_3NH^+B(OR)_4^-$, not the simple complexes, $R'_3N:B(OR)_3$.

Exchange of alkoxy groups in mixtures of borate esters has been shown to occur readily (19–20). Similar exchanges of alkoxy groups on boron with amino groups (20–21), alkyl groups (22), and alkylthio groups (22) have also been observed. In most cases the mixed products have not been isolated, only the symmetrical starting materials are obtained.

The reactions of borate esters with organometallic compounds, particularly Grignard reagents, are standard preparative routes to alkyl boranes (eq. 3).

$$(RO)_3B + R'MgBr \rightarrow R'B(OR)_2 + ROMgBr \qquad (3)$$

Reaction of trialkoxyboranes with metal hydrides can lead to boron–hydrogen bonded compounds, and this reaction is the basis for the current production of sodium borohydride.

Preparative Methods. The most common preparative method for trialkoxy- and triaryloxyboranes is the reaction of the appropriate alcohol or phenol with boric acid, an inexpensive and readily available boron source. The equilibrium shown in equation 4 normally lies far to the left. Usually the water formed is removed by azeotropic distillation with excess alcohol or a suitable solvent such as benzene, toluene, or various petroleum distillate fractions.

$$B(OH)_3 + 3 ROH \rightleftharpoons B(OR)_3 + 3 H_2O \qquad (4)$$

The procedure used depends on the specific ester desired. One complication encountered with the two low molecular weight esters, methyl borate and ethyl borate, is the formation of low boiling azeotropes between the borate and one molar equivalent of alcohol (see Azeotropic and extractive distillation).

The closely related reaction of alcohols with boric oxide has also been used as a preparative method. Use of three equivalents of alcohol requires a filtration step to remove the boric acid formed (eq. 5). This method avoids formation of the azeotropes of methyl and ethyl borates,

$$3 ROH + B_2O_3 \rightarrow B(OR)_3 + H_3BO_3 \qquad (5)$$

but only 50% of the boron in boric oxide is converted to the desired product. In a commercial process B_2O_3 can be recovered by dehydration of the boric acid formed. Use of six equivalents of alcohol (eq. 6) and removal of the water formed by azeotropic distillation in a manner equivalent to equation 4 is also an accepted method. Boric acid is usually preferred as a starting material

$$6 ROH + B_2O_3 \rightarrow 2 B(OR)_3 + 3 H_2O \qquad (6)$$

because it is significantly less expensive than boric oxide. Production rates, which depend on the ease of water removal, are generally faster with boric oxide. Molecular sieves (qv) have also been used to remove water in the alcohol–boric oxide reaction (see also Esterification).

Boric acid esters of higher molecular weight alcohols and phenols can be prepared by transesterification of a low boiling ester such as methyl borate (eq. 7). The lower boiling alcohol is removed

$$(CH_3O)_3B + 3 ROH \rightarrow (RO)_3B + 3 CH_3OH \qquad (7)$$

by distillation. A continuous process has been developed using these reactions (24). Another modification of this method has been reported where use of molecular sieves to absorb the low boiling alcohol is used rather than distillation (25). In commercial practice use of boric acid provides a more economical route to higher boiling boric acid esters, but transesterification is a convenient laboratory procedure.

Other less practical boron sources than boric acid have been used for the preparation of boric acid esters. These include boron halides, tri(amino)boranes and others. Again some of these methods, particularly starting from boron trichloride, may be convenient in the laboratory, but they are not considered in large-scale production.

Manufacture

There is essentially no published information on the processes used for commercial production of boric acid esters or their total annual production rates. The boric acid ester prepared in the largest quantities is methyl borate, and most of it is used captively in the production of sodium borohydride by Ventron Corporation. Other sources of methyl borate are Anderson Development Co., May and Baker Ltd., and Borax Consolidated Ltd.

Methyl borate production was studied in great detail during the 1950s and 1960s when it was proposed as a key intermediate for production of high energy fuels. In 1976 about 4500 metric tons of methyl borate were produced in the manufacture of sodium borohydride. Other uses may have added 500–1000 t giving a total market which may have approached 5500 t. Methyl borate is sold as the pure compound and also as the methanol azeotrope which consists of an approximately 1:1 molar ratio of methanol to methyl borate. Methyl and ethyl borates both form azeotropes with their respective alcohols, and the azeotropes are the lowest boiling components of the reaction mixtures obtained in their preparation. In the preparation of methyl borate, methanol is added to boric acid as indicated in equation 4 and the azeotrope distilled. The volatile azeotrope must be separated in a second step to produce the pure ester. A large number of methods have been described to separate the methanol–methyl borate azeotrope including continuous fractional distillation, sulfuric acid extraction, and treatment with calcium or lithium chloride. Other methods used are hydrocarbon washing, high-pressure fractional distillation, adding a compound that will form a lower boiling azeotrope with methanol, use of molecular sieves, and semipermeable membranes. The method of choice in large commercial operations is that involving lithium chloride (26–27). The major disadvantage is corrosivity of this inorganic chloride. For smaller scale production the sulfuric acid method is more convenient (28).

Other borate esters that have been offered commercially include ethyl borate, n-propyl borate, isopropyl borate, n-butyl borate, and cresyl borate (from a mixture of *meta* and *para* cresols). Each of these is prepared by the standard reaction of the appropriate alcohol or phenol with boric acid. An azeotropic separation step similar to that used in the production of methyl borate is used to produce pure ethyl borate. In each case the borate is purified by distillation.

Two boroxines, methoxy and isopropoxy, probably are produced by azeotropic removal of water in reactions similar to the preparation of trialkoxyboranes. Boroxines, derivatives of metaboric acid (HBO_2), are trimeric as indicated in equation 8; the monomer, ROBO, can not be isolated.

$$3 \text{ ROH} + 3 \text{ H}_3\text{BO}_3 \rightleftharpoons (\text{ROBO})_3 + 6 \text{ H}_2\text{O} \qquad (8)$$

They are difficult to purify and are very susceptible to hydrolysis.

Processes for glycol borate derivatives also are based on boric acid. The appropriate glycol and water are heated in the presence of a suitable hydrocarbon azeotroping agent, and water is removed continuously by using a condenser which allows continuous solvent return. The glycol borates (1) and (8) have been prepared in this manner as undistilled bottoms products, since distillation can result in decomposition. The biborate (2) has been prepared as an undistilled product and also purified by vacuum distillation.

Catechol borane (9) is prepared by the reaction of catechol with diborane in tetrahydrofuran (29).

$$\text{(8)}$$

$$\text{catechol} + BH_3 \cdot THF \longrightarrow \text{catecholborane} + THF + 2H_2 \qquad (9)$$

The reaction proceeds readily at 0–25°C and the product is purified by distillation. The related hexylene glycol derivative (**10**) [*10199-14-1*] is most conveniently prepared by the reaction of the hexylene glycol boric anhydride (**1**) with lithium aluminum hydride (30). Both compounds are specialty chemicals and have been produced in very small quantities.

$$\text{(10)}$$

Economic Aspects

Since none of the boric acid esters are sold as commodity chemicals, their prices remain relatively high, but there is no reason to believe that prices for the esters derived from low cost alcohols and phenols would not come down significantly with increasing volume. Prices usually quoted for most of the pure borate esters shown in Table 1 are in the range of $2.20–6.60/kg. Methyl borate azeotrope used in gas fluxing (see under Uses) is priced well below $2.20/kg, and even pure methyl borate is available in the range of $2.20/kg.

Handling and Shipping

Procedures for shipping of boric acid esters vary according to the particular compound being handled. Aryl borates produce phenols when contacted with water, and they must be labeled as corrosive chemicals and are subject to the shipping regulations governing such materials. Lower alkyl borates are flammable (flash points of methyl, ethyl, and butyl borates, are 0, 32, and 94°C, respectively) and must be stored in approved areas. All of the low boiling esters from methyl to butyl must also be labeled as flammable. Other compounds, such as hexylene glycol biborate (**1**), offer no hazard and may be shipped or stored in any convenient manner.

Organic boron–oxygen compounds are difficult to handle because of their propensity to hydrolyze. The more sensitive compounds should be stored and transferred in an inert atmosphere.

Normally small quantities of borate esters are shipped in cans or drums, and tank cars can be used for bulk items. The only metal corrosion problem would occur with esters that hydrolyzed to corrosive products, eg, halogenated alcohols.

Uses

Sodium Borohydride. Methyl borate is used as an intermediate in the commercial production of sodium borohydride:

$$(CH_3O)_3B + 4\ NaH \rightarrow NaBH_4 + 3\ NaOCH_3 \qquad (10)$$

This route to sodium borohydride is based on work done by Schlesinger and Brown under contract to the United States government (31).

Gas Fluxing. A major commercial outlet for boric acid esters is the use of methyl borate azeotrope as gaseous flux for welding (qv) and brazing (see Solders). The azeotrope, which acts as a volatile source of boric oxide, is introduced directly into the gas stream as a flux for the surfaces to be joined in the welding process. The major utilization of this process has been in the European automobile industry, although some methyl borate is also manufactured for this use in the United States.

Polymer Additives. The area of potential commercial application of borate esters that has probably been most thoroughly investigated has been additives for polymers. The earliest reference in this field appears to be a 1931 British patent (32) which claims the use of β-naphthyl borate [1259-70-7] as a stabilizer in rubber. Since that time the patent literature in this field has expanded rapidly, but the actual industrial usage of borate esters as polymer additives has not shown comparable growth. Small quantities of several borate esters are used as epoxy resin curing agents, others have been used in polymer stabilization, but up to 1977, no market of even moderate size has developed. There may be commercial uses of a proprietary nature in which the esters are produced captively, but there is certainly no indication in the open literature of significant quantities of borate esters being produced for polymer applications.

Although there have been some claims in which borate esters have been added to modify the resin structure, the major use of borate esters in epoxy systems has been as curing agents or hardeners. At least five specific borate esters, tricresyl borate, the hexylene glycol biborate (2), 2-[2-(dimethylamino)ethoxy]-4-methyl-1,3,2-dioxaborinane [7024-33-1] (11), and the methyl and isopropyl boroxines (5) have been sold for this use. Some of the borates are claimed to give rapid cures at low temperatures; others lead to extended pot lives by modifying the curing action of amine hardening agents (see Epoxy resins).

A few of the patents describing these applications are references 33–38.

Following the initial disclosure of β-naphthyl borate as a rubber stabilizer, borate esters have been claimed as stabilizers in a wide variety of polymer systems as shown in Table 2. In some cases stabilization is precisely defined, eg, stabilization to heat, ultraviolet radiation, oxygen or ozone (see Antioxidants; Uv stabilizers).

Table 2. Borate Esters as Polymer Stabilizers

Polymer system	Type of stabilization	Borate esters claimed	References
poly(vinyl chloride)	heat	$(R_3SnO)_3B$ tin–boron compounds	39
	heat	$(RO)_3B$, hindered phenolic borates	40–41
	oxidation	octyl borate	42
poly(phenylene oxide)	(?)		43
polyolefins	heat	hindered phenolics	44
	oxidation	$(RO)_3B$, hindered phenolics, glycol borates, tetrahedral borates	42, 45
	general stabilizer	hindered phenolics plus phosphites	46–47
rubbers	heat	triphenyl, tricyclohexyl borates	48
	oxidation	hindered phenolics, tricresyl borate	42, 49
poly(vinyl alcohol)	heat	hindered phenolics	50
polyesters	heat	$(RO)_3B$ + bisphenols	51
polyamides	oxidation, ultraviolet	stearyl borate	52
polyurethane foam	color	$(RO)_3B$, hindered phenolics	53

Borate esters have also been investigated as secondary plasticizers (qv) in polyvinyl chloride (54) and as gelation preventatives in polysulfones (55) and in polyesters (56).

Numerous patents claim the use of borate esters as catalysts for polymerization reactions. In most cases the borate is one constituent of a catalyst mixture containing several components. It is not clear whether any of these patented mixtures are being used commercially. Some examples of the types of polymers that can be prepared using borate ester catalysts or mixtures containing borate esters include polyacrylonitrile (57), poly(vinyl chloride) and copolymers (58), polyurethanes (59), polypropylene (60), polyesters (61), and polyisobutylene (62).

Many boron compounds have good flame retardant properties, and inorganic borates are commonly used as additives in flame retardant plastics, wood, and cellulosic products (see Flame retardants). Brominated alkyl borate esters have been claimed as effective flame retardants in polyurethane foams (63) and in polyolefins (64). The borate esters of mono-, di-, and triethanolamines have been patented as flame retardants in cotton (65) and glycerol borates as antismoke additives in polymeric systems (66), but borate esters have found little commercial application in this area.

Gasoline Additives. In the late 1950s and through the 1960s a series of boron containing compounds were developed, mainly by Standard Oil of Ohio (Sohio), as gasoline additives. These additives, which are claimed to reduce engine knocking and engine deposits, were manufactured in large quantities for use in the United States and in a number of other countries (see Gasoline). The most popular examples were borate esters from glycols of the anhydride and biborate types. The use of these compounds has decreased somewhat in the 1970s.

Hydraulic Fluids and Lubricants. The use of borate esters in hydraulic fluids (qv) and lubricants has been described in numerous patents (see Lubrication). Borate additives have been claimed an anticorrosion agents, antioxidants, and heat stabilizers. The antioxidant properties of a number of borate esters in various hydrocarbon fluids are summarized in reference 42. Improved high pressure properties and lubricating properties have also been claimed. It is not clear whether any of these patented applications have been committed to practice. One application which has developed commercially is the use of borates of alkylated glycols to give a hydraulic fluid with improved water sensitivity. These fluids consist predominantly of borates such as $[CH_3O(CH_2CH_2O)_n]_3B$ plus some unreacted glycols (67). The glycol borates have the property of absorbing significant quantities of water without affecting the critical properties of the fluid, particularly the boiling point.

Biocides. Most boron compounds have mildly antiseptic properties and at relatively high concentrations they affect certain microorganisms. Although many boric acid esters have been tested against microorganisms, the only commercial use that has developed is a rather specific application to fuel systems containing a contaminant water layer. Microorganisms growing in the water layer feed on the hydrocarbon fuel, multiply, and cause scum and precipitates which foul engines using the fuel (see Aviation and other gas turbine fuels). Biobor JF, US Borax & Chem Corp. is an effective microbiocide which can be added at low levels to the fuel portion of the water-contaminated fuel system. This product, which contains a mixture of the glycol borates (1) and (8), has been used successfully in jet fuels, heating oils and various diesel applications. Current sales are in the range of 100 metric tons per year mainly for use in aircraft and fuel tanks in ships (see Industrial antimicrobial agents).

Flame Retardant Cotton Batting. Investigators at the USDA Southern Regional Research Center discovered that treatment with methyl borate is an effective method to produce flame retardant cotton batting for mattresses. Impregnation of the cotton (qv) with a solution of boric acid in anhydrous methanol was effective (68). However, vaporization of methyl borate onto the cotton fibers with subsequent rapid hydrolysis to boric acid from water in the atmosphere and in the cotton is the preferred method (69–70) since no drying is necessary. This is accomplished by placing the cotton batts above a solution of methanol and boric acid which is in equilibrium with vapors of the low-boiling (54°C) methanol–methyl borate azeotrope. Commercial usage of this process depends on cost and effectiveness relative to alternative methods which involve direct application of boric acid.

Hydrocarbon Oxidation. It has been well established that oxidations of hydrocarbon and hydrocarbon derivatives can be significantly altered by boron compounds, and this property has been the basis of several large-scale commercial processes, eg, the oxidation of cyclohexane to a cyclohexanol–cyclohexanone mixture in nylon manufacture (see Hydrocarbon oxidation; Polyamides).

A number of patents have been issued on the use of borate esters and boroxines in hydrocarbon oxidation reactions, but commercial processes apparently use boric acid as the preferred boron source. However, all of the oxidations involve intermediate formations of borate esters which are hydrolyzed to the desired alcohols. This process is used to obtain alcohols selectively rather than the normal complex mixture of alcohols, carbonyl compounds, carboxylic acids, etc. The literature in this field has been covered through 1967 in a review which describes uses and reaction mechanisms (42). There have been numerous publications particularly in the Russian and Japanese literature since that time, but no significant new applications have been reported for borate esters.

Specialty Chemicals. Several borate esters and related compounds are sold in relatively small volumes for specific chemical applications. Catechol borane (9) is a hydroborating agent for the convenient synthesis of boronic acid derivatives (29), and the related dioxaborinane (11) is a weak hydroborating agent (30) as well as a blocking group for alcohols, amines, etc (71) (see Hydroboration).

$$B(O\overset{O}{\overset{\|}{C}}CF_3)_3$$
(12)

Tris(trifluoroacetoxy)borane [350-70-9] (12) can be used to remove a variety of protective groups employed in peptide synthesis (72). Reaction conditions are mild and (12) does not react with the peptide chain or with other substituents in the structure. The reagent is prepared by the reaction of boron tribromide with trifluoroacetic acid (73).

BIBLIOGRAPHY

"Boric Acid Esters" are treated in *ECT* 2nd ed. under "Boron Compounds, Boric Acid Esters," Vol. 3, pp. 653–673, by H. C. Newson, U.S. Borax Research Corporation.

1. J. J. Ebelman and M. Bouquet, *Ann. Chim. et Phys.* **17,** 54 (1846); *Ann.* **60,** 251 (1846).
2. "The Nomenclature of Boron Compounds," *Inorg. Chem.* **7,** 1945 (1968).
3. H. Steinberg, *Organoboron Chemistry, Vol 1 Boron–Oxygen and Boron–Sulfur Compounds*, Interscience Publishers, New York, 1964.
4. P. M. Christopher and W. H. Washington, *J. Chem. Eng. Data*, **14,** 437 (1969).
5. L. Maijs, *Latv. PSR Zinat. Akad. Vestis, Khim. Ser.* **247** (1972).
6. P. M. Christopher and G. V. Guerra, *J. Chem. Eng. Data* **16,** 468 (1971).
7. J. W. Wilson and J. T. F. Fenwick, *J. Chem. Thermodynamics* **5,** 341 (1973).
8. P. M. Christopher and A. Shilman, *J. Chem. Eng. Data* **12,** 333 (1967).
9. P. M. Christopher, *J. Chem. Eng. Data* **5,** 568 (1960).
10. S. Makishima, Y. Yoneda, and T. Tajima, *J. Phys. Chem.* **61,** 1618 (1957).
11. J. R. Anderson, K. G. O'Brien, and F. H. Reuter, *J. Appl. Chem.* **2,** 241 (1952).
12. U.S. Pat. 3,359,298 (Dec. 19, 1967), D. L. Hunter and H. Steinberg (to United States Borax & Chemical Corporation); U.S. Pat. 3,356,707 (Dec. 5, 1967), J. B. Hinkemp, J. D. Bartleson, and G. E. Irish (to Ethyl Corporation); U.S. Pat. 3,102,902 (Sept. 3, 1963), R. M. Washburn and C. F. Albright (to American Potash and Chemical Corp.); U.S. Pat. 3,347,793 (Oct. 17, 1967), R. M. Washburn and F. A. Billig (to American Potash and Chemical Corp.).
13. E. J. Mezey, P. R. Girardot, and W. E. Bissinger, *Advan. Chem. Ser.* **42,** 192 (1964).
14. R. J. Brotherton and H. Steinberg, *J. Org. Chem.* **26,** 4632 (1961).
15. K. Pilgrim and F. Korte, *Tetrahedron* **19,** 137 (1963).
16. S. H. Dandegaonker and A. S. Mane, *J. Indian Chem. Soc.* **50,** 622 (1973).
17. U.S. Pat. 2,948,751 (Aug. 9, 1960), R. J. Brotherton (to United States Borax & Chemical Corp.).
18. J. W. Wilson, *J. Chem. Soc. Dalton Trans.*, 1628 (1973).
19. Y. N. Bubnov and co-workers, *J. Gen. Chem. USSR (Engl. Transl.)* **42,** 1308 (1972).
20. B. M. Mikhailov and L. S. Vasil'ev, *Bull. Acad. Sci. USSR (Engl. Transl.)*, 1962 (1961); *Proc. Acad. Sci USSR* (Engl. Transl.), **139,** 693 (1961).
21. J. P. Laurent and co-workers *J. Chim. Phys. Physicochim Biol.* **69,** 1022 (1972).
22. H. C. Brown and S. K. Gupta, *J. Am. Chem. Soc.* **93,** 2802 (1971).
23. Ger. Pat. 2,209,047 (Sept. 14, 1972), and U.S. Pat. 3,853,941 (Dec. 10, 1974), W. V. Hough, C. R. Guibert, and G. T. Hefferan (to Mine Safety Appliance Co.).
24. Brit. Pat. 842,534 (July 27, 1960), United States Borax & Chemical Corporation.

25. Ger. Pat. 2,209,065 (Sept. 7, 1972), W. V. Hough (to Mine Safety Appliance Co.).
26. H. J. Becher, *Z. Physik. Chem.* **2**, 276 (1954).
27. H. I. Schlesinger and co-workers, *J. Am. Chem. Soc.* **75**, 213 (1953).
28. U.S. Pat. 2,217,354 (Oct. 8, 1940), F. J. Appel (to E. I. du Pont de Nemours & Co., Inc.).
29. H. C. Brown and S. K. Gupta, *J. Am. Chem. Soc.* **93**, 1816 (1971).
30. R. H. Fish, *J. Am. Chem. Soc.* **90**, 4435 (1968); W. G. Woods and P. L. Strong, *J. Am. Chem. Soc.* **88**, 4667 (1966).
31. U.S. Pat. 2,461,661 (Feb. 15, 1949), H. I. Schlesinger and H. C. Brown (to United States Atomic Energy Commission).
32. Brit. Pat. 363,483 (Dec. 24, 1931), H. M. Bunbury, J. S. H. Davies, and W. J. S. Nauntan (to Imperial Chemical Industries Ltd.).
33. Fr. Pat. 1,579,439 (Aug. 22, 1969), R. M. Moran and H. T. Blekicki (to Ciba Ltd.).
34. Ger. Pat. 2,163,143 (Aug. 3, 1972), J. F. Bosso and M. Wismer (to PPG Industries, Inc.).
35. U.S. Pat. 3,378,504 (April 16, 1968), H. L. Lee (to Callery Chemical Co.); U.S. Pat. 3,382,217 (May 7, 1968), L. C. Case.
36. U.S. Pat. 3,257,347 (June 21, 1966), W. G. Woods, W. D. English, and I. S. Bengelsdorf, U.S. Pat. 3,269,853 (Aug. 30, 1966), W. D. English, I. S. Bengelsdorf, and G. W. Willcockson, and Brit. Pat. 955,491 (April 15, 1964) (to United States Borax & Chemical Corporation).
37. Brit. Pat. 928,835 (June 19, 1963) (Westinghouse Electric Corp.).
38. U.S. Pat. 3,102,873 (Sept. 3, 1963), M. M. Lee (to General Electric Co.).
39. Jpn. Pat. 11,495 (June 29, 1967), T. Seki and Y. Kawakami and Jpn. Pat. 19,177 (Sept. 28, 1967), K. Suzuki, T. Seki, and Y. Kawakami (to Nitto Chemical Industrial Co., Ltd.); U.S. Pat. 3,928,285 (Dec. 23, 1975), R. G. Gough and F. J. Buescher (to Cincinnati Millicron, Inc.).
40. Jpn. Pat. 11,494 (June 29, 1967), T. Seki and Y. Kawakami (to Nitto Chemical Industry Co., Ltd.).
41. T. Morikawa and T. Amano, *Kobunshi Kagaku*, **30**, 479 (1973).
42. W. G. Woods and R. J. Brotherton, "Oxidations of Organic Substrates in the Presence of Boron Compounds," in R. J. Brotherton and H. Steinberg, eds., *Progress in Boron Chemistry*, Vol. 3, Pergamon Press, 1970.
43. U.S. Pat. 3,450,670 (June 17, 1969), K. E. Holoch, A. Katchman, and R. A. Schufelt, U.S. Pat. 3,792,121 (Feb. 12, 1974), V. Abolins, P. F. Erhardt, and K. E. Holoch (to General Electric Co.).
44. U.S. Pat. 3,345,326 (Oct. 3, 1967), L. S. Chang (to Allied Chemical Corp.).
45. U.S. Pat. 3,131,164 (April 28, 1964), M. E. Doyle and G. S. Jaffe (to Shell Oil Co.).
46. U.S. Pats. 3,367,996 (Feb. 6, 1968), 3,435,097 (March 25, 1969), and 3,526,679 (Sept. 1, 1970), J. Bottomley and R. Strauss (National Polychemicals); Ger. Pat. 1,149,527 (May 30, 1963), J. D. Shimm and W. H. Skoroszewski (to Shell International Research, Maatschappij N.V.).
47. U.S. Pats. 2,598,755 (June 3, 1952), 3,445,498 (May 20, 1969), 3,505,226 (April 7, 1970), 3,598,757 (Aug. 10, 1971), and 3,669,926 (June 13, 1972), H. A. Cyba (to Universal Oil Products Corp.).
48. U.S. Pat. 3,600,351 (Aug. 17, 1971), J. O. Hunt and R. P. Spitz (to Firestone Tire and Rubber Co.).
49. U.S. Pat. 3,177,267 (April 6, 1965), J. P. Luvisi (to Universal Oil Products).
50. Jpn. Pat. 73 27,905 (Aug. 27, 1973), O. Murai, K. Miyamoto, and M. Jsutsumi (to Mitsui Toatsu Chemicals, Inc.).
51. Jpn. Pat. 71 29,976 (Aug. 31, 1971), Y. Masai, T. Murayama, and Y. Kato (to Toyo Spinning Co., Ltd.).
52. Fr. Pat. 1,576,859 (Aug. 1, 1969), H. Eggensperger, V. Frazen, and G. Huhner (to Deutsche Advance Producktion G.m.b.H.).
53. Jpn. Pat. 72 06,755 (Feb. 25, 1972), K. Kumada, T. Okuda, and K. Kato (to Sanyo Chemical Industries, Ltd.).
54. S. P. Potnis and A. G. Khanolkar, *J. Polym. Sci.* **33**, 161 (1971).
55. U.S. Pat. 3,377,324 (April 9, 1968), S. Mostert (to Shell Oil Co.).
56. Belg. Pat. 656,114 (May 24, 1965) W. C. Dearing, R. J. Shrontz, and J. M. Howald (to American Cyanamid Co.).
57. Fr. Pat. 1,486,471 (June 20, 1976), G. Balitrand (to Societe des Usines Chimiques Rhone-Poulenc).
58. Brit. Pats. 919,198 (Feb. 20, 1963), 950,769 (Feb. 26, 1964) and 953,930 (April 2, 1964), (Dynamit Nobel A.G.); Fr. Pat. 1,390,143 (Feb. 19, 1965) (Monsanto Chemicals Ltd.).
59. Ger. Pat. 2,034,166 (Jan. 21, 1971), G. V. D. Teirs and M. G. Allen (to Minnesota Mining and Manufacturing Co.).
60. Jpn. Pat. 69 14,353 (June 26, 1969), K. Machida, H. Ikegami, and K. Kubo (to Tokuyama Soda Co., Ltd.).
61. Jpn. Pats. 71 32,432 (Sept. 21, 1971), 71 32,433 (Sept. 21, 1971), and 71 40,712 (Dec. 1, 1971), H. Kobayashi, K. Sasaguri, and A. Kawamoto (to Asahi Chemical Industries Co., Ltd.).

62. Jpn. Pat. 68 18,143 (Aug. 1, 1968), K. Miyoshi, Y. Araki, S. Uemura, and S. Koya (to Japan Petrochemical Co., Ltd.).
63. U.S. Pat. 3,189,565 (June 15, 1965) and U.S. Pat. 3,250,797 (May 10, 1966), W. G. Woods, D. Laruccia, and I. S. Bengelsdorf (to United States Borax & Chemical Corporation).
64. Ger. Pat. 2,238,243 (Feb. 15, 1973) and U.S. Pat. 3,907,857 (Sept. 23, 1975), H. Mayerhoefer, W. Mueller, and U. Sollberger (to Sandoz Ltd.).
65. R. Liepins, and co-workers *J. Appl. Polymer Sci.* **17,** 2523 (1973).
66. Jpn. Pat. 75 100,147 (August 8, 1975), Y. Ogawa, H. Hisada, H. Kimoto, N. Shiina, and T. Takenaka (to Marubishi Oil Chemical Co. Ltd.).
67. Can. Pat. 980,756 (Dec. 30, 1975) and Brit. Pat. 1,214,171 (Dec. 2, 1970), A. W. Sawyer and D. A. Csejka (to Olin Chemical Co.); Fr. Pat. 2,122,540 (Oct. 6, 1972), R. A. Cameron and C. J. Harrington (to Burmah Oil Trading Ltd.); U.S. Pat. 3,925,223 R. L. Coffman and R. W. Shiffler (to Union Carbide); Ger. Pat. 2,257,546 (June 7, 1973), D. A. Dalman (to Dow Chemical Co.); Belg. Pat. 818,595 (Feb 10, 1975) (to Chuo Kagaru Kogyo K.K.).
68. J. P. Madacsi, J. P. Neumeyer, and N. P. Knoepfler, *Ind. Eng. Chem. Prod. Res. Div.* **15,** 71 (1976).
69. U.S. Pat. 4,012,507 (March 15, 1977), N. B. Knoepfler, J. P. Madacsi, and J. P. Neumeyer (to the U.S. Government, Secretary of Agriculture.).
70. N. B. Knoepfler, J. P. Madacsi, and J. P. Neumeyer, *J. Coated Fabr.* **6,** 121 (1976).
71. R. Fish and R. J. Brotherton, *Abstracts of Papers Presented at the 155th National Meeting of the American Chemical Society,* San Francisco, California, March 1968, Abst. P-98.
72. J. Pless and W. Bower, *Angew Chem. Int. Ed.* **12,** 147 (1973).
73. W. Gerrard, M. F. Lappert, and R. Schafferman, *J. Chem. Soc.* 3648 (1958).

R. J. BROTHERTON
U.S. Borax Research Corp.

REFRACTORY BORON COMPOUNDS

Borides have metallic characteristics, with high electrical conductivity and positive coefficients of electrical resistivity. Many of them have high melting points, great hardness, low coefficients of thermal expansion, and good chemical stability, particularly the borides of metals of Subgroups IVA, VA, and VIA, the MB_6 compounds of Groups II and III, and the borides of aluminum and silicon.

Borides are inert toward nonoxidizing acids; however, a few such as Be_2B and MgB_2, react with aqueous acids to form boron hydrides. Most borides dissolve in oxidizing acids such as nitric or hot sulfuric acid. They are also readily attacked by hot alkaline salt melts or fused alkali peroxides and more stable borates are formed. In dry air, where a protective oxide film can be preserved, borides are relatively resistant to oxidation. For example, the borides of vanadium, niobium, tantalum, molybdenum, and tungsten do not oxidize appreciably in air up to temperatures of 1000–1200°C. Zirconium and titanium borides are fairly resistant up to 1400°C. Engineering and other properties of refractory metal borides are summarized in ref. 1.

Table 1 lists many metal borides and their observed melting points. Most metals form more than one boride phase and borides often form a continuous series of solid

Table 1. Metal Borides [a]

Boride	CAS Registry No.	mp, °C	Boride	CAS Registry No.	mp, °C
AlB$_2$	[12041-50-8]	975, d[b]	NdB$_6$	[12008-23-0]	2540
AlB$_{12}$	[12041-54-2]	2070	NiB	[12007-00-0]	1080
BaB$_6$	[12046-08-1]	2070	Ni$_2$B	[12007-01-1]	1230
BeB$_2$	[12228-40-9]	>1970	Ni$_2$B$_2$	[12007-00-0]	1160
BeB$_6$	[12228-40-9]	2070	Ni$_3$B	[12007-02-2]	1155
Be$_2$B	[12536-51-5]	1520	Pd$_3$B	[12429-53-7]	820, d
Be$_5$B	[12536-53-7]	1160	Pd$_5$B$_2$	[11130-91-9]	870, d
CaB$_6$	[12007-99-7]	2235	ReB$_2$	[12355-99-6]	2400
CeB	[12045-00-0]		RuB	[12523-59-0]	1600
CeB$_4$	[12007-52-2]	2380, d	RuB$_2$	[12360-00-8]	1600, d
CeB$_6$	[12008-02-5]	550	Ru$_2$B$_3$	[12356-00-2]	1600
CoB	[12006-77-8]	1460	ScB$_2$	[12007-34-0]	2250
Co$_2$B	[12045-01-1]	1285	ScB$_6$	[12785-49-8]	mp, u
Co$_3$B	[12006-78-9]	1125, d	SiB$_4$	[12007-81-7]	1870, d
CrB	[12006-79-0]	2060	SiB$_6$	[12008-30-9]	1980, d
CrB$_2$	[12007-16-8]	2130	SmB$_4$	[12007-82-8]	mp, u
Cr$_2$B	[12006-80-3]	1875	SmB$_6$	[12008-29-6]	2540
Cr$_3$B$_2$	[12045-40-8]	1960	SrB$_6$	[12046-08-1]	2235
Cr$_3$B$_4$	[12045-71-5]	1920	TaB	[12007-07-7]	2040
Cr$_4$B	[12006-81-4]	1680	TaB$_2$	[12007-35-1]	3100
FeB	[12006-84-7]	≈1550	Ta$_2$B	[12045-26-0]	1900
Fe$_2$B	[12006-86-9]	1390	Ta$_3$B$_4$	[12045-92-0]	2650
Fe$_3$B	[12006-86-9]		Ta$_3$B$_2$	[12045-92-0]	2450

Compound	CAS	mp	Compound	CAS	mp
GdB$_4$	[12007-54-4]		ThB$_4$	[12007-83-9]	2500
GdB$_6$	[12008-06-9]	u^c	ThB$_6$	[12229-63-9]	2195
HfB	[12228-27-2]	2100	TiB	[12007-08-8]	2600
HfB$_2$	[12007-23-7]	2100, d	TiB$_2$	[12045-63-5]	2900 ± 50
LaB$_4$	[12008-21-8]	3250	Ti$_2$B	[12305-68-9]	2200
LaB$_6$	[12008-21-8]	1800	Ti$_2$B$_5$	[12447-59-5]	2670
MgB$_2$	[12007-25-9]	2150	UB$_2$	[12007-36-2]	2385
MgB$_6$	[12008-22-9]	800, d	UB$_4$	[12007-84-0]	2495
MgB$_{12}$	[12230-32-9]	1100, d	UB$_6$		2235
MnB	[12045-15-7]	1300, d	VB	[12045-27-1]	2250
MnB$_2$	[12228-50-1]	1890	VB$_2$	[12007-77-3]	2450
MnB$_4$	[12271-96-4]	1900	V$_2$B$_3$	[12313-74-5]	2300
Mn$_2$B	[12045-16-8]	2160	V$_3$B$_2$	[12421-53-3]	2060
Mn$_3$B$_4$	[12229-02-6]	1580	WB	[12007-09-9]	2660
Mn$_4$B	[12260-22-9]	1750, d	W$_2$B	[12007-10-2]	2670
MoB	[12006-98-3]	1285, d	W$_2$B$_5$	[12007-98-6]	2365
MoB$_2$	[12007-27-1]	600	YB$_2$	[12429-58-2]	2100
Mo$_2$B	[12006-99-4]	2300, d	YB$_4$	[12045-95-3]	2800
Mo$_2$B$_2$	[12006-98-3]	2280	YB$_6$	[12008-32-1]	2600, d
Mo$_2$B$_5$	[12007-97-5]	2066	YB$_{12}$	[12046-90-1]	2200, d
NbB	[12045-19-1]	2100	ZrB	[12045-28-2]	2800
NbB$_2$	[12007-29-3]	2270	ZrB$_2$	[12045-64-6]	3040
		2900	ZrB$_{12}$	[12046-91-2]	2250, d

^{a} Ref. 2.
^{b} d = Decomposes.
^{c} u = Compound exists but melting point unknown.

solutions with one another at elevated temperatures; thus close composition control is necessary to achieve particular properties. The relatively small size of boron atoms facilitates their diffusion.

The structures of borides range from the isolated boron atoms in the M_2B borides through single chains in MB borides, double chains (M_3B_4), two-dimensional hexagonal nets (MB_2), cross-linked nets (MB_4), and interconnected B_6 octahedra (MB_6), to cages of 24 boron atoms surrounding the central metal atom in the MB_{12} borides (3–4). The three-dimensional frameworks of the boron-rich borides provide stable lattices through which the metal atoms may migrate at high temperatures, ca 1600°C; damaged surfaces may thereby be rejuvenated. These stable lattices also improve the chemical stability of such borides.

Preparation. The simplest method of preparation is a combination of the elements at a suitable temperature, usually in the range of 1100–2000°C. On a commercial scale, borides are prepared by the reduction of mixtures of metallic and boron oxides with aluminum, magnesium, carbon, boron, or boron carbide followed by purification.

Borides can also be synthesized by vapor-phase reaction or electrolysis.

To produce wear-resistant or hardened surfaces, thin layers of borides can be prepared on metal surfaces by reaction and diffusion. Boride powders can be formed into monolithic shapes by cold pressing and sintering, or by hot pressing.

Uses. In spite of their unique properties, there are few commercial applications for monolithic shapes of borides. They are used for resistance-heated boats (with boron nitride), for aluminum evaporation, and for sliding electrical contacts. There are a number of potential uses in the control and handling of molten metals and slags where corrosion and erosion resistance are important. Titanium diboride and zirconium diboride are potential cathodes for the aluminum Hall cells. Lanthanum hexaboride and cerium hexaboride are particularly useful as cathodes in electronic devices because of their high thermal emissivities, low work functions, and resistance to poisoning.

Boron Carbide

Boron and carbon form one compound, boron carbide [12069-32-8], B_4C, although excess boron may dissolve in boron carbide, and a small amount of boron may dissolve in graphite (5). Usually excess carbon appears as graphite, except for the special case of boron diffused into diamond at high pressures and temperatures (eg, 5 GPa or 50 kbar and 1500°C) where boron may occupy both interstitial and substitutional positions in the diamond lattice, a property utilized in synthetic diamonds (see Carbon, diamond, synthetic).

Properties. Boron carbide has a rhombohedral structure consisting of an array of nearly regular icosahedra, each with twelve boron atoms at the vertices and three carbon atoms in a linear chain outside the icosahedra (3–4,6–7). Thus a descriptive chemical formula would be $B_{12}C_3$ [12075-36-4]. Each boron atom is bonded to five others in the icosahedron as well as either to a carbon atom or to a boron atom in an adjacent icosahedron. The structure is very similar to that of rhombohedral boron. The theoretical density for $B_{12}C_3$ is 2.52 g/cm^3. The rigid framework of relatively closely bonded atoms corresponds to a high melting point, about 2400°C, appreciable electrical conductivity, great hardness (about 27 GPa or 270 kbar on the Knoop scale, diamond indenter), and a high compressive strength. Its brittleness limits its useful tensile strength to about 150 MPa (1.5 kbar) at 950°C; this, in combination with a moderate coefficient of thermal expansion, makes it sensitive to thermal shock. It is noticeably oxidized in air at 800–1000°C, and is a semiconductor with a room temperature resistivity of 0.001–0.1 Ω·m.

Boron carbide is resistant to most acids but is rapidly attacked by molten alkalies. It may be melted without decomposition in an atmosphere of carbon monoxide, but is slowly etched by hydrogen at 1200°C. It withstands metallic sodium fairly well at 500°C and steam at 300°C (8).

Preparation. Boron carbide is most commonly produced by the reduction of boric oxide with carbon in an electric furnace between 1400 and 2300°C. In the presence of carbon, magnesium reduces boric oxide to boron carbide at 1400–1800°C. The reaction is best carried out in a hydrogen atmosphere in a carbon tube furnace. By-product magnesium compounds are removed by acid treatment.

In general the purified boron carbide is ultimately obtained as a granular solid which subsequently may be molded or bonded into useful shapes. To achieve high density and strength, it is hot pressed at 1800–2400°C in graphite molds.

Uses. Major applications for boron carbide relate either to its hardness or its high neutron absorptivity (B^{10} isotope). Hot-pressed boron carbide finds use as wear parts, sandblast nozzles, seals, and ceramic armor plates; but in spite of its hardness, it finds little use as an abrasive. However, this property makes it particularly useful for dressing grinding wheels.

Boron carbide is used in the shielding and control of nuclear reactors because of its neutron absorptivity, chemical inertness, and radiation stability. For this application it may be molded, bonded, or the granular material may be packed by vibration (see Nuclear reactors).

Boron Nitride

Boron and nitrogen form one compound, boron nitride [10043-11-5], BN, which may exist in a hexagonal graphite-like form, with a layered structure and planar 6-membered rings of alternating boron and nitrogen atoms (3,9). On alternate sheets, boron atoms are directly over nitrogen atoms. A denser cubic form with a zinc blende lattice also exists, in addition to an equally dense hexagonal form with a wurtzite lattice. The latter two forms are, like diamond, thermodynamically stable only at high pressures, but persist at normal temperatures and pressures because of the slowness of the transition.

Properties. Under nitrogen pressure, hexagonal boron nitride melts at about 3000°C, but sublimes at about 2500°C at atmospheric pressure. Despite the high melting point, the substance is mechanically weak because of the relatively easy sliding of the planes of rings past one another (3). The theoretical density is 2.27 g/cm^3 and the resistivity is about 10^5 Ω·m.

Hexagonal boron nitride is relatively stable in oxygen or chlorine up to 700°C, probably because of a protective surface layer of boric oxide. It is rapidly attacked by hot alkali or fused alkali carbonates and slowly by many acids as well as alcohol, acetone, and carbon tetrachloride. It is not wetted by most molten metals and many molten glasses.

The chemical properties (10) of cubic and wurtzitic boron nitride are similar to those of the graphite form to which they revert at temperatures above ca 1700°C.

The cubic form resembles diamond in its crystal structure, and is almost as hard.

It is colorless and a good electrical insulator when pure; traces of impurities add color and make it a semiconductor. The theoretical density is 3.48 g/cm^3.

Preparation. Hexagonal boron nitride can be prepared by heating boric oxide with ammonia, or by heating boric oxide, boric acid, or its salts with ammonium chloride, alkali cyanides, or calcium cyanamide. Elemental nitrogen does not react with boric oxide even in the presence of carbon, although it does react with elemental boron at high temperatures. Boron nitride obtained from the reaction of boron trichloride or boron trifluoride with ammonia is easily purified.

The cubic zinc blende form of boron nitride is usually prepared from the hexagonal graphitic form at high pressures (4–6 GPa or 40–60 kbar) and temperatures (1400–1700°C). The reaction is accelerated by lithium or magnesium nitride catalysts (10). At higher pressures of 6–13 GPa (60–130 kbar) the cubic or wurtzite forms are obtained without catalyst (11).

The wurtzite form differs only slightly from the cubic form, but it is not quite as stable. It is most easily obtained by static or dynamic compression of the hexagonal graphitic form at high pressures (11). In the presence of a liquid catalyst at high pressures, the wurtzite form changes to the cubic form rapidly. The change occurs more slowly without a catalyst above 6 GPa (60 kbar) (12).

Uses. Hot-pressed hexagonal boron nitride is useful for high-temperature electrical or thermal insulation, vessels, etc, especially in inert or reducing atmospheres. Its low thermal expansion makes it resistant to thermal shock. The powder can be used as a mold release agent or as thermal insulation. Pyrolytically deposited boron nitride has greater chemical resistance and is impervious to helium but, of course, displays some anisotropy. Boron nitride is also available in fiber form for special applications (13).

The greatest use of cubic boron nitride is as an abrasive under the name Borazon, in the form of small crystals of 1 to about 300 μm in size. Usually these crystals are incorporated in abrasive wheels and used to grind ferrous and nickel-based alloys. The extreme hardness of the crystals and their resistance to attack by air and hot metal, make the wheels very durable, and close tolerances can be maintained on the work pieces, especially where harder alloys are employed.

The cubic boron nitride crystals may also be bonded together into strong bodies which make excellent cutting tools for hard steels and nickel-based alloys. Such tools will produce red-hot chips and now permit the wider use of tough, high-temperature alloys which would otherwise be prohibitively expensive to shape (14) (see Abrasives).

BIBLIOGRAPHY

"Boron Compounds (Refractory)" in *ECT* 2nd ed., Vol. 3, pp. 673–680, by James L. Boone, U.S. Borax Research Corporation.

1. Battelle Memorial Institute, *Engineering Properties of Selected Ceramic Materials*, American Ceramic Society, Columbus, Ohio, 1966.
2. W. G. Moffat, *Handbook of Binary Phase Diagrams*, General Electric Co., Schenectady, N.Y., 1976.
3. R. Thompson, *Endeavour*, 34 (Jan. 1970).
4. N. N. Greenwood, "Boron" in J. C. Bailar and co-eds., *Comprehensive Inorganic Chemistry*, Vol. 1, Pergamon Press, New York, 1973, pp. 665–993.
5. C. E. Lowell, *J. Am. Ceram. Soc.* **50,** 142 (1967).
6. R. R. Ridgeway, *Trans. Electrochem. Soc.* **66,** 117 (1934).
7. C. W. Tucker and P. Senio, *Acta Crystallogr.* **7,** 450 (1954).

8. *Handbook on Boron Carbide, Elemental Boron, and Other Stable, Boron-Rich Materials,* Norton Company, Worcester, Mass., 1955.
9. R. S. Pease, *Acta Crystallogr.* **5,** 356 (1952).
19. R. H. Wentorf, Jr., *J. Chem. Phys.* **34,** 809 (1961).
11. F. P. Bundy and R. H. Wentorf, Jr., *J. Chem. Phys.* **38,** 1144 (1963).
12. F. P. Bundy and F. R. Corrigan, *J. Chem. Phys.* **63,** 3812 (1975).
13. J. Economy and R. V. Anderson, *J. Polym. Sci. C*(19), 283 (1967).
14. L. E. Hibbs, Jr., and R. H. Wentorf, Jr., *High Temp. High Pressures* **6,** 409 (1974).

R. H. WENTORF, JR.
General Electric Co.

BORON HALIDES

Physical Properties

The important physical and thermochemical properties of the boron trihalides are given in Table 1. The specific heat and thermal conductivity for BCl_3 at various temperatures are tabulated in ref. 1. The thermal conductivity of BBr_3 at 20°C is 0.112 W/(m·K).

Boron trichloride and tribromide are soluble in CCl_4, $SiCl_4$, S_2Cl_2, SCl_2, and SO_2. For ^{11}B nmr data of the trihalides in methylcyclohexane, see ref. 2. In coordinating solvents, spectra characteristic of BX_3 solvent complexes are obtained. Raman and infrared data for BCl_3 (3–5), BBr_3 (3,6), and BI_3 (3) have been reported.

The boron trihalides are planar (sp^2) molecules with X–B–X angles of 120°. Orbital energy assignments have been made for the trihalides based on their photoelectron spectra (7–8).

Reactions

Much of the chemistry of the boron halides (9,15–18) is dominated by their Lewis acidity. The generally accepted order of acidity is $BI_3 \sim BBr_3 > BCl_3 > BF_3$, which is opposite to the order of the degree of π-bond interaction between X and B.

Formation of Donor–Acceptor Complexes. Boron trihalides form complexes with Lewis bases containing O, S, N, P, or As donor atoms. For example:

$$(CH_3)_3N + BCl_3 \rightarrow (CH_3)_3N{:}BCl_3$$

$$R_2O + BCl_3 \rightarrow [R_2O{:}BCl_3] \rightarrow ROBCl_2 + RCl$$

Some transition metal compounds may also act as Lewis bases, and the compounds $[(C_6H_5)_3P]_3Pt\cdot2BCl_3$ and $[(C_6H_5)_3P]_2Rh(CO)Cl\cdot BCl_3$ and $[(C_6H_5)_3P]_2Rh(CO)Cl\cdot BBr_3$ are known (19).

130 BORON COMPOUNDS (HALIDES)

Table 1. Physical Properties of the Boron Trihalides

Property	BCl$_3$ [10294-34-5]	BBr$_3$ [10294-33-4]	BI$_3$ [13517-10-7]	Ref. BCl$_3$	Ref. BBr$_3$	Ref. BI$_3$
mp, °C	−107	−46	−49.9	9	9	9
bp, °C	12.5	91.3	210	9	9	9
density^a (liq)	1.434$_4^0$	2.643$_4^{18}$	3.35	10	11	
	1.349$_4^{11}$					
crit. temp, °C	178.8	300		9	9	
crit. pressure, kPa^b	3901.0			1		
vapor pressure, kPa^{b,c}				1	11	12
−80°C	0.53					
−40°C	8.9					
0°C	63.5					
40°C	243					
80°C	689					
viscosity^d				10	11	
ΔH_f^0, kJ/mol gas^e	−403	−206	+18	13	13	13
ΔH_{vap}, kJ/mol^e	23.8	34.3		13	13	
C_p, J/(mol·°C), for gas at 25°C^e	62.8	67.78		1	1	
C_p, J/(mol·°C), for liquid at 25°C^e	121	128		1	1	
ΔH_{hydrol} kJ/mol liquid at 25°C^e	−289	−351		1	1	
ΔH_{fusion}, J/g at mp^e	18			1		
B–X bond energy, kJ/mol	443.9	368.2	266.5	14	14	14
B–X distance, nm	0.173	0.187	0.210	14	14	14

^{a} For BCl$_3$: ρ (g/mL) = 1.3730−2.159 × 10^{-3}°C − 8.377 × 10^{-7} °C; −44 to 5°C.
   For BBr$_3$: ρ (g/mL) = 2.698−2.996 × 10^{-3}°C; −20 to 90°C.
^{b} To convert kPa to mm Hg, multiply by 7.50.
^{c} For BBr$_3$: pressure (kPa) = 0.1333 a log [6.9792 − 1311/(°C + 230)]; 0–90°C.
^{d} For BCl$_3$: η (mPa·s or cP) = 0.34417/(1−6.9662 × 10^{-3}°C − 5.9013 × 10^{-6}°C); −40 to 10°C.
   For BBr$_3$: η (mPa·s or cP) = 0.100 a log [333 K^{-1} − 0.257]; 0–90°C.
^{e} To convert J to cal, divide by 4.184.

If the Lewis base is a halide ion, the tetrahaloborate anions are formed, the BCl$_4^-$, BBr$_4^-$, and BI$_4^-$ anions being more difficult to synthesize than their BF$_4^-$ analogue. In general, the stability of their salts depends on the size of the countercation (14) and they are usually isolated as pyridinium, tetraalkylammonium, tropenium, or triphenylcarbonium salts. The stability order with a given cation is MBCl$_4$ > MBBr$_4$ > MBI$_4$. All of these compounds are sensitive to hydrolysis.

Reactions with Donors Containing Active Hydrogen. The trihalides react with H$_2$O, lower alcohols, H$_2$S, alkyl mercaptans, NH$_3$, primary and secondary amines, PH$_3$, and AsH$_3$, liberating hydrogen halide in each case. For example:

$$BX_3 + 3 H_2O \rightarrow B(OH)_3 + 3 HX$$

In many cases a donor–acceptor complex may be isolated at low temperature. The trihalides react with tertiary alcohols to yield RX and B(OH)$_3$.

Exchange Reactions. The boron trihalides undergo a number of exchange reactions with trialkyl, triaryl, trialkoxy, or triaryloxy boranes, other boron halides, and diborane:

$$BX_3 + 2 BR_3 \rightarrow 3 R_2BX$$

$$2 BX_3 + B(OR)_3 \rightarrow 3 X_2BOR$$

$$BCl_3 + BBr_3 \xrightarrow{<30°C} BCl_2Br + BClBr_2$$

$$BX_3 + B_2H_6 \rightarrow BHX_2 + B_2H_5X$$

The halogen–halogen exchange reactions have received a good deal of attention. A study of ^{11}B nmr for the trihalide mixtures has shown that the reactions are all nearly thermoneutral (20). The mixed Br–I compounds have been isolated.

The exchange reactions between BBr_3 or BI_3 and metal chlorides or oxides are used for the preparation of anhydrous metal bromides and iodides (21).

Reductions. Boron halides are reduced by active metals, metal hydrides, and hydrogen. Reaction with the alkali and alkaline earth metals at elevated temperature yields elemental boron and the metal halide (14). Stoichiometric amounts of BCl_3 and NaH react at room temperature in diglyme to give good yields of diborane, B_2H_6 (22). Boron trichloride reacts with H_2 in an electric discharge (23) or over Al at 350–500°C (24) to produce B_2H_6. Boron fibers are formed by the decomposition of the trihalides at high temperatures (>1000°C) in the presence of H_2. These decompositions are usually carried out on hot tungsten wires or in plasma torches (25–26).

Preparation. A number of routes have been reported for the preparation of BCl_3:

References

(a) $B_2O_3 + 3\,C + 3\,Cl_2 \xrightarrow{>500°C} 2\,BCl_3 + 3\,CO$ 9, 27–36

(b) $B_2O_3 + 3\,C + 6\,HCl \xrightarrow{900-1400°C} 2\,BCl_3 + 3\,CO + 3\,H_2$ 37

(c) $2\,B_2O_3 + 3\,SiCl_4 \xrightarrow{800°C} 4\,BCl_3 + 3\,SiO_2$ 38

(d) $B_2O_3 + 3\,COCl_2 \xrightarrow[\text{AlCl}_3 \text{ bath}]{425°C} 2\,BCl_3 + 3\,CO_2$ 39

(e) $7\,B_2O_3 + 6\,NaCl \xrightarrow{800-1000°C} 2\,BCl_3 + 3\,Na_2O.2B_2O_3$

(f) $B + 3\,AgCl \xrightarrow[\text{vacuum}]{650°C} BCl_3 + 3\,Ag$ 40

(g) $2\,NaBF_4 + 3\,MgCl_2 \xrightarrow{500-1000°C} 2\,BCl_3 + 2\,NaF + 3\,MgF_2$ 41

(h) $BF_3 + AlCl_3 \xrightarrow{\Delta} BCl_3 + AlF_3$ 42

(i) $TiB_2 + 10\,HCl \xrightarrow{600-800°C} 2\,BCl_3 + TiCl_4 + 5\,H_2$ 43–44

Variations on these routes are possible, eg, in (a), $B(OH)_3$, metal borates, B_4C, or B may be chlorinated; in (b), metal borates may be used; in (c), CCl_4, PCl_3, PCl_5, and S_2Cl_2 may act as the chlorinating agents; in (e), other alkali metal halides are effective; in (g) and (h), KBF_4 may be used as the source of boron; and in (i), other borides may be used. Route (a) is used in the manufacture of BCl_3 whereas route (h) is the most convenient in the laboratory.

Commercial BCl_3 may contain phosgene ($COCl_2$) and Cl_2 as impurities at about 0.01–0.09 wt % each. Other possible impurities are HCl, $FeCl_3$, $SiCl_4$, $AsCl_3$, and SO_2. A number of purification methods have been reported including distillation (45–47), countercurrent crystallization (47), and passage of the impure gas through Al_2O_3 (48), molten Zn (49), Ti sponge (50), Cu or charcoal (51) at high temperatures.

Boron tribromide has been prepared by high temperature bromination of boron, mixtures of carbon and boron oxide, or B_4C, by the action of HBr on carbon–boron oxide mixtures (37) or CaB_6 (44), by the reaction of $NaBF_4$ with $MgBr_2$ (41), and by the halogen exchange reaction of BF_3 and $AlBr_3$ (42). Again, the direct high temperature halogenation of carbon–boron oxide mixtures or B_4C is the usual commercial route whereas the BF_3–$AlBr_3$ exchange reaction is the best laboratory procedure. Bromine often remains as an impurity in commercially prepared BBr_3, but is easily removed by reduction with elemental mercury.

Boron triiodide can be prepared by the halogen exchange reactions of BF_3 and AlI_3 (52) or BBr_3 and LiI (53). Good yields have been obtained by the reaction of $LiBH_4$ at 125°C, or $NaBH_4$ at 200°C, with I_2 (54). Iodine in BI_3 is removed by treatment with Zn in CS_2 solution, followed by solvent evaporation and sublimation of the solid BI_3.

Production, Economic Aspects, and Shipping

The only United States producer of commercial quantities of BCl_3 and BBr_3 is the Kerr-McGee Chemical Corporation with annual capacities of about 275 and 16 metric tons, respectively (55). Their processes are proprietary, but they almost certainly produce both compounds by the reaction of B_4C with the halogens.

Boron triiodide is produced only in small quantities, mainly for research purposes, and is available at 99.9% purity.

In 1976 United States consumption of BCl_3 and BBr_3 was about 225 and 11 t, respectively (55). Prices for commercial-grade BCl_3 and BBr_3 were about $7.70 and $44.00 per kilogram, respectively, fob plant, in early 1977 (55).

Boron trichloride is shipped in 0.90-, 45-, 590-, and 817-kg (net) steel cylinders. Boron tribromide is shipped in 0.45- and 2.3-kg (net) glass bottles and in 91 kg (net) Monel drums (1). Both compounds are classed as corrosive liquids. Shipments must be made by private carriers.

Analytical Methods, Specifications, and Safety

Samples of BCl_3, BBr_3, and BI_3 are first hydrolyzed and then analyzed by standard methods (1,56) for B, halide, free halogen, and Si. The concentrations of HCl and $COCl_2$ in BCl_3 are determined by infrared measurements in a gas cell. The HCl band at 3.38 µm and the $COCl_2$ band at 11.77 µm are generally used (1). Gas chromatography has also been employed for the determination of volatile impurities (57). Table 2 gives the specifications of the Kerr-McGee products (1).

Boron trihalides are highly toxic and contact should be avoided. They react vigorously (sometimes explosively) with water to yield hydrogen halides. At high temperatures they decompose to yield toxic halogen-containing fumes.

Uses

Approximately 96% of United States production of boron trichloride is used in the manufacture of boron fibers, the remainder as catalyst in various organic reactions (9,58). About 95% of the tribromide production is used for catalysis, the remainder for semiconductors (55).

Boron fibers produced from BCl_3 have found application in the aerospace and

Table 2. Specifications for BCl$_3$ and BBr$_3$ (Kerr-McGee).

Assay	BCl$_3$, % Specification	Typical	BBr$_3$[a], % Specification	Typical
BX$_3$	99.9	99.95	99.9	99.98
Br$_2$			0.05	0.01
Cl$_2$	0.01			
COCl$_2$	0.09			
Si	0.001			0.0002
P				0.002
Fe				0.003
Mg				0.0002

[a] Ultrahigh purity (99.9999%) BBr$_3$ is available (55).

sporting goods industries (59) (see Ablative materials). Boron trichloride may be used as a catalyst (1) for the copolymerization of indene and piperylene, for the polymerization of cyclopentadiene and styrene (see Hydrocarbon resins), and for the production of alkyl and aromatic halosilanes (see Silicon compounds). Additional applications for BCl$_3$ include its use in the preparation of boron compounds (ie, NaBH$_4$, B$_2$Cl$_4$, metal borides, etc), in the prevention of solid polymer formation in liquid SO$_3$ (1), in the refining of aluminum, and in the production of optical wave guides (60) (see Microwave technology).

Boron tribromide is used as a catalyst in polymerization, alkylation, and acylation (58). Ultra-pure BBr$_3$ is used to supply boron to dope semiconductors.

Boron triiodide has found little commercial application.

Boron Subhalides

Boron subhalides are compounds in which the halogen to boron ratio is less than three (9,14,61).

The most important subhalides are the compounds B$_2$X$_4$, diboron tetrachloride [13701-67-2], B$_2$Cl$_4$; diboron tetrabromide [14355-29-4], B$_2$Br$_4$; diboron tetraiodide [18015-86-6], B$_2$I$_4$; and diboron tetrafluoride [13965-73-6], B$_2$F$_4$. These compounds are all thermally unstable, and are not produced in commercial quantities. They may act as Lewis acids, forming 1:2 complexes with Lewis bases such as (CH$_3$)$_3$N. The most potentially useful reactions of the tetrahalides are their addition to unsaturated organic molecules (diboration). The analogous hydroboration reactions are well known in organic synthesis (62) (see Hydroboration). Difficulties in preparing and handling large quantities of the B$_2$X$_4$ compounds have impeded their application.

BIBLIOGRAPHY

"Boron Compounds, Boron Halides," in *ECT* 1st ed., Vol. 4, pp. 592–593, by M. H. Pickard; "Boron Compounds, Boron Halides," in *ECT* 2nd ed., Vol. 3, pp. 680–683, by M. L. Iverson, Atomics International, S. M. Dragonov, U.S. Borax Research Corp.

1. Kerr-McGee Chemical Corp., *Technical Bulletin 0211*, 1973.
2. J. Pedley and co-workers, *J. Chem. Soc. A*, 2426 (1971).
3. R. J. H. Clark and P. D. Mitchell, *J. Chem. Phys.* **56**, 2225 (1972).
4. A. Loewenschuss, *Spectrochim. Acta.* **31A**, 679 (1975).

5. O. Brievx de Mandirola, *Spectrochim. Acta.* **23A,** 767 (1967).
6. M. C. Pomposiello and O. Brievx de Mandirola, *J. Mol. Struct.* **11,** 191 (1972).
7. P. J. Bassett and D. R. Lloyd, *J. Chem. Soc. A,* 1551 (1971).
8. G. H. King and co-workers, *Faraday Disc. Chem. Soc.,* 70 (1972).
9. G. Urry in E. Muetterties, ed., *The Chemistry of Boron and Its Compounds,* John Wiley & Sons, Inc., New York, 1967, p. 325.
10. T. J. Ward, *J. Chem. Eng. Data Inc.* **14,** 167 (1969).
11. W. F. Barber, C. F. Boynton, and P. E. Gallagher, *J. Chem. Eng. Data* **9,** 137 (1964).
12. P. D. Ownby and R. D. Gretz, *Surface Sci.* **9,** 37 (1968).
13. A. Finch and P. J. Gardner in R. J. Brotherton and H. Steinberg, eds., *Progress in Boron Chemistry,* Pergamon Press, New York, 1970, p. 177.
14. N. N. Greenwood and B. S. Thomas in A. F. Trotman-Dickerson, ed., *Comprehensive Inorganic Chemistry,* Pergamon Press, Oxford, 1973, p. 956.
15. A. G. Massey in H. J. Emeleus and A. G. Sharpe, eds., *Inorganic and Radiochemistry,* Vol. 10, Academic Press, New York, 1967, p. 1.
16. W. Gerrard and M. F. Lappert, *Chem. Rev.* **58,** 1081 (1958).
17. D. R. Martin, *Chem. Rev.* **42,** 581 (1948).
18. D. R. Martin and J. M. Canon in G. Olah, ed., *Friedel-Crafts and Related Reactions,* Vol. 1, Interscience Publishers, a division of John Wiley & Sons, Inc., New York, 1963, p. 399.
19. M. Fishwick and co-workers, *Inorg. Chem.* **15,** 491 (1976).
20. M. F. Lappert and co-workers, *J. Chem. Soc. A,* 383 (1971).
21. P. M. Druce and M. F. Lappert, *J. Chem. Soc. A,* 3595 (1971).
22. H. C. Brown and P. A. Tierney, *J. Am. Chem. Soc.* **80,** 1552 (1958).
23. H. I. Schlesinger and A. B. Burg, *J. Am. Chem. Soc.* **53,** 4321 (1931).
24. D. T. Hurd, *J. Am. Chem. Soc.* **71,** 20 (1949).
25. B. A. Jacob, F. C. Douglas, and F. S. Galasso, *Am. Ceram. Soc. Bull.,* **52,** 896 (1973).
26. H. E. Hintermann, R. Bonatti, and H. Breiter, *Chem. Vap. Deposition, Int. Conf.,* 536 (1973).
27. T. Niemyski and Z. Olempska, *J. Less Common Met.* **4,** 235 (1962).
28. U.S. Pat. 3,000,705 (Sept. 19, 1961), P. R. Juckniess (to The Dow Chemical Co.).
29. N. Takamoto, T. Katagiri, and K. Fujii, *Nippon Kinzoku Gakkaishi* **30,** 915 (1966).
30. U.S. Pat. 3,019,089 (Jan. 30, 1962), J. B. O'Hara, (to Olin Mathieson Chemical Corp.).
31. U.S. Pat. 3,025,138 (March 13, 1962), R. G. Davis, J. N. Haimsohn, and J. T. Bashour (to Stauffer Chemical Co.).
32. U.S. Pat. 3,130,168 (April 21, 1964), L. J. Hov and R. Miller (to Stauffer Chemical Co.).
33. Czech. Pat. 99827 (June 15, 1961), (to J. Formanek).
34. U.S. Pat. 3,095,271 (June 25, 1963), G. H. McIntyre, N. T. Sprouse, and H. S. Haber (to U.S. Borax and Chemical Corp.).
35. U.S. Pat. 2,876,076 (March 3, 1959), C. W. Montgomery and W. A. Pardee (to Gulf Research and Development Co.).
36. U.S. Pat. 3,152,869 (Oct. 13, 1964), L. C. Bratt (to Stauffer Chemical Co.).
37. Brit. Pat. 944962 (Dec. 18, 1963), (to National Distillers and Chemical Corp.).
38. U.S. Pat. 2,983,583 (May 9, 1961), W. H. Schechter (to Callery Chemical Co.).
39. U.S. Pat. 4,024,221 (May 17, 1977), A. J. Becker and D. R. Careatti (to Aluminum Co. of America).
40. K. H. Lieser and co-workers, *Z. Anorg. Allg. Chem.* **313,** 193 (1961).
41. U.S. Pat. 2,762,691 (Sept. 11, 1956), E. Wainer (to Horizons, Inc.).
42. E. L. Gamble, *Inorg. Synth.* **3,** 27 (1950).
43. U.S. Pat. 3,144,306 (Aug. 11, 1974), F. H. May and J. L. Bradford (to American Potash and Chemical Corp.).
44. Ger. Pat. 2,113,591 (Nov. 30, 1972), G. Wiebke, G. Stohr, G. Vogt, and G. Kratel (to Electroschmelzwerke Kempten G.m.b.H).
45. N. Sevryugova and N. Zhavoronko, *Khim. Prom.* **46,** 433 (1970).
46. N. Sevryugova, *Metody, Poluch. Anal. Veshchestv. Osobi Chist. Tr. Vses. Knof.,* 64 (1970).
47. G. Devyatykh and co-workers, *Zh. Prikl Khim* **49,** 1280 (1976).
48. U.S. Pat. 3,037,337 (June 5, 1962), D. M. Gardner (to Thiokol Chemical Corp.).
49. U.S. Pat. 3,043,665 (July 10, 1962), J. R. Gould and D. M. Gardner (to Thiokol Chemical Corp.).
50. U.S. Pat. 3,207,581 (Sept. 21, 1965), D. R. Stern and W. W. Walker (to American Potash and Chemical Corp.).
51. U.S. Pat. 3,126,256 (March 24, 1964), J. N. Haimsohn, L. A. Smalheiser, and B. J. Luberoff (to Stauffer Chemical Co.).

52. E. L. Gamble, P. Gilmont, and J. Stiff, *J. Am. Chem. Soc.* **62,** 1257 (1940).
53. V. H. Tiensuu, *Diss. Abstr.* **23,** 1535 (1962).
54. W. C. Schumb, E. L. Gamble, and M. D. Banus, *J. Am. Chem. Soc.* **71,** 3225 (1949).
55. A. Ferguson and D. Treskon in *Chemical Economics Handbook,* Stanford Research Institute, Menlo Park, Calif., 1977, p. 717.1000A.
56. F. D. Snell and C. L. Hilton, eds., *Encyclopedia of Industrial Chemical Analysis,* Interscience Publishers, a division of John Wiley & Sons, Inc., New York, 1968, p. 332.
57. N. K. Agliulov, *Tr. Khim, Khim. Tekhnol.* **3,** 66 (1973).
58. G. Olah in ref. 18, p. 235–237.
59. *Bus. Week,* (Aug. 9, 1976).
60. W. G. French, G. W. Tasker, and J. R. Simpson, *Appl. Opt.* **15,** 1803 (1976).
61. P. L. Timms, *Acc. Chem. Res.* **6,** 118 (1973).
62. H. C. Brown, *Hydroboration,* W. A. Benjamin, New York, 1962.

LOREN D. LOWER
U.S. Borax Research Corp.

BORON HYDRIDES, HETEROBORANES, AND THEIR METALLO DERIVATIVES

Although reports of volatile hydrides of boron appeared before the turn of the century, as Sidgwick (1) aptly put it, "All statements about the hydrides of boron earlier than 1912, when Stock began to work upon them, . . . are untrue."

Through nearly a quarter century of classic synthetic work, Stock (2) fathered a family of toxic, air-sensitive, and volatile hydrides of general composition B_nH_{n+4} and B_nH_{n+6}. This feat is most remarkable because it required that he and his collaborators first develop most of the basic techniques of vacuum-line manipulation, synthesis, and characterization. Subsequently, Schlesinger initiated the American effort in boron hydride research with his students Burg and Brown sharing significant advances in the synthesis of lower boron hydrides. After World War II, activity in the area grew tremendously. Hydroboration (qv) procedures, now so important to synthetic organic chemists, were developed by Brown (3). Significant interest was generated also by the high-energy fuel projects HEF and ZIP (see Explosives and propellants).

Nearly all of the structural proposals for the boranes during the first half century were incorrect. It was not until the second half of the century that the structural and theoretical aspects of boron hydrides were incisively delineated by the work of Lipscomb and co-workers with x-ray diffraction studies, theoretical analyses of bonding, and predictions about structure and reaction chemistry. The 1976 Nobel Prize in Chemistry was awarded to Lipscomb for his definitive research in borane chemistry (4).

The emergence of a theoretical understanding of boranes and the residual momentum of the high-energy fuel program spawned a recent (since ca 1960) rapid rise in the discovery of significant new boranes and the elaboration of their chemistry. It was a half century from Stock's first report of highly reactive hydrides to the discovery of the most thermodynamically and kinetically stable class of boranes, the polyhedral anions, carboranes, and metalloboranes. Although it may be premature, certain investigators appear to be primarily responsible for these recent advances. The great preponderance of data on the polyhedral ions $[B_{10}H_{10}]^{2-}$ [12356-12-6] and $[B_{12}H_{12}]^{2-}$ [12356-13-7] comes from the extensive work of Muetterties and co-workers (5). Synthesis of isomeric icosahedral $C_2B_{10}H_{12}$ and an investigation of their organic derivative chemistry was reported almost simultaneously by American industrial chemists Bobinski, Fein, Cohen, Heying, Schroeder, Papetti, and the Russians, Zakharkin, and Stanko (6).

Hawthorne and co-workers (7) are responsible for the creative merger of transition metal and borane chemistry to give polyhedral metallo analogues of the boranes and heteroboranes.

Nomenclature

Because of the number and complexity of known boron hydride derivatives, their nomenclature is in a state of confusion; however, the IUPAC and ACS have made nomenclature recommendations (8). The following is a brief summary of what appears to be currently accepted practice.

The neutral boron hydrides are termed boranes. The molecule BH_3 is called borane or borane(3) [13283-31-3]. For more complex boranes, the number of boron atoms is indicated by the common prefixes (di-, tri-, tetra-, etc) and the number of hydrogens (substituents) given by an Arabic numeral in parentheses directly following the name. For example, B_5H_{11} is named pentaborane(11) [18433-84-6], $B_{20}H_{16}$ is named icosaborane(16) [12008-84-3], and $B_{10}H_{12}I_2$ is diiododecaborane(14). The position of substituents can be designated precisely from framework numbering conventions. The numbering conventions for most common polyhedra are given in Figure 1. Since borane polyhedra have both closed and open skeletons, it has become common practice to include the appropriate structural classification in the compound's name. Closed polyhedra with only triangular faces are termed *closo*, and the open structures are designated *nido*, *arachno*, or *hypho*, depending on other criteria to be discussed later. For instance, more complete names for the previous examples are *arachno*-pentaborane(11), *closo*-icosaborane(16), and 2,4-diiodo-*nido*-decaborane(14) [23835-60-1]. Borane anions are termed hydroborates with prefixes to designate the number of hydrogens and borons; the charge follows the name in parentheses. For example, $K_2[B_{10}H_{10}]$ is potassium decahydro-*closo*-decaborate(2-) [12447-89-1] and the 2,4-dichlorododecahydro-*nido*-decaborate(2-) [51668-03-2] ion is $[2,4-Cl_2-B_{10}-H_{12}]^{2-}$.

When a boron atom of a borane is replaced by a heteroelement, the compounds are called carbaboranes, azaboranes, phosphaboranes, thiaboranes, etc, by an adaptation of organic replacement nomenclature. The numbering of the skeleton in heteroboranes is such that the heteroelement is given the lowest possible number consistent with the convention in the parent borane. Thus $C_2B_3H_5$ is dicarba-*closo*-pentaborane(5) and could occur in the 1,2-, 2,3-, or 1,5-isomeric forms ([23777-70-0], [30396-61-3], and [20693-66-7], respectively). When different heteroelements occur in combination in a polyhedron, Chemical Abstracts Service gives priority by descending group number and increasing atomic number within a group, eg, 1,2-

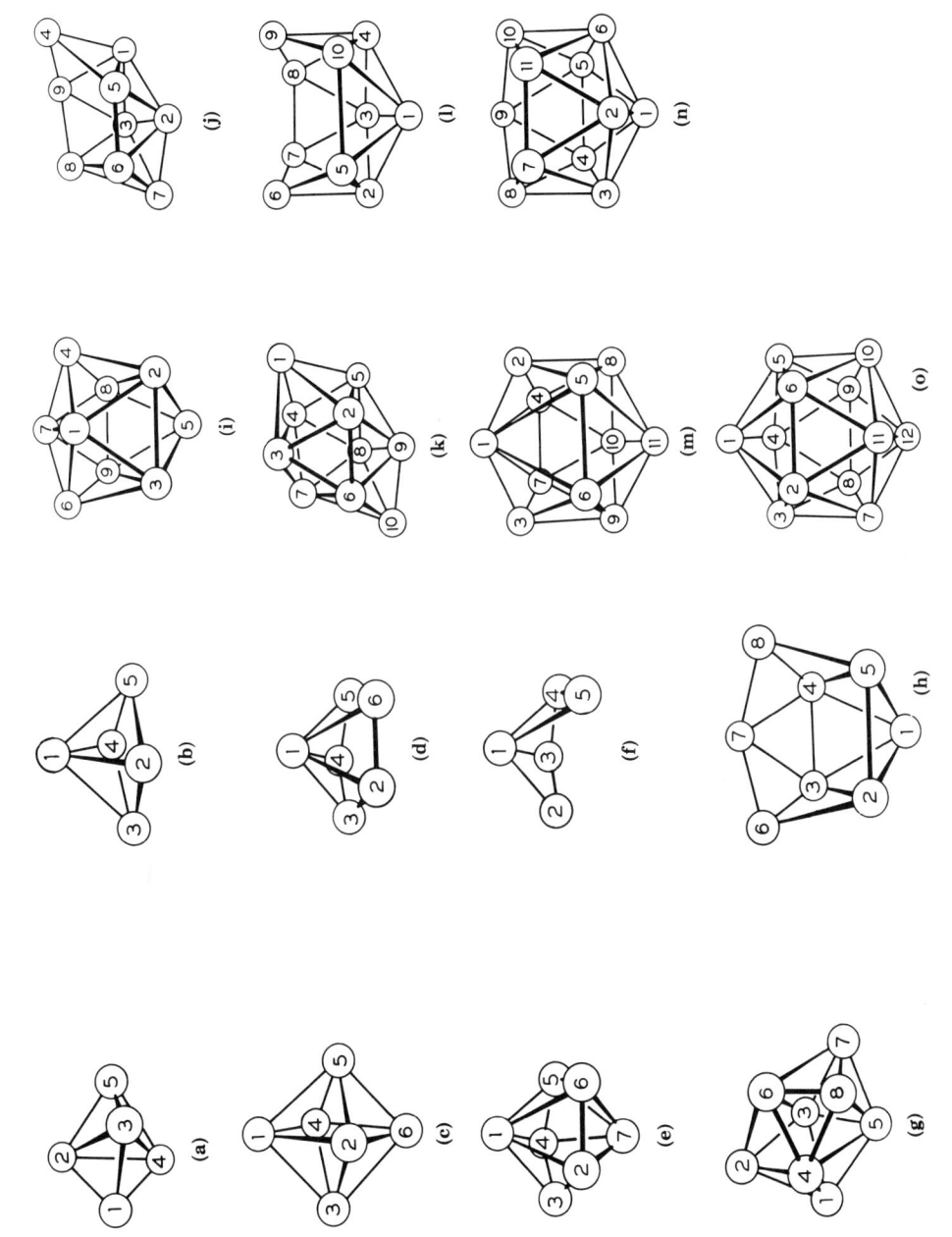

Figure 1. Commonly accepted numbering conventions for most borane polyhedra. The numberings shown for (**h**) and (**j**) are the more commonly used in literature. They can be numbered by *closo* rules (as shown) or by a different numbering system for *nido* structures. See IUPAC "Nomenclature of Inorganic Boron Compounds," *Pure Appl. Chem.* **30**(3–4), 683 (1972), Rules 3.221 and 3.25 and Figure 14.

137

PCB$_{10}$H$_{11}$ or 1-phospha-2-carba-*closo*-dodecaborane(*11*) [*30112-97-1*]; however, the hierarchy in the original literature often gives the lowest number to the element of lowest atomic number, eg, 1,2-CPB$_{10}$H$_{11}$ or 1-carba-2-phospha-*closo*-dodecaborane(*11*) [*30112-97-1*]. This convention carries over to metalloboranes and metalloheteroboranes when a metal occupies a polyhedral vertex. Examples are, 9,9-bis(triphenylphosphine)-6-thia-9-platina-*nido*-decaborane(*10*) [*52628-81-6*], 9,9-[P(C$_6$H$_5$)$_3$]$_2$-6,9-SPtB$_8$H$_{10}$, 3-η_5-cyclopentadienyl-1,2-dimethyl-1,2-dicarba-3-ferra-*closo*-dodecaborane(*1*) [*66750-82-1*], 3-(C$_5$H$_5$)-1,2-(CH$_3$)$_2$-3,1,2-FeC$_2$B$_9$H$_9$. In this review we will attempt to retain the numbering of the original literature. The Arabic numeral in parentheses following the name does not include exopolyhedral ligands bonded to the metal, only the total of hydrogens plus other substituents bonded to boron and main-group heteroelements. Examples of metalloborane anions are the 7,7'-*commo*-bis(dodecahydro-7-nickela-*nido*-undecaborate) dianion [*31388-28-0*], [7,7'-Ni(B$_{10}$H$_{12}$)$_2$]$^{2-}$, and the decahydro-2,11-bis(cyclopentadienyl)-2,11-dicobalta-1-carba-*closo*-dodecaborate(*1-*) [*59422-34-3*] ion, [2,11-(C$_5$H$_5$)-2,11,1-Co$_2$CB$_9$H$_{10}$]$^-$. The *commo* prefix indicates that the metal vertex is shared by two polyhedra.

Structural Systematics

The polyhedral skeletons described here can be accurately described as deltahedra (all faces triangular) or deltahedral fragments. The left-hand column of Figure 2 illustrates the deltahedra from $n = 4 - 12$ vertices: the tetrahedron, trigonal bipyramid, octahedron, pentagonal bipyramid, bisdisphenoid, symmetrically tricapped trigonal prism, bicapped square antiprism, octadecahedron, and icosahedron. In their regular form, all of these idealized structures are convex deltahedra (the plane of any face does not intersect the polyhedron) except for the octadecahedron, which is not a regular polyhedron as shown in Figure 2. The left-hand column of Figure 2 also constitutes the class of deltahedral *closo* molecules from which all the other idealized structures (deltahedral fragments) can be generated systematically. Any *nido* or *arachno* cluster can be generated from the appropriate deltahedron by ascending a diagonal from left to right. This progression generates the *nido* structure (center column) by removing the most highly connected vertex of the deltahedron and the *arachno* structure (right column) by removal of the most highly connected atom of the open (nontriangular) face of the *nido* cluster. The structural correlations shown in Figure 2 were first formulated by Williams (9) but were later embellished to include the new *hypho* class (10). The nomenclature *closo*, *nido*, *arachno*, and *hypho* is derived from Latin and Greek and implies a closed, nest-like, web-like, and net-like structure, respectively. Because of the embryonic nature of the *hypho* class, the known structures have not been included in Figure 2. In addition to boranes, the classifications *closo*, *nido*, *arachno*, and *hypho* apply to heteroboranes and their metallo analogues and are intimately connected to a quantity called the framework electron count. The partitioning of electrons into exopolyhedral and framework (endopolyhedral) classes allows for the accurate prediction of structure in most cases even though these systematics are not concerned explicitly with the assignment of localized bonds within the skeleton. That is, the lines depicting the skeletons of the structures illustrated are not electron-pair bonds. The lines merely join nearest neighbors and illustrate cluster geometry. However, exopolyhedral lines do represent the usual electron-pair bonds. Localized bond treatments will be discussed below.

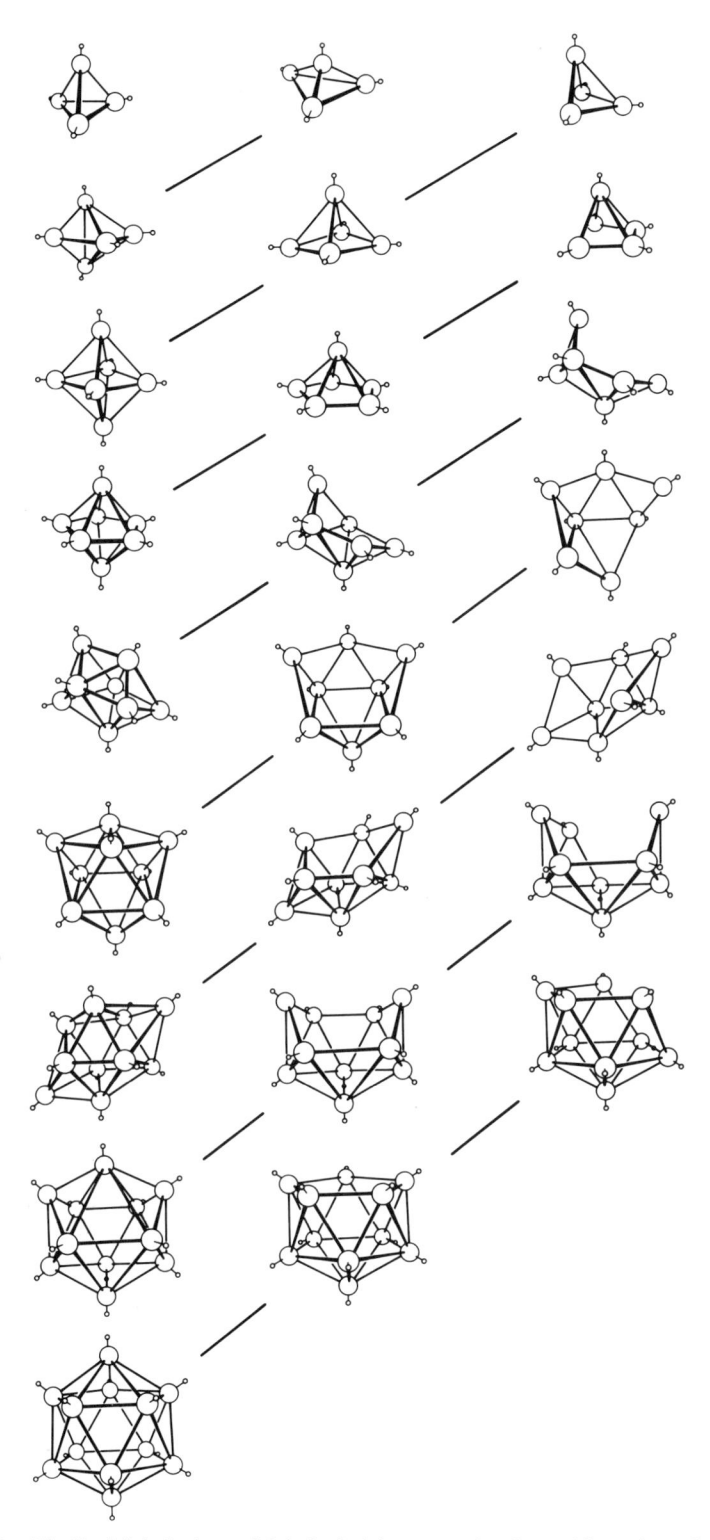

Figure 2. Idealized deltahedra and deltahedral fragments for *closo, nido,* and *arachno* boranes and heteroboranes. From left to right, the vertical columns give the basic *closo, nido,* and *arachno* frameworks; bridge hydrogens and BH$_2$ groups are not shown, but when appropriate they are placed around the open face of the framework (see text). The diagonal progression is described in the text as are the known members of the *hypho* class.

Proposal of a structure from Figure 2 for a given borane or heteroborane proceeds by: (1) selecting the row which corresponds to the number of framework atoms n, and (2) determining the number of electrons that can be reasonably assigned to the skeleton as opposed to exopolyhedral electrons. (Counts of $2n + 2$, $2n + 4$, and $2n + 6$ framework electrons give a *closo*, *nido*, or *arachno* classification, respectively, and suggest the structure corresponding to the appropriate column of Fig. 2.) Other empirical rules, to be mentioned later, refer to the preferred placement of heteroatoms and so-called extra hydrogens. The systematics also emphasize the oxidation–reduction nature of *closo–nido–arachno* interconversions for frameworks of the same size.

Closo Molecules ($2n + 2$ Systems). The usual assignment of valence electrons and factoring out of those in exopolyhedral bonds gives $2n$ framework electrons from n boron atoms, two electrons short of the $2n + 2$ *closo* count for a B_nH_n molecule. In fact, no neutral B_nH_n molecules are known; however, the $[B_nH_n]^{2-}$ anions ($n = 6$–12) and the isoelectronic $C_2B_{n-2}H_n$ dicarbaboranes ($n = 5$–12) are the best known examples of *closo* molecules. Thus, in these respective cases, the anion charge and the two carbon atoms with one more valence electron each than boron, supply the 2 electrons in excess of the $2n$ count. In general, the substitution of heteroatoms for boron presumably alters the number of framework electrons contributed by a given vertex, but as long as the total amounts to $2n + 2$, the molecule is classified as *closo*. Heteroatoms in the same group as boron contribute 2 framework electrons, those to the left in the periodic table contribute fewer, those to the right contribute more. As generalized in the form of an equation by Wade (11), the number of framework electrons contributed by a main group element is equal to $(v + x - 2)$ where $v =$ the number of valence shell electrons, and $x =$ the number of electrons from ligands (eg, for H, $x = 1$, and for Lewis bases, $x = 2$). Examples of $2n + 2$ boranes and heteroboranes and the framework electron contributions of the skeletal atoms are given in Table 1. Note that these considerations suggest an exodeltahedral lone pair in molecules such as :SB_9H_9 and :$PCB_{10}H_{11}$ as well as nominal bicapped square antiprism ($n = 10$) and icosahedral ($n = 12$) structures, respectively.

Nido Molecules ($2n + 4$ Systems). A number of *closo* heteroboranes add the two electrons characteristic of a *nido* molecule and undergo a concomitant structural distortion from a deltahedron to a deltahedral fragment. For instance, *closo*-$C_2B_9H_{11}$ [17764-89-0] ($2n + 2 = 24\ e^-$) is easily reduced to [*nido*-$C_2B_9H_{11}]^{2-}$ ($2n + 4 = 26\ e^-$) (12) and conversely, [7,9-$C_2B_9H_{11}]^{2-}$ [39469-99-3] is readily oxidized to $C_2B_9H_{11}$ (13). Reduction also occurs in effect upon addition of donors to such molecules, ie, the octahedron of the *closo* molecule $C_2B_4H_6$ opens to the *nido* pentagonal pyramid upon addition of :$NR_3 = L$ to give $C_2B_4H_6 \cdot L$ (14). Such addition of donors can formally be regarded as addition of H^-, ie, $C_2B_4H_6 \cdot L$ (14) is analogous to $C_2B_4H_7^-$. Other examples of known *nido* molecules are given in Table 1.

In molecules such as $C_2B_3H_7$ one can recognize that there are extra hydrogens, extra in the sense that there are more than necessary for an exopolyhedral hydrogen (substituent) for each framework atom. Rationales for regarding extra hydrogens as contributing to framework electron count can be offered. Extra hydrogens usually are found on a nontrigonal or so-called open face of the deltahedral-fragment skeleton in the form of bridging hydrogens or as the second hydrogen in a BH_2 group. Both of the latter hydrogen locations are reminiscent of framework positions in that the bridge positions usually reside close to a spheroidal extension of the skeletal surface and in that one hydrogen of a BH_2-group is usually *endo* (close to a framework extension)

Table 1. Representative Examples for the $2n + 2$, $2n + 4$, and $2n + 6$ Rules

Compound	Boron[a]	Carbon[a]	Heteroatom[a]	Charge	Total	Ref.
2n + 2 systems						
$B_3C_2H_5$	3 (2)	2 (3)		0	12	17
$[B_6H_6]^{2-}$	6 (2)			2	14	18
$B_4C_2H_6Ga(CH_3)$	4 (2)	2 (3)	1 (2)	0	16	19
$[B_8H_8]^{2-}$	8 (2)			2	18	17
$B_7C_2H_9$	7 (2)	2 (3)		0	20	20
$[B_9CH_{10}]^-$	9 (2)	1 (3)		1	22	21
$B_9C_2H_{11}$	9 (2)	2 (3)		0	24	22
$B_{11}H_{11}S$	11 (2)		1 (4)	0	26	23
$B_9C_2H_{11}Sn$	9 (2)	2 (3)	1 (2)	0	26	24
$B_{10}CH_{11}P$	10 (2)	1 (3)	1 (3)	0	26	25
2n + 4 systems						
$B_3C_2H_7$	3 (2)	2 (3)		2^b	14	26
$B_3C_3H_7$	3 (2)	3 (3)		1^b	16	27
$B_2C_4H_6$	2 (2)	4 (3)		0	16	28
$B_8H_{10}S$	8 (2)		1 (4)	2^b	22	23
$B_9H_{11}S$	9 (2)		1 (4)	2^b	24	23
$[B_9CH_{10}P]^{2-}$	9 (2)	1 (3)	1 (3)	2	26	25
2n + 6 systems						
B_5H_{11}	5 (2)			6^b	16	29
$B_7C_2H_{13}$	7 (2)	2 (3)		4^b	24	30
$[B_9H_{12}S]^-$	9 (2)		1 (4)	4^b	26	23

[a] Number of atoms multiplied by electrons contributed to the framework.
[b] In this case, charge is used in only a formal sense. Actually the "charge" is balanced by protons, often resulting in bridge hydrogens and BH_2 groups in $2n + 4$ and $2n + 6$ systems, namely:

$$[C_2B_9H_{11}]^{2-} \xrightarrow{H^+} [C_2B_9H_{12}]^- \xrightarrow{H^+} C_2B_9H_{13}.$$

and the other *exo*. Furthermore, extra hydrogens are generally acidic and can be removed to give anions without substantially altering framework geometry, ie, extra hydrogens conceptually amount to protonated framework electrons. Reasoning in this vein, it also follows that the addition of a lone-pair donor (conceptually hydride ion, H^-) to a framework adds two electrons and changes the molecule's classification accordingly.

Arachno Molecules ($2n + 6$ Systems). In comparison to the number of known *closo* and *nido* boranes and heteroboranes, there are relatively few *arachno* counterparts. Therefore, some of the empirically based structural types in Figure 2 may become exceptions as more *arachno* molecules are discovered or, expressed differently, *arachno* structures may prove to be less predictable than *closo* and *nido*. For example, two isomeric forms of B_9H_{15} are known, one with the *arachno* [19465-30-6] framework shown in Figure 2 (15), the other with a framework more reminiscent of that shown for the 9-atom *nido* classification (16). The electron count for *arachno* molecules of course involves the recognition of even more extra hydrogens than for *nido* molecules. Typical examples are shown in Table 1.

Hypho Molecules ($2n + 8$ Systems). Shore has prepared and structurally characterized (31) $B_5H_9[P(CH_3)_3]_2$ [39661-74-0], $[B_5H_{12}]^-$ [11056-98-7], and B_6H_{10}-$[P(CH_3)_3]_2$ [57034-29-4], three molecules that contain $2n + 8$ framework electrons and which represent the first well established members of the *hypho* class of boranes. This new class adopts structures that are as expected even more open than the *arachno* and *nido* counterparts. That of B$_5$H$_9$[P(CH$_3$)$_3$]$_2$ is illustrated in Figure 3. In this *hypho* molecule there are two nonbonding basal B–B distances (those not bridged by hydrogen). In *arachno*-B$_5$H$_{11}$ [18433-84-6] there is only one nonbonding basal B–B distance and in *nido*-B$_5$H$_9$ [19624-22-7] all basal distances of the pyramid are bonding (see the $n = 5$ horizontal row of Fig. 2). Because of the embryonic nature of the *hypho* class, the known structures have not been included in Figure 2. A complete range of *hypho* framework sizes eventually may be found. For instance, Kodama (32) recently described B$_4$H$_8$[P(CH$_3$)$_3$]$_2$ [66750-83-2].

Metalloboranes and Metalloheteroboranes. The correlation of skeletal electron count with structure appears to be applicable to metalloborane derivatives and other cluster molecules such as carbonium ions (33) and metal clusters (10–11). The method has been termed the PERC approach (Paradigm for the Electron Requirements of Clusters) (10).

Metalloboranes formed from a main group metal and a borane or heteroborane can be treated as heteroboranes by the PERC method. For example, tricarbaborane analogues are found for MC$_2$B$_9$H$_{11}$ (M = Ge [27071-52-9], Sn [23151-46-4], Pb [27071-51-8]) species which appear to be carbenoid in that the metal is a bare vertex (34). The metal appears to have a nonbonding lone pair of largely ns^2 character ($n = 4, 5, 6$) (35) and to contribute the 2 remaining valence electrons to the framework giving a 26-electron *closo* icosahedron.

In addition to the framework electron requirements of the cage, transition element metalloboranes and metalloheteroboranes generally adhere to the rule of eighteen, and therefore require a somewhat different PERC treatment. If one rather arbitrarily assumes that the metal vertex uses only three orbitals in cluster bonding, then 12 of the 18 electrons at a metal vertex are not involved in cluster bonding. The d-electrons in effect are not included as framework electrons. Mingos (36) has generalized such

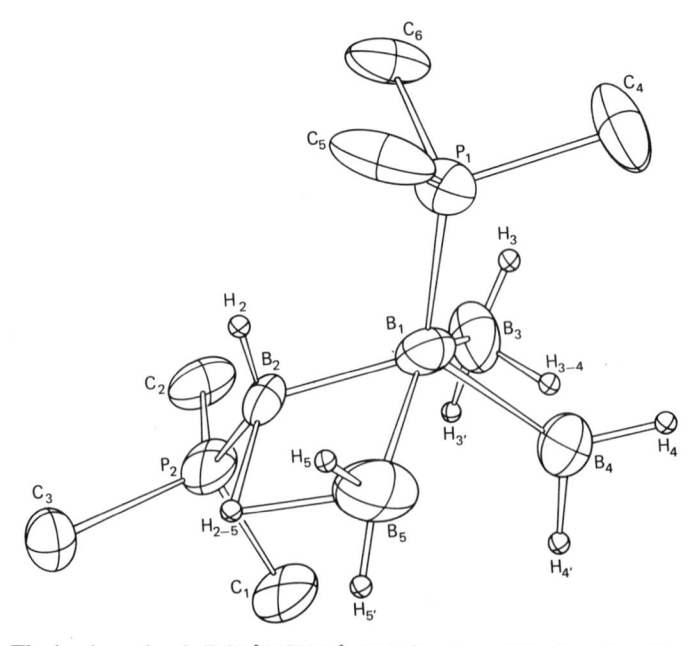

Figure 3. The *hypho* molecule B$_5$H$_9$[P(CH$_3$)$_3$]$_2$ (31). Courtesy of the American Chemical Society.

premises to give the number of skeletal electrons per metal vertex as $(v + x - 12)$ where v = the number of valence electrons of the metal, and x = the number of electrons donated by exocluster ligands and substituents. In this formalism moieties such as $Fe(CO)_3$ and $Co(\eta\text{-}C_5H_5)$ are analogous to a BH vertex and $Ni(\eta\text{-}C_5H_5)$ is effectively a CH vertex. Other examples of common vertex groups and their framework electron contributions are given in Table 2 (11) and Figure 4. The extension of these principles to organometallics is straightforward: $(C_5H_5)Mn(CO)_3$ and $(C_3H_5)Co(CO)_3$ are *nido* and *arachno* examples, respectively, also included in Figure 4 (see Organometallics).

Table 2. Examples of Framework Electron Contributions for Some Representative Transition Metal Moieties

V^a	Metal (M)	$M(CO)_2$, $x = 4$	$M(\eta\text{-}C_5H_5)$, $x = 5$	$M(CO)_3$, $x = 6$
6	Cr	(−2)	(−1)	0
7	Mn	(−1)	0	1
8	Fe	0	1	2
9	Co	1	2	3
10	Ni	2	3	4

[a] The general contribution for a metal with V valence electrons and exocluster ligands donating x electrons is $v + x - 12$ framework electrons.

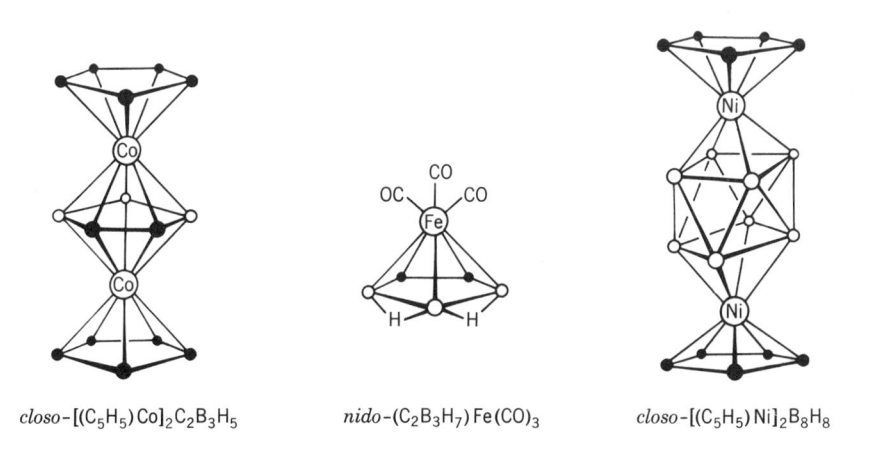

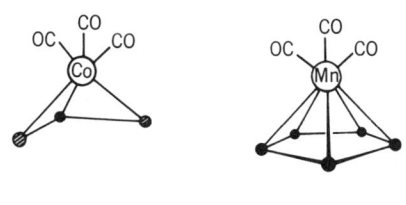

closo-[(C$_5$H$_5$)Co]$_2$C$_2$B$_3$H$_5$ *nido*-(C$_2$B$_3$H$_7$)Fe(CO)$_3$ *closo*-[(C$_5$H$_5$)Ni]$_2$B$_8$H$_8$

arachno-(C$_3$H$_5$)Co(CO)$_3$ *nido*-(C$_5$H$_5$)Mn(CO)$_3$

Figure 4. Metalloboranes and organometallics of the *closo*, *nido*, and *arachno* classification. ○, BH; ●, CH; ◐, CH$_2$.

Heteroatom Placement. Obviously many of the deltahedra and deltahedral fragments of Figure 2 have two or more nonequivalent vertices. Nonequivalent vertices are easily recognized as having a different order; ie, a different number of nearest neighbor vertices within the framework. It does appear that heteroatoms exhibit a positional preference which can be deduced on the basis of the electron richness of the heteroatom relative to boron, and the order of the polyhedral vertex. Electron-rich heteroatom groupings contribute more framework electrons than a :B—H moiety (2 framework electrons) and seem to prefer low-order vertices (those with fewer neighbors). For example, two of the three isomeric forms of $C_2B_8H_{10}$ can be isomerized thermally to the 1,10-isomer [23653-23-8] (20), the molecule with the carbons at the lowest order vertices. The pyrolysis of *nido*-6-SB_9H_{11} [59120-72-8] gives *closo*-1-SB_9H_9 [41646-56-4] (sulfur axial at lowest-order vertex) even though a least-motion mechanism would predict *closo*-2-SB_9H_9 (23) (see Fig. 1 for numbering conventions). When the heteroatom is in the same group as boron, it preferably adopts a high-order vertex; eg, $CH_3GaC_2B_4H_6$ [36607-02-2] (19). The most stable polyhedron with a heteroelement to the left of boron in the periodic chart presumably places that element at the highest-order vertex, although no definitive examples are known (in $R_3NBeC_2B_9H_{11}$ (37) all vertices are order five).

Transition metal moieties appear to occur predominately at high-order vertices. In a study of $(\eta^5\text{-}C_5H_5)CoC_2B_nH_{n+2}$, n = 6–10 *closo* molecules, Hawthorne and coworkers (38) concluded that the following empirical rules govern the thermal rearrangements: (*1*) the cobalt atom will occupy the highest-order vertex and remain there, (*2*) the carbons will not decrease their mutual separation, (*3*) carbon will migrate to the lowest-order vertices, and (*4*) carbon will migrate away from cobalt providing rules (*2*) and (*3*) are not violated. Exceptions to these rules have been found, but Grimes (39) has shown that at least some of the exceptions are due to kinetic rather than thermodynamic control of the rearrangement; eg, the formally *closo* 14-vertex cluster $(\eta^5\text{-}C_5H_5)Fe_2((CH_3)_4C_4B_8H_8)$ [60735-91-3] is isolated in isomeric forms with open (*nido*) skeletons. Upon heating, the 14-vertex framework isomerizes at 300°C to a *closo* form where previously nonadjacent carbons become adjacent and finally to the expected *closo* deltahedron (bicapped hexagonal antiprism) with the iron atoms at the highest order vertices and carbons nonadjacent (Fig. 5). Other exceptional cases involve electron-rich d^8 and d^9 metals. For example, the PERC approach predicts *closo* structures for 3-[$(C_2H_5)_2NCS_2$]-3,1,2-$AuC_2B_9H_{11}$ [62572-50-3] (40) and 8,8-[$(CH_3)_3P$]$_2$-7,8,10-$CPtCB_8H_{10}$ [58348-40-6] (41), but they have been found to have *nido* structures by x-ray crystallography. Although unlikely, it may be that all the hydrogens have not been located in these exceptional *nido* structures. For instance, a recent x-ray characterization of platinathiaboranes showed open skeletal structures but left an ambiguity regarding the formulation of the thiaborane ligand as SB_8H_8 or SB_8H_{10}. Spectroscopic evidence was found for the two extra hydrogens and, therefore, the framework electron count corresponds to the *nido* skeletal structure and cannot qualify as an exceptional case (42).

There are other examples of exceptional heteroboranes, eg, *nido*-1,2-$C_2B_3H_7$ [26249-71-8] presents a notable exception to the predilection of carbon for low-order vertices. It has been suggested that this exception is in effect related to the placement of bridge hydrogens on the open face (43).

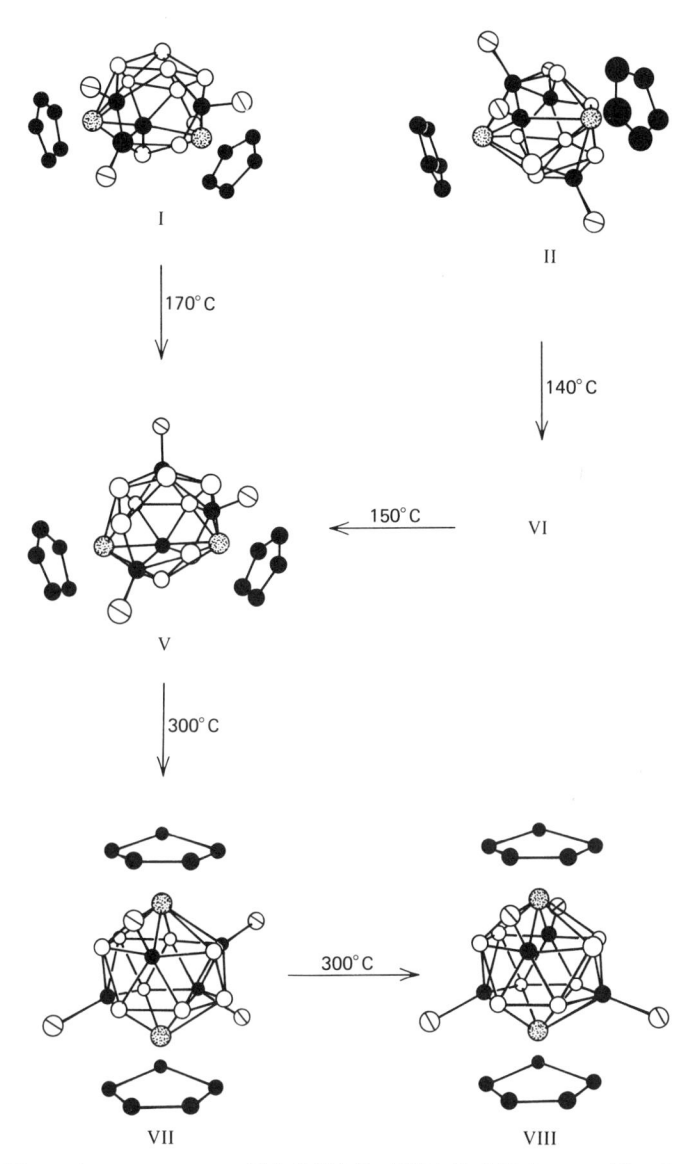

Figure 5. Thermal rearrangement of $(\eta^5\text{-}C_5H_5)_2Fe_2(CH_3)_4C_4B_8H_8$ isomers. Structures of I, II, V, and VII are established, and that of VIII is proposed from nmr data (39). ●, C or CH; ○, BH; ⊖, CH₃; ⊚, Fe. Courtesy of the American Chemical Society.

Placement of Extra Hydrogens. The placement of extra hydrogens is a moot point. In effect, their exact position sometimes appears to depend on the physical state of the molecule, eg, different bridge hydrogen placements for $[B_{10}H_{13}]^-$ *[36928-50-4]* in the crystal (44) and solution (45) can be inferred from the experimental evidence, but the solution data are also consistent with a dynamic process of bridge hydrogen tautomerism. A well documented example of fluxionality for bridge hydrogens is provided by B_6H_{10} (46). In spite of the polemics regarding hydrogen placement in boranes, some empirical rules are evident: (*1*) bridging occurs only between boron

146 BORON COMPOUNDS (B-H, METALLO DERIVATIVES)

atoms, usually an adjacent pair on the open (nontriangular) face of the skeleton (an edge), and only rarely does a hydrogen bridge a triangular array of borons (47); (2) when possible, the bridge termini are the low-order vertices of the open face; and (3) there is only one bridge per edge. Generally, BH$_2$ groups may be postulated as tautomeric intermediates in fluxional *nido* boranes, but they occur as ground state moieties in *arachno* molecules and then at vertices of order three or lower. In metalloboranes, hydrogens sometimes bridge between boron and the metal. Topological rules governing the balance between bridge hydrogens and BH$_2$ groups in boranes will be discussed below.

The contention has been made that bridge-hydrogen placement is the most important variable in the determination of relative isomeric stability, outranking placement of the heteroatom (43). It is true that in *nido*-1,2-C$_2$B$_3$H$_7$ one of the heteroatoms is at an unanticipated high-order vertex (48–49). Williams (43) suggests that there are other similar cases where the heteroatoms will adopt high-order vertices in deference to bridge-hydrogen placement at low-order vertices. However, it also could be argued that *nido*-1,2-C$_2$B$_3$H$_7$ is unique in that alternative C-atom placement completely eliminates the availability of a B—B edge suitable for a bridge hydrogen. Until there are sufficient thermodynamic data on the relative stabilities of isomeric heteroboranes, the question remains contestable.

M—H—B Bridges. Numerous metalloboranes and metalloheteroboranes now are known to contain hydrogens bridging between the metal and skeletal borons, but complexes containing covalently bound [BH$_4$]$^-$ [16971-29-2] constitute the prototypical class. Three modes of bonding are feasible, but only (**b**) and (**c**) of Figure 6 are well established (50). Metal tetrahydroborates have been reviewed recently by Marks and Kolb (51) and will be discussed below. Polyboranes bind through bridge hydrogens in a wide variety of manners as shown in (**d–f**) of Figure 6. Although the bonds involve one, two, and three hydrogens in bridging to the metal just as for the tetrahydroborates, the presence of the cage atoms affords distinctly different structural possibilities.

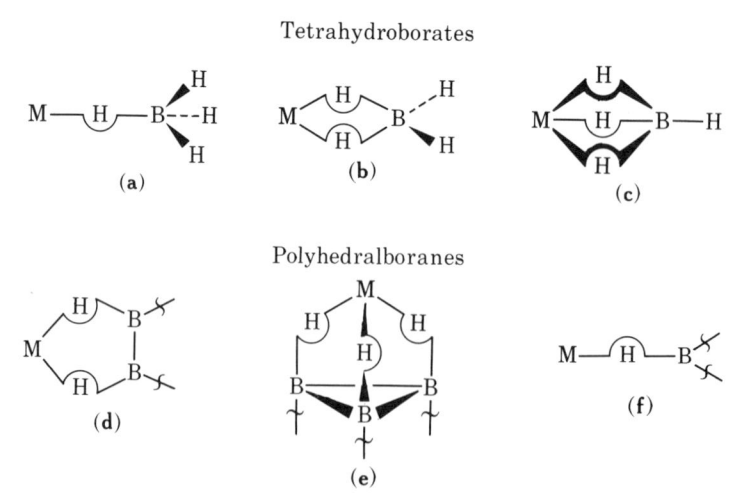

Figure 6. Possible modes of bonding a metal through bridge hydrogens.

Exceptions to Structural Systematics. Certain exceptions to the rule have been discussed above but others can be anticipated. Although $C_4B_2H_6$ [12403-20-2] has a pyramidal *nido* geometry, the structural evidence favors a planar form for $C_4B_2F_2H_4$ [20534-09-2] where B—H has been replaced by B—F (52). With fluorine and other halogens attached to boron, there is, of course, the possibility of π-bonding. It appears that the electron deficiency of boron can be ameliorated in some cases by back-donation rather than by the multicenter bonding afforded in a cage framework. Thus it has been suggested that PERC exceptions will occur where back-donation from the substituent to a cluster boron is possible. Salient examples are the distorted square antiprism B_8Cl_8 and the tetrahedral B_4Cl_4 (53–54).

B_4Cl_4 is not an exception in the same sense as the other halogenated boranes since the *closo* electron count for a tetrahedral cluster is unique among those considered here; B_4H_4 [27174-99-8] is the only case where a closed shell configuration is predicted for the neutral deltahedral borane rather than the dianion. The LUMO (Lowest Unoccupied Molecular Orbital) for B_4H_4 is degenerate (*e*) suggesting $[B_4H_4]^{4-}$ and more feasibly C_4H_4 as tetrahedral molecules with *nido* electron counts. Protonation of $[B_4H_4]^{4-}$ can give $[B_4H_7]^-$ of C_{3v} symmetry which has been proposed on the basis of simplified molecular orbital (MO) descriptions (55); there is nmr evidence consistent with C_{3v} [*nido*-B_4H_7]$^-$ (56). Of course, the butterfly structure (C_{2v}) for *arachno*-B_4H_{10} [18283-93-7] has been well established. Therefore, with the expected electron count and structures of lower symmetry than the *closo* counterpart, the tetraborane cluster appears to return to normalcy for the *nido* and *arachno* classifications. Nonetheless, MO theory suggests the possibility of tetrahedral skeletons for certain *nido* and *closo* heteroboranes or metalloheteroboranes. MO and localized bonding treatments for the boranes will be discussed in more detail after attention to some metalloheteroboranes that do not conform to the PERC formalism.

The symmetrical sandwich structure of the $((C_2B_9H_{11})_2M^{n+})^{n-4}$ complexes of d^6 metals, M = Fe(II) [51868-94-1], Co(III) [11078-84-5], Ni(IV) [57185-17-4], and Pd(IV) [22537-46-8], nicely fits the paradigm. The corresponding d^8 complexes with M = Cu(III) [15721-63-8] and Ni(II) [36733-09-2] could be expected to show an asymmetrically distorted *nido–closo* structure; however, a symmetrical slipped sandwich structure is observed, indicative of delocalization. Alternatively, the slipped structure can be explained in terms of a reduction of the *closo* molecule with concomitant distortion as observed for *closo* carboranes (52); in accord with these ideas, the d^9 copper complex is opened slightly more than the d^8 complex (57) and the distortion can be rationalized with Jahn-Teller arguments (58).

Since the electron counting paradigm incorporates the 18 electron rule when applied to transition metal complexes, exceptions are expected just as is the case for classical coordination complexes. Relatively minor exceptions are found in (η-$C_5H_5)_2Fe_2C_2B_6H_8$ [54854-86-3] (59) and $[Ni(B_{10}H_{12})_2]^{2-}$ [11141-32-5] (60). The former (2*n* electrons) is noticeably distorted from an idealized structure, and the latter is reminiscent of the d^8 and d^9 complexes discussed above. However, the extremely deficient count obtained for $[(C_2B_9H_{11})_2Cr(III)]^-$ [37036-06-9] presents a disconcerting situation in view of its clearly *closo* structural classification (61). In some of these cases, it is more satisfying to consider the borane as a multidentate ligand; eg, $[B_{10}H_{12}]^{2-}$ [12430-37-4] is effectively bidentate giving square planar and tetrahedral complexes, $[Ni(B_{10}H_{12})_2]^{2-}$ [31388-28-0], and $[Zn(B_{10}H_{12})_2]^{2-}$ [19154-53-1], respectively (60). Both the latter and the former are cases where the metal in effect occupies the position of a bridge hydrogen of the conjugate acid borane, a rather common occurrence which gives rise to another classification for metalloboranes and metalloheteroboranes. Wegner (62) and Lippard (63) have discussed in more detail situations where the metal vertex is equivalent to an H^+, BH^{2+}, or BH_2^+ moiety; however, in the latter formalism the *arachno* molecules $[(C_6H_5)_3P]_2CuB_3H_8$ [*12368-70-6*] and $[(OC)_4Cr(B_3H_8)]^-$ [*50725-42-3*] were incorrectly classified as *nido*.

Bonding

Localized Bonds. Since there are available more valence orbitals than valence electrons in the boron hydrides, they have been termed electron deficient molecules. This so-called electron deficiency is responsible, at least in part, for the great interest surrounding borane chemistry and molecular structure. Even the simplest boron hydride, diborane(6) [*19287-45-7*], B_2H_6, was sufficiently challenging that its structure was debated for many years before it was resolved beyond question (64) in favor of the hydrogen-bridged structure shown in Figure 7.

The structure of diborane(6) led to the description of a new bond type, the three-center bond, in which one electron pair is shared by three atomic centers (65). The delocalization of a bonding electron pair over a three-center bond allows for the utilization of all the available orbitals in an electron deficient system. This key point was recognized by Eberhardt, Crawford, and Lipscomb (66) who formulated a valence–bond description of the bonding in boron hydrides, sometimes termed a topological description. The valence structures of this topological approach give localized bonding descriptions which include delocalized 3-center bonds in the basis set of bond types. In addition to the B—H—B three-center bridge bond introduced previously, a B—B—B 3-center bond was introduced to describe bonding in the framework. In its present state of refinement, Lipscomb's valence theory (4) includes those 3-center bonds shown in Figure 8.

Of course, normal two-center bonds (B—B and B—H) are used also. For instance, one resonance structure of pentaborane(9) is given in projection in Figure 9.

From the valence structure of B_5H_9 it can be seen that the octet of electrons about each B atom is attained only if three-center bonds are used in addition to two-center bonds. In many cases it is easy to deduce the valence structure by a "back-of-the-envelope" accounting procedure. First, one determines the total number of orbitals

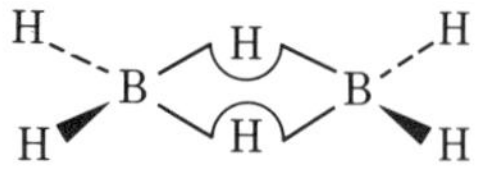

Figure 7. The structure of diborane(6).

Figure 8. Three center bonds: (a) three-center bridge-hydrogen bond; (b) central three-center bond.

Figure 9. The 4120 valence structure of B_5H_9.

and valence electrons available for bonding. Next, the B—H and B—H—B bonds are accounted for. Finally, the remaining orbitals and valence electrons are used in framework bonding. With a little reflection, it is obvious that alternative placements of hydrogen require different valence structures, eg, in B_5H_9 with two BH_2 groups and two bridge hydrogens the valence structure shown in Figure 10 is obtained.

Lipscomb has described precisely the possible number of valence structures for a given boron hydride with three general equations of balance. For a borane of composition $[B_p H_{q+p+c}]^c$ where c is the charge, the equations are:

$$s + x = q + c \tag{1}$$

$$s + t = p + c \tag{2}$$

$$t + y = p - c - \tfrac{1}{2} q \tag{3}$$

where p = the number of BH units; s = the number of B—H—B bonds; t = the number of B—B—B bonds; y = the number of B—B bonds; and x = the number of BH_2 groups.

Usually there are several possible solutions to the equations of balance which are differentiated by a so-called styx number, a four-digit number which gives the respective values of s, t, y, and x. The two dimensional representations of B_5H_9 given in Figures 9 and 10 have 4120 and 2302 styx numbers, respectively; other examples are given in Figure 11.

The correct styx number reflects molecular geometry, but the other styx structures may be transition states for fluxional molecules. Hexaborane(*10*) provides a well-documented example of bridge-hydrogen tautomerism which is detectable by nmr (46). The tautomerism probably proceeds through a 3311 valence structure in which a bridge hydrogen of the 4220 ground-state has been converted to a BH_2 group (Fig. 12).

It should be noted that the sum of the digits of the styx number gives one half

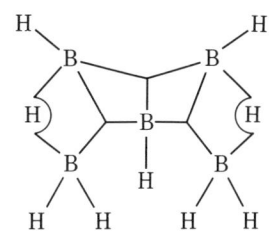

Figure 10. The 2302 valence structure of B_5H_9.

Figure 11. Valence structures of B_4H_{10}, B_5H_{11}, and B_6H_{10}.

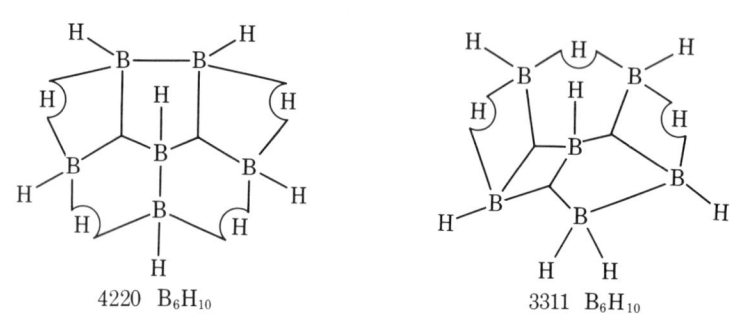

Figure 12. Valence tautomers of B_6H_{10}.

of the number of electrons involved in framework bonding, ie, the $2n + 2$, $2n + 4$, and $2n + 6$ framework-electron counts of the PERC formalism (10) and the topological description are intimately related. Therefore, after using PERC to arrive at a framework structure, a valence–bond description of the localized bonds can be deduced from the styx formalism (4). The simultaneous application of these two formalisms allows the prediction of rearrangement during certain reactions and the prediction of transition-state structure. Major skeletal rearrangement would not be anticipated as long as the number of skeletal electrons remains unchanged, as would be the case for both associative and dissociative electrophilic mechanisms since H^+ is the model electrophile. For the model nucleophile H^-, associative and dissociative nucleophilic mechanisms increase and decrease the framework count, respectively, and framework rearrangement is expected during the course of the reaction (67).

Molecular Orbital Descriptions. In addition to the localized bond descriptions just described, Lipscomb and co-workers have developed molecular orbital (MO) descriptions of bonding in the boranes and carboranes (4). In fact, their early work on boranes developed one of the most widely applicable approximate MO methods, the extended Hückel method. Molecular orbital descriptions are particularly useful for *closo* molecules where localized bond descriptions become cumbersome because of the large number of resonance structures which do not accurately reflect molecular symmetry. The bonding molecular orbitals and their symmetry designations are given for a number of representative boranes in Table 3. Such descriptions show nicely that the highest occupied MO(HOMO) is degenerate for most deltahedral B_nH_n molecules, but that a closed shell is obtained for the corresponding anions $[B_nH_n]^{2-}$. After accounting for the electrons in exopolyhedral bonds, $2n + 2$ electrons remain for

Table 3. Energy Levels and Symmetries in Selected Polyhedral Species (Approximate Calculation Using Only Framework Orbitals and Not Exopolyhedral Orbitals)

B_4H_4, T_d	$[B_5H_5]^{2-}$, D_{3h}	$[B_6H_6]^{2-}$, O_h	$[B_7H_7]^{2-}$, D_{5h}	$[B_{10}H_{10}]^{2-}$, D_{4d}	$[B_{10}H_{10}]^{2-}$, I_h
f_2 −0.849	e' −0.863	e_g −0.884	a_2'' −0.851	e_2 −0.884	f_{2u} −0.886
f_1 −0.556	a_2'' −0.858	f_{1u} −0.829	a_1' −0.847	e_3 −0.862	h_g −0.856
e 0.033	a_1' −0.823	f_{1g} −0.671	e_2' −0.845	b_2 −0.850	f_{1g} −0.782
f_2 0.685	e'' −0.649	f_{2u} −0.416	e_1' −0.840	a_1 −0.812	f_{1u} −0.773
a_1 2.094	a_2' −0.556	f_{2g} 0.493	a_2' −0.726	b_2 −0.812	h_u −0.678
	e' −0.301	f_{1u} 1.023	e_1' −0.669	e_1 −0.776	g_g −0.471
	e'' 0.335	a_{1g} 2.969	e_1'' −0.508	e_3 −0.747	g_u 0.518
	e' 0.758		e_2'' −0.371	a_2 −0.739	h_g 0.984
	a_2'' 1.108		e_2' 0.494	e_1 −0.615	f_{1u} 1.907
	a_1' 2.567		e_1'' 0.524	e_2 −0.598	a_g 4.163
			a_2'' 0.740	b_1 −0.590	
			e_1' 1.420	e_3 −0.327	
			a_1' 3.273	e_1 0.328	
				e_2 0.756	
				e_3 0.760	
				a_1 1.004	
				e_1 1.447	
				b_2 1.978	
				a_1 3.877	

framework bonding which gives some theoretical justification for the PERC formulation (10). Symmetry considerations such as those outlined by Pearson (68) also give justification for the opening of a *closo* borane (69) or carbaborane (49) upon addition of electrons.

This progressive opening of a given size of cluster with the addition of electrons (movement across any row of Fig. 2) might appear paradoxical for electron deficient molecules. However, as alluded to above, the addition of two electrons to the lowest unoccupied MOs(LUMOs) (e', t_{2u}, e_2'', e_3, and g_g, respectively) for the *closo* anions $[B_nH_n]^{2-}$ ($n = 5$–7, 10, 12) can result in orbital degeneracy and first-order Jahn-Teller instability. In those cases where degeneracy cannot be invoked ($n = 8, 9, 11$) there is reason to suspect second-order instability and distortion from the *closo* structure for the reduced anions (68–70). These effects were examined by mapping the energy of the B_5H_5 framework as a function of a symmetry-allowed deformation from the D_{3h} *closo* structure and as a function of electron count. Energy minima for the $2n + 2$, $2n + 4$, and $2n + 6$ systems occur at *closo*, *nido*, and *arachno* clusters, respectively. This correspondence of electron count with structure was also demonstrated for $C_2B_3H_5$ where the pyramidal (*nido*) isomers of $[C_2B_3H_5]^{2-}$ [54844-32-5] were found to be lower in energy than the corresponding trigonal bipyramidal isomers (Fig. 13) (49).

Boranes

Nido and Arachno Boranes. The *nido* and *arachno* boranes will be discussed before their *closo* relatives. They are in general much more reactive and thermally less stable than *closo* boranes. Chemical knowledge of boranes is based primarily on the rather readily available diborane(*6*), B_2H_6, and decaborane(*14*), $B_{10}H_{14}$, and to a lesser degree on the less readily available B_4H_{10}, B_5H_9, and B_6H_{10}. The topic has

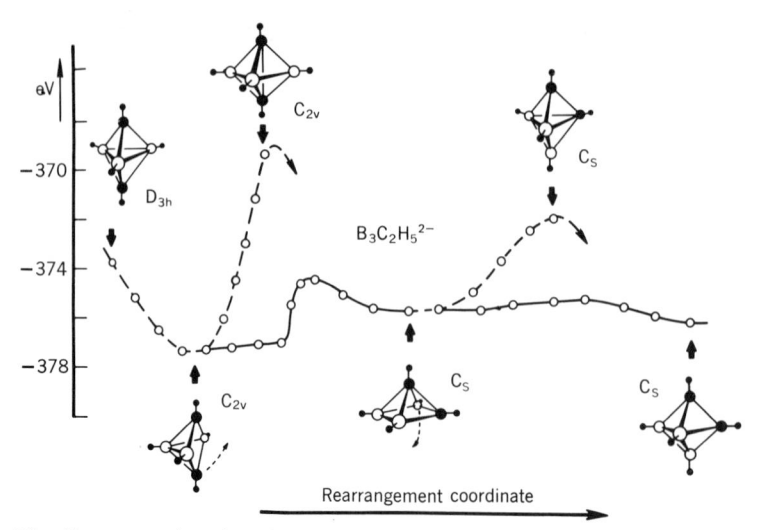

Figure 13. Energy as a function of geometry for $C_2B_3H_5^{2-}$. The solid line represents the interconversion of the isomeric forms of $C_2B_3H_5^{2-}$ via symmetry-allowed arcing movements of the designated atoms. The broken line designates the inconversion via a diamond–square–diamond mechanism.

been reviewed recently by Shore (64). The early work in the area was associated with the high-energy fuel program; it is covered principally in technical reports which have been summarized in *Production of Boranes and Related Research* (71).

The *nido* and *arachno* boranes smaller than $B_{10}H_{14}$ are very reactive to oxygen and moisture. Certain of their properties are summarized in Table 4.

Preparation. Diborane(6) is available from Callery Chemical Company, Callery, Pennsylvania. Convenient laboratory-scale preparations are given in equations 4–6 of which the last appears to afford the best method.

$$3\ Na[BH_4] + 4\ BF_3 \xrightarrow{diglyme} 2\ B_2H_6 + 3\ Na[BF_4] \quad (4)$$

$$[BH_4]^- + H^+(H_3PO_4\ or\ H_2SO_4) \rightarrow \tfrac{1}{2} B_2H_6 + H_2 \quad (5)$$

$$2\ Na[BH_4] + I_2 \xrightarrow{diglyme} 2\ NaI + B_2H_6 + H_2 \quad (6)$$

Tetraborane(10) can be obtained easily from salts of $B_3H_8^-$ [12429-74-2] which are also readily available (eqs. 7–8).

$$3\ Na[BH_4] + I_2 \xrightarrow[100°C]{diglyme} Na[B_3H_8] + 2\ H_2 + 2\ NaI \quad (7)$$

Table 4. Physical Properties of Some Boranes

Compound	Name	CAS Registry No.	mp, °C	bp, °C	ΔH_f° kJ/mol[a]	ΔG_f° kJ/mol[a]	S_{298}° J/(K·mol)[a]
B_2H_6	diborane(6)	[19287-45-7]	−164.9	−92.6	35.5	86.6	232.0
B_4H_{10}	tetraborane(10)	[18283-93-7]	−120	18	66.1		
B_5H_9	pentaborane(9)	[19624-22-7]	−46.6	48	73.2	174.0	275.8
B_5H_{11}	pentaborane(11)	[18433-84-6]	−123	63	103.0		
B_6H_{10}	hexaborane(10)	[23777-80-2]	−62.3	108	94.6		
$B_{10}H_{14}$	decaborane(14)	[17702-41-9]	99.7	213	31.5	216.1	353.0

[a] To convert J to cal, divide by 4.184.

$$4 \text{ HCl} + 4 \text{ Na}[B_3H_8] \rightarrow 3 \text{ B}_4H_{10} + 4 \text{ NaCl} + 3 \text{ H}_2 \tag{8}$$

Pentaborane(9) and decaborane(14) are available from Callery Chemical Company or can be prepared by pyrolysis of B_2H_6 under different conditions.

Reactions with Lewis Bases. Boranes which contain a BH_2 moiety (B_2H_6, B_4H_{10}, B_5H_{11}, B_6H_{12} [28375-94-2], B_9H_{15} [19465-30-6]) can generally be cleaved in two fashions by nucleophiles. The two fragmentation patterns have been termed symmetrical and unsymmetrical bridge cleavage by Parry (72). With neutral bases the two modes of cleavage lead to molecular and ionic fragments, respectively, as shown in Figure 14 for B_2H_6. Equations 9–14 give some examples.

$$B_2H_6 + 2 \text{ NH}_3 \rightarrow [H_2B(NH_3)_2]^+[BH_4]^- \tag{9}$$

$$B_2H_6 + 2 \text{ (CH}_3)_2S \rightarrow 2 \text{ (CH}_3)_2S.BH_3 \tag{10}$$

$$B_4H_{10} + 2 \text{ NH}_3 \rightarrow [H_2B(NH_3)_2]^+[B_3H_8]^- \tag{11}$$

$$B_4H_{10} + 2 \text{ N(CH}_3)_3 \rightarrow (CH_3)_3N.BH_3 + (CH_3)_3N.B_3H_7 \tag{12}$$

$$B_5H_{11} + 2 \text{ NH}_3 \rightarrow [H_2B(NH_3)_2]^+[B_4H_9]^- \tag{13}$$

$$B_5H_{11} + 2 \text{ CO} \rightarrow BH_3(CO) + B_4H_8(CO) \tag{14}$$

The mechanism of cleavage and how the nature of the base affects the type of cleavage are controversial; they have been discussed recently (64).

Certain base adducts of borane, BH_3, such as $(C_2H_5)_3N.BH_3$ [1722-26-5], $(CH_3)_2S.BH_3$ [13292-87-0], and tetrahydrofuranborane [14044-65-6], $C_4H_8O.BH_3$, are more easily and safely handled than B_2H_6 and are commercially available. They find wide use as reducing agents and in hydroboration reactions (50) (see Hydrides). A wide variety of borane reducing agents and hydroborating agents is available from Aldrich Chemical Co., Milwaukee, Wisc.

Base displacement reactions are used to convert one adduct to another. The relative stabilities of BH_3 adducts as a function of group V and VI donor atom are P > N; S > O which has sparked polemics because the trend opposes the normal order established by BF_3.

In the case of anionic nucleophiles, base displacement leads to ionic hydroborate adducts (eqs. 15–17).

$$(C_2H_5)_2O.BH_3 + KF \rightarrow K^+[BH_3F^-] + (C_2H_5)_2O \tag{15}$$

$$(C_2H_5)_2O.BH_3 + Na[SCN] \rightarrow Na^+[BH_3SCN]^- + (C_2H_5)_2O \tag{16}$$

Unsymmetrical cleavage of B_2H_6 by metal hydrides gives metal tetrahydroborate salts, sometimes termed metal borohydrides or hydroborates (see below).

$$2 \text{ MH} + B_2H_6 \rightarrow 2 \text{ M}[BH_4] \, (M = \text{Li, Na, K}) \tag{17}$$

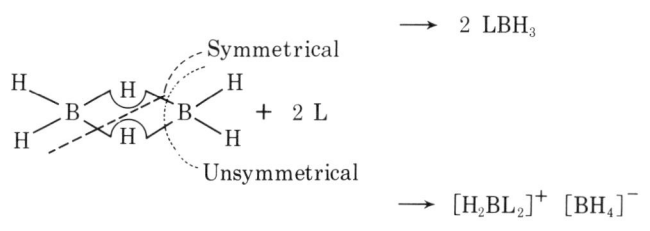

Figure 14. Symmetrical and unsymmetrical cleavage of diborane(6).

Except for $B_{10}H_{14}$, a more detailed consideration of the nature of the interaction of ligands with other boranes is beyond the scope of this review. Weak bases fail to deprotonate decaborane but do react with the evolution of H_2.

$$B_{10}H_{14} + 2L \rightarrow B_{10}H_{12}L_2 + H_2 \tag{18}$$

The ligands coordinate at the 6,9-positions of the decaborane skeleton as shown in Figure 15.

Base displacement reactions have established the relative basicities (73) of a number of ligands towards $B_{10}H_{12}$.

$(C_6H_5)_3P >$ pyridine $> (C_2H_5)_3N > CH_3CON(CH_3)_2 >$
$$HCON(CH_3)_2 > (C_2H_5)_2NCN > CH_3CN > (CH_3)_2S \tag{19}$$

$B_{10}H_{12}L_2$ species are pivotal intermediates in the synthesis of two key *closo* species ($[B_{10}H_{10}]^{2-}$ and $1,2-C_2B_{10}H_{12}$) which have a diverse chemistry (see below).

Proton Abstraction. Even though the exopolyhedral hydrogens of *nido* and *arachno* boranes are generally considered hydridic, the bridge hydrogens are acidic as first demonstrated by titration of $B_{10}H_{14}$ and deuterium exchange (74). The relative Brønsted acidities of several boranes have been determined and found to be essentially that predicted by Parry and Edwards (72) on the basis of framework size for boranes without BH_2 groups: $B_5H_9 < B_6H_{10} < B_{10}H_{14} < B_{16}H_{20} < B_{18}H_{22}$. However, the property is quite general to bridge hydrogens, and even boranes with BH_2 groups such as B_4H_{10} have acidic character, $B_6H_{10} < B_4H_{10} < B_{10}H_{14}$ (75). Some typical reactions are:

$$B_{10}H_{14} + NaOH \rightarrow Na[B_{10}H_{13}] + H_2O \tag{20}$$
$$B_{10}H_{14} + NaH \rightarrow Na[B_{10}H_{13}] + H_2 \tag{21}$$
$$B_5H_9 + n\text{-}C_4H_9Li \rightarrow Li[B_5H_8] + C_4H_{10} \tag{22}$$
$$B_4H_{10} + NaH \rightarrow Na[B_4H_9] + H_2 \tag{23}$$

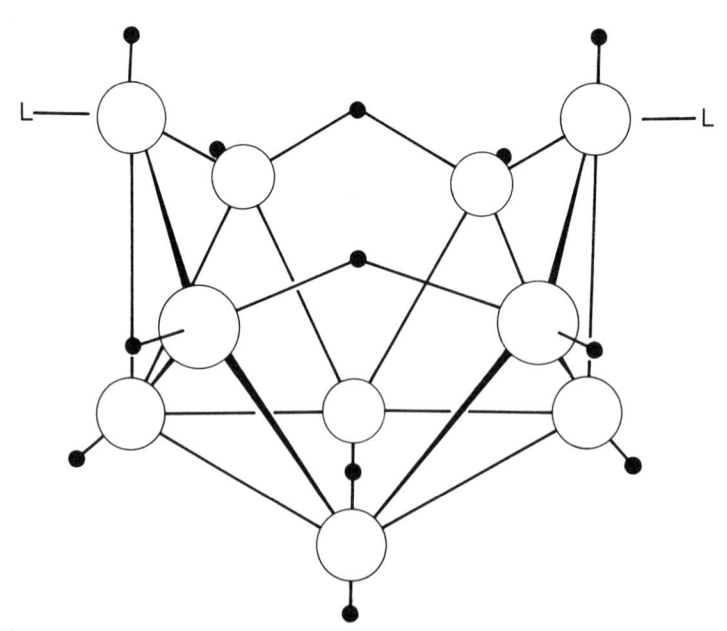

Figure 15. Structure of 6,9-L_2-$B_{10}H_{12}$, bis(ligand) adducts of decaborane(*12*).

$$B_6H_{10} + KH \rightarrow K[B_6H_9] + H_2 \tag{24}$$

The hydropolyborate ions formed by proton abstraction are useful intermediates for the preparation of metalloboranes and heteroboranes to be discussed later.

Polyhedral Expansion. The term polyhedral expansion is used to describe a host of reactions whereby the size of the polyhedron is increased by the addition of more atoms (boron, heteroelements and metals). In the case of the boranes, the pyrolysis of B_2H_6 has been used to obtain B_5H_9 and $B_{10}H_{14}$ industrially. However, the mechanism of such pyrolytic expansions is not completely established.

The expansion of $B_{10}H_{14}$ to $[B_{11}H_{14}]^-$ appears to proceed through $B_{10}H_{13}^-$ (76).

$$[BH_4]^- + B_{10}H_{14} \rightarrow [B_{11}H_{14}]^- + 2\,H_2 \tag{25}$$

The removal of H^+ from a bridge hydrogen in $B_{10}H_{14}$ gives a B—B bond in $B_{10}H_{13}^-$. Other expansion reactions between diborane and borane anions with a B—B edge bond have been observed (77) (eqs. 26–28).

$$\left.\begin{array}{l}[B_4H_9]^- \\ [B_5H_8]^- \\ [B_6H_9]^- \end{array}\right\} + \tfrac{1}{2}\,B_2H_6 \rightarrow \left\{\begin{array}{l}[B_5H_{12}]^- \quad (26) \\ [B_6H_{11}]^- \quad (27) \\ [B_7H_{12}]^- \quad (28) \end{array}\right.$$

Closo borane anions can be formed by similar means, but will be discussed later. Boron halides have also been shown to insert into the B—B bond and give an initial product with the new boryl moiety in a bridge position (78).

$$[B_5H_8]^- + (CH_3)_2BCl \rightarrow \mu\text{-}[(CH_3)_2B]\text{-}B_5H_8 + Cl^- \tag{29}$$

$$\mu\text{-}[(CH_3)_2B]\text{-}B_5H_8 \xrightarrow{(C_2H_5)_2O} 4,5\text{-}(CH_3)_2B_6H_8 \tag{30}$$

Electrophilic Attack. A variety of boranes, heteroboranes, and metalloboranes undergo electrophilic substitution. The latter two classes are discussed below. In some cases the site of substitution correlates with the ground-state charge distribution of the molecule obtained from MO-calculations. However, such correlations must be considered fortuitous because the activated-state charge distribution is not necessarily similar to that of the ground state.

The susceptibility of boranes to electrophilic attack is often detected by deuteron–proton exchange studies. For example, DCl–AlCl$_3$ with $B_{10}H_{14}$ in CS$_2$ gives initial deuteration at the 2,4-positions followed by the 3,4-positions (79) (Fig. 1l). The trend to increasingly positive sites in $B_{10}H_{14}$ is 2,4 < 3,4 < 5,7,8,10 < 6,9. The initial halogenation and alkylation of $B_{10}H_{14}$ also occur at the 2-position (80–81).

The electrophilic substitution of pentaborane(9) has also been observed to give halogenation, alkylation, and deuteration. The apical site (the more negative 1-position) is substituted initially, but the 1-isomer can be catalytically converted to the basally substituted isomer (2-position) (82) (Fig. 1b).

The basal B—B bond of B_6H_{10} can be protonated to give the only example of an isolable polyhedral borane cation, $[B_6H_{11}]^+$ (83) (Fig. 1d).

Closo Boranes. This class contains some very stable molecules which makes their chemistry atypical among the boranes. The molecules that have been characterized constitute a series of $[B_n H_n]^{2-}$ ($n = 6$–12) deltahedral anions as shown in Figure 2. Unlike their *nido* and *arachno* counterparts with bridge hydrogens, for practical purposes, the abstraction of H^+ does not occur in *closo* borane chemistry. Instead, acid catalysis plays an important role in their substitution chemistry. Both $[B_{10}H_{10}]^{2-}$ [12356-12-6] and $[B_{12}H_{12}]^{2-}$ [12356-13-7] are stable as their conjugate acids in aqueous solution with an acidity function similar to sulfuric acid and yet their chemistry is remarkably diverse.

Properties. Large unipositive cations, such as Tl^+, Cs^+, Rb^+, $[(CH_3)_4N]^+$, and $[(CH_3)_3S]^+$, yield relatively water insoluble salts of $B_{12}H_{12}{}^{2-}$ and $[B_{10}H_{10}]^{2-}$. The $[B_{10}H_{10}]^{2-}$ salts are usually more water soluble than their $[B_{12}H_{12}]^{2-}$ counterparts (84). Small unipositive cations and most dipositive cations, such as Ba^{2+} and Ca^{2+}, form water soluble salts which are strong electrolytes and give hydrates on evaporation. The divalent transition and rare earth elements also give soluble salts and hydrates, but their solubilities decrease when the water of the coordination sphere is replaced by ligands such as NH_3. Polarizing cations, such as Ag^+, Cu^+, and Hg^{2+}, form notably water-insoluble salts. The latter are unusual since there is no evidence for metal ion reduction (Ag^+ reduction was long a diagnostic test for boranes). Salts of nonreducible cations display exceptional thermal stabilities. Thus $Cs_2[B_{12}H_{12}]$ and $Cs_2[B_{10}H_{10}]$ can be heated to 810 and 600°C, respectively, in a sealed evacuated tube and recovered unchanged.

The salts of $[B_6H_6]^{2-}$ [12429-97-9], $[B_7H_7]^{2-}$ [12430-07-8], $[B_8H_8]^{2-}$ [12430-13-6], $[B_9H_9]^{2-}$ [12430-00-0], and $[B_{11}H_{11}]^{2-}$ [12430-44-3] are probably strong electrolytes but little definitive evidence is available. Although silver salts of $[B_6H_6]^{2-}$, $[B_9H_9]^{2-}$, and $[B_{11}H_{11}]^{2-}$ have been isolated, they are shock-, heat-, and light-sensitive (85–86). Anhydrous $Cs_2[B_6H_6]$, $Cs_2[B_8H_8]$, and $Cs_2[B_9H_9]$ are thermally stable to 600°C, but $Cs_2[B_{11}H_{11}]$ gives equal amounts of $Cs_2[B_{10}H_{10}]$ and $Cs_2[B_{12}H_{12}]$ at temperatures above 400°C (18).

Preparation and Polyhedral Expansion. The reaction of B_2H_6 with $NaBH_4$ or $[(C_2H_5)_3N]BH_3$ gives $B_{12}H_{12}{}^{2-}$ in nearly quantitative yield at 180°C in $N(C_2H_5)_3$ at high pressure (87).

$$2\, Na[BH_4] + 5\, B_2H_6 \rightarrow Na_2[B_{12}H_{12}] + 13\, H_2 \tag{31}$$

$$2\, [(C_2H_5)_3N]BH_3 + 5\, B_2H_6 \rightarrow [(C_2H_5)_3NH]_2[B_{12}H_{12}] + 11\, H_2 \tag{32}$$

In diglyme at 162°C, the same reactants give a 5–10% yield of $[B_6H_6]^{2-}$; at even lower temperature (85°C) and at atmospheric pressure $Na[B_3H_8]$ [12429-74-2] results (86,88). Pyrolysis of $Cs[B_3H_8]$ at 230°C gives $Cs_2[B_9H_9]$ (60%) along with some $Cs_2[B_{10}H_{10}]$, $Cs_2[B_{12}H_{12}]$, and $Cs[BH_4]$ (89). The sensitivity of polyhedral expansion reactions to solvent, temperature, and pressure is further exemplified by the results in dioxane at 120°C under pressure. To obtain the *closo* borane, the $NaB_{11}H_{14}$ is first converted to $Cs_2B_{11}H_{13}$ which can be pyrolyzed to give $Cs_2[B_{11}H_{11}]$ (86).

$$Na[BH_4] + 5\, B_2H_6 \xrightarrow{120°C} Na[B_{11}H_{14}] + 10\, H_2 \tag{33}$$

Pyrolysis of $[(C_2H_5)_4N][BH_4]$ at 185°C gives a 90% yield of $[(C_2H_5)_4N]_2[B_{10}H_{10}]$ (90).

If $B_{10}H_{14}$ is available, $[B_{12}H_{12}]^{2-}$ can be prepared by a reaction involving $(C_2H_5)_3NBH_3$ in an inert high boiling hydrocarbon at 190°C (91).

$$2\,[(C_2H_5)_3N]BH_3 + B_{10}H_{14} \xrightarrow{190°C} [(C_2H_5)_3NH]_2[B_{12}H_{12}] + 3\, H_2 \tag{34}$$

A base-promoted closure of $6{,}9\text{-}L_2\text{-}B_{10}H_{12}$ species gives $[B_{10}H_{10}]^{2-}$ in nearly quantitative amounts (92). The conversion is quite convenient with triethylamine.

$$B_{10}H_{14} + 2\, N(C_2H_5)_3 \xrightarrow{-H_2} 6{,}9\text{-}[(C_2H_5)_3N]_2B_{10}H_{12} \rightarrow [(C_2H_5)_3NH]_2[B_{10}H_{10}] \tag{35}$$

Reactions with Nucleophiles. The $[B_6H_6]^{2-}$, $[B_7H_7]^{2-}$, $[B_9H_9]^{2-}$, and $[B_{11}H_{11}]^{2-}$ ions are hydrolytically less stable than $[B_{10}H_{10}]^{2-}$ and $[B_{12}H_{12}]^{2-}$. All of these anions are more stable in basic than in acidic solution. $[B_7H_7]^{2-}$ is hydrolytically least stable and is degraded even in basic media. $[B_6H_6]^{2-}$, $[B_8H_8]^{2-}$, and $[B_9H_9]^{2-}$ are stable in neutral and alkaline solutions but react rapidly with aqueous acid. Strongly acidic solutions (> 2N HCl) are necessary for the hydrolysis of $[B_{11}H_{11}]^{2-}$; $[B_{12}H_{12}]^{2-}$ is the most hydrolytically stable borane anion, withstanding even 3N HCl at 95°C, conditions which slowly degrade $[B_{10}H_{10}]^{2-}$ (18,85).

Virtually nothing is known about the nucleophilic substitution chemistry of the $B_nH_n^{2-}$ ions except for $n = 10, 12$ where certain reactions have been described as acid-catalyzed nucleophilic substitution. Mechanistically, the steps appear to be:

$$[B_nH_n]^{2-} + H^+ \rightarrow [B_nH_{n+1}]^- \xrightarrow[-H_2]{L} [B_nH_{n-1}L]^- \xrightarrow{H+} B_nH_nL \xrightarrow[-H_2]{L} (B_nH_{n-2}L_2) [n = 10, 12]$$

When L is an anion the product remains charged. Most acid-catalyzed nucleophilic substitutions of $B_{10}H_{10}^{2-}$ give equatorial substituents (reactions with amides, ethers, and sulfones).

$$[B_nH_n]^{2-} + 2\,R_2O \xrightarrow{H+} [B_nH_{n-2}(OR)_2]^{2-} + 2\,RH \tag{36}$$

$$[B_nH_n]^{2-} + H^+ + HCON(CH_3)_2 \rightarrow [B_nH_{n-1}(OCH{=}N(CH_3)_2)]^- + H_2 \tag{37}$$

$$[B_nH_n]^{2-} + H^+ + R_2SO_2 \rightarrow [B_nH_{n-1}OS(O)R_2]^- + H_2 \tag{38}$$

These O-bonded substituents are easily cleaved with hydroxide ion to give the corresponding hydroxyl derivative, $B_nH_{n-1}(OH)^{2-}$ or $[B_nH_{n-2}(OH)_2]^{2-}$, $n = 10, 12$. Halogenation of $[B_{12}H_{12}]^{2-}$ by HCl and HF has been termed acid-catalyzed nucleophilic attack, but no stereochemical information was reported in these cases (93).

Reactions with Electrophiles. No simple pattern in site preference emerges for electrophilic substitution (93). The prediction that electrophilic substitution in $[B_{10}H_{10}]^{2-}$ should occur first at the apical positions (positions 1,10; see Fig. 1**k**) is not always borne out. Notable exceptions are benzoyl chloride, hydroxylamine-O-sulfonic acid, and the Vilsmeier reagent. The benzene diazonium cation and nitrous acid give exclusively apical substitution.

The reaction of $B_{10}H_{10}^{2-}$ with excess nitrous acid gives a very explosive intermediate which can be reduced to the nonexplosive bis inner diazonium salt 1,10-$(N_2)_2$-$B_{10}H_8$ [66750-86-5] (eq. 39).

$$[B_{10}H_{10}]^{2-} \xrightarrow{HONO} [\text{explosive intermediate}] \xrightarrow{Na[BH_4]} 1,10\text{-}(N_2)_2\text{-}B_{10}H_8 \tag{39}$$

The diazonium species is an important synthetic intermediate. Unfortunately, $[B_{12}H_{12}]^{2-}$ doesn't undergo the corresponding reaction. Nitrogen is the only ligand that can be displaced from the B_{10}-cluster by a variety of nucleophiles (L = ammonia, amines, nitriles, hydrogen sulfide, azide ion, hydroxide ion, and carbon monoxide (see eq. 40) (94).

$$1,10\text{-}(N_2)_2\text{-}B_{10}H_8 + 2\,L \rightarrow 1,10\text{-}L_2\text{-}B_{10}H_8 + 2\,N_2 \tag{40}$$

The dicarbonyl [12539-66-1] available from 1,10-$(N_2)_2$-$B_{10}H_8$ is another very important species because of the scope and flexibility of its chemistry. Carbonyls of $[B_{12}H_{12}]^{2-}$ can be formed from CO and the conjugate acid of $[B_{12}H_{12}]^{2-}$. The direct formation of a B_{10}-carbonyl from $B_{10}H_{10}^{2-}$ has not been reported. Almost without exception the B_{10}- and B_{12}-carbonyls exhibit the same reactivity (95). The carbonyls can be considered anhydrides of carboxylic acids and accordingly react with alcohols and amines to give esters and amides:

$$B_{10}H_{10}(CO)_2 \quad [12363\text{-}61\text{-}0] \quad \begin{cases} \xrightleftharpoons{H_2O} H_2[B_{10}H_8(COOH)_2] \quad [12541\text{-}38\text{-}7] & (41) \\ \xrightleftharpoons{ROH} H_2[B_{10}H_8(COOR)_2] & (42) \\ \xrightleftharpoons{R_2NH} [R_2NH_2]_2[B_{10}H_8(CONR_2)_2] & (43) \end{cases}$$

A further discussion of the extensive chemistry derived from the carbonyls is given in ref. 93.

The halogenation of $[B_nH_n]^{2-}$ ions has not been clearly defined as electrophilic. Halogenation of $[B_{12}H_{12}]^{2-}$ and $[B_{10}H_{10}]^{2-}$ occurs with elemental halogen in solvents such as water, alcohol, and tetrachloroethane. Initial rates are extremely high (approaching collisional rates) in all cases with $[B_{10}H_{10}]^{2-} > [B_{12}H_{12}]^{2-}$ and the kinetic order is: $F \geq Cl > Br > I$, but F_2 causes degradation (96). Exemplary products are $[B_{10}Cl_{10}]^{2-}$ [12430-33-0], $[B_{10}H_3Br_7]^{2-}$ [12360-16-6], $[B_{10}I_{10}]^{2-}$ [12430-43-2], $[B_{12}Cl_6H_6]^{2-}$ [12430-46-5], $[B_{12}H_3Br_6Cl_3]^{2-}$ [12536-79-7], and $[B_{12}I_{12}]^{2-}$ [12587-25-6]. In general, the alkali and alkaline earth metal salts of the B_{10}- and B_{12}-halogenated derivatives have excellent thermal, oxidative, and hydrolytic stabilities.

Oxidation–Reduction Reactions. Oxidative degradation of $[B_{10}H_{10}]^{2-}$ and $[B_{12}H_{12}]^{2-}$ to boric acid is extremely difficult and requires Kjeldahl digestion or neutral permanganate. The heat of reaction obtained from the permanganate degradation leads to a calculated heat of formation for $[B_{10}H_{10}]^{2-}$ (aq) of 92.5 ± 21.1 kJ/mol (22.1 ± 5.04 kcal/mol) (97). The $[B_8H_8]^-$ radical as detected by epr spectroscopy is obtained by air oxidation (18) of $[B_9H_9]^{2-}$. However, the oxidative coupling of both $[B_{10}H_{10}]^{2-}$ and $[B_{12}H_{12}]^{2-}$ has received the most attention (98).

$[B_{10}H_{10}]^{2-}$ can be oxidized chemically or electrochemically (eqs. 44–46) to give $[B_{20}H_{19}]^{3-}$ and $[B_{20}H_{18}]^{2-}$ [59724-35-5] which is two B_{10}-clusters through a double 3-center B—B—B bond as shown in Figure 16 (99).

$$[B_{10}H_{10}]^{2-} \rightleftharpoons [B_{10}H_{10}]^- + e^- \quad (44)$$

$$2\,[B_{10}H_{10}]^- \rightarrow [B_{20}H_{19}]^{3-} + H^+ \quad (45)$$

$$[B_{20}H_{19}]^{3-} \rightarrow [B_{20}H_{18}]^{2-} + H^+ + 2\,e^- \quad (46)$$

Reduction of $[B_{20}H_{18}]^{2-}$ by sodium metal in liquid ammonia gives $[B_{20}H_{18}]^{4-}$ [59724-36-6] in which the B_{10}-clusters are joined by a single 2-center B—B bond through the apical or equatorial positions. The nondestructive oxidation of $[B_{12}H_{12}]^{2-}$ has been reported only electrochemically and gives $[B_{24}H_{23}]^{2-}$. However, the derivative $[B_{24}H_{20}I_2]^{2-}$ is known (100).

Tetrahydroborates

Tetrahydroborates of virtually every metal in the periodic chart except the actinides (qv) have been reported. Their preparations have been reviewed by Wallbridge (50). However, the important commercial ones are the alkali metal tetrahydroborates. Some of their properties are given in Table 5.

Treatment of trimethyl borate with a metal hydride (eg, NaH) in the absence of

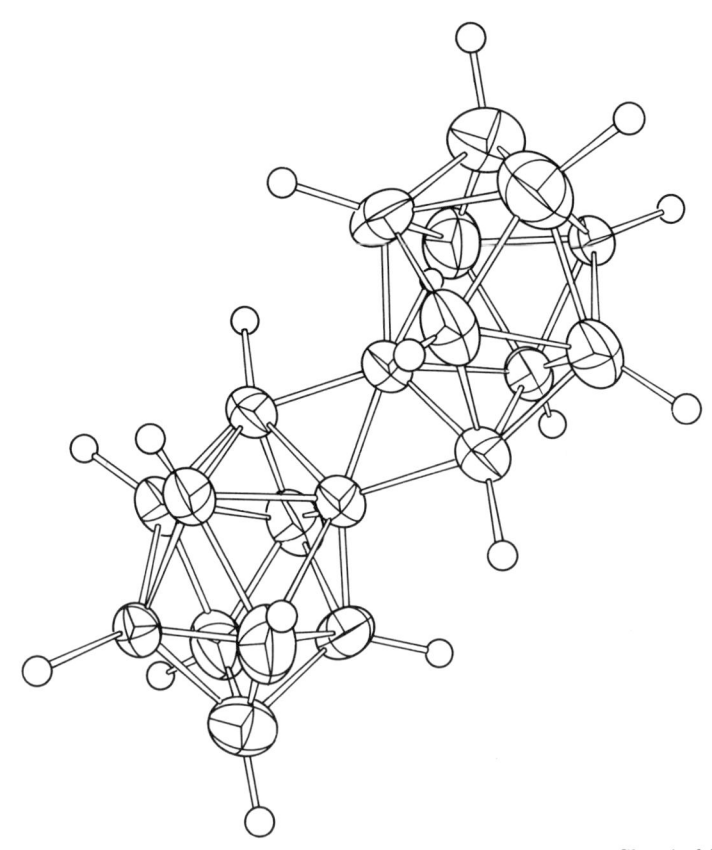

Figure 16. The structure of $[B_{20}H_{18}]^{2-}$ (99). Courtesy of the American Chemical Society.

Table 5. Some Properties of the Alkali Metal Tetrahydroborates

Property	Li[BH$_4$] [16949-15-8]	Na[BH$_4$] [16940-66-2]	K[BH$_4$] [13762-51-1]	Rb[BH$_4$] [20346-99-0]	Cs[BH$_4$] [19193-36-3]	Refs.
mp, °C	268	505	585			101–103
decomp. temp, °C	380	315	584	600	600	101–103
density, g/cm^3	0.68	1.08	1.17	1.71	2.40	102, 104–105
refractive index		1.547	1.490	1.487	1.498	106
lattice energy, kJ/mol[a]	792.0	697.5	657	648	630.1	102
ΔH_f°, kJ/mol[a]	−184	−183	−243	−246	−264	105, 107
S_{298}°, J/(mol·K)[a]	−128.7	−126.3	−161	−179	−192	108

[a] To convert J to cal, divide by 4.184.

a solvent yields sodium hydrotrimethoxyborate [16940-17-3] (eq. 47) which disproportionates in the presence of solvents such as tetrahydrofuran at 60–70°C (eq. 48) (109).

$$\text{MH} + \text{B(OCH}_3)_3 \rightarrow \text{M[BH(OCH}_3)_3]; \quad \text{M} = (\text{Li, Na, K}) \tag{47}$$

Addition of diborane under the latter conditions renders the production of Na[BH$_4$]

essentially continuous until all the original metal hydride is consumed since trimethyl borate is regenerated (eq. 49).

$$4 \text{ M[BH(OCH}_3)_3] \rightarrow \text{M[BH}_4] + 3 \text{ M[B(OCH}_3)_4] \tag{48}$$

$$3 \text{ M[B(OCH}_3)_4] + 2 \text{ B}_2\text{H}_6 \rightarrow 3 \text{ M[BH}_4] + 4 \text{ B(OCH}_3)_3 \tag{49}$$

This general method has been used for the commercial production of Na[BH$_4$], but is less satisfactory for Li[BH$_4$] and K[BH$_4$] (110). Some metathetic conversions are shown in equations 50–55.

$$\text{TlNO}_3 + \text{K[BH}_4] \xrightarrow{\text{H}_2\text{O}} \text{Tl[BH}_4] + \text{KNO}_3 \tag{50}$$

$$\text{Na[BH}_4] + \text{KOH} \xrightarrow{\text{H}_2\text{O}} \text{K[BH}_4] + \text{NaOH} \tag{51}$$

$$[(\text{C}_6\text{H}_5)_4\text{P]F} + \text{K[BH}_4] \xrightarrow{\text{H}_2\text{O}} [(\text{C}_6\text{H}_5)_4\text{P][BH}_4] + \text{KF} \tag{52}$$

$$\text{Na[BH}_4] + \text{LiCl} \xrightarrow{\text{isopropylamine}} \text{Li[BH}_4] + \text{NaCl} \tag{53}$$

$$\text{MgCl}_2 + 2 \text{ Na[BH}_4] \xrightarrow{\text{ethanol}} \text{Mg[BH}_4]_2 + 2 \text{ NaCl} \tag{54}$$

$$[\text{NH}_4]\text{F} + \text{Na[BH}_4] \xrightarrow{\text{ammonia}} [\text{NH}_4][\text{BH}_4] + \text{NaF} \tag{55}$$

The physical and chemical properties of the tetrahydroborates show more contrasts than the salts of nearly any other anion. The alkali metal salts are the most stable. In dry air, Na[BH$_4$] is stable at 300°C, and *in vacuo* to 400°C. The potassium and sodium salts have been sublimed in vacuo at over 400°C with only partial decomposition. In contrast, several tetrahydroborates, including the titanium, thallium, gallium, copper, and silver salts, are unstable at or slightly above ambient temperatures. The tetrahydroborates of the polyvalent metals are in many cases the most volatile derivatives of these metals known. Aluminum tris(tetrahydroborate) [*16962-07-5*] has a normal boiling point of 44.5°C and uranium bis(tetrahydroborate) [*33725-14-3*], U[BH$_4$]$_2$, has a vapor pressure of 530 Pa (4 torr) at 61°C.

The chemical and physical properties of the tetrahydroborates are closely related to molecular structure. Sodium tetrahydroborate, which is typical of the alkali metal tetrahydroborates (except for the lithium salt), has a fcc crystal lattice, which is essentially ionic and contains the tetrahedral BH$_4^-$ ion. The volatile tetrahydroborates, however, are relatively nonpolar, and the bonding resembles that of diborane. The structure of aluminum tris(tetrahydroborate) is shown in Figure 17.

The alkali metal tetrahydroborates are stable in dry air, and the sodium and potassium salts can be crystallized from a water solution. On the other hand, aluminum tris(tetrahydroborate) is hydrolyzed explosively and is violently pyrophoric in air.

The solubility of sodium tetrahydroborate in a variety of solvents is given in Table 6. It is appreciably soluble only in polar solvents of high dielectric constant and those which can solvate the metal ion. The rate of hydrolysis of NaBH$_4$ is increased by either lowering the pH or by increasing the temperature. It can be recrystallized from alkaline solutions. Dissolution in water results in a slow hydrolysis until the solution becomes akaline. A 1 M solution gives a pH of 10.5 and only very slow hydrolysis occurs at pH > 13.

The tetrahydroborates have been used as reducing agents for a variety of inorganic

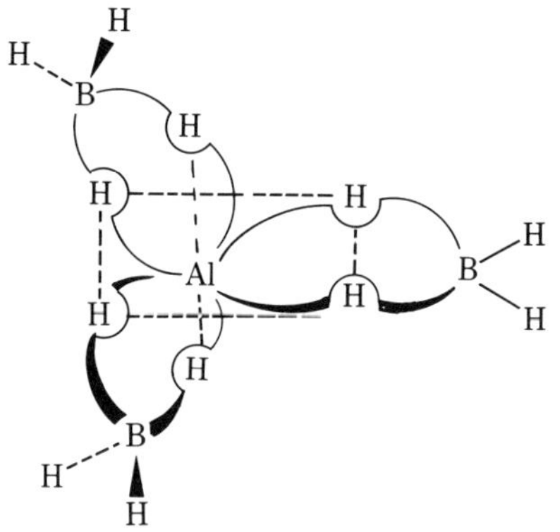

Figure 17. The structure of Al(BH$_4$)$_3$.

Table 6. Solubility of Sodium Hydroborate in Various Solvents[a]

Solvent	bp of solvent, °C	Temp, °C	Solubility, g/100 g solvent
water	100	0	25
		25	55
		60	88.5
liquid ammonia	−33.3	−33.3	104
methylamine	−6.3	−20	27.6
ethylamine	16.6	17	20.9
n-butylamine	77.8	28	4.9
pyridine	115.3	25	3.1
dimethylformamide	153	20	18.0
ethylenediamine	118	75	22
acetonitrile	80.1	28	2
ethanol	78.5	20	4 (reacts slowly)
ethylene glycol monomethyl ether	125	25	16.7
diethylene glycol monomethyl ether	193	25	16.3
dimethyl ether of ethylene glycol (monoglyme)	85	20	0.8
dimethyl ether of diethylene glycol (diglyme)	162	25	5.5
		45	8.0
		75	10.0
dimethyl ether of triethylene glycol (triglyme)	216	25	8.7
		50	8.5
		100	6.7
tetrahydrofuran	65	20	0.1
dimethyl sulfoxide	189	25	5.8

[a] Compiled by Ventron, Metal Hydrides Division, Beverly, Mass.

reductions. For example, tin is reduced from Sn^{2+} to SnH$_4$, germanium from GeO$_2$ to GeH$_4$, arsenic from arsenite to AsH$_3$, and antimony from its trivalent salts to SbH$_3$. Several such reactions have been utilized in quantitative analytical procedures. However, the use of hydroborates (and boranes) for organic processes has proven to be even more significant since the reduction reactions are highly selective and nearly quantitative. The reducing characteristics of the hydroborate may be varied by changing the associated cation and by changing the solvent.

Covalent metal tetrahydroborates derived from Ti, Zr, Co, Ni, and Rh have shown catalytic activity in hydrogenation, polymerization and isomerization reactions (51). The function of the $[BH_4]^-$ ligand in these reactions is not clear. It may (1) reduce the central metal to a lower oxidation state, (2) provide a source of hydrogen forming, in effect, a metal hydride, (3) provide a variable coordination sphere for the metal atom by a series of bidentate ⇌ tridentate interconversions which would induce coordinative unsaturation at the metal, or (4) function as a ligand which activates other moieties in the coordination sphere.

Heteroboranes

Relatively little is known about *arachno* heteroboranes in comparison with their *nido* and *closo* counterparts. This section will deal with all three structural classifications. Here a heteroelement is classified as a nonmetal. The heteroatoms known to form part of a borane polyhedron include C, N, Si, P, S, As, Se, and Sb either singly or in combination. Those heteroboranes with the greatest demonstrated scope and flexibility of chemistry are the carbaboranes and thiaboranes. Both carbaboranes and thiaboranes are also available in practical amounts.

Carbaboranes. The heteroborane which ushered in a new era of borane chemistry is 1,2-dicarba-*closo*-dodecaborane(12) [16872-09-6], 1,2-$C_2B_{10}H_{12}$, which is often termed *ortho*-carborane (*ortho*-barene in the Russian literature); it is prepared by the reaction of acetylene with 6,9-$[(CH_3)_2S]_2B_{10}H_{12}$ [28377-92-6] or other 6,9-L_2-$B_{10}H_{12}$ species.

$$6,9\text{-}L_2\text{-}B_{10}H_{12} + HC\equiv CH \longrightarrow HC\underset{B_{10}H_{10}}{\overset{}{\diagdown\hspace{-0.5em}O\hspace{-0.5em}\diagup}}CH + H_2 + 2\,L \tag{56}$$

A large variety of C-substituted *o*-carboranes (Table 7) can be obtained by utilizing substituted acetylenes, $RC\equiv CR'$. The symbols

$$HC\underset{B_{10}H_{10}}{\overset{}{\diagdown\hspace{-0.5em}O\hspace{-0.5em}\diagup}}CH$$

and $HCB_{10}H_{10}CH$ and $H\overline{CB_{10}H_{10}C}H$ are commonly used in the literature to represent 1,2-$C_2B_{10}H_{12}$ [16872-09-6], 1,7-$C_2B_{10}H_{12}$ [16986-21-6], and 1,12-$C_2B_{10}H_{12}$ [20644-12-6], respectively. The latter two isomers, sometimes called *m*-carborane and *p*-carborane, respectively, are obtained conveniently by thermal isomerization (eqs. 57–58) of 1,2-$C_2B_{10}H_{12}$ (see Fig. 1o for the numbering convention for these icosahedral species).

$$1,2\text{-}C_2B_{10}H_{12} \xrightarrow[1\text{-}2\text{ days}]{425\text{-}500°C} 1,7\text{-}C_2B_{10}H_{12}\ (\text{ca }100\%) \tag{57}$$

$$1,2\text{-}C_2B_{10}H_{12} \xrightarrow[N_2\text{ flow}]{700°C} 1,7\text{-}C_2B_{10}H_{12}\ (\text{ca }25\%) + 1,12\text{-}C_2B_{10}H_{12}\ (\text{ca }25\%) \tag{58}$$

C-substituted dicarbaboranes can also be obtained conveniently through the intermediacy of lithium reagents such as 1,2-Li_2-1,2-$C_2B_{10}H_{10}$ [22220-85-5] and 1,7-Li_2-1,7-$C_2B_{10}H_{10}$ [17217-89-9], which are readily prepared by treatment of *o*- or *m*-carborane with *n*-BuLi. The scope of this organic derivative chemistry has been reviewed by Grimes (6).

Table 7. Selected C-Substituted *o*-Carboranes

Substituent groups	mp, °C
Alkyl derivatives	
1-CH$_3$	218–219
1-*n*-C$_4$H$_9$	11.2–11.3
1-*n*-C$_6$H$_{13}$	bp 101–102 (67 Pa or 0.5 mm Hg)
1,2-(CH$_3$)$_2$	265
Haloalkyl derivatives	
1-CH$_2$Cl	91.5
1-CH$_2$Br	47–49
1-(CH$_2$)$_2$Br	110
Aryl derivatives	
1-C$_6$H$_8$	69.5–70
1-CH$_2$C$_6$H$_5$	57–58
1,2-(C$_6$H$_5$)$_2$	148–149
1-*o*-C$_6$H$_4$Br	104–104.5
1-*m*-C$_6$H$_4$Br	107–108
1-*p*-C$_6$H$_4$Br	134.5–134.7
Alkenyl and alkynyl derivatives	
1-CH=CH$_2$	78–79
1-CH$_3$-2-CH=CH$_2$	164–165
1-C≡CH	75–78
Carboxylic acids	
1-COOH	150–150.5
1-(CH$_2$)$_2$COOH	147–149
1,2-(CH$_2$COOH)$_2$	203–205
1-C$_6$H$_5$-2-CH$_2$COOH	196–198
Esters and acyl halides	
1-CH$_2$COOC$_2$H$_5$	10–11
1,2-(COOCH$_3$)$_2$	66–67
1,2-(CH$_2$COOCH$_3$)$_2$	47–48
1-COCl	39–41
1,2-(COCl)$_2$	69–70
Alcohols	
1-CH$_2$OH	229–230
1-(CH$_2$)$_2$OH	51–52
1-(CH$_2$)$_3$OH	53–53.5
1,2-(CH$_2$OH)$_2$	299–300
1-CH$_3$-2-CH$_2$OH	268–269
1-C$_6$H$_5$-2-CH$_2$OH	89–90
Ethers and epoxides	
1-CH$_2$CHCH$_2$ (epoxide, O bridging CH–CH$_2$)	68
1-C$_6$H$_5$-2-CH$_2$OC$_6$H$_5$	59–60
Aldehydes	
1-CHO	212–213
1-CH$_2$CHO	94–95
1-CH$_3$-2-CH$_2$CHO	140–142
Ketones	
1-COCH$_3$	54–55
1-COC$_6$H$_5$	57.5–58.0
1-CH$_3$-2-COC$_6$H$_5$	67

Table 7. (continued)

Substituent groups	mp, °C
Amines	
1-NH$_2$	304–305
1-CH$_2$N(C$_2$H$_5$)$_3$	33–35
1-CH$_3$-2-NH$_2$	303
1-C$_6$H$_5$-2-NH$_2$	98–99
Amides, azides, cyanides	
1-CONH$_2$	118.5–119
1-CH$_2$CONH$_2$	149–149.5
1-CH$_3$-2-CON$_3$	42–43
1-C$_6$H$_5$-2-CN	105–106

The nonicosahedral dicarbaboranes can be prepared from the icosahedral species by degradative procedures, or in some cases by reactions between boranes such as B$_4$H$_{10}$ and B$_5$H$_9$ with acetylenes. The degradative procedures for intermediate C$_2$B$_n$H$_{n+2}$ species (n = 6–9) have been described in detail by Dunks and Hawthorne (110) and are outlined schematically below.

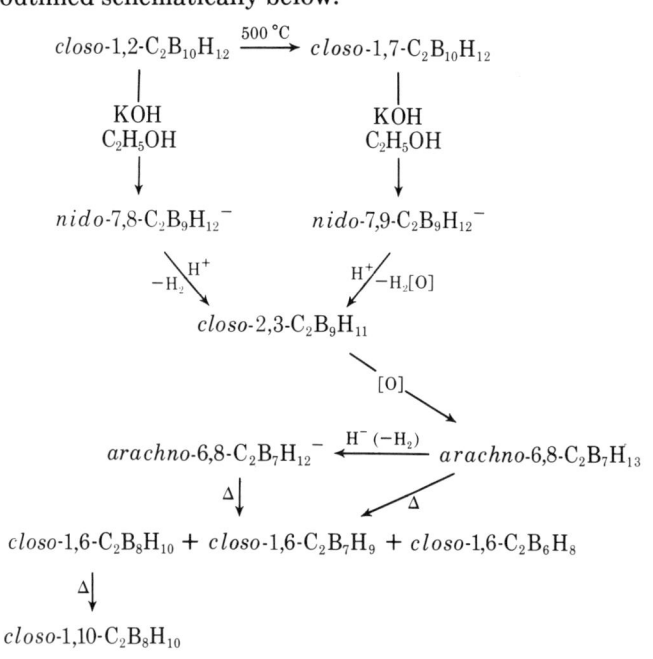

$$closo\text{-}1,2\text{-}C_2B_{10}H_{12} \xrightarrow{500\,°C} closo\text{-}1,7\text{-}C_2B_{10}H_{12}$$

$$\downarrow \text{KOH, C}_2\text{H}_5\text{OH} \qquad \downarrow \text{KOH, C}_2\text{H}_5\text{OH}$$

$$nido\text{-}7,8\text{-}C_2B_9H_{12}^- \qquad nido\text{-}7,9\text{-}C_2B_9H_{12}^-$$

$$\searrow_{-H_2}^{H^+} \qquad \swarrow_{-H_2[O]}^{H^+}$$

$$closo\text{-}2,3\text{-}C_2B_9H_{11}$$

$$\searrow [O]$$

$$arachno\text{-}6,8\text{-}C_2B_7H_{12}^- \xleftarrow{H^-\,(-H_2)} arachno\text{-}6,8\text{-}C_2B_7H_{13}$$

$$\Delta \downarrow \qquad \swarrow \Delta$$

$$closo\text{-}1,6\text{-}C_2B_8H_{10} + closo\text{-}1,6\text{-}C_2B_7H_9 + closo\text{-}1,6\text{-}C_2B_6H_8$$

$$\Delta \downarrow$$

$$closo\text{-}1,10\text{-}C_2B_8H_{10}$$

The small $closo$-C$_2$B$_n$H$_{n+2}$ species (n = 3–5) are obtained by the direct thermal reaction (500–600°C) of B$_5$H$_9$ and acetylene in a continuous-flow system. The combined yields approach 70% with the product distribution of 2,4-C$_2$B$_5$H$_7$ [20693-69-0] to 1,6-C$_2$B$_4$H$_6$ [20693-67-8] to 1,5-C$_2$B$_3$H$_5$ [20693-66-7] around 5:5:1 (111). A similar reaction involving the base catalyst 2,6-dimethylpyridine at ambient temperature gives $nido$-2,3-C$_2$B$_4$H$_8$ [21445-77-2] (112).

$$B_5H_9 + C_2H_2 \xrightarrow{\text{base}} 2,3\text{-}C_2B_4H_8 + \text{base}\cdot BH_3 \tag{59}$$

Pyrolysis of 2,3-C$_2$B$_4$H$_8$ is another, apparently more convenient route to 2,3-C$_2$B$_5$H$_7$ [30347-95-6], 1,2- [20693-68-9] and 1,6-C$_2$B$_4$H$_6$ [20693-67-8], and 1,5-C$_2$B$_3$H$_5$ [20693-66-7] (113).

Two *arachno* dicarbaboranes have been reported: 6,8-$C_2B_7H_{13}$ [17653-38-2] and 6,9-$C_2B_8H_{14}$ [38670-58-5] (30,114). Two of the extra hydrogens in these *arachno* carbaboranes are quite acidic and appear in CH_2 groups. The other two extra hydrogens bridge B—B edges.

It appears that most readily accessible carbaborane known is *nido*-7-(NH_3)-7-$CB_{10}H_{12}$ [12539-44-5], a zwitterionic species formally derived from $[CB_{10}H_{13}]^-$ by replacement of a H^- by NH_3. Knoth (115) found that the latter monocarbaborane is obtained in excellent yield by treatment of $B_{10}H_{14}$ with CN^- followed by passage through an acidic ion-exchange column (eq. 60).

$$B_{10}H_{14} + 2\,NaCN \xrightarrow[-HCN]{H_2O} Na_2[B_{10}H_{13}CN] \xrightarrow{H^+} CB_{10}H_{12}(NH_3) \quad (60)$$

Todd (116) has obtained the related mono-*N*-alkylated monocarbaboranes, 7-(NH_2R)-7-$CB_{10}H_{12}$, by treatment of decaborane(*14*) with alkyl isocyanides.

$$B_{10}H_{14} + RNC \rightarrow CB_{10}H_{12}(NH_2R) \quad (61)$$

The nitrogen of these aminocarboranes can be alkylated to give, eg, 7-[$N(CH_3)_3$]-7-$CB_{10}H_{12}$ [31117-16-5]. The latter upon treatment with Na followed by iodine oxidation gives [*closo*-2-$CB_{10}H_{11}$]$^-$ [38192-45-9], not [1-$CB_{10}H_{11}$]$^-$ [12539-21-8] as originally proposed on the basis of low-resolution 32.1 MHz ^{11}B nmr (117).

$$[CB_{10}H_{12}(NR_3)] \xrightarrow[THF]{Na} [CB_{10}H_{11}]^{3-} + NR_3 + \tfrac{1}{2}H_2 \xrightarrow{I_2} [CB_{10}H_{11}]^- + 2\,I^- \quad (62)$$

Other large monocarbaboranes include *nido*-6-(NR_3)-6-CB_9H_{11}, [*closo*-1-CB_9H_{10}]$^-$ [38192-43-7] and [*closo*-$CB_{11}H_{12}$]$^-$ [38192-46-0]. The *closo* monocarbaboranes can be functionalized at carbon through the intermediacy of lithium reagents just as for dicarbaboranes. Halogenation also has been reported. Metalloderivatives will be discussed below.

The small monocarbaboranes *closo*-1-CB_5H_7 [25301-90-0], *nido*-2-CB_5H_9 [12385-35-2], and a variety of their alkylated derivatives also are known (118–119).

Tetracarbaboranes were rare until Grimes' (120) discovery of a metallodicarbaborane-mediated synthesis of $(CH_3)_4C_4B_8H_8$ [58815-26-2]. (In this review, linked dicarbaborane clusters are not classified as tetracarbaboranes. A tetracarbaborane must have 4 carbons in a single skeleton.)

$$[2,3\text{-}(CH_3)_2\text{-}2,3\text{-}C_2B_4H_5]^- \xrightarrow[25\,°C]{CoCl_2} [(CH_3)_2C_2B_4H_4]_2CoH \xrightarrow{[O]} (CH_3)_4C_4B_8H_8 \xleftarrow{[O]} [(CH_3)_2C_2B_4H_4]_2FeH_2 \xleftarrow[-30\,°C]{FeCl_2} \quad (63)$$

The $C_4B_{n-4}H_n$ series of tetracarbaboranes is classified in the PERC formalism as *nido*. Therefore, these molecules are not expected to have closed structures even though extra hydrogens are absent. Microwave spectroscopy (121) has confirmed this expectation for 2,3,4,5-$C_4B_2H_6$ [28323-17-3]. One isomer of $(CH_3)_4C_4B_8H_8$ has the open nonicosahedral structure shown in Figure 18 and the other known isomer, the 1,2,3,8-tetramethyl compound [54387-54-1], is proposed to be even more open (122).

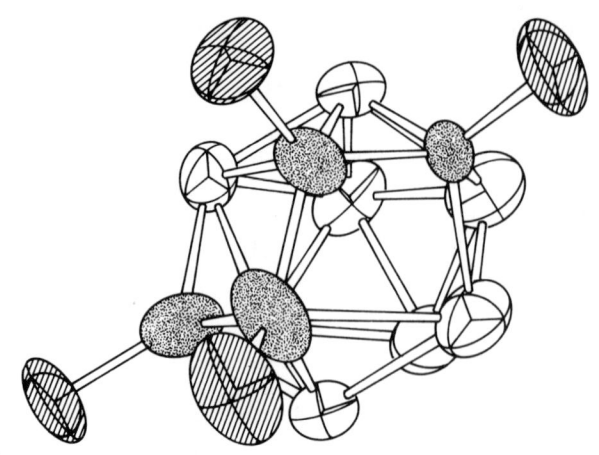

Figure 18. Structure of $(CH_3)_4C_4B_8H_8$ showing no hydrogen atoms (122). ○, BH; ⊘, CH_3; ●, C. Courtesy of the American Chemical Society.

Other 12-atom carborane cages with *nido* electron counts, eg, $[(C_6H_5)_2C_2B_{10}H_{11}]^-$ [*39322-84-4*], are also nonicosahedral (123). Theoretical considerations presented earlier in this article (10,49) predict that in general a 2-electron reduction of a *closo* molecule opens the skeleton. However, at least for $[C_2B_{10}H_{12}]^{2-}$ molecules, there appears to be some controversy regarding the latter contention vs the role of subsequent protonation in the opening of the icosahedron; ie, it is not known whether both $[C_2B_{10}H_{12}]^{2-}$ [*12542-25-5*] and $[C_2B_{10}H_{13}]^-$ [*56902-29-5*] are open.

Heteroboranes Other Than Carbaboranes. In principle, most heteroboranes could have a wide range of skeletal sizes; however, with the major exception of the carbaboranes, such a series is emerging only for the thiaboranes. Most of the other known heteroboranes have a 12- or 11-atom framework. The thiaboranes [*arachno*-6-$SB_9H_{12}]^-$ [*45979-10-0*] and *nido*-6-SB_9H_{11} [*59120-72-8*] were the first reported heteroboranes (other than carbaboranes) with 10-atom frameworks (124). The latter two thiadecaboranes appear to have icosahedral fragment structures, but they can be converted (eq. 64) to *closo*-1-thiadecaborane(9) [*41646-56-4*], 1-SB_9H_9, which was the first heteroborane with other than an icosahedral-fragment structure excluding the carbaboranes (23). The isolation of smaller thiaboranes 4-SB_8H_{10} [*59351-07-4*] and [4-$SB_8H_9]^-$ [*59366-23-3*] after degradation of 1-SB_9H_9 (eq. 65) presages the eventual preparation of an entire series of thiaboranes.

$$B_{10}H_{14} \xrightarrow[H_2O]{S_x^{2-}} [arachno\text{-}6\text{-}SB_9H_{12}]^- \xrightarrow[C_6H_6]{I_2} nido\text{-}6\text{-}SB_9H_{11} \xrightarrow{\Delta} closo\text{-}1\text{-}SB_9H_9 \quad (64)$$

$$1\text{-}SB_9H_9 \xrightarrow[(CH_3)OH]{KOH} [4\text{-}SB_8H_9]^- \underset{-H^+}{\overset{+H^+}{\rightleftharpoons}} 4\text{-}SB_8H_{10} \quad (65)$$

The larger thiaboranes are attainable by expansion reactions (eq. 66):

$$[6\text{-}SB_9H_{12}]^- \xrightarrow{\Delta, H^+} 7\text{-}SB_{10}H_{12} \xrightarrow{R_3NBH_3} SB_{11}H_{11} \quad (66)$$

Attempts to oxidize [*nido*-7-$SB_{10}H_{10}]^{2-}$ to *closo*-$SB_{10}H_{10}$ [*12421-68-0*] in aromatic solvents instead have led to a novel cage substitution reaction (eq. 67):

$$[nido\text{-}7\text{-}SB_{10}H_{10}]^{2-} + 2\,Ag^+ + ArH \rightarrow nido\text{-}2\text{-}Ar\text{-}7\text{-}SB_{10}H_{11} + 2\,Ag \quad (67)$$

Where ArH equals benzene, toluene, *n*-propylbenzene, and *tert*-butylbenzene. By contrast, oxidation of [6-$SB_9H_{12}]^-$ changes the framework electron count (eq. 64).

The thiaboranes 1-SB$_9$H$_9$ and SB$_{11}$H$_{11}$ [56464-75-6] undergo electrophilic alkylation, deuteration, and halogenation (125–126), but the site of substitution does not correlate with the ground-state charge distribution. Rearrangement plays an important role in the stereochemistry of substitution products. Under electrophilic conditions, initial attack appears to be at the lower belt (Fig. 19); however, the thermodynamically preferred axial isomer is obtained by intramolecular rearrangement if the workup or the reaction conditions are too strenuous.

$$1\text{-SB}_9\text{H}_9 + \text{EX} \xrightarrow{\text{AlCl}_3} 6\text{-E-1-SB}_9\text{H}_8 + \text{HX} \qquad (68)$$

E = X = Cl, Br, I; or E = D, CH$_3$, C$_2$H$_5$, and X = I. The thiaboranes [arachno-6-SB$_9$H$_{12}$]$^-$, closo-1-SB$_9$H$_9$, and closo-SB$_{11}$H$_{11}$ are quite resistant to moisture and air oxidation. The other thiaboranes mentioned above are moderately susceptible to air and moisture. Alcohols and wet solvents in which they are soluble effect a slow degradation of 1-SB$_9$H$_9$ and SB$_{11}$H$_{11}$. The substituted derivatives of 1-SB$_9$H$_9$ and SB$_{11}$H$_{11}$ are similar to their parents in reactivity towards air and moisture.

A reaction with potential for the preparation of a greater variety of organothiaboranes than that obtainable by electrophilic substitution appears to be the recently reported hydroboration reaction of 6-SB$_9$H$_{11}$ (127). Decaborane(14) and 6-SB$_9$H$_{11}$ are isoelectronic and similar in structure. As discussed above, B$_{10}$H$_{14}$ reacts with alkynes in donor solvents to yield C-substituted dicarbaboranes, C$_2$B$_{10}$H$_{10}$RR'. By contrast, 6-SB$_9$H$_{11}$ reacts rapidly and in high yield with alkynes in nondonor solvents to give B-substituted organothiaboranes (eq. 69).

$$6\text{-SB}_9\text{H}_{11} + \text{RC}\equiv\text{CR}' \rightarrow 9\text{-(RCH=CR')-6-SB}_9\text{H}_{10} \qquad (69)$$

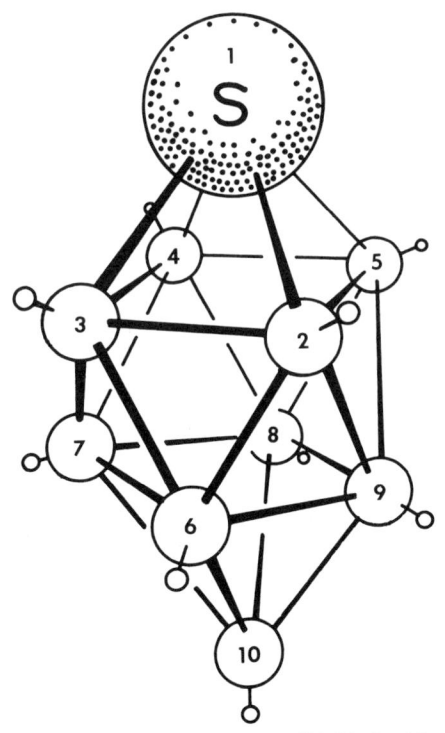

Figure 19. Structure and numbering convention for 1-SB$_9$H$_9$. Positions 2–5, 6–9, and 10 are designated upper-belt, lower-belt, and axial, respectively.

168   BORON COMPOUNDS (B-H, METALLO DERIVATIVES)

There is no evidence of insertion into the cage framework to give dicarbathiaboranes; the reaction is best termed a hydroboration which gives alkenylthiaboranes. Alkenes also undergo hydroboration to give alkylthiadecaboranes.

$$RR'C=CR''R''' + 6\text{-}SB_9H_{11} \rightarrow 9\text{-}(RR'CH-CR''R''')\text{-}6\text{-}SB_9H_{10} \qquad (70)$$

Of course, this means that certain alkenylthiadecaboranes undergo a second hydroboration to give bis(thiadecaboranyl)alkanes.

The pyrolysis of these *nido* thiadecaboranes leads to organo-substituted *closo* thiadecaboranes, R-1-SB$_9$H$_8$ (mixed isomers).

$$9\text{-}R\text{-}6\text{-}SB_9H_{10} \xrightarrow{\Delta} R\text{-}1\text{-}SB_9H_8 + H_2 \qquad (71)$$

It will be interesting to see what new dimensions hydroboration via thiaboranes brings to organic synthesis (see Hydroboration).

Table 8 is a selected list which gives an indication of the variety of heteroboranes other than carbaboranes that have been reported. Most of the entries shown in Table 8 form metalloderivatives, but except for the thiaboranes mentioned above, little is known about other potential substitution chemistry.

Metallo Analogues

In the early 1960s Hawthorne recognized the bonding similarities between the pentagonal face of the isomeric *nido*-C$_2$B$_9$H$_{11}$$^{2-}$ ions and the well-known cyclopentadienide ion [C$_5$H$_5$]$^-$ or Cp$^-$ (Fig. 20) (7).

Indeed, Hawthorne and co-workers have since established that [7,8-C$_2$B$_9$H$_{11}$]$^{2-}$ [*56902-43-3*], [7,9-C$_2$B$_9$H$_{11}$]$^{2-}$ [*39469-99-3*], and a variety of other *nido*-carbaborane anions form metallocene-like complexes with transition elements (see Organometallics). Other investigators naturally explored similar veins, and it now has been demonstrated that seemingly limitless varieties of metals and heteroatoms are capable of combination with boron in cluster compounds. In fact, the scope and variety of chemistry shown by the metalloderivatives of the boranes and heteroboranes probably exceeds that of the metallocenes. This review will give the reader selected examples of representative metalloderivatives. The clusters will be presented according to decreasing size with the metal(s) included in the count.

13- and 14-Atom Clusters.   Polyhedral expansion (reduction followed by metal

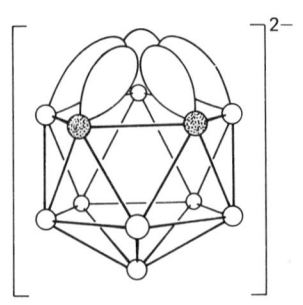

Figure 20.   Structural analogy between [7,8-C$_2$B$_9$H$_{12}$]$^{2-}$ and [C$_5$H$_5$]$^-$ (7). ○, BH; ●, CH. Courtesy of the American Chemical Society.

Table 8. A Selected List of Heteroboranes Including Elements Other than Carbon[a]

Heteroborane	CAS Registry Number	Ref.
Group V heteroatoms		
arachno-6-NB$_8$H$_{13}$	[58920-21-1]	128
nido-10-C$_6$H$_5$CH$_2$-7,8,10-C$_2$NB$_8$H$_{10}$	[58614-34-9]	128
nido-7-CH$_3$-7-PB$_{10}$H$_{12}$	[57108-87-9]	129
closo-C$_6$H$_5$PB$_{11}$H$_{11}$	[57139-68-1]	130
closo-1,7-CPB$_{10}$H$_{11}$	[17398-92-4]	130
[*nido*-7,8-CPB$_9$H$_{11}$]$^-$	[52110-38-0]	25
closo-1,2-As$_2$B$_{10}$H$_{10}$	[51292-90-1]	131
[*nido*-7-AsB$_{10}$H$_{12}$]$^-$	[51292-97-8]	131
closo-1,2-CAsB$_{10}$H$_{11}$	[23231-66-5]	132
Group VI heteroatoms		
[*arachno*-6-SB$_9$H$_{12}$]$^-$	[51358-27-1]	133
nido-4-SB$_8$H$_{10}$	[59351-07-4]	23
closo-1-SB$_9$H$_9$	[41646-56-4]	23
nido-7-SB$_{10}$H$_{12}$	[58984-44-4]	134
arachno-6,8-S$_2$B$_7$H$_9$	[63115-77-5]	135
arachno-6,8-CSB$_7$H$_{11}$	[63115-78-6]	135
nido-7-SeB$_{10}$H$_{12}$	[61649-90-9]	136
nido-7,8-Se$_2$B$_9$H$_9$	[61618-06-2]	136
nido-7-TeB$_{10}$H$_{12}$	[61649-91-0]	136

[a] The *closo*, *nido*, *arachno* classification is given on the basis of framework electron count and not structure.

insertion) of C$_2$B$_{10}$H$_{12}$ leads to supraicosahedral metallodicarbaboranes such as *closo*-13-Cp-13,11,7-CoC$_2$B$_{10}$H$_{12}$ [33340-90-8] (Fig. 21) (137). Further expansion of 13-vertex species leads to the 14-vertex cluster (CpCo)$_2$C$_2$B$_{10}$H$_{12}$ [52649-56-6] and [52649-57-7] (138). Similar 14-vertex species have been obtained from tetracarbaboranes (39) and show unusual structures (Fig. 5).

12-Atom Clusters. The first demonstration of the insertion of a transition metal into an open face of a borane was the reaction of iron(II) chloride with [7,8-C$_2$B$_9$H$_{11}$]$^{2-}$, which has sometimes been termed the dicarbollide ion [12541-50-3].

$$2\,[7,8\text{-}C_2B_9H_{11}]^{2-} + Fe^{2+} \rightarrow [(7,8\text{-}C_2B_9H_{11})_2Fe(II)]^{2-} \tag{72}$$

The product was readily air-oxidized to the complex containing a formal Fe(III), [(7,8-C$_2$B$_9$H$_{11}$)$_2$Fe(III)]$^-$ [12547-76-1]. The latter complexes as well as those formed from a variety of formally d^3, d^5, d^6, and d^7 transition metals (Table 9) have symmetrical sandwich structures (7) as shown in Figure 22. By contrast, with d^8 and d^9 metals the sandwich is slipped (Fig. 22). A theoretical rationale for these structural differences was discussed earlier.

The transition metals form a myriad of icosahedral dicarbollyl complexes as shown in part by equations 73 (139), 74 (140), and 75 (141):

$$[C_2B_9H_{11}]^{2-} + Cp^- + M^{2+} \rightarrow [CpM(II)C_2B_9H_{11}]^- \\ \downarrow [O] \\ CpM(III)C_2B_9H_{11} \tag{73}$$

$$M = Fe,\ Co,\ Ni,\ Cr$$

170   BORON COMPOUNDS (B-H, METALLO DERIVATIVES)

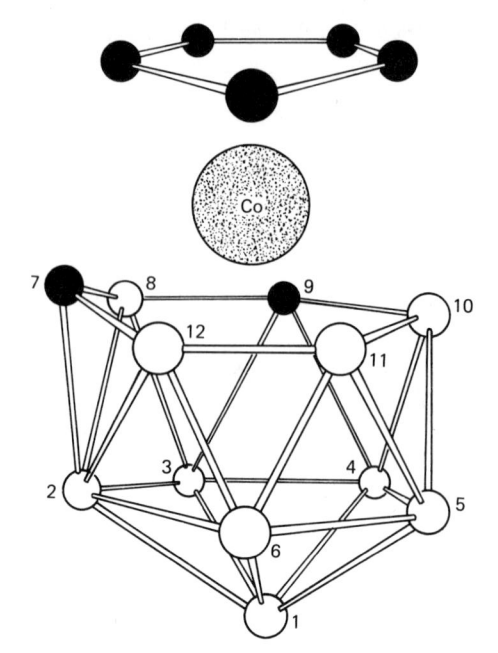

Figure 21. Structure of 13-(Cp)-13,7,9-CoC$_2$B$_{10}$H$_{12}$ (Cp = cyclopentadienyl) (137). ○, BH; ●, CH.

Table 9. Structure as a Function of *d* Electronic Configuration for [(C$_2$B$_9$H$_{11}$)$_2$M^{n+}]$^{n-4}$ Complexes

Symmetrical sandwich				Slipped sandwich	
d^3	d^5	d^6	d^7	d^8	d^9
Cr(III)	Fe(III)	Fe(II)	Co(II)	Cu(III)	Cu(II)
		Co(III)	Ni(III)	Ni(II)	Au(II)
		Ni(IV)	Pd(III)	Pd(II)	
		Pd(IV)		Au(III)	

$$[C_2B_9H_{11}]^{2-} + BrM(CO)_5 \xrightarrow{\Delta} [(C_2B_9H_{11})M(CO)_3]^- + Br^- + 2\ CO \quad (74)$$

$$M = Mn(I), Re(I)$$

$$[C_2B_9H_{11}]^{2-} + M(CO)_6 \xrightarrow{h\nu} 3\ CO + [(C_2B_9H_{11})M(CO)_3]^{2-} \quad (75)$$

$$M = Cr, Mo, W$$

Main-group metals also form *closo* complexes as shown in equations 76 (37), 77 (142), and 78 (34).

$$7,8\text{-}C_2B_9H_{13} + Be(CH_3)_2 \cdot 2(C_2H_5)_2O \rightarrow 3\text{-}[(C_2H_5)_2O]\text{-}3,1,2\text{-}BeC_2B_9H_{11} + O(C_2H_5)_2 + 2\ CH_4 \quad (76)$$

$$7,8\text{-}C_2B_9H_{13} + MR_3 \xrightarrow{\Delta} 3,1,2\text{-}MC_2B_9H_{11}R + 2\ RH$$

$$M = Al;\ R = CH_3, C_2H_5 \quad (77)$$

$$M = Ga;\ R = C_2H_5$$

$$[7,8\text{-}C_2B_9H_{11}]^{2-} + MX_2 \rightarrow 3,1,2\text{-}MC_2B_9H_{11} + 2\ X^- \quad (78)$$

$$M = Ge, Sn, Pb\ \text{with}\ X = I, Cl,\ \text{and}\ OAc^-,\ \text{respectively}$$

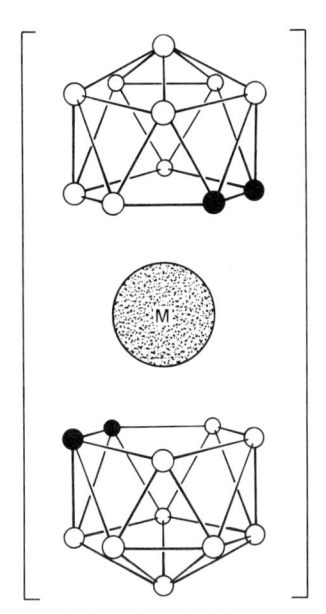

Figure 22. Generalized structure of slipped-sandwich derivatives. ○, BH; ●, CH. Courtesy of the American Chemical Society.

The isomeric germadicarbaborane [*27071-52-9*], 3,1,7-GeC$_2$B$_9$H$_{11}$, was obtained from GeI$_2$ and [7,9-C$_2$B$_9$H$_{11}$]$^{2-}$, but SnCl$_2$ effected an oxidative closure (13) rather than insertion (eq. 79).

$$[7,9\text{-}C_2B_9H_{11}]^{2-} + SnCl_2 \rightarrow Sn + 2,3\text{-}C_2B_9H_{11} + 2\,Cl^- \tag{79}$$

The group-IV metallodicarbaboranes are proposed to have a bare metal atom at one vertex of the icosahedron (35).

With the exception of those including Be, Al, and Ga, most of the *closo* metalloderivatives mentioned above show remarkable oxidative and hydrolytic stability.

Icosahedral metalloheteroboranes other than those based on the isomeric [C$_2$B$_9$H$_{11}$]$^{2-}$ ions are known. Ligands analogous to Cp$^-$ include [7-CB$_{10}$H$_{11}$]$^{3-}$, [7-PB$_{10}$H$_{11}$]$^{2-}$, 7,8- and [7,9-CEB$_9$H$_{10}$]$^{2-}$ (E = P, As), and [7-EB$_{10}$H$_{10}$]$^{2-}$ (E = S, Se, Te) which have been reported to form metalloderivatives with [Cr(III), Fe(III), Co(III), Co(IV), Ni(IV), Mn(IV), Mo(II), Ge(II)] (115,132), [Mn(I)] (129), [Fe(II), Fe(III), Co(II), Co(III), Mn(I), Ni(II)] (143), and [Fe(III), Co(III), Co(II), Re(I), Pt(II), Rh(III)] (124,144), respectively.

Nickelaboranes equivalent to [CB$_{11}$H$_{12}$]$^-$ and C$_2$B$_{10}$H$_{12}$ by the PERC formalism are known, *closo*-[CpNi(B$_{11}$H$_{11}$)]$^-$ and *closo*-1,2-(CpNi)$_2$B$_{10}$H$_{10}$ [*55266-88-1*], respectively. Although the latter complexes are the only reported examples of *closo* 12-atom metalloboranes (Fig. 23) they display remarkable hydrolytic, oxidative, and thermal stability, and other examples can be expected (145).

Perhaps the most interesting icosahedral metalloheteroboranes to be reported recently are 3,3-[P(C$_6$H$_5$)$_2$]$_2$-3-*H*-3,1,2-RhC$_2$B$_9$H$_{11}$ [*61250-52-0*] and 2,2-[P(C$_6$H$_5$)$_3$]$_2$-2-*H*-2,1-RhSB$_{10}$H$_{10}$ [*61796-42-7*] which have been demonstrated to function as homogeneous olefin isomerization and hydrogenation catalysts (144,146). Even when the rhodium carborane is immobilized by attachment to polystyrene, it retains appreciable catalytic activity (147–148). The catalytically active form of these

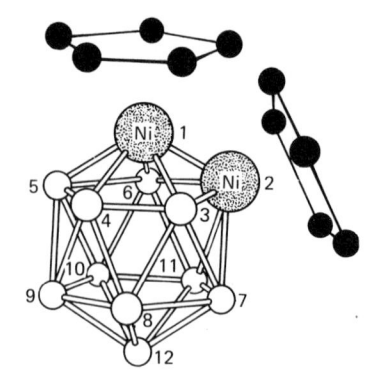

Figure 23. Proposed structure of 1,2-(CpNi)$_2$B$_{10}$H$_{10}$ (Cp = cyclopentadienyl) (149). ○, BH; ●, CH. Courtesy of the American Chemical Society.

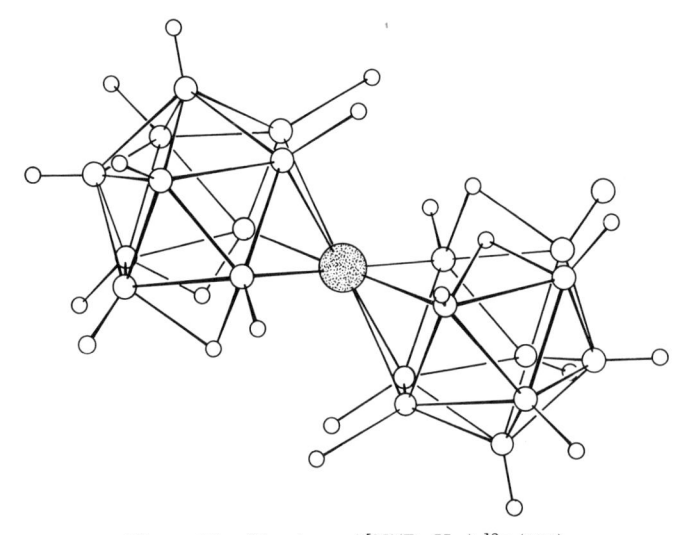

Figure 24. Structure of [Ni(B$_{10}$H$_{12}$)$_2$]$^{2-}$ (154).

complexes appears to require phosphine dissociation to give a coordinatively unsaturated rhodaheteroborane which is still immobilized by the polymer. The latter characteristic may be an important advantage of carborane-attached catalysts over phosphine-attached catalysts which leach metal to solution by phosphine dissociation (147–148).

11-Atom Clusters. A variety of heteroboranes are readily reduced from *closo* to *nido* molecules with a concomitant opening of the deltahedron (left column of Fig. 2). The open, nontriangular faces of the resulting *nido* anions appear to be ideally suited for coordination to metals. Since this sequence of reduction followed by metal insertion increases the size of the starting polyhedron by one vertex, it has been termed polyhedral expansion. The scheme (eqs. 80–81) has been reported most frequently for dicarbaboranes.

$$C_2B_nH_{n+2} + 2\,e^- \rightarrow [C_2B_nH_{n+2}]^{2-} \tag{80}$$
$$[C_2B_nH_{n+2}]^{2-} + M^x \rightarrow [(C_2B_nH_{n+2})_2M^x]^{x-4} \tag{81}$$

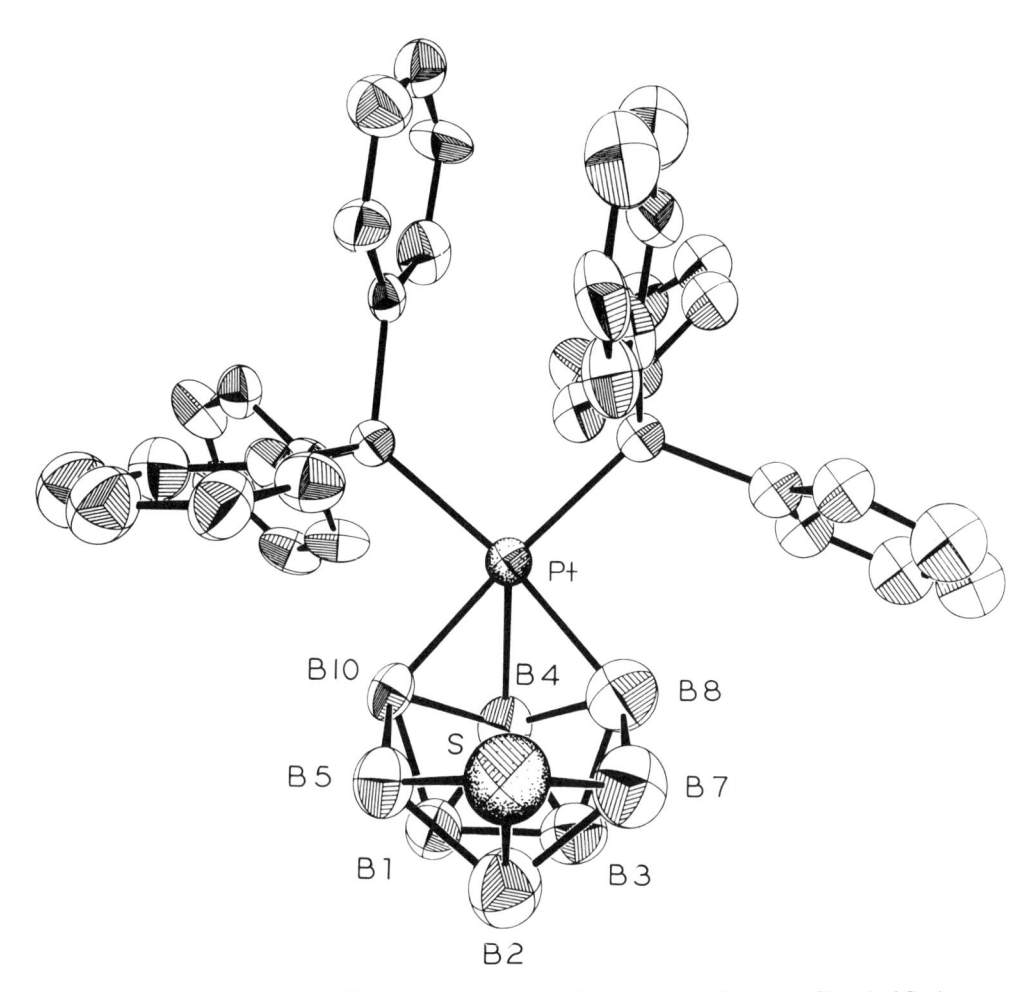

Figure 25. Structure of 9,9-[P(C$_6$H$_5$)$_3$]$_2$-6,9-SPtB$_8$H$_{10}$. Courtesy of the American Chemical Society.

However, it has also been reported for other heteroboranes and metallo analogues (to give multimetal heteroboranes). The preparation of 11-atom metallodicarbaboranes (149) illustrates the expansion synthetic strategy (eq. 82).

$$closo\text{-}1,6\text{-}C_2B_8H_{10} \xrightarrow[\text{2. CoCl}_2, \text{Cp}^-]{\text{1. Na/naphthalene}} closo\text{-}1\text{-}(CpCo)\text{-}2,3\text{-}C_2B_8H_{10} \quad (82)$$
$$[23704\text{-}81\text{-}6] \qquad\qquad\qquad [11072\text{-}14\text{-}3]$$

A variety of nido 11-atom clusters are known, including both main-group and transition metals. For instance, 7,7-(CH$_3$)$_2$-7-MB$_{10}$H$_{12}$ (M = Ge [34514-72-2], Sn [34514-71-1]) (150) and [7,7-(CH$_3$)$_2$-7-TlB$_{10}$H$_{12}$]$^-$ [34134-32-2] (151) have been reported. In most 11-atom *nido* metalloboranes it appears that the borane is effectively bidentate with the "bite" of the ligand coming from electron density between boron atoms 2,11 and 3,8 with the metal at position 7 (Fig. 1**n**). In the case of the [B$_{10}$H$_{12}$]$^{2-}$ ligand, the bridging hydrogens lie between positions 8,9 and 10,11. Typical complexes with [B$_{10}$H$_{12}$]$^{2-}$ include: [M(B$_{10}$H$_{12}$)$_2$]$^{2-}$ [M = Zn, Cd, Hg (152), Co, Ni, Pd, Pt (153)]; L$_2$M(B$_{10}$H$_{12}$) [M = Pd, Pt; L = PR$_3$ (153)]; and [L$_3$M(B$_{10}$H$_{12}$)]$^-$ [M = Co, Rh, Ir; L = CO, PR$_3$ (153)].

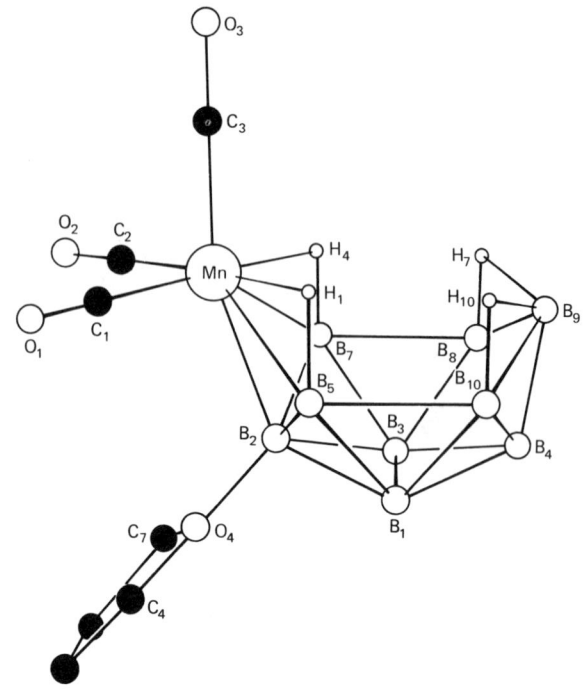

Figure 26. Structure of 2-(THF)-6,6,6-(CO)$_3$-6-MnB$_9$H$_{12}$, terminal hydrogens omitted (THF = tetrahydrofuran) (161). Courtesy of the American Chemical Society.

The x-ray structure (154) of [Ni(B$_{10}$H$_{12}$)$_2$]$^{2-}$ is shown in Figure 24. If the [B$_{10}$H$_{12}$]$^{2-}$ ligand is bidentate, then the coordination about Ni is effectively square planar. This effective geometry varies according to the metal and is tetrahedral about Zn in [Zn(B$_{10}$H$_{12}$)$_2$]$^{2-}$ (155).

The [B$_{10}$H$_{13}$]$^-$ ion is especially useful for the synthesis of 11-atom *nido* metalloboranes. The syntheses are influenced by: (*1*) the cation of the [B$_{10}$H$_{13}$]$^-$ [36928-50-4] salt; (2) the nature of other ligands about the metal; and (3) the availability of a proton trap for the reaction (153,156).

10-Atom Clusters. The precursor to several *closo* 10-atom cobaltadicarbaboranes is [*arachno*-6,8-C$_2$B$_7$H$_{11}$]$^{2-}$ [42319-46-0] which is obtained by deprotonation of 6,8-C$_2$B$_7$H$_{13}$ [17653-38-2]. When treated with excess CoCl$_2$ and cyclopentadienide ion, [6,8-C$_2$B$_7$H$_{11}$]$^{2-}$ gives *closo*-CpCo(C$_2$B$_7$H$_9$) [37381-23-0 and 51539-00-5] (eq. 83) which exists as both *closo*-2-(CpCo)-1,6-C$_2$B$_7$H$_9$ [41348-11-2] and *closo*-2-(CpCo)-1,10-C$_2$B$_7$H$_9$ [42808-86-6] (30).

$$[C_2B_7H_{11}]^{2-} + Cp^- + \tfrac{3}{2} Co^{2+} \rightarrow H_2 + \tfrac{1}{2} Co + CpCo(C_2B_7H_9) \qquad (83)$$

Other interesting 10-atom *closo* species include *closo*-10-(CpNi)-1-CB$_8$H$_9$ [52540-76-8] (157) and *closo*-[1-(CpNi)B$_9$H$_9$]$^-$ [57034-64-7] (158) in which cases the metal is bound to a B$_4$-face. In the latter case, the 2-isomer, where Ni is bound to 5 borons, is obtained initially and then thermally isomerized to the 1-position. In the case of another molecule with a *closo* electron count, the x-ray structure of 2,7-(CH$_3$)$_2$-9,9-[P(C$_2$H$_5$)$_3$]$_2$-2,7,9-C$_2$PtB$_7$H$_7$ [54203-62-2 and 58320-38-0] shows major distortions from the anticipated deltahedral structure (159) which are manifested

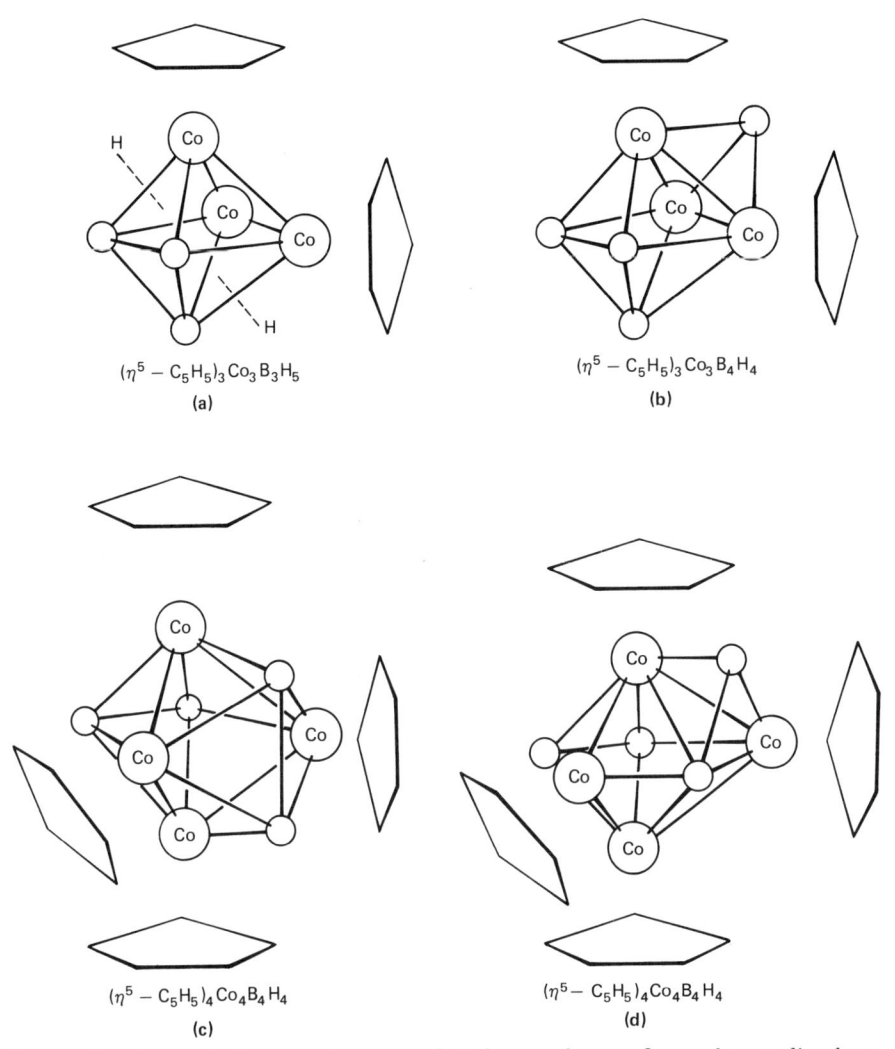

Figure 27. Proposed structures for some cobaltaborane clusters. One cyclopentadienyl group is omitted from (a) and (b) for clarity. Open circles are BH groups (166). Courtesy of the American Chemical Society.

primarily in a nonbonding Pt–C distance of 0.283 nm. Similar distortions are observed for other platinacarboranes.

Nido ten-atom metalloboranes appear to have the boat-like framework shown in Figures 25 and 26 with the metal at a prow, 6- or 9-, position. However, the nature of the bonding depends upon the metal. Examples include 6,6'-[(C$_2$H$_5$)$_3$P]$_2$-9-L-6-PtB$_9$H$_{11}$ (L = phosphine, amine, nitrile, or sulfide) (160) and 9,9'-(L)$_2$-6,9-SPtB$_8$H$_{10}$ [L = P(C$_6$H$_5$)$_3$, P(CH$_3$)$_2$C$_6$H$_5$, P(C$_2$H$_5$)$_3$] (42) where the Pt achieves effective square–planar coordination with a B—B—B edge acting as a η^3 bidentate (Fig. 25). In the case of complexes such as 2-(THF)-6,6,6-(CO)$_3$-6-MnB$_9$H$_{12}$ [51801-39-9] or 5-(THF)-6,6,6-(CO)$_3$-6,6,6-MnB$_9$H$_{12}$ [51801-38-8], where THF equals tetrahydrofuran (Fig. 26) the borane ligand is effectively tridentate, bound to Mn through two Mn—H—B bridges and a B—Mn bond (161).

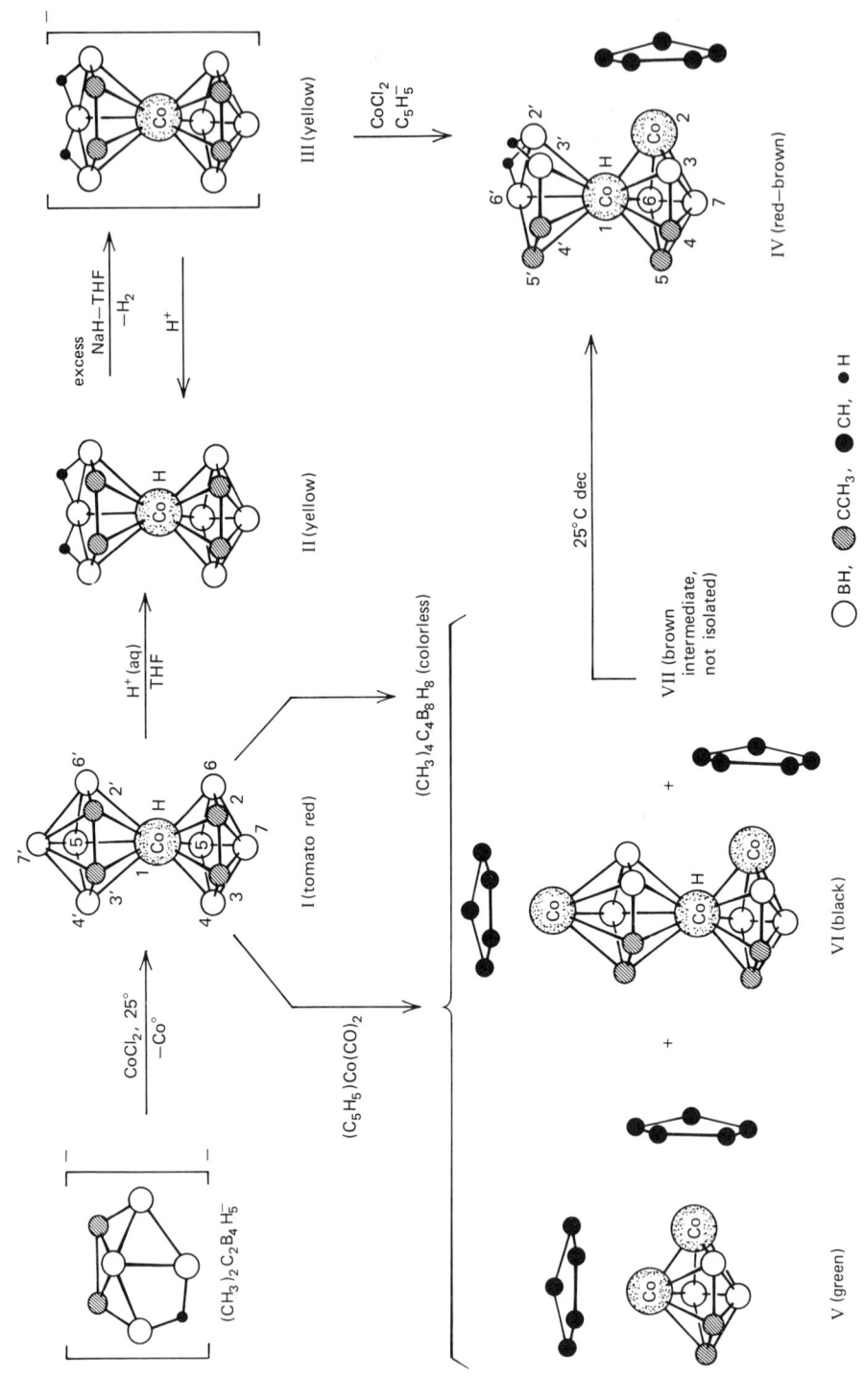

Figure 28. Reaction schemes and proposed structures of some smaller cobaltacarboranes (167). Courtesy of the American Chemical Society.

5, 6, 7, 8 and 9-Atom Clusters. Since a wide variety of cluster sizes is often obtained from the same reaction, and since there are fewer reports of the smaller metalloboranes, they will be treated in one section. The expected tricapped trigonal prism structure of a 9-atom *closo* cluster is observed for [2-(CO)$_3$-1,2,6-CMnCB$_6$H$_8$]$^-$ *[41267-49-6]* (162). Similar structures have since been proposed and found for (CpNi)$_2$C$_2$B$_5$H$_7$ *[51108-05-5]* (163) and 6,8-(CH$_3$)$_2$-1,1-[P(CH$_3$)$_3$]$_2$-1,6,8-PtC$_2$B$_6$H$_6$ *[55888-13-6]* (164), respectively.

Grimes has uncovered a number of novel products by careful analysis of the reaction mixture obtained when [B$_5$H$_8$]$^-$ *[31426-87-6]* is treated with CoCl$_2$ and Cp$^-$ in THF (165–166). The predominant product is *nido*-2-(CpCo)-B$_4$H$_8$ *[43061-99-0]*, and there are clusters containing up to 4 cobalt atoms, (CpCo)$_4$B$_4$H$_8$ *[59370-82-0]*. Characterization gives an unusual 2n framework electron count and favors some cluster geometries unprecedented for boranes. The geometries (Fig. 27) are reminiscent of strictly metallic clusters as might be expected (10,166).

The variety of structures possible for smaller metallodicarbaboranes and various synthetic strategies are shown in part in Figure 28 (167). Especially noteworthy is the occurrence of triple-decked sandwich compounds. The polyhedral expansion synthetic strategy can also be used with small dicarbaboranes (163) as shown in Figure 29. However, the reaction is not clean as evidenced by the formation of some naphthyl derivatives. Another technique used for small metalloboranes is a hot–cold reactor which can be used to prepare volatile species such as *closo*-1,1,1-(CO)$_3$-1,2,3-FeC$_2$B$_4$H$_6$

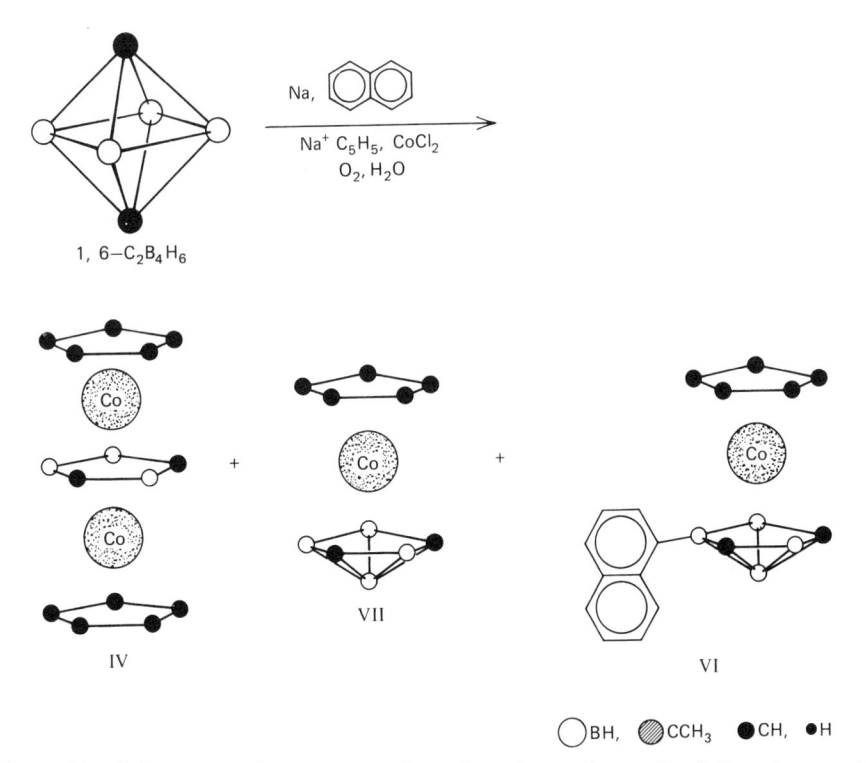

Figure 29. Other proposed structures and reaction schemes for small cobaltacarboranes (163). Courtesy of the American Chemical Society.

178 BORON COMPOUNDS (B-H, METALLO DERIVATIVES)

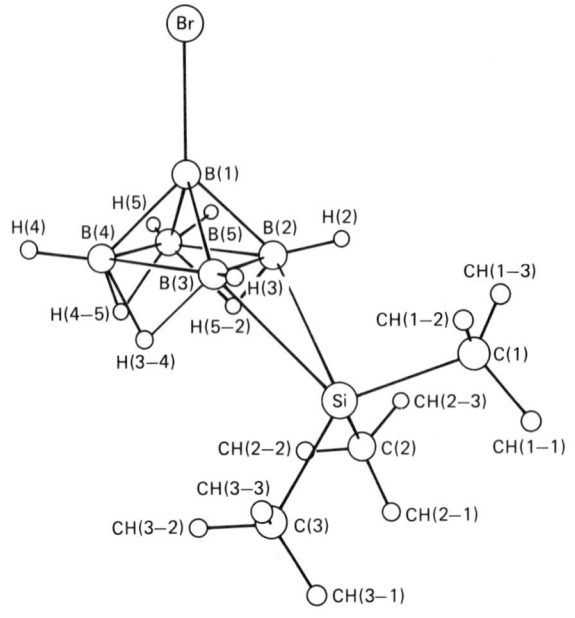

Figure 30. Structure of 1-Br-μ-[(CH$_3$)$_3$Si]B$_5$H$_7$ (171). Courtesy of the American Chemical Society.

[32761-40-3] (168) and *nido*-2,2,2-(CO)$_3$-2-FeB$_5$H$_9$ [61403-41-6] (169) from C$_2$B$_4$H$_8$ and B$_5$H$_9$, respectively, upon treatment with Fe(CO)$_5$.

The smaller boranes and dicarbaboranes have also been shown to form a wide variety of metalloderivatives in which the metal occupies a bridging position between two boron atoms, not a polyhedral vertex. An extensively studied system is μ-R$_3$MB$_5$H$_8$, where R = H, CH$_3$, C$_2$H$_5$, halogen, and M = Si, Ge, Sn, Pb (170). The structure of 1-Br-μ-[(CH$_3$)$_3$Si]-B$_5$H$_8$ [28323-19-5] is shown in Figure 30 (171).

Exopolyhedral Metalloboranes. Metalloboranes with an exopolyhedral M—B σ-bond have been prepared by nucleophilic displacement reactions (172) (eq. 84) and oxidative addition (173) of B—H and B—Br bonds to the metal center (eq. 85). The structure of an Ir complex is shown in Figure 31.

$$\text{Na[M(CO)}_5\text{]} + 1\text{- or 2-Cl-B}_5\text{H}_8 \rightarrow \text{NaCl} + 2\text{-[(CO)}_5\text{M]B}_5\text{H}_8 \tag{84}$$

M = Mn [7439-96-5], Re [7440-16-6]

$$\text{IrCl(CO)[P(CH}_3\text{)}_3\text{]}_2 + 1\text{- or 2-BrB}_5\text{H}_8 \rightarrow 2\text{-[IrBr}_2\text{(CO)[P(CH}_3\text{)}_3\text{]}_2\text{B}_5\text{H}_8$$
[53221-42-4] \hfill (85)

Other examples of oxidative addition to the B—H of *closo* heteroboranes include 3-[[P(C$_6$H$_5$)$_3$]$_2$IrHCl]-1,2-C$_2$B$_{10}$H$_{11}$ [54330-42-6] (174) and 2-[[P(C$_6$H$_5$)$_3$]$_2$IrHCl]-1-SB$_9$H$_8$ [62337-20-6] (175).

It has also been shown that metals can be bound to polyhedral boranes through exo B—H—M bonds as in the case of [[(C$_6$H$_5$)$_3$P]$_2$Cu]$_2$-μ-B$_{10}$H$_{10}$ [54020-26-7] which is shown in Figure 32 (63). In the latter case, the metal centers are bound through a bidentate Cu—H—B—B—H chelate ring.

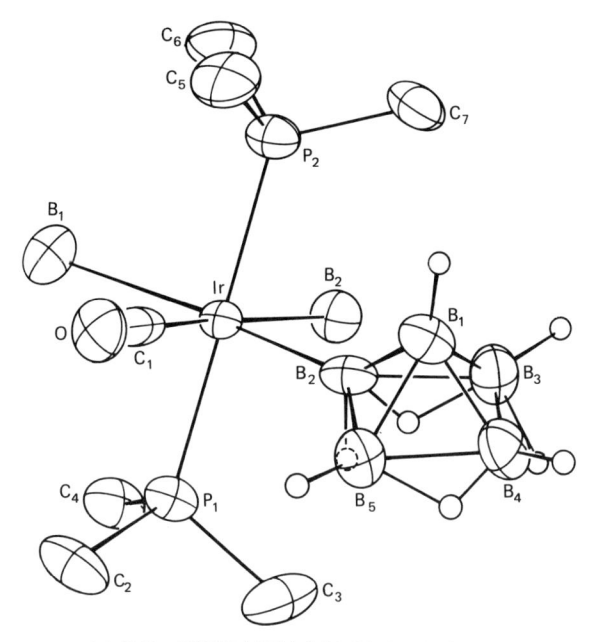

Figure 31. Structure of 2-[IrBr$_2$(CO)[P(CH$_3$)$_3$]$_2$]B$_5$H$_8$ (173). Courtesy of the American Chemical Society.

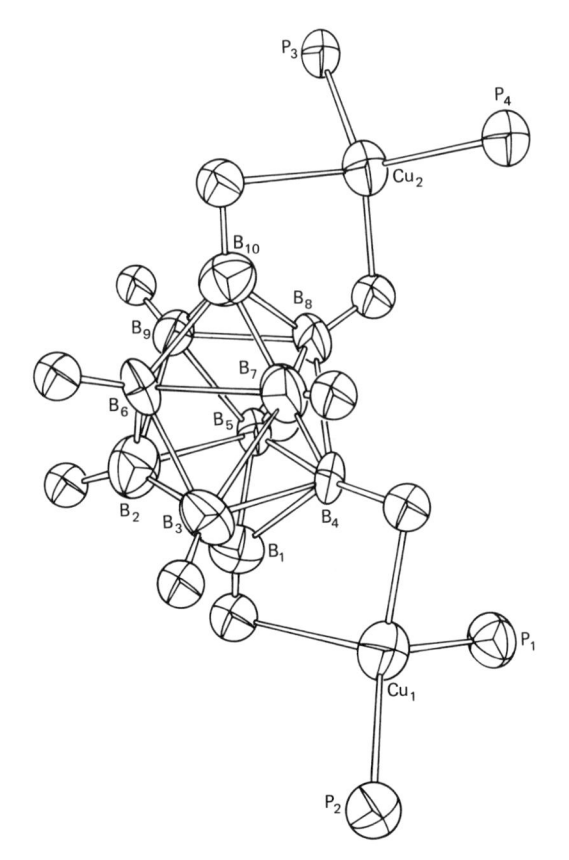

Figure 32. Structure of [[(C$_6$H$_5$)$_3$P]$_2$Cu]$_2$-μ-B$_{10}$H$_{10}$ (63). Courtesy of the American Chemical Society.

BIBLIOGRAPHY

"Boron Hydrides" under "Boron Compounds" in *ECT* 1st ed., Vol. 2, pp. 593–600, by S. H. Bauer, Cornell University; "Boron Hydrides and Related Compounds" under "Boron Compounds," Suppl. 1, pp. 103–130, by S. H. Bauer, Cornell University; "Diborane and Higher Boron Hydrides" under "Boron Compounds," Suppl. 2, pp. 109–113, by W. J. Shepherd and E. B. Ayres, Callery Chemical Company; "Boron Hydrides" under "Boron Compounds" in *ECT* 2nd ed., Vol. 3, pp. 684–706, by G. W. Campbell, Jr., U.S. Borax Research Corporation.

1. N. V. Sidgwick, *The Chemical Elements and Their Compounds*, Vol. I, Clarendon, London, Eng., 1950, p. 338.
2. A. Stock, *Hydrides of Boron and Silicon*, Cornell University Press, Ithaca, N. Y., 1933.
3. H. C. Brown, *Hydroboration*, W. A. Benjamin, Inc., New York, 1962.
4. W. N. Lipscomb, *Science* **196,** 1047 (1977).
5. E. L. Muetterties and W. H. Knoth, *Polyhedral Boranes,* Marcel Dekker, Inc., New York, 1968.
6. R. N. Grimes, *Carboranes,* Academic Press, Inc., New York, 1970.
7. M. F. Hawthorne, *Acc. Chem. Res.* **1,** 281 (1968).
8. *Pure Appl. Chem.* **30,** 683 (1972); *Inorg. Chem.* **7,** 1945 (1968).
9. R. E. Williams, *Inorg. Chem.* **10,** 210 (1971).
10. R. W. Rudolph, *Acc. Chem. Res.* **9,** 446 (1976).
11. K. Wade, *Adv. Inorg. Chem. Radiochem.* **18,** 1 (1976).
12. W. J. Evans, G. B. Dunks, and M. F. Hawthorne, *J. Am. Chem. Soc.* **95,** 4565 (1973).
13. R. W. Rudolph and co-workers, *J. Am. Chem. Soc.* **95,** 4560 (1973).
14. B. Lockman and T. Onak, *J. Am. Chem. Soc.* **94,** 7923 (1972).
15. W. N. Lipscomb and co-workers, *J. Chem. Phys.* **27,** 200 (1957).
16. P. C. Keller, *Inorg. Chem.* **9,** 75 (1970).
17. I. Shapiro, C. D. Good, and R. E. Williams, *J. Am. Chem. Soc.* **84,** 3837 (1962).
18. E. L. Muetterties and co-workers, *Inorg. Chem.* **6,** 1271 (1967).
19. R. N. Grimes and W. J. Rademaker, *J. Am. Chem. Soc.* **91,** 6498 (1969).
20. M. F. Hawthorne and co-workers, *Inorg. Chem.* **8,** 1907 (1969).
21. W. H. Knoth, *J. Am. Chem. Soc.* **89,** 1274 (1967).
22. F. N. Tebbe, P. M. Barret, and M. F. Hawthorne, *J. Am. Chem. Soc.* **86,** 4222 (1964).
23. W. R. Pretzer and R. W. Rudolph, *J. Am. Chem. Soc.* **98,** 1441 (1976).
24. R. L. Voorhees and R. W. Rudolph, *J. Am. Chem. Soc.* **91,** 2173 (1969).
25. L. J. Todd, J. L. Little, and H. T. Silverstein, *Inorg. Chem.* **8,** 1698 (1969).
26. D. A. Franz and R. N. Grimes, *J. Am. Chem. Soc.* **92,** 1438 (1970).
27. R. N. Grimes, C. L. Bramlett, and R. L. Vance, *Inorg. Chem.* **7,** 1066 (1968).
28. P. Binger, *Tetrahedron Lett.*, 2675 (1966).
29. K. Borer, A. B. Littlewood, and C. S. G. Phillips, *J. Inorg. Nucl. Chem.* **15,** 316 (1960).
30. P. M. Garrett, T. A. George, and M. F. Hawthorne, *Inorg. Chem.* **8,** 2008 (1969).
31. S. G. Shore and co-workers, *J. Am. Chem. Soc.* **98,** 449 (1976); **96,** 3014 (1974).
32. G. Kodama and A. R. Dodds, paper presented at meeting, Imeboron III, Ettal, Ger., July 6, 1976.
33. H. Hogeveen and P. W. Kwant, *Acc. Chem. Res.* **8,** 413 (1975).
34. R. W. Rudolph, R. L. Voorhees, and R. E. Cochoy, *J. Am. Chem. Soc.* **92,** 3351 (1970).
35. R. W. Rudolph and V. Chowdhry, *Inorg. Chem.* **13,** 248 (1974).
36. D. M. P. Mingos, *Nature Phys. Sci.* **236,** 99 (1972).
37. G. Popp and M. F. Hawthorne, *Inorg. Chem.* **10,** 391 (1971).
38. D. F. Dustin and co-workers, *J. Am. Chem. Soc.* **96,** 3085 (1974).
39. R. N. Grimes and co-workers, *J. Am. Chem. Soc.* **99,** 4016 (1977).
40. M. G. H. Wallbridge and co-workers, *J. Chem. Soc. Chem. Commun.*, 1019 (1976).
41. F. G. A. Stone and co-workers, *J. Chem. Soc. Chem. Commun.*, 804 (1975).
42. D. A. Thompson, T. K. Hilty, and R. W. Rudolph, *J. Am. Chem. Soc.* **99,** 6774 (1977).
43. R. E. Williams, *Adv. Inorg. Chem. Radiochem.* **18,** 67 (1976).
44. R. Schaeffer and co-workers, *J. Chem. Soc. Chem. Commun.*, 474 (1972).
45. A. R. Siedel, G. M. Bodner, and L. J. Todd, *J. Inorg. Nucl. Chem.* **33,** 3671 (1971).
46. V. T. Brice, H. D. Johnson, II, and S. G. Shore, *J. Am. Chem. Soc.* **95,** 6629 (1973).
47. R. A. Beaudet and co-workers, *J. Chem. Soc. Chem. Commun.*, 765 (1974).
48. D. A. Franz, V. R. Miller, and R. N. Grimes, *J. Am. Chem. Soc.* **94,** 412 (1972).
49. B. J. Meneghelli and R. W. Rudolph, *Inorg. Chem.* **14,** 1429 (1975).
50. M. G. H. Wallbridge, *Prog. Inorg. Chem.* **11,** 99 (1970).

51. T. J. Marks and J. R. Kolb, *Chem. Rev.* **77,** 263 (1977).
52. P. L. Timms, *J. Am. Chem. Soc.* **90,** 4585 (1968).
53. R. A. Jacobson and W. N. Lipscomb, *J. Chem. Phys.* **31,** 605 (1959).
54. M. Atoji and W. N. Lipscomb, *J. Chem. Phys.* **21,** 172 (1953).
55. W. N. Lipscomb, *Boron Hydrides,* W. A. Benjamin, Inc., New York, 1963.
56. G. Kodama and co-workers, *J. Am. Chem. Soc.* **94,** 407 (1972).
57. R. M. Wing, *J. Am. Chem. Soc.* **90,** 4828 (1968).
58. C. Glidewell, *J. Organometal. Chem.* **102,** 339 (1975).
59. M. F. Hawthorne and co-workers, *J. Am. Chem. Soc.* **97,** 296 (1975).
60. L. J. Guggenberger, *J. Am. Chem. Soc.* **94,** 114 (1972).
61. D. St. Clair, A. Zalkin, and D. H. Templeton, *Inorg. Chem.* **10,** 2587 (1971).
62. P. Wegner in E. Muetterties, ed., *Boron Hydride Chemistry,* Academic Press, Inc., 1975, Chapt. 12.
63. J. T. Gill and S. Lippard, *Inorg. Chem.* **14,** 751 (1975).
64. S. G. Shore in E. Muetterties, ed., *Boron Hydride Chemistry,* Academic Press, Inc., 1975, Chapt. 3.
65. H. C. Longuet-Higgins, *Q. Rev. Chem. Soc.* **11,** 121 (1957).
66. W. H. Eberhardt, B. L. Crawford, Jr, and W. N. Lipscomb, *J. Chem. Phys.* **22,** 989 (1954).
67. R. W. Rudolph and D. A. Thompson, *Inorg. Chem.* **13,** 2779 (1974).
68. R. G. Pearson, *J. Am. Chem. Soc.* **91,** 4947 (1969).
69. R. W. Rudolph and W. R. Pretzer, *Inorg. Chem.* **11,** 1974 (1972).
70. E. L. Muetterties and B. F. Beier, *Bull. Soc. Chim. Belg.* **84,** 397 (1975).
71. R. T. Holzmann, ed., *Production of Boranes and Related Research,* Academic Press, Inc., New York, 1967.
72. R. W. Parry and L. J. Edwards, *J. Am. Chem. Soc.* **81,** 3554 (1959).
73. R. J. Pace, J. Williams, and R. L. Williams, *J. Chem. Soc.,* 2196 (1961).
74. G. A. Guter and G. W. Schaeffer, *J. Am. Chem. Soc.* **78,** 3546 (1956); M. F. Hawthorne and co-workers, *J. Am. Chem. Soc.* **82,** 1825 (1960).
75. H. D. Johnson, II, and S. G. Shore, *J. Am. Chem. Soc.* **92,** 7586 (1970).
76. E. L. Muetterties and co-workers, *Inorg. Chem.* **1,** 734 (1962).
77. H. D. Johnson, II and S. G. Shore, *J. Am. Chem. Soc.* **93,** 3798 (1971).
78. D. Gaines and T. V. Iorns, *J. Am. Chem. Soc.* **92,** 4571 (1970).
79. J. A. Dupont and M. F. Hawthorne, *J. Am. Chem. Soc.* **84,** 1804 (1962).
80. L. I. Zakharkin and V. N. Kalinin, *J. Gen. Chem. USSR* **36,** 2154 (1966).
81. R. L. Williams, I. Dunstan, and N. J. Blay, *J. Chem. Soc.,* 5006 (1960).
82. D. Gaines and J. A. Martens, *Inorg. Chem.* **7,** 704 (1968).
83. S. G. Shore and co-workers, *J. Am. Chem. Soc.* **94,** 6711 (1972).
84. E. L. Muetterties and co-workers, *Inorg. Chem.* **3,** 444 (1964).
85. F. Klanberg and E. L. Muetterties, *Inorg. Chem.* **5,** 1955 (1966).
86. J. L. Boone, *J. Am. Chem. Soc.* **86,** 5036 (1964).
87. H. C. Miller, M. E. Miller, and E. L. Muetterties, *Inorg. Chem.* **3,** 1456 (1964).
88. E. L. Muetterties, ed., *Inorg. Synth.* **10,** 82 (1967).
89. W. L. Jolly, ed., *Inorg. Synth.* **11,** 27 (1968).
90. J. M. Makhlouf, W. V. Hough, and G. T. Hefferan, *Inorg. Chem.* **6,** 1196 (1967).
91. Ref. 88, p. 88.
92. S. Y. Tyree, Jr., ed., *Inorg. Synth.* **9,** 16 (1967).
93. E. L. Muetterties and W. H. Knoth, *Polyhedral Boranes,* Marcel Dekker, Inc., New York, 1968, p. 130.
94. W. H. Knoth, *J. Am. Chem. Soc.* **88,** 935 (1966).
95. W. H. Knoth and co-workers, *J. Am. Chem. Soc.* **89,** 4842 (1967).
96. W. H. Knoth and co-workers, *Inorg. Chem.* **3,** 159 (1964).
97. A. Kaczmarczyk and co-workers, *Inorg. Chem.* **7,** 1057 (1968).
98. M. F. Hawthorne and co-workers, *J. Am. Chem. Soc.* **85,** 3704 (1963); R. L. Middaugh and F. Farha, Jr., *J. Am. Chem. Soc.* **88,** 4147 (1966).
99. C. H. Schwalbe and W. N. Lipscomb, *Inorg. Chem.* **10,** 151 (1971).
100. R. J. Wiersema and R. L. Middaugh, *J. Am. Chem. Soc.* **92,** 223 (1970).
101. E. M. Fedneva, V. I. Alpatova, and V. I. Mikheeva, *Russ. J. Inorg. Chem.* **9,** 826 (1964).
102. V. I. Mikheeva and S. M. Arkhipev, *Russ. J. Inorg. Chem.* **11,** 805 (1966).
103. V. I. Mikeeva, M. S. Selivokhina, and G. N. Kryukova, *Russ. J. Inorg. Chem.* **1,** 838 (1962).
104. S. C. Abrahams and J. Kalnajs, *J. Chem. Phys.* **22,** 434 (1954).

105. W. D. Davis, L. S. Mason, and G. Stegman, *J. Am. Chem. Soc.* **71**, 2775 (1949).
106. M. D. Banus, R. W. Bragdon, and A. A. Hinckley, *J. Am. Chem. Soc.* **76**, 3848 (1954).
107. A. P. Altschuler, *J. Am. Chem. Soc.* **77**, 5455 (1955).
108. M. B. Smith and G. E. Bass, *J. Chem. Eng. Data* **8**, 342 (1963).
109. H. I. Schlesinger and co-workers, *J. Am. Chem. Soc.* **75**, 199 (1953).
110. G. B. Dunks and M. F. Hawthorne, *Acc. Chem. Res.* **6**, 124 (1973).
111. J. F. Ditter and co-workers, *Inorg. Chem.* **9**, 889 (1970).
112. T. Onak, F. J. Gerhart, and R. E. Williams, *J. Am. Chem. Soc.* **85**, 3378 (1963).
113. T. Onak, R. P. Drake, and G. B. Dunks, *Inorg. Chem.* **3**, 1686 (1964).
114. B. Stibr, J. Plesek, and S. Hermanek, *Coll. Czech. Chem. Commun.* **39**, 1805 (1974).
115. W. H. Knoth, *Inorg. Chem.* **10**, 598 (1971).
116. L. J. Todd and co-workers, *Inorg. Chem.* **6**, 2229 (1967).
117. R. J. Wiersema and M. F. Hawthorne, *Inorg. Chem.* **12**, 785 (1973).
118. G. B. Dunks and M. F. Hawthorne, *Inorg. Chem.* **8**, 2667 (1969).
119. T. Onak and co-workers, *Inorg. Chem.* **10**, 2770 (1971).
120. W. M. Maxwell, V. R. Miller, and R. N. Grimes, *J. Am. Chem. Soc.* **96**, 7116 (1974).
121. J. P. Pasinski and R. A. Beaudet, *J. Chem. Phys.* **61**, 683 (1974).
122. R. N. Grimes and co-workers, *Inorg. Chem.* **16**, 1847 (1977).
123. E. I. Tolpin and W. N. Lipscomb, *Inorg. Chem.* **12**, 2257 (1973).
124. W. R. Hertler, F. Klanberg, and E. L. Muetterties, *Inorg. Chem.* **6**, 1696 (1967).
125. R. W. Rudolph and co-workers, *Inorg. Chem.* **16**, 3008 (1977).
126. B. J. Meneghelli and R. W. Rudolph, *J. Organomet. Chem.* **133**, 139 (1977).
127. B. J. Meneghelli and R. W. Rudolph, *J. Am. Chem. Soc.* **100**, in press (1978).
128. J. Plesek and co-workers, *J. Chem. Soc. Chem. Commun.*, 934–935 (1975).
129. J. L. Little and A. C. Wong, *J. Am. Chem. Soc.* **93**, 522 (1971).
130. J. L. Little, J. T. Moran, and L. J. Todd, *J. Am. Chem. Soc.* **89**, 5495 (1967).
131. J. L. Little, S. S. Pao, and K. K. Sugathan, *Inorg. Chem.* **13**, 1752 (1974).
132. L. J. Todd and co-workers, *J. Am. Chem. Soc.* **91**, 3376 (1969).
133. W. R. Hertler, F. Klanberg, and E. L. Muetterties, *Inorg. Chem.* **6**, 1696 (1967).
134. W. R. Pretzer and R. W. Rudolph, *J. Am. Chem. Soc.* **95**, 931 (1973).
135. J. Plesek, S. Hermanek, and Z. Janousek, *Coll. Czech. Chem. Commun.* **42**, 785 (1977).
136. J. L. Little, G. D. Friesen, and L. J. Todd, *Inorg. Chem.* **16**, 869 (1977).
137. D. Dustin, G. B. Dunks, and M. F. Hawthorne, *J. Am. Chem. Soc.* **95**, 1109 (1973).
138. W. J. Evans and M. F. Hawthorne, *J. Chem. Soc. Chem. Commun.*, 38 (1974).
139. R. J. Wilson, L. F. Warren, Jr., and M. F. Hawthorne, *J. Am. Chem. Soc.* **91**, 758 (1969).
140. M. F. Hawthorne and T. D. Andrews, *J. Am. Chem. Soc.* **87**, 2496 (1965).
141. M. F. Hawthorne and H. W. Ruhle, *Inorg. Chem.* **8**, 176 (1969).
142. D. A. T. Young, R. J. Wiersema, and M. F. Hawthorne, *J. Am. Chem. Soc.* **93**, 5687 (1971).
143. L. J. Todd and co-workers, *Inorg. Chem.* **9**, 63 (1970).
144. D. A. Thompson and R. W. Rudolph, *J. Chem. Soc. Chem. Commun.*, 770 (1976).
145. B. P. Sullivan, R. N. Leyden, and M. F. Hawthorne, *J. Am. Chem. Soc.* **97**, 455 (1975).
146. T. E. Paxon and M. F. Hawthorne, *J. Am. Chem. Soc.* **96**, 4674 (1974).
147. M. F. Hawthorne and co-workers, *J. Am. Chem. Soc.* **99**, 6768 (1977).
148. E. S. Chandrasekaran, D. A. Thompson, and R. W. Rudolph, *Inorg. Chem.* **17**, 761 (1978).
149. W. J. Evans and M. F. Hawthorne, *J. Am. Chem. Soc.* **93**, 3063 (1971).
150. R. E. Loffredo and A. D. Norman, *J. Am. Chem. Soc.* **93**, 5587 (1971).
151. N. N. Greenwood, N. F. Travers, and D. W. Waite, *J. Chem. Soc. Chem. Commun.*, 1027 (1971).
152. N. N. Greenwood and N. F. Travers, *J. Chem. Soc. A*, 3257 (1971).
153. F. Klanberg, P. A. Wegner, G. W. Parshall, and E. L. Muetterties, *Inorg. Chem.* **7**, 2072 (1968).
154. L. J. Guggenberger, *J. Am. Chem. Soc.* **94**, 114 (1972).
155. N. N. Greenwood, J. A. McGinnety, and J. D. Owen, *J. Chem. Soc. A*, 809 (1971).
156. N. N. Greenwood and D. N. Sharrocks, *J. Chem. Soc. A*, 2334 (1969).
157. C. G. Salentine, R. R. Rietz, and M. F. Hawthorne, *Inorg. Chem.* **12**, 3025 (1974).
158. R. N. Leyden and M. F. Hawthorne, *J. Chem. Soc. Chem. Commun.*, 310 (1975).
159. A. J. Welch, *J. Chem. Soc. Dalton Trans.*, 2270 (1975).
160. A. R. Kane, L. J. Guggenberger, and E. L. Muetterties, *J. Am. Chem. Soc.* **92**, 2571 (1970).
161. J. W. Lott and D. F. Gaines, *Inorg. Chem.* **13**, 2261 (1974).
162. F. J. Hollander, D. H. Templeton, and A. Zalkin, *Inorg. Chem.* **12**, 2262 (1973).
163. R. N. Grimes and co-workers, *Inorg. Chem.* **13**, 1138 (1974).
164. A. J. Welch, *J. Chem. Soc. Dalton Trans.*, 225 (1976).

165. V. R. Miller and R. N. Grimes, *J. Am. Chem. Soc.* **95**, 5078 (1973).
166. V. R. Miller and R. N. Grimes, *J. Am. Chem. Soc.* **98**, 1600 (1976).
167. W. M. Maxwell, V. R. Miller, and R. N. Grimes, *J. Am. Chem. Soc.* **98**, 4818 (1976).
168. R. N. Grimes, *J. Am. Chem. Soc.* **93**, 261 (1971).
169. T. P. Fehlner and co-workers, *J. Am. Chem. Soc.* **98**, 7085 (1976).
170. D. F. Gaines and T. V. Iorns, *J. Am. Chem. Soc.* **90**, 6617 (1968).
171. J. C. Calabrese and L. F. Dahl, *J. Am. Chem. Soc.* **93**, 6042 (1971).
172. D. F. Gaines and T. V. Iorns, *Inorg. Chem.* **7**, 1041 (1968).
173. M. R. Churchill and co-workers, *J. Am. Chem. Soc.* **96**, 4041 (1974).
174. E. L. Hoel and M. F. Hawthorne, *J. Am. Chem. Soc.* **97**, 6389 (1975).
175. D. A. Thompson and R. W. Rudolph, *J. Chem. Soc. Chem. Commun.*, 770 (1976).

RALPH W. RUDOLPH
The University of Michigan

BORON HYDRIDES, HETEROBORANES AND THEIR METALLODERIVATIVES, COMMERCIAL ASPECTS

Even though several materials that can be classified as boron hydrides, heteroboranes, or metalloboranes, possess unique chemical and/or physical properties including polyhedral structures, thermal stability, etc, none of the polyhedral boranes and, with the exception of sodium tetrahydroborate [16940-66-2] and possibly certain of the amine boranes, none of the monoboron compounds presently enjoy significant commercial application. Generally, when a borane-containing material has provided enhanced properties for a particular application, the effect of those enhanced properties has not been sufficiently dramatic to displace less expensive materials. Thus, the virtual lack of commercial exploitation of boranes is largely owing to their cost, which in turn is a reflection of the complex procedures required for their synthesis. In order for boranes to graduate from the laboratory to the market place, their cost must be drastically reduced. The problem is cyclic in that the costs associated with the development of less expensive production processes cannot be justified until profitable uses are defined, however; consumers are reluctant to design applications until a ready supply of cheaper materials is available.

Boranes, or B–H-containing compounds, of proven commercial value will be discussed in this section, together with some materials and potential applications that have been cited in the literature but are presently not commercially important.

Tetrahydroborate, BH_4^-

Sodium tetrahydroborate, more commonly called sodium borohydride, is the only boron hydride used routinely on a commercially significant scale. Over 90% of the sodium tetrahydroborate produced each year (about 1000 metric tons) is manufactured by Ventron Corporation (U.S.) (1) with smaller quantities produced by Farbenfabrik Bayer A.G., Leverkusen, (FRG). The predominant form is a concentrated aqueous solution stabilized by sodium hydroxide (2). Solid sodium tetrahydroborate, a white powder, is available in ton quantities for $18–55/kg and is shipped in polyethylene bags in metal containers. It can be manipulated in air, does not ignite upon contact with moisture, and is not shock sensitive.

One of the earliest commercial and presently important uses of sodium tetrahydroborate is as a specific reducing agent in the synthesis of pharmaceuticals, vitamins, antibiotics, flavors, and other fine chemicals (3–4). A large and rapidly growing market for sodium tetrahydroborate is its use in wood pulp bleaching (5–7), clay bleaching (8), and textile dye reductions (9–11). Other applications include its use as a solid propellant additive (12–13), a blowing agent for the preparation of cellular plastics (14–16), a bleaching agent used in leather production (17–18), and an olefin polymerization catalyst (19), often in the form of a transition metal tetrahydroborate (20). The reducing capability of sodium tetrahydroborate can be used to control odor in alcohols (21), decolorize ketones (22) and polyols (23), stabilize olefins (24), to plate metals onto various substrates (including plastics) without the use of electricity (25–26) (see Electroless plating), and to provide a source of hydrogen for use in fuel cells (27) (see Batteries). An emerging application for sodium tetrahydroborate is in the removal of heavy metals such as lead and mercury from waste water streams (28). Similarly, sodium tetrahydroborate is used to recover precious metals such as silver from photographic solutions and rhodium from catalyst preparations (28). Some more esoteric applications of the tetrahydroborate ion include its use in hair waving preparations (29–30), and in its tritiated form, to analyze prostaglandins by radioassay (31).

Octahydrotriborate, $B_3H_8^-$

The octahydrotriborate(−1) ion [33055-82-2] is most often isolated as quarternary ammonium salts, which are white crystalline solids that are relatively stable in air (slow oxidation). Only research quantities are presently produced at $2–3/g.

Applications reported for the octahydrotriborate ion include its use as an ignition aid in solid propellants (32), as a fogging agent in photographic films (33–34), and as a specialty reducing agent in organic synthesis.

Mercaptoundecahydrododecaborate(2−), $B_{12}H_{11}SH^{2-}$

One of the most intriguing, albeit commercially insignificant, uses of boranes is in neutron capture therapy for the sterilization of malignant tissue in the treatment of brain tumors. This technique involves the localization of a boron compound in the affected tissue followed by irradiation with thermal neutrons. The particles released by the reaction [$^{10}B(\eta, \alpha)^7Li$], transfer energy to the malignant cells which are sterilized in the process. Among the problems associated with the procedure, is the necessity of obtaining a high boron concentration in the malignant tissue relative to the boron concentration in the surrounding healthy tissue. Loading of boron preferentially in the malignant cells may be accomplished by differential absorption (35). Brain tumors do not limit the transport of chemical agents from the blood (blood-brain barrier) as does normal brain tissue (36), thus an appropriate borane can be absorbed by malignant cells from the blood stream with much less incorporated by the surrounding healthy cells. Another approach is to utilize boron-labeled, tumor-specific antibodies to concentrate boron in brain tumors (37–38). Several boranes, eg, $B_{12}H_{11}SH^{2-}$ [12448-23-6] (39–40), have been studied in neutron-capture therapy with some positive results (35).

Borane Complexes

The adducts formed from alkyl amines, acting as Lewis bases, and the borane unit (BH$_3$), are known collectively as amine boranes. The stability of amine boranes depends on the substituent groups present on both the nitrogen and boron atoms; however, the most common trialkylamineboranes and dialkylamineboranes are stable in air and react only slowly with moisture.

Although many amine boranes are available in research quantities, their major commercial application is the use of an estimated several thousand kilograms per year of dimethylamineborane [74-94-2], (CH$_3$)$_2$NHBH$_3$, as the reducing agent in electroless plating baths (41–43) in the deposition of nickel, copper, and cobalt on the surface of metals and plastics.

Other uses cited for amine boranes are in photographic emulsions (44), insect chemosterilants (45), as a chemical stabilizer for organoisocyanates (46), as a polymerization accelerator for epoxy resins (47), and as reducing agents.

Borane complexes of tetrahydrofuran and dimethyl sulfide are commercially available and have found limited use as specialty reducing agents.

Heteroboranes

Heteroboranes that have carbon and boron atoms in the polyhedral structure are called carboranes. The neutral, closed-structure carboranes with the general formula C$_2$B$_n$H$_{n+2}$ are known where n = 3–10. The largest member of the series, C$_2$B$_{10}$H$_{12}$, has been incorporated into polymeric systems including polyvinyls, polyamides, polyesters, and carborane siloxanes (48). Carborane siloxanes have also been prepared with C$_2$B$_5$H$_7$ (49). Inclusion of carboranes into the backbone of such polymers imparts enhanced chemical and thermal stability, which was realized most dramatically in the carborane siloxanes (50-52); incorporation of the carborane raised the use temperature at least 50°C (53) over conventional polydimethylsiloxane. At present, carborane siloxane polymers that incorporate the *meta*-C$_2$B$_{10}$H$_{12}$ moiety are marketed under the trade name Dexsil and are commercially available for use as the stationary phase in gas–liquid chromatography columns (54) (see Analytical methods). Such columns can be operated in excess of 350°C and are capable of separating long-chain waxes, fatty acids, etc. Other suggested uses for carborane siloxane polymers are as an impregnant in mica paper used in high-temperature wire insulation (55) and as a component in a catalyst for the methylation of toluene, to *para*-xylene (56).

Derivatives of the C$_2$B$_{10}$H$_{12}$ carborane have been reported useful in optical switching devices (57), as a defoliant–desiccant which aids in harvesting crops (58), and as propellant additives (59). The separation by gas–liquid phase chromatography of several amino acid enantiomers has been accomplished using an optically active carborane dipeptide stationary phase (60).

Metalloboranes

Metalloboranes have not been shown to have commercial value although some potential uses have been cited. Among those uses are gasoline additives (61); a homogeneous hydrogenation, isomerization, hydrosilylation and hydrogen–deuterium exchange catalyst (62); and as a phase-transfer catalyst for the recovery of rubidium and cesium ions from aqueous solution by extraction into an immiscible organic phase (63) (see Catalysis, phase-transfer).

BIBLIOGRAPHY

"Boron Hydrides" under "Boron Compounds" in *ECT* 1st ed., Vol. 2, pp. 593–600, by S. H. Bauer, Cornell University; "Boron Hydrides and Related Compounds" under "Boron Compounds," Suppl. 1, pp. 103–130, by S. H Bauer, Cornell University; "Diborane and Higher Boron Hydrides" under "Boron Compounds," Suppl. 2, pp. 109–113, by J. W. Shepherd and E. B. Ayres, Callery Chemical Company; "Boron Hydrides" under "Boron Compounds" in *ECT* 2nd ed., Vol. 3, pp. 684–706, by G. W. Campbell, Jr., U.S. Borax Research Corporation.

1. A. Ferguson, D. J. Treskon, and R. E. Davenport, *Boron Minerals and Chemicals, Chemical Economics Handbook*, Stanford Research Institute, Menlo Park, Ca., 1977, p. 717.1000A.
2. *Private Communication,* Ventron Corporation, Beverly, Mass. (May 1977).
3. M. A. Kaplan, J. H. Lannon, and F. H. Buckwalter, *Appl. Micro.* **13,** 505 (1965).
4. U.S. Pat. 3,127,394 (Mar. 31, 1964), Y. G. Perron and L. B. Crast, Jr. (to Bristol-Meyers Company).
5. U.S. Pat. 3,017,316 (Jan. 16, 1962), W. H. Rapson (to Hooker Chemical Corp.).
6. U.S. Pat. 3,284,283 (Nov. 8, 1966), R. R. Kindron and L. P. Cartsunis (to FMC Corporation).
7. U.S. Pat. 3,981,765 (Sept. 21, 1976), E. Kruger and co-workers (to Vita Mayer and Company).
8. Belg. Pat. 653,535 (Jan. 18, 1965), F. R. Sheldon, W. H. Kibbel, Jr., and J. E. Kressbach (to FMC Corporation).
9. U.S. Pat. 2,745,788 (May 15, 1956), R. Stanley, M. Frohnsdorff, and J. Upton (to The Gillette Company).
10. U.S. Pat. 3,127,231 (March 31, 1964), C. E. Neale (to Southern Bleachery and Print Works, Inc.).
11. U.S. Pat. 3,195,974 (July 20, 1965), C. E. Neale (to Southern Bleachery and Print Works, Inc.).
12. U.S. Pat. 3,108,431 (Oct. 29, 1963), D. L. Armstrong and W. P. Knight (to Aero-jet General Corporation).
13. U.S. Pat. 3,703,359 (Nov. 21, 1972), S. Papetti (to Olin Mathieson Chemical Corporation).
14. U.S. Pat. 2,930,771 (Mar. 29, 1960), R. C. Wade (to Metal Hydrides Inc. (Ventron Corporation)).
15. U.S. Pat. 3,167,520 (Jan. 26, 1965), R. C. Wade (to Metal Hydrides Inc. (Ventron Corporation)).
16. U.S. Pat. 3,355,398 (Nov. 28, 1967), R. E. Kass (to The National Cash Register Company).
17. U.S. Pat. 3,347,751 (Oct. 17, 1967), T. C. Thorstensen (to Metal Hydrides Inc. (Ventron Corporation)).
18. U.S. Pat. 3,892,523 (July 1, 1975), G. H. Redlich and M. L. Alderman (to Rohm and Haas Company).
19. U.S. Pat. 3,306,886 (Feb. 28, 1967), F. Grosser, E. V. Hort, and A. Schwartz (to GAF Corporation).
20. U.S. Pat. 3,057,835 (Oct. 9, 1962), H. W. Coover, Jr. (to Eastman Kodak Company).
21. U.S. Pat. 3,860,520 (Jan. 14, 1975), L. D. Lindemuth, D. K. Whitehead, and C. Fronczek (to Continental Oil Company).
22. U.S. Pat. 3,865,880 (Feb. 11, 1975), J. R. Quelly and T. P. McNamara (to Exxon Research and Engineering).
23. U.S. Pat. 3,679,646 (July 25, 1972), J. E. Bristol (to E. I. du Pont de Nemours and Company, Inc.).
24. U.S. Pat. 3,527,821 (Sept. 7, 1970), C. W. Montgomery and C. M. Selwitz (to Gulf Research and Development Company).
25. U.S. Pat. 2,942,990 (June 28, 1960), E. A. Sullivan (to Metal Hydrides Inc. (Ventron Corporation)).
26. U.S. Pat. 3,917,885 (Nov. 4, 1975), K. D. Baker (to Englehard Minerals and Chemicals Corporation).
27. U.S. Pat. 3,374,121 (Mar. 19, 1968), J. N. Hogsett (to Union Carbide Corporation).
28. E. A. Sullivan, *Third International Meeting on Boron Chemistry,* Munich and Ettal, (FRG), July 5–9, 1976, Pergamon Press, 1977.
29. U.S. Pat. 2,766,760 (Oct. 16, 1956), H. Bogaty and A. E. Brown (to The Gillette Company).
30. U.S. Pat. 2,983,569 (May 9, 1961), R. Charle and G. Kalopisis (to Societé Monsavon-l'Ovéal).
31. U.S. Pat. 3,934,980 (Jan. 27, 1976), W. O. McClure (to Nelson Research and Development Company).
32. U.S. Pat. 3,563,818 (Feb. 16, 1971), C. L. Miller and R. Anderson (to the United States of America).
33. Fr. Pat. 1,548,122 (Nov. 29, 1968), C. C. Bard and H. W. Vogt (to Eastman Kodak Company).
34. Fr. Pat. 1,548,123 (Nov. 29, 1968), H. C. Baden and C. C. Bard (to Eastman Kodak Company).
35. H. Hatanaka, *J. Neurol.* **209,** 81 (1975).
36. H. Darson, "The Blood Brain Barrier" in G. H. Bourne, ed., *The Structure and Function of Nervous Tissue,* Vol. IV, Academic Press Inc., New York, 1972.
37. M. F. Hawthorne, R. J. Wiersema, and M. Takasugi, *J. Med. Chem.* **15,** 449 (1972).
38. H. S. Wong, E. I. Tolpin, and W. N. Lipscomb, *J. Med. Chem.* **17,** 785 (1974).

39. E. I. Tolpin and co-workers, *Oncology* **32,** 223 (1975).
40. A. H. Soloway, H. Hatanaka, and M. A. Davis, *J. Med. Chem.* **10,** 714 (1967).
41. U.S. Pat. 3,667,991, G. A. Miller (to Texas Instruments Inc.) (June 6, 1972).
42. U.S. Pat. 3,507,681 (Apr. 21, 1970), W. J. Cooper (to Mine Safety Appliance Company).
43. U.S. Pat. 3,431,120 (Mar. 4, 1969), L. M. Weisenberger (to Allied Research Products Incorporated).
44. U.S. Pat. 3,655,390 (Apr. 11, 1972), J. W. Overman (to E. I. du Pont de Nemours and Company, Inc.).
45. U.S. Pat. 3,463,851 (Aug. 26, 1969), A. B. Borkovec and J. A. Settepani (to the United States of America).
46. U.S. Pat. 3,479,393 (Nov. 18, 1969), R. L. Sandridge (to Mobay Chemical Company).
47. U.S. Pat. 3,347,827 (Oct. 17, 1967), H. L. Lee, Jr. (to Callery Chemical Company).
48. R. N. Grimes, *Carboranes,* Academic Press, Inc., New York, 1970.
49. R. E. Williams, *Pure Appl. Chem.* **29,** 569 (1972).
50. J. F. Ditter and A. J. Gotcher, *Office of Naval Research, AD-770625, 1973,* National Technical Information Service, U.S. Department of Commerce, Springfield, Va., 1973.
51. H. A. Schroeder and co-workers, *Rubber Chem. Technol.* **39,** 1184 (1966).
52. E. N. Peters and co-workers, *Rubber Chem. Technol.* **48,** 14 (1975).
53. E. N. Peters and co-workers, *J. Polym. Sci.* **15,** 723 (1977).
54. *Chem. Week* 37 (July 20, 1977).
55. U.S. Pat. 3,989,875 (Nov. 2, 1976), F. D. Bayles and M. A. Dudley (to Canada Wire and Cable Ltd.).
56. U.S. Pat. 3,965,210 (June 22, 1976), C.-C. Chu (to Mobil Oil Corp.).
57. U.S. Pat. 3,711,180 (Jan. 16, 1973), T. J. Klingen and J. R. Wright (to The University of Mississippi).
58. U.S. Pat. 3,539,330 (Nov. 10, 1970), D. C. Young (to Union Oil Company).
59. U.S. Pat. 3,764,417 (Oct. 9, 1973), W. E. Hill and L. R. Beason (to the United States of America).
60. R. Brazell and co-workers, *Chromatographia* **9,** 57 (1976).
61. U.S. Pat. 3,526,650 (Sept. 1, 1970), D. C. Young (to Union Oil Company).
62. G. E. Hardy and co-workers, *Acta Cryst.* **B32,** 264 (1976).
63. M. Kyrs and co-workers, *Third International Meeting on Boron Chemistry,* Munich and Ettal, (FRG), Pergamon Press, 1977, July 5–9, 1976.

GARY B. DUNKS
Union Carbide Corp.

ORGANIC BORON-NITROGEN COMPOUNDS

Organic boron–nitrogen compounds include all compounds in which there is a direct boron–nitrogen bond with at least one conventional organic group or moiety attached to boron, nitrogen, or both.

There are three types of boron–nitrogen bonds: (*1*) a coordinate covalent bond in which the nitrogen atom supplies both electrons in the B—N bond, generally represented by the formula R_3N-BR_3, sometimes by $R_3N \rightarrow BR_3$ or $R_3\overset{+}{N}-\overset{-}{B}R_3$ to show bond polarity (amine boranes); (*2*) a normal covalent bond in which an electron from each atom is shared. In such systems the hybridization of both boron and nitrogen is sp^2 resulting in a planar moiety capable of π-interaction by utilizing the nitrogen's free electron pair and boron's vacant *p* orbital; $R_2\overset{+}{N}=\overset{-}{B}R_2$ (aminoboranes); (*3*) formal analogues of aromatic carbon compounds in which the isoelectronic B—N linkage replaces the C—C bond.

(borazines)

The aforementioned classification of boron–nitrogen compounds according to the nature of the B—N bond was first presented by Wiberg (1) 30 years ago. Subsequent research in the field of boron–nitrogen chemistry has greatly increased the number of known organic compounds containing boron–nitrogen bonds. However, most of the B—N species can be placed in one of the three classes described above.

A fourth distinct class might be compounds in which an organonitrogen moiety is attached to a boron atom in a boron cluster compound (substituted boron hydrides, carboranes, metalloboranes, etc) via either a sigma or dative bond (2–3). The distinguishing feature of this class is that the boron atom of interest has a coordination number greater than four.

Amine Boranes

Table 1 lists representative examples of amine boranes and illustrates the wide variety of compound types known. The amine adducts of borane(*3*), BH_3, are relatively stable when pure and many are commercially available. The derivatives of haloboranes are quite sensitive to hydrolysis and are best prepared *in situ*.

Synthesis. *L.BH₃ (L = Amine).* The most convenient method for preparing amine boranes of the type $L.BH_3$ is illustrated by equation 1 (4).

$$MBH_4 + R_{3-n}H_nN.HCl \xrightarrow{\text{an ether}} H_2 + MCl + R_{3-n}H_nN.BH_3 \quad (1)$$

Other methods for preparing $L.BH_3$ compounds are illustrated in the following equations (5–8).

$$B_2H_6 + 2\,NRR'R'' \rightarrow 2\,R''R'RN.BH_3 \quad (2)$$

$$R_3N.Br'_3 + 3\,H_2 \xrightarrow[\text{pressure}]{\text{high}} R_3N.BH_3 + 3\,R'H \quad (3)$$

Table 1. Examples of Amine Boranes

Compound	CAS Registry No.	Mp, °C
$(C_2H_5)_3N.BH_3$	[25783-22-6]	−2
$(CH_3)_2NH.BH_3$	[74-94-2]	37
$tert\text{-}BuNH_2.BH_3$	[7337-45-3]	96
$C_6H_5(C_2H_5)_2N.BH_3$	[24144-20-5]	
$C_5H_5N.BH_3$	[110-51-0]	10–11
$(C_2H_5)_3N.BCl_3$	[2890-88-2]	92–93.5
$(p\text{-}BrC_6H_5)NH_2.BCl_3$	[1516-54-1]	300
$(CH_3)_3N.BBr_3$	[13423-86-4]	238
$(C_2H_5)NH_2.BF_3$	[75-23-0]	89
$(C_2H)_2NH.BF_3$	[372-56-5]	160
$(C_2H_5)_3N.BF_3$	[368-37-6]	29.5
$(CH_3)_2NH.BH_2Cl$	[52920-74-8]	18
$C_5H_5N.BH_2Cl$	[58178-53-3]	45
$(CH_3)_2NH.BH_2Br$	[66634-94-4]	5–6
$(CH_2)_2NH.BH_2I$	[52920-75-9]	
$(CH_3)_3N.BHBr_2$	[32805-31-5]	
$(CH_3)_3N.BHCl_2$	[25741-83-7]	145–146
$(C_2H_5)_2NH.B(CH_3)_3$	[30665-21-5]	27
$(CH_3)_2NH.B(tert\text{-}Bu)_3$	[66634-95-5]	
$CH_3CN.BF_3$	[21093-66-3]	
$p\text{-}NO_2C_6H_4CN.BCl_3$	[66634-96-9]	135
$C_6H_5CN.BBr_3$	[66675-86-3]	

$$B(OR)_3 + 3\,H_2 \xrightarrow{R'_3N} R'_3N.BH_3 + 3\,ROH \quad (4)$$

$$4\,R_3N + BX_3 + 3\,NaBH_4 \rightarrow 4\,R_3N.BH_3 + 3\,NaX \quad (5)$$

Amine boranes can also be produced via the reduction of amides with diborane(6), B_2H_6, as illustrated in equation 6.

$$B_2H_6\,(\text{excess}) + CH_3-\overset{\overset{O}{\|}}{C}-N(CH_3)_2 \rightarrow C_2H_5(CH_3)_2N\cdot BH_3 \quad (6)$$

Organic diamines form both 1:1 and 1:2 adducts depending on reaction conditions (9–10). Hydrazine boranes are also known and exist as mono- and bisboranes (11–12).

L.BX$_3$, L.BHX$_2$, and L.BH$_2$X. Interaction of a wide variety of amines with trihaloboranes in the vapor phase or in solution yields amine adducts of the type L.BX$_3$ (examples listed in Table 1) (13–14).

Compounds of the type L.BX$_2$H are prepared by reaction of the appropriate amine borane with a hydrogen halide (15–16).

$$R_3N.BH_3 + HX \rightarrow R_3N.BH_2X + H_2 \quad (7)$$

Further halogenation of amine monohaloborane adducts yields amine dihaloboranes (16–17).

$$(CH_3)_3N.BH_2X + X_2 \rightarrow (CH_3)_3N.BHX_2 + HX \quad (8)$$

L.BH$_{3-n}$R$_n$. A large number of amine organoboranes have been reported. Most are prepared by the reaction of the amine and the organoborane at low temperatures (18).

The reduction of B-trialkylboroxines, (RBO)$_3$, and arylhydroxyboranes with LiAlH$_4$ in the presence of an amine also produces amine organoboranes in high yields (19–20).

$$(RBO)_3 + 3\ NR'_3 \xrightarrow[\text{(C}_2\text{H}_5)_2\text{O, 35°C}]{\text{LiAlH}_4} 3\ RBH_2 \cdot NR'_3 \tag{9}$$

$$Ar_2BOR \xrightarrow[\text{Pyridine}]{\text{LiAlH}_4} Ar_2BH \cdot C_5H_5N \tag{10}$$

Unlike the formation of unsubstituted borane and haloborane adducts of amines, the organoborane–amine adduct formation is very sensitive to the nature of the organic substituent on boron. Steric hindrance prevents the formation of many amine organoboranes (21–22).

Both 1:1 and 2:1 complexes are reported formed when trialkylboranes react with ethylenebis(diethylamine) (C$_2$H$_5$)$_2$NCH$_2$CH$_2$N(C$_2$H$_5$)$_2$ (23).

L.BRX$_2$ and L.BR$_2$X Compounds. Amine adducts of organohaloboranes are rare. Two reported examples (CH$_3$)$_3$N.BF$_2$CH$_3$ [557-87-9] and (CH$_3$)$_3$N.BF(CH$_3$)$_2$ [373-62-6] were prepared by reaction of trimethylamine with the appropriate borane (24).

$$(CH_3)_3N + BF_{3-n}(CH_3)_n \xrightarrow{-80°C} (CH_3)_3N \cdot BF_{3-n}(CH_3)_n \tag{11}$$

Compounds Involving Internal B—N Coordination. Compounds containing amine and borane functionalities in the same molecule frequently exhibit internal adduct formation. Dimethyl(aminomethyl)borane [18494-94-5], (CH$_3$)$_2$BCH$_2$NH$_2$, has been studied and ^{1}H nmr evidence suggests that internal B—N coordination occurs (25). Five-membered heterocycles of the type

X = R, H, Cl, Br
Y = R, OR, NR$_2$, RS

have been prepared by hydroboration (qv) of allylamines, by the reaction of trialkylboranes with allylamines (26–27) and by the addition of HX to 1,2-azaboroles (28).

The esters formed from hydroxyboranes and aminoalcohols show unusual hydrolytic stability, a fact believed due to coordination of nitrogen to boron (29).

Amine–Boronium Cations. The first boronium salt to receive extensive scrutinization was the now famous diammoniate of diborane(6) [23777-63-1] (30–31) synthesized in the exothermic reaction of diborane and ammonia.

$$\left[\begin{array}{c}H\quad NH_3\\ \diagdown B \diagup \\ H\quad NH_3\end{array}\right]^+ \quad BH_4^-$$

Similar cationic systems have been reported in recent years and this area has been reviewed by Ryschkewitsch (32).

Properties and Reactions. Amine boranes are molecular species in which the strength of the B—N interaction is a function of the nature of the groups attached to boron and nitrogen. Conductivity studies show amine boranes to be essentially nonconductors (33). They are monomeric even in polar solvents. Molecular association, due to dipole–dipole interactions, is observed in benzene. Vibrational spectral data indicates that the B—N stretching frequency of amine boranes lies in the region 700–800 cm^{-1} (34). Most amine boranes are thermally stable when pure, exhibiting sharp melting points. Appreciable dissociation is observed in the vapor phase.

Characteristic reactions of amine boranes are illustrated in the following equations.

Displacement

$$\begin{cases} \text{(a) } R_3N \cdot BR'_3 + R''_3N \longrightarrow R''_3N \cdot BR'_3 + R_3N \\ \text{(b) } BR_3 + R'_3N \cdot BR''_3 \longrightarrow R''_3B + R'_3N \cdot BR_3 \end{cases} \qquad (12)$$

Redistribution

$$2\,(C_2H_5)_3N \cdot BH_3 + (C_2H_5)_3NBCl_3 \longrightarrow 3\,(C_2H_5)_3N \cdot BH_2Cl \qquad (13)$$
$$[13240\text{-}39\text{-}6]$$

Solvolysis

$$R_3N \cdot BH_3 + 3\,H_2O \longrightarrow R_3N + 3\,H_2 + B(OH)_3 \qquad (14)$$

Hydroboration

$$R_3N \cdot BH_3 + 3\;\;\mathrm{C{=}C} \longrightarrow R_3N + B(-\overset{|}{\underset{|}{C}}-\overset{|}{\underset{|}{C}}-H)_3 \qquad (15)$$

Selective reduction

$$R_3N \cdot BH_3 + R\overset{O}{\overset{\|}{-C-}}R' \xrightarrow{[R^+]} R\overset{OH}{\overset{|}{-CHR'}} \qquad (16)$$

Pyrolysis

$$R_2NH \cdot BR'_2H \xrightarrow{\Delta} R_2NBR'_2 + H_2 \qquad (17)$$

Aminoboranes

Aminoboranes are classified as mono-, bis-, or trisaminoboranes depending on the number of amine groups attached to the boron atom. Typical examples are listed in Table 2.

Synthesis. *Monoaminoboranes.* Several convenient routes to monoaminoboranes are available and are illustrated in the following equations (35–38).

$$R_2NH \cdot HCl + MBH_4 \rightarrow R_2NBH_2 + MCl + 2\,H_2 \qquad (18)$$

$$R_2NH + BR'_2X \xrightarrow{(C_2H_5)_3N} R_2NBR'_2 + (C_2H_5)_3N \cdot HX \qquad (19)$$

(R = alkyl, aryl or halide)

$$R_2NH \cdot BR'_2H \xrightarrow{\Delta} R_2NBR'_2 + H_2 \qquad (20)$$

$$B(NR_2)_3 + 2\ BR_3 \rightarrow 3\ R_2NBR_2 \tag{21}$$

In addition to the above methods, aminolysis of B—C, B—H, B—S, and B—O bonds yields aminoboranes. Some interesting and novel examples are illustrated below (39–42).

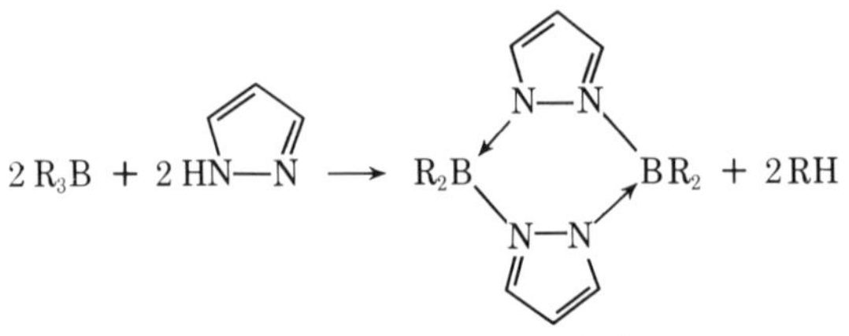

$$2\ R_3B + 2\ HN\text{—}N \longrightarrow R_2B(\text{pyrazabole})BR_2 + 2\ RH \tag{22}$$

(a pyrazabole)

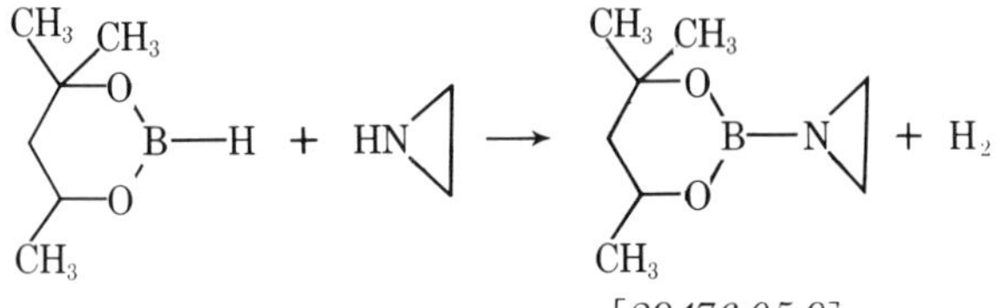

$$\text{(dimethyl dioxaborinane)B—H} + HN\triangleleft \longrightarrow \text{(dimethyl dioxaborinane)B—N}\triangleleft + H_2 \tag{23}$$

[29476-05-9]

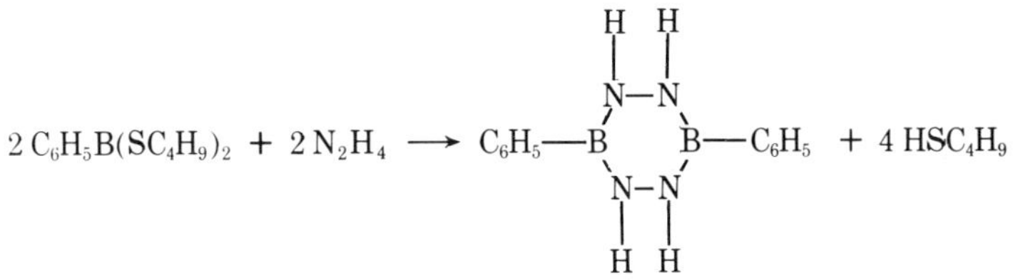

$$2\ C_6H_5B(SC_4H_9)_2 + 2\ N_2H_4 \longrightarrow C_6H_5\text{—B}(N_2H_4)_2\text{B—}C_6H_5 + 4\ HSC_4H_9 \tag{24}$$

$$\text{o-aminophenol} + C_6H_5B(OH)_2 \longrightarrow \text{benzoxazaborole—}C_6H_5 + 2\ H_2O \tag{25}$$

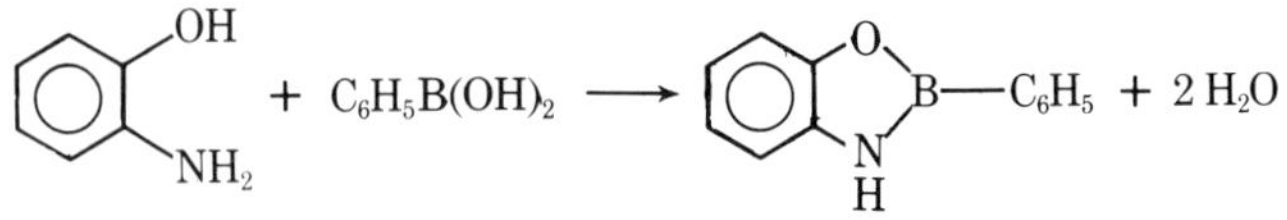

Bisaminoboranes. Ligand exchange of a trisaminoborane with a boron trihalide yields a bisaminohalogenoborane (43).

$$2\ B(NR_2)_3 + BX_3 \rightarrow 3\ (R_2N)_2BX \tag{26}$$

Subsequent reaction of the >B—X moiety with metal hydrides or metal alkyls affords the corresponding bisaminoborane and bisaminoalkylborane, respectively (44–45).

$$(R_2N)_2BX + MH \rightarrow (R_2N)_2BH + MX \tag{27}$$
$$(R_2N)_2BX + MR' \rightarrow (R_2N)_2BR' + MX \tag{28}$$

At elevated temperatures secondary amines react with trimethylamine adducts of alkylboranes yielding bisaminoboranes (46).

$$(CH_3)_3N \cdot BH_2R + 2\ HNR'_2 \xrightarrow{\Delta} (R'_2N)_2BR + 2\ H_2 + (CH_3)_3N \tag{29}$$

Table 2. Examples of Aminoboranes

Compound	CAS Registry No.	Bp, °C(kPa)^a
[(CH$_3$)$_2$N-BH]$_2$	[23884-11-9]	73^b
(CH$_3$)$_2$N-B(CH$_3$)$_2$	[11130-30-0]	49–51 (101)
H$_2$N-B(n-Bu)$_2$	[1767-37-9]	64–66 (0.23)
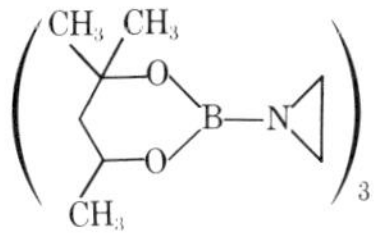 (structure with CH$_3$, O, B—N)	[66634-84-2]	82–83^b
[(C$_2$H$_5$)$_2$N]$_2$BH	[2386-99-4]	62–64 (2.1)
[(CH$_3$)$_2$N]$_2$BCH$_3$	[6914-63-2]	29–32 (2.0)
[(CH$_3$)$_2$NH]$_2$BC$_6$H$_5$	[19125-66-7]	106–107 (2.1)
B[N(tert-Bu)H]$_3$	[18379-73-2]	101–107 (3.2)
B[N(C$_2$H$_5$)$_2$]$_3$	[867-97-0]	220 (101)

^{a} To convert kPa to mm Hg, multiply by 7.50.
^{b} Melting point.

Trisaminoboranes. Trisaminoboranes are most conveniently prepared by adding boron trichloride to an excess of amine in an inert solvent at low temperatures (47).

$$BCl_3 + 6\ HNRR' \rightarrow B(NRR')_3 + 3\ HNRR'\cdot HCl \qquad (30)$$

The reaction of boron trifluoride with an amine in the presence of an active reductant such as a Grignard reagent or an alkali metal yields a trisaminoborane (48–49). Tris-(phenoxy)borane, B(OC$_6$H$_5$)$_3$, reacts with aluminum and hydrogen in the presence of diethylamine to form tris(diethylamino)borane [867-97-0] (50). The reaction is possibly applicable to the synthesis of other trisaminoboranes. Trialkylborates undergo aminolysis when treated with tris(dimethylamino)aluminum (51).

$$B(OR)_3 + Al[N(CH_3)_2]_3 \rightarrow B[N(CH_3)_2]_3 + Al(OR)_3 \qquad (31)$$
$$[4375\text{-}83\text{-}1]$$

Unsymmetrical trisaminoboranes may be prepared via a transamination reaction (52).

$$B(NR_2)_3 + 3\text{-}n\ R'_2NH \rightarrow (R_2N)_nB(NR'_2)_{3-n} + 3\text{-}n\ R_2NH \qquad (32)$$

Iminoboranes. Compounds containing an *sp*-hybridized nitrogen attached to boron, as shown below, are called iminoboranes.

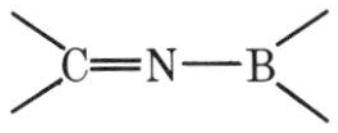

$$>\!\!C\!\!=\!\!N\!\!-\!\!B<$$

A recent x-ray study has verified the allene-like geometry proposed for iminoboranes (53).

Iminoboranes are conveniently prepared by the reaction of imines with (alkylthio)dialkylboranes (54)

$$R'C(R)\!\!=\!\!NH + R''SBR''_2 \rightarrow R''SH + R'C(R)\!\!=\!\!N\!\!-\!\!BR''_2 \qquad (33)$$

or via the reaction of imine hydrochlorides with aminoboranes (55).

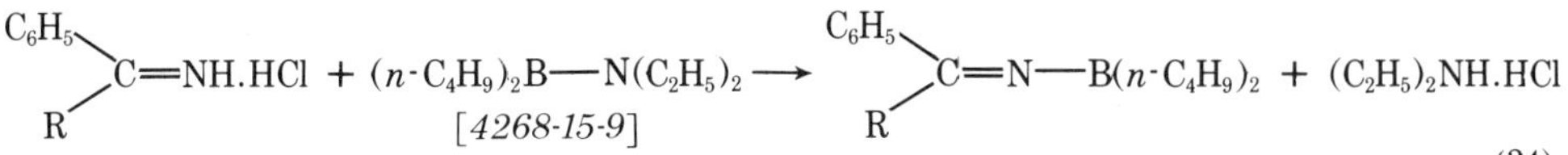

$$\begin{array}{c}C_6H_5\\R\end{array}\!\!\!>\!\!C\!\!=\!\!NH\cdot HCl + (n\text{-}C_4H_9)_2B\!\!-\!\!N(C_2H_5)_2 \longrightarrow \begin{array}{c}C_6H_5\\R\end{array}\!\!\!>\!\!C\!\!=\!\!N\!\!-\!\!B(n\text{-}C_4H_9)_2 + (C_2H_5)_2NH\cdot HCl$$
$$[4268\text{-}15\text{-}9] \qquad (34)$$

An interesting boronium cation is also produced as a by-product in the reaction.

$$\left[\left(\underset{R}{\overset{C_6H_5}{>}}C=NH\right)_2 B(n\text{-}C_4H_9)_2\right]^+ Cl^-$$

Iminoborane chemistry has been reviewed recently by Niedenzu (56).

Properties and Reactions. Monoaminoboranes readily undergo association, the extent of which depends principally on the steric requirements of the groups attached to boron and nitrogen. The monomers are generally liquids or low-melting solids whereas the dimers and trimers are crystalline solids. The monomers are readily hydrolyzed but the oligomers show appreciable hydrolytic stability except at high temperatures. Monoaminoboranes undergo extensive dissociation at elevated temperatures. This fact probably accounts for the hydrolytic instability at higher temperatures since nucleophilic attack on the tricoordinate boron is possible. Monoaminoboranes are soluble in a variety of organic solvents. The degree of solubility decreases in the order monomer > dimer > trimer.

Bisaminoboranes and trisaminoboranes are known only as monomers. They are less sensitive to hydrolysis than the monomeric monoaminoboranes. They are generally high-boiling liquids or crystalline solids.

Pyrolysis. Monoaminoboranes containing hydrogen attached to nitrogen and hydrogens or halogens bonded to boron undergo an internal elimination reaction on heating, yielding borazines (57).

$$3 \text{ RNH-BHR}' \xrightarrow{-H_2} (-BR'-NR-)_3 \tag{35}$$

Bisaminoboranes in general are thermally stable. When pyrolysis does occur a principal reaction product is a borazine as illustrated for bis(ethylamino)borane [7360-03-4] (58).

$$3 (C_2H_5NH)_2BH \rightarrow (-BH-NC_2H_5-)_3 + 3 C_2H_5NH_2 \tag{36}$$

Some bisaminoboranes undergo ligand exchange reactions when heated (59).

$$4 [(CH_3)_2N]_2BF \rightarrow 2 B[N(CH_3)_2]_3 + [(CH_3)_2NBF_2]_2 \tag{37}$$
$$[383\text{-}90\text{-}4] \quad [4375\text{-}83\text{-}1] \quad [6792\text{-}24\text{-}1]$$

Trisaminoboranes derived from secondary amines are thermally stable up to approximately 280°C. Decomposition to uncharacterized polymeric products occurs at higher temperatures (59). Trisaminoboranes derived from primary amines readily form borazines via amine elimination.

$$3 B(NHR)_3 \rightarrow [-B(NHR)-N(R)-]_3 + 3 RNH_2 \tag{38}$$

Representative Reactions. *Solvolysis.*

$$B(NR_2)_3 + 3 ROH \rightarrow B(OR)_3 + 3 R_2NH \tag{39}$$

Transamination.

$$B(NR_2)_3 + 3 R'_2NH \rightarrow B(NR'_2)_3 + 3 R_2NH \tag{40}$$

Conversion to haloboranes.

(a) $B(NR_2)_3 + HCl \text{ (excess)} \rightarrow R_2NH \cdot BCl_3 + 2 R_2NH \cdot HCl$ (41)

(b) $B(NR_2)_3 + HCl \text{ (excess)} \rightarrow [(R_2NH)_2BCl_2]^+Cl^- + R_2NH \cdot HCl$ (42)
(a boronium salt)

Addition to double bonds.

$$B(NR_2)_3 + 3\ C_6H_5N{=}C{=}O \rightarrow B(NC(O)NR_2)_3 \quad (43)$$
$$\underset{C_6H_5}{|}$$

Borazines

The largest and most extensively studied family of boron–nitrogen compounds is the borazines, characterized by a six-membered ring system containing alternating boron and nitrogen atoms. In contrast to the trimeric aminoboranes, $(-R'_2N-BR_2-)_3$, borazines are tricoordinate, planar, and have B—N bond distances intermediate between calculated single and double bond distances. Because borazines are isostructural and isoelectronic with benzenes, an inevitable comparison has evolved. The physical properties of borazines tend to confirm a resonance structure quite similar to that of benzenes. However, the chemical evidence indicates that the reactions of borazines are dominated by polarization of the B—N bonds.

Several brief review articles detailing various aspects of borazine chemistry as well as a comprehensive treatise on borazine chemistry are available (57,60–63).

Synthesis. The parent compound, borazine [6569-51-3], is best prepared by a two-step process as shown in the following equations,

$$3\ BCl_3 + 3\ NH_4Cl \rightarrow (-BCl-NH-)_3 + 9\ HCl \quad (44)$$

$$(-BCl-NH-)_3 + 3\ NaBH_4 \rightarrow (-BH-NH-)_3 + 3\ NaCl + 3/2\ B_2H_6 \quad (45)$$

These reactions have been studied in some detail by Hohnstedt and Haworth (64). The direct interaction of solid sodium tetrahydridoborate with solid ammonium chloride gives borazine in 20% yield (65).

The pyrolysis of ammonia borane is a commercially feasible method for the production of borazine, although the pyrolysis is reported to be difficult (66).

Symmetrically substituted borazines are most often prepared by condensing amines and boranes (57).

$$BX_3 + NH_2R \rightarrow BX_3 \cdot NH_2R \rightarrow BX_2NHR + HX \quad (46)$$

$$BX_2 \cdot NHR \rightarrow 1/3\ (-BX-NR-)_3 + HX \quad (47)$$

$$X = H, \text{alkyl or halogen};\ R = H, \text{alkyl or aryl}$$

When X = halogen, *B*-trihalogenoborazines are formed. These materials readily react with Grignard reagents yielding *B*-substituted trialkyl and triaryl borazines (67).

Other typical reactions which yield symmetrically substituted borazines are (64,68–72):

$$(-BX-NR-)_3 + 3\ NaBH_4 \rightarrow (-BH-NR-)_3 + 3/2\ B_2H_6 + 3\ NaX \quad (48)$$

$$4\ (-BCl-NR-)_3 + 3\ TiF_4 \rightarrow 4\ (-BF-NR-)_3 + 3\ TiCl_4 \quad (49)$$

$$3\ R'B(NHR)_2 \rightarrow (-BR'-NR-)_3 + 3\ NH_2R \tag{50}$$

$$3\ B(NHR)_3 \rightarrow [-B(NHR)-NR-]_3 + 3\ NH_2R \tag{51}$$

$$B(OC_6H_5)_3 + Al + RNH_2 \xrightarrow{H_2} \tfrac{1}{3}(-BH-NR-)_3 + Al(OC_6H_5)_3 + 1/2\ H_2 \tag{52}$$

$$3\ RB(SCH_3)_2 + 3\ R'NH_2 \rightarrow (-BR-NR'-)_3 + 6\ CH_3SH \tag{53}$$

Unsymmetrically substituted borazines can be prepared by the reaction of substituted borazines with Grignard reagents (57).

$$H_3B_3N_3R_3 + x\ R'MgX \rightarrow R'_xH_{3-x}B_3N_3R_3 + x\ MgXH \tag{54}$$
$$(x = 1\ or\ 2)$$

The treatment of B-trihalogenoborazines with the appropriate amount of Grignard reagent also yields unsymmetrically substituted borazines (57).

$$X_3B_3N_3R_3 + x\ R'MgY \rightarrow R'_xX_{3-x}B_3N_3R_3 + x\ MgXY \tag{55}$$

Unsymmetrical N-alkylborazines have been prepared by the reaction of lithium borohydride with ammonium chloride and an alkylammonium chloride, or two different alkylammonium chlorides (73–75).

Properties and Reactions. Symmetrically substituted borazines are generally crystalline solids and unsymmetrically substituted borazines are generally liquids or low melting solids. Borazines are planar molecules, although some puckering may be observed in highly substituted borazines owing to steric interactions.

Thermal Stability. Borazine itself shows negligible decomposition at 0–5°C. At ambient temperature 1–2% decomposition has been observed during the first month followed by an increase in rate thereafter (66). At higher temperatures appreciable decomposition occurs yielding condensed borazines (analogues of naphthalene, biphenyl, etc) and eventually boron nitride (66). Some substituted borazines exhibit outstanding thermal stabilities. When there is hydrogen attached to nitrogen and halogen attached to boron, the borazine decomposes rather rapidly at temperatures below 200°C. Organosubstituted borazines vary in thermal stability and apparently decompose via boron–carbon homolysis (76).

Hydrolysis. Borazine is hydrolyzed by water at ambient or higher temperatures.

$$H_3B_3N_3H_3 + 12\ H_2O \rightarrow 3\ H_3BO_3 + 3\ NH_4OH + 3\ H_2 \tag{56}$$

However, borazine is not soluble in water so that hydrolysis is retarded by the formation of a protective coating of boric acid. Substituted borazines are also fairly readily hydrolyzed with resultant ring cleavage (57). B-Trihaloborazines react vigorously with water at room temperature.

Addition Reactions. Unlike benzenes, borazines readily undergo addition reactions, much in the manner of olefins. At 0°C, instead of undergoing hydrolysis, borazine actually adds three moles of water forming a cyclotriborazane [67124-95-2] (77).

$$(-BH-NH-)_3 + 3\ H_2O \rightarrow \text{cyclotriborazane} \tag{57}$$

The addition of hydrogen halides to borazines at low temperatures yields the appropriate borazine addition product (a cyclotriborazane).

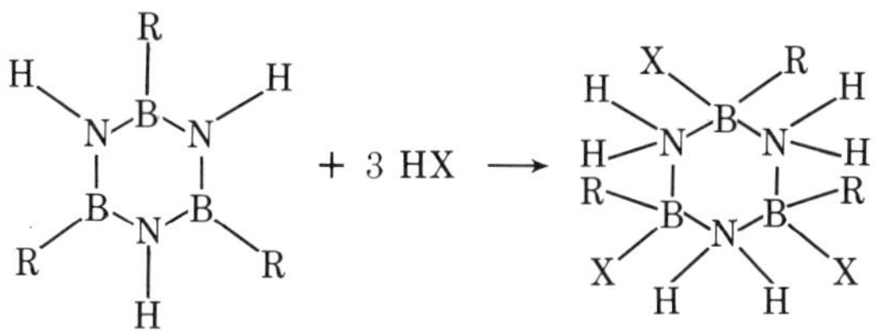

[reaction scheme: borazine + 3 HX → cyclotriborazane adduct]

Although borazine reacts readily with the halogens, only the 1:3 bromine adduct has been characterized (78). In general, polar molecules add to borazines yielding intermediate cyclotriborazanes which on warming undergo elimination reactions yielding borazines or decomposition products (57).

Substitution Reactions. Substitution reactions on borazines are confined mainly to substitution on boron—substituents on nitrogen are inert to most reagents. Treatment of B-haloborazines with metal–organic reagents (eg, alkyl lithium or Grignard reagents) results in the substitution of the halogen atom by the organic moiety without destruction of the borazine ring.

$$(-BX-NR-)_3 + R'M \rightarrow (-BR'-NR-)_3 + MX \tag{58}$$

Since B-haloborazines are relatively easy to prepare, substituted borazines are most conveniently prepared via the above reaction.

The reaction of a N-substituted borazine with a Grignard reagent results in a substitution of a boron-bonded hydrogen by an alkyl or aryl group.

$$(-BH-NR-)_3 + 3\,R'MgX \rightarrow (-BR'-NR-)_3 + 3\,MgHBr \tag{59}$$

Interaction of methyllithium with B-trimethylborazine [5314-85-2] is reported to yield B-trimethyl-N-lithioborazine which can be converted to N-monoalkyl-borazines via treatment with alkyl halides (79).

Other established substitution or exchange reactions are illustrated below (57,68,80).

$$(-BCl-NR-)_3 + 6\,R'_2NH \rightarrow [-B(NR'_2)-NR-]_3 + 3\,R'_2NH\cdot HCl \tag{60}$$

$$(-BCl-NR-)_3 + 3\,MOR' \rightarrow (-BOR'-NR-)_3 + 3\,MCl \tag{61}$$
$$(M = \text{alkali metal})$$

$$(-BH-NH-)_3 + 3\,DBF_2 \rightarrow (-BD-NH-)_3 + 3\,HBF_2 \tag{62}$$

$$4\,(-BCl-NR-)_3 + 3\,TiF_4 \rightarrow 4\,(-BF-NR-)_3 + 3\,TiCl_4 \tag{63}$$

Miscellaneous Reactions. *Photolysis.* Irradiation of borazine in the presence of various substrates using a mercury lamp is a convenient route to monosubstituted borazines (81–83). For example, the irradiation of chlorinated hydrocarbon–borazine mixtures yields B-monochloroborazines.

$$(-BH-NH-)_3 + CHCl_3 \xrightarrow{h\nu} ClH_2B_3N_3H_3 + CH_2Cl_2 \tag{64}$$

When gaseous borazine is irradiated the products include borazanaphthalene [253-18-9], and biborazinyl [18464-85-2] (84).

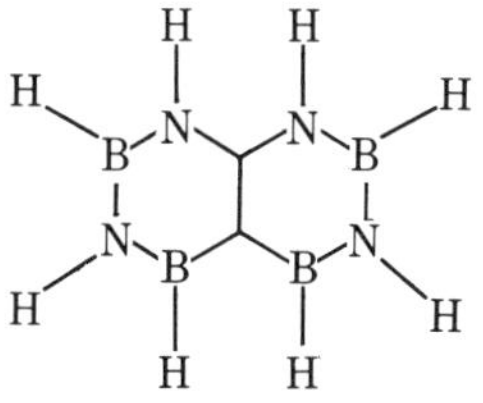

borazanaphthalene 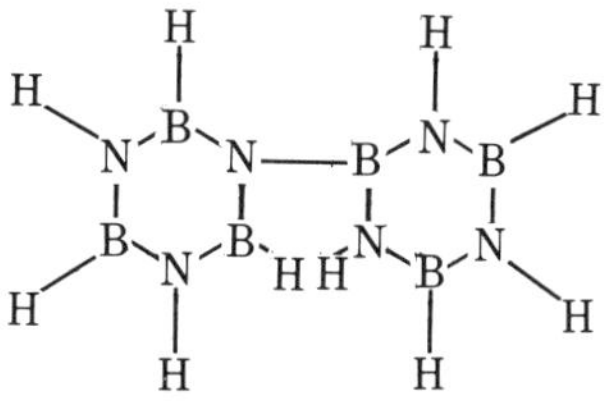   biborazinyl

Alkylborazines, in the presence of hydrogen, undergo photoinduced dimerization to C—C-bonded diborazinyl derivatives (85).

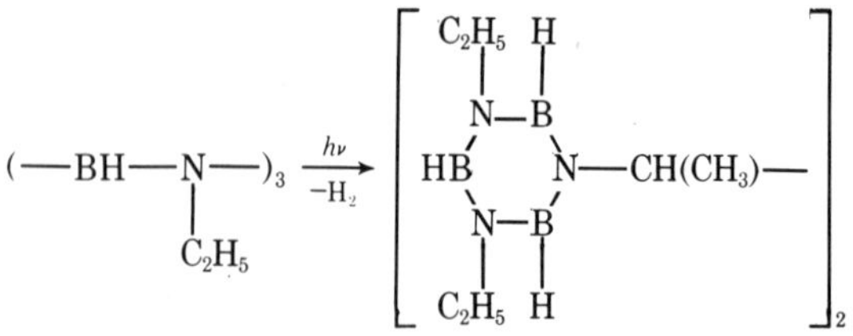

$$(-BH-N(C_2H_5)-)_3 \xrightarrow[-H_2]{h\nu} \left[HB \underset{\underset{C_2H_5\ H}{|\ \ |}}{\overset{\overset{C_2H_5\ H}{|\ \ |}}{\begin{array}{c} N-B \\ N-B \end{array}}} N-CH(CH_3)- \right]_2$$

Complexation. Chromium and molybdenum carbonyl react with alkylborazines to form π complexes analogous to aromatic π complexes (86–87).

$$Mo(CO)_6 + (-BR-NR'-)_3 \xrightarrow{h\nu} \pi\text{-}(-BR-NR'-)_3 Mo(CO)_3 + 3\ CO \qquad (65)$$

Some borazines form donor-acceptor complexes with organic bases such as pyridine and diglyme (dimethyl ether of diethylene glycol) (88).

$$(-BCl-NH-)_3 + n\ C_5H_5N \rightarrow (-BCl-NH-)_3 \cdot n C_5H_5N \qquad (66)$$
$$(n = 2\ \text{or}\ 4)$$

Other B—N Ring Systems

Several unusual ring systems containing only B—N linkages have been reported (89). These include the eight-membered borazocines (1) and the four-membered diazaboretanes (2) (90–93) as well as systems containing B_2N_3, B_3N_2, BN_4, B_2N_4, B_4N_2, etc, frameworks (89).

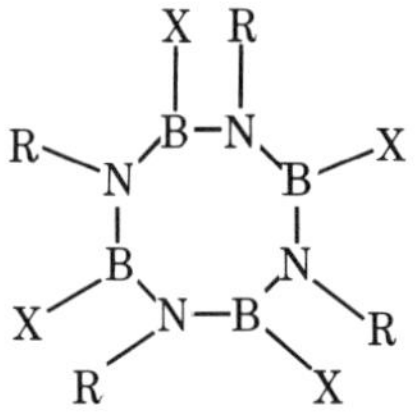

(1) 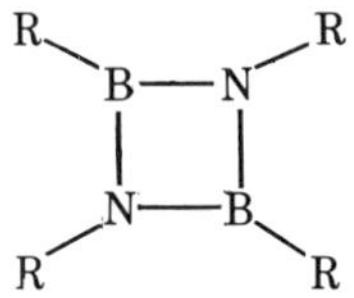   (2)

The incorporation of the B—N linkage in heteroaromatic systems of the type shown below has been reported by several researchers, most notably Dewar (94), and reviewed by Lappert (95).

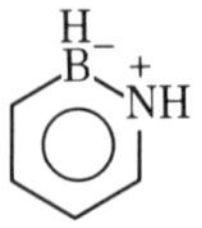

[6680-69-9]

Molecular orbital calculations (96–97) indicate that π-bonding and charge separations are small. Although much of the chemistry of heteroaromatic boron compounds is conveniently rationalized by invoking the concept of aromaticity. A typical synthesis is shown in equation 67 (98).

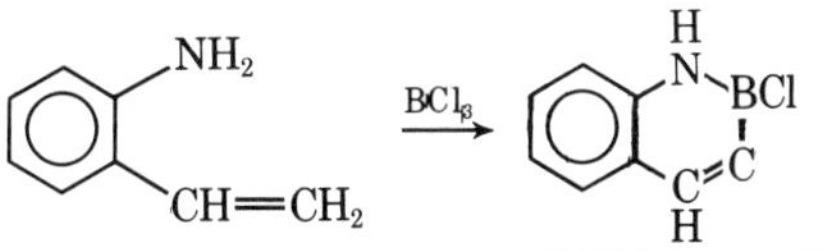

$$\text{(67)}$$

[67114-34-5]

Saturated CNB heterocycles have also been characterized. Typical examples are shown below (99–101).

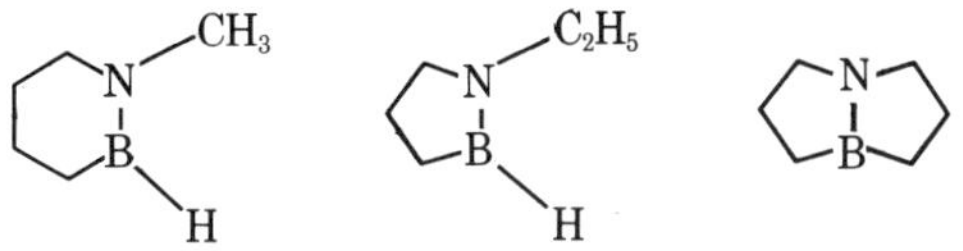

[38369-75-4]   [67114-35-6]   [16153-13-2]

Manufacture and Uses

Organic boron–nitrogen compounds have not found extensive usage and, therefore, very few are manufactured on a large scale. Research quantities of most stable amine boranes are available from Callery Chemical Company, Aldrich Chemical Company, and Ventron Corporation (Alfa). Amine boranes not listed in the respective catalogues can generally be obtained via a custom synthesis.

Callery Chemical Company appears to be the largest manufacture of amine boranes and borazines. The equipment and exact conditions used are proprietary information.

Amine boranes are principally used as reagents to produce synthetic intermediates in organic syntheses. The amine adducts of BH_3 are valuable reducing agents, both in organic chemistry and in metal plating (see Electroless plating). Borazines and aminoboranes have found limited use as polymerization catalysts or oxidation stabilizers (see Boron hydrides, commercial aspects).

BIBLIOGRAPHY

"Borazines" in *ECT* 1st ed., Suppl. 2, pp. 89–108, by S. F. Stafiej, American Cyanamid Company; "Boron Nitrogen" under "Organic Boron Compounds" under "Boron Compounds," in *ECT* 2nd ed., Vol. 3, pp. 728–737, by H. C. Newsom, U.S. Borax Research Corporation.

1. E. Wiberg, *Naturwissenschaften* **35**, 182, 212 (1948).
2. L. I. Zakharkin, V. N. Kalinin, and V. V. Gedymin, *J. Organometal. Chem.* **16**, 371 (1969).
3. M. L. Denniston, Ph.D. Thesis, Ohio State University, Columbus, 1970.
4. G. W. Schaeffer and E. R. Anderson, *J. Am. Chem. Soc.* **71**, 2150 (1949).
5. A. Stock and E. Pohland, *Ber. Dtsch. Chem. Ges.* **59B**, 2210, 2215, 2223 (1926).
6. R. Köster, *Angew. Chem.* **69**, 64 (1957).
7. E. C. Ashby and W. E. Forester, *J. Am. Chem. Soc.* **84**, 3407 (1962).
8. U.S. Pat. 3,037,985 (1962), K. Lang and F. Schubert.
9. A. R. Gatts and T. Wartik, *Inorg. Chem.* **5**, 329 (1966).
10. H. C. Kelly and J. O. Edwards, *Inorg. Chem.* **2**, 226 (1963).

11. Ger. Pat. 1,068,231 (1959), K. Lang.
12. R. J. Baumgarten and M. C. Henry, *J. Org. Chem.* **29,** 3400 (1964).
13. W. Gerrard, M. F. Lappert, and C. A. Pearce, *J. Chem. Soc.,* 381 (1957).
14. W. Gerrard and E. F. Mooney, *J. Chem. Soc.,* 4028 (1960).
15. H. Nöth and H. Beyer, *Chem. Ber.* **93,** 2251 (1960).
16. J. M. VanPaasschen and R. A. Geanangel, *J. Am. Chem. Soc.* **94,** 2680 (1972).
17. G. E. Ryschkewitsch and J. W. Wiggins, *Inorg. Syn.* **12,** 116 (1970).
18. W. Gerrard, *The Organic Chemistry of Boron,* Academic Press, Inc., New York, 1961.
19. M. F. Hawthorne, *J. Am. Chem. Soc.* **83,** 831 (1961).
20. *Ibid.,* **80,** 4291 and 4293 (1958).
21. H. C. Brown, *Rec. Chem. Progr.* **14,** 83 (1953).
22. M. F. Lappert, *Chem. Review* **56,** 959 (1956).
23. S. Schroeder and K. H. Thiele, *Anorg. Allg. Chem.* **428**(1), (1977).
24. A. Burg and A. A. Green, *J. Am. Chem. Soc.* **65,** 1838 (1943).
25. R. Schaeffer and L. J. Todd, *J. Am. Chem. Soc.* **87,** 488 (1965).
26. B. M. Mikhailov, V. A. Dorokhov, and N. V. Mostovoi, *Iz. Akad. Nauk. SSSR, Ser. Khim.,* 223 (1965).
27. N. V. Mostovoi, V. A. Dorohkov, and B. M. Mikhailov, *Iz. Akad. Nauk. SSSR, Ser. Khim.,* 90 (1966).
28. *Ibid.,* p. 70.
29. H. K. Zimmerman, *Advances in Chemistry,* Vol. 42, American Chemical Society, Washington, D.C., 1964, p. 183.
30. A. Stock and E. Pohland, *Ber. Deut. Chem. Ges.* **56,** 789 (1923).
31. D. R. Schultz and R. W. Parry, *J. Am. Chem. Soc.* **80,** 4 (1958).
32. G. E. Ryschkewitsch in E. L. Muetterties, ed., *Boron Hydride Chemistry,* Academic Press, New York, 1975.
33. W. Gerrard, M. F. Lappert, and C. A. Pearce, *J. Chem. Soc.,* 381 (1957).
34. R. C. Taylor, *Adv. Chem.* **42,** 59 (1964).
35. G. W. Schaeffer and E. R. Anderson, *J. Am. Chem. Soc.* **71,** 2143 (1949).
36. K. Niedenzu and co-workers, *J. Org. Chem.* **26,** 3037 (1961).
37. M. F. Hawthorne, *J. Am. Chem. Soc.* **83,** 267 (1961).
38. K. Niedenzu, J. W. Dawson, and H. Jenne, *Chem. Ber.* **96,** 1653 (1963).
39. S. Trofimenko, *J. Am. Chem. Soc.* **89,** 3903 (1967).
40. H. D. Smith and R. J. Brotherton, *Inorg. Chem.* **9,** 2443 (1970).
41. B. M. Mikhailov and T. K. Kozminskaya, *Iz. Akad. Nauk. SSSR, Ser. Khim.,* 439 (1965).
42. F. Umland, *Angew. Chem.* **79,** 583 (1967).
43. E. Wiberg and E. Amberger, *Hydrides of the Elements of Main Groups I-IV,* Elsevier Scientific Pub. Co., Amsterdam, 1971.
44. H. Nöth and co-workers, *Z. Anorg. Allg. Chem.* **318,** 293 (1962).
45. H. Nöth and D. Fritz, *Z. Anorg. Allg. Chem.* **322,** 297 (1963).
46. M. F. Hawthorne, *J. Am. Chem. Soc.* **83,** 2671 (1961).
47. W. Gerrard, M. F. Lappert, and C. A. Pearce, *J. Chem. Soc.,* 381 (1957).
48. A. Dornow and H. H. Gehrt, *Z. Anorg. Allg. Chem.* **294,** 81 (1958).
49. C. A. Kraus and E. H. Brown, *J. Am. Chem. Soc.* **52,** 4414 (1930).
50. R. A. Kovar, R. Culbertson, and E. C. Ashby, *Inorg. Chem.* **10,** 900 (1971).
51. J. K. Ruff, *J. Org. Chem.* **17,** 1020 (1962).
52. R. J. Brotherton and T. Buckman, *Inorg. Chem.* **2,** 424 (1963).
53. G. J. Bullen and K. Wade, *J. Chem. Soc. Chem. Commun.,* 1122 (1971).
54. B. M. Mikhailov, *Bull. Acad. Sci. USSR* **23,** 2307 (1975).
55. B. M. Mikhailov and co-workers, *Izv. Akad. Nauk. SSSR, Ser. Khim* **9,** 20536 (1976).
56. K. Niedenzu, *International Review of Science, Inorg. Chemistry Series Two,* Vol. 4, Butterworths and Co., London, 1975.
57. H. Steinberg and R. J. Brotherton, *Organoboron Chemistry,* Vol. 2, John Wiley & Sons, Inc., New York, 1966.
58. B. M. Mikhailov and V. A. Dorokhov, *J. Gen. Chem. USSR* **32,** 1497 (1962).
59. A. B. Burg and J. Banus, *J. Am. Chem. Soc.* **76,** 3903 (1954).
60. K. Niedenzu, *International Review of Science, Inorg. Chemistry Series Two,* Butterworth and Co. Ltd., Reading, Mass., 1975, p. 41.

61. D. F. Gaines and J. Borlin in E. L. Muetterties, ed., *Boron Hydride Chemistry,* Academic Press, Inc., New York, 1975.
62. K. Niedenzu and J. W. Dawson in E. L. Muetterties, ed., *The Chemistry of Boron and its Compounds,* John Wiley & Sons, Inc., New York, 1967.
63. N. N. Greenwood and B. S. Thomas in A. F. Trotman-Dickenson, ed., *Comprehensive Inorganic Chemistry,* Vol. 1, Pergamon Press Ltd., Oxford, Eng., 1973, 919.
64. L. F. Hohnstedt and D. T. Haworth, *J. Am. Chem. Soc.* **82,** 89 (1960).
65. L. P. Moisa and V. B. Spivakovskii, *Russ. J. Inorg. Chem.* **15,** 1510 (1971).
66. *Borazine,* Callery Chemical Company, Callery, Penn., 1977.
67. D. T. Haworth and L. F. Hohnstedt, *Chem. Ind.,* 559 (1960).
68. K. Niedenzu, H, Beyer, and H. Jenne, *Chem. Ber.* **96,** 2649 (1963).
69. B. M. Mikhailov and T. K. Kozyminskaya, *Dok. Akad. Nauk. SSSR* **121,** 656 (1958).
70. D. W. Aubrey and M. F. Lappert, *J. Chem. Soc.,* 2927 (1959).
71. E. C. Ashby and R. A. Kovar, *Inorg. Chem.* **10,** 1524 (1971).
72. D. Nölle and H. Nöth, *Angew. Chem. Intern. Ed.* **10,** 126 (1971).
73. C. S. G. Phillips, P. Powell, and J. A. Semlyen, *J. Chem. Soc.,* 1202 (1963).
74. C. S. G. Phillips and co-workers, *Z. Anal. Chem.* **197,** 202 (1963).
75. O. T. Beachley, *J. Am. Chem. Soc.* **94,** 4223 (1972).
76. H. C. Newsom and co-workers, *J. Am. Chem. Soc.* **83,** 4134 (1961).
77. W. E. Weibrecht, *Dissertation Abstr.* **26**(9), 5028 (1966) from Univ. Microfilms (Ann Arbor) Order No. 64-8762.
78. R. F. Riley and C. J. Schack, *Inorg. Chem.* **3,** 1651 (1964).
79. R. I. Wagner and J. L. Bradford, *Inorg. Chem.* **1,** 93 (1962).
80. A. J. DeStefano and R. F. Porter, *Inorg. Chem.* **15,** 2569 (1976).
81. M. Nadler and R. F. Porter, *Inorg. Chem.* **8,** 599 (1969).
82. G. H. Lee, II, and R. F. Porter, *Inorg. Chem.* **6,** 648 (1967).
83. M. Nadler and R. F. Porter, *Inorg. Chem.* **6,** 1739 (1967).
84. M. A. Ness and R. F. Porter, *J. Am. Chem. Soc.* **94,** 1438 (1972).
85. G. A. Kline and R. F. Porter, *Inorg. Chem.* **16,** 11 (1977).
86. H. Wernen, R. Prinz, and E. Deckelmann, *Chem. Ber.* **102,** 95 (1969).
87. J. L. Adcock and J. J. Lagowski, *Inorg. Chem.* **12,** 2533 (1973).
88. M. F. Lappert and G. Srivastava, *J. Chem. Soc. A,* 602 (1967).
89. A. Finch, J. B. Leach, and J. H. Morris, *Organomet. Chem. Rev. Sect. A* **4,** 1 (1969).
90. H. S. Turner and R. J. Warne, *Adv. Chem. Soc.* **42,** 290 (1964).
91. M. F. Lappert and M. K. Majumdar in ref. 90, p. 208.
92. P. Geymayer, E. G. Rochow, and U. Wannagat, *Angew. Chem. Intern. Ed. Eng.* **3,** 633 (1964).
93. P. I. Paetzold, *Z Anorg. Allg. Chem.* **326,** 53 (1963).
94. M. J. S. Dewar, *Adv. Chem.* **42,** 227 (1964).
95. M. F. Lappert in E. L. Muetterties, ed., *The Chemistry of Boron and Its Compounds,* John Wiley & Sons, Inc., New York, 1967.
96. M. J. S. Dewar in H. Steinberg and A. L. McCloskey, eds., *Progress in Boron Chemistry,* Vol. 1, Pergamon Press, Oxford, Eng., 1964.
97. J. J. Kaufman and J. R. Hamann, *Adv. Chem. Ser.* **42,** 273 (1964).
98. M. J. S. Dewar and R. Dietz, *J. Chem. Soc.,* 2728 (1959).
99. H. Willie and J. Goubeau, *Chem. Ber.* **105,** 2156 (1972).
100. U. A. Dorokhov, O. G. Boldyrena, and B. M. Mikhailov, *J. Gen. Chem. USSR* **40,** 1515 (1970).
101. H. Bock and W. Fuss, *Chem. Ber.* **104,** 1687 (1971).

H. D. SMITH, JR.
Virginia Polytechnic Institute and State University

BRAKE FLUIDS. See Hydraulic fluids.

BRAKE LININGS AND CLUTCH FACINGS

This review concerns the current state of friction materials technology with regard to types of friction materials, their applications, friction and wear characteristics, raw materials, manufacturing methods, evaluation and test methods, environmental considerations, and current trends.

During a stop, a brake converts the kinetic energy of the moving vehicle into heat, absorbs the heat, and gradually dissipates it into the atmosphere. The brake is a sliding friction couple consisting of a rotor connected to the wheel and a stator on which is mounted the friction material (pad, lining, or block). The friction material is considered to be the expendable portion of the brake couple which, over a long period of use, is converted to debris and gases (1).

During engagement a clutch transfers the kinetic energy of a rotating crankshaft (coupled to a power source) to the transmission and wheels. Any slippage results in the generation of heat, which is absorbed and eventually dissipated to the atmosphere by the clutch. Thus the clutch is basically a static friction couple which momentarily slides during gear shifts. The clutch friction material is considered to be expendable.

Brakes and clutches operate both dry and wet. In dry friction couples, the heat is removed by conduction to the surrounding air and structural members. Wet friction couples operate within a fluid (usually an oil) which absorbs the heat and maintains the couple at relatively low temperatures (below 200°C). The fluid also traps the wear debris (see Hydraulic fluids).

Friction materials serve in a variety of ways to control the acceleration and deceleration of a variety of vehicles and machines which may be as small as a clutch in a business machine or a brake on a bicycle to as large as jumbo-aircraft brakes. The brakes on bicycles may have friction couples of iron sliding against iron in the coaster brake or rubber-bound composite pads sliding against a chromium-plated wheel rim in hand-activated brakes. Passenger cars may have drum brakes or disk brakes exclusively or a combination of disk fronts and drum rears. The friction materials may be resin- or rubber-bound composites based on asbestos or metallic fibers. Trucks and off-highway vehicles usually have very large drum brakes; only a few now have disk brakes, with friction couples that usually operate at higher friction levels and temperatures than passenger cars. Large aircraft are equipped exclusively with disk brakes which contain multiple rotor and stator arrangements with the most popular friction couple consisting of a sintered friction material sliding against a high-temperature resistant steel. The newer aircraft brakes consist of carbon composites serving as both the rotor and the stator (see Carbon).

Friction and Wear

Friction. A qualitative analysis of frictional force suggests that a frictional force is likely to consist of several components such as adhesion-tearing, ploughing (or abrasion), elastic and plastic deformation, and asperity interlocking, all occurring at the sliding interface. These mechanisms presumably depend upon the temperature as well as the normal load and sliding speed, since material properties are known to be dependent upon these variables. In the case of automotive friction materials, the coefficient of friction is usually found to decrease with increasing unit pressure and sliding speed at a given temperature, contrary to Amontons' law (2–4). This decrease in friction is controlled by the composition and microstructure of friction materials. As the temperature of the sliding interface increases, the coefficient of friction varies. This variation is unpredictable, and there exists no general trend except that at extremely high temperatures the coefficient may become very low (less than 0.1). This temporary loss in friction is referred to as fade (5). Like automotive friction materials, aircraft cermet friction materials exhibit decreasing coefficient of friction with increasing unit pressure (6).

Terminology. *Effectiveness*, essentially a measure of the stopping efficiency, can be expressed in a number of different ways: as the coefficient of friction, ie, hydraulic or air line pressure required, torque developed, or distance required to stop a vehicle. The various temperatures are identified for passenger cars as *cold* (under 100°C), *normal* (150–250°C), or *hot* (above 300°C). Effectiveness can be measured *new* or *off-rack* (without any prior use), *preburnished* (after little prior use), *burnished* (after moderate use), and *faded* (after use at elevated temperatures). Although the same terms are used for all friction materials, for large aircraft materials the temperatures are: cold, under 300°C; normal, 400–600°C; and hot, above 700°C. The normal fade-free maximum operating temperatures of various friction materials are: drum linings and clutch friction materials, 250°C; Class A organic disk pads, 300°C; Class B organic materials and blocks, 350°C; semimetallics, 400°C; and ceremetallics and carbon composites, 700°C.

Friction peaking is an increase in friction known to occur during or after prior high temperature operation. *Unbalance* occurs when friction peaking or fade causes one wheel or axle to change in friction, yielding side-to-side or front-to-rear unbalance. *Friction stability* is the ability of the friction materials to produce similar friction or friction changes at all wheels through all duty-cycles and especially during a return to normal operation after a temporary severe duty. *Recovery* from fade is the ability of the friction material to return to its prefade friction level. *Speed sensitivity* is the ability to maintain effectiveness at varying surface or rubbing speeds. Most materials show losses in effectiveness at higher speeds, with semimetallics being the notable exception. *Load sensitivity* is the ability to maintain effectiveness at various weight loadings. The ability of a friction material to recover from loss in effectiveness due to exposure to water is called *water recovery*.

Wear. For a fixed amount of braking the amount of wear of automotive friction materials tends to increase slightly or remain practically constant with respect to brake temperature, but once the brake rotor temperature reaches about 204°C the wear of resin-bonded materials increases exponentially with increasing temperature (7–10). This exponential wear is due to thermal degradation of organic components. At low temperatures the practically constant wear rate is primarily controlled by abrasion and adhesion (11). The wear, W, of friction materials can best be described by the following wear equation (12–13):

$$W = KP^a V^b t^c$$

where K is the wear coefficient, P the normal load, V the sliding speed, t the sliding time, and a, b, and c are a set of parameters for a given friction material–rotor pair at a given temperature.

Wear is an economic consideration. Wear resistance generally (not always) is inversely related to friction level and other desirable performance characteristics within any class of friction material. The formulator's objective is to provide the highest level of wear resistance in the normal use temperature range, a controlled moderate increase at elevated temperatures, and a return to the original lower wear rate when temperatures again return to normal. Contrary to common belief, maximum wear life does not require maximum physical hardness.

There are four possible mechanisms of wear at sliding surfaces: (1) Adhesion of the two materials, followed by cohesion failure of one as the two slide past each other. (2) Ploughing or gouging of one material by a fragment of another (harder) material. (3) Thermal or mechanical fatigue or melting, which permits solid pieces to become detached from the surface. (4) Oxidation and pyrolysis, leading to gas formation (see Ablative materials).

In the low temperature range Class A organic materials and semimetallic materials wear at a substantially lower rate than Class B organic materials. At extremely high temperatures the wear rate of Class B organic materials is the best, that of semimetallic materials next-best, and that of Class A organic materials the worst. In the intermediate temperature range, semimetallic materials are the best, followed by Class B organic materials.

Cermet or carbon friction materials operate at substantially higher temperatures than normal automotive or truck friction materials. Still the wear rates of these materials increase with brake temperature. One unique feature of these two materials is the formation of a glazed friction layer which reduces the wear rate. Without this glazed layer the wear rate is usually very high.

Types of Friction Materials

Organic Materials. The most common type of friction materials used in brakes and clutches for normal duty are termed organics. These materials usually contain about 30–40 wt % of organic components (14).

The primary applications of organic frictional materials and their requirements are as follows: (1) primary drum brake linings must provide high and stable friction at all temperatures and pressures; (2) secondary drum brake linings must provide stable friction and wear resistance; (3) Class A disk pads provide nonabrasive friction and wear properties, quiet operation, and rotor compatibility; (4) Class B disk pads provide higher friction and high temperature wear resistance at the expense of some low temperature wear resistance, noise properties, and rotor compatibility; (5) Class C friction materials consist of both disk pads and block-type friction materials for extremely heavy-duty operations where high friction, minimal fade, and wear resistance are essential at the expense of other brake characteristics; and (6) clutch friction materials provide stable friction, good wear properties, quiet operation, and rotor compatibility combined with high strength properties.

The major constituent of practically all organic friction materials is asbestos fiber, although small quantities of other fibrous reinforcement may be used. Asbestos is chosen because of its thermal stability, its relatively high friction, and its reinforcing properties (see Asbestos; Composite materials; Laminated and reinforced plastics). Since asbestos alone does not offer all of the desired properties, other materials called property modifiers are added to provide desired levels of friction, wear, fade, recovery, noise, and rotor compatibility. A resin binder holds the other materials together. This binder is not completely inert and makes contributions to the frictional characteristics of the composite. More commonly used ingredients can be found in various patents (15–17); see also Table 1 (18).

Table 1. Organic Friction Materials, Wt %[a]

	Copolymers[b]	5K asbestos	Sulfur	Zinc oxide	Cardolite no. 753	Resin	Additive 1	Additive 2
X-1	22.00	49.00	2.00	4.00			11.50[c]	11.50[d]
X-2	11.00	63.00	1.00	2.00			11.50[c]	11.50[d]
X-3	17.00	54.30	1.60	3.10			11.50[c]	11.50[d]
								1.00[e]
X-4	15.00	57.80	1.48	2.72			11.50[c]	11.50[d]
X-5	15.00	57.80		2.72			12.98[c]	11.50[d]
X-6	15.00	78.80	1.48	2.72		2.00		
X-7	15.00	68.80	1.48	2.72	10.00	2.00		
X-8	15.00	68.80	1.48	4.72	10.00			
X-9	15.00	78.80	1.48	2.72				
X-10	15.00	68.80	1.48	2.72		2.00	10.00[f]	
X-11	13.30	55.17	1.31	2.42	8.85	1.75	17.20[g]	
X-12	13.00	64.20	1.30	3.00	10.00	6.50	1.00[h]	1.00[e]
X-13	15.00	15.00	0.30	6.00		5.00	40.00[i]	18.70[j]
X-14	15.00	68.80	1.48	2.72		2.00	10.00[k]	
X-15	15.00	68.80		4.20	10.00	2.00		
X-16	17.00	48.80		4.20	30.00			
X-17	17.00	64.80		4.20	10.00		4.0[l]	
X-18	17.00	68.80		4.20	10.00			

[a] Ref. 18.
[b] Acrylonitrile–butadiene copolymer.
[c] Barytes.
[d] Rottenstone.
[e] Alundum (600X).
[f] B-196 friction particle.
[g] Zinc dust.
[h] Paraformaldehyde (curing agent).
[i] Steel wool.
[j] Iron oxide.
[k] B-221 friction particle.
[l] Orlon fiber.

Metallic Materials. Gray cast iron is of reasonably low cost, provides good wear resistance and damping characteristics, and has been in long time use as a brake drum or disk material for passenger cars and trucks. However, under severe operating conditions, gray cast iron can undergo phase transformations leading to heat checking of brake drums or disks (19). Copper or aluminum rotors have been evaluated experimentally (20–22) and alloy steel rotors are being used for certain nonautomotive brakes, including aircraft and trains. In the case of carbon-composite aircraft brakes the rotor material is the same as the stator material.

Semimetallic Materials. Semimetallics were introduced in the late 1960s and gained widespread usage in the mid 1970s. These materials usually contain more than 50 wt % metallic components. They are primarily used as disk pads and blocks for heavy-duty operation (23).

The major constituent of practically all semimetallics is iron powder in conjunction with a small amount of steel fiber. Various property modifiers are added to enhance performance to desired levels, and a resin binder—necessary to hold the materials together in a mass—is also added. Semimetallic friction materials may therefore consist of metallic powder, sponge iron particles, ceramic powder, steel fiber, rubber particles, graphite powder, and phenolic resin (24).

Sintered Materials or Cermets. Heavy weights and high landing speeds of modern aircraft or high-speed trains require friction materials that are extremely stable thermally. Organic or semimetallic friction materials are frequently unsatisfactory for these applications. Cermet friction materials are metal-bonded ceramic compositions (25–26). The metal matrix may be copper or iron. Several typical compositions are given in Table 2 (27) (see Glassy metals; Ceramics).

Carbon Composites. Cermet friction materials tend to be heavy, thus making the brake system less energy-efficient. Compared with cermets, carbon (or graphite) is a thermally stable material of low density and reasonably high specific heat. A combination of these properties makes carbon attractive as a brake material. Currently several companies are actively developing and field-testing lighter brake materials based on carbon fiber-reinforced carbon-matrix composites, which are to be used primarily for aircraft brakes (see Carbon and artificial graphite; Ablative materials).

Raw Materials

Binders. In selecting a resin binder system the processing characteristics must be considered along with the final frictional and physical properties. Two types of systems are used. In wet processing, the binder is a viscous liquid (usually a resole) having characteristics suited to thermoplastic processing techniques (see Plastics technology). In dry processing, the binder is a powdered material (usually a novolac) that is mixed directly with the other materials; it does not cross-link until heat and pressure are applied.

Synthetic resins such as phenolic and cresylic resins are most commonly used

Table 2. Cermet Friction Materials, Wt % [a]

Material	1	2	3	4	5	6	7	8	9	10
Matrix										
copper	47.1	49.6	44.6	44.7	44.9	46.2	52.4	43.6	18.6	46.2
zinc										5.5
tin								7.4		3.3
nickel	7.1	7.1	7.1			7.0	6.6	6.6	7.1	
titanium	3.6	3.6	3.6			3.5	3.3	3.3	3.6	
brass chips									28.6	
iron				15.0	15.0					8.8
Total	57.8	60.3	55.3	59.7	59.9	56.7	62.3	60.9	57.9	63.8
Friction material										
calcined kyanite	26.1	26.1	26.1	24.9	19.9	25.6	24.0	24.2	26.1	22.0
silica	4.8	4.8	4.8	4.6	4.6	4.8	4.4	4.5	4.8	4.4
Total	30.9	30.9	30.9	29.5	24.5	30.4	28.4	28.7	30.9	26.4
Lubricant										
graphite	1.2	1.2	1.2	1.1	6.1	1.1	1.1	1.1	1.2	8.8
hard coal										
lead						2.0	3.7			
Total	1.2	1.2	1.2	1.1	6.1	3.1	4.8	1.1	1.2	8.8
Antioxidant										
molybdenum	10.0	7.5	12.5	9.5	9.5	9.8	4.6	9.3	10.0	1.0
Total	10.0	7.5	12.5	9.5	9.5	9.8	4.6	9.3	10.0	1.0

[a] Ref. 27.

friction material binders, and are usually modified with drying oils, rubber, cardanol, or an epoxy (see Phenolic resins). They are prepared by the condensation of the appropriate phenol with formaldehyde in the presence of an acidic or basic catalyst. Polymerization takes place at elevated temperatures. Other resin systems are based on elastomers, drying oils, or combinations of the above or other polymers.

Metals such as copper, iron, tin, and lead are used as binders of sintered friction materials where deformation under the high forming pressure is required to lock together the property modifiers within a matrix.

Fibrous Reinforcements. The asbestos usually used in friction materials is chrysotile (mined in Quebec and Vermont). Chrysotile is the principal mineral of the serpentine group ($3MgO.2SiO_2.2H_2O$) (28). Long-fiber asbestos (eg, Grade 5) is generally used when dry processing techniques are employed in manufacturing. Shorter-fiber asbestos (eg, Grade 7) is used with wet processing techniques. The two grades differ considerably in length of fiber, bulk, absorptiveness, cost, and reinforcing value. Longer fibers permit the bending of secondary linings from flat blanks to curved segments.

Steel fiber may have a variety of aspect ratios, cleanliness (straight versus curved fibers), surface roughness, and chemical compositions. Fibers with tight specifications in terms of cleanliness, chemical composition, and aspect ratio are necessary for sintered friction materials. The fibers are usually machined from larger metallic forms.

Carbon fibers are produced by graphitization of organic or pitch fibers by techniques that provide parallel alignment of the carbon chain to the fiber length for maximum tensile strength.

Organic clutch materials contain continuous-strand reinforcements in addition to fibrous reinforcements. These include cotton (primarily for processing) and asbestos yarn, brass wire, or copper wire for high burst strength.

Property Modifiers. Property modifiers can, in general, be divided into two classes: nonabrasive and abrasive. Nonabrasive modifiers can be further classified as high friction and low friction.

The most frequently used nonabrasive modifier is a cured resinous friction dust derived from cashew nutshell liquid. Ground rubber is used in particle sizes similar to or slightly coarser than those of the cashew friction dusts for noise, wear, and abrasion control. Carbon black, petroleum coke flour, natural and synthetic graphite, or other carbonaceous materials are used to control the friction and improve wear (when abrasives are used) or to reduce noise. The above-mentioned modifiers are primarily used in organic and semimetallic materials except for graphite which is used in all friction materials.

Abrasive modifiers are used in several types of friction materials. Very hard materials such as alumina, silicon carbide, and kyanite are used in fine particle sizes in organic, semimetallic, and cermet materials (generally less than 74 μm or 200 mesh). Particle size is limited by the fact that large particles of such hard materials would groove cast-iron mating surfaces. Larger particle sizes are possible for harder mating surfaces in the special steel rotors used with sintered materials.

Minerals are generally added to improve wear resistance at minimum cost. The most commonly used are ground limestone (whiting) and barytes, although various types of clay, finely divided silicas, and other inexpensive or abundant inorganic materials may also perform this function.

Metal or metal oxides may be added to perform specific functions. Brass chips and copper powder are frequently used in heavy-duty organics where they act as scavengers to break up undesirable surface films. Zinc and aluminum are also used. Zinc chips used in Class A organics contribute significantly to recovery of normal performance following fade. Most of these inorganic materials tend to detract from anti-noise properties and mating surface compatability.

Manufacturing

An important relation exists between composition, including performance requirements and ease of manufacture. Organic linings which must bend also require higher resin contents and longer fibers. Heavy-duty materials with reduced resin loading for improved performance require molding-to-shape. Sintered and carbon friction materials require high pressure forming and high temperature treatment in inert atmospheres. Woven and some clutch materials require special fiber-forming methods.

Linings. Most linings are produced from resin wet-mix by either an extrusion or a rolling process. Initially, the fibrous reinforcement and the friction modifiers are mixed with a liquid resin at approximately 50°C. The binder solvent serves as a plasticizer to yield a dense putty-like mass with good wet strength. In the extrusion process, the mix is heated to approximately 90°C and extruded at 13.8–27.6 MPa (2000–4000 psi) as a flat, pliable tape which is dried for 2 h at 80°C. In the rolling process, the mix is partially dried, sized into particles, then fed between two rolls of slightly different speeds to align the fibers in the flat, pliable tape or green lining which is formed.

The green lining is then cut to length, formed into an arcuate segment at 150°C, then placed in curved mold cavities and cured 4–8 h at 180–250°C. Final grinding produces the finished brake lining.

Linings for heavy-duty use such as for medium-sized trucks are produced by a dry-mix process. The fiber, modifiers, and a dry novolac resin are mixed in a Littleford mixer. The blend is then formed into about a 60 by 90 cm briquet at 2.8–4.1 MPa (400–600 psi). The briquets are hot-pressed for 3–10 min at 140–160°C and then cooled. The resin is only partially cured at this point and will be thermoplastic when subsequently reheated for bending. The hot-pressed briquets are then cut to desired size and bent at 170–190°C and cured in curved molds for 4–8 h at 220–280°C. Final grinding produces the finished segments.

Disk Pads. Organic and semimetallic disk pads are produced by somewhat similar processes after the mixes are formed. The mix for organics is prepared as for heavy-duty linings. The mix for semimetallics is produced in a less intensive blender. The mix is then formed into briquets at room temperature and 27.6–41.4 MPa (4000–6000 psi). These briquets are then hot-pressed at 160–180°C for 5–15 min at 27.6–55.2 MPa (4000–8000 psi). The pads are then cured at 220–300°C for 4–8 h and ground to produce the finished disk pads.

In many instances the friction material mix is integrally molded into holes within the backing plate or shoe. Painting of the final assembly is rare in the United States but it is the rule in Europe.

Blocks. The mix for organic blocks is prepared as for heavy-duty segments and the mix for semimetallic blocks is prepared as for semimetallic disk pads. Briquets are formed at 10.3–17.2 MPa (1500–2500 psi). To reduce blisters in hot pressing, the briquets may be heated to 90°C for 15–30 min. The blocks are formed at 130–150°C at 13.8–20.7 MPa (2000–3000 psi) for periods of 10–30 min. After slitting to width, the blocks undergo grinding of internal and external radii. Final cure may be in an unconfined form at temperatures as low as 180°C for 15 h or in confined form at temperatures as high as 280°C for 6 h. Grinding, drilling, and chamferring produce the final block.

Clutch Materials. Methods for producing clutch friction materials are concerned with the placement of the reinforcement strand or wire within the matrix. Processing includes: molding of mix without strands or wire; molding of mix around strand or wire preforms; and weaving curable preforms. In the first two cases, a dry mix is used. In the latter case, a wet mix is prepared and a strand is run through the premix to pick up the viscous mass along the strand which can be woven after drying. After hot-pressing and curing, the surface is ground to final shape.

For trucks and heavy off-the-road equipment, cermet friction materials (see below) are also used. Sintered cermet segments are attached to a metal plate to form the clutch facing.

Woven Bands. Woven bands for heavy-duty operation are produced by an expensive process which begins with an asbestos cord, which may be reinforced with wire, being passed through a wet-mix to pick up resin and modifiers. The saturated cord is then woven into tapes which pass through heated rolls to partially cure the resin. The material can be post-cured at low temperatures (ca 160°C) to remain as a flexible roll lining or post-cured at higher temperature (180–230°C) to form rigid segments. Such materials are used in large band brakes used to control large machinery.

Cermets. Cermet materials are manufactured using the powder metallurgy (qv) technique. Desired amounts of individual ingredients are weighed, mixed, compacted, sintered, and coined (or recompacted). The sintering is performed in a reducing or neutral atmosphere, and the sintering temperature has to be high enough so that the metal ingredients will adhere to each other.

Carbon Composites. In this class of materials, carbon or graphite fibers are embedded in a carbon or graphite matrix. The matrix can be formed by two methods: chemical vapor deposition and coking. In the case of chemical vapor deposition a hydrocarbon gas is introduced into a reaction chamber in which carbon formed from the decomposition of the gas condenses on the surface of carbon fibers. An alternative method is to mold a carbon fiber–resin mixture into shape and coke the resin precursor at high temperatures. In both of the methods the process has to be repeated until a desired density is obtained.

Evaluation Methods

Chemical, Physical, and Mechanical Tests. Manufactured friction materials are characterized by various chemical, physical, and mechanical tests in addition to friction and wear testing. The chemical tests include tga (thermogravimetric analysis), dta (differential thermal analysis), pgc (pyrolysis gas chromatography), acetone extraction, lc (liquid chromatography), ir analysis, and x-ray or sem (scanning electron microscope) analysis (see Analytical methods). Physical and mechanical tests determine properties such as thermal conductivity, specific heat, tensile or flexural strength, and hardness.

Dynamometer and Vehicle Testing. Friction materials are evaluated in the laboratory by a great variety of tests and equipment. Friction and wear characteristics can be evaluated with sample dynamometers such as the Chase machine. In the most reliable sample test machines the output torque is controlled so that different materials all do the same amount of work. One disadvantage of sample test machines is that the ratios of friction-material area to rotor area and friction-material mass to rotor mass are quite different from the ratios used on vehicles. The heat generation, storage, and rejection conditions are therefore quite different. A second disadvantage is that only one material is tested, whereas in vehicles with drum brakes two types of friction materials are used together and there are interaction effects. The advantage is mainly one of economics—more tests at less cost.

The full brake dynamometer, when properly instrumented and controlled, reflects with reasonable accuracy the actual brake performance in a vehicle. The high initial investment is recovered through operation independent of the climatic conditions and by a fully automatic operation for extended periods, thus minimizing manpower costs.

No attempt will be made here to detail the numerous vehicle test procedures used by different organizations. Performance tests are essentially designed to appraise initial effectiveness, burnish and normal effectiveness, fade and recovery, and final effectiveness. Side-to-side and front-to-rear balance can also be determined. Only vehicle tests can determine noise properties accurately. Wear measurements are generally made in accelerated performance tests, but the results are a reflection of high-temperature wear properties and are usually not valid for predicting normal driving wear. More valid predictions of normal wear life result from specifically designed extended road traffic wear measurement tests involving a great number of stops with restricted maximum temperatures.

Vehicle tests are considered the ultimate in friction material evaluation, but to be accurate they must be carefully designed to eliminate variations caused by changing weather. Controlled-temperature tests and parallel-test control vehicles normally perform the function satisfactorily, but at increased cost.

Environment and Health

Manufacturing. Organic friction materials have been in increasing jeopardy because OSHA regulations have limited the exposure of workers to airborne asbestos fibers. High performance friction materials can only be produced by the dry-mix process and this tends to be dustier than other processes. Prior to any regulations, typical fiber concentrations were in the 10–30 fibers/cm^3 range. In 1974 the time-weighted average was set at 5 fibers/cm^3 and in 1976 it was reduced to 2 fibers/cm^3. The cost of capital equipment to effect these improvements is extremely high. OSHA has been considering a further reduction to 0.5 or 0.1 fibers/cm^3.

Because of the wide variety of asbestos bundle sizes and even individual fibrils in commercial asbestos, the handling of asbestos causes degradation of the larger fiber bundles to fibers with diameters less than 2 µm which remain airborne for extended periods of time. These airborne fibers are prone to inhalation and lung entrapment which may lead to disabling diseases. The exact definition of harmful fibers and the mechanism by which they affect the body is not accurately known.

Friction materials are known to contain other harmful materials. Lead is found in secondary linings, Class A, B, and C organic disk pads, and other friction materials. At least one original equipment and after-market supplier is known to have a policy against incorporation of lead in its products.

Wear Products. Friction material and rotor emissions are generated by normal wear. Because of large-scale usage and the potential health hazard of asbestos, organic friction materials and their wear debris have been extensively studied (1,29–30). Below 250°C, abrasive and adhesive wear are considered to be the most important mechanisms and the wear rates are low. Above 250°C, organic friction materials begin to pyrolyze or oxidize such that both gases and particulates are released.

In order to define the extent of emissions from automotive brakes and clutches, a study was carried out in which specially designed wear debris collectors were built for the drum brake, the disk brake, and the clutch of a popular United States vehicle (1). The vehicle was driven through various test cycles to determine the extent and type of brake emissions generated under all driving conditions. Typical original equipment and after-market friction materials were evaluated. Brake relinings were made to simulate consumer practices. The wear debris was analyzed by a combination of optical and electron microscopy to ascertain the asbestos content and its particle size distribution. It was found that more than 99.7% of the asbestos was converted to a nonfibrous form and that only 3.2% of the total asbestos was emitted to the atmosphere. A second study of brake emissions adjacent to a city freeway exit ramp on the downwind side indicated that the asbestos emissions were so low they could not be distinguished from the background on the upwind side (30).

Future Prospects

The trend toward more energy-efficient passenger cars and trucks will put an increased demand on friction materials. There is already a trend toward smaller and lighter vehicles. More efficient vehicles have manual transmissions with smaller brakes. Organic friction materials will continue to serve the drum brake industry. In the past 10 years, more and more vehicles were equipped with ventilated disk brakes. In time these may be replaced by solid disk brakes for weight savings. When the brakes become smaller, thus producing higher brake temperatures, the Class A organic materials will become less suitable and Class B organic and semimetallic materials will fill the gap. Trucks and other heavy vehicles are moving toward more efficient disk brakes. More sintered friction materials will appear in the heavy-vehicle clutch market. At the same time, aircraft will move toward light carbon brakes.

Future brakes will be required to satisfy health standards. Some passenger car manufacturers are known to be in favor of removing all asbestos from passenger cars. One large automotive manufacturer has a policy for removal of asbestos-based friction materials by model year 1981. The only proven asbestos-free friction materials at this time are the semimetallics. In some cases, current vehicles already have front disk brakes equipped with semimetallics (15–20% of the original equipment market in model year 1977). Equipping rear disk brakes with semimetallic materials also would be one way to achieve a vehicle with asbestos-free friction materials before 1980.

Current friction materials are relatively noise-free. The major problem with new heavier-duty friction materials will be their noise properties. Noise-free friction materials will continue to be used in the heavy-truck and bus markets (see Noise pollution).

BIBLIOGRAPHY

"Brake Linings and Other Friction Facings" in *ECT* 1st ed., Vol. 2, pp. 622–628, by F. C. Stanley, The Raybestos Division, Raybestos-Manhattan, Inc.; "Friction Material" in *ECT* 2nd ed., Vol. 10 pp. 124–134, by Clyde S. Batchelor, The Raybestos Division, Raybestos-Manhattan, Inc.

1. M. G. Jacko, R. T. DuCharme, and J. H. Somers, *SAE Trans.* **82,** 1813 (1973).
2. S. K. Rhee, *SAE Trans.* **83,** 1575 (1974).
3. S. K. Rhee, *Wear* **28,** 277 (1974).
4. W. R. Tarr and S. K. Rhee, *Wear* **33,** 373 (1975).
5. J. M. Herring, *Mechanisms of Brake Fade in Organic Brake Linings, SAE Paper No. 670146,* SAE, New York, Jan. 1967.
6. N. A. Hooton, *Bendix Tech. J.* **2,** 55 (1969).
7. S. K. Rhee, *SAE Trans.* **80,** 992 (1971).
8. S. K. Rhee, *Wear* **29,** 391 (1974).
9. T. Liu and S. K. Rhee, *Wear* **37,** 291 (1976).
10. T. Liu and S. K. Rhee, "High-Temperature Wear of Semimetallic Disc Brake Pads," in K. C. Ludema, W. A. Glaeser, and S. K. Rhee, eds., *Wear of Materials—1977,* American Society of Mechanical Engineers, New York, 1977.
11. S. K. Rhee, *Wear* **23,** 261 (1973).
12. S. K. Rhee, *Wear* **16,** 431 (1970).
13. S. K. Rhee, *Wear* **18,** 471 (1971).
14. F. W. Aldrich and M. G. Jacko, *Bendix Tech. J.* **2**(1), 42 (Spring 1969).
15. U.S. Pat. 2,428,298 (Sept. 30, 1947), R. E. Spokes and E. C. Keller (to American Brake Shoe Co.).
16. U.S. Pat. 2,685,551 (Aug. 3, 1954), R. E. Spokes (to American Brake Shoe Co.).
17. U.S. Pat. 3,007,549 (Nov. 7, 1961), B. W. Klein (to Bendix Corporation).
18. U.S. Pat. 3,007,890 (Nov. 7, 1961), S. B. Twiss and E. J. Sydor (to Chrysler Corporation).
19. S. K. Rhee, *Characterization of Cast Iron Friction Surfaces, SAE Paper No. 720056* SAE, New York, Jan. 1972.
20. S. K. Rhee, R. M. Rusnak, and W. M. Spurgeon, *SAE Trans.* **78,** 1031 (1969).
21. S. K. Rhee, J. L. Turak, and W. M. Spurgeon, *SAE Trans.* **79,** 503 (1970).
22. S. K. Rhee and J. E. Byers, *SAE Trans.* **81,** 2085 (1972).
23. B. W. Klein, *Bendix Tech. J.* **2**(3), 109 (Autumn 1969).
24. U.S. Pat. 3,835,118 (Sept. 10, 1974), S. K. Rhee and J. P. Kwolek (to Bendix Corporation).
25. N. A. Hooton, *Bendix Tech. J.* **2,** 55 (1969).
26. K. Aoki and J. Shirotani, *Bendix Tech. J.* **6,** 1 (1973–1974).
27. U.S. Pat. 2,948,955 (Aug. 16, 1960), A. W. Allen and R. H. Herron.
28. A. A. Hodgson, "Fibrous Silicates," *Lecture Series No. 4,* Royal Institute of Chemistry, London, Eng., 1965.
29. A. E. Anderson and co-workers, "Asbestos Emissions from Brake Dynamometer Tests," *SAE Paper No. 730549,* SAE, New York, May 1973.
30. J. C. Murchio, W. C. Cooper, and A. DeLeon, "Asbestos Fibers in Ambient Air of California," *University of California (Riverside) Report EHS No. 73-2,* Mar. 1973.

M. G. JACKO
S. K. RHEE
Bendix Corporation

BRANDY. See Beverage spirits, distilled.

BRASS. See Copper alloys.

BREAD. See Bakery processes and leavening agents.

BRIGHTENERS, FLUORESCENT

The operation of bleaching or brightening is concerned with the preparation of fabrics whose commercial value is dependent on the highest possible whiteness. In bleaching, textile converters and paper manufacturers follow procedures that are concerned with the removal of colored impurities or their conversion into colorless substances. In chemical bleaching, impurities are oxidized or reduced to colorless products. Physical bleaching involves the introduction of a complementary color whereby the undesired color is made invisible to the eye in an optical manner. In the long-known operation of blueing, the yellow cast of substrates such as textiles, paper, sugar, etc, is eliminated by means of blue or blue-violet dyes. Through color compensation the treated product appears whiter to the eye; however, it is actually grayer than the untreated material.

With the aid of optical brighteners, also referred to as fluorescent whitening agents (FWAs) or fluorescent brightening agents, optical compensation of the yellow cast may be obtained. The yellow cast is produced by the absorption of short-wavelength light (violet-to-blue). With optical brighteners this lost light is in part replaced, thus a complete white is attained without loss of light. This additional light is produced by the brightener by means of fluorescence. Optical brightening agents absorb the invisible ultraviolet portion of the daylight spectrum and convert this energy into the longer-wavelength visible portion of the spectrum, ie, into blue to blue-violet light (see Fig. 1). Optical brightening, therefore, is based on the addition of light, whereas the blueing method achieves its white effect through the removal of light.

Two requirements are indispensable for an optical brightener: it should be optically colorless on the substrate, and it should not absorb in the visible part of the spectrum. In the application of optical brighteners it is possible not only to replace the light lost through absorption, thereby attaining a neutral, complete white but, through the use of excess brightener, to convert still more uv radiation into visible light,

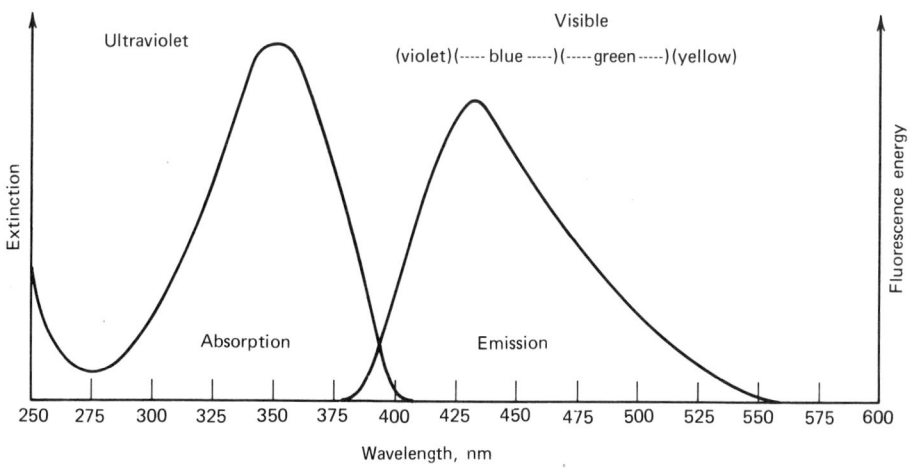

Figure 1. Absorption and emission spectra in solution of a compound of structure (1). (See Table 1.)

so that the whitest white is made still more sparkling. Since the fluorescent light of an optical brightener is itself colored, ie, blue-to-violet, the use of excess brightener always gives either a blue-to-violet or, sometimes, a bluish-green cast.

The principle of optical bleaching was described in 1929 by Krais (1), but the industrial use of optical brightening began about ten years later. Since that time optical brighteners have found increasing use in the most diverse fields (2–5). The toxicological properties of optical brighteners have been summarized (6). The commercial products investigated thus far have all been found to be completely harmless. More than 2000 patents for FWAs exist, there are several hundred commercial products, and approximately one hundred producers and distributors.

The 1975 worldwide consumption of FWAs has been estimated at over 45,000 metric tons of commercial product, of which approximately 15,000 t found use in the paper industry, 500 t in plastics, 20,000 t were used in washing powders, and roughly 10,000 t in the textile industry.

Many chemical compounds have been described in the literature as fluorescent, and since the 1950s, intensive research has yielded many more fluorescent compounds that provide a suitable whitening effect; however, only a small number of these compounds found practical uses. Collectively these compounds belong to the aromatic or heterocyclic series; many of them contain condensed ring systems. An important feature of these compounds is the presence of an uninterrupted chain of conjugated double bonds, the number of which is dependent on substituents as well as the planarity of the fluorescent part of the molecule. Almost all of these compounds are derivatives of stilbene or 4,4′-diaminostilbene, biphenyl, 5-membered heterocycles such as triazoles, oxazoles, imidazoles, etc, or 6-membered heterocycles (coumarins, naphthalimide, s-triazine, etc). The principle of classification used here is based largely on practical considerations.

Types of Brightening Agents

Stilbene Derivatives. *4,4′-Bis(triazin-2-ylamino)stilbene-2,2′-disulfonic Acids.* A considerable number of the commercial brighteners are bistriazinyl derivatives (1) of 4,4′-diaminostilbene-2,2′-disulfonic acid as shown in Table 1. The usual compounds are built up symmetrically, wherein X and X′, and Y and Y′ are the same. Their preparation is carried out by reaction of 2 moles of cyanuric chloride with 1 mole of 4,4′-diaminostilbene-2,2′-disulfonic acid. Asymmetric derivatives can be synthesized via 4-amino-4′-nitrostilbene-2,2′-disulfonic acid; however, their preparation is more expensive, and they show little advantage over the symmetrical compounds. The principal effects of these variations are changes in solubility, substrate affinity, acid fastness, etc. The bistriazinyl compounds are not stable toward hypochlorite; however, some compounds show a certain amount of fastness after application to the fiber. The bistriazinyl brighteners are employed principally on cellulosics, such as cotton or paper. Some products also show affinity for nylon at the weakly alkaline pH of most commercial detergents.

Mono(azol-2-yl)stilbenes. Mono(azol-2-yl)stilbenes arose from efforts to find hypochlorite-stable products with neutral fluorescence.

2-(Stilben-4-yl)naphthotriazoles (**2**) are prepared by diazotization of 4-aminostilbene-2-sulfonic acid or 4-amino-2-cyano-4′-chlorostilbene, coupling with an ortho-coupling naphthylamine derivative, and finally, oxidation to the triazole. With

Table 1. Examples of Symmetrical Derivatives[a]

(1) [structure: R'-substituted triazine–NH–phenyl(SO₃H)–CH=CH–phenyl(HO₃S)–NH–triazine-R',R]

CAS Registry No.	R	R'	References
(E) [17118-48-9]	—anilino	—methoxy	7
[31900-04-6], disodium salt	—anilino	—methylamino	8–9
(E) [17118-46-6]	—anilino	—N-methyl-N-hydroxyethylamino	10–11
(E) [17118-44-4]	—anilino	—bis(hydroxyethyl)amino	12–14
(E) [32466-46-9]	—anilino	—morpholino	15
(E) [17863-51-3]	—anilino	—anilino	16
[16470-24-9], tetrasodium salt	—sulfanilic acid	—bis(hydroxyethyl)amino	17
(E) [17118-40-0]	—metanilic acid	—bis(hydroxyethyl)amino	18–19
[41098-56-0]	—anilin-2,5-disulfonic acid	—diethylamino	20–21

[a] The groups R and R' are substituted or unsubstituted amino groups, substituted hydroxyl groups, etc.

water-solubilizing groups these types of compounds are suitable for brightening cellulosic materials or nylon from soap and detergent baths. In solution and on the substrate these compounds show good fastness to hypochlorite and to light. Water-insoluble derivatives of this family, eg, compounds having the nitrile group, are suitable for brightening synthetic fibers and resins.

2-(4-Phenylstilben-4-yl)benzoxazoles (3) are prepared by means of the anil synthesis from 2-(4-methylphenyl)benzoxazoles and 4-biphenylcarboxaldehyde anil and used for brightening polyester fibers (22–23).

Bis(azol-2-yl)stilbenes (24–25). 4,4'-Bis(triazol-2-yl)stilbene-2,2'-disulfonic acids. 4,4'-Dihydrazinostilbene-2,2'-disulfonic acid, obtained from the diamino compound, on treatment with 2 moles of oximinoacetophenone and subsequent ring closure leads to the formation of structure (4). Such compounds are used chiefly as washing powder additives for the brightening of cotton fabrics, and exhibit excellent light- and hypochlorite-stability.

Styryl Derivatives of Benzene and Biphenyl. A further group of compounds based on the styryl group was prepared to lengthen the conjugated system of stilbene.

1,4-Bis(styryl)benzenes. 1,4-Bis(styryl)benzenes (5) are obtained by the Horner modification of the Wittig reaction, eg, 1,4-bis(chloromethyl)benzene is treated with 2 moles of triethyl phosphite and the resulting phosphonate reacts with 2 moles of o-cyanobenzaldehyde to yield the bisstyryl compound (28). A strong brightening effect with a reddish cast is obtained on polyester fibers.

4,4'-Bis(styryl)biphenyls. 4,4'-Bis(styryl)biphenyls are also obtained by the Horner-Wittig reaction of the phosphonate derived from 1,4-bis(chloromethyl)biphenyl and triethyl phosphite with benzaldehyde-o-sulfonic acid, giving the corresponding bisstyrylbiphenyl (6) (29). They are used in washing powders for brightening cotton to a very high degree of whiteness with improved hypochlorite stability.

Pyrazolines. *1,3-Diphenyl-2-pyrazolines.* 1,3-Diphenyl-2-pyrazolines (7) (Table 2) are obtainable from appropriately substituted phenylhydrazines by the Knorr re-

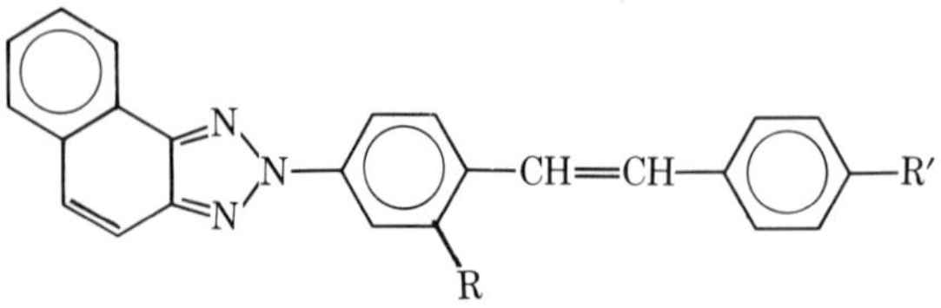

examples:
R = —SO₃H, R' = —H [4434-38-2] (26)
R = —C≡N, R' = —Cl [5516-20-1] (27)

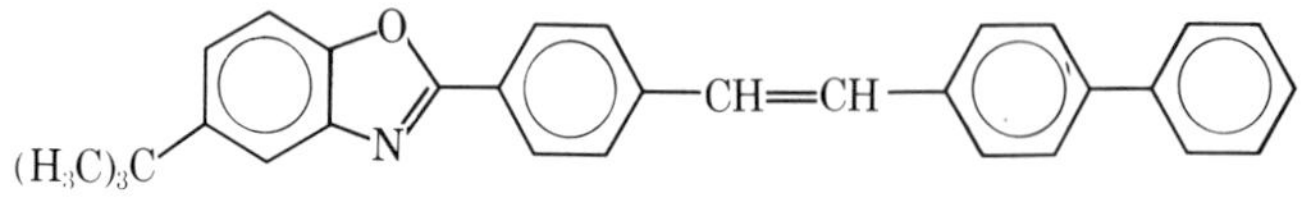

(3)

[16143-18-3]

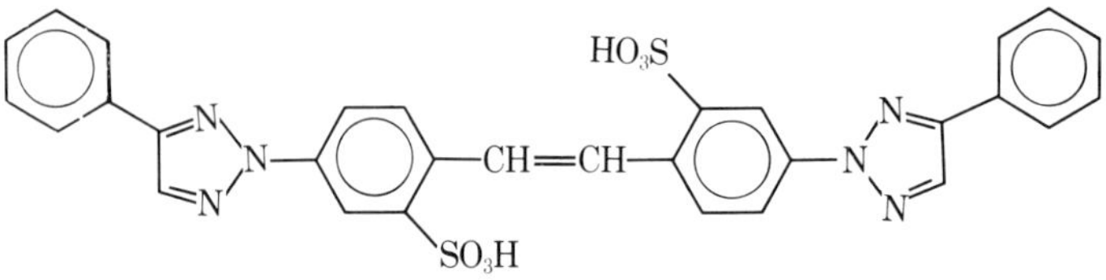

(4)

disodium salt [23743-28-4]

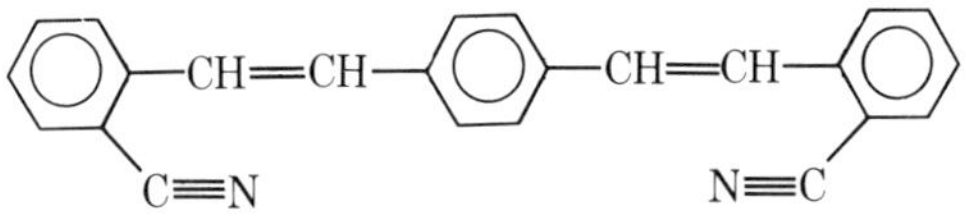

(5)

[13001-39-3]

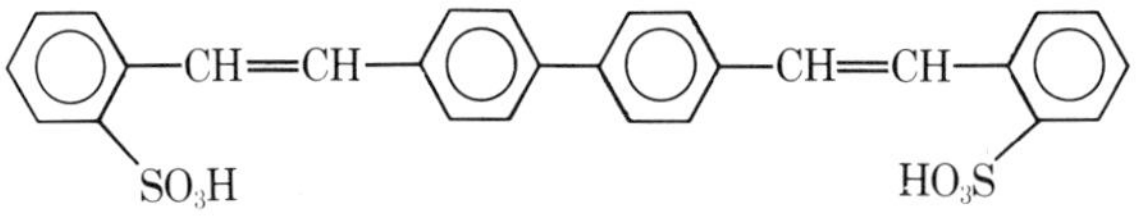

(6)

disodium salt [27344-41-8]

action with either β-chloro- or β-dimethylaminopropiophenones (30–31). They are employed for brightening synthetic fibers such as polyamides, cellulose acetates, and polyacrylonitriles.

Bis(benzazol-2-yl) Derivatives. *Bis(benzoxazol-2-yl) Derivatives.* Bis(benzoxazol-2-yl) derivatives (8) (Table 3) are prepared in most cases by treatment of dicarboxylic acid derivatives of the central nucleus X, (eg, stilbene-4,4'-dicarboxylic acid, naphthalene-1,4-dicarboxylic acid, thiophene-2,5-dicarboxylic acid, etc), with 2 moles of an appropriately substituted o-aminophenol, followed by a ring-closure reaction. These compounds are suitable for the brightening of plastics and synthetic fibers.

Table 2. 1,3-Diphenyl-2-pyrazoline Derivatives

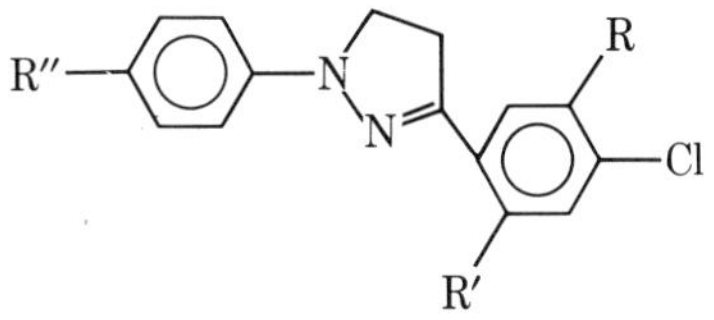

(7)

CAS Registry No.	R	R'	R''	References
[2697-84-9]	—H	—H	—SO$_3$H	32
[2744-49-2]	—H	—H	—SO$_2$NH$_2$	33–34
[38848-70-3]	—H	—H	—SO$_2$NHCH$_2$CH$_2$CH$_2$N(CH$_3$)$_3$ ⁻SO$_3$OCH$_3$	35
[27441-70-9]	—H	—H	—SO$_2$CH$_2$CH$_2$SO$_3$H, sodium salt	36

[6608-82-8]	—H	—H	—SO$_2$CH$_2$CH$_2$O—CH—CH$_2$—N(CH$_3$)$_2$
			                    |
			                    CH$_3$
				37

[35441-13-5]	—Cl	—CH$_3$	—SO$_2$CH$_2$CH$_2$SO$_3$H	38–39

Table 3. Bis(benzoxazol-2-yl) Derivatives

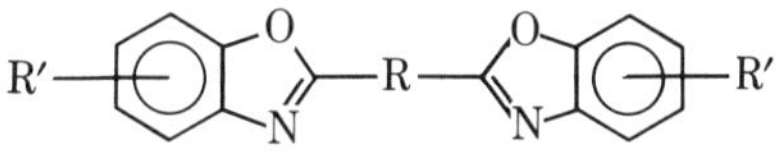

(8)

R[a]	R' =	CAS Registry No.	References
—CH=CH—	alkyl, 5-CH$_3$	[1041-00-5]	40–41
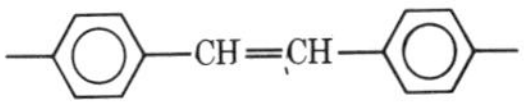 —⌬—CH=CH—⌬—	H alkyl	[1533-45-5]	42–43
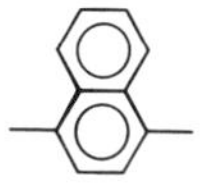 (naphthyl)	H	[5089-22-5]	44–46
(thienyl)	COO-alkyl, SO$_2$-alkyl H alkyl	[2866-43-5]	47

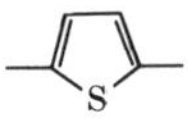

[a] R represents a conjugated system.

Bis(benzimidazol-2-yl) Derivatives. A large number of patents cover derivatives (9) of these structures. In structure (9) two benzimidazole radicals are joined by a conjugated bridge. Besides being effective on cotton, compounds of this type show good affinity for nylon.

2-(Benzofuran-2-yl)benzimidazoles. 2-(Benzofuran-2-yl)benzimidazoles (10) may be synthesized by reaction of substituted benzofuran-2-carboxylic acid chlorides with substituted o-phenylenediamines and ring closure of the resulting o-aminoamide, followed by quaternization (48–51). Such products are brighteners for synthetic fibers, in particular those of polyacrylonitrile.

Coumarins, Carbostyrils. *7-Hydroxy and 7-(Substituted amino)coumarins.* By treatment of flax with esculin, a glucoside of esculetin (11), a brightening effect was achieved, however this effect was not fast to washing and light. Later, the use of β-methylumbelliferone (12) and similar compounds as brighteners for textiles and soap was patented.

As an improvement over β-methylumbelliferone (55–57), 4-methyl-7-aminocoumarin derivatives (13) (58–61) were proposed. These compounds are used for brightening wool and nylon either in soap powders or detergents, or as salts under acid dyeing conditions. They are obtained by the Pechmann synthesis from appropriately substituted phenols and β-ketocarboxylic acid esters or nitriles in the presence of Lewis acid catalysts (see Coumarin).

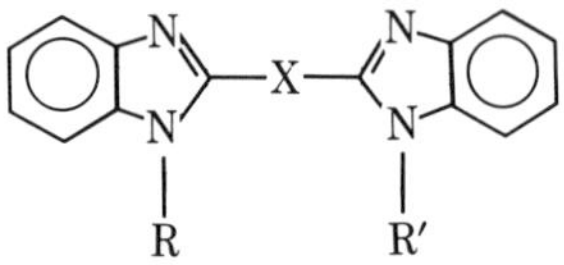

(9)

X = —CH=CH— (52–53)   X = —⟨⟩—CH=CH—⟨⟩— (54)

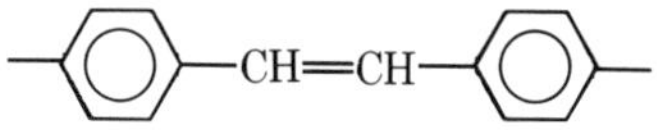

[95-34-1]
[2620-60-2]

[92-37-5]
[66675-85-2]

R = R' = H
R = R' = CH₃
R, R' = H, alkyl, hydroxyalkyl, etc

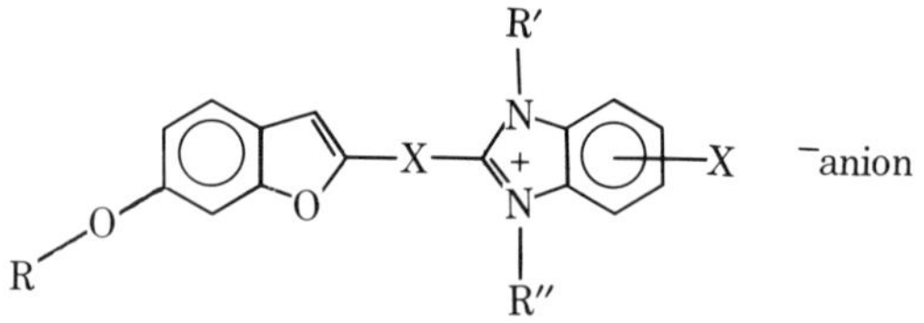

(10)

R = alkyl, alkylaryl
R', R" = X = H, colorless substituent

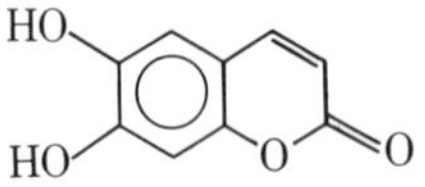

(11) esculetin [305-01-1]

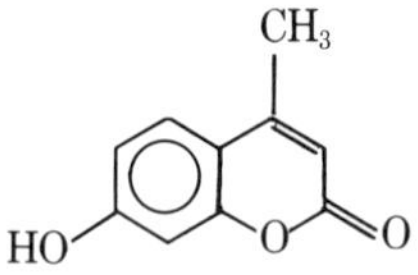

(12) β-methylumbelliferone [90-33-5]

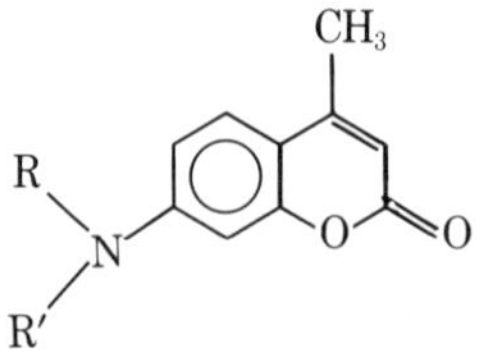

(13)
R, R' = H, alkyl, hydroxyalkyl, aryl, etc
R = R' = H [26093-31-2]
R = R' = CH₃ [87-01-4]

3-Phenyl-7-(triazin-2-ylamino)coumarins (14). A further development in the coumarin series is the use of derivatives of 3-phenyl-7-aminocoumarin as building blocks for a series of light-stable brighteners for various plastics and synthetic fibers, and, as the quaternized compounds, for brightening polyacrylonitrile (62).

3-Phenyl-7-aminocoumarin is obtained by a Knoevenagel reaction of substituted salicylaldehydes with phenylacetic acid or benzyl cyanide. Further synthesis of the individual end-products is carried out by usual procedures.

Other related substances are 3-phenyl-7-(azol-2-yl)coumarins (15) and 3,7-bis(azolyl)coumarins (16).

Carbostyrils. Carbostyrils (17) are prepared by the reaction of 2-alkylamino-4-nitrotoluene with ethyl glyoxalate in the presence of piperidine, reduction of the resulting 3-phenyl-7-nitrocarbostyril derivatives, and finally methylation of the corresponding amino compounds (68). They are brighteners for polyamides, wool, and cellulose acetates (see Cellulose acetate and triacetate fibers).

Other Heterocyclic Systems. *Naphthalimides.* Naphthalimides (18) are derivatives of 4-aminonaphthalimide (69) and are used in plastics. The alkoxynaphthalimides (19) are of particular interest as optical brighteners.

Naphthalimides are prepared from naphthalic anhydride, obtained from naph-

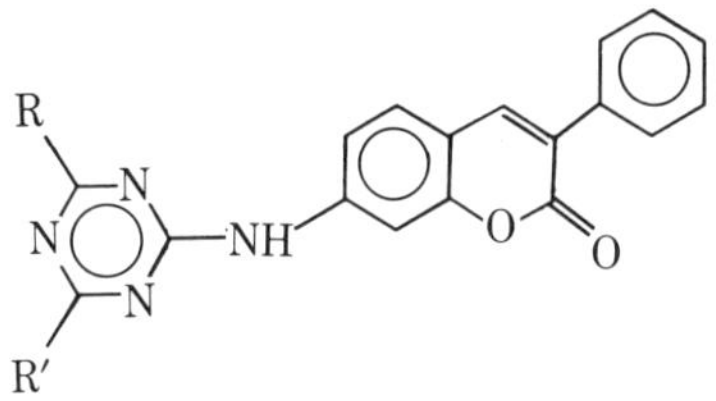

(14)
R, R' = Cl, substituted amines

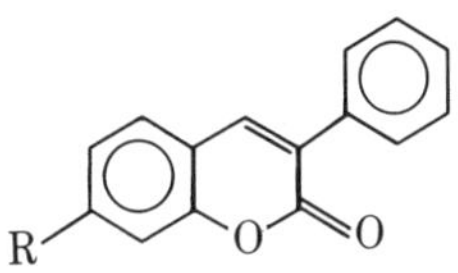

(15)
R = 3-methylpyrazol-1-yl [6025-18-9] (63);
4-methyl-5-phenyl-1,2,3-triazol-2-yl [19683-09-1] (64);
naphthotriazol-2-yl [3333-62-8] (65)

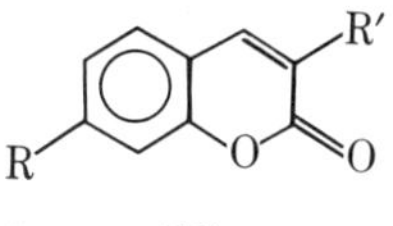

(16)
R = *vic*-triazol-2-yl-, R' = 4-alkyl-1,2,4-triazol-1-yl-, quat (66)
R = *vic*-triazol-2-yl-, R' = pyrazol-1-yl- [66634-92-2] (67)

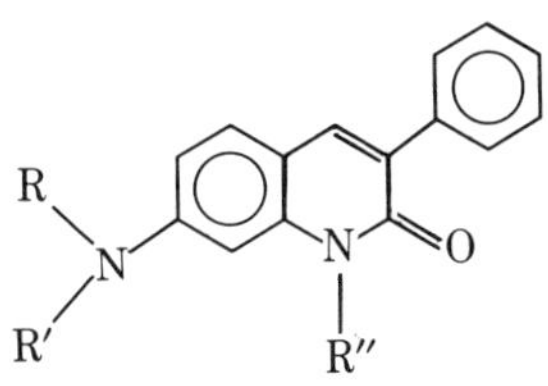

(17)
R, R', R" = alkyl, eg, —C$_2$H$_5$ [33934-60-0]

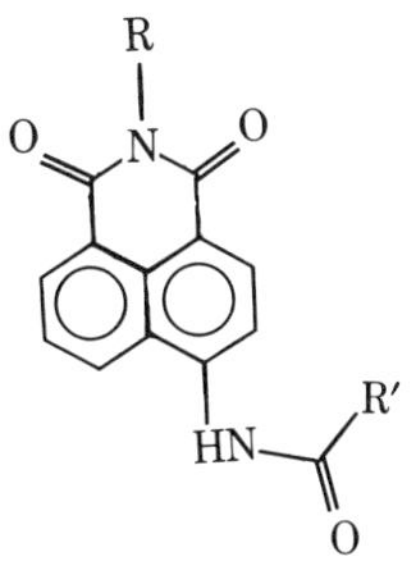

(18)
R = alkyl, —aryl
R' = alkyl, —NH—aryl

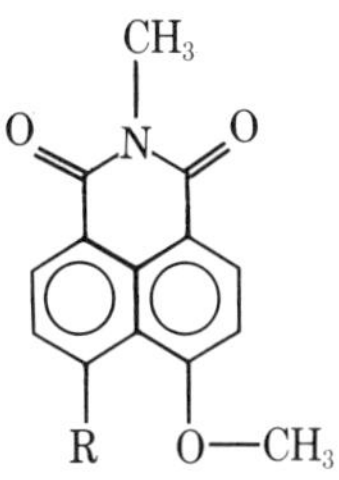

(19)
R = —H [3271-05-4] (70)
—OCH$_3$ [22330-42-3] (71)

thalene-1,8-dicarboxylic acid, the oxidation product of acenaphthene or its derivatives, by reaction with amines. They are utilized for the brightening of synthetic fibers such as polyesters.

Derivatives of Dibenzothiophene-5,5-dioxide. A further group of brighteners was found in the derivatives (20) of 3,7-diaminodibenzothiophene-2,8-disulfonic acid-5,5-dioxide. The preferred acyl groups are alkoxybenzoyl (72–74). These compounds give a greenish fluorescence and are relatively weak in comparison with stilbene derivatives on cotton; however, they show good stability to hypochlorite.

Pyrene Derivatives. Compound (21) is obtainable by the Friedel-Crafts reaction of pyrene with 2,4-dimethoxy-6-chloro-s-triazine, and is used for brightening polyester fibers (75).

Pyridotriazoles. Quaternized pyridotriazoles (22) can be used for brightening acrylic fibers (76).

Uses

Initially, fluorescent whitening agents (FWAs) were used exclusively in textile finishing. The detergent and paper industries followed thereafter, and today these products are in wide-spread use in fiber spinning-masses, plastics, and paints.

There are more than a thousand known products from 200 compounds based on ca 40 fundamental structures.

The approximate use distribution of whiteners is shown in Table 4.

Textile Applications. A 1971 estimate of world textile fiber consumption (2) showed that approximately 60% of textile goods are dyed and about 30% (6,690,000 metric tons) are whites. The proportion of white goods (>40%) is highest for cotton. Table 5 shows the proportions by individual fiber types (see Textiles).

Textile substrates of natural or synthetic fibers are contaminated in the raw state by substances of varying degrees of yellowness. Bleaching is required for the removal of the yellowish cast. Chemical bleaching agents destroy the yellow coloring matter in fibers. However, even if bleaching processes are carried to the technically acceptable

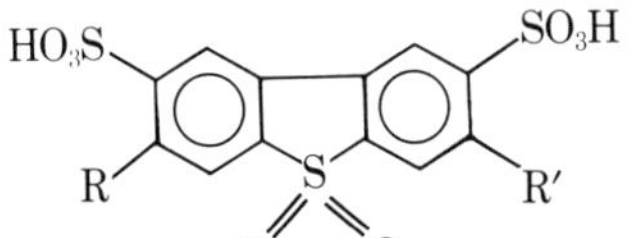

(20)

R, R' = acylamino

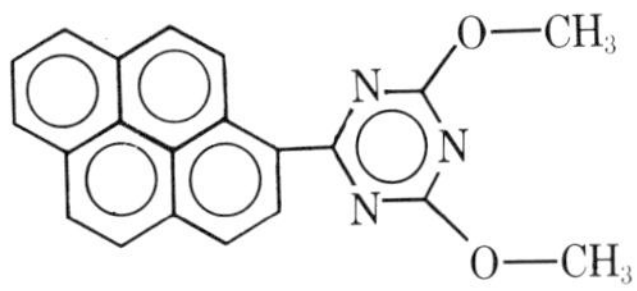

(21)

[3271-22-5]

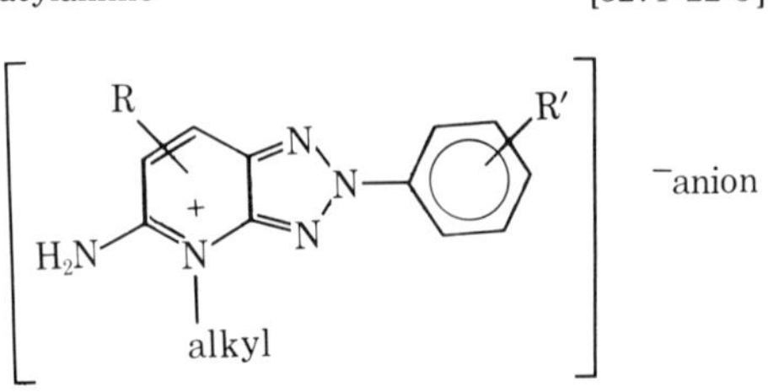

(22)

R = H, alkyl, hal, COOH, etc
R' = H, alkyl, alkoxy, phenyl

Table 4. Use of Fluorescent Whitening Agents

Industry	Proportion, %	Number of fundamental structures
textile	ca 20	>30
detergents	ca 45	ca 7
paper	ca 35	
synthetic fibers, plastics	ca 1	ca 9

Table 5. Fluorescent Whitening Agents in the Textile Industry [a]

Substrate	Proportion of whites, %	Most commonly used products (structure numbers)	Maximum light fastness (no international standard)
Natural fibers			
wool	10	(1) (6) (7) (13)	2–3
cotton	42	(1) (4) (6)	3
regenerated cellulose	28	(1) (4) (6)	3
Synthetic fibers	25		
secondary acetate		(7) (8) (13) (15) (18) (19) (21)	3–4
polyamide		(1) (4) (6) (7) (8) (13) (15)	5
polyester		(2) (3) (5) (6) (8) (10) (19) (21)	8
polyacrylonitrile		(7) (9) (10) (13) (19) (22)	4
polypropylene		(8)	3
poly(vinyl chloride)		(8)	4

[a] Ref. 2.

limits of damage to the fibers, they never succeed in completely removing this intrinsic color.

To produce the color white it is necessary to dye with a fluorescent whitener. FWAs used in textiles can be roughly divided into: products containing sulfonic acid groups, corresponding to acid dyes (for cotton, wool, and polyamides); cationic whiteners that behave in the same way as basic dyes (for polyacrylonitrile fibers); and whiteners containing no solubilizing groups, corresponding to disperse dyes (for polyester and secondary acetate fibers).

The above is not a strict division since nonionic FWAs are known that can whiten polyacrylonitrile and polyamide, and certain cationic FWAs produce effects on polyester (77). The second generation of synthetic fibers includes types that have been acid- or base-modified and consequently display different dyeing characteristics (see Fibers, man-made and synthetic). For dyeing fiber blends such as viscose–polyamide, polyamide–Spandex, or polyester–cotton, only compatible FWAs may be used that do not interfere with one another or have any detrimental effect on fastness properties.

In conjunction with the increasing use of synthetic fibers, and blends of these with natural fibers and the modernization of application processes which has taken place at the same time, the technique of textile whitening has been improved quite considerably within the last few years. Fixing the dye to a fiber can be performed by an exhaust procedure or a padding procedure.

In an exhaust procedure the fluorescent whitener is exhausted from a "long liquor" on to the substrate until an approximate equilibrium is reached between whitener in the bath and whitener on the substrate. In this procedure the equilibrium is biased primarily towards FWA on the substrate, ie, the highest possible degree of exhaustion is desired. Exhaust procedures are mainly used for loose stock, yarns, woven fabrics, and knitgoods which give poor or unsatisfactory results in padding processes, and for garments and garment parts.

Padding methods, ie, application from "short liquors", are becoming increasingly important for whitening piece-goods. Woven fabrics or knitgoods are passed in an unfolded, open-width state through a small trough charged with treatment liquor containing FWA and subsequently between squeeze rollers to express the liquor to a precisely defined liquor pick-up.

During the drying and, if required, heat treatment that follows, the fluorescent whitener is fixed on the substrate. FWAs and dyes used in padding procedures must have low substantivity during the padding operation. This is an important prerequisite for level whitening with no tailing.

In contrast to dyes, fluorescent whiteners are not applied exclusively in special processes, but often in combination with bleaching and finishing steps. Fluorescent whiteners used in such processes must be stable and should not interfere with the operation (see Dyes, application).

The most common chemical bleaching procedures are: hypochlorite bleach (cotton); hydrogen peroxide bleach (wool, cotton), sodium chlorite bleach (cotton, polyamide, polyester, polyacrylonitrile); and reductive bleaching with dithionite (wool, polyamide).

Whitening in combination with the finishing process is used primarily for woven fabrics of cellulosic fibers and their blends with synthetic fibers.

Detergent Applications. Laundering is characterized by repeated application to the same item. Fluorescent whiteners used in this repetitive process have to compensate for the reduction in whiteness and contribute towards prolongation of the useful life of the textile material. Approximately two-thirds of all goods laundered are whites (see Drycleaning and laundering).

In the case of textile whitening there is a practically unlimited degree of freedom in application; the processing method can be adapted to suit the optimum color behavior of the substrate. It is quite a different matter in the case of laundering. The condition of the application bath is predetermined by the composition of the detergent, and the soil loading of the laundry varies from batch to batch. The pH of the washbath is always neutral to alkaline. The bath temperature does not correspond to an optimum dyeing temperature; laundering temperatures vary from cold to boiling water (2). Detergent fluorescent whiteners must remain stable under the conditions that prevail during the washing, rinsing, drying, and ironing cycle (2), and they should possess light fastness. Detergent whiteners must not cause a shade change in the original, generally neutral white of the goods. Fluorescent whiteners that tend to accumulate in excessive amounts on the substrate, thus lending it a greenish cast, are not suitable for use in detergents. Detergent FWAs should also whiten the washing powder, or at least not discolor it.

In countries where laundry is done at the boil, sodium perborate is added to the detergent as a bleaching agent. In the United States, South Africa, and Australia, sodium hypochlorite is generally added to the washbath, and in various countries where laundry is done cold the goods are treated in rinsebaths containing hypochlorite. Only FWAs stable or fast to hypochlorite can be used successfully in these countries (2).

The types of FWAs used in the detergent industry today are listed in Table 6.

Table 6. Fluorescent Whitening Agents in the Detergent Industry

Substrate	FWAs used (structure numbers)
wool	(13)
cellulosics	(1) (2) (4) (6) (9)
secondary acetate	(7) (8) (13)
polyamide	(1) (2) (4) (6) (7) (8) (9) (13)
polyester[a]	(8)
polyacrylonitrile	

[a] Good effects can be obtained only in special processes with high washing and/or drying temperatures.

Paper Industry Applications. Derivatives of bistriazinylstilbene (1) are used in the paper industry.

Most papers are whitened by addition of FWA both to the pulp and to the surface coating. Roughly one-third of the FWA is added at the pulp stage and the remaining two-thirds are added to the preformed sheet to give surface whiteness. In the first operation, the FWA in aqueous solution is added to the stock at the Hollander or other beater. For this application the products must be inexpensive and readily soluble in water. Besides satisfactory exhaustion at low temperatures, good paper FWAs also require good acid and alum stability as well as compatibility with fillers (qv). Good affinity for pulp is also required, since any unabsorbed FWA is lost in the effluent (see Pulp; Paper).

The surface of various sorts of paper are treated with size coatings and pigment coatings to improve their writability and printability. Since the natural white of the raw paper is reduced by this surface coating, suitable FWAs (1) are added to compensate for the yellowish cast and improve the whiteness.

Synthetic Fiber and Plastics Industries. The whitener is not applied from solutions as in textiles, the substrate itself serves as the solvent. In the case of synthetic fibers such as polyamide and polyester which are produced by the melt-spinning process, FWAs can be added at the start or during the course of polymerization or polycondensation. However, FWAs can also be powdered onto the polymer chips prior to spinning. The above types of application place severe thermal and chemical demands on FWAs. They must not interfere with the polymerization reaction and must remain stable under spinning conditions.

In the case of solvent spinning (secondary acetate, polyacrylonitrile, poly(vinylchloride)), the FWA is added to the polymer solution. An exception is gel-whitening of polyacrylonitrile, where the wet tow is treated, after spinning, in a washbath containing FWA (see Acrylic and modacrylic fibers).

In the case of poly(vinyl chloride) plastics, the FWA is mixed dry with the PVC powder before processing or dissolved in the plasticizing agent (see Vinyl polymers). Polystyrene, acrylonitrile–butadiene–styrene (ABS), and polyolefin granulates are powdered with FWA prior to extrusion (2,78) (see Styrene plastics; Acrylonitrile polymers; Olefin polymers).

Table 7. Fluorescent Whitening Agents for Use in the Synthetic Fibers and Plastics Industries

Substrate	FWAs used (structure numbers)
polyamide	(1) (3) (5) (6)
polyester	(2) (6) (8) (15) (16)
secondary acetate	(8)
polyacrylonitrile (PAC)	(2) (7) (9)
PAC gel whitening	(7) (10) (13)
poly(vinyl chloride)	(2) (6) (8) (15)
polystyrene	(2) (6) (8) (15)
acrylonitrile–butadiene–styrene	(6) (8) (15)

The types of FWA used in the synthetic fibers and plastics industries are listed in Table 7 (see Plastics technology).

Measurement of Whiteness. The Ciba-Geigy Plastic White Scale is effective in the visual assessment of white effects (79). Its whiteness formula makes use of instrumental measurement of whiteness levels. As a complement to the whiteness level, shade is determined from the Ciba-Geigy Shade Formula (2,80–83).

BIBLIOGRAPHY

"Whitening Agents" in *ECT* 1st ed., Vol. 15, pp. 45–48, by H. B. Freyermuty, General Aniline & Film Corporation; "Brighteners, Optical" in *ECT* 2nd ed., Vol. 3, pp. 737–750, by R. Zweidler and H. Haüsermann, J. R. Geigy S.A.

1. P. Krais, *Melliand Textilber.* **10,** 468 (1929).
2. "Fluorescent Whitening Agents," in F. Coulston and F. Korte, eds., *Environmental Quality and Safety,* Suppl. Vol. 4, Georg Thieme Verlag, Stuttgart, Ger., and Academic Press, Inc., New York, London, 1975.
3. H. Gold, "Fluorescent Brightening Agents," in K. Venkataraman, ed., *The Chemistry of Synthetic Dyes,* Vol. 5, Academic Press, Inc., New York, 1971, pp. 535–679.
4. A. Dorlars, C. W. Schellhammer, and J. Schroeder, *Angew. Chem.* **87,** 693 (1975).
5. R. Zweidler, *Textilveredlung* **4,** 75 (1969).
6. R. Anliker and co-workers, *Fluorescent Whitening Agents, MVC-Report 2, Proceedings of a Symposium Held at the Royal Institute of Technology, Miljövardscentrum, Stockholm, Sweden, April 11, 1973.*
7. U.S. Pat. 2,713,046 (Oct. 9, 1951), W. W. Wilson and H. B. Freyermuth (to General Aniline & Film Corp.).
8. U.S. Pat. 2,612,501 (Sept. 30, 1952), R. H. Wilson (to Imperial Chemical Industries).
9. U.S. Pat. 2,763,650 (Sept. 18, 1956), F. Ackermann (to Ciba A.G.).
10. U.S. Pat. 2,762,801 (Sept. 11, 1956), H. Häusermann (to J.R. Geigy A.G.).
11. Brit. Pat. 1,116,619 (June 6, 1968), H. Häusermann (to J.R. Geigy A.G.).
12. Fr. Pat. 874,939 (Aug. 3, 1942), B. Wendt (to I.G. Farbenindustrie).
13. Ger. Pat. 752,677, B. Wendt (to I.G. Farbenindustrie).
14. U.S. Pat. 3,211,665 (Oct. 12, 1965), W. Allen and R. F. Gerard (to American Cyanamid).
15. U.S. Pat. 2,618,636 (Nov. 18, 1952), W. W. Williams and W. E. Wallace (to General Aniline & Film Corp.).
16. U.S. Pat. 2,171,427 (Aug. 29, 1939), J. Eggert and B. Wendt (to I.G. Farbenindustrie).
17. U.S. Pat. 3,025,242 (Mar. 13, 1962), R. C. Seyler (to E. I. du Pont de Nemours & Co., Inc.).
18. Ger. Pat. 1,090,168 (Oct. 6, 1960), J. Hegemann, A. Mitrowsky, and H. Roos (to Bayer A.G.).
19. Ger. Pat. 1,250,830 (Sept. 28, 1967), E. A. Kleinheidt and H. Gold (to Bayer A.G.).
20. East Ger. Pat. 55,668 (May 5, 1967), B. Noll and co-workers (to VEB Farbenfabrik).
21. U.S. Pat. 3,479,349 (Nov. 18, 1969), R/ C. Allison, F. Fischer, and H. Häusermann (to Geigy Chemical Corp. Ardsley).
22. U.S. Pat. 3,725,395 (Apr. 3, 1973), A. E. Siegrist and co-workers (to Ciba A.G.).
23. U.S. Pat. 3,781,278 (Dec. 25, 1973), A. E. Siegrist and co-workers (to Ciba A.G.).

24. U.S. Pat. 3,485,831 (Dec. 23, 1969), A. Dorlars, O. Neuner, and R. Putter (to Bayer A.G.).
25. U S. Pat. 3,666,758 (May 30, 1972), A. Dorlars and O. Neuner (to Bayer A.G.).
26. U.S. Pat. 2,784,183 (Mar. 5, 1957), E. Keller, R. Zweidler, and H. Häusermann (to J.R. Geigy A.G.).
27. U.S. Pat. 2,972,611 (Feb. 21, 1961), R. Zweidler and E. Keller (to J.R. Geigy A.G.).
28. U.S. Pat. 3,177,208 (Apr. 6, 1965), W. Stilz and H. Pommer (to Badische Anilin- und Soda-Fabrik).
29. U.S. Pat. 3,984,399 (Oct. 5, 1976), K. Weber and co-workers (to Ciba A.G.).
30. A. Wagner, C. W. Schellhammer, and S. Petersen, *Angew. Chem. Int. Ed. Engl.* **5,** 699 (1966).
31. U.S. Pat. 2,610,969 (Sept. 16, 1952), J. D. Kendall and G. F. Duffin (to Ilford Ltd.).
32. U.S. Pat. 2,640,056 (May 26, 1953), J. D. Kendall and G. F. Duffin (to Ilford Ltd.).
33. U.S. Pat. 2,639,990 (May 26, 1953), J. D. Kendall and G. F. Duffin (to Ilford Ltd.).
34. U.S. Pat. 3,135,742 (June 2, 1964), A. Wagner, A. Schlachter, and H. Marzolph (to Bayer A.G.).
35. U.S. Pat. 3,131,079 (Apr. 28, 1964), A. Wagner and S. Petersen (to Bayer A.G.).
36. U.S. Pat. 3,255,203 (June 7, 1966), E. Schinzel and K. H. Lebkücher (to Hoechst A.G.).
37. U.S. Pat. 3,560,485 (Feb. 2, 1971), E. Schinzel, S. Bildstein, and K. H. Lebkücher (to Hoechst A.G.).
38. Brit. Pat. 1,360,490 (Oct. 14, 1971), H. Mengler (to Hoechst A.G.).
39. U.S. Pat. 3,865,816 (Feb. 11, 1975), H. Mengler (to Hoechst A.G.).
40. U.S. Pat. 2,488,289 (Nov. 15, 1949), J. Meyer, Ch. Gränacher, and F. Ackermann (to Ciba A.G.).
41. U.S. Pat. 3,649,623 (Mar. 14, 1972), F. Ackermann, M. Dünnenberger, and A. E. Siegrist (to Ciba A.G.).
42. U.S. Pat. 3,260,715 (July 12, 1966), D. G. Saunders (to Eastman Kodak).
43. U.S. Pat. 3,322,680 (May 30, 1967), D. G. Hedberg, M. S. Bloom, and M. V. Otis (to Eastman Kodak).
44. Brit. Pat. 1,059,687 (Feb. 22, 1967), (to Hoechst A.G.).
45. U.S. Pat. 3,709,896 (Jan. 9, 1973), H. Frischkorn, U. Pinschovius, and H. Behrenbruch (to Hoechst A.G.).
46. U.S. Pat. 3,993,659 (Nov. 23, 1976), H. R. Meyer (to Ciba-Geigy A.G.).
47. U.S. Pat. 2,995,564 (Aug. 8, 1961), M. Dünnenberger, A. E. Siegrist, and E. Maeder (to Ciba A.G.).
48. U.S. Pat. 3,900,419 (Aug. 19, 1975), H. Schläpfer and G. Kabas (to Ciba-Geigy A.G.).
49. U S. Pat. 3,772,323 (Nov. 13, 1973), H. Schläpfer and G. Kabas (to Ciba-Geigy A.G.).
50. U.S. Pat. 3,940,417 (Feb. 24, 1976), H. Schläpfer (to Ciba-Geigy A.G.).
51. Swiss Pat. 560,277 (Mar. 27, 1955), H. Schläpfer (to Ciba-Geigy A.G.).
52. U.S. Pat. 2,488,094 (Nov. 15, 1949), Ch. Gränacher and F. Ackermann (to Ciba A.G.).
53. U.S. Pat. 2,488,289 (Nov. 15, 1949), J. Meyer, Ch. Gränacher, and F. Ackermann (to Ciba A.G.).
54. U.S. Pat. 2,838,504 (June 10, 1958), N. C. Crounse (to Sterling Drug Inc.).
55. Ger. Pat. 765,901, (to Hoffmanns Stärkefabriken A.G.).
56. U.S. Pat. 2,590,485 (Mar. 25, 1952), H. Meyer (to Lever Brothers Co.).
57. U.S. Pat. 2,424,778 (July 29, 1947), P. W. Tainish (to Lever Brothers Co.).
58. U.S. Pat. 2,610,152 (Sept. 9, 1952), F. Ackermann (to Ciba A.G.).
59. U.S. Pat. 2,600,375 (June 17, 1952), F. Ackermann (to Ciba A.G.).
60. U.S. Pat. 2,654,713 (Oct. 6, 1953), F. Fleck (to Sandoz A.G.).
61. U.S. Pat. 2,791,564 (May 7, 1957), F. Fleck (to Saul & Co.).
62. U.S. Pat. 2,945,033 (July 12, 1960), H. Häusermann (to J.R. Geigy A.G.).
63. U.S. Pat. 3,123,617 (Mar. 3, 1964), H. Häusermann (J.R. Geigy A.G.).
64. U.S. Pat. 3,646,052 (Feb. 29, 1972), O. Neuner and A. Dorlars (to Bayer A.G.).
65. U.S. Pat. 3,288,801 (Nov. 29, 1966), F. Fleck, H. Balzer, and H. Aebli (to Sandoz A.G.).
66. Brit. Pat. 1,201,759 (Aug. 12, 1970), (to Bayer A.G.).
67. U.S. Pat. 3,839,333 (Oct. 1, 1974), A. Dorlars and W. D. Wirth (to Bayer A.G.).
68. U.S. Pat. 3,420,835 (Jan. 7, 1969), W. D. Wirth and co-workers (to Bayer A.G.).
69. Brit. Pat. 741,798 (Dec. 14, 1955), E. Nold (to Badische Anilin- und Soda-Fabrik A.G.).
70. U.S. Pat. 3,310,564 (Mar. 21, 1967), T. Kasai.
71. Jpn. Pat. 71 13,953 (Apr. 14, 1971), T. Kasai.
72. U.S. Pat. 2,563,795 (Aug. 7, 1951), M. Scalera and D. R. Eberhart (to American Cyanamid Co.).
73. U.S. Pat. 2,573,652 (Oct. 30, 1951), M. Scalera and W. S. Forster (to American Cyanamid Co.).
74. U.S. Pat. 2,702,759 (Feb. 22, 1955), M. Scalera and D. E. Eberhart (to American Cyanamid Co.).
75. U.S. Pat. 3,157,651 (Nov. 17, 1964), J. R. Atkinson and S. Hartley (to Imperial Chemical Industries Ltd.).
76. U.S. Pats. 3,058,989 (Oct. 16, 1962), and 3,049,438 (Aug. 14, 1962), B. G. Buell and R. S. Long (to American Cyanamid).
77. R. Anliker and co-workers, *Textilveredlung* **11,** 369 (1976).

78. G. W. Broadhurst and A. Wieber, *Plast. Eng.*, 36 (1973).
79. Ciba-Geigy Review, *White*, 1973/1.
80. E. Ganz, *J. Color Appearance* **1**, 33 (1972).
81. G. Anders, *Textilveredlung* **9**, 10 (1974).
82. E. Ganz, *J. Appl. Optics* **15**, 2039 (1976).
83. R. Levene and A. Knoll, *J. Soc. Dyers Colour.* **94**, 144 (1978).

REINHARD ZWEIDLER
HEINZ HEFTI
Ciba-Geigy Ltd.

BUBBLE MEMORY. See Magnetic material.

BROMINE

Bromine [7726-95-6], a nonmetallic element of the halogen family, is a dark red-brown liquid at ordinary temperatures and pressures. It vaporizes readily and has a sharp, penetrating odor. The diatomic nature of bromine persists throughout the solid, liquid, and gaseous phases. Bromine's properties are between those of chlorine and iodine. The most common oxidation states of bromine are -1 and $+5$, but positive valences of 1, 3, and 7 are also observed.

Bromine was discovered in 1824 by Balard and independently in 1825 by Lowig. The first mineral to contain bromine, of which there are few, was discovered in 1841 by Berthier; the mineral was apparently bromyrite (silver bromide). A valuable dye containing bromine, Tyrian purple (6,6-dibromoindigo) is believed to have been used as early as 1600 BC (see Dyes, natural). The dye was extracted from several species of Mediterranean mollusks of the *Murex* family. The dye industry is still a significant consumer of bromine (see Dyes and dye intermediates).

The discovery of potash in the Stassfurt salt deposits in 1858 opened the way for the first production of bromine as a by-product in Germany. The initial production was used in photography and medicine. The use of bromine in the form of light-sensitive silver bromide in photography was introduced about 1840 (see Photography). Medicinal use was introduced by Lacock in 1857 for the treatment of epilepsy (see Hypnotics, sedatives, and anticonvulsants). Today, bromides are still marketed in nonprescription drugs, but they have been replaced in ethical drugs by newer and more effective compounds. The first bromine production in the United States was in 1846 at Freeport, Pa., by Alter (see Chemicals from brine).

Ethylene dibromide is an effective gasoline additive in preventing lead fouling from the tetraalkyl lead antiknock agents. In 1963, ethylene dibromide accounted for 77% of the bromine produced in the United States. This percentage is steadily decreasing because of EPA regulations requiring the reduction of lead emissions from automobiles.

Fire retardant chemicals are becoming a major outlet for bromine. Since the late

1960s there has been increased emphasis on reducing the flammability of both natural and synthetic polymeric materials; numerous organic bromine compounds are being produced as flame retardants (qv).

Physical Properties

Bromine is a dense, dark red, fuming liquid which is highly corrosive and lachrimatory. Table 1 summarizes the physical properties of the element.

The bromine molecule is diatomic and thermal dissociation does not occur below 600°C. It is the only liquid, nonmetallic element at ordinary temperatures. Its solubility in water (Table 2) is low, but aqueous solubility can be increased by bromide and chloride ions owing to complex formation (Table 3). Equilibrium constants for the formation of the tribromide and pentabromide ions at 25°C have been reported (10):

$$Br_2 + Br^- \rightleftharpoons Br_3^- \quad K = 16.85$$
$$Br_3^- + Br_2 \rightleftharpoons Br_5^- \quad K = 1.45$$

The partial pressure of bromine from aqueous solution is lowered in the presence of dissolved salts. For example, the partial pressures of bromine from 0.10% and 0.27% solutions of water free from other solutes at 20°C are 0.72 and 1.97 kPa (5.4 and 14.8 mm Hg), respectively. If the solution is a brine containing 10.1% NaCl, 8.1% each of KCl and $MgCl_2$, and 6.2% $MgSO_4$, the partial pressures at the same temperature become 0.33 and 0.8 kPA (2.5 and 6.0 mm Hg), respectively (5). The density of bromine solutions in the presence of bromides is high. A solution containing 408.3 g/L NaBr and 10.43 mol/L Br_2 has a density of 2.42 g/mL at 25°C. Kramer and Stenger (11) have investigated the solvent characteristics of bromine and applied the results to the separation of cesium bromide from other alkali bromides.

Bromine is miscible with many organic solvents such as carbon tetrachloride, ether, methanol, alkyl bromides, carbon disulfide, methylene chloride, and chloroform. Reactions can take place with some of these solvents under certain conditions. The solubilities of several bromides in bromine are shown in Table 4.

A convenient summary of the physical and thermodynamic properties of the halogens has recently been presented in graphical form (12). The spectroscopic properties of bromine (13), and bromine and its compounds (14), have been reviewed.

Chemical Properties

Many reactions of bromine are a result of its strong oxidizing properties. The tendency of bromine atoms to gain a single electron lies between chlorine and iodine, as do most other chemical properties. The standard potential for bromine in the aqueous system is $E° = +1.087$ V. The ionization potential for bromine is 11.8 eV and the electron affinity is 3.78 eV. The heat of dissociation of the Br_2 molecule is 192 kJ (46 kcal). Other reduction potentials for bromine and oxybromide anions in aqueous acid solutions at 25°C are (15):

$$Br_2 + 2\,e^- \rightarrow 2\,Br^- \quad E° = +1.07 \text{ V}$$
$$HOBr + H^+ + e^- \rightarrow \tfrac{1}{2}\,Br_2 + H_2O \quad E° = +1.59 \text{ V}$$
$$BrO_3^- + 5\,H^+ + 4\,e^- \rightarrow HOBr + 2\,H_2O \quad E° = +1.49 \text{ V}$$
$$BrCl + 2\,e^- \rightarrow Br^- + Cl^- \quad E° = +1.2 \text{ V}$$

Table 1. Physical Properties of Bromine[a]

Property	Value
stable isotopes	Br^{79}, 50.54%
	Br^{81}, 49.46%
mol wt	159.808
freezing point, °C	−7.25
bp, °C	58.8
density, g/cm³, 15°C	3.1396
20°C	3.1226
25°C	3.1055
30°C	3.0879
vapor density, g/L, 0°C, 101.3 kPa[b]	7.139
refractive index, 20°C	1.6083
25°C	1.6475
viscosity, mm²/s (= cSt), 20°C	0.314
30°C	0.288
40°C	0.264
50°C	0.245
surface tension, mN/m (= dyn/cm), 25°C	40.9
solubility parameter, 25°C	11.5
critical temperature, °C	311
critical pressure, MPa (atm)	10.3 (102)
thermal conductivity, W/(m·K)	0.123
specific conductivity, (Ω·cm)$^{-1}$	9.10×10^{-12}
dielectric constant, 25°C, 10⁵ Hz	3.33
electrical resistivity, 25°C, Ω·cm	6.5×10^{10}
expansion coefficient at 20–30°C, per °C	0.0011
compressibility, 20°C (0–10 MPa)[c]	62.5×10^{-6}
heat of vaporization, J/g[d], 50°C	187
heat of fusion, J/g[d], −7.25°C	66.11
heat capacity[e], J/mol[d], 15 K	7.217
30 K	22.443
60 K	36.33
240 K	57.94
solid, 265.9 K	61.64
liquid, 265.9 K	77.735
288.15 K	78.66
electronegativity	3.0
electron affinity, kJ[d]	330.5

[a] Refs. 1–4.
[b] To convert kPa to mm Hg, multiply by 7.50.
[c] To convert MPa to bar, multiply by 10.
[d] To convert J to cal, divide by 4.184.
[e] Ref. 5.

The reduction potentials at unit hydroxyl ion activity in basic solution (25°C) are:

$$½ Br_2 + e^- \rightarrow Br^- \quad E° = +1.07 \text{ V}$$
$$BrO^- + H_2O + 2\,e^- \rightarrow Br^- + 2\,OH^- \quad E° = +0.76 \text{ V}$$
$$BrO^- + H_2O + e^- \rightarrow ½ Br_2 + 2\,OH^- \quad E° = +0.45 \text{ V}$$
$$BrO_3^- + 2\,H_2O + 4\,e^- \rightarrow BrO^- + 4\,OH^- \quad E° = +0.54 \text{ V}$$
$$BrO_3^- + 3\,H_2O + 6\,e^- \rightarrow Br^- + 6\,OH^- \quad E° = +0.61 \text{ V}$$

Comparison with other oxidizing materials shows bromine to be a stronger oxidizing agent than dilute nitric acid and ferric ion, and a weaker agent than oxygen.

Table 2. Solubility of Bromine in Water[a]

Temp, °C	Solubility, g/100 g soln	Temp, °C	Solubility, g/100 g soln
0	2.31 (4.05 metastable)	20	3.41
3	3.08 (3.85 metastable)	25	3.35
5	3.54 (3.77 metastable)	40	3.33
10	3.60	53.6 (bp)	3.50

[a] Refs. 5–7.

Table 3. Solubility of Bromine in Various Aqueous Solutions at 25°C[a]

Solute	Solute, g/L soln	Bromine, g/L soln	Solute	Solute, g/L soln	Bromine, g/L soln
none		34.0	NaCl	58.5	55.9
KBr	11.9	49.3	NaCl[b]	118.0	86.4
KBr	59.5	119.0	KNO$_3$	101.2	29.0
KBr	119.0	216.0	NaNO$_3$	85.1	28.0
KBr	360.8	632.0	K$_2$SO$_4$	91.2	24.8
NaBr	92.6	99.2	Na$_2$SO$_4$	63.6	23.9
NaBr	206.0	248.0	H$_2$SO$_4$	49.0	29.3
NaBr	320.0	546.0	HBr[c]	33.7	108.8
KCl	74.6	57.4	HCl[c]	39.0	79.2

[a] Refs. 8–9.
[b] At 20.6°C.
[c] At 20.8°C.

Bromine reacts with many metals to form bromides. Sodium is stable to dry bromine, but sodium vapor reacts vigorously. Potassium and cesium react violently with bromine. Aluminum also reacts vigorously with bromine with the emission of light. Magnesium, silver, lead, and nickel become coated with their bromide which prevents further reaction. This protective coating makes lead a useful material for containers of bromine.

Moisture plays an important part in the corrosion of metals by bromine. This is probably caused by the hydrolysis of bromine to hydrobromic and hypobromous acids. At moisture contents below 40 ppm, nickel vessels can be used for transporting bromine (16). Mercury is attacked by bromine and should not be used in instruments and gages exposed to bromine vapors. Dry bromine reacts slowly with iron to form a protective layer of ferric bromide, but when wet, a mixture of hydrated iron bromides is formed which does not adhere to the surface (17). Other heavy metals such as copper, manganese, chromium, antimony, cadmium, cobalt, and bismuth react with bromine, although some require higher temperatures.

When bromine dissolves in water, some disproportionation occurs:

$$Br_2 + H_2O \rightleftharpoons HOBr + H^+ + Br^-$$

The equilibrium constant for this reaction at 25°C is 7.2×10^{-9} (18). The formation of hypobromous acid can be repressed by the addition of acid or bromide ion and en-

Table 4. Solubilities of Bromides in Liquid Bromine[a]

Cation	Solubility of bromide, g/100 g soln at 25°C
lithium	0.005
sodium	0.010
potassium	0.0185
rubidium	0.058
cesium	19.3
strontium	0.002
barium	0.005
calcium	0.003
magnesium	0.002
zinc	0.0146
aluminum	0.002, hydrated soluble, anhydrous
iron (III)	0.324, anhydrous
nickel	0.0003

[a] Ref. 3.

couraged by the addition of metal ions which form insoluble bromides. Hypobromous acid undergoes decomposition when exposed to light, producing hydrogen bromide and oxygen, which are responsible for the bleaching action of bromine. In the dark, decomposition occurs with the formation of bromic acid and bromine.

$$5\ HOBr \rightarrow HBrO_3 + 2\ Br_2 + 2\ H_2O$$

Bromic acid is also relatively unstable and decomposes slowly to give bromine and oxygen. Bromic and bromous acids exist only in solution.

$$4\ HBrO_3 \rightarrow 2\ Br_2 + 5\ O_2 + 2\ H_2O$$

In alkaline solution, bromine rapidly reacts to produce hypobromite:

$$Br_2 + 2\ OH^- \rightleftharpoons Br^- + BrO^- + H_2O \quad K = 2 \times 10^8$$

This reaction must be carried out below 0°C to minimize the disproportionation reaction of hypobromite to bromate and bromide:

$$3\ BrO^- \rightarrow 2\ Br^- + BrO_3^-$$

Because of their instability, hypobromites are usually prepared *in situ* for commercial uses such as textile bleaching and desizing (see Bleaching agents; Textiles). Bromate is formed quantitatively at 50–80°C by reaction of bromine in alkaline solution:

$$3\ Br_2 + 6\ OH^- \rightarrow 5\ Br^- + BrO_3^- + 3\ H_2O$$

Acidification of such a reaction mixture or combination of a 5:1 molar ratio of bromide ion and bromate under acid conditions produces bromine:

$$5\ Br^- + BrO_3^- + 6\ H^+ \rightarrow 3\ Br_2 + 3\ H_2O$$

Oxidation Reactions. Oxidation by bromine occurs under many conditions. In aqueous solutions, alkaline conditions are usually most favorable because the oxyhalogen systems are stronger oxidizing agents. The reaction of sulfur with bromine in water is exothermic and occurs rapidly above 70°C. The reaction can be regulated by controlled addition of either component. Several common reactions with inorganic sulfur compounds are:

$$H_2S + Br_2 \rightarrow S + 2\,HBr$$

$$S + 3\,Br_2 + 4\,H_2O \rightarrow H_2SO_4 + 6\,HBr$$

$$SO_2 + Br_2 + 2\,H_2O \rightarrow 2\,HBr + H_2SO_4$$

$$Na_2S + 4\,Br_2 + 8\,NaOH \rightarrow Na_2SO_4 + 8\,NaBr + 4\,H_2O$$

$$Na_2S_2O_3 + 4\,Br_2 + 5\,H_2O \rightarrow 2\,NaHSO_4 + 8\,HBr$$

Bromine also reacts with red phosphorus to produce phosphorous acid which is oxidized further by excess bromine. This reaction can be quite violent without proper control.

$$2\,P + 3\,Br_2 + 6\,H_2O \rightarrow 6\,HBr + 2\,H_3PO_3$$

$$H_3PO_3 + Br_2 + H_2O \rightarrow H_3PO_4 + 2\,HBr$$

$$H_3PO_3 + Br_2 + 3\,NaOH \rightarrow NaH_2PO_4 + 2\,NaBr + 2\,H_2O$$

Bromine is reduced by ammonia and hydrazine with the evolution of nitrogen. Nitrites, azides, and urea are also reducing agents relative to bromine.

$$2\,NH_3 + 3\,Br_2 + 6\,NaOH \rightarrow 6\,NaBr + N_2 + 6\,H_2O$$

$$8\,NH_4OH + 3\,Br_2 \rightarrow 6\,NH_4Br + N_2 + 8\,H_2O$$

$$N_2H_4 + 2\,Br_2 + 4\,NaOH \rightarrow 4\,NaBr + N_2 + 4\,H_2O$$

$$NaNO_2 + Br_2 + H_2O \rightarrow NaNO_3 + 2\,HBr$$

$$2\,NaN_3 + Br_2 \xrightarrow{H_2O} 2\,NaBr + 3\,N_2$$

$$NH_2CONH_2 + 3\,Br_2 + 6\,NaOH \rightarrow 6\,NaBr + N_2 + CO_2 + 5\,H_2O$$

In water, bromamine formation is greatly dependent on the pH and the ratio of ammonia to bromine (19).

Bromine also reacts with organic amides, imides, and amines to produce N-bromo derivatives. These reactions usually take place under basic conditions (20), but acidic conditions have also been used (21). In base, the temperature must be kept low to avoid the Hofmann reaction with amides and imides.

Carbon and carbon compounds are oxidized by bromine. Formic acid is oxidized by bromine but not by iodine, which is the basis for an analytical technique for bromine in the presence of iodine (22).

$$C + Br_2 + H_2O \rightarrow 2\,HBr + CO$$

$$HCHO + 2\,Br_2 + 4\,NaOH \rightarrow 4\,NaBr + CO_2 + 3\,H_2O$$

$$HCO_2H + Br_2 \xrightarrow{H_2O} CO_2 + 2\,HBr$$

Aqueous bromine is an effective oxidizing agent for numerous organic compounds including alcohols, aldehydes, ethers, and carbohydrates (23). Ethanol, eg, is oxidized to acetaldehyde, which can be further oxidized to acetic acid.

Elemental bromine readily adds to unsaturated compounds. The manufacture of ethylene dibromide by the addition of bromine to ethylene is by far the most important commercial example of this type of reaction.

$$Br_2 + CH_2{=}CH_2 \rightarrow BrCH_2CH_2Br$$

Addition of bromine to other unsaturated compounds is usually carried out in inert solvents and can be accelerated by ultraviolet and visible radiation. The rate of bromination of ethylenic compounds is dependent on several factors such as structure, solvent, salt effects, and concentrations. The influence of these parameters and polar and steric substituent effects have been recently studied by Dubois and co-workers (24). The importance of complex formation between bromine and olefins in the mechanism of these reactions is discussed by Downs and Adams (1). Bromine also readily adds to acetylenic compounds (see Acetylene-derived chemicals; Ethylene).

Substitution Reactions. Bromine undergoes electrophilic substitution reactions with aromatic substrates such as benzene. Depending on the degree and type of substitution on the aromatic ring, the reaction may involve only bromine or require a catalyst. The most common catalysts are the bromides or chlorides of iron or aluminum, although many other catalysts including iodine are effective (20,25) (see Friedel-Crafts reactions).

$$C_6H_6 + Br_2 \xrightarrow{FeBr_3} C_6H_5Br \text{ (bromobenzene)} + HBr$$

Activated aromatic systems are usually brominated without catalysts; however, as the degree of bromine substitution increases, steric effects begin to play a role in reducing reaction rates. Phenol, eg, reacts very rapidly with bromine in aqueous solution to produce 2,4,6-tribromophenol. Further bromination of phenol to the pentasubstituted product requires anhydrous conditions and an iron catalyst. Increasing alkyl substitution on the benzene ring has been observed to increase the rate of electrophilic bromination.

$$C_6H_5OH + Br_2 \longrightarrow \text{2,4,6-tribromophenol} \xrightarrow{Br_2, FeBr_3} \text{pentabromophenol}$$

The homolytic substitution of hydrogen by bromine is a useful reaction for the preparation of several bromine compounds. This reaction occurs most readily with alkyl aromatic compounds and is exemplified by the benzylic bromination of toluene to yield benzyl bromide. Increasing substitution at the α-position increases the rate of reaction so that the relative rates of α-bromination for toluene, ethylbenzene, and cumene have been reported to be 1, 17.2, and 37.0, respectively (26).

$$Br_2 \xrightarrow{h\nu} 2\ Br\cdot$$

$$Br\cdot + C_6H_5CH_3 \rightarrow C_6H_5CH_2\cdot + HBr$$

$$C_6H_5CH_2\cdot + Br_2 \rightarrow C_6H_5CH_2Br + Br\cdot$$

$$2\ Br\cdot \rightarrow Br_2$$

Bromine substitution of allylic hydrogens is also possible under special conditions to minimize the addition reaction. This type of bromination, however, is usually accomplished with N-bromosuccinimide and similar reagents. A free radical reaction mechanism related to benzylic bromination is involved (27). Bromine radicals are less reactive in hydrogen abstraction than chlorine. This can be advantageous where a high degree of specificity is required (28).

Other reactions of bromine with organic compounds include α-bromination of carbonyl compounds and carboxylic acids, the latter usually with the aid of phosphorus catalysts.

Occurrence

Bromine does not occur in nature as the free element, but is found only as the bromide. Bromyrite (AgBr), embolite (AgCl, Br), and iodobromite (AgBr, Cl, I) are some of the few bromine-containing minerals. The ionic radius of the bromide ion (0.196 nm) is so similar to chloride (0.181 nm), that it is capable of partially replacing chlorine in some minerals (29). Thus, bromine usually follows chlorine through the geochemical cycle. Crustal rocks usually contain 2–3 ppm bromine but 20 ppm was found in labradorite. Igneous rocks, sediments, and meteorites have the highest ratio of bromide to chloride, about 1:100 (30).

Although geochemically the halogens are considered lithophile elements, bromine also appears in variable amounts in the biosphere. The bromide concentration of terrestrial plants is ca 7 ppm, with 20 ppm in fresh water plants and higher amounts in marine algae and other sea plants, often found as organic bromine (31–35). The amounts of bromide in vegetables, fruits, and cereal grains vary from 2–14 ppm depending on the species and locality. Some of these compounds are found to have potential medicinal use and antibacterial activity. Dibromoindigo is obtained from the Mediterranean sea snail *Murex brandaris*; dibromotyrosine is found in the skeleton of corals. Animal tissues contain 1–9 ppm bromide; blood bromide ranges from 5–15 ppm (36).

The most readily recoverable form of bromine occurs in the hydrosphere as soluble bromide salts. The current accepted value of 65 mg/L for the bromide concentration in sea water was reported in 1871, although many determinations have since been carried out (36). Other extractable sources of bromine occur in salt lakes, salines, and inland seas. The Dead Sea is the richest source, containing nearly 4 g/L of bromide at the surface. Searles Lake in California is reported to contain 0.085% bromide.

Other important sources of bromine are brine wells, which are the principal sources of bromine in the United States. The richest brines are found in Arkansas and Michigan (see Chemicals from brine).

Manufacture

The concentration of bromine, in the form of bromide, in natural brine deposits, bitterns, and ocean water varies from 0.065 g/L in sea water to 12 g/L in effluent from Dead Sea potash production. The concentration of bromide in natural brines is commonly ca 1.3–5 g/L. The molar ratio of chloride ions to bromide ions in these solutions is usually 200:700. The key factors of bromine isolation processes are the selective separation of bromide from chloride and the removal of small concentrations of bromine from large volumes of aqueous solution. The selective oxidation of bromide in the presence of large amounts of chloride is possible because of the difference in their reduction potentials. In fact, chlorine is the most economical and convenient oxidant for bromide and is employed in all current methods of bromine production.

$$\tfrac{1}{2}\,Cl_2(g) + e^- \rightarrow Cl^- \quad E° = +1.356\ V$$

$$\tfrac{1}{2}\,Br_2(g) + e^- \rightarrow Br^- \quad E° = +1.065\ V$$

The bromine released by this reaction is, fortunately, quite volatile and can be driven out of the solution.

$$2\ Br^- + Cl_2 \rightarrow Br_2 + 2\ Cl^-$$

The four principal steps in bromine production are: (1) oxidation of bromide to bromine; (2) bromine vapor removal from solution; (3) isolation of bromine from the vapor; and (4) purification. For step (1), the use of chlorine has been mentioned, but other methods have been explored including oxidation by manganese dioxide, chlorates, bromates, hypochlorite, and electrochemical methods (37).

It is in step (2) that the two most significant developments in bromine manufacture have been made. These are the steaming-out process as modified by Kubierschky in 1909 and the blowing-out process (38) developed by Dow in 1889 (39). Steam is suitable when the raw brine contains more than one gram of bromine per liter; however, air is more economical when the source is as dilute as ocean water. When steam is used the vapor may be condensed directly; otherwise, the bromine must be trapped in an alkaline or reducing solution. In either case, purification is necessary to remove chlorine and possibly other impurities that may vary with the brine source.

Steaming-Out Process. The Kubierschky process is a modification of a process used earlier in Germany (37,40). This process is the basis for most current bromine production. A flow sheet is shown in Figure 1.

For efficient chlorine use, it is reported as being advantageous to introduce 10–25% chlorine with 90°C brine at the chlorinator, and to add the remaining chlorine at several positions in the steaming-out tower (41). The halogen-laden vapor passes from the top of the steaming-out tower into a condenser and then into a gravity separator. Some uncondensed chlorine and bromine are returned to the bottom of the chlorinator for halogen recovery. Saturated with bromine and chlorine, the water layer from the separator is returned to the steaming-out tower. Crude bromine collects at the bottom of the separator and passes through a trap to a distillation column where excess chlorine is removed. The bromine is then fractionated for final purification. The removal of chlorine from crude bromine by a countercurrent reaction with sodium bromide solution has been reported (42).

Although the raw brine is nearly neutral, it can contain small amounts of reducing substances that react with chlorine to produce hydrochloric acid. This acidity increases the efficiency of bromine liberation by preventing hydrolysis of bromine to hypobromous acid. If the brine contains ammonium salts or amines, it may be desirable to add excess acid to prevent haloamine formation (19,43). The use of sulfuric acid is not advisable if the brine contains calcium or strontium which may be precipitated as sulfates and foul the tower packing and heat exchangers. Sulfates will, however, be formed if the brine contains sulfides. It has been reported (44) that the passage of brine, supersaturated with calcium sulfate, through 4–16 cm layers of gypsum reduces subsequent crystallization.

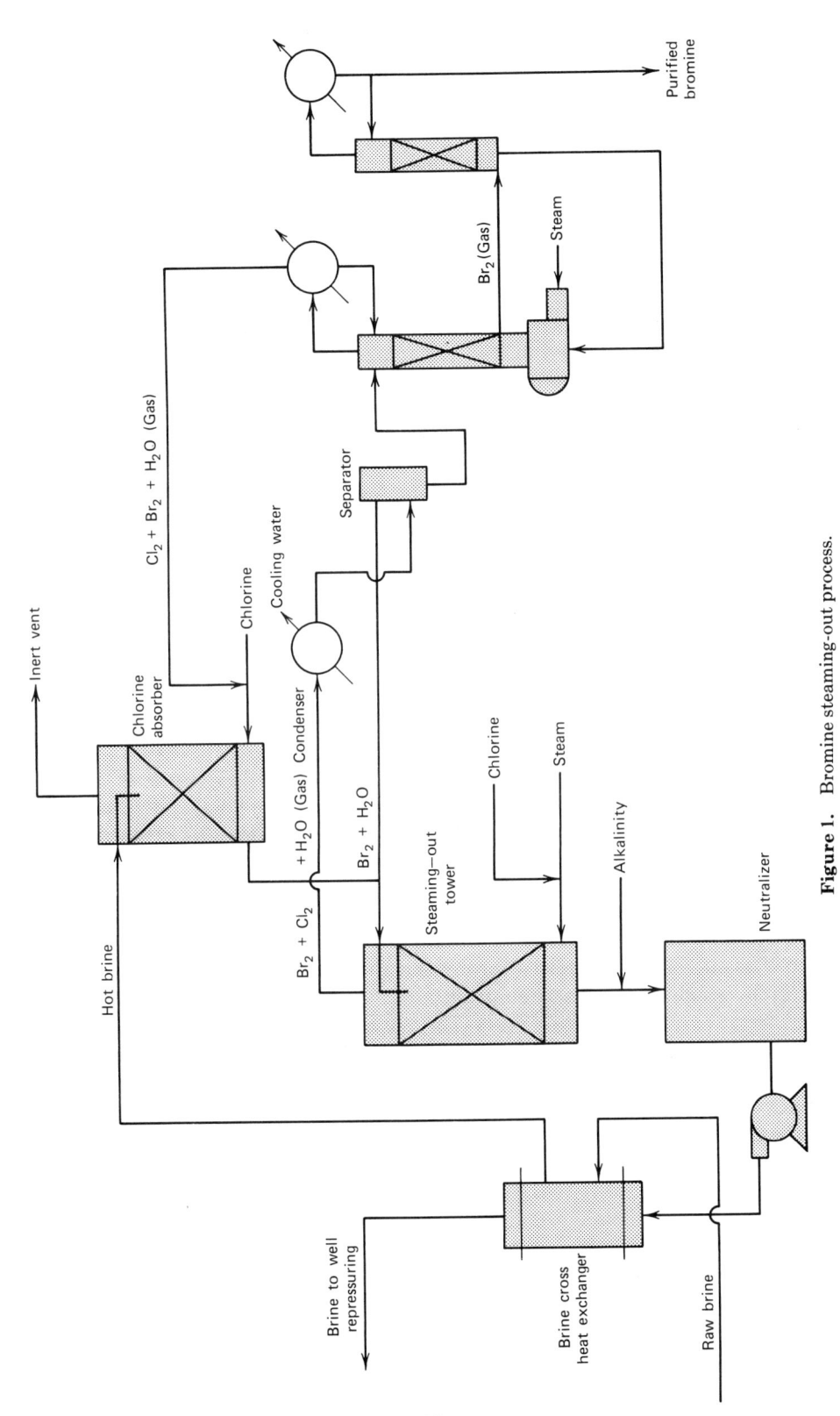

Figure 1. Bromine steaming-out process.

The steaming-out process is controlled by measurements of temperature and pressures within the towers, the temperature, pH, and halogen content of the out-going brine, and the oxidation potential of the halogen mixture that leaves the steaming-out tower. The oxidation potential furnishes a good measure of excess chlorine, because chlorine is a stronger oxidant. A moderate chlorine excess must be maintained for efficient operation; little is wasted and excess chlorine is removed and recycled. More than 95% of the bromine is recovered from the raw brine.

Equipment coming into contact with wet halogens must be made of corrosion-resistant materials. Glass and ceramics have been extensively used in the past, but new construction materials such as tantalum and fluoropolymers are coming into use. Dry liquid bromine is handled in lead, nickel, glass, and fluoropolymer-lined steel equipment. Under certain conditions steel alone can be used (45).

The Blowing-Out Process. The blowing-out process, in which air rather than steam is used to drive out bromine from oxidized brine, is the most important process for obtaining bromine from sources containing less than 1 g/L, such as ocean water. A large scale blowing-out plant that extracted bromine from ocean water was operated by The Dow Chemical Company at Freeport, Texas, for several years (46). Plants currently extracting bromine from ocean water are in operation in Japan and the United Kingdom (see Ocean raw materials).

In this process, ocean water is pumped to the tops of a number of blowing-out towers, with sulfuric acid and chlorine being added just above the pumps so that mixing takes place during the ascent. About 1.3 kg of 10% sulfuric acid per metric ton of water is required to neutralize the natural bicarbonates and reduce the pH; a 15% excess of chlorine over the theoretical requirement is used. The acidification is necessary to minimize bromine hydrolysis to hypobromous acid which dissolves in water to the extent of 33.0 g/L. Bromide ion, formed upon bromine hydrolysis, also increases the aqueous solubility of bromine by formation of the tribromide ion. Therefore, at pH 7 there is nearly no bromine liberated. Acidity of pH 3–3.5 is required to liberate nearly all the bromine.

A series of distributor troughs at the top of the blowing-out tower divides the bromine solution into several smaller streams. The streams then descend through parallel, packed chambers where the bromine is blown out by a countercurrent air stream. The outgoing air is passed through an absorption tower and the halogens are taken up by a sodium carbonate solution.

$$3\ Na_2CO_3 + 3\ Br_2 \rightarrow 5\ NaBr + NaBrO_3 + 3\ CO_2$$

The absorption tower is divided into several vertically parallel compartments, each having its own circulating system for absorption liquor. The air enters through a compartment where the liquor is nearly exhausted, then passes successively through the others until it is scrubbed with fresh solution. Each compartment contains a number of spray nozzles at the top through which the sodium carbonate solution is pumped and falls by gravity to the reservoir bottom. It is recycled until the alkalinity becomes low, at which time it is transferred to storage. The bromine is released or desorbed from the concentrated solution by treatment with sulfuric acid. The freed bromine is steamed out and condensed as in the steaming-out process or used directly in the production of ethylene dibromide.

$$NaBrO_3 + 5\ NaBr + 3\ H_2SO_4 \rightarrow 3\ Br_2 + 3\ Na_2SO_4 + 3\ H_2O$$

A modification of this process that employed sulfur dioxide to absorb the blown-out bromine was introduced in 1937.

$$Br_2 + SO_2 + 2\,H_2O \rightarrow 2\,HBr + H_2SO_4$$

The resulting bromide solution is reoxidized with chlorine and recovered in a steaming-out process.

Bromine produced by either the steaming-out or blowing-out process usually contains impurities such as chlorine, water, halogenated hydrocarbons, and halogen acids. Fractional distillation brings the purity into the 99.5–99.7% range of commercial bromine. Purification to the 99.9+% quality which is now available, may be done by a number of proprietary methods. Distillation with a small excess of potassium bromide eliminates free chlorine. Heating bromine vapor to 370–540°C decomposes chlorine and organic impurities to HCl and higher boiling compounds that can be removed by distillation (47–48). A fractional vaporization of bromine at 50°C lowers the chlorine content from 0.12 to 0.06% and moisture from 0.05 to 0.005% (49). Organic halogen compounds such as chloroform can be removed by passing predried bromine vapor and oxygen through a hot quartz-packed tube (50).

To avoid corrosion of most metal containers, the water content of bromine should be less than 30 ppm. Drying with strong sulfuric acid removes most of the moisture. Sulfuric acid has been used to remove both organic compounds and water (51) and drying with molecular sieves (qv) has been reported (52). The laboratory preparation of very pure bromine for physical property determinations has been described (53).

The electrochemical oxidation of bromine is similar to that of chlorine. Two electrochemical processes were developed in Germany. The Wunsche process was based on the use of diaphragms to provide good current efficiency. The carbon-stick cathodes were separated from the carbon anodes by porous clay cylinders which allowed ion diffusion; bromine was the anode product.

$$2\,Br^- \rightarrow Br_2 + 2\,e^-$$

The cathode reaction produced hydrogen and magnesium hydroxide which clogged the clay diaphragm pores.

$$Mg^{2+} + 2\,e^- + 2\,H_2O \rightarrow Mg(OH)_2 + H_2$$

The Kossuth electrochemical process used bipolar, diaphragmless, carbon-plate electrodes with one side serving as anode and the other side as cathode. To dislodge the cathodic magnesium hydroxide deposits, the current was periodically briefly reversed. Despite this advantage over the Wusche process, the Kossuth process had low current efficiency and could not compete with the newly developed Kubierschky process (37) (see Electrochemical processing).

Dow's electrolytic cell was similar to the Kossuth bipolar, carbon-plate electrodes with some modifications and simplification. His first blowing-out process consisted of burlap strips stretched across a wooden trough. The electrolyzed brine, containing free bromine, was dropped on the burlap while air was blown through counter to the brine. The air carried the bromine vapor into a tower packed with scrap iron where it reacted and was concentrated as iron bromide for refining (38).

Processes using aniline and phenol to react with bromine in oxidized brine have been unsuccessful.

Economic Aspects

Bromine plant locations are dictated primarily by the availability of natural brines or bitterns containing adequate amounts of bromine. Ocean water plants may be operated most economically in areas where the water temperature is high and power costs are low. A proposed site must ensure a supply of sea water undiluted by fresh river water or plant effluent.

Investment and maintenance costs are relatively high because of corrosion. The principal material cost is for chlorine, 1 kg is required for about every 2–2.2 kg of bromine produced. The major operational expenses are for pumping and air-blowing in ocean water plants and for pumping and steam in brine plants. A 1974 Bureau of Mines energy survey of United States bromine manufacturers revealed that about 40.2 MJ (11.17 kW·h or 38,140 Btu) was required to produce a kg of bromine.

The average price of bromine decreased regularly between 1965 and 1973, as quoted by the Bureau of Mines (54), from 44.38 to 35.38 ¢/kg. A sharp reversal of this trend was observed in 1974 when the average price was 60.05 ¢/kg. In 1977, delivered prices quoted in the *Chemical Marketing Reporter* were 121–137 ¢/kg (drums, car lots, truck loads) and 55–66 ¢/kg (bulk, 11,340 kg minimum).

Bromine production, exports and imports, and supply–demand relationships are summarized and reported annually by the U.S. Bureau of Mines. Production of the major producing countries, taken from that source (55) and others (54,56–57) is summarized in Table 5.

Little information is available on bromine production in the Union of Soviet Socialist Republics, but it was estimated to be 12,690 metric tons in 1973. Production in Spain and India for 1973 was estimated to be 399 and 249 t, respectively.

In 1974, the United States had ten bromine-producing plants in three states operated by seven companies. The Dow Chemical Company, with three plants, was the largest United States producer, followed by the Ethyl Corporation. The richest brines are in Arkansas where six plants are located.

Specifications and Standards

Typical, freshly prepared bromine from a modern plant is likely to be 99.9% pure. Usual impurities are water, chlorine, and organic materials at levels of less than 50 ppm each. Specifications of the USP and the ACS Committee on Analytical Reagents allow up to 0.05% chlorine, 0.001% iodine, 0.001% sulfur, and 0.005% nonvolatile matter.

Table 5. Annual Bromine Production (Metric Tons)

Year	United States	Israel	Japan	France	F. R. Germany	Italy	United Kingdom
1900	236						
1950	44,691		120	1,060		391	
1960	77,132[a]	2,400	1,950	2,009		1,278	
1965	124,578	4,220[a]	4,314	2,800	2,945	2,060	
1969	152,106	9,982	6,350		3,175	4,082	
1972	175,528	14,104	10,473[a]	13,558	2,720	4,519	30,395
1973	189,769	13,044	11,020[a]	14,060[a]	2,900	5,215[a]	30,595
1974	196,050	18,013	11,383	14,060[a]	3,400	4,989[a]	30,975[a]

[a] Estimates.

Analytical Methods

The analysis of bromine and bromine compounds has been reviewed by Stenger (58). Liquid bromine can be assayed by titrating the iodine liberated from an acidified sample in potassium iodide solution with standard sodium thiosulfate. The following reactions are involved:

$$2\,I^- + Br_2 \rightarrow I_2 + 2\,Br^-$$

$$I_2 + 2\,Na_2S_2O_3 \rightarrow Na_2S_4O_6 + 2\,NaI$$

Bromine vapor can be absorbed in aqueous potassium iodide solution and the liberated iodine titrated.

The van der Meulen method is used to analyze for small amounts of bromine. This method uses hypochlorite to oxidize bromine to bromate; excess hypochlorite is decomposed with formate and the bromate is determined iodometrically.

Bromine can be detected by several reagents but often other oxidizing agents such as chlorine can interfere. Bromine forms colored solutions with carbon tetrachloride and carbon disulfide which can be used as a simple qualitative test or in colorimetric methods for estimation. The conversion of fluorescein to tetrabromofluorescein has been used as a test for bromine where chlorine does not interfere. Filter paper soaked in fluorescein is used to detect bromine vapor by the formation of a yellow stain. Phenol red, methyl orange, o-tolidine, and rosaniline can be used in a similar manner for bromine detection.

Two titration techniques for bromide using silver nitrate are well known. The Mohr titration utilizes potassium chromate as an indicator. The Volhard titration involves the back titration of excess silver nitrate, in the presence of ferric ammonium sulfate, with standard potassium thiocyanate. The end point is indicated by formation of the red ferric thiocyanate.

A reliable method for the separation of iodide, bromide, and chloride ions in any mixture involves selective oxidation and distillation followed by potentiometric or turbidimetric determination. Other methods for analyzing mixtures of halide ions include chromatography and precipitation. A spectrophotometric method is useful for the rapid determination of bromide in natural brines and similar salt solutions. Chloride does not interfere and iodine can be removed in a preliminary extraction. The spectrophotometric and potentiometric titration methods are suitable for bromine as a minor constituent (see Analytical methods).

Various methods are used for the determination of impurities in bromine. The chlorine content can be determined by density or freezing point depression measurements. For amounts of chlorine in the order of 5–10%, the halogen mixture is reduced with sulfur dioxide to the halides and titrated potentiometrically. Pure bromine is transparent in the infrared so that water and organic impurities can be measured by the ir absorption spectrum. The residue after evaporation of bromine is determined by heating a sample in a tared container on a steam bath, drying at 105°C, and weighing. The residue is then used for analysis for heavy metals and sulfur.

Neutron activation and x-ray fluorescence are convenient methods for the determination of the bromine content in organic materials.

Health and Safety Factors

Liquid bromine or its vapors attack the skin and other tissues to produce irritation and necrosis. Comparatively low concentrations of vapor are quite painful and highly irritating to the eyes and respiratory tract. Excessive exposure to acutely dangerous concentrations will result in serious inflammation and edema, frequently followed by pneumonia, whereas lower concentrations will result in inflammatory eye and respiratory reactions. The maximum time-weighted average concentration of bromine considered safe for repeated 8 hour exposure is 0.1 ppm. This is the limit specified by OSHA (59). The American Conference of Governmental Industrial Hygienists has also established 0.1 ppm as the threshold limit value (TLV) for bromine. Air monitoring is necessary to detect bromine at this low concentration. At higher concentrations the characteristic odor of bromine will be noticed. A concentration of 1 or 2 ppm is highly disagreeable and 10 ppm of vapor in the air is practically intolerable.

Production facilities where bromine is manufactured or used should be designed to rapidly dispose of liquid bromine spills. Solutions or slurries of 10–50% potassium carbonate, 10–13% sodium carbonate, and 5–10% sodium bicarbonate or saturated "hypo" solution (prepared by dissolving 4 kg of technical-grade sodium thiosulfate in 9.5 L of water and adding 113 g of soda ash) are preferred neutralizing agents for liquid bromine spills (4). A 5% lime slurry or a 5% sodium hydroxide solution may be used, but heats of reaction are higher for these reagents. Ammonia solutions should not be applied to liquid spills because of the high heat of reaction and nitrogen evolution. Anhydrous ammonia gas is useful for neutralization of bromine fumes.

Full body protection constructed of resistant materials should be worn when handling bromine in significant quantities. Major manufacturers of bromine will furnish information on the safe handling procedures for bromine.

Uses

Elemental bromine is primarily used in the manufacture of bromine compounds (qv) which are mainly characterized by their chemical or biological activity, high density, or fire retarding and extinguishing ability (see Flame retardants). Bromine compounds are well represented in such use areas as gasoline additives, agricultural chemicals, flame retardants, dyes, photographic chemicals, pharmaceuticals, and others; the amount of total world bromine production in each of these areas has been estimated (1976) to be 55, 17, 9, 7, 3.5, 3.5, and 5%, respectively (57).

The high density characteristics of bromine compounds are advantageously applied in hydraulic fluids (qv), gage fluids, and ore flotation (qv). Calcium bromide in aqueous solution was introduced in 1972 as a dense fluid product for use as oil well packs and completion fluids (60) (see Drilling fluids).

The largest industrial use of bromine is as ethylene dibromide (EDB), a lead scavenger additive used in gasoline with tetraethyllead and tetramethyllead to reduce engine knocking. In 1970, 71% of United States bromine production was used for the manufacture of EDB. A decrease in production to 66% in 1974 was attributed partly to regulations issued in 1973 by the EPA to reduce the lead content of gasoline. Consumption of bromine as EDB will decrease at a rate dependent on the reduction and removal of lead from gasoline according to the EPA schedule. Several factors are affecting the rate of lead use reduction that make current predictions of the impact on

Table 6. Bromine Demand According to Use[a]

Use	Thousands of metric tons (%)				
	1966	1968	1970	1972	1974
ethylene dibromide	81.6 (69.8)	94.1 (70.0)	110.1 (71.0)	122.2 (75.2)	108.9 (66.2)
sanitary preparations	16.3 (14.0)	18.9 (14.0)	21.7 (14.0)	16.2 (10.0)	16.5 (10.0)
fire retardants	5.9 (5.0)	6.7 (5.0)	7.8 (5.0)	11.4 (7.0)	25.4 (15.4)
other	13.2 (11.2)	14.8 (11.0)	15.5 (10.0)	12.6 (7.7)	13.8 (8.4)
Total	117.0	134.5	155.1	162.4	164.6

[a] Ref. 55.

the bromine industry difficult (see Gasoline). The use pattern for bromine in the United States is shown in Table 6. Sanitary preparations include the use of bromine in water disinfection in swimming pools (see Water) and cooling towers for control of bacteria, algae, and odors, and other uses (see Industrial antimicrobial agents). This has generally been the second largest use area for bromine until 1974 when flame retardants showed a significant increase. The continued growth of brominated flame retardants will depend largely on the nature and timing of new flammability standards and regulations (61–62). The use pattern in other bromine-producing countries does not differ greatly from that in the United States.

Bromine is also used, either directly or indirectly, in the desizing of cotton, in bleaching of pulp and paper, air conditioning absorption fluids, hair waving compositions, and water disinfection (see Bleaching agents; Disinfectants; Hair preparations).

BIBLIOGRAPHY

"Bromine" in *ECT* 1st ed., Vol. 2, pp. 629–645, by V. A. Stenger, The Dow Chemical Company; "Bromine" in *ECT* 2nd ed., Vol. 3, pp. 750–766, by V. A. Stenger, The Dow Chemical Company.

1. A. J. Downs and C. J. Adams in A. F. Trotman-Dickenson, ed., *Comprehensive Inorganic Chemistry*, Pergamon Press, New York, 1973, p. 1107.
2. F. Yaron in Z. E. Jolles, ed., *Bromine and Its Compounds*, Ernest Benn Ltd., London, 1966, pp. 43–49.
3. V. A. Stenger, *Angew. Chem. Int. Ed.* **5**(3), 280 (1966).
4. Dow Chemical U.S.A., *Bromine—Unloading, Storing, Handling*, Form No. 101-2-76.
5. J. d'Ans and P. Hofer, *Angew. Chem.* **47**, 71 (1934).
6. F. H. Rhodes and C. H. Bascom, *Ind. Eng. Chem. Ind. Ed.* **19**, 480 (1927).
7. L. W. Winkler, *Chem. Ztg.* **23**, 687 (1899).
8. A. Seidell in W. F. Linke, ed., *Solubilities of Inorganic and Organic Compounds* 3rd ed., Vol. 1, Van Nostrand, Princeton, N. J., 1940, pp. 207–209.
9. *Ibid*, 4th ed., Vol. 1, 1958, pp. 440–444.
10. A. I. Popov in V. Guttman, ed., *Halogen Chemistry*, Vol. 1, Academic Press, New York, 1967, p. 225.
11. U.S. Pat. 2,481,455 (Sept. 6, 1949), W. R. Kramer and V. A. Stenger (to The Dow Chemical Company).
12. H. S. N. Setty, J. D. Smith, and C. L. Yaws, *Chem. Eng.*, 70 (June 10, 1974).
13. J. J. Turner in V. Guttman, ed., *MTP International Review of Science, Inorganic Chemistry*, Series One, Vol., 3, 1972, pp. 253–291.
14. E. F. Mooney and M. Goldstein in ref. 2, pp. 802–869.
15. W. M. Latimer, *The Oxidation States of the Elements and their Potentials in Aqueous Solutions*, 2nd ed., Prentice-Hall, New York, 1952.
16. D. E. Lake and A. A. Gunkler, *Chem. Eng.* **67**, 136 (1960).
17. N. W. Gregory and B. A. Thackery, *J. Am. Chem. Soc.* **72**, 3176 (1950).

18. Ref. 1, p. 1191.
19. J. D. Johnson and R. Overby, *Proceedings of the National Specialty Conference on Disinfection,* American Society of Civil Engineers, New York, July 8–10, 1970, pp. 37–60.
20. Z. E. Jolles, A. H. Oxtoby, and P. J. M. Radford in ref. 2, pp. 351–407.
21. U.S. Pat. 2,971,959 (Feb. 14, 1961), T. D. Waugh and R. C. Waugh.
22. L. Spitzer, *Ind. Eng. Chem. Anal. Ed.* **8,** 465 (1936).
23. I. R. L. Barker, *Chem. Ind.,* 1936 (1964).
24. D. Grosjean, G. Mouvier, and J. E. Dubois, *J. Org. Chem.* **41,** 3869, 3872 (1976).
25. G. A. Olah, *Friedel-Crafts Chemistry,* John Wiley & Sons, Inc., New York, 1973, p. 126.
26. G. Russel and C. De Boer, *J. Am. Chem. Soc.,* 3136 (1963).
27. M. D. Johnson in ref. 2, p. 266.
28. E. S. Huyser in S. Patai, ed., *The Chemistry of the Carbon Halogen Bond,* Part 1, John Wiley & Sons, Inc., New York, 1973, p. 553.
29. K. Rankama and Th. G. Sahama, *Geochemistry,* The University of Chicago Press, Chicago, Ill., 1950.
30. W. T. Halser, *Bromide Geochemistry of Salt Rocks, Second Symposium on Salt,* Vol. 1, No. 2, 1968.
31. B. J. Burreson, R. E. Moore, and P. O. Roller, *J. Agric. Food Chem.* **24**(4), 856 (1976).
32. J. A. Pettus, Jr., R. M. Wing, and J. J. Sims, *Tetrahedron Lett.* (1), 41 (1977).
33. R. Kazlauskas and co-workers, *Tetrahedron Lett.* (1), 37 (1977).
34. G. R. Pettit and co-workers, *J. Am. Chem. Soc.* **99**(1), 262 (1977).
35. H. Farkas-Himsley and G. I. Jolles in ref. 2, p. 491.
36. J. P. Riley and G. Skirrow, *Chemical Oceanography,* Vol. I, Academic Press, New York, 1965.
37. F. Yaron in ref. 2, pp. 3–42.
38. D. Whitehead, *The Dow Story,* McGraw-Hill Book Company, New York, 1968, p. 22.
39. U.S. Pat. 460,370 (Sept. 27, 1891), H. H. Dow.
40. Ger. Pat. 194657 (1906), K. Kubierschky, *Z. Chem. Apparatenk.* **3,** 212 (1908).
41. U.S.S.R. Pat. 447,108 (July 15, 1975), I. T. Zhukov and co-workers.
42. N. Kalyanam, T. N. Balakrishnam, and S. Vasudevan, *Chem. Age India* **26**(4), 335 (1975).
43. U.S. Pat. 3,442,614 (May 6, 1969), C. D. Frazee and H. R. Greeb (to Proctor and Gamble Co.).
44. M. A. Schenker, A. M. Ponizovskii, and Z. S. Chernova, *Zh. Prikl. Khim.* **44,** 2601 (1971).
45. M. R. Block and co-workers, *Corros. Sci.* **11,** 453 (1971).
46. L. C. Stewart, *Ind. Eng. Chem. Ind. Ed.* **26,** 361 (1934).
47. U.S. Pat. 3,314,762 (Apr. 18, 1967), L. H. Hahn (to Michigan Chemical Corp.).
48. U.S. Pat. 3,642,447 (Feb. 15, 1972), L. H. Hahn and C. W. Dunbar.
49. U.S. Pat. 3,615,265 (Oct. 26, 1971), R. Gartner (to Veb Kombinat Kali).
50. U.S. Pat. 2,929,686 (Mar. 23, 1960), M. Codell and E. Norwitz.
51. Ger. Pat. 2,034,473 (Feb. 3, 1972), R. Sell and O. Braun (to Kali und Salz G.m.b.H.).
52. Ger. Pat. 2,034,038 (Jan. 20, 1972), R. Sell and G. Budan (to Kali und Salz G.m.b.H.).
53. M. S. Chao and V. A. Stenger, *Talanta* **11,** 271 (1964).
54. C. L. Klingman, *Mineral Facts and Problems, Bulletin 667,* U.S. Dept. of Interior, Bureau of Mines, Washington, D.C., 1975.
55. *Minerals Yearbook,* U.S. Dept. of Interior, Bureau of Mines, U.S. Govt. Printing Office, Washington, D.C., 1975 and earlier editions.
56. *Statistical Summary of the Mineral Industry,* Institute of Geological Sciences, Mineral Resources Div., Her Majesty's Stationery Office, London, 1971, and earlier editions.
57. M. Wilsker, *European Chemical News, Large Plants Supplement,* 1976, p. 27.
58. V. A. Stenger in F. D. Snell and C. L. Hilton, eds., *Encyclopedia of Industrial Chemical Analysis,* Vol. 8, John Wiley & Sons, Inc., New York, 1969, pp. 1–29.
59. *Occupational Safety and Health Administration Regulations,* Subpart G, paragraph 1910.93, Table G-1.
60. Dow Chemical U.S.A., *Well Pack Fluid-14,* Form No. 101-22-74.
61. R. C. Kidder, *Plast. Eng.,* 36 (Feb. 1977).
62. *Mod. Plast.,* 40 (Sept. 1976).

General References

Z. E. Jolles, ed., *Bromine and Its Compounds,* Academic Press, New York, 1966, pp. 940.
A. J. Downs and C. J. Adams, *Chlorine, Bromine, Iodine and Astatine,* in A. F. Trotman-Dickenson, ed., *Comprehensive Inorganic Chemistry,* Vol. 2, Chapt. 26, Pergamon Press, New York, 1973, pp. 1107–1573.
V. Gutmann, ed., *Halogen Chemistry,* Vol. 1, Academic Press, New York, 1967.
J. W. Mellor, *Comprehensive Treatise on Inorganic and Theoretical Chemistry,* Vol. II, Suppl. I, John Wiley & Sons, Inc., New York, 1956.
S. Patai, ed., *The Chemistry of the Carbon-Halogen Bond,* Parts 1 and 2, John Wiley & Sons, Inc., New York, 1973, 1187 pp.

CHARLES E. REINEKE
Dow Chemical U.S.A.

BROMINE COMPOUNDS

The most important compound of bromine is ethylene bromide, used mainly in antiknock gasoline. During 1976 it represented about 69% of world bromine output. Fumigants other than ethylene bromide accounted for 10% and fire retardants 8%. Only about 5% went into inorganic products, eg, hydrobromic acid, alkali bromides, and bromates.

INORGANIC COMPOUNDS

Bromamines

These extremely unstable compounds are likely to explode violently, even at low temperature, if isolated. They are usually prepared, observed, or used in quite dilute solutions (1). Traces may be formed when water containing a little bromide and ammonia is chlorinated.

Monobromamine [14519-10-9], NH_2Br, (bromamide) was originally prepared by adding a dilute solution of bromine in anhydrous ether to a 50% excess of ammonia, also in dry ether, at $-60°C$ (2):

$$2 NH_3 + Br_2 \rightarrow NH_2Br + NH_4Br$$

After removal of ammonium bromide and excess ammonia, a pale straw-colored solution of the product remains. Clemens and co-workers (3) carried out the same reaction in the gas phase, under nitrogen in a Sisler generator, and studied the further reactions of the bromamine vapor with substituted phosphines to form addition products. If bromamine is isolated at low temperature, it decomposes with violence upon being allowed to reach $-70°C$ (4). In ether, bromamine reacts with Grignard reagents yielding primary amines, ammonia, and nitrogen. The formation and stability of bromamine in dilute aqueous solutions, its spectrophotometric detection, and its effect on bacteria have been reported (1,5–6). Its $\lambda_{max}^{H_2O}$ is 278 nm.

Dibromamine [*14519-03-0*], NHBr$_2$, (bromimide) may be prepared by adding an ether solution of ammonia to an excess of bromine in ether at −50°C until the color of the latter changes from red to yellow (2,7):

$$3\,NH_3 + 2\,Br_2 \rightarrow NHBr_2 + 2\,NH_4Br$$

An ether solution, after removal of ammonium bromide, is stable for several hours at −72°C. On treatment with a Grignard reagent, both primary and secondary amines are formed along with ammonia and nitrogen. The disinfection efficiency of dibromamine in water exceeds that of dichloramine (1). Its λ_{max} is 232 nm (6).

Tribromamine [*15162-90-0*], NBr$_3$, (nitrogen tribromide) can be obtained as an ammoniate by the reaction of bromine and ammonia below −70°C (4). Tribromamine predominates in water below pH 8 when 2 to 3 moles of bromine are added per mole of ammonia (6); it can be extracted from water by chloroform. Its $\lambda_{max}^{H_2O}$ is ca 260 nm, but interference is caused by hypobromite or tribromide ion. Nitrogen tribromide is fairly stable in dilute aqueous solution if an excess of bromine is present (1).

Bromides

Alkali and alkaline earth bromides are prepared most simply by neutralizing a solution of the corresponding hydroxide or carbonate with hydrobromic acid, evaporating excessive water, and crystallizing. In place of acid, bromine and a reducing agent such as ammonia are used in the van der Meulen process (8):

$$3\,K_2CO_3 + 3\,Br_2 + 2\,NH_3 \rightarrow 6\,KBr + N_2 \uparrow + 3\,CO_2 + 3\,H_2O$$

Classically, alkali bromides and bromates are formed simultaneously by absorbing bromine in a carbonate solution. After removal of much of the bromate by crystallization, the remainder is reduced with iron and the solution filtered and evaporated further to yield the bromide. Ammonium bromide [*12124-9-9*] is prepared commercially by the direct reaction of bromine with aqueous ammonia:

$$3\,Br_2 + 8\,NH_4OH \rightarrow 6\,NH_4Br + N_2 + 8\,H_2O$$

Laboratory preparations of bromide solutions have been based upon a metathesis reaction of the appropriate sulfate or carbonate with barium bromide.

Considerable amounts of sodium bromide [*7647-15-6*] and potassium bromide [*7758-02-3*] are employed in the preparation of light-sensitive silver bromide [*7785-23-1*] emulsions for photography. Various inorganic bromides, chiefly those of the alkalies, alkaline earths, and ammonium ion, are prescribed in medicine. Their sedative action is of value in the treatment of nervous disorders. Lithium bromide [*7550-35-8*] and calcium bromide [*7789-41-5*] are effective desiccants used in the industrial drying of air (see Drying agents). A more recent application of calcium bromide is in a dense packing fluid for oil well completion (see Drilling fluids). A nominal 53% solution with sp gr 1.70–1.72 at 25°C is used. A saturated solution at 20°C contains 58.8% CaBr$_2$ and has a density of 1820 kg/m^3. Anhydrous aluminum bromide is a good catalyst for some types of bromination reactions. It is prepared directly from the elements with precautions to allow for the high heat of reaction. For further information on specific inorganic bromides, see articles dealing with compounds of the metals in question.

Hydrogen bromide [*10035-10-6*], HBr, (hydrobromic acid), is a very irritating, colorless gas that fumes strongly in moist air; mp, −86°C (9); liquid density 2152 kg/m^3 bp −67°C; heat capacity, J/(kg·K) [cal/(kg·K)], solid at −91°C, 636 [152], liquid, 737 [176], gas at 27°C, 356 [85]; heat of fusion at mp, 29.8 kJ/kg (7.12 kcal/kg); heat of vaporization at −66.7°C, 218 kJ/kg (52 kcal/kg); critical temperature, 89.8°C; critical pressure, 8510 kPa (84 atm). The gas is highly soluble in water, forming azeotropic mixtures whose compositions at various pressures have been determined by Bonner, Bonner, and Gurney (10). At normal atmospheric pressure the boiling point is 124.3°C and the HBr content is 47.63%. This mixture freezes at about −11°C and has a density of 1478.6 kg/m^3 at 22°C. At very low temperatures hydrogen bromide forms crystalline hydrates with 1, 2, 3, and 4 moles of water (11). Work with liquid hydrogen bromide under pressure in glass at or above room temperature should be avoided, as chemical attack, possibly intergranular and not obvious, can lead to unexpected shattering.

Hydrobromic acid resembles hydrochloric acid but is a more effective solvent for some ore minerals because of its higher boiling point and stronger reducing action. Certain higher oxides such as ceric oxide are readily dissolved. In respect to hydrogen ion activity, hydrobromic acid is one of the strongest acids. With the bromides of several metals it forms complexes (HFeBr$_4$ [*19567-68-1*], amber; HCuBr$_3$ [*31415-59-5*], violet; etc). Bromine is very soluble in strong aqueous hydrobromic acid. Hot 48% acid cleaves alkoxy and phenoxy compounds almost as rapidly as does hydriodic acid.

Hydrogen bromide gas is prepared commercially by burning a mixture of hydrogen and bromine vapor. Processes involving the use of platinized asbestos as a catalyst are also available (9,12). The vapor is passed through hot, activated charcoal to remove free bromine (13), and is either liquefied by cooling for shipment in cylinders, or is absorbed in water or other solvent such as acetic acid. Aqueous solutions may also be prepared by the reaction of bromine with sulfur or phosphorus and water (14–15) (see Bromine), or on a laboratory scale by distillation from potassium bromide and dilute sulfuric acid (16). Another laboratory method suitable for preparing the dilute acid involves passing an alkali bromide solution through the hydrogen form of a cation-exchange resin.

Technical 48% and 62% acids are colorless to light-yellow liquids available in carboys, drums, or tank quantities. They are classified under DOT Regulation 173.262 as corrosive materials; a black and white label is required for rail or truck shipment. A medicinal grade 48% acid is furnished in carboys or small bottles. Anhydrous hydrogen bromide is available in 6.8-kg and 68-kg cylinders, under its vapor pressure of approximately 2.4 MPa (350 psi) at 25°C. Classified as a nonflammable gas, it requires a green label for rail or truck shipment.

Hydrobromic acid resembles hydrochloric acid in its toxic effects but may be somewhat less potent. Water solutions are irritating to the skin. The gas and fumes are almost as painful and irritating to the eyes and respiratory passages as hydrochloric acid.

A fair amount of hydrobromic acid is consumed in the manufacture of inorganic bromides, as well as in the synthesis of alkyl bromides from alcohols. The acid can also be used to hydrobrominate olefinic linkages directly. In the petroleum industry hydrogen bromide can serve as an alkylation catalyst, and its functions as a catalyst in the controlled oxidation of aliphatic and alicyclic hydrocarbons to ketones, acids, and peroxides have also been reported (17–18). Applications of HBr with NH$_4$Br (19) or with H$_2$S and HCl (20) as promoters in the dehydrogenation of butene to butadiene have been described, and either HBr or HCl can be used in the vapor-phase ortho methylation of phenol with methanol over alumina (21). Various patents dealing with catalytic activity of HCl also cover the use of HBr. An important reaction of HBr in organic syntheses is the replacement of aliphatic chlorine by bromine in the presence of an aluminum catalyst (22). Small quantities of hydrobromic acid are employed in medicine and for analytical purposes.

Bromine Halides

Bromine and chlorine react reversibly in the liquid or vapor states to form bromine chloride [13863-41-7]. The heat of reaction is ca −975 J/mol (−233 cal/mol) for the formation of gaseous bromine chloride (23), and the equilibrium constant, $K = \sqrt{(Br_2)(Cl_2)}/(BrCl)$, from spectroscopic data, has values of 0.338 at 25°C and 0.374 at 127°C (24). It follows that in an equimolar mixture of bromine and chlorine vapors at room temperature, about 60 mol % of the halogen is present as BrCl (25). This proportion does not change greatly with temperature. A similar but somewhat more stable compound, iodine bromide [7789-33-5], IBr, is formed from bromine and iodine; the heat of formation is about −7830 J/mol (−1871 cal/mol) (vapor at 127°C) and K ranges from 4.95×10^{-2} at 25°C to 9.04×10^{-2} at 127°C (24,26). There is also evidence for the existence of a tribromo complex, IBr_3 [7789-58-4]. These compounds are soluble in carbon tetrachloride or acetic acid and are chiefly of interest as halogenating agents for organic substances (27–28).

Bromine chloride is useful in both addition and substitution reactions. In the former, olefins are converted to bromochloro compounds and, in the latter, hydrogen is replaced by bromine yielding an organic bromide and hydrogen chloride. Use of bromine chloride rather than chlorine as a disinfectant in wastewater treatment has been recommended (29) because chlorine reacts with ammonia in the water to produce chloramines, for which high fish toxicity has been reported (30). The bromamines formed by bromine chloride are less stable, hence residuals are lower. Presumably the larger molecule is also less likely to be absorbed through the gill system. The corrosivities of bromine chloride to iron, nickel, and various alloys have been studied (31) and, at corresponding low moisture levels, are considerably less than those of liquid bromine.

Fluorine reacts violently with bromine forming various fluorides (BrF [13863-59-7], BrF_3 [7787-71-5], and BrF_5 [7789-30-2]) depending upon the proportions taken and other conditions (see Fluorine compounds, inorganic). The tri- and pentafluorides are commercially available. As strong fluorinating agents they are useful in organic syntheses (32–34) and in forming uranium fluorides, both for isotope enrichment (35–36) and for fuel element reprocessing (37). With alkali fluorides they form salts $MBrF_4$ and $MBrF_6$ analogous (in regard to the bromine valence state) to bromites and bromates.

Because of their reactivity, the bromine fluorides must be handled in apparatus and tubing of resistant materials such as nickel, Monel metal, or Teflon plastics, with all due precautions against operator exposure. For shipping, bromine trifluoride and pentafluoride are classified as oxidizers under DOT Regulation 173.246. The trifluoride also requires a poison label.

Bromine Oxides, Acids, and Salts

Oxides. At least some of the oxides of bromine may be considered as anhydrides of the oxygen-containing bromo acids described below. They are all unstable at ordinary temperature and for that reason none was isolated before 1929. Bromine monoxide [21308-80-5], Br_2O, is a dark brown solid, stable below $-40°C$ and melting with decomposition at $-17.5°C$. It has been reported as formed by thermal decomposition of the dioxide [21255-83-4], BrO_2 or Br_2O_4 (38), but is most often used as a solution in carbon tetrachloride or fluorotrichloromethane, prepared in the dark by reaction in the solvent under anhydrous conditions (39–40):

$$HgO + 2\, Br_2 \rightarrow Br_2O + HgBr_2$$

Mercuric bromide and excess mercuric oxide are removed by filtration. The solution can be used for brominations.

A trioxide, Br_2O_3 [32062-14-9], is reported (41) to be formed during thermal decomposition of bromine dioxide, Br_2O_4. The former has the structure OBrOBrO and the latter may be O_2BrBrO_2, lacking an oxygen bridge. The dioxide is prepared by ozonization of a solution of bromine in Freon 11 at $-50°C$ (42). The solvent is evaporated at low temperature, leaving a yellow–orange solid that decomposes at $0°C$. A white higher oxide stable below $-80°C$ formed by reaction of ozone or active oxygen with bromine, has been reported as BrO_3, Br_2O_5, or $(Br_3O_8)_n$ (43–46). According to Pascal and co-workers (47) the material obtained at -28 to $+4°C$ is 80% Br_2O_5 and 20% Br_2O_4; Raman spectra indicate that Br_2O_5 is analogous to iodine pentoxide.

In the decomposition of bromine oxides or in collisions between bromine and oxygen atoms, free radicals such as BrO [15656-19-6] can be observed spectrally (48–49). These exist very briefly and are of interest mainly in studies of reaction mechanisms.

Acids and Salts. The oxygen acids of bromine are unstable strong oxidants capable of existing at ordinary temperatures only in solution. An aqueous solution of hypobromous acid [13517-11-8] may be prepared by treating bromine water with silver nitrate (50) or mercuric oxide (40,50):

$$Br_2 + H_2O + AgNO_3 \rightarrow HOBr + AgBr + HNO_3$$

$$2\, Br_2 + H_2O + HgO \rightarrow 2\, HOBr + HgBr_2$$

In the former method a weak solution was distilled in vacuum after filtration. To prepare a more concentrated solution, the mercuric oxide reaction can be carried out in Freon 11 without water, yielding a solution of bromine monoxide which is filtered and hydrolyzed. Hypobromous acid is very slightly ionized; its dissociation constant at $25°C$ is 1.6×10^{-9}.

Hypobromites, the salts of hypobromous acid, disproportionate gradually to bromide and bromate, thus are not well suited for shipping or storage. Solutions are prepared as needed from bromine and alkali (see Bromine) with cooling and without removal of the simultaneously formed bromide. Because disproportionation is catalyzed by cobalt, nickel, and copper (51), care should be taken to avoid these impurities. Solid alkaline earth hypobromites, or more properly bromide hypobromites such as CaBr(OBr) [67530-61-4], have been known since the early days of bromine chemistry, but the pure crystalline hydrates $NaOBr.5H_2O$ [13824-96-9] and $KOBr.3H_2O$ [13824-97-0] were not described until the work of Scholder and Kraus (52) in 1952. Hypobromites are strong bleaching agents, similar to hypochlorites, and have been incorporated in scouring formulations (53).

Bromous acid [7486-26-2], HOBrO, is formed in low concentrations when excess bromine is present in the reactions shown above (54–56). It seems quite likely that a stronger aqueous solution stable at least for short periods might be obtained by mixing cold solutions of sulfuric acid and barium bromite. Crystalline barium bromite [20161-18-6], Ba(BrO$_2$)$_2$.H$_2$O, was first reported by Kircher and Periat (57); methods for preparing this salt and strontium bromite dihydrate [20161-17-5] have been patented by Meybeck and co-workers (58). The appropriate hypobromite is treated with bromine at 0°C and pH 11.2. After slow evaporation the compounds crystallize. Under vacuum at 0°C they are stable for at least six months.

Sodium bromite [7486-26-2], available in the United States as an approximately 10% solution, is used as a desizing agent in the textile industry. A solid trihydrate, NaBrO$_2$.3H$_2$O, has been commercialized in Europe (59). The enthalpy of formation of NaBrO$_2$ in aqueous solution is −273.7 kJ (−65.4 kcal) at 25°C (60). Methods for the analysis of hypobromite–bromite mixtures have been reviewed and improved by Andersen and Madsen (61). A review of the structures and thermal stabilities of metal bromites has been compiled (62).

Bromic acid [7789-31-3], HBrO$_3$, was originally prepared in solution by the reaction of barium bromate with sulfuric acid. A more convenient procedure involves passage of an alkali bromate solution, about 0.5 M, through a cation-exchange resin such as Dowex 50X8 in hydrogen form (63–64). The resulting acid can be concentrated *in vacuo* to about 50% HBrO$_3$. Stronger solutions are not stable and the 50% solution begins to decompose around 40°C. Preparation of bromic acid solutions by the electrolysis of bromine in bromic acid solution, with platinum or lead dioxide anodes, has been reported (65).

As an acid HBrO$_3$ is quite strong, its pK value of 0.7 being close to that for the second dissociation step of sulfuric acid. It is also a strong oxidant, with a normal potential of 1.5 V for the reaction HBrO$_3$ → ½ Br$_2$ in acid medium.

Bromates are stable under ordinary conditions and have various applications based upon their oxidizing properties. The alkali bromates are usually prepared by electrolysis of bromide solutions in the presence of a depolarizing ion such as dichromate (66). Cathodes may be stainless steel or copper, and anodes graphite, platinum-clad copper or titanium, or lead dioxide on stainless steel or graphite, with current densities of 0.10–0.25 A/cm^2. The bromate is crystallized by cooling the hot solution after removal from the cell, and the mother liquor is treated with more bromide and recycled. Another method of manufacture involves addition of bromine to the appropriate alkali carbonate or alkaline earth hydroxide to form hypobromite, which further reacts to yield bromate and bromide (see Bromine). The products are separable by virtue of the lower solubilities of bromates.

As a source of active oxygen when heated, subjected to shock, or acidified, bromates represent a potential fire and explosion hazard. They should be kept from contact with reactive organic matter, including paper and wood. Lots of 11.3 and 90.7 kg are packed in fiber drums with polyethylene liners (either bonded or bags). Metal drums have also been used. Laboratory quantities are supplied in glass bottles. For shipment a yellow oxidizer label is required under DOT Regulations 173.153-4.

Barium bromate [10326-26-8], Ba(BrO$_3$)$_2$.H$_2$O, forms colorless, monoclinic crystals, sp gr, 3.99; it decomposes at 260°C. The solubility in water is 0.79 g/100 g soln at 25°C, and 3.52 g at 80°C. It is a poisonous substance, used mainly for the preparation of other bromates.

Potassium bromate [7758-01-2], KBrO$_3$, forms white or colorless granules or crystals, sp gr, 3.27; it decomposes at about 370°C. The solubility in water is 7.53 g/100 g soln at 25°C, and 25.4 at 80°C. Grades available are: pure (not less than 99.5%), cp, and reagent (not less than 99.8%). The major use for potassium bromate is in flour treatment. Wheat flour containing 5–10 ppm, or more if soybean proteins are also present, shows considerable improvement over untreated flour in baking characteristics. Some potassium bromate is used as a primary standard as well as a brominating agent in analytical chemistry.

Sodium bromate [7789-38-0], NaBrO$_3$, is a white, crystalline powder, sp gr 3.40, n_D^{20} 1.594; it decomposes at 381°C. The solubility in water is 28.3 g/100 g soln at 25°C, 43.1 at 80°C. Pure and reagent grades (not less than 99.5 and 99.7%, respectively) are available. An important outlet for sodium bromate is as a neutralizer or oxidizer in certain hair-wave preparations. Formerly mixtures of sodium bromate with sodium bromide were sold as mining salts or bromine salts and were used as a source of bromine (released upon acidification in aqueous solution).

Perbromates and perbromic acid [19445-25-1], HBrO$_4$, were long thought to be incapable of existence because attempts to prepare them had been unsuccessful. In 1968, however, E. H. Appelman at the Argonne National Laboratory found that a small amount of perbromate could be obtained by oxidizing a bromate either electrolytically or with xenon difluoride (67). The following year he reported a fluorination procedure (68) suitable for making somewhat larger quantities.

Perbromic acid solution is prepared by passing the sodium salt solution through a cation-exchange resin in hydrogen form. It can be concentrated to nearly 6 M by evaporation under a heat lamp. Stronger solutions are not stable in air. Perbromic is similar to perchloric acid, having little oxidizing activity in dilute solution but capable of violent reactivity when concentrated. A 3 M solution attacks stainless steel. The acid can be used to prepare perbromate salts. Sodium perbromate [33497-30-2] is quite soluble in water, potassium perbromate [22207-96-1] only to the extent of about 0.2 M at room temperature. The reduction of the potassium salt by hydrochloric acid and by potassium bromide and iodide has been studied (69). Appelman and co-workers have reported the thermodynamic properties of perbromate and bromate ions (70) as well as the effects of photolysis and pulse-radiolysis on perbromate in aqueous solutions (71–72). The normal oxidation potential of the BrO$_4^-$/BrO$_3^-$ couple in acid solution is 1.74 V.

ORGANIC COMPOUNDS

Organic compounds of bromine usually resemble their chlorine analogues but have higher densities and lower vapor pressures. The bromo compounds are more reactive toward alkalies and metals; brominated solvents should generally be kept from contact with active metals such as aluminum. On the other hand, they present less fire hazard: one bromine atom per molecule reduces flammability about as much as two chlorine atoms. Although a great many bromine compounds are described in the literature, those considered here are mainly the ones that have some commercial application. Roughly in order of their sales volumes, they are: ethylene bromide, methyl bromide, bromochloropropanes, methylene chlorobromide, ethyl bromide, miscellaneous organic bromine compounds, pharmaceuticals, and dyes and indicators. Bromine derivatives of compounds such as acetic acid and camphor are described in the sections on Derivatives in the articles on these compounds. For compounds used especially as fire retardants, see Flame retardants.

Ethylene Bromide

Ethylene bromide [106-93-4], (ethylene dibromide, 1,2-dibromoethane), CH$_2$BrCH$_2$Br, is a clear, colorless liquid with a characteristic sweet odor. Following are its properties: mp, 9.9°C; bp, 131.4°C; density, 2179.2 kg/m^3 at 20°C (73); n_D^{20}, 1.5380 (74). Vapor pressure, 1.13 (8.5), 15.98 (119.8), and 38.03 kPa (285.2 mm Hg) at 20, 75, and 100°C, respectively. Viscosity at 20°C, 1.727 mPa·s (= cP). Heat capacity, J/(kg·K): solid at 15.3°C, 519 (124 cal/kg·K); liquid at 21.3°C, 724 (173 cal/kg·K). Heat of fusion at 9.9°C, 53.4 kJ/kg (12.76 kcal/kg); heat of vaporization at bp, 191 kJ/kg (45.7 kcal/kg); heat of transition at −23.6°C, 10.34 kJ/kg (2.47 kcal/kg). Critical temperature, 309.8°C; critical pressure, 7154 kPa (70.6 atm). Expansion coefficient at 15–30°C, 0.000958/K (see ref. 75); dielectric constant at 20.5°C (0.1 MHz), 4.77. The liquid is completely miscible with carbon tetrachloride, benzene, gasoline, ether, and anhydrous alcohols at 25°C. Its solubility in water at 20°C is 0.404 g/100 g sol.

Ethylene bromide is nonflammable and under ordinary conditions is quite stable. Slight decomposition may occur on exposure to light. The heat of combustion in an oxygen bomb is 6647 J/g (1589 cal/g). At 340–370°C in a glass vessel, ethylene bromide decomposes into vinyl bromide and hydrogen bromide. Vinyl bromide is also formed slowly on contact with a warm alkaline solution. Ethylene glycol is produced by high temperature hydrolysis under pressure, and the reaction with zinc in alcohol yields ethylene and zinc bromide. If a mixture of liquid ammonia and ethylene bromide is allowed to reach room temperature, an explosion may result with formation of ethylenediamine and higher homologues.

In the manufacture of ethylene bromide, gaseous ethylene is brought into contact with bromine by various methods, allowing for dissipation of the heat of reaction (76–78). In one of these the reaction is carried out in ethylene bromide as a solvent, promoted by a little water. Free acids are neutralized and the product may be fractionally distilled for purification. Typical specifications call for a clear liquid of sp gr (25/25°C), 2.165–2.185; initial boiling point not less than 126.0°C, and distillation range within 2.0°C, including 131.4°C, for the 90% distilled between 5 and 95%; APHA color, 60 max; water-extractable free acidity, max 1.2 cm^3 of 0.01N per 100 cm^3. Shipments are made in 208-L (55-gal) drums (450 kg net) or tank cars (ca 65,800 kg net) under DOT regulation 173.620; the hazard class is ORM-A.

The vapor of ethylene bromide is toxic and exposure to a time-weighted average of 20 ppm by volume in air should be avoided. Even moderately prolonged contact of liquid with the skin can cause lesions. In case of accidental exposure, injury is usually averted by prompt removal of contaminated apparel, thorough wiping of the skin, and opportunity for unhampered evaporation. That ethylene bromide may be a carcinogen has been reported from feeding studies on rodents. During 1977 its use as an agricultural fumigant was being examined by EPA.

Ethylene bromide is one of the cheapest organic compounds of bromine, priced during 1977 at 77–81¢/kg in tank lots or carloads of drums, freight equalized. World production in 1976 was estimated at 200,000 metric tons. The principal use as an additive in leaded gasoline is declining because of regulation of the amount of lead in automotive fuel. However, owing to a number of factors, the decline is not as rapid as had been predicted. Ethylene bromide permits vaporization of lead from engine deposits as lead bromide. The other important use is as an ingredient of soil fumigants against wireworms and nematodes, and of grain fumigant formulations for insect control (under Food Additive Regulations 121.1020, 121.1133, and 131.1152). Minor uses are as an intermediate in syntheses and as a nonflammable solvent for resins, gums, and waxes.

Methyl Bromide

Methyl bromide [74-80-9], CH_3Br, (bromomethane) is a colorless liquid or gas with practically no odor. Its physical properties are: mp, $-93.7°C$; bp, $3.56°C$; liquid d_4, 1730 kg/m^3; vapor d^{20}, 3.974 kg/m^3; n_D^{-20}, 1.4432; vapor pressure at 20°C, 189.3 kPa (1420 mm Hg); viscosity at -20, 0, and 25°C: 0.475, 0.397, and 0.324 mPa·s (= cP), respectively. Heat capacity, J/(kg·K) [cal/(kg·K)]: liquid at $-13.0°C$, 824 [197]; vapor at 25°C, 448 [107]; heat of vaporization at 3.6°C, 252 kJ/kg (60.2 kcal/kg); critical temperature (calculated), 194°C; expansion coefficient, -15 to 3°C, 0.00163/K; dielectric constant at 0°C and 0.001–0.10 MHz, 9.77. The liquid is miscible with most organic solvents and forms a bulky, crystalline hydrate with cold water. The solubility in water varies with pressure; at normal atmospheric pressure (methyl bromide plus water vapor) the solubility is 1.75 g/100 g soln (20°C).

Hydrolysis of methyl bromide to methanol and hydrobromic acid proceeds slowly in aqueous solution but more rapidly in dilute alkalies. Methyl bromide is an effective methylating agent; it reacts with amines, particularly the more basic ones, to form methylammonium bromide derivatives. It also reacts with sulfur compounds under alkaline conditions to give mercaptans, thio ethers, and disulfides. Most structural metals, other than aluminum, are inert toward pure, dry methyl bromide, but surface reactions take place on zinc, tin, and iron in the presence of impurities such as alcohol or moisture. Methyl bromide is nonflammable over a wide range of concentrations in air at atmospheric pressure and offers practically no fire hazard; with an intense source of ignition, flame propagation within the narrow range from 13.5 to 14.5% by volume has been reported. The material has no flash point. Thermal decomposition in a glass vessel begins somewhat above 400°C.

Commercial and laboratory methods of manufacturing methyl bromide are generally similar and are based primarily upon the reaction of hydrobromic acid with methanol. For generation of the acid in contact with the alcohol, the best known process is its liberation from an alkali bromide with sulfuric acid (79). Other methods involve direct use of hydrobromic acid or the treatment of bromine with a reducing agent, such as sulfur dioxide or phosphorus, in the presence of water (80–83). In the sulfuric acid process, the acid is added to a strong solution of sodium bromide and methanol in a reactor and methyl bromide is distilled through a reflux column which returns methanol and hydrobromic acid to the reactor. The product is passed through sulfuric acid for drying and is condensed, then fractionated to remove impurities. More recently proposed processes involve the reaction of hydrogen bromide with excess methyl chloride at 400–500°C (84) and the thermal or photochemical reaction of bromine chloride with methane (85).

Methyl bromide is sold both as the essentially pure compound, 99.8% minimum, with not more than 0.005% water and 0.001% acidity as HBr, and in formulations.

Formulations include mixtures with ethylene bromide or other fumigants, with or without chloropicrin to serve as a warning agent, and hydrocarbons as inert diluents. During 1977 methyl bromide in 63.5-t tanks was priced at 90¢/kg, freight allowed. Cylinders containing 4.5–170 kg are also available; these are equipped with siphon tubes to facilitate emptying without inversion. The natural pressure may be augmented with nitrogen or carbon dioxide before shipment to permit rapid ejection at low temperatures. Methyl bromide can also be furnished in 0.45 kg or 0.68 kg cans. Alone or in formulations, it is classified as a poison, class B, and requires a poison label. DOT Regulation 173.353 applies to packaging.

Exposure to methyl bromide in either the liquid or vapor state should be avoided. Contact of liquid with the skin for more than a few seconds may cause itching and blisters; leather and rubber clothing, particularly the latter, fail to give protection since they absorb methyl bromide. The upper safe limit for daily 8-h exposure to the vapor in air is considered as 15 ppm by volume, or about 0.06 mg/L. Repeated exposure to slightly higher concentrations may cause disorders of the central nervous system, recovery from which is generally complete some time after the exposure has been discontinued. Acute exposure (over 1000 ppm for one hour, or different concentrations in roughly inverse proportions in relation to time) may cause double vision, dizziness, headache, and nausea, followed in more severe cases by convulsions, muscular tremors, and even death. With observance of the proper precautions, methyl bromide may be transported and used safely. Precautions for handling can be obtained from manufacturers or the Manufacturing Chemists' Association (86). A positive-pressure hose gas mask or one equipped with a canister approved by the Bureau of Mines for organic vapors may be used if needed. A canister should be replaced after twenty minutes' exposure to a concentration of 10–40 mg/L.

The major use for methyl bromide is in the extermination of insect and rodent pests (see Poisons, economic). The material is suitable for the fumigation of food commodities and the facilities in which they are processed or stored, as well as for tobacco and many kinds of nursery stock. Dosages of about 16 mg/L (1 lb/1000 ft^3) for 8 h or more at 20°C, are effective against insects, spiders, and mites in any stage of development. Vacuum fumigations of shorter duration and atmospheric fumigations at lower temperatures may be carried out with higher concentrations. Residue studies (87–88) have shown that all of the fumigant disappears from fumigated foods within a few days, leaving only inorganic bromide residues for which the FDA establishes tolerances. In the past, methyl bromide had some use as a fire extinguisher for airplane engines. Small amounts are used in organic syntheses, for methylations.

Other Bromomethanes. Bromochloromethane [74-97-5], CH_2BrCl, (methylene chlorobromide) is a clear, colorless liquid with a characteristic sweet odor and very low freezing point, −88.0°C. Its properties include: bp, 68.1°C; d_4^{25}, 1922.9 kg/m^3; n_D^{25}, 1.4808; heat of vaporization at bp, 232 J/kg (55.4 cal/kg). The liquid is completely miscible with common organic solvents and soluble in water to the extent of about 0.9 g per 100 g at 25°C. Common methods of preparation involve the partial replacement of chloride in methylene chloride (dichloromethane) by reaction with anhydrous aluminum bromide, treatment with bromine and aluminum (89), or by reaction with hydrogen bromide in the presence of an aluminum halide catalyst (90), followed by water washing and distillation. The major outlet for bromochloromethane is as a fire-extinguisher fluid; its effectiveness per unit weight suits it for use in aircraft and portable extinguishers (see Fire extinguishing agents). Its toxicity is also lower than that of many bromine compounds (91). Prices quoted in 1977 were $1.57–1.61/kg in tanks or drum carlots, fob equalized.

Dibromomethane [74-95-3], CH$_2$Br$_2$, (methylene bromide) is a similar liquid, mp −52.7°C, bp 96.9°C, d$_4^{20}$ 2497.0 kg/m^3, n$_D^{20}$ 1.542. It is prepared by the same methods, allowing the reaction to proceed to completion. For a laboratory preparation, see ref. 92. The compound has limited uses in syntheses, as a solvent, and in gage fluids. Both of these dihalomethanes have potential uses as dense, readily volatile media for mineral and salt separations (see Gravity concentration).

Tribromomethane [75-25-2], CHBr$_3$, (bromoform) is usually sold mixed with 3–4% ethanol as a stabilizer. The pure liquid has mp, 7.7°C; bp, 149.5°C; d$_4^{20}$, 2891.2 kg/m^3; n$_D^{19}$ 1.5980 (93). Water solubility is about 0.3 g/100 mL at 25°C. Bromoform is prepared from chloroform by the replacement procedures above (94). The classical method of preparation involves reaction of acetone and sodium hypobromite; the latter may be generated from sodium hypochlorite and a bromide (95). Uses have been found in syntheses, in pharmacy as a sedative and antitussive, in gage fluids, and as a dense liquid. Traces of bromoform and bromochloroforms are likely to be present in municipal waters and wastes as a result of chlorination in the presence of naturally occurring bromide ions and humic substances (96). Removal can be accomplished by adsorption on activated charcoal.

Tetrabromomethane [558-13-4], CBr$_4$, (carbon tetrabromide) is a white to brownish powder, mp, 90.1°C; bp, 189.5°C; d$_4^{20}$, 3240 kg/m^3; n$_D^{99.5}$, 1.600 (97). The compound is monoclinic at room temperature with a transition to cubic at 46.9°C. It is prepared by replacement of chlorine in carbon tetrachloride with bromine using hydrogen bromide and an aluminum halide catalyst (90). It can also be prepared from bromoform (or acetone) and sodium hypobromite containing excess bromine, by an extension of the haloform reaction (98). Carbon tetrabromide is capable of direct addition to olefins; with ethylene it forms 1,1,1,3-tetrabromopropane [62127-50-8] (99). It is also light-sensitive. A number of patents have covered possible uses in photography and photo-duplicating systems.

Various bromofluoromethanes have been described and proposed for use as fire extinguishing agents (qv). Two that have been recommended highly for this purpose are CBr$_2$F$_2$ [75-61-6] and CBrF$_3$ [75-63-8] (100). These and similar substituted methanes, some containing chlorine as well as bromine and fluorine, are potentially useful for the synthesis of other halo-fluoro compounds.

Ethyl Bromide

Ethyl bromide [74-96-4], CH$_3$CH$_2$Br, (bromoethane) is a volatile, clear, colorless liquid of mp, −119.3°C; bp, 38.4°C; d$_4^{25}$, 1449.2 kg/m^3; n$_D^{25}$, 1.421 (73). It is completely miscible with most neutral or acidic organic solvents but may react with bases. The solubility in water is ca 0.9 g/100 g at 25°C. Ordinarily ethyl bromide is prepared by refluxing ethanol with hydrobromic acid, or with an alkali bromide and sulfuric acid (101). The reaction of ethane with sulfur trioxide and potassium bromide at 300–325°C is reported to produce ethyl bromide with a yield of 91% based on ethane consumed (102). A process involving the reaction of ethylene and hydrogen bromide induced by γ-radiation (103) was employed successfully for several years but became uneconomical for the small scale of production required during the 1970s. Ethyl bromide is used mainly as an ethylating agent in syntheses, particularly of pharmaceuticals. Some has been employed as a solvent or refrigerant, and occasionally as a local anesthetic applied as a spray. Its toxicity is markedly lower than that of methyl bromide. Technical-grade (98%) product in carloads of 208-L (55-gal) drums was quoted in 1977 at $1.34–1.37/kg, freight allowed east of the Mississippi River.

Bromochloropropanes

1-Bromo-3-chloropropane [109-70-6], CH$_2$BrCH$_2$CH$_2$Cl, (trimethylene chlorobromide) is a clear, colorless liquid, mp, −58.9°C; bp, 143.4°C; d_4^{25}, 1589 kg/m^3; n_D^{25}, 1.484 (73). It is prepared by addition of hydrogen bromide to allyl chloride (104–106) and used for the synthesis of cyclopropane. It can also be made from ethylene and methylene chlorobromide (107).

1,2-Dibromo-3-chloropropane [96-12-8] (DBCP) is also a clear, colorless liquid, mp, 6°C; d_4^{25}, 2081 kg/m^3; n_D^{25}, 1.552. At the reduced pressure of 40 kPa (300 mm Hg) it boils at 164.5°C; higher temperatures cause decomposition. The compound is prepared by brominating allyl chloride below 25°C, with provision for removing heat from the vigorous reaction. Employed either alone or in formulations under trade names such as Nemagon and Fumazone, it had become one of the most useful fumigants or fungicides for control of nematodes, root-knot disease, etc (108–111). Applications were carried out either by direct injection into the soil or by addition to irrigation water (the solubility of about 300 ppm in water is roughly ten times the recommended dosage). During the summer of 1977 sterility problems were encountered among male employees in production facilities and production was suspended pending further study. The compound has also been reported as carcinogenic in feeding studies on rats. OSHA has issued an emergency temporary standard limiting exposure to 10 ppb for an 8-h workday and 50 ppb for any 15-min period. Previously a limit of 1 ppm had been considered safe.

Dyes and Indicators

Among the dyes that contain bromine, those of the indigo group have been in greatest demand. Compounds with from one to eight atoms of bromine per molecule can be prepared. The reaction may be carried out dry or in a solvent such as nitrobenzene, carbon tetrachloride, or acetic acid. Dibromo- [19201-53-7] and tetrabromoindigo [2475-31-2], are brighter in color and show better light stability and covering power than indigo. A closely related compound, Ciba Bordeaux B [6371-14-8] (CI 73325) is a distinctive red dye obtained by brominating thioindigo (see Dyes, natural).

Anthraquinone dyes comprise a large group, members of which are capable of wide variation in dyeing characteristics depending upon their substituent groups. One of these, useful in dyeing wool, is Acid Blue 78 [6424-75-5] or Alizarin Pure Blue B (CI 62105). It is prepared by brominating 1-aminoanthraquinone in aqueous sulfuric acid to form the 2,4-dibromo compound, condensing the product with p-toluidine, and sulfonating. If the aminoanthraquinone is sulfonated before bromination, another useful intermediate results. Termed bromamine acid [116-81-4], it is 1-amino-4-bromoanthraquinone-2-sulfonic acid. A number of bromo anthraquinone, pyranthrone, and benzanthrone dyes are known (CI 59105, 59705, 59805, 60005, 62060).

Other bromo compounds employed as intermediates for the preparation of dyes include 6-bromo-2,4-dinitroaniline [1817-73-8], made by brominating the parent compound suspended in water at 40–60°C, and 5,7-dibromoisatin [6374-91-0], from bromination of isatin in sulfuric acid or the degradation of indigo with bromine in water. Bromodinitroaniline is used in preparing azo dyes by condensation, and dibromoisatin serves in making Alizarin Indigos B, 3R, and G (CI 73885, 73815, and 73820) (see Azo dyes).

Eosin [15086-94-9] (tetrabromofluorescein), made by the bromination of fluorescein in alcohol, is known both as a dye and as an adsorption indicator. Eosin Y [17372-87-1], the disodium salt, is used as a biological stain. Substances of interest mainly as acid–base indicators are bromophenol blue [115-39-9], bromcresol green [76-60-8], bromophenol red [2800-80-8], bromcresol purple [115-40-2], bromothymol blue [76-59-5], and dibromoxylenol blue. The compound 2,6-dibromo-N-chloro-p-benzoquinone imine [537-45-1] has been used as a color-test reagent for phenol, whereas the reaction product 2,6-dibromoindophenol [2582-33-4] is an oxidation-reduction indicator. Several bromine-containing substances, including p-bromophenacyl bromide [99-73-0], p-nitrobenzyl bromide [100-11-8], and p-bromoaniline [591-19-5], are employed as reagents in qualitative organic analysis.

Bromine as a substituent in dye or indicator molecules causes absorption of light at longer wavelengths. This is probably due to increased polarization caused by its electronegativity. Indicators show increased dissociation of phenolic hydroxyl groups for the same reason. Bromine is also likely to lower the solubility of a compound. For a more detailed discussion see ref. 112.

Pharmaceuticals

Historically, bromine compounds, particularly the alkali and some alkaline earth bromides, have been employed as sedatives, but this use has fallen into disfavor. Although a number of inorganic and organic bromine compounds have appeared in earlier editions of various pharmacopias, few remain in the more recent editions. Bromine may be significant in medicinal chemistry in several ways: bromide released from a compound may have a sedative effect; the presence of one or more bromine atoms in a molecule may impart desirable physical or chemical properties; and bromine compounds may be useful synthetic intermediates.

The general anesthetic halothane [151-67-6], $CF_3CHBrCl$, (2-bromo-2-chloro-1,1,1-trifluoroethane) exemplifies a compound in which a particular combination of halogens provides suitable physical properties and chemical stability (see Anesthetics). With a boiling point of 50.2°C it is easily administered by inhalation, and low water solubility allows ready removal through the respiratory system when exposure is discontinued. It is also nonflammable. Introduced in 1954 (113), it is listed in *USP XIX* and is widely used. Several methods of preparation are available. Madai and Muller (114) brominate 2-chloro-1,1,2-trifluoroethylene to form $CBrF_2CBrClF$ [354-51-8], isomerize the product with anhydrous aluminum chloride to CF_3CBr_2Cl [754-17-6], and reduce the latter with hydrogen (catalyzed by platinum). Patented processes include reduction of the same dibromo compound with steel turnings in hydrochloric acid (115), the antimony chloride-catalyzed reaction of HF with $CHBrClCBrCl_2$ [3749-38-7] (116), the exothermic rearrangement of $CBrF_2CHClF$ [354-51-8] with aluminum chloride (117), and the originally proposed chlorination of CH_2BrCF_3 [421-06-7] or bromination of CH_2ClCF_3 (118).

Table 1. Miscellaneous Bromine Compounds

Compound	CAS Registry No.	Formula	mp, °C	bp, °C[a]	d[4][20]	n[D][20]	Method of manufacture[b]	Uses[d], miscellaneous properties	Refs.
acetylene tetrabromide[c] (1,1,2,2-tetrabromoethane)	[79-27-6]	CHBr$_2$CHBr$_2$	0.1	151$^{5.9}$ 243.5	2.9638	1.6380	I	C, D, E, G, M, solv, syn, properties, prep	120–121
acetyl bromide[c]	[506-96-7]	CH$_3$COBr	−96	76	1.529		X	syn, prep	122
allyl bromide[c] (3-bromopropene)	[106-95-6]	CH$_2$=CHCH$_2$Br	−119.4	71.3	1.398		IV, VIII(a)	F, syn, prep, fumigations	123–124
benzyl bromide[c]	[100-39-0]	C$_6$H$_5$CH$_2$Br	−3.9	198–199	1.440	1.4655	II (uv)	L, syn, prep	125
bromacil	[314-40-9]	C$_9$H$_{13}$BrN$_2$O$_2$	157.5–160					H, prep, use	126–128
bromoacetic acid[c]	[79-08-3]	CH$_2$BrCOOH	50	208	1.934		II (in CS$_2$), III	syn, prep	129
bromoacetone[c]	[598-31-2]	CH$_3$COCH$_2$Br	−36.5	137	1.634^{23}	1.4697	II	L, syn, prep	
bromobenzene[c]	[108-86-1]	C$_6$H$_5$Br	−30.6	156.1	1.4951	1.5600	II(a)	solv, syn, flash pt, 51°C; fire pt, 155°C	130–131
2-bromo-4-tert-butylphenol	[2198-66-5]	(CH$_3$)$_3$CC$_6$H$_3$(OH)Br	<−20	120$^{0.67}$ (approx)	1.333^{25}	1.550^{25}	II (in C$_2$H$_4$Cl$_2$)	syn, monohydrate mp, 51–52°C	
p-bromochlorobenzene	[106-39-8]	C$_6$H$_4$BrCl	67.4	196				syn	
bromodichloromethane	[75-27-4]	CHBrCl$_2$	−57.1	90.1	1.9945	1.4985	III	E, solv, syn, properties	75
2-bromo-1,4-diethylbenzene	[52076-43-4]	BrC$_6$H$_3$(C$_2$H$_5$)$_2$	<−70	236–239	1.255^{25}	1.538^{25}	II(a)	G, solv, isomers	
2-bromodiphenyl (2-bromobiphenyl)	[2052-07-5]	C$_6$H$_5$C$_6$H$_4$Br	<−20	296–298	1.217$^{5 26}$		II(a) in (C$_6$H$_6$)	T, syn, prep	132–134
4-bromodiphenyl (4-bromobiphenyl)	[92-66-0]	C$_6$H$_5$C$_6$H$_4$Br	89.5–90	310–311			II(a) in (C$_6$H$_6$)	syn, prep, flash pt, 144°C	135–136
4-bromoethylbenzene	[1585-07-5]	BrC$_6$H$_4$C$_2$H$_5$	−65	200–203	1.335^{25}	1.454^{25}	II(a)	G, T, solv (Alkazene 40), isomers	137–139
p-bromoisopropylbenzene (4-bromocumene)	[586-61-8]	BrC$_6$H$_4$CH(CH$_3$)$_2$	−22.4	219.0	1.278^{25}	1.533$^{8 25}$	II(a)	G, T, solv, isomers	140–141
2-bromo-p-cymene	[4478-10-8]	Br(CH$_3$)C$_6$H$_3$CH(CH$_3$)$_2$	<−20	234.3	1.269^{17}	1.535^{26}	II(a)	G, T, syn, prep, and properties	142
1-bromonaphthalene	[90-11-9]	C$_{10}$H$_7$Br	6.2	281.1	1.4875	1.6582	II (in H$_2$O), II(a) (in CCl$_4$)	M, syn, prep	143–145
p-bromophenol	[106-41-2]	BrC$_6$H$_4$OH	64	237–238	1.840^{15}		II (in CCl$_4$)	A, syn, dissoc const, prep.	146–147
2-bromo-4-phenylphenol	[92-03-5]	C$_6$H$_5$C$_6$H$_3$(OH)Br	94	157$^{0.53}$	1.536^{25}		II (in CCl$_4$)	A (Dowicide 5), prep	148
2-bromopropionic acid	[598-72-1]	CH$_3$CHBrCOOH	25.7 (dl)	203.5	1.671	1.475	II(b) (100–130°C)	syn, prep	149
3-bromopyridine	[626-55-1]	BrC$_5$H$_4$N		173–174	1.632^{10}	1.566^{25}	II	syn (nicotinic acid), prep	150–151
5-bromosalicylic acid	[89-55-4]	Br(OH)C$_6$H$_3$COOH	166		2.098		II (in C$_2$H$_4$Cl$_2$)	syn (5-bromo-aspirin), prep	152
N-bromosuccinimide	[128-05-5]	(CH$_2$CO)$_2$NBr	173–175 (dec)				II (cold, alk)	syn (bromination), prep	153–154
2-bromothiophene	[1003-09-4]	BrC$_4$H$_3$S		149–151	1.649^{23}	1.522^{25}	II	syn, prep	155
o-bromotoluene	[95-46-5]	CH$_3$C$_6$H$_4$Br	−27	181.4	1.422	1.549	II(a)	solv, syn, prep, vapor pressure	156–157
p-bromotoluene	[106-38-7]	CH$_3$C$_6$H$_4$Br	28	184	1.390		II(a)	solv, syn, flash pt, 85°C	
bromoxynil (3,5-dibromo-4-hydroxylbenzonitrile)	[1689-84-5]	C$_7$H$_3$Br$_2$NO					II (alk)	H, prep, use	158–159
n-butylbromide[c] (1-bromobutane)	[109-65-9]	CH$_3$(CH$_2$)$_2$CH$_2$Br	−112.7	100.5	1.268$^{7 25}$	1.4398	IV	syn, properties, prep	160–161
cyanogen bromide[c]	[506-68-3]	CNBr	52	61.3^{100}	2.015		Br$_2$ + KCN	F, L, syn, very poisonous	162
o-dibromobenzene	[583-53-9]	C$_6$H$_4$Br$_2$	1.8	225	1.965	1.6081	II(a) (poor yield)	D, T, solv, syn	
p-dibromobenzene	[106-37-6]	C$_6$H$_4$Br$_2$	87.3	220.4	2.261^{17}	1.574$^{3 99}$	II(a)	syn, prep	
4,4'-dibromodiphenyl	[92-86-4]	BrC$_6$H$_4$C$_6$H$_4$Br	163–164	355–360	1.897		II(a)	syn, prep	136, 163–164
dibromoethylbenzene	[93-52-7]	Br$_2$C$_6$H$_3$C$_2$H$_5$	<−70	258–261	1.727^{25}	1.587^{25}	II(a)	G, T, (Alkazene 42), prep	165
diphenylbromomethane[c]	[776-74-9]	(C$_6$H$_5$)$_2$CHBr	45	193$^{3.5}$	1.491^{13}		II (in glass)	syn (pharmaceuticals), prep	166
ethyl bromoacetate	[105-36-2]	CH$_2$BrCOOC$_2$H$_5$		159	1.480^{25}	1.451	V	L, syn, prep	167–168
ethylene chlorobromide (1-bromo-2-chloroethane)	[107-04-0]	CH$_2$ClCH$_2$Br	−16.5	106.8	1.7392	1.4917	VI	F, solv, syn, thermal prop	169
ethylidene bromide (1,1-dibromoethane)	[557-91-5]	CH$_3$CHBr$_2$		108–110	2.06	1.5122	VII(b)	D, syn, prep	170
isopropyl bromide (2-bromopropane)	[75-26-3]	(CH$_3$)$_2$CHBr	−89.0	59.3	1.3138	1.4254	IV	solv, syn, prep, prop	161, 171–172

256

lauryl bromide (1-bromododecane)	[143-15-7]	CH₃(CH₂)₁₀CH₂Br	−5.5	175–180⁶	1.0382	1.4581	IV	syn (p), prep, prop	101, 173–174
metobromuron [3-(p-bromophenyl)-1-methoxy-1-methylurea]	[3060-89-7]	C₉H₁₁BrN₂O₂	95–96				II	H, prep, use, and prop	175–176
pentabromochlorocyclohexane	[87-84-3]	C₆H₆Br₅Cl	203.5				II (photochem)	E, prep	177–178
n-propyl bromide (1-bromopropane)	[106-94-5]	CH₃CH₂CH₂Br	−110	70.9	1.3514	1.4341	IV	solv, syn, prep	161, 179
propylene bromide (1,2-dibromopropane)	[78-75-1]	CH₃CHBrCH₂Br	−55.3	140	1.9333	1.5194	I	solv, syn, prep	180
styrene bromide (1,2-dibromoethylbenzene)	[93-52-7]	C₆H₅CHBrCH₂Br	74–75	131²·⁷			I (in CCl₄)	C (polym), syn F	181–182
tetrabromobisphenol A (4,4′-isopropylidenebis-2,6-dibromophenol)	[79-94-7]	(C₆H₂Br₂OH)₂C(CH₃)₂	181	316 (dec)			II (in alcohol)	E, prep	183
tetrabromo-o-cresol	[576-55-6]	CH₃C₆H₂OH	206–207	dec			II(a) (in CCl₄)	A, prep	184
2,4,6-tribromophenol	[118-79-6]	Br₃C₆H₂OH	96	subl	2.55		II (in H₂O)	A, E, dissoc const	185–187
1,2,3-tribromopropane	[96-11-7]	CH₂BrCHBrCH₂Br	16	220	2.4076²⁵	1.5835²⁵	I (from allyl bromide)	D, f, syn	188
trimethylene bromide (1,3-dibromopropane)	[109-64-8]	CH₂BrCH₂CH₂Br	−34.2	167.3	1.9790	1.5232	IV	syn (cyclopropane prep)	101
vinyl bromide (bromoethylene)	[593-60-2]	CH₂=CHBr	−137.8	15.8	1.5152	1.4462	VII(c), VIII(a,b)	in copolymers, prep, properties	189–192
xylyl bromides^c									
o-xylyl bromide	[89-92-9]	o-CH₃C₆H₄CH₂Br	21	223–224	1381²³	1.573²⁷	II	L, syn, prep	193
p-xylyl bromide	[104-81-4]	p-CH₃C₆H₄CH₂Br	38	218–220⁹⁸·⁷			II	L, syn, prep	194
1,2-bis(bromomethyl)benzene	[91-13-4]	o-C₆H₄(CH₂Br)₂	93–94				II	L, syn, prep	195
1,4-bis(bromomethyl)benzene	[623-24-5]	p-C₆H₄(CH₂Br)₂	143.5	245			II	L, syn, prep	194

^a At 101 kPa unless noted. To convert kPa to mm Hg, multiply by 7.5.

^b I Direct combination of bromine and the parent unsaturated compound, in a solvent if so indicated. The product may be washed to remove acids, dried, and fractionated, under vacuum if necessary.

II Replacement of hydrogen in parent compound by bromine: (a) in the presence of iron; (b) in the presence of phosphorus trichloride. Hydrogen bromide is allowed to escape, preferably being washed by incoming raw material. In some cases a method for separating isomeric compounds is necessary; otherwise the product, if a brominated hydrocarbon, is purified as in I. Phenols may be dissolved in alkali, precipitated with acid, and crystallized directly from the solvent in which they were formed.

III Treatment of the chlorine analogue with slightly more than the stoichiometric quantity of anhydrous aluminum bromide, or with hydrogen bromide in the presence of an aluminum halide catalyst. After replacement of chlorine by bromine, the mixture is washed with cold water to remove the inorganic materials and the product is distilled.

IV Treatment of the parent hydroxyl compound with a bromide and sulfuric acid, as in the preparation of methyl bromide. If the parent compound is unstable toward sulfuric acid, hydrobromic acid is passed in and alkyl bromide is flashed over from the hot mixture. The product is condensed, neutralized, and fractionated. In modifications of these procedures, bromine may be used along with a reducing agent such as sulfur, sulfur dioxide, phosphorus, or sodium borohydride.

V Esterification of the parent bromo acid with alcohol and sulfuric acid. After reaction the acid is washed out and the product distilled, under reduced pressure if necessary.

VI Direct combination of an unsaturated compound with an equimolar mixture of chlorine and bromine, followed by purification as in I. Some dichloro and dibromo compounds are formed at the same time.

VII Hydrobromination of unsaturated parent compound (a hydrocarbon or one of its chloro or bromo substitution products): (a) in the presence of a peroxide; (b) with exclusion of peroxides; (c) photochemically at 0°C. Thus vinyl bromide may be prepared from acetylene, and ethylidene bromide from vinyl bromide.

VIII Partial dehydrobromination of a dibromide to form an unsaturated monobromide: (a) by cracking at high temperature; (b) by hydrolyzing with alkali.

IX Bromomethylation of an aromatic ring by reaction with hydrogen bromide and an aldehyde in the presence of zinc chloride or bromide. In this way α-bromomethylnaphthalene may be prepared starting with naphthalene and paraformaldehyde (197).

X Treatment of an acid or acid anhydride with phosphorus tribromide.

^c Covered specifically in shipping regulations (196).

^d A, as an antiseptic, germicide, or fungicide.
C, as a catalyst.
D, as a dense liquid for solid separations based upon differences in specific gravity.
E, as an ingredient of fire-extinguishing fluids or as a fire retardant.
F, as a fumigant (if quite volatile) or as a contact poison.
G, as a gage fluid.
H, as an herbicide.
L, as a lacrimator or warning agent.
M, in microscopy or refractometry.
P, as an ingredient of pharmaceutical or medicinal products.
prep, preparation method described.
solv, as a solvent, generally for fats, waxes, or resins, but in a few cases as a medium for carrying out reactions.
syn, as an intermediate in the synthesis of other compounds.
T, as an ingredient of heat-transfer media or transformer oils.

258 BROMINE COMPOUNDS

Various quaternary ammonium compounds or amines are used pharmaceutically in the form of bromides or hydrobromides because these salts offer suitable solubility, preferred crystalline character, or are the most readily prepared. Among those in *USP XIX* are the cholinergics neostigmine bromide [114-87-7] and pyridostigmine bromide [101-26-8] and the anticholinergics propantheline bromide [50-34-0], scopolamine hydrobromide [114-49-8], and homatropine hydrobromide [51-56-9], the last being an ophthalmic drug. A number of others are listed in *NF XIV*, including demecarium bromide [56-94-0], homatropine methylbromide [80-49-9], and hyoscyamine hydrobromide [306-03-6]. Hexafluorenium bromide [317-52-2] is a muscle relaxant, hydroxyamphetamine hydrobromide [306-21-8] is an adrenergic, and dextromethorphan hydrobromide [125-69-9] is an antitussive. Few compounds in *NF XIV* contain bromine on a benzene ring; two that do are the antihistaminics bromodiphenhydramine hydrochloride [108-12-4] and brompheniramine maleate [980-71-2]. Dexbrompheniramine maleate [132-20-7] is the active dextrorotary form of the latter. A compound containing an alkyl bromide is pipobroman [54-91-1], 1,4-bis(3-bromopropionyl)piperazine, an antineoplastic.

Sulfobromophthalein sodium [71-67-0] is the *USP XIX* listing for the disodium salt of an indicator with four bromine atoms on a benzofuran moiety. It is used as a diagnostic aid in testing for liver function.

Merbromin [129-16-8] (mercurochrome, the disodium salt of 2,7-dibromo-4-hydroxymercurifluorescein) was once widely employed as an antiseptic but is no longer recommended because the structure includes organic mercury. This and a number of other proposed or formerly used bromine compounds such as carbromal [77-65-6], bromisovalum [496-67-3] (bromural), α-bromoisovaleryl-p-phenetidine [52964-39-3], bromazepam [1812-30-2], bromanylpromide [332-69-4], β-tribromoethanol [75-80-9], and 2,4,6-tribromo-m-cresol [4619-74-3] are described in the *Merck Index* (119). A more complete account of medicinal applications of bromine compounds is given in Jolles (see General references) including the subject of disinfection by bromine-releasing agents.

Miscellaneous Organic Bromine Compounds

A number of bromine compounds, most of which have some industrial importance, are listed in Table 1. In order to present condensed information on preparation methods, properties, and uses, abbreviations have been used in the table as explained in the footnotes.

Shipping and Precautions. A compound with footnote *c* in Table 1 indicates that shipping of the material is covered specifically under Hazardous Materials Regulations (196). Other compounds included in the table may usually be shipped under the classification of chemicals, NOIBN, without special requirements unless from their nature they would fall under a category such as combustible liquid, compressed gas, corrosive liquid (or solid), disinfectant liquid (or solid), drug, dye intermediate (liquid), fire extinguisher, flammable gas (liquid, or solid), insecticide, medicine, oxidizer or oxidizing material, poisonous liquid (gas, or solid), solvent, or tear gas. Specific provisions apply to each of these categories and appropriate packaging and labeling are required.

In general, exposure to bromine compounds should be avoided. This applies to inhalation, ingestion, and skin contact, except under medical supervision. In case of skin contact, the affected parts should be washed thoroughly first with water, then soap and water. Following any significant exposure, medical treatment should be obtained promptly.

BIBLIOGRAPHY

"Bromine Compounds" in *ECT* 1st ed., Vol. 2, pp. 645–660, by V. A. Stenger and G. J. Atchison, The Dow Chemical Company; "Bromine Compounds" in *ECT* 2nd ed., Vol. 3, pp. 766–783, by V. A. Stenger and G. J. Atchison, The Dow Chemical Company.

1. J. D. Johnson and R. Overby, *J. Sanit. Eng. Div. Am. Soc. Civil Eng.* **97**(SA5), 617 (1971).
2. G. H. Coleman, H. Soroos, and C. B. Yager, *J. Am. Chem. Soc.* **55**, 2075 (1933); **56**, 965 (1934).
3. D. F. Clemens and co-workers, *Inorg. Chem.* **8**, 998 (1969).
4. J. Jander and E. Kurzbach, *Z. Anorg. Allgem. Chem.* **296**, 117 (1958); J. Jander and C. Lafrenz, *Z. Anorg. Allgem. Chem.* **349**, 57 (1966).
5. J. K. Johannesson, *Analyst*, **83**, 155 (1958); *J. Chem. Soc.*, 2998 (1959); *Am. J. Public Health*, **50**, 1731 (1960).
6. H. Galal-Gorchev and J. C. Morris, *Inorg. Chem.* **4**, 899 (1965).
7. G. H. Coleman and G. E. Goheen, in H. Booth, ed., *Inorganic Syntheses*, Vol. I, McGraw-Hill Book Co., Inc., New York, 1939, p. 62.
8. U.S. Pat. 1,775,598 (Sept. 8, 1930), J. H. van der Meulen.
9. C. P. Smyth and C. S. Hitchcock, *J. Am. Chem. Soc.* **55**, 1830 (1933).
10. W. D. Bonner, L. G. Bonner, and F. J. Gurney, *J. Am. Chem. Soc.* **55**, 1406 (1933).
11. J. O. Lundgren and I. Olovsson, *J. Chem. Phys.* **49**, 1068 (1968); J. O. Lundgren, *Acta Crystallogr.* **B 26**, 1893 (1970).
12. J. M. Schneider and W. C. Johnson in ref. 7, p. 152.
13. U.S. Pat. 2,070,263 (Feb. 9, 1937), G. F. Dressel and O. C. Ross (to The Dow Chemical Company).
14. Ger. Pat. 701,073 (Dec. 5, 1940), C. J. Hansen (to Byk Guldemoerke Chemische Fabrik A.G.).
15. U.S. Pat. 2,342,465 (Feb. 22, 1944), F. Goldschmidt and F. Deutsch.
16. G. B. Heisig and E. Amdur, *J. Chem. Ed.* **14**, 187 (1937).
17. U.S. Pats. 2,369,182 (Feb. 13, 1945), 2,380,675 (July 31, 1945), 2,383,919 (Aug. 28, 1945), 2,391,740 (Dec. 25, 1945), F. F. Rush and co-workers (to Shell Development Co.).
18. Ger. Offen. 1,960,558 (June 9, 1971), H. Jenkner and O. Rabe (to Chemische Fabrik Kalk G.m.b.H.).
19. U.S. Pat. 3,426,093 (Feb. 4, 1969), O. C. Karkalits, Jr., and C. A. Leatherwood (to Petro-Tex Chem. Corp.).
20. M. Vadekar and I. S. Pasternak, *Can. J. Chem. Eng.* **48**, 216 (1970).
21. U.S. Pat. 3,426,358 (Feb. 4, 1969), A. S. Barbopoulos and W. H. Prahl (to Hooker Chem. Corp.).
22. U.S. Pat. 2,553,518 (May 15, 1951), D. E. Lake and A. A. Asadorian (to The Dow Chemical Company).
23. C. M. Beeson and D. M. Yost, *J. Am. Chem. Soc.* **61**, 1432 (1939).
24. L. G. Cole and G. W. Elverum, *J. Chem. Phys.* **20**, 1543 (1952).
25. H. G. Vesper and G. K. Rollefson, *J. Am. Chem. Soc.* **56**, 620 (1934).
26. H. G. Zeise, *Z. Elektrochem.* **41**, 267 (1935).
27. W. Militzer, *J. Am. Chem. Soc.* **60**, 256 (1938).
28. E. P. White and P. W. Robertson, *J. Chem. Soc.* **142**, 1509 (1939).
29. J. F. Mills, *Disinfectants: Water Wastewater*, 113 (1975).
30. D. R. Grothe and J. W. Eaton, *Trans. Am. Fish. Soc.* **104**, 800 (1975).
31. J. F. Mills and B. D. Oakes, *Chem. Eng.* **80**(18), 102 (1973).
32. U.S. Pats. 2,432,997 (Dec. 23, 1947); 2,471,831 (May 31, 1949); 2,480,080 (Aug. 23, 1949); 2,489,969 (Nov. 29, 1949), W. B. Ligett, E. T. McBee and V. V. Lindgren (to Purdue Research Foundation).
33. Brit. Pat. 1,059,234 (Feb. 15, 1967), R. A. Davis and E. R. Larsen (to The Dow Chemical Company).
34. R. A. Davis and E. R. Larsen, *J. Org. Chem.* **32**, 3478 (1967).
35. G. Strickland, F. L. Horn, and R. Johnson, *U.S. Atomic Energy Comm. BNL-471 (T-107)*, 1957.
36. J. F. Ellis, *U.K. At. Energy Authority, Ind. Group*, 8092D, 8111D, 1959.

37. U.S. Pat. 3,294,493 (Dec. 27, 1966), A. A. Jonke, R. K. Stennenberg, and R. C. Vogel (to U.S. Atomic Energy Comm.).
38. R. Schwarz and H. Wiele, *Naturwissenschaften* **26,** 742 (1938); *J. Prakt. Chem.* **152,** 157 (1939).
39. W. Brenschede and J. J. Schumacher, *Z. Anorg. Allgem. Chem.* **226,** 370 (1936).
40. K. Ody, A. Neshvatal, and J. M. Tedder, *J. Chem. Soc. Perkin Trans. 2,* 521, (1976).
41. J. L. Pascal, A. C. Pavia, and A. Potier, *Compt. Rend. Ser. C* **279,** 43 (1974); **280,** 661 (1975).
42. M. Schmeisser and K. Jorger, *Angew. Chem.* **71,** 523 (1959).
43. A. J. Arvia, P. J. Aymonino, and H. J. Schumacher, *Z. Anorg. Allgem. Chem.* **298,** 1 (1959).
44. R. Mungen and J. W. T. Spinks, *Can. J. Res.* **B18,** 363 (1940).
45. B. Lewis and H. J. Schumacher, *Z. Anorg. Allgem. Chem.* **182,** 182 (1929).
46. A. Pflugmacher, H. J. Robben, and H. Dahmen, *Z. Anorg. Allgem. Chem.* **279,** 313 (1955).
47. J. L. Pascal and co-workers, *Compt. Rend. Ser. C* **282,** 53 (1976).
48. D. A. Ramsay, *Ann. N.Y. Acad. Sci.* **67,** 485 (1957).
49. D. S. A. G. Radlein, J. C. Whitehead, and R. Grice, *Mol. Phys.* **29,** 1813 (1975).
50. F. Pollak and E. Z. Doktar, *Z. Anorg. Allgem. Chem.* **196,** 89 (1931).
51. A. Luneckas and A. Prokopcikas, *Liet. TSR Mokslu Akad. Darbei,* **B 1960** (3), 53; *Chem. Abstr.* **56,** 53 (1962).
52. R. Scholder and K. Krauss, *Z. Anorg. Allgem. Chem.* **268,** 279 (1952).
53. U.S. Pat. 3,575,865 (Apr. 20, 1971), R. L. Burke, L. T. Murray, and W. Chirash (to Colgate-Palmolive Co.).
54. A. H. Richards, *J. Soc. Chem. Ind.* **25,** 4 (1906).
55. M. L. Josien, *Compt. Rend.* **208,** 348 (1939).
56. G. Sourisseau, *Compt. Rend.* **226,** 1605 (1948).
57. R. Kircher and R. Periat, *Ind. Chim. Belge, Suppl.,* 877 (1959).
58. U.S. Pat. 3,178,262 (Apr. 13, 1965), J. Meybeck, R. Kircher, and J. Breiss (to Soc. d'Etudes Chim. Ind. Agr.).
59. Ger. Pat. 1,069,023 (Nov. 19, 1959), J. Meybeck, J. Leclerc, and R. Kircher (to Soc. d'Etudes Chim. Ind. Agr.).
60. M. B. Kennedy and M. W. Lister, *Thermochim. Acta* **7,** 249 (1973).
61. T. Andersen and H. E. L. Madsen, *Anal. Chem.* **37,** 49 (1965).
62. F. Solymosi, *Kem. Kozlem* **40,** 113 (1974).
63. V. I. Bogatyrev and A. I. Vulikh, *J. Appl. Chem. U.S.S.R.* **36,** 205 (1963).
64. U.S. Pat. 3,187,044 (June 1, 1965), D. Robertson (to Arapahoe Chemicals, Inc.).
65. U.S.S.R. Pats. 254,490 (Oct. 17, 1969), 297,576 (Mar. 11, 1971), D. P. Semchenko, V. I. Lyubushkin, and S. S. Khadirov.
66. D. T. Ewing and H. W. Schmidt, *Trans. Electrochem. Soc.* **47,** 117 (1925).
67. E. H. Appelman, *J. Am. Chem. Soc.* **90,** 1900 (1968).
68. E. H. Appelman, *Inorg. Chem.* **8,** 223 (1969); E. H. Appelman in F. A. Cotton, ed., *Inorganic Syntheses,* Vol. 13, McGraw-Hill Book Company, Inc., New York, 1972, p. 1.
69. S. K. Shukla, S. Fanali, and L. Ossicini, *J. Chromatogr.* **114,** 451 (1975).
70. G. K. Johnson and co-workers, *Inorg. Chem.* **9,** 119 (1970).
71. V. K. Klaening, K. J. Olsen, and E. H. Appelman, *J. Chem. Soc. Faraday Trans. 1* **71,** 473 (1975).
72. K. J. Olsen, K. Sehested, and E. H. Appelman, *Chem. Phys. Letters* **19,** 213 (1973).
73. R. R. Dreisbach, *Physical Properties of Chemical Compounds,* Vol. 2, American Chemical Society, Washington, D.C., 1959.
74. M. S. Kharasch, M. C. McNab, and F. K. Mayo, *J. Am. Chem. Soc.* **55,** 2530 (1933).
75. J. Timmermans and F. Martin, *J. Chim. Phys.* **23,** 747 (1926).
76. U.S. Pat. 2,746,999 (May 22, 1956), A. A. Gunkler, D. E. Lake, and B. C. Potts (to The Dow Chemical Co.).
77. Brit. Pats. 804,995 and 804,996 (Nov. 26, 1958), W. J. Read, D. B. Clapp, D. R. Stephens, and J. E. Russell (to Associated Ethyl Co., Ltd.).
78. U.S. Pat. 2,921,967 (Jan. 19, 1960), F. Yaron (to Dead Sea Bromine Co., Ltd.).
79. N. Weiner in A. H. Blatt, ed., *Organic Syntheses,* Collective Vol. II, John Wiley & Sons, Inc., New York, 1943, p. 280.
80. R. H. Goshorn, T. Boyd, and E. F. Degering in H. Gilman and A. H. Blatt, eds., *Organic Syntheses,* Collective Vol. I, John Wiley & Sons, Inc., New York, 1941, p. 36.
81. U.S. Pat. 2,173,133 (Sept. 19, 1939), D. J. Pye and H. H. Purcell (to The Dow Chemical Company).
82. U.S. Pat. 2,244,324 (June 3, 1941), H. Bender (to The Dow Chemical Company).
83. U.S. Pat. 2,359,828 (Oct. 10, 1944), H. Davies.

84. Neth. Appl. 6,400,652 (July 30, 1964), (to Soc. Chim Dell' Aniene S.p.A.).
85. Fr. Pat. 1,392,045 (Mar. 12, 1965), (to Soc. Chim. Dell' Aniene S.p.A.).
86. *Chemical Safety Data Sheet SD-35, Methyl Bromide,* Manufacturing Chemists' Association, Washington, D.C., 1949.
87. S. A. Shrader, A. W. Beshgetoor, and V. A. Stenger, *Ind. Eng. Chem. Anal. Ed.* **14,** 1 (1942).
88. K. A. Scudamore and S. G. Heuser, *Pestic. Sci.* **1,** 14 (1970).
89. Ger. Pat. 727,690 (Oct. 8, 1942), O. Scherer, F. Dostal, and K. Dachlauer (to I.G. Farbenindustrie A.G.); U.S. Pat. 2,347,000 (Apr. 18, 1944) (to Alien Property Custodian).
90. U.S. Pat. 2,553,518 (May 15, 1951), D. E. Lake and A. A. Asadorian (to The Dow Chemical Company).
91. T. R. Torkelson, F. Oyen, and V. K. Rowe, *Am. Ind. Hyg. Assoc. J.* **21,** 275 (1960).
92. W. W. Hartman and E. E. Dreger in ref. 79, p. 357.
93. A. Sherman and J. Sherman, *J. Am. Chem. Soc.* **50,** 1119 (1928).
94. U.S. Pat. 1,891,415 (Dec. 20, 1932), I. F. Harlow and O. C. Ross (to The Dow Chemical Company).
95. A. Kergomard, *Bull. Soc. Chim.* 2360, (1961).
96. J. J. Rook, *Water Treat. Exam.* **33**(2), 234 (1974).
97. K. J. Frederick and J. H. Hildebrand, *J. Am. Chem. Soc.* **61,** 1555 (1939).
98. W. H. Hunter and D. E. Edgar, *J. Am. Chem. Soc.* **54,** 2025 (1932).
99. M. S. Kharasch, E. V. Jensen, and W. H. Urry, *J. Am. Chem. Soc.* **68,** 154 (1946).
100. D. L. Engibous and T. R. Torkelson, *Wright Air Dev. Center Tech. Report 59-463,* Dayton, Ohio, Jan. 1960.
101. O. Kamm and C. S. Marvel in ref. 80, p. 29.
102. U.S. Pat. 2,698,347 (Dec. 28, 1954), A. P. Giraitis (to Ethyl Corp.).
103. D. E. Harmer and J. S. Beale, Jr., *Nucleonics,* **21**(9), 76 (1963).
104. U.S. Pat. 2,255,605 (Sept. 9, 1941), R. E. Windecker and A. Schormuller.
105. U.S. Pat. 2,303,549 (Dec. 1, 1942), A. G. Horney (to Air Reduction Co.).
106. U.S. Pat. 2,352,782 (July 4, 1944), C. B. Gardenier (to Thomas A. Edison, Inc.).
107. U.S.S.R. Pat. 187,749 (Feb. 9, 1965), I. B. Afanas'ev and T. N. Eremina (to State Sci.-Res. and Design Inst.).
108. C. H. McBeth and G. B. Bergeson, *Plant Disease Reptr.* **39,** 223 (1955).
109. U.S. Pat. 2,937,936 (May 24, 1960), C. T. Schmidt (to Pineapple Res. Inst. of Hawaii).
110. U.S. Pat. 3,049,472 (Aug. 14, 1962), A. W. Swezey (to The Dow Chemical Company).
111. L. A. Lider, A. N. Kasimatis and R. V. Schmitt, *Am. J. Enol. Vitic.* **18,** 55 (1967).
112. G. Booth, in Z. E. Jolles, ed., *Bromine and Its Compounds,* Academic Press, New York and London, 1966, pp 531–543.
113. C. R. Stephen and D. M. Little, *Halothane (Fluothane),* Williams and Wilkins, Baltimore, Md., 1961.
114. H. Madai and R. Muller, *J. Prakt. Chem.* **19,** 83 (1963).
115. U.S. Pat. 3,082,263 (Mar. 19, 1963), R. L. McGinty (to Imperial Chemical Industries).
116. Brit. Pat. 805,764 (Dec. 10, 1958); U.S. Pat. 2,921,099 (Jan. 12, 1960), J. Chapman and R. L. McGinty (to Imperial Chemical Industries, Ltd.).
117. Ger. Pat. 1,041,937 (Oct. 30, 1958); U.S. Pat. 2,959,624 (Nov. 8, 1960), O. Scherer and H. Kuhn (to Farbwerke Hoechst A.G.).
118. Brit. Pat. 767,779 (Feb. 6, 1957); U.S. Pat. 2,849,502 (Aug. 26, 1958), C. W. Suckling and J. Raventos (to Imperial Chemical Industries Ltd.).
119. M. Windholz, ed., *The Merck Index,* 9th ed., Merck & Co., Inc., Rahway, N. J., 1976.
120. F. E. Bartell and A. D. Wooley, *J. Am. Chem. Soc.* **55,** 3521 (1933).
121. Brit. Pat. 889,649 (Feb. 21, 1962), W. J. Read and co-workers (to Associated Ethyl Co., Ltd.).
122. T. M. Burton and E. F. Degering, *J. Am. Chem. Soc.* **62,** 227 (1940).
123. M. S. Kharasch and F. K. Mayo, *J. Am. Chem. Soc.* **55,** 2495 (1933); O. Kamm and C. S. Marvel in ref. 80, p. 27.
124. H. D. Young and R. T. Cotton, *J. Econ. Entomol.* **36,** 796 (1943).
125. A. H. Oxtoby in ref. 112, p. 356.
126. U.S. Pat. 3,235,357 (Feb. 15, 1966), H. M. Loux (to E. I. du Pont de Nemours & Co., Inc.).
127. A. C. Chapham, S. Kenyon, and R. S. Bell, *Proc. Northeast. Weed Control Conf.* **21,** 395 (1967).
128. J. W. Shriver and S. W. Bingham, *Weed Sci.* **21,** 212 (1973); F. A. Peevy, *Weed Sci.* **21,** 54 (1973).
129. U.S. Pat. 3,130,222 (Apr. 21, 1964), A. A. Asadorian and G. A. Burk (to The Dow Chemical Company).
130. U.S. Pat. 2,452,154 (Oct. 26, 1948), J. Ross (to Colgate-Palmolive-Peet Co.).

131. U.S. Pat. 2,752,341 (June 26, 1956), B. J. Magerlein (to Upjohn Co.).
132. T. M. Berry, J. E. Copenhaver, and E. E. Reid, *J. Am. Chem. Soc.* **49,** 3146, 3162 (1927).
133. H. R. Snyder, R. R. Adams, and A. V. McIntosh, *J. Am. Chem. Soc.* **63,** 3280 (1941).
134. S. H. Zaheer and S. A. Faseeh, *J. Indian Chem. Soc.* **21,** 27 (1944).
135. M. Gomberg and W. E. Bachmann in ref. 80, p. 113.
136. U.S. Pat. 1,835,754 (Dec. 8, 1931), E. C. Britton and W. C. Stoesser (to The Dow Chemical Company).
137. J. H. Brown and C. S. Marvel, *J. Am. Chem. Soc.* **59,** 1176 (1937).
138. G. W. Pope and M. T. Bogert, *J. Org. Chem.* **2,** 276 (1937).
139. P. S. Varma, V. Sahay, and B. R. Subramonium, *J. Indian Chem. Soc.* **14,** 157 (1937).
140. G. F. Hennion and V. R. Pieronek, *J. Am. Chem. Soc.* **64,** 2751 (1942).
141. E. C. Sterling and M. T. Bogert, *J. Org. Chem.* **4,** 20 (1939).
142. K. A. Kobe and T. S. Okabi, *J. Am. Chem. Soc.* **63,** 3251 (1941).
143. W. Militzer, *J. Am. Chem. Soc.* **60,** 256 (1938).
144. H. T. Clarke and M. R. Brethen in ref. 80, p. 121.
145. J. F. Suyver and J. P. Wibaut, *Rec. Trav. Chim.* **64,** 65 (1945).
146. G. M. Bennett, G. L. Brooks, and S. Glasstone, *J. Chem. Soc.* **138,** 1821 (1935).
147. Ger. Offen. 2,005,259 (Aug. 29, 1971), O. Rabe (to Chemische Fabrik Kalk G.m.b.H.).
148. S. E. Hazlet, G. Alligher, and R. Tiede, *J. Am. Chem. Soc.* **61,** 1447 (1939).
149. M. Weinig, *Ann. Chem.* **280,** 247 (1894).
150. U.S. Pat. 1,977,662 (Oct. 23, 1934), J. P. Wibaut and H. J. den Hertog, Jr., (to The Dow Chemical Company).
151. S. M. McElvain and M. A. Goese, *J. Am. Chem. Soc.* **65,** 2227 (1943).
152. N. W. Hirwe and B. V. Patil, *Proc. Indian Acad. Sci.* **A5,** 321 (1937).
153. J. S. Pizey, *Synthetic Reagents,* Vol. 2, John Wiley & Sons, Inc., New York, 1974, pp. 1–63.
154. E. Campaigne and B. F. Tullar, in C. C. Price, ed., *Organic Syntheses,* Vol. 33, John Wiley & Sons, Inc., New York, 1953, p. 97.
155. L. Gatterman, *Ann. Chem.* **393,** 230 (1912).
156. U.S. Pat. 3,077,503 (Feb. 12, 1963); J. W. Crump (to The Dow Chemical Company); J. W. Crump and G. A. Gornowicz, *J. Org. Chem.* **28,** 949 (1963).
157. J. M. Stuckey and J. H. Saylor, *J. Am. Chem. Soc.* **62,** 2922 (1940).
158. U.S. Pat. 3,349,111 (Oct. 24, 1967); R. W. Luckenbaugh (to E. I. du Pont de Nemours & Co., Inc.).
159. R. T. Leonard and R. C. Wakefield, *Proc. Northeast Weed Contr. Conf.* **21,** 278 (1967); also H. P. Wilson and co-workers, p. 294.
160. R. F. Deese, Jr., *J. Am. Chem. Soc.* **53,** 3673 (1931).
161. J. F. Norris, *Am. Chem. J.* **38,** 639 (1907).
162. L. N. Ferguson, A. Y. Garner, and J. L. Mack, *J. Am. Chem. Soc.* **76,** 1250 (1954).
163. K. Alder and H. F. Rickert, *Ber. Deut. Chem. Ges.* **B71,** 373 (1938).
164. R. E. Buckles and N. G. Wheeler in N. Rabjohn, ed., *Organic Syntheses,* Collective Volume II, John Wiley & Sons, Inc., New York, 1963, p. 256.
165. U.S. Pat. 2,257,903 (Oct. 7, 1941), R. R. Dreisbach (to The Dow Chemical Company).
166. W. Schlenk and H. Bergmann, *Ann. Chem.* **463,** 196 (1928).
167. H. J. Ziegler, L. Walgraeve, and F. Binon, *Synthesis* **1,** 39 (1969).
168. B. Y. Eryshev and B. P. Yatsenko, *Tr. Mosk. Khim. Tekhnol. Inst.* **66,** 50 (1970).
169. W. E. Railing, *J. Am. Chem. Soc.* **61,** 3349 (1939).
170. M. S. Kharasch, M. C. McNab, and F. K. Mayo, *J. Am. Chem. Soc.* **55,** 2530 (1933).
171. U.S. Pat. 2,058,465 (Oct. 7, 1936), M. S. Kharasch (to E. I. du Pont de Nemours & Co.).
172. E. L. Skau and R. McCullough, *J. Am. Chem. Soc.* **57,** 2439 (1935).
173. Belg. Pat. 444,538 (Aug. 31, 1942), (to N. V. Chemische Fabriek "Naarden").
174. A. I. Vogel, *J. Chem. Soc.* **146,** 636 (1943).
175. Swiss Pat. 405,821 (July 29, 1966), H. Martin, H. Aebi, and L. Ebner (to CIBA Ltd.).
176. G. D. Hill, *Proc. Annu. Calif. Weed Conf.* **23,** 158, 171 (1971).
177. Brit. Pat. 867,468 (May 1, 1961), (to The Dow Chemical Company).
178. G. Emschwiller and J. L. Sacconey, *Bull. Soc. Chim. Fr.,* 118 (1949).
179. U.S. Pat. 2,058,466 (Oct. 27, 1936), M. S. Kharasch (to E. I. du Pont de Nemours & Co., Inc.).
180. U.S.S.R. Pat. 196,805 (May 31, 1967), M. Guseinov, S. M. Mamedov, O. M. Salmanova, and Sh. S. Akhnazarova.
181. G. W. Barber, *J. Econ. Entomol.* **36,** 330 (1943).

182. M. C. Swingle, A. M. Phillips, and J. B. Gahan, *U.S.D.A., Bur. Entomol. Plant Quarantine,* **E-621,** 1944.
183. U.S. Pat. 3,029,291 (Apr. 10, 1962), A. J. Dietzler (to The Dow Chemical Company).
184. U.S. Pat. 2,319,960 (May 25, 1943), C. S. Treacy (to Merck & Co.).
185. E. Jeney and T. Zsolnai, *Zentralbl. Bakteriol. Parasitenkd. Infectionskr. Hyg. Abt. I Orig.* **202**(4), 547 (1967); *Chem. Abstr.* **67,** 41234f (1967).
186. Ger. Offen. 1,957,030 (June 9, 1971), W. Helmus (to Chemische Fabrik Kalk G.m.b.H.).
187. A. G. Ogston, *J. Chem. Soc.* **139,** 1713 (1936).
188. U.S. Pat. 3,003,914 (Oct. 30, 1959), C. R. Youngson and C. A. I. Goring (to The Dow Chemical Company).
189. U.S. Pat. 2,361,504 (Oct. 31, 1944), W. Scott and R. B. Seymour (to Wingfoot Corp.).
190. Ger. Offen. 2,041,303 (Apr. 1, 1971), R. A. Davis (to The Dow Chemical Company).
191. Fr. Pat. 1,441,233 (June 3, 1966), (to Shell Int. Res. Maatschappij N.V.).
192. A. Guyer, H. Schütze, and M. Weidemann, *Helv. Chim. Acta* **20,** 936 (1937).
193. S. Dev, *J. Indian Chem. Soc.* **32,** 408 (1955).
194. E. F. Atkinson and J. F. Thorpe, *J. Chem. Soc.* **91,** 1695 (1907).
195. E. F. M. Stephenson in ref. 164, p. 984.
196. *Tariff No. 31, Hazardous Materials Regulations of The Department of Transportation,* R. M. Graziano, Washington, D.C., 1977.
197. A. Darzens and G. Levy, *Compt. Rend.* **202,** 83 (1936).

General References

Z. E. Jolles, ed., *Bromine and Its Compounds,* Academic Press, New York and London, 1966.
A. G. Sharpe and co-workers, in J. W. Mellor, ed., *Comprehensive Treatise on Inorganic and Theoretical Chemistry,* Vol. II. Supplement II, John Wiley & Sons, Inc., New York and London, 1956, pp. 689–812.
A. J. Downs and C. J. Adams in A. F. Trotman-Dickenson, ed., *Comprehensive Inorganic Chemistry,* Vol. II, Pergamon Press, Oxford and New York, 1973, pp. 1107–1594.
V. Gutmann, ed., *Halogen Chemistry,* Vol. I–III, Academic Press, London and New York, 1967.
P. Pascal, *Nouveau Traité de Chimie Minérale,* Vol. 16, Masson et Cie, Paris, 1960, pp. 337–447.

V. A. STENGER
Dow Chemical, U.S.A.

BTX PROCESSING

Benzene [71-43-2] (B), toluene [108-88-3] (T), and xylenes (X) are familiar chemicals to those even slightly acquainted with the chemical industry; BTX, however, is less familiar as the major contributor to high-performance gasolines. Processing for benzene, toluene, xylene, and the enormous gasoline industry are inextricably linked. If benzene, toluene, or xylenes are removed as basic chemicals, the gasoline reservoir or pool has to be compensated for the loss of its more important high-octane components (see Benzene; Gasoline; Petroleum; Toluene; Xylenes and ethylbenzene).

A refinery must maintain a pool of gasoline with sufficiently high octane values to supply transportation demands. Removal of BTX depletes the gasoline pool. Addition of tetraethyl- or tetramethyllead raises the octane rating, however, government regulations restrict the amount of lead that may be used. Thus additional processing is needed to replace those aromatics removed for chemical purposes, resulting in higher costs.

The crude oil received by the refiner contains a fraction in the 65–175°C boiling range. It is separated by distillation, but is of little use in gasoline or for chemical purposes because the concentration of aromatics and thus the octane value are low (see Feedstocks). The process that increases the aromatics concentration is called reforming. The balance of the crude oil is further distilled to recover fuels for diesel or jet engines or for heating purposes. The even higher-boiling fractions are broken down or cracked to bring them into the more useful boiling range.

In the United States, benzene, toluene, and the xylenes are largely petroleum based except for 10% of the benzene which is derived from coal tar distillates (see Coal). Benzene, toluene, and the xylenes are put to a great variety of uses and eventually enter the modern household in hundreds of different forms. Figure 1 illustrates the transformation of BTX to fabrics, resins, molded products, etc; Table 1 gives the United States and world consumptions in the manufacture of various chemicals.

BTX From Petroleum

Recovery of BTX from petroleum is accomplished through a series of steps which, although differing in detail and sequence, are common throughout the industry. A typical aromatics processing loop is illustrated in Figure 2.

The 65–175°C fraction obtained directly or by cracking does not contain sufficient aromatics to be valuable. It is reformed in a catalytic reformer (See Unit A, Fig. 2). Materials boiling higher or lower than those required for a specific purpose are rejected in a pair of distillation columns (Fig. 2, Units B and C). The fraction with the desired boiling range is referred to as the heart cut.

The heart cut may still contain a large proportion of nonaromatics which are removed by an extraction process (Unit D). The extract contains benzene, toluene, C_8 aromatics, and probably a small amount of C_{9+} aromatics. The aromatics are preferentially dissolved in a solvent and separated from the nonaromatic compounds, now referred to as the raffinate. The aromatics are then recovered by distillation.

The o-xylene [95-47-6] is generally recovered by a two-stage distillation. First it is separated (or "split") from m-xylene [108-38-3] in a distillation column, the xylene splitter, with about 150 trays (Unit H). The bottoms, a mixture of o-xylene and C_{9+} aromatics, is redistilled (or "rerun") in Unit I to recover o-xylene of 96+% purity and

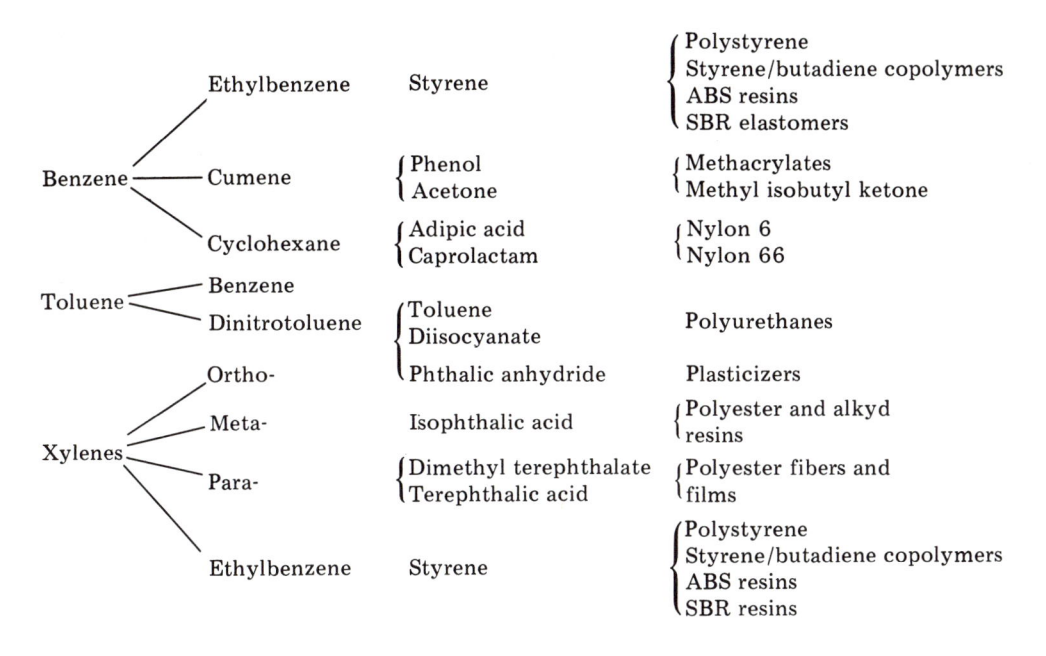

```
                                                        ┌ Polystyrene
                                                        │ Styrene/butadiene copolymers
            Ethylbenzene ───── Styrene                  │ ABS resins
           ╱                                            └ SBR elastomers
          ╱                         ┌ Phenol              ┌ Methacrylates
Benzene ──────── Cumene             └ Acetone             └ Methyl isobutyl ketone
          ╲
           ╲                        ┌ Adipic acid         ┌ Nylon 6
            Cyclohexane             └ Caprolactam         └ Nylon 66
           ┌── Benzene
Toluene ───┤
           └── Dinitrotoluene       ┌ Toluene
                                    │ Diisocyanate        Polyurethanes
          ┌─ Ortho-                 └ Phthalic anhydride  Plasticizers
          │                                               ┌ Polyester and alkyd
          ├─ Meta-                    Isophthalic acid    └ resins
Xylenes ──┤                         ┌ Dimethyl terephthalate  ┌ Polyester fibers and
          ├─ Para-                  └ Terephthalic acid       └ films
          │                                               ┌ Polystyrene
          │                                               │ Styrene/butadiene copolymers
          └─ Ethylbenzene ───── Styrene                   │ ABS resins
                                                          └ SBR resins
```

Figure 1. Uses of BTX compounds.

Table 1. 1976 Consumption of BTX in the Manufacture of Chemicals, 1000 Metric Tons Per Year

Product	Worldwide	United States
benzene		
cumene	3,346	1,310
cyclohexane	2,778	954
ethylbenzene	6,758	2,797
phenol	146	97
total	*13,028*	*5,158*
toluene		
caprolactam	107	
phenol	155	31
terephthalic acid	24	
toluene diisocyanate	419	155
total	*705*	*186*
o-xylene		
phthalic anhydride	1,722	334
m-xylene		
isophthalic acid	79	32
p-xylene		
dimethyl terephthalate	2,530	1,011
terephthalic acid	749	287
total	*3,279*	*1,298*

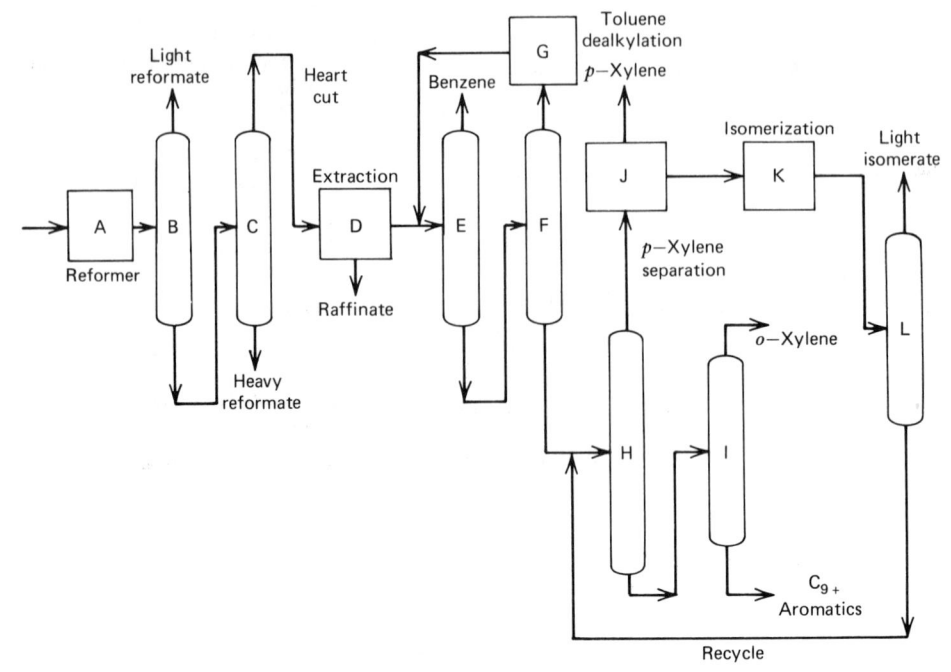

Figure 2. General BTX processing sequence.

C_{9+} aromatics. The latter are used as solvents or as blending components for gasoline.

The distillate (overhead), containing ethylbenzene, p-xylene [106-42-3], m-xylene, and o-xylene becomes the feed for the p-xylene separation process (Unit J), usually a crystallization or an adsorption plant (see below). The mother liquor obtained from the crystallization, or the raffinate after removal by adsorption, is isomerized to convert m-xylene to the p- and p-isomers (Unit K). In addition, some processes convert ethylbenzene to a mixture of xylenes.

In the isomerization step, light and heavy impurities are generated including saturated hydrocarbons, benzene, toluene, and C_{9+} aromatics. These are removed by recycling the isomerate through the xylene splitter and the light isomerate column (Unit L). If the light isomerate contains much benzene or toluene, it could be recycled to Unit G.

m-Xylene may be recovered between the xylene splitter and the p-xylene separation plant.

There are many variations of the basic processing loop shown in Figure 2. Processing for BT only is common, often in conjunction with a toluene-to-benzene demethylation unit. If benzene and toluene are not to be recovered, Column B may be used to remove toluene and lighter components. In that case, Units E, F, and G would be eliminated.

To recover p-xylene only, the xylene splitter is reduced in size and is used to split the o-xylene and C_{9+} aromatics. The o-xylene rerun still is then eliminated.

Gasoline refining affects this general process. For example, a gasoline refinery may produce 16,000 m³ per day (100,000 42-gal barrels per day, BPD) of gasoline and only 800 m³ (5000 BPD) of aromatic chemicals. Hence, the gasoline requirements take precedence over the needs of BTX processing.

Upstream Treatment. Crude oil is designated by the location of its sources which are spread all over the world. Some crude oil properties are listed in Table 2.

A crude oil is a complex mixture of chemical compounds and may contain all the possible saturated hydrocarbons and their even more numerous nitrogen and sulfur derivatives.

Native crude is distilled into different fractions in atmospheric or vacuum distillation columns. The BTX precursors are generally in the 50–175°C boiling range, representing less than 20% of the total crude (straight-run naphtha). The highest boiling stocks may be used as feeds to other refining processes, eg, fluid catalytic cracking, thermal cracking, or hydrocracking, which break the heavy stocks down to lighter fractions, making them usable as BTX sources (see Petroleum).

In the cracking process, n-paraffins form isoparaffins, and because ethyl groups crack more readily than methyl groups, the cycloparaffins (naphthenes) contain fewer ethyl side chains than those in straight-run naphthas. Thus, the BTX precursors, whether from straight-run or cracked stocks, are concentrated in materials with a 50–175°C boiling range.

Another classification of crude oils is the PCA (paraffin, cycloparaffins, aromatics) content of the straight-run portion (see Table 2). Middle Eastern crudes are higher in paraffins than most continental United States crudes. Since the key to BTX processing is the catalytic reforming step (which simplified, may be considered a P → C → A conversion), the PCA content of the straight-run naphtha can be vitally important because it determines the degree of processing needed to achieve the required aromatics content.

The sulfur and nitrogen content of a given crude affects the preparation of a reformer feed. In catalytic reforming the feed is treated with a platinum catalyst which is adversely affected by nitrogen, sulfur, and oxygen compounds. These elements are removed by pretreating the reformer feed over a nickel–molybdenum or a cobalt–molybdenum catalyst in the presence of hydrogen. They are thereby converted to ammonia, hydrogen sulfide, and water, respectively, and readily removed by distillation.

Catalytic Reforming. The three general classes of reforming processes are semi-regenerative, regenerative, and those where the catalyst is regenerated continuously. The main process variables are pressure (790–3550 kPa; 100–500 psig), temperature (ca 480°C), space velocity, and catalyst. An excess of hydrogen (2–8 moles per mole of hydrogen) is usually employed. Aromatics formation is favored by low pressures, but this is uneconomical because of catalyst fouling. However, with the advent of improved catalysts, fouling, ie, coking became much less of a problem. Today semi-regenerative reformers of 962 kPa (125 psig) are commercially operable. Figure 3 shows the effect of pressure on yield.

The semiregenerative process is represented by Houdriforming (1), Magnaforming (2), Platforming (3–4), Powerforming (5–8), Rheniforming (9–10), and Ultraforming (11). Figure 4 is a diagram of a Rheniformer (12) where the hydrotreated feed is heat-exchanged with the reformer product. Further heating is achieved by passage through a furnace. The feed then enters the first of several reactors that contain the platinum–rhenium on alumina catalyst. To replenish the heat of reaction lost by the conversion of cycloparaffins to aromatics, semiregenerative processes may have three or four sets of alternating furnaces and reactors.

Depending on feed and conditions, hydrogen production is in the range of 90–320

Table 2. Properties of Crude Oil Straight-Run Fractions

Property	Louisiana U.S. Empire Mix	California Huntington Beach	Venezuela Leona	Norway Ekofisk	U.K. Ninian	Libya Sarir	Iran Light	Kuwait	Abu Dhabi
vol %	9.5	12.2	11.9	19.2	19.8	15.0	18.7	17.6	21.9
gravity, °API	59.0	58.0	61.5	63.0	66.0	66.5	65.8	69.7	66.0
octane, F-1 +3[a]	82.0	89.6	86.0	80.9	78.0	76.5	79.0	75.3	70.0
paraffins, %	38	43	59	54	57	64	55	78	73
cycloparaffins, %	43	42	20	32	34	33	35	16	17
aromatics, %	9	15	21	14	9	3	10	6	10

[a] F-1 +3 = a measure of octane rating after addition of 0.79 g tetraethyllead per liter.

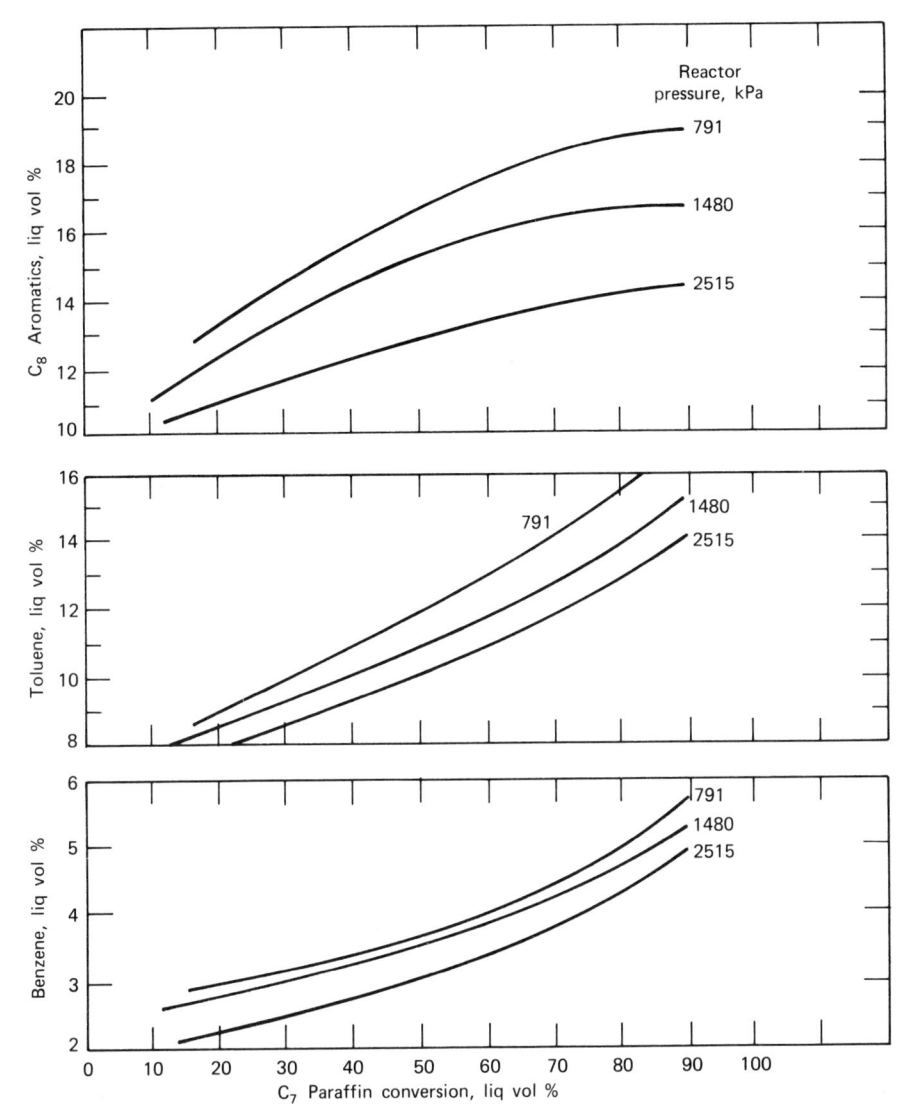

Figure 3. Higher yields from lower pressures; data for Arabian Naphtha, 54–154°C fraction. To convert kPa to psi, divide by 6.895.

m^3/m^3 feed (500–1800 SCF/bbl); it is partially recirculated to the reactors. The C$_1$–C$_4$ products are recovered to be used as fuels.

Catalyst coking is a problem with higher-boiling feeds and higher-severity operations, and time must be taken to burn off the coke during a several-day regeneration. Redistribution of platinum is also carried out at this time. A typical temperature range from start to end-of-run is about 40°C, and a cycle length of 6 to 12 months is acceptable to most refiners.

Magnaforming (2), Powerforming (5–8), and Ultraforming (11) have regenerative versions. The latter has an extra or swing reactor, that can be taken out of service for regeneration. These processes operate with hydrogen-to-hydrocarbon ratios of 2–4 to 1 and may run to higher severity than the semiregenerative processes. The regen-

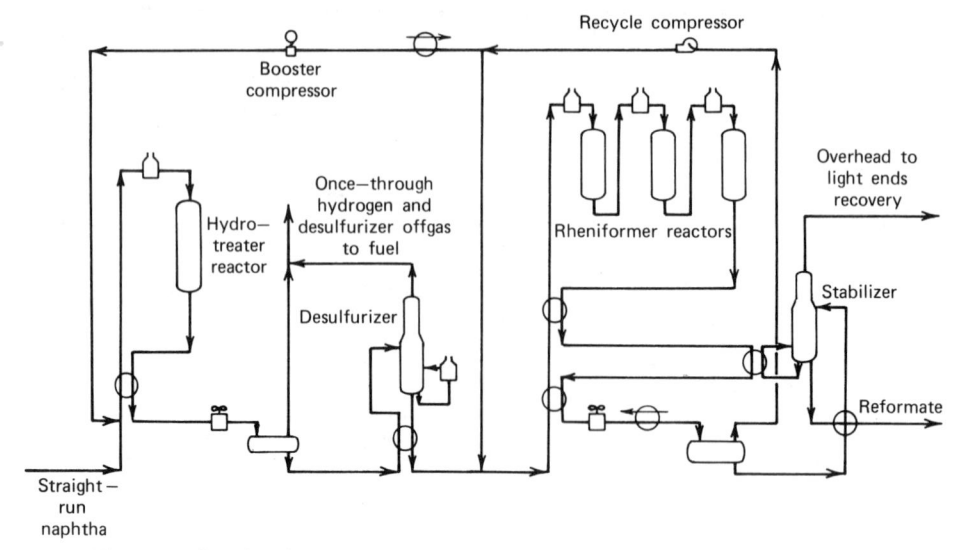

Figure 4. Simplified process flow diagram of a Naphtha Hydrotreater and Rheniformer.

erative process unit usually has four to six reactors, two of which may be regenerated daily. Thus the catalyst has a higher average activity than that used in semiregenerative processes.

Continuous regeneration, similar in principle to the regenerative process, is offered by UOP and the IFP Aromizing process (13). A portion of the catalyst is continuously removed for regeneration and fresh catalyst is returned.

The choice of reforming process depends on the product desired. If aromatics are to be a coproduct, the refiner would probably select a semiregenerative process. Severity is usually lower, and the important factor is the C_{5+} yield of gasoline. More recently it was shown that the yield for some combinations of feeds and catalyst improves with time (14). For high BTX yields the swing reactor or continuous regenerative process might be the choice because BTX yields are highest at higher severities.

Chemistry of Reforming. The octane rating of the various classes of reformer feeds is in the order of aromatics > cycloparaffins > isoparaffins > n-paraffins. Thus it is the objective of gasoline refiner and BTX producer to increase the proportion of aromatics. Figure 5 shows an approximate relationship between the octane number and the aromatics content of reformates.

The main reactions occurring in a reformer are shown in Figure 6 (12,15–17); most are reversible indicating the importance of reaction equilibrium.

In the alkylcyclohexane to aromatic equilibrium, ACH $\rightleftharpoons$ Ar, aromatics are favored by high temperatures and low pressures. Reforming conditions promote rapid ACH dehydrogenation with high conversion to aromatics.

The alkylcyclopentane (ACP) to aromatics process is less efficient than ACH dehydrogenation (ACP $\rightleftharpoons$ ACH $\rightleftharpoons$ Ar), owing to the slowness of the first step and ACP ring opening, and it requires an acidic platinum catalyst. Cyclohexane is converted to benzene in close to 100% efficiency, whereas only 50–75% of methylcyclopentane is converted to benzene.

Reformer feeds of high aromatics-plus-cycloparaffins content produce high-octane

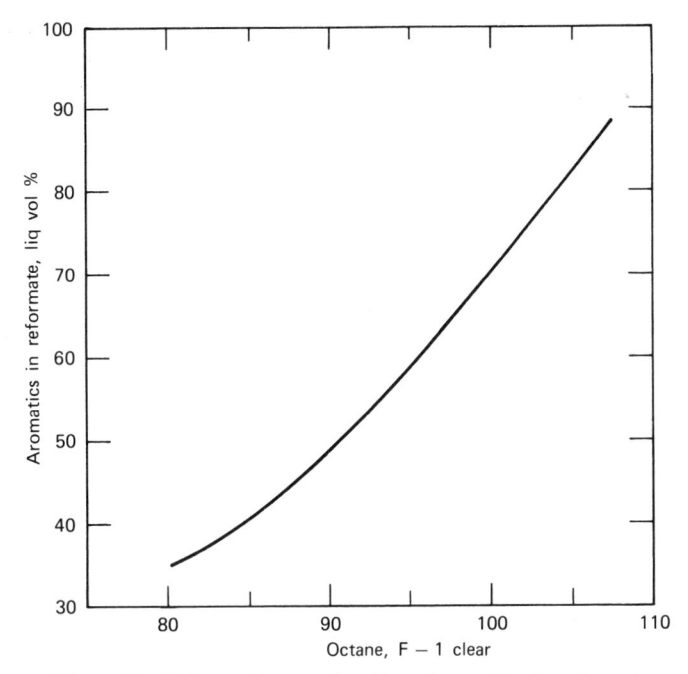

Figure 5. Octane rating as a function of aromatics in reformate.

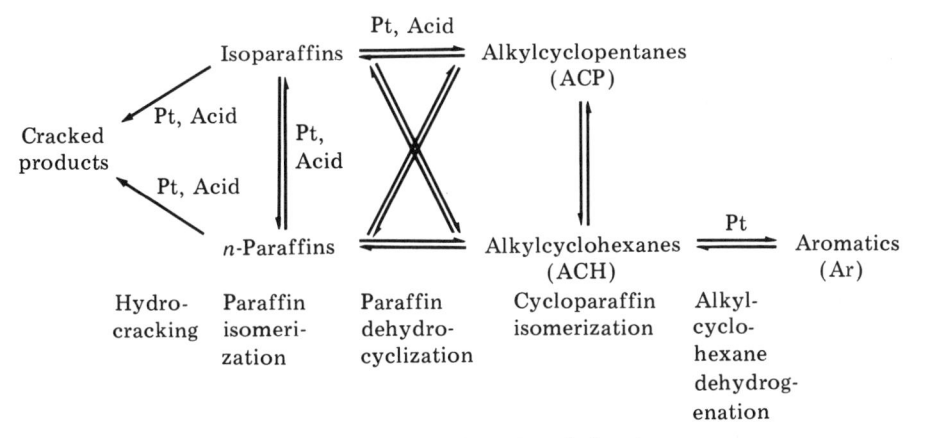

Figure 6. Main reactions of catalytic reforming.

reformates using only the cycloparaffins-to-aromatics reactions. Cyclization of paraffins is generally more difficult.

Comparing the reforming under the same mild conditions of two straight-run fractions with different PCA, we can see that the feed with the high C + A content achieves a much higher octane rating (and aromatics content) than the one with a low C + A content. The latter must be treated more severely to achieve this high octane rating, and as cracking increases, the yield of C_{5+} product decreases.

272   BTX PROCESSING

Reforming for Specific BTX Compounds.   Most benzene is formed from cyclohexane and methylcyclopentane. In order to obtain a good yield the severity must be raised and, if possible, low pressures should be used. The use of broad-range feeds is not advisable because of higher catalyst coking rates.

Toluene is less valuable as a chemical, but it is an important octane contributor. Toluene demand may increase if a Mitsubishi patent (18) related to earlier DuPont work (19) finds commercial application as an alternative route to polyesters via *p*-tolualdehyde.

Reforming for xylenes is very common. Examination of Figure 7 reveals a big overlap of C_8 with C_7 and C_9 hydrocarbons. Thus, even a sharply cut 90–150°C fraction may contain only 40–50% C_8 hydrocarbons. A reformate from such a feed will contain only ~30% C_8 aromatics.

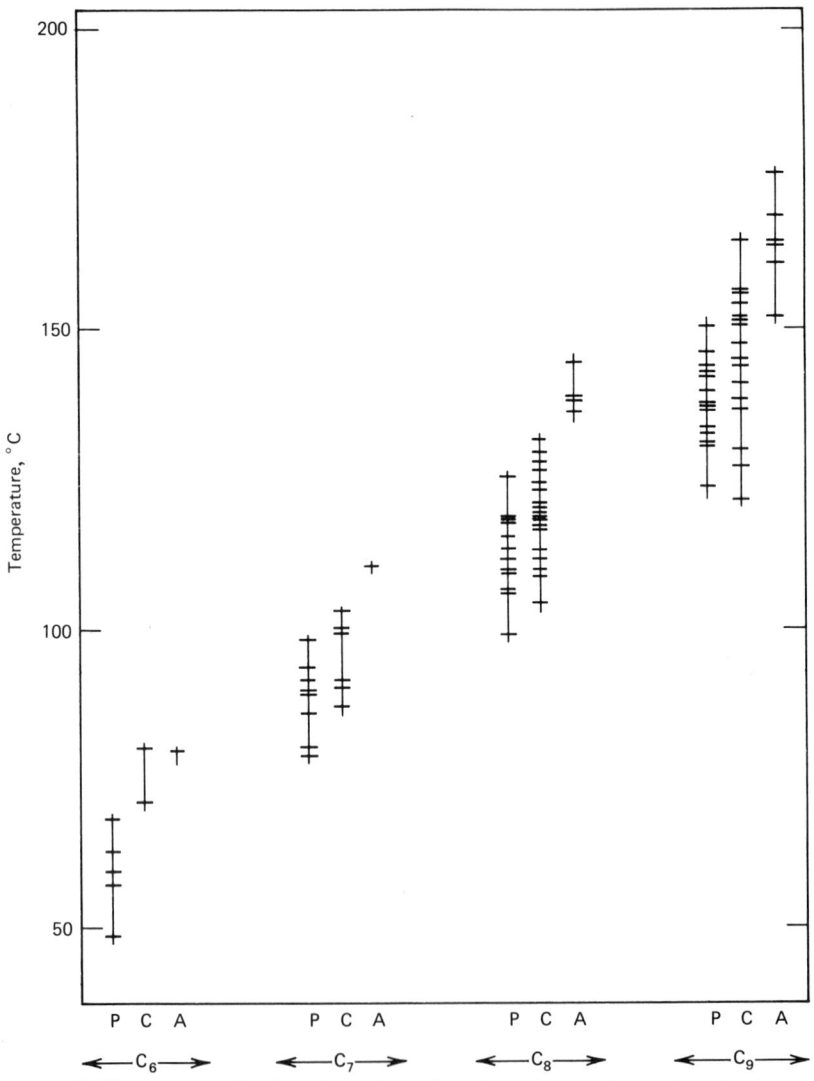

Figure 7.   Boiling points of C_6–C_9 hydrocarbons. P = Iso and normal paraffins; C = C_5- and C_6-cycloparaffins; and A = aromatics.

Recovery of xylenes, which are always accompanied by ethylbenzene, utilizes the heart-cut fraction described earlier. If the reformate was made under severe conditions, most of the C_9 paraffins, which boil in the xylene range, will be removed. Then, if the heart-cut fraction is sharp, the xylenes may contain very little nonaromatics and extraction is not necessary; another advantage of high-severity reforming, particularly as practiced by the regenerative or continuous reforming processes.

Extraction. If the heart-cut reformate contains appreciable nonaromatics, extraction may be necessary to prevent the accumulation of nonaromatics in the processing loop. Some isomerization processes are capable of cracking the paraffins but at the expense of additional hydrogen consumption. These processes do, however, eliminate the need for extraction of the xylenes and heavier compounds.

The choice of extraction solvent depends upon boiling point, polarity, thermal stability, selectivity, and high aromatics capacity (see Extraction).

Capacity, defined by the distribution factor, ie, the ratio of aromatics concentration in solvent and raffinate phases, must be balanced against selectivity, the ratio of the aromatics to the nonaromatics distribution. Most high-capacity solvents have low selectivity. The ultimate choice of solvent is determined by costs.

The most important extraction processes use sulfolane or glycols (20–21). Sulfolane, used in Shell's process, offers good thermal and hydrolytic stability, high density and boiling point, and a good balance of solvent properties. A diagram of a sulfolane extraction unit is shown in Figure 8. Fresh feed enters the extractor and flows countercurrent to the down-flowing solvent. The raffinate is withdrawn at the top of the extractor and leaves the system after water washing. The solvent, now rich in aromatics, is sent to the top of the extractive stripper where the nonaromatic hydrocarbons are removed. Aromatics and sulfolane are separated in the recovery column. The lean solvent is recycled to the extractor and the aromatics are washed with water and removed.

The recovery of benzene and toluene is usually 99+%; of C_8 aromatics 97%; and of C_{9+} aromatics 75–90%.

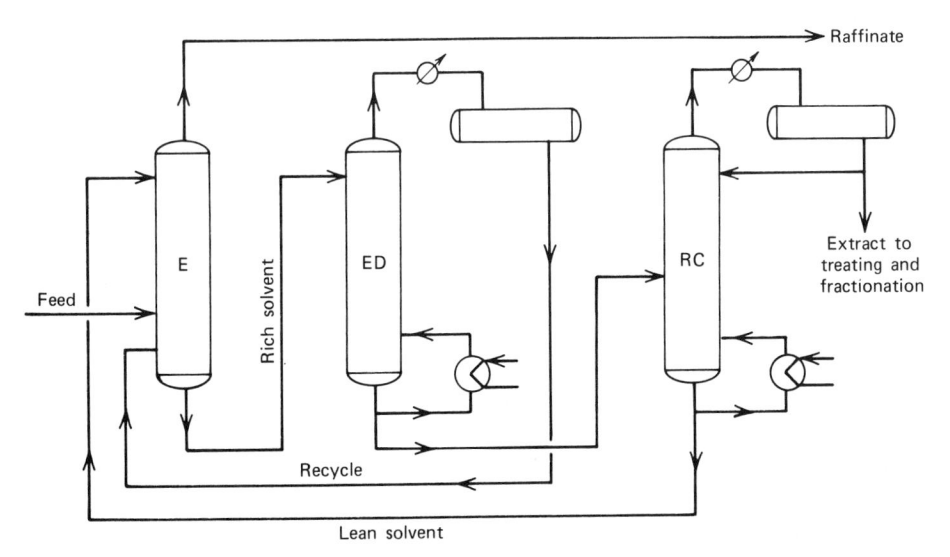

Figure 8. Shell sulfolane extraction process. E = Extraction; ED = extractive distillation; and RC = recovery column. (Presented at UOP Inc. 1975 Technology Conference, courtesy of UOP Inc.)

The widely employed Udex process uses a glycol solvent. Diethylene glycol was used in early versions of the process; however, increased capacity was obtained by adding dipropylene glycol or, in some cases, a change was made to triethylene glycol. More recently, further improvement has been made by using tetraethylene glycol (22–24).

Other extraction processes include the Lurgi Arosolvan (N-methyl-2-pyrrolidinone) (25–26), the Institut Français du Pétrôle (dimethylsulfoxide) (27), SNAM Progetti's Formex (N-formylmorpholine) (28), Howe-Baker's Aromex (diglycolamine) (29), and Koppers (N-methylmorpholine) (30).

Downstream Processes. Dealkylation of toluene provides benzene. Temperatures in the 540–810°C range are required, and an excess of hydrogen is used. Nonaromatic materials or higher aromatics lead to greater hydrogen consumption, coking problems, and poorer economics.

Transalkylation of toluene or toluene and C_{9+} aromatics has not been widely used in the past but is likely to become more important as feed preparation becomes more expensive. Transalkylation of toluene alone leads mainly to benzene and xylenes, whereas transalkylation of toluene and C_{9+} aromatics gives mainly xylenes. These reactions are equilibrium controlled, and unreacted feed must be recycled after distillation of the desired products.

p-Xylene separation is accomplished by crystallization or adsorption. When a typical reformate-derived C_8 aromatic mixture is cooled, p-xylene crystallizes first. Most plants operate at −60 to −75°C, depending on feed composition. The process is limited by a eutectic temperature below which o- or m-xylene also crystallizes. The solubility of p-xylene in the remaining C_8 aromatic mixture over the −60 to −75°C range is 9.6 to 6.2%.

The Parex adsorption process for p-xylene separation was developed by UOP. Toray has a similar process called Aromax. These processes take advantage of the fact that p-xylene is adsorbed more easily than the other C_8 aromatics by a suitable molecular sieve. The p-xylene is desorbed by either a lighter or heavier hydrocarbon which is readily separated by distillation. p-Xylene is recovered in about 90% yield (see Adsorptive separation).

m-Xylene separation is less important and is based on selective sulfonation followed by regeneration or via complex formation with HF/BF_3.

Isomerization of xylenes or the C_8 aromatics increases the recovery of the required product from a given amount of feed. The processes may isomerize xylenes only; isomerize ethylbenzene to xylenes; or isomerize xylenes and selectively crack ethylbenzene (recently developed by Mobil). The xylene isomerization requires an acidic catalyst, whereas ethylbenzene isomerization requires a hydrogenation catalyst, usually platinum (see Xylenes and ethylbenzene).

BTX by Pyrolysis

Ethylene is made throughout the world on an enormous scale. An ethylene plant with an annual production of 450,000 metric tons is very common and considered to be close to the economically optimum size. Side products may include large quantities of BTX, depending on the feed. Table 3 shows the relationship of feed type to ethylene and BTX yields (31). Low-octane stocks, often in excess in a refinery, are favored feeds to an ethylene cracker (see Ethylene).

Table 3. BTX Yields From Various Pyrolysis Feeds[a,b]

Feed	Ethane	Propane	n-Butane	Mid-crude naphtha	C$_6$–C$_8$ Raffinate	Gas oil, distilled Atmospheric	Gas oil, distilled Vacuum
feed rate, 1000 t/yr	583	1080	1135	1349	1535	1750	2213
ethylene once through yield	48.2	34.2	35.8	30.0	26	23	18
benzene, 1000 t/yr	5	26.6	34.3	90.0	72.7	105.5	82.5
toluene, 1000 t/yr	0.7	5.8	9.4	45.1	41.5	50.8	64.3
C$_8$ aromatics, 1000 t/yr			4.0	23.8	18.4	38.0	41.4

[a] Ethylene plant of an annual production of 450,000 metric tons; ethane recycled to extinction.
[b] Ref. 31.

Pyrolysis gasoline contains a large proportion of olefins which must be hydrogenated before BTX recovery. The C_{5-} and C_{9+} fractions are usually separated by distillation prior to hydrogenation in order to save hydrogen. The C_6–C_8 heart cut is hydrotreated to desulfurize and saturate the olefins. The aromatics are recovered by extraction as discussed earlier. The raffinate from the aromatic extraction step has a high cycloparaffin content and is an excellent reformer feed.

The Role of the Computer in BTX Processing

Most refineries have an overall computer program (32), including models of the various units in the refinery, which provides a material balance. By inserting operating costs of each unit, values for each stream, and specific product requirements, an optimum economic operating mode can be reached. It should not be expected that the model representing any given unit will be very detailed. In fact, hydrotreaters, hydrocrackers, or reformers may all be lumped together as one or costs may be coupled to feed rate, but without consideration of severity. The computer, in a reiterative scheme, such as the BTX processing scheme shown in Fig. 2, may be used to predict the effects of a change or to optimize a variable.

Computer simulation of an individual plant is also possible. Each of the tanks or vessels will be modeled, and heat and material balances included. In addition, the chemical reactions are simulated for highly complex processes based on empirical observations or on the kinetics and thermodynamis of the system. Thus the performance of different feed types or operating modes can be predicted.

The computer is used widely for optimizing process control by providing a complex feedback or feedforward equipment control. It is a flexible sensor-based system and can respond to many different forms of signals (see Instrumentation and Control).

Environmental Considerations

The petroleum industry has to operate in a highly competitive area within the framework of a continuously changing set of rules, many of which are intended to protect the environment. For example, if the permissible lead content in gasoline is reduced by government regulation, the refiner must run his reformers to higher severity at higher cost to compensate for the loss of octane rating. One alternative is the installation of paraffin isomerization units to upgrade the octane value of pentane and hexane streams; another would be the production of vehicles requiring lower octane fuels (see Air pollution; Exhaust control, automotive).

Limits are also imposed in the permissible level of sulfur in heavy fuel oils. The BTX industry uses fuel oil which must now be either treated or replaced by an acceptable and more expensive fuel.

Finally, there is an ever-present problem in handling large volumes of low boiling materials. Our greater awareness of the need to protect the environment has led to a great deal of expense in ways of monitoring the operation and reducing emissions (see Industrial hygiene and toxicology).

BIBLIOGRAPHY

1. *Oil Gas J.* **69,** 45 (Dec. 20, 1971).
2. *Ibid.* p. 48.

3. M. J. Sterba and co-workers, *Oil Gas J.* **66,** 140 (1968).
4. Ref. 1, p. 51.
5. R. R. Cecil and co-workers, *Oil Gas J.* **70,** 50 (1972).
6. W. F. Taylor and A. R. Welty, Jr., *Oil Gas J.* **61,** 142 (1963).
7. J. Bernstein and L. Daurer, *Oil Gas J.* **66,** 163 (1968).
8. Ref. 1, p. 54.
9. C. S. McCoy and P. Munk, *Chem. Eng. Prog.* **67**(10), 78 (Oct. 10, 1971).
10. Ref. 1, p. 57.
11. Ref. 1, p. 58.
12. E. M. Blue and G. D. Gould, *Paper Presented to the Chemical/Petrochemical Processing Equipment and Technology Technical Sales Seminar,* Oct./Nov. 1975.
13. *Oil Gas J.* **74,** 48 (1976).
14. T. R. Hughes and co-workers, *Hydrocarbon Process.,* 75 (May 1976).
15. J. Henningsen and M. Bundgaard-Nielson, *Brit. Chem. Eng.* **15,** 1433 (Nov. 1970).
16. W. S. Kmak and A. N. Stuckey, *American Institute for Chemical Engineers Paper 56A,* New Orleans, March 14, 1973.
17. A. M. Kueelman, *Hydrocarbon Process.,* 95 (Jan. 1976).
18. U.S. Pat. 3,948,998 (Dec. 5, 1974), F. Fujiyama and co-workers (to Mitsubishi Gas Chemical Co.).
19. U.S. Pat. 2,485,237 (Nov. 8, 1966), W. F. Gresham and G. E. Taber (to E. I. du Pont de Nemours Co., Inc.); U.S. Pat. 2,284,508 (Oct. 18, 1949), E. R. Gray and co-workers (to E. I. du Pont de Nemours Co., Inc.).
20. *Oil Gas J.* **63,** 128 (Apr. 5, 1965).
21. D. Read, *Oil Gas J.* **51**(7), 82 (1952).
22. G. S. Somekh and B. I. Friedlander, *Hydrocarbon Process.,* 127 (Dec. 1969).
23. G. S. Somekh, *Hydrocarbon Process. Pet. Refiner,* (July, Aug., Sept., Oct. 1963).
24. T. S. Hoover in ref. 22, p. 131.
25. E. Muller, *Chem. Ind.,* 518 (June 2, 1973).
26. E. Muller, *7th World Petrol. Congr.* **16**(2), (Apr. 29, 1967).
27. B. Choffe and co-workers *Hydrocarbon Process.* **45**(5), 188 (May 1966).
28. E. Cinelli and co-workers, *Hydrocarbon Process.,* 141 (Apr. 1971).
29. W. T. Jones and V. Payne, *Hydrocarbon Process.,* 91 (March 1973).
30. M. Stein, *Hydrocarbon Process.,* 139 (Apr. 1973).
31. J. G. Freiling and A. A. Simone, *Oil Gas J.* **71,** 25 (1973).
32. P. Dyer quoted in *Oil Gas J.,* 48 (Jan. 24, 1977).

Derek L. Ransley
Chevron Research Company

BURNER TECHNOLOGY

Most familiar and important combustion processes involve diffusion flames in which fuel and oxidant are separately introduced, and in which the rate of the overall process is determined mainly by mixing, eg, by laminar or turbulent diffusion, or by some other physical process. Flames of candles, matches, gaseous fuel jets, oil sprays, and large fires, accidental or otherwise, are essentially of this type; but many others of practical interest are premixed flames, propagating in homogeneous mixtures of reactants. Much of the knowledge of flames has been derived from study of the latter type, and many of the generally applicable concepts and quantities are readily and clearly defined only for flames or explosives in uniform premixtures. A general description of such a flame is thus a prerequisite for subsequent discussion of more complex combustion processes as well as the operation of premixing torches, burners, etc.

In addition to the journals devoted to the subject and the texts listed under General References, the sixteen volumes of papers given at the biennial International Symposia on Combustion, sponsored by the Combustion Institute, contain many reviews of special topics in the field as well as important current research reports. There are also specialized texts devoted to related specific technologies, eg, power generation (qv) (1), furnaces (qv) (2), reciprocating engines (3), gas turbines (see Aviation and other gas turbine fuels) (4), fuel chemistry and processing (see Fuels) (5), condensed explosives (see Explosives) (6), fire, explosion, and detonation phenomena (7), and incineration (see Incinerators) (8). As noted, most of these are reviewed elsewhere in the Encyclopedia. Therefore, this review is directed mainly to combustion of fossil fuels with air (or oxygen).

Definitions, Terminology, and Basic Ideas

The overall composition of a mixture (whether uniform or not) of fuel with air or other oxidant is often specified as the ratio of fuel to oxidant: the weight ratio (m_f/m_o), or for gaseous reactants, the molar or the volume ratio (V_f/V_o). In either case, it can be normalized to the stoichiometric composition by defining the equivalence ratio:

$$\phi = (m_f/m_o)/(m_f/m_o)_{stoich} = (V_f/V_o)/(V_f/V_o)_{stoich} \qquad (1)$$

The denominator describes the composition that would yield just CO_2, H_2O, and N_2 on complete combustion, eg, in a calorimeter if the initial mixture contained only C, H, O, and N (sometimes called perfect combustion). A mixture of $\phi < 1$ is said to be lean, and its burned gas contains unreacted or excess O_2; if $\phi > 1$, some of the C and H necessarily appears as CO and H_2 in the burned gas and the mixture is said to be rich. ϕ can be readily calculated from other composition variables in common use, eg:

$$\% \text{ excess air} = 100(1 - \phi)/\phi \qquad (2)$$

This is usually applied to overall lean mixtures although it may also be applied to rich mixtures; negative values indicate a deficiency instead of excess. If the volume fraction (= mol frac) x_f of a gaseous fuel in a mixture is given $(x_f = V_f/(V_f + V_o))$:

$$\frac{1}{\phi} = \left(\frac{V_f}{V_o}\right)\left(\frac{(1 - x_f)}{x_f}\right) \qquad (3)$$

In mixtures of a given fuel (gas or vapor) with oxidant, there are usually two values of ϕ: the lean or lower limit (ϕ_L) and the rich or upper limit (ϕ_U) of flammability between which the mixtures are said to be flammable (ie, explosive), or can propagate a flame as a steady wave from some local ignition source, such as a spark, at a velocity that depends on ϕ; it is normally a maximum near $\phi = 1$, and near the limits tends to lower values. The limits most often measured and quoted refer to fuel–air mixtures initially at ambient conditions, ignited with a spark or small flame at the bottom of a vertical tube of diameter ≥ 10 cm. Many measurements have also been made at other temperatures and pressures on mixtures of fuels with O_2 and other O_2/N_2 oxidant atmospheres (9). ϕ_U increases and ϕ_L decreases markedly with increasing O_2/N_2; but lean limits expressed as volume or mole % fuel in the whole mixture (100 x_f) are about the same in O_2 and in air (Fig. 1), because O_2 and N_2 have nearly the same molar heat capacity, as is also true of air diluted or vitiated with N_2 (though not of air vitiated with CO_2, Ar, etc).

For hydrocarbons of molecular weight >30, ϕ_U is found to be considerably larger when the flame propagates upward from the ignition source than it is for downward propagation; and the lean limit, ϕ_L, for H_2 is much lower with upward than with downward propagation. Such nonisotropic flame propagation is observed when the deficient reactant diffuses faster than does the reactant in excess (eg, H_2 vs O_2 in lean flames); it presumably is due to the complex interaction of convection and diffusion in the propagating flame and often results in only partial combustion or cellular flames. Nevertheless, the mixture must be considered flammable, and the limits to be considered, especially in questions of safety, should always be the more conservative value (9).

There is an oxygen content of the atmosphere, sometimes called the oxygen index (OI) of the fuel, below which no mixture with the fuel is flammable. Near the OI, both ϕ_U and ϕ_L approach unity, and only an approximately stoichiometric mixture will be flammable. ϕ_L and OI may be considered roughly equivalent as measures of flammability, eg, in hazard evaluation; both decrease slowly with increasing mixture temperature (about 10% for 100°C increase) while ϕ_U increases. The flammability region widens with increasing temperature, as it usually does also with elevated pressure, though the effect of pressure is less predictable. Figure 1 is an example of some typical limit data, derived mainly from values given in ref. 9 for propane/O_2/N_2 mixtures.

A closely related measure of liquid flammability is the flash point. It is primarily a measure of volatility, and is ideally the temperature at which the vapor pressure of the liquid is such that the mole fraction of its vapor in the atmosphere above the liquid ($x_f \simeq$ vapor pressure in atmospheres) is that of the lower limit mixture of the fuel in air at that temperature (10); ie, it will just propagate a flame. The measured value may depend slightly on the method used, especially for liquid mixtures, because the composition of the vapor evolved can vary with the heating rate, but it is an important hazard criterion. (The fire point temperature, though sometimes also measured, is normally of less interest; its absolute value should be and usually is about 7% higher than the absolute flash point.) For a pure substance of known elemental composition or combustion stoichiometry, the flash point may be estimated quite well from its boiling point and a rough estimate of either its OI or ϕ_L (11).

The flammability limits and OI of a few typical mixtures of fuels with air and oxygen are shown in Table 1, with other properties to be discussed. As rough rules,

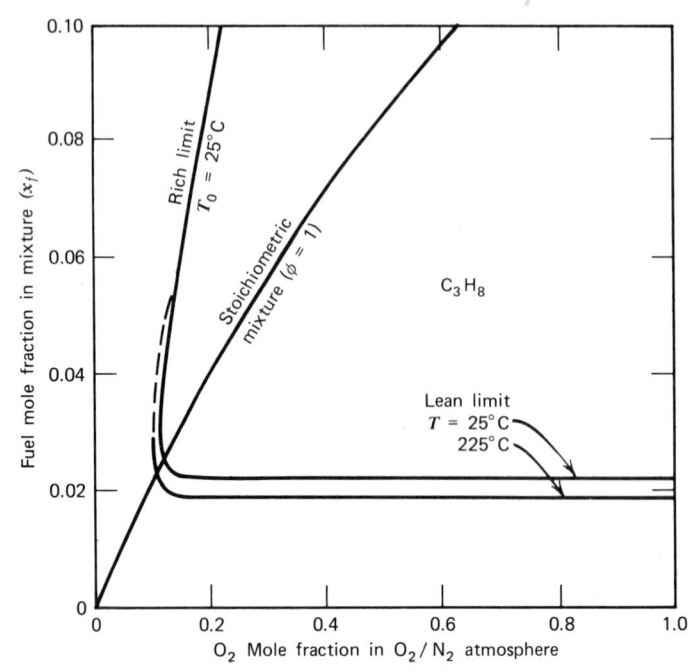

Figure 1. Flammability limit diagram for propane–O_2/N_2 mixtures at 101.3 kPa (1 atm).

for most hydrocarbons, the oxygen index is in the range of 12–14%, the lower limit in mixtures with air is at ϕ_L ca 0.5; and the flame temperatures of their limit mixtures are about the same (1500–1700 K).

Thermochemical Quantities. The higher heating value of a fuel is its heat of combustion at constant pressure and temperature (usually ambient), from calorimetric measurements in which any water formed by combustion is completely condensed. The lower heating value is the similarly measured or defined heat of combustion if there was no condensation of water. The lower value is most often used in combustion and flame calculations, since the water is usually steam in most processes of practical interest; the combustion efficiency of the process is then the ratio of the heat actually released (enthalpy of the burned gas) to this ideal heat release.

The lower heating value is numerically equal to the corresponding enthalpy decrease; eg, in the complete combustion of methane with O_2 (or with air, since the N_2 is not significantly involved in the flame reactions) at 25°C (298 K):

$$CH_{4(g)} + 2\ O_{2(g)} \rightarrow CO_{2(g)} + 2\ H_2O_{(g)} \tag{4}$$

The enthalpy change as written for 1 mole of fuel is:

$$\Delta H^C_{298} = -Q_p = -802.1 \text{ kJ/mol } (-191.7 \text{ kcal/mol}) \tag{5}$$

Or, in other commonly used units (atmospheric pressure):

$$Q_p = 32.78 \text{ J/cm}^3 \text{ (CH}_4 \text{ gas, 25°C) or 7.83 cal/cm}^3 \text{ or 886.1 Btu/ft}^3 \tag{6}$$

The heat of vaporization of water at 298 K is 44.01 kJ/mol (10.52 kcal/mol), thus the higher heat of combustion is 802.1 + 2(44.0) = 890.1 kJ/mol of CH_4.

Similarly for n-octane (vapor):

$$C_8H_{18(g)} + 12\ O_{2(g)} \rightarrow 8\ CO_{2(g)} + 9\ H_2O_{(g)} \tag{7}$$

$$Q_p = -\Delta H^C_{298} = 5124.3 \text{ kJ/mol (1224.7 kcal/mol)} \tag{8}$$

Table 1. Some Calculated and Experimental Data for Typical Explosive Mixtures at Ambient Temperature and Pressure

Fuel	Q_p, kJ/mol	Oxidant	$(V_f/V_o)_{\phi=1}$	$(m_f/m_o)_{\phi=1}$	$q_p, \phi=1$, J/cm^3	$\phi_L{}^a$	$\phi_U{}^a$	OI^a, %	$T_{ig}{}^b$ (AIT), K	T_a, K	S_u^0 cm/s	d_q, cm	E_{min}, mJ ($\phi=1$)	T_a^v, K ($\phi=1$)	r_p, (calc.)^c ($\phi=1$)	r_p, (meas.)^d ($\phi=1$)
CH$_4$	802.1	air	0.1050	0.0582	3.10	0.53	1.6	12.1	810	2220	40	0.22	0.45	2580	8.8	
		O$_2$	0.5000	0.2469	10.96	0.027	1.3			3050	350	0.035	0.004	3540	14.9	
C$_3$H$_8$	2044.4	air	0.0420	0.0641	3.37	0.54	2.5	11.4	740	2260	45	0.19	0.28	2630	9.3	6.5
		O$_2$	0.200	0.2756	13.91	0.118	6.1			3090	390	0.02	0.002	3630	18.3	
n-C$_8$H$_{18}$ (V)	5124.3	air	0.0168	0.0664	3.46	0.60	2.7	12	480	2270	45		0.5	2640	9.5	6.5
		O$_2$	0.0800	0.285	15.50					3100				3670	20.1	
C$_2$H$_4$	1322.5	air	0.0700	0.0675	3.54	0.455	6.75	10.0	760	2370	75			2730	9.4	6.8
		O$_2$	0.3333	0.2922	13.52	0.093	12.0			3170	550			3730	16.9	
C$_3$H$_6$ (propene)	1926.2	air	0.0466	0.0675	3.52	0.53	2.5	11.5	730	2340	50					
		O$_2$	0.2222	0.2922	14.33	0.097	5.1			3150						
C$_2$H$_2$	1257.0	air	0.0840	0.0752	3.96	0.30	48	6.0	575	2540	160	0.07		2920	9.8	8.5
		O$_2$	0.4000	0.3254	14.56	0.06				3400	1100	0.02		3980	17.1	
C$_6$H$_6$	3168.5	air	0.0280	0.0752	3.53	0.51	2.73	11.2	830	2300	45		0.5			6.5
		O$_2$	0.1333	0.3254	15.21	0.10	10.0			3100						
H$_2$	242.1	air	0.4202	0.0292	2.91	0.092	7.2	5.0	790	2400	230	0.06	0.02			7.0
		O$_2$	2.0000	0.1258	6.57	0.021	7.8			3100	1100	0.019				
CO (+1% H$_2$ or H$_2$O)	284.5	air	0.4202	0.0871	3.44	0.34	6.8	5.6	880	2400	45					7.0
		O$_2$	2.0000	1.751	7.75	0.088	7.8			3000	120					
C$_2$H$_5$OH (V)	1277.2	air	0.0700	0.1111	3.42	0.642	3.35		670	2250	ca 40					6.2
		O$_2$	0.3333	0.4792	13.0	0.13		13		3000						
Suspensions																
aluminum dust	826 (=30.6 kJ/g)	air		0.262	9.6	0.133		3	900	2500	20–100		50			ca 4
bituminous coal dust	30 kJ/g	air		0.09	3.2	0.35		15	800	2100	ca 30		40			ca 4
kerosene mist, (CH$_2$)$_x$	42 kJ/g	air		0.068	3.4	0.5	ca 3	14	450	2200	45					5.8

^{a} For upward propagation, mainly from data in refs 9–10.
^{b} Approximate for ignition delay ca seconds.
^{c} Ref. 13.
^{d} Mostly from ref. 14, other data collected from various sources.

= 209.5 J/cm³ (vapor, 25°, 101.3 kPa) or 50.1 cal/cm³ (1 atm) (9)

For liquid n-octane, the corresponding value is less by its heat of vaporization at 298 K (41.5 kJ/mol). The lower heating value of the liquid is thus 5028.8 kJ/mol, and in units commonly given for substances normally solid or liquid, it would be: 5028/114.2 = 44.50 kJ/g (= 10.63 kcal/g = 19,140 Btu/lb). On a weight basis, fuels of similar type and C:H ratio, eg, higher paraffin hydrocarbons of gross composition, roughly $(CH_2)_x$, all have roughly this heating value.

The evaluation of ΔH_{298} for use in thermochemical calculations is more generally and conveniently done by the use of tables of self-consistent values of heats of formation from the elements (ΔH^f_{298}) (12) of reactants and products, known or assumed, for the reaction in question.

In general, for any reaction:

$$\Delta H_{298} = \sum_{\text{products}} \Delta H^f_{298} - \sum_{\text{reactants}} \Delta H^f_{298} \quad (10)$$

For example, if the process were to involve burning CH_4 at $\phi = 1.5$ according to:

$$CH_{4(g)} + 3/2\, O_{2(g)} \rightarrow CO_{(g)} + 1/2\, CO_{2(g)} + H_2O_{(g)} + H_{2(g)} \quad (11)$$

$$\Delta H_{298} = (\Delta H^f_{298})_{CO} + 1/2\,(\Delta H^f_{298})_{CO_2} + (\Delta H^f_{298})_{H_2O} + (\Delta H^f_{298})_{H_2}$$
$$- (\Delta H^f_{298})_{CH_4} - (\Delta H^f_{298})_{O_2} \quad (12)$$

$$= -110.5 + 1/2\,(-394.8) - 242.1 + 0 - (-74.85 + 0) =$$
$$-475.15 \text{ kJ/mol } (113.56 \text{ kcal/mol}) \quad (13)$$

In Table 1, the quantity $q_{p,\phi=1}$, or the heat evolution on combustion of the stoichiometric mixture, is simply derived from Q_p and the corresponding stoichiometry with air and with O_2. For gaseous fuels with air, $q_{p,\phi=1}$ does not vary greatly; in effect, the increase in molar or volumetric heating value with increasing molecular weight of the fuel is accompanied by a nearly proportional increase in volumetric air requirement for complete combustion.

Adiabatic Flame Temperatures. If the products of any combustion process or of any chemical reaction are at equilibrium and uniform with respect to composition, pressure, and temperature (whether or not the reactant mixture is uniform), and if there are no heat losses to the surroundings in the process, the thermodynamically attainable temperature is calculable as the adiabatic flame temperature T_a. In addition, if the process occurs at constant pressure, the overall enthalpy change is zero, and T_a is the temperature that satisfies the condition. For a uniform initial temperature T_o from the first law of thermodynamics for such a reaction:

$$\Delta H_{T_o} + \sum_{\text{products at } T_a} n_i (H_{T_a} - H_{T_o})_i = 0 \quad (14)$$

The first term is the enthalpy change at T_o per mole of fuel for a combustion reaction written with products that correspond to thermodynamic equilibrium at T_a. The sum is the enthalpy of the burned gas mixture at equilibrium at T_a relative to its value at T_o. The number of moles of a species of the burned gas produced per mole of fuel is n_i; its molar enthalpy may be evaluated as $\int_{T_o}^{T_a} C_p dT$ or found directly from tables (12). (The assumption of a uniform T_o of the reactants is not essential; with only slight modification the same calculation procedure can be applied to arbitrarily different states of the reactants.)

If the n_is do not depend on T_a, eg, if the products can be assumed to be only CO_2,

H_2O, and N_2, $\Delta H_{298} = \Delta H^c_{298}$, the value of T_a that satisfies the equality is easily found from the enthalpy tables, and generally this can be done if T_a is less than ca 1700 K and $\phi < 1$. On the other hand, a rich hydrocarbon flame, whatever T_a may be, always yields some CO and H_2, and their partial pressures must accord with the water–gas equilibrium ($K_{WG} = p_{CO}p_{H_2O}/p_{CO_2}p_{H_2}$). K_{WG} varies with temperature, although slowly (ΔH = 41 kJ/mol), and therefore so does the gas composition and ΔH_{298}. If T_a turns out to be much higher than 1700 K for any input composition, rich or lean, in effect the endothermic dissociation of, eg, CO_2 and H_2O, leads to significant equilibrium partial pressures of other species (OH, H, O, O_2, H_2, and from a carbon-containing flame, also CO) that must simultaneously satisfy the equilibria: $H_2O \rightleftharpoons OH + \frac{1}{2} H_2$, $H_2O \rightleftharpoons H_2 + \frac{1}{2} O_2$, $CO_2 \rightleftharpoons CO + \frac{1}{2} O_2$, $\frac{1}{2} O_2 \rightleftharpoons O$ and $\frac{1}{2} H_2 \rightleftharpoons H$, or an equivalent set. The five equilibrium constants for these reactions, eg, $K_p = p_{OH}p_{H_2}^{1/2}/p_{H_2O}$, etc, which vary rapidly with temperature, must then be combined with the atom-conservation equations (for C, H, O, and N if all are present) to give a set of nine simultaneous equations, many of which are nonlinear, to be solved for the partial pressures of mole fractions of the nine species (including N_2), or the equilibrium composition of the gas. (The dissociation of N_2 can generally be ignored if the temperature is below 4000 K.) This calculation is made for many temperatures in the range of interest and the calculated n_is at some temperature are used to determine ΔH_{T_o} (eq. 10). Its magnitude, which will always be less than $(-\Delta H^c_{T_o})$ and depends on the temperature, is compared with the enthalpy sum calculated for that composition to test the equality (eq. 14). If they do not agree, a new temperature and its corresponding equilibrium composition are used to give a new value of ΔH_{T_o} to compare with the new enthalpy sum for that composition, and the process is iterated to converge eventually on the temperature that satisfies the equality.

Computers and available programs (15) have made routine these otherwise very tedious iterations, so that rapid calculation of equilibrium gas compositions and their thermodynamic properties can be done for any fuel–oxidizer system for which the enthalpies and heats of formation are available or can be estimated. The programming techniques can easily be extended to include other minor species and equilibria of interest, eg, of air pollutants or emissions such as: $NO \rightleftharpoons \frac{1}{2} NO_2 + \frac{1}{2} O_2$; and in some very rich hydrocarbon flames, the precipitation of solid carbon and the Boudouard equilibrium: $C_{(s)} + CO_2 \rightleftharpoons 2 CO$. See Table 2 for compositions calculated for a few typical adiabatic flames. The temperatures given in Tables 1 and 2 and Figures 2 and 3 are theoretical values. With increasing T_o, the increase of T_a is less and often much less than the increment in T_o, owing to the increase in heat capacity and to dissociation at the higher temperature. Figure 3 illustrates the effect of drastic preheating of the combustion air on T_a of a $\phi = 1$ CH_4–air mixture to attain flame temperatures in the range required for open-cycle MHD power generation (16) (see Coal, magneto-hydrodynamics). Air temperatures resulting from adiabatic compression in most engine cycles are much lower.

The effect of pressure and of oxygen enrichment is also illustrated in Figure 3; increasing pressure tends to suppress dissociation of the products and thus generally raises T_a. A rough but sometimes useful rule for O_2-enrichment effects can be seen if T_a for ambient conditions, $\phi = 1$, mixtures of most common fuels is plotted vs the ratio N_2/O_2: between $N_2/O_2 = 0$ and 3.76 (air), such a plot happens to be nearly linear, T_a decreasing by about 220 K per unit increase of N_2/O_2.

Adiabatic flame temperatures so calculated agree reasonably well with values measured, eg, by optical techniques, when the combustion is essentially complete and

284 BURNER TECHNOLOGY

Table 2. Equilibrium Burned-Gas Composition of Some Adiabatic Flames; 101.3 kPa (1 atm); $T_o = 298$ K

	T_a, K	N_2[a]	CO_2	CO	H_2O	H_2	O_2	OH	O	H	NO
H_2–air; $\phi = 1$	2400	0.643			0.322	0.0164	0.0053	0.0068	0.007	0.0020	0.0028
H_2–O_2; $\phi = 1$	3100				0.566	0.158	0.052	0.115	0.034	0.075	
CH_4–air; $\phi = 0.8$	1995	0.7270	0.0768	0.0005	0.1536	0.0002	0.0369	0.0016	0.0001	0.0000	0.0032
CH_4–air; $\phi = 1$	2225	0.7091	0.0853	0.0089	0.1831	0.0036	0.0045	0.0029	0.0002	0.0004	0.0020
C_3H_8–air; $\phi = 1$	2265	0.7213	0.1026	0.0124	0.1481	0.0033	0.0058	0.0032	0.0003	0.0005	0.0024
CH_4–air; $\phi = 1.5$	1900	0.6262	0.0408	0.0838	0.1668	0.0822	0.0000	0.0000	0.0000	0.0002	0.0000
CH_4–O_2; $\phi = 1$	3050		0.1133	0.1551	0.3907	0.0732	0.0834	0.0958	0.0388	0.0499	

[a] Includes argon (air is 0.93% Ar).

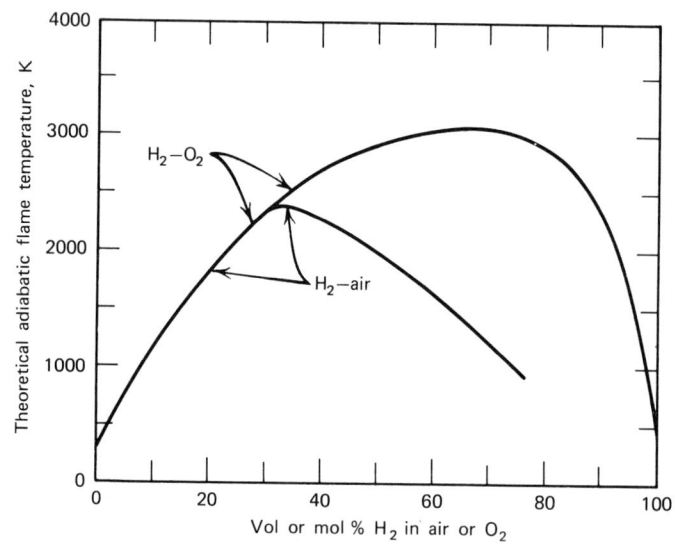

Figure 2. Calculated adiabatic flame temperatures T_a of hydrogen–air and of hydrogen–oxygen mixtures at 101.3 kPa (1 atm).

when losses are known to be relatively small. Calculated temperatures and gas compositions thus are extremely useful or even essential for assessment of combustion processes and prediction of effects of variation of process parameters. Direct and accurate measurement of gas temperatures, especially above 1200°C, is usually difficult or may involve large corrections. Among the optical techniques that can be applied to high temperature flame gases, the most common is sodium line-reversal (17). It has often been used with laboratory flames when there are heat losses of unknown magnitude and to verify calculated values (18).

If the process takes place, not at constant pressure, but at constant volume, the

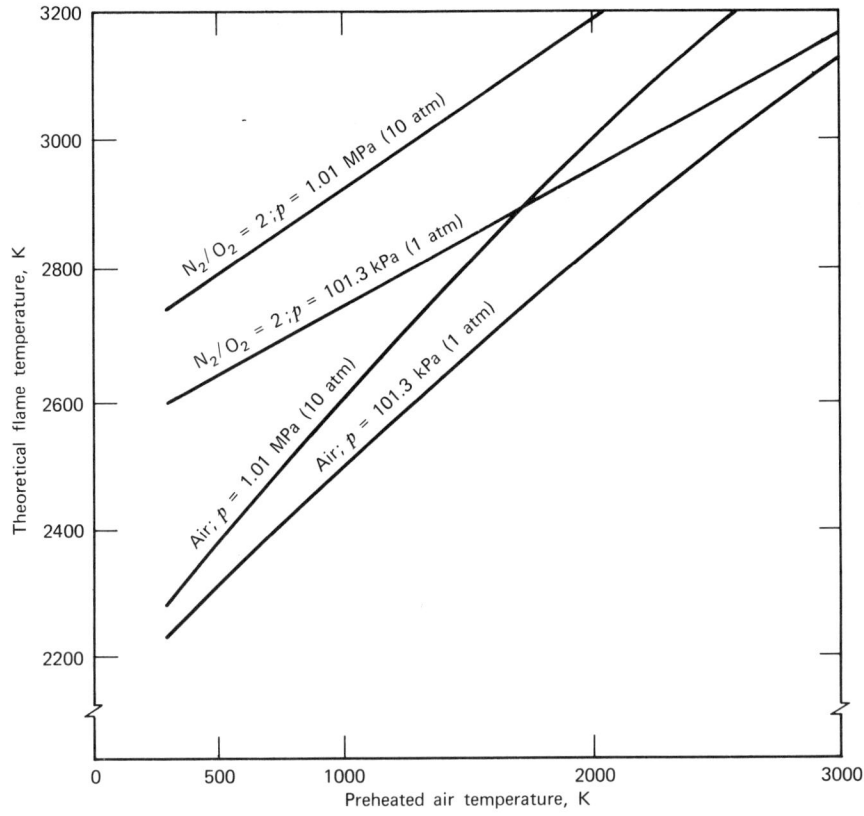

Figure 3. Flame temperatures of stoichiometric mixtures of methane with preheated atmospheres of air and of $O_2 + 2\,N_2$ at two pressures.

overall change in internal energy is zero, and the analogous adiabatic flame temperature calculation is made by equating $(-\Delta E_{T_o})$ to:

$$\sum_{products} n_i \cdot (E_{T_a^v} - E_{T_o}) = n_i \cdot \int_{T_o}^{T_a^v} C_v dT \qquad (15)$$

Because no flow-work is done on or by the surroundings, the temperature T_a^v is of course higher than T_a for the same mixture and initial condition. The pressure increase at constant volume and its effect on dissociation tends also to raise the temperature and affects the n_is. The iteration procedures for the temperature must in effect be extended to iteration for the pressure. The pressure ratio (p_b/p_o) is of practical interest, eg, in assessing explosion hazards, and is commonly in the range of 5–10. Theoretically, it can be calculated from the gas law (for moderate pressures):

$$r_p \equiv \rho_b/\rho_o = (\overline{M}_o/\overline{M}_b)(T_a^v/T_o) \qquad (16)$$

where $\overline{M}_o$ and $\overline{M}_b$ are, respectively, the average molecular weights of initial and burned gas mixtures. $\overline{M}_o/\overline{M}_b$ of course varies with the stoichiometry and with dissociation of product gas; it may be <1 as in H_2/O_2 explosions but is usually about unity for most fuel–air mixtures. See Table 1 for calculated values of T_a^v for a few mixtures of $\phi = 1$, with the corresponding values of r_p (13) and some measured pressure ratios in closed bomb explosions of the same mixtures (14). The latter will tend always to be made smaller by heat losses from any vessel of finite volume.

Burning Velocity of Premixed Flames. A propagating flame is a quite thin wave in which the temperature rises abruptly though not discontinuously to the flame temperature. The thickness of the wave depends on ϕ and on pressure; in a typical hydrocarbon–air mixture of $\phi = 1$ at 101.3 kPa (1 atm) the entire rise from T_o to T_a occurs in ca 0.1 cm, and the gradients of concentration and temperature are accordingly very steep. An adiabatic combustion wave is usually considered to consist of a preheat zone in which the unburned mixture is heated by conduction down the gradient from the burned gas and a reaction zone near T_a in which the complex exothermic reactions occur.

The reaction zone of hydrogen flames is practically invisible, but in hydrocarbon combustion the position of the flame is identifiable by the nonthermal radiation or chemiluminescence (qv) from transient, incidental species in extremely low concentrations, mainly excited C_2 and CH, which are significant only in the production of some ions (19). The rate of the overall flame reaction and its apparent temperature dependence (global activation energy, ca 150–250 kJ/mol) are such that the flame reactions occur mainly at or near T_a. In most of the transit time of a gas element through the wave, typically in ms, it is simply being heated to T_a. There is thus a coupling of the chemical reaction or heat evolution rate and the heat transfer by conduction against the flow of unburned mixture, and an observer moving with the wave would see a steady laminar flow of unburned gas at a uniform velocity, S_u^o, into the stationary wave or flame. S_u^o is thus defined as the normal burning velocity of the mixture. If its unburned density is ρ_o, the mass velocity, $\rho_o S_u^o$, is then the rate of mass consumption of the mixture per unit area of the wave. S_u^o as conventionally used implies measurement at the ambient ρ_o [101.3 kPa (1 atm), 298 K]; though it is given, eg, as cm/s, it actually refers to a mass velocity, with that value of ρ_o and should be so understood (see Mass transfer).

S_u^o has in fact been approximated with flames stabilized by a steady, uniform flow of unburned gas from porous metal diaphragms or other flow straighteners (20). However, this situation is generally unattainable in practice and S_u^o is usually determined less directly from the speed and area of transient flames in tubes (21), closed vessels (22), and soap bubbles blown with the mixture (23), but most commonly from the shape of steady Bunsen flames (24). The observed speed of a transient flame usually differs markedly from S_u^o. For example, it may be calculated and is confirmed by observation that a flame spreads from a central ignition point in an unconfined explosive mixture such as a soap bubble at a speed: $(\rho_o/\rho_b)S_u^o$, in which the density ratio across the flame is typically 5–10. In this and in many other situations the expansion of the burning gas imparts a considerable velocity to the unburned mixture, and the observed speed will be the sum of this velocity and S_u^o.

Although S_u^o should be calculable from other properties of the mixture, they are rarely known well enough to yield useful results; experimental values are abundant, however, and some are shown in Table 1 for $\phi = 1$ mixtures at ambient conditions. S_u^o typically varies with ϕ as shown in Figure 4. The maximum is usually at $\phi =$ ca 1.05, though for some fuels, eg, H_2, C_2H_2, the mixture of maximum burning velocity is considerably richer. Around ϕ_L and ϕ_U, S_u^o of course becomes small, though perhaps not zero (25).

S_u^o increases with increasing T_o or preheating because T_a and the rates of flame

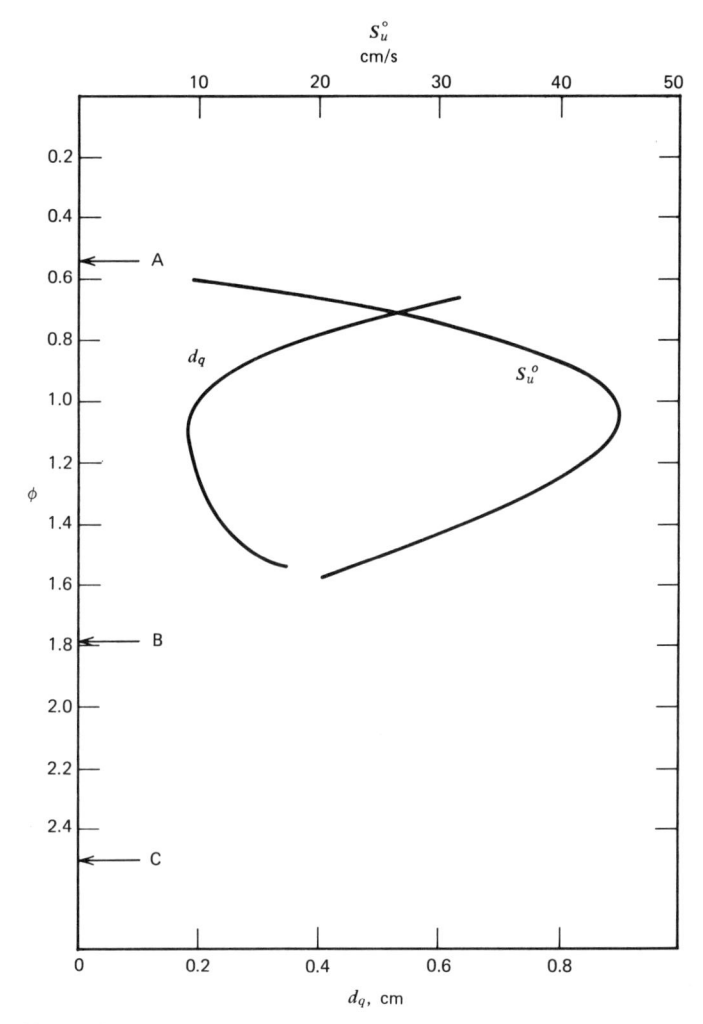

Figure 4. Normal burning velocity S_u^o and quenching distance d_q of propane–air mixtures. A, lean limit; B, rich limit (downward); C, rich limit (upward).

reactions are increased, and the temperature dependence usually has an Arrhenius form, $S_u^o = \text{constant} \cdot e^{-B/T}$, in which flame temperature T is the variable. The constant B appears to be typically ca 12,000 K, and is assumed to be related to the apparent or global activation energy of the flame process. A similar temperature dependence is also found with flames in which T_o is reduced, or there is heat loss to an upstream flameholder (26), so that the actual burned gas temperature is less than T_a (nonadiabatic flames, Fig. 5f).

With increasing pressure, S_u^o decreases (for some usually low S_u^o) mixtures and for some it varies little if at all; it increases with pressure for most fast-burning mixtures, eg, H_2 or C_2H_2 in air or oxygen, or generally those mixtures in which a flame may become a detonation (27). Often loosely applied to a variety of explosion processes, the term detonation should refer to a specific, constant velocity and violent mode of flame propagation which usually takes place only in mixtures of high S_u^o. Nevertheless,

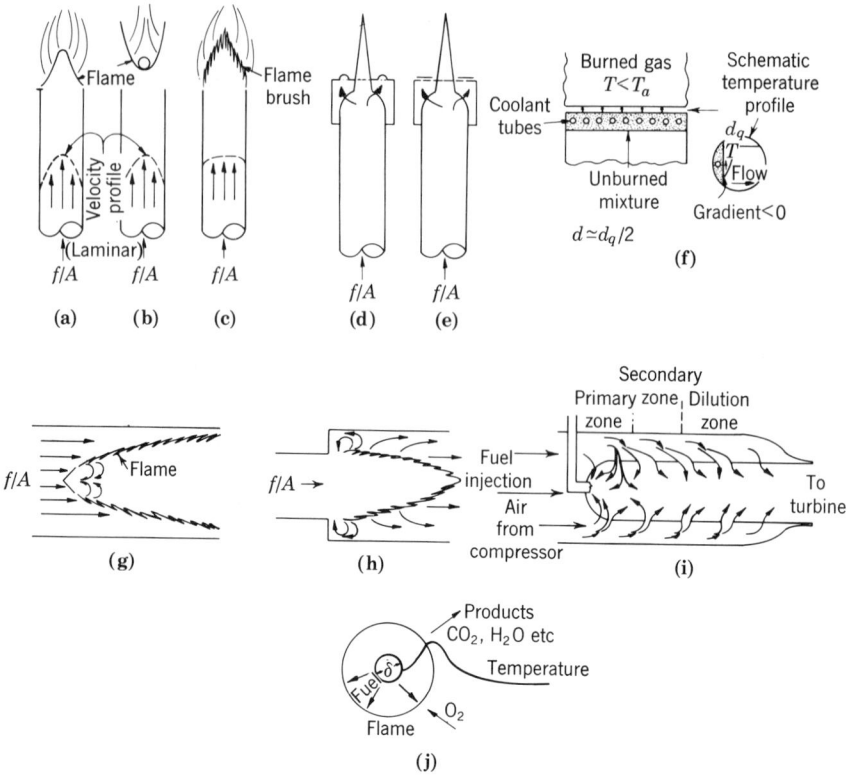

Figure 5. Schematic illustrations of various types of flames and flame stabilizers. (**a**), Bunsen; (**b**), rod stabilizer; (**c**), turbulent Bunsen; (**d**), holes; (**e**), screen; (**f**), nonadiabatic flat flame; (**g**), V-gutter in duct; (**h**), corner in duct; (**i**), can in duct (gas turbine); (**j**), ideal droplet diffusion flame, small δ (Reynolds number ca 2).

much of the transition process can also occur in much weaker mixtures, resulting in the unexpected violence of accidental explosions.

Detonation in Gases. Owing to the volume increase behind the propagating flame, ignition of any confined explosive mixture will adiabatically compress and heat the unburned gas ahead of the flame. Friction at the pipe wall complicates the transient motion of the unburned gas, but always causes some curvature and increase in area of the flame, which will be further enhanced if the flow becomes turbulent. Altogether these changes tend to increase the flame speed relative to the tube or pipe container, resulting in a highly unsteady and unstable condition in which the flame speed may eventually reach the velocity of sound in the mixture. The initially weak pressure waves ahead of the rapidly accelerating flame coalesce to form a shock wave, which discontinuously heats the mixture further, possibly to ignition. When this occurs, the shock and the flame unite, at least momentarily, and the instantaneous wave velocity may be many km/s and the maximum local pressure may be fifty to one hundred times the initial pressure (see also Acetylene).

The distance or elapsed time for this complex sequence of events will be less the larger is S_u^o and its pressure dependence, but the process can and does occur in a relatively weak explosive mixture in a large and sufficiently long pipe, eg, in natural gas–air mixtures inadvertently formed in pipelines. Sometimes called pressure piling (28), the condition, though very transient, can be astonishingly destructive.

If the combustion rate is appropriately high, the process leads directly to the formation of a true detonation wave, a steady state in which the shock and combustion behind it remain coupled and continue indefinitely as a nearly plane wave of constant and reproducible velocity. Thus within certain limits narrower than their flammability ranges, fuels such as H_2 and C_2H_2 mixed with air, and most fuels with oxygen can be readily detonated. The decomposition flame of pure acetylene $C_2H_2 \rightarrow 2\,C_{(s)} + H_2$ can also become a violent detonation at elevated initial pressure (29). The distance required for the formation process as described increases with the pipe diameter and varies considerably with initial conditions; typically it is in the range of 30–100 pipe diameters.

The steady wave velocity D, unlike S_u^o, can be precisely computed from the mixture properties and hydrodynamics if an additional, well-verified assumption is made (27). D is in effect the sum of the sound velocity a of the burned gas and its velocity v relative to the pipe; or relative to the wave front, the burned gas flows out of the wave at its sound velocity at that temperature, which is 10% or so higher than T_a for the mixture. D is typically 1–3 km/s or three to ten times the sound velocity in the unburned gas; the pressure ratio across the wave is ca 15:20, but if the wave meets or is reflected by an obstacle, eg, the closed end of a pipe, the burned gas (at velocity v) is stopped and the pressure abruptly rises by another factor of ca 2. Through a complex sequence of expansions and rarefactions, the pressure in the pipe, if still intact, is eventually equalized.

Other Properties of Flammable Mixtures. The quenching distance d_q is the limiting dimension of passage, containing an essentially quiescent flammable mixture, through which a flame can propagate. d_q is roughly related to the heat loss rate that a flame can experience without extinction, and can be precisely defined and measured in various ways, each with slightly varying results and meaning, eg, with parallel plates, tubes, slots, etc (30). For a wide range of fuels and mixture compositions, d_q can be correlated with S_u^o and the thermal diffusivity, α, of the mixture: $S_u^o d_q/\alpha \simeq$ constant (31); d_q varies approximately inversely with absolute pressure, as does α (Fig. 4).

A flame arrester, exemplified by the grid in the Davy Safety lamp, is essentially a porous or perforated barrier that allows the passage of a gas mixture that may be explosive, but quenches a propagating flame should the mixture be ignited on one side of the arrester. It may consist of one or more screens, a perforated plate or block, rolled-up corrugated sheet, sintered metal or the like. In practice, the allowable size of any opening in the barrier is almost always smaller than d_q, and depends on: whether there is a high gas velocity ahead of the flame to be quenched; whether there is a steady flame at the arrester against a low mixture flow rate even though the transient flame is stopped, etc (32). Thus the design and effectiveness of a flame arrester often must be determined empirically although the quenching distance may be a useful guide as an upper limit to opening size.

Many of the values of d_q shown in Table 1 were obtained from parallel plate measurements in which the minimum spark ignition energy was also determined (33). E_{min} is the minimum energy that must be imparted virtually instantaneously to a very small volume of the mixture ($<d_q^3$) to raise its temperature to the level required (ca 1500 K) for initiation of a self-propagating flame while overcoming heat losses. It is of the order of mJ for hydrocarbon–air mixtures and μJ for hydrocarbon–oxygen mixtures and can be a useful measure of sensitivity of explosive mixtures, though the spark energy required for reliable ignition of burners may be several orders of magnitude larger, especially at high gas velocity.

Autoignition

If a homogeneous fuel–oxidant mixture is uniformly and instantaneously heated to some temperature (≥ 500 K), the mixture may spontaneously self-ignite after a time delay or induction period τ_{ig}. Ideally, the temperature at which such self-ignition is observed is called the autoignition temperature, T_{ig}, corresponding to that value of the time delay or ignition lag (34). The τ_{ig} is an inverse though complex measure of overall reaction rate and decreases with increasing temperature; conversely T_{ig} decreases with increasing τ_{ig}, and varies appreciably with the heating time imposed by the test methods (35). In the range of T_i usually of interest (500–1000 K), τ_{ig} is typically 10^{-4}–10 s. Though plots of log τ_{ig} vs $1/T_{ig}$ often are not linear, the temperature dependence is usually equivalent to an overall activation energy of ca 150 kJ/mol (36 kcal/mol).

Although thermal ignition has been exhaustively studied, the complex nature of the process permits only a few useful generalities, even with regard to its gross features. The ignition lag decreases with increasing pressure, varying about as p^{-1} or p^{-2}, though the dependence appears to vary with the range of temperature. The τ_{ig} varies slowly with O_2 content of the mixture and is usually somewhat less in O_2 than in air at the same equivalence ratio. It varies with mixture composition ϕ in a complicated way, usually decreasing with increasing fuel concentration. There are no definite composition limits for thermal ignition, but in any case the range is much wider than the range of flammability at ambient temperature. By preheating to a sufficiently high temperature, any fuel–air mixture will, of course, react exothermically in some characteristic time.

Autoignition is to be avoided in a spark-ignition engine but is essential to the compression-ignition or Diesel engine, in which the air, after compression to ca 5 MPa (50 atm) is at ca 1000 K. Liquid fuel must be injected and ignited at this condition in about 1 ms. To study this process various kinds of rapid adiabatic compression apparatus are generally used (36), including suitably instrumented shock tubes (37). However, many commonly tabulated values of T_{ig} or AIT, as in Table 1, are obtained on all kinds of substances including dusts (38) by methods at ambient pressure that involve τ_{ig}s of many seconds. They are meant to be useful though approximate criteria of potential hazard in fuel–air mixtures (34–35). The measurement may involve nonuniform temperature and composition owing to evaporation and mixing of cold liquid vapor with heated air. Ignition may then occur at hot spots and the flames may propagate in local mixtures of various compositions rather than by homogeneous reaction. Combustion of this kind following autoignition is often observed in the Diesel and other piston engines, with transient flames at very high velocity leading to abnormally high peak pressures.

The order of increasing T_{ig} (AIT) among hydrocarbons of various types is very roughly: higher straight-chain paraffins (n-hexane to cetane, etc) < lower paraffins < branched-chain paraffins < aromatics and other cyclic hydrocarbons (eg, benzene, decalin). The range from cetane (n-hexadecane) to decalin (decahydronaphthalene) is in effect the basis for the cetane number for rating diesel fuels (38) where ease of autoignition is sought. In inverse fashion, the range from highly branched paraffins (eg, isooctane) to n-heptane is the basis for octane number or the antiknock rating of spark-engine fuels, where freedom from autoignition at high compression ratio is desired.

Thermal Ignition and Flame Chemistry

At low temperature or large τ_{ig} most of the ignition delay involves relatively very slow, nearly isothermal initiation reactions. A fairly definite distinction between the initiation and explosion periods within τ_{ig} is often made, eg, in Diesel ignition (34,39). With increasing T_{ig} (or decreasing τ_{ig}) for a given mixture, these times become more nearly comparable. At high temperature, eg, 1500 K or approaching a typical flame temperature, both periods may be ca 10^{-4} s or virtually simultaneous and, like the reaction zone of a flame in the mixture, would ordinarily be observed as one unresolved event. In effect, the rate determining step in low temperature autoignition may not be rate controlling and thus the usual values of T_{ig} often do not correlate well with other flame properties of the mixture. Nevertheless, the following highly simplified, even speculative, qualitative description may be useful in thinking about flames.

Early in the ignition or induction period, the attack of fuel by O_2 is initially slow, but generates free radicals (OH, H, O, HO_2, hydrocarbons) and other intermediate species (CO, H_2, and partial oxidation and decomposition products of hydrocarbons, if present). For some time, there may be little or no temperature rise, the energy essentially being stored in the free radicals to be released later. In this stage the reactions may be similar to those in the very slow and nearly isothermal oxidation and cool flames observed in very rich hydrocarbon–O_2 mixtures, usually at lower temperatures and pressures. These often stop at the production of stable oxidation and decomposition products, such as aldehydes, peroxides, lower hydrocarbons, etc, but under some conditions may lead to explosion (see Hydrocarbon oxidation). In a variety of chain reactions, the fuel, any intermediates, etc, are rapidly attacked by radicals, some of which are also undergoing very fast branching-chain reactions, known from studies of the mechanism of H_2/O_2 and similar explosions (40–41). Those free radicals are thus enormously multiplied and attain high transient concentrations; they are eventually consumed to form stable species by three-body recombination reactions in which most of the heat release occurs, eg, if M represents any atom or molecule in the gas, by:

$$H + OH + M \rightarrow H_2O + M; \Delta H = -498 \text{ kJ/mol} (-119 \text{ kcal/mol}) \qquad (17)$$

along with other reactions of this type (including a sequence involving HO_2) the relative importance of which depends on mixture composition (41–42). The rates of these termolecular processes increase with pressure but have little or no temperature dependence.

When the partial pressures of radicals become high, their homogeneous recombination becomes fast, the heat evolution much exceeds losses, eg, to any walls, and the temperature rise accelerates any uncompleted consumption of fuel to produce more radicals. Around the maximum temperature, recombination exhausts the radical supply and the heat evolution rate may not compensate radiation losses. Thus the final approach to thermodynamic equilibrium by recombination of OH, H, and O (at concentrations still many times equilibrium) is often observed to occur over many milliseconds after the maximum temperature is attained, especially in flame products of relatively low (<2000 K) burned gas temperature (41,43–44).

Because the radicals and other components are all connected by various equilibria, their decay by recombination occurs in effect as one process on which the complete conversion of CO to CO_2 therefore also depends (41,45). For this reason, the hot, otherwise completely burned gas of any lean hydrocarbon flame typically has a higher-than-equilibrium CO content (41), slowly decreasing toward equilibrium (CO afterburning) along with the radicals, so that the oxidation of CO is actually a radical recombination process (45).

In most hydrocarbon flames, there are many radical consumption reactions that in effect interfere with the branching chain, and the peak radical concentrations are therefore always lower (41) than in analogous flames of H_2, H_2/CO, or moist CO, and the overall reaction rate for a given temperature as reflected, eg, in S_u^o should also be lower. Similar but more drastic reduction of radical concentrations and interference with the flame chemistry no doubt accounts for inhibition of flames by various additives (41,46), notably halogen-containing substances, which can narrow flammability limits, reduce S_u^o, increase d_q, etc, even when they cause little or no reduction in T_a. Those containing bromine are particularly effective, and many brominated organic compounds are extensively used to suppress unwanted diffusion flames or fires as well as premixed flames in extinguishing devices, in commercial plastics as flammability inhibitors, etc (47) (see Flame retardants).

Although most of the chemistry summarized very briefly here has evolved from study of flames and explosions in premixtures, it also applies at least semiquantitively to many nonpremixed systems as well. It is generally consistent with results obtained in studies of such systems by the quite different approach of the well-stirred reactor (48–51), as well as with observations made by shock-heating of premixtures (37). Combustion in the well-stirred reactor resembles that in some types of combustion chambers in common use, and some aspects of combustion of practical as well as fundamental interest can be experimentally examined with such a device.

Reactor Models. In the adiabatic well-stirred reactor, the mixing of product gas with reactants, introduced as very high velocity jets of fuel and oxidizer or of premixture, is so intense that the steady-state reactor content ideally is uniform in composition and temperature. How closely the burned gas condition approaches that expected from the input composition is determined by the residence time (computed from the steady total flow rate and the reactor volume) which is thus an inverse measure of overall reaction rate at that temperature. Detailed analysis of the exit products at various pressures, flow rates and input compositions can yield kinetic data at conditions common to flames; but also, within constraints imposed by relative heat losses, intensity of mixing, etc, steady and nearly complete combustion of mixtures not ordinarily flammable can be maintained. With appropriately long residence times, such mixtures can be burned at temperatures much lower than ordinarily required for flame propagation. The overall temperature dependence or activation energy (typically ca 150 kJ/mol or 36 kcal/mol), as well as the observable details of the chemistry, agree fairly well with those deduced from flame studies (48). The data and basic ideas may be used as a guide in estimating required residence times or volume for nearly complete combustion, eg, in furnaces, incinerators (qv), exhaust gas afterburners, etc, where the maximum temperature may be limited by materials of construction or by the nature of the process.

As the mass flow rate is increased to a well-stirred reactor or combustor, the temperature (ideally T_a at long residence time) falls continuously until a limiting residence time is reached, the combustion is no longer self-sustaining and the flame abruptly blows out, usually at about 0.8 T_a. Measurements at this condition (of low combustion efficiency, ca 80%) can be related indirectly to the overall kinetics (48) but they also provide a measure of the maximum volumetric heat release rate or the limit of attainable space rate or combustion intensity, terms often used to describe loading of combustors, furnaces, etc. For a typical stoichiometric hydrocarbon–air mixture at 101 kPa (1 atm) it is ca 10^6 kJ/(s·m^3) [ca 10^8 Btu/(h·ft^3)], and it is the same order of magnitude as that estimated from the reaction zone thickness and burning velocity of a steady adiabatic premixed flame of the same composition and pressure.

Although in practice, this order of magnitude of intensity is seldom approached, it may be considered an asymptotic maximum, attainable if the fuel supply rate is not limited by physical processes, such as atomization and evaporation in the case of liquid fuels, and particle devolatization or reaction in the case of solids, and if unlimited pressure drop is available to provide infinite mixing rate. Moreover, there is usually some practical limit on allowable reactor temperature, and higher combustion efficiency (>99% rather than 80%) and therefore longer residence times are required. For these reasons, intensities approaching the maximum are reached only in some special cases, eg, in rocket engine combustion chambers, and the space rate attainable with hydrocarbon–air combustion in most furnaces and combustion chambers is very much lower. A typical domestic gas or oil-burning furnace, eg, may be designed for a space rate as low as 10^2 kJ/(s·m^3) and bears little resemblance to a well-stirred reactor.

On the other hand, there are processes that necessarily involve high intensity ($\simeq 10^5$) and are often usefully modeled by a stirred reactor, at least in part. In a typical steady flow gas turbine combustor, the primary combustion zone can be viewed as a reacting volume of this kind, formed and stabilized by mixing of air jets, fuel, and recirculated burned gas, at the highest velocities consistent with the allowable pressure drop. The mixing or stirring action is derived almost entirely from the warm air flowing from the compressor discharge condition [0.5–2 MPa (5–20 atm) depending on the engine cycle] through holes in the combustor (can, annular passage, etc), and so distributed over its length that a stable zone of combustion with the fuel, injected near the upstream end, can exist at any required operating condition. The air must also provide cooling of the combustor wall as well as uniform reduction of the gas temperature by dilution to a tolerable turbine-inlet condition (the firing temperature of the turbine). To avoid serious cycle losses, the pressure drop thus incurred should usually not exceed 4 or 5% of the combustor absolute pressure, and the possible mixing intensity is thus limited by the available energy or velocity of the air jets (Fig. 5i).

The primary zone must in general be maintained at an overall equivalence ratio of 1:1.5 (though it is seldom accurately known) with a burned-gas temperature of usually 2000 K or higher; this is the region of intense luminosity, arising from unoxidized soot particles. The soot is formed in local regions much richer than the average, and incidentally confirms that the stirring is less than perfect, since premixed hydrocarbon flame products at the same equivalence ratio would be practically nonluminous. Making the primary zone somewhat rich in effect accounts for imperfection in the mixing process so that the actual gas-phase reactions can take place near $\phi = 1$ for maximum temperature, space rate, and stability of the primary zone. It is followed by similarly rapid dilution and cooling to the desired firing temperature, typically 1200–1600 K, depending on the sophistication of the cooling and the cycle. Some of the dilution air added just down stream of the primary zone should complete the combustion of any unburned products (including soot, though often some survives as smoke). It may be called secondary air although most of the air subsequently added simply cools the gas. In any case, the low equivalence ratio and temperature of the mixture at the turbine-inlet condition makes necessary the indicated division of the combustor flow into primary and dilution regions. The division is determined only by the internal gas dynamics of the combustor, and the energy available for primary zone mixing is severely limited; nevertheless, the approximation of the primary zone as a well-stirred reactor has often proved useful for description or modeling of the time–temperature history of the gas, eg, as it influences emissions.

Flame Chemistry and Emissions

From combustion with air of fuels containing only C, H, and O (clean fuels) the usual air pollutants or emissions of interest are carbon monoxide, unburned hydrocarbons, and oxides of nitrogen (NO_x); the interaction of the last two in the atmosphere produces photochemical smog (see Air pollution). NO_x, the sum of NO and NO_2, is formed almost entirely as NO in the products of flames; typically 5 or 10% of it is subsequently converted to NO_2 at low temperatures. Occasionally conditions in a combustion system may lead to a much larger fraction of NO_2 and undesirable visibility of, eg, a very large exhaust plume.

NO is formed to some extent from N_2 and O_2 in flame products when N atoms are somehow produced at a significant rate. Above ca 1700 K, the important step in the much-studied Zeldovitch (thermal or hot-air) mechanism is the production of N atoms by (41):

$$O + N_2 \rightarrow NO + N \qquad (18)$$

This is followed by a very fast reaction.

$$N + O_2 \rightarrow NO + O \qquad (19)$$

When $[NO] \ll [NO]_{eq}$, as is usually true in practice, its formation is essentially irreversible, and its rate is proportional to $[O][N_2]$ with a large temperature dependence (activation energy of ca 316 kJ/mol or 75 kcal/mol). Unfortunately, the rate becomes appreciable just in the range of typical hydrocarbon–air flame conditions. If it is also assumed that $[O] = [O]_{eq}$, the observed rate in most lean-flame products in which N_2 is roughly 75 mol % of the gas can be approximated by (52–53):

$$d[NO]/dt = (3.3 \times 10^{18}/T) \exp(-68{,}700/T)(x_{O_2})^{1/2} \qquad (20)$$

Here x_{O_2} is the mol fraction of O_2 in the products at T, and the rate is given in ppm/ms. The exponential implies an enormous effective activation energy of 570 kJ/mol (136 kcal/mol), the sum of that for the $O-N_2$ reaction and half the dissociation energy of O_2. The line in Figure 6 is a plot of equation 20 for $x_{O_2} = 0.02$ (or ϕ = ca 0.9). Thus in typical hydrocarbon–air flames, the rate is about 8 ppm/ms, or in a 10 ms residence time the thermal NO would be about 80 ppm. If preheating of the mixture were to raise T by 100 K, eg, by precompression in an engine cycle, the rate would be nearly tripled, making [NO] probably unacceptable. Conversely, the rate can be reduced by the same factor with a 100 K reduction of the temperature by precooling or some equivalent heat abstraction from the flame itself (26,53), or by dilution of the mixture with excess air, steam, or other inert gas such as recirculated, relatively cool exhaust gas as is often done in piston engines (3).

Control of thermal NO_x thus involves reduction of the maximum attainable temperature, or the residence time at high temperature, or both. In nonaircraft gas turbines, eg, steam or water injection to the primary combustion zone is commonly done to minimize NO_x (see Power generation). In aircraft gas turbines, the analogous approach necessarily involves the use of air for the cooling, thus a lean primary zone. Such measures, however, always entail some compromise in stability and control, and

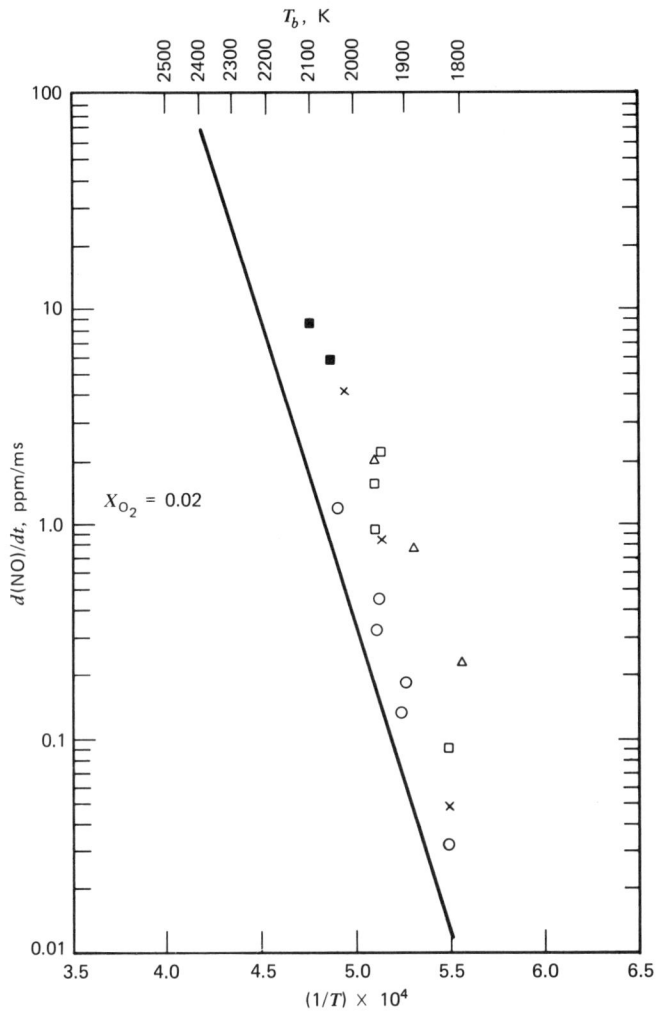

Figure 6. Arrhenius plot of approximate rate of formation of nitric oxide in lean ($\phi \simeq 0.9$) products of combustion with air. Calculated line corresponds to thermal or Zeldovitch mechanism, assuming equilibrium O atom concentration at T (eq. 20). Data shown are from measurements on nonadiabatic flame products ($T_b < T_a$) of some typical hydrocarbons (53); 101 kPa (1 atm) ○, propane; □, heptane; △, toluene, ×, gasoline A; ■, gasoline B.

possibly also in the efficiency of the combustion process. The afterburning of CO tends to be quenched by rapid temperature reduction (54), and the resulting increase in the emission of CO must be balanced against the desired NO_x reduction.

Heat abstraction or cooling of the flame must always occur to some extent by radiation from the highly luminous flames of pulverized coal or oil, typical of boilers and similar furnaces. When the rejected heat is taken away, eg, by a boiler fluid, and is not returned or recuperated to the unburned mixture, the maximum temperature and thermal NO_x formation will be reduced by the heat transfer. An extension of this effect has been applied to achieve low NO_x emissions in some furnaces and boilers in which combustion occurs in a very rich, relatively low temperature primary stage, followed by heat abstraction by convection as well as radiation to reduce the gas enthalpy (55). Secondary and any excess air is then introduced to complete the combustion and, owing to the previous heat transfer, the maximum temperature attainable in that stage will never approach the adiabatic flame temperature. Much soot, which is responsible for the radiative heat loss, may be present in the rich primary flame products. To avoid smoke from such two-stage processes, care must be taken to assure its oxidation in the second stage.

$[O]/[O]_{eq}$ is seldom unity as assumed. Though [O] is decreasing in the burned gas, its average value may be several times $[O]_{eq}$ and NO formation may be correspondingly higher as illustrated by the data shown in Figure 6. Similarly, the very high radical concentration (eg, [O]) in the reaction zone of a flame often leads to some practically instantaneous NO production even though the temperature is still relatively low and the time is short (52–53). Other fast reactions involving transient flame species producing N atoms (eg, $CH + N_2 \rightarrow HCN + N$, notably in rich flames) can also contribute some NO. In any case, what is inevitably formed almost simultaneously with the very fast flame reactions may be called prompt NO (56). The total is not large, 10–50 ppm depending on the composition of the flame, but is significant if very low NO_x emission is sought.

A different and often more serious source of NO_x is chemically bound nitrogen or fuel-N in any form, NH_3, amines, nitrites, pyridine, etc, because some of it is almost certain to appear as NO. Most coals contain at least 1% N, of which half or more may also appear in gaseous or liquid fuels derived from coal (see Fuels, synthetic). The N-content of distillate and gaseous petroleum fuels from most sources is usually very low, but it can be 0.5% and if, eg, such a fuel were burned with air at stoichiometric or lean conditions, the conversion of its fuel-N to NO in the flame would be essentially direct, very fast, and nearly quantitative, yielding ca 400 ppm of NO in the burned gas.

NO_x emissions from fuel-N and control measures as well as the mechanism of the conversion are known from correlations of data on the conversion in flames and combustors (57). Formation occurs in the flame reaction zone by OH radical oxidation of intermediate species formed by decomposition of the fuel-N (eg, NH_2, NH, N) which if unoxidized would in a short time simply form stable N_2. Whether NO or N_2 formation prevails depends on the flame conditions as well as on the concentrations of intermediates. At high levels of fuel-N, NO can also be converted directly to N_2. In general, the yield (mol of NO/mol of fuel-N) is much higher in flames of $\phi \leq 1$ than in rich flames, and asymptotically approaches unity at low fuel-N. It decreases with increasing fuel-N, and at some level that depends on ϕ, the NO concentration becomes constant (but in general rather high) and any further additions of fuel-N are converted to N_2. In sufficiently rich flames ($\phi > 1.5$) nearly all of the fuel-N tends to be converted rapidly to N_2, from which NO can then be formed only by the relatively slow thermal process with oxygen.

Thus if combustion can be effected in two stages, with or without the intermediate heat rejection for thermal NO_x control discussed above, the conversion of fuel-N to NO can be largely circumvented by: (1) a primary stage at ϕ ca 1.5–2 with a modest residence time to allow formation of N_2 in the hot primary products, followed by (2) rapid addition of secondary air to complete the combustion at an effective $\phi \simeq 0.8$. There will of course be a maximum in the temperature near $\phi = 1$ in the course of the secondary air addition, but if time at that condition is minimized, the production of thermal NO will also be minimized.

These ideas form the basis of most approaches to NO_x control with N-containing fuels. In principle they should be readily applicable to modification of a gas turbine combustor in which qualitatively the desired division of the combustion process already exists for other reasons. Although such improvements have been demonstrated, it is difficult in practice to make the required revisions of air and fuel distribution without adverse effects on other emissions or on performance. It has also been found that when steam is used to reduce thermal NO_x, the formation of NO_x from fuel-N is enhanced, or the reduction is less than otherwise expected.

Sulfur Dioxide. When present in the fuel as inorganic sulfides or organic compounds, sulfur is always converted practically quantitatively to SO_2 in the products of complete combustion. There are no known techniques for prevention of this conversion process in flames, and emission control measures necessarily involve either desulfurization of the fuel or removal of the SO_2 in the exhaust or stack gases. A little of the SO_2 (usually 5 or 10% of it) is oxidized to SO_3 at lower temperatures in most combustion processes, and the total sulfur oxides emission is often given as the sum of SO_2 and SO_3, or SO_x. In the inevitable presence of H_2O in the combustion products, at low temperatures, any SO_3 forms H_2SO_4 which can be both seriously corrosive to heat exchange surfaces and a highly undesirable stack emission.

Smoke and Unburned Hydrocarbons (UHC). Emitted smoke from clean (ash-free) fuels consists of unoxidized and aggregated particles of soot (sometimes referred to as carbon, though it is actually a hydrocarbon). Typically, the particles are of submicrometer size, and are initially formed by pyrolysis or partial oxidation of hydrocarbons in very rich but hot regions of hydrocarbon flames; conditions that cause smoke will usually also tend to produce unburned hydrocarbons and their potential contribution to smog formation. Both may be objectionable, though for different reasons, at concentration levels equivalent to only 0.01–0.1% of the initial fuel; although their effect on combustion efficiency would be practically negligible, it may nevertheless be important to reduce such emission levels, with measures based on the known chemistry of very rich soot-forming flames (58).

Neither soot nor unburned hydrocarbons are ever found in the products of a lean or stoichiometric premixed flame with air or O_2, although hydrocarbons may be formed or survive unburned in lean flames partially quenched, eg, by the cold wall of a combustion chamber (59). A moderately rich flame also should yield only the water–gas equilibrium products (and N_2); but with increasing equivalence ratio, at some ϕ still well below the upper flammability limit, the burned gas becomes faintly luminous with precipitated soot particles which increase in number density and luminous intensity with further increase in ϕ (60). The appearance of the condensed phase (soot) is connected with appreciable nonequilibrium concentrations of fairly stable low molecular weight hydrocarbons, notably acetylene and some related homologues, from decomposition of unoxidized fuel; these polymerize (condense with elimination of H_2) to form high molecular weight products, mostly with ring structures (61). If these intermediates are not consumed, eg, by OH, in simultaneous and competing oxidation reactions, they grow until nucleation and eventually precipitation occurs as visibly radiating soot, typically over a period of several ms after the main flame reaction zone. The particle growth and competing oxidation then continue in the burned gas.

The composition, properties, and size of soot particles collected from flame products vary considerably with flame conditions and growth time. Typically the C–H atom ratio ranges from 2 to 5 and a particle consists of irregular chains or clusters of tiny spheres 10–40 nm in diameter with overall dimensions of perhaps 200 nm (61) although some may agglomerate further to much larger sizes.

Whether soot particles form at all and their growth rate thus depend on ϕ, the fuel types and other variables, eg, the growth, is easier and faster at a given ϕ with fuels of high C–H ratio and at elevated pressure (62). Given sufficient residence time to attain equilibrium at the burned gas condition, soot (and the hydrocarbons) would eventually be consumed. In practice, their rapid oxidation occurs in a secondary flame in which the very hot primary products are burned with the required excess air, added by diffusion or by more intensive mixing. However, if a large excess is added too rapidly, the cooling can in effect quench the oxidation of both unburned hydrocarbons and the accompanying soot which would then persist as visible smoke. The blackness of the smoke depends on the size and number density of particles when quenched, their further aggregation, etc. Some may also survive as much smaller invisible particles, or condensation nuclei.

In a pure diffusion flame, the flame reactions can be thought of as occurring mainly in that region where the local composition is roughly stoichiometric and the temperature is the highest (near T_a for $\phi = 1$). The fuel diffusing from its side of that region is also heated and pyrolyzed to high concentrations of soot-forming precursors similar to those of the premixed flame. But the precursors are produced in the presence of little or no oxidizer, and soot tends to precipitate before they can diffuse to the region where they can be oxidized. Thus a diffusion flame of a hydrocarbon in air tends to form soot more copiously and to be much more luminous than is a rich (but flammable) premixture of the same fuel. Although the structure of the diffusion flame is more complex and difficult to analyze, the same basic description of soot formation and oxidation should apply and has proved to be a useful guide in understanding nonpremixed systems as well.

Based on these concepts, combustor modifications for smoke elimination in gas turbines have been devised and successfully applied, using rough though reasonable models of the primary combustion process and flow through the combustor. It may be expected, however, that compromises are inevitable if NO_x, UHC, CO, and smoke are to be simultaneously minimized; experience with emissions control in various types of furnaces, engines, etc, has generally confirmed that those measures or techniques effective for NO_x (control of the maximum temperature and residence time at high temperature) tend to make control of one or more of the other three more difficult (3,63). The magnitude of the problem varies with other constraints of the system such as size limitations, variability of operating conditions, etc. For example, in gas turbines, a combustor redesign that reduces NO_x to acceptable levels at all engine operating conditions usually results in excessive CO at low engine power (64).

A more intractable source of particulate matter or smoke is the inorganic or ash content of some fuels; it usually is converted quantitatively to oxides in the combustion process that persist in the burned gas, or condenses from vapor on subsequent cooling of the combustion products as very finely divided solids. Like sulfur oxides, smoke of this kind is not susceptible to control through combustion chemistry, and usually can be reduced only by control of the original fuel composition or by its removal from the exhaust or stack gas (65). The ash content of a petroleum residual oil is typically ca 0.1% (1), but coals range in ash content from a few % to 25% or more. While a relatively small fraction of it is carried into the stack gas when coal is burned on grates, most of it tends to appear as fly-ash from pulverized-coal furnaces, requiring elaborate stack-gas cleaning procedures and equipment (65) (see Air pollution control methods). In some types of furnace, much of the combustion-product ash is withdrawn as molten slag, thus considerably reducing the emitted ash (66).

Stabilization of Premixed Flames

The Bunsen Flame. In any useful burner or torch, some mechanism must exist, or some device (a flameholder or pilot) must be provided to stabilize the flame against a variable flow of unburned mixture and to fix the position of the flame at the burner port. The volume flow rate of mixture $\dot{V}_{mix}$ into the steady flame must in general be such that the average mixture velocity is many times S_u^o. Although burners vary greatly in form and complexity, in many the stabilization mechanism is fundamentally the same; in some small fraction of the mixture flow, its linear velocity is maintained $\leq S_u^o$ to form a steady pilot flame from which flame spreads to consume the main flow at much higher velocity. Although the shape of the resulting flame may be complex, the area of the steady flame, A_f, will be such that $A_f \cdot S_u^o = \dot{V}_{mix}$, to which the heat input rate $\dot{Q}$ is proportional ($\dot{Q} = q_p \cdot \dot{V}_{mix}$).

The basic idea may be illustrated by the simple Bunsen flame on a tube of circular cross-section in which the stabilization depends on the inevitable velocity variation in the flow emerging from the tube. If the flow is laminar (parabolic velocity profile) in a tube of radius R, the velocity at radius r is: $v_r = \text{const}\,(R^2 - r^2)$; the maximum velocity at the axis is twice the average, whereas at the wall, $v_r = 0$ (Fig. 5a).

Although V_{av} may be several times S_u^o, there is a thin annulus of the flow (of thickness ca $d_q/2$) at the wall where $0 < v_r < S_u^o$. In general, flame cannot propagate upstream in that region because it would be quenched (67), or alternatively, the cooling effect of the wall locally decreases the burning velocity of the mixture. The cooling effect of the wall is essential to the stabilization; but it also follows that the wall is heated and that the heat must somehow be dissipated. Ordinarily the heat rejection occurs adventitiously by conduction, convection, or radiation to the surroundings with no great rise of the wall temperature. But if that heat loss is seriously impeded, the presumed quenching or local reduction of burning velocity may not be obtained, and upstream propagation of flame may result.

In any case, there should be some radius, depending on $\dot{V}_{mix}$, at which $v_r = S_u^o$ so that the flame can be stationary or stable against the flow in a very thin annular region near the rim. This annulus then serves as a pilot, from which the main flow of mixture is progressively ignited, and flame spreads toward the center. Most of the mixture flow, in which $v_r > S_u^o$, is thus consumed in the familiar, roughly conical, and steady flame surface. Its height and area of course increase with mixture flow, and measurement of the area of a stable Bunsen flame is incidentally the basis for the method most commonly used to determine S_u^o.

If the mixture is fed through a converging nozzle instead of a straight tube, V_r is still zero at the wall, but the profile may be made nearly flat or uniform rather than parabolic. A Bunsen flame in such a flow has a smaller range of stability but the mechanism is essentially the same, and the flame very closely approximates a cone. If its apex angle is θ, the equality: $\dot{V}_{mix} = A_f S_u^o$ leads to:

$$S_u^o = v_r \sin(\theta/2) \qquad (21)$$

v_r is the (constant) mixture velocity at the nozzle exit or port, and has often been used to determine S_u^o (68). The ordinary Bunsen flame on a straight tube is not really a cone because v_r is not a constant and the corresponding measurement may be more tedious and uncertain.

If the tube diameter is appreciably larger than d_q, there will be some mixture flow below which S_u^o will exceed v_r at some value of r in the profile where the flame would be little affected by the cooling or quenching effect of the wall. The flame would then propagate down the tube as far as there is mixture to consume, and this undesirable condition is called flashback. There is also some mixture flow above which v_r exceeds S_u^o everywhere and the flame lifts from the port and blows off. These stability limits have been extensively studied for a wide variety of mixtures and conditions and shown to be well-correlated in terms of the critical velocity gradient at the tube wall (when it is known from the nature of the flow) (69). The gradient at the wall is: $g_w \equiv (dv/dr)_{r=R}$. For example, in laminar or Poiseuille flow in a round tube of radius R: $g_w = 4 g_{av}/R$, and for a given primary mixture composition, flashback (or blowoff) will occur at the same value of g_w in tubes of various sizes, whereas the corresponding average velocity at flashback (or blowoff) is proportional to R. Some measured values fairly typical of stoichiometric hydrocarbon–air flames are (69): for a $\phi = 1$ CH_4–air mixture g_w at flashback is about 400 s^{-1}, and at blowoff it is ca 2000 s^{-1}. Thus if it is burned on a 1-cm dia tube, the average velocity at flashback is $400 \times 0.5/4 = 50$ cm/s and at blowoff: $v_{av} = 250$ cm/s, or the range of stability would be roughly one and a half to seven times S_u^o.

The variations of measured values of g_w at flashback with ϕ and with the oxygen content of the atmosphere are similar to but somewhat faster than the corresponding variation of S_u^0. For example, g_w at flashback is about one hundred times larger for a stoichiometric fuel–O_2 mixture than for a fuel–air mixture of the same ϕ, and the ratio of S_u^o values is usually about 10.

g_w at flashback is a maximum around $\phi = 1$, as is also g_w at blowoff if the burner is operated in a surrounding inert atmosphere. However, as normally used with air surrounding the burner, the behavior of rich mixtures is complicated by the entrainment of air at the burner port which sustains combustion of the hot rich products of the primary flame near the port. g_w, or the blowoff velocity, is then found to increase continuously with ϕ, or richer mixtures are more stable with respect to blowoff. Together with the lesser tendency toward flashback in rich mixtures (relative to stoichiometric), it follows that a Bunsen burner has much more latitude for stable operation if the primary mixture is rich. For this reason many appliance burners that involve assemblies of such flames are routinely adjusted by first making the primary mixture so rich that soot just forms in the burned gas (yellow-tipping), and then increasing the air until the yellow luminosity disappears (70). The primary equivalence ratio is then perhaps 1.5 or more; the rich products of that primary flame are burned in the secondary diffusion flame in the surrounding air, or the faintly luminous, so-called outer mantle of a Bunsen flame.

Atmospheric or Induced-Air-Flow Burners. A great variety of gas–air premixed flame burners are basically laminar-flow Bunsen burners. Most of these are atmospheric, ie, the primary air is induced from the atmosphere by the fuel flow with which it mixes in the burner passage or pipe leading to the burner port or ports where the mixture is ignited and the flame stabilized. The induced air flow is determined by the fuel flow through momentum exchange (71–72) and by the position of a shutter or throttle at the air inlet; the air flow is thus a function of the fuel velocity as it issues from its orifice or nozzle, or of the fuel supply pressure at the orifice. With a fixed fuel flow, the equivalence ratio is adjusted by the shutter, and the resulting induced air flow also determines the total mixture flow, since the desired air–fuel volume ratio is usually 7 or more, depending on the stoichiometry.

Among burners of this general type are many multiple-port burners for domestic furnaces, heaters, and stoves, as well as for industrial use (73). In many the flame stabilizing ports are not round, but may be slots of various shapes to conform to the heating task. The familiar laboratory Meker burner is also a multiple-port burner in which the ports are the square openings in a grid; a large number of relatively short Bunsen flames for better distribution of the hot flame gas in certain heating operations are thus produced.

Atmospheric industrial burners are made for a heat release capacity of up to 50 kJ/s. Although they may differ considerably in design details and geometry and often bear little resemblance to Bunsen burners, the stabilization principle is basically the same. In some, the mixture is fed through a fairly thick-walled pipe or casting of appropriate shape for the application and the desired distribution of flame. The mixture issues from many small and closely spaced drilled holes, typically 1- or 2-mm dia, and burns as a row or many rows of small Bunsen flames. It may be ignited with a small pilot flame, spark, or heated wire, usually located near the first holes to avoid accumulation of unburned mixture before ignition. The rated total heat release for a given fuel–air mixture can be scaled with the size and number of the holes: eg, for 2-mm dia holes it would be 10–100 J/(s·hole) or in general ca 0.3–3 kJ/(s·cm^2) of port area, depending on the fuel. The ports may also be narrow slots, sometimes packed with corrugated metal strips to improve the flow distribution and lessen the tendency to flashback.

Burners with Independent Air Supply. The mixture may also be prepared from separately metered or controlled supplies of fuel and air if they are available at somewhat elevated pressures. Torches and burners that require such supplies are usually intended for much higher mixture velocity or heating intensity, as is often the case in industrial applications, and the stabilization against blowoff must therefore be enhanced. In one type of single-port burner (Fig. 5d) this is done by surrounding the main port with a number of pilot ports. The main port is typically a nozzle so that the upstream pressure is high enough to feed a small fraction of the mixture flow to the surrounding pilot chamber through metering holes, and it burns at low velocity in Bunsen flames at the pilot ports, or flame retention pilots, which in turn continuously ignite or pilot the high velocity main flow. In this way, most of the mixture can be burned at a port velocity as high as 100 S_u^o to produce a long pencil-like flame, suitable for operations requiring a high local heat flux. The retention pilots in some torches of this kind consist of metal screens rather than an array of holes (72). The pilot mixture flow is rendered more or less uniform by the flow resistance of the screen, from which it emerges at a superficial velocity $<S_u^o$. When ignited, the pilot flow burns as a sheet of flame distributed over the screen, instead of as discrete small flames, but otherwise the piloting of the main flame is the same (Fig. 5e). There are many variations in the detailed design of the retention pilots as well as in the form of the burner structures but most are elaborations of the basic forms shown in Figure 5d and e.

Stabilization of such a pilot flame or any similar flat flame on a porous diaphragm, screen, etc, involves reduction of the mixture burning velocity or S_u^o to the imposed superficial velocity, with consequent reduction of the burned gas temperature (26) and therefore uniform heat rejection to the surface (Fig. 5f). The screen must in turn dispose of this steady heat flux to attain a stable situation for reasons similar to those discussed with regard to the process at the rim of a Bunsen burner, but applied to the entire surface of the screen and to all of the gas flow through it. The required heat dissipation depends on the superficial velocity, reaching a maximum at about 40% of S_u^o; and it may be shown that a $\phi = 1$ CH_4–air flame would thus reject a maximum of about 10 W/cm^2 to the surface, equivalent to about 10% of the total enthalpy flux of the flame at that velocity (53,74). If radiation through the transparent burned gas were the only mode of heat loss, the surface temperature would rise to about 1100 K. However, other losses make the temperature attained more reasonable, though such screens do sometimes become quite hot (and marginally effective in their incidental role of flame arrester). On the other hand, a $\phi = 1$ H_2–air flame would reject a maximum of about 60 W/cm^2 to such a screen, which could not be so dissipated at any reasonable surface temperature. Flashback through the screen is virtually certain with mixtures of such high burning velocity.

The corresponding heat dissipation required in most Bunsen flame devices is relatively small, since only a small fraction of the total flow is usually involved, but assemblies of such flames can approximate a flat flame as the ports are made smaller and more closely spaced, as in the Meker burner, and the metal structure of the burner in the limit may have to dissipate a comparable heat flux. Stabilized nonadiabatic flames of this general type have been applied in various burners for special applications. Some are radiant refractory diaphragms, either uniformly porous or with a great many small holes (75), in which most or practically all of the stabilizing heat rejection occurs by radiation (76), and some are independently cooled with another fluid, eg, water (77).

High local heat flux can also be obtained with Bunsen flames using mixtures of high burning velocity, as in H_2/O_2 and C_2H_2/O_2 torches; the stabilization mechanism is essentially the same as it is with slower burning mixtures but the port or nozzle of the torch is usually much smaller, in part to avoid turbulence in the mixture flow as mentioned below. The consequences of flashback are also much more severe since most such mixtures are detonable, and the premixing chamber or tube must accordingly be more rugged. For this reason, many large hydrogen–oxygen and hydrocarbon–oxygen flames used (eg, in quartz working flames and similar very high temperature processing) are not premixed flames. They actually consist of assemblies of closely spaced diffusion flames, produced from separately fed but contiguous fuel and oxidizer flows. In such surface-mixing burners, the surface is an array of very small and closely packed alternating fuel and oxidizer ports. The arrangement and number of the ports and the complexity of the required manifolding of the reactant passages vary with the application and the desired geometry of the burned gas flow. With a very fast-burning fuel–oxidizer combination, the individual diffusion flames may be so short that the assembly approximates a large flat premixed flame, as is also the case with some rocket engine injectors that are similarly constructed for basically the same reasons.

Interchangeability and Substitution of Gaseous Fuels. It is often desired to substitute directly some more readily available fuel for the gas for which a premixed burner or torch and its associated feed system were designed. Satisfactory behavior with respect to flashback, blowoff, and heating capability, or the local enthalpy flux to the work, generally requires reproduction as nearly as possible of the maximum temperature and velocity of the burned gas, and of the shape or height of the flame cone. Often this must be done precisely, and with no changes in orifices or adjustments in the feed system.

If the substitute fuel is of the same general type, eg, propane for methane, the problem reduces to control of the primary equivalence ratio. For nonaspirating burners, ie, those in which the air and fuel supplies are essentially independent, it is further reduced to control of the fuel flow, since the air flow is usually most of the mass flow and is fixed. With a given fuel supply pressure and fixed flow resistance of the feed system, the volume flow rate of fuel is inversely proportional to $\sqrt{\rho_f}$. The same total heat input rate or enthalpy flow of the flame simply requires satisfactory reproduction of the product of the lower heating value of the fuel and its flow rate, or that: $WI \equiv Q_p/\sqrt{\rho_f}$ should be the same on substitution.

WI is the Wobbe Index of the fuel gas, and is a commonly used criterion for interchangeability in adjusting the composition of a substitute fuel. The units of WI are variously given but, if used consistently, are unimportant since only ratios of Wobbe Indexes are ordinarily of interest. Sometimes ρ_f is taken as the specific gravity relative to some reference gas, eg, air, or average molecular weights may be used.

The Wobbe Index criterion also applies to substitution with aspirating or atmospheric burners in which the volume flow of primary air induced by momentum exchange with the fuel increases with $\sqrt{\rho_f}$. Because the volumetric air requirement for a given ϕ is nearly proportional to the heating value of the fuel (Table 1), an adjustment of Q_p and ρ_f to the same WI results in about the same stoichiometry of the primary flame. For example, if propane is an available substitute for methane or natural gas, it is common practice to prepare a mixture approximately 60% propane–40% air (which of course is well above the upper flammability limit) to use as the fuel supply; though the heating value of the mixture is 1.53 times that of CH_4, its density or molecular weight is 2.36 times as large, and $1.53/\sqrt{2.36} = 1.0$ or its Wobbe Index is the same. Though there would be slight differences in the stoichiometry of the flame (arising from the air mixed with the propane) and in S_u^o of the final mixture, substitution with the same supply conditions would be quite satisfactory. On the other hand, if a mixture of the same heating value as that of methane were used (39.2% propane in air) at the same supply pressure, the flame would be much leaner and generally unsatisfactory.

There are direct substitutions of possible interest that would not be feasible without drastic changes in the feed system or pressure. Thus if the available substitute for natural gas is, eg, a manufactured gas containing much CO, there would almost always be a bad mismatch of WI unless the fuel could be further modified by mixing with some other gaseous fuel of high volumetric heating value (propane, butane, or possibly vaporized fuel oil, etc). Moreover, if there are substantial differences in S_u^o, eg, due to the presence of considerable H_2 as well as CO in the substitute gas, the variation in flame height and flashback tendency can also make the substitution unsatisfactory for some purposes, even if WI is reproduced. Refinements and additional criteria are occasionally applied to measure these and other effects in more complex substitution problems (71,78).

Turbulence in the flow of a premixture flattens the velocity profile and increases the effective burning velocity of the mixture; eg, at a pipe-Reynolds number of 40,000 the turbulent burning velocity will be several times the laminar burning velocity and perhaps fifty times larger at very high Reynolds numbers (79). A turbulent flame is always somewhat noisy, the apparent flame surface becomes diffuse (Fig. 5c) owing to the fluctuations of the actual or flame surface about its average position, and its stability tends to be less predictable. The instantaneous flame surface may be thought of as wrinkled by velocity variations in turbulent flow, or on the average distributed over a greater thickness (or time). Although the resulting enhancement of the mixture consumption rate may be considerable, turbulence is often considered undesirable in Bunsen-type flames. For this among other reasons, a large number of burner ports of small characteristic dimension, rather than a single large port, are frequently used to assure laminar flow to the individual flames. However, turbulence has an essential role in facilitating the mixing of fuel, oxidizer and flame products, and in the function of various types of flame-stabilizers of practical importance.

Other Types of Flame Stabilizers. Flames in high velocity, highly turbulent streams of premixtures in large ducts, etc, are often stabilized with obstructions or flameholders of various kinds, such as bars, grids, rods, V-gutters, abrupt changes in duct cross section, etc (80). To some extent, local reduction in velocity and establishment of a piloting region analogous to that of the Bunsen flame may be involved, as can be observed with such obstructions in low velocity flows as well (81) (Fig. 5**b**).

Most often, however, such devices involve extensive recirculation of hot burned gas to the region at or just behind the flameholder, and intensive local mixing of hot combustion products with unburned mixture. Although such recirculation of mostly inert gas tends to dilute or vitiate the mixture and thus to reduce the combustion rate, the effect is more than offset by the temperature increase which produces an exponential increase of chemical reaction rate and makes possible a locally stable zone. The region just behind the flameholder may then approximate a steady well-stirred reactor at quite high temperature, from which flame spreads to the rest of the flow (Fig. 5**g** and **h**).

Combustion of Nonflammable Mixtures (Catalytic Burners)

The effluent gas from process equipment, engines, etc, often contains environmentally objectionable organic vapor too dilute to be either economically recoverable or flammable *per se*. In principle it can be consumed in a homogeneous well-stirred reactor of sufficient volume or residence time (49), and this is in fact one approach sometimes applied in engine exhaust treatment. However, the most straightforward method is incineration in a conventional flame in the stack or exhaust of the process if fuel can be added to render the mixture flammable and if it contains sufficient oxygen (if not, air may also be added).

However, it is sometimes possible and more economical to consume the mixture directly by passing it through a bed of catalyst, especially if the effluent gas is at a somewhat elevated temperature (ca 200–300°C) and if the gas is free of potential catalyst poisons.

The catalyst structures most commonly used consist of platinum or palladium deposited on some porous support material of low flow resistance, such as screens, a ceramic honeycomb structure, etc, which are commercially available from vendors of noble metals. The effective space rates are considerably lower than those of ordinary flames at typical flame temperatures, but the rates are much higher than would be estimated, eg, by extrapolation of stirred-reactor data to the low operating temperature typical of such weak mixtures and of interest in catalytic reactors.

In general, the maximum temperature will be and should be low (ca 1000°C or less) to avoid damage to the catalyst or to the support. Moreover, if the mixture should become flammable at the entry conditions (ie, with $T_a > 1300°C$) the catalyst will act as an ignitor and flashback may occur. The required bed depth is determined by the mixture velocity which is usually kept low to minimize pressure drop, and by the contact time for complete oxidation. The latter is typically a few seconds, although it varies considerably with fuel and inlet conditions, and it has a temperature dependence suggesting an activation energy of 100–150 kJ/mol (comparable with that of homogeneous combustion processes) (see Exhaust control).

An interesting special case is the catalytic recombination of the radiolytic off-gas (mainly a stoichiometric H_2–O_2 mixture) that separates from the steam in the turbine condenser of a boiling-water nuclear reactor. The flow of detonable mixture can be directly and safely burned in a steady flame on burners of special design (77). But in current practice, the mixture composition is adjusted with low pressure steam so that it is just nonflammable (ca 4% H_2) and passed at low velocity through a large bed of supported noble-metal catalyst. The diluted off-gas is completely burned or recombined with a temperature rise of about 300°C.

Combustion of Particle Suspensions of Liquids and Solids

In the design of efficient, high intensity burners for liquid and solid fuels, a basic objective is to minimize limitations on the overall combustion rate imposed by surface-limited processes, eg, evaporation, as well as by other transport processes such as mixing of fuel with oxidizer and hot products. With less volatile distillate liquid fuels, such as kerosene or no. 2 fuel oil, it is sometimes possible at least in principle to circumvent part of the problem by preheating to complete evaporation of the fuel. Its combustion would then simulate that of a gaseous fuel.

The vapor may be produced directly in some type of heat exchanger, eg, immersed in the hot combustion products, or it may be heated and vaporized in a stream of hot but unreactive carrier gas, eg, steam if it is available, and burned as a turbulent diffusion flame. It may also be vaporized in a flow of preheated air, usually to prepare a very rich premixture. However, a much larger fraction of the air flow may be used to make a $\phi = 1$ mixture, resembling carburetion of a more volatile fuel as in a gasoline engine. In some proposed applications, eg, those of catalytic combustion, the entire air flow would be hot and used to prepare a very lean premixture. If the air temperature is much over 200°C, autoignition of the hydrocarbon vapor in such hot air-fed vaporizers is always an undesirable possibility, depending on the boiling range of the fuel, its AIT and the residence time of the vapor–air mixture (Table 1). Moreover, in most vaporizers, carbonization of the heated fuel can also be a problem. Nevertheless, some degree of prevaporization of less volatile distillate fuels (eg, no. 2 fuel oil, kerosene, or JP-4) is often incorporated in the design and does occur in many burners and combustors, even though some or most of the fuel may burn in suspension.

At ordinary temperatures, in the limit of very finely dispersed fuel particles with dimensions ca 1 μm or less, a mist of liquid (82) or a suspension of solid dust of appropriate overall composition can exhibit many characteristics of a premixed flammable gas mixture, as illustrated in Table 1. However, in most fuel suspensions produced by practical methods of particle size reduction, at least some of the fuel must burn or evaporate as discrete particles or droplets, each with a surrounding diffusion flame fed with fuel diffusing from the particle surface and with oxygen from the gas (Fig. 5j).

The mass or volume rate of consumption of a particle in such a diffusion flame is proportional to the first power of the particle diameter; the observed consumption of a particle of initial diameter δ_o is often expressed in an equivalent form:

$$\delta^2 = \delta_o{}^2 - \lambda t \qquad (22)$$

δ is the diameter at time t and λ is its evaporation constant and the time for complete consumption of the droplet is $\delta_o{}^2/\lambda$ (83). Hence if λ = ca mm²/s (typical of most common liquids), the combustion or evaporation time of a 100 μm droplet is about 10 ms, and a 10 μm droplet would evaporate completely in 1 ms. The latter is comparable to a typical chemical reaction time in a flame.

Theoretically, the evaporation constant depends on the thermal and transport properties of the gases involved (83), and on the temperature difference between the flame and the droplet surface, though it varies only slowly with the latter. In fact, λ varies little among most liquid fuels. Particles of solids, eg, most coals, which evolve volatile fuel vapors or which pyrolyze to gaseous fuels when strongly heated, often appear to burn in a similar fashion with observed λ of the same order of magnitude (84), even though the details of the actual mechanism may differ significantly from that of liquid droplet combustion (85). In large turbulent flames of atomized oil, pulverized coal, etc, of practical interest, the process is much more complex. Among other things, interactions among neighboring particles must modify considerably this simple picture (86). Extensive studies have been made of these and of many other aspects, theoretical and experimental, of the combustion of fuel sprays, suspensions of solid, and other complex processes involving turbulent diffusion flames; some examples of this work can be found in refs. 87–95.

The most important features of practical burner and combustion design are the mechanisms provided for particle size reduction or maximizing the surface:mass ratio of the fuel, and for its suspension, rapid evaporation (of liquids), and mixing of the suspension with additional air or with hot burned gas. In effect, they limit the degree of approximation to the ideal of the well-stirred reactor which in practice is seldom approached.

In the process of combustion of such heterogeneous and nonuniform mixtures, these steps always occur simultaneously and are not easily separable or amenable to analysis and prediction. The development of burners and combustors still tends in part to be empirical, as is reflected in the wide variety of detailed mechanical designs and techniques, even when applied to the same general purpose. Descriptions of some of these as applied to solid fuels (coals) can be found in ref. 5, and for liquid fuels as well as coal in ref. 96. Here it is possible only to emphasize a few of the common and essential elements of techniques commonly used in the combustion of condensed phase fuels, with particular reference to liquids, although some of it applies to the combustion of pulverized coal as well (see Furnaces, fuel fired).

For distillate fuels of moderate viscosity (ca 30 mm²/s or 30 cSt) at ordinary temperatures, simple pressure atomization with some type of spray nozzle is most commonly used. Operating typically with fuel pressure (or pressure drop) of ca 700–1000 kPa (7–10 atm) such a nozzle produces a distribution of droplet diameters from ca 10–150 μm. They range in design capacity of ca 0.5–10 or more cm³/s and the pumping power dissipated is generally less than 1% of the corresponding heat release rate. A typical domestic oil burner nozzle uses about 0.8 cm³/s of no. 2 fuel oil at the design pressure. Although pressure-atomizing nozzles are usually equipped with filters, the very small internal passages and orifices of the smallest tend to be easily plugged, even with clean fuels. With decreasing fuel pressure the atomization becomes progressively less satisfactory. Much higher pressures often are used, especially in engine applications, to produce higher velocity of liquid relative to the surrounding air and accordingly smaller droplets and evaporation times. Other mechanical atomization techniques for production of more nearly monodisperse sprays or smaller average droplet size (spinning disk, ultrasonic atomizers, etc) are sometimes useful in burners for special purposes and may eventually have more general application, especially for small flows (97) (see Ultrasonics).

Conventional spray nozzles are relatively ineffective for atomization of fuels of high viscosity, such as no. 6 or residual oil (Bunker C) and other viscous dirty fuels. In order to transfer and pump no. 6 oil, it must usually be heated to about 100°C, at which its viscosity is still typically ca 40 mm^2/s (40 cSt). Relatively large nozzle passages and orifices are necessary for the possible suspended solids. Atomization of such fuels is often accomplished or at least assisted by atomizing air, pumped at high velocity through adjacent passages in or around the liquid injection ports. Much of the relative velocity required to shear the liquid and form droplets is thus provided by the atomizing air; its mass flow is usually comparable with the fuel flow and thus a small fraction of the stoichiometric combustion air (although it is sometimes called primary air). A typical high pressure, air-atomizing nozzle designed for injecting residual oil in a gas turbine combustor is illustrated in Figure 7 (98). The air for this purpose is usually supplied by an auxiliary compressor with a power absorption about 1% of the combustion heat release rate. Dry steam, if available, may also be used in a similar way, as is common practice in the furnaces of power plant boilers using residual oil.

Air atomization with low pressure, relatively low velocity air is also used in some burners for low viscosity distillate oils; and in most aircraft gas turbines some or even a large part of the atomization is done in this way by a small fraction of the warm, compressed combustion air supplied in swirling flow around the fuel nozzles. Imparting swirl to at least some of the air flow around fuel injectors of all types is a common

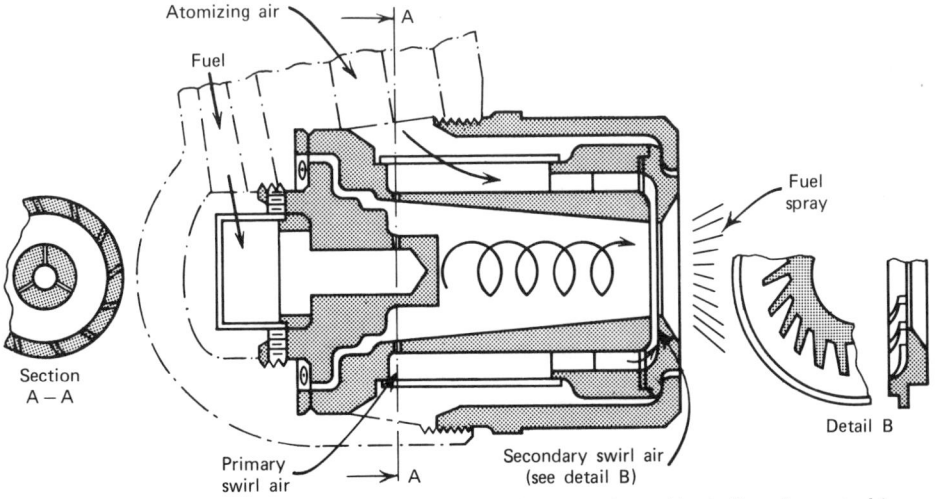

Figure 7. Typical high pressure air-atomizing fuel injector for residual oil use in gas turbine combustor. Courtesy of General Electric Company.

feature of many burners and combustors; in some, swirl is introduced on a larger scale in all of the primary combustion air. The velocity gradients or shear in the resulting vortex-like flow promote mass transfer or mixing, including the recirculation of hot products to the rich mixture or suspension in the low pressure core that contributes to stabilization of the primary combustion zone (99). The angular or swirl velocity imparted to the air or the strength of such flows is of course limited by the available pressure drop; eg, in gas turbine combustors the allowable pressure loss is usually <4% of the absolute pressure.

The cyclone combustor, as applied mainly to coal in the boilers of many large power plants (100), involves similar swirling flow of the reactants, usually with somewhat higher pressure loss (ca 7%). The combustion mechanism in such combustors, although surface-limited, differs physically from that of pulverized coal burners: most of the coal, crushed typically to ca 5 mm pieces, is stuck in a slowly moving layer of slag or molten ash at the cylindrical wall of the water-cooled, refractory-lined vessel. In the steady state, 85–90% of the coal ash is thus removed as slag. There most of the coal burns in contact with tangentially introduced high-velocity air, usually at ca 300 m/s and preheated to ca 400°C, some of which carries coal into the chamber. The gas flow in a combustor of this kind is quite complex (99) and the design is largely empirical; in any case it must be operated slagging, ie, with temperatures high enough to keep the ash above its fusion temperature (usually > 1700 K), and so the overall coal–air ratio generally is near stoichiometric (see Reactor technology).

The attainable combustion rate in the cyclone is determined mainly by the exposed fuel surface or roughly the wall area. The intensity or volumetric rate accordingly decreases with increasing combustor size for constant aspect ratio of the cylinder, as has been found to be desirable (length to diameter ratio of 1–1.5 seems to be optimal), so individual cyclone combustors in practice are generally limited in diameter to about 3 m. The rate, referred to the wall area, may reach ca 2×10^3 kJ/(s·m^2) [7×10^5 Btu/(h·ft^2)], which is about an order of magnitude higher than is found or estimated for coal particles of this size burning in the relatively quiescent conditions appropriate to Figure 5(j) and equation 22. Most of the difference is no doubt due to the high relative velocity of air and coal in this type of combustor, as was presumably an objective in its development and to some degree is sought in the design of any large, high intensity combustor or burner for condensed-phase fuels.

Nomenclature

A_f	= area of steady flame, cm^2
C_p	= molar heat capacity at constant pressure, J/(mol·K)
C_v	= molar heat capacity at constant volume, J/(mol·K)
d_q	= quenching distance, cm
E_{min}	= minimum ignition energy, J
g_w	= boundary or wall velocity gradient; s^{-1}
H_T	= gas enthalpy at temperature T, kJ/mol
ΔH_{298}	= enthalpy change of reaction at 25°C, kJ/mol$_f$
ΔH_{298}^f	= standard enthalpy or heat of formation at 25°C, kJ/mol$_f$
ΔH_{298}^c	= enthalpy change on complete combustion at 25°C, kJ/mol
m	= mass (weight)
$\overline{M}$	= average molecular weight
OI	= oxygen index
p	= pressure, kPa
q_p	= heat of combustion of mixture unit volume; J/cm^3

Q	= heat input rate; watts
Q_p	= $-\Delta H^c_{298}$, lower heat of combustion at constant pressure, kJ/mol$_f$
$R; r$	= radius, cm
r_p	= pressure ratio in adiabatic constant volume combustion
S_u^o	= normal burning velocity, cm/s
T_a	= adiabatic flame temperature at constant pressure, K
$T_a{}^v$	= adiabatic flame temperature at constant volume, K
T_{ig} = AIT	= autoignition temp, K
T_o	= initial temperature, K
v	= gas velocity, cm/s
V	= volume flow rate, cm^3/s
$\dot{V}_{mix}$	= volume flow rate of mixture, cm^3/s
WI	= Wobbe Index
x	= mol fraction
ρ	= density
τ	= time delay, s
ϕ	= equivalence ratio
ϕ_L	= lean, lower limit equivalence ratio
ϕ_U	= rich, upper limit equivalence ratio

Subscripts

a	= adiabatic
b	= burned gas
f	= fuel
ig	= autoignition
o	= oxidant; initial condition
w	= wall
av	= average

BIBLIOGRAPHY

"Fuels (Combustion Calculations)," *ECT* 1st ed., Vol. 6, pp. 913–935, by Henry R. Linden, Institute of Gas Technology; "Fuels (Combustion Calculations)," *ECT* 2nd ed., Vol. 10, pp. 191–220, by D. M. Himmelblau, The University of Texas.

1. *Steam, Its Generation and Use*; 38th ed., Babcock and Wilcox Co., New York, 1972.
2. M. W. Thring, *The Science of Flames and Furnaces,* John Wiley & Sons, Inc., New York, 1962.
3. D. J. Patterson and N. A. Henein, *Emissions from Combustion Engines and Their Control,* Ann Arbor Science Publishers, Ann Arbor, Mich., 1972.
4. J. Hodge, *Gas Turbine Cycles and Performance Estimation,* Butterworths, London, Eng., 1955; D. B. Spalding, *Some Combustion Fundamentals,* Academic Press, Inc., New York, 1955.
5. H. H. Lowry, ed., *Chemistry of Coal Utilization,* John Wiley & Sons, Inc., New York, 1963.
6. M. A. Cook, *The Science of High Explosives,* Reinhold, New York, 1958.
7. S. S. Penner and B. P. Mullins, *Explosions, Detonations, Flammability, and Ignition, AGARD Monograph,* Pergamon Press, Inc., New York, 1959.
8. R. C. Corey, *Principles and Practices of Incineration,* Wiley-Interscience, New York, 1969.
9. H. F. Coward and G. W. Jones, *Limits of Flammability of Gases and Vapors, Bulletin 503,* U.S. Bureau of Mines, Government Printing Office, Washington, D.C., 1952.
10. M. G. Zabetakis, *Flammability Characteristics of Combustible Gases and Vapors, Bulletin 627,* U.S. Bureau of Mines, Government Printing Office, Washington, D.C., 1965, p. 43.
11. A. F. Roberts, *15th International Symposium on Combustion,* The Combustion Institute, Pittsburgh, Pa., 1974, p. 305; G. E. Moore, *Report No. 76CRD179,* General Electric Corporate Research and Development, Schenectady, N. Y., 1976.
12. D. R. Stull and H. Prophet, eds., "JANAF Thermochemical Tables," *Nat Stand. Ref. Data Ser. Nat. Bur. Stand.* (1971).
13. R. J. Steffenson, J. T. Agnew, and R. A. Olsen, *Tables for Adiabatic Gas Temperatures and Equ. Gas Composition of 6 Hydrocarbons, Engineering Bulletin #122,* Purdue University, Lafayette, Ind., May 1966.

14. *National Fire Codes,* Vol. 2, pp. 68–35ff (tables of explosion data from Eastman Kodak Co. closed bomb tests).
15. D. R. Cruise, *J. Phys. Chem.* **68,** 3797 (1964).
16. L. P. Harris and G. E. Moore, *Combustion—MHD Power Generation for Central Stations, Paper 71TP79-PWR, IEEE Meeting,* New York, Jan. 1971.
17. R. M. Fristrom and A. A. Westenberg, *Flame Structure,* McGraw-Hill, New York, 1965, p. 153ff.
18. B. Lewis and G. von Elbe, *Combustion, Flames and Explosions of Gases,* 2nd ed., Academic Press, Inc., New York, pp. 628–638.
19. W. J. Miller, *14th International Symposium on Combustion,* The Combustion Institute, Pittsburgh, Pa., 1973, pp. 307–320.
20. A. C. Egerton and co-workers, *Proc. Roy. Soc. London Ser. A* **211,** 445 (1952), **228,** 297 (1955).
21. M. Gerstein, *7th International Symposium on Combustion,* The Combustion Institute, Pittsburgh, Pa., 1959, p. 903.
22. Ref. 18, p. 367ff.
23. Ref. 18, p. 394.
24. Ref. 18, p. 381ff.
25. G. Dixon-Lewis and G. L. Isles in ref. 21, p. 473.
26. W. E. Kaskan, *6th International Symposium on Combustion,* The Combustion Institute, Pittsburgh, Pa., 1957, p. 134.
27. Ref. 18, Chapt. 8, pp. 511–554.
28. D. J. Rasbash, *Symp. Chem. Proc. Hazards, Inst. Chem. Eng. (UK),* 58 (1960); *Gas Engineers Handbook,* Industrial Press, New York, 1966, p. 2/82.
29. J. W. Reppe, *Chemistry and Technology of Acetylene Pressure Reactions* (in Ger.), 2nd ed., Verlag Chemie, Weinheim, Ger., 1952, pp. 3–29; J. A. Niewland and R. R. Vogt, *ACS Monogr.* **99,** 16 (1945).
30. Ref. 18, pp. 228, 323–340.
31. A. E. Potter, *Prog. Combust. Sci. Technol.* **1,** 166 (1960).
32. K. N. Palmer in ref. 27, p. 50.
33. Ref. 18, pp. 323–340.
34. B. P. Mullins, *The Spontaneous Ignition of Liquid Fuels,* AGARD Monograph, Butterworths, London, Eng., 1955.
35. W. Jost, *Explosions—und Verbrennungsvorgänge in Gasen,* Springer, Berlin, Ger., 1939, p. 30ff; Engl. trans., H. O. Croft, McGraw-Hill, New York, 1946, pp. 32–45.
36. W. Jost, *3rd International Symposium on Combustion,* The Combustion Institute, Pittsburgh, Pa., 1949, p. 424.
37. A. G. Gaydon and I. R. Hurle, *The Shock-tube in High Temperature Chemistry and Physics,* Reinhold, New York, 1963.
38. J. Nagy and co-workers, *U.S. Bur. Min. Rep. Invest.,* 7132 (1968); 5971 (1962); Baumeister and Marks, *Standard Handbook for Mechanical Engineers,* 7th ed., McGraw-Hill, New York, 1967, p. 7–40.
39. Ref. 18, p. 146.
40. Ref. 18, Chapt. II.
41. C. P. Fenimore, *Chemistry in Premixed Flames,* Pergamon Press, Inc., New York, 1964, Chap. 14; R. M. Fristrom and A. A. Westenberg, *Flame Structure,* McGraw-Hill, New York, 1965.
42. C. P. Fenimore and G. W. Jones, *10th International Symposium on Combustion,* The Combustion Institute, Pittsburgh, Pa., 1965, p. 489.
43. E. M. Bulewicz, C. G. James, and T. M. Sugden, *Proc. Roy. Soc. London Ser. A* **235,** 89 (1956).
44. W. E. Kaskan, *Combust. Flame* **2,** 229, 286 (1958).
45. W. E. Kaskan, *Combust. Flame* **3,** 49 (1959); T. Singh and R. F. Sawyer, *13th International Symposium on Combustion,* The Combustion Institute, Pittsburgh, Pa., 1971, p. 403.
46. M. J. Day and co-workers, *13th International Symposium on Combustion,* The Combustion Institute, Pittsburgh, Pa., 1971, p. 705; G. Lask and H. G. Wagner, *8th International Symposium on Combustion,* The Combustion Institute, Pittsburgh, Pa., 1962, p. 432.
47. C. P. Fenimore and F. J. Martin, *Combust. Flame* **10,** 135 (1966); C. P. Fenimore and G. W. Jones, *Combust. Flame* **10,** 295 (1966).
48. J. P. Longwell and M. A. Weiss, *Ind. Eng. Chem.* **47,** 1634 (1955); **50,** 257 (1958).
49. P. H. Kydd in ref. 42, p. 101.
50. H. C. Hottel in ref. 42, p. 111; *12th International Symposium on Combustion,* The Combustion Institute, Pittsburgh, Pa., 1969, p. 913.

51. H. B. Palmer and J. M. Beér, eds., *Combustion Technology, Some Modern Developments*, Academic Press, Inc., New York, 1974, pp. 117–125, 374–415.
52. G. C. Williams, A. F. Sarofim, and N. Lambert, *Symposium on Emissions from Continuous Combustion Systems*, Plenum Press, New York, 1972, p. 141.
53. G. E. Moore and B. E. Gans, *Report No. 73-CRD-209*, General Electric, Corp., Schenectady, N. Y., 1973.
54. C. P. Fenimore, *Combust. Flame* **22**, 343 (1974).
55. B. P. Breen and co-workers, *13th International Symposium on Combustion*, The Combustion Institute, Pittsburgh, Pa., 1971, p. 391.
56. C. P. Fenimore, in ref. 55, p. 373.
57. C. P. Fenimore, *Combust. Flame* **19**, 289 (1972).
58. K. H. Hohmann, *Combust. Flame* **11**, 265 (1967); *15th International Symposium on Combustion*, The Combustion Institute, Pittsburgh, Pa., 1975, pp. 1415–1470 (5 papers).
59. J. B. Heywood, *15th International Symposium on Combustion*, The Combustion Institute, Pittsburgh, Pa., 1975, p. 1191; W. A. Daniel in ref. 26, p. 886.
60. J. C. Street and A. Thomas, *Fuel* **34**, 4 (1955).
61. U. Bonne, K. H. Hohmann, and H. G. Wagner in ref. 42, p. 503.
62. J. J. Macfarlane, F. H. Holderness, and F. S. Whitcher, *Combust. Flame* **8**, 215 (1964).
63. A. H. Lefebvre in ref. 59, p. 1169; J. B. Heywood in ref. 59, p. 1191.
64. A. M. Mellor, *Prog. Energy Combust. Sci.* **1**, 111 (1976).
65. A. C. Stern, ed., *Air Pollution*, Academic Press, Inc., New York, 1968, pp. 319–519.
66. H. H. Lowry, ed., *Chemistry of Coal Utilization*, John Wiley & Sons, Inc., New York, 1963, p. 825.
67. Ref. 18, pp. 228–265.
68. K. Bartholomé, *Z. Electrochem.* **53**, 191 (1949).
69. G. von Elbe and B. Lewis in ref. 36, pp. 68, 80; A. Putnam and R. Jensen in ref. 36, p. 89.
70. *Gas Engineers Handbook*, Industrial Press, New York, 1966, Chapt. 12a, pp. 12/193ff.
71. Ref. 70, Chapt. 14; Ref. 18, pp. 490–510.
72. H. K. Richardson, *Ind. Gas (Duluth)* (4), 8 (1949).
73. Ref. 70, Chapt. 13, p. 12/211.
74. J. O. Hirschfelder, *9th International Symposium on Combustion*, The Combustion Institute, Pittsburgh, Pa., 1963, p. 555.
75. Ref. 70, p. 12/225.
76. W. A. Bone and D. T. A. Townend, *Flame and Combustion in Gases*, Longmans, Green, London, Eng., 1927, p. 471.
77. M. Siegler and G. E. Moore, *AIChE Symp. Ser.* (104), 1 (1970).
78. E. R. Weaver, *J. Res. Nat. Bur. Stand.* **46**, 213 (1951).
79. G. E. Andrews, D. Bradley, and S. B. Lwakabamba, in ref. 59, p. 655.
80. F. H. Wright and E. E. Zukoski, *8th International Symposium on Combustion*, The Combustion Institute, Pittsburgh, Pa., 1961, p. 933 (and other papers in this symposium; pp. 944–980).
81. Ref. 18, p. 224ff.
82. J. A. Browning and W. G. Krall, *5th International Symposium on Combustion*, The Combustion Institute, Pittsburgh, Pa., 1955, p. 159; T. H. Pierce and J. A. Nicholls in ref. 19, p. 1277.
83. G. A. E. Godsave, *4th International Symposium on Combustion*, The Combustion Institute, Pittsburgh, Pa., 1953, p. 818.
84. Ref. 2, p. 197.
85. J. B. Howard and R. H. Essenhigh, *11th International Symposium on Combustion*, The Combustion Institute, Pittsburgh, Pa., 1967, p. 399; R. H. Essenhigh and J. Csaba in ref. 74, p. 111.
86. F. A. Williams in ref. 80, p. 50; D. B. Spalding, *Selected Combustion Problems (AGARD)*, Butterworths, London, Eng. 1954, p. 340.
87. Ref. 51, Chapts. 3, 4, and 14.
88. H. W. Emmons in ref. 55, p. 1.
89. J. Swithenbank and co-workers in ref. 19, p. 627.
90. G. S. Canada and G. M. Faeth in ref. 19, p. 1345.
91. C. G. McCreath and N. A. Chigier in ref. 19, p. 1355.
92. R. T. Waibel and R. H. Essenhigh in ref. 19, p. 1413.
93. P. Wolanski and S. Wojacki in ref. 59, p. 1229.
94. D. J. Pratt in ref. 59, p. 1339.
95. J. Odgers in ref. 59, p. 1321.
96. Ref. 1, Chapts. 7, 9, and 10.

97. Ref. 51, pp. 417 ff.
98. R. B. Schiefer, *Fuel Nozzles and Combustion Systems, Ger-2191-D,* General Electric Gas Turbine Reference Library, Schenectady, N. Y., 1966, p. 8.
99. J. M. Beér and N. A. Chigier, *Combustion Aerodynamics,* Scientific Publishing Co., Inc., New York, 1972; N. Syred and J. M. Beér in ref. 19, p. 537.
100. Ref. 1, Chapt. 10.

General References

References 1–8 are good general references.
B. Lewis and G. von Elbe, *Combustion, Flames and Explosions of Gases,* 2nd ed., Academic Press, Inc., New York, 1961.
A. G. Gaydon and H. G. Wolfhard, *Flames, Their Structure, Radiation, and Temperature,* 2nd ed., Chapman & Hall, London, Eng., 1960.
R. M. Fristrom and A. A. Westenberg, *Flame Structure,* McGraw-Hill, New York, 1965.
C. P. Fenimore, *Chemistry in Premixed Flames,* Pergamon Press, New York, 1964.
G. J. Minkoff and C. F. H. Tipper, *The Chemistry of Combustion Reactions,* Butterworths, London, Eng., 1962.
F. A. Williams, *Combustion Theory,* Addison-Wesley, Reading, Mass., 1965.
J. M. Beér and N. A. Chigier, *Combustion Aerodynamics,* Elsevier Scientific Publishing Co., Inc., New York, 1972.
F. J. Weinberg, *Optics of Flames,* Butterworths, London, Eng., 1963.
Y. B. Zeldovitch and A. S. Kompaneets, *Theory of Detonation,* Academic Press, Inc., New York, 1960.
H. B. Palmer and J. M. Beér, eds., *Combustion Technology, Some Modern Developments,* Academic Press, Inc., New York, 1974.
American Gas Association, *Gas Engineers Handbook,* The Industrial Press, New York, 1966.
W. Jost, *Explosions und Verbrennungsvorgange in Gasen,* Springer, Berlin, Ger., 1939; Engl. trans., H. O. Croft, McGraw-Hill, New York, 1945.
L. A. Vulis, *Thermal Regines of Combustion,* Engl. trans., M. D. Friedman and G. C. Williams, McGraw-Hill, New York, 1961.
B. Lewis, R. N. Pease and H. S. Taylor, eds., *Combustion Processes* (High Speed Aerodynamics and Jet Propulsion, Vol. 2), Princeton University Press, Princeton, N. J., 1956.

GEORGE E. MOORE
Consultant

BUTADIENE

Butadiene [106-99-0] (1,3-butadiene), $CH_2\!\!=\!\!CHCH\!\!=\!\!CH_2$, is a major commodity product of the petrochemical industry. Annual U.S. production has been well over a million metric tons since 1963 and about 1.5 million metric tons since 1971. A principal use is the manufacture of SBR (butadiene–styrene copolymer) elastomer, of which more than 60% is used for tires. The first commercial process employed ethanol as a raw material, and it is still used in some countries, but no longer in the United States, Western Europe, or Japan. In general, butadiene is made by one or more of the following processes: catalytic dehydrogenation of n-butane or n-butene; oxidative dehydrogenation of n-butene; and as a coproduct in steam cracking of petroleum fractions, usually for the primary production of ethylene. The isomeric 1,2-butadiene [590-19-2], $CH_3CH\!\!=\!\!C\!\!=\!\!CH_2$, is sometimes found as a contaminant of 1,3-butadiene, but has been of little commercial interest although its use and recovery is described in recent patents (1–2).

Properties

1,3-Butadiene is a colorless gas, slightly soluble in water, somewhat more soluble in methanol and ethanol, but readily soluble in common organic solvents. It forms azeotropes with ammonia, methyl amine, acetaldehyde, n-butene, and 2-butene (3). Its physical properties are given in Table 1.

Thermochemical data show that 1,3-butadiene is more stable (ie, has a lower energy content) than a diene with isolated or nonconjugated double bonds. On the

Table 1. Physical Properties of 1,3-Butadiene[a]

Constant	Value
mol wt	54.09
bp, °C	−4.41
freezing point, °C	−108.9
critical temperature, °C	152
critical pressure, MPa (kg/cm^2)	4.32 (44.1)
critical volume, mL/mol	221
critical density, g/mL	0.245
density, g/mL	
20°C	0.6211
25°C	0.6149
50°C	0.5818
heat of fusion, J/g (cal/g)	147.6 (35.28)
heat of vaporization, J/g (cal/g), 25°C	389 (93)
bp	418 (100)
heat of formation, kJ/mol (kcal/mol) 25°C	
gas	110.2 (26.33)
liquid	88.7 (21.21)
free energy of formation, kJ/mol (kcal/mol), 25°C, gas	150.7 (36.01)
explosive limits, vol % butadiene in air	
lower	2.0
upper	11.5

[a] Refs. 4–9.

basis of electron diffraction studies, the central C–C bond in 1,3-butadiene has a bond distance (0.148 nm) intermediate between that of a double bond (0.133 nm) and a single bond (0.154 nm), indicating the bond has some double-bond character. Like other acyclic conjugated dienes, 1,3-butadiene shows an ultraviolet absorption in the 210–220 nm range, ie, at a much longer wavelength than hydrocarbons with isolated double bonds. These and the chemical properties of butadiene have been interpreted on the basis of valence bond, molecular orbital, and other considerations (10–19). In terms of valence bond theory, butadiene is considered a resonance hybrid with a resonance energy of about 15 kJ/mol (3.5 kcal/mol):

$$\overset{+}{C}H_2-CH=CH-\overset{-}{C}H_2 \leftrightarrow CH_2=CH-CH=CH_2 \leftrightarrow \overset{-}{C}H_2-CH=CH-\overset{+}{C}H_2$$

The butadiene molecule has a planar configuration; two conformations are possible, namely the cis and trans, which are in equilibrium:

trans ⇌ cis

The cis form predominates at about −75°C, the trans form at room temperature (12–13). However, in some reactions of butadiene, the cis orientation is required by stereochemical considerations (20–22).

Reactions

The versatility of butadiene as a chemically raw material is reflected in a large variety of reactions, especially 1,4 additions with formation of a double bond in the 2,3 position.

Diels-Alder Reactions. Butadiene undergoes the Diels-Alder reaction (20,23–26) with a wide variety of dienophiles to form six-membered ring compounds (20,25–27). The second-order reaction is exothermic and reversible and proceeds in the gas phase as well as in polar and nonpolar solvents. Catalysts are generally not required and reactions often proceed at ca 100–200°C and autogeneous pressure. The rate of reaction with strongly electrophilic dienophiles is accelerated by aluminum chloride and other Friedel-Crafts catalysts (28–31). However, the tendency of butadiene to undergo polymerization and other reactions, as well as the instability of some dienophiles in the presence of Lewis acids, limits the general utility of such catalysts (32).

Acrolein and butadiene give 1,2,3,6-tetrahydrobenzaldehyde in 90+% yields when the reaction is carried out over a short period of time (~½ h) at elevated temperatures up to about 200°C (33–34). In the presence of an aluminum–titanium catalyst, the reaction also takes place at room temperature (28).

Reactions with unactivated olefins (eg, ethylene) require more severe conditions (35–38). Acetylenic compounds are normally less reactive than the corresponding olefins (39–40).

The reaction of 1,3-butadiene with maleic anhydride, which forms 1,2,3,6-tetrahydrophthalic anhydride, has been used as a basis for the determination of butadiene (41). The reaction of butadiene and maleic anhydride is catalyzed by aluminum silicates (42). The Diels-Alder reaction is stereospecific, the addition is cis in regard to butadiene and the dienophile (43). Thus, the adduct of maleic anhydride is the cyclic *cis*-1,2-dicarboxylic acid anhydride, whereas the adduct of fumaric acid is the cyclic *trans*-1,2-dicarboxylic acid (44).

With styrene, 4-phenylcyclohexene (45) is obtained; with quinones, cycloolefinic ketones can be formed containing two or more fused rings. Thus 1,4-naphthoquinone gives tetrahydroanthraquinone (34,46), whereas 1,4-benzoquinone adds either 1 or 2 moles of butadiene, depending upon the reaction conditions (47–48) to give:

The reaction of butadiene with acrylonitrile, CH_2=CHCN, yields 1,2,3,6-tetrahydrobenzonitrile, which on subsequent fusion with sodium hydroxide, followed by acidification, gives pimelic acid, $HOOC(CH_2)_5COOH$ (49–51).

4-Vinylcyclohexene (52) has been prepared by heating butadiene to 425°C under 1.3 MPa (13 atm) pressure in the presence of silicon carbide (53), or at 110–150°C under 4.0–100 MPa (40–1000 atm) pressure with catalytic amounts of copper or chromium naphthenates or resinates (54). It is also formed during the prolonged storage of butadiene and is easily isolated by distillation. The dimer has been homo- and copolymerized thermally and in the presence of free radicals, Lewis acids, mineral acids, and metal–alkyl catalysts. Further autocondensation leads to 1,2,3,6,1',2',-3',6'-octahydrobiphenyl (45). Butadiene also condenses with other dienes (eg, isoprene or piperylene) at elevated temperatures (55–56).

Catalytic Oligomerization and Related Reactions. Butadiene can be converted into several different cyclic or open chain dimers and trimers depending upon the reaction conditions and catalyst used. An efficient route to 1,5,9-cyclodecatriene led to the large scale manufacture of dodecandioic acid used in the manufacture of DuPont's Qiana.

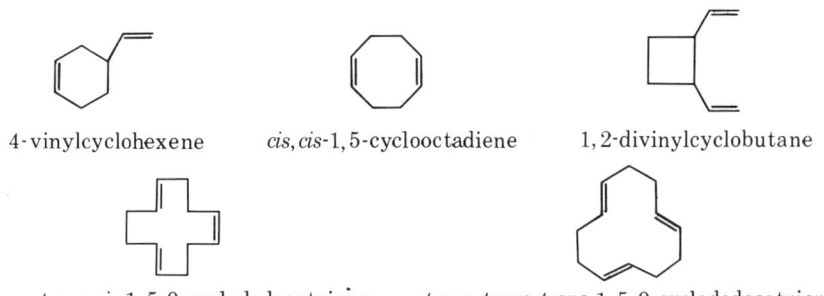

4-vinylcyclohexene   *cis,cis*-1,5-cyclooctadiene   1,2-divinylcyclobutane

trans,trans,cis-1,5,9-cyclododecatriene   *trans,trans,trans*-1,5,9-cyclododecatriene

Nickel catalysts are used extensively for the cyclodimerization and cyclotrimerization of butadiene (57–63). In addition to low-valent nickel compositions, complexes of zero-valent nickel are particularly effective for the cyclodimerization reaction when used in the presence of electron donor ligands, such as certain phosphites and phosphines. Catalysts prepared by the reaction of bisacetylacetonate nickel(II) with an alkyl aluminum cyclodimerize butadiene in the presence of catalytic amounts of specific phosphine ligands (59,64). Electrolytic reduction has been used to prepare nickel catalysts for the cyclodimerization reaction (65). Vinylcyclohexene and 1,5-cyclooctadiene are usually the predominant products; however, 1,2-divinylcyclobutane is also formed in high selectivity with proper ligand and suitable reaction conditions (59). In the presence of a controlled amount of alcohol, several nickel catalyst systems convert butadiene selectively to 1-vinyl-2-methylenecyclopentane (66–67). The 1-vinyl-3-methylenecyclopentane isomer is a by-product (68). Complexes of iron (69–71), copper(I), zeolite (38), and other compositions (72–73) also promote cyclodimerization often giving cyclooctadiene as the principal product (see Catalysis).

In the cyclotrimerization to 1,5,9-cyclododecatriene with nickel(0) complexes, the distribution of isomers depends on the temperature, extent of conversion, and the nature of additional ligands (58,62). Other catalysts include compositions containing chromium, vanadium, titanium, or manganese (74–76).

The principal open-chain dimers are the n-octatrienes formed by 1,4- and 1,2-additions of the isomers depending upon the catalyst system and ligand. Dimerization to 1,3,7-octatriene is catalyzed by a number of palladium complexes (64,77–80). Other catalysts promoting open-chain polymerization include complexes of iron (81–84), rhodium and ruthenium (85), cobalt (68,86–91), and nickel (68,92–94). The dimerization rate is enhanced by CO_2 when using palladium(0)-tertiary phosphine compositions (78) or a tetrakis(triphenylphosphine)platinum or palladium composition. In the absence of the CO_2, 4-vinylcyclohexane predominates (79). Further dimerization of octatrienes produces hexadecapentanes. With other catalysts, branched dimers, linear trimers and higher oligomers are also obtained (64,83,88,91,93).

With nickel(0)-ligand catalysts, butadiene and ethylene or alkynes give *cis,-trans*-1,5-cyclodecadiene (95) or *cis,cis,trans*-1,4,7-cyclodecatriene (96–97), respectively. Reaction with allene yields the 8- and 9-methylene-*cis,trans*-1,5-cyclodecadienes (98). Ethylene adds readily to butadiene in the presence of rhodium(III) chloride to give *trans*-1,4-hexadiene with high selectivity (85). Other suitable codimerization catalysts include complexes of iron (81,99), nickel (100–101), and cobalt (73,102–105).

Polymerization. Butadiene is readily polymerized under a variety of conditions to yield structures corresponding to 1,4 and 1,2 addition of the monomer. The 1,4 addition results in an internal double bond which, as a site of geometric isomerism, can have a cis or a trans configuration. These forms are readily distinguishable by their infrared spectra (106–107). The *cis*-1,4 structure leads to absorption bands centering at about 13.5 μ. The 1,2 addition produces a structure with a tertiary carbon that serves as a site of steric isomerism. Hence 1,2-polybutadienes can be of the isotactic or syndiotactic type.

Several synthetic elastomers are based on butadiene, the three most important being styrene–butadiene, (SBR), polybutadiene (BR), and acrylonitrile–butadiene (NBR) (see Elastomers, synthetic).

The most widely used is SBR, and its manufacture accounts for about half the butadiene consumption in the United States. Most of it is made by an emulsion process with a free-radical initiator giving a product with a styrene–butadiene ratio of about 23:77. Termination is mainly due to chain transfer, the usual transfer agent being a mercaptan. The monomer units are distributed randomly and the butadiene configuration comprises about 65% *trans*-1,4; 18% *cis*-1,4; and 17% vinyl or 1,2 structure. Acrylonitrile–butadiene rubber is also made by an emulsion process. Oil resistance is perhaps the most important property of the nitrile rubber. Butadiene does not polymerize readily in solution (or bulk) via a free-radical process; both termination and branching reactions are too fast for producing a satisfactory product. Emulsion polymerization techniques avoid these problems by localizing the growing chain radical.

Solution polymerization, by alkali metal catalysts such as lithium, is readily carried out commercially (108). Random copolymers formed in these processes resemble emulsion SBR but have a narrower mol wt, higher cis-content, and less chain branching; the polymers also have some improved properties (109). The solution process is also used to produce block copolymers having the general structure, S–B–S, where S represents a styrene segment which is glassy or crystalline in use but fluid at higher temperatures, and B represents a butadiene segment which is elastomeric at use temperatures. These polymers tend to be thermoplastic elastomers (see Styrene plastics).

The four types of stereoregular polybutadiene (*cis*-1,4; *trans*-1,4; 1,2 isotactic; and 1,2 syndiotactic) have been prepared using anionic catalysts based on transition metals (110). The *trans*-1,4 polybutadiene exists in two crystalline modifications with a transition at about 70–75°C and a melting point of the high temperature modification at 145°C. The isotactic 1,2 polybutadiene has a crystalline melting point of 128°C and the syndiotactic modification melts at 156°C (110). The *cis*-1,4 polybutadiene is an elastomer with highly satisfactory properties and is produced commercially in high volume. In 1977, there were five plants in the United States with a total capacity for *cis*-1,4 polybutadiene of close to 380,000 metric tons per year (111). Catalyst systems that promote a high selectivity to *cis*-1,4 polybutadiene include titanium-based compositions, eg, AlR_3–$TiCl_4$ or TiI_4 (112–120); cobalt catalysts, eg, AlR_2Cl–Co compounds (121–129); nickel-based catalysts, and various other transition metal compositions (130). Commercial-grade polymers having a 94–98% *cis*-1,4 content are produced by solution polymerization processes with titanium-, cobalt-, and nickel-based catalyst systems in a hydrocarbon solvent. A lower *cis*-1,4 content is obtained with an alkyl lithium catalyst (131–132). There is also some commercial interest in vinyl polybutadiene, ie, a polymer with a comparatively high 1,2 content. These polymers are made by solution polymerization of butadiene in the presence of an alkyl lithium or sodium catalyst and an anionic complexing agent (133). Cobalt catalysts have been used in preparing highly stereospecific 1,2-polybutadiene with an almost exclusively syndiotactic structure (134).

The *trans*-1,4-polybutadiene was first produced using an aluminum triethyl–$\alpha TiCl_3$ catalyst system (135). Other catalyst systems include aluminum trialkyl–VCl_3 and related compositions (136,110), vanadium compositions (136), rhodium salts (137–138), etc.

Hydrogenation. A variety of heterogeneous catalysts are used for the hydrogenation of butadiene to give butenes and butane. The selectivities to the butenes vs butane and the ratio between hydrogen addition in the 1,2- and 1,4-positions depend upon type of catalyst, its method of manufacture, and process parameters including the presence of promoters and inhibitors. Catalysts include platinum, palladium, iridium, ruthenium, nickel, cobalt, vanadium, titanium, copper chromite, and other compositions (139–150). In a series of experiments with Group VIII metal catalysts on various supports, selectivities to n-butenes of 99+ mol % and to 1-butene of about 80 mol % were demonstrated with a 0.1% platinum on alumina catalyst at 80–90°C in the presence of carbon monoxide which served to suppress the isomerization of 1-butene to 2-butene (144).

Telomerization. The reaction of butadiene with an active hydrogen compound in the presence of a palladium catalyst, eg, a *tert* phosphine complex, yields a 2,7-octadienyl derivative. Under appropriate conditions butadiene reacts with alcohols, particularly methanol, to form 1-alkoxy-2,7-octadiene and smaller amounts of 3-alkoxy-1,7-octadiene (151). Primary and secondary amines produce secondary and tertiary amines with 2,7-octadienyl groups (152–153). Similar derivatives have been obtained with carboxylic acids (151b–156), phenols (152,157), ethanol, and water (158). Phenol, in the presence of a catalytic amount of phenoxide ion, yielded 1-phenoxy-2,7-octadiene as a major product (159). Reaction with trimethyl silane proceeds somewhat differently to yield 1-trimethylsilyl-2,6-octadiene (160). In the presence of an aldehyde, palladium complexes catalyze the cocyclization of 2 moles of butadiene with the heteropolar bond to yield a mixture of isomeric divinyltetrahydropyrans and a 1-substituted 2-vinyl-4,6-heptadien-1-ol (161–164). Isocyanates react similarly (165).

In the presence of palladium and platinum catalysts, active methylene compounds react with butadiene to form 1:2 adducts, principally 2,7-octadienyl derivatives (166–167). A catalyst has been prepared by mixing dichloro–bis(triphenylphosphine)palladium and a basic compound such as sodium phenoxide (167). However, when a bidentate phosphine ligand is used, eg, ethylenebis(diphenylphosphine), the 1:1 adducts, rather than telomer products, are obtained (168). When butadiene reacts with a nitroalkane in the presence of a triphenylphosphine–palladium complex at about 50°C, the α-hydrogens of the nitroalkane are displaced with 2,7-octadienyl groups (169).

Addition Reactions. Butadiene reacts easily and rapidly with both electrophilic and free radical reagents via 1,2- and 1,4-mechanisms, depending upon reagent and reaction conditions.

The addition of hydrogen chloride in acetic acid at ambient temperature gives mostly the 1,2 addition product, with the yield being kinetically controlled (170). However, the 1,4 adduct is more stable thermodynamically, thus isomerization to an equilibrium mixture leads to the 1,4 adduct as the main product (171). In the absence of oxygen and peroxides, hydrogen bromide adds at low temperatures (−78°C) to give mainly 3-bromo-1-butene by a 1,2 addition. At higher temperatures, under the influence of hydrogen bromide (with or without peroxide), this product rearranges to an equilibrium mixture containing about 80% 1-bromo-2-butene. In the presence of air or peroxide the principal product is the 1,4 addition product. Hydrogen iodide addition at 20°C gives 52% 1-iodo-2-butene (172–173). In the absence of halogens or hydrogen halides, hypohalide adds to butadiene, predominantly in the 1,2 position (160,174). In an aqueous medium, chlorine and butadiene produce a mixture of the 1,2- and 1,4-chlorohydrins which, upon further reaction, led to an 84% yield of butadiene dichlorohydrin, based on chlorine consumed (175). Alkyl hypochlorites also added principally in the 1,2 position to give the analogous alkoxyhalobutenes or haloethers (171,176).

Halogenation of butadiene has been studied in solution and in the vapor phase (177–190). Chlorination of butadiene in chloroform or carbon disulfide below 0°C gives a mixture of 3,4-dichloro-1-butene and 1,4-dichloro-2-butene, in a ratio of about 2:1 (191–192). Tetrachlorides are also formed. In solution, bromination or iodination are more selective giving mostly a 1,4-dihalo-2-butene product (193–195). The oxidative chlorination of butadiene with HCl and O_2 or air has been described (182,196–197).

The vapor-phase chlorination of butadiene has been used as a basis for commercial routes to adiponitrile, $NCCH_2CH_2CH_2CH_2CN$ (198–199), chloroprene (see Chlorocarbons and chlorohydrocarbons), and 1,4-butanediol (see Acetylene-derived chemicals). Adiponitrile can be the precursor of hexamethylenediamine (see Diamines) and adipic acid by hydrolysis in the manufacture of Nylon 66 (see Adipic acid; Polyamides).

Chloroprene is manufactured by vapor-phase chlorination of butadiene at 300°C to a mixture of 3,4-dichloro-1-butene and 1,4-dichloro-2-butene (190,200) involves liquid-phase catalytic rearrangement of 1,4-dichloro-2-butene to 3,4-dichloro-1-butene which is dehydrohalogenated by heating with aqueous sodium hydroxide with exclusion of oxygen, followed by distillation.

$$\underset{\text{3,4-dichloro-1-butene}}{CH_2-CH-CH=CH_2} \xrightarrow{\text{NaOH}} \underset{\text{chloroprene}}{CH_2=C-CH=CH_2} + NaCl$$

The direct chlorination of butadiene to chloroprene has also been studied (188,201).

In 1971, a process for the manufacture of 1,4-butanediol via chlorination of butadiene was developed and commercialized in Japan utilizing vapor-phase chlorination at 260–300°C to 1,4-dichloro-2-butene (with the 3,4-dichloro-1-butene also obtained used for chloroprene manufacture) followed by hydrolysis with aqueous alkali to yield the 2-butene-1,4-diol. Care is taken to minimize by-product 3-butene-1,2-diol and its readily formed polymer. The 1,4-diol is hydrogenated to 1,4-butanediol.

In another method, 1,4-diacetoxy-2-butene is the intermediate obtained from 1,4-dichloro-2-butene and acetate (202). In a preferred route, butadiene reacts readily with oxygen or air and acetic acid in the presence of a suitable catalyst containing, eg, palladium (203–207), platinum (208), or manganese (209) at about 80–100°C to produce 1,4-diacetoxy-2-butene in high yield and some 3,4-diacetoxy-1-butene. The former is reduced to 1,4-diacetoxybutane with a nickel (210) or other suitable catalyst and then hydrolyzed to produce 1,4-butanediol. As noted previously, acetoxyoctadienes are formed when butadiene and acetic acid react in the presence of a palladium complex catalyst, eg, a palladium acetylacetonate catalyst with phosphine or phosphite ligand (211,155–156).

Palladium is a good catalyst for the carbonylation of butadiene. For example, with $PdCl_2$ or $PdBr_2$ as the catalyst, butadiene reacts with carbon monoxide in alcohol to give 3-pentenoate (212–213). However, with a halide-free palladium–phosphine catalyst, 2 moles butadiene react with 1 mole carbon monoxide in alcohol at 75–80°C to yield 3,8-nonadienoate (214). The reaction is a one-step carbonylation–dimerization

$$2\,CH_2=CHCH=CH_2 + CO + C_2H_5OH \xrightarrow[(C_6H_5)_3P]{\substack{Pd \\ \text{acetylacetonate}}}$$

leading directly to the product (214a). The cobalt carbonyl catalyzed reaction of butadiene with carbon monoxide in methanol in the presence of a pyridine base has been studied (215), as has hydroformylation in the presence of cobalt (216–217) and rhodium (218) catalysts. A patent (219) describes a direct synthesis of adipic acid from butadiene by reaction under pressure with carbon monoxide and water (about 220°C) in the presence of a rhodium catalyst (see Oxo process).

$$H_2C=CHCH=CH_2 + 2\ CO + 2\ H_2O \xrightarrow[CH_3I]{RhCl_3} HOOC-(CH_2)_4-COOH$$

Aromatic compounds are readily alkylated with butadiene under a variety of conditions, catalyzed by acids and Friedel-Crafts compositions (220–227). The addition of butadiene to toluene in the presence of metallic sodium at 90°C yields products containing 1, 2, 3, and 4 butadienes per toluene. All give benzoic acid upon oxidation, thus appearing to be monoalkenyl-substituted benzenes (228). In the presence of benzyl lithium, butadiene forms linear dimers, trimers, and tetramers to which the reagent adds at the terminal positions (229). Butadiene reacts with a dispersion of metallic sodium at −20 to −50°C in an ether solvent in the presence of catalytic amounts of polycyclic hydrocarbons to yield disodiooctadiene. The disodiooctadiene can be carbonated by reaction with carbon dioxide to yield a mixture of the sodium salts of ten-carbon-atom unsaturated dibasic acids (230–231). Hydrogenation and acidification lead to a mixture containing about 6–10% sebacic acid, 12–18% 2,5-diethyladipic acid, and 72–80% 2-ethylsuberic acid, a mixture marketed as isosebacic acid.

$$2\ CH_2=CHCH=CH_2 + 2\ Na \rightarrow Na^{+-}(CH_2CH=CHCH_2CH_2CH=CHCH_2)^-Na^+$$

Butadiene reacts with ammonia (232), cyanogen (233), aliphatic diazo compounds (234), and atomic sulfur (235), and condenses with aniline (236) and α-picoline (237). The reaction of butadiene with secondary amines initiated by n-butyl lithium can be carried out to produce 1-dialkylamino-cis-2-butene derivatives selectively (238). Similarly, secondary amines add to butadiene in the presence of alkali metal hydrides and amides to yield a simple butenyl compound; however, higher molecular weight products are formed with primary amines. The interaction of methylamine and butadiene (1:1 mol ratio) in the presence of a catalytic amount of sodium hydride produces a mixture of butenyl- and dibutenylmethylamines, as well as some higher boiling amines (239). The addition of organometallic compounds to butadiene has been studied by Ziegler (229,240–241) and Morton (242–243), and hydroboration (qv) of butadiene with diborane has been studied by H. C. Brown and co-workers (244–245) as well as by others (246).

The 1,4 cyclization of butadiene with reactants containing sulfur, nitrogen, or phosphorus (247–249) can lead to five- and six-membered heterocyclic compounds. Sulfur dioxide reacts almost quantitatively with butadiene to give 3-sulfolene (250–252). The reaction is reversible and is used for purification of dienes (253–254). Sulfur gives thiophene at 350–450°C in low yields (255–257). A similar reaction has been observed with selenium (258). Hydrogen sulfide reacts with butadiene over pyrites or iron oxide on activated alumina at 600°C to give thiophene in 30–60% yields (232,259). Singlet state methylene adds almost exclusively in the 1,2-position to give vinylcyclopropane and some cyclopentene (260–263). Vinyldihalocyclopropanes are synthesized by the reaction of dihalocarbenes with butadiene (264–266).

$$CHCl_3 + t\text{-BuOK} \rightarrow :CCl_2 \xrightarrow{butadiene} \underset{CCl_2}{CH_2-CH-CH=CH_2}$$

The addition of other carbenes (267) and of atomic and triatomic carbon (268–269) has been reported. A number of cyclobutane derivatives have been prepared by the thermally-induced addition of tetrafluorethylene (270), dichlorodifluoroethylene (271), ketenes (272), etc, or sensitized uv-promoted addition of acrylonitrile (273) to butadiene.

Butadiene combines with free radicals to form new free radicals which either dimerize, or react further (274–290). Among the compounds that add to butadiene via a free-radical reaction are carbon tetrachloride (285), diazonium halides (286), thiophenol, and thioacetic acid (287), thiocyanogen (288), nitrogen tetroxide (289), etc. Butenyl radicals are also obtained and dimerize or react with other radical species present (290).

Oxidation Reactions. The vapor-phase air oxidation in the presence of vanadium or molybdenum oxide catalysts at 250–400°C produces maleic anhydride (qv) or maleic acid (291–300).

Butadiene and oxygen (301–306) give polymeric peroxides which can be explosive, especially when concentrated, and are sensitive to shock (301–306). The polyperoxide, formed in liquid butadiene at 50°C, is composed of equal amounts of 1,4- and 1,2-butadiene units separated by peroxide units, and has a mol wt corresponding to about 35 $C_4H_6O_2$ units. Oxygen can also promote "popcorn" polymerization of butadiene (307–311).

Oxidation by a peroxy acid (eg, perbenzoic acid) yields butadiene monoxide, $CH_2=CHCHCH_2O$ (312–313), which can also be prepared by hydrochlorination followed by dehydrochlorination (314). Butadiene dioxide has also been prepared (315). Ozonolysis of butadiene in petroleum ether yields the slightly soluble, white 1,3-adduct, butadiene monoozonide. Ozonolysis in chloroform gives the soluble, unstable butadiene 1:2,3:4-diozonide (316). Butadiene treated with aqueous sodium hexafluoroplatinate (317) or palladium(II) chloride (318) forms crotonaldehyde.

Substitution Reactions. The replacement of a hydrogen in butadiene by a substituent, is readily effected by a number of reagents. With nitric acid at −30°C, nitrobutadiene is obtained (319). Butadiene monoacetate is made from butadiene, palladous chloride, disodium hydrogen phosphate, and glacial acetic acid in isooctane (320). Butadiene is methylated to 1,3-pentadiene (80% trans, 20% cis) with dimethyl sulfoxide in the presence of a base such as potassium t-butoxide (321). Alkylation with tertiary alkyl halides to produce substituted butadienes (322), phosphonation via reaction with PCl_5 (323), and the use of pyridine–sulfur trioxide as sulfonating agent to produce 1,3-butadiene-1-sulfonic acid (324) have been investigated.

Manufacture

Butadiene was first obtained over 100 years ago by various pyrolysis and red-hot tube reactions and was prepared in 1886 by pyrolysis of petroleum hydrocarbons (325). In the mid-1970s, principal processes for butadiene in the United States were distributed as follows: steam cracking of naphtha and gas oil fractions (35%); catalytic dehydrogenation of n-butene (30%) and of n-butane (20%); and oxidative dehydrogenation of n-butene (15%). The distribution is expected to change as additional steam cracking plants are built to meet a growing demand for ethylene and other lower olefins. In Western Europe and Japan, butadiene is produced almost exclusively from the C_4 stream produced in naphtha steam cracking operations for ethylene manufacture. (For investigations of butadiene preparation, see refs. 326–331.)

Production from Ethyl Alcohol. Through World War II, the catalytic conversion of ethyl alcohol was considered a valuable route to butadiene. After World War II, this route was no longer cost competitive with hydrocarbon conversion processes. However, alcohol-based plants require less elaborate equipment, hence they are cheaper to build. In areas where ethanol may be obtained fairly cheaply, there could be advantages for its use as a raw material.

The Lebedev process is a one-step vapor phase process where ethanol is simultaneously dehydrogenated and dehydrated over an oxide catalyst at temperatures of about 400–450°C. The reaction is highly endothermic and free-energy data have been published (332a). A variety of catalytic oxide mixtures have been investigated including single, binary, and ternary systems (327,330b,332–336). Application of the fluidized-bed technique to the process has been described (332b).

$$2 \, C_2H_5OH \rightarrow H_2C{=}CHCH{=}CH_2 + H_2 + 2 \, H_2O$$

The other alcohol conversion process originated with Ostromislensky (337) and was used in an improved form extensively in the United States during World War II and the immediate postwar period. It is a two-step process where ethyl alcohol is first dehydrogenated to acetaldehyde, and then the acetaldehyde and ethyl alcohol mixture is converted to butadiene. Copper-containing catalysts are suitable for the dehydrogenation, and silica gel, containing a small amount of tantalum oxide, is an effective catalyst for the second step carried out at about 325–350°C (338–343).

Catalytic Dehydrogenation of Butenes. Large amounts of butenes are available in petroleum refineries from the production of gasoline by catalytic cracking of middle and heavy distillates and also from steam cracking operations for the production of ethylene. The butenes are recovered in dilute form as a C_4 fraction by conventional refinery operations. A C_4 fraction from catalytic cracking may typically contain about 44 wt % butanes, 13 wt % 1-butene, 11 wt % *cis*-2-butene, 13 wt % *trans*-2-butene, 18 wt % butylene, and small amounts (~0.5 wt %) of C_3s and C_5s (344). The over-all yield depends upon composition of the feed, type of catalyst used (345), and operating conditions.

For an economical dehydrogenation process, the butene fraction must be concentrated to at least 70% and preferably 80–95% *n*-butenes. The isobutylene is generally removed from the C_4 fraction by a selective extraction–reaction process that takes advantage of differences in hydration rates of the C_4 olefins. The mixed *n*-butenes in the raffinate are further purified by separation from the butanes by extractive distillation with acetonitrile or furfural (346b).

A C_4 fraction from a naphtha steam cracking operation at medium severity may typically contain: 6.5 wt % butanes, 16 wt % 1-butene, 5.3 wt % *cis*-2-butene, 6.6 wt % *trans*-2-butene, 27.3 wt % isobutylene, 37 wt % butadiene, 0.5 wt % acetylenes, as well as about 0.3 wt % C_3s and 0.5 wt % C_5s (344,347b).

The catalytic dehydrogenation of the *n*-butenes to 1,3-butadiene is carried out in the presence of superheated steam as a diluent and a heating medium. The dehydrogenation reactions are reversible in the presence of the catalyst used. They are also endothermic and hence, from thermodynamic considerations, the equilibrium is more favorable at higher temperatures. The effects of temperature and pressure on the equilibrium conversion of butene to butadiene are shown in Table 2.

$$\begin{array}{c} CH_2{=}CHCH_2CH_3 \rightarrow \\ CH_3CH{=}CHCH_3 \rightarrow \end{array} CH_2{=}CHCH{=}CH_2 + H_2$$

Table 2. Dehydrogenation of 1-Butene at Equilibrium[a]

Temperature, °C	Conversion at equilibrium, %	
	At 101 kPa (1 atm)	At 10 kPa (0.1 atm)
500	12	36
600	31	70
700	60	93

[a] Ref. 348.

To obtain useful butene conversions, the dehydrogenation reactor has to be operated above 620°C and at the lowest feasible partial pressure of butene. The use of superheated steam allows adiabatic high-temperature reactor operation, keeps down the partial pressure of butenes, and lowers coke formation. The process can be cyclic, with n-butene and steam passed over a catalyst at 620–675°C followed by a regeneration cycle with air and steam to remove coke and oxidize the catalyst. By-products include hydrogen and low concentrations of carbon dioxide, carbon monoxide, methane, ethane, ethylene, propane, and propylene. Commercial processes differ in the catalyst (349), mode of operation (cyclic vs continuous), and conversion/selectivity relationship. A chromium-promoted calcium nickel phosphate catalyst–Dow Type B, has been used extensively (350) in commercial operation achieving a selectivity to butadiene of about 90% of theoretical at a butene conversion of about 35–45% per pass. A steam/butene mole ratio of about 20:1 is used, with a reactor pressure of about 20 kPa (0.2 atm). Another highly suitable catalyst is Shell 205 which comprises a mixture of about 70% Fe_2O_3, 27% K_2O, and 3% Cr_2O_3 (349).

Dehydrogenation of n-Butane. The dehydrogenation of n-butane both to butenes and to butadiene is endothermic, and temperatures above 500°C must be maintained to obtain commercially feasible conversions. As noted previously, a reduction in pressure favors the formation of butadiene. The effects of both pressure and temperature on the equilibrium between butadiene and the three isomeric butenes and data for various butane–butenes and butenes–diolefin equilibrium relationships have been reported by Kearby (348). Some typical equilibrium conversion data are shown in Tables 3 and 4. In commercial operations, preferred temperatures are in the range of 600°C, but once-through conversions are kept low by short contact times to minimize undesirable degradation reactions and coke formation.

The two processes introduced in the United States during World War II were based upon Houdry and Phillips technology. In the Houdry process, the two reactions take place in one reactor and n-butane is dehydrogenated to butadiene in a single-stage adiabatic operation. A 95% or better n-butane feed is dehydrogenated over an aluminum oxide–chromium oxide catalyst at about 550–650°C and space rates of about

Table 3. Dehydrogenation of n-Butane to 1-Butene[a]

Temperature, °C	Conversion at equilibrium, %	
	At 101 kPa (1 atm)	At 10 kPa (0.1 atm)
500	17	50
600	50	84
670	70	95

[a] Ref. 348.

324   BUTADIENE

Table 4. Thermodynamic Conversion of n-Butane[a]

	Conversion, mol %					
	101 kPa (1 atm) at °C			17 kPa (0.17 atm) at °C		
Products	600	700	750	600	700	750
1,3-butadiene	6	27.5	45	27.5	69	82
1-butene	22.5	26	22.5	23	13	7.5
cis-2-butene	16	16	13	16.5	7.5	4
trans-2-butene	24	23	18	25	11	6

[a] Ref. 348.

1–3 volume charge per volume catalyst per hour (351–354). A mixture of butenes and butadiene is formed. Hot effluent from the reactors is cooled in a quench tower by direct contact with circulating oil and then passed to compression, cooling, and to an absorption and stabilization system. The butadiene is extracted and the butenes are returned with the n-butane stream to the charge accumulator. The process is cyclic and the catalyst is regenerated periodically by burning off coke deposits with preheated air at atmospheric pressue. A catalyst life of 18–24 months is normal, and in excess of four years has also been obtained (355). Table 5 gives data on yields.

In the Phillips process n-butenes are produced in the first stage over an aluminum oxide–chromium oxide catalyst. The recovered n-butenes are then dehydrogenated to butadiene with a catalyst composition of iron, magnesium, potassium, and chromium oxides (349,356–357). Changes have been made, and in 1970 Phillips developed a commercial process based on oxidative dehydrogenation of n-butenes for the second stage. In this process a mixture of compressed air and steam is heated in a furnace, mixed with the butenes feed, and passed over an oxidative dehydrogenation catalyst in a continuous reactor. Heat is recovered from the reactor effluent and used to generate process steam. Following heat recovery, the reactor effluent is quenched, cooled, washed, and the C_4 components are recovered in an oil absorber. The C_4 components are stripped from the oil with steam and then purified. Unconverted butenes are recycled (358).

Table 5. Butadiene Yields for Houdry Process[a]

		Products, % based on feed		
Component	Fresh feed, wt %	Butadiene	Fuel gas	Other
dry gas			16.3	
isobutane	1.0			
isobutylene			0.5	
n-butenes		0.3	2.5	
n-butane	99.0		1.8	
butadiene		64.4	0.6	
C_{5+}, coke, H_2, CO, etc			1.6	12.0
total	100.0	64.7	23.3	12.0

[a] Ref. 355. Courtesy of Gulf Publishing Co., Houston.

Oxidative Dehydrogenation of n-Butene. The catalytic dehydrogenation of n-butenes to butadiene is highly selective, but yields per pass are restricted by thermodynamic considerations. In oxidative dehydrogenation, the reaction is irreversible and high yields per pass are feasible. Oxygen may be used to remove the hydrogen as water. The reaction is exothermic and temperature control is important. When air is used as oxidant, the nitrogen serves to replace steam as diluent. Suitable catalysts are mixtures of the oxides of tin, bismuth, and boron with phosphoric acid, and compositions based on the oxides of tin and antimony, etc (349,357,359).

Table 6 gives a comparison of conditions for the two types of dehydrogenation.

Halogens, particularly iodine, can also be used to remove hydrogen (360–363); mixtures of H_2S and oxygen act similarly (364).

Thermal Cracking. Thermal cracking of hydrocarbons in the presence of steam at 700–900°C is a principal source of ethylene and other olefins and diolefins. Residence times are short, and steam-hydrocarbon weight ratios are generally in the range of 0.2–0.8. Conditions depend upon the hydrocarbon composition of the feed stock and on the severity of operation desired (269b,344). Ethylene can be produced from a wide range of hydrocarbon feed stocks including ethane, propane, butane, naphthas (ie, fractions boiling up to about 230°C), gas oils (ie, fractions boiling in the range of about 315–480°C), etc. With naphthas and heavier feedstocks, the C_4 fraction of the product contains appreciable quantities of butadiene. Yield data for butadiene obtained by the steam cracking of various feedstocks at high severity, with recycle ethane cracking to extinction, are shown in Table 7.

Other Processes. Butadiene has also been commercially produced by chlorination of butene (365). A small plant using this process was in operation in the United States during World War II (366).

The aldol route to butadiene, which is based largely on acetaldehyde (qv) derived from the hydration of acetylene, was used on a commercial scale in Europe (330c,367). The final step involves the dehydraton of 1,3-butanediol formed by hydrogenation of the aldol (366). A butadiene plant designed to utilize acetaldehyde derived from the oxidation of propane and butane was built in the United States but did not come on stream. The manufacture of butadiene by dehydration of 1,4-butanediol formed via the reaction of acetylene and 2 moles of formaldehyde to give 2-butyne-1,4-diol, $HOCH_2C{\equiv}CCH_2OH$, was developed and used in Germany during World War II. The

Table 6. Butene Conversion Processes[a]

Conditions	Dehydrogenation, Dow Type B[b]	Oxidative dehydrogenation[c]
temperature, °C	620–675	400–450
feed composition		
steam diluent, volumes	18–20	2–4[d]
butene/oxygen ratio		~1
conversion per pass, %	35–45	55–70
selectivity, %	90	>80%

[a] Ref. 359. Courtesy of the American Chemical Society.
[b] Catalyst.
[c] Typical feed composition: 13 mol % C_4 (about 80% n-butenes), 60 mol % air, and 27 mol % steam.
[d] In addition to nitrogen from air.

Table 7. Butadiene Yields From Steam Cracking at High Severity[a]

Feed stock	Once-through ethylene yield, wt %	Ratio[b]
ethane	48.2	2.5
propane	34.5	7.2
n-butane	35.8	8.7
medium-range naphtha	30.0	13.6
atmospheric gas oil	23.0	17.6
light vacuum gas oil	18.0	26.2

[a] Ref. 347b.
[b] kg butadiene per 100 kg ethylene.

overall yield based on the consumed acetylene was about 70% (367–369) (see Acetylene-derived chemicals).

Other potential routes to butadiene have been investigated (24). Air oxidation of a mixed butene feed to produce 1,3-butadiene and methacrolein (371–372), and, the ammoxidation of mixed butenes to produce methacrylonitrile and butadiene (372–374) have been considered (see Methacrylic acid).

Separation and Purification. The hydrocarbon conversion processes yield a crude C_4 fraction containing butadiene and other closely boiling hydrocarbons (see Table 8). For synthetic rubber manufacture a 99.0 wt % minimum purity butadiene is needed with acetylenes in the ppm range. Acetylenes are particularly undesirable because they can polymerize, contributing to equipment fouling and foaming problems.

Commerically, two basic methods have been used for the separation and purification of 1,3-butadiene. One is selective extraction with aqueous cuprous ammonium acetate, the CAA process (376–382), which produces high purity butadiene with a recovery higher than 98%. In the other method, butadiene is extractively distilled with selective solvents (375), including acetonitrile (383–384), furfural (385), dimethylformamide (386–389), N,N-dimethylacetamide (390–392), N-methylpyrrolidinone (393–395), β-methoxypropionitrile (396) and others (397). Azeotropic distillation has also been investigated using ammonia (398) or amines (399) (see Azeotropic and extractive distillation). Butadiene can also be separated by complexing with solid cuprous chloride (400–402). Attempts have been made to evaluate the selectivity of solvents for extractive distillation separations in terms of basic properties (403–406).

Table 8. Composition[a] of a Crude Butadiene Fraction[b]

Component	bp °C	Vol %
C_3 hydrocarbons		0.9
isobutylene	−6.9	27.7
1-butene	−6.3	17.2
1,3-butadiene	−4.4	39.1
n-butane	−0.5	4.1
$trans$-2-butene	+0.9	6.0
cis-2-butene	+3.7	4.5
C_4 acetylenes	+5.1	0.2
1,2-butadiene	+10.9	<0.1
C_5 hydrocarbons		0.1

[a] Varies depending on process and conditions.
[b] Ref. 375.

Both single-stage and two-stage extractive distillation processes are employed. In a two-stage system, butadiene and acetylenes are first separated from the butenes and butanes, and then butadiene is separated from the acetylenic compounds. In a single-stage extractive distillation system the acetylenes are separated by a conventional fractional distillation. Alternatively, the butadiene-containing fractions can be treated before purification by selective hydrogenation in the gaseous or liquid phase. Suitable catalysts include various metal and metal oxide composition, eg, palladium on alumina or silica, copper oxide–chromium oxide compositions, etc (407–410).

Specification and Standards

The bulk of the butadiene manufactured in the United States is sold under the specifications for rubber-grade butadiene shown in Table 9. Several other grades of butadiene are available, ie, research grade of 99.86 mol % purity; a special purity 99.5 mol % grade; and commercial grade 98 mol % purity.

Butadiene can dimerize during handling and storage. The rate of formation is not apparently affected by the presence of peroxides, but is a function of temperature, as shown in Table 10. Consequently, butadiene should be stored at low temperature. The dimer, vinylcyclohexene, is miscible with butadiene in all proportions and can be removed by distillation.

Table 9. Butadiene Specification and Test Methods

Material	Specification	ASTM procedure
1,3-butadiene, wt %, min	99.0	D-1717
trace impurities, max		
acetylenes (as vinylacetylenes), ppm	500	D-1020
hydrocarbons (C_5), wt %	0.05	D-2593
butadiene dimer, wt %	0.2	D-2426
nonvolatile materials, wt %	0.1	D-1025
peroxides (as H_2O_2), ppm	10	D-1022
sulfur (as H_2S), ppm	10	D-1266
carbonyl (as acetaldehyde), ppm	50	D-1089
inhibitor (*t*-butylcatechol), ppm	100	

Table 10. Effect of Temperature on Dimerization Rate

Temperature, °C	Wt % diolefin dimerized per hour
20	0.00015
40	0.0014
60	0.013
80	0.12
100	1.1

328 BUTADIENE

Table 11. U.S. and Japanese Butadiene Production and Price Data [a]

Year	1000 Metric tons U.S.	1000 Metric tons Japan	Unit value (U.S.) ¢/kg
1960	856		28.4
1965	1,221		22.7
1970	1,410	496	8.5
1972	1,603	643	17.2
1974	1,674	640	32.1
1976	1,594	590	38.7
1977	1,550 (est)		

[a] Refs. 417–420.

Table 12. Approximate Use Pattern for Butadiene (Mid-1970s)

End-use	Percentage of total	
Synthetic elastomers		78
styrene–butadiene rubber	50	
polybutadiene rubber	18	
polychloroprene (Neoprene)	7	
nitrile rubber	3	
other polymers and resins		10
ABS (acrylonitrile–butadiene–styrene)	5	
styrene–butadiene copolymers	5	
chemicals and other specialties		12

Safety, Health, and Shipping

1,3-Butadiene is a noncorrosive, colorless, flammable gas with a mildly aromatic odor. The principal hazards arise from its high flammability and chemical reactivity. The explosive limits are 2–11.5% by volume of butadiene in air; the autoignition temperature is 450°C (411–412).

Butadiene forms explosive peroxides in contact with air. Peroxides can be dangerous, particularly when concentrated and heated, as during distillation. Furthermore, they promote polymerization. Peroxide formation is best prevented by the exclusion of air or by the addition of an inhibitor, eg, t-butylcatechol, hydroquinone, di-n-butylamine, etc. However, the inhibitor has no affect on butadiene in contact with oxygen in the vapor phase; moreover the inhibitor can be used up by frequent or prolonged exposure to air and its content should be checked at appropriate intervals.

The physiological effect of 1,3-butadiene may vary individually. Exposure at high concentrations has a narcotic effect. It is irritating to skin, eyes, and upper respiratory passages. Butadiene is shipped as liquid via tank cars, tank trucks, and in DOT approved cylinders. It must contain an inhibitor and have a red-gas label. In handling butadiene, the liquid should not be permitted to come into contact with the skin or clothing. Rapid evaporation can cause a burn or frostbite (413–416).

Uses and Economic Aspects

Table 11 summarizes U.S. and Japanese production data for the 1960–1977 period. About two-thirds of U.S. production was manufactured by dehydrogenation, the rest was obtained as coproduct in ethylene manufacture. A significant increase in coproduct capacity is expected by the early 1980s. In Western Europe, butadiene production was in excess of 1,400,000 metric tons in 1976, and is expected to be 1,800,000 metric tons by 1980 (421). In both Western Europe and Japan, butadiene is produced almost exclusively by steam cracking operations in ethylene manufacture (422–424).

The main use for butadiene is in the manufacture of synthetic elastomers, principally styrene–butadiene rubber and poly(*cis*-1,4-butadiene). The use pattern for butadiene in the United States in the mid-1970s is shown in Table 12 (see Elastomers, synthetic).

Butadiene is the raw material for many important chemicals, such as adiponitrile (for conversion to hexamethylenediamine, a nylon 66 precursor); 1,4 hexadiene (a monomer for ethylene–propylene terpolymer, EPDM); 1,5,9-cyclodecatriene (converted to dodecanedioic acid); and many others.

BIBLIOGRAPHY

"Butadiene" in *ECT* 1st ed., Vol. 1, pp. 669–674, by A. A. Dolnick and Max Potash, Publicker Industries, Inc.; "Manufacture of Butadiene" under "Rubber, Synthetic", in *ECT* 1st ed., Vol. 11, pp. 857–870, by C. E. Morrell, Esso Laboratories, Standard Oil Development Company; "Butadiene" in *ECT* 2nd ed., Vol. 3, pp. 784–815 by I. Kirshenbaum and R. P. Cahn, Exxon Research and Engineering Company.

1. Brit. Pat. 1,327,594 (Aug. 22, 1973), J. Lefebvre and G. Sartori (to Exxon Research and Engineering Co.).
2. Ger. Pat. 2,331,547 (Jan. 17, 1974), R. W. Andrews and R. E. Carpani (Polysar Ltd.).
3. L. H. Horsely, *Azeotropic Data No. 116* in *Advances in Chemistry Series,* American Chemical Society, Washington, D.C., 1973.
4. *Selected Values of Physical and Thermodynamic Properties of Hydrocarbons and Related Compounds,* American Petroleum Institute Research Project 44, Carnegie Press, 1953.
5. J. E. Kilpatrick and co-workers, *J. Res. Natl. Bur. Stand.* **42,** 225 (1949).
6. B. J. Zwolinski and R. C. Wilhoit, *Handbook of Vapor Pressures and Heats of Vaporization of Hydrocarbons and Related Compounds,* Publications in Science and Engineering, No. 101, API44-TRC, 1971.
7. W. Braker and A. L. Mossman, *Matheson Gas Data Book,* 5th ed., 1971.
8. C. L. Yaws, *Chem. Eng.* **83**(5), 107 (Mar. 1, 1976).
9. C. H. Meyers, C. S. Cragoe, and E. F. Mueller, *J. Res. Natl. Bur. Stand.* **39,** 507 (1947).
10. V. Schomaker and L. Pauling, *J. Am. Chem. Soc.* **61,** 1769 (1939).
11. L. C. Jones, Jr. and L. W. Taylor, *Anal. Chem.* **27,** 228 (1955).
12. R. S. Rasmussen, D. D. Tunnicliff, and R. R. Braittain, *J. Chem. Phys.* **11,** 432 (1943).
13. L. C. Pauling, *The Nature of the Chemical Bond and the Structure of Molecules and Crystals,* 3rd ed., Cornell University Press, Ithaca, N.Y., 1960.
14. D. J. Marais, N. Sheperd, and B. Stoicheff, *Tetrahedron* **17,** 163 (1962).
15. A. Almenningen, O. Bastiansen, and M. Traetteberg, *Acta Chem. Scand.* **12,** 1221 (1958).
16. G. W. Wheland, *Resonance in Organic Chemistry,* John Wiley & Sons, Inc., New York, 1955.
17. R. S. Mulliken, *J. Chem. Phys.* **7,** 1121 (1939).
18. M. J. S. Dewar and H. N. Schmeising, *Tetrahedron* **5,** 166 (1959); **11,** 97 (1960).
19. R. S. Mulliken, *Tetrahedron* **6,** 68 (1959).
20. R. B. Woodward and T. R. Katz, *Tetrahedron* **5,** 70 (1959); *Tetrahedron Lett.*(5), 19 (1959).
21. J. G. Martin and R. K. Hill, *Chem. Rev.* **61,** 537 (1961).
22. K. Alder, *Ann.* **571,** 157 (1951).
23. J. A. Norton, *Chem. Rev.* **31,** 319 (1942).
24. W. J. Bailey, "Butadiene", in E. C. Leonard, ed., *Vinyl and Diene Monomers* Part II, Chapt. 4, Wiley-Interscience, 1971.

25. J. C. Little, *J. Am. Chem. Soc.* **87,** 4020 (1965).
26. J. A. Berson and A. Remanick, *J. Am. Chem. Soc.* **83,** 4947 (1961).
27. S. Seltzer, *J. Am. Chem. Soc.* **87,** 1534 (1965); *Tetrahedron Lett.*(11), 457 (1962).
28. Brit. Pat. 835,840 (May 25, 1960), R. Robertson and G. I. Fray (to Shell Research Ltd.); G. I. Fray and R. Robertson, *J. Am. Chem. Soc.* **83,** 249 (1961).
29. T. Inukai and M. Kasai, *J. Organomet. Chem.* **30,** 3567 (1965).
30. T. Kojima and T. Inukai, *J. Organomet. Chem.* **35,** 1342 (1970).
31. H. W. Thompson and D. G. Melillo, *J. Am. Chem. Soc.* **92,** 3218 (1970).
32. S. Miyajima and T. Inukai, *Bull. Soc. Jpn.* **45,** 1553 (1972).
33. O. Diels and K. Alder, *Ann.* **460,** 98 (1928).
34. N. A. Chayanov, *J. Gen. Chem. USSR* **8,** 460 (1938).
35. U.S. Pat. 2,473,472 (June 14, 1949), E. Gorin and A. G. Oblad (to Socony-Vacuum Oil Co.).
36. U.S. Pat. 2,662,102 (Dec. 8, 1953), G. M. Whitman (to E. I. du Pont de Nemours & Co.).
37. J. L. Skinner and C. M. Sliepevich, *Ind. Eng. Chem. Fundam.* **2,** 168 (1963).
38. D. Rowley and H. Steiner, *Discuss. Faraday Soc.* **10,** 198 (1951).
39. H. D. Scharf and G. Zoche, *Chem. Ber.* **102,** 2478 (1969).
40. H. Reimlinger, E. H. de Ruiter, and U. K. Krüerke, *Chem. Ber.* **103,** 2317 (1970); U.S. Pat. 3,444,253 (May 13, 1969), (to Union Carbide Corporation).
41. H. Tropsch and W. Mattox, *Ind. Eng. Chem. Anal. Ed.* **6,** 104 (1934).
42. U.S. Pat. 3,359,285 (Dec. 19, 1967), P. S. Landis (to Mobil Oil Corp.).
43. J. G. Martin and R. K. Hill, *Chem. Rev.* **61,** 537 (1961).
44. A. Korolev and V. Mur, *Dokl. Akad. Nauk SSSR* **59,** 71 (1948).
45. K. Alder and H. F. Rickert, *Ber.* **71B,** 373 (1938).
46. Ger. Pat. 494,433 (May 8, 1928), (to I. G. Farbenindustrie A.G.).
47. K. Alder and G. Stein, *Ann.* **501,** 247 (1933).
48. K. Alder and G. Stein, *Angew. Chem.* **50,** 510 (1937).
49. U.S. Pat. 2,217,632 (Oct. 8, 1940), W. D. Wolfe (to Wingfoot Corp.).
50. A. A. Petrov and N. P. Sopov, *J. Gen. Chem. USSR Engl. Transl.* **17,** 2228 (1948).
51. H. J. Pistor and H. Plieninger, *Ann.* **562,** 239 (1949).
52. S. Lebedev and N. Skavronskaya, *J. Russ. Phys. Chem. Soc.* **43,** 1124 (1911).
53. U.S. Pat. 2,468,432 (Apr. 26, 1949), H. L. Johnson (to Sun Oil Co.).
54. U.S. Pat. 2,544,808 (Mar. 13, 1951), E. E. Stahley (to Koppers Co., Inc.).
55. A. A. Petrov and R. A. Shlyakhter, *Dokl. Akad. Nauk SSSR* **75,** 703 (1950).
56. A. A. Petrov, *Usp. Khim.* **22**(8), 905 (1953).
57. P. Heimbach, P. W. Jolly, and G. Wilke, *Adv. Organomet. Chem.* **8,** 29 (1970).
58. G. Wilke, *Angew. Chem., Int. Ed. Engl.* **2,** 105 (1963).
59. W. Brenner and co-workers, *Ann. Chem.* **727,** 161 (1969).
60. P. W. Jolly, I. Tkatchenko, and G. Wilke, *Angew. Chem., Int. Ed. Engl.* **10,** 329 (1971).
61. H. W. B. Reed, *J. Chem. Soc.,* 1931 (1954).
62. B. Bogdanovic and co-workers, *Ann. Chem.* **727,** 143 (1969).
63. M. F. Semmelhack, *Organic Reactions,* John Wiley & Sons, Inc., New York, 1971, Vol. 19, Chapt. 2.
64. G. Wilke and co-workers, *Angew. Chem. Int. Ed. Engl.* **5,** 151 (1966).
65. W. B. Hughes, *J. Organomet. Chem.* **36,** 4073 (1971).
66. J. Kiji, K. Masui, and J. Furukawa, *Bull. Chem. Soc. Jpn.* **44,** 1956 (1971).
67. J. Kiji and co-workers, *Bull. Chem. Soc. Jpn.* **46,** 1791 (1973).
68. H. Mueller and co-workers, *Angew. Chem. Int. Ed. Engl.* **4,** 327 (1965).
69. A. Yamamoto and co-workers, *J. Am. Chem. Soc.* **87,** 4652 (1965); **90,** 1878 (1968).
70. C. Y. Wu and H. E. Swift, *J. Catal.* **24,** 510 (1972).
71. J. P. Candlin and W. H. Janes, *J. Chem. Soc.* **C,** 1856 (1968).
72. P. Heimbach, *Angew. Chem. Int. Ed. Engl.* **12,** 975 (1973).
73. R. Baker, *Chem. Rev.* **73,** 487 (1973).
74. A. Carbonaro, *Chim. Ind. Milan* **55,** 244 (1973).
75. H. Breil and co-workers, *Makromol. Chem.* **69,** 18 (1963).
76. V. M. Akhemedov and co-workers, *Chem. Commun.* 777 (1974).
77. S. Takahashi, T. Shibano, and N. Hagihara, *Tetrahedron Lett.,* 2451 (1967); *Bull. Chem. Soc. Jpn.* **41,** 454 (1968).
78. A. Musco and A. Silvani, *J. Organomet. Chem.* **88,** C41 (1975).
79. J. F. Kohnle, L. H. Slaugh, and K. L. Nakamaye, *J. Am. Chem. Soc.* **91,** 5904 (1969).

80. T. Arakawa and H. Miyake, *Kogyo Kagaku Zasshi* **74**, 1143 (1971).
81. A. Carbonaro, A. Greco, and G. Dall'Asta, *Tetrahedron Lett.* 2037 (1967).
82. H. Takahasi, S. Tai, and M. Yamaguchi, *J. Organomet. Chem.* **30**, 1661 (1965).
83. H. Hidai, Y. Uchida, and A. Misono, *Bull. Chem. Soc. Jpn.* **38**, 1243 (1965).
84. A. Carbonaro and A. Greco, *J. Organomet. Chem.* **25**, 477 (1970).
85. T. Alderson, E. L. Jenner, and R. V. Lindsey, Jr., *J. Am. Chem. Soc.* **87**, 5638 (1965).
86. S. Otsuka, T. Kikuchi, and T. Taketomi, *J. Am. Chem. Soc.* **85**, 3709 (1963).
87. J. Beger and C. Duschek, *Z. Chem.* **12**(1), 18 (1972).
88. T. Saito, Y. Uchida, and A. Misono, *Bull. Chem. Soc. Jpn.* **37**, 105 (1964).
89. U.S. Pat. 3,393,245 (July 16, 1968), E. A. Zuech (to Phillips Petroleum Co.).
90. H. Matschiner, H. J. Kerrinnes, and K. Issleib, *Z. Anorg. Chem.* **380** (1971).
91. D. Young and M. L. H. Green, *J. Appl. Chem. Biotechnol.* **25**, 641 (1975).
92. N. Yamazaki and S. Murai, *Chem. Commun.*, 147 (1968).
93. P. Heimbach, *Angew. Chem. Int. Ed. Engl.* **7**, 882 (1968).
94. T. Ohta, K. Ebina, and N. Yamazaki, *Bull. Chem. Soc. Jpn.* **44**, 1321 (1971).
95. P. Heimbach and G. Wilke, *Ann. Chem.* **727**, 183 (1969).
96. W. Brenner, P. Heimbach, and G. Wilke, in ref. 95, p. 194.
97. W. Brenner and co-workers, *Angew. Chem. Int. Ed. Engl.* **8**, 753 (1969).
98. P. Heimbach, H. Selbeck, and E. Troxler, *Angew. Chem. Int. Ed. Engl.* **10**, 659 (1971).
99. G. Hata and D. Hoki, *J. Organomet. Chem.* **32**, 3754 (1967); G. Hata, *J. Am. Chem. Soc.* **86**, 3903 (1964).
100. R. G. Miller, T. J. Kealy, and A. L. Barney, *J. Am. Chem. Soc.* **89**, 3756 (1967).
101. N. Kawata and co-workers, *Bull. Chem. Soc. Jpn.* **47**, 2003 (1974).
102. D. Wittenberg, *Angew. Chem. Int. Ed. Engl.* **3**, 153 (1964).
103. M. Iwamoto and S. Yuguchi, *Chem. Commun.* **28**, (1968).
104. Y. Inoue, T. Kagawa, and H. Hashimoto, *Tetrahedron Lett.*, 1099 (1970).
105. T. Kagawa, Y. Inoue, and H. Hashimoto, *Bull. Chem. Soc. Jpn.* **43**, 1250 (1970).
106. R. S. Silas, J. Yates, and V. Thornton, *Anal. Chem.* **31**, 529 (1959).
107. D. Morero and co-workers, *Chim. Ind.* **41**, 758 (1959).
108. L. E. Forman in J. P. Kennedy and E. G. M. Tornqvist, eds., *Polymer Chemistry of Synthetic Elastomers*, Wiley-Interscience, New York, 1969, Chapt. 6.
109. W. M. Saltman in M. Morton, ed., *Rubber Technology*, 2nd ed., Van Nostrand Reinhold Company, New York, 1973.
110. G. Natta and L. Porri in J. P. Kenney and E. G. M. Tornqvist, eds., *Polymer Chemistry of Synthetic Elastomers*, Wiley-Interscience, New York, 1969, Chapt. 7.
111. C. F. Ruebensaal, *The Rubber Industry Statistical Report*, International Institute of Synthetic Rubber Producers, Inc., 1976.
112. U.S. Pat. 3,050,513 (Aug. 21, 1962), R. P. Zelinski and D. R. Smith (to Phillips Petroleum Company).
113. G. Natta and co-workers, *Chim. Ind.* **41**, 398 (1959).
114. N. G. Gaylord, T. W. Kwei, and H. F. Mark, *J. Poly. Sci.* **42**, 417 (1960).
115. P. H. Moyer and M. H. Lehr, *J. Poly. Sci. A* **3**, 217 (1965).
116. W. M. Saltman and T. H. Link, *Ind. Eng. Chem. Prod. Res. Dev.* **3**, 199 (1964).
117. J. F. Henderson, *J. Poly. Sci. C* **4**, 233 (1964).
118. W. Marconi and co-workers, *J. Polym. Sci. A* **3**, 735 (1965).
119. A. Mazzei and co-workers, *J. Polym. Sci. A* **3**, 753 (1965).
120. W. Marconi and co-workers, *Chim. Ind.* **51**, 1084 (1969).
121. C. Longrave, R. Castelli, and G. R. Croce, *Chim. Ind.* **43**, 625 (1961).
122. M. Gippin, *Ind. Eng. Chem. Prod. Res. Dev.* **1**, 32 (1962); **4**, 160 (1965).
123. A. I. Draconescu and S. S. Medvedev, *J. Polym. Sci. A* **3**, 31 (1965).
124. C. E. H. Bawn, *Rubber Plast. Age* **43**, 510 (1965).
125. J. G. Balas, H. E. DeLaMare, and D. O. Schissler, *J. Polym. Sci. A* **3**, 2243 (1965).
126. H. Scott and co-workers, *J. Polym. Sci.* **2**, 3233 (1964).
127. D. E. O'Reilly and co-workers, *J. Polym. Sci.* **2**, 3257 (1964).
128. W. Cooper, *Rubber Plast. Age*, **44**, 44 (1963).
129. A. Takahashi and S. Kambara, *J. Polym. Sci. B* **3**, 279 (1965).
130. F. Dawans and P. Teyssie, *Ind. Eng. Chem. Prod. Res. Dev.* **10**(3), 261 (1971).
131. R. S. Hanmer and H. E. Railsback in M. Morton, ed., *Rubber Technology*, 2nd ed., Van Nostrand Reinhold Company, New York, 1973.

132. J. A. Collingwood, *Rubber J.* **67,** (Sept. 1970).
133. *Europ. Rubber J.* **25,** (Nov. 1973).
134. E. Susa, *J. Polym. Sci. C* **4,** 399 (1963).
135. Ital. Pat. 536,631 (Dec. 7, 1955), G. Natta, L. Porri, and M. Mazzanti (to Montecatini S.p.A.).
136. G. Natta and co-workers, *Chim. Ind.* **40,** 362 (1958); **41,** 116 (1959).
137. R. E. Rinehart and co-workers, *J. Am. Chem. Soc.* **83,** 4864 (1961); **84,** 4145 (1962).
138. U.S. Pat. 3,025,286 (Mar. 13, 1962), H. P. Smith and G. Wilkinson (to U.S. Rubber Company).
139. W. G. Young and co-workers, *J. Am. Chem. Soc.* **69,** 2046 (1947).
140. U.S. Pat. 1,982,536 (Nov. 27, 1935), G. A. Perkins (to Union Carbide and Carbon Chemicals Corp.).
141. U.S. Pat. 3,408,415 (Oct. 29, 1968), F. S. Dovell and H. Greenfield (to Uniroyal Inc.).
142. Y. Furukawa and co-workers, *Bull. Jpn. Petrol. Inst.* **15**(1), 56 (1973).
143. *Ibid*, p. 64.
144. *Ibid*, p. 71.
145. U.S. Pat. 3,478,123 (Nov. 11, 1969), J. F. Brennen (to Universal Oil Products Co.).
146. U.S. Pat. 3,481,999 (Dec. 2, 1969), M. U. Reich (to Chemische Werke Huls).
147. F. W. Kirsch and S. E. Shull, *Ind. Eng. Chem., Prod. Res. Dev.* **2**(1), (1963).
148. Y. Tajima and E. Kunioka, *J. Org. Chem.* **23,** 1689 (1968).
149. J. Kwiatek and co-workers, *Adv. Chem.* (37), 201 (1963).
150. M. G. Burnett, P. J. Connolly, and C. Kemball, *J. Chem. Soc. A,* 991 (1968).
151. (a) S. Takahashi, H. Yamazaki, and N. Hagihara, *Bull. Chem. Soc. Jpn.* **41,** 254 (1968); (b) S. Takahashi, T. Shibano, and N. Hagihara, *Tetrahedron Lett.,* 2451 (1967).
152. S. Takahashi, T. Shibano, and N. Hagihara, *Bull. Chem. Soc. Jpn.* **41,** 454 (1968).
153. W. E. Walker and co-workers, *Tetrahedron Lett.,* 3817 (1970).
154. U.S. Pat. 3,534,088 (Oct. 13, 1970), D. R. Bryant and J. E. McKeon (to Union Carbide Corp.).
155. D. Rose and H. Lepper, *J. Organomet. Chem.* **49,** 473 (1973).
156. W. E. Walker and co-workers, *Tetrahedron Lett.,* 3817 (1970).
157. U.S. Pat. 3,518,315 (June 30, 1970), E. J. Smutny (to Shell Oil Company).
158. K. E. Atkins, W. E. Walker, and R. M. Manyik, *Chem. Commun.,* 330 (1971).
159. T. C. Shields and W. E. Walker, *Chem. Commun.,* 193 (1971).
160. S. Takahashi, T. Shibano, and N. Hagihara, *Chem. Commun.,* 161 (1969).
161. P. Haynes, *Chem. Commun.,* 3687 (1970).
162. R. M. Manyik and co-workers, *Chem. Commun.,* 3813 (1970).
163. K. Ohno, T. Mitsuyasu, and J. Tsuji, *Chem. Commun.,* 67 (1971).
164. K. Ohno, T. Mitsuyasu, and J. Tsuji, *Tetrahedron* **28,** 3705 (1972).
165. K. Ohno and J. Tsuji, *Chem. Commun.,* 247 (1971).
166. G. Hata, K. Takahashi, and A. Miyake, *Chem. Ind. London,* 1836 (1969).
167. G. Hata, K. Takahashi, and A. Miyake, *J. Org. Chem.* **36,** 2116 (1971).
168. K. Takahashi, A. Miyake, and G. Hata, *Bull. Chem. Soc. Jpn.* **45,** 1183 (1972).
169. T. Mitsuyasu and J. Tsuji, *Tetrahedron* **30,** 831 (1974).
170. M. S. Kharasch, J. Kritchevsky, and F. R. Mayo, *J. Org. Chem.* **2,** 489 (1937).
171. U.S. Pat. 2,123,504 (July 12, 1938), H. B. Dykstra (to E. I. du Pont de Nemours & Co.).
172. R. Voigt, *J. Prakt. Chem.* **151,** 3071 (1938).
173. H. Kubota, *Rev. Phys. Chem. Jpn.* **37**(1), 25 (1967).
174. A. A. Petrov, *Zh. Obsch. Khim.* **8,** 131 (1938).
175. U.S. Pat. 3,093,690 (June 11, 1963), P. H. Moss (to Jefferson Chemical Co.).
176. A. A. Petrov, *Zh. Obsch, Khim* **19,** 1046 (1949).
177. R. F. Taylor and G. H. Morey, *Ind. Eng. Chem.* **40,** 432 (1948).
178. U.S. Pat. 2,484,042 (Oct. 11, 1949), P. Mahler (to Publicker Industries, Inc.).
179. U.S. Pat. 2,299,477 (Oct. 20, 1943), G. W. Heurne and D. S. LaFrance (to Shell Development Co.).
180. U.S. Pat. 2,453,089 (Nov. 2, 1948), G. H. Morey and R. F. Taylor (to Commercial Solvents).
181. U.S. Patent 2,581,929 (Jan. 8, 1952), K. C. Eberly and R. J. Reid (to Firestone Tire and Rubber Co.).
182. U.S. Pat. 3,050,568 (Aug. 21, 1962), R. P. Arganbright (to Monsanto Chemical).
183. Brit. Pat. 669,338 (Apr. 2, 1952), R. C. Chuffart (to Imperial Chemical Industries).
184. Brit. Pat. 676,691 (July 30, 1952), R. C. Chuffart (to Imperial Chemical Industries).
185. Brit. Pat. 798,027 (July 16, 1958), H. P. Crocker, C. W. Capp, and B. J. Bellringer (to Distillers Co. Ltd.).
186. Brit. Pat. 798,393 (July 23, 1958), C. W. Capp and co-workers, (to Distillers Co. Ltd.).

187. Brit. Pat. 800,787 (Sept. 2, 1958), F. J. Bellringer and H. P. Crocker (to Distillers Co. Ltd.).
188. Ger. Pat. 1,115,236 (Oct. 19, 1961), M. Minsinger (to BASF).
189. Ger. Pat. 1,118,189 (Nov. 30, 1961), N. W. Luft, K. Essler, and H. Waider (to Hans J. Zimmer Verfahrenstechnik).
190. A. J. Besozzi and W. H. Taylor, *Commercial Production of Chloroprene via Butadiene* paper presented before Division of Petroleum Chemistry, ACS Meeting, Fall, 1972.
191. I. E. Musket and H. E. Northrup, *J. Am. Chem. Soc.* **52**, 4043 (1930).
192. K. Mislow and H. M. Hellman, *J. Am. Chem. Soc.* **73**, 244 (1951).
193. E. H. Farmer, C. D. Lawrence, and J. C. Thorpe, *J. Chem. Soc.*, 729 (1928).
194. N. G. V'yunova, *Izv. Akad. Nauk. SSSR, Ser. Khim* (3), 467 (1964).
195. H. Kubota, *Revs. Phys. Chem. Jpn.* **37**(1), 32 (1967).
196. Brit. Pat. 1,007,077 (Oct. 13, 1965), (to Shell International Research).
197. Brit. Pat. 1,139,516 (Jan. 8, 1969), A. Lambert and G. K. Makinson (to Imperial Chemicals Industries Ltd.).
198. J. C. Hillyer and P. S. Stallings, *Petrol. Refiner.* **35**(12), 157 (1956).
199. R. W. Moncrieff, *Man-Made Fibres,* 6th ed., John Wiley & Sons, Inc., New York, 1975, p. 341.
200. J. H. Prescott, *Chem. Eng.* **47,** (Feb. 8, 1971).
201. U.S. Pat. 3,406,215 (Oct. 15, 1968), H. E. Hanquist (to E. I. du Pont de Nemours & Co.).
202. Y. Tsutsumi, *Chem. Econ. Eng. Rev. Jpn.* **8**(5), 45 (1976).
203. U.S. Pat. 3,922,300 (Nov. 25, 1975), T. Onoda and co-workers, (to Mitsubishi Chemical Industries Ltd.).
204. Ger. Pat. 2,414,341 (Oct. 23, 1975), H. M. Weitz and J. Hartig (to Badische Anilin & Soda Fabrik).
205. Belg. Pat. 832,254 (Feb. 9, 1976), (to Badische Anilin & Soda Fabrik).
206. U.S. Pat. 3,755,423 (Aug. 38,1973), T. Onoda and J. Haji (to Mitsubishi Chemical Industries Ltd.).
207. U.S. Pat. 3,872,163 (Mar. 18, 1975), T. Shimizu, T. Yasui, and S. Nakamura (to Kuraray Co. Ltd.).
208. Belg. Pat. 827,837 (Oct. 13, 1975), (to Badische Anilin & Soda Fabrik).
209. Ger. Pat. 2,354,218 (May 15, 1975), H. Mueller, G. Daumiller, and H. Hoffman, (to Badische Anilin & Soda Fabrik).
210. Ger. Pat. 2,345,160 (Mar. 21, 1974), T. Onoda, A. Ohno, and K. Shiraga (to Mitsubishi Chemical Industries Ltd.).
211. J. Tsuji, *Acc. Chem. Res.* **6,** 8 (1973).
212. J. Tsuji, J. Kiji, and S. Hosaka, *Tetrahedron Lett.* 605 (1964).
213. S. Hosaka and J. Tsuji, *Tetrahedron* **27,** 3821 (1971).
214. (a) W. E. Billups, W. E. Walker, and T. C. Shields, *Chem. Commun.,* 1067 (1971); (b) J. Tsuji, Y. Mori, and M. Hara, *Tetrahedron* **28,** 3721 (1972).
215. A. Matsuda, *Bull. Chem. Soc. Jpn.* **46,** 524 (1973).
216. (a) H. Adkins and J. L. R. Williams, *J. Org. Chem.* **17,** 980 (1952); (b) U.S. Pat. 2,517,383 (Aug. 1, 1950), R. E. Brooks (to E. I. du Pont de Nemours & Co.).
217. L. F. Hatch, *The Chemistry of Petrochemical Reactions,* Gulf Publishing Co., Houston, Texas, 1955.
218. B. Fell and W. Rupilius, *Tetrahedron Lett.*, 2721 (1969).
219. U.S. Pat. 3,876,695 (Apr. 8, 1975), N. von Kutepow (to Badische Anilin & Soda Fabrik).
220. T. Inukai, *J. Org. Chem.* **31,** 24 (1966).
221. V. N. Ipatieff, H. Pines, and R. E. Schaad, *J. Am. Chem. Soc.* **66,** 816 (1944).
222. W. Proell, *J. Org. Chem.* **16,** 178 (1951).
223. U.S. Pat. 2,382,260 (Aug. 14, 1945), R. E. Schaad (to Universal Oil Products Co.).
224. V. N. Ipatieff and co-workers, *J. Am. Chem. Soc.* **67,** 1060 (1945).
225. U.S. Pat. 2,403, 963 (July 16, 1946), W. N. Axe (to Phillips Petroleum Co.).
226. H. Pines and co-workers, *J. Am. Chem. Soc.* **73,** 5173 (1951).
227. U.S. Pat. 3,160,674 (Dec. 8, 1964), L. G. Cannell and G. Holzman (to Shell Oil Co.).
228. Brit. Pat. 315,312 (July 11, 1928), (to I. G. Farbenindustrie).
229. K. Ziegler, F. Dersch, and H. Wollthan, *Ann.* **511,** 13 (1934).
230. U.S. Pat. 2,352,461 (June 27, 1944), J. F. Walker (to E. I. du Pont de Nemours & Co.).
231. *Chem. Week* **60,** (Dec. 10, 1955).
232. G. G. Schneider, H. Bock, and H. Hausser, *Ber.* **70B,** 425 (1937).
233. G. J. Janz, R. G. Ascah, and A. G. Keenan, *Can. J. Res.* **25B,** 272 (1947).
234. E. Muller and O. Roser, *J. Prakt. Chem.* **133,** 291 (1932).
235. K. S. Sidhu and co-workers, *J. Am. Chem. Soc.* **88,** 254 (1966).
236. W. J. Hickenbottom, *J. Chem. Soc.*, 1981 (1934).

237. R. Wegler and G. Pieper, *Ber.* **83,** 6 (1950).
238. T. Narita, N. Imai, and T. Tsuruta, *Bull. Chem. Soc. Jpn.* **46,** 1242 (1973).
239. E. A. Zuech, R. F. Kleinschmidt, and J. E. Mahan, *J. Org. Chem.* **31,** 3713 (1966).
240. K. Ziegler and L. Jacob, *Ann.* **511,** 45 (1934).
241. K. Ziegler and co-workers, *Ann.* **567,** 43 (1950).
242. A. A. Morton and co-workers, *J. Am. Chem. Soc.* **68,** 93 (1946).
243. A. A. Morton, M. L. Brown, and E. Magat, *J. Am. Chem. Soc.* **69,** 161 (1947).
244. G. Zweifel, K. Nagase, and H. C. Brown, *J. Am. Chem. Soc.* **84,** 183 (1962).
245. H. C. Brown, E. Negishi, and S. K. Gupta, *J. Am. Chem. Soc.* **92,** 2460 (1970).
246. R. Koster, *Angew, Chem.* **71,** 520 (1959); **72,** 626 (1960).
247. V. Hasserodt, K. Hunger, and F. Korte, *Tetrahedron* **19,** 1563 (1963); **20,** 1593 (1964).
248. U.S. Pat. 1,663,736-7 (Dec. 22, 1953), W. B. McCormack (to E. I. du Pont de Nemours & Co.).
249. N. A. Razumova and A. A. Petrov, *Zh. Obshch. Khim.* **33,** 783 (1963).
250. H. Staudinger and B. Rilzenthaler, *Ber.* **68B,** 455 (1935).
251. U.S. Pat. 2,420,834 (May 20, 1947), R. C. Morris and H. de V. Finch (to Shell Oil Co.).
252. L. R. Drake, S. C. Stowe, and A. M. Partansky, *J. Am. Chem. Soc.* **68,** 252 (1946).
253. D. Craig, *J. Am. Chem. Soc.* **65,** 1006 (1943).
254. O. Grummitt, A. E. Ardis, and J. Fick, *J. Am. Chem. Soc.* **72,** 5167 (1950).
255. A. F. Shepard, A. L. Henne, and T. J. Midgley, *J. Am. Chem. Soc.* **56,** 1355 (1934).
256. H. E. Rasmussen, A. L. Hansford, and A. N. Sachanen, *Ind. Eng. Chem.* **38,** 376 (1946).
257. U.S. Pat. 2,410,401 (Oct. 29, 1946), D. D. Coffman (to E. I. du Pont de Nemours & Co.).
258. B. A. Arbuzov and E. G. Katiev, *Dokl. Akad. Nauk SSSR* **96,** 983 (1954).
259. Brit. Pat. 603,103 (June 9, 1948), B. S. Greensfelder and H. J. Moore (to Shell Oil Co.).
260. V. Franzen, *Chem. Ber.* **95,** 571 (1962).
261. B. Grzybowska, J. H. Knox, and A. F. Trotman-Dickenson, *J. Chem. Soc.,* 4402 (1961).
262. H. M. Frey, *Trans. Faraday Soc.* **58,** 516 (1962).
263. J. Kiji and M. Iwamoto, *Tetrahedron Lett,* 2749 (1966).
264. R. C. Woodworth and P. S. Skell, *J. Am. Chem. Soc.* **79,** 2542 (1957).
265. M. Orchin and E. C. Herrick, *J. Org. Chem.* **24,** 139.
266. P. S. Skell and A. Y. Gardner, *J. Am. Chem. Soc.* **78,** 5430 (1956).
267. E. Ciganek, *J. Am. Chem. Soc.* **88,** 1979 (1966).
268. P. S. Skell and R. R. Engel, *J. Am. Chem. Soc.* **88,** 3749 (1966).
269. P. S. Skell and co-workers, *J. Am. Chem. Soc.* **87,** 2829 (1965).
270. D. D. Coffman and co-workers, *J. Am. Chem. Soc.* **71,** 490 (1949).
271. P. D. Bartlett, L. K. Montgomery, and B. Seidel, *J. Am. Chem. Soc.* **86,** 616 (1964); **86,** 628 (1964).
272. (a) E. Vogel and K. Muller, *Ann. Chem.* **615,** 29 (1958); (b) J. C. Martin and co-workers, *J. Org. Chem.* **30,** 4175 (1965).
273. W. L. Dilling and R. D. Kroening, *Tetrahedron Lett.* 695 (1970).
274. M. S. Kharasch and M. Sage, *J. Org. Chem.* **14,** 537 (1949).
275. M. S. Kharasch and co-workers, *Science* **113,** 392 (1951).
276. M. S. Kharasch, F. S. Arimoto, and W. Nudenberg, *J. Org. Chem.* **16,** 1556 (1951).
277. (a) M. S. Kharasch, P. Pauson, and W. Nudenberg, *J. Org. Chem.* **18,** 322 (1953); (b) p. 328.
278. M. S. Kharasch and W. Nudenberg, *J. Org. Chem.* **19,** 1921 (1954).
279. M. S. Kharasch, F. Kawahara, and W. Nudenberg, *J. Org. Chem.* **19,** 1977 (1954).
280. M. S. Kharasch, W. Nudenberg, and F. Kawahara, *J. Org. Chem.* **20,** 1550 (1955).
281. C. J. Albisetti and co-workers, *J. Am. Chem. Soc.* **81,** 1489 (1959).
282. J. B. Conant and B. F. Chow, *J. Am. Chem. Soc.* **55,** 3475 (1933).
283. R. Klein and M. D. Scheer, *J. Phys. Chem.* **67,** 1874 (1963).
284. C. S. H. Chen and R. F. Stamm, *J. Org. Chem.* **28,** 1580 (1963).
285. M. Asscher and D. Vofsi, *J. Chem. Soc.,* 1887 (1963).
286. A. V. Dombrovskii and co-workers, *Zh. Obshch. Khim.* **26,** 2776 (1956); **27,** 2000 (1956); **31,** 1284 (1961).
287. A. A. Oswald and co-workers, *Am. Chem. Soc. Div. Petrol. Chem. Prep.* **7**(3), 139 (1962).
288. E. Muller and A. Freytag, *J. Prakt. Chem.* **146,** 58 (1936).
289. C. R. Porter and B. Wood, *J. Inst. Petrol.* **38,** 877 (1952).
290. R. V. Lindsey, Jr. and M. L. Peterson, *J. Am. Chem. Soc.* **81,** 2073 (1959).
291. U.S. Pat. 2,097,094 (Nov. 2, 1937), C. H. Walters (to Union Carbide Corp.).
292. M. Matsumoto, T. Ikawa, and N. Nagasako, *Kogyo Kagaku Zasshi* **63,** 734 (1960).
293. R. H. Bretton, Shen-Wu Wan, and B. F. Dodge, *Ind. Eng. Chem* **44,** 594 (1952).

294. U.S. Pat. 2,260,409 (Oct. 28, 1941), O. C. Slotterbeck and S. W. Tribit (to Standard Oil Development Co.).
295. U.S. Pat. 2,097,904 (Nov. 2, 1937), C. H. Walters (to Carbide and Carbon Chemicals Corp.).
296. M. Akimoto and Echigoya, *J. Catal.* **29,** 191 (1973).
297. M. Akimoto and E. Echigoya, *Chem. Lett.,* 305 (1972).
298. M. Ai, *Kogyo Kagaku Zasshi* **73,** 950 (1970).
299. M. Ai and S. Suzuki, *J. Catal.* **26,** 202 (1972).
300. M. Ai, *Bull. Chem. Soc. Jpn.* **43,** 3490 (1970).
301. D. G. Hendry and co-workers, *Ind. Eng. Chem. Prod. Res. Dev.* **7,** 145 (1968).
302. R. F. Robey, H. K. Wiese, and C. E. Morrell, *Ind. Eng. Chem.* **36,** 3 (1944).
303. D. S. Alexander, *Ind. Eng. Chem.* **51,** 733 (1959).
304. D. G. Hendry, F. R. Mayo, and D. Schuetzle, *Ind. Eng. Chem. Prod. Res. Dev.* **7,** 136 (1968).
305. D. G. Hendry in N. M. Bikales, ed., *Encyclopedia of Polymer Science and Technology,* Interscience, New York, Vol. 9, p. 807, 1968.
306. C. T. Handy and H. S. Rothrock, *J. Am. Chem. Soc.* **80,** 5306 (1958); U.S. Pat, 2,898,377 (Aug. 4, 1959), (to E. I. du Pont de Nemours & Co.).
307. G. S. Whitby, *Ind. Eng. Chem.* **47,** 806 (1955).
308. E. H. Immergut, *Makromol. Chem.* **10,** 93 (1953).
309. G. H. Miller, R. L. Alumbaugh, and R. J. Brotherton, *J. Polym. Sci.* **9,** 453 (1952).
310. G. H. Miller, V. R. Larson, and G. O. Pritchard, *J. Polym. Sci.* **61,** 475 (1962).
311. G. H. Miller, R. R. Eliason, and G. O. Pritchard, *J. Polym. Sci.* **C**(4), 1109 (1963).
312. U.S. Pat. 2,976,223 (Mar. 21, 1961), M. Kovach and W. H. Rideout (to Columbia-Southern Chem. Corp.).
313. U.S. Pat. 2,785,185 (Mar. 12, 1957), B. Phillips and P. S. Starcher (to Union Carbide and Carbon Corp.).
314. R. G. Kadesch, *J. Am. Chem. Soc.* **68,** 41 (1946).
315. U.S. Pat, 2,861,084 (Nov. 18, 1958), P. S. Starcher, O. L. MacPeek, and B. Phillips (to Union Carbide Corp.).
316. C. C. Spencer and co-workers, *J. Org. Chem.* **5,** 610 (1940).
317. R. D. W. Kemmitt and D. W. A. Sharp, *J. Chem. Soc.* 2567 (1963).
318. J. Smidt, *Chem. Ind.* **54,** (1962).
319. U.S. Pat. 2,478,243 (Aug. 9, 1949), C. S. Coe and T. F. Doumani (to Union Oil Co.).
320. U.S. Pat. 3,479,392 (Nov. 18, 1969), E. W. Stern and M. L. Spector (to W. R. Grace and Co.).
321. P. A. Argabright and co-workers, *J. Org. Chem.* **30,** 3233 (1965).
322. Z. N. Kolyaskina and A. A. Petrov, *Zh. Obshch. Khim* **32,** 1089 (1961).
323. U.S. Pat. 2,486,657 (Nov. 1, 1949), G. M. Kosolapoff (to Monsanto Chemical Co.).
324. A. P. Terentev and A. N. Dombrovskii, *Zh. Obshch. Khim.* **21,** 704 (1951); *Doklady Akad. Nauk SSSR* **67,** 859 (1949).
325. H. E. Armstrong and A. K. Miller, *J. Chem. Soc.* **49,** 80 (1886).
326. B. G. Egloff and G. Hulla, *Chem. Rev.* **35,** 279 (1944).
327. G. Egloff and G. Hulla, *Oil Gas J.* **41**(31), 45; (32), 36 (1942).
328. A. A. Appleton, *J. Inst. Petrol.* **46,** 367 (1960).
329. J. A. R. Bennett, *Chem. & Ind. London,* 410 (1961).
330. (a) C. E. Morrell in G. S. Whitby, ed., *Synthetic Rubber,* John Wiley & Sons, Inc., New York, 1954, Chapt. 3 p. 56; (b) W. J. Toussaint and J. L. Marsh, Chapt. 4, p. 86; (c) H. L. Fisher, Chapt. 5 p 105.
331. J. C. Reidel, *Oil Gas J.* **55**(45) 166; (48), 87; (49), 114; (50), 110; (51), 74 (1957).
332. (a) S. K. Bhattacharyya and N. D. Ganguly, *J. Appl. Chem.* **12**(3), 97 (1962); (b) S. K. Bhattacharyya and B. N. Avasthi, *Ind. Eng. Chem. Proc. Design Dev.* **2**(1), 45 (1963).
333. Brit. Pat. 331,482 (June 30, 1930), S. V. Lebedev.
334. *BIOS (British Intelligence Objectives Subcommittee) Report 1060,* Feb. 28, 1947.
335. A. Talalay and L. Talalay, *Rubber Chem. Technol.* **15,** 403 (1942).
336. U.S. Pat. 2,357,855 (Sept. 12, 1944), W. Szukiewicz.
337. I. I. Ostromislensky, *J. Russ. Phys. Chem. Soc.* **47,** 1472 (1915).
338. B. B. Corson and co-workers, *Ind. Eng. Chem.* **41,** 1012 (1949).
339. *Ibid.,* **42,** 359 (1950).
340. W. J. Toussaint, J. T. Dunn, and D. R. Jackson, *Ind. Eng. Chem.* **39,** 120 (1949).
341. P. M. Kampmeyer and E. E. Stahly, *Ind. Eng. Chem.* **41,** 550 (1949).
342. W. M. Quattlebaum, W. J. Toussainte, and J. T. Dunn, *J. Am. Chem. Soc.* **69,** 593 (1947).

343. H. E. Jones, E. E. Stahly, and B. B. Corson, *J. Am. Chem. Soc.* **71**, 1822 (1949).
344. J. A. Philpot, *Euro. Chem. News, Large Plants Suppl.* **27**(707), 15 (Oct. 17, 1975).
345. R. H. Ebel, *Oil and Gas J.* **116,** (Apr. 1, 1968).
346. (a) J. P. Kennedy and I. Kirshenbaum in E. C. Leonard, ed., *Vinyl and Diene Monomers,* Part 2, Wiley-Interscience, New York, 1971, Chapt. 3, p. 691; (b) R. E. Haney in E. C. Leonard, ed., *Vinyl and Diene Monomers,* Part 2, Wiley-Interscience, New York, 1971, Chapt. 2, p. 577.
347. (a) J. F. Freiling and A. A. Simone, *Pet. Petrochem. Int.* **13**(2), 30 (1973); (b) B. Schleppinghoff, *Erdol u Kohle* **27**(5), 240 (1974).
348. K. Kearby, in B. T. Brooks and co-workers, eds., *The Chemistry of Petroleum Hydrocarbons,* Vol. 2, Chapt. 30, Reinhold, New York, 1955, p. 224.
349. C. L. Thomas, *Catalytic Processes and Proven Catalysts,* Academic Press, New York, 1970.
350. H. E. Swift, H. Beuther, and R. J. Rennard, Jr., *Ind. Eng. Chem. Prod. Res. Dev.* **15**(2), 131 (1976).
351. S. Carra and L. Formi, *Catal. Rev.* **5**(1), 159 (1971).
352. G. D. Lyubarskii, S. K. Merilyainen, and S. Y. Psliezhetskii, *Zh. Fiz. Khim.* **28**, 1272 (1954).
353. J. Happel, H. Blanck, and T. D. Hamill, *Ind. Eng. Chem. Fundam.* **5**(3), 289 (1966).
354. S. Noda, R. R. Hudgins, and P. L. Silveston, *Can. J. Chem. Eng.* **45**, 294 (1967).
355. *Hydrocarbon Process.* **50**(11), 136 (1971).
356. (a) K. Kearby in P. H. Emmett, ed., *Catalysis,* Reinhold, New York, 1955, Vol 3, Chapt. 10; (b) K. Kearby, *Ind. Eng. Chem.* **42,** 295 (1950).
357. U.S. Pat. 3,501,547 (Mar. 17, 1970), G. J. Nolan, R. J. Hogan, and F. Farha, Jr. (to Phillips Petroleum).
358. P. C. Husen, K. R. Deel, and W. D. Peters, *Oil Gas J.* **69,** 60 (Aug. 2, 1971).
359. F. C. Newman, *Ind. Eng. Chem.* **62**(5), 42 (1970).
360. R. W. King, *Hydrocarbon Process.* **45**(11), 189 (1966).
361. U.S. Pat. 2,890,253 (May 16, 1956), R. D. Mullineaux and J. H. Raley (to Shell Development Co.).
362. Brit. Pat. 973,564 (Oct. 14, 1960) F. Wattimena and W. F. Engel (to Shell International Research).
363. U.S. Pats. 3,359,343 (Dec. 19, 1967) and 3,374,283 (Mar. 17, 1968), L. Bajars (to Petro-Tex Chemical Corp.).
364. M. Vadekar and I. S. Pasternak, *Can. J. Chem. Eng.* **48,** 664 (1970).
365. D. V. Tishchenko, *J. Gen. Chem. USSR Engl. Transl.* **17,** 460 (1947).
366. E. R. Gilliland and H. M. Lavender, Jr., *India Rubber World III,* (1), 67 (Oct. 1944).
367. B. B. Randall, *Inst. Petrol. Rev.* **2,** 107 (1948).
368. S. A. Miller, *Acetylene: Its Properties, Manufacture and Uses,* Vol. 1, Academic Press, New York, 1965.
369. A. E. Hanford and D. L. Fuller, *Ind. Eng. Chem.* **40,** 1171 (1948).
370. Neth. Pat. 74/7755 (Dec. 13, 1974) H. Tshii and co-workers (to Mitsubishi Rayon Co.).
371. Ger. Pat. 2,554,648 (June 24, 1976) M. Kawakami, N. Audon, and A. Toi (to Japan Synthetic Rubber).
372. Ger. Pat. 2,342,328 (Mar. 21, 1974) S. Umemura and co-workers, (to Ubbe Industries Ltd.).
373. Ger. Pat. 2,258,821 (July 5, 1973) W. G. Shaw and R. K. Grasselli (to Standard Oil Co. Ohio).
374. U.S. Pat. 4,000,176 (Dec. 18, 1976) T. Yoshino and co-workers (to Nitto Chemical Industries Co. Ltd.).
375. T. Reis, *Chem. Proc. Eng.* **51**(3) 65, 1970).
376. C. E. Starr, Jr. and W. F. Ratcliff, *Ind. Eng. Chem.* **38**(10) 1020 (1946).
377. C. E. Morrell and co-workers, *Trans. Am. Inst. Chem. Eng.* **42,** 473 (1946).
378. (a) U.S. Pat. 2,369,559 (Feb. 13, 1945), E. R. Gilliland (to Jasco, Inc.); (b) U.S. Pat. 2,429,134 (Oct. 14, 1947), C. E. Morrell and M. W. Swaney (to Jasco, Inc.).
379. U.S. Pat. 2,788,378 (Apr. 9, 1957), W. S. Cotton, P. C. Kupa, and G. W. Taylor (to Polymer Corp.).
380. U.S. Pat. 2,847,487 (Aug. 12, 1958), W. N. Kestner, W. C. Kohfeldt, and S. N. Eubank (to Esso Research and Engineering Company).
381. U.S. Pat. 3,192,282 (June 29, 1965), C. E. Porter (to Esso Research and Engineering Company).
382. U.S. Pat. 2,985,697 (May 23, 1961), R. P. Cahn (to Esso Research and Engineering Company).
383. H. D. Evans and D. H. Sarno, *Proc. 7th World Petrol. Congr.* **5,** 259 (1967).
384. S. Y. Pavlov and co-workers, *Soviet Chem. Ind.* (11), 9 (1970).
385. J. W. Hobbs, R. S. Josephson, and H. L. Rogers, *Oil Gas J.* **73,** 76 (Jan. 20, 1975).
386. S. Takao, *Hydrocarbon Process.* **45**(11), 151 (1966).
387. U.S. Pats. 3,436,436 and 3,436,438 (April 1, 1969), S. Takao and H. Hokari (to Japanese Gen Co.).
388. T. Yoshimura, *Chem. Eng.* **73**(10), 134 (1966).

389. N. Bushin and co-workers, *Soviet Chem. Ind.* (1), 8 (1971).
390. Brit. Pat. 1,158,566 (July 16, 1969), R. R. Bannister, G. W. Harris, and J. F. Boston (to Union Carbide).
391. W. W. Coogler, Jr., *Oil Gas J.* **65**(21), 99 (1967); *Chem. Eng.* **74**(16) 70 (1967); *Hydrocarbon Process.* **46**(5), 166 (1967).
392. R. R. Bannister and E. Buck, *Chem. Eng. Prog.* **65**(9), 65 (1969).
393. U. Wagner and H. M. Weitz, *Ind. Eng. Chem.* **62**(4), 43 (1970).
394. H. Kroeker and H. M. Weitz, *Oil Gas J.* **65**(2), 98 (1967).
395. P. Ellwood, *Chem. Eng.* **75**(20), 172 (1968).
396. *Oil Gas J.* **67**(42), 86 (1969).
397. U.S. Pat. 3,328,480 (June 27, 1967), J. W. Begley, L. C. Kahre, and R. W. Carney (to Phillips Petroleum Co.).
398. N. Poffenberger and co-workers, *Trans. Am. Inst. Chem. Engr.* **42**, 815 (1946).
399. W. Hunsmann, *Chem. Ing. Tech.* **33**(8), 537 (1961).
400. R. B. Long, *Chem. Eng. Progr. Symp. Ser.* **66**(103), 82 (1972).
401. E. R. Gilliland, H. L. Bliss, and C. E. Kip, *J. Am. Chem. Soc.* **63**, 2088 (1941).
402. U.S. Pat. 3,206,521 (Sept. 14, 1965), R. B. Long (to Esso Research and Engineering Company).
403. H. J. Bittrich, *Z. Chem.* **10**(6), 201 (1970).
404. D. Tassios, *Chem. Eng.* **76**(3), 118 (1969).
405. D. C. Cronauer and L. D. Moore, *Ind. Eng. Chem. Fundam.*, **8**(4), 734 (1969).
406. B. S. Rawat, K. L. Mallik, and I. B. Gulati, *J. Appl. Chem. Biotechnol.* **22**, 1001 (1972).
407. U.S. Pat. 3,274,286 (Sept. 20, 1966), M. Reich (to Chemische Werke Huels).
408. Ref. 349, Chapt. 12.
409. Brit. Pat. 993,515 (May 26, 1965), (to Shell Internationale Research).
410. U.S. Pat. 3,293,316 (Dec. 20, 1966), H. Clay (to Phillips Petroleum Co.).
411. J. Osugi, H. Kubota, and K. Ueba, *Rev. Phys. Chem. Jpn.* **35**(1), 38 (1965).
412. J. H. Buehler and co-workers, *Chem. Eng.* **77**(19), 77 (Sept. 7, 1970); S. Griffith and R. G. Keister, *Hydrocarbon Process.* **49**(9), 323 (1970).
413. W. Braker and A. L. Mossman, *Matheson Gas Data Book,* 5th ed., Matheson Gas Products, New Jersey, 1971.
414. *Storing and Handling Liquified Olefins and Diolefins,* Exxon Chemical Company, Houston, 1974.
415. *Hazardous Chemical Data* Department of Transportation, U.S. Coast Guard Manual CG-446-2, January 1974.
416. N. V. Steere, ed., *Handbook of Laboratory Safety,* 2nd ed., The Chemical Rubber Co., 1971.
417. *Synthetic Organic Chemicals, U.S. Production and Sales,* United States International Trade Commission, Washington; *Chem. Eng. News* 34, (Dec. 19, 1977).
418. R. L. Ericsson, *Chem. Eng. Prog.* **68**(10), 80 (1972).
419. J. H. Prescott, *Chem. Eng.* **83**(16), 46 (Aug. 2, 1976).
420. *Chem. Week,* 70 (Sept. 8, 1976); *Chem. Eng. News,* 11 (Sept. 13, 1976).
421. *Chem. Age,* 8 (Feb. 25, 1977).
422. *Chem. Tech.* **23**, 7, 443 (1971); *Chem. Age,* 13 (Mar. 19, 1971); *Chem. Ind.* **25**, 703 (1973); *Eur. Chem. News,* 8 (June 7, 1974).
423. *Petrol Times,* 57 (Sept. 3, 1976); *Eur. Chem. News,* 40 (Oct 15, 1976); *Chem. Market. Report.,* 29 (May 19, 1974).
424. *Chem. Econ. Eng. Rev. Jpn.* **8**(7–8), 21 (1976).

ISIDOR KIRSHENBAUM
Exxon Research and Engineering Co.

BUTANEDIOLS, $C_4H_8(OH)_2$. See Glycols.

BUTANES. See Hydrocarbons, C_1–C_6.

BUTTER. See Milk products.

BUTYL ACETATE, $CH_3OOC_4H_9$. See Esters, organic.

BUTYL ALCOHOLS

There are four isomeric, 4-carbon alcohols of molecular formula $C_4H_{10}O$; two are primary, one is an unsymmetrical secondary alcohol, and one is tertiary (Table 1).

The butyl alcohols were discovered and their structures established between 1852 and 1871. Wurtz believed he had isolated n-butyl alcohol from fusel oil. Later, Erlenmeyer and Markovnikov showed that it was really isobutyl alcohol since it yielded isobutyric acid upon oxidation. In 1863, de Luynes isolated an alkyl iodide from the reaction of hydrogen iodide and erythritol, $CH_2OHCHOHCHOHCH_2OH$ (see Alcohols, polyhydric). This alkyl iodide was treated with silver acetate and the resultant ester was hydrolyzed and then oxidized. Since ethyl methyl ketone was isolated, the intermediate alcohol had to be sec-butyl alcohol. In 1864, Butlerow synthesized tert-butyl alcohol from acetyl chloride and dimethyl zinc using a procedure similar to the Grignard reaction. This was the first tertiary alcohol to be prepared although Kolbe had predicted their existence. Finally, in 1871, Lieben and Rossi prepared n-butanol by reduction of n-butyraldehyde.

Table 1. Physical Properties of the Butyl Alcohols

Properties	1-Butanol [71-36-3]	2-Methyl-propanol [78-83-1]	2-Butanol [78-92-2]	2-Methyl-2-propanol [75-65-0]
formula	$CH_3(CH_2)_2CH_2OH$	$(CH_3)_2CHCH_2OH$	$CH_3CH_2CH(OH)CH_3$	$(CH_3)_3COH$
alternative name	n-butyl alcohol	isobutyl alcohol	sec-butyl alcohol	tert-butyl alcohol
classification	primary	primary	secondary	tertiary
mp, °C	−90.2	−108	−114.7	25.5
bp, °C	117.7	108.1	99.5	82.5
density, d_4^t, g/mL	0.8133[15]	0.8057[15]	0.8108[15]	0.7762[30]
refractive index, n_D^t	1.3971[20]	1.3976[15]	1.3944[15]	1.3811[25]
flash pt, °C	35.0	27.5	24.4	8.9
viscosity, mPa·s[t] (=cP[t])	33.79[15]	47.03[15]	42.10[15]	33.16[30]
heat of vaporization, J/g[a]	591.2	578.4	562.4	535.4
sp heat, J/(g·K)[a]	2.33[20]	2.38[20]	2.73[20]	3.04[27]
heat of fusion, J/g[a]	125			91.6
heat of combustion, J/g[a]	2674	2670		2633
critical temp, °C	287	265	265	235
critical pressure, kPa[b]	4890	4850		
vapor pressure, kPa[b]	0.628[20]	1.173[20]	4.132[32]	4.079[20]
electrical conductivity, $(\Omega \cdot cm)^{-1}$	9.12×10^{-9}	8×10^{-8}		
dipole moment, C·m[c]	1.66×10^{-18}	5.97×10^{-30}		5.54×10^{-30}
dielectric constant, ϵ^t	17.7[17.2]	17.95[25]	15.5[19]	11.4[19]
solubility at 30°C, wt %				
in water	7.08	7.5	18	miscible
of water	20.62	17.3	36.5	miscible

[a] To convert J to cal, divide by 4.184.
[b] To convert kPa to mm Hg, multiply by 7.5.
[c] To convert C·m to debye (D), divide by 3.336×10^{-30}.

Physical and Chemical Properties

The butyl alcohols are colorless, clear liquids with characteristic odors. The straight-chain primary alcohol is the highest boiling and the highly branched tertiary alcohol is lowest boiling. With the exception of *tert*-butyl alcohol (mp 25°C) the butyl alcohols have very low melting points. Their relative solubilities in water also correspond to their molecular structures; *n*-butyl alcohol has a solubility of approximately 8%, whereas the tertiary alcohol is miscible. The four butyl alcohols are miscible with most common organic solvents.

Physical constants of the butyl alcohols are given in Table 1. These alcohols are known to form a number of azeotropic mixtures, the most common of which are given in Table 2 (1–2). Constants for the optically pure stereoisomers of 2-butanol have been reported: *d*-2-butanol [*4221-99-2*], n_D^{23} 1.3955, d_{30} 0.799, α_D^{27} 13.28° (3); and *l*-2-butanol [*14898-79-4*], n_D^{20} 1.3975, α_D^{20} −11.8° (4).

The chemical properties of the butyl alcohols are primarily a function of the hydroxyl group, and consequently their most important reactions are dehydration, dehydrogenation or oxidation, and esterification. The corresponding butylenes may be obtained by passing the alcohols over various dehydration catalysts at elevated tem-

Table 2. Azeotropic Mixtures of the Butyl Alcohols

Components	Weight %	bp of mixture, °C
Binary azeotropes		
n-butyl alcohol		
water	42.4	92.6
n-butyl acetate	32.8	117.6
n-butyl formate	76.3	105.8
methyl isovalerate	67	113
cyclohexane	90	79.8
tetrachloroethylene	68	110.0
ethyl isobutyrate	83	109.2
toluene	68	105.5
isobutyl alcohol		
water	33	90
cyclohexane	86	78.1
benzene	91	79.8
toluene	56	101.1
sec-butyl alcohol		
water	32	88.5
sec-butyl acetate	13.7	99.6
Ternary azeotropes		
n-butyl alcohol	10	83.6
n-butyl formate	68.7	
water	21.3	
n-butyl alcohol	27.4	89.4
n-butyl acetate	35.3	
water	37.3	

peratures. Thus butanol gives a mixture of 1- and 2-butenes at 175–400°C in the presence of such catalysts as alumina, tungsten oxide, and magnesium chloride. In general, 2-butanol gives predominately 2-butene, whereas 2-methyl-1-propanol and 2-methyl-2-propanol yield 2-methylpropene (isobutylene). When butanol is heated at 400°C over alumina, the olefinic mixture produced contains 78% 1-butene, 15% *cis*-2-butene, and 7% *trans*-2-butene. When the temperature is increased to 450°C, the yields are 50, 30, and 20%, respectively. When *sec*-butyl alcohol is similarly dehydrated, the yields are 35, 40, and 25%, respectively, and remain unchanged with temperature variation (5). When butanol is heated with dilute hydrochloric acid at 300°C, a 66% yield of butenes is obtained; *sec*-butyl alcohol gives 80% butenes and *tert*-butyl alcohol yields 84% isobutylene (6). The ease of dehydration increases from primary alcohol to tertiary alcohol. In the latter case, acids such as dilute sulfuric are effective dehydration agents even in the liquid phase at low temperatures (see Butylenes). Ether formation occurs when the other butyl alcohols are treated with sulfuric acid. When *tert*-butyl alcohol is cracked at 850–1050°C using steam as a diluent, propyne (methylacetylene) and allene ($CH_2{=}C{=}CH_2$) are produced (7).

Except for *tert*-butyl alcohol, the butyl alcohols may be dehydrogenated to their corresponding carbonyl compounds when passed over hot copper- or silver-containing catalysts. Butyraldehyde is obtained from butanol, ethyl methyl ketone (2-butanone) from *sec*-butyl alcohol, and isobutyraldehyde from isobutyl alcohol in the presence of copper at 300°C. The yields are generally very good. The same products are formed from the catalytic oxidation of the alcohol with air over oxides of copper or zinc at 250–400°C. On continued oxidation of the primary alcohols the corresponding butyric acids may be isolated whereas the secondary alcohol is degraded to acids of shorter chain length. Chemical oxidation with acid dichromate, alkaline permanganate, hydrogen peroxide or sodium perchlorate give analogous results. In some instances it is possible to isolate the ester from the acid and unreacted alcohol. The oxidation of *tert*-butyl alcohol is much more difficult than that of its isomers. Although *tert*-butyl alcohol is resistant to dilute dichromic acid, with increasing strength of sulfuric acid oxidative degradation to smaller fragments such as acetone, acetic acid, methanol, formaldehyde, and oxides of carbon takes place. Presumably, *tert*-butoxy radicals, $(CH_3)_3CO\cdot$, are formed which break down into acetone and methyl radicals. The methyl radicals then react with oxygen to give the one-carbon products (8). *Tert*-butyl alcohol has been oxidized in 50% aqueous hydrogen peroxide solution in the presence of silicotungstic acid to give high yields of *tert*-butyl hydroperoxide (9).

Esterification (qv) of the butyl alcohols with organic acids may be carried out in the usual manner with traces of mineral acids to catalyze the reaction. The alcohols can also be converted to the butyl chlorides or bromides by treatment with mineral acids, thionyl chloride, or phosphorus halides. The rate of esterification is fastest with the tertiary alcohol and slowest with the primary alcohols, although considerable dehydration occurs for the tertiary alcohol. Whereas *tert*-butyl alcohol reacts with phosphorus triiodide in carbon disulfide to give only 20% *tert*-butyl iodide (10), an 88% yield can be obtained by resorting to a mixture of phosphoric acid and potassium iodide (11).

Treatment of benzene in the Friedel-Crafts reaction (qv) with either *n*- or *sec*-butyl alcohols gives *sec*-butylbenzene and *o*- and *p*-di(*sec*-butyl)benzene; with isobutyl or *tert*-butyl alcohol the analogous *tert*-butyl isomers are produced. Alkylation of phenols with these alcohols gives rise to their corresponding butylphenols. The reaction of butanol with concentrated alkali hydroxide at 275°C yields hydrogen, butyric acid, and 2-ethylhexanoic acid, $CH_3CH_2CH_2CH(C_2H_5)CH_2COOH$; with *sec*-butyl alcohol the products are hydrogen, 5-methyl-3-heptanol, 5-ethyl-7-methyl-3-nonanol, and small amounts of formic, oxalic, and propionic acids. This convertion of *sec*-butyl alcohol to higher molecular weight alcohols is known as the Guerbet reaction, and may be carried out under less alkaline conditions by using potassium carbonate, magnesium oxide, and copper chromite at 245°C (12). Chlorination of butanol without cooling leads to the formation of dichlorobutyraldehyde dibutyl acetal. With isobutyl alcohol chlorination also causes simultaneous oxidation and substitution leading to the formation of mono- and dichloroisobutyraldehyde along with other chlorinated oxidation products. The chlorination of *tert*-butyl alcohol in the presence of sunlight gives *tert*-butyl chloride, 1,2-dichloro-2-methyl-propane, and other chlorinated products. Whereas *n*-, *sec*-, and isobutyl alcohols react with boron trichloride at −10°C in pentane to produce the corresponding borates, $(C_4H_9O)_3B$, in better than 90% yields, *tert*-butyl alcohol gives a 90% yield of *tert*-butyl chloride (13). The same type of products result with silicon tetrachloride.

Butanol and isobutyl alcohol may be aminated with ammonia over alumina at 300–350°C to give the corresponding butyl amine, dibutyl amine, and tributyl amine in the ratio of approximately 4:3:1; no detectable amine formation is observed with *tert*-butyl alcohol (14). Butyronitrile, $CH_3CH_2CH_2CN$, is produced when butanol is treated with ammonia at 460–470°C over Zn_3P_2 or $ZnSe$ (15). The vapor-phase nitration of butanol gives a mixture of nitromethane, nitroethane, nitropropane, and 2-nitro-2-methyl-1-propanol (16). In addition, *n*-butyl alcohol reacts with hydrogen sulfide at 180°C to give 5% *n*-butyl thiol (mercaptan) and 33% dibutyl sulfide; *tert*-butyl alcohol, under the same conditions, gives 34% *tert*-butyl thiol (17) (see Sulfur compounds).

Analysis

The qualitative determination of the butyl alcohols is carried out by the standard procedures of qualitative organic analysis for distinguishing primary, secondary, and tertiary alcohols and by preparation of the usual derivatives. The quantitative determination of aqueous fermentation butanol where it occurs along with acetone and ethanol, has been described (18). ASTM describes the butanol specification and testing methods (ASTM D304-58) (19).

Manufacture

***n*-Butanol.** Butanol can be manufactured by a synthetic process based on aldol, by the oxo process, or by the selective bacterial fermentation of carbohydrate-containing materials such as molasses and grains. It was first produced on a large scale during World War I as a by-product in the formation of acetone by fermentation of cornstarch (see Fermentation). The expanding lacquer industry created a great demand for butanol, and acetone soon became the by-product. The first fermentation process to find practical use in the United States was that developed by Weizmann in 1919. Production by this process was dominant in the United States until shortly after 1930 when synthetic production began and fermentation processes using molasses instead of grain were developed. Using either corn or molasses, several interesting by-products are produced which have commercial value, including acetone, ethanol, carbon dioxide, livestock feed, hydrogen, and riboflavin.

The most widespread means of producing butanol today is the oxo process (qv). Propylene is treated with carbon monoxide and hydrogen in the presence of an appropriate catalyst to give a mixture of n- and isobutyraldehydes (see Butyraldehyde). The alcohols are formed by reduction and separated, or the aldehydes are separated then reduced to the corresponding alcohols (20).

The aldol process uses acetaldehyde (qv) as the precursor. Acetaldehyde is most commonly produced by the Wacker process, which involves direct ethylene oxidation, and by dehydrogenation of ethanol (21). When acetaldehyde is treated with base at low temperatures, 5–25°C, the resulting product is acetaldol (1), via aldol condensation.

$$2\ CH_3CHO \xrightarrow{base} CH_3CH(OH)CH_2CHO$$
$$(1)$$

The reaction mixture is then acidified and distilled to dehydrate the acetaldol to crotonaldehyde (2) (2-butenal). The butanol is produced by catalytic hydrogenation of this unsaturated aldehyde (22).

$$(1) \xrightarrow{-H_2O} CH_3CH=CHCHO \xrightarrow{2\ H_2} CH_3CH_2CH_2CH_2OH$$
$$(2)$$

A number of other synthetic methods have been described in the patent literature. Ethyl alcohol may be converted directly to 1-butanol at 325°C and 13 MPa (128 atm) over magnesium oxide–copper oxide (23). A mixture of butanol, hexyl and ocytl alcohols, acetaldehyde, butyraldehyde, and crotonaldehyde is obtained when ethanol and hydrogen are passed over magnesium oxide at 200°C and 10 MPa (99 atm) (24–25). Butyl bromide can be hydrolyzed at 130–180°C at 350–700 kPa (3.5–6.9 atm) to give a mixture of butanol and dibutyl ether; the dibutyl ether can be converted to 81% butanol by heating with 48% aqueous hydrobromic acid in an autoclave at 150°C (26). An 82% yield of 1-butanol can be obtained from a low temperature reduction of n-butyraldehyde with sodium borohydride (27). At 200–300°C and 10 MPa (99 atm), furan has been reduced in the presence of copper chromite–barium chromite catalyst to butanol in 70% yields (28). Other interesting routes to butanol are described in reference 29.

The United States production of butanol in 1976 was 248,100 metric tons (30), at $0.48/kg, and the worldwide production was about 540,000 t.

Other Butyl Alcohols. Isobutyl alcohol (2-methyl-1-propanol) is now produced as a by-product of the oxo process. The normal process yields about 1 kg of isobutyl alcohol for every 3 kg of n-butanol (31), although modification of the cobalt catalyst or the use of a rhodium catalyst will substantially increase the ratio of normal to branched chain product (32–33). The United States production was about 90,000 metric tons at ca $0.44/kg in 1976 (30). The world production was about 170,000 t.

Hydration, in the presence of sulfuric acid, of 1-butene and isobutylene derived from the cracking of petroleum leads to the production of 2-butanol and tert-butyl alcohol, respectively. Tert-butanol is also produced as a by-product from the isobutane oxidation process for manufacturing propylene oxide (qv) (34). In general, the process for making 2-butanol is carried out in a continuous system involving the absorption of 1-butene in sulfuric acid, hydrolysis at elevated temperatures on dilution with water, and purification of the product by distillation or extraction. A great number of different conditions can be used, as indicated in the patent literature. Thus the absorption of butene may be carried out in 65% sulfuric acid at 30–45°C (35), in 75% acid at 10–15°C (36), or in 90–100% acid at 15°C or below (37). Isolation of the product by benzene extraction has been described (38). The hydration to 2-butanol may also be carried out in the vapor phase by passage of 1-butene and steam over a solid catalyst containing phosphoric acid and oxides of certain metals at 240°C and 1000 kPa (9.9 atm) (39). Brown and Zweifel describe the synthesis of optically pure 2-butanol by the hydroboration of cis-2-butene followed by oxidation with alkaline hydrogen peroxide (4) (see Hydroboration). The United States production of 2-butanol is estimated at 250,000 metric tons in 1976 at $0.46/kg. Most of the product is used for the manufacture of methyl ethyl ketone.

Toxicology

In general, the histological effects of the butyl alcohols are similar; they cause fatty accumulations in the liver, heart, and kidneys of experimental animals. The secondary and tertiary alcohols are more strongly narcotic than the primary alcohols. This stronger effect is probably related to the higher vapor pressures of the secondary and tertiary alcohols. The results of animal experiments are summarized in Table 3 (40). A complete discussion of the toxicity of the butyl alcohols has been prepared (41).

Uses

The largest use for butanol and its derivatives is in the coatings industry for the formulation of nitrocellulose lacquers. Butanol acts synergistically with butyl acetate and other solvents to increase solvency, without being a nitrocellulose solvent itself (21). An emerging market for butanol is latexes, in the form of butyl acrylate, used in the coatings industry (42). The more important derivatives of butanol are: butyl acetate, butyl glycol ether, and plasticizers such as dibutyl phthalate (43). At one time, 75–85% of the butanol produced was converted to the acetate, which found application as a solvent in the preparation of artificial leather-coated paper and textiles, and plastics; as an extraction solvent for oils, drugs, and perfumes; and as an ingredient in perfumes and flavors (44). With the growth of urea–formaldehyde finishing materials, increasing amounts of butanol have been used as a solvent in their manufacture with a resulting substantial reduction in the proportion of butanol converted to the

Table 3. Butyl Alcohols Toxicology

	LD_{50}[a]	Exposure limits TWA[b]
n-butanol	790 mg/kg	100 ppm
isobutyl alcohol	2460 mg/kg	100 ppm
2-butanol	6480 mg/kg	150 ppm
tert-butanol	3500 mg/kg	100 ppm

[a] Rat, acute oral.
[b] Time-weighted average, 8 h.

ester. Butanol has been used in alkyd resin coatings. Butanol also provides an excellent diluent for formulating brake fluids suitable for use in passenger cars, and is used as an extractant in the manufacture of antibiotics, vitamins, and hormones.

Practically the entire production of *sec*-butyl alcohol is converted to methyl ethyl ketone which has found extensive use as a solvent where a ketone of higher boiling point than acetone is required (see Ketones). In addition, *sec*-butyl alcohol employed in conjunction with aromatic hydrocarbons forms a good solvent mixture for alkyd enamels and lacquers based on ethyl cellulose (see Alkyd resins). The fact that 2-butanol has solvency for water and oils makes it a useful coupling agent for application in hydraulic brake fluids (see Hydraulic fluids), industrial cleaning compounds, and paint removers. It is used in the manufacture of *sec*-butyl acetate, a nitrocellulose solvent, and in the production of a xanthate derivative which is a collector in ore flotation (see Flotation). It has also been used in the manufacture of fruit essences, perfumes, dyestuffs, and wetting agents.

Tert-butyl alcohol is used in the manufacture of *tert*-butyl chloride and in the manufacture of *tert*-butylphenol. In the field of synthetic perfumes, *tert*-butyl alcohol is an important raw material for the preparation of artificial musk. It is an authorized denaturant for use in proprietary ethanol mixtures as well as in several specially denatured alcohols. Catalytic dehydration of *tert*-butanol is a means of obtaining isobutylene and it has been patented for use as a gasoline antiknock agent (45).

Isobutyl alcohol can be used in practically all of the applications that have been mentioned for butanol. Since butanol has been commercially available for a much longer time, it is more widely used. However, isobutyl alcohol can frequently be used interchangeably and has consistently been less expensive than butanol since it first became available in large commercial quantities from the oxo process in 1951. As in the case of butanol, the major use for isobutyl alcohol is in the production of its acetate ester which is widely used in the lacquer industry.

BIBLIOGRAPHY

"Butyl Alcohols" in *ECT* 1st ed., Vol. 2, pp. 674–680, by C. L. Gabriel and A. A. Dolnick, Publicker Industries, Inc.; "Butyl Alcohols" in *ECT* 2nd ed., Vol. 3, pp. 822–830.

1. L. H. Horsley, *Azeotropic Data—III,* American Chemical Society, Washington, D.C., 1973.
2. L. H. Horsley, *Anal. Chem.* **19,** 588 (1947).
3. S. W. Kantor and C. R. Hauser, *J. Am. Chem. Soc.* **75,** 1744 (1953).
4. H. C. Brown and G. Zweifel, *J. Am. Chem. Soc.* **83,** 486 (1961).
5. M. R. Musaev and V. G. Zizin, *Zh. Priklady Khim.* (*Leningrad*) **29,** 803 (1956).
6. Brit. Pat. 576,480 (Apr. 5, 1946), (to Universal Oil Products Co.).
7. U.S. Pat. 2,752,405 (June 26, 1956), J. Happel and C. J. Marsel.
8. C. F. Cullis and E. A. Warwicker, *Proc. Roy. Soc. A* **264,** 392 (1961).
9. Brit. Pat. 682,424 (Dec. 12, 1952), (to V. N. de Bataafsche Petroleum Maatschappij).
10. M. C. Berlak and W. Gerrard, *J. Chem. Soc.,* 2309 (1949).
11. H. Stone and H. Schechter, *J. Org. Chem.* **15,** 491 (1950).
12. M. N. Dvornikoff and M. R. Farrar, *J. Org. Chem.* **22,** 540 (1957).
13. W. Gerrard and M. F. Lappert, *J. Chem. Soc.,* 2545 (1951).
14. N. S. Kozlov and N. I. Panova, *J. Gen. Chem. USSR* (Engl. Transl.) **26,** 2901 (1956).
15. Swiss Pat. 324,670 (Nov. 30, 1957), R. Jacob (to Lonza Electrizitatswerke und Chemische Fabriken A.G.).
16. H. B. Hass and D. E. Hudgen, *J. Am. Chem. Soc.* **76,** 2692 (1954).
17. T. L. Cairns, A. W. Larchar, and B. C. McKusick, *J. Org. Chem.* **18,** 748 (1953).
18. L. M. Christensen and E. I. Fulmer, *Ind. Eng. Chem. Anal. Ed.* **7,** 180 (1935).

19. *1977 Annual Book of ASTM Standards,* Part 29, American Society for Testing and Materials, Philadelphia, Pa., 1977.
20. J. Falbe, *Carbon Monoxide in Organic Synthesis,* Springer-Verlag New York, Inc., New York, 1970, pp. 4–14.
21. A. M. Brownstein, *Trends in Petrochemical Technology,* Petroleum Publishing Co., Tulsa, Okla., 1976, pp. 200–201.
22. Fr. Pat. 1,500,656 (Nov. 3, 1967), (to Chemische Werke Huels, A.G.).
23. J. A. Monick, *Alcohols, Their Chemistry Properties and Manufacture,* Reinhold Book Corp., New York, 1968, pp. 125–136.
24. V. Nagarajan, *Chem. Process. Eng. (Bombay)* **4**(11), 29 (1970).
25. Brit. Pat. 381,185 (Sept. 26, 1932), (to Deutsche Gold-und Silber-Scheideanstalt vorm. Roessler).
26. Brit. Pat. 866,563 (Apr. 26, 1961), E. J. Gasson and D. J. Hadley (to Distillers Co., Ltd.).
27. U.S. Pat. 2,683,721 (July 13, 1954), H. I. Schlessinger and H. C. Brown.
28. Brit. Pat. 586,222 (Mar. 11, 1947), J. G. M. Bremner (to Imperial Chemical Industries Ltd.).
29. I. T. Harrison and S. Harrison, *Compendium of Organic Synthetic Methods,* Vol. 1, Wiley-Interscience, a division of John Wiley & Sons, Inc., New York, 1971, pp. 75–131; I. T. Harrison and S. Harrison, *Compendium of Organic Synthetic Methods,* Vol. 2, Wiley-Interscience, a division of John Wiley & Sons, Inc., New York, 1974, pp. 28–52.
30. *Synthetic Organic Chemicals, U.S. Production and Sales, 1976,* U.S. Tariff Commission, U.S. Govt. Printing Office, Washington, D.C.
31. B. Cornils, R. Payer, and K. C. Traenckner, *Hydrocarbon Process.,* 83 (1975).
32. U.S. Pat. 3,239,571 (1963), L. H. Slaugh, R. D. Mullineaux, and L. G. Cannel (to Shell Oil Co.).
33. U.S. Pat. 3,527,809 (Sept. 8, 1970), R. L. Pruett and J. A. Smith (to Union Carbide Corp.).
34. D. E. Winkler and co-workers, *Ind. Eng. Chem.* **53**(8), 65 (1961).
35. U.S. Pat. 2,271,092 (Jan. 27, 1942), G. A. Perkins and J. A. Davies (to Union Carbide and Carbon Corp.).
36. U.S. Pat. 1,408,320 (Feb. 28, 1922), C. Weizmann and A. A. Legg.
37. U.S. Pat. 1,948,286 (Feb. 29, 1934), B. T. Brooks (to Standard Alcohol Co.).
38. U.S. Pat. 2,196,177 (Apr. 9, 1940), R. E. Burk and H. A. Lankelma (to Standard Oil Co., Cleveland).
39. U.S. Pat. 2,052,095 (Aug. 25, 1936), W. P. Joshua, H. M. Stanley, and J. P. Dymock.
40. H. E. Christenson and co-workers, *Registry of Toxic Effects of Chemical Substances,* 1976 ed., U.S. Dept. of Health, Education, and Welfare, Rockville, Md., June 1976.
41. E. Browning, *Toxicity and Metabolism of Industrial Solvents,* Elsevier Publishing Co., New York, 1965, pp. 342–355.
42. A. M. Brownstein, *Trends in Petrochemical Technology,* Petroleum Publishing Co., Tulsa, Okla., 1976, 189 pp.
43. *Chem. Mark. Rep.,* (Nov. 10, 1975).
44. I. Mellan, *Industrial Solvents,* Reinhold Publishing Corp., New York, 1939, Chapts. 10 and 15.
45. Belg. Pat. 618,629 (Dec. 6, 1962), W. Muenster, W. Wolf, and G. Notres (to Badische Anilin- und Soda-Fabrik A.G.).

PAUL DWIGHT SHERMAN, JR.
Union Carbide Corporation

BUTYLENE OXIDES, C_4H_8OO. See Epoxides.

BUTYLENES

The four C_4H_8 mono-olefins—1 butene, cis-2-butene, trans-2-butene, and 2-methylpropene—collectively are called butylenes. These four isomers are treated as a group because, with only minor exception, they are obtained as a mixture from the C_4 fraction from processes that crack petroleum fractions and natural gas. This C_4 fraction is commonly known as the B–B stream because it contains butanes as well as butylenes. The name isobutylene for 2-methylpropene is firmly established by usage and will be used in this review. The three linear isomers will be referred to collectively as butenes.

Physical Properties

The structures and important physical properties for the four butylenes are summarized in Table 1. Other thermodynamic and transport properties are in reference 1.

Table 1. Physical Properties of the Butylenes

Properties	1-Butene [106-98-9] CH CH CH=CH$_2$	cis-2-Butene [590-18-1]	trans-2-Butene [624-64-6]	Isobutylene [115-11-7]
mp, °C	−185.35	−138.922	−105.533	−140.337
bp, °C	−6.25	+3.718	+0.88	−6.896
density of liquid at 25°C, g/L	588.8	615.4	598.4	587.9
vapor pressure (Antoine equation constants)[a,c]				
temp range, °C	−82 to +13	−73.4 to +23	−76 to +20	−82 to +12
A	7.7180	7.74436	7.74462	7.71644
B	926.1	960.1	960.8	932.2
C	240.0	237.0	240.0	240.0
heat of vaporization, J/g[b]				
at 25°C	359.21	359.08	380.87	367.46
at boiling point	391.18	416.74	406.18	394.78
critical temperature, °C	146.4	162.43	155.48	144.75
critical pressure, kPa[c]	4022.6	4205.0	4103.6	4000.3
critical volume, L/mol	0.240	0.234	0.238	0.239
heat of combustion at constant pressures and 25°C, J/mol[b]	2.524×10^6	2.515×10^6	2.512×10^6	2.507×10^6
isobaric specific heat at 25°C, J/(kg·K)[b]				
gas in ideal state	1528.9	1408.7	1567.5	1590.5
liquid at 101.3 kPa[c]	2300.3	2251.7	2277.3	2338.0
surface tension at 20°C, mN/m (= dyn/cm)	0.0125	0.01507	0.01343	0.01242

[a] $\log_{10} P = A - B/(t + C)$, where P is in kPa and t in °C.
[b] To convert J to cal, divide by 4.184.
[c] To convert kPa to mm Hg, multiply by 7.5.

Chemical Properties

Butylenes behave as typical olefins. In addition to the reactions listed in the literature under the individual butylene isomers, the many reactions studied with the more easily handled C_5–C_{10} olefins may be expected to apply to butylenes. The double bonds are centers of high electron density and reactions take place by electrophilic, metallation, and free-radical mechanisms. Discussions and reviews of oxidation (2) and of alkylation (qv) of aromatics as well as other reactions catalyzed by Friedel-Crafts catalysts (3) (qv) are available. The main reactions exhibited by butylenes are addition reactions, isomerization, and polymerization.

Addition Reactions. For a given reaction, rates for each butylene isomer differ from those of the other isomers because the arrangement of substituent groups around the double bond results in a different electron density, polarity, and steric effect. The ^{13}C nmr shifts (4) that provide a measure of the first two effects are summarized in Table 2. High shifts for a carbon atom indicate a high electron density, and larger differences in shifts between carbon atoms indicate a higher polarity (see Analytical methods). Both these data and the calculated activation energies for hydration in which carbenium ion formation (proton addition) is assumed to be the rate controlling step (5) are consistent with experimental facts. Thus simple electrophilic additions of HX (H_2O, ROH, HCl) to the butylenes yield the products of Markovnikov addition (H^+ attack on the least substituted, most negative carbon atom) and the relative rates of reaction are: isobutylene $\gg$ 1-butene > 2-butenes. Likewise, in the polymerization of the butylene isomers, the relative rates of reaction are: isobutylene $\gg$ 1-butene > cis-2-butene > trans-2-butene. This difference in reactivity permits selective reaction of isobutylene by electrophilic attack and forms the basis for separation of isobutylene from the remaining isomers by sulfuric acid extraction, eg, under typical extraction conditions at 30°C, isobutylene reacts with 45–60% H_2SO_4 about one thousand times as fast as 1- and 2-butene (6).

Although the rates of addition of HX to 1-butene and 2-butene differ, as indicated by activation energies in Table 2, polarity indicates the products will be the same, viz, 2-substituted butanes. Thus sec-butyl alcohol is produced by hydration of the 1- and 2-butene mixture typically remaining after removal of isobutylene.

Table 2. ^{13}C Nmr Shifts and Hydration Rates for Butylenes

Butylene	^{13}C nmr[a] ppm shift at A	B	E_a, kJ/mol[b], for hydration at A	B
C_A=C_B—C—C	80.4	53.5	188	145
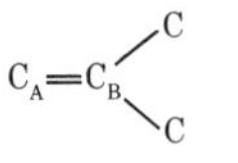 C_A=C_B(C)(C)	82.0[c]	50.4[c]	711	108
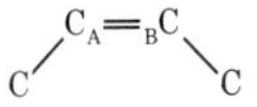 C—C_A=$_B$C—C (cis)	69.1	69.1	203	203
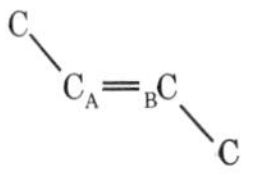 C—C_A=$_B$C—C (trans)	67.7	67.7	203	203

[a] Relative to external CS_2 reference.
[b] To convert kJ to kcal, divide by 4.184.
[c] These values obtained by interpolation.

Alkylation of aromatics with butylenes may be considered as an addition reaction:

$$C_6H_6 + CH_3CH=CHCH_3 \xrightarrow{H^+} C_6H_5-CH(CH_3)CH_2CH_3$$

and it follows the same general rules with regard to relative rates and product structure. Thus 1- and 2-butenes yield *sec*-butyl derivatives and isobutylene yields *tert*-butyl derivatives.

Electrophilic addition to the double bond generally proceeds by a step-wise addition as shown by bromination in the presence of methanol (7) or acetic acid (8) to yield the bromomethoxy and bromoacetoxy derivatives as well as the dibromides.

Because the addition is stereospecifically trans and the intermediate bromonium ion acts as a rigid, cyclic structure, *cis*- and *trans*-2-butenes yield different optical

$$CH_3CH=CHCH_3 + Br_2 \longrightarrow CH_3\overset{Br^+}{CH}-CHCH_3 + Br^- \longrightarrow CH_3\overset{Br}{C}H\overset{Br}{C}HCH_3$$

$$\xrightarrow{CH_3OH} CH_3\overset{Br}{C}H\overset{OCH_3}{C}HCH_3 + HBr$$

$$\xrightarrow{CH_3COOH} CH_3\overset{Br}{C}H\overset{OCOCH_3}{C}HCH_3 + HBr$$

isomers (9–10).

Although electrophilic addition of HX proceeds normally with isobutylene, addition of chlorine does not take place to a significant extent and the reaction intermediate stabilizes itself by loss of a proton rather than by reaction with a nucleophile or an anion.

$$CH_3-C(CH_3)=CH_2 + Cl_2 \longrightarrow CH_3-\overset{Cl^+}{C}(CH_3)-CH_2 + Cl^- \longrightarrow CH_2=C(CH_3)-CH_2Cl + HCl$$

This mode of reaction is attributed to the high degree of steric hindrance against approach by the anion. Reaction is very rapid, even at ambient temperature (11).

Metallation proceeds by a simple four-centered transition state. Since the metallic entity is electron-seeking, attachment is primarily to the terminal carbon in both 1-butene and isobutylene (see Hydroboration; Organometallics).

$$CH_3CH_2\overset{\delta^+}{CH}=\overset{\delta^-}{CH_2} + \overset{\delta^-\delta^+}{HB} \longrightarrow CH_3CH_2CH\cdots CH_2 \longrightarrow CH_3CH_2CH_2CH_2B$$

Butylene isomers also can be expected to show significant differences in reaction rates for metallation reactions such as hydroboration and hydroformylation (addition of HCo(CO)$_4$). For example, the rate for addition of di(*sec*-isoamyl)borane to *cis*-2-butene is about six times that for addition to *trans*-2-butene (12). For hydroformylation of typical 1-olefins, 2-olefins, and 2-methyl-1-olefins, specific rate constants are in the ratio 100:31:1, respectively.

The composition of the products of reactions involving intermediates formed by metallation depends on whether the measured composition results from kinetic control or from thermodynamic control. Thus the addition of diborane to 2-butene initially yields tri-*sec*-butylborane. If heated or allowed to react further, this product isomerizes about 93% to the tributylborane, the product initially obtained from 1-butene (12). Similar effects are observed during hydroformylation reactions; however, interpretation is more complicated because the relative rates of isomerization and of carbonylation of the reaction intermediate depend on temperature and on hydrogen and carbon monoxide pressures (13).

These reactions are also quite sensitive to steric factors, as shown by the fact that if 1-butene reacts with di(*sec*-isoamyl)borane the initially formed product is 99% substituted in the 1-position (12) compared to 93% for unsubstituted borane. Similarly, the product obtained from hydroformylation of isobutylene is about 97% isoamyl alcohol and 3% neopentyl alcohol (14). Also, reaction of isobutylene with aluminum hydride yields only triisobutyl aluminum.

Free radical additions to 1-butene, as in the case of HBr, RSH, and H$_2$S to other olefins (15–17), can be expected to yield terminally substituted derivatives.

Selectivity among butene isomers also occurs in vapor phase heterogeneous catalysis, at least in the case of dehydrogenation of butenes to butadiene, where maximum yields can be obtained by employing slightly different conditions for each isomer (18). In practice, mixtures of isomers are used and an average set of conditions is employed.

Isomerization. Isomerization of any of the butylene isomers to increase supply of another isomer is not practiced commercially; however, their isomerization has been studied extensively because: formation and isomerization accompanies many refinery processes; maximization of 2-butene content maximizes octane number when isobutane is alkylated with butene streams using HF as catalyst; and isomerization of high concentrations of 1-butene to 2-butene in mixtures with isobutylene could simplify subsequent separations (19a). One plant (Phillips) is now being operated for this latter purpose (19b,c). The general topic of isomerization is covered in detail by several reference books (20–22). Isomer distribution at the thermodynamic equilibrium in the range 300–1000 K is summarized in Table 3 (23).

The three isomerizations, *cis*-2-butene $\rightleftharpoons$ *trans*-2-butene, 1-butene $\rightleftharpoons$ 2-butene, and butenes $\rightleftharpoons$ isobutylene require increasingly severe reaction conditions. When the position of the double bond is shifted, cis–trans isomerization also occurs, and when the carbon skeleton is rearranged, mixtures of butenes result. However, during isomerization of 1-butene to 2-butene, with solid catalysts, the cis isomer is preferentially formed initially even though it is the thermodynamically less-favored isomer.

An extremely wide variety of catalysts, Lewis acids, Brønsted acids, metal oxides, molecular sieves, dispersed sodium and potassium, and light, are effective. Typical

350 BUTYLENES

Table 3. Equilibrium Butylenes Distribution, Ideal Gas, 101.3 kPa[a,b]

Temp, K	1-Butene	cis-2-Butene	trans-2-Butene	Total butenes	Isobutylene
			Mole %		
300	0.4	3.8	11.8	16.0	84.0
400	1.9	8.3	18.0	28.2	71.8
500	4.5	11.9	21.6	38.0	62.0
600	7.5	14.4	23.4	45.3	54.7
700	10.8	15.8	24.2	50.8	49.2
800	14.0	16.6	24.5	55.1	44.9
900	16.9	17.0	24.6	58.6	41.4
1000	19.6	17.2	24.4	61.2	38.8

[a] 101.3 kPa = 1 atm.
[b] Ref. 20.

Table 4. Isomerization of Butylenes

Catalyst	Temp, °C	Ref.
cis-2-Butene ⇌ trans-2-butene		
Co[60] γ-rays	ambient	24
mercury arc	ambient	25
SO_2 adsorbed on NaX and KL zeolite, Na mordenite and porous Vycor	24.5	26
bis(acetylacetonato)Pd–SiO_2	61.5	27
1-Butene ⇌ 2-butene		
Na–Al_2O_3	−60	28
Pt–SiO_2–Al_2O_3	−10	29
$SnCl_2$–[$(C_6H_5)_3P]_2NiX_2$	0	30
$RhCl_3$–$SnCl_2$–CH_3OH	ambient	31
$Cl_2(Bu_3P)_2Ni$–$(C_2H_5)_2AlCl$–SiO_2	ambient	32
BF_3–Al_2O_3	ambient	33
NaX and HY zeolite, Al_2O_3 and SiO_2	24.5	26
BF_3–H_2O	25	34
60% H_2SO_4	73	35
85% H_3PO_4	73	35
F–Al_2O_3	200	36
Ga_2O_3	190–330	37
iodine	200–250	38
KOH–Al_2O_3	350	39
$ZnCrFeO_4$	465	40
ZrO_2–η-Al_2O_3	540	41
Butene ⇌ isobutylene		
HCl–γ-Al_2O_3	350–450	39
α-Al_2O_3	450	39
Pt–Al_2O_3–SiO_2	475	42
SiO_2	520	43

examples and their use are summarized in Table 4. Generally, acidic catalysts are required for skeletal isomerization and reaction is accompanied by polymerization, cracking, and hydrogen transfer, typical of carbenium ion intermediates. Double bond shift is accomplished with high selectivity by the basic and metallic catalysts.

Polymerization. Polymerization reactions are used to produce the major products formed directly from butylenes: butyl elastomers, polybutylenes, and polyisobutylene (see Elastomers, synthetic; Olefin polymers).

Polymerization is induced by acidic substances, Ziegler-Natta catalysts (qv), and various other initiators (qv). The acidic initiators are also called cationic initiators because they induce the polymerization by creation of carbocation intermediates. Review articles (44–46) explain the mechanism of cationic initiation and the fundamentals of cationic polymerization. Many Brønsted acids, such as sulfuric acid, phosphoric acid, dihydroxyfluoroboric acid, and molybdic acid, initiate the oligomerization of butylenes but cannot give high molecular weight polymers. More versatile initiators are the Lewis acids (BF_3, $AlCl_3$, $AlBr_3$, $TiCl_4$, $SnCl_4$, etc) and related electron acceptors such as trialkylaluminums and alkylaluminum halides. Most of the Lewis acids need a co-initiator such as water, methanol, acetic acid, hydrochloric acid, hydrobromic acid, certain alkyl halides, or halogens. Many solids having acidic surfaces can also initiate the polymerization of butylenes. Examples are natural clays, synthetic zeolites (see Molecular sieves), silica–alumina, active carbon, and the reaction product of phosphoric acid with kieselguhr (47). A few of the solid catalysts can produce polyisobutylenes with a number average molecular weight of a few thousand, but most of them give only oligomers (48). The Lewis acids are most important because they can produce high molecular weight polyisobutylenes. Considerable effort has been made to increase either the polymerization rate or the polymer molecular weight (at a given temperature) by combination of the initiators and coinitiators or by modifications of the initiator system. Reference 49 is an excellent review of the related literature.

The second group of initiators encompasses the Ziegler-Natta coordination catalysts. Although it may be possible to obtain polymers from all butylene isomers with Ziegler-Natta catalysts, only the 1-butene polymerization is practical and significant. The high molecular weight, stereoregular poly(1-butene) obtained finds application as a premium-grade plastic (50).

The many other types of initiation reported for butylenes include thermal initiation, which yields only oligomers (51), high energy radiation (52–53), and uv irradiation of the charge transfer complexes of isobutylene (54).

The heat of the exothermic polymerization for the butylene monomers is in the 50–57 kJ/mol (12–13.6 kcal/mol) range. The thermodynamic aspects of the cationic polymerization have been reviewed (55–57).

Relatively little is known about the rates of cationic polymerizations (58). The rate increases and simultaneously the polymer molecular weight decreases with increasing polymerization temperature. For limited temperature regions, quantitative Arrhenius correlations have been established between the logarithm of the molecular weight and the reciprocal of the polymerization temperature.

Isobutylene is the most reactive of the butylene isomers. All the initiators listed above can induce this monomer to polymerize, usually at very low concentrations. Extremely fast polymerizations are achieved with the more active initiators. The molecular weight range of the isobutylene polymers can be varied from the dimer to a few millions by the appropriate selection of the initiator and the polymerization temperature. 1-Butene gives a high molecular weight thermoplastic product with Ziegler-Natta catalysts but with cationic initiators only oligomers or viscous liquids can be obtained. The cationic polymerization of 1-butene is much slower than that of isobutylene. The *cis*-2-butene isomer is even less reactive and gives only oligomers with Lewis acids. It cannot be directly polymerized with Ziegler-Natta catalysts, but its isomerization to 1-butene is effected at a slow rate and poly(1-butene) forms (59). *trans*-2-Butene does not polymerize with cationic initiators. However, similarly to *cis*-2-butene, it can give poly(1-butene) with Ziegler-Natta catalysts (59).

Isobutylene has no tendency to form alternating copolymers by cationic polymerization (see Copolymers). It can form copolymers with numerous monoolefins and dienes (eg, propene, butenes, styrene, α-methylstyrene, β-pinene, isoprene, etc) (60–61). The copolymer molecular weight is much lower than that of the isobutylene homopolymer. With free radical initiators, isobutylene can be easily copolymerized with vinyl chloride, vinyl acetate, acrylonitrile, acrylic esters, maleic anhydride, etc (60).

The chemical structure of the butylene homopolymers has been extensively studied. Polyisobutylenes have a regular skeleton (see below) but minor deviations from the regular structure can occur (62).

$$\left(-CH_2-\underset{\underset{CH_3}{|}}{\overset{\overset{CH_3}{|}}{C}}- \right)_n$$

The polymer contains one double bond per molecule. The nature of the double bond depends on the initiator and the polymerization conditions (62). The polymers of 1-butene and *cis*-2-butene formed by cationic polymerization have very irregular structures. They are the products of isomerization polymerizations involving intramolecular hydride shifts during propagation (63). The 1-butene polymers have a tendency to lose unsaturation during polymerization, presumably by intramolecular alkylation of a tertiary carbon atom by the polymer double bond (63).

Manufacture

Figure 1 shows how the manufacture and use of butylenes fits into the general United States refinery–petrochemicals system (see Feedstocks; Petroleum).

Almost all commercially produced butylenes are obtained as by-products from two principal processes: catalytic or thermal cracking, refinery processes which upgrade high boiling petroleum fractions to gasoline; and steam cracking, which produces light olefins for chemical feedstocks by pyrolysis of saturated hydrocarbons derived from natural gas or crude oil. Table 5 shows that the combined 1976 production in the United States, Western Europe, and Japan amounted to 15.7 million metric tons. Catalytic cracking accounts for about 95% of the total in the United States. However, production is divided about equally between catalytic and steam cracking in Europe and Japan, where refinery operation emphasizes all fuels rather than just gasolines.

The yields of butylenes vary according to the type of feed and the specific processing scheme. As shown in Table 6, the largest yields generally come from the catalytic cracking of gas oil, the crude fraction boiling in the range of 343–565°C. Catalytic or thermal conversion of resid (the undistillable crude fraction boiling above 565°C) offers a potentially large source of butylenes, but such processes are not yet widely used.

Other commercial processes that are sometimes used to produce specific isomers

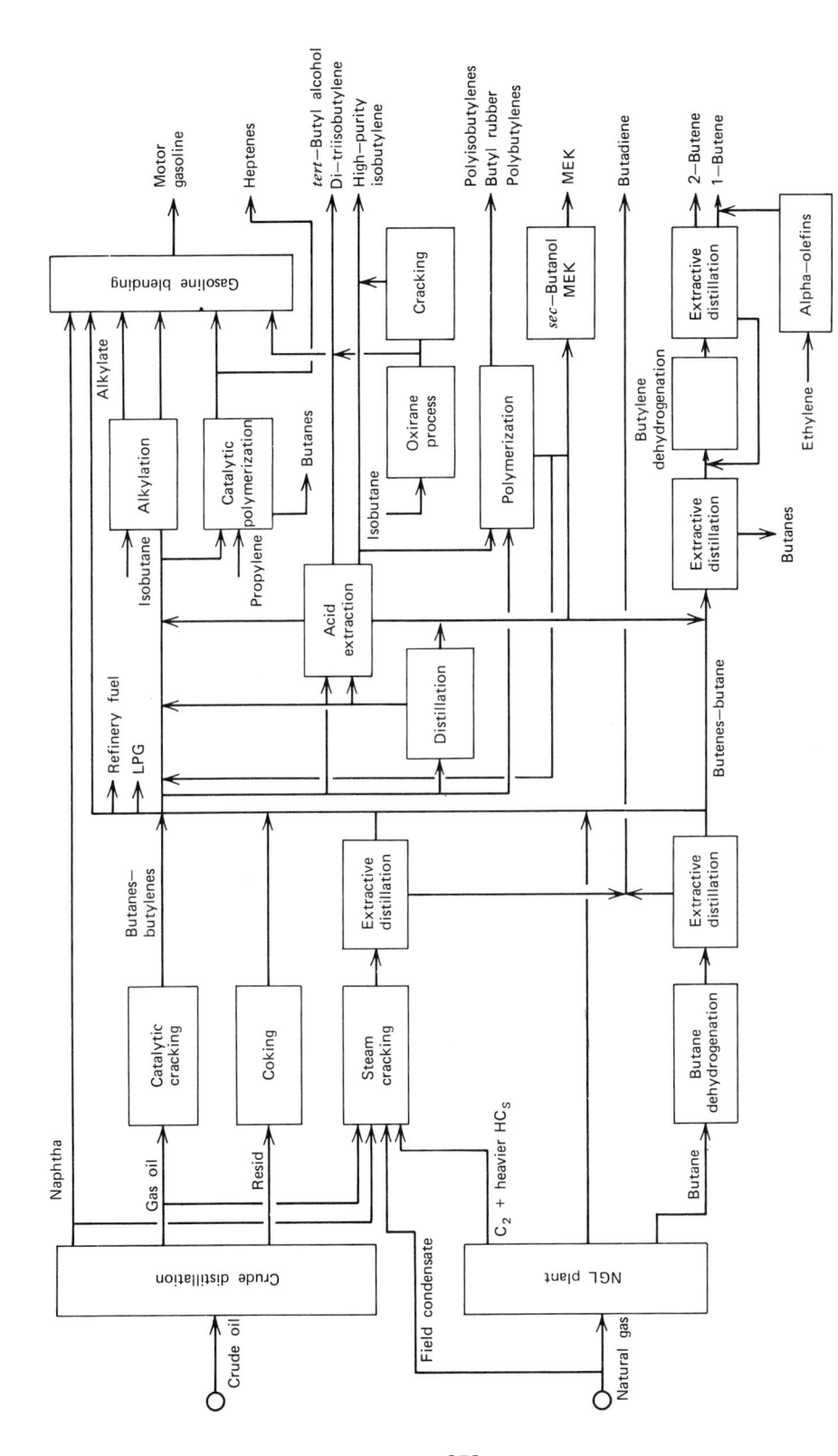

Figure 1. General United States refinery–petrochemical system. NGL, natural gas liquid; HCs, hydrocarbons; LPG, liquefied petroleum gas; MEK, methyl ethyl ketone.

Table 5. Butylene Production in 1976, 1000 Metric Tons Per Year

Country	Refinery	Steam crackers	Total
United States	11,400	490	11,890
Western Europe	1,450	1,200	2,650
Japan	600	600	1,200
Total	13,450	2,290	15,740

or mixtures of butenes or both, either directly or as by-products, include: the Oxirane process for making propylene oxide; the dehydrogenation of butane and isobutane; the disproportionation of olefins; and the oligomerization of ethylene. All or any of them may become useful feedstock sources should the need arise.

Catalytic and Thermal Cracking. Catalytic and thermal cracking are refinery processes which produce butylenes in admixture with butane, isobutane, and traces of butadiene. Because the butylenes react further, as shown in Figure 2, catalyst type and operating severity determine the specific distribution and yield of C_4s (65). Yield and stream compositions given in Table 6 are therefore only approximate.

The most prominent process is fluid catalytic cracking of gas oil. In this process, partially vaporized gas oil contacts a hot flowing stream of catalyst at reaction temperatures of 450–565°C, contact times of 5 s to 2 min, depending on the design of the individual process unit, and pressures in the reaction zone of about 250–400 kPa (2.5–4 atm). The current generation of cracking catalysts comprises alumino-silicate molecular sieves (either refined natural clay or synthetic) with controlled amounts of acidic sites. These catalysts strongly promote such secondary reactions as carbenium-ion cracking, isomerization–disporportionation of lighter olefin fragments, and hydrogen transfer reactions between lighter and heavier fragments that convert olefins to their corresponding paraffins.

The butane–butylene (B–B) fraction is separated from C_3 and C_5 hydrocarbons by conventional pressure distillation. Depending on the sharpness of fractionation, the B–B stream may contain significant concentrations of C_3 and C_5 olefins and paraffins.

Although still not widely practiced, the conversion of resid by catalytic (66) and thermal cracking should increase in the future, both because of more favorable economics and because of the greater use of heavier, foreign crude oils containing more gas oil and resid.

As shown in Table 6, two thermal cracking processes—delayed coking and Flexicoking (64, 67)—show different selectivities for the production of butylenes and butanes. Flexicoking produces more butylenes because it uses higher reaction temperatures (ca 550°C vs 450°C for delayed coking). Both processes operate at relatively low pressures (300–600 kPa or 3–6 atm). Because the free-radical, thermal cracking mechanism does not show high selectivity for the production of branched-olefin and saturated products, which result from the carbenium-ion mechanism characteristics of catalytic processes, 1-butene is the principal butylene isomer resulting from thermal processes (see Petroleum refinery processes).

Table 6. Typical Yields and Compositions of C$_4$ Fractions from Cracking Operations

Butylene yield	Catalytic cracking Gas oil		Residue	Thermal cracking of residue Delayed coking		Flexicoking[a]		Steam cracking of naphtha and light gas oil	
wt % on crude	0.5–5		1.5–3	0.1–0.6		0.15–0.8		0.4–5	
wt % on feed	3–10		3–5[a]	1–1.5		1.5–2			
C$_4$ composition	Total	Olefin	Total	Total	Olefin	Total	Olefin	Total	Olefin
butane	7–13		7	47		14–23		2–5	
isobutane	28–52		18–14	12		5		1.5–0.6	
isobutylene	26–8	40–23	79	16	40	13	20–18	27.4–22.0	48
1-butene	8–7	12–20	79	13	31	17	26–24	16.0–14.0	30
cis-2-butene	31–20	48–57	75–79	5	12	35–42	54–58	5.5–4.8	10
trans-2-butene	31–20	48–57	79	7	17	35–42	54–58	6.5–5.8	12
1,3-butadiene	0.1–0.5			0.5		7–9		37.0–47.5	

[a] Ref. 64.

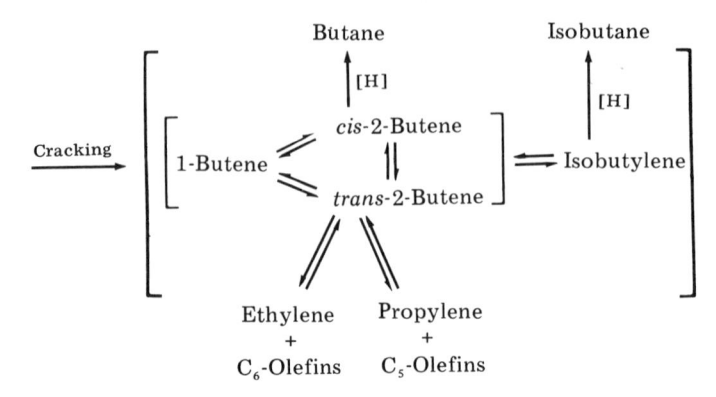

$$\text{Cracking} \longrightarrow \begin{bmatrix} \begin{bmatrix} & & \text{Butane} & & & & \text{Isobutane} \\ & & \uparrow [\text{H}] & & & & \uparrow [\text{H}] \\ & & \text{cis-2-Butene} & & & & \\ \text{1-Butene} & \rightleftarrows & \updownarrow & & \rightleftarrows & \text{Isobutylene} \\ & & \text{trans-2-Butene} & & & & \\ & \swarrow & & \searrow & & & \\ \text{Ethylene} & & & \text{Propylene} & & & \\ + & & & + & & & \\ \text{C}_6\text{-Olefins} & & & \text{C}_5\text{-Olefins} & & & \end{bmatrix} \end{bmatrix}$$

Figure 2. Reactions of butylenes in catalytic cracking.

Steam Cracking. In steam cracking, reaction conditions are selected to maximize production of light olefins. The type of feedstocks and the reaction conditions determine the amount of butylenes produced. Typical wt % yields for different feedstocks are: ethane, 0.4; propane, 1.4; butane, 1.6; light naphtha, 4.1; light gas oil, 5.0. Steam crackers operating on light paraffins (ethane and propane), as is the case for roughly 60% of the total United States steam cracker capacity, do not produce enough butylenes to make recovery worthwhile. Composition of the butylenes produced, shown in Table 6, is substantially the same from all feedstocks. However, the amounts of isobutylene and 1-butene in butylenes from steam cracking are much greater than those in butylenes from catalytic cracking.

Cracking conditions are selected to maximize production of light olefins. Typically, cracking is practiced at a weight ratio of 0.3:1.0 of steam to hydrocarbon with the reactor coil outlet at 760–870°C, and slightly above 100 kPa (atmospheric) pressure. In older plants cracking ethane and propane, the residence time was in the order of 1.0 s; in newer plants where heavier feedstocks are used, shorter residence times of 0.1–0.4 s are used.

During steam cracking, the cracked gases emerging from the reactors are rapidly quenched to arrest undesirable secondary reactions which destroy light olefins. Both direct and indirect heat exchange are used to recover heat while cooling the cracked gases to about 50°C. These cooled gases are then compressed and separated for recovery of olefins. Generally, one of the two flow schemes (front-end demethanizer or front-end depropanizer), shown in Figure 3, is selected. The choice depends on the feedstock and is purely an economic decision.

The mixed C_4 stream, which contains butylenes, is separated in a debutanizer as an overhead product, where a sharp separation is performed between C_4 and C_5 hydrocarbons. Ethylene and propylene are recovered as the principal products, and butylenes are obtained in an approximately 50–50 mixture with butadiene.

Because butadiene commands a higher price, some steam cracking units include a satellite unit for recovering butadiene from butylenes by extractive distillation using a polar solvent such as acetonitrile (68), dimethylformamide (69), N-methylpyrrolidinone (70), or dimethylacetamide (71). These processes efficiently recover polymerization-grade butadiene and leave a butylene raffinate with less than 1% butadiene residue (see Butadiene; Elastomers, synthetic).

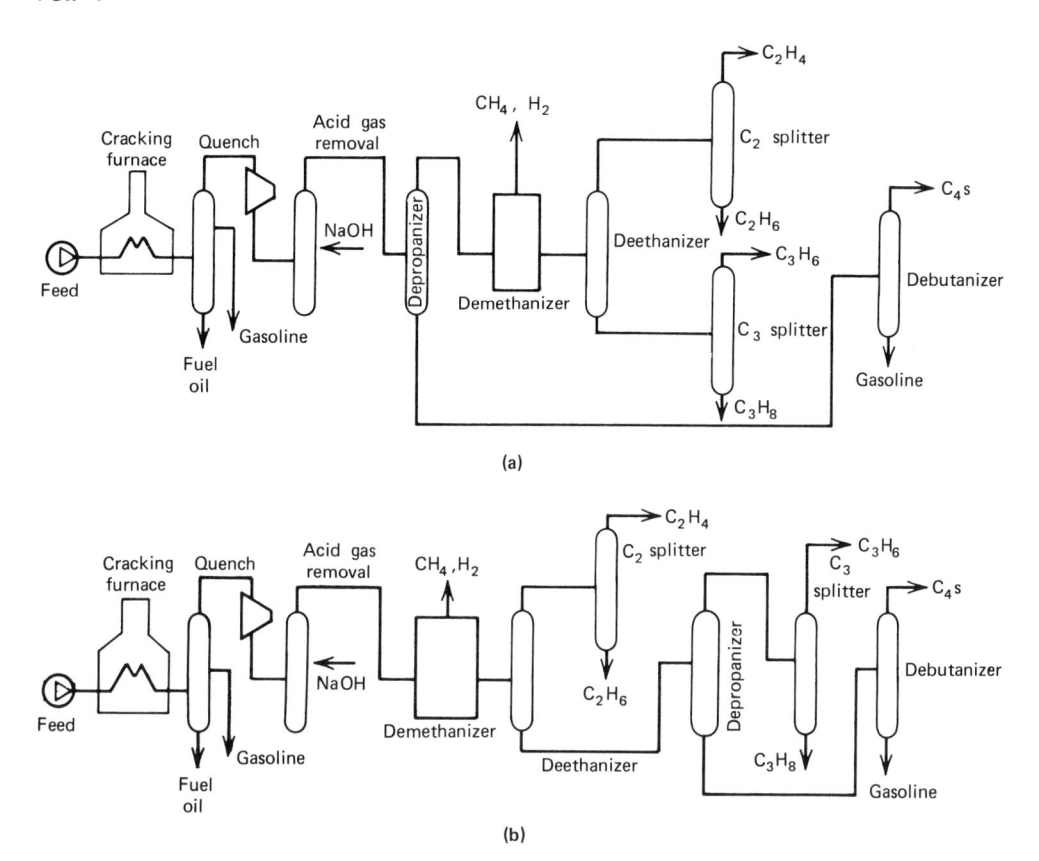

Figure 3. Steam cracking (**a**) front-end depropanizer and (**b**) front-end demethanizer.

Oxirane Process. The production of propylene oxide by the Oxirane process produces *tert*-butyl alcohol as a by-product.

The alcohol from current production could easily be dehydrated to provide a potential of 240,000 metric tons per year of high-purity isobutylene, an amount approximately equal to that now being used for polyisobutylene, butyl elastomers, and alkylation of phenols and cresols. However, most of the current production of the alcohol is being added to the gasoline pool as a high octane blending component.

Dehydrogenation of Butane and Isobutane. Mixtures of 1- and 2-butenes are prepared commercially as intermediates in the Phillips two-step process for dehydrogenation of butane to butadiene. 1-Butene and 2-butene are generated from butane and then are either purified and separated or are fed to a second stage for conversion to butadiene. Butenes are also produced in the Houdry single-stage dehydrogenation process, and could be separated from the butane–butene recycle stream. The butenes are produced in the ratio of about 2 parts each of 1-butene and *cis*-2-butene to 3 parts of *trans*-2-butene. If steam crackers proliferate to the extent that they can supply total butadiene demand, then current dehydrogenation capacity could be used to supply amounts of butenes equivalent at least to those of butadiene derived from butane.

The Coastal States process for thermal dehydrogenation of isobutane (72–73) to isobutylene at about 1400°C and 100–200 kPa (1–2 atm) appears to be commercially feasible, but has not been commercialized, possibly because it is relatively nonselective and produces nearly equal quantities of propylene along with the isobutylene.

Disproportionation of Olefins. Butenes, particularly 2-butene, can be readily prepared by disproportionation (metathesis) of propylene (74–76).

$$2\ CH_3-CH=CH_2 \rightleftharpoons CH_2=CH_2 + CH_3CH=CH-CH_3$$

A large amount of research has been conducted on this general reaction since about 1964. A commercial unit of about 29,500 t/yr has been operated by Shawinigan Chemical Ltd (77) but is currently shut down, probably because economics still favor the direct production of ethylene by steam cracking with its concomitant production of propylene (already abundant) and butylenes.

Oligomerization of Ethylene. A small amount of 1-butene results from ethylene oligomerization processes which yield a spectrum of linear, terminal olefins ranging from C_4 to C_{24}. 1-Butene is 6–20% of the product. However, because C_{16} and higher olefins are the most desired for detergent manufacture, the production of 1-butene is usually minimized. Total linear, terminal olefins capacity (Ethyl, Shell, and Gulf) is about 395,000 t/yr (1977) from which about 20,000 t/yr of 1-butene is produced. This amount can be varied depending on the product distribution of the plants.

Separation and Purification. As shown by the volatility data in Table 7, butylene isomers cannot be separated by simple fractionation. For example, isobutylene and 1-butene usually behave as a single component. Consequently, commercial processes rely on other means of separation such as extraction with sulfuric acid or selective adsorption.

Figure 4 shows the flow diagram for a typical process in which isobutylene is separated from butylenes by sulfuric acid extraction (78). At low acid concentration, only isobutylene reacts and is hydrated to *tert*-butyl alcohol (TBA). At higher acid concentrations, formation of isobutylene dimers and trimer increases. Generally, the butylenes are contacted with 45–65% sulfuric acid in a two-stage reactor. The extract containing the TBA and oligomers (dimers, trimers, etc) is flashed to remove entrained hydrocarbon impurities and steam stripped in a regenerator to recover isobutylene, TBA, and oligomers from the acid. Unconverted TBA is recycled. Oligomers can be recovered as such or may be cracked to recover monomers. Isobutylene with 99+% purity can be produced in this process.

A recent advent is selective adsorption, typified by the Union Carbide OlefinSiv process (79). When vaporized butylenes are passed through a bed of Type 5A molecular sieves (qv), butenes are selectively adsorbed. Unadsorbed isobutylene is distilled and recovered at >99% purity. Adsorbed butenes are periodically desorbed by a purge stream (see Adsorptive separation).

Further separation of butenes into isomers after isobutylene has been removed

Table 7. Volatility of Butylenes and Butanes

Compound	Boiling point at 101.3 kPa[a], °C	Vapor pressure at 25°C, kPa[a]	Ideal relative volatility
isobutane	−11.72	358	1.14
isobutylene	−6.90	300	1.0
1-butene	−6.26	300	0.9945
butane	−0.55	245	0.8139
trans-2-butene	0.88	228	0.7855
cis-2-butene	3.72	213	0.7183

[a] To convert kPa to mm Hg, multiply by 7.5.

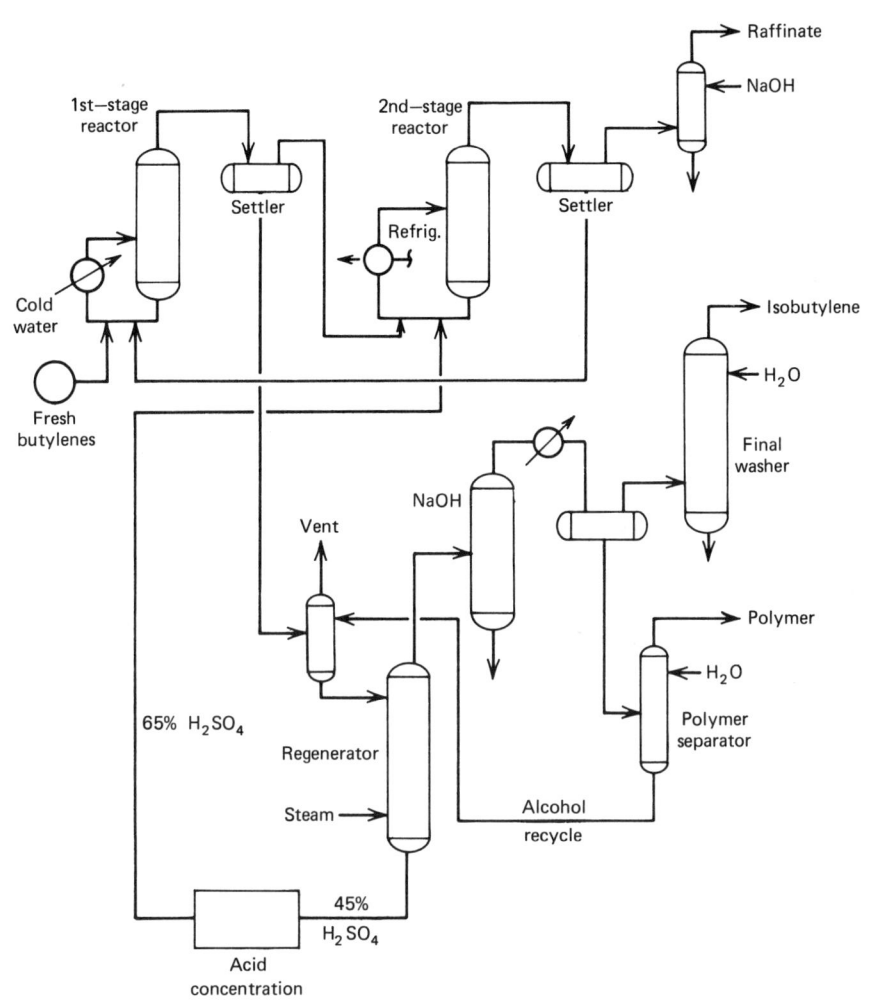

Figure 4. Isobutylene extraction process.

by sulfuric acid extraction or selective adsorption depends on the purity of individual isomers required. 1-Butene can be separated from 2-butene by distillation. However, low-boiling contaminants, such as butanes, butadiene, and C_4 acetylenes, will distribute themselves largely with 1-butene in the distillation process. Recovery of high purity 1-butene thus requires additional steps to remove these contaminants. Butadiene is most conveniently removed by selective hydrogenation employing a noble metal catalyst. Because the catalyst used for this purpose usually causes some butenes isomerization, the hydrogenation is best done before distillation. If butanes are undesirable, they may be removed by additional extractive distillation (see Azeotropic and extractive distillation). Residual isobutylene can be removed by dimerization. All of this additional processing adds to the cost of the isomer produced, so that the purity desired and the cost of achieving it must be carefully balanced.

Recovery of 1-butene may be considerably simplified by the new UOP liquid-phase process in which only 1-butene is selectively adsorbed from a butane–butylene

(B–B) stream (80). Butadiene, if present, will contaminate the product (see Adsorptive separation, liquids). There is no large commercial use for either *cis*- or *trans*-2-butene. These isomers can be separated by fractionation, but the process is inefficient.

Handling and Analysis

Storage and Transportation. In the range of 0–40°C butylenes are liquids at a pressure of about 100–400 kPa (1–4 atm), lower than the 500–1500 kPa (5–15 atm) for LPG (see Liquefied petroleum gas). Therefore, they are quite easily stored in tanks and underground caverns and transported as liquids in pipelines, tank wagons, and tank cars in much the same manner as is LPG.

Health and Safety. Butylenes are not regarded as toxic. However, because their physiological properties are not adequately known, they should be handled with caution. Some of the isomers may have narcotizing action if inhaled. All of them are asphyxiants.

The butylenes are extremely flammable. Their flash points are in the range of −80 to −73°C, and their autoignition temperatures in the range of 324–465°C. The fire hazards should be minimized by appropriate preventive measures. In case of fire, both water spray and carbon dioxide extinguishers are applicable. Butylenes also form explosive mixtures with air and oxygen. In air the explosive limits are 1.7–9.7% butylene.

Analysis. Measurement of the individual butylene isomers in a hydrocarbon mixture and identification and determination of the impurities in butylenes are the most frequent analytical requirements. Gas–liquid chromatography and mass spectrometry are the most commonly used analytical tools. The "Petroleum Products and Lubricants" section in the *1975 Book of ASTM Standards* describes several routine analyses applicable to butylenes; among them, Method D-1717 is used for the gas chromatographic analysis of butane–butylene mixtures.

Commercial Utilization

The general use pattern for butylenes in the United States, Western Europe, and Japan is summarized in Table 8. The primary factor influencing the general use pattern is the demand for butylenes to produce alkylate for motor fuels.

In the United States, about 87% of the total butylenes used in motor fuels is converted to alkylate. As a result, the butylene price in the United States is set by its value in alkylate. In Western Europe and Japan, however, the refineries are designed to maximize fuel oil production rather than motor gasoline, and alkylation accounts for a negligible fraction of the gasoline supply. Also, steam cracking operations in these countries have always depended on naphtha as the chief feedstock, thus butylenes from steam cracking contribute significantly more to the total supply than in the United States. Not having an outlet as alkylation feed, the butylenes have always been in over-supply and their prices are traditionally set by their use as fuel. The fraction

Table 8. Butylenes Consumption in 1976, % of Supply

Use	United States	Western Europe	Japan
motor gasoline	86.8	41.5	15.0
chemical uses	13.2	15.5	13.0
as fuel	0	43.0	72.0

consumed as chemical feedstocks is nearly the same in all three areas. The detailed use pattern for the United States is shown in Table 9.

Fuels. *Alkylate.* The largest use of butylenes is as feed to alkylation units within refineries and this reaction has been extensively studied (82–83). Alkylation (qv) converts C_3–C_5 olefins and isobutane to highly branched, high octane C_5–C_{12} paraffins for use in gasoline. The process utilizes a series of acid-catalyzed, hydride transfer and carbenium-ion chain reactions to produce the desired branched, high octane paraffins. The principal product from reaction of isobutane and butylenes is 2,2,4-trimethylpentane (isooctane).

$$\overset{+}{CH_3}CCH_2 + CH_3CH=CHCH_3 \longrightarrow \left[\begin{array}{c} CH_3 \\ | \\ CH_3C-\overset{+}{C}HCHCH_3 \\ | \ | \\ CH_3 \ CH_3 \end{array} \right] \longrightarrow \begin{array}{c} CH_3 \\ | \\ CH_3\overset{+}{C}CH_2\overset{+}{C}CH_3 \\ | \ | \\ CH_3 \ CH_3 \end{array} \longrightarrow$$

$$\begin{array}{c} CH_3 \\ | \\ CH_3\overset{+}{C}CH_2\overset{+}{C}CH_3 \\ | \ | \\ CH_3 \ CH_3 \end{array} + \begin{array}{c} H \\ | \\ CH_3\overset{\cdot}{C}CH_3 \\ | \\ CH_3 \end{array} \longrightarrow \begin{array}{c} CH_3 \\ | \\ CH_3CCH_2CHCH_3 \\ | \ | \\ CH_3 \ CH_3 \end{array} + \begin{array}{c} \overset{+}{C}H_3CCH_3 \\ | \\ CH_3 \end{array}$$

Table 9. Commercial Uses of Butylenes, United States, 1977[a]

Product	1000 Metric tons	Isomers used
Gasoline and fuels		
alkylate/polymer gasoline	11,750	mixed butylenes
direct blending and LPG	3,000	mixed butylenes
Total	14,750	
Chemicals		
butadiene	570	butenes
sec-butyl alcohol	200	butenes
di–triisobutylene	48	isobutylene
alkylated phenols and cresols	20	isobutylene
heptenes	13	mixed butylenes
tert-butylamine	8.6	isobutylene
tert-butyl alcohol	5.0	isobutylene
primary amyl alcohol	3.8	butenes
tert-butyl mercaptan	3.4	isobutylene
butylene oxide	2.8	butenes
methallyl chloride	2.1	isobutylene
p-tert-butyltoluene	1.7	isobutylene
neopentanoic (pivalic) acid	1.7	isobutylene
di–triisobutyl aluminum	0.9	isobutylene
ethylene–butylene copolymer	0.7	butenes
butylated hydroxyanisole	0.3	isobutylene
Total	882	
Polymers		
butyl elastomers	186	isobutylene
polybutenes	187	mixed butylenes
polyisobutylene	33	isobutylene
poly(1-butene)	10	butene-1
Total	416	
Total consumption	16 048	

[a] Ref. 81.

This compound makes up from 35 to 60% of butylene alkylate depending on the process used. Other compounds are produced from the isooctane by isomerization and disproportionation side reactions. Both commercial processes, sulfuric acid- and hydrofluoric acid-catalyzed alkylation, involve two-phase, liquid–liquid contacting in a reaction zone held at subambient temperatures, followed by settling and separation of the dense acid phase from the alkylate (see Petroleum refinery processes).

Polymer Gasoline. Polymer gasoline or polygas results from the copolymerization of propane–propene and butane–butylene streams, to yield mainly branched C_6–C_{10} olefins. These oligomers have good octane ratings and blending characteristics and are blended directly into gasoline. Polygas units have been largely replaced by alkylation units because the latter produce higher octane gasoline, but they are still used where insufficient isobutene is available for alkylation. The main process used for manufacture is that developed by UOP (47) which employs phosphoric acid on kieselguhr in a fixed bed.

Gasoline Blending and LPG. Because the vapor pressures and octane values of butylenes and butanes are very similar, butylenes can be used in place of butanes in motor gasoline for vapor pressure control. The heating values of butylenes are also only slightly less than those of butanes, and they may be blended with LPG for bottled gases (84–85).

These uses, however, are limited by temperature. In cold climates and seasons, the amount of C_4 in motor gasoline is increased to enhance volatility and ease of starting. For warm climates and seasons the amount is reduced to prevent excess vaporization and vapor-lock in automotive carburetors. Only limited amounts of C_4 hydrocarbons can be used in LPG (bottled gas) because the vapor pressure is not sufficiently high to provide adequate pressure at low temperatures.

In the United States, blending and LPG is not a major use because paraffinic light hydrocarbons are readily available from natural gas liquids and butenes have a higher value in alkylate. In Europe where refineries are the major LPG source, olefins are used.

Chemicals. Although the amount of butylenes produced in the United States is roughly equal to the amounts of ethylene and propylene produced, the amount consumed for use in chemicals is considerably less. Thus, as shown in Table 10, the utilization of either ethylene or propylene for each of at least six major chemical derivatives is about the same or greater than utilization of butenes for butadiene, their

Table 10. Utilization of Ethylene, Propylene, and Butylenes for Production of Chemicals in 1977, 1000 Metric Tons

Ethylene[a]		Propylene[a]		Butylenes[b]	
polyethylene, low density	2941	polypropylene	1248	butadiene	570
polyethylene, high density	1660	isopropanol	595	sec-butyl alcohol	200
vinyl chloride	1183	cumene	645	polybutenes	190
ethylene oxide	1278	propylene oxide	625	butyl rubber	186
acetaldehyde	500	acrylonitrile	590	dimers and trimers	48
ethanol	360	butyraldehyde	544	polyisobutylene	33
vinyl acetate	226			butylated phenols and cresols	20

[a] Based on product data (80), assumes 100% conversion of olefin to product.
[b] Ref. 81.

main use. (This production is only about one third of the total—the remaining two thirds is derived directly from butane.) The underlying reasons are poorer price–performance compared to derivatives of ethylene and propylene and the lack of applications of butylene derivatives. Some of the C_4 products are more easily derived from 1-, 2-, and 3-carbon atom species, eg, butanol (1), 1,4-butanediol (2), and isobutyl alcohol (3).

$$2\ CH_2{=}CH_2 \xrightarrow{[O]} 2\ CH_3CHO \rightarrow CH_3CH{=}CHCHO \xrightarrow{H_2} CH_3CH_2CH_2CH_2OH \quad (1)$$

$$HC{\equiv}CH + 2\ CH_2O \xrightarrow{H_2} HOCH_2CH_2CH_2CH_2OH \quad (2)$$

$$CH_3CH{=}CH_2 \xrightarrow[H_2]{CO} CH_3CH_2CH_2CH_2OH + (CH_3)_2CHCH_2OH$$
$$\qquad\qquad\qquad\qquad\qquad (1) \qquad\qquad\qquad (3)$$

The value of butylenes in the United States is determined by their value in alkylation of isobutane to high-octane gasoline and is about 9–11¢/kg (1976). This value is now considerably lower than that for ethylene, 24–26¢/kg (1976), and propylene, 20–22¢/kg (1976) (86). Cost of isomer separation for butylenes, if practiced on a large scale, should not increase the cost beyond that of ethylene or propylene. Since much chemical-products usage is determined on a price–performance basis, a shift to development of butylene-based technology may occur.

The major chemical products derived from butylenes are shown in Table 9. Their uses and manufacture are discussed below.

Butadiene. Although manufacture of butadiene is the single largest consumer of butylenes for chemicals use, only about one third of the capacity (1,590,000 t/yr in 1977) is based on processes in which butenes are the feedstock. About half of the remaining two thirds is derived from direct dehydrogenation of butane (Houdry Process) and the other half is coproduct from ethylene units. These processes are described in detail elsewhere in this Encyclopedia (see Butadiene) and review articles (87).

Although the production of butadiene is expected to increase, consumption of butanes and butenes for this use is expected to decrease because of the increase in butadiene production from the steam cracking of heavy feedstocks (88–89). Butadiene yields range from 1.5% for plants operating on ethane to more than 10% for plants operating on heavy feedstocks (see Feedstocks).

Sec-Butyl Alcohol. sec-Butyl alcohol is produced entirely from butenes using indirect hydration with sulfuric acid. Nearly all of the sec-butyl alcohol produced is converted to methyl ethyl ketone (MEK) by catalytic dehydrogenation.

Typical feed to the commercial process is refinery or steam cracker B–B (butane–butylene) from which butadiene and isobutylene have been removed, to leave a mixture of 1-butene, 2-butene, butane, and isobutane. This feed is extracted with 75–85% sulfuric acid at 35–50°C to yield butyl hydrogen sulfate. This ester is then diluted with water and stripped with steam to yield the alcohol. (Both 1-butene and 2-butene give sec-butyl alcohol.) The sulfuric acid is generally concentrated and recycled (90).

$$\begin{array}{c} CH_3CH_2CH{=}CH_2 \\ CH_3CH{=}CHCH_3 \end{array} \xrightarrow{H_2SO_4} CH_3CH_2CHCH_3 \xrightarrow{H_2O} CH_3CH_2CHCH_3 + H_2SO_4$$
$$\qquad\qquad\qquad\qquad\qquad\quad |\qquad\qquad\qquad\qquad |$$
$$\qquad\qquad\qquad\qquad\qquad OSO_3H \qquad\qquad\qquad OH$$

Di- and Triisobutylenes. Di- and triisobutylenes are prepared by heating the sulfuric acid extract of isobutylene from a separation process to about 90°C. A 90% yield containing about 80% dimers and 20% trimers results. Use centers on the dimer, a mixture of mainly 2,4,4-trimethylpentene-1 and -2. The main use is for alkylation of phenol to yield octylphenol which then reacts with ethylene oxide to produce a surfactant (see Alkylation).

$$C_8H_{16} + C_6H_5OH \rightarrow (C_8H_{17})C_6H_4OH \xrightarrow{nC_2H_4O} (C_8H_{17})\text{-}C_6H_4\text{-}(OCH_2CH_2)_nOH$$

Other uses of dimer are amination to octylamine and octyldiphenylamine, used in rubber processing; hydroformylation to nonyl alcohol for phthalate plasticizer production; and carboxylation via the Koch synthesis to yield acids used in formulating paint driers (see Driers).

Butylated Phenols and Cresols. Butylated phenols and cresols, used primarily as oxidation inhibitors and chain terminators, are manufactured by direct alkylation of the phenol using a wide variety of conditions and acid catalysts, including sulfuric acid, *p*-toluene sulfonic acid, and sulfonic acid ion-exchange resins (3,91). By use of a small amount of catalyst and short reaction times, the first-formed, ortho-alkylated products can be made to predominate. For the preparation of the 2,6-substituted products, aluminum phenoxides generated *in situ* from the phenol being alkylated are used as catalyst. Reaction conditions are controlled to minimize formation of the thermodynamically favored 4-substituted products (see Alkylphenols). The most commonly used compounds and their principal uses are: *p-tert*-Butylphenol for manufacture of phenolic resins. The *tert*-butyl group leaves only two rather than three active sites for condensation with formaldehyde and thus modifies the characteristics of the resin.

2,6-Di-*tert*-butylphenol as an antioxidant.

2,6-Di-*tert*-butyl-4-methylphenol (di-*tert*-butyl-*p*-cresol or butylated hydroxytoluene (BHT)) is most commonly used as an antioxidant in plastics and rubber. Use in food is decreasing because of legislation and it is being replaced by butylated hydroxy anisole (BHA) (see Antioxidants; Food additives).

p-tert-Butyl catechol as an inhibitor for styrene monomer.

Heptenes. Heptenes are used for the preparation of isooctyl alcohol by hydroformylation (see Oxo process).

$$\underset{R'}{R-C=CH_2} \xrightarrow{H_2/CO} \underset{R'}{R-CHCH_2CHO} \xrightarrow{H_2} \underset{R'}{R-CHCH_2CH_2OH}$$

The heptenes are prepared by very carefully controlled fractionation of poly gas. Specifications generally call for >99.9% C_7 content (including some paraffin which is also formed) to simplify processing.

About 32 isomers are present in these heptenes. Dominant carbon skeletons are:

5-15%   15-20%   20-35%   40-50%

The amounts of these heptenes used commercially are decreasing because the butenes are being diverted from polygas manufacturing to alkylation which produces a better motor fuel, and because expansion in production of the phthalate esters used as plasticizers for poly(vinyl chloride) has been based mostly on C_6–C_{10} linear alcohols.

Tert-Butylamine. *tert*-Butylamine is used as an intermediate in the manufacture of lubricating oil additives and miscellaneous chemicals. It is manufactured using the Ritter reaction. Isobutylene first reacts with sulfuric acid and then HCN to yield *tert*-butylformamide. Hydrolysis yields the amine.

$$(CH_3)_2C{=}CH_2 \xrightarrow{H_2SO_4} (CH_3)_3COSO_3H \xrightarrow{HCN} (CH_3)_3CN{=}CHOSO_3H$$

$$\xrightarrow{H_2O} (CH_3)_3CNHCHO \xrightarrow{OH^-} (CH_3)_3CNH_2$$

Tert-Butyl Alcohol. Until recently *tert*-butyl alcohol was a small volume (ca 5000 t/yr) chemical used primarily as a solvent. It was manufactured by the hydrolysis of the sulfuric acid extract obtained during the separation of pure isobutylene from mixed butane-butylene streams. However, with the introduction of the Oxirane process for manufacture of propylene oxide (92), by-product *tert*-butyl alcohol will be available in quantity. An estimated 332,000 t/yr will be available from the projected propylene oxide capacity of 253,000 t/yr. It is derived from isobutane which is the oxygen carrier for the process:

$$(CH_3)_3CH \xrightarrow{[O]} (CH_3)_3COOH$$

$$(CH_3)_3COOH + CH_3CH{=}CH_2 \longrightarrow CH_3\overset{O}{\overset{|}{CH}}{-}CH_2 + (CH_3)_3COH$$

Currently most of it is blended directly into automotive fuel and some is converted to high purity isobutylene by dehydration.

Tert-Butyl Mercaptan. *tert*-Butyl mercaptan is used primarily as an odorant at <30 ppm for natural gas so that leaks can be readily detected. It is manufactured by the reaction of isobutylene and hydrogen sulfide in the presence of acid catalysts (93).

Primary Amyl Alcohols. Primary amyl alcohols are manufactured by hydroformylation of mixed butenes, followed by hydrogenation (94). Both 1-butene and 2-butene yield the same products although in slightly different ratios depending on the catalyst and conditions. Some catalysts and conditions produce the alcohols in a single step. By modifying the catalyst, typically a cobalt carbonyl, with phosphorus derivatives, such as tri(*n*-butyl)phosphine, the linear alcohol can be the major product from 1-butene.

$$\begin{array}{c} CH_3CH{=}CHCH_3 \\ \text{or} \\ CH_3CH_2CH{=}CH_2 \end{array} \xrightarrow[\text{catalyst}]{H_2/CO} \begin{array}{c} \overset{CHO}{\underset{|}{CH_3CH_2CHCH_3}} \\ + \\ CH_3CH_2CH_2CH_2CHO \end{array} \xrightarrow{H_2} \begin{array}{c} \overset{CH_2OH}{\underset{|}{CH_3CH_2CHCH_3}} \\ + \\ CH_3CH_2CH_2CH_2CH_2OH \end{array}$$

The main use of the amyl alcohols is as esters such as acetates for solvents.

Di- and Triisobutylaluminums. Triisobutylaluminum is prepared by reaction of isobutylene with aluminum at 80°C and 20.3 MPa (200 atm) of hydrogen (95). It is used as catalyst for ethylene oligomerization to prepare even-numbered, linear 1-olefins. Use of stoichiometric quantities of triisobutylaluminum followed by oxidation of the resulting mixed, long-chain aluminum alkyls yields even-numbered, terminal primary alcohols in the plasticizer and detergent range (96). Conoco uses that process in the United States to manufacture plasticizer (C_6–C_{10}) and detergent (C_{16}–C_{22}) range alcohols (see Alcohols, synthetic).

Triisobutylaluminum is converted to diisobutylaluminum chloride and diisobutylaluminum hydroxide which are used as cocatalysts for Ziegler polymerization systems. Corresponding ethyl compounds are prepared via the reaction of the triisobutylaluminum with ethylene (see Organometallics; Ziegler-Natta catalysts).

Butylene Oxide. Butylene oxides are prepared on a small scale (Dow) by chlorohydrin technology. There appears to be no technical reason why they could not be prepared by the propylene oxide Oxirane process (see Chlorohydrin).

A major use of butylene oxide is as an acid scavenger for chlorine-containing materials such as trichloroethylene. Inclusion of about 0.25–0.5% of butylene oxide, based on the solvent weight, during preparation of vinyl chloride and copolymer resin solutions minimizes container corrosion which may be detrimental to resin color and properties.

p-Tert-Butyltoluene. p-tert-Butyltoluene, prepared by acid-catalyzed alkylation of toluene with isobutylene under mild conditions (97–98) is an intermediate in the production of p-tert-butyl benzoic acid. This acid is used as a chain-length control agent in the preparation of unsaturated polyester resins. Solubility characteristics offer some advantage over benzoic acid.

Neopentanoic (Pivalic) Acid. Neopentanoic acid is prepared using the Koch technology in which isobutylene reacts with carbon monoxide in the presence of strong acids such as H_2SO_4, HF, and $BF_3 \cdot H_2O$ (99–102). General reaction conditions are 2–10 MPa (ca 20–100 atm) of CO and 40–150°C.

$$(CH_3)_2C{=}CH_2 \xrightarrow{H_2SO_4} [(CH_3)_3C]^+ OSO_3H^- \xrightarrow{CO} (CH_3)_3C\overset{O}{\overset{\|}{C}}OSO_3H \xrightarrow{H_2O} (CH_3)_3C\overset{O}{\overset{\|}{C}}COH + H_2SO_4$$

The acids are converted to peroxy esters for use as polymerization initiators. The metal salts are used as driers in paint formulations (see Carboxylic acids).

Methallyl Chloride. Methallyl chloride is the major product when isobutylene and chlorine react over a wide range of temperature (11). Very little addition takes

$$\underset{\underset{CH_3}{|}}{CH_3C}{=}CH_2 + Cl_2 \longrightarrow \underset{\underset{CH_3}{|}}{CH_2{=}CCH_2Cl} + HCl$$

place.

It is a chemical intermediate for various specialty products, but it has no single significant commercial use.

Butylated Hydroxy Anisole (BHA). This material is an oxidation inhibitor and has been accepted for use in foods where the use of butylated hydroxy toluene (BHT) is restricted (see Food additives). It is manufactured by the alkylation of 4-hydroxyanisole alkylation with isobutylene which yields the mixture of 2- and 3-*tert*-butyl isomers used as product (103).

$$\text{4-hydroxyanisole} + C_4H_8 \longrightarrow \underbrace{\text{2-tert-butyl isomer} + \text{3-tert-butyl isomer}}_{\text{BHA}}$$

Potential Use. Processes using butylenes as feedstocks have been developed for a group of industrial chemicals that are not currently produced by these processes or are produced only on a relatively small scale. Such chemicals are: isoprene, methyl methacrylate, maleic anhydride, methyl *tert*-butyl ether, and acetic acid. These processes are of interest because they may emerge as the dominant processes with suitable improvements, changes in product values, or development of new markets.

Isoprene. Although current United States synthetic capacity for isoprene (**4**) is based entirely on dehydrogenation of refinery isoamylenes and demethanation of propylene dimer, there has been a large amount of work on alternative processes, particularly those using the Prins reaction (104–107).

$$(CH_3)_2C{=}CH_2 + CH_2O \xrightarrow[H^+]{H_2O} (CH_3)_2CCH_2CH_2(OH)(OH) \xrightarrow[-H_2O]{CH_2O} \text{(dioxane intermediate)}$$

$$\xrightarrow{-2H_2O} \quad CH_2{=}C(CH_3)CH{=}CH_2 \quad \xleftarrow[-H_2O]{-CH_2O}$$

(**4**)

Processes have been developed both in the Union of Soviet Socialist Republics and by the Institute Francais de Petrol. A significant advantage of the Prins reaction process is that isobutylene need not first be separated from the butane–butylene (B–B) fraction because formaldehyde reacts selectively with the isobutylene.

A second route based on olefins disproportionation was developed by Phillips Petroleum (77). Here isobutylene reacts with propylene to form isoamylenes, which are dehydrogenated to isoprene.

$$CH_3{-}C(CH_3){=}CH_2 + CH_3{-}CH{=}CH_2 \longrightarrow CH_3{-}C(CH_3){=}CH{-}CH_3 + CH_2{=}CH_2 \xrightarrow{-H_2} (\mathbf{4})$$

2-Butene can be used in place of propylene since it also yields isoamylene and the co-product propylene can be recycled. Use of mixed butylenes causes the formation of pentenes, giving piperylene which contaminates the isoprene.

Although the availability of butane–butylene streams containing high concentrations of isobutylene from steam crackers will increase and possibly make these technologies more attractive, these same steam crackers also produce recoverable amounts of isoprene directly. Currently the projected isoprene production from the increasing numbers of steam crackers, which will be using heavier feeds and therefore producing higher yields of isoprene, is expected to supply the slowly increasing demand for isoprene (108).

Methacrylates. Currently the production of methacrylate esters, in the range of 340,000 t/yr in the United States, is accomplished entirely from acetone by reaction with HCN and subsequent conversion to methyl ester and by-product ammonium bisulfate.

$$CH_3\overset{O}{\overset{\|}{C}}CH_3 \xrightarrow{HCN} CH_3\underset{CN}{\overset{OH}{\overset{|}{C}}}CH_3 \xrightarrow[CH_3OH]{H_2SO_4} CH_2{=}\underset{CO_2CH_3}{\overset{|}{C}}CH_3 + NH_4HSO_4$$

There are two technologies for production from isobutylene: ammoxidation to methacrylonitrile (Sohio) which is then solvolyzed, similarly to acetone cyanohydrin, to methyl methacrylate; and direct oxidation of isobutylene in two stages via methacrolein to methacrylic acid which is then esterfied (109a). Although there is a large body of technical literature related to both processes, neither is in current commercial use. Since direct oxidation bypasses the need for NH_3 and HCN and eliminates disposal of ammonium bisulfate and handling of toxic materials, it appears to be the preferred process, and recent announcements indicate plants based on this technology will come on stream in 1981 (Oxirane) and possibly 1985 (Rohm and Haas) (109b). The Oxirane plant will use by-product *tert*-butyl alcohol directly rather than first converting it to isobutylene (see Methacrylic acid).

Maleic anhydride. Until recently all maleic anhydride production in the United States was based on benzene as feedstock, even though there exists a substantial literature on use of butenes (2,110–111). However, the rapidly increasing demand and price for benzene have made butenes appear to be a better feedstock. Not only are theoretical yields much better, 1.75 kg/kg butenes compared to 1.26 kg/kg benzene, but less oxygen is required and the oxidation produces less heat, which is critical in reactor design.

However, at the same time benzene prices are escalating, need for butenes for use in alkylates for motor fuel has also increased and butenes prices have also escalated. As a result technology based on butane has been developed and Amoco Chemicals Corporation has recently (1977) brought a 27,000 t/yr plant on stream (112–113). Commercialization of this technology will probably preclude significant development of butenes-based technology, at least in the United States. In Europe and particularly in Japan, where butane is much less abundant and needs for butenes in alkylate are much less, butenes may become the dominant feedstock (see Maleic anhydride).

Methyl tert-butyl ether. Recently much effort has been spent looking for new components for automotive fuels that can maintain octane ratings as lead is removed, and which have burning characteristics that minimize exhaust emission problems. Methyl *tert*-butyl ether appears to be a possible candidate (114–116a). It can be prepared by acid catalyzed addition of methyl alcohol to isobutylene using commercially available sulfonated styrene–divinylbenzene resin catalysts. The reaction has a particular advantage in that not only can a total butane–butylene (B–B) stream be used with a selective isobutylene reaction, but because the isobutylene is removed, the spent stream produces a higher octane, more valuable alkylate from isobutane.

Although methyl *tert*-butyl ether is being produced in Europe for use as automotive fuel (116b), first large-scale production in the United States will probably be from a 150,000 t/yr plant being constructed by Arco Chemicals and due to start up in 1980 (116c). Several other plants are being planned (116d).

Acetic acid. A process to convert butenes to acetic acid has been developed by Farbenfabriken Bayer AG (117). This process could be of particular interest in Europe and Japan where butylenes have only fuel value. In this process a butane–butylene stream from which isobutylene and butadiene have been removed reacts with acetic acid in the presence of an acid ion-exchange resin at 100–120°C and 1500–2500 kPa (ca 15–25 atm) (see Acetic acid). Both butenes react to yield *sec*-butyl acetate (5) which is then oxidized at about 200°C and 6 MPa (ca 60 atm) without catalyst to yield acetic acid.

$$CH_3CH=CHCH_3 + CH_3CH_2CH=CH_2 + CH_3COOH \longrightarrow CH_3CH_2CH(OCOCH_3)CH_3 \xrightarrow{[O]} 3\ CH_3COOH$$
$$(5)$$

This process may be competitive with butane oxidation (see Hydrocarbon oxidation) which produces a spectrum of products (118), but neither process appears competitive with the newer process from synthesis gas practiced by Monsanto (119) and BASF (120a) which have been used in 90% of the new acetic acid capacity added during the past five years (120b).

$$2H_2 + CO \longrightarrow CH_3OH \xrightarrow{CO} CH_3C(O)-OH$$

Polymers. Polymers account for about 3–4% of the total United States butylene consumption and about 30% of the nonfuels use.

Homopolymerization of butylene isomers is relatively unimportant commercially. Only stereoregular poly(1-butene) and a small volume of polyisobutylene are produced in this manner. High molecular weight polyisobutylenes have found limited use because they cannot be vulcanized. To overcome this deficiency a butyl rubber copolymer of isobutylene with isoprene has been developed. Low molecular weight viscous liquid polymers of isobutylene are not manufactured because of the high price of purified isobutylene. Copolymerization from relatively inexpensive refinery butane–butylene fractions containing all the butylene isomers yields a range of viscous polymers which satisfy most commercial needs.

Three of the commercially important polymers derived from butylenes, poly(1-butene), polyisobutylene, and butyl elastomer, are discussed only briefly here (see Olefin polymers; Elastomers, synthetic).

Polybutylenes. In the trade the viscous butylene polymers are incorrectly called polybutenes since this name implies that they exclude isobutylene. They should be called polybutylenes. Viscous polybutylene production was started by Chevron Chemicals in 1940. Currently, Amoco Chemicals has nearly a two thirds share of United States polybutylene production. The other producers are Chevron Chemicals, Cosden Oil, Lubrizol, and Exxon Chemicals. Canada, Western Europe, and Japan also have production facilities. In 1975 polybutylene prices in tank car quantities were in the range of 33–49¢/kg, the higher molecular weight polymers representing the high end of the price range.

The feed for the polybutylene plants is the C_4 fraction from either catalytic cracking units or steam cracking units. Steam cracking units provide fractions with high isobutylene concentration which are preferred feeds. The butane and isobutane in these feeds act as solvents for the polymerization.

Only Cosden Oil's process has been described (121). The process used by the other producers can only be inferred from patent descriptions (122–126). However, the processes are basically very similar and all are continuous-flow processes. Sulfur and nitrogen compounds must first be removed from the feed. In addition, moisture must also be removed, typically by chemical drying agents (qv) (bauxite, calcium chloride, etc) or azeotropic distillation (qv). The feed is often cooled before its continuous introduction into the reactor. The reactors are agitated cooled vessels, and two or three reactors may be joined into series. They are designed to operate in the range of -10 to $+80°C$ and up to 2 MPa (20 atm). The cooling is provided either by direct heat exchange or indirectly by refluxing the monomers and solvents. The initiator (catalyst) appears to be $AlCl_3$ in all the processes, occasionally co-initiators such as HCl or $CHCl_3$ are used. The initiator(s) are introduced continuously into the reactor, either in solution or as a slurry. From the effluent leaving the reactor continuously, the $AlCl_3$ is removed by settling and/or reaction with ammonia, amines, methanol, water or aqueous sodium hydroxide. The stream is usually passed through a clay or bauxite column to remove catalyst residues and color. Then, in a distillation section, the unreacted C_4 hydrocarbons are flashed in one column, and the light (low molecular weight) polymers are taken overhead in other columns.

The polymer viscosity and molecular weight are regulated by the selection of the polymerization temperature and catalyst concentration. All the isobutylene in the feed is usually converted to polymer; however, the conversion of the butenes is low. The unconverted butenes can serve as an excellent feed source for isobutane alkylation if an alkylation plant is on the site. A process has also been described to convert the butenes to low molecular weight polymers in a separate polymerization stage (127).

The viscous polybutylenes are colorless, odorless liquids. They are soluble in hydrocarbons and chlorinated hydrocarbons. At room temperature those having low molecular weight flow freely but the higher molecular weight grades are quite viscous and usually require heating for transfer and pumping. Table 11 lists the properties of several commercial grades. The heavier polymers are excellent lubricants, as indicated by their viscosity index. Polybutylenes are also excellent electrical insulators. Chemically the polybutylenes are monoolefins. They are nondrying, ie, have no tendency to cross-link. They are thermally stable to ca 280–300°C. Under the combined

Table 11. Properties of Several Commercial Polybutylene Grades[a]

Property[b]	ASTM test method	L-14E[d]	L-50E[d]	H-35	H-300E[d]	H-1900
viscosity	D-445					
at 37.8°C, mm²/s (= cSt)		27–33	106–112			
at 98.9°C, mm²/s (= cSt)				74–79	627–675	4069–4382
flash point COC[e], min, °C	D-92	138	149	166	227	243
API gravity at 15.6°C	D-287	36–39	23–36	28–31	25–28	23–26
color	APHA					
haze free, max		70	70	70	70	70
haze, max		15	15	15	15	15
average mol wt	D-2503	320	420	660	1290	2300
viscosity index	D-2270	69	90	100	117	122
ref. index, N$_D$	D-1218	1.4680	1.4758	1.4872	1.4970	1.5042
acidity, mg KOH/g	D-974	0.03	0.02	0.04	0.01	0.01
pour point, °C	D-97	−51	−40	−15	+2	+18
power factor, % max, 100 tan Δ	D-924	0.05	0.05		0.03	
power factor after aging	D-924; D-1934	0.07	0.07			
resistivity at 100°C, 10¹² Ω·cm	D-1169		150			
dielectric breakdown at −3.9°C, kV	D-877	35	35		30	
water content, ppm	D-1533	20	45		45	
organic chlorides, max, ppm		40	40		40	
total sulfur, max, wt %	D-1552	0.01	0.01		0.02	

[a] Abstracted from ref. 128.
[b] Viscosity, flash point, API gravity, and color are included in the specification.
[c] Selected grades for illustration. Other grades are also available.
[d] The letter E indicates electrical grade polymer.
[e] COC = Cleveland open cup.

influence of an acid and heat, isobutylene homopolymers degrade more easily than the butylene copolymers and the butene polymers. Empirical techniques such as iodine number or bromine number determinations have been used to measure the polymer unsaturation but they have proved unreliable. Hydrogenation or reaction with m-chloroperbenzoic acid gives accurate measurement. In Amoco's H-100 and H-300 grade polymers, the unsaturation is nearly one double bond per molecule (62). Because of their olefin content, the very low molecular weight polybutylenes can become oxidized to a small extent if exposed to air at room temperature, but this does not occur with the more viscous polymers because they are impermeable to gases. Nevertheless, as a precautionary measure, polybutylenes are usually blanketed with an inert gas during storage. Polybutylenes having either little unsaturation or a high degree of unsaturation are also produced; the former by hydrogenation of polybutylenes, the latter by copolymerization with a diene.

Viscous polybutylenes find a great variety of uses. By far the largest quantity is used in the manufacture of additives for motor oils. Although some polybutylenes may be blended directly into lubricating oils, they usually serve as chemical intermediates for additive synthesis. The double bond provides the functionality for the syntheses. Other major uses are: in formulations of sealants, caulks, coatings, adhesives, and laminating agents; in high voltage electrical cables as impregnating oils and pipe oils; and as industrial lubricants, eg, for sheet metal cutting and as compressor oil.

Polybutylenes are nontoxic, nonirritating, and thus find applications in paper

Polyisobutylene. The manufacture of high molecular weight polyisobutylene differs from the manufacture of polybutylenes in two important aspects: the feed is pure isobutylene in an inert solvent; and the polymerization temperature is much lower (−10 to −100°C). The molecular weight is controlled by the polymerization temperature. Polyisobutylenes are amorphous in the unstretched state but they can crystallize if stretched. The lower molecular weight grades are used as viscosity index improvers for lube oils, the higher molecular weight products as adhesives, caulks, sealants, and polymer additives. Exxon Chemical Company and Lubrizol Corporation are the domestic producers.

Butyl Elastomers. Butyl rubbers are manufactured in the same way as the high molecular weight polyisobutylenes, but about 1–3% isoprene is added to the feed. The butyl rubbers are amorphous and have no tendency to crystallize. Having low permeability to gases, they can be used for the manufacture of automotive inner tubes and inner liners. The butyl rubbers can be easily halogenated. The manufacture of chlorinated and brominated butyl rubbers has helped to extend the areas of utility. The halogenated rubbers vulcanize faster and are more compatible with other polymers. In recent years, small volumes of lower molecular weight butyl rubbers have also been produced mostly for caulk and sealant applications. Exxon Chemical Company and Cities Service Company are the domestic producers (see Elastomers, synthetic).

Poly(1-butene). Manufacture of poly(1-butene) is similar to that of high density polyethylene and polypropylene using Ziegler-Natta catalysts (qv). The commercial products have isotactic structure. They crystallize slowly to 50–55% crystallinity. The polymers excel in resistance to creep and environmental stress cracking. Their fabrication by molding is trouble-free because their slow rate of crystallization prevents buildup of stress. They retain their mechanical properties up to their softening point (94°C for the regular, partially crystallized polymer). They are particularly suitable for pipe fabrication. Films are other potential areas for use. Shell Chemicals is the only present commercial manufacturer. Currently efforts are concentrated on developing markets for this recently introduced product for which prospects for growth are good. The price will be considerably higher than prices for polyethylene or polypropylene and will be greatly influenced by the size of the market.

BIBLIOGRAPHY

"Butylenes" in *ECT* 2nd ed., Vol. 3, pp. 830–865, by C. E. Morrell, Esso Research and Engineering Company.

1. *Technical Data Book—Petroleum Refining,* American Petroleum Institute, New York, 1970; L. N. Canjor and F. S. Manning, *Thermodynamic Properties and Reduced Correlations for Gases,* Gulf, 1967; R. B. Scott, W. J. Ferguson, and F. G. Brickwedde, *J. Res. Nat. Bur. Stand.* **33,** 1 (1944); C. H. Marron and co-workers, *J. Chem. Eng. Data.* **1,** 394 (1962); R. R. Dreisbach, *Physical Properties of Chemical Compounds II,* American Chemical Society, Washington, D. C., 1959.
2. D. J. Hucknall, *Selected Oxidations of Hydrocarbons,* Academic Press, Inc., New York, 1974, p. 104.
3. G. A. Olah, *Friedel-Crafts and Related Reactions,* Vol. II, Interscience Publishers, a division of John Wiley & Sons, Inc., New York, 1964, Part 1.
4. D. E. Dorman, M. Jantelot, and J. D. Roberts, *J. Org. Chem.* **36,** 2157 (1971).
5. I. Hirano, O. Kikuchi, and K. Suzuki, *Bull. Chem. Soc. Jpn.* **49,** 3321 (1976).

6. H. Kroper and co-workers, *Hydrocarbon Process.* **48**(9), 195 (1969).
7. J. Chetron, M. Henant, and G. Marnier, *Bull. Chim. Soc. Fr.,* 1966 (1969).
8. J. H. Polstar and K. Yates, *J. Am. Chem. Soc.* **91**, 1469 (1969).
9. J. B. Hendrickson, O. J. Cram, and G. S. Hammond, *Organic Chemistry,* McGraw-Hill Book Co., New York, 1970, p. 614.
10. M. C. Hoff, K. W. Greenlee, and C. E. Boord, *J. Am. Chem. Soc.* **73**, 3329 (1951).
11. J. Burgin and co-workers, *Ind. Eng. Chem.* **31**, 1413 (1939).
12. H: C. Brown, *Hydroboration,* W. A. Benjamin Inc., New York, 1962, pp. 114, 192, 200.
13. I. Wender and co-workers, *J. Am. Chem. Soc.* **78**, 5401 (1956).
14. *Ibid.,* **77**, 5760 (1955).
15. L. F. Fieser and M. Fieser, *Advanced Organic Chemistry,* Reinhold Publishing Co., New York, 1961, p. 162.
16. U.S. Pat. 2,392,294 (Jan. 1, 1946), W. E. Vaughn and F. W. Rust (to Shell Dev. Co.).
17. U.S. Pat. 2,925,443 (Feb. 1, 1960), W. L. Walsh (to Gulf Research and Development Co.).
18. U.S. Pat. 2,555,054 (May 22, 1951), J. Owen (to Phillips Pet. Co.).
19. (a) V. J. Guerico, *Oil Gas J.,* 68 (Feb. 21, 1977); (b) *Chem. Eng.* 62, (Feb. 13, 1978); (c) *Chem. Week* 49, (Nov. 16, 1977).
20. J. E. Germain, *Catalytic Conversion of Hydrocarbons,* Academic Press, Inc., New York, 1969.
21. T. Brooks and co-workers, *The Chemistry of Petroleum Hydrocarbons;* Reinhold Publishing Corp., 1955.
22. G. Egloff, G. Hulla, and V. I. Komarewski, *Isomerization of Pure Hydrocarbons,* Reinhold Publishing Corp., New York, 1942.
23. D. R. Stuhl, E. F. Westrum, and G. C. Sinke, *The Chemical Thermodynamics of Organic Compounds,* John Wiley & Sons, Inc., New York, 1969.
24. W. G. Burns and co-workers, *Trans. Faraday Soc.* **64**, 129 (1968).
25. R. D. Cundall and T. F. Palmer, *Trans. Faraday Soc.* **56**, 1211 (1960).
26. K. Otsuka and A. Morikawa, *Chem. Commun.* (6), 218 (1975).
27. Y. Misono, Y. Saito, and Y. Yoneda, *J. Catal.* **10**(2), 200 (1968).
28. T. M. O'Grady, R. M. Alm, and M. C. Hoff, *Am. Chem. Soc. Div. Pet Prepr.* **4**(4), (1959).
29. R. Nicolova and co-workers, *C. R. Acad. Sci. Ser. C* **265**(8), 468 (1967).
30. H. Kanai, *Chem. Commun.* 203 (1972).
31. K. Tanaka, *Sci. Pap. Inst. Phys. Chem. Res. Jpn.* **69**(2), 50 (1975).
32. U.S. Patent 3,641,184 (Feb. 8, 1972), C. E. Smith and B. J. White (to Phillips Petroleum Co.).
33. K. Matsuura, A. Suzuki, and M. Itoh, *J. Catal.* **23**, 396 (1971).
34. A. M. Eastham, *J. Am. Chem. Soc.* **78**, 6040 (1956).
35. W. B. Smith and W. Y. Watson, *J. Am. Chem. Soc.* **84**, 3174 (1962).
36. D. I. Epshtein, M. I. Farberov, and G. A. Stozhkova, *Osnoun. Org. Sint. Neftekhim* **2**, 45 (1975); *Chem. Abstr.* **83** 178175n (1976).
37. T. A. Gilmore and J. J. Rooney, *Chem. Commun.,* 219 (1975).
38. K. W. Egger and S. W. Benson, *J. Am. Chem. Soc.* **88**, 236 (1966).
39. G. M. Panchenkov, G. S. Maksimova, and Y. M. Zhorov, *Tr. Mosk. Inst. Neftekhim. Gasoz. Prom.* (86), 191 (1969); *Chem. Abstr.* **71** 83225s, (1969).
40. U.S. Pat. 3,527,834 (Sept. 8, 1970), W. L. Kehl, R. J. Rennard, Jr., and H. E. Swift (to Gulf Research and Development Co.).
41. U.S. Pat. 3,642,933 (Sept. 8, 1971), L. F. Heckelsberg (to Phillips Petroleum Co.).
42. J. Dubien, L. DeMourgues, and Y. Trambouze, *Bull. Soc. Chem. Fr.* (1), 108 (1967).
43. U.S. Pat. 3,479,415 (Nov. 18, 1969), E. Shull (to Air Products and Chemicals, Inc.).
44. D. C. Pepper, *Quart. Rev.* **8**, 88 (1954).
45. Idem, *Proceedings of the Intl. Symp. on Macromolecular Chemistry, Prague, Czech., Sept. 1957; Tetrahedron* Supplement No. 2 (1957).
46. G. Heublein and B. Adelt, *Zeitschr. Chem.* **11**, 321 (1971).
47. P. C. Weinert and G. Egloff, *Pet. Process.* **3**, 585 (1948).
48. R. A. Rhein, paper presented at the 97th American Chemical Society Meeting, Division of Rubber Chemistry, Washington, D.C., May 5–8, 1970, paper no. 10.
49. J. P. Kennedy, *Cationic Polymerization of Olefins: A Critical Inventory,* John Wiley & Sons, Inc., New York, 1975.
50. I. D. Rubin, "Poly-1-Butene, Its Preparation and Properties," in H. Morawetz, ed., *Polymer Monographs Series,* Gordon and Breach Science Publishers, New York, 1968.
51. F. M. Seger, H. G. Doherty, and A. N. Sachanen, *Ind. Eng. Chem.* **42**, 2446 (1950).

52. S. H. Pinner in P. H. Plesch, ed., *The Chemistry of Cationic Polymerization,* The Macmillan Company, New York, 1963, p. 611.
53. F. Williams, A. Shinkawa, and J. P. Kennedy, *J. Polym. Sci. Polym. Symp.,* **56,** 421 (1976); *J. Polym. Sci. A-1* **9,** 1551 (1971).
54. U.S. Pat. 3,897,322 (July 29, 1975), M. Marek, L. Toman, and J. Pecka (to the Czechoslovakian Academy of Sciences).
55. H. Sawada, *J. Macromol. Sci. Revs. Macromol. Chem.* **7,** 161 (1972).
56. H. Sawada, *Thermodynamics of Polymerization,* Marcel Dekker, Inc., New York, 1976.
57. R. M. Joshi and B. J. Zwolinski "Vinyl Polymerization, Part I," in *Kinetics and Mechanisms of Polymerization,* Vol. I, Marcel Dekker, Inc., New York, 1967, p. 44.
58. P. H. Plesch, *Adv. Polym. Sci.* **8,** 137 (1971).
59. J. P. Kennedy and T. Otsu, *Adv. Polym. Sci.* **7,** 369 (1970).
60. H. Gueterbock, *Polyisobutylene und Isobutylen—Mischopolymerizate,* Springer Verlag, Berlin, Ger., 1959, p. 108.
61. Y. T. Eidus and B. K. Nefedov, *Usp. Khim.* **29,** 833 (1960).
62. I. Puskas, E. M. Banas, and A. G. Nerhein, *J. Polym. Sci. Polym. Symp.,* **56,** 191 (1976).
63. I. Puskas and co-workers, unpublished.
64. "Flexicoking" in J. J. McKetta, ed., *Encyclopedia of Chemical Process and Design,* Marcel Dekker, Inc., New York, in press.
65. E. G. Wollaston and co-workers, *Hydrocarbon Process.* **54**(9), 93 (1975).
66. J. A. Finneran, J. R. Murphy, and E. L. Whittington, *Oil Gas J.* **72,** 52 (Jan. 14, 1974).
67. *Oil Gas J.* **73,** 53 (Mar. 10, 1975).
68. H. D. Evans and D. H. Sarno, *Seventh World Petroleum Congress* **5,** 259 (1967).
69. S. Tukao, *Hydrocarbon Process.* **45**(11), 151 (1966).
70. H. Klein and H. M. Weitz, *Hydrocarbon Process.* **47**(11), 135 (1968).
71. W. W. Coogler, *Hydrocarbon Process.* **46**(5), 166 (1967).
72. *Hydrocarbon Process.* **48**(11), 190 (1969).
73. B. W. Struth, *Oil Gas J.* **67**(11), 79 (1969).
74. R. L. Banks, *Pet. Div. Prepr. ACS Meeting, New York, Aug. 27–Sept. 1, 1972.*
75. D. L. Crain and R. E. Reusser, "Synthesis of Olefins Via Disproportionation," *Pet. Div. Prepr., ACS Meeting, New York, Aug. 27–Sept. 1, 1972.*
76. K. L. Anderson and T. D. Brown, "New Routes to Petrochemicals by Olefin Disproportionation," *81st AIChE National Meeting, Apr. 11–14, 1976, Kansas City, Mo.,* paper no. 270.
77. K. L. Anderson and T. D. Brown, *Hydrocarbon Process.* 119 (Aug. 1976).
78. G. P. Baumann and M. R. Smith, *Oil Gas J.,* 71 (Sept. 27, 1954).
79. J. R. Barber, J. J. Collins, and T. C. Sayer, "Separation of n-Butenes from i-Butene," *68th National Meeting AIChE,* Houston, Tex. Feb. 28, 1971; *Hydrocarbon Process.* **55** (9), 226 (1976).
80. *Chem. Eng. News* 48 (May 12, 1978).
81. "Butylenes," *Chemical Economics Handbook,* SRI International, Menlo Park, Calif.; Chem Systems Inc., 747 Third Ave., New York, N. Y. 10017.
82. C. R. Cupit, J. E. Gwyn, and E. C. Jernigan, *Petr. Manage.* **33** (12), 203 (Dec. 1961); C. R. Cupit, J. E. Gwyn, and E. C. Jernigan, *Petr. Manage.* **34** (1), 207 (Jan. 1962).
83. J. H. Gary and G. E. Handwerk, *Petroleum Refining,* Marcel Dekker, Inc., New York, 1975, p. 142.
84. *Ibid.,* p. 7.
85. G. D. Hobson and W. Pohl, *Modern Petroleum Technology,* 4th ed., John Wiley & Sons, Inc., New York, 1973, p. 517.
86. *Chem. Mark. Rep.,* (Dec. 27, 1976).
87. R. Stobaugh, *Hydrocarbon Process.* **46**(6), 141 (1967).
88. *Chem. Week,* 70 (Sept. 8, 1976).
89. *Chem. Eng. News,* 11 (Sept. 13, 1976).
90. A. L. Waddams, *Chemicals from Petroleums,* 2nd ed., John Murray, London, Eng., 1968, p. 128.
91. P. Wiseman *Industrial Organic Chemistry,* Wiley-Interscience, New York, 1972, p. 169.
92. *Ibid.,* p. 93.
93. A. V. Hahn, *The Petrochemical Industry,* McGraw Hill Book Co., New York, 1970, p. 591.
94. Ref. 91, p. 219.
95. C. E. Coates, M. L. H. Green, and K. Wade, *Organometallic Compounds,* Methnew & Co., Ltd., London, Eng., Vol. I., p. 299.
96. Ref. 91, p. 245.
97. Ref. 3, p. 153.

98. Ref. 93, p. 526.
99. *Chem. Eng. News,* 46 (Aug. 5, 1963).
100. U.S. Pat. 3,296,286 (1967), (to Esso Research and Engineering).
101. Brit. Pat. 1,174,209 (Dec. 17, 1969), A. Kiwantes and B. Stouthamer (to Shell Internationale Research Maalschappij N.V.).
102. Jpn. 73 23,413 (July 13, 1973), Y. Komatsu, T. Tamura, and H. Okayama (to Maruzen Oil Co.).
103. Ref. 3, p. 93.
104. T. Reis, *Chem. Process. Eng. (Bombay)* (2), 68 (1972).
105. R. B. Stobaugh, *Hydrocarbon Process.* **46**(7), 149 (1967).
106. *Chem. Week,* 39 (Mar. 24, 1971).
107. *Chem. Process. Eng. (Bombay)* (3), 70 (1971).
108. N. P. Chopey, *Chem. Eng. N.Y.* 90 (Dec. 14, 1970).
109. (a) Y. Oda and co-workers, *Hydrocarbon Process* (10), 115 (1975); (b) *Chem. Week,* 39 (May 31, 1978); *Chem. Mark. Rep.,* 3 (May 8, 1978); *Chem. Mark. Rep.,* 5 (May 1, 1978).
110. S. Ushio, *Chem. Eng. N.Y.,* 107 (Sept. 20, 1971).
111. *Eur. Chem. News,* 32 (Apr. 9, 1971).
112. *Chem. Week,* 79 (Oct. 13, 1976).
113. U.S. Pat. 3,862,146 (Jan. 21, 1975), E. M. Boghosian (to Standard Oil Co. (Indiana)).
114. *Chem. Week,* 39 (Dec. 1, 1976).
115. *Oil Gas J.* **73,** 50 (June 16, 1975).
116. (a) R. Csikos and co-workers, *Hydrocarbon Process.* (7), 121 (1976); (b) *Hydrocarbon Process.,* 98 (Dec. 1977); (c) *Chem. Mark. Rep.,* 5 (July 24, 1978); (d) *Chem. Week,* 47 (Feb. 15, 1978).
117. W. A. Schwerdtel, *Hydrocarbon Process.* (11), 117 (1970).
118. *Pet. Ref.* **36**(11), 233 (1954).
119. *J. Catal.* **13**(1), 105 (1969).
120. *Eur. Chem. News* **5,** 148 (1969).
121. D. Mark and A. R. Orr, *Pet. Ref.* **35**(12), 185 (1956); R. A. Labine, *Chem. Eng.,* 98 (Aug. 8, 1960).
122. U.S. Pat. 3,121,125 (Feb. 11, 1964), G. Nichols (to Standard Oil (Indiana)).
123. U.S. Pat. 3,119,884 (July 28, 1964), J. R. Allen and D. M. Krausse (to Cosden Oil and Chemical Company).
124. U.S. Pat. 2,484,384 (Oct. 11, 1949), I. F. Lavine and L. T. Folsom (to Chevron Chemicals).
125. U.S. Pat. 2,677,000 (Apr. 27, 1954), N. Fragen (to Standard Oil (Indiana)).
126. U.S. Pat. 2,677,001 (Apr. 27, 1954), L. W. Russum (to Standard Oil (Indiana)).
127. U.S. Pat. 3,501,551 (Mar. 17, 1970), J. C. Heidler and R. J. Lee (to Standard Oil (Indiana)).
128. *Bulletin 12-H,* Amoco Chemicals Corporation, Naperville, Ill.

MELVERN C. HOFF
UN K. IM
WILLIAM F. HAUSCHILDT
IMRE PUSKAS
Amoco Chemicals Corporation

BUTYL RUBBER. See Elastomers, synthetic.

BUTYRALDEHYDE

Both n-butyraldehyde and isobutyraldehyde occur naturally in small quantities. The former has been isolated in small quantities in the essential oils of several plants. It has also been detected in oil of lavender, the oil of the Eucalyptus globulus of California, in tobacco smoke, in tea leaves, and in other leaves. Isobutyraldehyde is found in trace quantities in Jeffrey pine oil and tea leaves.

n-Butyraldehyde (1), $CH_3CH_2CH_2CHO$, is a colorless, flammable liquid with a characteristic aldehydic odor. It is used chiefly as an intermediate in the production of synthetic resins, rubber accelerators, solvents, and plasticizers. Because of the large number of condensation and addition reactions it can undergo, it is a useful starting material in the production of a wide variety of compounds containing at least 6 to 8 carbon atoms. Butyraldehyde became a commercial chemical in the decade following World War II. It was discovered shortly after 1860 and was prepared by reduction of crotonaldehyde as early as 1880.

Isobutyraldehyde (2), $(CH_3)_2CHCHO$, is also a colorless liquid with a pungent odor and is available commercially. Like n-butyraldehyde, it can be used in the manufacture of resins and rubber chemicals as well as in the synthesis of isobutyric acid, acetals, mercaptals, and derivatives for use as corrosion inhibitors, insecticides, and amino acids. Isobutyraldehyde was first synthesized by Linnemann and von Zitta in 1872 by the hydrolysis of 1,2-dibromo-2-methylpropane at 150–160°C.

Physical Properties

The four-carbon aldehydes are highly flammable liquids with physical properties as shown in Table 1. They are colorless and have pungent odors which are characteristic of the lower carbon aldehydes. Both butyraldehyde and isobutyraldehyde are slightly soluble in water (see Table 2) and miscible with most organic solvents, eg, ethanol, ethyl ether, benzene, toluene, and acetone. The known azeotropes of butyraldehyde are shown in Table 3.

Reactions

The butyraldehydes contain a terminal carbonyl group and undergo reactions that are characteristic of aldehydes (qv), ie, oxidation, reduction, and condensation. For reactions of commercial importance, see the section Uses. Photochemical and thermal decomposition of n-butyraldehyde results mainly in propane and carbon monoxide. The thermal decomposition of isobutyraldehyde in the presence of copper or palladium, at elevated temperatures, gives propylene, carbon monoxide, and hydrogen; the photoinduced chain decomposition between 20 and 501°C gives carbon monoxide, hydrogen, methane, ethylene, propylene, propane, and 2,3-dimethylbutane (1).

Aldehydes are intermediate in the sequence:

$$RCH_2OH \underset{\text{reduction}}{\overset{\text{oxidation}}{\rightleftarrows}} RCHO \underset{\text{reduction}}{\overset{\text{oxidation}}{\rightleftarrows}} RCOOH$$
$$\text{alcohol} \qquad \text{aldehyde} \qquad \text{acid}$$

Table 1. Physical Properties

Properties	Butanal [123-72-8] CH$_3$CH$_2$CH$_2$CHO (1) (butyraldehyde)	2-Methylpropanal [78-84-2] (CH$_3$)$_2$CHCHO (2) (isobutyraldehyde)
mp, °C	−99.0	−65.9
bp, °C	75.7	64.5
density, d_{20}^{20}	0.8048	0.7938
vapor density (air = 1)	2.48	
refractive index, n_D^{20}	1.3843	1.3730
flash point, °C[a]	−9.4	−10.6
viscosity (20°C), mPa·s (= cP)	0.433	
heat of formation, kJ/mol[b]	240.3	
specific heat, J/(kg·K)[b]	2121	2544
heat of vaporization at bp, J/g[a]	436	409
heat of combustion, kJ/mol[b]	2478.7	2510.0
dipole moment (vapor), C·m[c]	9.07 × 10^{-30}	
surface tension, mN/m (= dyn/cm), 24°C	29.9	
vapor pressure, kPa[d] (20°C)	12.2	18.4

[a] Tag open cup, ASTM D 1310.
[b] To convert J to cal, divide by 4.184.
[c] To convert C·m to debye, divide by 3.336 × 10^{-30}.
[d] To convert kPa to mm Hg, multiply by 7.5.

Table 2. Butyraldehyde–Water Solubilities

| Butyraldehyde in water || Water in butyraldehyde ||
Temperature, °C	Butyraldehyde, wt %	Temperature, °C	Water, wt %
0	8.7	9	3.01
10	7.9	10	3.08
20	7.1	20	3.17
30	6.3	30	3.27
40	5.4	40	3.39

Thus butanol and 2-methylpropanol (isobutyl alcohol) can be obtained by hydrogenation of the corresponding aldehydes, and butyric acid and isobutyric acid can be produced by oxidation. The aldehyde reduction can be accomplished in the presence of an activated nickel catalyst or by sodium amalgam in dilute sulfuric acid. Hydrogenation of butyraldehyde over a NiO–SiO$_2$–Al$_2$O$_3$ catalyst or Raney copper gives practically quantitative yields of butanol (2–3). At relatively high pressures, the reduction of butyraldehyde over Raney nickel under alkaline conditions gives 2-ethylhexanol (4). Sodium borohydride and diborane (B$_2$H$_6$) have been used in the conversion of n-butyraldehyde to butanol (5–6) (see Hydroboration). Isobutyl alcohol

Table 3. Butyraldehyde Azeotropes

Other component(s)	Wt %	bp, °C
homogeneous binary mixtures		
ethanol	60.6	70.7
methanol	51	62.6
heterogeneous binary mixture		
water	12	68.0[a]
heterogeneous ternary mixtures		
ethanol	11	67.2[b]
water	9	

[a] The upper layer (94 vol %) contains 3.5 wt % water. The lower layer (6 vol %) contains 91.8 wt % water.
[b] The upper layer (97.8 vol %) contains 11 wt % ethanol and 7 wt % water.

can be formed by reduction of isobutyraldehyde over Raney nickel or sodium amalgam and by electrolytic reduction in aqueous sulfuric acid.

The pure aldehydes are easily oxidized by treatment with air. In fact, the unpleasant odor often associated with lower molecular weight aldehydes may be due to the corresponding carboxylic acids. The oxidation can be catalyzed by manganese or cobalt bromides (7). Selenium dioxide, SeO_2, oxidizes butyraldehyde at the α-carbon atom to give 2-oxobutyraldehyde (ethylglyoxal), CH_3CH_2COCHO (8). Isobutyraldehyde can be oxidized in air at 30–50°C to give isobutyric acid in 95% yield (9). Catalytic oxidation of isobutyraldehyde with air using a solution of cobalt isobutyrate as catalyst leads to considerable α-oxidation, as detected by the acetone formed (9).

Several species of bacteria under suitable conditions cause n-butyraldehyde to undergo the Cannizzaro reaction (simultaneous oxidation and reduction to butyric acid and butanol, respectively) (10); this reaction can also be catalyzed by Raney nickel with or without sodium hydroxide (11–12). The direct formation of butyl butyrate or isobutyl isobutyrate (Tishchenko reaction) from the corresponding aldehyde takes place rapidly with aluminum ethylate or aluminum butyrate as catalyst (13).

In the presence of traces of oxygen, butyraldehyde (**1**) polymerizes (1.2 GPa; ca 12,000 atm) to an amorphous mass without strength or elasticity, which reverts to the aldehyde on standing at room temperature and pressure. Hydrogen chloride or a few drops of concentrated sulfuric acid catalyze the formation, at room temperature, of the trimer parabutyraldehyde (**3**) (2,4,6-tripropyl-1,3,5-trioxane), bp 210–220°C, d^{21} 0.917. The reaction is reversed by heating the parabutyraldehyde in the presence of acid. In the presence of a base, dilute sodium or potassium hydroxide, butyraldehyde (**1**) undergoes the aldol condensation to form 2-ethyl-3-hydroxyhexanal (**4**) (butyraldol), bp 103–105°C (at 1.9 kPa or 14 mm Hg), d_4^{20} 0.9397, n_D^{20} 1.4409 (14). Butyraldol can be dehydrated to 2-ethylhexenal by heating in the presence of acid. The hydrogenated product, 2-ethylhexanol, is a major commercial chemical used in plasticizers, ie, di(2-ethylhexyl) phthalate, and coatings, ie, 2-ethylhexyl acrylate. Butyraldol may be hydrogenated directly to produce 2-ethylhexane-1,3-diol, another commercial chemical used as an insect repellent and solvent.

Isobutyraldehyde (2) undergoes an aldol condensation in the presence of base to form isobutyraldol (2,2,4-trimethyl-3-hydroxypentanal), although the product is isolated as a trimer, 2,6-diisopropyl-5,5-dimethyl-4-hydroxy-1,3-dioxane (5), bp 110–111°C (at 1.1 kPa or 8 mm Hg), n_D^{25} 1.4461, d_{25}^{25} 0.9670.

The isobutyraldol can be isolated by treatment of the trimer with dilute acid. A continuous process for making isobutyraldol has been patented (15). The other trimer of isobutyraldehyde, paraisobutyraldehyde (6) (2,4,6-triisopropyl-1,3,5-trioxane) is formed on standing or with catalysis by ultraviolet light, chlorine, bromine, and mineral acids. Paraisobutyraldehyde crystallizes from solution as long white needles, mp 59–60°C. Like parabutyraldehyde, the isomeric trimer is degraded by heating.

The aldol condensation of butyraldehydes will also occur with other aldehydes. Although these crossed aldol reactions are not generally useful (they produce many products) some commercial processes are presently used. For example, the reaction of acetaldehyde and butyraldehyde in base produces, after dehydration, 2-butenal (crotonaldehyde), 2-ethyl-2-butenal, 2-hexenal and 2-ethyl-2-hexenal (16). The products may be hydrogenated to the corresponding saturated aldehydes, or more completely to the alcohols. Acetone and n-butyraldehyde, in the presence of dilute sodium hydroxide, have been reported to give 4-hydroxy-2-heptanone, bp (at 1.6 kPa or 12 mm Hg) 92–94°C, which readily loses water to give 3-heptene-2-one, bp (at 2 kPa or 15 mm Hg) 70°C, bp (at 101.3 kPa or 760 mm Hg) 156–157°C (17).

Crossed Tishchenko reactions with other aldehydes have been studied; equimolar amounts of acetaldehyde and butyraldehyde in the presence of aluminum butyrate are said to give a mixture of the four possible esters (ethyl and n-butyl acetates, ethyl and n-butyl butyrates); acetaldehyde in excess gives mainly n-butyl acetate (18).

Equimolar amounts of n-butyraldehyde and benzaldehyde under Tishchenko conditions give only benzyl benzoate, whereas butyl butyrate, benzyl butyrate, and butyl benzoate are formed when excess benzaldehyde is used (19). Isobutyl isobutyrate is formed in 90% yield by adding isobutyraldehyde to aluminum isobutoxide (9). n-Butyraldehyde reacts with excess isobutyraldehyde in refluxing aqueous potassium carbonate solution to give a high yield of 2,2-dimethyl-3-hydroxyhexanal (20).

Acetic anhydride reacts with n-butyraldehyde (Perkin reaction) to give 1,1-diacetoxybutane and 1-butenyl acetate (the acetate of the enol form of butyraldehyde), $CH_3COOCH=CHCH_2CH_3$; the reaction appears to be catalyzed by traces of mineral acid or greater than 2% acetic acid (21). Butyraldehyde (1) reacts with ketene, $CH_2=C=O$, to give β-caprolactone (7), and with 2-bromoethyl acetate under Reformatsky conditions (mercuric chloride in tetrahydrofuran) to give a 35% yield of 2-hexenol (22). In the presence of bases such as pyridine, n-butyraldehyde reacts with malonic acid (Knoevenagel reaction) to give hexenoic acids (23); isobutyraldehyde gives 4-methyl-2-pentenoic acid. Isobutyraldehyde also undergoes the Mannich reaction with formaldehyde and primary amines to give good yields of secondary amines, the Perkin reaction with propionic anhydride to give 2,4-dimethyl-2-pentenoic acid, the Reformatsky reaction with β-bromoesters, and the Stobbe condensation with diethyl succinate. β-Isocaprolactone (8) is formed in the reaction of isobutyraldehyde (2) with ketene.

$$(1) + CH_2=C=O \longrightarrow \text{(7)}$$

$$(2) + CH_2=C=O \longrightarrow \text{(8)}$$

When butyraldehydes are treated with alcohols in the presence of a mineral acid or calcium chloride, an acetal of the carbonyl group is formed.

$$(1) + 2\,ROH \xrightarrow{H^+} CH_3CH_2CH_2CH(OR)_2 + H_2O$$

Table 4 gives the properties of butyraldehyde acetals. Ethylene glycol forms a cyclic acetal (9) (2-propyl-1,3-dioxolane, bp 130°C), and other 1,2-glycols form the corresponding dioxolanes. Equimolar quantities of phenol and n-butyraldehyde in acetic acid solution, with hydrogen chloride passed into the reaction mixture, result in a polymer that yields 4-butylphenol on slow pyrolysis (24); n-butyraldehyde on treatment with phenol in the presence of hydrogen chloride has been found to yield 1,1-bis(p-hydroxyphenyl)butane (10) and a brittle resin that is converted almost completely to structure (10) on vacuum distillation (25). Mercaptals have been formed in the reaction of these butyraldehydes with mercaptans in the presence of zinc chloride. These thiobutyrals have been known since 1889. Ethyl mercaptan and isobutyraldehyde react to form isobutyraldehyde diethyl mercaptal, bp (at 101.3 kPa or 1 atm) 210°C, which is oxidized with potassium permanganate or hydrogen peroxide

Table 4. Properties of Butyraldehyde Acetals

Compounds	bp, °C	d_4^t	n_D^t
butyraldehyde acetal			
dimethyl	114	0.847[21]	1.3900[21]
diethyl	143	0.8417[25]	
di-n-propyl	182		
di-n-butyl	213	0.8578[17.5]	1.4211[17.5]
isobutyraldehyde acetal			
dimethyl	103		
diethyl	135–136	0.8295[20]	1.3885[20]
	99.3 kPa (745 mm Hg)		
di-n-propyl	132	0.834[15]	1.4066[20]
di-n-butyl	81		1.4155[20]
	0.8 kPa (6 mm Hg)		

to the disulfone (11), mp 94°C.

(9) (10) (11)

Butyraldehyde reacts with aqueous ammonia solutions at 0°C to give butyraldehyde–ammonia; with alcoholic ammonia at 150°C or with liquid ammonia, 2-propyl-3,5-diethyl-pyridine is obtained (26–27). Butyraldehyde and alcoholic ammonia hydrogenated in the presence of a nickel catalyst under pressure at 90–125°C yield up to 80% of butylamine, with smaller amounts of di- and tributylamines (28) (see also Amines, lower aliphatic). Butyraldehyde and aniline in neutral media form a monomeric Schiff base or anil, N-butylideneaniline, $C_6H_5N{=}CHCH_2CH_2CH_3$, which gives a crystalline dimer on standing; traces of acid cause the formation of a yellow oil, the anil of 2-ethyl-2-hexenal (29).

Isobutyraldehyde is unique among the aliphatic aldehydes in that the formation of the aldol is reversed under alkaline conditions at 60°C so that good yields of primary products can be obtained. Isobutylamine can thus be prepared in 70% yield by reductive amination of the aldehyde. Diisobutylamine is prepared by the catalytic reduction of the Schiff base from the aldehyde and ammonia. Under Leuckart reaction conditions, isobutyraldehyde reacts with ammonium formate and formic acid at 90–130°C to give triisobutylamine. Ammonia and isobutyraldehyde react in the presence of chromic oxide on alumina to give isobutyronitrile.

Hydroxylamine reacts with these aldehydes to form the corresponding butyraldoximes (liquids), whereas phenylhydrazines and semicarbazides give the usual crystalline compounds with characteristic melting points.

Acetylene reacts with these aldehydes in the presence of copper acetylide. n-Butyraldehyde gives a 45% yield of 1-hexyn-3-ol, bp 140–141°C, and 13% 1,4-dipropyl-2-butyn-1,4-diol, bp 110–115°C (at 1.3 kPa or 10 mm Hg) (30). In the presence of anhydrous KOH and sodium amide the yield of the hexynol is 60–70% (31). An 82% yield of 1-hexyn-3-ol has been reported when the reaction is carried out in the presence of KOH in ethylene glycol dimethyl ether (32) (see also Acetylene-derived chemicals).

n-Butyraldehyde reacts with benzene in the presence of aluminum chloride (Friedel-Crafts reaction (qv)) to give 1,1-diphenyl-butane and some butylbenzene (33), whereas isobutyraldehyde gives a mixture of 2-methyl-1,1-diphenylpropane and isobutyrophenone. Styrene and n-butyraldehyde in acetic acid give a 14% yield of 1-phenyl-1,3-hexadiene and 39% of 1-phenyl-1,3-hexanediol monoacetate in the presence of sulfuric acid (34). The photochemically induced reaction of n-butyraldehyde with 1-octene and bromochloroform gives 4-dodecanone (35). When isobutyraldehyde reacts with simple olefins or unsaturated ethers in the presence of sulfuric acid a m-dioxane is formed; ethoxyethylene reacts to give 2,6-diisopropyl-4-ethoxy-m-dioxane (12) (9,36).

Copper is attacked fairly rapidly by n-butyraldehyde in the presence of air, forming a greenish-blue butyraldehyde solution of cupric butyrate which deposits metallic copper on refluxing or on being heated in a sealed tube (37).

Hydrocyanic acid forms a cyanohydrin with n-butyraldehyde which hydrolyzes to 2-hydroxyvaleric acid; under the same conditions isobutyraldehyde forms a cyanohydrin which hydrolyzes and condenses to a hydroxylactone (13). Grignard reagents react with these aldehydes to produce alcohols upon hydrolysis of the reaction intermediate; n-propylmagnesium bromide reacts with n-butyraldehyde to give a product from which 4-heptanol can be obtained. Benzylmagnesium chloride reacts with n-butyraldehyde to give the expected 1-phenyl-2-pentanol and an abnormal product, 1-(α-hydroxybutyl)-2-(β-hydroxypentyl)benzene (14), resulting from ortho alkylation of the major product (38). Butyraldehyde reacts with phosphorus pentachloride to yield 1,1-dichlorobutane, whereas 1,2-dichloro-3-methylpropane is formed with iso-

(12)   (13)   (14)

butyraldehyde.

n-Butyraldehyde reacts with 1-nitropropane in alkaline solution to give an 88% yield of 3-nitro-4-heptanol (39); its isomer undergoes the same reaction to give lower yields of 2-methyl-4-nitro-3-hexanol. 2-Chlorobutyraldehyde, bp 106–107°C (at 98.6 kPa or 740 mm Hg), n_D^{25} 1.441, is prepared in 80% yield by chlorination of the aldehyde with sulfuryl chloride (40). The α-bromo derivatives of n-butyraldehyde and isobutyraldehyde, prepared by treatment of the aldehyde with dioxane–dibromide (orange crystalline solid from equimolar amounts of bromine and dioxane), have bp 68–70°C (at 4.7–5.3 kPa or 35–40 mm Hg) and normal bp 112–113°C, respectively (41).

The direct sulfonation of isobutyraldehyde with sulfur trioxide in dioxane gives a 41% yield of 2-formyl-2-propanesulfonic acid, $(CH_3)_2C(CHO)SO_3H$ (42).

Manufacture

The most widely used manufacturing technique for both butyraldehyde and isobutyraldehyde is the Oxo process, in which propylene, carbon monoxide, and hydrogen are combined with a suitable catalyst, usually a cobalt compound, at about 130–160°C and 10–20 MPa (100–200 atm) pressure (43–44)

$$CH_3CH=CH_2 + CO + H_2 \rightarrow (CH_3)_2CHCHO + CH_3CH_2CH_2CHO$$

The ratio of the products is variable, depending on temperature, pressure, and reactant concentrations, but the usual ratio is about 3 kg of butyraldehyde to 1 kg of isobutyraldehyde. The use of rhodium complexes as catalysts for the Oxo reaction became commercial in early 1976. (45–46). The advantages of the rhodium process for making butyraldehydes include low temperatures (80–120°C), low pressure operation (0.7–3 MPa (7–30 atm)), higher efficiency to the more valuable normal butyraldehyde and less by-product formation (46) (see Oxo process).

Butyraldehyde can also be produced from 2-butenal (crotonaldehyde) formed by the aldol condensation of acetaldehyde.

$$2\ CH_3CHO \xrightarrow{OH^-} CH_3\overset{OH}{\underset{|}{CH}}-CH_2CHO \xrightarrow{-H_2O} CH_3CH=CHCHO \xrightarrow{H_2} (1)$$

This process was a major source of butyraldehyde until about 1970.

In 1975, butyraldehyde production was 242,000 metric tons (47) priced at $0.53/kg (48). Isobutyraldehyde production was 174,000 t (47) and the price was $0.49/kg (48).

Analysis

The butyraldehydes can be identified by the crystalline compounds they form with hydrazines and semicarbazides; eg, the 2,4-dinitrophenylhydrazone (three modifications) and 4-phenylsemicarbazone derivatives of n-butyraldehyde melt at 120–122°C (49) and 134–136°C, respectively; the semicarbazone and 2,4-dinitrophenylhydrazone of the isomeric aldehyde melt at 125 and 182°C, respectively. The methone (**15**) (dimedone, 5,5-dimethyl-1,3-cyclohexanedione) derivative of n-butyraldehyde melts at 134–135°C (50). In the absence of ketones and other aldehydes, n-butyraldehyde can be determined by addition of sodium bisulfite; the excess bisulfite is titrated with iodine or thiosulfate (51).

(**15**)

The nmr spectrum of butyraldehyde has a fine-structured triplet centered at 0.97 ppm downfield from the tetramethylsilane (TMS) standard, a multiplet at 1.67, a triplet at 2.42, and a triplet at 9.74 ppm (52). The ultraviolet spectrum contains a broad maximum centered at 283 nm, attributed to the excitation of the electrons of the carbonyl group. The infrared absorption of the carbonyl group is at 1725 cm^{-1}.

Health and Safety Factors

Although tests have shown that n-butyraldehyde exhibits some adverse physiological effects, there is no danger to health in normal plant practice. No threshold limit value has been assigned for either butyraldehyde or isobutyraldehyde. The LD_{50} (acute oral ingestion, rat) for butyraldehyde is 2490 mg/kg and for isobutyraldehyde 2810 mg/kg (53). Breathing of the vapors of the aldehydes and contact of the liquid with the skin should be avoided. Injury may result if the aldehydes are splashed in the eye. The flash points for the aldehydes are well below room temperature. Thus precautions must be taken to avoid heat, sparks, or open flame. The aldehydes should be stored in cool, well ventilated places. Since the aldehydes are easily oxidized by air to the corresponding butyric acids, precautions associated with these carboxylic acids must also be noted (54).

Storage and Handling

Stainless steel, baked phenolic-lined steel, or aluminum are often used for storage and handling of butyraldehyde and isobutyraldehyde. The aldehydes are flammable and reactive; on exposure to air they are easily oxidized, and in contact with acid or bases they will undergo an exothermic condensation reaction. Storage of the aldehydes under nitrogen will avoid these problems and maintain the integrity of the material (54). The dry aldehydes will undergo some polymerization—to form parabutyraldehyde (3) and paraisobutyraldehyde (6). There is some evidence that the presence of water will inhibit this reaction.

Uses

The butyraldenydes, produced via the Oxo process, are hydrogenated to form the corresponding butyl alcohol (qv). Butyraldehyde reacts with poly(vinyl alcohol) to form poly(vinyl butyral) (Butacite, Butvar, Mowital B, Saflex, vinyl butyral resins) which is used as an interlayer for safety glass as well as for coating fabrics and for injection-molding compositions (see Vinyl polymers).

Oil-soluble resins and resins suitable for use in molding powders and coating compositions can be prepared by the condensation of butyraldehyde with phenol in the presence of hydrogen chloride (55–57) or sodium hydroxide (58); formaldehyde has been added after the main reaction (59), and butyraldehyde–formaldehyde mixtures have also been used (60). Alcohol-soluble resins for use in baking finishes have been obtained by pressure reaction between urea and butyraldehyde (61) (see Amino resins).

Butyraldehyde in the presence of dilute alkali aldolizes to butyraldol which splits off water readily to form 2-ethyl-2-hexenal, which is reduced to 2-ethyl-1-hexanol, a useful solvent, defoaming, dispersing, and wetting agent. Esters such as di(2-ethylhexyl) phthalate (dioctyl phthalate) are valuable plasticizers for vinyl resins, cellulose esters, and synthetic rubbers. Mixtures of acetaldehyde and butyraldehyde similarly aldolized are used to produce both 2-ethyl-1-butanol and 1-hexanol.

Butyraldehyde is a raw material for the maufacture of butyric acid and butyric anhydride (see Carboxylic acids).

Isobutyraldehyde is the starting material in the manufacture of pantothenic acid (16), vitamin B_5, which is used in poultry feeds as a chick antidermatitis factor (see

Vitamins). DL-Valine, $(CH_3)_2CH_2CH(NH_2)COOH$, and DL-leucine, $(CH_3)_2$-$CHCH_2CH(NH_2)COOH$ (see Amino acids), are also prepared from isobutyraldehyde. These essential amino acids are useful where there is a known deficiency in basal diet. Isobutyraldehyde derivatives have been found useful as repellents, insecticides, mold inhibitors, and similar agents. 2,2,4-Trimethyl-1,3-pentanediol is particularly good as a repellent for mosquitoes, chiggers, and the like (62) (see Repellents). Isobutyraldehyde has also been used in the synthesis of perfumes and flavoring agents, plasticizers, resins, and gasoline additives.

(16)

BIBLIOGRAPHY

"Butyraldehydes" in *ECT* 1st ed., Vol. 2, pp. 684–693, by M. S. W. Small, Shawinigan Chemicals, Ltd., and P. R. Rector, Carbide and Carbon Chemicals Corporation; "Butyraldehydes" in *ECT* 2nd ed., Vol. 3, pp. 865–877, by A. P. Lurie, Eastman Kodak Company.

1. J. A. Kerr and A. F. Trotman-Dickinson, *Trans. Faraday Soc.* **55,** 921 (1959).
2. Ger. Pat. 1,115,232 (July 18, 1958), W. Rottig (to Ruhrchemie A.G.).
3. J. Jadot and R. Braine, *Bull. Soc. Roy. Sci. Liege* **25,** 62 (1956).
4. Jpn. Pat. 116 (Jan. 18, 1952), K. Sagawa (to Kao Soap Co.).
5. S. W. Chaikin and W. G. Brown, *J. Am. Chem. Soc.* **71,** 122 (1949).
6. H. C. Brown and B. C. Subba Rao, *J. Am. Chem. Soc.* **82,** 681 (1960).
7. Brit. Pat. 824,116 (Nov. 25, 1959), E. T. Crisp and G. H. Whitfield (to Imperial Chemical Industries, Ltd.).
8. J. Vene, *Bull. Soc. Chim. Fr.* **12**(5), 506 (1945).
9. H. J. Hagemeyer, Jr., and G. C. DeCroes, *The Chemistry of Isobutyraldehyde and its Derivatives,* Eastman Kodak Co., 1953,
10. C. Neuberg and F. Windisch, *Biochem. Z.* **166,** 454 (1925).
11. M. Delepine and C. Hanegraaff, *Bull. Soc. Chim. Fr.* **4**(5), 2087 (1937).
12. M. Delepine and A. Horean, *Bull. Soc. Chem. Fr.* **4**(5), 1524 (1937).
13. H. S. Kulpinski and F. F. Nord, *J. Org. Chem.* **8,** 256 (1943).
14. A. T. Nielson and W. J. Houlihan in A. C. Cope, ed., *Organic Reaction,* Vol. 16, John Wiley & Sons, Inc., New York, 1968.
15. U.S. Pat. 2,829,169 (Apr. 1, 1958), (to Eastman Kodak Company).
16. Jpn. Pat. 74 24891 (June 26, 1974), A. Matsukuma, I. Takakiski, and K. Yoshida (to Mitsubishi Chemical Industries Company).
17. S. G. Powell and D. A. Ballard, *J. Am. Chem. Soc.* **60,** 1916 (1938).
18. U.S. Pat. 1,700,103 (Jan. 22, 1929), R. H. Van Schaak, Jr., (to Van Schaak Bros. Chemical Works, Inc.).
19. I. Lin and A. R. Day, *J. Am. Chem. Soc.* **74,** 5133 (1952).
20. U.S. Pat. 2,811,562 (Oct. 29, 1957), H. J. Hagemeyer, Jr., (to Eastman Kodak Company).
21. M. Crawford and W. T. Little, *J. Chem. Soc.,* 722 (1959).
22. R. E. Miller and F. F. Nord, *J. Org. Chem.* **16,** 728 (1951).
23. S. E. Boxer and R. P. Linstead, *J. Chem. Soc.,* 740 (1931).
24. J. B. Niederl, and co-workers, *J. Am. Chem. Soc.* **59,** 1113 (1937).
25. L. H. Baekeland and H. L. Bender, *Ind. Eng. Chem.* **17,** 225 (1925).
26. L. Haskelberg, *J. Soc. Chem. Ind.* **54,** 261, 534 (1935).
27. H. H. Strain, *J. Am. Chem. Soc.* **54,** 1221 (1932).

28. U.S. Pat. 2,411,802 (Nov. 26, 1946), J. F. Olin (to Sharples Chemicals, Inc.).
29. M. S. Kharasch, I. Richlin, and F. R. Mayo, *J. Am. Chem. Soc.* **62,** 494 (1940).
30. Brit. Pat. 508,062 (June 20, 1939), (to Farbenindustrie, A.G.).
31. P. Seguin, *Bull. Soc. Chem. France* **54,** 1221 (1932).
32. H. A. Stansbury, Jr., and W. R. Proops, *J. Org. Chem.* **27,** 279 (1962).
33. K. Bodendorf, *J. Prakt. Chem.* **129,** 337 (1931).
34. W. S. Emerson, *J. Org. Chem.* **10,** 464 (1945).
35. M. S. Kharasch, W. H. Urry, and B. M. Kuderna, *J. Org. Chem.* **14,** 248 (1940).
36. U.S. Pat. 2,628,257 (Feb. 10, 1953), R. I. Hoaglin and D. H. Hirsh (to Union Carbide and Carbon Corporation).
37. T. L. Davis and W. P. Green, Jr., *J. Am. Chem. Soc.* **62,** 3015 (1940).
38. S. Siegal, W. M. Boyer, and R. A. Jay, *J. Am. Chem. Soc.* **73,** 3237 (1951).
39. J. E. Bourland and H. B. Hass, *J. Org. Chem.* **12,** 704 (1947).
40. H. C. Brown and A. B. Ash, *J. Am. Chem. Soc.* **77,** 4019 (1958).
41. L. A. Lanovskaya and A. P. Terent'ev, *J. Gen. Chem.* **22,** 1598 (1957).
42. W. E. Truce and C. E. Alfieri, *J. Am. Chem. Soc.* **72,** 2740 (1950).
43. J. Falbe, *Carbon Monoxide in Organic Synthesis,* Springer-Verlag, New York, 1970, pp. 3–77.
44. B. Cornils, R. Payer, and K. C. Traenckner, *Hydrocarbon Process.* **54,** 83 (1975); H. Weber, W. Dimmling, and A. M. Desai, *Hydrocarbon Process.* **55,** 127 (1976).
45. U.S. Pat. 3,527,809 (Sept. 8, 1970), R. L. Pruett and J. A. Smith (to Union Carbide Corporation); R. L. Pruett and J. A. Smith, *J. Org. Chem.* **34,** 327 (1969).
46. R. Fowler, H. Connor, and R. A. Baehl, *Chemtech* **6**(12), 772 (1976); E. A. V. Brewester, *Chem. Eng.* **83,** 90 (1976).
47. U.S. International Trade Commission, *Synthetic Organic Chemicals, U.S. Production and Sales,* U.S. Gov. Printing Office, Washington, D.C., 1976.
48. *Chem. Mark. Rep.,* (Aug. 1976).
49. G. L. Clark, W. I. Kaye, and T. D. Parks, *Ind. Eng. Chem. Anal. Ed.* **18,** 310 (1946).
50. E. C. Horning and M. G. Horning, *J. Org. Chem.* **11,** 95 (1946).
51. A. E. Parkinson and E. C. Wagner, *Ind. Eng. Chem. Anal. Ed.* **6,** 433 (1934).
52. *NMR Spectra Catalog,* Vol. 1, Varian Associates, 1962, No. 78.
53. H. E. Christenson and co-workers, *Registry of Toxic Effects of Chemical Substances,* 1976 ed., U.S. Department of Health, Education, and Welfare, Rockville, Md., 1976, pp. 263, 632.
54. *Aldehydes,* Union Carbide Corporation, 270 Park Ave., New York, 1974.
55. E. F. Andruva, *Org. Chem. Ind.* **5,** 513 (1938).
56. I. P. Losev, V. M. Kotrolov, and A. Y. Fegina, *Org. Chem. Ind.* **3,** 558 (1937).
57. S. N. Ushakov and E. N. Freidberg, *Khim. Referat. Zhur.* **3,** 107 (1940).
58. U.S. Pat. 2,176,951 (Oct. 24, 1939), W. J. Bannister (to Resinox Corp.).
59. U.S. Pat. 2,231,860 (Feb. 11, 1941), L. C. Swallen (to Monsanto Chemical Co.).
60. T. S. Carswell, ed., *Phenoplasts,* Vol. VII, in H. Mark, C. S. Marvel, and P. J. Flory, eds., *High Polymers,* Interscience Publishers, New York, 1947, Chapt. 3.
61. U.S. Pat. 2,124,151 (July 19, 1938), H. S. Rothrock (to E. I. du Pont de Nemours and Co., Inc.).
62. U.S. Pat. 2,407,205 (Sept. 3, 1946), B. C. Wilkes (to Carbide and Carbon Chemicals Corporation).

PAUL DWIGHT SHERMAN, JR.
Union Carbide Corporation

BUTYRIC ACID AND BUTYRIC ANHYDRIDE. See Carboxylic acids.

BUTYROLACTONE. See Acetylene-derived chemicals.

C

CABLE COVERINGS. See Insulation, electric.

CACAO. See Chocolate and cocoa.

CADMIUM AND CADMIUM ALLOYS

Cadmium [7440-43-9], Cd, a Group-IIB element (between zinc and mercury) is a soft, ductile, silvery-white metal with a distorted hexagonal close-packed structure ($a = 0.29793$ nm, $c = 0.56181$ nm). It was discovered by Stromeyer in 1817 from an impurity in zinc carbonate. The crustal abundance of cadmium is somewhere between 0.1 and 0.5 ppm, and although several cadmium minerals have been identified—the most common one being greenockite, CdS—the element is generally encountered in zinc ores, zinc-bearing lead ores, or complex copper–lead–zinc ores, where it forms an isomorphic impurity in the zinc mineral sphalerite, ZnS, usually in concentrations of 0.1–0.5% cadmium. For this reason, cadmium is almost invariably recovered as a by-product from the processing of zinc, lead and copper ores.

Properties

Some typical physical properties of cadmium are listed in Table 1. Its electronic structure is: $1s^2 2s^2 2p^6 3s^2 3p^6 3d^{10} 4s^2 4p^6 4d^{10} 5s^2$, and its oxidation state in almost all of its compounds is +2, although a few compounds have been reported (1) in which cadmium exists in the oxidation state of +1. There are eight natural isotopes:

Table 1. Physical Properties of Cadmium

Property	Value	
atomic weight		112.40
melting point, °C		321.1
boiling point, °C		767
latent heat of fusion, kJ/mol[a]		6.2
latent heat of vaporization, kJ/mol[a]		99.7
specific heat, J/(mol·K)[a]		
20°C		25.9
321–700°C		29.7
coefficient of linear expansion at 20°C, μcm/(cm·°C)		31.3
electrical resistivity, μΩ·cm		
22°C		7.27
400°C		34.1
600°C		34.8
700°C		35.8
electrical conductivity, % IACS[b]		25
density, kg/m^3 (temp, °C)	8642 (26)	8020 (330 liq)
	7930 (400)	7720 (600)
volume change on fusion, % increase		4.74
thermal conductivity, W/(m·K) (temp, K)		98 (273)
	95 (373)	89 (573)
vapor pressure, kPa[c] (temp, °C)	0.1013 (382)	1.013 (473)
	10.13 (595)	101.3 (767)
surface tension, mN/m (=dyn/cm) (temp, °C)	564 (330)	598 (420)
	611 (450)	
viscosity, mPa·s (=cP) (temp, °C)	2.37 (340)	2.16 (400)
	1.54 (600)	1.84 (500)
molar magnetic susceptibility, cm^3/mol (=emu/mol)		-19.8×10^{-6}
Brinell hardness, kg/mm^2		16–23
tensile strength, MPa[d]		71
elongation, %		50
Poisson's ratio		0.33
modulus of elasticity, GPa[e]		49.9
shear modulus, GPa[e]		19.2
thermal neutron capture cross-section at 2200 m/s, m^2/atom		$2450 \pm 50 \times 10^{-28}$

[a] To convert J to cal, divide by 4.184.
[b] International Annealed Copper Standard.
[c] To convert kPa to mm Hg, multiply by 7.5.
[d] To convert MPa to psi, multiply by 145.
[e] To convert GPa to psi, multiply by 145,000.

Mass	Relative abundance, %	Mass	Relative abundance, %
106	1.22	112	24.07
108	0.88	113	12.26
110	12.39	114	28.86
111	12.75	116	7.58

Although it is only slowly oxidized in moist air at ambient temperature, cadmium forms a fume of brown-colored cadmium oxide, CdO, when heated in air. Other elements which react readily with cadmium metal upon heating include the halogens, phosphorus, selenium, sulfur, and tellurium.

The standard reduction potential for $Cd^{2+} + 2\,e \rightarrow Cd$ is 0.402 V at 25°C (2), and therefore cadmium is only slowly attacked by warm dilute hydrochloric or sulfuric acid with the evolution of hydrogen. Because of its position in the electromotive series of elements, cadmium is displaced from solution by more electropositive metals such as zinc or aluminum.

Cadmium is rapidly oxidized by hot dilute nitric acid with the simultaneous generation of various oxides of nitrogen. Unlike the zinc ion, the cadmium ion is not markedly amphoteric, and therefore its hydroxide, $Cd(OH)_2$, is virtually insoluble in alkaline media. However, the cadmium ion forms stable complexes with ammonia as well as cyanide and halide ions. The metal is not attacked by aqueous solutions of alkali hydroxides.

Production

Cadmium occurs primarily as sulfide minerals in zinc, lead–zinc, and copper–lead–zinc ores. Beneficiation of these minerals, usually by flotation (qv) or heavy-media separation, yields concentrates which are then processed for the recovery of the contained metal values. Cadmium follows the zinc with which it is so closely associated.

The zinc concentrate is first roasted in a fluid-bed, flash, or multiple-hearth roaster to convert the zinc sulfide to the oxide and, in the case of electrolytic zinc plants, some sulfate. Normally, roasting is carried out with an excess of oxygen below 1000°C so that comparatively little cadmium is eliminated from the calcined material in this operation (3). Further treatment determines the form in which the cadmium is concentrated for further processing.

Air pollution problems and labor costs have led to the closing of older pyrometallurgical plants, and to increased electrolytic production. On a worldwide basis, 56% of total zinc production in 1970 was by the electrolytic process (4). In electrolytic zinc plants, the calcined material is dissolved in an aqueous sulfuric acid, usually spent electrolyte from the electrolytic cells. The filtered leach solution is treated with zinc dust to remove cadmium and other impurities.

The processing (5) of the cadmium-bearing precipitate generally follows the flow sheet shown in Figure 1. The precipitate, containing four to twenty-nine times more zinc than cadmium as well as other impurities, notably residual copper, is dissolved at 45–82°C in a mixture of spent electrolyte from the zinc plant, sulfuric acid, and spent cadmium electrolyte. The copper is removed by galvanic precipitation with a small amount of zinc dust. After filtering the copper cake, cadmium is reprecipitated in two stages, usually at pH 5.2 and with 0.6–2 kg zinc dust per kg cadmium, so that the product will contain about 80% cadmium and less than 5% zinc. Steam oxidation of this sponge is optional. It is then dissolved at 45–82°C in spent cadmium electrolyte and make-up sulfuric acid to give a solution of about 200 g Cd/L, and is mixed with recirculating spent electrolyte to form the cell electrolyte. The electrowinning is carried out at 21–25°C in cells equipped with silver–lead anodes and aluminum cathodes at a current density varying between 26 and 240 A/m^2, and cell voltages of 2.5–2.8 V. Glue is added at rates of 0–2.5 kg per metric ton of cadmium deposited. Cathode deposits are stripped from the aluminum blanks every six to twenty-four hours, depending on the current density. They are washed, dried, and melted at 380–400°C under sodium hydroxide, which not only acts as a flux to prevent oxidation but also effectively removes any zinc or arsenic which may still be present. Finally, the metal is cast into commercial shapes, ie, slabs, balls, ingots, rods, splatters, and powder.

Precautions have to be taken during the dissolution of cadmium precipitates or

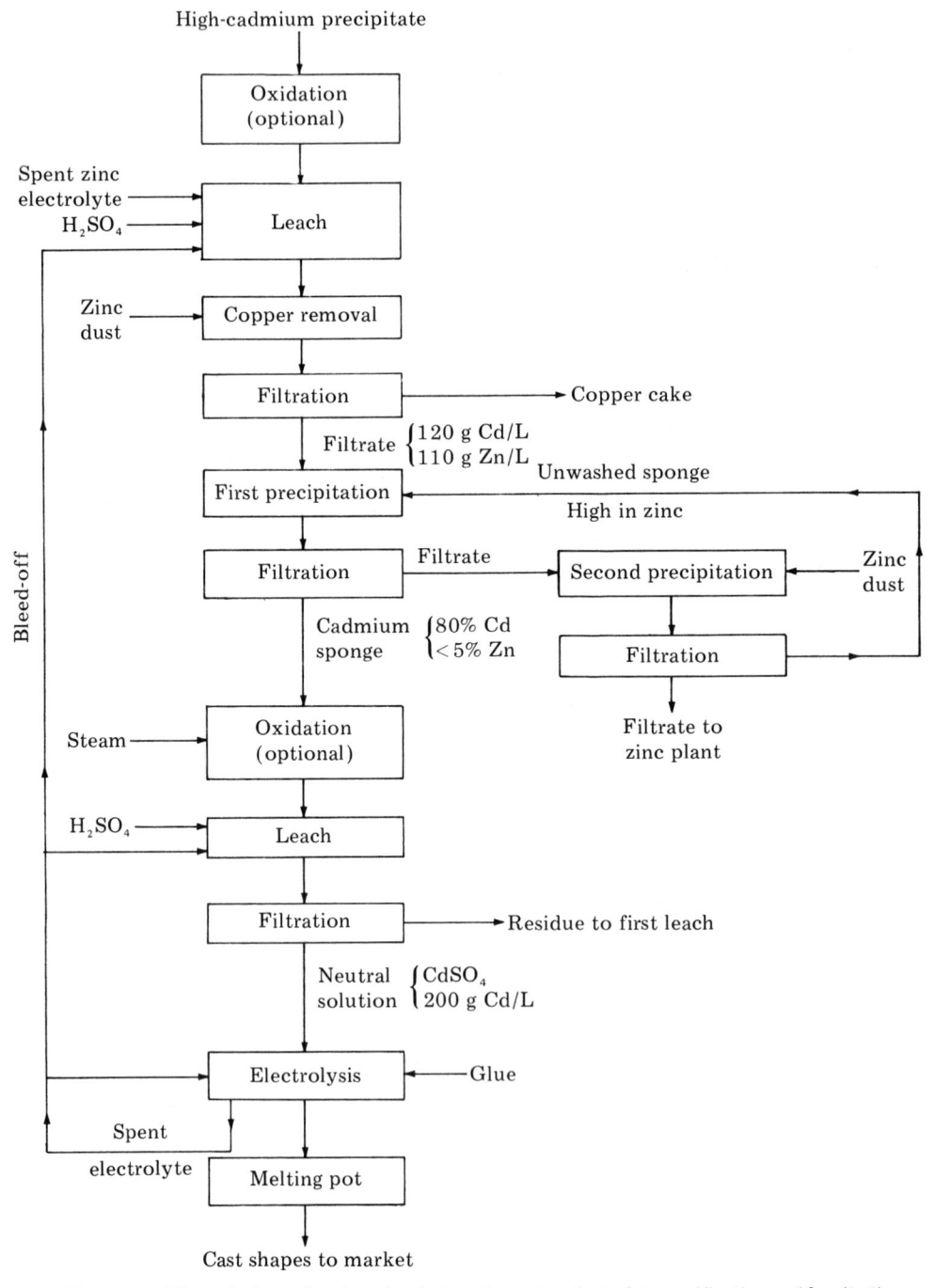

Figure 1. Electrolytic production of cadmium from zinc electrolyte purification residue (5–6).

the galvanic precipitation of cadmium with zinc to remove possible mist and toxic gases such as arsine. Suitable exhaust hoods and scrubbers must be provided. The fume which may be formed during cathode melting must be removed similarly.

When the calcined product from a roaster is first subjected to sintering with coal or coke, a binder may be added. The briquettes formed are then fired with air at about 1200°C. This procedure results in considerable volatilization of cadmium and lead compounds, enhanced by the presence of chloride, leading to 90–99% recovery of cadmium. The fume and dust from the sintering machine are collected in a baghouse (7–8). Cadmium not removed during sintering and subsequent operations follows the zinc metal and often is recovered during zinc metal purification by distillation.

The cadmium content in the feed to lead and copper smelters is lower than that generally encountered in zinc plants, and this necessitates upgrading of the initial cadmium level in the fume by one or more refuming steps in a kiln or reverberatory furnace so that the final fume may contain as much as 45% cadmium. In general, the composition of these fumes as well as those obtained from zinc sintering vary with respect to cadmium content and impurities. They usually require more processing and purification steps for cadmium recovery than is the case with purification residues from electrolytic zinc plants. Galvanic precipitation is the most frequently adopted method for the final recovery of cadmium in pyrometallurgical plants, but electrowinning may also be used (see Extractive metallurgy).

The flow sheet in Figure 2 illustrates cadmium recovery from cadmium-bearing fumes. Depending on its composition, the fume may have to be roasted with or without sulfuric acid or oxidized with sodium chlorate or chlorine in order to convert cadmium into a water- or acid-soluble form and to eliminate volatile constituents. However the leach solution is obtained, it must generally be purified to remove arsenic, iron, copper, thallium, and lead, by various treatments as shown in Figure 2, for the recovery of cadmium from baghouse fume. The cadmium may also be galvanically precipitated from the leach solution and then redissolved (see alternative 1 in Fig. 2).

In the recovery of cadmium from fumes evolved in the Imperial Smelting Process for the treatment of lead–zinc concentrates, cadmium is separated from arsenic with a cation exchange resin such as Zeocarb 225 or Amberlite 120 (9–10). Cadmium is absorbed on the resin and eluted with a brine solution. The cadmium may then be recovered directly by galvanic precipitation.

Alternative 2 in Figure 2 indicates the most common method for the recovery of cadmium from purified leach solution by galvanic displacement with zinc in the form of dust, sheets, or even rods or rectangular anodes. The final processing depends on the grade of zinc. In most cases, the pH for galvanic precipitation is below 2, although one plant operates at pH 6.2. Temperatures range from ambient to 70°C and precipitation times vary from 30 min to 18 h, depending on temperature and aggregation of the zinc. The weight of zinc required to precipitate one kilogram of cadmium varies between 0.65–0.95 kg. In most plants, the final cadmium sponge is washed to remove soluble impurities, and then compacted by briquetting. The briquettes may be melted under a flux of sodium hydroxide or ammonium chloride or be distilled for final purification.

In alternative 3 (Fig. 2), the electrolysis may be operated on a semicontinuous basis with the cadmium eventually being stripped completely from the electrolyte, which is then discarded after suitable treatment. Instead of the usual silver–lead anodes, high silicon–iron anodes, such as Duriron, are commonly used.

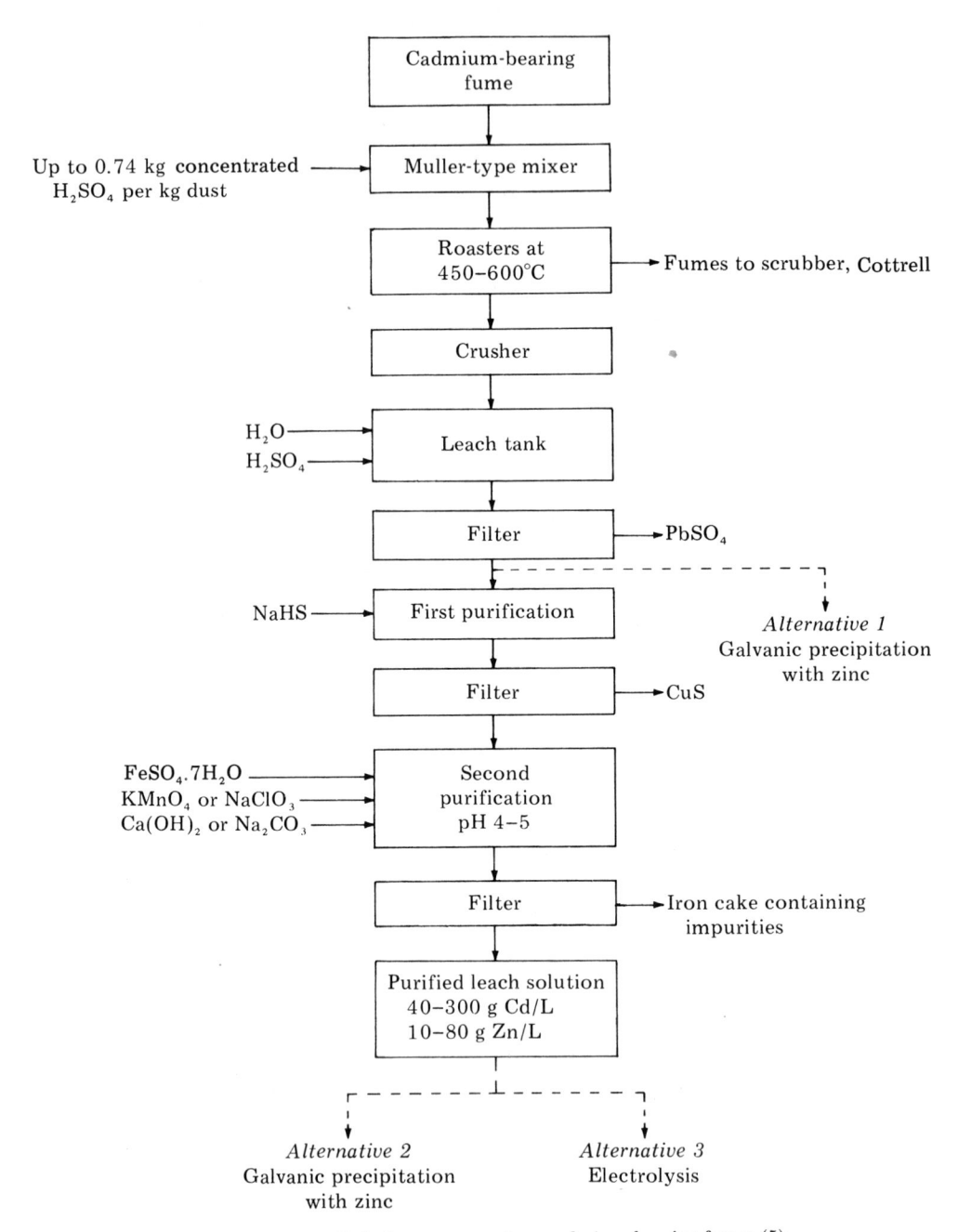

Figure 2. Cadmium recovery from cadmium-bearing fumes (5).

Economic Aspects, Specifications

Cadmium production is dependent on the processing of zinc ores. As the United States demand for cadmium normally exceeds the domestic supply, the United States is dependent on imports.

Recent United States production has dropped sharply owing to a cut-back in zinc production (see Table 2). The outlook is influenced by reduced cadmium levels of new ore sources and changes in environmental regulations. Before 1974, the United States production of cadmium had been in the range of 3179–4539 t/yr, which has represented approximately 25% of the non-Communist world production. About 30 countries are cadmium producers, led by the United States, Canada, Mexico, Australia, the Federal Republic of Germany, Belgium and Luxembourg, France, Italy, and Japan.

The apparent United States consumption of cadmium in 1977 was estimated to be 4226 t, of which 2333 t was imported, with over 75% coming from Canada, Australia, Mexico, Belgium and Luxembourg, and Yugoslavia.

The major domestic producers and suppliers of cadmium metal are AMAX Lead and Zinc, Inc.; ASARCO, Inc.; The Bunker Hill Company; National Zinc Company; The New Jersey Zinc Company; and St. Joe Minerals Corp.

The ASTM specification for cadmium metal is for a minimum purity of 99.90%. It is also supplied to a purity of 99.999% in minimum quantities of 22.5 kg (at $22.00/kg, 1977 price).

Table 2. United States Cadmium Statistics, t[a]

Year	Production	Imports	Apparent consumption	Price, $/kg
1972	3762	1099	5731	6.60
1973	3405	1768	5689	8.25
1974	3026	1801	5492	9.35
1975	1990	2376	3033	4.40
1976	2048	3096	5524	6.60
1977[b]	2040	2333	4226	6.60

[a] Ref. 11.
[b] Estimated.

Analysis

Although the most sensitive line for cadmium in the arc or spark spectrum is at 228.8 nm, the line at 326.1 nm is more convenient to use for spectroscopic detection. The limit of detection at this wavelength amounts to 0.001% cadmium with ordinary techniques and 0.00001% with specialized methods. Determination in concentrations up to 10% is accomplished by solubilization of the sample followed by atomic absorption measurement. This range can be extended to still higher cadmium levels provided that a relative error of 0.5% is acceptable. Another quantitative method is by titration at pH 10 with a standard solution of ethylenediaminetetraacetic acid (EDTA) and Eriochrome Black T indicator. Zinc interferes and therefore must first be removed.

Safety, Handling

Cadmium is classified as a toxic metal. Most cases of cadmium poisoning result from the inhalation of dust or fumes affecting the respiratory tract. Protection should be provided by a properly designed exhaust ventilation system or, for some intermittent exposures, by a suitable individual filtered or air-supplied respirator (12). Fumes are formed at vaporization temperatures in welding or brazing. Immediate medical attention should be obtained after severe exposure.

The current OSHA atmospheric limit (8-h time-weighted average standard) is 100 $\mu g/m^3$ of air for cadmium fume and 200 $\mu g/m^3$ of air for cadmium dust.

Cadmium wastes should be disposed of according to existing standards and regulations.

Controls proposed by the EPA for enactment in the near future may tend to reduce usage of cadmium for some purposes.

Uses

Elemental cadmium is used principally as an electroplated coating on fabricated steel and cast iron parts for corrosion protection (see Corrosion). The advantages of cadmium metal for this purpose are: ease of electroplating and high rate of deposition (high throwing power, ie, ability to deposit uniformly on intricate objects); good corrosion resistance to alkali and salt water; high ductility (parts plated can be stamped or otherwise formed); good solderability; and high retention of silvery-white luster for extended periods.

Cadmium is usually plated from a cyanide bath which consists of an aqueous solution of cadmium oxide (35 g/L) and sodium cyanide (75 g/L). An additive and brightener are used to produce smooth, fine-grain deposits. Current density ranges from 1.4 to 3.7 A/m^2, depending on the concentration of cadmium cations in the electrolyte.

It may also be applied by vacuum deposition, dipping, spraying, or mechanical plating with cadmium powder. The future of the cadmium plating industry depends on its ability to meet tighter waste water restrictions.

Mechanical plating with cadmium powder using glass shot is a substitute for electroplating.

Batteries. Cadmium is the negative electrode in rechargeable nickel–cadmium and silver–cadmium batteries. Their major use is in small electric appliances and in flashlights, and also in vehicular batteries (see Batteries and electric cells, secondary).

Alloys

Cadmium is an important component in brazing and low-melting alloys, used in bearings, solders, and nuclear reactor control rods, and as a hardener for copper (see Bearing materials).

Of interest are two brazing alloys, 20% Ag–45% Cu–30% Zn–5% Cd (mp 615°C; flow point 815°C), and ASTM Ag 2 which is 35% Ag–26% Cu–21% Zn–18% Cd (mp 607°C; flow point 702°C). Other useful brazing compositions are also available (13–14) (see Solders and brazing alloys).

The commonly used low-melting or fusible alloys are: AsarcoLo 158 or Cerrobend

Table 3. Low-Melting Alloys

System	Composition, %					mp, °C
	Bi	Cd	In	Sn	Pb	
Sn–Pb–Cd		18		51	31	145
Bi–Cd	60	40				144
Cd–In		25	75			123
Bi–Cd–Sn	54	20		26		103
In–Cd–Sn		14	44	42		93
Bi–Cd–Pb	52	8			40	92
In–Bi–Cd	30	8	62			62

(50% Bi; 26.7% Pb; 13.3% Sn; 10% Cd; mp 70°C) for bending pipes and thin sections, glass-lens grinding blocks, and fire protection devices; AsarcoLo 158–190 or Cerrosafe (42.5% Bi; 37.7% Pb; 11.3% Sn; 8.5% Cd; mp 70–87°C) for foundry patterns, spotting fixtures, solder, and proof-casting molds; and AsarcoLo 117 or Cerrolow 117 (44.7% Bi; 22.6% Pb; 8.3% Sn; 5.3% Cd; 19.1% In; mp 47°C) for fusible cores, soldering and sealing, holding irregular pieces for machining, and plastic lens grinding blocks. Other low-melting alloys (15) are listed in Table 3.

For soldering aluminum, combinations of cadmium and zinc are widely used, the most satisfactory being the 60% Cd–40% Zn alloy, in addition to a 95% Cd–5% Ag solder.

In high speed and high temperature applications, which are too severe for tin or lead bearings, SAE 18, containing 1% nickel and 99% cadmium, and SAE 180, containing 0.7% silver, 0.6% copper, and 98.7% cadmium are employed.

Additions of cadmium (0.05–1.3%) to copper raise the recrystallization temperature and improve the mechanical properties, especially in cold-worked conditions, with relatively little reduction in conductivity. Copper containing 0.07% cadmium is used in automotive cooling fins, heavy-duty radiators, motor commutators, and electrical terminals.

An alloy containing 80% Ag, 15% In, and 5% Cd is used in control rods in nuclear reactors since it has a high neutron cross-section and good mechanical strength (see Nuclear reactors).

Electrical contacts. Silver containing 2.5–15% cadmium oxide is used extensively for producing electrical contacts.

Semiconductor uses. The intermetallic compounds with Group-VI elements including CdS, CdSe, and CdTe, have interesting semiconductor properties for photoconductors, photovoltaic cells, and ir windows. Cadmium sulfide is widely used as a phosphor in television tubes.

BIBLIOGRAPHY

"Cadmium and Cadmium Alloys" in *ECT* 1st ed., Vol. 2, pp. 716–732, by S. J. Dickinson, American Smelting and Refining Company; "Cadmium and Cadmium Alloys" in *ECT* 2nd ed., Vol. 3, pp. 884–899, by H. E. Howe, American Smelting and Refining Company.

1. I. M. Kolthoff and P. J. Elving, *Treatise on Analytical Chemistry,* Part II, Vol. 3, Interscience Publishers, New York, 1961, p. 178.
2. Ref. 1, p. 177.

3. *Technical and Microeconomic Analysis of Cadmium and its Compounds,* EPA 560/3-75-005, National Technical Information Service, Springfield, Va., 1975, p. 38.
4. Ref. 3, p. 35.
5. R. E. Lund and R. E. Sheppard, *J. Metals* **16,** 724 (1964).
6. G. D. Van Arsdale, *Hydrometallurgy of Base Metals,* McGraw-Hill Book Company, Inc., New York, 1953, pp. 239–240, 299–307.
7. R. L. Nauert, *J. Metals* **18,** 15 (1966).
8. Ref. 3, p. 43.
9. F. H. Baker and J. G. Munro, *J. Metals* **17,** 255 (1965).
10. W. Ryan, *Non-Ferrous Extractive Metallurgy in the United Kingdom,* The Institution of Mining and Metallurgy, London, Eng., 1968, pp. 31–32.
11. *Cadmium Reports for 1972–78,* Mineral Industry Surveys, U.S. Bureau of Mines, Washington, D.C., 1977.
12. *Cadmium Data Sheet 312. Revision A, (Extensive),* National Safety Council Chemical Division, 425 N. Michigan Ave., Chicago, Ill., 1970.
13. A. Butts and C. D. Coxe, eds., *Silver Economics, Metallurgy and Use,* D. Van Nostrand Co., Inc., Princeton, N.J., 1967, p. 387.
14. *Brazing Alloy Handbook,* ASARCO Incorporated, New York, 1968.
15. T. Lyman and co-eds., *Metals Handbook,* 8th ed., Vol. 1, American Society for Metals, Metals Park, Ohio, 1976, p. 864.

General References

F. Prince, "A Study of Industrial Exposure to Cadmium," *J. Ind. Hyg. Toxicol.* **25,** 5 (1947).

D. M. Chizhikov, *Cadmium,* translated by D. E. Hayler, Pergamon Press, Inc., New York, 1966.

M. C. Sneed and R. C. Brasted, eds., *Comprehensive Inorganic Chemistry,* Vol. IV, D. Van Nostrand Co., Inc., Princeton, N.J., 1955, pp. 65–90.

M. L. HOLLANDER
S. C. CARAPELLA, JR.
ASARCO Incorporated

CADMIUM COMPOUNDS

The only naturally occurring cadmium compound of significance, the sulfide greenockite [1317-58-4], CdS, which is fairly rare, is almost always associated with the polymetallic sulfide ores of zinc, lead, and to a lesser extent, copper at grades up to ca 0.5%. An oxycarbonate, otavite is also known (1). Hence, cadmium compounds are prepared from metallic cadmium obtained from lead–zinc production. The cadmium can be converted to the oxide which is a more convenient starting material for many compounds.

Properties

Cadmium, a member of Group II-B (Hg, Cd, Zn) of the Periodic Table, exhibits almost entirely a +2 valence state in its compounds. Under special conditions, dimeric formulations of the type X–Cd–Cd–X have been isolated as, eg, from the reaction of liquid cadmium with $CdCl_2$ in certain molten salt mixtures. In these rare instances, cadmium resembles mercury in its behavior but generally cadmium compounds exhibit properties very similar to those of the corresponding zinc compounds. Table 1 lists some properties of the major cadmium compounds. Cadmium shows a marked tendency to form aqueous complexes in which it binds from one to four ligands (see under Cadmium complexes). The most important are the cyanide, amine, and various halide complexes.

Economic Aspects, Uses

Technically and commercially important cadmium compounds include the oxide, sulfide, selenide, chloride, sulfate, nitrate, hydroxide, and various organic cadmium salts such as the stearate and benzoate. Some of the main areas of application are listed below according to decreasing consumption of cadmium.

Metal Finishing. This is primarily a process where thin metallic cadmium coatings from cadmium cyanide electrolytes are used for their special corrosion protection qualities.

Pigment Manufacture. Cadmium sulfide and sulfoselenide and lithopone pigments are used for colors ranging from yellow, orange, and red through deep maroon.

Batteries, Cells. Cadmium hydroxide serves as the active anode material in Ag–Cd and Ni–Cd batteries; cadmium sulfide-based solar cells are being rapidly developed.

Stabilizers. Organocadmium salts are employed as heat and light stabilizers in plastics and are used to retard discoloration, particularly in PVC.

Electronics Applications. Included are silver–cadmium oxide contacts, cadmium chalcogenide electroluminescent and photoconductive devices, and phosphors and semiconductors of several compositions.

Catalysts. Cadmium dialkyls and many inorganic cadmium salts are widely used as catalysts, especially in organic polymerization reactions.

In the United States, the present cadmium demand is 6000–7000 metric tons per year. This is growing at a rate of 4% per year and is projected to be 9560 metric tons by 1985 (12). However, because of the highly toxic nature of many cadmium forms, tighter governmental control over industrial cadmium usage is anticipated.

Table 1. Physical and Chemical Properties of Selected Cadmium Compounds[a]

Cadmium compound	CAS Registry No.	$\Delta H^\circ_{f,298}$ kJ/mol[b]	$\Delta G^\circ_{f,298}$ kJ/mol[b]	S°_{298} J/mol·deg[b]	Density, g/cm³	C°_p J/mol·deg[b]	mp, °C	ΔH_{fus}, J/mol[b]	bp, °C	ΔH_{vap}, kJ/mol[b] at bp	Solubility in H₂O g/100 g H₂O	Crystal structure	Unit cell dimensions, nm
antimonide CdSb	[12050-27-0]	−14.4	−13.0	92.9	6.92		456	32,050		138		ortho-rhomb	a = 0.6471, b = 0.8253, c = 8526
bromide CdBr₂	[7789-42-6]	−316	−296	137.2	5.192	76.7	568	20,920	963	dissoc 113	95/18°C	hex	a = 0.395, c = 1.867
carbonate CdCO₃	[513-78-0]	−751	−669	92.5	4.26		332 decomp				2.8 × 10⁻⁶	rhomb	a = 0.61306
chloride CdCl₂	[10108-64-2]	−391	−344	115.3	4.05	74.7	568	22,176	980	125	128.6/30°C	hex	a = 0.3854, c = 1.746
fluoride CdF₂	[7790-96-6]	−700	−648	77.4	6.33		1110	22,594	1747	234	4.35/25°C	cubic	a = 0.53880
hydroxide Cd(OH)₂	[21041-95-2]	−561	−474	96.2	4.79		150 decomp				2.6 × 10⁻⁴	hex	a = 0.3475, c = 0.467
iodide CdI₂, α-form	[7790-80-9]	−203	−201	161.1	5.67	80.0	387	33,472	790	106	86/25°C	hex (hex)[c]	a = 0.424, c = 0.684 (c = 1.367)[c]
nitrate Cd(NO₃)₂	[10325-94-7]	−456	−255	197.9			350				109/0°C	cubic	a = 0.756
nitrate, hyd Cd(NO₃)₂·4H₂O	[10022-68-1]	−1649			2.455		59.4	32,636	132		132/0°C	ortho-rhomb	a = 0.583, b = 2.575, c = 1.099

compound	CAS Reg. No.									
oxide CdO	[1306-19-0]	−258		54.8	8.2	43.4	1540 subl	243,509 subl	cubic	$a = 0.46953$
selenide CdSe, α-form	[1306-24-7]	−136	−228	96.2	5.81		680	305,307	hex	$a = 0.4309$, $c = 0.7021$
			−100				dissoc 1252	dissoc	(cubic)[c]	$(a = 0.605)$[c]
m-silicate CdSiO₃	[13477-19-5]	−1189	−1105	97.5	4.928	88.6			monoclinic	$a = 1.504$, $b = 0.710$, $c = 0.696$
sulfate CdSO₄	[10124-36-4]	−933	−823	123.0	4.691	99.6	1000	20,084	ortho-rhomb	$a = 0.4717$, $b = 0.6559$ $c = 0.4701$
sulfate, hyd CdSO₄·H₂O	[7790-84-3]	−1240	−1069	154.0	3.79	134.6	105 trans		monoclinic	$a = 7.607$, $b = 7.541$ $c = 8.186$
sulfate, hyd 3CdSO₄·8H₂O	[22465-18-5]	−1729	−1465	229.6	3.09	213.3	80 trans		monoclinic	$a = 0.947$, $b = 1.184$ $c = 1.635$
sulfide CdS, α-form	[1306-23-6]	−162	−156	64.9	4.82 (4.50)[c]		980 subl in N₂ 1045	201,669 subl	hex	$a = 0.41348$, $c = 0.6749$
									(cubic)[c]	$(a = 0.5818)$[c]
telluride CdTe	[1306-25-8]	−92	−92	100.4	6.20				hex	$a = 0.457$, $c = 0.747$
									(cubic)[c]	$(a = 0.6480)$[c]

[a] Refs. 2–8 and 10–11.
[b] To convert J to cal, divide by 4.184.
[c] β-form.

399

Analysis

The determination of cadmium at low concentrations (13-14) has become increasingly important because of environmental problems. Atomic absorption spectrophotometry has, to a considerable extent, displaced the polarographic and emission spectroscopic methods which were previously considered satisfactory for low-level cadmium analyses. Relative detection limits of 5 µg/L in aqueous solution have been reported (15). Extraction with ammonium pyrrolidine dithiocarbamate in MIBK is useful for concentrating cadmium and for separation of interfering materials such as sodium chloride. Other widely used methods include dithizone colorimetry, neutron activation analysis, and anodic stripping voltammetry (16).

At higher levels cadmium may be gravimetrically determined after isolation from interfering elements by standard sulfide precipitation and separation techniques (17). Cadmium may be weighed as metal following dissolution of the sulfide in a cyanide electrolyte and electrolytic deposition onto a stationary platinum cathode. A rotating cylindrical or disk-shaped cathode is used for faster analyses.

Other gravimetric methods include weighing as the β-naphthoquinoline complex (18) and as the ammonium phosphate monohydrate. In volumetric methods, the purified cadmium sulfide precipitate may be dissolved in acid and the liberated H_2S titrated with iodine. If zinc is a contaminant, the cadmium is precipitated with sodium diethyldithiocarbamate. The precipitate is redissolved and the cadmium titrated with a standard solution of disodium ethylenediamine tetraacetate using Eriochrome Black-T as indicator (19).

Toxicity and Environmental Aspects

Unlike zinc, cadmium shows no indication of being an essential trace element in biological processes. On the contrary, cadmium is toxic and poisoning occurs through inhalation and ingestion (20). Only about 6% of the estimated 40–50 µg/d of ingested cadmium is absorbed by the body, whereas 25–50% of the ca 2–10 µg/d cadmium content of inhaled dust and fume is absorbed (21). Hence, an increase in airborne cadmium is more dangerous than a similar rise in foodstuff content. Cadmium is also present in cigarette smoke (20). Deaths from acute cadmium poisoning have resulted from inhalation of cadmium oxide fumes, usually owing to welding of cadmium plated steels in areas without adequate ventilation.

Acute inhalation of cadmium oxide fumes and cadmium chloride aerosols, lethal dose = 6 mg/m^3 over an 8-h period (22), causes severe lung damage in the form of pneumonitis or pulmonary edema accompanied by abdominal pain and nausea. Furthermore, cadmium compounds are powerful emetics. Exposure to 1 mg/m^3 over an 8-h period is immediately dangerous. One minute inhalation of 2500 mg/m^3 is fatal (23).

At present, atmospheric time-weighted average exposure limits are 200 µg/m^3 over 8 h for cadmium metal dusts and dusts of soluble cadmium salts, and 100 µg/m^3 over 8 h for cadmium metal and cadmium oxide fumes (24). Acceptable OSHA 15-min ceiling concentrations are 600 µg/m^3 for dusts and 3000 µg/m^3 for fume (25). In 1977, NIOSH proposed standards that will limit the 8-h exposure level to 40 µg/m^3 for all forms of airborne cadmium, this will take effect when published in the Federal Register. The ceiling concentrations will be lowered to 200 µg/m^3 for dusts and 100 µg/m^3 for fumes (26).

Biological halftime for cadmium in the human body is between 10 and 30 years (21). The critical organ is the kidney where one-third of body cadmium accumulates (27). Renal tubular dysfunction and resultant protein urea occur after chronic exposure giving rise to renal cortex cadmium concentrations in excess of 200 ppm (28). Other results of chronic exposure include bone effects (osteomalacia), anemia, and liver dysfunction (20). Chronic inhalation of cadmium oxide has caused emphysema.

Cadmium discharges to air and water are decreasing as primary zinc producers convert to electrolytic plants and as more efficient water and air pollution control technologies take effect (29). Most of the cadmium compounds released to the environment are contained in solid wastes such as coal ash, sewage sludge (75 ppm), flue dust, and fertilizers (2–20 ppm).

Although effluent discharge regulations have allowed in some instances up to 15 ppm cadmium in waste waters, a limit of 1.0 ppm (av 0.5 ppm) became law in 1978 (30). In general, tighter standards can be anticipated as the Federal Environmental Protection Agency issues regulations concerning 65 toxic pollutants one of which is cadmium. By 1981, the EPA will expect industry to control cadmium discharges by the best available technology which will include membrane processes in addition to precipitation and filtration techniques (31).

The effect of lower effluent limits is strongly felt in the metal finishing industry where electroplating accounts for over half of the United States cadmium usage. Cadmium can poison biological sewage treatment operations such as anaerobic bacteriological sludge digestion. Recovery by evaporation of cadmium cyanide rinses used within plate shops involves capital investment but may be one solution to the problem. Stringent disposal restrictions may influence consumption significantly. In Japan, where several areas have had serious cadmium pollution problems (16, 28), plating accounted for a large portion of cadmium consumption (271 t in 1969) until it was recognized, in 1970, as a pollution source and reduced by about 90% (32) (see Electroplating).

Inorganic Compounds

Cadmium Arsenides, Antimonides, and Phosphides. Cadmium forms two arsenides with the formulas Cd_3As_2 and $CdAs_2$. The former forms grey tetragonal crystals (a = 0.8945 nm, c = 1.265 nm), d = 6.21 g/cm^3, mp 721°C (3); $\Delta H^°_{f,\,298}$ = −41.84 kJ/mol (−10.0 kcal/mol) (2). It is an n-type semiconductor and has a high electron mobility (see Semiconductors); its electron concentration is 3×10^{18} per cm^3 and the activation energy at 298 K is 0.13 eV (8).

Cadmium arsenide [12006-15-4], Cd_3As_2, is prepared by heating stoichiometric amounts of cadmium and arsenic in an inert atmosphere or by precipitation from an ammoniacal cadmium sulfate solution with arsine. Thin films of n-type Cd_3As_2 on glass, mica, or salts are used in low-temperature semiconductor applications such as ultrasonic multipliers, photodetectors, thermodetectors, and thin-film Hall generators (33).

The other cadmium arsenide [12044-40-5], $CdAs_2$, mp 621°C, also forms tetragonal crystals (a = 0.465 nm, c = 0.793 nm) (7); $\Delta H^°_{f,\,298}$ = −17.5 kJ/mol (−4.2 kcal/mol) (2), and upon heating, decomposes to Cd_3As_2 and arsenic. It has a high optical absorption and transparency to a wide range of wavelengths, and is mainly used in research related to its electronic properties.

Cadmium antimonide CdSb, is formed by direct union of the elements. Some of its properties are listed in Table 1. Monocrystals of CdSb exhibit hole-type conductivity and the concentration of current carriers (8) is 1×10^{16}/cm^3. Cadmium antimonide is a thermoelectric generator (see Thermoelectric energy conversion). The width of its forbidden zone is 0.46 eV.

The phosphides, Cd$_3$P$_2$, CdP$_2$, and CdP$_4$, are prepared by direct union of cadmium with phosphorus in the desired ratio. Cadmium phosphide [12014-28-7], Cd$_3$P$_2$, can also be produced by the action of phosphine on solutions of cadmium salts. It forms grey tetragonal crystals ($a = 0.8746$ nm, $c = 1.228$ nm, d = 5.60 g/cm^3). It is an n-type semiconductor and the first compound of the A$_3$(II)B$_2$(V) group in which laser action was observed (33) (see Lasers).

Cadmium Borates. Cadmium borates with the general formula nCdO·mB$_2$O$_3$ have been prepared and are used as phosphors. Cadmium borate [20571-45-3], CdO·3B$_2$O$_3$, d = 3.76 g/cm^3, shows green cathodoluminescence like 2CdO·3B$_2$O$_3$ and 3CdO·B$_2$O$_3$. When activated by trace amounts of manganous ion, 2CdO·B$_2$O$_3$ fluoresces strongly orange under the influence of electromagnetic radiation.

Cadmium borotungstate [1306-26-9] (2CdO·B$_2$O$_3$·WO$_3$·18H$_2$O) solutions with densities up to 3.28 are used for heavy media separation of minerals (8).

A cadmium fluoborate [14486-19-2] bath is used for electrodeposition of cadmium on high strength steels to avoid the problem of hydrogen embrittlement inherent in cyanide plating. The 1977 price of a 44% Cd(BF$_4$)$_2$ solution was $6.60 per kilogram of contained cadmium.

Cadmium Carbonate. A hydrated, amorphous, basic carbonate is precipitated from cadmium salt solutions upon addition of sodium or potassium carbonate. It is converted to a crystalline form by heating with ammonium chloride at 150–180°C in the absence of oxygen. The normal cadmium carbonate, CdCO$_3$, is obtained by adding an excess of ammonium carbonate to a solution of cadmium chloride; the precipitate is dried at 100°C (see Table 1). Cadmium oxide slowly absorbs carbon dioxide to form the carbonate. For the decomposition reaction, CdCO$_3$ → CdO + CO$_2$, the carbon dioxide partial pressure reaches 101 kPa (1 atm) at 357°C.

The carbonates are soluble in acids and form complexes in solutions of cyanides and ammonium salts. They are used as catalysts in organic reactions and as a source of cadmium in the generation of other compounds. Phosphor-grade cadmium carbonate sells for $12.54/kg in one-ton lots.

Cadmium Complexes. The aqueous complexes of Cd^{2+} have been studied extensively. Cadmium does not bind more than 4 anionic ligands and solutions of the cadmium complexes are generally colorless. Thermodynamic and stability constant data for some of the more important complexes are shown in Table 2.

Other inorganic complexes of cadmium include the selenocyanate, Cd(SeCN)$_4{}^{2-}$ [4701-28-4], log K_4 = 0.24; hydrazine, Cd(N$_2$H$_4$)$_4{}^{2-}$ [302-01-2], log K_4 = 1.11; nitrite, CdNO$_2{}^+$ [7790-83-2], log K_1 = 1.80; pyrophosphate, CdP$_2$O$_7{}^{2-}$ [15600-62-1], log K_1 = 8.7; chlorate, CdClO$_3{}^+$ [22750-54-5], log K_1 = −0.30; and bromate, CdBrO$_3{}^+$ [14518-94-6], log K_1 = 0.06. Many complexes with organic ligands are also known such as with methylamine, thiourea, oxalic acid, tartaric acid, dimethylglyoxime, pyridine, acetic acid, EDTA, and glycolic acid (see ref. 34 for stability constants).

The outstanding application for cadmium complexes is in the metal finishing industry. Electroplating baths containing complexed cadmium have better throwing power and yield deposits with finer grains. By complexing the cation, higher cathodic

Table 2. Thermodynamic and Stability Constant Data for Selected Aqueous Cadmium Complexes[a,b]

Complex ion	CAS Registry No.	$\Delta H°_{f, 298}$ kJ/mol[c]	$\Delta G°_{f}, 298$ kJ/mol[c]	Stability constant
CdCl+	[14457-58-0]	−240.5	−224.4	log K_1 = 1.32
CdCl$_3$−	[21439-35-0]	−561.0	−487.0	log K_3 = 0.09
Cd(CN)$_4$$^{2-}$	[16041-14-8]	428.0	507.5	log K_4 = 3.58
Cd(NH$_3$)$_2$$^{2+}$	[47942-20-1]	−266.1	−159.0	log K_2 = 2.24
Cd(NH$_3$)$_4$$^{2+}$	[18373-05-2]	−450.2	−226.4	log K_4 = 1.18
CdBr+	[15691-37-9]	−200.8	−193.9	log K_1 = 1.97
CdBr$_3$−	[21439-36-1]		−407.5	log K_3 = 0.24
CdI+	[15691-38-0]	−141.0	−141.4	log K_1 = 2.08
CdI$_3$−	[15691-42-6]		−259.4	log K_3 = 2.09
CdI$_4$$^{-2}$	[15975-72-1]	−341.8	−315.9	log K_4 = 1.59
CdSCN+	[18194-99-5]		7.5	log K_1 = 1.90
Cd(SCN)$_4$$^{-2}$	[19438-35-8]			log K_4 = ca 0.1
Cd(N$_3$)$_4$$^{-2}$	[16408-27-8]		1,295.0	log K_4 = 0.76

[a] Standard state, M = 1.
[b] Refs. 2 and 34.
[c] To convert J to cal, divide by 4.184.

concentration overvoltages can be achieved, thus the cathode potentials are changed to regions where other metals codeposit with cadmium.

In commercial operations the cyanide bath (cadmium is complexed as Cd(CN)$_4$$^{2-}$) predominates because almost every plating characteristic of consequence favors this system. However, other electrolytes including the cadmium sulfate, sulfamate, chloride, fluoborate, and pyrophosphate baths do not have the disadvantages of cyanide plating, namely, hydrogen embrittlement of high strength steels and strict cyanide effluent limitations.

Cadmium Ferrite. The formation of cadmium ferrite [12013-98-8], CdFe$_2$O$_4$, from cadmium oxide and ferric oxide proceeds rapidly at 750°C. The crystal form is cubic (8) (a = 0.8684 nm), of the spinel type, d = 5.76 g/cm^3. Many other mixed oxides of cadmium are known, such as with molybdenum, tungsten, antimony, arsenic, niobium, tantalum, and titanium oxides.

Cadmium Halides. Cadmium fluoride, CDF$_2$, with an ionic structure, behaves differently from the other cadmium halides which are all characterized by partly covalent bonds. The fluoride is generally less soluble but dissolves in mineral acids. The other halides are soluble in alcohols, ethers, and liquid ammonia, and have higher water solubilities than the fluoride (see Table 1).

Aqueous solutions of the halides have low conductivities which is, to a large extent, because of complex ions (eg, CdX+, CdX$_3$−, CdX$_4$$^{2-}$). Numerous double halides, such as K$_2$CdCl$_4$ [20648-91-3], with alkali and alkaline earth halides, are known. The cadmium halides find use in pyrotechnics (qv) where they color flame blue, and in the manufacture of photographic film. The iodide and bromide are used in lithography and engraving. All of the cadmium halides are used as catalysts in a wide variety of organic reactions.

Cadmium Fluoride. A solution of CdF_2 is prepared by the dissolution of $CdCO_3$ in 40% hydrofluoric acid. The pure fluoride is obtained by evaporating the solution followed by vacuum drying at 150°C. Reaction of gaseous fluorine or HF with cadmium metal or compounds such as the oxide, sulfide, and chloride also yields the fluoride. Cadmium fluoride is fluorescent and is used in certain phosphors. Research applications, eg, in solid ionic transport studies (35–36), provide its predominant use.

Cadmium Chloride. In addition to the anhydrous compound listed in Table 1, cadmium forms hydrated chlorides with 1, 2, 2.5, and 4 molecules of water. Under normal conditions of temperature and humidity, dicadmium chloride pentahydrate [*34330-64-0*] $2CdCl_2 \cdot 5H_2O$ is the stable compound forming colorless, monoclinic crystals, d = 3.33 g/cm³, $\Delta H°_{f,298}$ = -1132 kJ/mol (-270.54 kcal/mol) (2); from the anhydrous salt and water, $\Delta H°_{298}$ = 26.3 kJ/mol (6/3 kcal/mol) (8); for the monohydrate, $\Delta H°_{f,298}$ = -688.4 kJ/mol (-164.54 kcal/mol).

Cadmium chloride is formed in solution by the reaction of hydrochloric acid with cadmium metal, carbonate, sulfide, oxide, or hydroxide. Upon evaporation, a hydrated salt crystallizes. To prepare the anhydrous salt, the hydrate is refluxed with thionyl chloride, or calcined in a hydrogen chloride atmosphere. It may also be obtained by (a) addition of dry cadmium acetate to a mixture of glacial acetic acid and acetyl chloride; (b) distillation from a mixture of $Cd(NO_3)_2 \cdot 4H_2O$ in hot concentrated hydrochloric acid; and (c) by reaction of HCl or chlorine gas with cadmium metal.

Cadmium chloride is generally sold as the pentahydrate at $6.20/kg (technical grade) in 45-kg lots. All of the hydrates are efflorescent in air. The compound finds use in photocopying, printing, and dyeing. The cadmium chloride metal finishing bath, introduced in 1969, provides good throwing power and a bright deposit and may displace cadmium cyanide plating where cyanide effluent control is a particular problem.

Cadmium chloride aerosols are among the most toxic cadmium compounds and are usually included with cadmium oxide fumes as severe hazards.

Cadmium Bromide. Anhydrous crystalline cadmium bromide (see Table 1) is a yellow compound prepared by (a) direct combination of the elements at elevated temperature; (b) mixing dry cadmium acetate with glacial acetic acid and acetyl bromide, and (c) calcining the tetrahydrate at 200°C.

The white cadmium bromide tetrahydrate [*13464-92-1*], $CdBr_2 \cdot 4H_2O$, $\Delta H°_{f,298}$ = -1492.55 kJ/mol (-356.73 kcal/mol) is the form stable up to 36°C at which point it decomposes to the monohydrate. It is prepared by dissolving cadmium oxide in bromine water. It is only sold as a reagent grade in small quantities for $132/kg.

Cadmium Iodide. Besides the normal white α-form (see Table 1), a brownish β-CdI_2 crystal can be obtained by slow crystallization from solutions or from fused salt mixtures (7).

Cadmium iodide is prepared by dissolving cadmium metal, oxide, hydroxide, or carbonate in hydroiodic acid or by heating the elements together in the absence of air. Reagent-grade cadmium iodide sold in 1977 for $62.30/kg.

Cadmium Hydroxide. Crystalline cadmium hydroxide, $Cd(OH)_2$, can be prepared by the addition of a solution of cadmium nitrate to boiling sodium or potassium hydroxide. It is soluble in acids, and in strongly complexing systems such as solutions of ammonium salts, ferric chloride, alkali halides, cyanides, and thiocyanates.

Compared to zinc hydroxide, $Cd(OH)_2$ is much more basic. It is soluble in 5 N NaOH solution at 0.13 g/100 mL (3) as the anionic complex $Cd(OH)_4^{2-}$ (aq) [*26214-93-7*] (37).

Cadmium hydroxide is used as a starting material in the manufacture of other cadmium compounds. Technical grade was $10.45/kg in 1977. The important application for cadmium hydroxide is as the active anode material in alkaline nickel–cadmium and silver–cadmium secondary storage batteries (38). These batteries accounted for over 20% of annual United States cadmium consumption in 1976 (39). They are used particularly in long-life heavy-duty applications and in rechargeable tools, appliances, and instruments (see Batteries).

Cadmium Nitrate. In addition to anhydrous cadmium nitrate, $Cd(NO_3)_2$, hydrated nitrates with 2 and 4 molecules of water are formed (see Table 1). The tetrahydrate is obtained by nitric acid digestion of cadmium metal, oxide, hydroxide, or carbonate followed by crystallization.

Cadmium nitrate is the preferred starting material for $Cd(OH)_2$, the active anode material in sintered plate nickel–cadmium, and silver–cadmium alkaline storage batteries where the anode matrix is saturated with a nitrate solution containing 480–500 g Cd/L. Cadmium hydroxide is formed by a standardized electrolysis and drying procedure (40). Uses for the nitrate include colorants for the ceramics industry and, in combination with magnesium, as a flash powder in photography. Technical-grade cadmium nitrate sold for $4.62/kg in 180-kg lots in 1977.

Cadmium Nitride. Cadmium nitride [4215-29-3], $Cd(N_3)_2$, prepared by heating the amide, $Cd(NH_2)_2$, at 180°C, forms black, cubic crystals (a = 1.079 nm, d = 7.67 g/cm^3), $\Delta H^\circ_{f,298}$ = 38.7 kcal/mol (3)). Cadmium nitride is unstable in air and it decomposes in moist environments and in acids and bases (see Nitrides).

Cadmium Oxide. Rust-brown crystalline cadmium oxide, CdO, dissolves in simple acids but is insoluble in water and alkali. It forms a variety of soluble complexes. An amorphous cadmium oxide is also known (41), in addition to cadmium peroxide [12139-22-9], CdO_2.

Cadmium oxide is an n-type semiconductor with an energy gap of 222 kJ/mol (53 kcal/mol). It is reduced to the metal at temperatures ranging from 300–670°C by carbon, carbon monoxide, hydrogen, and methane (see Semiconductors).

Commercially, cadmium oxide is prepared by the reaction of cadmium metal vapor with air. Pure cadmium metal is melted in a cast iron or steel kettle and pumped to a heated chamber where it is vaporized. The vapor is conducted to a reactor and air is blown through oxidizing the cadmium and carrying the reaction product into a baghouse. Finer and coarser particles are produced, depending on the ratio of air to cadmium vapor. The oxide can be calcined at a low red heat to ensure uniform physical properties. Annual U.S. production exceeeds 1000 metric tons. The 1977 price was $6.60/kg. Other possible high-temperature routes include oxidation of the sulfide and thermal decomposition of cadmium carbonate, nitrate, sulfate, or hydroxide. Pyrolysis of the formate [4464-23-7] or oxalate [814-88-0] results in a very active finely divided form of CdO.

Cadmium oxide is used as a starting material for PVC heat stabilizers (qv) (see under Organocadmium Compounds) and inorganic cadmium compounds. Silver–cadmium oxide contacts containing up to 30% CdO (42) are used in electrical devices. High purity CdO is used as a second depolarizer (in addition to silver oxide) in silver–zinc storage batteries (43) (see Batteries). Cadmium oxide dissolved in excess sodium cyanide solution is widely used in electroplating baths. The use of CdO in nitrile rubbers improves heat resistance, and added to plastics such as Teflon, it improves their high-temperature properties.

Cadmium Selenide and Telluride. All of the cadmium chalcogenides are n-type semiconductors and the energy barriers (3) for cadmium sulfide, selenide, and telluride (CdS, CdSe, and CdTe) are 230 (45), 180 (41), and 155 kJ/mol (37.0 kcal/mol), respectively (see Table 1 for other properties).

Both the selenide and telluride are made at high temperature by direct combination of the elements, from solutions of cadmium salts by treating with H_2Se or H_2Te gas or by addition of alkali selenides or tellurides. Cadmium selenide and telluride are luminescent photoconductive compounds; however, because their fundamental absorption edges lie outside the visible spectrum they have received less attention than CdS (44).

Cadmium selenide and sulfide form a series of pigments known as cadmium sulfoselenides ranging in color from orange through maroon. CdSe is also used in photocells, rectifiers, luminous paints, and in glass manufacture as a ruby colorant.

Cadmium telluride is used in infrared optics (45), phosphors, electroluminescent devices, photocells, and as a detector for nuclear radiation (46) (see Nuclear reactors).

Cadmium Silicates. Cadmium o-silicate [15857-59-2], Cd_2SiO_4, (mp 1246°C, d = 5.83), and m-silicate, $CdSiO_3$, are known (see Table 1). They are prepared by the reaction of cadmium oxide with amorphous silicon or silica at 390°C and 30.4 MPa (300 atm), or at 900°C and atmospheric pressure with steam catalyst.

Cadmium silicates are fluorescent and phosphorescent and, when activated by trace amounts of manganous ion, they are used as phosphors. They are also used as catalysts in organic reactions.

Cadmium Sulfate. The principal cadmium sulfates, $CdSO_4$, $CdSO_4 \cdot H_2O$, and $CdSO_4 \cdot 8/3 H_2O$, are listed in Table 1. They are crystallized from cadmium sulfate solutions or may be precipitated by addition of alcohol. Anhydrous cadmium sulfate is prepared by oxidation of the sulfide or sulfite at elevated temperatures, or by the action of dimethyl sulfate on finely powdered cadmium nitrate, halides, oxide, or carbonate. Solutions are prepared by dissolving cadmium metal, oxide, sulfide, hydroxide, or carbonate in sulfuric acid. Cadmium sulfate solution is the electrolyte in standard cells such as the Weston cell and as electrolyte is also used industrially as an alternative to the cadmium cyanide electroplating bath. Technical grade $CdSO_4 8/3H_2O$ sold in 1977 for $8.18/kg in 22.7-kg drums.

Cadmium Sulfide. The dimorphic sulfide CdS is the most widely used cadmium compound (see Table 1). β-CdS can be transformed to α-CdS by heating at 750°C in a sulfur atmosphere. Cadmium sulfide is oxidized by air to the normal sulfate, basic sulfate, and finally the oxide upon heating from 300 to 700°C. Air oxidation is promoted by the presence of moisture (8).

Both α and β forms may be prepared in colors ranging from lemon yellow through orange and red, depending on their preparation and particle size. Cadmium sulfide may be prepared by the reaction between H_2S and cadmium vapor at 800°C, or by heating a mixture of cadmium or cadmium oxide with sulfur. Usually, however, the sulfides are precipitated from aqueous solutions of cadmium salts by adding H_2S or a soluble sulfide such as Na_2S. Yellow crystals are precipitated at room temperature from cadmium sulfate, nitrate, and chloride solutions in the acidic to neutral pH range. When the same solutions are precipitated at boiling, yellow sulfides are obtained in the absence of acid and red sulfides are formed from acidified solutions. A yellow sulfide precipitates from acidified cadmium acetate solutions and a red modification precipitates from ammoniacal solution. The hexagonal crystal form is precipitated from halide solutions but the cubic crystals are generally prepared from sulfate solution and acidified nitrate systems. Passing H_2S into a warm, acidified solution of cadmium perchlorate also leads to the precipitation of cubic CdS (3).

The main use of cadmium sulfide is for pigments. The consumption of cadmium colorants by the plastics industry from 1972 to 1976 is shown below (47).

Year	*Cadmium colorant consumption (metric tons)*
1972	2300
1974	2750
1976	2600

The contained cadmium was over 1000 t/yr which represents 75% of all cadmium pigments used. Pure yellow cadmium sulfides are formulated with red cadmium selenides in varying proportions to give C.P. toners ranging from yellows and oranges with low selenium content to high selenium reds and maroons (see Colorants for plastics).

In general, industrial pigment manufacture can be summarized as follows: (a) cadmium sticks are dissolved in sulfuric acid to form cadmium sulfate solutions; (b) sodium sulfide is mixed with varying amounts of soluble selenides; and (c) the precipitated cadmium sulfide or sulfoselenide pigments are filtered, washed, dried, and finally calcined at 700°C to obtain a uniform product.

Cadmium lithopones, ZnS–CdS (or CdSe)–$BaSO_4$ are produced by treating a mixture of zinc and cadmium sulfates with barium rather than sodium sulfide. These pigments are generally yellow shades; however, reds and maroons are obtained with additions of selenium. The 1977 prices for the various pigments are listed in Table 3. Cadmium–mercury lithopones are produced in similar colors for slightly lower cost than the corresponding selenium lithopones. Cadmium pigments are resistant to hydrogen sulfide, sulfur dioxide, light, heat, and other atmospheric conditions. They are dense, heavy colorants with good covering power and they provide bright, deep shades.

In addition, cadmium colorants find application in paints (especially artists colors such as cadmium yellow), soaps, rubber, paper, glass, printing inks, ceramic glazes, textiles, and for blue color in fireworks. Red and yellow cadmium sulfide–zinc sulfide fluorescent and phosphorescent pigments are also produced (see Pigments).

Cadmium sulfide, an intrinsic n-type semiconductor, is also used in the conversion

Table 3. 1977 Prices of Cadmium Pigments

Pigment	$/kg
C.P. toners	
yellows	13.30
oranges	21.80
reds	23.40
maroons	27.10
lithopones	
yellows	5.60
oranges	7.15
reds	10.74
maroons	12.70

of solar energy to electrical power (48–49) (see Photovoltaic cells; Solar energy). Its photoconductive and electroluminescent properties have been studied (44) and applied (49–50), not only to photocells, but also in a wide variety of phosphors, light amplifiers, radiation detectors, thin film transistors and diodes, electron beam-pumped lasers, and household smoke detectors.

Cadmium Tungstate. Cadmium tungstate [7790-85-4] forms white or yellowish fluorescent monoclinic crystals, insoluble in water and dilute acids (a = 0.5029 nm, b = 0.5859 nm, c = 0.5074 nm; d = 8.033 g/cm^3).

It is used in x-ray screens, phosphors, and as a catalyst in organic reactions (51).

Organic Compounds

A wide variety of organocadmium compounds have been synthesized since Wanklyn first isolated diethylcadmium in 1856 but, until recently, they have been of little industrial importance (see Organometallics). Alkyl and aryl cadmiums are now employed as polymerization catalysts (52), and cadmium salts of organic acids are used as heat and light stabilizers in plastics (see Heat stabilizers; Uv absorbers).

The dialkyls are prepared by the reaction of anhydrous cadmium halides with Grignard reagents followed by distillation; however, only dimethyl cadmium is stable for extended periods (diethyl cadmium is only moderately stable). The others are, in general, decomposed at room temperature and by light (the thermal stability of organocadmium compounds is lower than that of organozinc or organomercury compounds). They react with air and water and, in some cases ignite (52). For manufacturing purposes the more unstable cadmium dialkyls are generated *in-situ*, and used without separation, eg, in the conversion of acid chlorides to specialty ketones:

$$2\ RCOCl + CdR'_2 \longrightarrow 2\ RCOR' + CdCl_2$$

Physical properties of some dialkyl cadmium compounds are given in Table 4 (2,53–54).

Diethylcadmium is a polymerization catalyst for vinyl chloride (55), vinyl acetate, and methyl methacrylate (52). With admixture of TiCl$_4$ it catalyzes the polymerization of polyethylene and a highly crystalline polypropylene suitable for filaments, textiles, glues, and coatings (52). A catalyst containing more than 50% TiCl$_4$ is used for the

Table 4. Properties of Dialkyl Cadmium Compounds

Compound	CAS Registry No.	Formula	mp, °C	bp, °C (kPa)[a]	Density g/cm^3
dimethylcadmium[b]	[506-82-1]	(CH$_3$)$_2$Cd	−4.5	105.5 (101.3)	1.9846
diethylcadmium	[592-02-9]	(C$_2$H$_5$)$_2$Cd	−21	64 (2.6)	1.6564
dipropylcadmium	[5905-48-6]	(C$_3$H$_7$)$_2$Cd	−83	84 (2.8)	1.4184
dibutylcadmium	[3431-67-2]	(C$_4$H$_9$)$_2$Cd	−48	103.5 (1.6)	1.3054
diisobutylcadmium	[3431-67-2]	(C$_4$H$_9$)$_2$Cd	−37	90.5 (2.6)	1.2674
diisoamylcadmium	[35061-27-9]	(C$_5$H$_{11}$)$_2$Cd	−115	121.5 (2.0)	1.2184

[a] To convert kPa to mm Hg, multiply by 7.5.
[b] $\Delta H°_{f,\ 298}$ = 63.6 kJ/mol (15.2 kcal/mol), $\Delta G°_{f,\ 298}$ = 139.3 kJ/mol (33.3 kcal/mol), and $S°_{298}$ = 201.88 J/(mol·K) [48.25 cal/(mol·K)].

polymerization of dienes. Diethyl cadmium is used as an intermediate ethylating agent in the production of tetraethyllead by one company (see Lead compounds).

Cadmium alkyl halides (RCdX, $R_2Cd \cdot CdX_2$) have been prepared but have no industrial application. Cadmium allyls are made by Grignard reactions and are reactive toward many organic and inorganic compounds (56). There are, however, no present commercial applications for these compounds.

Cadmium aryl compounds are prepared by Grignard synthesis in THF or by direct metal substitution reactions. They are less important than the corresponding mercury compounds. Diphenyl cadmium [2674-04-6] (mp = 174°C) is generated *in situ* by the reaction: $(C_6H_5)_2Hg + Cd = (C_6H_5)_2Cd + Hg$ and has also been used as a polymerization catalyst.

Cadmium Acetate. Cadmium forms acetates $Cd(CH_3.COO)_2 \cdot nH_2O$ (n = 1–3); anhydrous cadmium acetate [1543-90-8], d = 2.34 g/cm^3, melts at 256°C (3). The dihydrate is obtained by dissolving cadmium metal or oxide in acetic acid followed by crystallization. Calcination can be controlled to yield cadmium acetate monohydrate [543-90-8] and the anhydrous acetate. The acetates are very soluble in water. Aqueous complexes of Cd^{2+} with up to four acetate ligands are known.

Cadmium acetate is the starting material for cadmium halides and is a colorant in glass, ceramics, and textiles (see Colorants for ceramics). In 5.45-kg lots, reagent-grade cadmium acetate dihydrate [5743-04-4] sold for $55.90/kg in 1977.

Organocadmium Soaps. Cadmium salts of organic acids find widespread use as heat and light stabilizers in plastics such as poly(vinyl chloride). Temperatures during the plastic molding operation reach 150°C for flexible products and 230°C for rigid compositions (little plasticizer). Without stabilizers, PVC loses HCl at 95°C resulting in discoloration (yellowing). Cadmium stabilizers only prevent early discoloration and are generally used in combination with barium organic soaps which help in preventing long-term yellowing (57). The Ba–Cd stabilizers at $0.65–1.30/kg combine good performance with low cost (58). Cadmium soaps are marketed as solid and liquid. The former include the cadmium salts of lauric [2605-44-9], stearic [2223-93-0], myristic, and palmitic acids. They are produced by the addition of the sodium salt of the preferred organic acid, to a solution of cadmium chloride. The cadmium salt precipitates from solution and is filtered, washed, and dried.

Most stabilizers are used in liquid form including cadmium octoate [2191-10-8] phenolate [18991-05-4], decanoate [2847-16-7], naphthenate, and benzoate [3026-22-0]. They are prepared from a suspension of cadmium oxide in a mixture of the organic acid and an inert organic solvent. An acid–base reaction slowly takes place and water formed during the reaction is driven off by heating. A clear solution of the cadmium soap in the organic medium is used in plastics manufacture. The liquid stabilizers impart better physical characteristics to PVC and are more economical. The United States consumption of Ba–Cd stabilizers between 1969 and 1976 is shown below (47):

Year	Metric tons
1969	16,300
1972	18,800
1974	22,300
1976	22,000

The use of Ba–Cd stabilizers in plastics that come in contact with food is not permitted by the FDA because of the high toxicity of cadmium. The FDA is considering an extension of this ban to other food-oriented plastic products and increased substitution by calcium–zinc stabilizers may be anticipated.

BIBLIOGRAPHY

"Cadmium Compounds" in *ECT* 1st ed., Vol. 2, pp. 732–738, by Gerald U. Greene, Fenn College; "Cadmium Compounds" in *ECT* 2nd ed., Vol. 3, pp. 899–911, by Gerald U. Greene, New Mexico Institute of Mining and Technology.

1. Bureau of Mines, *Mineral Facts and Problems,* U.S. Dept. of the Interior, 1970, p. 516.
2. D. D. Wagman and co-workers, *Selected Values of Chemical Thermodynamic Properties,* National Bureau of Standards Technical Note No. 270-3, Washington, D.C., 1968, p. 248.
3. B. J. Aylett in J. C. Bailar and A. F. Trotman-Dickenson, eds., "Group IIB" in *Comprehensive Inorganic Chemistry,* Pergamon Press, Oxford, 1973, pp. 187, 190, and 258–272.
4. K. E. Almin, *Acta-Chem. Scand.* **2,** 400 (1948).
5. F. D. Rossini and co-workers, *Selected Values of Chemical Thermodynamic Properties,* National Bureau of Standards Circular No. 500, Washington, D.C., 1952.
6. H. M. Cyr "Cadmium" in M. C. Sneed and R. C. Brasted eds., *Comprehensive Inorganic Chemistry,* Vol. IV, Van Nostrand, Princeton, N. J., 1955, p. 71.
7. R. W. G. Wyckoff, *Crystal Structures,* 2nd ed., Vols. I–V, Interscience Publishers, Inc., a division of John Wiley & Sons, Inc. New York, New York, 1963–1965.
8. D. M. Chizhikov, *Cadmium,* trans. D. E. Hayler, Pergamon Press, Oxford, 1966, pp. 10–48, 61, 63, and 68–70.
9. H. M. Haendler and W. S. Bernard, *J. Am. Chem. Soc.* **73,** 5218 (1951).
10. P. Goldfinger and M. Jeunehomme, *Trans. Faraday Soc.* **59,** 2851 (1963).
11. O. Kubaschewski and E. L. Evans, *Metallurgical Thermochemistry,* Academic Press, New York, 1951, p. 268.
12. Versar Inc., *Technical and Microeconomic Analysis of Cadmium and Its Compounds,* Final Report EPA Contract 68-01-2926, Task 1, Environmental Protection Administration, Washington, D.C., 1976, pp. 12 and 157.
13. W. Fulkerson and H. E. Goeller, eds., *Cadmium: The Dissipated Element,* Oak Ridge National Laboratory, Oak Ridge, Tenn., 1973.
14. National Environmental Research Center, *Scientific and Technical Assessment Report on Cadmium,* U.S. N.T.I.S., Springfield, Va., 1975, Sect. 4.2.
15. W. Slavin, *Atomic Absorption Spectroscopy,* Wiley-Interscience, New York, 1968, pp. 59–61, 74, 86, 201.
16. L. Friberg and co-workers, *Cadmium in the Environment,* 2nd ed., CRC Press, Cleveland, Ohio, 1974, Chapt. 2.
17. W. W. Scott in N. H. Furman, ed., *Standard Methods of Chemical Analysis,* Vol. I, 5th ed, Van Nostrand, Princeton, N.J., 1945, pp. 197–204.
18. ASTM Standards, *Part 12—Chemical Analysis of Metals and Metal Bearing Ores,* Std. E40-58, 1974, pp. 192–193.
19. *Ibid.,* Std. E56-63, p. 295.
20. H. A. Schroeder and J. J. Balassa, *J. Chronic Dis.* **14,** 236 (1961); The *New York Times,* June 15, 1978.
21. Ref. 16, pp. 88 and 204.
22. H. M. Barrett and B. Y. Card, *J. Ind. Hyg. Toxicol.* **29,** 286 (1947).
23. Ref. 8, p. 17.
24. AGGIH, *TLV's for Chemical Substances and Physical Agents in the Workroom Environment with Intended Changes for 1973,* American Conference of Governmental and Industrial Hygienists, Cincinnati, Ohio, 1973.
25. OSHA, *General Industry Standards and Interpretation,* Vol 1, U.S. Dept. of Labor, Washington, D.C., 1975, p. 642.6.
26. N. Molinelli, *Am. Met. Mark.* **84**(54), 10 (3/18/77).
27. J. P. Smith and co-workers, *J. Pathol. Bacteriol.* **80,** 287 (1960).
28. L. Friberg and co-workers, Stockholm, *Cadmium in the Environment II,* U.S. Environmental Protection Agency Report R2-73-190, Washington, D.C., 1973.
29. Ref. 12, p. 56.

30. F. Haflich, *Am. Met. Mark.* **84**(54), 13 (3/18/77).
31. E. Suriani to author, personal communication, NPDES Branch, Region 2, U.S. Environmental Protection Administration.
32. T. Furukawa, *Am. Met. Mark.* **84**(54), 15 (3/18/77).
33. W. Zdanowicz and L. Zdanowicz, *Ann. Rev. Mater. Sci.* **5**, 301 (1975).
34. L. G. Sillén and A. E. Martell, eds., *Stability Constants of Metal-Ion Complexes,* Chemical Society Special Publication No. 17, Suppl. 1, London, 1971.
35. M. Cyris, P. Mueller, and J. Teltow, *J. Phys. (Paris) Colloq.* **9**, 63 (1973).
36. M. Bancie-Grillot and E. Grillot, *Proceedings of the 10th Rare Earth Research Conference, Vol. 2,* U.S. N.T.I.S. CONF-730402-P2, Springfield, Va., 1973, p 950.
37. R. D. Armstrong, K. Edmonson, and G. D. West, *Electrochemistry* **4**, 18 (1974).
38. S. U. Falk and A. J. Salkind, *Alkaline Storage Batteries,* John Wiley & Sons, Inc., New York, 1969.
39. C. F. Baker, *Eng. Min. J.* **178**(3), 180 (1977).
40. Ref. 38, p. 132.
41. Ref. 8, p. 11.
42. L. E. Antonelli, *Am. Met. Mark.* **84**(43), 22 (3/3/77).
43. L. Hajdu and J. Zahoran, *Acta Tech. (Budapest)* **73**(1–2), 117 (1972).
44. M. Aven and J. S. Prener, eds., *Physics and Chemistry of II–VI Compounds,* North-Holland Publishing Co., Amsterdam, 1967, pp. 76, 104, and 106.
45. C. L. Gupta and R. C. Tyagi, *Def. Sci. J.* **24**(2), 71 (1974).
46. A. J. Strauss, *Proceedings of the International Symposium on Cadmium Telluride as Material for Gamma-Ray Detectors,* p. I-1 (1972).
47. *Mod. Plast.* **48**, 9 (1971); **49**, 9 (1972); **50**, 9 (1973); **51**, 9 (1974); **52**, 9 (1975); and **53**, 9 (1976).
48. R. J. Mytton, *Sol. Energy* **16**(1), 33 (1974).
49. A. G. Stanley, *Appl. Solid State Sci.* **5**, 251 (1975).
50. E. S. Turner, *Cadmium Sulfide—Preparation, Optical and Electrical Properties and Applications,* Bibliography No. 83, Bell Labs, Inc., November 1965.
51. A. Karl, *Compt. Rend.* **196**, 1403 (1933).
52. J. H. Harwood, *Industrial Applications of the Organometallic Compounds,* Reinhold Publishing Co., New York, 1963, p. 59.
53. N. Hagihara, M. Kumada, and R. Okawara, *Handbook of Organometallic Compounds,* Benjamin Publishing Co., New York, 1968, pp. 777–803.
54. Ref. 3, p. 272.
55. J. Furukawa, *J. Polym. Sci.* **28**, pp. 234, 450 (1958).
56. G. Curtois and L. Miginiac, *J. Organometal. Chem.* **69**(1), (1974).
57. S. Donald Brilliant, *Mod. Plast. Encycl.* **49**(10A), 403 (1972).
58. Ref. 12, p 81.

General References

Bureau of Mines, *Mineral Facts and Problems, Bulletin 667,* U.S. Dept. of the Interior, Washington, D.C., 1975, pp. 195, 914.
F. Wagenknecht and R. Juza in G. Braver, ed., *Handbook of Preparative Inorganic Chemistry,* 2nd ed., Vol. 2, Academic Press, New York, 1965, pp. 1092–1108.

P. D. PARKER
AMAX Base Metals Research & Development, Inc.

CAFFEIN. See Alkaloids.

CAFFEINE. See Alkaloids.

CALCIUM AND CALCIUM ALLOYS

Calcium [7440-70-2], Ca, a member of Group II of the periodic table (between magnesium and strontium) is classified, together with barium and strontium, as an alkaline earth metal and is the lightest of the three. Calcium metal does not occur free in nature; however, in the form of numerous compounds, it is the fifth most abundant element constituting 3.63% of the earth's crust.

Calcium in the form of its oxide has been known since prehistoric times. The Romans used large quantities of calcium oxide or lime as mortar in construction. Indeed, the word calcium is derived from calx, the Latin word for lime. Calcium compounds are very stable, and it was not until 1808 that Davy produced elemental calcium as a mercury amalgam by electrolysis of calcium chloride in the presence of a mercury cathode. However, he and others were only marginally successful in their attempts to isolate the pure metal by distilling the mercury.

Matthiessen produced calcium metal in 1855 by electrolysis of a mixture of calcium, strontium, and ammonium chlorides, but his product was highly contaminated with chlorides (1). This process was slowly improved, and by 1904 Rathenau obtained fairly large quantities of calcium by the electrolysis of molten calcium chloride held at a temperature above the melting point of the salt but below the melting point of calcium metal (2–3). An iron cathode just touched the surface of the bath and was raised slowly as the relatively chloride-free calcium solidified on its end. This process became the basis for commercial production of calcium metal until World War II.

Prior to 1939 calcium was manufactured exclusively in France and Germany. However, with the outbreak of war, an electrolytic calcium plant was constructed in the United States at Sault Ste. Marie, Michigan, by the Electro Metallurgical Corporation. Large amounts of calcium were required as the reducing agent for uranium production. In addition, the U.S. Army Signal Corps used calcium to produce calcium hydride, which could easily be transported to remote areas and used as a source of hydrogen for meteorological balloons. To satisfy these increased requirements, the aluminothermal reduction process was developed whereby calcium oxide is reduced by aluminum in vacuum at high temperature.

Calcium is mainly used as a reducing agent for many reactive, less common metals; to remove bismuth from lead; as a desulfurizer and deoxidizer for ferrous metals and alloys; and as an alloying agent for aluminum, silicon, and lead. Smaller amounts are used as a dehydrating agent for organic solvents, and as a purifying agent for removal of nitrogen and other impurities from argon and other rare gases.

Physical Properties

Pure calcium is a bright silvery-white metal, although under normal atmospheric conditions freshly exposed surfaces of calcium quickly become covered with an oxide layer. The metal is extremely soft and ductile with a hardness between that of sodium and aluminum. It can be readily rolled into plate or extruded into wire. Calcium can also be melted under an inert atmosphere and cast into ingots or billets. It can be work-hardened to some degree by mechanical processing. Although its density is low, calcium's usefulness as a structural material is limited by low tensile strength and high chemical reactivity (4).

Calcium has a face-centered cubic crystal structure (a = 0.5582 nm) at room temperature, but transforms into a body-centered cubic ($a = 0.4477$ nm) form at 428 ± 2°C (5). Earlier work had suggested additional allotropic forms but these were only associated with impurities (6–7). Some of the more important physical properties of calcium are given in Table 1. (For additional physical properties, see refs. 11–16.) Measurements of the physical properties of calcium are usually somewhat uncertain owing to the effects which small levels of impurities can exert.

Chemical Properties

Calcium has a valence electron configuration of $4s^2$ and characteristically forms divalent compounds. It is very reactive and reacts vigorously with water, liberating hydrogen and forming calcium hydroxide, $Ca(OH)_2$. Calcium does not readily oxidize in dry air at room temperature, but is quickly oxidized in moist air or in dry oxygen at about 300°C. The oxide layer is nonprotective and complete oxidation of a massive piece of calcium will eventually occur. Calcium reacts with fluorine at room temperature and with the other halogens at 400°C. When heated to 900°C, calcium reacts with nitrogen to form calcium nitride, Ca_3N_2. The metal becomes incandescent when heated to 400–500°C in an atmosphere of hydrogen with the formation of calcium hydride, CaH_2, which reacts with water to give hydrogen:

$$CaH_2 + 2\ H_2O \rightarrow Ca(OH)_2 + 2\ H_2$$

Thus the hydride is a very efficient carrier of hydrogen. Upon heating, calcium reacts with boron, sulfur, carbon, and phosphorus to form the corresponding binary compounds, and with carbon dioxide to form CaC_2, calcium carbide, and CaO.

Table 1. Physical Properties of Calcium[a]

Property	Value
atomic weight ($^{12}C = 12.000$)	40.08
electron configuration	2-8-8-2
stable isotopes	40 42 43 44 46 48
natural abundance, %	96.947 0.646 0.135 2.083 0.186 0.18
specific gravity at 20°C, kg/m³	1.55×10^3
melting point, °C	839 ± 2
boiling point, °C	1484
heat of fusion, kJ/mol[b]	9.2
heat of vaporization, kJ/mol[b]	161.5
heat of combustion kJ/mol[b]	634.3
vapor pressure data	
pressure, kPa (mm Hg)	0.133 (1) 1.33 (10) 13.3 (100) 53.3 (400) 101.3 (760)
temperature, °C	800 970 1200 1390 1484
specific heat at 25°C, J/g·K[b]	0.653
coefficient of thermal expansion, 0–400°C, m/(m·K)	22.3×10^{-6}
electrical resistivity at 0°C, $\mu\Omega$·cm	3.91
electron work function, eV	2.24
tensile strength (annealed), MPa (psi)	48 (6960)
yield strength (annealed), MPa (psi)	13.7 (1990)
modulus of elasticity, GPa (psi)	22.1–26.2 ($3.2–3.8 \times 10^6$)
hardness (as cast)	
HB	16–18
HRC B	36–40

[a] Refs. 8–10.
[b] To convert J to cal, divide by 4.184.

Calcium is an excellent reducing agent and is widely used for this purpose. At elevated temperatures it reacts with the oxides or halides of almost all metallic elements to form the corresponding metal. It also combines with many metals forming a wide range of alloys and intermetallic compounds. Among the phase systems that have been better characterized are those with Ag, Al, Au, Bi, Cd, Co, Cu, Hg, Li, Na, Ni, Pb, Sb, Si, Sn, Tl, Zn, and the other Group IIA metals (17).

Commercially produced calcium metal is analyzed for metallic impurities by emission spectroscopy. Carbon content is determined by combustion, whereas nitrogen is measured by Kjeldahl determination.

Manufacture

Electrolysis. Although the electrolytic method has now been completely replaced by the aluminothermal process, it is still of interest since this method accounted for virtually all calcium production from 1904 to 1940 (18). The basic cell is a graphite-lined steel vessel filled with partially molten calcium chloride. Temperature of the salt bath is maintained at 780–800°C by the direct electrolyzing current. This temperature is slightly above the melting point of $CaCl_2$ but below the melting point of calcium metal. Calcium metal forms as a solid deposit at the end of the water-cooled anode, which is raised slowly keeping the end of the growing deposit in contact with the salt bath. A stick of calcium metal is produced (referred to as a carrot because of its characteristic shape). The carrot may contain 15–25% entrapped salts and has to be remelted to reduce impurity levels. However, appreciable amounts of chlorine and nitrogen remain in the purified metal. The process is further complicated by the requirement of fairly precise temperature and current control. Power consumption of a cell of this type is about 33–55 kW/kg of metal for a current efficiency of about 60%.

Aluminothermal Method. At present, all calcium metal is produced by high-temperature vacuum reduction of calcium oxide with aluminum. The method was tried on a laboratory scale in 1922 (19–20), but only low yields of calcium were obtained. The process was improved and commercialized by the New England Lime Co. in Canaan, Connecticut (now a division of Pfizer Inc.) (21). Depending on initial stoichiometry, the following reactions appear to occur (22–27):

$$6 \ CaO + 2 \ Al \rightarrow 3CaO \cdot Al_2O_3 + 3 \ Ca \ (gas)$$

$$33 \ CaO + 14 \ Al \rightarrow 12CaO \cdot 7Al_2O_3 + 21 \ Ca \ (gas)$$

$$4 \ CaO + 2 \ Al \rightarrow CaO \cdot Al_2O_3 + 3 \ Ca \ (gas)$$

These reactions are thermodynamically unfavorable at temperatures below 2000°C. However, in the range of 1000–1200°C a small but finite equilibrium pressure of calcium vapor is established at the reaction site. The calcium vapor is then transferred with a vacuum pump to a cooled region of the reactor where condensation takes place. This shifts the equilibrium at the reaction site and allows more calcium vapor to be formed.

A typical flow sheet for the process is given in Figure 1. High calcium limestone, $CaCO_3$, is quarried and calcined to form calcium oxide. The calcium oxide is ground to a small particle size and dry-blended with the desired amount of finely divided aluminum. This mixture is then compacted into briquettes to ensure good contact of reactants. The briquettes are placed in horizontal metal tubes (retorts) made of

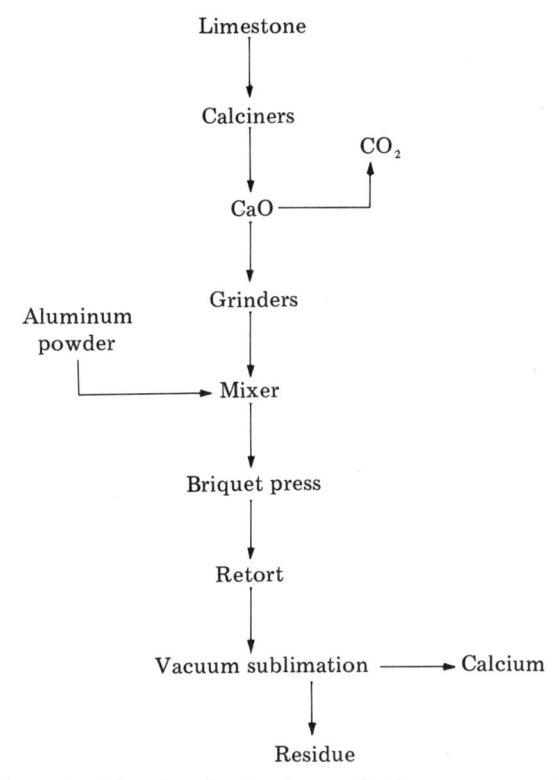

```
                    Limestone
                        │
                        ▼
                    Calciners
                                    CO₂
                        │            ▲
                        ▼            │
                       CaO ──────────┘
                        │
                        ▼
                     Grinders
   Aluminum             │
   powder               ▼
       └──────────▶   Mixer
                        │
                        ▼
                   Briquet press
                        │
                        ▼
                      Retort
                        │
                        ▼
              Vacuum sublimation ──▶ Calcium
                        │
                        ▼
                     Residue
```

Figure 1. Flow sheet for aluminum reduction process.

heat-resistant steel and heated in a furnace to 1100–1200°C. The open ends of the retorts protrude from the furnace and are cooled by water jackets to condense the calcium vapor. The retorts are then sealed and evacuated to a pressure of less than 13 Pa (0.1 mm Hg). After the reaction has been allowed to proceed for about 24 h the vacuum is broken with argon, and the condensed blocks of ca 99% pure calcium metal (known as crowns) and calcium aluminate residue are removed. Large amounts of energy are required by this method, partially because of the high temperatures of the process itself and partially because of the energy-intensive raw materials employed (calcined CaO and electrolytically produced aluminum).

The calcium crowns can be sold as such for certain applications. However, further processing may be required, and the crowns can be reduced in size to pieces of about 25 cm or nodules of about 3 mm. They can also be melted under a protective atmosphere of argon and cast into billets or ingots. Calcium wire can be made by extrusion, and calcium turnings are produced as lathe cuttings from cast billets.

Redistillation. For certain applications (especially those involving reduction of other metal compounds), a better than 99% purity is required. This can be achieved by redistillation, a process developed at the Ames, Iowa, Laboratory of the DOE (formerly the Atomic Energy Commission) (28). Crude calcium is placed in the bottom of a large vertical retort made of heat-resistant steel equipped with a water-cooled condenser at the top. The retort is sealed and evacuated to a pressure of less than 6.6 Pa (0.05 mm Hg) while the bottom is heated to 900–925°C. Under these conditions calcium quickly distills to the condensing section leaving behind the bulk of the less volatile impurities. Similar equipment is used by Pfizer Inc. for the commercial production of high-purity, redistilled calcium. Subsequent processing must take place under exclusion of moisture to avoid oxidation.

Redistillation does not greatly reduce the impurity level of volatile materials such as magnesium, and several investigators have studied the use of fractional distillation for its removal (6,28–31). Volatile alkali metals can be separated from calcium by passing the vapors over refractory oxides such as TiO_2, ZrO_2, or Cr_2O_3 to form nonvolatile Na_2O and K_2O (32). More recent purification techniques employ reactive distillation (33), growth of crystals from the melt (34), and combined crystal growth and distillation techniques (35).

Shipment

Because of its extreme chemical reactivity, calcium metal must be carefully packaged for shipment and storage. The United States producer packages the metal in sealed argon-filled containers. Calcium is classed as a flammable solid and is nonmailable. Sealed quantities of calcium should be stored in a dry, well-ventilated area so as to remove any hydrogen formed by reaction with moisture.

Specifications and Standards

The purity of commercial grade calcium depends to a large extent on the purity of the calcium oxide used in its production. Impurities such as magnesium oxide, or other alkaline earth or alkali metal compounds will be reduced together with the calcium oxide and can contaminate the calcium. In addition, small amounts of aluminum reducing agent may distill with the calcium vapor, and small amounts of calcium nitride may be produced by reaction with atmospheric nitrogen.

Redistilled-grade calcium still contains magnesium, which codistills with the calcium. For most applications the magnesium content is not detrimental. For the exceptions, however, the United States producer supplies a grade known as center cut, consisting of the central portions of redistilled crowns. It is very low in magnesium, which tends to concentrate in the first metal fraction condensed, and less-volatile impurities, which tend to concentrate in the last fraction. The typical compositions of United States commercial- and redistilled-grade calcium are given in Table 2.

Table 2. Typical Analysis of Commercial- and Redistilled-Grade Calcium

Component	Commercial, %	Redistilled, %
Ca and Mg	99.5	99.9
Mg	0.70	0.70
Al	0.14	<0.0014
N	0.04	<0.0070
Fe	<0.034	<0.0017
Mn	<0.0088	<0.0016
Co		<0.0002
Li		<0.0001
Be		<0.0001
Cr		<0.0002
B		<0.0001

Economic Aspects

Calcium metal is produced in the United States by Pfizer Inc., Canaan, Connecticut, and in Canada by Chromasco Corporation, Ltd., Montreal, Quebec (plant at Haley, Ontario). In France it is produced by Planet Wattohm S.A., a subsidiary of Compagnie de Mokta. It is also produced in the Union of Soviet Socialist Republics, but no details are available. Both Pfizer and Chromasco supply the commercial and redistilled grades in a variety of sizes and forms (full and broken crowns, nodules, billets, ingots, and turnings). In addition, Pfizer supplies an 80% Ca–20% Mg alloy and a steel-clad calcium wire for use in deoxidation of steel and other metals.

United States imports of calcium metal fluctuate greatly. Virtually all imports during the past ten years have been from Canada, although the 1974 imports of about 50 metric tons include approximately 11 t from France and about 1.8 t from the Union of Soviet Socialist Republics (36). Annual worldwide production is probably on the order of 1000–2000 metric tons.

The price of calcium metal has increased from an average of $2.44/kg since 1974 because of large increases in fuel and aluminum costs. As of January 1977, prices quoted by Pfizer Inc. for commercial grade uncast calcium ranged about $3.30–8.95/kg, depending on size and quantity. Prices for cast or cast and machined forms ranged about $5.65–19.70/kg. Redistilled-grade prices ranged about $7.05–17.45/kg. Canadian prices are similar.

Calcium-containing alloys are produced in the United States by Union Carbide Corp., Niagara Falls, New York, and by Foote Mineral Company, Exton, Pennsylvania, for metallurgical uses. Union Carbide produces Ca–Si alloy (28–32% Ca, 60–65% Si, ca 5% Fe) and two proprietary Ca–Si alloys containing barium and barium and aluminum. Prices quoted for 18 t lots or more are $1.12/kg of alloy for Ca–Si and Ca–Si–Ba alloys, and $1.36/kg of alloy for the Ca–Si–Ba–Al alloy. Foote Mineral produces a number of complex proprietary alloys containing 1.5–7% calcium. Considerable amounts of Ca–Si alloy are imported into the United States from a number of foreign producers (37).

Health and Safety

Calcium metal and most calcium compounds are nontoxic. In massive pieces the metal does not spontaneously burn in air. Calcium can be touched with dry bare hands without harm. Care must be taken, however, to avoid contact with water owing to the exothermic liberation of hydrogen and the resulting explosion hazard. Calcium must always be kept dry and preferably sealed in the shipping containers.

Calcium Alloys

Calcium alloys are produced commercially by various techniques. For example, direct alloying of the pure metals is used in the production of 80% calcium–magnesium alloys. Others are formed directly during chemical reduction of one or more of the components. An example of this technique is the electrolysis of a mixture of calcium chloride, calcium fluoride, and aluminum fluoride to produce a 50% calcium–aluminum alloy (38). Similar methods can be used for certain calcium–lithium, calcium–magnesium, and calcium–beryllium alloys (39). An alternative method for preparation of calcium–aluminum alloys is reduction of calcium oxide with excess aluminum at high temperatures in an inert atmosphere (40–42).

Lead alloys containing small amounts of calcium are formed by adding sodium metal to a molten lead bath in contact with a fused mixture of NaCl and $CaCl_2$. Under these conditions sodium will reduce calcium chloride with the formation of the stable intermetallic compound trilead calcium [12049-58-0], Pb_3Ca. Alternatively, calcium carbide can be added to the lead bath to form Pb_3Ca and carbon by thermal decomposition of the CaC_2. A flux of NaCl–$CaCl_2$ can then be added to remove the carbon and any residual CaO.

Alloys of calcium with silicon are used in ferrous metallurgy and are generally produced in an electric furnace from CaO (or CaC_2), SiO_2, and a carbonaceous reducing agent (43). The resulting alloy, calcium disilicide [12013-56-8], is nominally of composition $CaSi_2$ and has the following typical wt % analysis: 30–33% Ca, 60–65% Si, 1.5–3% Fe (44). Proprietary Ca–Si alloys containing other elements such as Ba, Al, Ti, or Mn are sometimes produced by a combination of carbothermic ore reduction followed by direct alloying. In general, the chemical reactivity of calcium is greatly reduced when it is present in an alloyed state.

Uses

Calcium metal is an excellent reducing agent for production of the less-common metals because of the large free energy of formation of its oxides and halides. The following metals have been prepared by the reduction of their oxides or fluorides with calcium: hafnium (45), plutonium (46), scandium (47), thorium (48), tungsten (49), uranium (50–51), vanadium (52), yttrium (53), zirconium (45,54), and most of the rare earth metals (55).

For some processes, calcium metal first reacts with hydrogen to form calcium hydride which is then used as the actual reducing agent. Oxides of uranium, vanadium, titanium, and niobium have been reduced in this way. Recently this technique has been used to reduce samarium oxide in the presence of cobalt powder to directly form the intermetallic compound samarium pentacobalt, $SmCo_5$, a material which is finding increased use in the manufacture of extremely powerful permanent magnets (56) (see Magnetic materials). Additional amounts of calcium metal are converted into calcium hydride for use as a portable source of hydrogen gas (see Hydrides).

Calcium metal is also used in strip form as the anode material in thermal batteries (see Batteries), which are used as the power source in artillery fuses (57).

Metallurgical industries use large amounts of calcium and calcium-containing alloys for a variety of purposes. In ferrous metallurgy, calcium and certain of its alloys are used extensively as addition agents to deoxidize, desulfurize, and degas steel and cast iron; to control the type and distribution of nonmetallic inclusions in steel and to promote a uniform microstructure in gray iron (58–60). Because of its extreme reactivity, addition of pure calcium to molten steel can be difficult. Using alloys such as calcium–silicon may overcome this problem but introduces unwanted alloying elements into the iron or steel. In a new and promising technique a steel-clad calcium wire is quickly fed into molten steel (61).

Calcium is used in the lead industry as a refining agent and as an alloying ingredient. When added to molten lead during refining, calcium metal removes bismuth impurities through formation of the insoluble intermetallic compound Bi_2Ca_3 [*11056-25-0*] (62). Bismuth levels can be reduced to less than 0.05% by this method. As an alloying ingredient calcium is used at levels of 0.04–0.065% to increase the strength, creep resistance, corrosion resistance, and formability of lead (63) (see Lead). Because of their corrosion resistance to sulfuric acid (64), calcium alloys are being used as grids in lead–acid storage batteries and have enabled development of the sealed maintenance-free auto battery (see Batteries).

The mechanical and electrical properties of aluminum alloys are improved through additions of small amounts of calcium (65). Such calcium-containing alloys are used for die casting of automobile trim. The eutectic alloy $Al-Al_4Ca$ (7.6 wt % Ca) has been found to exhibit superplastic behavior (66). Other metallurgical uses of calcium include deoxidation of copper, magnesium, and tantalum (67). In the case of magnesium-base alloys, calcium also improves corrosion resistance and acts as a grain refiner.

A number of relatively minor uses for calcium depend upon its high chemical reactivity. These include the dehydration of certain organic solvents, the desulfurization of petroleum, and the removal of nitrogen in the purification of argon gas. In addition, a number of calcium alloys have been patented for a variety of uses but apparently never commercialized. Among these are Ca–Li–Na and Ca–Li–Ba alloys for use as water-reactive solid fuels (68–69), a Ca–Li alloy for catalysis of the polymerization of conjugated diolefins (70), a Ca–Ag alloy for catalyzing the oxidation of ethylene to ethylene oxide (qv) (71), and Ca–Ge alloys for use as rectifier materials (72). Calcium–silicon alloys have been proposed for removing heavy metals from waste waters and brine solutions. (73).

It has been noted that calcium's extremely low density and low electrical resistivity would make it suitable as a building material and electrical conductor if suitably protected from corrosion (74). Whether or not these applications develop, the amount and variety of calcium metal usage will undoubtably increase in the future.

BIBLIOGRAPHY

"Calcium and Calcium Alloys" treated in *ECT* 1st ed. under "Alkaline Earth Metals and Alkaline Earth Metal Alloys," Vol. 1, pp. 458–463, by C. L. Mantell, Consulting Chemical Engineer; "Calcium and Calcium Alloys" in *ECT* 2nd ed., pp. 917–927, by O. N. Carlson and J. A. Haffling, Ames Laboratory, United States Atomic Energy Commission.

1. J. W. Mellor, *Comprehensive Treatise on Inorganic and Theoretical Chemistry,* Vol. 3, Longmans, Green & Co., Inc., New York, 1923, pp. 619–631.
2. W. Rathenau, *Z. Elektrochem.* **10,** 508 (1904).
3. W. Rathenau, *Proc. Am. Philos. Soc.* **43,** 381 (1904).
4. W. Hodge, R. I. Jafee, and B. W. Gonser, *RAND Corp. Report R-123,* Battelle Memorial Institute, Santa Monica, Calif., Jan. 1, 1949.
5. J. Katerberg and co-workers, *J. Phys. F* **5,** L74 (1975).
6. J. F. Smith, O. N. Carlson, and R. W. Vest, *J. Electrochem. Soc.* **103,** 409 (1956).
7. J. F. Smith and B. T. Bernstein, *J. Electrochem. Soc.* **106,** 448 (1959).
8. R. C. Weast, ed., *Handbook of Chemistry and Physics,* 57th ed., CRC Press, Cleveland, Ohio, 1976, pp. B1, B3, B12, B276, D62, D165, D211, E81, and F170.
9. F. Emley, "Calcium" in D. M. Considine, ed., *Chemical and Process Technology Encyclopedia,* McGraw-Hill, New York, 1974, p. 192.
10. C. L. Mantell, "The Alkaline Earth Metals—Calcium, Barium, and Strontium" in C. A. Hampel, ed., *Rare Metals Handbook,* 2nd ed., Reinhold Publishing Corp., London, Eng., 1961, pp. 20–21.

11. F. X. Kayser and S. D. Sonderquist, *J. Phys. Chem. Solids* **28**, 2343 (1967).
12. G. de Maria and V. Piacente, *J. Chem. Thermodyn.* **6**, 1 (1974).
13. K. L. Agarwal and J. O. Betterton, Jr., *J. Low Temp. Phys.* **17**, 509 (1974).
14. E. Schuermann, P. Fünders, and H. Litterscheidt, *Arch. Eisenhuettenwes.* **45**, 433 (1974).
15. J. G. Cook, M. J. Laubitz, and M. P. Van der Meer, *Can. J. Phys.* **53**, 486 (1975).
16. E. Schurman and R. Schmid, *Arch. Eisenhuettenwes.* **46**, 773 (1975).
17. M. Hansen, *Constitution of Binary Alloys*, McGraw-Hill Book Co., Inc., New York, 1958, pp. 11–13, 75–77, 190, 302–303, 394–414.
18. C. L. Mantell and C. Hardy, *Calcium Metallurgy and Technology*, Reinhold Publishing Corp., New York, 1945, pp. 21–34.
19. W. Biltz and G. Hoborst, *Z. Anorg. Chem.* **121**, 1 (1922).
20. W. Biltz and W. Wagner, *Z. Anorg. Chem.* **124**, 1 (1924).
21. C. C. Loomis, *Trans. Electrochem. Soc.* **89**, 207 (1946).
22. L. M. Pidgeon and J. T. N. Atkinson, *Can. Min. Met. Bull.* **41**, 14 (1948).
23. P. Vignial and J. L. Andrieux, *Compt. Rend.* **242**, 709 (1956).
24. V. V. Zhukovetskii, *Tr. Sev. Kavk. Gornometall. Inst.* (15), 210 (1957); *Chem. Abstr.* **54**, 8514h (1960).
25. V. A. Pazukhin and A. Ya. Fisher, *Sb. Nauchn. Tr. Mosk. Inst. Tsvetn. Met. Zolata Vses. Nauchno Inzhener. Tekh. Obshchestvo Tsvetn. Met.* (26), 172 (1957); *Chem. Abstr.* **54**, 17175e (1960).
26. J. D. Pedregal, *Electron Fis. Apl.* **9**, 461 (1966).
27. D. Goncalves de Oliviera and co-workers, *Met ABM (Ass. Brasil Metais)* **25**, 817 (1969); *Chem. Abstr.* **72**, 81739j (1970).
28. H. A. Wilhelm and O. N. Carlson, *J. Met.* **16**, 170 (1964).
29. I. I. Betcherman and L. M. Pidgeon, *Can. Min. Met. Bull.* **44**, 253 (1951).
30. E. Fujita, H. Yokomizo, and Y. Kurosaki, *J. Electrochem. Soc. Jpn.* **19**, 196 (1951).
31. W. J. McCreary, *J. Met.* **10**, 615 (1958).
32. U.S. Pat. 2,375,198 (May 8, 1945), P. P. Alexander (to Metal Hydrides, Inc.).
33. J. Evans and co-workers, *J. Less Common Met.* **30**, 83 (1973).
34. A. V. Vakhobov, V. G. Khudaiberdiev, and M. K. Nasyrova, *Izv. Akad. Nauk SSSR Met.* (4), 162 (1974).
35. A. V. Vakhobov, V. N. Vigdorovich, and V. G. Khudaiberdiev, *Izv. Vyssh. Uchebn Zavad. Tsvetn. Metall.* (4), 115 (1973); *Chem. Abstr.* **80**, 62397v (1974).
36. A. H. Reed, "Calcium and Calcium Compounds" in *Minerals Yearbook, 1974*, Vol. 1, U.S. Department of the Interior, Bureau of Mines, Washington, D.C., 1974, pp. 247–249.
37. C. O. Babcock, "Calcium and Calcium Compounds" in *Minerals Yearbook 1963*, Vol. 1, U.S. Department of the Interior, Bureau of Mines, Washington, D.C., 1964, pp. 341–346.
38. U.S. Pat. 2,829,092 (Apr. 1, 1958), J. L. Andrieux and E. Bonnier (to Société d'Electrochimie, d'Electrométallurgie et des Aciéries Electriques d'Ugine).
39. G. Boisde and co-workers, *Met. Corros. Ind.* **47**, 205 (1972).
40. U.S. Pat. 2,190,290 (Feb. 13, 1940), G. N. Kirsebom (to Calloy Ltd.).
41. U.S. Pat. 2,257,988 (Oct. 7, 1941), R. Suchy and H. Seliger (to Walther H. Duisberg).
42. U.S. Pat. 2,955,936 (Oct. 11, 1960), A. J. Deyrup (to E. I. du Pont de Nemours & Co., Inc.).
43. H. Walter, "Calcium–Silicon" in G. Volkert, ed., *Metallurgy of Ferroalloys*, Springer, Berlin, Ger., 1972, pp. 570–581.
44. C. L. Mantell, "Calcium" in C. A. Hampel, ed., *The Encyclopedia of the Chemical Elements*, Reinhold Book Corp., New York, 1968, p. 102.
45. O. N. Carlson, F. A. Schmidt, and H. A. Wilhelm, *J. Electrochem. Soc.* **104**, 51 (1957).
46. W. Z. Wade and T. Wolf, *J. Nucl. Sci. Technol.* **6**, 402 (1969).
47. F. H. Spedding and co-workers, *Trans. Met. Soc. AIME* **218**, 608 (1960).
48. H. A. Wilhelm, *The Metal Thorium*, American Society for Metals, Novelty, Ohio, 1958, pp. 78–103.
49. U.S. Pat. 2,763,542 (Sept. 18, 1956), C. H. Winter, Jr. (to E. I. du Pont de Nemours & Co., Inc.).
50. H. A. Wilhelm, *J. Chem. Ed.* **37**, 56 (1960).
51. T. Oki and J. Tanikawa, *Nippon Kinzoku Gakkaishi* **31**, 1048 (1967).
52. R. K. McKechnie and A. U. Seybolt, *J. Electrochem. Soc.* **97**, 311 (1950).
53. O. N. Carlson and co-workers, *J. Electrochem. Soc.* **107**, 540 (1960).
54. C. J. DeCroly, J. Gerard, and D. Tytgat, *Rev. Met. (Paris)* **56**, 143 (1959).
55. F. H. Spedding and A. H. Daane, *Met. Rev.* **5**(19), 297 (1960).
56. R. E. Cech, *J. Met.* **26**, 32 (1974).
57. R. P. Clarke and K. R. Grothaus, *J. Electrochem. Soc.* **118**, 1680 (1971).

58. F. J. Shortsleeve and D. C. Hilty, "Calcium in Iron and Steel" in *Boron, Calcium, Columbium, and Zirconium in Iron and Steel,* John Wiley & Sons, Inc., New York, 1957, pp. 61–101.
59. E. Rudielka, *Radex Rundsch.* (2), 483 (1965).
60. D. C. Hilty and J. W. Farrell, *Iron Steelmaker* **2,** 17 (1975).
61. E. J. Dunn, D. W. P. Lynch, and S. B. Gluck, *Electric Furnace Proceedings, St. Louis, 1976,* Vol. 34, American Institute of Mining Metallurgical and Petroleum Engineers, New York, 1977, pp. 248–253.
62. U.S. Pat. 1,428,041 (Sept. 11, 1922), W. Kroll.
63. M. V. Rose and J. A. Young, *5th International Lead Conference, Paris, Fr., Nov. 1974,* preprints, (1974).
64. A. A. Abdul Azim and K. M. El-Sobki, *Corros. Sci.* **12,** 371 (1972).
65. U.S. Pat. Reissue 15,407 (July 11, 1922), F. C. Frary (to Aluminum Company of America).
66. G. Piatti, G. Pellegrini, and R. Trippodo, *J. Mater. Sci.* **11,** 186 (1976).
67. J. H. Swisher, O. Fuchs, and W. M. Baldwin, *J. Met.* **23,** 34 (1971).
68. U.S. Pat. 2,978,304 (Apr. 4, 1961), R. B. Cox (to Aerojet-General Corp.).
69. U.S. Pat. 3,000,732 (Sept. 19, 1961), R. B. Cox (to Aerojet-General Corp.).
70. U.S. Pat. 2,908,672 (Oct. 13, 1959), H. R. Jackson (to E. I. du Pont de Nemours & Co., Inc.).
71. U.S. Pat. 2,562,858 (July 31, 1951), A. Cambron and F. L. W. McKim (to National Research Council, Ottawa).
72. U.S. Pat. 2,588,253 (Mar. 4, 1952), K. Lark-Horovitz and R. M. Whaley (to Purdue Research Foundation).
73. J. P. McKaveney, W. P. Fassinger, and D. A. Stivers, *Environ. Sci. Technol.* **6,** 1109 (1972).
74. C. Gentaz and F. Pruvost, *Metal Bulletin Monthly,* (1), 37 (1975).

CHARLES J. KUNESH
Pfizer Inc.

CALCIUM COMPOUNDS

Survey, 421
Calcium carbonate, 427
Calcium chloride, 432
Calcium sulfate, 437.

SURVEY

Calcium carbonate [*1317-65-3*] in the forms of limestone and marble has been widely used since antiquity for buildings and monuments and as the precursor of lime [*1305-78-8*], CaO. Among the first industries of the American colonies was lime-burning, accomplished in kilns dug out of the sides of hills.

Calcium is the fifth most abundant element and the third most abundant metal, amounting to about 4% of the earth's crust (1). Over 90% of the upper ten miles is composed of igneous rocks (eg, basalt, granite); the rest is mainly shale, sandstone, and limestone. The feldspars, $M_2O \cdot Al_2O_3 \cdot 6SiO_2$ (M = K, Na) and $CaO \cdot Al_2O_3 \cdot 2SiO_2$ are the most abundant minerals making up about 60% of the igneous rocks. Lime feldspar [1302-54-1], CaAl$_2$Si$_2$O$_8$ (anorthite), accounts for about half of the feldspars (2).

About 7% of the earth's crust is calcite [3397-26-7], CaCO$_3$, in the forms of limestone, marble, and chalk (3). Some other important calcium minerals are: gypsum [13397-24-5] or selenite [10101-41-4], CaSO$_4$.2H$_2$O; anhydrite [7778-18-9], CaSO$_4$; fluorite [7789-75-5] or fluorspar, CaF$_2$; and apatite [1306-05-4], Ca$_5$FP$_3$O$_{12}$ (4).

Inorganic Compounds

Calcium Carbonate. Limestone is the most widely used of all rocks since calcium carbonate is an indispensable chemical in industry, either as such, or as the precursor of lime and hydrated lime.

Lime (CaO) and Hydrated Lime [1305-62-0] **(Ca(OH)$_2$).** In industrial countries, lime is second only to sulfuric acid in tonnage consumed (5–6). It is produced by calcination of calcium carbonate in various forms including limestone, marble, chalk, oyster shells, and dolomite. Lime is one of the few chemical compounds having a negative temperature coefficient of solubility.

Mortar. Mortar, principally slaked lime and sand, sets by evaporation of water and absorption of water by the bricks or cement blocks, followed by hardening due to action of CO$_2$:

$$Ca(OH)_2 + CO_2 \rightarrow CaCO_3 + H_2O$$

Metallurgy. Calcium oxide reacts readily with acid anhydrides:

$$CaO + SO_3 \rightarrow CaSO_4$$

Reactions of this type are important in high-temperature metallurgical processes in which CaO (which may be produced by decomposition of CaCO$_3$) reacts with and removes acidic impurities, eg, in the pig-iron blast furnace (7).

$$CaO \text{ (from limestone)} + SiO_2 \rightarrow CaSiO_3 \text{ (blast furnace slag)}$$

In modern steel manufacture, pebble quicklime is used as a flux in the basic oxygen, basic open hearth, basic Bessemer, and basic electric furnaces.

Treatment of Industrial Wastes. Neutralization of acid waste liquors is usually accomplished by use of lime, the cheapest alkali (7). Examples of such liquors are: waste pickle liquids from steel plants; wastes from metal plating operations (eg, chrome plating, copper plating); acid wastes from chemical and explosives plants; acid mine waste waters.

Great quantitites of stack gases containing acidic substances such as H$_2$S and SO$_2$ are produced in the smelting and refining of nonferrous metals such as copper, zinc, and lead obtained from sulfide ores. These can be removed by stack gas scrubbers utilizing Ca(OH)$_2$ suspensions.

Control of emission of acid gases (eg, SO$_2$) from fossil fuel power plants is also accomplished with Ca(OH)$_2$ scrubbers (see Air pollution control methods). An important process now under development is fluidized-bed combustion of coal mixed with limestone which traps SO$_2$ before it reaches the exhaust stacks (8).

Treatment of Municipal and Industrial Water Supplies. Hard water contains dissolved solids, chiefly calcium and magnesium salts (7). Hardness in water causes great economic losses. It necessitates boiler feedwater treatment and softening of water used in dyeing and other textile processing operations and in laundering (see Drycleaning and laundering) (9). Slaked lime is the water softener used in the greatest quantity in precipitation of calcium and magnesium ions from water:

$$Ca(HCO_3)_2 + Ca(OH)_2 \rightarrow 2\ CaCO_3 \downarrow + 2\ H_2O$$

$$MgX_2 + Ca(OH)_2 \rightarrow Mg(OH)_2 \downarrow + CaX_2$$

$$(X = Cl,\ NO_3,\ HCO_3,\ {\textstyle\frac{1}{2}}\ SO_4^{2},\ {\textstyle\frac{1}{2}}\ CO_3^{2})$$

Other Applications. Among other industrial uses of lime are the following: causticizing agent in kraft (sulfate) paper plants; recovering of NH_3 from NH_4Cl (Solvay process); recovery of magnesium (qv) from seawater and brines (precipitation of $Mg(OH)_2$); production of pesticides such as calcium arsenate (eg, from arsenic acid), lime–sulfur sprays, and Bordeaux mixture (copper sulfate and lime in water); and neutralizing acid soils (5,7,10).

Halogen Compounds. Halides. Calcium halides are easily made by reaction of the acid with $CaCO_3$, CaO, or $Ca(OH)_2$ (4).

Fluorspar, CaF_2, is used as a flux in metallurgical processes such as production of steel in the open hearth furnace. It is the source of HF (by reaction with H_2SO_4) and fluorine.

Hypochlorites. Common dry forms of chlorine (used as bleaches) are chloride of lime (available chlorine ca 35%) and high test calcium hypochlorite (70% available chlorine) (4,7). Both are made by reaction of gaseous chlorine with high calcium hydrated lime:

$$Ca(OH)_2 + Cl_2 \rightarrow CaCl(OCl).H_2O\ (\text{bleaching powder})$$

Calcium hypochlorite [7778-54-3], $Ca(OCl)_2$, can be made by salting it out (with NaCl) from a solution of bleaching powder. In contrast with bleaching powder, calcium hypochlorite does not decompose on standing (see Bleaching agents).

Sulfates and Sulfites. Calcium sulfate occurs in large deposits as $CaSO_4$ (anhydrite) and $CaSO_4.2H_2O$ (gypsum) (4,5).

Calcium sulfite [10257-55-3] and acid sulfite may be prepared by reaction of SO_2 with hydrated lime or limestone. Calcium acid sulfite [13780-03-5], $Ca(HSO_3)_2$, has been used to remove lignin from wood pulp in paper manufacture (7) (see Paper; Pulp).

Phosphates. The major constituent of phosphate rock (eg, large deposits are found in Florida and Morocco) is a fluorapatite $Ca_5FP_3O_{12}$, from which industrial phosphates are obtained. Among these are the phosphate fertilizers, phosphoric acid and calcium phosphates (4) (see Phosphoric acid; Fertilizers).

Superphosphate, Triple Superphosphate, Phosphoric Acid. Since phosphate rock is too insoluble to be very useful as a fertilizer, it is converted to superphosphate [12431-88-8] ($Ca(H_2PO_4)_2 + 2\ CaSO_4$) by H_2SO_4 and triple superphosphate [7758-23-8] ($Ca(H_2PO_4)_2$) by H_3PO_4 (11). Phosphoric acid may also be produced from phosphate rock by reaction with H_2SO_4.

Calcium Metaphosphate [13477-39-9]. This is made by reaction of P_2O_5 in HPO_3 with rock phosphate. The product, polymeric and insoluble, has to hydrolyze to act as a fertilizer (11):

$$[Ca(PO_3)_2]_x + 2\ H_2O \rightarrow Ca(H_2PO_4)_2$$

Baking Powder. One of the baking acids used in baking powder to release CO_2 from $NaHCO_3$ is monocalcium phosphate [10031-30-8], $Ca(H_2PO_4)_2.H_2O$ (4). Monocalcium phosphate is crystallized from a hot reaction mixture of concentrated (electric furnace) phosphoric acid and lime, or is made by spray drying a slurry of the product of reaction of lime and phosphoric acid (11) (see Bakery processes).

Silicates. Glass. Ordinary glass (qv) (soda-lime glass) is a complex mixture of silicates, chiefly those of sodium and calcium [1344-95-2] (5,7).

Portland Cement. Portland cement is obtained by calcining a mixture of substances to produce an appropriate ratio of the oxides CaO, MgO, Al_2O_3, Fe_2O_3, and SiO_2 (4,5) (see Cement).

Whitewares (Earthenware, China, Porcelain). The chief raw materials of ceramics manufacture are clay, feldspar, and sand (5). All of the three common types of feldspars are used, ie, soda, potash, and lime ($M_2O.Al_2O_3.6SiO_2$, M = Na, K) and $CaAl_2Si_2O_8$ [1327-39-5] (see Ceramics; Enamels). Clays are hydrated aluminum silicates such as kaolinite, $Al_2O_3.2SiO_2.2H_2O$, formed by the weathering of igneous rocks such as feldspars (7) (see Clay). Fluxing agents include the calcium compounds apatite, fluorspar, and calcined bones (mainly apatite). Special refractory ingredients may be lime, limestone, or dolomite [17069-72-6].

Calcium Silicate Brick. Sand–lime brick is used in masonry in the same way as common clay brick. The bricks, molded from a wet mixture of sand and high calcium hydrated lime, are heated under pressure in a steam atmosphere. Complex hydrosilicates are formed which give the bricks high dimensional stability (7).

Coordination Chemistry of Calcium

Calcium ion is essential in the sequence of reactions resulting in coagulation of blood (see Blood, coagulants). Blood samples can thus be protected against coagulation by substances that mask Ca^{2+}. One of the most effective agents, citrate, reduces the Ca^{2+} concentration by coordination to a value below that needed for coagulation.

Ethylenediaminetetraacetate (EDTA, Versene, Sequestrene):

$$\begin{array}{c} ^-OOC \\ ^-OOC \end{array} NCH_2CH_2N \begin{array}{c} COO^- \\ COO^- \end{array}$$

forms a complex with Ca^{2+} with an extremely large formation constant (4). EDTA is potentially hexadentate, although it may act as a tetradentate ligand utilizing only the four carboxyl groups. The calcium complex may have the composition $[CaY]^{2-}$.

Calcium halides form addition complexes with NH_3 such as $CaCl_2.nNH_3$ (n = 2, 4, 8) and $CaBr_2.nNH_3$ (n = 2, 4, 6, 8). Hexammine calcium [12133-31-2], $Ca(NH_3)_6$, is formed by reaction of calcium metal with anhydrous NH_3 (4) (see Coordination compounds).

Organic Chemistry of Calcium

Calcium Carbide and its Derivatives. Probably the most important organic calcium compound is calcium carbide [75-20-7], produced in the electric furnace (12):

$$CaO + 3C \rightarrow CaC_2 + CO$$

Acetylene is produced from the CaC_2 (by reaction with H_2O) from which other organic compounds (eg, ethanol, acetaldehyde) may be obtained (7) (see Acetylene-derived chemicals; Carbides).

The calcium carbide may also be converted to calcium cyanamide:

$$CaC_2 + N_2 \rightarrow CaCN_2 + C$$

Calcium cyanamide [156-62-7] (lime nitrogen) has been used as a fertilizer (12). It hydrolyzes in the moist soil to produce ammonia:

$$CaCN_2 + 3 H_2O \rightarrow CaCO_3 + 2 NH_3$$

Calcium cyanamide can also be converted to calcium cyanide [592-01-8], used in cyanidation of metallic ores and production of sodium cyanide and ferrocyanides (4) (see Cyanamide).

Salts of Organic Acids; Calcium Sucrate. Calcium salts of organic acids may be prepared by reaction of the carbonate hydroxide with the acid (5). Calcium lactate [814-80-2] is an intermediate in the purification of lactic acid from fermentation of molasses. Calcium soaps (salts of fatty acids), soluble in hydrocarbons, are useful as waterproofing agents and constituents of greases (5).

In production of sugar, the juice extracted from the sugar cane or sugar beets is treated with a suspension of $Ca(OH)_2$ which neutralizes the syrup acidity and precipitates calcium sucrate, leaving impurities in the solution. This is filtered and the calcium sucrate is converted to sugar and $CaCO_3$ by reaction with CO_2.

Reagents in Synthesis. Calcium borohydride [17068-95-0], $Ca(BH_4)_2$ (reaction of $NaBH_4$ with $CaCl_2$), has been reported as superior to $NaBH_4$ and $LiAlH_4$ for some reductions (see Hydrides; Boron compounds). Hexammine calcium, $Ca(NH_3)_6$, (prepared by passing NH_3 into an ether suspension of calcium) reduces polycyclic aromatic compounds leaving one isolated aromatic ring. Calcium hydride [7789-78-8], CaH_2, and anhydrous calcium sulfate (Drierite), $CaSO_4$, are useful as drying agents qv (13).

Organometallic Chemistry. Only a few organocalcium compounds have been reported. Alkyl calcium halides have been prepared by reaction of the halides with calcium in tetrahydrofuran (13) (see Organometallics).

Physiological Role of Calcium

Calcium accounts for about 2% of body weight; about 99% is found in bones and teeth (10,14,15,16). The major mineral component of these structures is a hydroxyapatite [1306-06-5], $Ca(OH)_2 \cdot 3Ca_3(PO_4)_2$ (16). Calcium of bones appears to be present in two forms: the less soluble crystalline hydroxyapatite; and the more soluble intercrystalline calcium salts (14,16). In addition to its major functions of formation and maintenance of the skeletal structure, calcium is also essential in the following ways (15,17): blood clotting; control of muscle and nerve cell response; influencing permeability of cell membranes; functioning of certain enzyme systems; and milk formation.

The normal concentration of calcium in blood is 10–11 mg/100 mL. Approximately half of the serum calcium is ionic. The rest is bound to blood proteins, and a little is present as a citrate complex.

Maintenance of the appropriate blood calcium level involves the bones, kidneys, intestine, and parathyroid glands (14). Most of the normal calcium requirement of the blood is taken care of by the equilibrium between blood Ca^{2+} and the more soluble intercrystalline calcium salts of the bones (16). The rest is accounted for by a feedback mechanism, involving the less soluble crystalline hydroxyapatite, under the control of the parathyroid glands.

In vitamin D deficiency, transport of calcium from bones to blood is inhibited. This vitamin is involved in biosynthesis of a calcium transport protein in the mucosal cells of the intestine (see Vitamins).

The recommended daily allowances of calcium are (mg): children to ten years of age, 360–800; teenage children, 1200; adults, 800 (increasing to 1200 during pregnancy and lactation). Milk supplies ca 1.27 g/L of calcium in available form (see Mineral nutrients).

BIBLIOGRAPHY

"Calcium Compounds" are treated in *ECT* 1st ed. under "Calcium Compounds," Vol. 1, pp. 747–779, by C. R. Hough and in *ECT* 2nd ed. under "Calcium Compounds" Vol. 4, pp. 1–7, by C. L. Rollinson, University of Maryland.

1. H. Ahrens, *Distribution of the Elements in Our Planet*, McGraw-Hill Book Company, New York, 1965, p. 97.
2. *Ibid*, p. 29.
3. *Chem. and Eng. News* 33 (April 18, 1977).
4. R. D. Goodenough and V. A. Stenger, "Magnesium, Calcium, Strontium, Barium and Radium" in *Comprehensive Inorganic Chemistry*, Vol. 1, Pergamon Press, Oxford, England, 1973, pp. 591–664.
5. R. Norris Shreve, *Chemical Process Industries*, McGraw-Hill Book Company, New York, 1967, pp. 143–190.
6. *Chem. and Eng. News* 12 (May 16, 1977).
7. *Chemical Lime Facts, Bulletin 214,* National Lime Association, Washington, D.C., 1973, pp. 9–24.
8. *Annual Report,* Pennsylvania Power and Light Company, 1976, p. 8.
9. Ref. 5, p. 41.
10. *A Handbook for the AG Lime Salesman,* National Limestone Institute, Inc., Washington, D.C., 1973, p. 21; *100 Questions and Answers on Liming Land,* National Lime Association, Washington, D.C., 1967.
11. Ref. 5, p. 265.
12. Ref. 5, pp. 262, 301.
13. M. Fieser and L. F. Fieser, *Reagents for Organic Synthesis,* John Wiley & Sons, Inc., New York, Vol. I, 1967, p. 103, Vol. III, 1969, p. 4, Vol. V, 1975, p. 89.
14. J. M. Orten and O. W. Neuhaus, *Human Biochemistry,* The C. V. Mosby Co., St. Louis, 1975, p. 665.
15. J. B. Peterson, "The Role of Agricultural Limestone in National Health," *Limestone,* Fall, 1968.
16. Ref. 14, p. 533.
17. Ref. 14, p. 430.

General References

Limestone Purifies Water, National Limestone Institute, Inc., Fairfax, Virginia, May, 1977; valuable up-to-date discussion of limestone including statistics of production and use, application in agriculture, neutralization of acidic wastes and acidic rain, effects on natural waters, etc (92 references).
Lime, Bulletin 213, National Lime Association, Washington, D.C., 1976; useful practical treatise on handling, application and storage; includes much chemical and engineering data.
Water Supply and Treatment, National Lime Association, Washington, D.C., 1976; comprehensive treatise on all aspects of treatment of municipal and boiler feedwaters (softening, coagulation, fluoridation, sterilization; includes methods of water analysis).

CARL L. ROLLINSON
University of Maryland

CALCIUM CARBONATE

Calcium carbonate [471-34-1], $CaCO_3$, mol wt 100.09, is the major constituent of limestone which is used in the manufacture of quicklime or hydrated lime; these in turn are the sources of most calcium compounds including precipitated calcium carbonate. Calcium carbonate occurs naturally in the form of marble, chalk, and coral.

Powdered calcium carbonate is produced by either chemical methods or by the mechanical treatment of the natural materials. The term, precipitated calcium carbonate, applies to the commercial types of the compound produced chemically in a precipitation process. The precipitated products are distinguished by a finer, more uniform particle size, a narrower size range, and a higher degree of chemical purity.

Precipitated calcium carbonate dates back to 1850 when J & E Sturge Ltd., Birmingham, England, started production using calcium chloride and sodium carbonate as the reactants. Commercial production in the United States was started in ca 1913 by the West Virginia Pulp and Paper Company (Westvaco), Luke, Md. In 1928, production was stopped at this plant and manufacture began in Covington, Va.; both plants employed the direct carbonation of milk-of-lime.

Within the next few years production was underway by two of the present U.S. producers of precipitated calcium carbonate: the Mississippi Lime Company, St. Genevieve, Mo., where production was started in 1928 by the Peerless Lime Company and the Minerals, Pigments and Metals Division of Pfizer Inc., Adams, Mass., where manufacture was begun in 1933 by the New England Lime Company.

Precipitated calcium carbonate is one of the most versatile mineral fillers and is consumed in a wide range of products including paper, paint, plastics, rubber, textiles, putties, caulks, sealants (qv), adhesives (qv), and printing ink (see Fillers). USP grades are used in dentifrices (qv), cosmetics, foods, and pharmaceuticals. Precipitated calcium carbonate is produced in a number of grades for these applications.

Properties

Calcium carbonate occurs naturally in two crystal structures, calcite [13397-26-7] and aragonite [14791-73-2]. Calcite is thermodynamically stable at all investigated pressures and temperatures (1). The aragonite polymorph is metastable and irreversibly changes to calcite when heated in dry air to about 400°C, the rate increasing with temperature. The transformation is much more rapid when in contact with water or solutions containing calcium carbonate and may take place at room temperature.

The crystal forms of calcite are in the hexagonal system. There are more than 600 reported crystal habits for calcite in contrast to 10–15 for other isostructural carbonates.

Aragonite is in the orthorhombic system. The usual crystal habits are acicular or elongated prismatic. In the commercial forms of precipitated calcium carbonate where aragonite predominates, crystals have parallel sides and large length-to-width ratios. Rapid precipitation, high concentration of reactants, high temperatures, and the presence of divalent cations increase the tendency to produce aragonite (2).

Most commercial grades of precipitated calcium carbonate have a dry brightness in excess of 98% and have a minimum purity of 98%, the major contaminants being magnesium carbonate and silica. Products are available with average particle sizes ranging from submicrometer (ca 0.03 µm) to coarse (ca 5 µm).

The essential properties of the two crystal polymorphs are shown below.

Property	Calcite	Aragonite
refractive index		
α		1.530
β		1.681
γ		1.685
ϵ	1.4864	
ω	1.6583	
density, kg/m³	2710	2930
bp (dec), °C	898	825
solubility, g/100 cm³ H$_2$O		
at 25°C	0.0014	0.00153
at 75°C	0.0018	0.00190

Manufacturing and Processing

Precipitated calcium carbonate can be produced by several methods but only the carbonation process is commercially used in the United States today. This is the simplest and most direct process, using the most readily available and lowest cost raw materials.

Limestone is calcined in a kiln to obtain carbon dioxide and quicklime. Generally, these products are purified separately before recombining. The quicklime is mixed with water to produce either a milk-of-lime or dry hydrated lime; both are essentially all calcium hydroxide. When dry hydrate is used in the process, water is added to produce a milk-of-lime slurry.

In the carbonation process, the cooled and purified carbon dioxide-bearing kiln gas is bubbled through the milk-of-lime in a reactor known as a carbonator. Gasing continues until all the calcium hydroxide has been converted to the carbonate. The end point can be monitored by pH or by chemical measurements.

The reactions involved in this production method are:

$$\text{calcination, } CaCO_3 \rightarrow CaO + CO_2$$
$$\text{hydration or slaking, } CaO + H_2O \rightarrow Ca(OH)_2$$
$$\text{carbonation, } Ca(OH)_2 + CO_2 \rightarrow CaCO_3 + H_2O$$

Reaction conditions determine the type of crystal, the size of particles, and the size distribution produced. The process variables include starting temperature, temperature during carbonation, rate of mixing, pH, concentration of reactants, and the presence or absence of chemical additives.

Following carbonation, the product can be further purified by screening as the impurities in the milk-of-lime remain as coarse particles in comparison to the micrometer-sized, precipitated calcium carbonate. This screening, also used to control the maximum size of the product, is followed by dewatering. Rotary vacuum filters, pressure filters, or centrifuges are used in the mechanical removal of water. Washing of the filter cake is unnecessary as water is the only by-product of carbonation.

The filter cake solids are generally 25–60% $CaCO_3$, depending, to a large degree, on the particle size of the precipitated carbonate. Final drying is accomplished in either a rotary, tunnel, spray, or flash dryer (see Drying). This dryer product is usually disintegrated in a micropulverizer. The milled material is conveyed to large storage bins for bulk loading or packing in bags.

Some coated grades are available for special applications. The precipitated calcium carbonate is coated to improve flow properties, processing, and the physical properties of the final product. Fatty acids, resins, and wetting agents used as coating materials are applied before or after drying.

The dried calcium carbonate is normally shipped in bulk or in 22.7-kg multiwall bags. The product is sometimes supplied to the paper industry in repulpable paper bags to minimize the cost and problem of bag disposal.

Convenience, cost, and energy consumption factors prompted the investigation of methods for supplying the product in slurry form. A considerable volume of both precipitated and natural ground carbonates is now being shipped in this manner.

Prior to 1977, a significant quantity of the precipitated calcium carbonate was produced by manufacturers of synthetic soda ash. The calcium carbonate plants were located adjacent to the Solvay process soda ash works which supplied the two raw materials. A solution of soda ash reacts with a solution of purified calcium chloride yielding calcium carbonate and a sodium chloride by-product. The sodium chloride was difficult to wash from the carbonate filter cake and to remove in waste treatment facilities. The closing of the Solvay plants for either economic or ecological reasons necessitated the shutdown of these precipitated carbonate operations (see Alkali and chlorine products).

Economic Aspects

In general, the pricing structure of precipitated calcium carbonate is related to particle size; ultrafine grades are priced higher than the coarser grades. Specially controlled physical properties, chemical purity, and the presence of a coating agent are other price governing factors, eg, current prices range from 8.2¢/kg for a paper-filling grade to 27.6¢/kg for a coated ultrafine grade.

The closing in 1975 and 1976 of Solvay process-connected plants caused a shortage of precipitated calcium carbonate. Production facilities were expanded by the remaining manufacturers.

Current U.S. producers of precipitated calcium carbonate include the Minerals, Pigments and Metals Division of Pfizer Inc., New York, and the Mississippi Lime Company, Alton, Ill. Production in 1978 is estimated to be approximately 163,000 metric tons; capacity will approximate 181,000 metric tons.

Specifications, Standards, and Quality Control

The American Society for Testing and Materials (ASTM) and the Technical Association of the Pulp and Paper Industry (TAPPI) have issued specifications and methods of testing technical grades of precipitated calcium carbonate. ASTM Standard Specification D-1199-69 has a section covering precipitated carbonate for filler use (3). Beyond a 2% maximum limit on moisture, other properties and limits are "to be agreed upon by purchaser and seller." TAPPI method T660 covers procedures for measuring moisture, water absorption, and 44 μm (325 mesh) residue. Other TAPPI methods cover sampling (T657), measuring pH (T658), oil absorption (T658), and brightness (T646) (4). Federal specifications include MIL-C-151898-A and JAN-C-293.

Food and pharmaceutical grades are covered in the *United States Pharmacopeia* (5) and the *Food Chemical Codex* (6). These publications include both purity limits and methods of testing.

Differences in grades are often distinguished by physical properties. Tapped density is determined by measuring the volume occupied by a given weight of dry material after compacting by a standard dropping method; it is presently expressed within the United States by lb/ft^3 (multiply by 16.0 to convert to kg/m^3). Dry brightness is a measure of the whiteness or reflecting power of the dry product compared with that of a white standard. Oil absorption is determined by measuring the amount of oil required to form a paste of specified consistency using a prescribed weight of dry calcium carbonate. (This test is important where the product is used as an extender pigment.) Residue is the measured amount of material remaining on a screen of specified mesh after wet sieving; a 325 mesh (44 μm) screen is usually specified.

Health and Safety Factors

Precipitated calcium carbonate is listed as a nutrient and dietary-supplement food additive (7) (see Mineral nutrients). Thus, there are minimal environmental concerns in the handling of the material. The dust is classified as a nuisance particulate and present threshold limits are 10 mg/m^3 (30 million particles per cubic foot of air or mppcf). This limit, for a normal working day, does not apply to brief exposures at higher concentrations.

Uses

The manufacture of paper is, by far, the largest use of precipitated calcium carbonate. It is estimated that the United States paper industry consumes more than 75% of the total production (8). In paper coating, precipitated calcium carbonate is used in combination with kaolin; kaolin provides the necessary gloss and the carbonate adds brightness, opacity, ink receptivity, and smoothness. Increasing the amount of carbonate reduces the gloss to obtain the matte or dull finish that is in growing demand for educational textbooks. The percentage of carbonate may vary from 5 to 50% depending upon the desired coating properties and economic factors.

Precipitated calcium carbonate is also used as a filler in paper and its use is increasing in printing and writing papers. Development of sizing materials that are effective at neutral or alkaline pHs permit the use of precipitated calcium carbonate as a filler. Compared to low-pH sizing, alkaline sizing provides increased paper strength, increased brightness, better retention of strength and brightness properties on long-term aging, less corrosion, cleaner machine operation, and the opportunity to use precipitated calcium carbonate to reduce filler costs (9–10). Precipitated calcium carbonate continues to be the primary filler in cigarette papers, where particle size is critical in order to control the burning rate and porosity (11) (see Fillers; Paper).

The plastics industry is a rapidly growing consumer of precipitated calcium carbonate for cost reduction. A variety of both coated and uncoated grades that cover a range of particle sizes is used.

Calcium carbonate is usually categorized as a nonreinforcing filler. However, when used in resins in concentrations greater than 10%, calcium carbonate improves physical properties such as heat resistance, dimensional stability, stiffness, hardness, and processability (12–13).

The coarser uncoated grades of precipitated calcium carbonate have found wide application in latex and alkyd flat wall paints to impart uniform flatting as well as good color and sheen uniformity on recoat and overlap. These low water- and oil-demand products permit high loading levels (14). The use of ultrafine grades in colored enamels eliminates flooding, floating, streaking, sagging, and excessive drip-off.

Ultrafine precipitated calcium carbonate is used in rubber to provide reinforcement of white and light colored products. Coated grades are used to improve the dispersibility in mixing and to lower the modulus of the rubber.

Other miscellaneous uses for the technical grades of precipitated calcium carbonate include printing ink, putties, caulks, sealants, and adhesives. In all of these applications the carbonate provides body and some degree of reinforcement.

Food, drug, and pharmaceutical uses include direct use as an antacid, a mild abrasive agent in dentifrices, a source of calcium for calcium enrichment, a source of calcium and alkalinity in antibiotics manufacture, a constituent in chewing gum, and a filler in cosmetics, to name only a few.

BIBLIOGRAPHY

"Calcium Carbonate" under "Calcium Compounds" in *ECT* 1st ed., Vol. 2, pp. 750–759, by R. H. Buckie, West Virginia Pulp and Paper Company; "Calcium Carbonate" under "Calcium Compounds" in *ECT* 2nd ed., Vol. 4, pp. 7–11, by Robert F. Armstrong, Diamond Alkali Company.

1. C. Palache, H. Berman, and C. Frondel, *Dana's System of Mineralogy,* 7th ed., Vol. II, John Wiley & Sons, Inc., New York, 1951, p. 151.
2. R. W. Hagemeyer in R. W. Hagemeyer, ed., *Paper Coating Pigments,* Monograph No. 38, TAPPI, Atlanta, Ga., 1976, p. 39.
3. *1976 Annual Book of ASTM Standards,* Part 28, ASTM, Philadelphia, Pa., 1976, pp. 237–238.
4. *TAPPI Testing Procedures,* TAPPI, Atlanta, Ga., 1976.
5. *The United States Pharmacopeia,* 19th revision, United States Pharmacopeial Convention, Inc., Rockville, Md., 1975, pp. 60–61.
6. *Food Chemicals Codex,* 2nd ed., National Academy of Sciences, Washington, 1972, pp. 121–123.
7. N. I. Sax, *Dangerous Properties of Industrial Materials,* 4th ed., Van Nostrand Reinhold Co., New York, 1975, p. 509.
8. *Ind. Miner.* **55,** 9 (1972).
9. S. H. Watkins, *Paper Trade J.* **157**(36), 28 (1973).
10. J. J. Guerrier, *Pulp Paper* **46**(10), 131 (1972).
11. R. K. Mays, *J. Tech. Assoc. Pulp Paper Ind.* **53,** 2116 (1970).
12. J. Agranoff, *Modern Plastics Encyclopedia, 1976–1977,* Vol. 53, McGraw-Hill, New York, 1976, p. 176.
13. C. A. Harper, *Handbook of Plastics and Elastomers,* McGraw-Hill, New York, 1975, pp. 46–47.

14. P. F. Woerner in T. C. Patton, ed., *Pigment Handbook*, Vol. I, John Wiley & Sons, Inc., New York, 1973, p. 123.

General References

R. L. McCleary in R. L. Myers and J. S. Long, eds., *Pigments, Part I*, Marcel Dekker, New York, pp. 159–259.
A. P. Wilson, ed., *Precipitated Calcium Carbonate—History, Manufacture, and Standardization*, J and E Sturge, Birmingham, England, 1948.

RICHARD H. LEPLEY
Pfizer Inc.

CALCIUM CHLORIDE

Calcium chloride was discovered in the 15th century, but received little attention until late in the 18th century. It is an extremely soluble salt which forms many hydrates with properties as shown in Table 1. Although calcium chloride is highly soluble in water at ordinary temperatures, solid phase separation will occur under certain temperature–concentration conditions as shown in the phase diagram in Figure 1.

Table 1. Properties of Calcium Chloride Hydrates[a]

Property	$CaCl_2 \cdot 6H_2O$ [7774-34-7]	$CaCl_2 \cdot 4H_2O$ [25094-02-4]	$CaCl_2 \cdot 2H_2O$ [10035-04-8]	$CaCl_2 \cdot H_2O$ [22691-02-7]	$CaCl_2$ [10043-52-4]
composition, % $CaCl_2$	50.66	60.63	75.49	86.03	100.00
mol wt	219.09	183.05	147.02	129.00	110.99
mp[b], °C	29.9	45.3	176	187	772
bp, °C			175[c]	181[c]	1935
density, d_4^{25}	1.71	1.83	1.85	2.24	2.16
heat of fusion, J/g (Btu/lb)	209 (90)	163 (70)	88 (38)	134 (58)	257 (111)
heat of soln (to infinite diln) in H_2O, J/g (Btu/lb)	72 (31)	−59.4 (−25.6)	−304.6 (−131.1)	−405 (−174.3)	−737.2 (−317.2)
heat of formation[d], kJ/mol[e], at 25°C	−2608	−2010	−1404	−1111	−795.4
heat capacity, $J/(g \cdot K)$[e], at 25°C	1.4	1.4	1.2	0.84	0.67

[a] Courtesy of Dow Chemical U.S.A.
[b] Incongruent mp for hydrates.
[c] Temperature where dissociation pressure is 101.3 kPa (1 atm).
[d] Negative sign means heat is evolved (exothermicity).
[e] To convert J to cal, divide by 4.184.

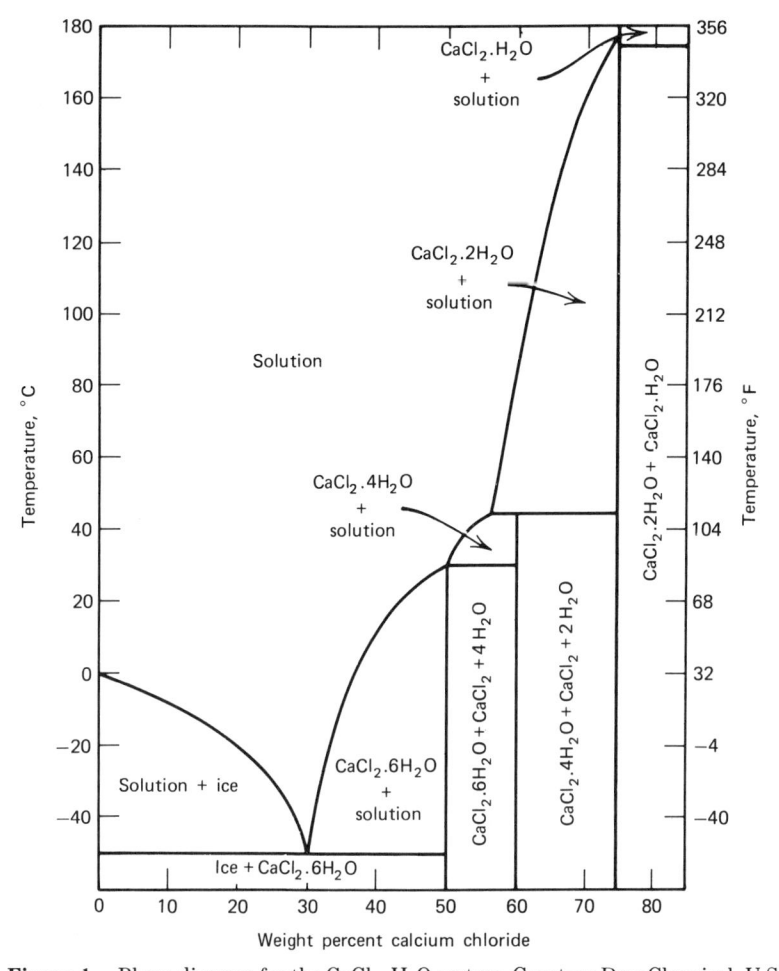

Figure 1. Phase diagram for the CaCl$_2$–H$_2$O system. Courtesy Dow Chemical, U.S.A.

Crystallization points of brines made from commercial calcium chloride closely approximate the heavy black line in the diagram. The crystallization point is the temperature at which crystals form from solution. For example, a 20% solution has a crystallization point of about −20°C, at which point ice crystals form. The solution will increase in concentration as more ice separates with continued cooling and may appear solid. The mixture of calcium chloride hexahydrate and ice will completely solidify when the temperature reaches about −51°C.

The eutectic temperature for a calcium chloride solution is variously given as −50 to −55°C owing to other ingredients in commercial products.

Calcium chloride and its solutions will absorb moisture from the air at various rates depending upon the vapor pressure of water in the air with respect to that over the calcium chloride, the rate at which the air circulates over the calcium chloride, and the surface area of the calcium chloride exposed to the air. At 25°C and 40% relative humidity (rh) approximately one gram of water is absorbed per gram of dihydrate calcium chloride until the system reaches equilibrium. At 25°C and 95% rh, one gram of dihydrate will adsorb approximately 14 g of water. Under similar conditions one gram of the anhydrous product (technical grade) will absorb about 1.4 g of water at 40% rh and 17 g of water at 95% rh. The theoretical energy requirement to evaporate water as steam from a 20% calcium chloride solution to an anhydrous state is about 10.5 MJ/kg (4500 Btu/lb) of $CaCl_2$.

Manufacture

Commercial production of calcium chloride was developed about 1860 as a result of the successful manufacture of soda ash by the ammonia–soda process which basically consists of the reaction of sodium chloride with calcium carbonate to form sodium carbonate (soda ash) and calcium chloride after the introduction of ammonia in various stages of the decomposition operation (1–2) (see Alkali and chlorine products). At present the greatest volume of calcium chloride is derived from evaporation of underground brines (3) (see Chemicals from brine). Calcium chloride is produced in the laboratory and in small-scale operations by several different methods, but most commonly by the action of hydrochloric acid on calcium carbonate (limestone, sea shells) followed by crystallization and dehydration. It also occurs in surface waters—seas, lakes, shallow streams—, and as a constituent in some natural mineral deposits.

Economic Aspects

Total domestic production of calcium chloride in 1973 reported in the *U.S. Bureau of Mines Minerals Handbook* (1975) was 778,400 metric tons, 100% $CaCl_2$ basis. Solid forms of $CaCl_2$ are sold in bags and drums and in bulk hopper cars, hopper trucks, and sparger cars. Liquid forms are sold in tank trucks and tank cars. Examples of price quotations are shown in Table 2.

Specifications

The grading specification most generally used is Grade A with the following limits specified: 100% passing a 9.5 mm sieve, 80–100% passing no. 4 (4.76 mm) and 0–5% passing no. 30 (590 μm). Other grades are defined for special conditions. (See current ASTM D98 Specification.)

Table 2. Price Quotations for Calcium Chloride [a] (per Metric Ton)

Form	1976
flake or pellet, 94–97%[b]	$117
flake, 77–80%[b]	$93.1
powdered, 77% min[b]	$88.2
liquor, 40%[c]	$26.5
granulated, USP[d]	$860

[a] Ref. 4.
[b] Paper bags (36.4 kg), carload lots, plant, freight equalized.
[c] Tank cars and trucks, freight equalized.
[d] 102-kg drums, freight equalized.

Uses

Major markets for calcium chloride are in deicing, dust control, road stabilization, and production of concrete products, as well as in oil well drilling and completion operations because of stability and consistency at various temperatures, and the advantage of using a clear heavy solution completely free of solids to minimize or prevent the destruction of formation permeability.

As a deicer, calcium chloride is the most effective of the commonly used chemicals because of its ability to perform quickly and at low subfreezing temperatures. Mixtures of rock salt (basically, sodium chloride which has an eutectic temperature of approximately −21°C) and calcium chloride have been found to be efficient in large scale deicing operations owing to the economics of using mined rock salt in combination with manufactured calcium chloride. Some observers have reported that by using such mixtures less total chemicals are required to attain desired deicing objectives (5).

The deliquescent property of calcium chloride makes it uniquely effective in dust control and stabilization operations. It dissolves readily by attracting moisture from the air and other sources. Also, in solution it retards the rate of moisture evaporation, and acts as a compaction aid. Consequently, it retains the fine particles in the unpaved surface and base and contributes to the stability of the structure.

Calcium chloride is used at the rate of 1–2% (by weight of cement) in ready-mix concrete and manufacture of other concrete products to accelerate set time and early strength development. In winter construction its use permits a reduction in time required to protect concrete exposed to low or freezing temperatures. The amount of calcium chloride used will lower the freezing point of water in the concrete only to an insignificant extent. Therefore, it is not considered an antifreeze material. Calcium chloride does not adversely affect corrosion of the usual reinforcement in concrete if adequate concrete cover is provided over the steel. However, it should not be used in prestressed concrete because of possible stress corrosion of the prestressing steel or in regular concrete containing imbedded aluminum or galvanized metals because of aggravated corrosion which may occur. Other uses are shown below.

Other uses of calcium chloride:

In	Use
adhesives	humectant; lowers gel temperature
cement	reduces alkalies, eg, Na_2O and K_2O
herbicides	controls growth of vegetation
hydrocarbons (l) and (s)	desiccant
liquid feed supplements for beef and dairy cattle	calcium source
plastics	controls particle size development; minimizes coalescence
pulp and paper processing	drainage aid
refrigeration	heat transfer medium
selected organic compounds	fire retardant
steel	controls scaffolding in blast furnaces by reducing alkalies
tractor tire weighting	improves traction and draw bar pull

Food grade calcium chloride (meeting requirements of the FDA and other governmental groups) is used in processing of cheese and other milk products, fruits, vegetables, and in preparation of pharmaceuticals.

Environmental Factors and Toxicity

Most calcium chloride environmental concerns are about its effects on automobiles, concrete and asphalt pavements, vegetation, and drinking water supplies. Monitoring run-off from normal use of deicing salts, including calcium chloride, at rates of coverage standard for ice control operations has shown chloride concentration levels in public water supplies to be less than 10% of the 250 ppm level established as the upper limit for water for public consumption by the U.S. Public Health Service (6).

Calcium chloride has been determined to be low in acute oral toxicity, similar to common table salt. The LD_{50} for rats is greater than 2500 mg/kg. From these animal data, the estimated lethal dose for a man of medium weight (ca 68 kg) would be about 150 g (single accidental swallowing) (7).

BIBLIOGRAPHY

The Calcium Chlorides are treated in *ECT* 1st ed., under "Calcium Compounds (Halides)," Vol. 2, pp. 759–761, by G. H. Kimber, and in *ECT* 2nd ed., under "Calcium Compounds—Calcium Chloride," Vol. 4, pp. 11–14, by Robert F. Armstrong, Diamond Alkali Company.

1. *Calcium Chloride, Tech. Bull. 16,* Allied Chemical Co., 1951.
2. *Calcium Chloride Handbook,* technical handbook, The Dow Chemical Co., 1966.
3. *Chem. Process. (Chicago),* **39**(5), 71 (1976).
4. *Chem. Mark. Rep.* **210,** 26 (1976).
5. *Minimizing Deicing Chemical Use,* National Cooperative Highway Research Program Practice Report no. 24, Transportation Research Board, National Research Council, 1974.
6. *Effects of Deicing Salts on Water Quality and Biota, Report 91,* National Cooperative Highway Research Program, p. 30.
7. W. S. Spector, ed, *Handbook of Toxicology,* Vol. 1, Saunders, Philadelphia, Pa., 1954, pp. 58–59.

WALKER L. SHEARER
Dow Chemical, U.S.A.

CALCIUM SULFATE

Mineral calcium sulfate is commonly called anhydrite, and occurs in many parts of the world. The mineral gypsum, calcium sulfate dihydrate, is widely distributed and is of much more economic importance. About fifty-five million metric tons of gypsum are consumed annually; about half is processed to calcium sulfate hemihydrate. The hemihydrate is also called plaster of Paris [26499-65-0]. Smaller amounts are dehydrated to hexagonal calcium sulfate, or soluble anhydrite, and orthorhombic calcium sulfate, identical to the mineral anhydrite. The latter is also called dead-burnt gypsum. Both anhydrite and gypsum are produced in large quantities as a by-product of various chemical operations.

Pure anhydrite and the hydrates have the following percent compositions:

Common name	CAS Registry No.	Molecular formula	Lime (CaO)	Sulfur trioxide (SO_3)	Combined water (H_2O)
anhydrite	[7778-18-9]	($CaSO_4$)	41.2	58.8	
gypsum	[10101-41-4]	($CaSO_4 \cdot 2H_2O$)	32.6	46.5	20.9
hemihydrate	[10034-76-1]	($CaSO_4 \cdot \tfrac{1}{2}H_2O$)	38.6	55.2	6.2

Major uses of gypsum are in construction, portland cement, and agriculture. There are many smaller but economically significant industrial uses of gypsum and anhydrite.

Gypsum has been used by artists and builders for more than five thousand years. Certain Egyptian pyramids contain alabaster ornamental and utilitarian objects, the walls of the tombs were plastered with gypsum and the masonry mortar contained burnt gypsum. In the United States, gypsum processing to the hemihydrate began around 1835 using ore imported from the Canadian Maritime Provinces. Rather precise set control (rehydration time) of the hemihydrate was developed by the end of the nineteenth century. This development led to the rise of gypsum plaster, blocks, and wallboard as the primary wallcladding materials in the United States. Probably an equally outstanding property of gypsum for building is its fire resistance.

Gypsum is added to portland cement to regulate set. This use accounts for over 40% of the world consumption of gypsum (see Cement).

The other major use of gypsum is in agriculture as a soil conditioner. In the United States about 1,200,000 metric tons are used annually for this purpose.

Gypsum and anhydrite are used to some extent in the production of heavy chemicals. Lower priced alternatives have resulted in a steady erosion of gypsum volume for these purposes.

Physical Properties

Gypsum, $CaSO_4 \cdot 2H_2O$, is the most useful form of $CaSO_4$. It is useful primarily because a controlled, modest amount of heat will convert gypsum to hemihydrate $CaSO_4 \cdot \tfrac{1}{2}H_2O$. This intermediate, relatively stable phase of calcium sulfate is the basis for over 90% of the commercial value of all calcium sulfate products sold in the United States.

There are two types of gypsum: a natural mineral and a synthetic product of chemical reaction from a variety of industries. The natural mineral is quarried or mined in many areas of North America. Major producing areas are Nova Scotia, Mexico, Michigan, Texas, California, Iowa, and Oklahoma. France, the Union of Soviet Socialist Republics, the United Kingdom, Spain, Italy, and Germany also have significant deposits of natural gypsum. Deposits vary in both ease of extraction and purity. Surface quarrying is practiced at many locations with little or no overburden removal required. Other deposits require underground mining techniques hundreds of meters below the earth's surface.

Natural gypsum is seldom found in the pure form. The anhydrous form and the dihydrate are commonly found together. Other impurities found in gypsum deposits are calcium carbonate, magnesium carbonates, silica, clay minerals and a variety of soluble salts. Most gypsum that is commercially used is a minimum of 80% pure, although some deposits require beneficiation or selective mining practices to achieve this level of purity.

There are forms of natural gypsum that have minimal commercial value. Alabaster, a fine-grained, relatively soft, pure rock gypsum, is occasionally found in deposits. Colorado is the only commercial source in the United States. Virtually all of its production is used by sculptors. Satinspar, also a pure form of crystalline gypsum, is fibrous in nature. In dense form it is translucent. Selenite is a monoclinic pure form of gypsum. Sheets of selenite can occur in dimensions up to several meters and it is sometimes mistaken for mica because of its transparency and parallelogram shape.

Synthetic gypsum, although available in North America in very large quantities, is little used commercially owing to objectionable impurities. These synthetic gypsums often are by-products from chemical processes such as stack gas scrubbing and syntheses of phosphoric acid, titanium dioxide, and citric acid. Owing to the lack or depletion of natural gypsum reserves, some countries use large quantities of synthetic gypsum. Technology exists to reduce significantly the quantity of objectionable impurities in most synthetic gypsum. The added capital investment and production costs associated with impurity removal deter its application in areas where natural gypsum is readily available. Japan, a country with few good natural gypsum deposits, has adopted purification process technology to permit commercial use of millions of metric tons of by-product gypsum from its phosphoric acid industry.

World supply of by-product gypsum (nonpurified) from phosphoric acid production alone is estimated to be 35–45 million metric tons per year.

Anhydrite, the anhydrous form of calcium sulfate, $CaSO_4$, occurs in a natural mineral form. In its natural dense, massive state it frequently can be visually differentiated from gypsum by its bluish-grey color.

In addition to occurring in a natural state, anhydrite can be manufactured by dehydration of gypsum, or by precipitation.

Physical properties	*Gypsum*	*Hemihydrate*	*Anhydrite*
mol wt	172.17	145.15	136.14
mp, °C	128 ($-1\frac{1}{2}\,H_2O$)	163 ($-\frac{1}{2}\,H_2O$)	1360 (dec)
	163 ($-2\,H_2O$)		
sp gr	2.32		2.96
Mohs hardness	1.5–2.0		3.0–3.5
soly in 100 g of H_2O at 25°C, g	0.24	0.30	0.20

Decomposition of Gypsum

The thermodynamic properties of the system:

$$CaSO_4 \cdot 2H_2O \xrightarrow{\Delta} CaSO_4 \cdot \tfrac{1}{2}H_2O + 1\tfrac{1}{2}\,H_2O \uparrow$$

$$CaSO_4 \cdot \tfrac{1}{2}H_2O \xrightarrow{\Delta} CaSO_4 + \tfrac{1}{2}\,H_2O \uparrow$$

have been the subject of much theoretical and practical study. Two forms of hemihydrate, α and β, were identified (1). The beta form was obtained when the dihydrate was partly dehydrated in a vacuum at 100°C or under conditions where a nearly saturated steam atmosphere did not prevail. The alpha form was prepared by dehydration of gypsum in water at temperatures above 97°C and by dissociation in an atmosphere of saturated steam. The β-hemihydrate has a higher energy content and a higher solubility than the α-hemihydrate.

The terms α and β are often used to differentiate two generally accepted, yet controversial forms of hemihydrate. Practically speaking, alpha is distinguishable from beta in that its particles disintegrate very little when mixed with water. It requires far less mixing water to form a workable slurry; consequently, it has the ability to produce denser and higher compressive-strength casts and less excess water, beyond that required for recrystallization, has to be removed after hydration is complete.

Anhydrite also has several common classifications. Anhydrite I designates the natural rock form. Anhydrite II identifies a relatively insoluble form of $CaSO_4$ prepared by high temperature thermal decomposition of dihydrate. It has an orthorhombic lattice. Anhydrite III denotes a relatively soluble form made by lower temperature decomposition of dihydrate, which is quite unstable (converts to hemihydrate easily with exposure to water or free moisture) and has the same crystal lattice as the hemihydrate phase. Soluble anhydrite is readily made from gypsum by dehydration at temperatures of 140–200°C. Insoluble anhydrite can be made by heating the dihydrate, hemihydrate, or soluble anhydrite for about 1 h at 900°C. Conversion can also be achieved at lower temperatures; however, longer times are necessary.

Manufacture

Natural Gypsum. Typical crude natural gypsum processing and calcination is shown in Figure 1. Gypsum rock from the mine or quarry is crushed and sized to meet the requirements of future processing or direct marketing of the dihydrate.

Fine ground dihydrate is commonly called landplaster, regardless of its intended use. The degree of fine grinding is dictated by the ultimate use. The majority of fine ground dihydrate is used as feed to calcination processes for conversion to hemihydrate.

The dehydration of gypsum (dihydrate), commonly referred to as calcination in the gypsum industry, is used to prepare hemihydrate or anhydrite. Hemihydrate is generally called stucco in North America and plaster in many other continents. In North America, plaster is differentiated from hemihydrate or stucco by the inclusion of additives to control intended use properties; eg, dehydration time, density, coverage, strength, and viscosity.

Kettle calcination continues to be the most commonly used method of producing

CALCIUM COMPOUNDS (CALCIUM SULFATE)

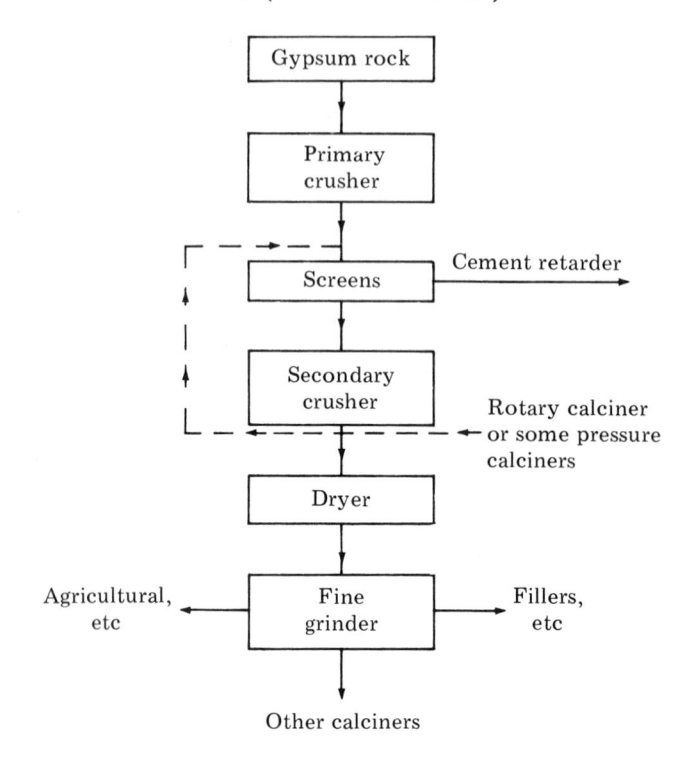

Figure 1. Mineral dihydrate flow diagram.

beta hemihydrate. The kettle can be operated on either a batch or continuous basis. Its construction is shown in Figure 2. The kettle is a cylindrical steel vessel enclosed in a refractory shell with a plenum between. The steel vessel is suspended above a fire box from which heated air flows up and into the plenum surrounding the steel vessel and through multiple horizontal flues which completely penetrate the vessel. The plenum and flues provide heat transfer to the kettle contents before the heated air is exhausted. An agitator with horizontal arms penetrates the depth of the kettle and is driven from above. Landplaster, usually ground 85–95% to 149 µm is fed from the top. In batch operation, with an 18.1 metric ton capacity kettle, filling takes 20–30 min. Another 90–120 min are usually required to convert the dihydrate to hemihydrate. The steam released from the dehydration reaction is vented from the kettle top. When conversion to hemihydrate is complete (usually determined by temperature measurement of the kettle contents), the stucco is discharged by gravity through the quick-opening gate located at the periphery and bottom of the steel vessel. A typical temperature pattern for the kettle contents is shown in Figure 3. Approximately 1 GJ (950,000 Btu) is required per metric ton of hemihydrate in a modern well-designed kettle.

During the fill portion of a kettle cycle, firing rate is usually controlled to maintain the kettle contents at a temperature of approximately 104°C. When the fill is complete, the firing rate is increased to a level dictated by the desired stucco properties. The mass boils at a temperature of 115–120°C. The boil or drag continues for about 1 h, then subsides. Heating continues for a short period to allow moisture release and the mass temperature increases to approximately 150–155°C if the hemihydrate form is desired,

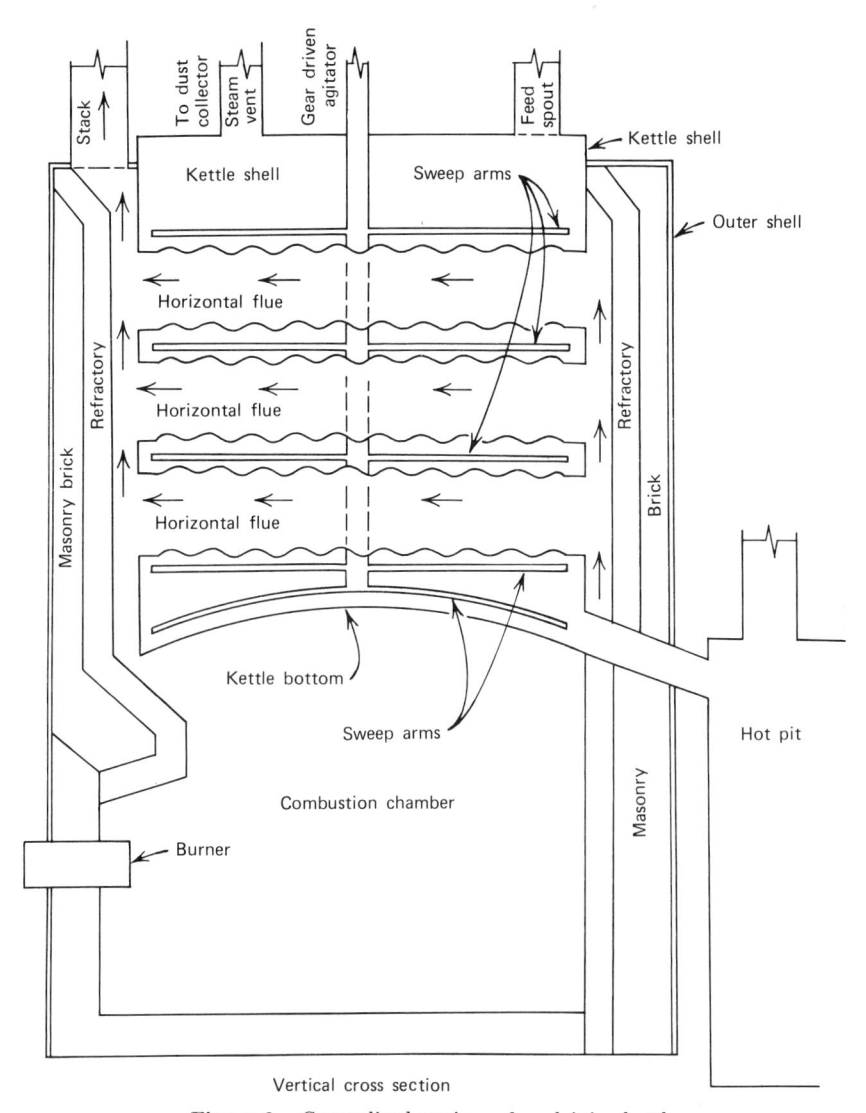

Figure 2. Generalized sections of a calcining kettle.

after which firing is reduced and the contents dumped. In practice, owing to the inability to heat all particles of gypsum adequately, the discharged mass will often contain small percentages of dihydrate, soluble anhydrite, and at times insoluble anhydrite.

If soluble anhydrite is desired, firing will be maintained until a second boil occurs accompanied by a second temperature plateau at about 190°C. Virtually all the water of crystallization has been removed at 215°C.

Soluble salts are impurities that increase the vapor pressure within the kettle. Aridized stucco refers to kettle-calcined hemihydrate which has been made with the intentional addition of 0.55–1.1 kg of NaCl or $CaCl_2$ per metric ton of landplaster. The stucco characteristic of lower water demand permits higher density and higher strength casts. The hygroscopic nature of such salts prevents the use of aridized stucco for some applications.

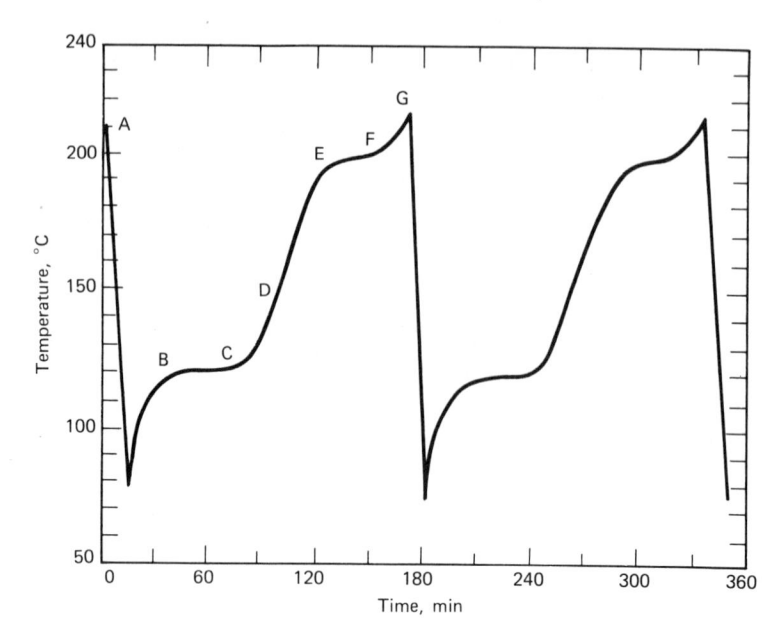

Figure 3. Time–temperature profile for kettle calcination. Points A–G are the fill period; B–C, the boil or drag; C–D, falling rate or cook-off; D, discharge for hemihydrate. Points D–E show firing rate to second boil; E–F, second boil; F–G second cook-off; G, second settle discharge.

In another process liquid water is introduced into the hot calcined gypsum mass in a kettle to reduce a portion of the mass below the boiling point of water, and then the mass is reheated (2). Stabilized setting and water demand properties were claimed plus water demand levels below those attainable through aridizing.

There is a technique that permits continuous calcination with kettles (3). Knauf (4) designed a continuous perforated grate charged with a single layer or multiple layers of sized rock, with the bed passing through a machine wherein hot gases are drawn through the bed. The material is cooled by air at a selected point to control the degree of dehydration. In 1976 Keller and Spitz (5) were issued a Canadian patent describing a flash calcination method for hemihydrate.

As with the calcination of beta hemihydrate described above, many processing innovations have evolved for producing alpha hemihydrate. In the 1930s Randel, Dailey, and McNeil (6) described charging lump gypsum rock 1.3–5 cm in size into a vertical retort, sealing it, and applying steam at a pressure of 117 kPa (17 psi) and a temperature of about 123°C. After calcination under these conditions for 5–7 h the hot moist rock is quickly dried and pulverized.

In 1957 Hoggatt (7) related a method of producing hemihydrate having low water demand by heating dihydrate in a water solution containing a metallic salt such as $CaCl_2$ at pressures not exceeding atmospheric.

In 1967 Cafferata (8) disclosed preparing very low water-demand alpha hemihydrate by autoclaving powdered gypsum in a slurry process. A crystal-modifying substance such as succinic or malic acid is added to the slurry in the autoclave to produce large squat crystals.

All of the three cited alpha hemihydrate processing methods are known to be practiced today.

In addition to the kettle calcination method described earlier, soluble anhydrite is commercially manufactured in a variety of forms, from fine powders to granules 4.76 mm (4 mesh) in size.

Insoluble anhydrite is manufactured commercially by several methods. Where large rock gypsum is the starting material, beehive kilns are used and 24-h processing time is not unusual. Rotary calciners or traveling grates are often used for small rock feed. Fine ground gypsum is calcined to the insoluble form in flash calciners. Temperature control is somewhat critical in all methods; low temperatures result in soluble anhydrite being present and high temperatures dissociate the $CaSO_4$ into CaO and oxides of sulfur.

Synthetic Gypsum. Phosphogypsum [13397-24-5], the by-product of phosphoric acid manufacture, is the major source of synthetic gypsum. The $CaSO_4$ can be produced in either the dihydrate or hemihydrate form. Waste sulfuric acid from the sulfate TiO_2 process using ilmenite may be used to make the dihydrate:

$$H_2SO_4 + CaCO_3 \text{ or } Ca(OH)_2 \rightarrow CaSO_4 \cdot 2H_2O$$

Stack-gas scrubbing of SO_2 from power plants and other industries burning high sulfur fuel is an increasing source of synthetic gypsum:

$$SO_2 + Ca(OH)_2 \rightarrow CaSO_3 \cdot \tfrac{1}{2}H_2O$$

$$2CaSO_3 \cdot \tfrac{1}{2}H_2O + O_2 + 3H_2O \rightarrow 2CaSO_4 \cdot 2H_2O$$

Because these synthetic gypsums are by-products, most contain objectionable quantities of impurities that relate to the process and raw materials from which they originate.

In the 1960s production of the alpha hemihydrate from synthetic gypsum resulting from the wet phosphoric acid process was developed in Japan using the CGC process and in Germany (9).

Shipment

Major gypsum and anhydrite products are bulky and low cost. In the United States, the normal economic shipping range for these materials is about 500 km. Specialized flat cars move most rail shipments of unitized wallboard products. In the last decade there has been a pronounced shift from rail to flat-bed truck transportation of board products. Most truck shipments are made direct to the job site.

Gypsum and anhydrite for the portland cement industry are bulk-shipped in closed hopper cars by rail or tank trucks. Agricultural gypsum is also shipped in bulk by similar methods. However, considerable tonnage is packaged in three-ply paper bags, including all landplaster for the home lawn and garden market.

Most plaster products are shipped in three- or four-ply paper bags. The bag construction often includes a vapor-proof liner to protect the contents from aging (change in set time and water demand resulting from moisture absorption).

Economic Aspects

Crude gypsum rock is the only form of calcium sulfate that has a significant movement in international trade, although certain special products move across international borders. Widespread location of gypsum deposits plus the relatively low ratio of manufacturing cost to product weight (shipping cost) are the reasons for the limited trade. Canada is the largest exporter of gypsum rock with most of it being shipped to eastern seaboard facilities in the United States (Table 1).

France is the largest exporter of gypsum in Europe, although the Federal Republic of Germany, Poland, and Austria also export significant quantities.

The United States imports more gypsum than any other nation. Table 2 shows where the imports originate. Table 3 shows the tonnage distribution by major uses in the United States for 1976. Table 4 shows gypsum production in the United States since 1962. The crude gypsum prices shown in Table 4 also indicate the commodity status of the business. Typical gypsum wallboard prices in 1976 were similar to those in 1961.

Table 1. World Production of Gypsum and Anhydrite, Thousands of Metric Tons[a]

Country	1970	1974[b]
Austria[c]	630	901
Bulgaria	169	187
Czechoslovakia	487	535
France[c]	6,087	6,168
GDR	289	366
FRG	1,473	1,764
Greece	308	420
Ireland	295	393
Italy	3,300	3,605
Poland[d]	850	850
Spain	4,228	4,082
USSR[d]	4,716	4,716
UK[c]	4,275	3,628
Yugoslavia	250	288
Argentina	422	523
Brazil[d]	290	290
Canada[c]	5,731	7,256
Jamaica	283	367
Mexico	1,291	1,559
United States	8,558	11,286
South Africa	410	499
Egypt	500[d]	486
Australia	845	1,028
China[d]	550	599
India	921	910
Iran	2,100	2,476
Japan	539	389
Thailand	144	243
Turkey	320	383
other countries	1,661	3,388
Total	51,652	59,585

[a] Excludes by-product gypsum (10).
[b] 1974 figures have been obtained for Canada, the United States, France, Spain, and the United Kingdom, and were estimated for other countries.
[c] Includes anhydrite.
[d] Estimate.

Table 2. Crude Gypsum Imported for Consumption in the United States, 1976[a]

Country of origin	Amount, t
Canada	4,053,340
Mexico	1,192,960
Jamaica	323,230
Dominican Republic	75,790
Brazil	7,000
Italy	44
Australia	9

[a] Ref. 10.

Table 3. Gypsum Sold or Used by Producers in the United States, 1976[a]

Use	Amount, t
Uncalcined[b]	
portland cement	2,985,600
agriculture[c]	1,288,200
other	289,850
Total	*4,563,650*
Calcined	
board products[d]	10,968,000
building plaster	445,700
industrial plaster	277,000
Total	*11,690,700*

[a] In cooperation with the Gypsum Association (10).
[b] Some data are estimated from quarterly reports.
[c] Includes by-product gypsum.
[d] Includes weight of paper and other materials.

Table 4. Crude Gypsum in the United States, Thousands of Metric Tons[a]

Year	Consumption[b]	Production[c]	Imports[d]	Exports[d]	Inventories[e]	Representative price, $/t[f]
1962	13,940	9,044	4,918	18	700	4.02
1966	13,700	8,752	4,970	19	680	4.08
1970	14,110	8,560	5,559	9	708	4.10
1974	17,610	10,885	6,735	13	870	4.86
1976	16,625	10,990	5,653	19	940	5.29

[a] Excludes by-product gypsum.
[b] Includes production plus imports, minus exports.
[c] From the quarterly and annual gypsum canvasses and from data furnished by the Gypsum Association. Includes sold or used by producers.
[d] From Bureau of Census.
[e] Estimated from consumption.
[f] Company-reported value per metric ton, fob mine or plant.

Specifications

Table 5 shows ASTM specifications relative to gypsum and plaster products.

Table 5. ASTM Specifications of Gypsum and Plaster Products

C 22–50[a] (1974)	gypsum
C 28–68[a] (1973)	gypsum plasters
C 35–70[a] (1975)	inorganic aggregates for use in plaster
C 52–54[a] (1972)	gypsum partition tile or block
C 59–73[a]	gypsum casting and molding plaster
C 61–64[a] (1975)	Keene's cement
C 265–64[a] (1970)	test for calcium sulfate in hydrated portland cement mortar
C 317–641[a] (1975)	gypsum concrete
C 377–66 (1972)	precast reinforced gypsum slabs
C 471–75	chemical analysis of gypsum and gypsum products
C 472–73	physical testing of gypsum plasters and gypsum concrete
C 473–76	physical testing of gypsum board products, etc
C 587–68 (1973)	gypsum veneer plaster

[a] Approved as American National Standard by the American National Standards Institute.

Uses

Uncalcined Gypsum and Anhydrite. Calcium sulfate, generally in the form of gypsum, is added to portland cement clinker to stop the rapid reaction of calcium aluminates (flash set). Also, gypsum accelerates strength development. For this reason, gypsum is more properly termed a set regulator, rather than a retarder, for portland cement. Used in proper amounts it also minimizes volume change. Normal gypsum addition to clinker is 5–6%.

Another large volume use of gypsum is in agriculture and for this use it is finely ground. Generally, successful use is associated with soil type rather than plants.

Very finely ground white gypsum, terra alba, has many accepted uses as a filler or an inert dilutent. It is also used in water clarification and animal feed. Food and pharmaceutical-grade gypsum is used as a source of calcium in foods and complies with the Federal Pure Food and Drug laws. Calcium sulfate is added to brewing water to burtonize it (see Beer).

Glass batch gypsum, a specially sized product, is added to the ingredients used to make glass as an oxidizing and fining agent and scum remover.

In Germany, ground anhydrite chemically accelerated to hydrate is used to construct temporary roof support walls in longwall mining of coal. Estimated yearly use for this purpose is two to three hundred thousand metric tons.

Hemihydrate. The ability of plaster of Paris to readily revert to the dihydrate form and harden when mixed with water is the basis for its many uses. However, certain other properties of plaster are important in its uses. Of equal significance is the ability to control the time of rehydration rather precisely in the range of four minutes to over eight hours through additions of retarders and accelerators. Other favorable properties include its fire resistance, excellent thermal and hydrometric dimensional stability, good compressive strength, and neutral pH. Upon setting, gypsum expands slightly and this property can be used to reproduce the finest detail, down to ca 1 μm, as is done in certain dental and jewelry castings employing the lost wax process. Normal linear expansion upon setting of gypsum plaster is 0.2–0.3% but with additives it may be controlled for special uses from 0.03 to over 1.2%.

The calcination procedures and processing techniques previously described produce a family of base plasters best described by the amount of water, in wt % of the plaster, which must be added when mixing to obtain standard fluidity. The range of fluidity permits casting neat plaster in the dry range of specific gravity of about 0.85–1.8 and consequent dry compressive strength of about 3.5–>70 MPa (35–>700 atm). Frequently these plasters are formulated with set and expansion control additives as well as many other materials to meet the needs of a particular application. Properties which limit gypsum plaster usage include plastic flow under load (which is increased under humid conditions), strength loss in a humid atmosphere, and dissolution and erosion in water. Thus gypsum is not normally used for permanent performance structurally and in exposed, exterior locations. To prevent long-term calcination gypsum products should not be used where temperatures exceed 45°C.

The largest single use of plaster in the United States is in the production of gypsum board. Gypsum wallboard replaced plaster in the United States about twenty years ago as the premier wall-cladding material.

Molding plasters have been used for centuries to form cornices, columns, decorative moldings, and other building interior features. Molding plaster is a good utility plaster where expansion control, high hardness, and strength are not needed. Its miscellaneous uses are numerous.

High quality atmospheric-calcined pottery plaster is used by the ceramic industry in the production of dishes, sanitary ware, art ware, stone ware, and related products. The major tonnage is used in the production of working molds which serve to shape the clay and absorb water from the clay until it is dry enough to be self-supporting.

Art plasters are essentially molding plasters modified to increase surface hardness, chip resistance and reduce paint absorption of casts made from this material.

Orthopedic plasters are used by hospitals and clinics for all types of orthopedic cast work.

A moderate amount of plaster is used in making impressions and casting molds for bridges, etc, by dental laboratories. Both alpha and beta plasters are used by the dental trade (see Dental materials).

The alpha plasters are tailored to meet the needs of modern industrial tooling, where they are used for master patterns, models, mock-ups, working patterns, match plates, etc. They are the accepted material for many of these applications because their use results in great time and labor savings, and excellent accuracy and stability of cast dimensions. Also the material is adaptable to intricate, irregular shapes, complex intersections and quick modification.

An important use for plaster is casting nonferrous metals in specially formulated plaster molds.

Calcined Anhydrite. Soluble anhydrite, or second settle stucco, has somewhat similar physical properties to those of gypsum plaster. It hydrates to the dihydrate rapidly in water. Its outstanding property is its extreme affinity for any moisture, which makes it a very efficient drying agent (see Drying agents). In ambient air it readily hydrates to hemihydrate. Soluble anhydrite, under the trade name Drierite, is widely used as a desiccant in the laboratory and in industry. A small amount is also used as an insecticide carrier. Small amounts of soluble anhydrite are unintentionally produced in certain calciners, such as rotary kilns and imp mills, during hemihydrate production.

Dead-burnt gypsum when finely ground is formulated with chemical accelerators to produce Keene's cement. Gauged with lime, Keene's cement is a common finish for use over cement plaster. Finely ground dead-burnt gypsum is also used in industry for many of the same applications as terra alba. A food and pharmaceutical grade is available in the United States.

BIBLIOGRAPHY

"Calcium Sulfate" treated in *ECT* 1st ed. under "Calcium Compounds," Vol. 2, pp. 767–779, by W. A. Hammond, W. A. Hammond Drierite Company; "Calcium Sulfate" treated in *ECT* 2nd ed. under "Calcium Compounds," Vol. 4, pp. 14–27, by W. A. Hammond, W. A. Hammond Drierite Company.

1. K. K. Kelly, J. C. Southard, and C. T. Anderson, *U.S. Bur. Min. Tech. Papers* **625**, (1941).
2. U.S. Pat. 3,415,910 (Dec. 10, 1968), W. A. Kinkade and R. E. McCleary (to United States Gypsum Company).
3. Belg. Pat. 624,555 (Feb. 28, 1963), R. C. Blair (to British Plaster Board Ltd.).
4. Brit. Pat. 886,602 (Jan. 10, 1962), A. N. Knauf (to Gebr. Knauf Saar-Gipswerke).
5. Can. Pat. 986,145 (Mar. 23, 1976), J. A. Keller and R. T. Spitz (to National Gypsum Company).
6. U.S Pats. 1,979,704 (Nov. 6, 1934); 2,074,937 (Mar. 23, 1937), W. S. Randel, M. C. Dailey, and W. M. McNeil (to United States Gypsum Company).
7. U.S. Pat. 2,616,789 (Nov. 4, 1952), G. A. Hoggatt (to Certain-Teed Products Corporation).
8. Brit. Pat. 1,079,502 (Aug. 16, 1967), G. W. Cafferata (to BPB Industries Ltd.).
9. U.S. Pat. 3,337,298 (Aug. 22, 1967), H. Ruter, E. Cherdon, and F. Fässle (to Gebruder Giulini).
10. *Mineral Industry Surveys*, U.S. Department of Interior, Bureau of Mines, Mar. 1977.

ROBERT J. WENK
PAUL L. HENKELS
United States Gypsum Co.

CALKING AND SEALING COMPOSITIONS. See Sealants; Chemical grouts.

CALORIMETRY

The measurement of amounts of heat (calorimetry) in its early history consisted essentially of the comparison of thermal effects with some reference thermal effect, but the meaning of heat was obscure. With the development of the first law of thermodynamics in the nineteenth century, as heat became more clearly defined, calorimetry assumed its present meaning: a procedure for establishing increments in the internal energy of a system by work done on the system plus energy transferred into it as a result of temperature gradients (heat flow) (see Thermodynamics). However, the early practices left terminology and concepts that have persisted somewhat incongruously up to the present. From the time of the earliest quantitative heat measurements the reference heat effect usually adopted was the amount of heat needed to raise the temperature of unit mass of water one temperature unit, leading to a heat scale on which the specific heat capacity of water is unity (1). This custom, which was a practical approach before the equivalence of thermal and mechanical energies was established by Joule (2), led to established units of heat, the calorie (cal) and the British thermal unit (Btu), different from those of other forms of energy. The continued use of these units and other derived units, which has persisted for many years after refinements in the work of Joule rendered them unnecessary, has tended to obscure the equivalence. The unit of energy derived from mechanical work, the newton–meter, was named the joule. This SI unit is properly applicable to heat. Table 1 lists some commonly encountered units used for heat, and their values in joules.

In common parlance heat is treated as if it were a material substance, probably a vestige of the theory of heat as caloric. With this obsolete principle, a hotter substance contains more heat than it had when it was colder. Care should be used in differentiating the loose terminology of common parlance from the more strict thermodynamic

Table 1. Common Units of Heat

Heat unit	Symbol	Energy value[a]
calorie (international tables)	cal_{IT}	4.186800 J[b]
calorie (thermochemical)	cal_{th}	4.184000 J[b]
calorie (15°C)	$cal_{15°C}$	4.18580 J
calorie (20°C)	$cal_{20°C}$	4.18190 J
calorie (mean)	cal_{mean}	4.19002 J
calorie (nutritional calorie)	Cal	4.184 kJ[c]
ton (of TNT) (10^9 cal_{th})[d]	T	4.184 GJ
British thermal unit (international tables)	Btu_{IT}	1.055056 kJ
British thermal unit (39°F)	$Btu_{39°F}$	1.05967 kJ
British thermal unit (59°F)	$Btu_{50°F}$	1.05480 kJ
British thermal unit (60°F)	$Btu_{60°F}$	1.05468 kJ
British thermal unit (mean)	Btu_{mean}	1.05587 kJ
therm (100,000 Btu)[e]		105.5056 MJ
barrel of oil (5.8×10^6 Btu)[f]	bbl	6.1 GJ

[a] Except where otherwise noted, energy values are from ref. 3.
[b] Defined exactly.
[c] Various values: 4.2 kJ (4); 4.184 kJ (4); 1 kcal (15–16°C) (5).
[d] According to Ornellas (6), T is 1093 δ $cal_{th} \cdot g^{-1}$ or 4.573 GJ per metric ton.
[e] Ref. 7.
[f] Ref. 8.

terminology; although the treatment of heat in common parlance so permeates the language that it is easier to use it than to avoid it.

In thermodynamic terminology (first law), heat is observable only in transfer, the transfer being the result of a temperature gradient. A material has a greater internal energy when hot than when cold. However, it is not correct to say that a substance contains more heat when molten than when solid, for the solid and liquid can coexist in contact indefinitely without any flow of heat between them if they are at the same temperature. Increase of temperature of a substance, melting of a solid, or vaporization of a condensed phase can be caused by doing work on the system as well as by the flow of heat into the system. Work can easily be done on the system mechanically, as by stirring, or by electrical work done on a resistor imbedded in the system. Thus measurements of heat and work are complementary; internal energy is a proper descriptor of the form of the energy, which heat and work change as they enter or leave the system. A calorimeter is an appropriate device for measuring these changes.

Calorimetry first became a quantitative measurement procedure in the late eighteenth century. It has found a considerable range of applications of both scientific and technological importance, and many complex instruments and subtle techniques have been applied to it. Calorimetry offers a convenient means of studying the energetics of a variety of processes. Unfortunately, the difficulty of completely characterizing temperature gradients somewhat offsets this convenience.

Principles

First Law of Thermodynamics.

At equilibrium at a given temperature, pressure, composition, and values of other intensive and extensive variables, the value of the internal energy U is unique for a system and independent of the path by which the state was reached. The first law states that increments in U are the sum of heat kq, flowing into the system and work, w, done on the system:

$$\Delta U = q + w \tag{1}$$

For an isobaric system,

$$\Delta H = q + w' \tag{2}$$

where w' is work other than that due to $P\Delta V$. The convention of the sign of w in statements of the first law is not universally the same, so care is needed in interpreting statements of the work done.

Both q and w' (or w) can be measured. Three mechanisms of heat flow are radiation, conduction; and convection, each follows a different law (9). For small temperature differences between separated surfaces, in an instrument of fixed configuration, the time rate of heat flow is proportional to the first power of the temperature difference, and is given approximately by the following:

$$\frac{dq}{dt} \sim k(\theta_2 - \theta_1) \tag{3}$$

where k is a constant of the system (but a function of the temperature), and θ_2 and θ_1 are the temperatures of the two surfaces. Equation 3 is often called Newton's law of heat transfer.

Mechanical or electrical work is a measure of w. Aside from pressure–volume work

($P\Delta V$ in an isobaric system), mechanical stirring and other frictional effects driven from outside the system are included in mechanical work done on the system. Electrical work, w_{el}, is easily done and easily measured.

$$w_{el} = EIt = I^2Rt = \frac{E^2}{R}t \tag{4}$$

E is the emf that drives a current, I, through a resistance, R, in the system for an elapsed time, t. Because w_{el} is easily measured it is customary to relate calorimetric experiments ultimately to electrical measurements of work.

Calorimeter Isolation. To control or measure q, two extreme procedures are used in practice. In one extreme procedure the calorimeter is isolated as fully as possible from its surroundings so that heat transfer is minimized; work added, or energy converted by a chemical process, causes a change in temperature of the calorimeter and its contents. Electrical work and the thermodynamic properties of the materials (heat capacity, enthalpy, and enthalpy or internal energy of reaction) are related to one another in terms of the temperature changes that occur in the system and the effective heat capacity of the calorimeter system. Heat flow is measured only in order to make small corrections. In the extreme case of ideal adiabatic calorimetry q is zero. The second extreme procedure has a good, well-defined path for heat flow between the calorimeter system and its surroundings, and a means by which q is measured accurately. This is a conduction-type calorimeter, and q must be measured with the full accuracy expected of the experiment.

Examples of the temperature–time responses of calorimeters of these two extreme types to an exothermic process of short duration are given in Figure 1.

In the nearly isolated calorimeter system (Fig. 1a), the calorimeter jacket is at a temperature θ_j higher than the temperature of the calorimeter except when the calorimeter vessel approaches θ_∞. Before the exothermic process starts, the calorimeter vessel is in a rating period during which its temperature approaches θ_∞ exponentially, but at such a slow rate that there can be a linear approximation. The temperature is disturbed by the heat released by the process, which begins at t_b and ends quickly. The temperature rises in a pattern characteristic of the calorimeter and the reaction until the heat of the reaction has been dissipated uniformly in the calorimeter vessel at t_e at which time the calorimeter enters a second rating period; its temperature drifting as before exponentially to θ_∞. The rating periods are used to establish the rate of energy transfer to the calorimeter from the jacket plus the rate of work done by stirring and by the thermometer. From these and the course of the time–temperature curve in the interval t_b to t_e, the increment in internal energy from extraneous effects can be calculated and the adiabatic temperature rise of the calorimeter (due to the reaction energy release) can be deduced. The geometric construction shows a time, t_m, at which the hatched areas are equal. Extrapolation of the rating period lines to t_m provides a convenient method of calculating the adiabatic (corrected) temperature rise.

In a conduction-type calorimeter in which temperature gradients are indicated by thermoelectric devices (Fig. 1b), an exothermic process that is nearly instantaneous, or lasts a few seconds at most, generates a response which follows a curve dependent upon the heatflow characteristics of the assembly, the thermoelectric constants of the sensors, and the response characteristics of the amplifier–recorder system. The rapid rise of sensor voltage begins at the time of occurrence of the process. The total energy

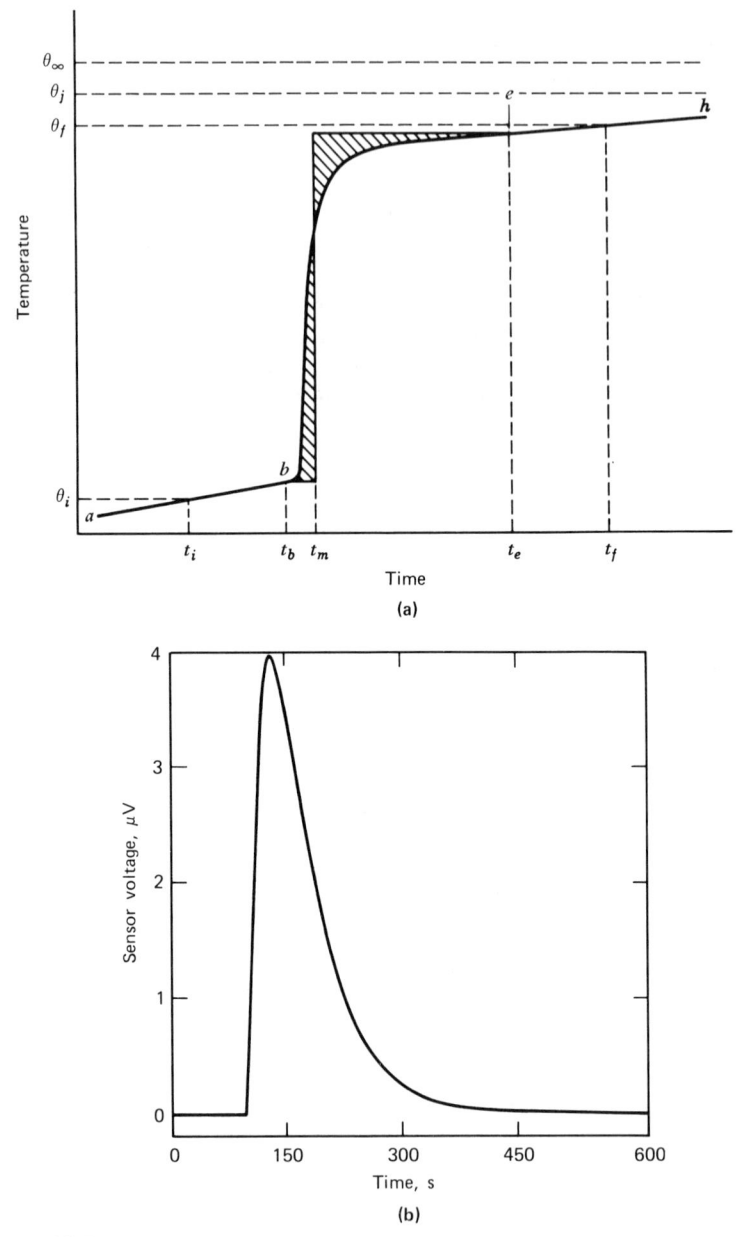

Figure 1. (a) Time–temperature plot for the calorimeter vessel of a stirred-water, isoperibol reaction calorimeter (10). Courtesy of NBS. (b) Heat burst in a conduction calorimeter (11). Courtesy of NBS.

of the process is measured by the integral of the peak above the base line. For a microcalorimeter, which is typically a conduction calorimeter, the total energy measured can be quite small, eg, a few millijoules.

Temperature Measurement. As a dominant parameter of calorimetry, temperature is measured for several purposes: (*1*) to establish the temperature with which the process or property measured is to be associated; (2) to measure a temperature increment caused by work done or by a chemical reaction in an isolated calorimeter or by heat flow to or from the calorimeter; (3) to measure temperature differentials that cause heat flow (eq. 3).

The thermodynamic properties of materials are functions of temperature, and their values are unambiguous only when temperatures are assigned to them. The appropriate temperature scale is the thermodynamic temperature scale, for which the kelvin is now internationally accepted as the unit of temperature or temperature difference. This scale is realized as the International Practical Temperature Scale of 1968 (IPTS-1968) (12). The instruments in terms of which interpolated temperatures on IPTS-1968 are stated up to the gold point temperature are the platinum-resistance thermometer, and the (platinum–10% rhodium)-platinum thermocouple. These instruments have been calibrated at defining fixed points. Above the gold point temperature (1337.58 K) the radiation law is specified and measurements are made by the radiation pyrometer. Figure 2 (13) shows a circuit for accurate platinum-resistance thermometry (on the right). The thermometer is a four-terminal resistance element, and the Mueller Bridge is a d-c Wheatstone bridge with one adjustable arm which consists of 6 or 7 decades of 10- or 11-digit resistors. The thermometer forms the arm of the bridge that is matched by the adjustable resistors. The Mueller bridge allows the current and voltage leads of the thermometer to be interchanged, and a mathematical manipulation of the two readings thus obtained allows the lead resistances to be eliminated (14). A-c resistance bridges of adequate accuracy and stability for accurate thermometry are now available (Automatic Systems Ltd., UK).

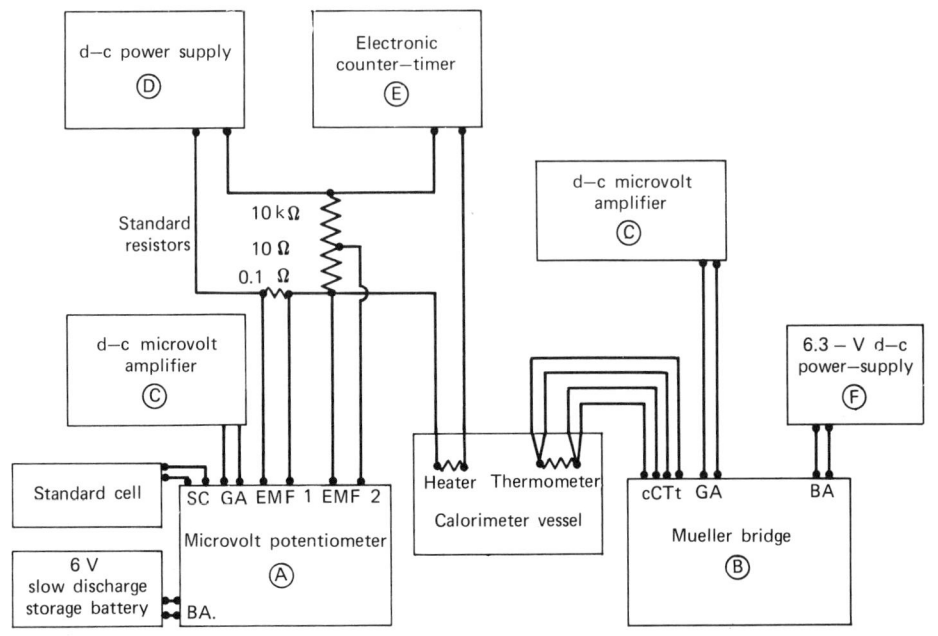

Figure 2. Calorimeter heater and thermometer circuits (13). BA = battery; GA = galvanometer; SC = standard cell; cCTt = current and potential leads for platinum resistance thermometer. Courtesy of NBS.

Some properties such as reaction energies are not strong functions of temperature, and alternative instruments to measure the temperature are suitable, even for work of high calorimetric accuracy. For calorimetry of such processes, temperature increments must still be measured accurately. High sensitivity and stability are the primary considerations for such an instrument. Examples of optional instruments are the quartz-crystal-oscillator thermometer (15–16), mercury-in-glass thermometers (17–20), and, for limited accuracy, the thermistor (17). Temperature differences needed to measure heat flow are most frequently measured by multiple-element thermocouples (21–25).

Measurement of Work Done. In a typical calorimetric experiment, work is done on the calorimeter electrically. The work done (eq. 4, p. 451) is converted to a thermal effect in the calorimeter. Figure 2 shows a circuit (on the left) for accurate measurement of electrical work. It is quite independent of the thermometer circuit shown in the same figure.

The electrical energy to establish the energy equivalent of the calorimeter is provided by a d-c power supply. The electrical power that is dissipated in the calorimeter is measured potentiometrically by measuring the emf across two standard resistors. The emf across a 0.1-Ω resistor is used to measure the heater current; the emf across a 10-Ω resistor, which forms part of a voltage divider of ratio about 1:1000 in parallel with the heater, is used to determine the emf across the heater terminals. The time of heating is measured electronically by counting cycles of a reference frequency, using the appearance and disappearance of the power supply voltage itself to trigger the electronic counter.

The power source may be an electronic power supply operable in constant current mode or constant voltage mode, or it may be a large capacity accumulator battery. The electronic counter-timer can be replaced by a stop watch without seriously impairing the overall accuracy of the measurement, if the heating periods are not too short.

Applications

Calorimetry is used to determine the thermodynamic properties of materials and also to measure thermal effects that can be derived from more or less complex devices or physical and chemical processes without special regard to the properties of materials.

Measurement of Properties. With respect to properties of materials, calorimetry is used to measure increments of internal energy and enthalpy ($U_2 - U_1$ and $H_2 - H_1$), and quantities derivable from them, such as heat capacity at constant volume (C_v), heat capacity at constant pressure (C_p), and increments in entropy ($S_2 - S_1$) and Gibbs energy ($G_2 - G_1$). The subscripts 1 and 2 refer to two different states of the material, representing differences in temperature, pressure, physical state, chemical composition, or other parameters necessary to characterize the thermodynamic state of the system. These quantities can be measured for pure materials or mixtures and for changes in the materials such as result from the formation of solutions or from chemical reactions (9,26). Table 2 shows values of some representative thermodynamic quantities. From a technological point of view these properties are valuable in choosing practical manufacturing processes in the chemical-process and metallurgical industries, predicting and optimizing yields of reaction products, making energy balances, choosing materials for particular applications, controlling or monitoring effluent

Table 2. Representative Calorimetrically Determined Thermodynamic Quantities for Substances [a,b]

Substance	Formula[c]	ΔH_f°(298.15 K), kJ/mol	S°(298.15 K), J/(mol·K)	H°(298.15 K) $-H^\circ$(T = 0), kJ/mol
water	$H_2O(l)$	−285.830 ± 0.042	69.950 ± 0.080	13.293 ± 0.021
fluoride	$F^-(aq)$	−335.35 ± 0.65	−13.18 ± 0.54	
hydrogen fluoride	$HF(g)$	−273.30 ± 0.70	173.665 ± 0.035	8.599 ± 0.004
sulfur dioxide	$SO_2(g)$	−296.81 ± 0.20	(248.11 ± 0.06)	(10.548 ± 0.013)
carbon (graphite)	$C(c)$	0.00	5.74 ± 0.12	1.050 ± 0.020
silicon dioxide	$SiO_2(c,\alpha)$	−910.7 ± 1.0	41.46 ± 0.20	6.916 ± 0.020
aluminum	$Al(c)$	0.00	28.35 ± 0.08	4.565 ± 0.010
aluminum oxide	$Al_2O_3(c,\alpha)$	−1675.7 ± 1.3	50.92 ± 0.10	10.016 ± 0.020
zinc oxide	$ZnO(c)$	−350.46 ± 0.27	43.64 ± 0.40	6.933 ± 0.040

[a] Ref. 27.
[b] Additional collections of thermodynamic properties of pure substances are found in refs. 28–34.
[c] (c) = crystal; (c,α) = crystal, α-form.

concentrations, and in other ways (28,34–36). Calorimetry is also used directly to determine technologically important properties of complex poorly-defined materials. Of principal importance in this area are the heating values of fuels, coal, and coke (37), petroleum products (38), gaseous fuels (7), and others (39–40). In recent years the heating values of incinerator refuse and refuse-derived fuels have become of interest because of their potential use in integrated utilities systems (41) (see Fuels from waste). Calorimetry of hazardous and explosive materials has extremely important industrial consequences for the safe handling and transport of chemicals (42–45). The calorific value of fuels is an important factor in comparative prices. In the United States, gaseous fuel delivered to consumers is sold, to a large extent, on the basis of its total heating value. In some countries, a laboratory inspection service guarantees the accuracy of coal heating values. Less extensively, calorimetry is also used to establish the energy values of foods and other agricultural products for the study or control of human and animal nutrition (46). The routine determination of heating values of fuels (Table 3) is perhaps the major application of commercial calorimetry.

Calorimetry as a Diagnostic Procedure. Calorimetry for the determination of the properties of materials is often carried out with the utmost precision attainable. This approach tends to make the procedures relatively slow. In recent years, interest in the use of thermal effects as a diagnostic tool has caused increasing development of procedures allowing rapid determinations, but with some loss of precision and definition. The procedures of differential thermal analysis (dta) and differential scanning calorimetry (dsc) fall into this category (58–60).

Other modes of rapid calorimetry are also used for assay and analysis. Enthalpimetric titrations fall in this area (61). Berger (62), Brown (63), and Goldberg and co-workers (64) have applied calorimetry to clinical laboratory analysis.

Calorimetry is applied to some processes that are not directly associated with the thermal properties of specific materials. Studies of the metabolism of living organisms attracted early attention (65) and have been made recently on organisms ranging in size from a single cell (25,66) to human beings (67–68). Calorimetric determination of processes occurring in electrolytic power cells is a useful diagnostic procedure. Recent interest has developed in the use of calorimetry for quality control of miniature power cells in critical applications areas such as in cardiac pacemakers (69). Calo-

456 CALORIMETRY

Table 3. Typical Heating Values of Materials of Commerce Used for Energy [a]

Material	Q_v(gross) [b] Btu/lb	MJ/kg	Ref.
anthracite coal, dry mineral free (high) [c]	16,000	37.2	47
dry mineral free (low) [c]	14,750	34.3	47
bituminous coal, dry mineral free (high) [c]	16,000	37.2	47
moist mineral free (low) [c]	10,500	24.4	48
subbituminous coal, moist mineral free (high) [c]	11,500	26.7	48
moist mineral free (low) [c]	8,300	19.3	48
lignite, moist mineral free (high) [c]	8,300	19.3	48
moist mineral free (low) [c]	<6,300	<14.6	48
petroleum, crude, and fuel oils (S, H_2O, ash free) of sp gr (15.5°C/15.5°C), 1.00			
1.00	18,540	43.1	49
0.90	19,260	44.8	49
0.80	19,900	46.2	49
kerosene (Jet A) (mean) [c]	19,900	46.2	50
aircraft fuel JP5 (mean) [c]	19,600	45.5	50
aircraft fuel JP4 (mean) [c]	19,900	46.2	50
aviation gasoline 115/145 (mean) [c]	20,400	47.4	50
natural gas [d] (dry basis)	1,014	37.78	51
wood, oak (quercus) air dried (C, 50.16%; H, 6.02%; O, 43.36%; N, 0.09%; ash 0.37%)	8,300	19.3	52
wood, pine (pinus) air dried (C, 50.31%; H, 6.20%; O, 43.08%; N, 0.04%; ash, 0.3%)	9,150	21.3	52
lignin (hard wood)	10,619	24.7	53
lignin (soft wood)	11,350	26.4	54
cellulose (from cotton linters)	7,446	17.30	55
sulfite liquor (spent, 55% solids, nonneutralized) (mean value)	8,212	19.1	56
incinerator refuse (municipal, St. Louis) } high	7,593	17.6	57
magnetic and heavy materials removed } low	2,293	5.33	57
sucrose (weighed in vacuum) [e]	7,085	16.46	46
animal fat (beef)	17,100	39.7	46
corn oil	16,704	38.8	46
starch	7,560	17.6	46
protein (beef)	10,170	23.6	46
milk (dried)	9,786	22.7	46

[a] More details are found in the original references, in most cases.
[b] Heat of combustion at constant volume with products: gaseous CO_2, N_2, SO_2, and liquid H_2O.
[c] High and low values represent approximate upper and lower limits of reported information; mean is the mean of a series of representative values.
[d] Methane, 15.5°C, 101.560 kPa.
[e] National Bureau of Standards Certificate, Standard Sample 17, July 1936 (reported value 16,476 International kJ/g).

rimetry is used in the determination of purities of materials, to determine the energy in electromagnetic radiation, and to observe radioactive decay energies and the energies of beams of particles.

Purity Determination. Purities of substances may be determined calorimetrically from measurements of heat capacity in the melting region. The heat capacity–temperature relationship for a pure substance shows a sharp discontinuity at the beginning of melting, whereas for a substance containing impurities it will exhibit a gradual change. An estimate of the impurity content may be obtained by analysis of the shape of the heat capacity–temperature relationships near the melting temperature in terms of thermodynamic relationships of phases in mixtures (70–71).

Laser Energy. The measurement of laser power and energy is complicated by the wide range of wavelengths as well as the range of powers and energies involved. In a calorimetric procedure laser energy absorbed is compared to work done by electrical energy in a specially designed calorimeter (isoperibol or conduction). Depending on the means of absorbing the laser energy, such calorimeters may be classified as disk, cone, hollow-sphere, volume-absorption, and partial-absorption type. The limits of systematic error for a reference laser calorimeter are reported to be ±1% of the laser energy measured by the calorimeter (72). Precision calorimetry for laser energy measurements (72–74), and volume-absorbing (75) and circulated-liquid (76) calorimeters have been reviewed (see Lasers).

Other calorimetric measurements of electromagnetic energy include the measurement of microwave and radiofrequency radiant energy (77). The thermopile measurement of the energy radiated from a star is essentially a calorimetric measurement.

Applications in Nuclear Science. The energies of high energy particles are converted to a thermal effect in an appropriately designed calorimeter. As a result, calorimetry has been applied for several purposes in the investigation of nuclear phenomena (78–79). Most calorimeters used in nuclear science and technology may be divided into four general categories:

Type	*Measurement*
radionuclide calorimeter	the power of a radioactive source contained within the calorimeter
beam calorimeter	total power in a collimated beam of photons or charged particles from a source external to the calorimeter
local absorbed-dose calorimeter	the power deposited in a specimen of material located in the isotopic gamma field of a source external to the calorimeter
in-reactor calorimeter	the various effects due to reactor radiations within a nuclear reactor

A receiving Bunsen ice calorimeter has been specially adapted for high-precision power measurements on radioactive heat sources (80). The power of a radioactive source is a measure of the rate of decay of the radioactive elements present, when properly weighted in terms of their individual radioactive decay energies and their proportions in the source. Properly designed calorimetric experiments can give specfic radioactive decay energies or half-lives (81) of elements, or can be used to assay sources for particular elements. These procedures are now used in the technology of nuclear power plants (82) (see Nuclear reactors).

Commercialization. A striking feature of calorimetry is the diversity of styles of instruments and procedures that either must be used or have been found desirable to use in making the variety of thermal measurements that are required. The large number of instrument types and the comparatively small number of experimental users led to a situation in which for many years a calorimeter was designed and built by the individual who would use it (or it was built to order); and few styles were commercially available. Fuel calorimetry was a major exception for many years; bomb calorimeters (39–40) and gas-flow calorimeters (40) were available for these measurements. In recent years more instruments have become commercially available

(83–84). Microcalorimeters of several types and macroscopic solution calorimeters are now available. The major growth of commercial instrumentation has occurred where applications in routine analysis and in rapid characterization of materials have been found. However, it can hardly be said that there is a large market or a large calorimeter manufacturing industry. Almost all the commercially available calorimeters deal with reacting systems or with physical properties of nonchemical systems as described above. There are few, if any, commercially available instruments for accurately determining thermal properties of nonreacting substances over a wide temperature range.

Types of Calorimetry

General Considerations. Nonreacting and reacting systems are treated differently in determination of the properties of materials. The properties determined are often differentiated as thermodynamic (or thermophysical) and thermochemical properties. The boundary between nonreacting and reacting systems may not be sharp. Calorimetry of a typical nonreacting system involves determining its heat capacity and enthalpy over a range of temperatures. This may very well include heating a substance through several phase transitions—a change in solid crystal form, fusion, and vaporization. The changes are usually reversible and the substance is still the same substance identified as a single component according to Gibbs' phase rule. The performance of mixtures may follow essentially the same pattern. The calorimetry of reacting systems is most unambiguous when a chemical reaction occurs in an irreversible way, leading to products that are not readily regained merely by reversing the temperature path. Reaction processes include mixing of gases, mixing of liquids, dissolving gases, liquids, or solids in a liquid, etc.

Calorimetric procedures are not uniquely different or suited for only one kind of process or the other, though certain types of instruments and procedures have been specifically developed and optimized in design to carry out a certain type of process. Modifications of detail rather than of principle may be sufficient to make a calorimetric system suitable for a quite different process.

Calorimeters for Nonreacting Systems. Most major classes of the calorimeters for nonreacting systems are discussed briefly here with special emphasis on those that have been developed since 1965. Recent advances have been in the high temperature areas. Improvements in receiving calorimeters have extended the accurate measurement of enthalpy to temperatures above 2000 K. Development of levitation calorimeters, which are special types of receiving calorimeters, have extended the measurements to high temperatures as well as to substances in their liquid phase. Also, accurate nonsteady-state techniques (pulse calorimetry) have been perfected for measurements on electrically conducting materials to several thousand K.

Isothermal Calorimeters. In an isothermal calorimeter there is no temperature change during the experiment. The use of the term isothermal implies constancy of temperature with time of any part of the calorimeter, rather than the uniformity of temperature over the calorimeter at any given time. A convenient way of establishing and maintaining an isothermal system is to use a substance that undergoes a phase change at a convenient temperature; the constant temperature of the phase change and the enthalpy of the phase changes are used. The quantity of heat is measured by the amount of isothermal phase change it produces in the calorimeter material. The solid–liquid phase change is used in the ice calorimeter (85) and the resulting volume change is used as a measure of the heat added or removed. The liquid–vapor-phase change also has been used (86–87). Other methods that have been used to maintain constant temperature during the experiment utilize electric heating to compensate for heat removal and thermoelectric cooling to compensate for heat added (see Thermoelectric energy conversion). Removal or addition of heat can also be achieved by passing a fluid through the calorimeter. The isothermal calorimeter is often operated as a receiving calorimeter.

Isoperibol Calorimeters. In an isoperibol calorimeter, the outer shield is maintained at a constant temperature throughout the experiment while the temperature of the specimen container is changing. This type of calorimeter is sometimes referred to as an isothermal shield (or jacket) calorimeter, and sometimes is erroneously classified as an isothermal calorimeter. A detailed description of typical isoperibol calorimeters for measurements in the range 10–300 K is given by Stout (90).

The calorimeter vessel includes an electric heater, a temperature measuring device, and the specimen to be studied. Figure 1a shows the course of temperature with time in such a calorimeter, including the effect of heat exchange. Heat exchange between the calorimeter vessel and the shield is minimized by evacuating the space between them, by minimizing in the space between them the amount of solid material used for mechanical support and electrical connections, and by coating the relevant surfaces with material of high reflectivity.

Adiabatic Calorimeters. In an idealized adiabatic calorimeter there is no heat transfer across the calorimeter boundary during the experiment.

Adiabatic calorimeters have been used for measurements of heat capacities of solids and liquids from cryogenic to moderately high temperatures (about 1500 K). Operation is based on the determination of the temperature rise in the specimen accompanying the input of a measured quantity of electrical energy. A detailed discussion of adiabatic calorimeters operating in the range 10–350 K is given by Westrum and co-workers (91).

Application of adiabatic calorimetry to measurements in the range 300–800 K and a survey (in tabular form) of adiabatic calorimeters is discussed by West and Westrum (92). Although, in principle, an adiabatic calorimeter for use above room temperature should not differ from that used below room temperature, there is a considerable practical difference. Low temperature calorimeters are operated in vacuum, which eliminates heat transfer by gas conduction and convection; however, this advantage is nullified at higher temperatures by the increased heat transfer by radiation. The larger coefficient for radiant heat exchange between the calorimeter and the adiabatic shield requires correspondingly more precise control of the temperature difference for comparable uncertainty in the heat capacity data. A schematic representation of an adiabatic shield temperature control is presented in Figure 3.

A sensor TC detects the temperature difference between the calorimeter vessel and the adiabatic shield. By way of a feedback loop and a current-adjusting control unit, electric power is supplied to the adiabatic-shield heater to minimize the temperature difference.

Receiving Calorimeters. A receiving calorimeter (also referred to as a drop calorimeter) is a combination of a furnace and a calorimeter vessel using principles described earlier. Receiving calorimeters are used primarily for moderate and high temperature enthalpy measurements. Their operation is based on heating the specimen

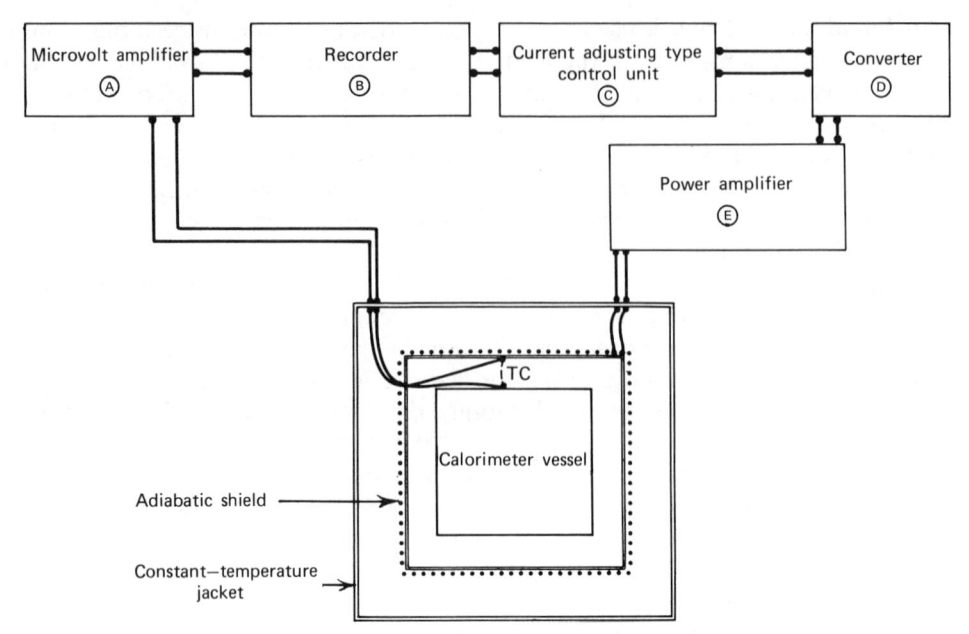

Figure 3. Adiabatic shield temperature control (13). Courtesy of NBS.

to a steady-state temperature in a furnace and transferring it quickly to the calorimeter vessel for measurement of the amount of energy released as the specimen cools to the calorimeter temperature, which generally is near room temperature (see Furnaces, electric).

A schematic representation of a receiving calorimeter is given in Figure 4. The specimen is suspended in the furnace and allowed to reach a steady temperature, then it is dropped into the well of the calorimeter vessel which may be isothermal, isoperibol or adiabatic (93).

There is a widespread use of receiving calorimeters for measurements above room temperature. An example is a Bunsen ice calorimeter (94–95). The wire-wound furnace can be operated up to near 1200 K (limited by the melting point of silver used as the furnace core). Silver is used to improve the temperature uniformity in the furnace and to minimize corrosion. The specimen is initially suspended in the isothermal zone of the furnace. After it reaches a constant temperature, it is dropped into the ice calorimeter. The heat transferred to the calorimeter vessel as the specimen cools to the calorimeter temperature (very nearly 273.15 K) melts a portion of the calorimeter ice. The resulting change in volume of the calorimeter's confined ice-water system is measured with a mercury dilatometer.

A second is a receiving adiabatic calorimeter which consists of an induction-heated, graphite-tube furnace and an adiabatic calorimeter placed above the furnace (96–97). The system is designed for operation in the range 1200–2600 K. The specimen (in a container) is initially suspended (with a thin tungsten wire) in the furnace until a constant temperature is reached. The specimen temperature in the furnace is measured with an automatic optical pyrometer. Then, the specimen is rapidly lifted into the calorimeter. The energy liberated by the specimen is determined from the measurement of temperature rise in the adiabatic calorimeter and the total heat capacity.

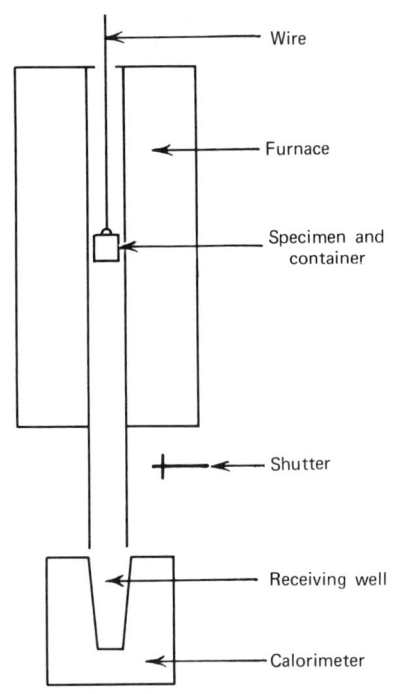

Figure 4. Schematic diagram of a receiving calorimeter. Courtesy of NBS.

Levitation Calorimeters. A levitation calorimeter is a receiving calorimeter, but is discussed separately because it has attracted particular attention recently. Its technique uses a combination of electromagnetic levitation heating of the specimen and energy measurement by a conventional calorimeter. This eliminates the use of any container for the specimen and thus minimizes specimen contamination, which makes it attractive for measurements at high temperatures on refractory materials, especially in their liquid phase.

The specimen is levitated in a coil and is heated to high temperatures using radiofrequency induction. The most commonly used receiving calorimeter is the isoperibol type. Two of the extensively used levitation calorimeters for measurements on refractory metals (solids and liquid phase) are described in refs. 98–99.

Pulse Calorimeters. The relatively long high-temperature exposure times of most of the previously-described techniques (minutes to hours) in conjunction with the rapid increase with temperature of various phenomena, such as heat transfer, chemical reactions, evaporation, loss of mechanical strength, limit the application of the conventional techniques to temperatures below about 2500 K. Accurate measurements above this limit use techniques in which the contribution of these undesirable phenomena can be made negligibly small by exposing the specimen to high temperatures for only a very short time (less than a second).

The advantages of pulse techniques have been realized over 50 years in which pulse calorimeters were constructed and used for measurements of heat capacity at moderate and high temperatures. In almost all cases the progress did not extend beyond the preliminary stage. This can be attributed largely to the lack of proper instrumentation and to the difficulties in accurate transient measurement techniques.

Increasing demand for properties of materials at high temperatures and rapid advances in the electronics field have stimulated new efforts in pulse calorimetry.

A pulse calorimeter system consists of an electrical power pulsing circuit and associated high-speed measuring circuits. The pulsing circuit includes the specimen in series with a power source, an adjustable resistance, a standard resistance, and a fast-acting switch (Fig. 5).

In general, power imparted to the specimen is obtained from measurements of current through the specimen and potential difference across the specimen as functions of time. Temperature measurements are made using resistive, thermoelectric thermocouples, or optical means. The experiment chamber of an accurate millisecond-resolution pulse calorimeter (100) is shown in Figure 6.

Historical developments of pulse calorimetry are given in detail for solids (101) and liquids (102). The needs for accurate data on solids up to their melting temperatures (up to about 4000 K) and on liquids over 5000 K have stimulated recent effort, eg, millisecond-resolution pulse calorimetry for solids (100,103) and the preliminary work on microsecond-resolution calorimetry for liquids (104–105). Because of its inherent advantages, pulse calorimetry has also been applied to measurements at very high pressures, up to 10^4 MPa (ca 10^5 atm) (106).

Modulation Calorimeters. A modulation calorimeter utilizes a periodically varying heating scheme for determining heat capacities of materials. The technique, more than 60 years old, is based on the measurement of the amplitude of temperature fluctuations in a wire when heated by alternating current. Temperature fluctuations are measured either optically or resistively. It has been used for heat capacity measurements of refractory metals at high temperatures approaching the melting point (107). The ap-

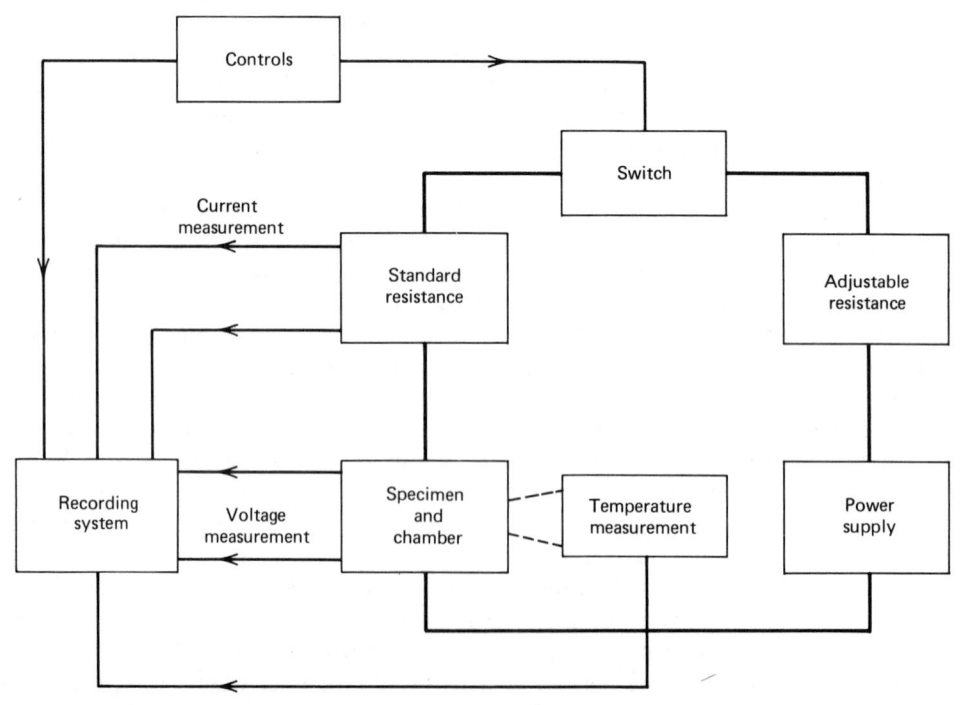

Figure 5. Simplified block diagram of a pulse calorimeter system. Courtesy of NBS.

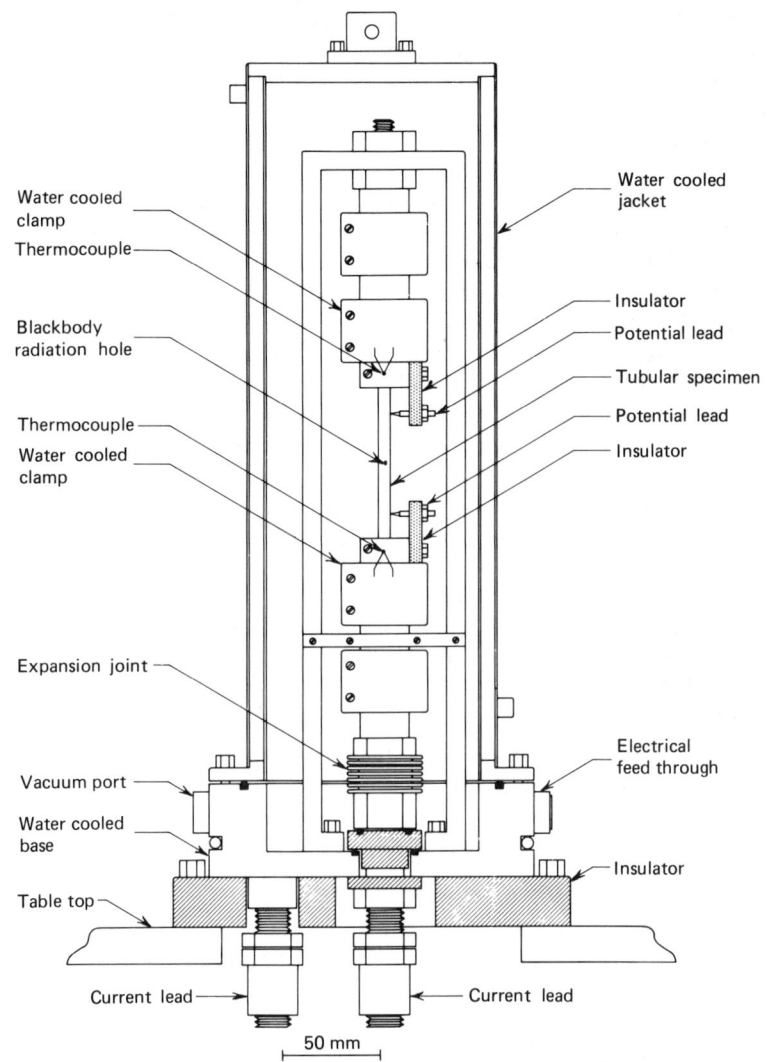

Figure 6. Experiment chamber of a millisecond resolution pulse calorimeter (100). Courtesy of NBS.

plication of the modulation method to nonconductors, as well as conductors, by heating the specimen externally using electron bombardment has been described (108).

Calorimeters for Reacting Systems. *Bomb Combustion Calorimeters.* The most commonly used type of reaction calorimetry is bomb calorimetry using oxygen as the oxidizer. The calorimeter bomb was devised by Berthelot in 1885. The widespread use of this technique is due to the fact that practically all organic materials, including foodstuffs, as well as solid and liquid fuels, burn almost completely when ignited in excess oxygen under a pressure of about 3 MPa (ca 30 atm) and lead to readily identifiable products in a form suitable for analysis. The procedure can be modified to allow the use of other gaseous oxidizers such as fluorine; and many inorganic substances can be burned or formed by suitable choice of reactant and oxidizer (26).

The common commercial calorimeter used for determining heating values of coal and other solid or liquid fuels is of this type, and may be operated in the isoperibol or the adiabatic mode as a stationary instrument (18–20,109). Figure 7 is a traditional stirred-water isoperibol bomb calorimeter. A is an optional air-bath enclosure to reduce the effects of room-air circulation and short-term room-temperature fluctuations. B is a stirred-water, controlled-temperature, isothermal jacket. Separating the interior surface, C, of the jacket and the exterior surface, D, of the calorimeter vessel is an air space, F, about 1 cm thick. Located close to surface, D, is a thermometer, E, which may use a platinum-resistance, a quartz-oscillator, a thermistor, or a mercury-in-glass sensing element. The reaction chamber is bomb, G. If electrical energy is to be used for calibration, it should be introduced as close to the bomb as possible. H represents such an electrical heater (optional) which can be used either for calibration or to adjust the initial temperature of the calorimeter vessel. Arrows, I, represent the circulatory flow of water in the calorimeter vessel. This flow, forced by propeller, L, is guided by devices such as a cylindrical channel, J, into a uniform flow pattern. The jacket temperature regulation also involves a careful placement, relative to one another of the sensor, K, for the temperature regulator, the propeller, M, and the jacket heater, N. All connections to the calorimeter are tempered in the jacket and are of as low a thermal conductance as feasible. Rotating bomb calorimeters analogous to this are also in use (111). Bomb calorimeters without stirred water (aneroid) are also in use.

The aneroid, adiabatic, rotating-bomb calorimeter is a cylindrically shaped bomb mounted on axial pins inside an adiabatic shield (112,113). This shield in turn is mounted with a geared belt around its equator between two axles projecting in from opposite sides of the exterior box. The biaxial rotation that is thus possible causes the small amount (1–10 cm^3) of aqueous solution in the bomb to wash the whole interior surface of the bomb after combustion, and thus to make the aqueous phase homogeneous and of a well-defined thermodynamic state.

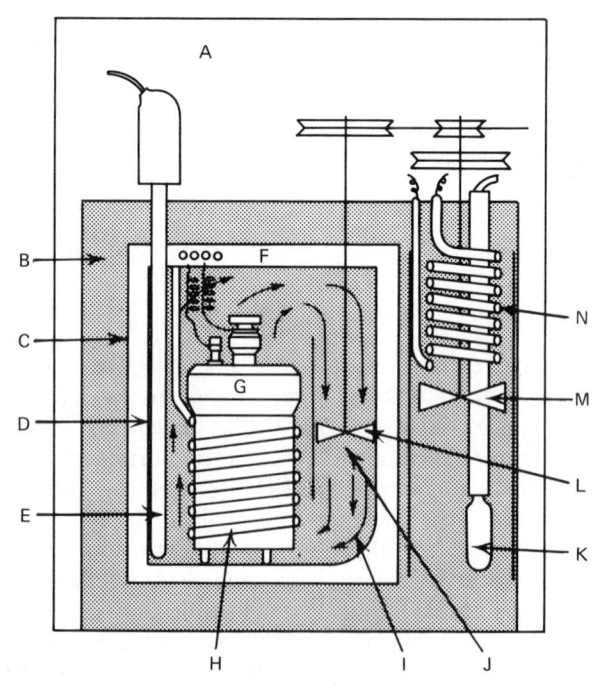

Figure 7. Bomb calorimeter assembly (see text for interpretation) (110). Courtesy of NBS.

The overall accuracy of the stirred-water calorimeter is better than that of the aneroid calorimeter for electrical-chemical comparisons, but the reproducibility of repetitive measurements is better with the aneroid calorimeter (113).

Gas-Flow Combustion Calorimeters. Because combustion of gaseous mixtures in a bomb is not recommended, a flow calorimeter is preferable to carry out the combustion calorimetry of a gaseous material. This type of calorimeter is more widely used industrially than as a scientific research instrument. Rossini (114) devised a reaction vessel in which gaseous oxidizer and reducer are introduced into the combustion area, one as an atmosphere (in excess), and the other as a fuel. The calorimetry differs little from bomb calorimetry, the difference lying principally in the reaction vessel. As with bomb calorimetry, various oxidizers or reducers can be used (115), the principal feature of the system being that a homogeneous, rather than a heterogeneous mixture is burning and the reactants and products pass through the calorimeter rather than being confined to it. The experiment can be initiated upon introduction of the fuel (substance not in excess), and ignited by a high voltage spark. The reaction is entirely under the control of the operator, occurring at a rate that is determined by the rate of introduction of fuel (and atmosphere) and lasts until the desired amount of reaction has occurred. This is in contrast to the bomb calorimetry process, which when initiated proceeds to completion at a rate dependent upon the materials and initial conditions.

For applications to commercial gaseous fuels various flow calorimeters (Junkers, Boys, and others) were used for many years (7,116–117). However, the tediousness of this test and its inadequacy for gaseous fuels of perhaps variable composition has led to an almost complete displacement of these manually-operated, timed-duration instruments by continuous-flow automatic recording calorimeters. The best known such calorimeter instrument in the United States, widely used by public utilities corporations, gas transmission companies, and gas producers, is the Cutler-Hammer (formerly Thomas) recording calorimeter.

In the continuous recording calorimeter (118–119), gas is burned at a constant rate and the heat developed is absorbed by a stream of air. The rates of flow of the gas, air for combustion, and the heat-absorbing air, are regulated by metering devices similar in construction to ordinary wet gas meters. These metering devices are geared so that the ratio of the rates of flow of gas and heat-absorbing air is constant. The products of combustion are kept separate from the heat-absorbing air and are cooled very nearly to the initial temperature of the air. The water formed is condensed, thus the rise in temperature of the heat-absorbing air is proportional to the total heating value of the gas.

Calorimeters of this kind operate with an imprecision of about 0.3% (120). A reference gas, with a known heating value established by a series of comparisons ultimately referred to the National Bureau of Standards, is used to standardize the calorimeter.

Solution Calorimeters. For reactions involving the mixing of liquids or the solution of a solid in a liquid, the reaction vessel may adequately serve also as the calorimeter vessel. The stirring of the liquid required in order to dissolve or mix the reactants also serves to bring the whole reaction system rapidly to uniform temperature. The smaller amount of reaction energy needs to heat only a relatively small calorimeter mass, and so the observable temperature rise is quite adequate for many reactions. The widely-used solution calorimeter developed by Sunner and Wadso, and other solution calorimeters are commercially available (83). An adiabatic vacuum-jacket solution calorimeter (13) has been used in numerous accurate studies such as certification of reference materials for solution calorimetry.

A somewhat different principle has found application in analytical chemistry (121–122) by which heat released during a reaction is detected by a thermoelement immersed in the reaction mixture at the point of injection of one of the reactants. In the instruments used, an enthalpimetric titration can be made.

The reaction vessel is commonly a Dewar-type glass vessel fitted with an insert allowing controlled delivery of one of the reactant solutions, together with a thermistor or other temperature sensor, provision for stirring, and an electrical heater (Fig. 8). Inside from left to right are a reservoir of titrant, electrical heater, sample bulb breaker, motor-driven pipette capillary, and thermistor. The titrant is brought to the isothermal bath temperature by passing through plastic tubing in a coil above the lid. The temperature changes of the system can be followed as a function of the addition of reactant. This technique has been variously termed enthalpy titration, direct injection enthalpimetry (die), thermometric enthalpy titration (tet), calorimetric titration, and thermometric titration (61,123–124). In thermometric enthalpy titration, continuous-flow enthalpimetry is a further variant and is indistinguishable from continuous-flow microcalorimetry.

Microcalorimeters. Although a microcalorimeter measures small amounts of heat, the calorimeter itself is not usually small.

The principal development of the modern microcalorimeter stems from the work of Tian and Calvet (24). Figure 9 shows the principle of such a conduction calorimeter and its realization as an instrument. In Figure 9a the essential components of a microcalorimeter are sketched in an appropriate arrangement. The hatched areas are thermoelectric sensors. These may be n- and p-type semiconductors such as bismuth selenide, bismuth telluride, or bismuth antimonide, with thermoelectric power in the range of 400 μV/K (25) (see Semiconductors). The output of the thermoelectric sensors ΔT is directly proportional to the rate of heat flow from the heat source to the isothermal heat sink; and its integral over an elapsed time $t_2 - t_1$ measures the energy q transferred in that time. In Figure 9b the cylindrically symmetrical heat sink or block (having an octagonal cross section) is mounted with axial supports in an enclosure which isolates it thermally. The electrical leads are brought out through the axle, which also has a handle to rotate the block and mix liquids in the heat source region. Openings in the block and enclosure permit specimens to be introduced without disassembling the calorimeter.

In this kind of device a reaction may be started by mixing solutions in a container in the region labeled heat source. Upon mixing the solutions, a pulse of heat is generated by the reaction, which causes the voltage to rise and decay as shown in Figure 1b. The procedure is also applicable to measurement of continuous heat sources, such as metabolic activity, or the decay period of a radioactive source.

Important factors determining the lower limit of heat detection by the microcalorimeter are noise level of the detector system and the emf response characteristic of the thermoelectric element. The number of elements is most important in achieving a proper coverage of the surface of the heat source and has little or no effect in obtaining greater sensitivity. Proper coverage assures that the measured emf appropriately sums

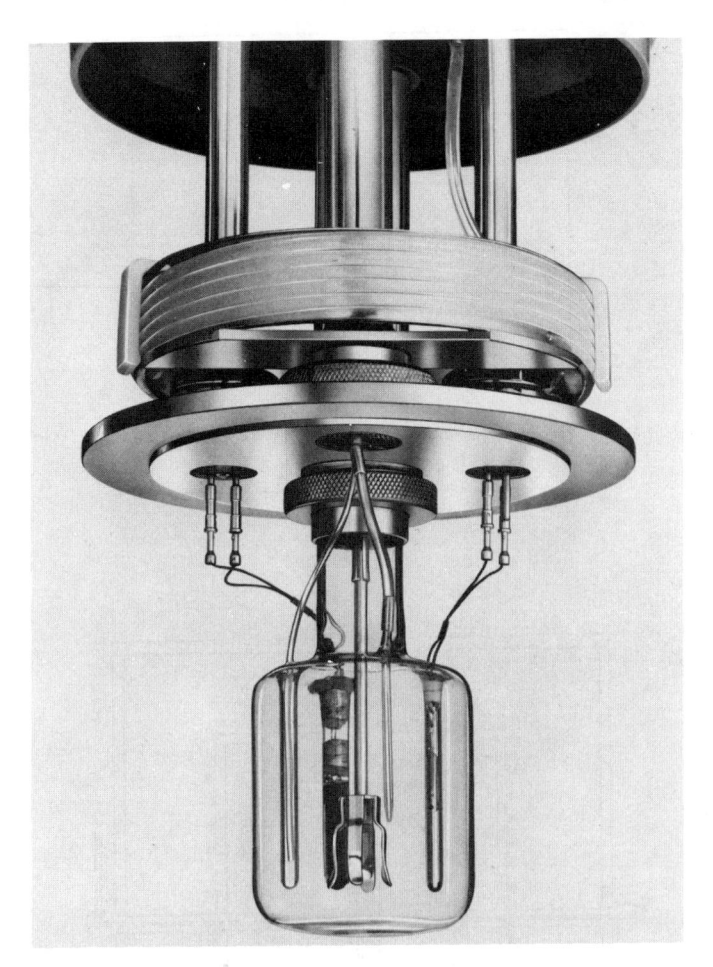

Figure 8. Solution-calorimeter reaction vessel for enthalpimetric titration. Courtesy LKB Instruments, Inc., Rockville, Md.

the total heat conducted to the heat sink. However, increased emf resulting from an increased number of junctions, n, is just balanced by an increased conduction path:

$$\frac{E}{q} = \frac{n \cdot \alpha \cdot \Delta T}{n \cdot \lambda \cdot \Delta T} \qquad (5)$$

Thus the ratio E/q is a characteristic of the system, determined by the emf characteristic α and the thermal conductance, λ, of the thermoelectric material.

Early microcalorimeters incorporated large numbers of thermal elements (22,24). Recent improvements in microcalorimeter instrument design (25) have depended largely on materials of high thermoelectric response (eg, BiTe) and the emplacement of the elements in orderly arrays in sheets convenient for fabricating into instruments.

468 CALORIMETRY

$$\frac{\partial q}{\partial t} = k\Delta T$$

$$q = k\int_{t_1}^{t_2} \Delta T\, dt$$

(a)

(b)

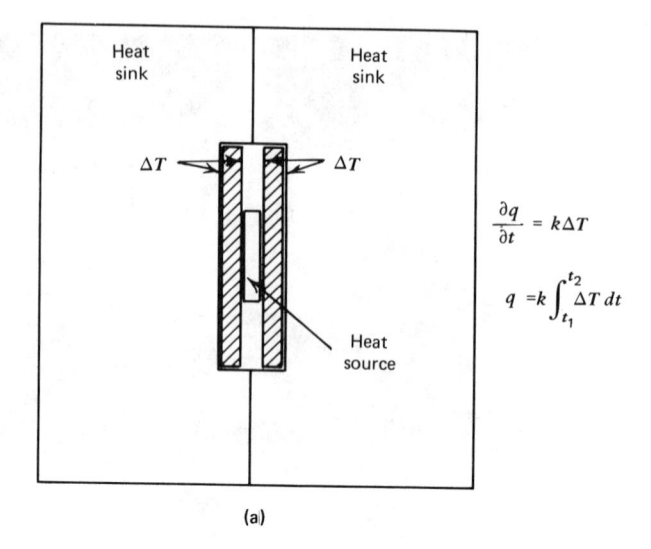

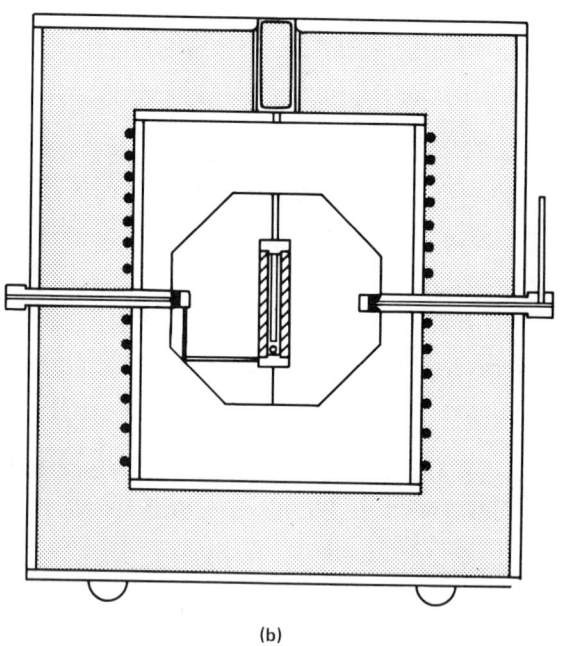

Figure 9. (a) Cross section and principle of operation of a simple conduction-type calorimeter (25). (b) An NBS microcalorimeter.

Many instruments depend upon a twin-cell design in which two identical reaction chambers are embedded in the heat sink (22,125). In one no reaction occurs and the observed measurement is the difference in the responses of the two cells. This eliminates small temperature inhomogeneities owing to temperature drift of the heat sink, and reduces the temperature control requirements for the heat sink. However, successful microcalorimeters have been built with a single cell (25). Microcalorimeters have been built for batch operation (22,25,125), continuous flow of fluids (126), and stopped-flow operation (127).

Trends in Calorimetry

Differential scanning calorimeters and similar devices are becoming more accurate. They are supplemented by other dynamic procedures that utilize rapid data-acquisition devices and convey the benefits of the rapid determination of properties. For instance, rapid dynamic procedures seem to offer the only feasible way, to obtain information about the properties of condensed-phase materials in the extremely high-temperature region, where containers are of little value because of extreme chemical reactivity of materials and melting-temperature limitations.

The availability of continuous or very rapid data-acquisition procedures combined with programmed temperature control is encouraging attempts to measure kinetic as well as thermodynamic information in the same procedure.

One effect of the increased interest in machine reduction of data, a real necessity for continuously recorded information, has been a trend to the use of auxiliary instrumentation that caters to this need. This has given a strong impetus to use of instruments that interface with a digital logging device. The quartz-oscillator thermometer is a device that has had a good reception because the technology has been available to digitize the output more readily than for platinum resistance thermometers that use d-c resistance-measurement technology (see Digital displays). Figure 10 shows instrumentation for a digital data-logging system for a microcalorimeter. A microcalorimeter is a typical apparatus for which quasi-continuous records are a valuable form of output.

In the traditional fields of calorimetry there seems to be little current trend toward greater accuracy. In traditional reaction calorimetry and for calorimetry of nonreacting

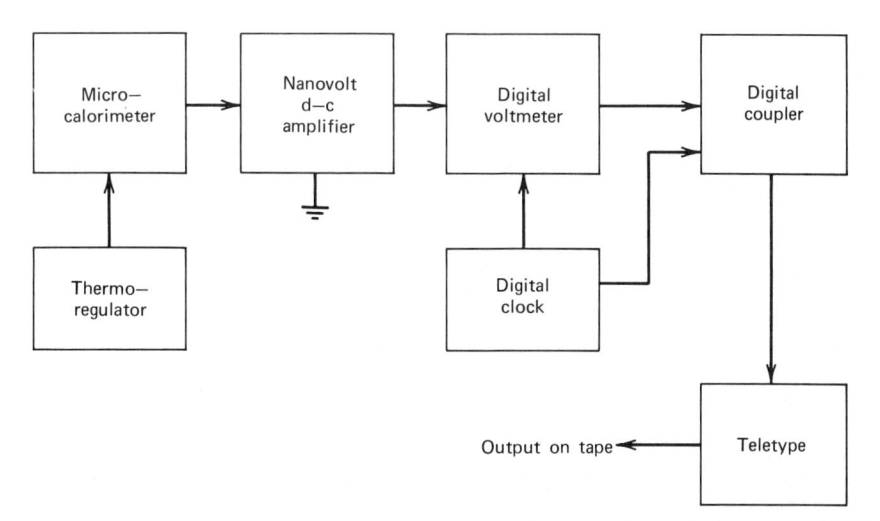

Figure 10. Data-logging system for a conduction calorimeter (25,128). Courtesy of NBS.

systems, it is not the calorimetry that limits the accuracy or meaning of the observations; rather the limiting factor is the precise identification of the process that has occurred.

Another trend is the increasing emphasis on biological materials and processes as subjects of calorimetric study. Microcalorimetry is especially apt for biological systems because only relatively small amounts of materials usually are available for investigation, and because microcalorimetry is applicable to continuous weak sources of heat.

Finally, it is reasonable to expect that a revitalization of calorimetric studies will be one response to an increasing awareness of the need for thrift in the use of energy, the desirability of developing new forms of convenient fuels, and an increasing need to utilize material resources that are either completely different or are of lower quality than have been used in the past (see Fuels).

Nomenclature

C_p = heat capacity at constant pressure
C_v = heat capacity at constant volume
E = emf
G = Gibbs free energy
H = enthalpy
I = current
q = heat flow
R = resistance
S = entropy
t = time
U = internal energy
w = work
w_{el} = electrical work
α = emf characteristic
θ = temperature of a surface
λ = thermal conductance

BIBLIOGRAPHY

"Calorimetry" in *ECT* 1st ed., Vol. 2, pp. 793–808, by J. G. Aston, Saul Isserow, and G. J. Szasz, Pennsylvania State University; "Calorimetry" in *ECT* 2nd ed., Vol. 4, pp. 35–53, by Edward S. J. Tomezsko and John G. Aston, The Pennsylvania State University.

1. G. T. Armstrong, *J. Chem. Ed.* **41,** 297 (1964).
2. J. P. Joule, *Philos. Mag.* **23,** 263, 347, 435 (1843).
3. *Standard for Metric Practice, ASTM Designation E 380-76; IEEE Std. 268-1975; American National Standard Z210.1,* American Society for Testing and Materials, Philadelphia, Pa., 1976.
4. *National Academy of Sciences Recommended Dietary Allowances,* 8th ed., National Academy of Science, Washington, D.C., 1974, p. 25.
5. A. White, P. Handler, and E. L. Smith, *Principles of Biochemistry,* 4th ed., McGraw Hill Book Company, New York, 1968, p. 292.
6. D. L. Ornellas, *J. Phys. Chem.* **72,** 2390 (1968).
7. C. G. Hyde and M. W. Jones, *Gas Calorimetry,* 2nd ed., Ernest Benn, Ltd., London, Eng., 1960, p. 18.

8. National Petroleum Council, *U.S. Energy Outlook,* National Petroleum Council, Washington, D.C., 1972, p. 134.
9. J. P. McCullough and D. W. Scott, eds., *Experimental Thermodynamics,* Vol. 1, Butterworths, London, Eng., 1968, Chapters 4 and 6.
10. J. Coops, R. S. Jessup, and K. van Nes in F. D. Rossini, ed., *Experimental Thermochemistry, Measurement of Heats of Reaction,* Vol. I, Interscience Publishers, New York, 1956, pp. 27–58.
11. R. N. Goldberg and G. T. Armstrong, *Med. Instrum. (Baltimore)* **8,** 30 (1974).
12. C. H. Page and P. Vigoreux, *Nat. Bur. Stand. (U.S.) Spec. Publ.* **330,** 38 (1974).
13. E. J. Prosen and M. V. Kilday, *J. Res. Nat. Bur. Stand.* **77A,** 179 (1972).
14. H. F. Stimson, D. R. Lovejoy, and J. R. Clement in ref. 9, Chap. 2, pp. 15–57.
15. P. S. Engel and co-workers, *J. Am. Chem. Soc.* **96,** 2381 (1974).
16. A. P. Brunetti, E. J. Prosen, and R. N. Goldberg, *J. Res. Nat. Bur. Stand.* **77A,** 599 (1973).
17. H. A. Skinner, J. M. Sturtevant, and S. Sunner in H. A. Skinner, ed., *Experimental Thermochemistry, Measurement of Heats of Reaction,* Vol. II, Interscience Publishers, New York, 1962, Chapt. 9.
18. *Annual Book of ASTM Standards, ASTM Designation D 2015,* Pt. 26, American Society for Testing and Materials, Philadelphia, Pa., 1976.
19. *ASTM Designation D 3286,* in ref. 18, Pt. 26.
20. *ASTM Designation D 240,* in ref. 18, Pt. 23.
21. E. F. Westrum, Jr., G. T. Furukawa, and J. P. McCullough, in ref. 9, Chapt. 5.
22. T. H. Benzinger and C. Kitzinger in J. D. Hardy, ed., *Temperature: Its Measurement and Control in Science and Industry,* Vol. 3, Pt. 3, Reinhold, New York, 1963, pp. 43–60.
23. H. D. Brown, ed., *Biochemical Microcalorimetry,* Academic Press, Inc., New York, 1969.
24. E. Calvet and H. Prat, *Microcalorimétrie: Applications, Physicochimiques et Biologiques,* Masson et cie., Paris, Fr., 1956; E. Calvet and H. Prat (transl. by H. A. Skinner), *Recent Progress in Microcalorimetry,* MacMillan Company, New York, 1963.
25. E. J. Prosen and co-workers in H. Kambe and P. D. Garn, eds., *Thermal Analysis: Comparative Studies on Materials,* John Wiley & Sons, Inc., New York, 1974, pp. 253–289.
26. F. D. Rossini, ed., *Experimental Thermochemistry, Measurement of Heats of Reaction,* Vol. I, 1956; H. A. Skinner, ed., Vol. II, 1962; S. Sunner, ed., Vol. III, 1977, Interscience Publishers, New York.
27. CODATA Task Group on Key Values for Thermodynamics, *CODATA Bulletin No. 17,* International Council of Scientific Unions, CODATA Secretariat, Paris, France, Jan. 1976.
28. O. Kubaschewski, E. Ll. Evans, and C. B. Alcock, *Metallurgical Thermochemistry,* 4th ed., Pergamon Press, London, Eng., 1967.
29. J. D. Cox and G. Pilcher, *Thermochemistry of Organic and Organometallic Compounds,* Academic Press, London, Eng., 1970, p. 140 ff.
30. D. R. Stull, E. F. Westrum, Jr., and G. C. Sinke, *The Chemical Thermodynamics of Organic Compounds,* John Wiley & Sons, Inc., New York, 1969, pp. 243 ff.
31. D. D. Wagman and co-workers, *Nat. Bur. Stand. (U.S.) Tech. Note* **270,** (1964 and later years).
32. *JANAF Thermochemical Tables,* 2nd ed., and supplements *NSRDS-NBS 37,* U.S. Government Printing Office, Washington, D.C., 1971.
33. R. Hultgren and co-workers, *Selected Values of Thermodynamic Properties of Metals and Alloys,* John Wiley & Sons, Inc., New York, 1963.
34. D. R. Stull, E. F. Westrum, Jr., and G. C. Sinke, *The Chemical Thermodynamics of Organic Compounds,* John Wiley & Sons, Inc., New York, 1969.
35. G. R. Fitterer, ed., *Applications of Fundamental Thermodynamics to Metallurgical Processes,* Gordon and Breach, New York, 1967.
36. T. Rosenqvist, *Principles of Extractive Metallurgy,* McGraw Hill Book Company, New York, 1974.
37. H. H. Lowry, ed., *Chemistry of Coal Utilization,* Vol. 1, John Wiley & Sons, Inc., New York, 1945, Chapters 2 and 4.
38. K. Boldt and B. R. Hall, eds., *Significance of Tests for Petroleum Products,* American Society for Testing and Materials, Philadelphia, Pa., 1977 pp. 60, 103, 121–124.
39. S. W. Parr, *The Analysis of Fuel, Gas, Water, and Lubricants,* 4th ed., McGraw Hill Book Company, Inc., New York, 1932, Chapt. 4; *Instruments* **3**(2) 71 (1930). Calorimeters developed by S. W. Parr are now manufactured by Parr Instrument Co., Moline, Ill.
40. B. Pugh, *Fuel Calorimetry,* Butterworths, London, Eng., 1966; ref. 7 and 116.

41. N. J. Weinstein and R. F. Toro, *Thermal Processing of Municipal Solid Waste for Resource and Energy Recovery*, Ann Arbor Science Publishers, Ann Arbor, Mich., 1976.
42. D. Gross and A. F. Robertson, *J. Res. Nat. Bur. Stand.* **61,** 413 (1958).
43. A. A. Duswalt, "Analysis of Highly Exothermic Reactions by DSC" in R. S. Porter and J. F. Johnson, eds., *Analytical Calorimetry,* Plenum Press, New York, 1968, p. 313.
44. D. I. Townsend, "Principles of Self-Accelerating Reactions, Hazard Evaluation, and Reaction Kinetics," *AIChE 83rd National Meeting, Houston, Texas, Mar. 1977, Session 44.*
45. Report by H. M. Factory Inspectorate, Health and Safety Executive, *The Explosion at The Dow Chemical Factory, King's Lynn, 27 June 1976,* H. M. Stationery Office, London, Eng., Mar. 1977, pp. 13–19.
46. A. L. Merrill and B. K. Watt, *Energy Value of Food, U.S. Department of Agriculture, Handbook No. 74,* U.S. Goverment Printing Office, Washington, D.C., Mar. 1955.
47. R. A. Mott, *Coal Assessment,* The Institute of Fuel London, Eng., 1948, pp. 27, 34.
48. *Annual Book of ASTM Standards, ASTM D 388-66,* Part 26, American Society for Testing and Materials, Philadelphia, Pa.
49. *Nat. Bur. Stand. (U.S.) Misc. Publ.* **97,** (1929).
50. R. L. Nuttall and G. T. Armstrong, *Nat. Bur. Stand (U.S.) Tech. Note* **937** (Apr. 1977).
51. G. T. Armstrong, E. S. Domalski, and J. I. Minor, Jr., *Am. Gas Assoc. Oper. Sect. Proc.,* D-74 (1972).
52. E. Gottlieb, *J. Prakt. Chem.* **28,** 385 (1883).
53. S. A. Rydholm, *Pulp Pap. Mag. Can.,* T2 (Jan. 1967).
54. J. Gullickson, *Proc. Symp. Recovery of Pulping Chemicals, Helsinki, Finland, 1968,* 221 (1969).
55. R. S. Jessup and E. J. Prosen, *J. Res. Nat. Bur. Stand.* **44,** 387 (1950).
56. B. Holden, *Proc. Symp. Recovery of Pulping Chemicals, Helsinki, Finland, 1968,* 387 (1969).
57. F. E. Wisely, G. W. Sutterfield, and D. L. Klumb in E. Eleskin, ed., *Proceedings of the Fourth Mineral Waste Utilization Symposium,* Illinois Institute of Technology Research Institute, Chicago, Ill., 1974, pp. 191–195.
58. P. D. Garn, *Thermoanalytic Methods of Investigation,* Academic Press, New York, 1965.
59. H. Kambe and P. D. Garn, eds., *Thermal Analysis: Comparative Studies on Materials,* John Wiley & Sons, Inc., New York, 1974.
60. R. S. Porter and J. F. Johnson, *Analytical Calorimetry,* Vol. I, 1968; Vol. 4, 1977, Plenum Press, New York.
61. J. Barthel, *Thermometric Titrations,* John Wiley & Sons, Inc., New York, 1975.
62. R. L. Berger in ref. 23, pp. 275–289.
63. H. D. Brown in ref. 23, pp. 291–296.
64. R. N. Goldberg and co-workers *Anal. Biochem.* **64,** 68 (1975).
65. A. L. Lavoisier and P. S. de Laplace, *Mém. Acad. R. Sci.* **1780,** 355 (1784); E. Mendelsohn, *Heat and Life,* Harvard University Press, Cambridge, Mass., 1964, p. 147 ff.
66. A. E. Beezer, "Microcalorimetric Investigators of Microorganisms" in I. Lamprecht and B. Schaarschmidt, eds., *Applications of Calorimetry in Life Sciences,* de Gruyter, Berlin, Ger., 1977, pp. 109–118.
67. E. Jequier in ref. 66, pp. 261–278.
68. T. H. Benzinger and C. Kitzinger, "Gradient Layer Calorimetry and Human Calorimetry" in ref. 22, pp. 87–109.
69. E. J. Prosen and J. C. Colbert, *Semiconductor Measurement Technology: Reliability Technology for Cardiac Pacemakers II—A Workshop Report, Nat. Bur. Stand. Spec. Publ. SP 400-42,* Aug. 1977, pp. 16–18.
70. A. R. Glasgow, Jr., and co-workers, *Anal. Chim. Acta* **17,** 54 (1957).
71. J. G. Aston, M. R. Cines, and H. L. Fink, *J. Am. Chem. Soc.* **69,** 1532 (1947).
72. E. D. West and co-workers, *J. Res. Nat. Bur. Stand.* **76A,** 13 (1972).
73. S. R. Gunn, *J. Phys. E* **6,** 105 (1973).
74. M. M. Birky, *Appl. Opt.* **10,** 132 (1971).
75. S. R. Gunn, *Rev. Sci. Instrum.* **45,** 936 (1975).
76. G. A. Fisk and M. A. Gusinow, *Rev. Sci. Instrum.* **48,** 118 (1977).
77. A. Y. Rumfelt and L. B. Elwell, *Proc. IEEE* **55,** 837 (1967); *Nat. Bur. Stand. (U.S.) Spec. Publ. 300* **4,** (1970).
78. S. R. Gunn, *Nucl. Instrum. Meth.* **29,** 1 (1964).
79. S. R. Gunn, *Nucl. Instrum. Meth.* **85,** 285 (1970).
80. D. A. Ditmars, *Int. J. Applied Radiat. Isot.* **27,** 469 (1976).

81. D. C. Ginnings, A. F. Ball, and D. T. Vier, *J. Res. Nat. Bur. Stand.* **50,** 75 (1953).
82. *American National Standard ANSI N15.22-1975, Calibration Techniques for the Calorimetric Assay of Plutonium-Bearing Solids Applied to Nuclear Materials Control,* American National Standards Institute, New York, 1975.
83. S. Sunner and co-workers, *Sci. Tools* **13**(1), (April 1966); *Acta Chem. Scand.* **13,** 97 (1959); *Acta Chem. Scand.* **22,** 1842 (1968); *Thermochemie, Colloques Internationaux CNRS no. 201, Marseille, 1971,* 1972, pp. 129–133 (calorimeters manufactured by LKB Produkter AB, Bromma, Sweden).
84. M. Laffitte in *Thermochemie, Colloques Internationaux du CNRS no. 201, Marseille, 1971,* 1972, pp. 135–151 (calorimeter manufactured by Seteram, Lyon, France); J. Christensen and co-workers, *Rev. Sci. Instr.* **36,** 779 (1965); **39,** 1356 (1968) (calorimeters manufactured by Tronac, Inc., Orem, Utah); P. Picker and co-workers, *Can. Res. Dev.* 11 (Jan.–Feb. 1974); *J. Chem. Thermodyn.* **1,** 469 (1969); **3,** 631 (1971); *Colloques Internationaux du CNRS no. 201, Marseille, 1971,* 1972, pp. 161–164 (calorimeters marketed by Techneurop, Inc., Montreal, Quebec, Canada); see refs. 118–120 (calorimeters manufactured by Cutler-Hammer, Inc., Milwaukee, Wisconsin); J. Jorden in I. M. Kolthoff, P. J. Elving, and E. B. Sandel, eds., *Treatise on Analytical Chemistry,* Pt. 1, vol. 8, Interscience Publishers, New York, 1968, Chapt. 91 (calorimetric device manufactured by American Instrument Co., Silver Spring, Maryland); R. C. Wilhoit, *J. Chem. Ed.* **44,** A571, A629, A685 (1967) (several commercial calorimeters are discussed); E. D. West and co-workers, *J. Res. Nat. Bur. Stand.* **76A,** 13 (1972); *J. Opt. Soc. Am.* **65,** 573 (1975) (laser power calorimeters manufactured by Calorimetrics, Boulder, Colorado).
85. D. C. Ginnings, T. B. Douglas, and A. F. Ball, *J. Res. Nat. Bur. Stand.* **45,** 23 (1950).
86. C. A. Kraus, and J. A. Ridderhof, *J. Am. Chem. Soc.* **56,** 79 (1934).
87. L. K. J. Tong and W. O. Kenyon, *J. Am. Chem. Soc.* **67,** 1278 (1945).
88. W. B. Mann, *J. Res. Nat. Bur. Stand.* **52,** 177 (1954).
89. Ref. 10, pp. 239, 297.
90. J. W. Stout in ref. 9, pp. 215–216.
91. E. F. Westrum, Jr., G. T. Furukawa, and J. P. McCullough in ref. 9, pp. 133–214.
92. E. D. West and E. F. Westrum, Jr., in ref. 9, pp. 333–367.
93. T. B. Douglas and E. G. King in ref. 9, pp. 293–331.
94. D. A. Ditmars and T. B. Douglas, *J. Res. Nat. Bur. Stand.* **75A,** 401 (1971).
95. G. T. Furukawa and co-workers, *J. Res. Nat. Bur. Stand.* **57,** 67 (1956).
96. E. D. West and S. Ishihara in S. Gratch, ed., *Advances in Thermophysical Properties at Extreme Temperatures and Pressures,* American Society of Mechanical Engineers, New York, 1965, pp. 146–151.
97. S. Ishihara and E. D. West, *J. Res. Nat. Bur. Stand.* **80A,** 65 (1976).
98. A. K. Chaudhuri and co-workers, *High Temp. Sci.* **2,** 203 (1970).
99. V. Ya. Chekhovskoi, A. E. Sheindlin, and B. Ya. Berezin, *High Temp. High Pressures* **2,** 301 (1970).
100. A. Cezairliyan, *J. Res. Nat. Bur. Stand.* **75C,** 7 (1971).
101. A. Cezairliyan and C. W. Beckett in H. A. Skinner, ed., *Thermochemistry and Thermodynamics, MTP International Review of Science, Physical Chemistry,* Series I. Vol. 10, Butterworths, London, Eng., 1972, pp. 159–175.
102. A. Cezairliyan and C. W. Beckett, "Electrical Discharge Techniques for Measurements of Thermodynamic Properties of Fluids at High Temperatures" in B. LeNeindre and B. Vodar, eds., *Experimental Thermodynamics of Non-Reacting Fluids,* Butterworths, London, Eng., 1975, pp. 1161–1192.
103. F. Righini, A. Rosson, and G. Ruffino, *High Temp. High Pressures* **4,** 597 (1972).
104. I. Ya. Dikhter and S. V. Lebedev, *High Temp. (USSR)* **9,** 845 (1971).
105. J. W. Shaner, G. R. Gathers, and C. Minichino, *High Temp. High Pressures* **8,** 425 (1976).
106. A. Cezairliyan and C. W. Beckett, in H. A. Skinner, ed., *Thermochemistry and Thermodynamics, MTP International Review of Science, Physical Chemistry,* Series Two, Vol. 10, Butterworths, London, Eng., 1975, pp. 247–260.
107. Ya. A. Kraftmakher and V. L. Tonaevskii, *Phys. Status Solids A* **9,** 573 (1972).
108. L. P. Filippov, *Int. J. Heat Mass Transfer* **9,** 681 (1966).
109. *ASTM Designation D-2382* in ref. 18, Pt 24.
110. R. S. Jessup, *Nat. Bur. Stand. (U.S.) Monogr.* **7,** 23 pp. (1960); K. L. Churney and G. T. Armstrong, *Nat. Bur. Stand. (U.S.) Report* **9626,** 132 pp. (1967).
111. G. Waddington, S. Sunner, and W. N. Hubbard in ref. 10, pp. 154–166.

112. G. T. Armstrong, *Calorimetry, Thermometry and Thermal Analysis,* *(Japan)* Sect. 6, 51 (1973); G. T. Armstrong and W. H. Johnson, *Third International Conference on Chemical Thermodynamics,* Baden, Austria 1973, Vol. 8, pp. 61–72. Gistel, Vienna, 1973.
113. E. J. Prosen and W. H. Johnson, NBS, private communication, 1977.
114. F. D. Rossini in ref. 10, Chapt. 4.
115. G. T. Armstrong in ref 17, pp. 129–145; G. T. Armstrong and R. C. King in S. Sunner, ed., in ref. 26, Chapter 15.
116. C. W. Waidner and E. L. Meuller, *Technol. Pap. Nat. Bur. Stand. (U.S.)* **36,** (1914); *Nat. Bur. Stand. (U.S.) Circ.* **48** (1916).
117. *ASTM Designation D-900* in ref. 18, Pt. 26, 1974.
118. R. S. Jessup, *J. Res. Nat. Bur. Stand.* **10,** 99 (1933).
119. J. H. Eiseman and E. A. Potter, *J. Res. Nat. Bur. Stand.* **58,** 213 (1957).
120. *ASTM Designation D 1826* in ref. 18, Pt. 26.
121. D. J. Eatough, J. J. Christensen, and R. M. Izatt, *J. Chem. Thermodyn.* **7,** 417 (1975).
122. J. Jordon, ed., *New Developments in Titrimetry,* Marcel Dekker, Inc., New York, 1974.
123. H. J. V. Tyrrell and A. E. Beezer, *Thermometric Titrimetry,* Chapman and Hall, Ltd., London, Eng., 1968.
124. G. A. Vaughn, *Thermometric and Enthalpimetric Titrimetry,* Van Nostrand-Reinhold Company, London, Eng., 1973.
125. I. Wadso, *Acta. Chem. Scand.* **22,** 927 (1968).
126. P. Monk and I. Wadso, *Acta Chem. Scand.* **22,** 1842 (1968).
127. R. L. Berger in ref. 23, pp. 275–289.
128. G. T. Armstrong, in M. Laffitte, ed., *Thermochimie, Colloques Internationaux du CNRS No. 201,* Centre National de la Recherche Scientifique, Paris, Fr., 1972, pp. 77–96.

General References

E. Calvet, ed., *Microcalorimétrie et Thermogènese, Colloques Internationaux du CNRS No. 156,* Centre National de la Recherche Scientifique, Paris, Fr., 1967.
I. M. Kolthoff, P. J. Elving, and E. B. Sandel, eds., *Treatise on Analytical Chemistry,* Pt. 1, Vol. 8, Interscience Publishers, New York, 1968, especially Chapters 90, 91, and 93.
D. C. Ginnings, ed., *Precision Measurement and Calibration—Heat, National Bureau of Standards Special Publication 300,* Vol. 6, U.S. Government Printing Office, Washington, D.C., 1970.
M. L. McGlashan, ed., *Chemical Thermodynamics,* Vol. 1, The Chemical Society, London, Eng., 1973, especially Chapters 3, 4, 6, and 9 (a review of recent techniques and measurements to 1971).
J. M. Sturtevant, "Calorimetry," in A. Weissberger, ed., *Physical Methods of Chemistry,* John Wiley & Sons, Inc., New York, 1971.
W. Swietoslawski, *Microcalorimetry,* Reinhold Publishing Corporation, New York, 1946.
W. P. White, *The Modern Calorimeter,* ACS Monograph No. 42, Reinhold Publishing Corporation, New York, 1928.

GEORGE T. ARMSTRONG
ARED CEZAIRLIYAN
National Bureau of Standards

CALORIZING. See Metallic coatings.

CAMPHOR. See Terpenoids.

CANCER CHEMOTHERAPY. See Chemotherapeutics, antimitotic.

CANDLES. See Waxes.

CAPACITOR FLUIDS. See Heat exchange technology.

CAPRIC ACID, CH$_3$(CH$_2$)$_8$COOH. See Carboxylic acids.

CAPROIC ACID, CH₃(CH₂)₄COOH. See Carboxylic acids.

CAPRYLIC ACID, CH₃(CH₂)₆COOH. See Carboxylic acids.

CARAMEL COLORS. See Colorants for foods, drugs and cosmetics.

CARBAMIC ACID

Carbamic acid [463-77-4], NH₂COOH, is the hydrated form of isocyanic acid, H—N=C=O. It is not known in the free state; hydrolysis rapidly gives ammonia and carbon dioxide.

$$H—N=C=O + H_2O \rightarrow [H_2NCOOH] \rightarrow CO_2 + NH_3$$

Carbamic acid is the monoamide of carbonic acid; the diamide is the well-known compound urea, (NH₂)₂CO, also called carbamide (see Urea). Guanidine HN=C(NH₂)₂, could be regarded as the amidine of carbamic acid (see Cyanamides). The acid chloride (chloroformamide, "urea chloride"), NH₂COCl, and its salts have been prepared. Ammonium carbamate, NH₂CO₂NH₄, can be obtained as a white crystalline solid by reaction of dry carbon dioxide and ammonia. It is an impurity in commercial ammonium carbonate (see Ammonium compounds). Esters of carbamic acid are quite stable. The best known is the ethyl ester usually called urethane.

```
       O(S)                    O(S)    O(S)
        ‖                        ‖       ‖
H₂N—C—NH₂              H₂N—C—N—C—NH₂
    urea                         |
  (thiourea)                     H
                              biuret
                          (2,4dithiobiuret)
```

Alkylcarbamates (urethanes) are formed from reaction of alcohols with isocyanic acid or urea (see Urethane polymers).

```
                          O                    O
                          ‖                    ‖
H—N=C=O + C₂H₅OH → NH₂—C—OC₂H₅ ⇌ H₂N—C—NH₂ + C₂H₅OH
                                  Δ
                       urethane
```

With excess isocyanic acid, stable allophanates are formed (See Cyanuric acid and isocyanuric acids).

```
                      O              O    O
                      ‖              ‖    ‖
HN=C=O + HC—OC₂H₅ → H₂N—C—N—C—OC₂H₅
                                |
                                H
                      ethyl allophanate
```

Salts of N-substituted dithiocarbamic acids are used as fungicides (qv) and rubber vulcanization accelerators (see Rubber chemicals).

CARBIDES

Survey, 476
Cemented carbides, 483
Industrial heavy-metal carbides, 490
Calcium carbide, 505
Silicon carbide, 520

SURVEY

Although the element carbon is comparatively inert at room temperature, at higher temperatures it forms carbides with most other elements, particularly metals and metal-like elements. Some of these carbides are extremely important in technology, for instance, calcium carbide, CaC_2, formerly the principal source of acetylene; the abrasives silicon carbide, SiC, and boron carbide, B_4C and tungsten carbide, WC, titanium carbide, TiC, and tantalum (niobium) carbide, TaC(NbC), the basic carbides of modern cemented carbides. Cementite, Fe_3C, should also be mentioned, as well as the numerous complexes such as $(Co,W)_6C$, $(Cr,Fe,Mo)_{23}C_6$, and $(Cr,Fe)_7C_3$. These are responsible for hardness, wear resistance, and excellent cutting performance in tool steels and Stellite-type alloys. (See Boron Compounds; Tungsten and tungsten alloys; Tantalum and tantalum compounds; Niobium and niobium compounds; Titanium and titanium compounds.)

Table 1 gives a survey of the most important and well-known binary compounds of carbon, according to their position in the periodic system. They are divided into four main groups: the saltlike, metallic, diamondlike, and volatile compounds of carbon (ie, those with hydrogen, nitrogen, oxygen, and sulfur). Some further subdivisions have been introduced into Table 1 in order to characterize the technically important carbides. However, this division is not rigid and there are certain transitional cases. Thus, some of the properties of Be_2C, eg, the very high degree of hardness, appertain to diamondlike carbides. Whereas some monocarbides of Group IIIB, eg, scandium carbide, ScC, and uranium carbide, UC, as well as thorium carbide, ThC, have comparatively pronounced metallic characteristics.

The important industrial carbides are all stable at high temperatures and thus can be prepared by the direct reaction of carbon with metals or metallike materials at high temperatures. This does not apply to the acetylides and the alkali metal–graphite compounds, which should by definition be included in the class of carbon compounds rather than carbides. From the technological and metallurgical point of view it is of interest that a series of elements (eg, Re, the platinum metals, and the metals of the Cu group), in the liquid state and at high temperatures, have a definite though generally very low dissolving power for carbon. The dissolved carbon separates on cooling in the form of graphite.

Saltlike Carbides. This group comprises almost all carbides of the Groups I, II, and III of the periodic system. Beryllium carbide and Al_4C_3 may be considered as derivatives of methane (C^{4-} anion) and most carbides with C_2 groups (chiefly C_2^{2-} anions) as derivatives of acetylene. This is supported to some extent by the following hydrolysis reactions:

Table 1. Binary Compounds of Carbon and Their Position in the Periodic System

	IA	IIA	IIIB	IVB	VB	VIB	VIIB		VIII		IB	IIB	IIIA	IVA	VA	VIA	VIIA	O
LEGEND:																	H$_x$C H$_x$C$_y$	He
(1a) Saltlike carbides	H$_2$C$_2$	Be$_2$C											B$_4$C	C	(NC)$_2$	OC O$_2$C	F$_4$C	Ne
(1b) Acetylides	Li$_2$C$_2$	Mg$_2$C$_3$ MgC$_2$	Sc$_4$C$_3$ ScC$_{1-x}$ Sc$_{15}$C$_{19}$	TiC	V$_2$C^{xxx} V$_4$C$_3$ V$_6$C$_5$ V$_8$C$_7$ VC	Cr$_{22}$C$_6$ Cr$_7$C$_3$ Cr$_3$C$_2$	Mn$_{23}$C$_6$ Mn$_{15}$C$_4$ Mn$_3$C Mn$_5$C$_2$ Mn$_7$C$_3$	Fe$_3$C Fe$_7$C$_3$ Fe$_2$C	Co$_3$C	Ni$_3$C	Cu$_2$C$_2$	ZnC$_2$	Al$_4$C$_3$	SiC	P$_2$C$_6$	S$_2$C	Cl$_4$C	Ar
(1c) Metal graphite compounds	Na$_2$C$_2$	CaC$_2$	YC$_{1-x}$ Y$_{15}$C$_{19}$ YC$_2$	ZrC$_{1-x}$	Nb$_2$C^{xxx} Nb$_3$C$_2$ NbC Nb$_4$C$_3$	Mo$_2$C^{xxx} MoC$_3$C$_2$ MoC$_{1-x}$ MoC	Tc?	RuC(?)	Rh	Pd	Ag$_2$C$_2$	CdC$_2$	Ga	Ge	As$_2$C$_6$	Se$_2$C	Br$_4$C	Kr
(2a) Metallic carbides of metals belonging to Groups IVB–VIB	K$_2$C$_2$	SrC$_2$	La$_2$C$_3$ LaC$_2$	HfC$_{1-x}$	Ta$_3$C^{xxx} Ta$_3$C$_2$ TaC Ta$_4$C$_3$	W$_2$C^{xxx} WC$_{1-x}$ WC	Re$_2$–$_4$C ReC	OsC (?)	Ir	Pt	Au$_2$C$_2$	HgC$_2$	Tl	Sn	Sb	Te$_2$C	I$_4$C	Xe
(2b) Metallic carbides of the iron metals, including Mn	Rb$_2$C$_2$	BaC$_2$											In	Pb	Bi	Po?	At?	Rn
(2c) Metallic carbides of Group VB	Cs$_2$C$_2$	Ra?	Ac?	xxx several modifications														
	Fr?																	
	NaC$_8$ NaC$_{16}$																	
	KC$_8$ KC$_{16}$ KC$_{60}$																	
	RbC$_8$ RbC$_{16}$ RbC$_{60}$																	
	CsC$_8$ CsC$_{16}$																	

(3) Diamondlike carbides
(4) Volatile nonmetallic carbides
E No carbide formation (in the case of the element (E) shown); however, some solubility of carbon in the melt (except the gases of Group O)
E? Carbide possible but still unknown

*IIIA Lanthanide Series

| Ce$_2$C$_3$ CeC$_2$ | PrC Pr$_2$C$_3$ PrC$_2$ | Nd$_2$C$_3$ NdC$_2$ | PmC$_2$ | Sm$_3$C Sm$_2$C$_3$ SmC$_2$ | EuC$_2$ | Gd$_3$C Gd$_2$C$_3$ GdC$_2$ | Tb$_3$C Tb$_2$C$_3$ TbC$_2$ | Dy$_3$C Dy$_2$C Dy$_2$C$_3$ DyC$_2$ | Ho$_3$C Ho$_2$C Ho$_2$C$_3$ HoC$_2$ | Er$_3$C Er$_2$C Er$_2$C$_3$ ErC$_2$ | Tm$_3$C Tm$_2$C$_3$ TmC$_2$ | Yb$_3$C Yb$_2$C$_3$ YbC$_2$ | Lu$_3$C Lu$_2$C$_3$ LuC$_2$ |

**Actinide Series

| ThC ThC$_2$ | PaC PaC$_2$ | UC U$_2$C$_3$ UC$_2$ | NpC Np$_2$C$_3$ NpC$_2$ | PuC Pu$_2$C$_3$ PuC$_2$ | Am? | Cm? | Bk? | Cf? | Es? | Fm? | Mv? | No? | Lw? |

$$Be_2C + 4\ HOH \rightarrow 2\ Be(OH)_2 + CH_4$$
$$Al_4C_3 + 12\ HOH \rightarrow 4\ Al(OH)_3 + 3\ CH_4$$
$$CaC_2 + 2\ HOH \rightarrow Ca(OH)_2 + C_2H_2$$

Propyne is obtained from Mg_2C_3 (probably C_3^{4-} anions), formed by the thermal decomposition of MgC_2, with separation of graphite:

$$Mg_2C_3 + 4\ HOH \rightarrow 2\ Mg(OH)_2 + CH_3C \equiv CH$$

In their pure state the carbides of Groups I and II are characterized by their transparency and lack of electroconductivity. The carbides of Group IIIB (Sc, Y, the lanthanides, and the actinides) are opaque. Some of them—depending on their composition—show metallic luster and electroconductivity. The M^{2+} cation may exist in the MC_2 phases of this group and the third valence electron apparently imparts to these compounds their partly metallic character.

Uranium monocarbide, which is important in reactor technology, is completely miscible with some of the metallic carbides of Groups IVB and VB and also with ThC. Methane, ethylene, and hydrogen, as well as acetylene, are formed during the hydrolysis of the Group IIIB carbides of varying composition (M_3C, MC, M_2C_3, MC_2). The term acetylides, in the stricter sense, applies to "carbides" precipitated from aqueous solutions or from solutions in aqueous ammonia with acetylene. These compounds are metastable-like acetylides of Cu, Ag, Au, Na, K, Rb, Cs, Zn, Cd, Hg, Pd, Os, Ce, Al, Mg, etc, and require additives such as H_2, H_2O, NH_3, C_2H_2, and metal salts for stabilization. Thus it is doubtful whether they can be considered as pure metal–carbon compounds and described as carbides.

The *alkali metal–graphite compounds* formed by graphite absorption of the fused metals Na, K, Rb, and Cs, represent a special type of metal–carbon compound. Compounds MC_8 (brown), MC_{16} (gray), and MC_{60} (strongly graphitic) are known.

Metallic Carbides. This class of compounds comprises the interstitial carbides of the transition metals of Groups IVB, VB, and VIB (for properties see Industrial heavy-metal carbides, Tables 1 to 3), and the carbides of the Group VIIB and Group VIII metals. The metalloconductive carbides P_2C_6 and As_2C_6 are also included.

IVB–VIB Metals. The term *interstitial carbides* (according to Hägg) (1) derives from the fact that not only the small atom carbon but also the elements H, N, and O can be deposited in the octahedral interstices of the parent lattice of the transition metal. The completion of the octahedral interstices by carbon atoms gives the metal–carbon phases as, eg, TiC, ZrC, HfC, VC, NbC, Tac, UC, PuC. Tungsten carbide (WC) and the isotypic phases MoC, (Mo, W)(C, N) are hexagonal. The hexagonal isotypic and closely related M_2C carbides, such as V_2C, Nb_2C, Ta_2C, Mo_2C, and W_2C, are obtained if only 50% of the available octahedral voids are occupied. A series of phases MC_2 exists in which both octahedral and tetrahedral interstices are filled. Neutron diffraction studies have shown that chromium carbide, Cr_3C_2, has a more complex structure.

Hägg's original hypothesis, that the existence and stability of interstitial compounds is closely related to the upper limit of the ratio of the radii $r_X/r_M < 0.59$ (X = C, H, N, O), is supported by more recent views on the electronic structure of the components and the interaction of the available outer electrons. Carbides have good electrical conductivity, mainly as a result of the metallic bonding. Rundle (2) has pointed out that the high melting points of carbides are also due to the additional strong metal–metalloid bonds and that their high degree of hardness is due to directed forces. This is in agreement with Pauling's resonance concepts, ie, it arises through directed half–bonds in this class of compounds (3).

Quantum mechanical calculations and electron–spectroscopic results (esca method) (4), as well as x–ray absorption and emission studies, have provided deeper knowledge of the electronic structure, particularly of the M^+C^- charge distribution in the hard carbides (5) (see Analytical methods). Thus, in these compounds the metal atoms are positively and the carbon atoms negatively charged and an electron transfer is taking the place, eg, of the titanium $3d$ electrons into the carbon $2p$ level. The main contribution to the mixed bond is furnished by the ionic part, which increases within the homogeneity range of the IVB carbides with increasing carbon content. The electron density at the Fermi level is lower in the carbides than in the parent metals and there is no gap between the valency and conducting bands. The valency band is determined mainly by the $2s$ and $2p$ states of the carbon atom.

Group VIIB and Group VIII Metals. The ratio of the radii in the carbides of Mn, Fe, Ni, and Co, as well as in chromium carbide, Cr_3C_2, is $r_X/r_M > 0.59$. As a result their structures deviate from the typical interstitial principle. Consequently, they are more complex. They also differ markedly in chemical and physical behavior and from carbides with interstitial structures. Hardness values and melting points are considerably lower and chemical stability to mineral acids is also no longer apparent. The chromium carbides also occupy an intermediate position in this respect.

This group is of technological importance; eg, cementite Fe_3C, and the wear-resistant complex carbides formed with, eg, Cr, Mo, and W are responsible for the hardness and cutting performance of heat-resistant steels, high-speed tool steels and alloys of the Stellite type. Manganese carbide, Mn_3C, fused in an electric furnace, is sometimes used as a cheap Mn master alloy. Powdered $(Mn,Cr,Fe)_7C_3$ is often used for sintered alloyed steel parts.

Crystal Structure. The lattice structure of metallic carbides is determined mainly by the ratio of metal and nonmetal atomic radii (Hägg's rule). The filling of the octahedral interstices of the transition metal lattice by the small carbon atoms results in close–packed structures with different layer sequences. Simple geometrical considerations permit a determination of the degree of interstice filling and thus of the formula type.

The complete octahedral filling, eg, in a cubic body-centered metal lattice, results in the cubic face-centered sodium chloride structure (B1–type) of the IVB and VB metal monocarbides. Thus, the characteristic structure unit is the M_6C octahedron, such as in complex carbides. The B1 monocarbides have a strong tendency to defect structure formation (partial nonoccupation of available lattice holes) which leads to broad homogeneity ranges. The most stable phase with the highest melting point, greatest hardness, and largest lattice constant then often occurs far below the stoichiometric composition. In the group VB metal-carbon systems, ordered phases also occur in the substoichiometric range.

In the hexagonal carbide phases of densest packing, the octahedral interstices are half filled in M_2C (eg, W_2C, Mo_2C, or Ta_2C) or completely filled as in MC (eg, WC or MoC). The carbon atom, surrounded by six metal atoms, is in the center of a trigonal prism, similar to the NiAs structure. The close relationship between the hexagonal and the cubic B1–carbides is well correlated with their mutual solubility.

If too few interstices between the metal atoms are available for the carbon atoms, they form pairs, resulting in carbides of the formula type MC_2 with tetragonally deformed NaCl structure (CaC_2–type). This structure is typical for the dicarbides of the

Table 2. Physical Properties of Diamondlike Carbides and Nonmetallic Hard Materials

Compound	Density, g/cm³	mp, °C	Micro hardness[a]	Transverse rupture strength, N/mm²[b]	Compression strength, N/mm²[b]	Modulus of elasticity, N/mm²[b]	Heat conductivity, W/(cm·K)	Coefficient of thermal expansion, $\beta \times 10^{-6}$	Electrical resistivity, $\mu\Omega\cdot cm$
diamond, C	3.52	3,800 dec	7600	~300	~2,000	~900,000	1.14	0.9–1.18	10^{18}
boron carbide, B$_4$C	2.52	2,450	2940	500	1,800	450,000	0.27	6.0	10^4
silicon carbide, SiC	3.2	2,300 dec	2580	<400[d]	1,400	480,000	0.15	5.7	10^3
beryllium carbide, Be$_2$C	2.42	2,300	2690		740	350,000	0.21	7.4	10^3
sintered alumina, Al$_2$O$_3$	3.9	2,050	2080	<700[d]	3,000	400,000	0.19	7.8	10^{18}
boron nitride, cubic, BN	3.45	2,730[c]	4700			600,000			10^{16}
aluminum nitride, AlN	3.26	2,250	1230			350,000			10^{13}
silicon nitride, Si$_3$N$_4$	3.44	1,900	1700	<750[d]		210,000	0.18	2.4	10^{16}
silicon boride, SiB$_6$	2.43	1,950	2300	~100		330,000		6.3	10^5

[a] See Hardness.
[b] To convert N/mm² to psi, multiply by 145 × 10⁻¹².
[c] Transition from cubic to hexagonal ~1650°C.
[d] Hot-pressed.

Table 3. Alphabetical List of Carbides Referred to in the Text

Carbide	CAS Registry Number	Formula
aluminum carbide (4:3)	[1299-86-1]	Al_4C_3
arsenic carbide (2:6)		As_2C_6
beryllium carbide	[57788-94-0]	Be_2C
boron carbide (4:1)	[12069-32-8]	B_4C
calcium carbide (2:1)	[75-20-7]	CaC_2
chromium carbide	[12011-60-8]	CrC
chromium carbide (3:2)	[12012-35-0]	Cr_3C_2
chromium carbide (4:1)	[12075-40-7]	Cr_4C
chromium carbide (7:3)	[12075-40-0]	Cr_7C_3
chromium carbide (23:6)	[12105-81-6]	$Cr_{23}C_6$
cobalt carbide (3:1)	[12011-59-5]	Co_3C
cobalt tungsten carbide (6:6:1)	[12538-07-7]	Co_6W_6C
hafnium carbide	[12069-85-1]	HfC
iron carbide	[12069-60-2]	FeC
iron carbide (2:1)	[12011-66-4]	Fe_2C
iron carbide (3:1)	[12011-67-5, 12169-32-3]	Fe_3C
iron carbide (5:2)	[12127-45-6]	Fe_5C_2
iron carbide (7:3)	[12075-42-2]	Fe_7C_3
iron carbide (23:6)	[12012-72-5]	$Fe_{23}C_6$
lanthanum carbide (1:2)	[12071-15-7]	LaC_2
manganese carbide (3:1)	[12121-90-3]	Mn_3C
manganese carbide (23:6)	[12266-65-8]	$Mn_{23}C_6$
magnesium carbide (1:2)	[12122-46-2]	MgC_2
magnesium carbide (2:3)	[12151-74-5]	Mg_2C_3
molybdenum carbide	[12011-97-1]	MoC
molybdenum carbide (2:1)	[12069-89-5]	Mo_2C
molybdenum carbide (23:6)	[12152-15-7]	$Mo_{23}C_6$
nickel carbide	[12167-08-7]	NiC
nickel carbide (3:1)	[12012-02-1]	Ni_3C
niobium carbide	[12069-94-2]	NbC
niobium carbide (2:1)	[12011-99-3]	Nb_2C
plutonium carbide	[12070-03-0]	PuC
plutonium carbide (2:3)	[12076-56-1]	Pu_2C_3
phosphorus carbide (2:6)		P_2C_6
scandium carbide	[12012-14-5]	ScC
silicon carbide	[409-21-2]	SiC
tantalum carbide	[12070-06-3]	TaC
tantalum carbide (2:1)	[12070-07-4]	Nb_2C
thorium carbide	[12012-16-6]	ThC
thorium carbide (1:2)	[12071-31-7]	ThC_2
titanium carbide	[12070-08-5]	TiC
tungsten carbide	[12070-12-1]	WC
tungsten carbide (2:1)	[12070-13-2]	W_2C
uranium carbide	[12170-09-6]	UC
uranium carbide (1:2)	[12071-33-9]	UC_2
uranium carbide (2:3)	[12076-62-9]	U_2C_3
vanadium carbide	[12070-10-9]	VC
vanadium carbide (2:1)	[12012-17-8]	V_2C
zirconium carbide	[12020-14-3]	ZrC

lanthanides and actinides, eg, UC_2 or LaC_2, which no longer exhibit pronounced metallic nature. C_2-groups are also found in the cubic body-center M_2C_3-carbides of the actinides, eg, U_2C_3 or Pu_2C_3.

If the critical ratio value is close to 0.59 or above, eg, in the group of iron metals and in chromium, interstitial structures, eg, having the formula $M_{23}C_6$ ($Cr_{23}C_6$, $Mn_{23}C_6$) are only formed at lower carbon contents. In the iron–carbon system, austenite, the solid solution of carbon in γ iron, is also an interstitial structure. If the ratio of 0.59 is exceeded and at higher carbon concentrations, complex structures with a lower degree of symmetry are formed. This applies particularly to the formula M_7C_3 (hexagonal), M_3C_2 (rhombic), and M_3C (rhombic). Corresponding compounds are formed with chromium and especially with the iron metals. In Cr_3C_2, neutron diffraction studies have shown that the carbon is in the center of a trigonal prism, similar to WC. In cementite, Fe_3C, the carbon is not in pronounced interstitial sites. Therefore, it cannot be considered as an interstitial structure.

Diamondlike Carbides. Silicon and boron carbides belong to this group; beryllium carbide, with a high degree of hardness, can be included. Some of its properties are related to reuse of silicon carbide (see under Silicon Carbide). Diamond itself may be considered a "carbide of carbon" because of its chemical structure. (See Boron compounds; Beryllium compounds.)

Table 2 summarizes the properties of the so-called nonmetallic hard materials, including diamond, the diamondlike carbides B_4C, SiC, and Be_2C. Also included in this category are corundum, Al_2O_3, cubic boron nitride, BN, aluminum nitride, AlN, silicon nitride, Si_3N_4, and silicon boride, SiB_6 (6).

BIBLIOGRAPHY

"Carbides" in *ECT* 1st ed., pp. 827–830 by C. R. Hough, Polytechnic Institute of Brooklyn; "Carbides" in *ECT* 2nd ed., Vol. 4, pp. 70–75, by R. Kieffer, University of Vienna and F. Benesovsky, Metallwerk Plansee A.G., Reutle, Austria.

1. G. Hägg, *Z. Physik. Chem.* **B6**, 221 (1929); **B12**, 33 (1931).
2. R. E. Rundle, *Acta Cryst.* **1**, 180 (1948).
3. H. Krebs, *Acta Cryst.* **9**, 95 (1956).
4. L. Ramquist, *Jernkontorets Ann.* **153**, 159 (1969).
5. H. Nowotny, A. Neckel, *J. Inst. Met.* **97**, 1961 (1969).
6. F. Binder, *Radex Rdsch.* 531 (1975).

RICHARD KIEFFER
Technical University, Vienna

FRIEDRICH BENESOVSKY
Metallwerke Plansee A.G., Reutle/Tyrol

CEMENTED CARBIDES

The term cemented carbides denotes powder-metallurgical products consisting of a carbide of a Group IVB to VIB metal in a matrix of a metal, usually cobalt or nickel. Various scientific and technical advances led to their development, particularly (1) discovery of the great hardness of cast tungsten carbide; (2) preparation of finely divided tungsten monocarbide, WC, by reaction of the elements or by carburizing with hydrocarbons; (3) application of ceramic sintering technology to carbides; (4) lowering of the high sinter temperature of pure carbides by the use of a liquid phase of eutectic alloys of the iron-group metals; (5) discovery of the alloy properties of cobalt and the great toughness of WC–Co alloys; and (6) application of wet-grinding methods, and the ceramic double sintering process.

The cemented carbides were developed in Germany during World War I and WC–Co patents were first applied by Krupp under the trademark Widia, and then in the United States by General Electric as Carboloy. At present, cemented carbides are produced in practically every industrially developed country.

Later, WC-free cemented carbides were developed and based on various compositions with W, Ti, and Ta, followed by the important WC–TaC and WC–TiC alloys, WC plus WC–Mo$_2$C–TiC solid solutions, and the WC–TiC–TaC alloys. A major breakthrough in the manufacture of these compositions was provided by P. M. McKenna in the mid 1930s. He developed "menstruum," composed of WC–TiC, WC–TaC, and similar solid solutions (see p. 491).

Efforts to replace cobalt by nickel, iron, or Stellite-type alloys for WC-based compositions were not very successful. However, the technique of pressure sintering or hot pressing, was applied, particularly in the preparation of nonporous dies, rolls, and projectile cores. During World War II, ammunition cores of WC–Co were made on both sides in quantities comparable to the present world output of all cemented carbides. The war also saw a resurgence of WC-free carbides in the form of TiC–VC–Ni–Fe alloys. Also among the WC-free carbides are the modern heat- and oxidation-resistant cemented carbides based on titanium carbide and chromium carbide with Ni–Co–Cr binder, and the heat-resistant cemented carbides based on TiC with steel binder. Later developments include the acid-resistant cemented carbides, either of WC with Pt or Ni–Cr binder, or Cr$_3$C$_2$–Ni, also cutting tools based on TiC–ZrB$_2$(TiB$_2$).

In the mid 1970s, a resurgence of WC-free alloys took place, originating in work done at the Ford Motor Co., and leading to a significant number of carbide cutting tool inserts based upon Ford technology and patents for the TiC–Mo$_2$C–Ni compositions. In addition, TiC–Mo–Ni and (Ti,Mo,V,Nb)C$_{1-x}$ alloys with Ni–Mo-binder and the first carbonitride alloys based on Ti(C,N) with Ni–Mo-binder and addition of up to 50% free carbides or solid solutions (1) appeared on the market. Furthermore, the spinoidal decomposition of (Ti,Mo)(C,N) based alloys with excellent cutting properties and high hardness, in combination with outstanding toughness, was discovered (2).

In 1976, molybdenum monocarbide (MoC with WC structure) stabilized with 20–40% WC, was introduced to the cutting field (3), and in 1977, the carbonitride (Mo,W)(C,N) showed promising properties (4). The coating of throw-away tips with thin layers (3–8 μm) of TiC, TiN, Ti(C,N), HfN, and Al$_2$O$_3$ prolonged the lifetime of the tools from 2–5 times (5). Furthermore, hot isostatic pressing and compression removal of all macro- and micropores of the finished, sintered tips, dies, and Sendzimir mill rolls with Ar, He, and similar gas mixtures improved the toughness. Cemented carbide tips based on WC–Co with WC powder (< 1 μm), are superior to standard tips, especially when machining superalloys. Today, coated tools constitute about 20–40% of production.

Figure 1 is a flow diagram of horizontal CVD equipment employed for the chemical vapor deposition (CVD) on throw-away tips. Nitrides, carbides, and carbonitride of Groups IVB and VB are used. The liquid chlorides (eg, $TiCl_4$ and VCl_4) are introduced into the system by bubbling a H_2/N_2 mixture through the chlorides. For the deposition of carbonitrides, corresponding amounts of CH_4 are added to the reaction gas. With a gas stream of H_2/CH_4 and increasing the temperature of the tips in the furnace from 1000–1100°C to 1050–1150°C pure carbides were deposited. The solid chlorides were inserted into a ceramic boat and directly into the furnace for evaporation. The total pressure in the system is 101.3 kPa (1 atm), but can be lower for TiC coatings. For mass production, vertical furnaces are used, especially when applying multilayers, eg, TiN on Ti(C,N), or TiC on cemented carbides or depositing HfN (6). For the coating of oxides, eg, Al_2O_3 on carbide tips, $AlCl_3$ vapor is applied together with H_2 containing small amounts of CO_2 or H_2O (7–8).

Preparation

The preparation of cemented carbides takes place in the following stages, as illustrated in Figure 2 and as follows: (*1*) preparation of the starting materials (metals, oxides, carbon black); (*2*) preparation of the carbides, and of the carbide solid solutions; (*3*) wet-ball milling of the carbides with powdered cobalt; (*4*) pressing the carbide–cobalt mixtures with hydraulic or mechanical presses; (*5*) forming of presintered shapes (double-sintering process) or pressing to the finished shape; and (*6*) high temperature sintering under a protective gas or *in vacuo;* (*7*) hot-pressing of finished powder batches, or isostatic hot-pressing to remove the last traces of porosity for applications where perfection is essential and opening of pores is unacceptable, such as in drawing dies and Sendzimir mill rolls; (*8*) inspection.

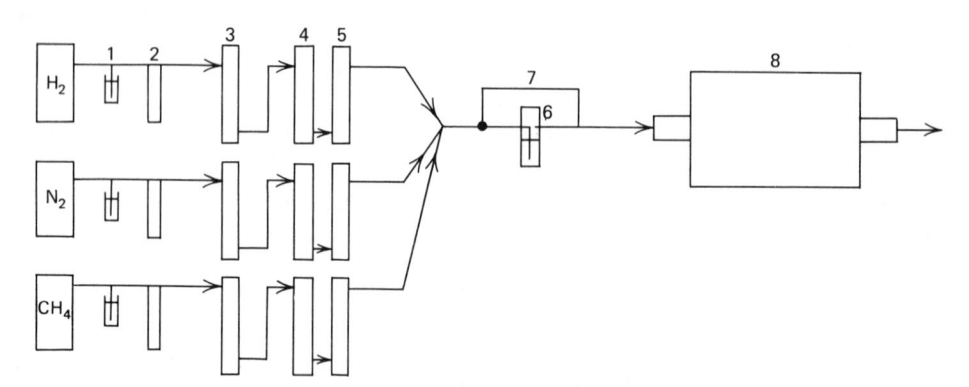

Figure 1. Coating system. (1 = safety valve, 2 = flow meter, 3 = gas purifying unit (BTS-catalyst, BASF), 4 = gas drying unit, 5 = gas drying unit, 6 = bottle for liquid chlorides, 7 = by-pass, and 8 = furnace (tube, 27 mm).)

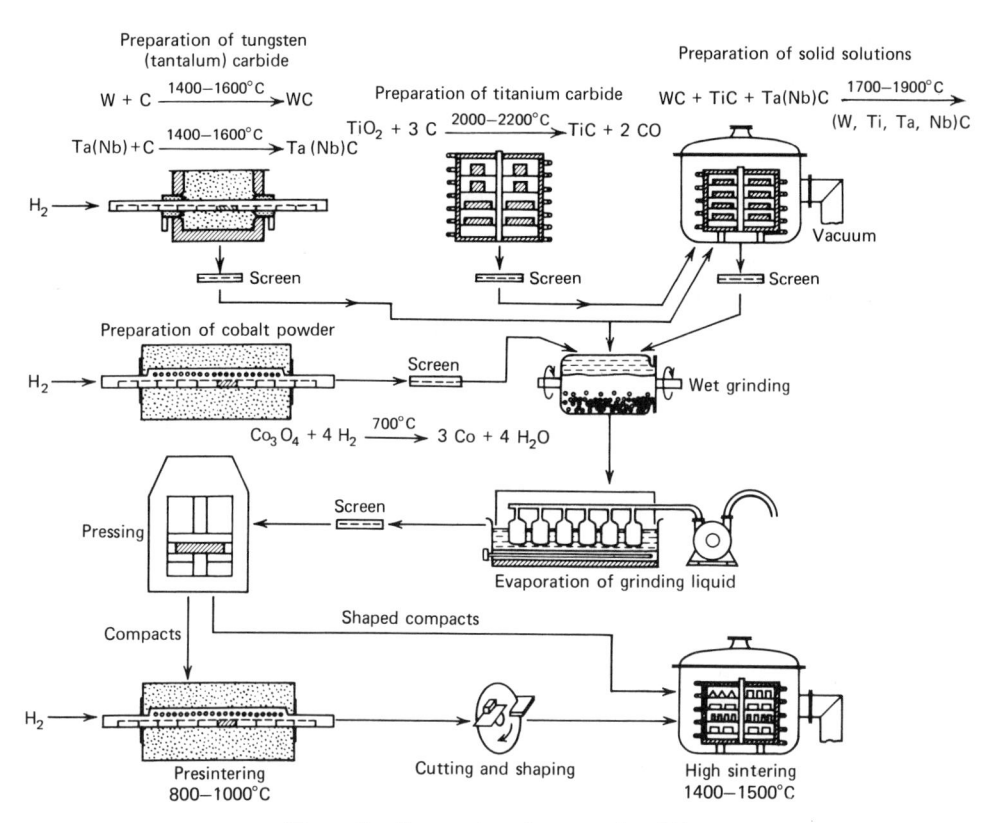

Figure 2. Preparation of cemented carbides.

Processing. The preparation of tungsten, tantalum, and niobium powder is described under Carbides, industrial heavy-metal. Most matrix cobalt is purchased as an extra, fine powder made by Metallurgie Hoboken (Belgium), by a proprietary process. Some matrix cobalt is produced by heating the oxide or oxalate in hydrogen in a push-type furnace or by other methods. Titanium dioxide and carbon black are usually purchased, as well as tungsten, WC, vacuum-cleaned solid solutions, and cobalt powder. The carbides and carbide solid solutions are prepared in continuous carbon tube furnaces, or alumina tube furnaces with molybdenum heating elements, or in graphite crucibles heated by induction, using protective gas or vacuum (see under Carbides, industrial heavy-metal). First the carbide pieces are coarsely ground in a jawcrusher or hammer mill, and then finely ground in a ball mill and screened. The carbide–cobalt mixture is wet-balled in a rotating or oscillating stainless steel ball mill, using cemented carbide balls, or in mills lined with a Stellite or cemented carbide. Attritors (stirring ball-mills) have replaced the conventional types. The solvents are usually isopropyl or ethyl alcohol, acetone, hexane, heptane, or tetrahydronaphthalene. The wet pulp is vacuum-dried or spray-dried and screened; cobalt oxide, which may be present, is reduced with hydrogen.

The finely divided, if necessary granulated, material is then pressed in hydraulic or mechanical presses, with or without an additive such as wax or camphor in ether. For special forms, the double sinter process may be used; large blocks are formed which, after a presintering at 450–900°C, are machined or ground to the final form with silicon carbide or diamond abrasive wheels.

Thin-walled tubes or small rods (eg, spikes for automobile snow tires) can be extruded employing plasticizing additives. However, the organic additives must be carefully removed before the final sintering.

High temperature sintering takes place at 1310–1650°C in carbon- or alumina-tube furnaces with molybdenum heating elements in hydrogen for pure WC–Co alloys or in vacuum induction furnaces for oxidation-susceptible WC–TiC–TaC(NbC)–Co alloys. In continuous furnaces, the cemented carbide tips are placed in graphite boats, bedded in carbon black or aluminum oxide; in vacuum furnaces the pieces are placed on graphite plates, and stacked. A horizontal continuous vacuum furnace with carbon rod heaters has been introduced with good results.

Hot-pressing is employed for the preparation of drawing dies and molds, or nonporous rolls. The cemented carbide compacts are sintered in graphite molds and heated by resistance or inductively, under a pressure of 6.9–14.7 MPa (1000–2100 psi). To save graphite dies and to prevent carbon reactions, powerful hot isostatic presses have been introduced that produce nearly pore-free material and an increase of 10–30% in transverse rupture strength.

Test Methods

In addition to chemical analysis, numerous physical test methods are employed by hard metal producers and consumers for process control and quality testing. The density is determined according to ISO DIS 3369, porosity according to ASTM B276-54 (1972). The transverse rupture strength provides maximum information on ductility (ISO DIS 3327). The hardness, in combination with the transverse rupture strength, is the most important property determining the performance of a dense cemented carbide grade. Rockwell A hardness (ASTM B294-76) is utilized and, preferably, the Vickers hardness with test loads of 9.8–490 N (1–50 kg) is determined on a polished specimen. The coercive force (ISO DIS 3326) is an indication of the grain size of the tungsten carbide and the carbon balance; the magnetic saturation (content of η-phase) is also determined. For special applications the elastic properties are of interest (ISO DIS 3312).

Micrographic examination provides information on porosity, grain size, and distribution of the carbides as carbide solid solutions in the binder metal (ISO DIS 4505, ISO DIS 4499). Owing to the very fine grain of cemented carbides, special grinding, polishing, and etching methods (electrolytic polishing, heat tinting), and high magnification are necessary. For electron microscopy, replica methods and shadowing techniques are used.

The machining performance is particularly important to the consumer and is characterized by the service life vs cutting speed. Short-time performance tests are usually under extreme conditions carried out on steel or cast iron bars. To determine friction wear, numerous test methods are available for special applications (bearings, sealings, and corrosion-resistant parts).

Classification

The classification of cemented carbides is a controversial subject (9). It is based on application, that is (*1*) cutting (including machining steel, nonferrous metals, or even wood), and (*2*) wear parts (including mining tools).

The classification system for wear parts is entirely different from that for cutting tools (10). For the former, a system can be worked out based on the composition of the cemented tungsten carbide material which would be difficult to develop for wear parts. Basically, all carbide manufacturers produce essentially the same composition (Co 6%, W 94%) with slight differences in addition to variation in grain size and additives. Thus, it is possible but not necessarily desirable to develop a carbide classification system for wear parts based on composition. Any attempt to devise a classification system for cutting tools based on composition has been unsuccessful because of the importance of physical properties.

The following classifications are generally accepted: (1) The ISO Recommendation R 513, *Application of Carbides for Machining by Chip Removal* (11); (2) the British Hard Metal Association, *A System of Calssification of Hard Metal Grades for Machining* (12); and (3) The unofficial C Classification System of the United States (9).

Special Grades. The standard grades of cemented carbides account for about 99% of the total, but the special grades fulfill a small, yet important function, and have considerable growth potential.

The acid resistance of the WC–Co carbides is unsatisfactory for some applications, and WC can be replaced by TiC or Cr_3C_2, and Co by Ni–Cr, Co–Cr, or Pt, including 90% WC, 8% Ni–2% Cr; 83% Cr_3C_2–2% W–15% Ni; 70% Cr_3C_2–15% TiC–15% Ni; 92% WC–8% Pt; 91% WC–8% Co–1% Re; 50% TiC–10% WC–32% Ni–8% Cr.

The TiC based alloys were developed for resistance to heat and oxidation.

The Ferro-TiC tool materials are based on the well-known heat resistance of cemented carbides with a high content of binder metal. They contain 35–50% TiC in a low- or high-alloy steel matrix. When annealed to Rockwell C hardness 40–45, these materials can be machined, for example, to make molds which can then be improved after quenching and heat treatment to Rockwell C hardness 60–70.

For high cutting speeds, and fine roughing, special alloys and carbide ceramics are available, such as cemented carbides of the compositions 72% TiC–18% Mo_2C–10% Ni; 82% [Ti:Mo (4:1)] C_{1-x} + 14% Ni + 4% Mo; 72% [Ti:Mo (3:1)] C_{1-x} + 12% VC–NbC + 12% Ni + 4% Mo; 90% TiC–8% Ni–2% Cr; 70% TiC–30% TiB_2; and composite ceramics such as 95% Al_2O_3–5% TiC; 85% Al_2O_3–15% (Mo_2C + WC); and 85% Al_2O_3 + 12% TiC + 3% Mo_2C. New compositions are based on carbonitride alloys, eg, 84% [TiC–TiN (4:1)] + 16% Ni–Mo; 44% [TiC–TiN (3:1)] + 40% [TiC–NbC–Mo_2C(2:1:1) + 16% Ni–Mo; and spinoidal compositions like the SD_3 alloy 83.5% [Ti:Mo (2.7:1)]-$(C,N)_{1-x}$ + 12.7% Ni + 3.8% Mo.

Recycling of Scrap

Recycling of cemented carbide scrap is an important operation. Formerly, the cleaned scrap was crushed in carbide-lined molds and ball-milled to fine powders. For embrittlement, the scrap can be heated to 1700–1800°C ("blowing up"), cooled with nitrogen, and the broken tips crushed by blowing them with a cold nitrogen jet against a carbide plate (Coldstream process). Furthermore, 20–40% powdered scrap is added to cheap wear-resistant parts like spikes or ammunition cores.

Chemical methods were applied to impure or TiC–TaC alloyed scrap. The tips were oxidized and transformed into a mixture of oxides to be chemically separated.

Recently, an ingenious method was initiated by the U.S. Bureau of Mines and developed by Teledyne, Wah Chang Division. The cleaned scrap is heated with molten zinc which attacks and embrittles the cobalt phase. The zinc is distilled *in vacuum* and reclaimed. The treated carbide pieces have the strength of a presintered block and can easily be broken, ball-milled, annealed, and worked (zinc content is less than 20 ppm).

Toxicity

Toxic effects by inhalation of dust can occur with extremely fine carbide and cobalt powders. Therefore, efficient exhaust devices, dust filters, and protective masks are essential; regular medical examinations of workers are recommended.

Uses

About 70% of the production is used for cutting tools (45% for long-chip, 25% for short-chip materials) and 30% for wear-resisting materials, principally mining tools for chipless shaping (grinding) operations.

In general, for cutting tools, the carbide tip is brazed to a steel shaft. The tool edges must be finished with silicon carbide or diamond. Tools with clamped tips have been developed, avoiding the disadvantages of brazing (such as grinding and brazing cracks). Throw-away tips can be rotated or inverted in such a way as to utilize several cutting edges on the same tip; when the tip is blunted, it is discarded. Some of these tips are coated with a 6–10 μm layer of TiC, TiN, Ti(C,N), HfN, or with an even thinner layer of Al_2O_3 to prolong the lifetime two to three times.

The wear resistance of cemented carbides is important for drawing dies (wire and tubes), mining tools (rotary drills, percussion drills, coal cutters, hard-surface deep-drilling tools), projectile cores, as well as machine and instrument parts such as stamping and deep-drawing tools, rolls, nozzles, guides, balls for bearings, valves, extrusion dyes and press dyes, calkins and, lately, spikes for snow tires.

Parts made of cemented carbides coated with gold-yellow TiN and red-gold Ti(C,N) are used increasingly for scratch-free watch blocks in the jewelry industry.

The application of cemented carbides in the chemical industry (corrosion-resistant parts) and for high temperature applications (hot-pressing dies, balls and points for hot-hardness testers, clamping devices for high temperature testing machines, valve seats and valve balls, etc) is, by comparison, not yet very extensive. Cemented carbide-lined cylinders for organic high pressure synthesis, and lately, also thrust collars and pistons have been found indispensable for the production of synthetic diamonds as well as cubic BN (Borazon).

BIBLIOGRAPHY

"Cemented Carbides" under "Carbides" in *ECT* 2nd ed., Vol. 4, pp. 92–100, by Richard Kieffer, University of Vienna, Friedrich Benesovsky, Metallwerke Plansee A.-G.

1. R. Kieffer, N. Reiter, and D. Fister, *BISRA-ISI Conference on Materials for Metal-Cutting,* Scarborough, Eng., 1970, p. 126; Fr. Pat. 2,064,842 (1970), (to Ugine-Carbone).
2. E. Rudy, S. Worcester, and W. Elkington, *High Temp. High Pressures* **6,** 447 (1974).
3. J. Schuster, E. Rudy, and H. Nowotny, *Montash. Chem.* **107,** 1167 (1976).
4. P. Ettmayer and R. Kieffer, patents pending.
5. R. Kieffer, D. Fister, and E. Heidler, *Metall.* **25,** 128 (1972).

6. E. Rudy, B. F. Kieffer, and E. Baroch, *Plansee Seminar Paper 37,* Reutte, Austria, 1977.
7. B. Lux and H. Schachner, *9 Plansee Seminar,* Reutte, Austria, 1977.
8. H. S. Kalish, *Technical Paper,* Society of Manufacturing Engineers, 1978.
9. H. S. Kalish, *Some Plain Talk About Carbides,* Manufacturing Engineering and Management, July 1973.
10. K. J. A. Brooks, *Metalworking Production's Guide to Hardmetals for Machining,* International Organization for Standardization, 1977.
11. *ISO Recommendation R-513, Application of Carbides for Machining by Chip Removal,* 1st ed., International Organization for Standardization, Nov. 1966.
12. *A System of Classification of Hard Metal Grades for Machining, Technical Publication No. 1967,* British Hard Metal Association, London, Eng.

General References

R. Kieffer and F. Benesovsky, *Hartstoffe,* Springer-Verlag, Vienna, Austria, 1963.
R. Kieffer and F. Benesovsky, *Hartmetalle,* Springer-Verlag, Vienna, Austria, 1965.
H. E. Exner and J. Gurland, *Powder Metall. Int.* **2,** 59–63, 104–105 (1970).
K. J. A. Brooks, *World Directory and Handbook of Hardmetals,* Engineers Digest Publ., London, Eng., 1976.
E. Browning, *Toxicity of Industrial Metals,* Butterworths, London, Eng., 1969.

RICHARD KIEFFER
Technical University, Vienna

FRIEDRICH BENESOVSKY
Metallwerke Plansee A.G., Reutle, Tyrol

INDUSTRIAL HEAVY-METAL CARBIDES

The four most important carbides for the production of hard metals are tungsten carbide, WC, titanium carbide, TiC, tantalum carbide, TaC, and niobium carbide, NbC. Chromium carbide, Cr_3C_2, molybdenum carbide, Mo_2C or MoC, vanadium carbide, VC, hafnium carbide, HfC, and zirconium carbide, ZrC, are of minor importance. The binary and ternary solid solutions of these carbides are, however, of great importance, especially the solid solutions WC–TiC and WC–TiC–TaC(NbC), which are discussed later. Carbides and their solid solutions are generally combined with cobalt, and used in the form of cemented carbides (see below). The carbides of the actinides Th, U, Pu, and Np have recently gained importance in reactor technology as nuclear fuels (see Nuclear reactors).

Preparation

In general the carbides of metals of Groups IVB–VIB are prepared by the action of elementary carbon and hydrocarbons on metals and metal compounds at sufficiently high temperatures. The process may be carried out in the presence of a protective gas, under vacuum, or in the presence of an auxiliary metal (menstruum).

Fusion. This method is still used for the preparation of tungsten carbide for the mining industry (ie, coarse-grained powder or castings for welding onto oil drills and wear-resistant parts).

$$3\,W + 2\,C \xrightarrow[H_2]{2800°C} W_2C\text{–}WC \quad \text{(eutectic cast carbide)}$$

$$3\,Cr_2O_3 + 6\,Al + 4\,C \xrightarrow[air]{2000°C} 2\,Cr_3C_2 + 3\,Al_2O_3 \quad \text{(slag)}$$

Carburization. The two types of carburization are (1) of powdered metal or metal hydrides; and (2) of oxides with solid carbon (loose or compacted mixtures), the presence of gases that yield carbon is optional.

$$(1) \quad W + C \xrightarrow[H_2]{1400\text{–}1600°C} WC$$

$$Ta(H) + C \xrightarrow[H_2,\,vacuum]{1400\text{–}1500°C} TaC + (H)$$

$$W + CH_4 \xrightarrow[(H_2)]{1400\text{–}1600°C} WC + 2\,H_2$$

$$(2) \quad WO_3 + 4\,C \xrightarrow[H_2,\,CO]{1400\text{–}1600°C} WC + 3\,CO$$

$$TiO_2 + 3\,C \xrightarrow[H_2,\,CO,\,vacuum]{1800\text{–}2000°C} TiC + 2\,CO$$

$$Nb_2O_5 + 6\,C + CH_4 \xrightarrow[(H_2)]{1400\text{–}1700°C} 2\,NbC + 5\,CO + 2\,H_2$$

Most cemented carbides based on WC, TiC, TaC, and NbC are prepared by either this method or a modified method. The direct or indirect carburization of halides is of importance for the metals of Group VB, especially when they have been processed by chlorination of the ores.

Reduction. Reduction of halides with hydrogen–hydrocarbon mixtures, is sometimes done in the presence of a graphite carrier or of metals with high melting points (ie, van Arkel gas deposition method) (1). If a plasma gun is employed, finely powdered carbides (< 1 μm) are obtained (2) (see Plasma technology). Chemical vapor deposition (CVD process) is used as a carrier for TiC, TiN, and Ti(C,N) coatings on cemented carbide tips (see under Cemented carbides).

$$2\ TaCl_5 + C_2H_2 + 4\ H_2 \xrightarrow[\text{graphite}]{1300-1500°C} 2\ TaC + 10\ HCl$$

$$HfCl_4 + CH_4 \xrightarrow[H_2,\ W\text{-wire}]{2400-2800°C} HfC + 4\ HCl$$

$$NbCl_5 + 5/2\ H_2 \xrightarrow{500-600°C} Nb + 5\ HCl \qquad Nb + C \xrightarrow[CH_4,\ CO,\ H_2]{1300-1500°C} NbC$$

Chemical Separation. Chemical separation occurs from carbon-saturated ferroalloys or metal baths (menstruum process) (3). When high-melting metals are carburized, they form carbides that can easily be isolated by treatment with acids.

$$Ti + (Fe) + C \xrightarrow{1600-1800°C} TiC + (Fe)$$

$$Nb,Ta + C + (Fe) \xrightarrow{1600-1800°C} (Nb,Ta)C + (Fe)$$

$$Ta_2O_4 + C + (Al) \xrightarrow{2000°C} TaC + (Al) + CO$$

This method is used for the preparation of TiC from pure titanium scrap, for the preparation of TaC, and especially for oxygen- and nitrogen-free solid solutions of various carbides (see under Solid Solutions).

Tungsten Carbide

On an industrial scale, tungsten monocarbide (annual world production approximately 15,000–18,000 metric tons) is prepared by the direct reaction of tungsten metal powders and low-ash carbon black in the presence of hydrogen at 1400–1600°C. Charcoal and reactor-grade pure graphite are sometimes used. Continuous carbon-tube furnaces, push-type furnaces with alumina muffles and molybdenum resistors, as well as high-frequency furnaces (protective gas or vacuum) are used for the carburization. The tungsten powder is prepared from tungstic oxide (WO_3), tungstic acid (H_2WO_4), or ammonium p-tungstate ($5(NH_4)_2O.12WO_3.5H_2O$) of >99.9% purity. Reduction with hydrogen is generally carried out in continuous push-type furnaces at approximately 800–900°C. Furnaces with ceramic or scale-resistant metal tubes or muffles, gas-heated multiple-tube furnaces, or walking-beam furnaces are also used. For very fine powders, rotary kilns may be employed (see Tungsten and tungsten alloys).

Purity, particle shape, and grain size of the starting material and the conditions employed for the reduction and carburization determine the properties—especially the grain size—of the final product. The course of the reaction $WO_3 \rightarrow W \rightarrow WC$ is dependent on the temperature which affects the coarsening of the grain; selection of suitable tungsten-containing raw materials and modification of the reduction and carburization conditions permit, therefore, the preparation of the WC powder in various grain sizes.

The following examples are illustrative: (1) Very pure tungstic acid of fine grain size (<0.1 µm) is reduced in dry hydrogen at approximately 800°C; the fine tungsten powder (<0.5 µm) obtained gives very fine grained WC (<1 µm) on carburization at 1350–1400°C. (2) Calcined coarse WO_3 (<2 µm), after reduction with hydrogen in the presence of water vapor at 900–950°C, gives tungsten powder (<6 µm) from which coarse crystalline WC powder (<10 µm) can be prepared by carburization at 1600°C.

The WC leaving the furnace is light-gray with a bluish tinge. It is generally caked and must be broken up, milled, and screened before use. It should contain about 6.1–6.25% total C, of which 0.03–0.15% is in the free, unbound state (theoretical C-content, 6.13%).

Finely powdered tungsten carbide (grain size 0.01–0.1 µm) is of growing interest for the production of very fine-grain cemented carbides. Usually the plasma gun technique or the reduction of very fine WO_3 with CO and CH_4 is applied (4) (Axel-Johnson process).

The properties of WC are listed in Table 1. Tungsten carbide and the fused W_2C–WC eutectic (the so-called cast tungsten carbide, see Fig. 1) are important in the mining industry and for wear-resistant weldings (eg, for oil drills).

Titanium Carbide

On an industrial scale, titanium carbide (annual world production 1200–1500 metric tons) is mainly prepared by the reaction of TiO_2 with carbon black. Smaller quantities can be obtained from titanium scrap or by recovery from metal baths (utilization of ferrotitanium or titanium alloy scrap), especially when it is prepared in conjunction with solid solutions of carbides.

Table 1. Physical Properties of the important Cemented Carbides WC, TiC, TaC, and NbC

Property	WC [12070-12-1]	TiC [12070-08-5]	TaC [12070-06-3]	NbC [12069-94-2]
mol wt	195.87	59.91	192.96	104.92
carbon, wt %	6.13	20.05	6.23	11.45
crystal structure	hex, Bh	fcc, B1	fcc, B1	fcc, B1
lattice constants, nm	$a = 0.29065$ $c = 0.28366$	0.43305	0.4454	0.4470
density, g/cm³	15.7	4.93	14.48	7.78
mp, °C	2720	2940	3825	3613
microhardness	1200–2500	3000	1800	2000
modulus of elasticity, N/mm² [a]	696,000	451,000	285,000	338,000
transverse rupture strength, N/mm² [a]	550–600	240–400	350–450	300–400
coefficient of thermal expansion per K	$a = 5.2 \times 10^{-6}$ $c = 7.3 \times 10^{-6}$	7.74×10^{-6}	6.29×10^{-6}	6.65×10^{-6}
thermal conductivity, W/(m·K)	121	21	22	14
heat of formation, ΔH_{298}, kJ/mol [b]	−40.2	−183.4	−146.5	−140.7
specific heat, J/(mol·K) [b]	39.8	47.7	36.4	36.8
electrical resistivity, µΩ·cm	19	68	25	35
superconductive below K	1.28	1.15	9.7	11.1
Hall constant, cm³/(A·s)	-21.8×10^{-4}	-15.0×10^{-4}	-1.1×10^{-4}	-1.3×10^{-4}
magnetic susceptibility	+10	+6.7	+9.3	+15.3

[a] To convert N/mm² to psi, multiply by 145.
[b] To convert J to cal, divide by 4.184.

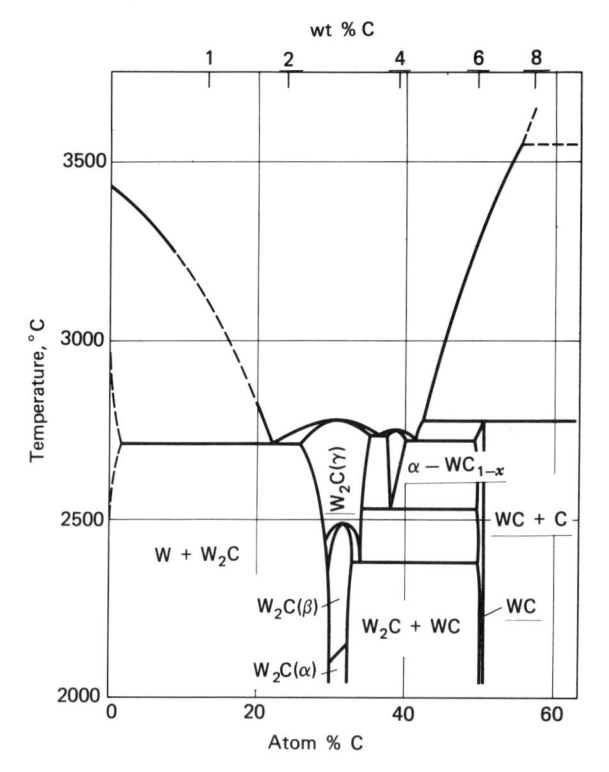

Figure 1. Phase diagram W–C.

Pure titanium oxide (99.8%, with minor impurities of Si, Fe, S, P, and alkalies) in the dry or wet state is mixed in 68.5:31.5 ratio with carbon black or finely milled low-ash graphite. The dry mixture is pressed into blocks which are heated in a horizontal or vertical carbon-tube furnace at 1900–2300°C; hydrogen (free of oxygen and nitrogen) serves as protective gas, or in the vertical push-type furnaces, the liberated CO has that function.

Titanium carbide is generally obtained in the form of gray, well-sintered lumps that are broken up in jaw-crushers and fine-milled in ball mills. Technical-grade contains 0.5–1.5% graphite, in addition to 0.5–1% oxygen and nitrogen and 0.1% impurities, such as Fe, Si, S, and P. If the graphite content is too high, carbon and titanium oxide may be added, followed by annealing. The oxygen and nitrogen content may be reduced considerably by heating the crude carbides under high vacuum for several hours at 2000–2500°C (eg, to 0.1–0.3% ($N_2 + O_2$)). This can also be achieved in a less costly way by formation of solid solutions with WC, with or without the use of Mo_2C or Cr_3C_2 (see under Molybdenum Carbide and Chromium Carbide, respectively).

Properties of titanium carbide are listed in Table 1; the phase diagram Ti–C is shown in Figure 2.

Titanium carbide is an important component of sintered cemented carbide because of its hardness and solvent action for other carbides. Standard cemented carbides for steel cutting contains 5–25%, and the special grades, 30–85% TiC.

Next to Cr_3C_2, TiC is the major component for heat- and oxidation-resistant

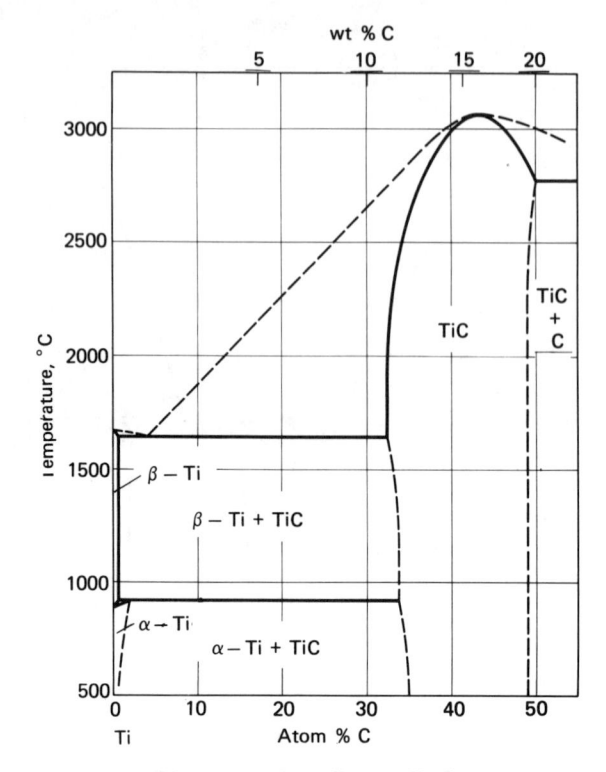

Figure 2. Phase diagram Ti–C.

cemented carbides. TiC-based boats, containing AlN, BN and, TiB$_2$, have been found satisfactory for the evaporation of metals.

Tantalum Carbide

On an industrial scale, TaC is prepared from tantalum metal or tantalum hydride powder, from tantalum pentoxide (Ta$_2$O$_5$), high-purity scrap obtained in the preparation of ductile Ta, or ferrotantalum–niobium. The chemical and metallurgical industry uses tantalites, niobites, Ta–Nb-containing tin slags and, sometimes, the rare ores microlite and pyrochlorite for the preparation of tantalum. The ores are decomposed and separated by fractional crystallization of the double fluorides or by fraction distillation of the pentachlorides, or by solvent extraction of the HF-containing solutions. The metals are prepared mainly by alkali metal reduction or fusion electrolysis of the halides.

The carbide is produced by carburization of the element or the oxide with carbon (see above), similar to the preparation of WC or of TiC. Final carburization in a vacuum gives a golden-yellow carbide, free of oxygen and nitrogen, that contains 6.1–6.3% C and 0–0.2% graphite.

The McKenna menstruum process (3), giving satisfactory yields, is based on the following principle: equal amounts of aluminum and Ta$_2$O$_5$ are heated in a graphite crucible to ca 2000°C, and pure graphite lumps are added. The excess aluminum and aluminum carbide are removed from the cooled, pulverized reaction product by extraction with mineral acids, and the remaining golden, highly pure TaC crystals are freed from the adhering graphite by flotation.

Various metal-bath reduction processes are used in plants where ferrotantalum–niobium or Ta–Nb-containing crude carbides are processed to TiC–TaC(NbC) or WC–TiC–TaC(NbC) solid solutions (see below).

The properties of TaC are given in Table 1; the phase diagram Ta–C is shown in Figure 3.

WC–TiC–Co steel grades for long-chip materials have been largely replaced by WC–TiC–TaC(NbC)–Co grades that show higher hot-hardness and better cutting performance. This, as well as the addition of small quantities (0.5–2%) of TaC to straight WC–Co alloys to prevent grain growth, has created an increasing market for TaC. Annual world consumption is estimated at 350–450 tons TaC(NbC) and it is believed that TaC will be used in increasing quantities in order to raise the (TaC + NbC) content in cemented carbides used for steel cutting from 3–5% to 7–18%.

Because of its high melting point (3000°C), small quantities of TaC are used in high pressure lamps and as barriers between tungsten and graphite in rocket jet nozzles.

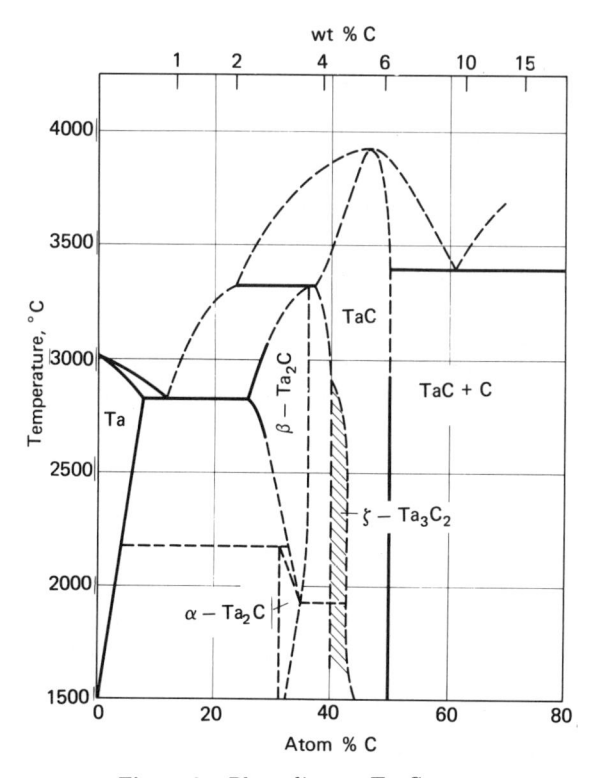

Figure 3. Phase diagram Ta–C.

Niobium Carbide

The preparation of Nb metal and Nb_2O_5, and the extractive metallurgy (qv) of niobium are very similar to that of tantalum.

Niobium carbide is most often prepared by the carburization of Nb_2O_5 with carbon black, and less frequently, by reaction of the two elements. The preparation of NbC has special importance because it is used on an industrial scale as a reducing agent to prepare niobium metal. In this process, finely divided mixtures of Nb_2O_5 and carbon black are pressed into cylindrical blocks in large hydraulic presses and converted to NbC in high-frequency furnaces in the presence of hydrogen or under vacuum at 1600–1800°C. The comminuted carbide may then be mixed with further quantities of the oxide in induction-heated vacuum furnaces and processed to technical-grade metal (1–3% O_2, 0.5–1% free graphite) which can be used directly for the preparation of cemented carbides.

The properties of NbC are given in Table 1.

The grayish-brown NbC powder is used in increasing quantities in sintered cemented carbides in order to replace TaC by the less expensive NbC. TaC–NbC solid solutions (3:1, 2:1, 1:1, and 1:2) and the corresponding ternary and quaternary solid solutions with TiC and WC are used (see under Solid Solutions).

Chromium Carbide

Mixtures of Cr_3C_2 and Cr_7C_3 (9–10% carbon) are sometimes fused in arc furnaces and used for the preparation of Stellite-type alloys (Co–Cr–W–(C–Si)) alloys. In the carbide series Cr_4C [12011-63-1], $Cr_{23}C_6$ [12105-81-6], Cr_7C_3 [12075-40-0], and Cr_3C_2 [12012-35-0], only the latter (which has the highest carbon content and the highest degree of hardness) is of interest for cemented carbides. It has not been possible to isolate a monocarbide CrC [12011-60-8] (analogous to WC) or to stabilize it by means of a second carbide.

Chromium carbide can be best prepared from pure Cr_2O_3. Compact materials containing 74% Cr_2O_3 and 26% carbon black can, for instance, be heated in carbon-tube furnaces at 1600°C in the presence of hydrogen, giving a carbide containing 13–13.3% total C and 0.1–0.3% free C.

The properties of Cr_3C_2 are listed in Table 2.

Chromium carbide is used in small quantities in oxidation-resistant hard alloys based on TiC (1–5%). As the main component (60–85% Cr_3C_2) with Ni or Ni–Cu binders, it forms acid- and wear-resistant alloys which have found application in the chemical industry in valve parts. Cast Cr_7C_3 or the eutectic Cr_7C_3–Cr_3C_2 powders are used in rods for hard facing.

Molybdenum Carbide

Mo_2C can be prepared by the carburization of MoO_3 or MoO_2 with carbon black or, more conveniently, by the reaction of molybdenum powder (93.4%) and carbon black or charcoal (ca 6.6%) at 1350–1500°C, in the presence of hydrogen. Thus the carbide formed contains 5.9–6.1% total C and 0.05–0.25% free C. The physical properties are listed in Table 2. There are two molybdenum carbides with higher carbon contents, ie, the cubic MoC_{1-x}, a high temperature phase often described as Mo_3C_{2+x}, and the hexagonal MoC, a low temperature phase isotypic with WC. Both carbides tend

Table 2. Physical Properties of a Few Less Important Carbides

Property	Cr$_3$C$_2$ [12012-35-0]	β-Mo$_2$C [12069-89-5]	η-MoC [12011-97-1]	VC [12070-10-9]	HfC [12069-85-1]	ZrC [12020-14-3]
mol wt	180.05	203.91	107.96	62.96	190.51	103.23
carbon, wt %	13.33	5.89	11.3	19.08	6.30	11.64
crystal structure	rhom D5$_{10}$	hex L′3	hex L′3	fcc, B1	fcc, B1	fcc, B1
lattice constants, nm	$a = 1.147$ $b = 0.554$ $c = 0.283$	$a = 0.300$ $c = 0.4734$	$a = 0.298$ $c = 0.281$	0.4165	0.4648	0.4698
density, g/cm^3	6.68	9.18	9.15	5.36	12.3	6.46
microhardness	1350	1500	2200	2900	2600	2700
modulus of elasticity, N/mm^2 [a]	373,000	533,000		422,000	352,000	348,000
mp, °C	1810	2520	2600	2684	3820	3420
coefficient of thermal expansion per K	10.3×10^{-6}	7.8×10^{-6}		7.2×10^{-6}	6.59×10^{-6}	6.73×10^{-6}
heat of formation ΔH_{298}, kJ/mol [b]	−94.2	−49		−124.8	−230.3	−196.8
specific heat, J/mol K [b]	32.7	30.3		32.3	37.4	37.8
electrical resistivity, μΩ·cm	75	71		60	37	42
superconductive below K	<1.2	2.78		<1.2	<1.2	<1.2
Hall constant, cm^3/(A·s)	-0.47×10^{-4}	-0.85×10^{-4}		-0.48×10^{-4}	-12.4×10^{-4}	-9.42×10^{-4}
magnetic susceptibility				+28	−25.2	−23

[a] To convert N/mm^2 to psi, multiply by 145.
[b] To convert J to cal, divide by 4.184.

to decompose to graphite and Mo$_2$C. The latter can be stabilized by W (5), N (6), or W + N (7) so that (Mo,W)C (8) or (Mo,W)(C,N) (7) mixed crystals are formed. These new phases may be future competitors for WC in the cemented carbide industry.

Molybdenum carbide is used in special tungsten-free alloys based on TiC–Mo$_2$C–Ni(Mo) or in the newly developed spinoidal alloys based on (Ti,Mo)(C,N)$_{1-x}$-Ni(Mo). During sintering, the latter alloy shows a spinoidal decomposition to an intimate mixture of the phases (Ti,Mo)C$_{1-x}$ and Ti(C,N)$_{1-x}$ (6,9).

Vanadium Carbide

Vanadium pentoxide or vanadium trioxide are the most satisfactory oxides for the preparation of VC. Vanadium pentoxide is best prepared by igniting chemically pure ammonium vanadate in the presence of moist oxygen to avoid reaction with nitrogen; V$_2$O$_3$ is obtained by reduction of V$_2$O$_5$ with hydrogen.

Vanadium carbide, with the highest content of chemically bound carbon, ie, 18.5–18.9% C, is prepared by the reaction of the elements under vacuum. In this process, V$_2$O$_5$ is reduced at a high temperature with calcium to reguli, which are melted in an arc furnace in the presence of argon, giving a 99.9% pure product. Vanadium powder of equal purity may be prepared by hydriding and crushing vanadium turnings. Vacuum carburization removes nitrogen and oxygen, which generally occur at 0.5–1% in carbides obtained from vanadium oxides.

The properties of VC are given in Table 2.

Although VC is very hard, it is very brittle and has, therefore, been used only in special cemented carbides. For instance, some grades of straight WC–Co alloys used in Europe contained ca 0.5% VC as a grain growth inhibitor; it has now been replaced by TaC. TiC–VC–Ni–Fe cemented carbides were used in Germany during World War II as tungsten carbide-free cutting-tool alloys.

Hafnium Carbide. The large-scale separation of hafnium from zirconium for the production of pure zirconium for nuclear reactors made it possible to obtain sufficient quantities of HfO_2 or Hf metal sponge to produce HfC for use in cemented carbides.

Hafnium carbide can be prepared industrially from hydrided hafnium sponge at 1500–1700°C or from HfO_2 at 2000–2200°C by carburization *in vacuo* in the presence of hydrogen. The resulting carbide contains almost the theoretical quantity of carbon (6.30% C) of which a maximum of 0.1% is free.

The properties of HfC are listed in Table 2. The distinctively high melting point (ca 3900°C) resembles that of TaC.

Addition of HfC as NbC–HfC or TaC–HfC solid solutions to WC–TiC–Co alloys, improves the properties (10).

Zirconium Carbide

ZrC may be prepared by igniting a mixture of 78.8% annealed ZrO_2 and 21.3% charcoal in a graphite crucible in a carbon-tube furnace at 2400°C in the presence of hydrogen. The carbide obtained has the following composition: 11.3% chemically bound C, traces of free C, 88.3% Zr, and 0.3% ($O_2 + N_2$).

Alternatively, a pressed mixture of ZrO_2 and carbon black reacts in an induction-heated graphite crucible in the presence of H_2 at 1800°C, and is then comminuted and annealed at 1700–1900°C, under vacuum, after addition of 1–2% carbon black. The product contains 11.8% C (ca 0.5% free C). Sintering under pressure at 2200°C yields a carbide that is substantially free of oxygen, and contains almost the theoretical amount of carbon (11.64% C).

The physical properties of ZrC are listed in Table 2.

Zirconium carbide is not as important as TiC for cemented carbides. To obtain the machining properties of WC–TiC–Co alloys, two parts of the more expensive ZrC must be added instead of one part TiC. ZrC is important as an intermediate formed during the preparation of zirconium from its ores.

Carbides of the Actinides Uranium and Thorium

Since World War II, the carbides of ^{235}U and thorium have gained importance as nuclear fuels and breeder materials for gas-cooled, graphite-moderated reactors.

The actinide carbides are prepared by the reaction of metal or metal hydride powders with carbon or better by the reduction of the oxides UO_2, U_3O_8, or ThO_2 at 1800–2200°C in carbon-tube furnaces in the presence of hydrogen or in vacuum furnaces. It is very difficult to obtain the monocarbides free of higher carbides.

Hot pressing and arc melting are very suitable methods for the preparation of homogeneous compacts, especially if followed by heat treatment in a tungsten-tube furnace in the presence of argon.

The properties of the uranium and thorium carbides are given in Table 3; the phase diagram for the system U–C is shown in Figure 4.

Uranium carbide is comparatively stable whereas UC_2 (especially in powder form) hydrolyzes rapidly in moist air. The latter (more or less enriched in ^{235}U) is used in the form of pellets or annealed spherical particles, coated with pyrographite, as a nuclear fuel for high-temperature reactors. For breeder reactors using thorium, it should be noted that UC and ThC form a continuous series of solid solutions, whereas UC_2 and ThC_2 have limited mutual solubility. Furthermore, UC can be stabilized with regard to its carbon content, even at high temperatures, by the formation of solid solutions with ZrC, HfC, NbC, or TaC so that no higher carbides are formed (11). Uranium carbides and plutonium carbides show a high degree of mutual miscibility (see Actinides).

The Carbides of the Iron Group Metals

The carbides of iron, nickel, cobalt, and manganese are not classified with the hard metallic materials having lower melting points and hardness, and different structures. Nonetheless, these carbides, particularly iron carbide and the double carbides with other transition metals, are of great technical importance as hardening components of alloy steels and cast iron.

The iron–carbon system contains the orthorhombic iron carbide [12011-67-5], (3:1), Fe_3C, which melts congruently and represents the cementite in steel metallurgy. The existence of other carbides, eg, Fe_2C [12011-66-4], Fe_5C_2 [12127-45-6], and Fe_7C_3 [12075-42-2], although they give x-ray patterns, is doubtful (see Steel).

Iron carbide, Fe_3C, mol wt 179.56, carbon 6.69%, density 7.64 g/cm^3, mp 1650°C, is obtained as a dark gray air-sensitive powder by anodic isolation with hydrochloric acid from high-carbon iron melts. In the microstructure of steels, cementite appears

Table 3. Physical Properties of Uranium and Thorium Carbides

Property	UC [12070-09-6]	UC_2 [12071-33-9]	ThC [12012-16-7]	ThC_2 [12071-31-7]
mol wt	250.08	262.09	244.06	256.07
carbon, wt %	4.8	9.16	4.92	9.37
crystal structure[a]	fcc, B1	tetr C11a	fcc, B1	mon
lattice constants, nm	0.49597	$a = 0.3524$	0.5346	$a = 0.6691$
		$c = 0.5996$		$b = 0.4231$
				$c = 0.6744$
				$\beta = 103°50'$
density, g/cm^3	13.63	11.86	10.64	8.65
microhardness	920	620	850	600
mp, °C	2560	ca 2500	2625	2655
coefficient of thermal expansion per K	$9.1 \cdot 10^{-6}$			
thermal conductivity, W/(m·K)	25			
heat of formation ΔH_{298}, kJ/mol[b]	−97.1	−96.3	−29.3	−125.2
specific heat, J/(mol·K)[b]	50.2	58.6		
electrical resistivity, $\mu\Omega$·cm	40	90	25	30
magnetic susceptibility	+3.15	+3.40		

[a] tetr = tetragonal; mon = monoclinic cubic.
[b] To convert J to cal, divide by 4.184.

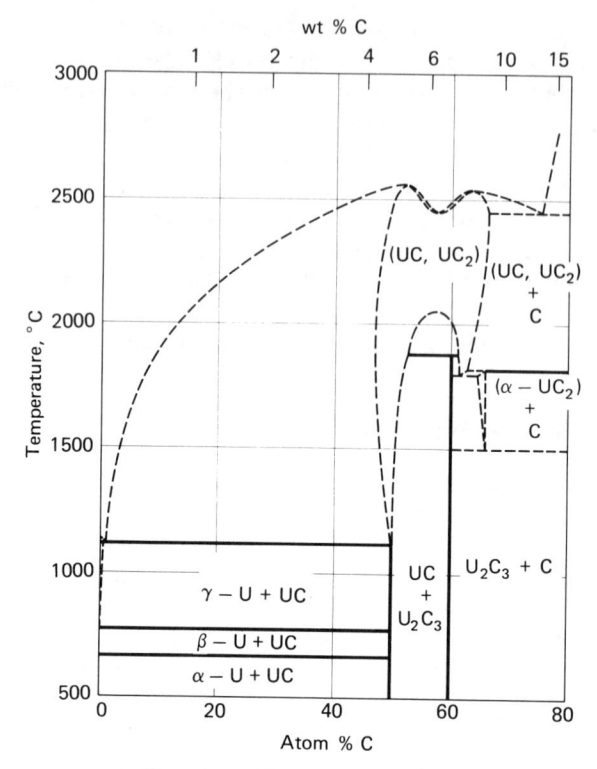

Figure 4. Phase diagram U–C.

in the form of etch-resistant grain borders, needles, or lamellae. When Fe_3C powder is sintered with binder metals, a change occurs so that cemented alloys cannot be produced. The hard components in alloy steels, such as chromium steels, are double carbides of the formulas $(Cr,Fe)_{23}C_6$, $(Fe,Cr)_7C_3$, or $(Fe,Cr)_3C_2$. They derive from the binary chromium carbides and can also contain tungsten or molybdenum. These double carbides are related to η-carbides, ternary compounds of the general formula $M_3M_3'C$ (M = iron metal; M′ = refractory transition metal) (see also under Complex Carbides). In cemented carbide technology, the very hard and brittle η-carbide Co_3W_3C is particularly undesirable.

The complex iron carbonitride is the hard component in steels that have been annealed with ammonia (nitrided steels). Complex carbonitrides with iron metals are also present in superalloys in the form of precipitates (see under Mixed Phases).

In the nickel–carbon and cobalt–carbon systems, the nickel carbide [*12012-02-1*], Ni_3C, and cobalt carbide [*12011-59-5*], Co_3C, are present. Both are isomorphous with Fe_3C and exist only at low temperatures. The manganese–carbon system contains manganese carbide [*12121-90-3*], Mn_3C (isomorphous with Fe_3C), and $Mn_{23}C_6$ [*12266-65-8*] isomorphous with $Cr_{23}C_6$ [*12105-81-6*]. These binary carbides occur frequently in the form of carbide solid solutions or double carbides with other transition metals in alloy steels, superalloys, and special hard metals.

Solid Solutions

Pure WC is of paramount importance in straight WC–Co alloys for short-chip materials. Fine-grain size straight WC inserts perform better than alloy carbides when cutting superalloys even though the latter produce long chips. In all other grades of cemented carbides, binary and ternary solid solutions of the type WC–TiC, WC–TaC(NbC), TiC–TaC(NbC), WC–TiC–TaC(NbC), WC–MoC, and (W,Mo)(C,N) occur besides the carbide component WC.

In tungsten-free alloys, the solid solutions $(Ti,Mo)C_{1-x}$, $(Ti,Mo,V,Nb)C_{1-x}$, and $(Ti,Mo)(C,N)_{1-x}$ are of growing interest.

The solid solutions are harder and more heat resistant than the pure components, and are substantially free from oxygen, nitrogen, and graphite as they are subjected to a type of autopurification during the diffusion annealing of the solid solutions. The vacuum-purified carbide solid solutions of the metals of Groups IVB, VB, and VIB are wetted more readily by cobalt than are the unalloyed carbides; they are also tougher than the latter.

The solid solutions of the carbides are generally prepared like the pure carbides employing the following methods, which can also be modified or combined.

Carburization of metal oxide mixtures with carbon black. Additives such as Co, Ni, Fe, or Cr (0.5–1%) may be used to promote diffusion.

$$WO_3 + TiO_2 + C \xrightarrow[H_2, CO]{1600-1800°C} (W,Ti)C + CO$$

$$WO_3 + TiO_2 + Ta_2O_5 + C \xrightarrow[H_2]{1600-1800°C} (W,Ti,Ta)C + CO$$

$$MoO_3 + WO_3 + C + (Fe,Ni,Co) \xrightarrow[CO, H_2, CH_4]{1000-1200°C} (Mo,W)C + CO + (Fe,Ni,Co)$$

Carburization of metal powder mixtures with carbon black.

$$W + Ta + C \xrightarrow[H_2, vacuum]{1500-1600°C} (W,Ta)C$$

$$Hf + Ti + C \xrightarrow[H_2, vacuum]{1600-1800°C} (Hf,Ti)C$$

Diffusion annealing of mixed preformed carbides at temperatures giving solid solutions. Additives, such as Co, Ni, Fe, or Cr (0.5–1%) promote diffusion.

$$WC + TiC \xrightarrow[H_2, vacuum]{2600-1900°C} (W,Ti)C$$

$$WC + TiC + TaC(NbC) \xrightarrow[H_2, vacuum]{1600-1800°C} (W,Ti,Ta,Nb)C$$

$$Mo_2C + TiC \xrightarrow[H_2]{1500-1700°C} (Mo_2Ti)C_{1-x}$$

$$HfC + TiC + WC \xrightarrow[H_2, vacuum]{1600-1900°C} (Hf,Ti,W)C$$

Separation of solid solutions of carbides from metal melts (metallic menstruum).

$$Ta + Nb + (Fe) + C \xrightarrow{1600-1800°C} (Ta,Nb)C + (Fe)$$

$$Ti\,(ferroalloy) + W\,(ferroalloy) + C \xrightarrow[H_2, CO, CH_4]{1800-2000°C} (Ti,W)C + (Fe)$$

$$WC + TiC + TaC(NbC) + (Ni + C) \xrightarrow{1600-1800°C} (W,Ti,Nb)C + (Ni)$$

502   CARBIDES (INDUSTRIAL HEAVY-METAL)

Reduction of oxides in the presence of a metal or carbide.

$$TiO_2 + (W,WC) + C \xrightarrow{1800°C} (Ti,W)C + CO$$

The monocarbides of Groups IVB and VB metals are completely miscible except for ZrC–VC and HfC–VC (see Fig. 5). At 1400–1500°C (cobalt sintering temperature), WC is soluble in the carbides of Groups IVB and VB 25–60% by wt, and at higher temperatures (1800–2400°C) even up to 90% by wt (12–13). WC itself, like the other carbides of Group VIB, shows practically no solvent power for face-centered cubic carbides. The cubic solid solutions which are saturated at higher temperatures and contain a high percentage of WC are very stable. Only sintering with liquid phase causes WC to separate readily. The hexagonal WC, however, forms complete series of solid solutions with the hexagonal MoC.

The pseudoternary system WC–TiC–TaC is especially important in the metallurgy of cemented carbides (14). Figure 6 shows the phase distribution and thus the solubility ratios at 1450°C (cobalt sintering temperature), at 2200°C (preferred

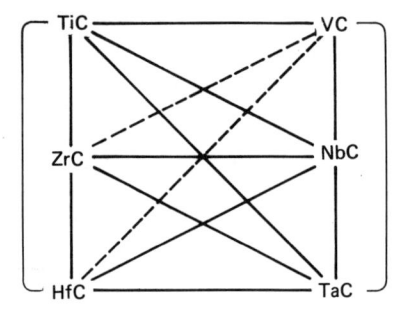

Figure 5. Solid solubility between carbides, schematic. Solid line = complete solubility; dotted line = limited solubility.

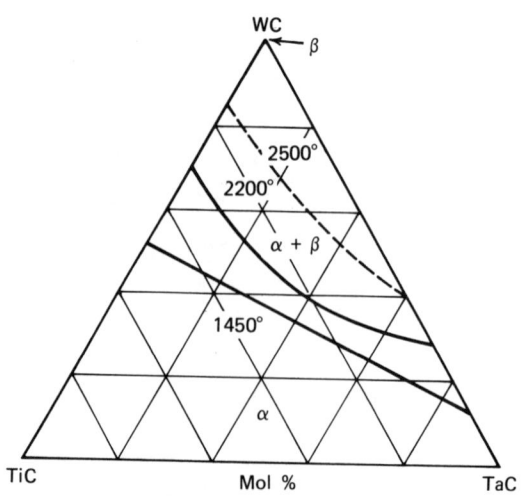

Figure 6. Phase distribution in the system WC–TiC–TaC at 1450, 2200, and 2500°C.

temperature for the formation of pure solid solutions with high WC content), and at 2500°C (hypothetical curve, maximum solubility). It is obvious that TiC is a better solvent for WC than TaC, ie, TaC additives in ternary solid solutions reduce the solvent power for WC. In industrial grades of cemented carbides containing 5–25% TiC, 3–15% TaC, 52–86% WC, and 6–13% Co, the amount of WC in solid solution is the same as TiC + TaC (8–40%); the remainder of the WC exists in the free form. The more spherical TiC–TaC–WC solid solutions can be readily distinguished from the angular WC in the microstructure.

Mixed Phases with Nitrogen, Boron, and Silicon

The chemistry of carbonitrides is attracting increasing attention. Ti(C,N) coatings appear to compete with TiC, TiN, HfN, and Al_2O_3 coatings (15). The Ni–Mo cemented carbonitrides Ti(C,N) and $(Ti,Mo,V,Nb,Ta)(C,N)_{1-x}$ show good cutting properties (16). The spinoidal $(Ti,Mo)(C,N)_{1-x}$ carbonitrides (9) are superior to the (Ti,Mo)-C_{1-x}-based alloys. Solid solutions of (Mo,W)C and (Mo,W)(C,N) may partially replace pure WC.

Complex Carbides

Complex carbides are ternary or quaternary intermetallic phases containing carbon and two or more metals. One metal can be a refractory transition metal and the second also a transition metal or from the iron or A-groups. Nonmetals can also be incorporated.

Complex carbides are very numerous. Particularly through the work of Nowotny and co-workers, many new compounds of this class have been discovered and their structure elucidated (17). The octahedron M_6C is typical with the metals arranged around a central carbon atom. The octahedra may be connected via corners, edges, or faces. Trigonal prismatic polyhedra also occur. Thus the complex carbides can be classified as follows (T = transition metal, M = metal or main group nonmetal): (1) T_3M_2C (filled β-manganese structure), eg, Mo_3Al_2C, W_3Re_2C; (2) T_2MC, H-phases, eg, Cr_2AlC, Ta_2GaC, Ti_2SC; (3) T_3MC, perovskite carbides (filled Cu_3Au-structure), eg, Ti_3AlC, VRu_3C, URh_3C; (4) T_3M_3C, T_2M_4C, η-carbides (filled Ti_2Ni-structure), eg, Co_3W_3C, Ni_2Mo_4C; and (5) κ-carbides, eg, $W_9Co_3C_4$, $Mo_{12}Cu_3Al_{11}C_6$. The preferred method for synthesis of complex carbides is the powder metallurgy technique. Hot-pressed powder mixtures must be subjected to prolonged annealing treatments; if low-melting or volatile components are present, autoclaves are used.

The η-carbides are not specifically synthesized, but are of technical importance, occurring in alloy steels, stellites, or as embrittling phases in cemented carbides. Other complex carbides in the form of precipitates may form in multicomponent alloys or in high temperature reactor fuels by reaction between the fission products and the moderator graphite (pyrographite-coated fuel kernels).

Test Methods and Quality Control

Chemical and physical examinations are needed during the preparation of cemented carbides. The carbon content is of primary interest, especially the free graphite carbon. The determination is carried out in a current of oxygen with addition of high-melting compounds that liberate oxygen, and the CO_2 formed is determined volumetrically or conductometrically. In order to determine the free carbon, the carbide is treated with HNO_3–HF and the graphite is separated. Decomposition of some carbides is troublesome (SiC, B_4C, Cr_3C_2). Nitrogen and oxygen determinations are carried out by the Kjeldahl method and by vacuum heat extraction.

For x-ray investigations, the Debye-Scherrer powder method is generally used. The lattice constants indicate purity or composition of solid solutions; the rapid counting-tube goniometric method can be used for plant control. The rotating-crystal and neutron diffraction methods are sometimes used for structure elucidation.

The hardness of carbides cannot be determined by the usual macro tests because of brittleness, but only by micro methods.

The extremely high melting points of the carbides are not readily determined by the usual methods. In the so-called Pirani hole method, a small hollow rod is placed between two electrodes and heated by direct current until a liquid drop appears in the cavity. The temperature is determined pyrometrically. When high-temperature tungsten tube furnaces are used the melting point can readily be estimated by the Seger-type cone method. The sample may also be fused in a Kroll arc furnace and the solidification temperature determined.

The eddy-current test (Sigma test, Förster) has been most satisfactory for the determination of electroconductivity, requiring only comparatively small, dense samples.

Special grinding and polishing methods are necessary for the metallographic examination of cemented carbides, since bort (low-grade diamond) must be employed; moreover, the cemented carbides show very high chemical stability so that vigorous etching agents (HNO_3, HF, H_2O_2) must be used. The Reinacher polishing and etching methods have proved to be most satisfactory (18).

BIBLIOGRAPHY

"Heavy-Metal Carbides" under "Carbides" in *ECT* 1st ed., Vol. 2, pp. 846–854, by P. M. McKenna and J. C. Redmond, Kennametal Inc.; (Industrial, Heavy-Metal Carbides) under "Carbides" in *ECT* 2nd ed., Vol. 4, pp. 75–92, by Richard Kieffer, University of Vienna, and Friedrich Benesovsky, Metallwerk Plansee A.-G.

1. A. E. van Arkel and J. H. de Boer, *Physica* **4**, 286 (1924); *Z. Anorg. Allg. Chem.* **148**, 345 (1925).
2. E. Neuenschwander, *J. Less Common Met.* **11**, 365 (1966).
3. U.S. Pats. 2,113,352; 2,113,353; 2,113,356 (1937), P. M. McKenna.
4. L. Ramquist, in H. H. Hausner, ed., *Modern Developments in Powder Metallurgy*, Vol. 4, Plenum Press, New York, 1970, pp. 75–84.
5. W. Dawihl, *Z. Anorg. Chemie* **262**, 212 (1950).
6. R. Kieffer and co-workers, *Monatsh. Chem.* **101**, 65 (1970); P. Ettmayer, *Monatsh. Chem.* **101**, 1720 (1970).
7. P. Ettmayer and R. Kieffer, Austrian and other foreign patents pending.
8. J. Schuster, E. Rudy, and H. Nowotny, *Monatsh. Chem.* **107**, 1167 (1976).
9. E. Rudy, S. Worcester, and W. Elkington, *High Temp. High Pressures* **6**, 447 (1974).
10. R. Kieffer, N. Reiter, and D. Fister, *BISRA-ISI Conference on Materials for Metal-Cutting, Scarborough, Eng.*, 1970, p. 126; Fr. Pat. 2,064,842 (1970), R. Kieffer (to Ugine-Carbone).
11. F. Benesovsky and E. Rudy, *Planseeber. Pulvermet.* **9**, 65 (1961); *Monatsh. Chem.* **94**, 204 (1963).
12. R. Kieffer and H. Nowotny, *Metallforschung* **2**, 257 (1947).
13. J. Norton and A. L. Mowry, *Trans. Am. Inst. Min. Met. Eng.* **185**, 133 (1949).
14. H. Nowotny, R. Kieffer, and O. Knotek, *Berg. Hüttenmänn. Monatsh.* **96**, 6 (1951).
15. Ger. Pat. 1,056,449 (1954), (to Metallgesellschaft A.G.); U.S. Pat. 3,717,496 (1970), (to Deutsche

Edelstahlwerke A.G.); Austrian Pat. 3,129,952 (1972), (to Metallwerk Plansee A.G.); U.S. Pat. 3,836,392 (Sept. 14, 1974), (to Sandvik A.B.); U.S. Pat. 3,736,107 (May 26, 1971), (to General Electric).
16. R. Kieffer, P. Ettmayer, and M. Freudhofmeier, *Metall.* **25**, 1335 (1971).
17. H. Nowotny, *Ang. Chem.* **84**, 973 (1972).
18. G. Reinacher, *Z. Metallk.* **48**, 162 (1947).

General References

R. Kieffer and F. Benesovsky, *Hartstoffe,* Springer-Verlag, Vienna, Austria, 1963, 1965.
E. K. Storms, *The Refractory Carbides,* Academic Press, New York, 1967.
L. Toth, *Transition Metal Carbides and Nitrides,* Plenum Press, New York, 1971.
H. J. Goldschmidt, *Interstitial Alloys,* Butterworth, London, Eng., 1967.
T. J. Kossolapova, *Carbides, Properties, Production and Applications,* Plenum Press, New York, 1971.
E. Fromm and E. Gebhardt, eds., *Gase und Kohlenstoff in Metallen,* Springer-Verlag, Berlin, Heidelberg, New York, 1976.
K. J. A. Brooks, *World Directory and Handbook of Hardmetals,* Engineering Digest Publ., London, Eng., 1976.

RICHARD KIEFFER
Technical University, Vienna

FRIEDRICH BENESOVSKY
Metallwerke Plansee A.G., Reutle, Tyrol

CALCIUM CARBIDE

Calcium carbide [75-20-7], is a transparent colorless solid. The pure material can be prepared only by very special laboratory techniques. Commercial calcium carbide is not a pure chemical compound but is composed of calcium carbide (CaC_2), lime (CaO), and other impurities occurring in the coke and lime used in its manufacture. Its calcium carbide content varies, and the commercial material is sold based upon a minimum acetylene yield as specified by U.S. Government regulations. Industrial-grade calcium carbide, sold for generation of acetylene gas, contains about 80% calcium carbide, the remainder being CaO with 2–5% other impurities.

Calcium carbide was made first in the laboratory by early workers such as Hare and Wöhler. Commercial production by the electric furnace method was developed about 1892 by Moissan in France and independently by Wilson in the United States. Development of the carbide industry began in 1895 and expanded rapidly.

During the First World War, acetylene generated from calcium carbide was recognized as a source of valuable organic chemicals. Plants built during this period were expanded after the war. Industrial calcium carbide was produced in very large quantities during the period 1940–1965 mainly as a source of acetylene as raw material for a wide variety of organic chemicals, resins and plastics, including production of calcium cyanamide. Annual production of calcium carbide in the United States rose from 335,000 metric tons in 1940 to 607,000 t in 1950, and reached a maximum of 943,000 t in 1960. Production remained at this level until after 1965 when many major plants were closed. Annual production decreased to 716,000 t in 1970 and dropped to 308,000 t by 1975. This rapid decline was due to the replacement of acetylene as the major building block for organic chemicals by ethylene produced by thermal cracking of hydrocarbons. Furthermore, a considerable amount of acetylene obtained as a co-product with ethylene (qv) in petrochemical production has displaced acetylene produced from calcium carbide.

The use of calcium carbide-based acetylene is now almost completely confined to the oxyacetylene welding and metal-cutting markets. Calcium carbide, which is safe and convenient to store and ship, provides an economical supply of acetylene to the metalworking industry. Minor quantities of calcium carbide are used for the desulfurization of steel and iron, a use expected to grow rapidly over the next few years.

Properties

Table 1 lists the more important physical properties of calcium carbide. Figure 1 (2) gives the phase diagram calcium carbide–calcium oxide for pure and technical grades.

Table 1. Physical Properties of Calcium Carbide[a]

Property	Value
molecular weight	64.10
melting point, °C	2300
crystal structure	
phase I	face centered tetragonal, 25–447°C
phase II	triclinic, below 25°C
phase III	monoclinic, metastable
phase IV	fcc, above 450°C
commercial	grain structure, 7–120 μm
specific gravity, coml grade	
at 15°C	2.34
at 2000°C (liquid)	1.84
electrical conductivity, tech grade, $(\Omega \cdot cm)^{-1}$	
at 25°C	3000–10,000[b]
at 1000°C	200–1000
at 1700°C (liquid)	0.36–0.47
at 1900°C (liquid)	0.075–0.078
viscosity at 1900°C, Pa·s(P)	
CaC_2 50%	6.0 (60)
CaC_2 87%	1.7 (17)
specific heat[c,d], 0–2000°C, J/(mol·K)	74.9
heat of formation[c,d], ΔH_{298}, kJ/mol	-59 ± 8
latent heat of fusion[c,d], H_f, kJ/mol	32

[a] For a recent and comprehensive compilation of physical and chemical properties, see ref. 1.
[b] Depending on impurities.
[c] 100% CaC_2.
[d] To convert J to cal, divide by 4.184.

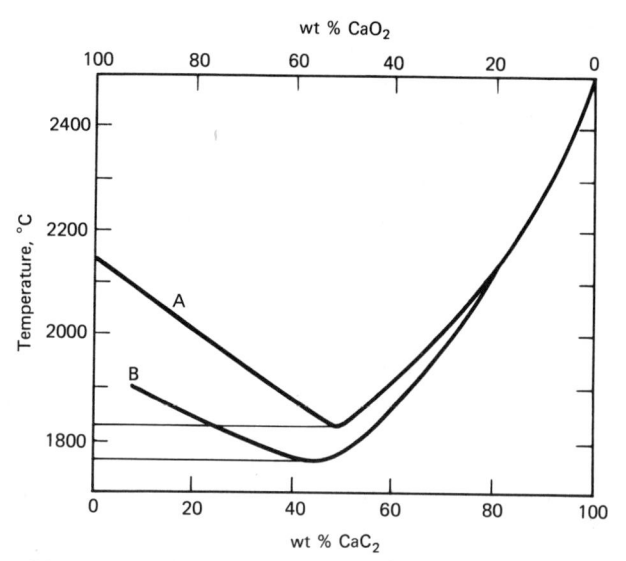

Figure 1. A, melting–freezing diagram of CaC$_2$–CaO with CaC$_2$. B, melting–freezing diagram of CaC$_2$–CaO with technical CaC$_2$.

Reactions

With Water. The highly exothermic reaction of calcium carbide and water to give acetylene is the basis of the most important industrial use of calcium carbide.

$$CaC_2 + 2\,H_2O \rightarrow C_2H_2 + Ca(OH)_2 \qquad \Delta H = 130 \text{ kJ/mol (31.0 kcal/mol)}$$

Commercially, 1 kg technical calcium carbide (80%) yields 0.31 m^3 acetylene (101 kPa, 20°C), consumes 0.59 kg water and leaves 1.18 kg of residue, mostly Ca(OH)$_2$. (This corresponds to 1 lb (80%) CaC$_2$ yielding 4.92 ft^3 acetylene at 1 atm.) An excess of water must be present in order that the heat of reaction is safely absorbed. Explosive conditions occur when calcium carbide reacts with limited quantities of water.

Carbide-to-water generators, operating at a temperature below 100°C use a large excess of cooling water (about 8 L water per kg carbide) to absorb the heat of reaction. So-called dry generators use only a slight excess of water but operate at temperatures above the boiling point of water with the reaction heat being absorbed by vaporization (see Acetylene). The Ca(OH)$_2$ obtained as coproduct is a free-flowing powder with 2–3% water. With a deficiency of water or in the presence of partially slaked carbide the following reaction can occur.

$$CaC_2 + Ca(OH)_2 \rightarrow C_2H_2 + 2\,CaO$$

This reaction proceeds slowly at room temperatures, appreciably faster at 100–120°C, and can occur in crushed carbide containing air-slaked material.

With Nitrogen. The second major industrial use of calcium carbide is the production of calcium cyanamide.

$$CaC_2 + N_2 \rightarrow CaCN_2 + C \qquad \Delta H = 295 \text{ kJ/mol (70.5 kcal/mol)}$$

The product contains about 10% free carbon. In the United States, the reaction is carried out by passing nitrogen through finely crushed carbide in a refractory oven at 1000–1200°C. To initiate the reaction, the carbide is heated electrically using a graphite-pencil electrode located in the center of the mass. Since the reaction is strongly exothermic, it proceeds autogenously.

European and Japanese producers have developed continuous nitrogenation furnaces designed on the basis of a rotary kiln into which powdered carbide and nitrogen gas are fed; the product is removed as a granular product (See Cyanamides).

With Sulfur. An important usage of calcium carbide is developing in the iron and steel industry where it has been found to be a very effective desulfurizing agent for blast furnace iron. Calcium carbide reacts with sulfur present in the molten metal as follows:

$$CaC_2 + S \rightarrow CaS + 2 C$$

Sulfur was controlled in the past in the iron and steel industry by careful selection of raw materials. Since the availability of high-grade raw materials is declining, and in an attempt to maximize production rates, producers are now shifting toward external desulfurization using additives such as calcium carbide in a separate step in the reduction process. Desulfurization by injection of calcium carbide powders in a "ladle" (3) or "torpedo car" is fast and efficient. It reduces sulfur content more than the traditional method of desulfurization under a basic reducing slag. In the cast iron and ductile iron industry, metal (4) is desulfurized in a ladle or in the electric furnace by injection of finely ground calcium carbide conveyed in fluidized state using nitrogen gas injection through a refractory nozzle inserted below the metal surface. With this method, sulfur content can be reduced from 0.1 to 0.01%. This technique is particularly valuable in the production of nodular iron, in which the use of calcium carbide promotes further savings in the amount of nodularizing element addition required (see Iron).

Manufacture

Calcium carbide is produced commercially by the reaction of high-purity quicklime with coke in an electric furnace at 2000–2200°C.

$$CaO + 3 C \rightarrow CaC_2 + CO \qquad \Delta H = 466 \text{ kJ/mol (111.3 kcal/mol)}$$

Calcium carbide, approximately 80% CaC_2, forms in a liquid state, the other impurities are mostly CaO. Carbon monoxide is usually collected and used as a fuel for lime production or the drying of the coke used in the process. The calcium carbide is drained or tapped from the furnace into cooling molds.

Raw Materials. The basic raw materials are limestone and coal or coke. The lime may be burned in rotary or vertical-shaft kilns. A limestone of high quality should be used, with a minimum of 95–97% $CaCO_3$, a maximum of 1% MgO, 1–1.5% SiO_2, 1% Fe_2O_3 plus Al_2O_3, 0.006% phosphorus, and 0.1% sulfur. The lime is screened to eliminate fines which interfere with the evolution of carbon monoxide during the smelting process.

If acetylene is produced on site, recycle of the lime hydrate from this operation can provide considerable savings. The usual route includes centrifuging the wet generator sludge, calcining the centrifuge cake, and briquetting the resulting CaO. Dry generator lime hydrate may be briquetted and charged directly to a lime kiln.

Up to 33% of the total lime charge can be replaced by recycle material. This is limited, however, by the fact that impurities eventually build up to a level where furnace operation is impeded.

Lime fines obtained by screening the charge material can be successfully briquetted with modern hydraulically-loaded briquette presses. These briquettes can be used as furnace charge (see Pelleting and briquetting). Metallurgical coke, petroleum coke, or anthracite, depending upon price and impurity, may be used as a source of carbon. Coke is the most common, whereas petroleum coke is desirable because of its low ash content and high resistivity. The presence of volatile compounds in petroleum coke can create difficulties in furnace operation. In some European plants a 3:1 mixture of coke and anthracite is used. Complete replacement of coke by anthracite is usually not successful because of low reactivity of the anthracite carbon. The reactivity of petroleum coke is also lower than that of metallurgical coke which is generally the preferred feed.

The presence of large quantities of fines in the furnace charge encourages the formation of crusts and reduces the porosity of the furnace burden impeding the flow of CO gas up through the furnace burden and causing furnace blows, which can be hazardous to the operators and destructive to the furnace equipment.

Mixed fines can be fed into the furnace, conveyed in a stream of recycle furnace gas or nitrogen, down through a vertical channel created in the center of each of the Soderberg electrodes, by the installation of a continuous length of 10-cm steel pipe. Although the steel pipe melts and disappears in the current-carrying zone of the electrode, it remains intact long enough to create a continuous channel. Fines delivered to this zone through the pipe react quickly. In addition to the economic gain of utilizing otherwise waste material and eliminating the disposal problem, savings in usage of electrode carbon and overall furnace power costs are also claimed (6).

Silica is usually the main impurity in the raw materials. In furnace operation it may be partly volatilized as silicon and later reoxidized in the cooler parts of the furnace; some is reduced and combines with the iron present to form a ferrosilicon alloy, with carbon to form silicon carbide, or with lime to form calcium silicate. Alumina forms a soluble calcium aluminate, whereas 80–90% of the magnesia is reduced to metallic magnesium which is volatilized and flushed from the smelting zone by the evolved carbon monoxide. Sulfur and phosphorus in the charge largely remain with the carbide as calcium sulfide and calcium phosphide.

The carbide impurities consume power in the smelting process and tend to distill and be reoxidized and form crusts near the top of the charge or around the cooler parts of the reaction crucibles. These crusts can cause trouble in furnace operation. Large amounts of dissolved calcium silicate and aluminate may form a viscous melt and impede the tapping process. Ferrosilicon is commonly removed from the crushed and screened carbide by electromagnets to avoid later blocking of the screens of wet acetylene generators.

Material and Energy Requirements. Material requirements per ton of carbide vary within moderate limits. On the basis of 95% available CaO in the lime and 88% fixed carbon in the coke, the coke-to-lime ratio required technically to produce calcium carbide (80%) is about 0.57, or about 865 kg of lime and 494 kg of coke per metric ton of carbide.

Electrode consumption varies from 14 to 26.8 kg/t of carbide. Theoretical power requirements per metric ton of calcium carbide are about 2200 kW·h but because of heat losses, about 2800–3100 kW·h is required (5,7). For every metric ton of 80% carbide produced, about 280 m^3 (10,000 ft^3 at 15°C) of furnace gas is evolved analyzing 75–85% carbon monoxide, 5–12% hydrogen, and the remainder N_2, O_2, CO_2, and CH_4.

510 CARBIDES (CALCIUM CARBIDE)

Furnace Design. Modern single calcium carbide furnaces provide capacities from 45,000 t (20 MW) to 180,000 t (70 MW) per year. The calcium carbide furnace (5) brought into operation in 1969 for Airco Alloys and Carbide, Louisville, Kentucky designed by Electrokemisk has a production capacity of 300 t per day, giving a carbide of about 0.30 m³ acetylene/kg (4.80–5.00 ft³/lb) capacity.

A cross-section of a three-phase covered carbide furnace is shown in Figure 2. The furnace shell is of simple construction, consisting of reinforced steel side walls

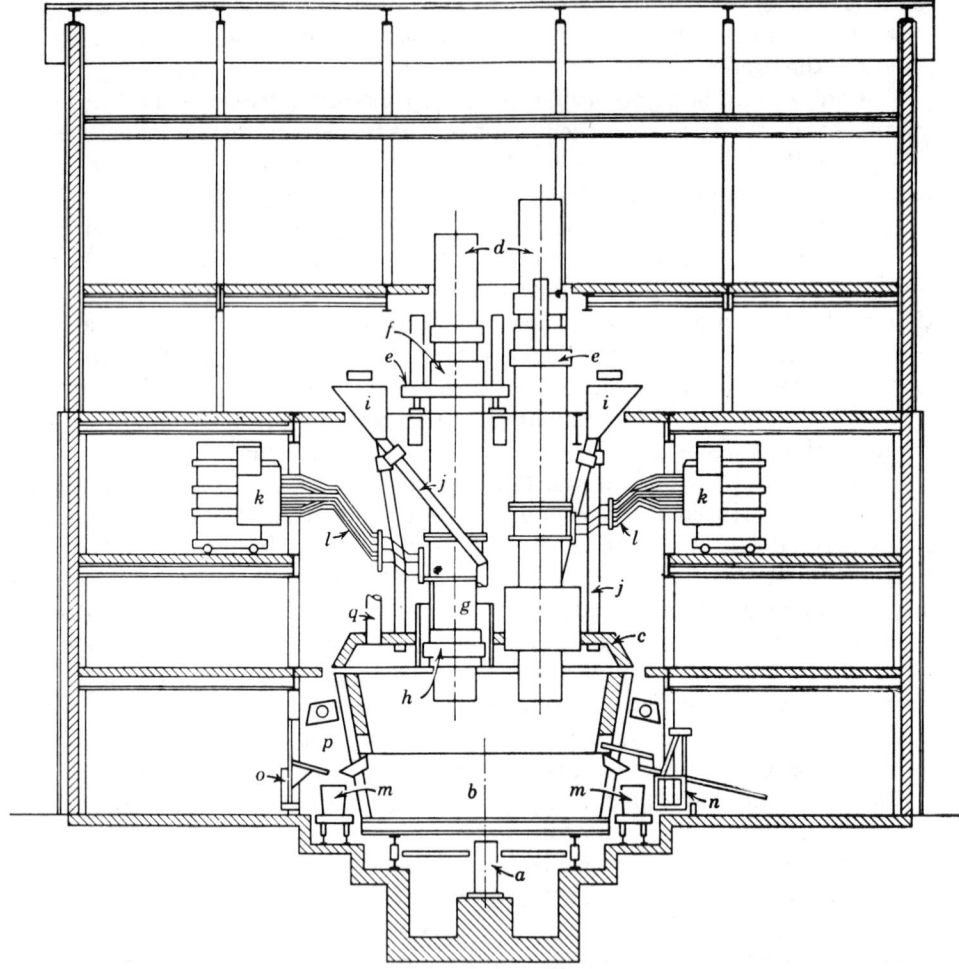

Figure 2. A 250 ton per day carbide furnace. The furnace rests on the furnace rotating mechanism a; the crucible b is built of brickwork and is closed with the furnace cover c. The Soderberg electrodes d can be moved with a hydraulic lifting device e. They are fitted with an electrode slipping device f. The contact shoes g provide the power connection to the electrodes. They are, contrary to conventional design (contact shoes outside of the cover), designed as low-position shoes; this affords some electrical advantages. The Soderberg electrodes are supported by a clamp ring h. The raw materials from the feed bunker i are conveyed to the furnace by feed chutes j. The single-phase transformers k may be in delta arrangement thereby giving a more favorable cos ϕ. The secondary leads l are connected to the electrodes. In place of the tapping receivers m, tapping channels or cooling drums may be used. During tapping with the tapping device n the working crew is protected by the tapping screens o. Fumes are removed by the hood p, and gases from the furnace are taken out through the tubes q; it is not necessary to erect a stack.

and bottom. Shell diameter for a furnace capable of producing 300 t/d is about 9 m. The shape is triangular or circular and the height-to-diameter ratio is shallow (0.25:1.0). The side walls are lined with mullite refractory brick (65 cm thick). The bottom consists of a 1-m layer of mullite refractory brick topped by a 1.6-m layer of rammed lining-grade Soderberg paste. The bottom layer can also be constructed from prebaked carbon blocks (see Carbon). The steel shell is usually supported on concrete piers rather than on a slab to allow cooling air to circulate under the bottom of the furnace.

Tapholes through which the liquid carbide can be removed are usually provided at or slightly above the level of the bottom. The cover consists of three separate pie-shaped segments of hollow-wall steel construction to allow the cooling water to circulate.

The furnace cavity is completely closed and relatively gas tight. The electrodes enter the furnace through water-cooled gland fittings or charge funnels. Carbon monoxide is removed through water-cooled pipe connections attached to the furnace cover. The gas pressure under the cover is usually slightly below atmospheric and is controlled by a pressure regulator.

Electrodes. Continuous self-baking Soderberg electrodes are generally used in modern carbide furnaces. These electrodes are positioned in a triangular layout, and consist of a light-gage steel casing which is suspended vertically above the furnace (see Fig. 2). The electrode casing, which is usually provided with internal steel reinforcing fins (see Figs. 3 and 4) is filled with Soderberg paste, which is a specially formulated mixture of electrically calcined anthracite and coal tar pitch usually supplied in the form of precast blocks. The Soderberg paste blocks become plastic at a temperature of 70°C, and the electrodes are maintained at this temperature with hot air jackets.

Baking of the Soderberg paste to a relatively dense amorphous carbon mass commences in the area just above the electrode clamp. Below the clamps the casing usually disappears as a result of melting and oxidation leaving the baked Soderberg paste as a monolithic carbon cylinder of considerable strength and high electrical conductivity, capable of carrying the furnace current to the charge melting zone.

When operating the furnace, carbon in the electrode tip is consumed requiring the electrode to be lowered continuously into the furnace to maintain a constant penetration of the electrode into the charge-melting zone. The degree of penetration controls the electrical resistance of each electrode circuit. As the electrode is lowered, additional sections of steel casing are welded on at the top of the electrode column. These sections are then filled with blocks of Soderberg paste, providing a continuous self-baking electrode.

In modern carbide furnaces, electrodes are usually suspended from a pair of hydraulic cylinders whose elevation is controlled in response to the electrode current signal. Slipping of the electrode, as it is consumed, is usually carried out with a pair of hydraulically tightened slipping bands mounted one above the other. The electrode is held by the lower set of bands, allowing the upper set of bands to be raised to grip the electrode casing in a new position. On release of the lower bands the electrode can be lowered safely using the hydraulic cylinders provided for the upper bands. Normally the electrode is held in the stable position by the action of both bands. If operated in reverse, the electrode can be raised out of the furnace by this procedure. Depending on the current carrying capacity required, Soderberg electrodes can be up to 152 cm in diameter and to 15 m in casing length. Normal electrode consumption is about 16 kg of electrode per ton of calcium carbide produced and normal slipping rates average about 2.5 cm of electrode per hour.

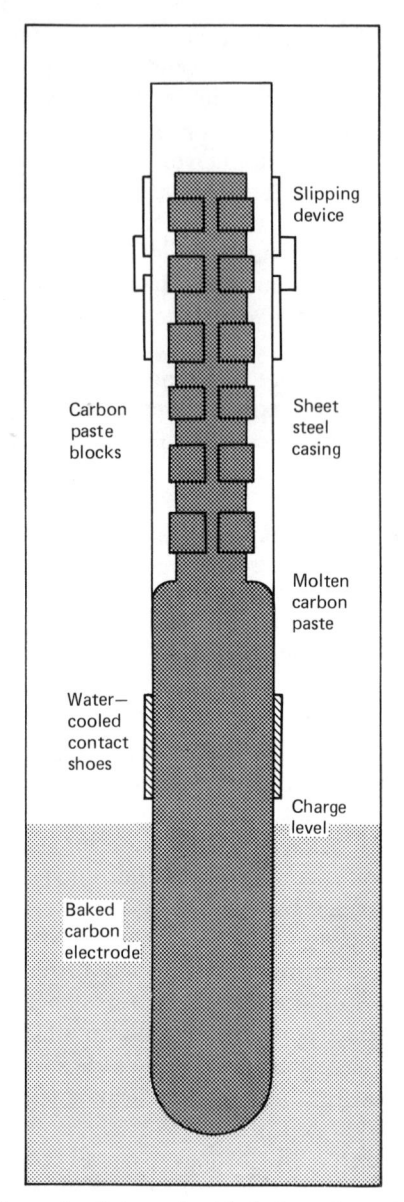

Figure 3. Schematic of Soderberg electrode.

Electrical Connections. Electric current is brought from the transformers by means of air-cooled copper busbars. Close to the individual electrodes, water-cooled copper pipe and flexible cables are used to connect to the water-cooled copper contact shoes which are held against the electrode surface by a hydraulic mechanism. The busbar is normally "interlaced" to allow cancelling the magnetic fields associated with individual conductors.

Usually three single-phase transformers are employed to step down from the high voltage supply to an operating voltage of 150–300 V. Power factor correction is carried out on the high voltage side, using banks of capacitors. Provision of a Wye/Delta

Figure 4. Top view of Soderberg electrode showing reinforcing fins and paste blocks stacked in casing.

switching system on the high voltage side of the transformer allows the operating voltage to be reduced for start-ups, etc.

Furnace Operation. A crew of five is needed for a typical large carbide installation; one man for raw material control, two for furnace operation, and two for tapping. The lime and coke are weighed separately on automatic continuous weighing scales which dump onto conveyors feeding to the furnace area. Coke is usually dried to a 1% max moisture content. The operation requires 56 parts calcium oxide to 36 parts carbon, but because of impurities approximately 20% excess lime is used. The charge is delivered to the furnace continuously by the feed conveyors which automatically maintain the charge levels in the feed pipes on the furnace cover or in the funnels surrounding the electrode. As the charge enters, it is heated in the upper zones by radiation from the electrodes and by heat exchange from the carbon monoxide leaving the reaction area. The lime is melted in the reaction zones close to the electrode tip. Generally the reaction zones are limited to the areas close to the electrodes. Charge material outside of this zone does not react and serves as a crucible liner providing the refractory walls in which the reaction zone is confined.

At the upper surface of the mix level, coke and lime are relatively cold and not capable of passing any current. As the mix moves down into the furnace it becomes hotter and its conductivity increases. At a distance of about 30 cm below the surface, the mix is hot enough to carry an appreciable amount of the current flowing between the electrodes. Penetration of the electrodes into the furnace is usually 90–125 cm. Mix at the level of the electrode tips may reach 1600°C. At this temperature its conductivity is good but usually not sufficient to cause melting of the lime.

Further down, about 75 cm below the electrode tips, the mix temperature is hot enough to allow the lime to melt (2200–2500°C). Since the coke does not melt, the liquid lime percolates downward through the relatively fixed coke bed forming calcium carbide which is also a liquid at this temperature. Both liquids erode the coke particles as they flow downward. The "weak carbide" formed first is converted to "richer" material by continued contact and reaction with the coke particles. The carbon monoxide gas produced in this area must be released by flowing back up through the charge into the active zone. This process continues down to the taphole level. Material in this area consists of solid coke wetted with liquid lime, and liquid calcium carbide pool being formed at the furnace bottom.

The ease with which carbon monoxide can escape from the reaction zone has an important bearing on the smooth operation of the furnace. The normal furnace charge, consisting of large particles, has good porosity, allowing a reasonably constant gas flow. Extremely fine raw materials or crusts of condensed impurities or of semimelted material impede the escape of gas from the reaction area, allowing pressure to develop and eventually result in "blows" in which hot mix and liquid carbide can be explosively ejected [7].

Metallurgical coke gives rise to ferrosilicon which, in the liquid phase, is more dense than calcium carbide and tends to settle and penetrate the bottom of the furnace. After a lengthy operating period, it may extend 30 cm or more below the taphole, eventually reaching the steel shell of the furnace where it will give rise to hot spots requiring repair and replacement of the furnace refractories.

An evenly operating furnace is essential for an efficient process, as indicated by (a) steady electrode penetration of the charge measured by the distance of the electrode tip above the taphole; (b) regular descent of the mix through all charging chutes; and (c) regular tapping of carbide equivalent to the power input to the furnace. These conditions are attained by maintaining standard operating procedures which include (a) frequent and adequate tapping of carbide; (b) constant electrical conditions and hence constant power input; and (c) a constant coke-to-lime ratio in the charge mix. Average rate of descent of mix in a modern high-capacity furnace is 90 cm per hour. Escape velocity for the carbon monoxide formed in the reaction is in the range of 15 cm/s (0.5 fps).

Tapping System. Traditionally, carbide furnaces were tapped intermittently but recently continuous tapping is being used. Most modern large carbide furnaces are provided with tapholes leading to the center of each electrode zone. Each electrode is tapped in turn, usually at 20–40 min intervals making the tapping operation an almost continuous process. This practice ensures uniform removal of accumulated liquid carbide from each electrode zone in turn. The taphole channel is opened by burning through the taphole with a tapping electrode. When the tip of the tapping electrode is applied to the hot solidified carbide in the taphole, a circuit is completed with the electrodes in the furnace and sufficient heat quickly develops to melt the carbide and establish a flow of liquid carbide from the furnace. Intermittently tapped furnaces should be drained to a uniform level, which usually takes from five to ten minutes.

Carbide is tapped from the furnace in a fluid stream at a temperature of 1900–2100°C. Its very low thermal conductivity makes it possible to tap directly into cast-iron chill cars, even though the melting point of cast-iron is lower than that of the carbide. The liquid product may be (*a*) cast into chills yielding a pig weighing up to 4.5 t; (*b*) It may be cast into smaller chills placed on a tapping wheel giving pigs of about 90 kg; (*c*) Or, it may be tapped into self-discharging tapping conveyors, consisting of shallow metal pans mounted on a continuous chain conveyor sufficiently long that the small ingot formed in each pan is solidified on reaching the discharge point, (*d*) Or it may be tapped continuously into a slightly inclined water-cooled rotating cylinder.

In methods *c* and *d* the carbide is ready for crushing upon discharge from the conveyor or cylinder; in method (*a*) the pigs must be cooled up to several hours before removal from the chills, and then cooled an additional 24–30 h before being crushed. All methods are used, but the continuously tapped, self-discharge conveyor is favored by the larger U.S. producers.

Crushing, Screening, Packing. The solidified carbide is sometimes first broken up by dropping the pig on a specially reinforced breaking table installed at the mouth of a jaw crusher. Gyratory crushers capable of handling a large pig in one piece are also used. Crushed carbide is fed to a screening plant where the carbide is screened for packing according to preestablished screen sizes. Recrushing of fractions may be required with recycle depending on the size ranges required. Gyratory screens are most successful for product screening. Magnetic separators are used to remove ferrosilicon in order to prevent damage to acetylene generators.

Previously, carbide was packed in nonreturnable light-gage sheet metal drums. Currently, returnable containers are preferred such as 170-L (45-gal) open-head gasketed drums or in 2.25–4.5 t steel bulk containers suitable for lift trucks and unloading conveyors. Oiling of granular carbide with a light lubricating oil decreases the rate of reaction when exposed to moist air and also decreases dust formation. A considerable amount of commercial carbide is now shipped oiled.

Computer Control. A large U.S. manufacturer now uses a digital computer to control the operation of two 23.5 MW carbide furnaces (8). Operations directly under computer control are: electrode position and slipping control; mix batching control; mix distribution; carbide gas (acetylene) yield; tapping intervals; CO gas collection and distribution; power control vs maximum demand; cooling water system; and dust collector system.

Computer control is reported to improve electrical usage by 5%, on time or operating time by 2.5% overall production by 10%, and product uniformity and quality significantly.

Environmental Considerations. The major environmental problem is prevention of particulate dust emission which can be handled with cloth filtration equipment (see Air pollution control methods). In addition, serious technical problems are encountered in the cleanup of the CO gas evolved and in the treatment of taphole fumes. Although the dust created in materials handling equipment is of relatively large particle size, this is easily handled in cloth filtration. Treatment of the furnace gas stream is complicated by the high temperature of the gas, its explosiveness, toxicity, the dust concentration, and particle size. Filtration of taphole fumes consisting entirely of submicrometer particles, which rapidly clog the filtration media, has been found to be difficult and expensive.

Typical carbide furnace gas contains about 80% CO with small amounts of CH_4, H_2, and CO_2, and some N_2 from the air leaked into the system. Dust concentration depending on the raw material varies from 3500–7000 grains/m^3 (100–200 grains/ft^3). The gas temperature at the point of discharge from the furnace cover is usually in the range of 650–900°C.

In larger U.S. installations, two methods are used for aspirating the gas from the furnace for delivery to the point of use as fuel. Wet handling combines pumping, cooling, and scrubbing. This method is carried out in a blower fitted with water sprays. Dry handling avoids cooling and uses water-cooled piping and specially designed high-temperature fans (5) to aspirate the gas (see Fans and blowers).

In general, the gas stream is used to fuel lime kilns associated with the carbide process. Particulate dust is finally removed by conventional cloth filtration located on the stacks of the lime kilns.

For the wet-scrubbing operation considerable quantities of water are required. The effluent water from the scrubber contains cyanides in sufficient quantities to create an expensive and difficult disposal problem. Dry handling methods are generally preferred. Shaker type cloth–tube dust collectors operating at low air-to-cloth ratios have been used successfully in recent years to filter taphole fumes containing large concentrations of submicrometer particulates.

Specifications, Shipping

Specifications and test methods vary from country to country. The U.S. Government has set standards for calcium carbide (Federal Specification O-C-101a, July 21, 1949) for government agencies, which are also used in the carbide trade. The British Standards Institution established a specification (B.S. 642-1951) to govern trade in the sales of calcium carbide. Other published specifications for carbide have been issued in France, Germany, Switzerland, and Japan.

Screen Size and Acetylene Yield. The U.S. standard of calcium carbide are given in Table 2, including the yield of acetylene at 15°C and 101 kPa (1 atm).

The U.S. specification states that the acetylene evolved shall contain not more than 0.05% by volume phosphine, whereas the British specify a maximum of 0.06% phosphine, 0.15% hydrogen sulfide, and 0.001% arsine by volume. The U.S. specification states that, unless otherwise specified, calcium carbide for domestic shipment shall be packed in industrial wide-mouth, screw-cover, 45.5-kg 26-gage metal drums. The drums must be marked "Calcium Carbide–Dangerous If Not Kept Dry." Calcium carbide is now classed as a hazardous chemical under regulations issued by the U.S. Department of Transportation (1977).

In the carbide trade, contracts are usually based on a size and gas-yield specification, with penalties for carbide which fails to meet specified gas yields. In general, gas yields range from 0.287–0.300 m^3/kg (4.60–4.80 ft^3/lb), depending on the screen size of the carbide.

Analytical and Test Methods

Test methods are specified in the U.S. and British specifications. The gas yield test is the most important, consisting of standard sampling, sample preparation, slaking in specified equipment, and collecting and measuring the volume of evolved

Table 2. U.S. Calcium Carbide Specifications

Designation[a]	Screen size, square opening, cm To pass	Screen size, square opening, cm Retained	Minimum average volume acetylene evolved at 15°C, 101 kPa, m³/kg
		81	
lump	10.8		0.28 (4.5)[d]
egg	5.1	0.95	0.28 (4.5)
nut	2.7	0.635	0.28 (4.5)
½ × ¼[b] (Miners)	1.27	0.635	0.28 (4.5)
¼ × 1/12[b]	0.635	0.17	0.28 (4.5)
rice	0.34	0.084	0.27 (4.3)
14 mesh ND[c]	0.17	0.042	0.27 (4.3)

[a] For all sizes except ½ × ¼ and ¼ × 1/12[b] not more than 5% by wt of carbide of other than the normal size shall be present.

[b] Note that ½ by ¼ is actually 1.27 by 0.635 cm and ¼ by 1/12 is 0.635 by 0.17 cm.

[c] ND = no dust; 14 mesh is ca 1500 µm.

[d] 0.28 and 0.27 m³/kg are 4.5 and 4.3 ft³/lb, respectively, taken at 60°F and 1 atm.

acetylene. This volume is then calculated to standard conditions. The phosphorus, sulfur, and arsenic content of the calcium carbide is checked by determining the phosphine, hydrogen sulfide, and arsine content of the evolved acetylene according to the specified procedures.

Phosphine may be determined by absorption in iodine solution followed by precipitation of the phosphomolybdate complex. Sulfur and arsenic are determined by absorption in sodium hypochlorite solution followed by precipitation of barium sulfate in the case of sulfur, and by acidification of the solution and volatilization of arsine by the Gutzeit procedure in the case of arsenic.

Health and Safety

No special health or safety factors are involved in the manufacture of calcium carbide. The usual precautions must be observed around the high-tension electrical equipment supplying power. The carbon monoxide formed, if collected in closed furnaces, is usually handled through blowers, scrubbers, and thence to a pipe transmission system. As calcium carbide exposed to water readily generates acetylene, the numerous cooling sections required in the high-temperature furnace must be constantly monitored for leaks. When acetylene is generated, proper precautions must be taken to prevent admixture with air owing to the explosibility of air–acetylene mixtures over a wide range of concentrations (from 2.5 to 82% acetylene by volume), and the flammability of 82–100% mixtures under certain conditions.

Although acetylene is considered to be a material with very low toxicity, a threshold limit value (TLV) of 2500 ppm has recently been set by NIOSH. In the presence of small amounts of water, carbide may become incandescent and ignite the evolved acetylene–air mixture. Nonsparking tools should be used when working in the area of acetylene-generating equipment.

Table 3. World Production of Calcium Carbide and Manufacturing Capacity, Thousand Metric Tons

Continent and country	1936	1959	1964	1975
Canada	209	317[a]		
United States	145	925	925	308
Czechoslovakia	25	70		
France	125	347	532	
Federal Republic of Germany	712	1887	953	
Great Britain		174	268	
Italy	156	321	30	
Norway	58	62	732	
Poland	42	254	40€	
Rumania	4	141	92	
Spain	15	81	100	
Sweden	36	79	75	
Switzerland	20		75	
U.S.S.R.	109	499		
Yugoslavia	33	66	97	
China		210		
Taiwan		91	150	
Japan	327	91	1602	
Korea	150			
Africa	15	80	60	
Australia	7	10	8	
Grand total	3389	5930	5373	

[a] 1958.

Economic Aspects

World production of calcium carbide is given in Table 3. Following a period of rapid expansion in the late 1950s, the calcium carbide industry is now in a state of decline. Calcium carbide was the sole source of acetylene and/or major source of all organic chemicals in the United States until 1955. In 1975 only 30% of the acetylene production was calcium carbide-derived. The total U.S. consumption of acetylene fell from a maximum of over 450,000 metric tons per year to less than half this figure by 1975. This decline is due to the replacement of acetylene by ethylene in many processes.

From 1967–1974 23 U.S. plants with a consuming capacity for 450,000 tons of acetylene were closed and replaced by facilities using other raw materials. Similar trends are apparent in other countries.

The 1976 U.S. market price of calcium carbide ranged from $181 to 226/t, depending on quantity and grade. Despite the pessimistic outlook for the carbide industry, the construction of a new carbide-based acetylene plant with a capacity of 44,000 metric tons per year was announced in South Africa in 1975.

Uses

The current largest use for calcium carbide is in the production of acetylene for oxyacetylene welding (qv) and cutting (see Acetylene). Companies producing compressed acetylene gas, the largest users of carbide, are strategically located near the user plants to minimize freight costs on the gas cylinders. In Canada and other countries, the production of calcium cyanamide from calcium carbide continues. A major and growing outlet for calcium carbide is the desulfurization of blast furnace metal for the production of steel and low-sulfur nodular cast iron. These uses are expected to increase over the next decade as the availability of high purity raw materials for the iron and steel industry decreases.

BIBLIOGRAPHY

Calcium Carbide treated in *ECT* 1st ed., under "Carbides (Calcium)," Vol. 2, pp. 834–846, by A. J. Abbott, Shawinigan Chemicals Ltd.; Calcium Carbide treated in *ECT* 2nd ed., under "Carbides, Calcium," Vol. 4, pp. 100–114, A. G. Scobie, Shawinigan Chemicals Ltd.

1. R. Juza and H. V. Schuster, *Z. Anorg. Chem.* **311,** 62 (1961).
2. S. A. Miller, *Acetylene, its Properties, Manufacture and Uses,* Vol. 1, Academic Press, New York, 1965, p. 475.
3. M. A. Palmer, and J. S. Beeker, *Desulfurization in the Transfer Ladle,* 34th Iron Making Conference Toronto, Apr. 1975, Inland Steel Co., Ind.
4. S. I. Karsay, *Ductile Iron Production Practices,* American Foundryman's Society, 1975, p. 188.
5. J. W. Frye, *50 M.W. Calcium Carbide Furnace Operation,* Electric Furnace Proceedings, Airco Alloys & Carbide, Louisville, Ky., 1970, p. 163.
6. U.S. Pat. 2,996,360; D. E. Hamby, *Hollow Electrode System for Calcium Carbide Furnaces,* Electric Furnace Proceedings, Union Carbide Corp., 1966, p. 208.
7. Kaess and Vogel, *Chem. Ing. Tech* **28,** 759, 1956.
8. G. E. Healy, *Why A Carbide Furnace Erupts,* Electric Furnace Proceedings, Penn State University, 1965, p. 62.
9. W. W. Wilbern, *Computer Control of Submerged Arc Ferro Alloy Furnace Operations,* Electric Furnace Proceedings, 1974, p. 101.

Noel B. Shine
Shawinigan Products Dept.
Gulf Oil Chemicals

SILICON CARBIDE

Silicon carbide [*409-21-2*] SiC, is a crystalline material, with a color that varies from nearly clear through pale yellow or green to black, depending upon the amount of impurities. It occurs naturally only as the mineral Moissanite [*12125-94-9*] in the meteoric iron of Cañon Diablo, Arizona. The commercial product, which is made in an electric furnace, is usually obtained as an aggregate of iridescent crystals. The iridescence is caused by a thin layer of silica produced by superficial oxidation of the carbide. The loose black or green grain of commerce is prepared from the manufactured product by crushing and grading for size.

The metallurgical, abrasive, and refractory industries are the largest users of silicon carbide. It is also used for heating elements in electric furnaces, in electronic devices, and in applications where its resistance to nuclear radiation damage is advantageous.

In 1891, Acheson (1) produced a small amount of silicon carbide by passing a strong electric current from a carbon electrode through a mixture of clay and coke contained in an iron bowl which served as the second electrode. He recognized the formula and the abrasive value of the crystals obtained (2) and founded The Carborundum Company in 1891. Even earlier, Colson and Schutzenberg (3) reported tetratomic radicals of silicon ($Si_2C_2O_2$, Si_2C_2N), and were also successful in obtaining SiC.

Properties

The properties of silicon carbide (4–6) depend upon purity, polytype, and method of formation. The measurements made on commercial, polycrystalline silicon carbide products should not be interpreted as being representative of single-crystal silicon carbide. The new sintered silicon carbides, being essentially single phase, fine grained, and polycrystalline, have distinct properties from both single crystals and direct-bonded silicon carbide refractories. Table 1 lists the properties of fully compacted, high-purity silicon carbide.

Silicon carbide is well known as a hard material occupying a relative position on Mohs scale between alumina at 9 and diamond at 10 (see Hardness). The average values for hardness under a load of 100 g are listed below:

Material	*Knoop hardness*
sapphire	2013
SiC, dense, direct-bonded	2740
SiC, sintered alpha	2800
SiC, black single crystal	2839
SiC, green single crystal	2875
boron carbide	3491

Because of its high thermal conductivity and low thermal expansion, silicon carbide is very resistant to thermal shock as compared to other refractory materials.

Table 1. Physical Properties

Property	Value	References
mol wt	40.10	
decomposition temp[a], °C		7
α-form	2825 ± 40	
β-form	2985	
sp gr, g/cm^3 at 20°C		
β-form	3.210	
6H polytype[b]	3.211 (3.208)	
commercial	3.1	
refractive index		
β-form	2.48	8
α-form	ϵ ω	4
4H[b]	2.712 2.659	
6H	2.69 2.647	
15R	2.687 2.650	
free energy of formation, $\Delta G°$, kJ/mol[c]		9
α-form	504.1	
β-form	506.2	
heat of formation, $\Delta H°_{298}$, kJ/mol[c]		10
α-form	−25.73 ± 0.63	
β-form	−28.03 ± 2	
thermal conductivity[d], W/(m·K)		11
commercial, high density	4.60	
80% density refractory	24.3	
emissivity		12–16
spectral (3–5 μm)	0.8	
total (0–1600°C)	0.8	
coefficient of thermal expansion[e], per °C		16–22
25–200°C	2.97×10^{-6}	
25–600°C	4.27×10^{-6}	
700–1500°C	6.08×10^{-6}	
elasticity		
Young's modulus, GPa[f]		23–24
α-form, hot-pressed	480	
α-form, sintered	410	25
β-form, sintered	410	23–24
shear modulus, GPa[f]		
reaction-bonded	167.3	26
α-form, sintered	177	27
β-form, sintered	140–190	23
sublimed	19	28

[a] The decomposition products are Si, Si$_2$C, Si$_2$, SiC, and Si$_3$.
[b] H = hexagonal, R = rhombohedral.
[c] To convert J to cal, divide by 4.184.
[d] To convert W/(m·K) to cal/(s·m·°C) divide by 4.184. Figure 1 shows the relation of thermal conductivity vs temperature.
[e] Sintered α-form.
[f] To convert GPa to psi, multiply by 145,000.

Crystal Structure. Silicon carbide may crystallize in the cubic, hexagonal, or rhombohedral structure. There is a broad temperature range where these structures may form. The hexagonal and rhombohedral structure designated as the alpha (noncubic) form may crystallize in a large number of polytypes.

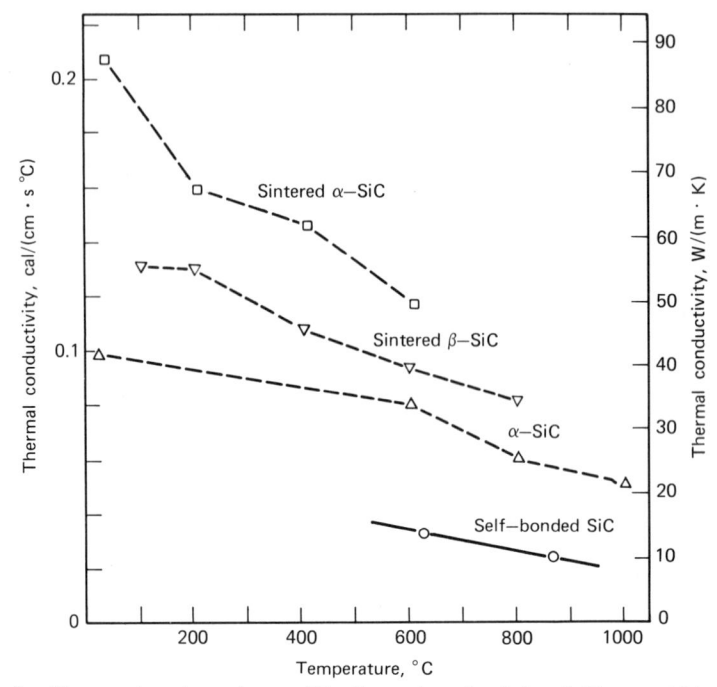

Figure 1. Temperature dependence of the thermal conductivity of silicon carbide.

Ramsdell suggested the now most generally used designation as most descriptive of the relation between types (29). Since there are three possible arrangements of atoms in a layer of SiC crystal, each type has the same layers but a different stacking sequence (30). Designation (29) is by the number of layers in the sequence, followed by H, R, or C to indicate whether the type belongs to the hexagonal, rhombohedral, or cubic class.

A number of theories have been put forth to explain the mechanism of polytype formation (30–36), such as the generation of steps by screw dislocations on single crystal surfaces which could account for the large number of polytypes formed (29,35–36).

At present, the growth of crystals via the vapor phase is believed to occur by surface nucleation and ledge movement by face specific reactions (37). The solid-state transformation from one polytype to another is believed to occur by a layer-displacement mechanism (38) caused by nucleation and expansion of stacking faults in close-packed double layers (of Si and C).

A progressive etching technique (39–40), combined with x-ray diffraction analysis, revealed the presence of a number of alpha polytypes within a single crystal of silicon carbide. More recent work, using lattice imaging techniques via transmission electron microscopy, has shown that alpha silicon carbide formed by transformation from the beta (cubic) phase can consist of a number of the alpha polytypes in a syntactic array (41).

A phase diagram for the carbon–silicon system and for the relationship between temperature and solubility of carbon in silicon has been determined (42).

Electrical Properties. The electrical properties of silicon carbide are highly sensitive to purity, density, and even to the electrical and thermal history of the sample.

Resistivity. The temperature coefficient of electrical resistivity of commercial silicon carbide at room temperature is negative. No data can be given for refractory brick since resistivity is greatly influenced by the manufacturing method and the amount and type of bond. For specific information, the manufacturer should be consulted.

The resistivity of silicon carbide heating elements varies with purity, porosity, raw material grain size, etc. The curve in Figure 2 shows the relation of resistivity vs temperature. This indicates the effect of reducing the total power input used to heat the furnace up to temperature while preventing it from overheating after the operating temperature has been reached.

Resistivity measurements of doped, alpha silicon carbide single crystals from −195 to 725°C, showed a negative coefficient of resistivity below room temperature which gradually changed to positive above room temperature (43). The temperature at which the change-over occurred increased as the ionization of the donor impurity increased. This is believed to be caused by a change in conduction mechanism.

Semiconducting Properties. Silicon carbide is a semiconductor, with a conductivity between that of metals and insulators or dielectrics (4,13,44–45). Because of the thermal stability of its electronic structure, silicon carbide has been studied for uses at high temperature (>500°C) (see Semiconductors).

The Hall mobility in silicon carbide is a function of polytype (46–47), temperature (41–48), impurity, and concentration (47). In n-type crystals, activation energy for ionization of nitrogen impurity varies with polytype (48–49).

Resistivity is strongly dependent on crystal structure and impurities. It is generally accepted that boron and nitrogen can substitute for carbon in the SiC lattice (50–51) and aluminum is thought to substitute for silicon (47) (see Boron, refractory boron compounds; Nitrides).

Optical absorption measurements give band-gap data for cubic silicon carbide as 2.2 eV and for alpha as 2.86 eV at 300 K (52). In the region of low absorption coefficients, optical transitions are indirect whereas direct transitions predominate for quantum energies above 6 eV. The electron affinity is about 4 eV.

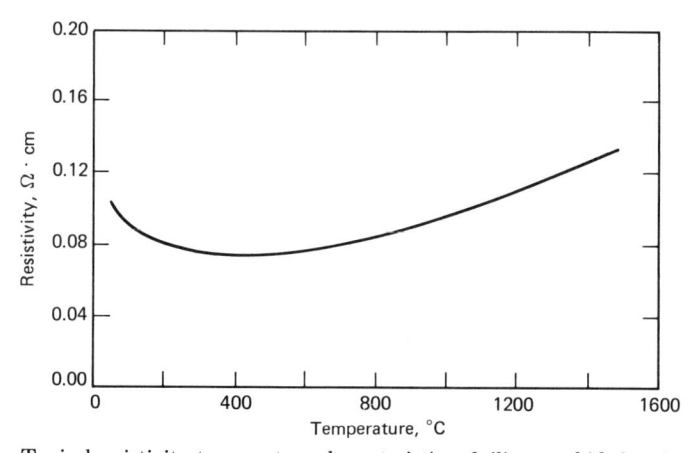

Figure 2. Typical resistivity-temperature characteristics of silicon carbide heating elements.

524 CARBIDES (SILICON CARBIDE)

The electronic bonding in silicon carbide is considered to be predominately covalent in nature, but with some ionic character (52).

In a Raman scattering study of valley-orbit transitions in 6H-silicon carbide, three electron transitions were observed, one for each of the inequivalent nitrogen donor sites in the silicon carbide lattice (53). The donor ionization energy for the three sites had values of 0.105, 0.140, and 0.143 eV (54).

Radiation Effects. Alpha silicon carbide exhibits a small degree of anisotropy in radiation-induced expansions along the optical axis and perpendicular to it (55). When diodes of silicon carbide were compared with silicon diodes in exposure to irradiation with fast neutrons (56), an increase in forward resistance was noted only at a flux about ten times that at which the increase occurs in a silicon diode. In general, it appears that silicon carbide with the more tightly bound lattice is less damaged by radiation than silicon.

Reactions

Silicon carbide is comparatively stable with the only violent reaction occurring when it is heated with a mixture of potassium dichromate and lead chromate. Chemical reactions do take place between silicon carbide and a variety of compounds at relatively high temperatures. Sodium silicate attacks it above 1300°C, and it reacts with calcium and magnesium oxides above 1000°C and copper oxide at 800°C to form the metal silicide. Silicon carbide decomposes in fused alkalies such as potassium chromate or sodium chromate and in fused borax or cryolite and reacts with carbon dioxide, hydrogen, air, and steam. Silicon carbide is resistant to chlorine below 700°C but forms carbon and silicon tetrachloride at high temperature. It dissociates in molten iron and the silicon reacts with oxides present in the melt, a reaction of use in the metallurgy of iron and steel. The dense self-bonded type has good resistance to aluminum up to about 800°C, to bismuth and zinc at 600°C, and to tin up to 400°C; a new silicon nitride bonded type exhibits improved resistance to cryolite.

In a study of oxidation resistance over the range 1200–1500°C an activation energy of 276 kJ/mol (66 kcal/mol) was determined (57). The rate law is of the form $\delta^2 = kT + C$; the rate-controlling step is probably the diffusion of oxygen inward to the SiC–SiO$_2$ interface while CO diffuses outwards. The oxidation rate of granular silicon carbide in dry oxygen at 900–1600°C was studied and an equation for the effect of particle size was derived (58). Small changes in impurity content did not affect this rate but the presence of water vapor and changes in partial pressure of oxygen were critical (58–59). Steam and various impurities and binders also affected the oxidation of silicon carbide (60). In the interaction of oxygen with silicon carbide a difference was observed in oxygen adsorption on the different crystal faces (61).

At high temperature, silicon carbide exhibits either active or passive oxidation behavior depending upon the ambient oxygen potential (62–63). When the partial pressure of oxygen is high, passive oxidation occurs and a protective layer of SiO$_2$ is formed on the surface.

$$2\,\text{SiC}_{(s)} + 3\,\text{O}_{2(g)} \rightarrow 2\,\text{SiO}_{2(s)} + 2\,\text{CO}_{(g)}$$

Active oxidation occurs where the oxygen partial pressure is low and gaseous oxidation products are formed.

$$\text{SiC}_{(s)} + \text{O}_{2(g)} \rightarrow \text{SiO}_{(g)} + \text{CO}_{(g)}$$
$$\text{SiC}_{(s)} + 2\,\text{SiO}_{2(s)} \rightarrow 3\,\text{SiO}_{(g)} + \text{CO}_{(g)}$$

A fresh surface of silicon carbide is thus constantly being exposed to the oxidizing atmosphere.

Active oxidation takes place at and below approximately 30 Pa (0.23 mm Hg) oxygen pressure at 1400°C (63). Passive oxidation is determined primarily by the nature and concentration of impurities (64).

Crystal Growth

Considerable effort has been expended to grow silicon carbide crystals of increased size and controlled purity for semiconductor use and to understand the mechanism of growth. Methods under examination are growth by sublimation from the vapor phase, thermal decomposition or reduction (gaseous cracking), and crystallization from a metal melt. Control of temperature at the crystal-growing face, the species, and the proportion of the constituents is essential, in addition to measures to reduce excess nucleation. Sintering of silicon carbide is discussed below under Uses.

Sublimation. In a modification of the Kroll graphite tube furnace, single crystals are grown from the vapor phase using commercial silicon carbide as a source (65). Purification takes place owing to a distillation effect. Argon or hydrogen may be used as the ambient gas in a furnace equipped with special temperature controls (66). Crystals are also produced in a graphite capsule in a furnace adapted to admit silicon vapor into a reaction gas and giving free-falling, silicon carbide crystallite "snow" (67).

Electronic-grade, hexagonal silicon carbide crystals with two nearly parallel faces (0.001–0.25 cm thick, and 0.16–1.3 cm long) were grown in a crucible-type electric furnace charged with silicon, the carbon being supplied by the furnace walls which acted as the substrate. These crystals can be doped for n- and p-type conductivity.

Thermal Decomposition or Reduction. A number of compounds providing a source for silicon and carbon have been heated in various reactors with some degree of success. Starting materials include silicon chloride with toluene, and methyl trichlorosilane or trimethyl chlorosilane alone, or with hydrogen. By growing crystals from the gas phase, cubic silicon carbide, silicon carbide whiskers, and microcrystalline silicon carbide were obtained (68). During the course of this work, the 2H or wurtzite type of silicon carbide was identified.

Growth From Supersaturated Metal Melts. Yellow crystals are grown ("pulled") on a seed-crystal rod from a carbon-supersaturated silicon melt. Blue crystals, containing 0.4% Fe, are grown from a carbon saturated, iron–silicon alloy melt (69). The crystals forming at the melt surface were mixed dendrites, needles, plates, and cubes, some 0.8 cm long. Silicon-enriched iron and nickel solutions containing carbon give n-type alpha crystals, whereas silicon solutions containing only carbon yielded beta-type crystals (70). The larger crystals appeared at the meniscus.

Manufacture

Silicon carbide is commercially produced by the electrochemical reaction of high grade silica sand (quartz) and carbon in an electric resistance furnace (71). The carbon is in the form of petroleum coke (pitch coke) or anthracite coal. The overall reaction is $SiO_2 + 3\,C \rightarrow SiC + 2\,CO$. Sawdust may be added to increase the porosity of the furnace mix (charge), thus increasing the circulation of the reacting gases and facili-

526   CARBIDES (SILICON CARBIDE)

tating the removal of CO. Lack of sufficient porosity may create furnace blowouts (see Silica).

Modern methods for preparing the furnace charge include a system of bins, automatic weighing, mixing, and charge delivery conveyor belts or large hoppers. The furnace is a trough, 12–24 m long, 3 m or more wide and 2–3 m high with a holding capacity of up to 125 metric tons. The sides are made of refractory bricks held in an inclined position by a steel frame. Each end of the trough is an electrode through which electric power is applied to a graphite core in the center of the furnace charge at a rate of up to 5000 kW, depending on the size of the furnace. As the silicon carbide forms, the conductivity of the charge increases and power to the furnace is adjusted by lowering the voltage. The core heats up to about 2600°C and then the temperature falls to a fairly constant 2040°C. The outer edges of the furnace mix remain at about 1370°C because of the burning of gases at the surface (72). When the heating cycle is completed, the furnace is cooled for several days. The side walls are then removed, the loose, unreacted mix taken away, and the remaining silicon carbide cylinder is raked to remove the crust, about 4 cm thick (see Fig. 3). This crust contains 30–50% silicon carbide as well as some condensed metallic oxides. The cylinder is then transported in sections to a cleaning room, where a further partially reacted layer (about 70% silicon carbide) is chipped away. The remaining oval cylinder constitutes high-grade silicon carbide with the exception of the central graphite core which is recovered for reuse.

A yield of 11.3 t of black silicon carbide can be expected from a 75 t furnace charge (70). Weight losses owing to handling and to escaping CO are over 16 t. The remaining

Figure 3. Two silicon carbide furnaces with sides and unreacted mix removed, thus showing the cylinders of silicon carbide between the electrodes.

mix, which contains about 10% silicon carbide, can be reused in the black silicon carbide (high aluminum) runs. For preparing a higher purity, green silicon carbide, only fresh high purity, low aluminum raw materials can be used. The processing, packing, and shipping of silicon carbide products are represented by the simplified flow sheet in Figure 4. The flow sheet can vary, depending upon the manufacturer, application, or even the particular users requirements.

Other developments include substitution of anthracite coal for some or all of the coke (73); introduction of chlorine gas into the reaction zone to remove impurities from the product (74); and addition of small amounts of boron, titanium, or zircon to the furnace charge in order to reduce product sensitivity to oxidation at 900–1100°C (75). Ultrafine silicon carbide is produced continuously in the electric arc from consumable anodes of silica and graphite (76). Fine crystalline silicon carbide was produced from waste rice hulls and iron oxide in a process developed at the University of Utah (77). The early patent literature has been reviewed (71).

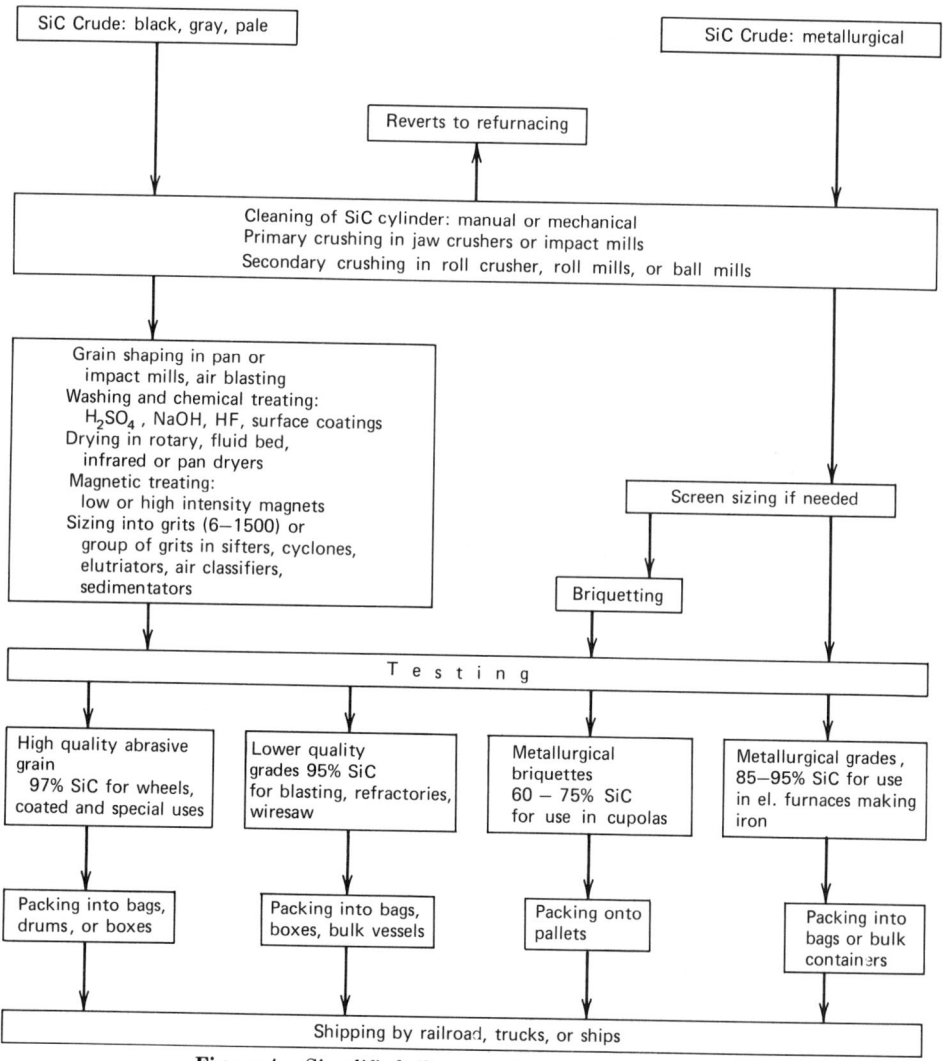

Figure 4. Simplified silicon carbide process flow sheet.

528   CARBIDES (SILICON CARBIDE)

Energy Requirements. The theoretical power consumption for the production of silicon carbide is 5.75 kW·h/kg of product (78). In practice it takes 6–12 kWh to produce 1 kg of crude, depending on the grade and recovery from a cylinder. Approximately 1.5 kW·h/kg is required to crush the crude into useable grain. Improvements in the energy efficiency of production are virtually impossible without radical change in the method of manufacture, which, however, is being investigated by some producers.

Economic Aspects

Silicon carbide was first manufactured on a large scale in 1892, but production did not reach 9000 t/yr until 1918. Including the usual fluctuations in the economy, production increased until the decline in the 1970s (see Tables 2 and 3).

In 1977, six firms in the United States and Canada were producing crude silicon carbide under various trade names, including American Metallurgical Products Co., Inc.; The Carborundum Company; Electro-Refractories and Abrasives, Ltd.; The Exolon Co.; General Abrasive Co.; and Norton Co. Silicon carbide is also produced in Argentina, Brazil, Czechoslovakia, France, GDR and FR Germany, Italy, Japan Norway, Spain, Switzerland, and the U.S.S.R. A silicon carbide plant is being built in Mexico.

Most plants are located in areas where electrical power is, or was at one time,

Table 2. World Production of Silicon Carbide and Corresponding Capacities for 1975 [a]

Region	Production, metric tons	Capacity, %
United States and Canada	121,560	70
Western Europe	156,040	90
Japan	72,570	95
other, including U.S.S.R.	107,050	83
total	*457,220*	*83*

[a] Ref. 79.

Table 3. Silicon Carbide Production in the United States and Canada

Year	Metric tons produced	Value, 1000$	Capacity, %	Amount used in non-abrasives, %	References
1891	0.023				80
1901	1,741				80
1911	4,706				81
1941	40,789				82
1955	67,860	11,028			83
1960	120,850	20,636			83
1965	125,520	19,963			83
1970	151,450	24,038	93	50	83
1972	150,590	24,690	85	43	83
1974	147,870	33,872	86	62	79, 84
1976	144,240	44,396	81	63	79

available at relatively low rates. Other considerations are availability of labor, reasonable air and water pollution standards, future expansion potential, and proximity of raw materials and markets.

One of the first important uses of silicon carbide was as an abrasive (see Abrasives). Other markets in refractories, electrical devices, and metallurgy have since been developed. Recent trends indicate a rapidly growing market for metallurgical applications. In the beginning of 1977, the silicon carbide market was estimated at 45% metallurgical, 35% abrasive, and 20% other.

Specifications and Standards

In the United States, the test methods, specifications, and standards for abrasive grain products are established by the Abrasive Grain Association (AGA) (85), the Grinding Wheel Institute (GWI) (86), and the Coated Abrasive Manufacturing Institute (CAMI) (87). These associations publish standards and specifications through the American National Standard Institute, Inc. (ANSI), New York. In Europe, the Federation of European Producers of Abrasive Products (FEPA) (88) issues standards for grain, microgrits, wheels, and coated products in cooperation with corresponding associations in Austria, France, Germany, Great Britain, Italy, Norway, Spain, Sweden, and Switzerland. In addition, major silicon carbide producers have their own tests and processes and product specifications for internal use, usually for products not covered by industry standards.

Analytical and Test Methods

The analysis of silicon carbide is classified into identification, chemical analysis, and physical testing. For identification, x-ray diffraction (89) and petrographic microscopy (90–91) are used. These techniques may also be used to distinguish the various polytypes.

Chemical analysis of abrasive grain and crude in the United States is carried out by using a standard analysis scheme approved by the AGA (85) and issued by the ANSI (92). Grain is usually analyzed for silicon carbide content, free silicon, free carbon, free silica, calcium oxide, magnesium oxide, and oxides of iron, titanium, and aluminum. A wet chemical analysis, using a combination of gravimetric, volumetric, and colorimetric techniques, is the standard method. The reference material, NBS SRM-112, which formerly served as a standard, is no longer available. In Europe, a standard analysis scheme, similar to the AGA scheme described above, has been developed by the FEPA (93).

Other methods include the determination of metals by spectrographic analysis in an a-c arc (94), mass spectrometry, or activation analysis (95); total silicon by atomic absorption, compared with the usual gravimetric analysis (96); and free silicon and free or combined silica by spectrophotometry after selective acid dissolution (97). Carbon is usually determined by combustion to carbon dioxide, and may be measured by weighing an absorption bulb (85,92) or by reading the electrical or thermal conductivity (97).

In the United States, a number of physical tests are performed on silicon carbide using standard AGA-approved methods, including particle size (sieve) analysis, bulk density, capillarity (wettability), friability, and sedimentation. Specifications for particle size depend on the use; for example, coated abrasive requirements (87) are different from the requirements for general industrial abrasives. In Europe, requirements are again set by FEPA. Standards for industrial grain are approximately the same as in the United States, but sizing standards are different for both coated abrasives and powders.

Toxicity

Sax (98) has described silicon carbide as having a slight toxicity by acute or chronic inhalation and an unknown toxicity with respect to acute or chronic systemic reaction. The threshold limit value for silicon carbide in the atmosphere is 10 mg/m^3 or 30 million particles per cubic foot (99).

Uses

Abrasives. Silicon carbide is used in loose form for lapping; mixed with a vehicle to form abrasive pastes or sticks; mixed with organic or inorganic binders; shaped and cured to form abrasive wheels, rubs, or tumbling nuggets; bonded to paper or cloth backings to form abrasive sheets, disks, or belts; or incorporated with the fibrous backing material before sheeting. Silicon carbide is harder yet more brittle than abrasives such as aluminum oxide. Since the grains fracture readily and maintain a sharp cutting action, silicon carbide abrasives are generally used for grinding hard, low tensile-strength materials such as chilled iron, marble, and granite, and materials that need sharp cutting action such as fiber, rubber, leather, or copper (see Abrasives).

Wear Surfaces. The extreme hardness of silicon carbide leads to its use where wear resistance is important as in brake linings (qv) or electrical contacts, and for nonslip applications such as floor or stair treads, terrazzo tile, deck-paint formulations, and in road surfaces.

Refractories. Its low coefficient of expansion, high thermal conductivity, and general chemical and physical stability make silicon carbide a valuable material for refractory use. Suitable applications for silicon carbide refractory shapes include boiler furnace walls, checker bricks, muffles, kiln furniture, furnace skid rails, trays for zinc purification plants, etc (see Refractories).

Electrical. Heating elements made from recrystallized silicon carbide, used in electric furnaces, operate up to about 1600°C and represent a major electrical use of silicon carbide. Heating elements are also used as a source of infrared radiation for drying operations, a light source for mineral determinations, and an ignition source for oil- or gas-fired burners. A more recent use is in ignition devices for gas clothes dryers and cooking ranges.

The semiconducting properties of silicon carbide have led to its use in thermistors (temperature-sensitive devices) and in varistors (voltage-sensitive devices). Thermistors are used for measuring and controlling temperature, as compensating devices for induction coils in electronic circuitry, and for time-delay applications. Varistors protect the coils or contacts in relays and solenoids against high-voltage surges, limit the inductive kick in oil burner ignition circuits, and stabilize circuits supplied by rectifiers.

A high-temperature thermocouple uses silicon carbide (100). It is also used in lightning arresters to protect high-tension power lines, and in high-loss-factor microwave attenuators (101).

Considerable interest in the solid-state physics of silicon carbide, that is, the relation between its semiconductor characteristics and the crystal growth reported above, has resulted from the expectation that it would be useful as a high-temperature-resistant semiconductor in devices such as point-contact diodes (102), rectifiers (103), and transistors (104–105) for use at temperatures above those where silicon or germanium metals fail (see Semiconductors).

Other solid-state applications of silicon carbide include its use as an electroluminescent diode for use in sound recording equipment and photomultipliers and controllers. It has been studied as a reflective surface for lasers.

Metallurgical. Silicon carbide is used extensively in ferrous metallurgy. When added to molten iron, a vigorous exothermic reaction takes place decomposing the silicon carbide and resulting in a hotter melt. The effect is to deoxidize and cleanse the metal and promote fluidity. Thus, a more desirable random distribution of the graphite flakes is achieved and a more machinable product obtained. Present practice is to add the silicon carbide as briquettes to the cupola or in loose granular form to induction furnaces when producing cast iron (see Iron). When added as granules to molten steel in the ladle, it reduces the number of undesirable inclusions and leads to better physical properties in the product. When added as granules to steel in a basic oxygen furnace it extends the capacity of the furnace to melt more scrap as a result of the exothermic reaction.

Other Uses. The special characteristics of silicon carbide give rise to a wide range of applications, including catalyst-carrier nuggets, tower packing, and pebbles for pebble-bed heaters or fluidized-bed reactors. It is used as a raw material for the production of silicon tetrachloride; in welding-rod compositions; as a filler in elastomers (see Fillers); as an additive in other ceramic materials to increase high-temperature resistance; and as an ingredient of red glaze (see Ceramics). Silicon carbide has been tested as a diluting agent in the coal gasification process.

The ultra-fine silicon carbide produced in an electric arc is used as insulation in cryogenic applications (106) (see Cryogenics). It greatly increases the wear resistance of the paint film when added to paint formulations.

Coatings of dense silicon carbide have been applied to materials such as graphite or silicon by vacuum evaporation with an electric gun. The coatings increase the oxidation and erosion resistance of the support. Graphite cathode supports may also be thus coated. A silicon carbide coating will protect the silicon metal rod which is to be purified by zone melting (qv).

Silicon carbide's relatively low neutron cross section and good resistance to radiation damage make it useful in some of its new forms (see below) in nuclear reactors. Silicon carbide temperature-sensing devices and structural shapes fabricated from the new dense types are expected to have increased stability. Silicon carbide coatings may be applied to nuclear fuel elements, especially those of pebble-bed reactors, or silicon carbide may be incorporated as a matrix in these elements (107–108) (see Nuclear reactors).

New Forms of Silicon Carbide. A number of new developments have appeared in recent years. The General Electric Co. and The Carborundum Co. have developed techniques for pressureless sintering of silicon carbide. Silicon carbide powder may be formed into complex shapes having bulk densities greater than 98% of theoretical and flexural strengths greater than 550 MPa (80,000 psi). These sintered parts retain their structural integrity above 1500°C. Both sintered alpha and beta silicon carbide are prime candidates as component materials for use in the hot zone of gas turbines (109–110), diesel engines, and other more advanced heat engines.

The Carborundum Company has a commercial process for the production of complex shapes made from sintered alpha silicon carbide. A $5.5 million high-performance silicon carbide technical center will include a pilot production line to produce prototypes and preproduction quantities by the end of 1978. Initial commercial production will include components for use in wear and corrosion areas (eg, paper foils, seal faces, valve trim, pump plungers, and textile guides). Components for diesel engines, turbochargers, gas turbines, and heat exchangers are also being considered for production as the markets develop.

The self-bonded silicon carbide referred to above, is prepared by siliconizing a shape of silicon carbide and carbon particles bonded temporarily with carbonaceous material. The product has zero porosity and small inclusions of uncombined silicon and carbon. The shapes produced have excellent abrasion resistance, corrosion resistance, high thermal conductivity and compressive strength, and can be prepared in complex shapes with smooth wear surfaces. These shapes are suitable for mold liners, hotpressing dies, spray and rocket nozzles, and orifices for handling abrasive and corrosive materials and for suction-box covers for Fourdrinier papermaking machines (111–112).

Hot-pressed silicon carbide of high hot strength and a density of up to 99+% of theoretical may be prepared under pressure of 69 MPa (10,000 psi) and 2000–2560°C. Small additions of other elements such as 1% aluminum assist in compaction and permit use of lower hot pressing temperatures.

A silicon carbide-bonded graphite material in which graphite particles are distributed through the silicon carbide matrix has even higher thermal shock resistance and is suitable for applications including rocket nose cones and nozzles and other severe thermal shock environments (113) (see Ablative materials).

A material made of silicon nitride or silicon oxynitride bonded to self-bonded silicon carbide has high corrosion resistance and may be used for pump parts, acid spray nozzles, and in aluminum reduction cells (114–117). A very porous silicon carbide foam has been considered for surface combustion burner plates and filter media.

BIBLIOGRAPHY

"Silicon Carbide" treated in *ECT* 1st ed. under "Carbides (Silicon)", Vol. 2, pp. 854–866, by M. Constance Parche, The Carborundum Company; "Silicon Carbide" treated in *ECT* 2nd ed. under "Carbides (Silicon)", Vol. 4, pp. 114–132, by M. Constance Parche, The Carborundum Company.

1. E. G. Acheson, *J. Franklin Inst.* **136**, 193, 279 (1893).
2. U.S. Pat. 492,767 (Feb. 28, 1893), E. G. Acheson (to The Carborundum Co.).
3. A. Colson, *Compt. Rend.* **94**, 1316 (1882).
4. R. C. Marshall and co-workers, eds., *Silicon Carbide—1973,* University of South Carolina Press, Columbia, S.C., 1974.
5. M. L. Torti, *Powder Metall. Int.* **6**, 186 (1974).
6. I. N. Frantsevich, *Silicon Carbide,* trans. from Russian, Consultants Bureau, New York, 1970, pp. 276.
7. J. Drowart and G. DeMaria in J. F. O'Connor and J. Smiltens, eds., *Silicon Carbide, A High Temperature Semiconductor,* Pergamon Press, Inc., New York, 1960, pp. 16–23.
8. N. W. Thibault, *Am. Mineral.* **29**, 327 (1944).

9. P. Grieveson and C. B. Alcock in P. Popper, ed., *Special Ceramics,* Vol. 5, 1972, p. 183.
10. G. L. Humphrey and co-workers, *Report of the U.S. Bureau of Mines,* Investigation No. 4888, 1952.
11. C. H. McMurtry and co-workers in ref. 4, pp. 411–419.
12. A. Goldsmith and co-workers, *Handbook of Thermophysical Properties of Solid Materials,* Vol. 3, rev. ed., The Macmillan Co., New York, 1961, pp. 923–936.
13. J. R. O'Connor and J. Smiltens, eds., *Silicon Carbide, A High Temperature Semiconductor,* Pergamon Press, Inc., New York, 1960.
14. L. Patrick and W. L. Choyke, *J. Appl. Phys.* **30,** 236 (1959).
15. J. A. Lely and F. A. Kroger, *Z. Krist* **109,** 514, 525 (1957).
16. S. D. Mark, Jr. and R. C. Emanuelson, *Bull. Am. Ceram. Soc.* **37,** 193 (1958).
17. A. H. Falter, *Internal Report,* The Carborundum Co., Res. & Dev. Lab., Niagara Falls, N.Y., 1953.
18. A. Taylor and R. M. Jones in ref. 7, pp. 147–154.
19. C. J. Engberg and E. H. Zehms, *J. Am. Ceram. Soc.* **42,** 300 (1959).
20. E. L. Kern and co-workers, in H. K. Henisch and R. Roy, eds., *Silicon Carbide—1968,* spec. publ. of *Materials Research Bulletin,* Vol 4, 1969, pp. 525–532.
21. E. H. Kraft, *Internal Report,* The Carborundum Co., Res. & Dev. Lab., Niagara Falls, N.Y., 1976.
22. S. Prochazka and co-workers, *Naval Air Systems Command Contract N62269-74-C-0255,* Nov. 1974, pp. 46, 71.
23. W. S. Coblenz, *J. Am. Ceram. Soc.* **58,** 530 (1975).
24. G. Q. Weaver and B. A. Olson in ref. 4, pp. 367–74.
25. E. H. Kraft and G. I. Dooher, *Mechanical Response of High Performance Silicon Carbides,* pres. at the *Second International Conference on Mechanical Behavior of Materials,* Aug. 16–20, 1976, Boston, Mass., to be published.
26. D. P. H. Hasselman, *Tables for the Computation of the Shear Modulus and Young's Modulus of Elasticity from the Resonant Frequencies of Rectangular Prisms,* The Carborundum Co., Res. & Dev. Lab., Niagara Falls, N.Y., 1961.
27. E. H. Kraft, *Carborundum Company Internal Report,* The Carborundum Co., Niagara Falls, N.Y., 1977.
28. P. T. B. Shaffer and C. K. Jun, *Mat. Res. Bull.* **7,** 63 (1972).
29. L. S. Ramsdell, *Am. Mineral.* **32,** 64 (1947).
30. P. T. B. Shaffer, *Acta Cryst.* **B25,** 477 (1969).
31. V. Vand, *Nature, London* **168,** 783 (1951).
32. V. Vand, *Philos. Mag.* **42,** 1384 (1951).
33. S. Amelinck, *Nature, London* **167,** 939 (1951).
34. F. C. Frank, *Philos. Mag.* **42,** 1014 (1951).
35. P. Krishna and A. R. Verma, *Z. Kristallogr.* **121,** 36 (1965).
36. A. R. Verma and P. Krishna, *Polymorphism and Polytypism in Crystals,* John Wiley & Sons, Inc., New York, 1966.
37. W. F. Knippenberg and co-workers in ref. 4, pp. 92–101.
38. D. Pandey and P. Krishna in ref. 4, pp. 198–205.
39. T. Nishida, *Mineral. J.* **6,** 216 (1971).
40. U. S. Ram and co-workers in ref. 4, p. 184–190.
41. G. Thomas and co-workers, *Proceedings of the 6th International Materials Symposium on Ceramic Microstructures,* Berkeley, Ca., Aug. 24–27, 1976, Westview Press, Boulder, Colorado, 1977.
42. R. I. Scace and G. A. Slack, *J. Chem. Phys.* **30,** 1551 (1959).
43. H. Kang and R. B. Hilborn, Jr. in ref. 4, pp. 493–499.
44. H. K. Henisch and R. Roy, eds., *Silicon Carbide-1968,* spec. publ. of *Mat. Res. Bull.* **4,** 525 (1969).
45. H. H. Woodbury and G. W. Ludwig, *Phys. Rev.* **124,** 1083 (1961); W. J. Choyke and co-workers, *Phys. Rev.* **133,** A1163 (1964); R. L. Hartman and P. J. Dean, *Phys. Rev.* **B2,** 951 (1970); L. A. Hemstreet and C. Y. Fong, *Solid State Commun.* **9,** 643 (1971); W. J. Choyke and L. Patrick, *Phys. Rev.* **187,** 1041 (1969); L. A. Hemstreet and C. Y. Fong, *Phys. Rev.* **B6,** 1464 (1972).
46. J. A. Van Vechten, *Phys. Rev.* **182,** 891 (1969).
47. W. J. Choyke and L. Patrick in ref. 4, pp. 261–283.
48. B. W. Wessels and H. C. Gatos, *J. Phys. Chem. Solids* **38,** 345 (1977).
49. H. O. Pritchard and H. A. Skinner, *Chem. Rev.* **55,** 145 (1955).
50. L. Pauling, *Nature of the Chemical Bond,* 3rd ed., Cornell University Press, Ithaca, N.Y., 1960.
51. M. Alexander, *Phys. Rev.* **172,** 331 (1968).
52. H. R. Philipp and E. A. Taft in ref. 7, pp. 366–370.

53. P. J. Colwell and M. V. Klein, *Phys. Rev.* **B6,** 498 (1972).
54. P. J. Dean and R. L. Hartman, *Phys. Rev.* **B5,** 4911 (1973); B. Ellis and T. S. Moss, *Proc. R. Soc. London,* **A299,** 393 (1967).
55. W. Primak in ref 7, pp. 385–387.
56. L. W. Aukerman and co-workers, in ref. 7, pp. 388–394.
57. R. F. Adamsky, *J. Phys. Chem.* **63,** 305 (1959).
58. P. J. Jorgenson and co-workers, *J. Am. Ceram. Soc.* **42,** 613 (1959); **43,** 209 (1960).
59. G. Wiebke, *Ber. Deut. Keram. Ges.* **37,** 219 (1960).
60. H. Suzuki, *Yogyo Kyokai Shi* **67,** 157 (1959).
61. J. A. Dillon, Jr., in ref. 7, pp. 235–240.
62. J. E. Antill and J. B. Warburton, *Corrosion Sci.* **11,** 337 (1971).
63. E. A. Gulbransen and S. A. Jansson, *Oxid. Met.* **4,** 181 (1972).
64. S. C. Singhal, *J. Mater. Sci.* **11,** 1246 (1976).
65. J. A. Lely, *Ber. Deut. Keram. Ges.* **32,** 229 (1955).
66. D. R. Hamilton in ref 7, pp. 43–52.
67. A. H. Smith in ref 7, pp. 53–59.
68. K. M. Merz in ref. 7, pp. 73–83.
69. F. A. Halden in ref. 7, pp. 115–123.
70. R. C. Ellis, Jr. in ref. 7, pp. 124–129.
71. J. C. McMullen, *J. Electrochem. Soc.* **104,** 462 (1957).
72. Abrasive Grain Association, *History of Abrasive Grain, Booklet 1,* Cleveland, Ohio, 1973.
73. A. V. Alterov, *Abrasivy* **13**(15), (1956).
74. U.S. Pat. 2,913,313 (Nov. 17, 1959), F. Schroll (to Electroschmelzwerk Kempten G.M.C.H.).
75. U.S. Pat. 2,908,553 (Oct. 13, 1959), H. Frank and E. Wilkendorf.
76. W. E. Kuhn, *J. Electrochem. Soc.* **110,** 298 (1963).
77. *Chem. Eng. News,* **50,** 13 (Sept. 18, 1973).
78. G. R. Finlay, *Chem. Canada* **14**(2), 25 (1952).
79. *Verbal Communications* with W. T. Adams, U.S. Bureau of Mines, Washington, D.C., July 1977.
80. *The Mineral Industry,* Vol. XVI, Hill Publishing Co., New York, 1907, pp. 149–145.
81. *The Mineral Industry,* Vol. XX, McGraw-Hill Book Co., New York, 1911, p. 668.
82. *Minerals Yearbook,* U.S. Bureau of Mines, Washington, D.C., 1921, 1931, 1941, and 1951.
83. *Minerals Yearbook,* U.S. Bureau of Mines, Washington, D.C., 1960, 1965, 1969, 1970, 1971, 1972, and 1973.
84. *Mineral Industry Surveys: Abrasive Materials in 1975, Minerals Yearbook,* U.S. Bureau of Mines, Washington, D.C., 1975.
85. Abrasive Grain Association, Cleveland, Ohio.
86. Grinding Wheel Institute, Cleveland, Ohio.
87. Coated Abrasives Manufacturers' Institute (CAMI), Cleveland, Ohio.
88. Federation of European Producers of Abrasive Products, Paris, France.
89. *Powder Diffraction File, Inorganic Section,* Joint Commission on Powder Diffraction Standards, Swarthmore, Pa.
90. A. N. Winchell and H. Winchell, *The Microscopical Characters of Artificial Inorganic Solid Substances: Optical Properties of Artificial Minerals,* 3rd ed., Academic Press, New York, 1964.
91. W. C. McCrone and J. G. Delly, *The Particle Atlas,* 2nd ed., Ann Arbor Science Publishers, 1973, pp. 407–408.
92. *ANSI Std. B74.15-1971,* American National Standards Institute, New York, N.Y.
93. *Method of Surface Chemical Analysis of Silicon Carbide for Abrasive Purposes,* Abrasive Industries Assoc., Victoria Station, London, England.
94. R. L. Pevzner, M. E. Zaionchik, and E. A. Shpital'naya, *Prikl. Spektrosk., Mater. Soveshch.* **16**(1), 389, 1965.
95. P. N. Kuin in ref. 44, pp. S273–S283.
96. K. Kato, *At. Absorpt. Newsl.* **15**(1), 4 (1976).
97. G. Serrini and W. Leyendecker, *Metal. Ital.* **64**(4) 129 (1972).
98. N. I. Sax, *Dangerous Properties of Industrial Materials,* Reinhold Book Corp., New York, 1968, p. 1089.
99. *Threshold Limit Values for Chemical Substances and Physical Agents in the Workroom Environment,* American Conference of Governmental Industrial Hygienists, 1976, p. 52.
100. M. T. Minamoto in ref. 7, pp. 443–446.
101. U.S. Pat. 2,857,338 (Oct. 2, 1958), J. C. Rolfs and L. Brecht (to Sperry Rand Corp.).

102. A. L. Hopkins, Jr. in ref. 7, pp. 482–495.
103. C. Goldberg and J. W. Ostroski in ref. 7, pp. 453–461.
104. H. Chang and co-workers in ref. 7, pp. 496–507.
105. R. N. Hall, *J. Appl. Phys.* **29,** 914 (1958).
106. W. E. Kuhn, *J. Electrochem. Soc.* **110,** 298 (1963).
107. G. M. Butler, *Am. Ceram. Soc. Bull.* **39,** 402 (1960).
108. K. M. Taylor and C. H. McMurtry, *Materials in Nuclear Applications,* Vol. 276, American Society of Testing Materials, Special Technical Publ., 1960, pp. 311–317.
109. J. A. Coppola and C. H. McMurtry, *Substitution of Ceramics for Ductile Materials in Design,* pres. National Symposium on Ceramics in the Service of Man, Carnegie Institution, Washington, D.C., June 7, 1976.
110. S. Prochazka in J. J. Burke and co-workers, eds., *Ceramics for High Performance Applications,* Brook Hill Publishing Co., Chestnut Hill, Mass., 1974, pp. 239–252.
111. R. E. Dial and G. E. Mangsen, *Corrosion* **17,** 35t (1961).
112. M. F. Kiachif, *Corrosion and Wear Resistant Ceramics, Papers,* American Institute of Chemical Engineers, 44th National Meeting, New Orleans, La., Feb. 1961.
113. J. F. Lynch, J. F. Quirk, and W. H. Duckworth, *Am. Ceram. Soc. Bull.* **37,** 443 (1958).
114. R. W. Brown and C. R. Landback, *Am. Ceram. Soc. Bull.* **38,** 352 (1959).
115. R. W. Brown and H. G. Noble, *Corrosion* **15**(10), 92 (1959).
116. R. W. Brown and R. F. Nering, *Ind. Eng. Chem.* **52,** 381 (1960).
117. M. E. Washburn and R. W. Lowe, *Am. Ceram. Soc. Bull.* **41,** 447 (1962).

RICHARD H. SMOAK
T. M. KORZEKWA
S. M. KUNZ
E. D. HOWELL
The Carborundum Co.

CARBOHYDRATES

Carbohydrates are the most abundant class of organic compounds found in living matter. They constitute three-fourths of the dry weight of the plant world and are widely distributed in other life forms (1). In plants and animals, carbohydrates mainly serve as structural elements and food reserves. Plant carbohydrates, in particular, represent an enormous store of energy, either as human and animal food or after transformation in the geological past, as coal and peat. Large industries process carbohydrates such as sucrose, starch, cellulose, pectin, and certain seaweed polysaccharides. Some carbohydrates and their derivatives have been examined as chemotherapeutic drugs for various pathological conditions such as cancer (2) (see Antibiotics; Chemotherapeutics). Derivatives have been successfully used in biochemical analysis (see Biopolymers; Diagnostic chemicals).

The term carbohydrate originated from the belief that these compounds were hydrates of carbon since elemental analysis of common carbohydrates such as lactose, sucrose, starch, and cellulose led to the empirical formula $C_x(H_2O)_y$. Although the formula represents the majority of carbohydrates, many have compositions that do not fit such a simplified generalization. Although it is not possible to give a simple, yet comprehensive definition of such a broad group of compounds, one fairly good definition describes them as compounds of carbon, hydrogen, and oxygen containing the saccharose group,

$$-\underset{\underset{\text{OH}}{|}}{\text{CH}}-\underset{\underset{\text{O}}{\|}}{\text{C}}-,$$

or its first reaction product, and which usually contain hydrogen and oxygen in the ratio found in water.

All monosaccharides and many oligosaccharides are called sugars. Frequently the monosaccharides are called simple sugars. All sugars are readily soluble in water but vary greatly in sweetness. The sweetest sugar, D-fructose [57-48-7], is about 1.7 times as sweet as table sugar, sucrose [57-50-1], and about 3 times as sweet as D-glucose, corn sugar. Other sugars are less or barely noticeably sweet; polysaccharides are not sweet, they have only a bland taste. Starch, if kept in the mouth will develop some sweetness after a time as a consequence of enzymatic breakdown to sweeter oligosaccharides and D-glucose.

Structure and Classification

Monosaccharides. Monosaccharides are the simplest carbohydrates. They are polyhydroxy aldehydes, ketones, or derivatives.

Classification is based on the number of carbon atoms in the chain and whether an aldehyde or ketone group is present. A monosaccharide containing three carbon atoms is a triose. A sugar containing four, five, or six carbon atoms is a tetrose, pentose, or hexose and is a monosaccharide rather than a polymer. If an aldehyde group is present, the sugar is an aldose; if a ketone group is present, the sugar is a ketose. Thus, a four carbon sugar containing an aldehyde group is an aldotetrose and a six carbon sugar containing a ketone group is a ketohexose.

Although glycolaldehyde [141-46-8], HOCH$_2$CHO, is the simplest carbohydrate structurally, D-glyceraldehyde, an aldotriose, is the simplest biologically important sugar. It is also the simplest aldose with a chiral, optically active, carbon atom. Dihydroxyacetone [96-26-4], is the simplest ketosugar.

$$\begin{array}{ccc}
\text{HC}=\text{O} \dots\dots 1 \dots\dots \text{CH}_2\text{OH} \\
| & & | \\
\text{H}-\text{C}-\text{OH} \dots 2 \dots \text{C}=\text{O} \\
| & & | \\
\text{CH}_2\text{OH} \dots\dots 3 \dots\dots \text{CH}_2\text{OH}
\end{array}$$

D-glyceraldehyde   dihydroxyacetone

The D-configuration for glyceraldehyde designates that the hydroxyl group attached on the second carbon (or highest-numbered chiral carbon in a sugar) lies to the right. In the enantiomorph, the mirror image, the hydroxyl group on the second carbon lies to the left and the sugar is L-glyceraldehyde [497-09-6].

For numbering carbon atoms, the sugar structure is written vertically with the aldehyde or ketone group at the top and the carbons numbered from top to bottom.

Structures for the most common aldoses can be derived from the structure of glyceraldehyde simply by adding carbon atoms in a chain lengthening process. Figure 1 illustrates the structures of D-aldoses containing up to six carbon atoms. The sugars belong to the D-series since the hydroxyl on highest-numbered chiral carbon atom in the chain is depicted to the right. It is number 2 in glyceraldehyde, number 3 in the tetroses, number 4 in the pentoses, and number 5 in the hexoses. The structures shown are drawn according to a shorthand known as the Rosanoff notation in which the circle denotes an aldehyde group, the vertical line represents the carbon chain, and the horizontal lines represent secondary hydroxyl groups (3).

In Figure 1, the tetroses, pentoses, and hexoses are shown grouped into discrete diastereomeric pairs. Such pairs of monosaccharides, differing only in the position of the hydroxyl group at carbon 2, are known as epimers. D-Glucose and D-mannose represent an important epimeric pair of aldohexoses.

Some important L-sugars include L-arabinose, L-rhamnose [3615-41-6] (6-deoxy-L-mannose), and L-fucose [2438-80-4] (6-deoxy-L-galactose). Their structures are shown in Figure 2.

In solution, the open chain structures exist in equilibrium with the ring forms which constitute the major population of molecules.

These forms arise by interaction between the carbonyl function and the C-4 or

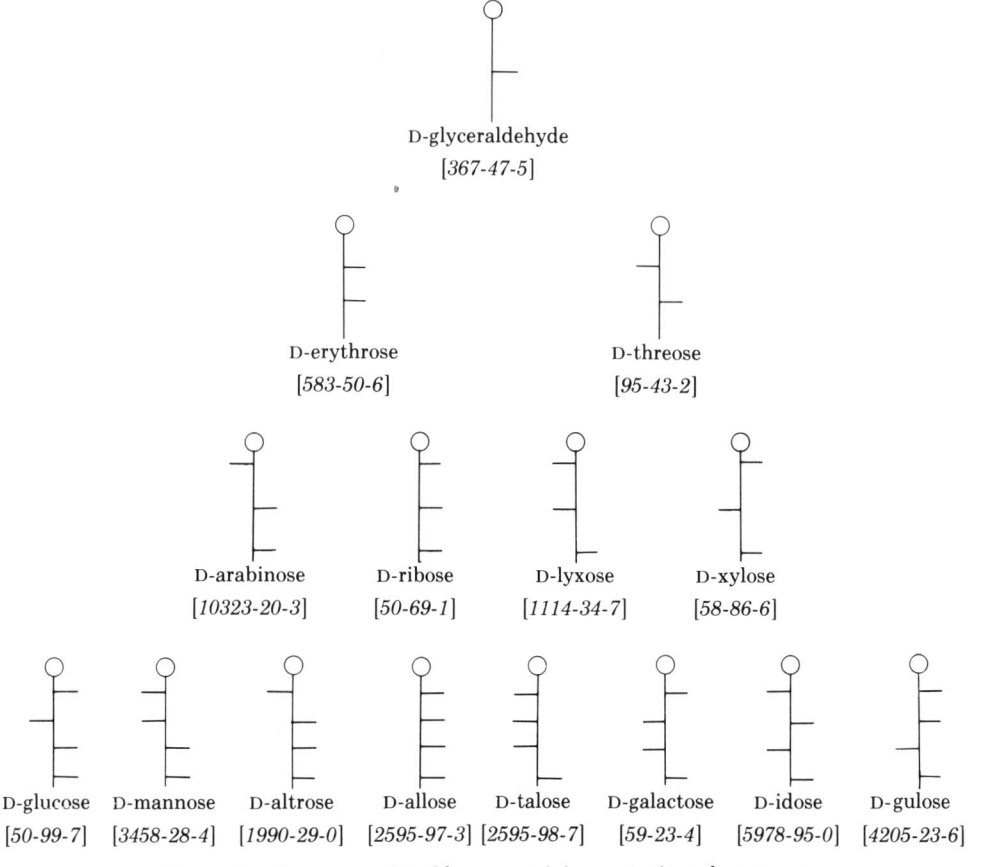

Figure 1. Structures of D-aldoses containing up to six carbon atoms.

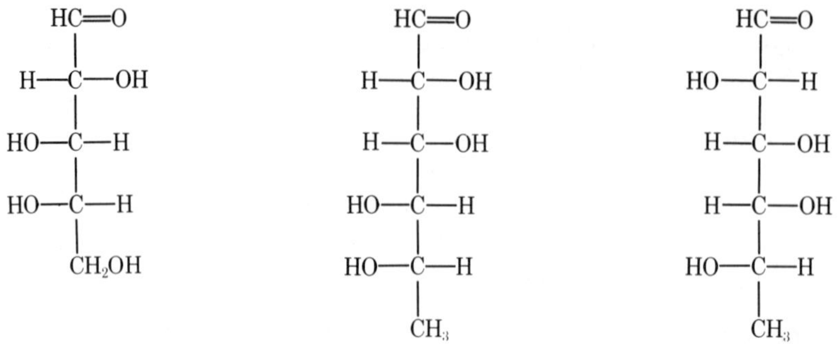

538   CARBOHYDRATES

```
    HC=O              HC=O              HC=O
     |                 |                 |
H—C—OH            H—C—OH            HO—C—H
     |                 |                 |
HO—C—H            H—C—OH            H—C—OH
     |                 |                 |
HO—C—H            HO—C—H            H—C—OH
     |                 |                 |
   CH₂OH           HO—C—H            HO—C—H
                       |                 |
                      CH₃               CH₃

 L-arabinose       L-rhamnose         L-fucose
 [5328-37-0]      [3615-41-6]       [2438-80-4]
```

Figure 2. The structures of L-arabinose, L-rhamnose, and L-fucose.

C-5 hydroxyl group to produce a relatively more stable hemiacetal. The oxygen from the C-4 or C-5 hydroxyl group participates in the ring formation and becomes the ring heteroatom (Fig. 3). These structures are drawn according to Fischer projections (4). The ring structures are more accurately represented in the Haworth cyclic convention (5) (Fig. 4) where (4) corresponds to (1) in Figure 3 and (5) corresponds to (3). In this convention, the bottom horizontal line of the ring between C-2 and C-3 is viewed as projected out of the plane of the paper. Groups attached to the ring are referred to as being either above or below the plane of the ring. Structures (4) and (5) differ only in the position of the hydroxyl group about the anomeric carbon C-1. Such diastereomeric pairs are known as anomers. The hydroxyl group at C-1 (4), is below the ring plane and is known as the α-D-anomer. The hydroxyl group (5) is above the ring plane and is referred to as the β-D-anomer. These structures are most accurately represented by stereochemical drawings (Fig. 5) in which hydrogens are axial, and hydroxyls and hydroxymethyl groups in β-D-glucose (7) are equatorial.

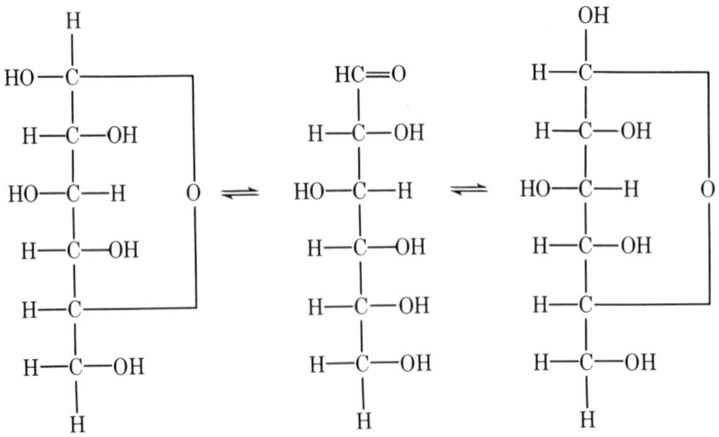

Figure 3. Oxygen from C-4 or C-5 hydroxyl group participates in ring formation and becomes the ring heteroatom.

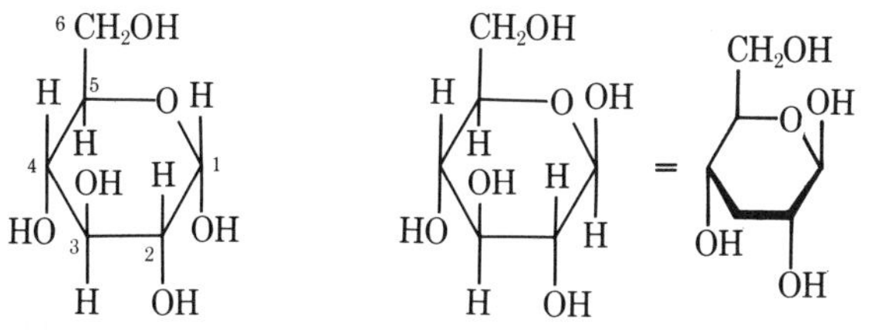

Figure 4. Haworth cyclic convention of α-D-glucose and β-D-glucose.

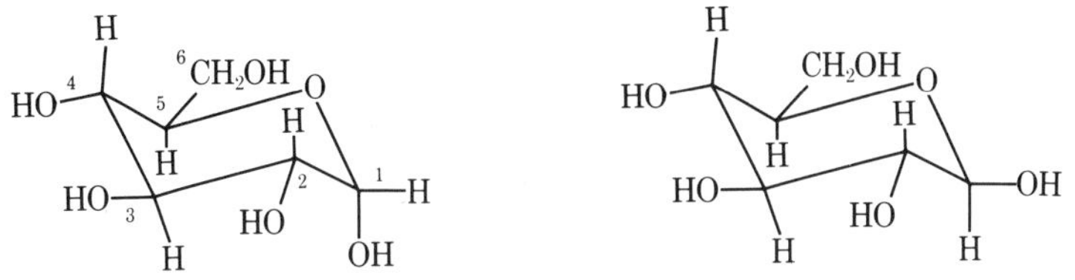

(6) α-D-glucose [492-62-6]

(7) β-D-glucose [492-61-5]

Figure 5. Stereochemical drawings of α-D-glucose and β-D-glucose.

Stereochemically, the ring structures (6) and (7) resemble the chair conformation of cyclohexane. This is also designated by the $^4C_1$ conformation indicating that C-4 is above and C-1 is below the central plane of 4 atoms.

Five and six member ring sugars are commonly called furanose and pyranose, respectively. Names are based on the structures of the parent compounds furan (8) and pyran (9).

(8) furan

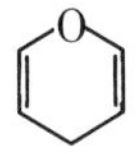

(9) pyran

Glycosides, Di- and Oligosaccharides. An important characteristic of monosaccharides is their ability to undergo glycoside formation in order to yield a mixed acetal. The attached group termed aglycone, may be a sugar or nonsugar. Treatment of D-glucose with methanol and an acid catalyst results in the formation of methyl glycosides. These are commonly methyl α- (10) or methyl β-D-glucopyranoside (11).

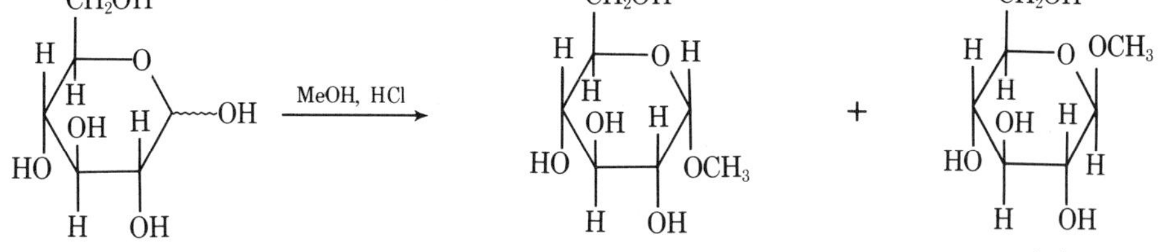

α- or β-D-glucopyranose [2280-44-6]

methyl α-D-glucopyranoside [97-30-3]

methyl β-D-glucopyranoside [709-50-2]

Instead of reacting with a monohydroxylic compound to form a glycoside, a sugar hemiacetal may react with a hydroxyl group of another sugar to form a disaccharide, and the reaction can be repeated to yield polysaccharides. Sucrose is a common disaccharide in which the anomeric carbon of each monosaccharide participates in a joint glycosidic bond (12), α-D-glycopyranosyl-(1 → 2)-β-D-fructofuranoside. The numbers in parenthesis indicate which carbon atom from each sugar is involved in the glycosidic bond. Carbon 1 of the α-D-glucose moiety is attached to C-2 of D-fructose via the glycosidic oxygen. Because this sugar has no potentially free aldehyde group to react with a mild oxidizing agent, it is said to be a nonreducing sugar.

(12)
sucrose (table sugar)

Maltose [69-74-4] (13), α-D-glucopyranosyl-(1 → 4)-α-D-glucopyranose, is a disaccharide produced by partial hydrolysis of starch. The glycosidic bond links carbon 1 of one D-glucose unit to carbon 4 of the other D-glucose unit. Since the D-glucose molecule attached at C-4 has a free hydroxyl at C-1, as indicated by the -ose ending, it can give rise to the aldehyde function. Because this aldehyde can be oxidized easily, the sugar is said to be reducing.

(13)
maltose (α-anomer) (α-D-glucopyranosyl-(1 → 4)-α-D-glucopyranose)

Disaccharides may be composed of two identical monosaccharides as in maltose, or may be composed of two dissimilar monosaccharides as in sucrose or in lactose [63-42-3] (milk sugar), which is made up of a D-galactose and a D-glucose unit.

Oligosaccharides are water-soluble polymers consisting of 2–10 monosaccharide units. Such polymers can be further classified as homopolymers containing only one type of monosaccharide, or heteropolymers containing several different kinds of monosaccharides. Examples of oligosaccharides are stachyose [10094-58-3], a naturally occurring tetrasaccharide, maltopentaose [34620-76-3], a five monosaccharide unit oligosaccharide, and cyclomaltohexaose [7585-39-9], a cyclic polysugar (Schardinger dextrin), consisting of six monosaccharide units.

Polysaccharides. Polysaccharides are large polymers of monosaccharides in a branched or unbranched chain. Common nonbranching polysaccharides such as cellulose [9004-34-6] and amylose [9005-82-7] consist of D-glucopyranose units linked by 1 → 4 glycosidic bonds. The glycosidic bond may either be α-D- as in amylose or β-D- as in cellulose. Amylopectin [9037-22-3] is a branched polysaccharide of α-D-glucose units in 1 → 4 linkages in the main chain and 1 → 6 glycosidic linkages at branch points (Fig. 6).

Cellulose, amylose, and amylopectin are examples of homoglycans, polysaccharides that contain one kind of monosaccharide. Polysaccharides with more than one kind of monosaccharide are known as heteroglycans, those with two different monosaccharide units are called diheteroglycans, and those with three different kinds of sugar units are known as triheteroglycans. Not more than six different monosaccharide units have been found in natural heteroglycans (6). The main chain of a natural heteroglycan is composed of no more than two different monosaccharide units. Whenever more than two monosaccharide units are found in a polysaccharide, the carbohydrate is bush-like in structure.

Linear 1 → 4 homoglycans are insoluble in water, or at best, are only slightly soluble because of their tendency to associate intermolecularly in an attempt to crystallize. They form ordered regions, often large enough to be discernible by x-ray diffraction as crystalline regions. Heteroglycans (homoglycans with 1 → 6 glycosidic linkages) and glycans with a mixture of different linkages are soluble. Branched polysaccharides are always soluble since they can not form strong intermolecular associations. Polysaccharides bearing anionic groups are soluble. Among these are pectates and alginates containing uronic acid carboxyl groups and carrageenans containing half-ester sulfate groups.

Oxidation and Reduction of Carbohydrates

Oxidation. Hydroxyl groups of carbohydrates are readily oxidized to acids, aldehydes, and ketones. The mechanism of oxidation is quite complex and only with lead tetraacetate and periodic acid is the reaction specific. In these cases α-glycol groups are cleaved to give carbonyl groups. Glycol groups containing a primary alcohol

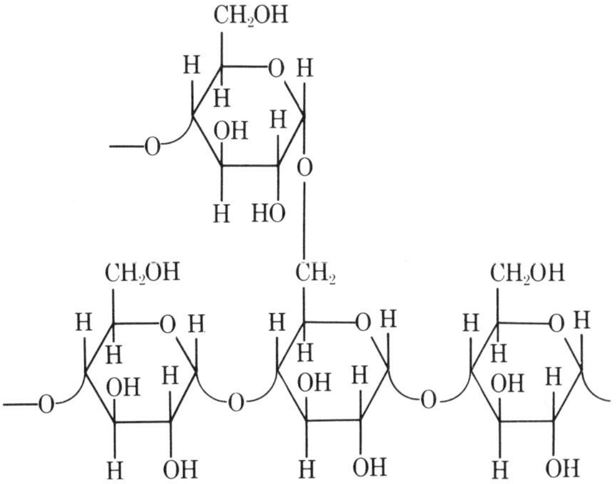

Figure 6. Branched polysaccharide of amylopectin.

group give rise to formaldehyde, whereas three vicinal secondary hydroxyl groups give formic acid. α-Hydroxycarboxylic acids react more slowly to produce carbon dioxide. Both lead tetraacetate and periodic acid are important quantitative reagents used for the structural determination of carbohydrates.

Aldoses undergo oxidation by several reagents. The most important of these include Tollens' and Fehling's reagents, bromine water, and nitric acid.

Sugars containing carbonyl groups ($-\overset{\overset{\displaystyle O}{\|}}{C}-$) are capable of undergoing oxidation by Tollens' ammoniacal silver nitrate or Fehling's alkaline copper reagent and they are known as reducing sugars. All monosaccharides and most disaccharides are reducing sugars. Sucrose has no carbonyl group and is the most important nonreducing disaccharide.

Oxidation of aldols by bromine and water yield glyconic acids where the aldehyde group is oxidized to a carboxyl group. Alternatively, nitric acid oxidizes aldols to glycaric acids in which both the aldehyde group and the primary alcohol are oxidized to carboxyl groups. Reagent specific oxidation of the primary alcohol group has been employed for the synthesis of glycuronic acids.

Reduction. Reduction of aldoses by hydrogen and nickel converts these sugars to polyhydroxy alcohols. Reduction of glucose in this manner yields sorbitol, an effective bacteriostat and food additive (see Alcohols, polyhydric).

Industrial Sugars

Sucrose is the most important industrial sugar, followed by D-glucose, and other much less important sugars, eg, lactose, maltose, and fructose. A few sugars are used in the synthesis of medicinal products (see Sugar).

Sucrose is obtained from sugar cane and sugar beets. World production of sucrose in 1976 was 342.4×10^6 metric tons of which 45–50 million tons moved in the global export market according to the U.S. Department of Agriculture.

Cane Sugar. Sugar cane is crushed between rolls to squeeze out the sugar-bearing juice. The dark cloudy juice is neutralized with lime, then filtered, concentrated, crystallized, and centrifuged to produce raw sugar which averages more than 97% sucrose. This raw sugar is refined by being washed, dissolved, passed over bone char, filtered, concentrated, crystallized, centrifuged, and dried to produce refined white sugar, which is 99.96% sucrose (7). Raw sugar from American refiners comes principally from Mexico, the Dominican Republic, Puerto Rico, Hawaii, the Phillipines, Louisiana, and Florida.

The mother liquors from both raw and refined sugar processes are reworked to produce additional yields of the respective sugars. When the concentration of impurities becomes so high that it is uneconomical to produce more sugar, the residue is a dark viscous syrup called molasses. The final mother liquor is blackstrap or cane final molasses. Refineries produce refiner's syrup, sugar-house molasses, or treacle. These products are fed to animals or are fermented to alcohol or pharmaceutical preparations (eg, citric acid (qv), penicillin) (7).

Intermediate stages of recovery produce brown sugars in which relatively small crystals of pure sucrose are coated with layers of colored molasses.

Beet Sugar. Beet sugar produced by counter-current extraction from beet slices called cossettes. The extract is limed, filtered, concentrated, passed over bone char, crystallized, centrifuged, and dried to produce a refined sugar. Raw beet sugar is not produced in the United States. Beet molasses contains much less invert sugar than cane molasses; hence, recovery of additional amounts of sugar by precipitation with alkaline earth hydroxides is possible. In the Steffen process, lime is added to molasses to precipitate sucrose as a calcium sucrate from which the sugar is released by carbon dioxide (7).

Liquid Sugar. Liquid sugar is an aqueous solution of sucrose made from raw sugar by first washing molasses film from crude crystals with a saturated sugar liquor. The washed raw sugar is then lime clarified, passed over char and vegetable or granular carbon, filtered, and concentrated if necessary (8). For the production of premium liquid sucrose (0.005% ash) (9), a single-unit ion-exchange process is used to remove remaining traces of color and ash and yield a water-white, essentially ash-free liquor. The sucrose solution is heat sterilized and concentrated in an evaporator to 67% solids (10).

Invert Sugars. Invert syrups are prepared by using a variety of inverting acids, the most common being tartaric and hydrochloric (37%) acids. Granulated sugar is heated with a specific volume of water to 100°C and mixed with a measured amount of acid. The mixture is gradually cooled, and a sodium bicarbonate solution is added with constant stirring (11).

Invert liquid sugars may contain 10–90% invert sugar. Since the presence of invert sugar minimizes the danger of crystallization, these liquid sugars can be shipped at higher densities than sucrose. Invert liquid sugars are used in carbonated beverages (qv), glacé fruits, canning, and bread baking.

Lactose. Lactose is present in the milk of all mammals at 2–8% concentration. It is produced from whey that has been treated with acid and heat to remove proteins. The filtrate is decolorized with carbon and concentrated to produce crystalline lactose (12). The commercial product is the monohydrate of the alpha-form (mp 201.6°C). If lactose is crystallized at temperatures above 93.3°C, the anhydrous beta-form is produced (mp 252.2°C). Lactose is much less soluble in water than sucrose.

Maltose. Maltose or malt sugar does not often occur naturally. Produced in sprouted grain, it is prevalent in the malting stage of the brewing process (see Beer). High maltose corn syrups can be produced commercially by hydrolyzing starch with beta-amylase. High maltose corn syrup, used extensively by preservers, contains about 10% D-glucose, 56% maltose, and 44% or more oligosaccharides. β-Maltose crystallizes from water as the monohydrate (mp 102–103°C). It is ca one-third as sweet as sucrose.

Fructose. D-Fructose is produced by isomerization of D-glucose with the enzyme isomerase. This process gives a mixture of 54% D-glucose and 46% D-fructose. The mixture is equivalent in sweetness to invert sucrose, the 50–50 mixture of D-glucose and D-fructose obtained on hydrolysis of sucrose by acid or the enzyme invertase. The isomerase-produced mixture is commercially called high fructose syrup and, because of its low cost, is taking over a significant part of the sweetener market formerly held by sucrose. The United States now produces more than 2.3×10^6 metric tons of isomerized syrup annually, and increased quantities are planned. In 1976 corn sweeteners accounted for 26% of the United States sweetener market or 15 kg per capita. High fructose syrup was consumed to the equivalent of 4.3 kg per capita. In the production of high fructose syrup, corn starch is pasted in a steam jet cooker. The paste, sometimes thinned with diastase, is passed over a fixed column of glucoamylase to produce high quality D-glucose. This syrup is passed through a column of fixed isomerase enzyme to give the equilibrium mixture of D-glucose and D-fructose (see Enzymes, immobilized). The mixture is usually sold as a syrup (liquid sugar). By using calcium ion-exchange columns, the D-fructose can be concentrated to higher levels or can be obtained pure, and crystallized for specialty uses as a sweetener, 1.7 times as sweet as sucrose (see Sweeteners).

Water Soluble Polysaccharide Gums

Gums, in the form of natural, biosynthetic, or modified polysaccharides, are industrially used in tremendous quantities. Over 450,000 metric tons are consumed annually within the United States where the growth of gum usage exceeds 8–10% per year. Gums seldom constitute an entire finished product. They are mainly used as additives to improve or control the properties of a commodity. The extent of their use results from the low cost of many gums and the important properties they contribute to products in low concentrations. Some gums exert their effect at very low concentrations indeed. Okra gum at 25 parts per million in water reduces friction in fluid flow by 80% (see Cellulose derivatives; Gums).

Commercial gums are water-soluble or water-dispersible hydrocolloids. Their aqueous dispersions usually possess suspending, dispersion, and stabilizing properties; or the gums may act as emulsifiers, have gelling characteristics, or be either adhesive or mucilaginous. Some act as coagulants, binders, lubricants, and film formers.

In practical terms, gums are either hydrophobic or hydrophilic high molecular weight molecules. They have usually colloidal properties that, in an appropriate solvent or swelling agent, produce gels, highly viscous suspensions, or solutions with low dry-substance content. Thus, the term gum applies to a wide variety of substances with "gummy" characteristics. Most commonly, however, the term gum as used in industry refers to plant or microbial polysaccharides or to their derivatives that are dispersible in either cold or hot water to produce viscous mixtures or solutions. Thus, modern usage includes the water-soluble or water-swellable derivatives of cellulose and the derivatives and modifications of other polysaccharides that are insoluble in the natural form. This definition does not require that gums have the property of tackiness, and therefore includes as gums those polysaccharides and derivatives that are slimy or mucilaginous. Some authors have separately categorized these slimy substances from plants as mucilages (see Table 1).

Cellulose

Cellulose, the most abundant polysaccharide, constitutes approximately one-third the weight of annual plants and one-half the weight of perennial plants (see Cellulose). It is a high molecular weight, stereoregular, linear polymer of repeating β-D-glucopyranose units. Owing to its availability from rich, continuously replenished sources and its stereoregularity, cellulose is a relatively low-cost polymer with valuable physical and chemical properties. Its largest use, approximately 45×10^6 metric tons annually, is as wood fiber in the manufacture of paper and paper products, including paper board (see Paper). Over 400,000 t are used annually in textile fibers. In spite of the increasing use of synthetic fibers, cellulose rayons (see Rayon) and cotton (qv) account for over 70% of textile production.

High quality cellulose is measured for its content of alpha-cellulose, the portion

Table 1. Important Gums of Commerce

Name	Source	U.S. production (thousands of metric tons)
corn starch	extracted from corn	5500
agar	extracted from algae, *Gelidium sp.*	0.32
arabic	exuded from the Acacia tree	14.0
alginate	extracted from brown seaweed, *Macrocystic pyrifera*	5.4
carrageenan	extracted from red seaweed, *Chondrus crispus*	4.2
furcellan	extracted from the *Furcellaria fastigiata*	0.14
ghatti	exuded from the tree *Anogeissus latifolia*	5.0
guar	milled endosperm from guar seed	27–29.5
karaya	exuded of the *Sterculia urens* tree	3.6
locust bean	milled endosperm from the locust bean seed	5.4
pectin	extracted from citrus fruit peel, lime, lemon	5.4
methyl cellulose	derivative of cotton linter or wood pulp methylated cellulose	25.0
carboxymethyl cellulose	derivative of cotton linter or wood pulp	51.0
tragacanth	exuded of the *Astragalus* tree	0.68
xanthan	fermentation product of *Xanthomonas compestris*	3.6

insoluble in 18% alkali. Beta-cellulose is the portion that dissolves in 18% alkali but precipitates when the solution is neutralized. Gamma-cellulose, which remains soluble after neutralization of the 18% alkali, usually consists of molecules with 10 or fewer D-glucopyranosyl units.

The purest natural cellulose is the cotton fiber which (on a dry basis) consists of about 98% cellulose, 1% protein, 0.45% wax, 0.65% pectic substance, and 0.15% mineral matter. Cotton that has been dewaxed by sequential extraction with solvents such as chloroform and ethanol, followed by boiling in oxygen-free alkali (kier boiling), has been historically taken as a standard for purity (13). However, as techniques for isolation and purification have improved, it has been recognized that chemical celluloses derived from wood and other sources can be fully equivalent to cotton cellulose in purity and performance.

Cellulose is a uniform chain of β-D-glycopyranosyl units joined uniformly by (1 → 4)-links. Because of the uniform nature and linear structure of the molecules, they can fit together over at least part of their length to form crystalline regions that give plant cell walls their great strength and rigidity. Structure (14) gives an example of cellulose showing a cellobiose disaccharide unit.

(14)

Cellulose can be swelled in alkaline solution or in a great variety of salt solutions and mixtures of organic reagents. In alkaline solution, swelling begins rapidly and passes through a maximum as the concentration of sodium hydroxide solution increases. At 25°C, maximum swelling occurs in a sodium hydroxide concentration range of 8.5–12%, depending on the source and history of the preparation of cellulose. At lower temperatures, lower alkali concentrations are required to effect maximum swelling. Different alkalis have different swelling powers, and the concentration at which maximum swelling occurs also varies. For example, at 25°C a maximum swelling of ramie cellulose occurs in 8.5–9% lithium hydroxide solution, in 11–12% sodium hydroxide solution, and in 15–17% potassium hydroxide solution.

The process of swelling, when carried to the extreme, results in disintegration of the well-ordered regions, and leads to complete solution. Many solvents for cellulose are based upon metal complex formation. One of these solvents which was widely used industrially, is the coordination compound $Cu(NH_3)_4(OH)_2$ in aqueous ammonia. This reagent (cuam), discovered by Schweitzer in the late 1800s, enjoyed extensive industrial success in the manufacture of industrial rayon. Cellulose dissolved in cuam was spun into an acid precipitation bath. Dissolution of the cellulose resulted from the ability of the hydroxyl groups to act in alkaline medium as ligands toward cupric ion. The exact structure of the complex is yet to be elucidated; the molar ratio of copper to anhydro–glucose units is found to be 1:1 (14). Currently, cellulose is also dissolved in other alkaline media through formation of metal ligands. Such solutions are widely used for viscosity measurements. Among these are copper complexes with ethylenediamine (cuen), and the newer solvent cadoxen, an aqueous solution of cadmium ions and ethylenediamine (15). Cadoxen is colorless; the cuen solutions are deep blue. Another solvent contains iron–sodium tartrate complex. Other solvents contain cobalt, nickel, and zinc.

Some tetraalkylammonium bases are strong swelling agents for cellulose. Of these dibenzyldimethylammonium hydroxide, $(CH_3)_2(C_6H_5CH_2)_2NOH$, actually dissolves cellulose.

Cellulose may also be dissolved in concentrated solutions of strong mineral acids, eg, 72% sulfuric acid, 40% hydrochloric acid, or 85% phosphoric acid. It undergoes rapid hydrolysis in mineral acid solutions at room temperature, but the hydrolysis is slower at low temperatures. Concentrated aqueous solutions of some salts, such as 75% zinc chloride, swell cellulose and may dissolve low molecular weight portions. The most modern solvents for cellulose, and those now recommended as possible solvents for cellulose to be spun into rayon, are a mixture of dinitrogen tetroxide and dimethylformamide or a mixture of paraformaldehyde and dimethyl sulfoxide. With the former, cellulose dissolves first as the nitrite ester and is regenerated by acid, and, with the latter, as the hydroxylmethyl ether or hemiacetal and is regenerated by alkali, but particularly by ammonia.

Derivatives. Since all D-glucose units in the chain, except a single unit at each end, have hydroxyl groups available at carbons C-2, C-3, and C-6, cellulose is trifunctional in alcohol groups that may be etherified or esterified (see Cellulose derivatives). When all the hydroxyl groups are substituted, the cellulose is said to have a degree of substitution of 3 (DS = 3); if only two hydroxyls are substituted, the cellulose has a DS of 2. A commercial cellulose acetate that has a DS of 2.3 has had the number of acetyl groups determined and the average number of groups per glucose unit in the molecule calculated as 2.3.

Cellulose xanthate [9032-37-5] made by the reaction of alkaline cellulose with carbon disulfide, is a soluble derivative of cellulose. Commercially, this viscose is spun into an acid bath where the cellulose xanthate decomposes to regenerate cellulose and carbon disulfide, thereby reforming cellulose into fibers or cellophane sheets (see Rayon).

Cellulose acetate [9004-35-7] (qv) is prepared commercially by the reaction of cellulose with acetic anhydride, acetic acid , and sulfuric acid. Cellulose acetate is spun into fibers by dissolving it in acetone and spinning the solution into a column of warm air which evaporates the acetone from the forming fiber. Cellulose acetate is also shaped into a variety of other plastic products, and its solutions are used as coating dopes. Cellulose acetate butyrate [9004-36-8], is made from cellulose, acetic anhydride, and butyric anhydride in the presence of sulfuric acid. Such products form high shock-resistant plastics.

Cellulose trinitrate [9046-47-3], made from cellulose and a mixture of nitric and sulfuric acids, is called gun cotton, and is used in explosives. Nitrates of lower DS find some application as coatings or adhesives.

Cellulose sulfate [9032-43-3], made by the reaction of cellulose with chlorosulfonic acid and dimethylformamide or with sulfur trioxide and methyl sulfoxide, is presently being test marketed (14) as an antistatic agent, photographic film additive, and solution thickener.

Carboxymethyl cellulose [9000-11-7] (CMC) is made in large quantity. Material of low DS is used in washing powders; it is adsorbed on fabrics during washing, thereby giving them a negative charge and preventing soil particles from redepositing (see Surfactants). Material of higher DS is used as water-soluble thickeners and dispersants (see Dispersants). Many other water soluble derivatives of cellulose are produced, among them methyl cellulose and hydroxyethyl cellulose.

In recent years, efforts have been made to modify the properties of cellulose by attaching (grafting) chains of synthetic polymers (see Copolymers). Usually a cellulose is mixed with a monomer, and chain growth is initiated in the mixture by additives that decompose into free radicals, radiation that produces free radicals, or by redox agents. The most successful of the latter have been ceric ions, which abstract hydrogen atoms from the hydroxyl groups of cellulose molecules, producing free radical sites (16). Initiation of the chain reaction leads to attachment of monomers that continue to grow into long-chain polymer molecules.

Hydrolysis. Cellulose molecules are hydrolyzed under acidic conditions with rapid depolymerization occurring in strong mineral acids. The β-D-glycopyranosidic bonds are as susceptible to hydrolysis as they are in simple glycosides. Hydrolysis proceeds by protonation of the glucosidic oxygen, followed by either by a displacement on C-1 through attack by a hydroxyl ion, or by simple cleavage of the protonated glucosidic bond to give a carbonium ion at C-1. The carbonium ion takes up hydroxyl ion from the aqueous medium to become a reducing end of the chain fragment.

Heterogeneous acid hydrolysis results in a rapid reduction in tensile strength and viscosity because of preferential attacks on those cellulose chains in the more accessible amorphous regions. Following this initial rapid reaction, the chain length approaches a leveling off or limiting value. Depending on the history of the sample, usually the leveling off value is approximately 250 D-glucose units per chain. Comprehensive reviews on acid hydrolysis are available (17).

Recently, efforts have been made to hydrolyze cellulose enzymatically with microbial cellulases. Such hydrolysis to D-glucose, if economically practical, could represent a new, large volume method of processing cellulose from woods or annual plants (see Enzymes).

The β-D-glucosidic bonds of cellulose are quite resistant to alkaline degradation. However, chains of cellulose are gradually shortened in an endwise degradation that

548   CARBOHYDRATES

proceeds by a β-elimination mechanism (18). If the cellulose has been oxidized at any point so that a carbonyl or carboxyl function is present, alkaline degradation will also proceed from there with chain rupture. Under more severe alkaline conditions, such as in the presence of one normal alkali at 170°C, hydrolysis of the glucosidic linkages occurs (19).

Oxidation. Many oxidants attack the cellulose chain to produce chain cleavage or to insert carbonyl functions. One such oxidant previously discussed is the periodate ion which attacks vicinal hydroxyl groups. This specific type of oxidation is also effected by lead tetraacetate in nonaqueous medium. Nonspecific oxidations are induced by permanganate, chromic acid, and hypochlorites. The hypochlorites are the most common bleaching agents industrially used. They cause a random attack on the polymer or molecule over most of the pH range (see Bleaching agents; Pulp).

Starches

Starches are widely distributed throughout the plant world as reserve polysaccharides. They are stored principally in seeds, fruits, tubers, roots, and stem pith (see Starch). They usually occur as discrete particles or granules of 2–150 μm in diameter. The physical appearance and properties of granules vary widely from one plant to another, and may be used to classify original starches. Some granules are round, some elliptical and some polygonal; many have a spot termed the hilum which is the intersection of two or more lines or creases. The granules are anisotropic and show strong birefringence. Two dark extinction lines extend from edge to edge of the granule intersecting at the hilum. Some granules show a series of striations arranged concentrically around the hilum.

Although starches hydrolyze only to D-glucose, they are not single substances. Except in very rare instances, they are mixtures of two structurally different glucans. Most starches contain 22–26% amylose and 74–78% amylopectin. Amylose, is a linear polymer (**15**) of D-glucose units joined by (1 → 4)-α-D-links, and the component amylopectin, is a bush-shaped structure (**16**) of (1 → 4) linked α-D-glucosyl units. (1 → 6)-α-D-Links occur at the branch points, about every 25–27 sugar units. Amylose can be selectively precipitated from a hot starch dispersion by the addition of butanol, fatty acids, various phenols, and nitro compounds such as nitropropane and nitrobenzene. On cooling slowly, the amylose combines with the fractionating agent to form a complex that separates as microscopic crystals. These complexes break down when they are dissolved in hot water or extracted with ethanol.

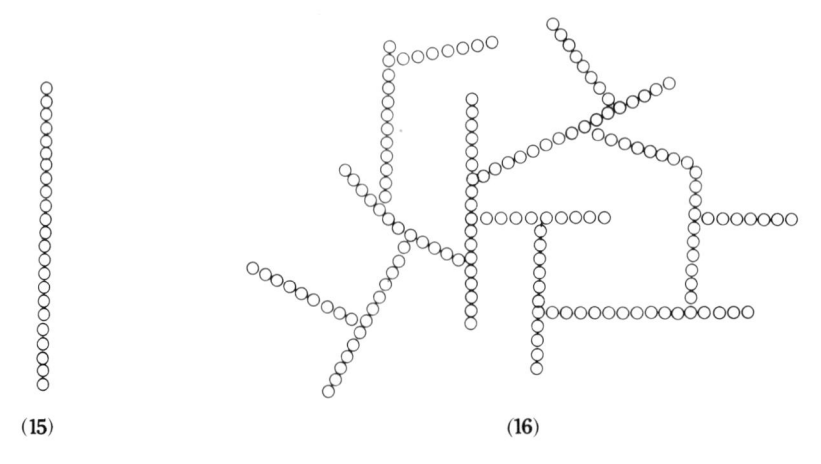

(15)   (16)

Corn starch is the most important of starches manufactured in the United States. Approximately 10.6×10^6 m^3 (300 million bushels) of corn are processed annually. Most waxy starches are entirely free of amylose and, therefore, consist solely of branched amylopectin molecules.

Starch of excellent quality can be prepared from white potatoes. Commercial wheat starch is usually prepared in small quantity in the United States but in large quantities elsewhere. Starches from other sources (cassava, tapioca, etc) are prepared in a manner similar to that for corn or potato starch.

X-ray diffraction analysis reveals several patterns for the various starches. Native cereal starches occur as the A form and tuber starches as the B form. Intermediates are designated as C types. In general, the B type is obtained by evaporation of pastes at room temperature, or by precipitation, by freezing or retrogradation; the C type is produced by evaporation of pastes at higher temperatures, and the A type at temperatures as high as 80–90°C. Precipitation of starch pastes by alcohols or some other precipitating agents gives a V type of starch with a more symmetrical arrangement of molecules.

One of the most important properties of starch granules is their behavior on heating with water. Water is at first slowly and reversibly taken up and limited swelling occurs without any perceptible changes in viscosity and birefringence. At a temperature characteristic for the type of starch, the granules undergo an irreversible rapid swelling, losing their birefringence, and with rapid increase in the viscosity of the suspension. Finally, at higher temperatures starch diffuses from some granules and others are ruptured leaving formless sacs. Swelling can be induced at room temperature by numerous chemicals such as formamide, formic acid, chloral, strong bases, and metallic salts. Sodium sulfate and to a lesser degree sodium chloride impede gelatinization.

On standing, an aqueous starch solution becomes opalescent and finally undergoes precipitation, known as retrogradation, to give a starch with a B x-ray pattern. Amylose molecules precipitate more readily than do amylopectin molecules. Generally, retrogradation proceeds faster as the starch concentration is increased and as the temperature is decreased toward 0°C.

Starch solutions have a high positive optical rotation because of the presence of α-D-glucosidic linkages. The values range from +180 to +220° for aqueous alkaline solutions. Osmotic pressure measurements on acetylated amyloses and amylopectins from several starches lead to molecular weights of the unacetylated components of 100,000–210,000 for amyloses and 1,000,000–6,000,000 for amylopectins.

Starches behave as polyhydroxy alcohols and, in the presence of an impelling agent, are capable of ether formation with alkyl and acyl halides and alkyl sulfates, and of ester formation with both inorganic and organic acids. For the preparation of acyl derivatives, the impelling agent such as pyridine or an acid anhydride absorbs the water formed during acylation. Alkalies, especially sodium hydroxide, are normally used as impelling agents for etherification. The fully methylated ethers of amylose and amylopectin have been used extensively in structural determinations.

For many industrial applications the properties of natural starches are changed by various treatments. These include the action of enzymes, acids or oxidizing agents on an aqueous suspension of the starch, or by heating essentially dry starch with or without small quantities of acids or alkalies. "Thin-boiling" starches are made by treating starches with acid at temperatures below the gelatinization point. Oxidized starches are usually prepared by the action of hypochlorite or peroxide. In the former case sodium or calcium hypochlorite is added to a slightly alkaline starch slurry, and the reaction allowed to take place at 30–50°C until the desired degree of oxidation is reached. Excess oxidizing agent is neutralized by the addition of sodium bisulfite. The oxidized starch retains its granular structure, is colored by iodine, gives suspensions of greater clarity than the parent starch, and has a shorter cooking time, lower viscosity, increased adhesiveness, and a lower rate of congealing. The major use of oxidized starches is in textiles and paper coatings.

British gums or Torrefection dextrins are prepared by high temperature roasting of starches. This treatment causes cross-linking and an increase in end groups. Addition of a small amount of alkali or acid prior to heating causes a more rapid change and yields products of a different nature. The major use of these starches is in adhesive applications for packages.

Hemicelluloses

Hemicelluloses are a large group of well-characterized polysaccharides found in the primary and secondary cell walls of all land and fresh water plants, and in some seaweeds. Hemicelluloses are made up of a relatively limited number of sugar residues, the principal ones are D-xylose, D-mannose, D-glucose, D-galactose, L-arabinose, D-glucuronic acid [6556-12-3], 4-O-methyl-D-glucuronic acid [4120-73-4], D-galacturonic acid [685-73-4], and to a lesser extent, L-rhamnose, L-fucose, and various O-methylated neutral sugars. Annual plants and woods contain about 20–35% hemicelluloses. A number of hemicelluloses are neutral molecules, but by far the most abundant have a backbone of (1 → 4)-linked β-D-xylopyranosyl units. The chain may be linear but is often branched and usually has other glycosidically bound sugar units. Some xylan chains have D-glucopyranosyluronic acid [26635-70-1] units attached, but the most important acidic hemicelluloses are O-acetyl-4-O-methyl-D-glucuronoxylans and L-arabino(4-O-methyl-D-glucurono)xylans. In the former polymers, which are the preponderant hemicellulose of woody angiosperms, the 4-O-methyl-D-glucopyranosyluronic acid [18462-68-5] units are joined to D-xylopyranose [2460-44-8] chain units by α-D-(1 → 2) linkages. High yields of 2-O-(4-O-methyl-α-D-glucopyranosyluronic acid)-D-xylose [57-86-6] are always obtained on extended acid hydrolysis. This aldobiouronic acid is quite stable to acid hydrolysis. The number of 4-O-methyl-D-glucuronic acid groups along the chain varies considerably. Most hardwood xylans have approximately one acid side chain per ten D-xylose units. The distribution of the groups along the xylan chain is not fully known, but the groups do not occur on adjacent D-xylose units (see Table 2).

Although the D-glucuronic acid [6556-12-3] and 4-O-methyl-D-glucuronic acid are most often linked to position C-2 of the D-xylose units, linkage to position C-3 has also been observed. Hemicellulose of sunflower heads yields 3-O-(α-D-glucopyranosyluronic acid)-D-xylose [6634-88-6] (20), and the hemicellulose of Monterey pine (*Pinus radiata*) and possibly those of maritime pine (*Pinus pinaster*) (21), and wheat straw (22) yield 3-O-(4-O-methyl-α-D-glucopyranosyluronic acid)-D-xylose [7382-52-7]. Jute (39) has been reported to yield a similar aldobiouronic acid, except that the methyl group is at position C-3. A hemicellulose from plum leaf gives 2-O-methyl-D-xylose [7434-28-8] on hydrolysis (40) the first known methylated aldopentose in nature.

Acetyl groups occur to the extent of 3–17% of wood and are at highest content

```
D-Xylp 1 ⟶ 4  D-Xylp 1 ⟶ 4  D-Xylp 1 ⟶ 4  D-Xylp 1 ⟶ 4  D-Xylp
                2                             3
                ↑                             ↑
                1                             1
        4-O-CH₃-D-GlcpUA                   L-Araf
```

Figure 7. Section of oat straw hemicellulose.

```
D-Xylp 1 ⟶ 4  D-Xylp 1 ⟶ 4  D-Xylp 1 ⟶ 4  D-Xylp 1 ⟶ 4  D-Xylp
                2
                ↑
                1
        4-O-CH₃-D-GlcpUA
```

Figure 8. Section of European beechwood hemicellulose A.

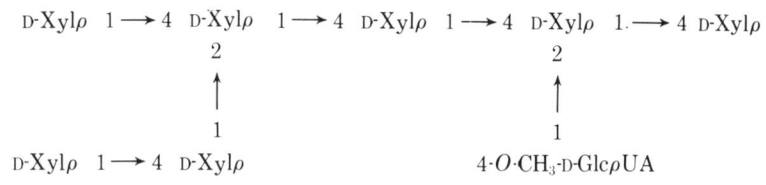

```
D-Xylp 1 ⟶ 4  D-Xylp 1 ⟶ 4  D-Xylp 1 ⟶ 4  D-Xylp 1 ⟶ 4  D-Xylp
                2                             2
                ↑                             ↑
                1                             1
D-Xylp 1 ⟶ 4  D-Xylp                   4-O-CH₃-D-GlcpUA
```

Figure 9. Section of American beechwood hemicellulose.

in hardwoods. Dimethyl sulfoxide extraction of angiosperm woods yields hemicelluloses with 16.9% acetate groups corresponding on the average to 7.1 ester groups per ten D-xylose units (41). Most of the O-acetyl groups are attached to C-3 and the remainder to C-2 of the D-xylose residues. These acetylated hemicelluloses are soluble in water and in solvents such as dimethyl sulfoxide, formamide, and N,N-dimethylformamide.

The acidic hemicelluloses of hardwoods are apparently devoid of L-arabinose.

Hemicelluloses are partially extractable with water from their natural cell wall sites, but they are usually removed by extraction with alkaline solutions. To prevent contamination with the coextractable lignin, the plant material is usually defatted with ethanol–benzene (azeotropic mixture) and delignified by conversion to holocellulose (42). The alkaline extract may be neutralized to precipitate the more linear and less acidic hemicelluloses, called the A portion. The more acidic or branched remainder, called the B portion, is precipitated with ethanol. Further fractionation and purification can be accomplished by fractional precipitation with ethanol, fractionation with cetyltrimethylammonium halide (42), fractionation with alkaline copper salts, by electrophoresis, or by several other procedures. Dimethyl sulfoxide extracts hemicelluloses without saponification of any acetate groups.

L-Arabino-(4-O-methyl-D-glucurono)xylans are found in softwoods and annual plants. The L-arabinose units are most often in furanose ring forms, although in some instances the more stable pyranose ring is present. Often the L-arabinosyl units occur as single-unit side chains, but sometimes they form branches several sugar units in length. Such branches may terminate with a 4-O-methyl-D-glucopyranosyluronic acid unit or perhaps even with a D-xylopyranosyl unit.

Table 2. Distribution of Groups Along the Xylan Chain

Name	Source	Comments
L-arabino-(4-O-methyl-D-glucourono)xylans	oat straw	linear chain of 45–50 (1 → 4) linked β-D-xylopyranose (D-Xylp) units; one 4-O-methyl-D-glucopyranosyluronic (4-O-Me-GlcpUA) acid at C-2; and one L-arabinoformosyl (L-Ara$f$) unit per 32 D-xylose units[a]
	flax straw	obtained by alkali extraction from delignified straw; approx 135 unit xylan chain with approx 15 4-O-Me-GlupUA and two L-rhamnose, one terminal[b]
	maize seed coat	obtained by lime water extraction[c]; highly branched xylan chain with 3-O-α-D-xylopyranosyl-L-arabinose [51755-05-6] and L-galactopyranosyl-(1 → 4)-D-xylopyranosyl-(1 → 2)-L-arabinose [23259-90-7] side chains; D- and L-galactose [59-23-4] [2438-80-4] at chain ends (nonreducing); GlcpUA at C-2 of D-xylose
	corn (maize) cobs	alkaline extraction from holocellulose followed by neutralization yields of 75% linear xylan with perhaps one GlcpUA per molecule; the polymer homologous series from xylobiose [6860-47-5] to xyloheptaose [20197-43-7] was first isolated from this xylan[d]
	European beechwood (*Fagus sylvatica*)	linear chain of about 70 (1 → 4) linked D-Xylp with one 4-O-Me-GlcpUA at C-2 of every[e] tenth unit (95% 4-O-Me-GlcpUA)
	American beechwood (*F. grandifolia*)	water soluble extracted from chlorite holocellulose[f] approx 45 D-Xylp units in Y shaped chain; branch at C-2 and five single unit 4-O-Me-α-D-GlcpUA (1 → 2) linked side chains
	Norway Spruce (*Picea abies*)	alkali extraction from hypochlorite holocellulose[g] 80–85 unit xylan chain with one 4-O-Me-GlcpUA at C-2 of every fifth unit; similar to beechwood xylan

D-gluco-D-mannans	cell walls of hard and soft woods	linear chains β-$(1 \rightarrow 4)$ linked units[h] in random orders; hardwood hemicelluloses are pure copolymers of D-glucose and D-mannose; softwoods have 1:3 glucose–mannose ratio and some D-galactose as single unit side chains at C-6 of mannose and possible additional branching at C-3 of some glucose units
D-galacto-D-gluco-D-mannans		by extraction with potassium hydroxide or dimethyl sulfoxide and precipitation as barium complex 1:1:3, D-galactose; D-glucose; D-mannose; main chain linear β-$(1 \rightarrow 4)$ linked D-glucose and D-mannose in random order with single side chain units of D-galactose at C-6 of both main chain units[i]
L-arabino-D-galactans	conifers (esp. Laris) and sap of sugar maple	water soluble, highly branched chain of L-arabinose and D-galactose[j]; extracted with hot water and precipitated with alcohol

[a] Ref. 23, also see, Figure 7, p. 551.
[b] Ref. 24.
[c] Refs. 25–29.
[d] Ref. 30.
[e] Ref. 31, also see Figure 8, p. 551.
[f] Ref. 32, also see Figure 9, p. 551.
[g] Ref. 33.
[h] Refs. 34–36.
[i] Ref. 37.
[j] Ref. 38.

554 CARBOHYDRATES

To give some idea of the structure of hemicelluloses, a few are described in Table 2. It should be kept in mind that many hemicelluloses of similar structure have been found in the cell walls of numerous plants.

Another subgroup of hemicellulose is composed of those containing large amounts of D-mannopyranosyl units or those containing large amounts of D-galactopyranosyl units. Both are found in the cell walls of wood. D-Gluco-D-mannans comprise from 3–5% of the wood of angiosperms and from 3–12% of the wood of gymnosperms where they occur in the cell wall in close association with cellulose and D-xylans. Similar polysaccharides are also found in the tubers of *Amorphophallus* species and in the bulbs and seeds of several plants where they may function as food reserve material. Their removal from the plant material is effected by extraction with potassium hydroxide solution, usually in the presence of borate, which complexes with the 2,3-*cis*-hydroxyl groups rendering the polysaccharide more soluble (43).

Although insoluble in water in the native state, the D-gluco-D-mannans are frequently water soluble after isolation.

Uses. Hemicelluloses have many properties similar to the exudate gums. Because of the great abundance of hemicelluloses and the continued price escalation of exudate gums, forced by increases in labor rates for harvesting, hemicelluloses are expected to enter the industrial market in growing volume.

BIBLIOGRAPHY

"Carbohydrates" in *ECT* 1st ed., Vol. 2, pp. 867–881, by Charles D. Hurd, Northwestern University; "Carbohydrates" in *ECT* 2nd ed., Vol. 4, pp. 132–148, by Charles D. Hurd, Northwestern University.

1. R. L. Whistler and W. M. Corbett in G. L. Clark and G. G. Hawley, eds., *The Encyclopedia of Chemistry,* Reinhold Publishing Corp., New York, 1966, pp. 161–164.
2. R. L. Whistler and co-workers *Adv. Carbohyd. Chem. Biochem.* **32,** 235 (1976).
3. C. S. Hudson, *Adv. Carbohydr. Chem.* **3,** 12 (1948).
4. C. S. Hudson, *J. Chem. Ed.* **18,** 353 (1941).
5. W. N. Haworth, *The Constitution of Sugars,* Edward Arnold and Co., London, 1928.
6. I. Danishefsky and R. L. Whistler in W. Pigman and D. Horton, eds., *The Carbohydrates,* Vol. II A, Academic Press, New York, 1970, p. 377.
7. J. L. Hickson in ref. 1, p. 1020.
8. P. X. Hoynak in A. H. Johnson and M. S. Peterson, eds., *Encyclopedia of Food Technology,* The AVI Publishing Company, Inc., Westport, Conn., 1974, p. 871.
9. W. R. Junk and H. M. Pancoast, *Handbook of Sugars,* The AVI Publishing Company, Inc., Westport, Conn., 1973, p. 19.
10. Ref. 8, p. 872.
11. Ref. 9, p. 53.
12. R. L. Whistler in ref. 8, p. 155.
13. W. M. Corbett in R. L. Whistler, ed., *Methods in Carbohydrate Chemistry,* Vol. III, Academic Press, New York, 1968, p. 3.
14. R. L. Whistler and co-workers, *Arch. Biochem. Biophys.* **121**(2), 59 (1966).
15. G. Jayme, and K. Neuschaffer, *Naturwissenschaften* **44,** 62 (1957).
16. R. J. E Cumberbirch and V. R. Holker, *J. Soc. Dyer Colour.* **82**(2), 59 (1966).
17. M. S. Feather and J. F. Harris, *J. Org. Chem.* **30,** 153 (1965).
18. R. L. Whistler and J. N. BeMiller in M. L. Wolfrom, ed., *Advances in Carbohydrate Chemistry,* Vol. 13, Academic Press, New York, 1958, p. 289.
19. E. Dryselius, B. Lindberg, and O. Theander, *Acta. Chem. Scand.* **12,** 340 (1958).
20. C. T. Bishop, *Can. J. Chem.* **33,** 1521 (1955).
21. A. Roudier and L. Eberhard, *Tappi* **38**(9), 156A (1955).
22. A. Rondier, *Assoc. Tech. Ind. Papet. Bull.* **2,** 53 (1954).
23. G. O. Aspinall and K. C. B. Wilkie, *J. Chem. Soc.,* 1072 (1956).

24. J. D. Geerdes and F. Smith, *J. Am. Chem. Soc.* **77,** 3569 (1955).
25. R. L. Whistler and J. N. BeMiller, *J. Am. Chem. Soc.* **78,** 1163 (1956).
26. R. L. Whistler and W. M. Corbett, *J. Org. Chem.* **21,** 694 (1956).
27. R. L. Whistler and W. M. Corbett, *J. Am. Chem. Soc.* **77,** 6328 (1955).
28. R. Montgomery, F. Smith, and H. C. Srivastava, *J. Am. Chem. Soc.* **79,** 698 (1957).
29. R. Montgomery and F. Smith, *J. Am. Chem. Soc.* **79,** 695 (1957).
30. R. L. Whistler and C. C. Tu, *J. Am. Chem. Soc.* **74,** 3609 (1952).
31. G. O. Aspinnall, E. L. Hirst, and R. S. Mahomed, *J. Chem. Soc.,* 1734 (1954).
32. G. A. Adams, *Can. J. Chem.* **35,** 556 (1957).
33. G. O. Aspinnall and M. E. Carter, *J. Chem. Soc.,* 3744 (1956).
34. G. P. Aspinall, R. Begbie, and J. E. McKay, *J. Chem. Soc.,* 214 (1962).
35. P. Kooiman and G. A. Adams, *Can. J. Chem.* **39,** 889 (1961).
36. H. Meier, *Acta Chem. Scand.* **12,** 1911 (1958).
37. E. C. A. Schwarz and J. E. Timell, *Can. J. Chem.* **41,** 1381 (1963).
38. M. J. Adams and C. Douglas, *Tappi* **46,** 544 (1963).
39. P. C. Das Gupta and P. B. Sarkar, *Text. Res. J.* **24,** 705 (1954).
40. P. Andrews and L. Hough, *Chem. Ind. London,* 1278 (1956).
41. H. O. Bouveng, P. J. Garegg, and B. Lindberg, *Chem. Ind. London,* 1727 (1958).
42. R. L. Whistler and M. L. Wolfrom eds., *Methods in Carbohydrates Chemistry,* Vol. V., Academic Press, New York, 1964.
43. T. E. Timell, *Sven. Papperstidn.* **63,** 472 (1960).

General References

R. L. Whistler, ed., *Industrial Gums,* 2nd ed., Academic Press, Inc., New York, 1973.
K. W. Britt, ed., *Handbook of Pulp and Paper Technology,* 2nd ed., Van Nostrand Reinhold Company, New York, 1970.
G. O. Aspinall, *Polysaccharides,* Pergamon Press, New York, 1970.
L. P. Miller, ed., *Phytochemistry,* Vol. I, Van Nostrand Reinhold Company, New York, 1973.
G. T. Dutton, ed., *Glucuronic Acid, Free and Combined,* Academic Press, New York, 1966.

ROY L. WHISTLER
JOHN R. ZYSK
Purdue University

CARBON

Carbon and artificial graphite, 556
Carbon black, 631
Diamond, natural, 666
Diamond, synthetic, 676
Natural graphite, 689

CARBON AND ARTIFICIAL GRAPHITE

Structure, terminology, and history, 556
Activated carbon, 561
Baked and graphitized carbon, 570
Processing of baked and graphitized carbon, 576
Properties of manufactured graphite, 589
Applications of baked and graphitized carbon, 596
Carbon fibers and fabrics, 622
Other forms of carbon and graphite, 628

STRUCTURE, TERMINOLOGY, AND HISTORY

Carbon [7440-44-0] occurs widely in its elemental form as crystalline or amorphous solids. Coal, lignite, and gilsonite are examples of amorphous forms; graphite and diamond are crystalline forms. Carbon forms chemical bonds with other elements, but is capable of forming compounds in which carbon atoms are bound to carbon atoms and, as such, is the chemical element that is the basis of organic chemistry. Carbon and graphite can be manufactured in a wide variety of products with exceptional electrical, thermal, and physical properties. They are a unique family of materials in that properties of the final product can be controlled by changes in the manufacturing processes and by raw material selection.

Crystallographic Structure

Elemental carbon exists in nature as two crystalline allotropes: diamond [7782-40-3] and graphite [7782-42-5]. The diamond crystal structure (see page 683) is face-centered cubic with interatomic distances of 0.154 nm. Each atom is covalently bonded to four other carbon atoms in the form of a tetrahedron. Diamond is transformed to graphite in the absence of air at temperatures above 1500°C.

The accepted ideal structure for graphite was proposed by Bernal in 1924 (1). This structure is described as infinite layers of atoms of carbon which are arranged in the form of hexagons lying in planes (see page 691). The stacking arrangement is ABAB so that the atoms in alternate planes align with each other. The spacing between the layers is 0.3354 nm, the interatomic distance within the planes 0.1415 nm, and the crystal density 2.266 g/cm^3. A less frequently occurring structure is rhombohedral with a stacking arrangement of ABCABC in which the atoms of every fourth layer align with each other (2). This rhombohedral form, which occurs only in conjunction with the hexagonal form, is less stable, and converts to the hexagonal form at 1300°C.

There are six electrons in the carbon atom with four electrons in the outer shell available for chemical bonding. Three of the four electrons form strong covalent bonds with the adjacent in-plane carbon atoms. The fourth electron forms a less-strong bond of the van der Waals type between the planes. Bond energy between planes is 17 kJ/mol (4 kcal/mol) (3) and within planes 477 kJ/mol (114 kcal/mol) (4). The weak forces between planes account for such properties of graphite as good electrical conductivity, lubricity, and the ability to form interstitial compounds.

Terminology

Although the terms carbon and graphite are frequently used interchangeably in the literature, the two are not synonymous. The terms carbon, formed carbon, manufactured carbon, amorphous carbon, or baked carbon refer to products that result from the process of mixing carbonaceous filler materials such as petroleum coke, carbon blacks, or anthracite coal with binder materials of coal tar or petroleum pitch, forming these mixtures by molding or extrusion, and baking the mixtures in furnaces at temperatures from 800–1400°C. The term carbon–graphite designates a formed, composite product in which the filler materials, in addition to a carbon, such as petroleum coke, contain a graphitized carbon that has been heat-treated to high temperatures. The filler is material that makes up the body of the finished product (see Fillers). Green carbon refers to formed carbonaceous material that has not been baked.

Graphite, also called synthetic or artificial graphite, electrographite, manufactured graphite, or graphitized carbon refers to a carbon product that has been further heat-treated at a temperature exceeding 2400°C. This process of graphitization, described in a later section, changes not only the crystallographic structure, but also the physical and chemical properties.

With the development of nuclear and aerospace technologies, several new forms of carbon and graphite are being commercially produced. Products that are deposited on a heated graphite substrate by vapor phase decomposition of gaseous hydrocarbons, usually methane, at 1800–2300°C, are termed pyrolytic carbons (5). Chemical vapor deposited (CVD) carbon or pyrocarbon are other terms used to designate pyrolytic carbons. Pyrolytic graphite is a product resulting from high temperature annealing, and has a crystallite interlayer spacing similar to that of ideal graphite (6) (see Ablative materials; Nuclear reactors).

The term polymeric carbon is a generic term for products that result when high polymers with some degree of cross-linking are heated in an inert atmosphere. These organic starting materials do not coke; instead, they result in chars that represent a distinct group of materials possessing a graphite ribbon-network structure rather than extensive graphite sheets (7). Vitreous or glassy carbon prepared by carbonization of cellulose and phenolic or polyfurfuryl resins is an example of a polymeric carbon. Carbonization of acrylic fibers to produce carbon fibers is another example.

Because of the diversity of carbon and graphite forms, British, French, and German carbon groups have agreed to cooperate in improving the characterization of carbon solids. The ultimate aim is to prescribe standard primary and secondary methods of characterization as well as making available standard samples (8). In the United States, the ASTM has issued standard definitions of terms relating to manufactured carbon and graphite (9).

There is no standard industry-wide system for designating the various grades of carbon and graphite that are commercially available. Each of the more than 20 manufacturers has its own nomenclature to describe grades, sizes, and shapes available for specific purposes. The *Directory of Graphite Availability* (10) characterizes many of the materials that were available in 1967, but the catalogues and technical literature issued by carbon and graphite manufacturers must be consulted for current grade and property data. The data cited in the following sections are representative average values for commercially available materials.

History of the Industry

Graphite as it is found in nature has been known for several centuries for its use in making clay–graphite crucibles and its lubricating properties. The first known use was for drawing or writing, and it was because of this attribute that the German mineralogist A. G. Werner named graphite after the Greek word *graphein,* which means to write (11).

Manufacture of artificial graphite did not come about until the end of the 19th century. Its manufacture was preceded by developments mainly in the fabrication

Table 1. Applications for Manufactured Carbon and Graphite[a]

Aerospace
  nozzles
  nose cones
  motor cases
  leading edges
  control vanes
  blast tubes
  exit cones
  thermal insulation

Chemical
  heat exchangers and centrifugal pumps
  electrolytic anodes for the production of chlorine, aluminum, and other electrochemical products
  electric furnace electrodes for making elemental phosphorus
  activated carbon
  porous carbon and graphite
  reaction towers and accessories

Electrical
  brushes for electrical motors and generators
  anodes, grids, and baffles for mercury arc power rectifiers
  electronic tube anodes and parts
  telephone equipment products
  rheostat disks and plates
  welding and gouging carbons
  electrodes in fuel cells and batteries
  contacts for circuit breakers and relays
  electric discharge machining

Metallurgical
  electric furnace electrodes for the production of iron and steel, ferroalloys, and nonferrous metals
  furnace linings for blast furnaces, ferroalloy furnaces, and cupolas
  aluminum pot liners and extrusion tables
  run-out troughs for molten iron from blast furnaces and cupolas
  metal fluxing and inoculation tubes for aluminum and ferrous furnaces
  ingot molds for steel, iron, copper, and brass
  extrusion dies for copper and aluminum

Nuclear
  moderators
  reflectors
  thermal columns
  shields
  control rods
  fuel elements

Other
  motion picture projector carbons
  turbine and compressor packing and seal rings
  spectroscopic electrodes and powders for spectrographic analyses
  structural members in applications requiring high strength-to-weight ratios

[a] Ref. 15.

and processing of carbon electrodes. H. Davy is credited with using the first fabricated carbon in his experiments on the electric arc in the early 1800s. During the 19th century, several researchers received patents on various improvements in carbon electrodes. The invention of the dynamo and its application to electric current production in 1876 in Cleveland, Ohio, by C. F. Brush, provided a market for carbon products in the form of arc-carbons for street lighting. The work of a Frenchman, F. Carré, in the late 19th century, established the industrial processes of mixing, forming, and baking necessary for the production of carbon and graphite (12).

A significant development occurred when E. G. Acheson patented an electric resistance furnace capable of reaching approximately 3000°C, the temperature necessary for graphitization (13). This development was the beginning of a new industry where improved carbon and graphite products were used in the production of alkalies, chlorine, aluminum, calcium and silicon carbide, and for electric furnace production of steel and ferroalloys. In 1942 a new application for graphite was found when it was used as a moderator by E. Fermi in the first self-sustaining nuclear chain reaction (14). This nuclear application and subsequent use in the developing aerospace industries opened new fields of research and new markets for carbon and graphite. Carbon and graphite fibers are an example of a new form and a new industry. A list of major applications is shown in Table 1.

BIBLIOGRAPHY

"Active Carbon" under "Carbon" in *ECT* 1st ed., Vol. 2, pp. 881–899, by J. W. Hassler, Nuclear Active Carbon Division, West Virginia Pulp and Paper Company, and J. W. Goetz, Carbide and Carbon Chemicals Corporation; "Arc Carbon" under "Carbon" in *ECT* 1st ed., Vol. 2, pp. 899–915, by W. C. Kalb, National Carbon Company, Inc.; "Baked and Graphitized Products" under "Carbon" in *ECT* 1st ed., Vol. 3, pp. 1–34, by H. W. Abbott, Speer Carbon Company; "Carbon Black" under "Carbon" in *ECT* 1st ed., Vol. 3, pp. 34–65, and Suppl. 1, pp. 130–144, by W. R. Smith, Godfrey L. Cabot, Inc.; "Acetylene Black" under "Carbon" in *ECT* 1st ed., Vol. 3, pp. 66–69, by B. P. Buckley; Shawinigan Chemicals Ltd.; "Diamond" under "Carbon" in *ECT* 1st ed., Vol. 3, pp. 69–80, by C. V. R.; "Lampblack" under "Carbon" in *ECT* 1st ed., Vol. 3, pp. 80–84, by W. J. Colvin and W. E. Drown; Monsanto Chemical Company; "Natural Graphite" under "Carbon" in *ECT* 1st ed., Vol. 3, pp. 84–104, by S. B. Seeley and E. Emendorfer, Joseph Dixon Crucible Co.; "Structural and Specialty Carbon" under "Carbon" in *ECT* 1st ed., Vol. 3, pp. 104–112, by F. J. Vosburgh, National Carbon Company; "Activated Carbon" under "Carbon" in *ECT* 2nd ed., Vol. 4, pp. 149–158, by E. G. Doying, Union Carbide Corporation; "Baked and Graphitized Products, Manufacture" under "Carbon" in *ECT* 2nd ed., Vol. 4, pp. 158–202, by L. M. Liggett, Speer Carbon Company; "Baked and Graphitized Products, Uses" under "Carbon" in *ECT* 2nd ed., Vol. 4, pp. 202–243, by W. M. Gaylord, Union Carbide Corporation; "Carbon Black" under "Carbon" in *ECT* 2nd ed., Vol. 4, pp. 243–282, by W. R. Smith, Cabot Corporation, and D. C. Bean, Shawinigan Chemicals Ltd.; "Diamond, Natural" under "Carbon" in *ECT* 2nd ed., Vol. 4, pp. 283–294, by H. C. Miller, Super-Cut, Inc.; "Diamond, Synthetic" under "Carbon" in *ECT* 2nd ed., Vol. 4, pp. 294–303, by R. H. Wentorf, Jr., General Electric Research Laboratry; "Natural Graphite" under "Carbon" in *ECT* 2nd ed., Vol. 4, pp. 304–335, by S. B. Seeley, The Joseph Dixon Crucible Company.

1. J. D. Bernal, *Proc. R. Soc. (London) Ser. A* **106,** 749 (1924).
2. H. Lipson and A. R. Stokes, *Proc. R. Soc. (London) Ser. A* **181,** 101 (1942).
3. G. J. Dienes, *J. Appl. Phys.* **23,** 1194 (1952).
4. M. A. Kanter, *Phys. Rev.* **107,** 655 (1957).
5. D. B. Fischbach, *Chem. Phys. Carbon* **7,** 28 (1971).
6. A. W. Moore, *Chem. Phys. Carbon* **8,** 71 (1973).
7. G. M. Jenkins and K. Kawamura, *Polymeric Carbons—Carbon Fibre, Glass and Char,* Cambridge University Press, New York, 1976.
8. International Cooperation on Characterisation and Nomenclature of Carbon and Graphite, *Carbon* **13,** 251 (1975).

9. *Standard Definitions of Terms Relating to Manufacturing Carbon and Graphite, ASTM Standard C 709-75,* American Society for Testing and Materials, Philadelphia, Pa.
10. J. Glasser and W. J. Glasser (Chemical and Metallurgical Research, Inc.), *Directory of Graphite Availability, Report AFML-TR-67-113,* 2nd ed., U.S. Air Force Materials Laboratory, 1967.
11. F. Cirkel, *Graphite; its Properties, Occurrence, Refining and Use,* Department of Mines, Montreal, Canada, 1906.
12. F. Jehl, *The Manufacture of Carbons for Electric Lighting and Other Purposes,* "The Electrician" Printing and Publishing Co., Ltd., London, Eng., 1899.
13. U.S. Pat. 568,323 (Sept. 28, 1896), E. G. Acheson.
14. E. Fermi, *Collected Papers of Enrico Fermi,* Vol. 2, University of Chicago Press, Chicago, Ill., 1965.
15. E. L. Piper, *Preprint Number 73-H-14,* Society of Mining Engineers of AIME, 1973.

General References

Periodicals

Carbon, Pergamon Press, New York, 1963.
Tanso (Carbons), Tanso Zairyo Kenkyukai, Tokyo, Japan, 1949.

Conferences

Proceedings of the First (1953) and Second (1955) Conferences on Carbon, Waverly Press, Baltimore, Md., 1956.
Proceedings of the Third Conference (1957) on Carbon, Pergamon Press, New York, 1959.
Proceedings of the Fourth Conference (1959) on Carbon, Pergamon Press, New York, 1961.
Proceedings of the Fifth Conference (1961) on Carbon, Vol. 1, 1962, and Vol. 2, 1963, Pergamon Press, New York.
M. L. Deviney and T. M. O'Grady, *Petroleum Derived Carbons,* American Chemical Society, Washington, D.C., 1976.
Carbon Society of Japan, *Symposium on Carbon, July 20–23, 1964,* Carbon Society of Japan, Tokyo, 1964.
Symposium on Carbonization and Graphitization, 1968, Societé de Chimie Physique, Paris, Fr., 1968.
Industrial Carbon and Graphite, 1st, 1957, Society of Chemical Industry, London, Eng., 1958.
Industrial Carbon and Graphite, 2nd, 1965, Society of Chemical Industry, London, Eng., 1966.
Industrial Carbons and Graphite, 3rd, 1970, Society of Chemical Industry, London, Eng., 1971.
Carbon and Graphite Conference, 4th, 1974, Society of Chemical Industry, London, Eng., 1976.
Carbon '72, 1972, Deutsche Keramische Gesellschaft, Badhonnef, Ger., 1972.
Carbon '76, 1976, Deutsche Keramische Gesellschaft, Badhonnef, Ger., 1976.

Books

L. C. F. Blackman, *Modern Aspects of Graphite Technology,* Academic Press, New York, 1970.
R. L. Bond, *Porous Carbon Solids,* Academic Press, New York, 1967.
H. W. Davidson and co-workers, *Manufactured Carbon,* Pergamon Press, New York, 1968.
Carbones, Par le Groupe Francais d'Etude des Carbones, Masson et Cie, Paris, Fr., 1965.
C. L. Mantell, *Carbon and Graphite Handbook,* 3rd ed., Interscience Publishers, a division of John Wiley & Sons, Inc., New York, 1968.
Nouveau Traite de Chimie Minerale, Vol. 8, Part 1, Masson et Cie, Paris, Fr., 1968.
A. R. Ubbelohde and F. A. Lewis, *Graphite and its Crystal Compounds,* The Clarendon Press, Oxford, 1960.
P. L. Walker, Jr., ed., *Chemistry and Physics of Carbon; a Series of Advances,* Vol. 1, Marcel Dekker, New York.

J. C. LONG
Union Carbide Corporation

ACTIVATED CARBON

Activated carbon, a microcrystalline, nongraphitic form of carbon, has been processed to develop internal porosity. Activated carbons are characterized by a large specific surface area of 300–2500 m^2/g which allows the physical adsorption of gases and vapors from gases and dissolved or dispersed substances from liquids. Commercial grades of activated carbon are designated as either gas-phase or liquid-phase adsorbents. Liquid-phase carbons are generally powdered or granular in form; gas-phase, vapor-adsorbent carbons are hard granules or hard, relatively dust-free pellets. Activated carbons are widely used to remove impurities from liquids and gases and to recover valuable substances from gas streams (see Adsorptive separation).

Physical Properties

Surface area is the most important physical property of activated carbon. For specific applications, the surface area available for adsorption depends upon the molecular size of the adsorbate and the pore diameter of the activated carbon. Generally, liquid-phase carbons are characterized as having a majority of pores of 3 nm diameter and larger (1), whereas most of the pores of gas-phase adsorbents are 3 nm in diameter and smaller. Liquid-phase adsorbents require larger pores because of the need for rapid diffusion in the liquid and because of the large size of many dissolved adsorbates. Methods of testing adsorbency employ substances having a range of molecular sizes. Liquid-phase carbons are usually characterized by phenol, iodine, and molasses numbers. Gas-phase carbons are characterized by carbon tetrachloride and benzene adsorption activities.

The bulk density or apparent density of an activated carbon, together with its specific adsorptive capacity for a given substance, can be used to determine bed capacity in the design of an adsorption system or to determine grades of carbon required for an existing system.

The range of particle sizes of activated carbon is important: the rate of adsorption has been shown to depend inversely upon the particle size, (small particles having the fastest rates); however, in fixed beds pressure drop increases as particle size decreases.

The mechanical strength or hardness and the attrition resistance of the particles are important where pressure drop and carbon losses are a concern.

The kindling point of the carbon must be high enough to prevent excessive carbon oxidation in gas-phase adsorption where high heats of adsorption, particularly of ketones, are involved.

Chemical Properties

The most important chemical properties of activated carbon are the ash content, ash composition, and pH of the carbon. Discrepancies between the expected performance of an activated carbon, based upon surface area and pore-size distribution data, and actual adsorptive capacity can often be explained by oxygen-containing groups on the surfaces of the carbon. The pH or pK_a of the carbon, as a measure of surface acidity or basicity of the oxygen-containing groups, assists in predicting hydrophilicity and anionic or cationic adsorptive preferences of the carbon (2–5).

Manufacture and Processing

Almost any carbonaceous material of animal, plant, or mineral origin can be converted to activated carbon if properly treated. Activated carbon has been prepared from the blood, flesh, and bones of animals; it has been made from materials of plant origin, such as hardwood and softwood, corncobs, kelp, coffee beans, rice hulls, fruit pits, nutshells, and wastes such as bagasse and lignin. Activated carbon has also been made from peat, lignite, soft and hard coals, tars and pitches, asphalt, petroleum residues, and carbon black. However, for economic reasons, lignite, coal, bones, wood, peat, and paper mill waste (lignin) are most often used for the manufacture of liquid-phase or decolorizing carbons, and coconut shells, coal, and petroleum residues are used for the manufacture of gas-adsorbent carbons.

Activation of the raw material is accomplished by two basic processes, depending upon the starting material and whether a low or high density, powdered or granular carbon is desired: (1) Chemical activation depends upon the action of inorganic chemical compounds, either naturally present or added to the raw material, to degrade or dehydrate the organic molecules during carbonization or calcination. (2) Gas activation depends upon selective oxidation of the carbonaceous matter with air at low temperature, or steam, carbon dioxide, or flue gas at high temperature. The oxidation is usually preceded by a primary carbonization of the raw material.

Decolorizing carbons are coal- and lignite-based granules, or light, fluffy powders derived from low density starting materials such as sawdust or peat. Many decolorizing carbons are prepared by chemical activation. Some raw materials, such as bones, contain inorganic salts that impart some degree of activity to the carbon when the raw material is simply carbonized or heated in an inert atmosphere (2–3). Decolorizing carbons are usually prepared by admixing or impregnating the raw material with chemicals that yield oxidizing gases when heated or that degrade the organic molecules by dehydration. Compounds used successfully are alkali metal hydroxides, carbonates, sulfides, and sulfates; alkaline earth carbonates, chlorides, sulfates, and phosphates; zinc chloride; sulfuric acid; and phosphoric acid (2).

Gas- and vapor-adsorbing carbons may be prepared by the chemical activation process by using sawdust or peat as raw material and phosphoric acid, zinc chloride, potassium sulfide, or potassium thiocyanate as the activator (2–3). In some cases, the chemically activated carbon is given a second activation with steam to impart physical properties not developed by chemical activation.

Processes involving selective oxidation of the raw material with air or gases are also used to make both decolorizing- and gas-adsorbing carbons. In both instances, the raw material is activated in granular form. The raw material is carbonized first at 400–500°C to eliminate the bulk of the volatile matter and then oxidized with gas at 800–1000°C to develop the porosity and surface area. Some decolorizing carbons are oxidized with air at low temperature; however, because this reaction is exothermic, it is difficult to control and is suitable only for low activity carbon. The high temperature oxidation process with steam, carbon dioxide, or flue gas is endothermic, easier to control, and generally used more often (6–7).

Some gas-adsorbing carbons are made from hard, dense starting materials such as nutshells and fruit pits. These are carbonized, crushed to size, and activated directly to give hard, dense granules of carbon. In other cases, it is advantageous to grind the charcoal, coal, or coke to a powder, form it into briquettes or pellets with a tar or pitch binder, crush to size, calcine at 500–700°C, and then activate with steam or flue gas at 850–950°C. This method gives more easily activated particles because they possess more entry channels or macropores for the oxidizing gases to enter and for the reaction products to leave from the center of the particles.

The production of activated carbon from granular coal (8) as well as the activation of briquetted coal (9–10) has been described. In commercial plants, the raw material is carbonized in horizontal tunnel kilns, vertical retorts, or horizontal rotary kilns. Activation is accomplished in continuous internally or externally fired rotary retorts, or in large, cylindrical, multiple hearth furnaces where the charge is stirred and moved from one hearth to the next lower one by rotating rabble arms, or in large vertical retorts where the charge cascades over triangular ceramic forms as it moves downward through the furnace.

Fibrous activated carbon has been made by carbonizing infusible, cured phenolic fibers, air oxidizing, and then steam activating the fibers. The activated carbon fibers of 300–1000 m^2/g surface area are relatively strong and flexible with tensile strengths of at least 103 MPa (ca 15,000 psi). The manufacture of the activated fibers and of activated quilted fabrics has been reported (11).

Antipollution laws have increased the sales of activated carbon for control of air and water pollution (see Air pollution control methods). Pollution control regulations also affect the manufacture of activated carbons. Chemical activating agents have a large potential for the emission of corrosive acid gases during the activation process. If the emissions from chemical activation processes cannot be economically controlled, alternative selective oxidation methods will dominate. Selection of raw materials favoring low-sulfur materials will depend upon the local standards for permissible emission levels of sulfur oxides.

Economic Aspects

Activated carbon usage in the United States in 1976 was ca 90,000 metric tons valued at 9×10^7 dollars, or $0.99/kg. Of the total activated carbon usage, 45,000 t was powdered carbon at an average price of $0.62/kg and 45,000 t was granular carbon at an average price of $1.32/kg (12–13). The capacities of activated carbon production of United States companies in 1976 are shown below (14–16):

Company	Location	Capacity, metric tons
Westvaco Corp.	Covington, Ky.	38,500
ICI United States	Marshall, Tex.	34,000
Calgon Corp.	Catlettsburg, Ky.	39,000
	Pittsburgh, Pa.	
Husky Industries	Romeo, Fla.	10,000
Barnebey-Cheney Co.	Columbus, Ohio	5,500
Union Carbide	Fostoria, Ohio	2,250
Witco Chemical	Petrolia, Pa.	2,250
Total		131,500

The 1976 production of gas- and vapor-adsorbent carbon in the United States is estimated to be 14,000 t. Gas-phase carbon consumption is much lower than liquid-phase carbon consumption, as there are fewer large volume applications and, in most applications, the carbon is regenerated repeatedly, remaining in service for several years. Prices were $1.10–4.08/kg with special grades running as high as $5.50/kg.

Since activated carbon manufacture is energy-intensive, prices increase with increases in the costs of fuels used to supply heat for calcination and activation of the carbon.

Regeneration of activated carbons is a major factor in the cost-effectiveness of the use of carbon (13). Liquid treatment costs can be minimized best by using the lowest effective carbon dosage of a carbon that retains its adsorptive capacity and mechanical strength after many thermal regeneration cycles (17).

Gas-phase activated carbons used for recovery of valuable process gases and solvents are responsible for the continued cost-effectiveness of many industrial processes, particularly solvent-based fiber and tape manufacture, dry cleaning, and rotogravure printing. Regeneration methods used are steam, thermal-vacuum, and pressure-swing desorption. Regeneration, pressure drop through carbon beds, carbon life, and system capital cost are the principal economic factors in gas and vapor recovery by activated carbon (18–19).

Specifications

Activated carbon is supplied on the basis of minimum values (eg, bulk density), maximum values (eg, moisture and ash content), and ranges of properties. In many cases, setting a narrow specification range for a property may considerably increase the price of the carbon. Specifications are based upon estimates and tests that indicate those properties of the carbon yielding the lowest cost of use or the maximum return on investment.

Table 1 shows some typical physical properties of activated carbons. These are approximate values; properties vary by grade and manufacturer within each raw material type.

Analytical and Test Methods

The American Society for Testing and Materials (ASTM), Committee D-28 on activated carbon, has defined several common property tests (20). These standard tests are listed in Table 2.

Many industrial activated carbon tests have not yet been standardized by ASTM but are in common use. The following are standard industrial tests: shaking a sample with several different sized steel balls for a standard time period reveals its hardness or strength; the percent hardness reported is based upon the decrease in average

Table 1. Physical Properties of Typical Activated Carbon Grades

Property	Liquid-phase carbon			Gas-phase carbon		
	Lignite base	Wood base	Bituminous coal base	Granular coal	Pelleted coke	Granular coconut
mesh (Tyler)	−100	−100	8–30	−4, +10	−6, +8	−6, +14
(mm)	(0.15)	(0.15)	(2.38–0.59)	(4.76, 1.70)	(3.36, 2.38)	(3.36, 1.18)
CCl_4 activity, % min	30	40	50	60	60	60
iodine no., min	500	700	950	1000	1000	1000
bulk density, g/mL, min	0.48	0.25	0.50	0.50	0.52	0.53
ash, % max	18	7	8	8	2	4

Table 2. ASTM Test Methods for Activated Carbon

Test	ASTM test no.
definition of terms relating to activated carbon	D2652
apparent density of activated carbon	D2854
liquid-phase evaluation of activated carbon	D2355
moisture in activated carbon	D2867
particle size distribution of granular activated carbon	D2862
total ash content of activated carbon	D2866

particle size of the carbon calculated from screen analyses before and after shaking with the steel balls.

Gas adsorptive properties of activated carbon may be statically determined by exposure to the desired substance at constant temperature and humidity levels in a bell jar. Adsorptive capacity is reported as weight percent pickup of the adsorbate by the carbon adsorbent. Gas adsorptive properties can be determined dynamically by blowing a test gas through a carbon bed or tube of a standardized depth and diameter. The equilibrium amount of adsorbate per unit weight of carbon is determined when the carbon bed becomes saturated with adsorbate. The most commonly applied gas adsorption tests are carbon tetrachloride and benzene activity tests. The carbon tetrachloride activity of a sample is its weight percent pickup, at equilibrium, of carbon tetrachloride from a dry air stream saturated with carbon tetrachloride. Benzene activity is based on the same principle, except that the benzene-saturated air stream is usually diluted to 10 vol % saturation (21). Another common gas-phase carbon test is retentivity. Both benzene and carbon tetrachloride retentivities measure the weight percent of adsorbate retained by a carbon sample first saturated with adsorbate and then subjected to air blowing for 6 h (2).

Activated carbon adsorption tests are useful in classifying carbons for general application areas and in manufacturing quality control. Although actual process fluids might appear more useful for testing the adsorptive powers of activated carbon than standard gas or liquid methods, process stream adsorbate and other impurity concentrations often vary within such wide limits that process fluid tests may be less reliable and less convenient than standard gas and liquid tests (2,22).

Storage

Activated carbon packaging includes bags, drums, cartons, railroad cars, and tank trucks. To avoid contamination and possible loss of properties, gas-phase activated carbon should not be exposed to vapors or moisture. Liquid-phase activated carbon may be stored as a water slurry. For storage of specially impregnated grades, activated carbon may be supplied in vapor-barrier drums.

Health and Safety Factors

Although activated carbon is not an easily combustible material (kindling points in pure oxygen are usually above 370°C), stored carbon should be kept away from heat, electricity, and flames. Dry carbon should not be allowed to contact strong oxidizing agents. Electrical controls used on or near activated carbon should be dust-tight and explosion-proof, and electrical motors should be totally enclosed. Foam or a fine water spray can extinguish activated carbon fires. When organic gases or vapors are adsorbed from an air stream, the carbon should be wet, steamed, or purged with inert gas prior to the first adsorption cycle since the initial heat of adsorption, especially for ketones, is sufficiently high to represent a fire hazard.

Use of activated carbon in protective masks does not protect the wearer from an oxygen depleted atmosphere's asphyxiation hazard (23). Also, an asphyxiation hazard exists inside activated carbon vessels because of adsorption of oxygen from the air onto the carbon. Proper ventilation and breathing equipment should be used by persons entering activated carbon vessels. Since flammable solvent vapors may be present, only explosion-proof lamps and spark-proof tools should be used inside the adsorber vessels. Dust-free loading of activated carbon into adsorbers may be accomplished by using special dry-loading equipment or by slurrying the carbon and wet-loading the adsorbers.

Uses

Liquid-Phase Carbon. Approximately 60% of the activated carbon manufactured for liquid-phase applications is used in powdered form. The two principal uses of powdered activated carbon are: (*1*) removal from solution of color, odor, taste, or other objectionable impurities such as those causing foaming or retarding crystallization, and (*2*) concentration or recovery from solution of a solute. Usually liquid treatment with powder is a batch process. The liquid to be treated is mixed with an appropriate amount of carbon predetermined by laboratory tests, heated if necessary to reduce viscosity, and agitated for 10–60 min. When the batch process is complete, the carbon is separated from the liquid by settling or filtration and is either discarded or eluted. The use of powdered carbons has been declining because granular carbons offer advantages in improved handling and reduced need for final filtering. The possibility of regenerating the carbon makes granular carbon more cost effective than powdered carbon. Increased exports to sugar-producing countries in Latin America mask the decline in domestic use of powdered carbon.

Granular carbon can be used in a continuous process where the liquid is slowly percolated through fixed beds of carbon until the carbon becomes saturated with adsorbate. The liquid stream is then diverted to a second adsorber, allowing the carbon in the first to be regenerated by hot gas or by solvent extraction. In applications where modest amounts of carbon are used, carbon saturated with adsorbed impurities is sent back to the manufacturer for regeneration in a multiple-hearth furnace.

Activated carbon is widely used to remove color from sugar (qv) (2). The use of carbon in glucose manufacture removes protein and hydroxymethylfurfural from the syrup, rendering it colorless and stable.

Although there are many chemical, physical, and biological methods of treating water contaminated with industrial and municipal wastes to produce safe and palatable drinking water, none have the potential of activated carbon treatment (24). Most methods require careful control for the effective removal of taste and odor contaminants. Activated carbon is a broad-spectrum agent that effectively removes toxic or biorefractive substances (25). Insecticides, herbicides, chlorinated hydrocarbons, and phenols, typically present in many water supplies, can be reduced to acceptable levels by activated carbon treatment. Powdered activated carbon is most effectively used in smaller water treatment systems where the equipment investment required for granular carbon would be prohibitive. Granular carbon systems may consist of moving beds, beds in series, beds in parallel, or expanded beds (26). Expanded-bed systems require water flow rates sufficient to expand the bed at least to 115% of static bed volume. Removal of organics by expanded-bed systems is comparable to fixed beds without the need for frequent backwashing required by fixed beds (27). Granular water-treatment carbons are thermally regenerated for reuse, thus reducing the cost of treatment.

Industrial and municipal waste water treatment will be an expanding market for liquid-phase activated carbons (14,28). *The 1972 Federal Water Pollution Control Act* expanded the authority of the EPA to conduct research for the improvement of municipal and industrial water treatment (29). Thus government and industry may cooperate through an EPA cost-sharing grant system in the improvement and application of activated carbon for effective, economical waste water treatment. The EPA program for advanced industrial waste water treatment will be especially concerned with activated carbon regeneration technologies (16) (see Water).

The dry cleaning industry uses powdered and granular carbons for reclaiming liquid solvent that has become contaminated with dyes and rancid extracted oils and grease. Carbon use is cheaper than distillation, and carbon more completely removes odor-causing substances.

Activated carbon has many applications in food and pharmaceutical manufacturing. Such uses include purification and color removal during the processing of fruit juices, honey, maple syrup, candy, soft drinks, and alcoholic beverages (2). Pharmaceutical uses include removal of pyrogens from solutions for injections, vitamin decolorizing and deodorizing, and insulin purification. Activated carbon (drug-pure grade) can be administered orally to poison victims (30). Research is being conducted for the development of activated carbon filters for artificial kidney devices and for the dialysis (qv) of poisons and drugs (31–32).

Activated carbon is used in electroplating (qv) to remove organic impurities from the bath. Carbon may also be used to allow the recovery of gold from cyanide-leach pulps (33). Gold may be recovered by burning the carbon or by eluting the gold (34). Activated carbon is not commonly used for the purification of inorganic chemicals (1,35).

Miscellaneous uses for liquid-phase carbons include laboratory uses, reaction catalysis, and aquarium water filters.

Gas-Phase Carbon. Gas-phase carbons typically have a surface area of 1000–2000 m^2/g and are made in larger particle sizes of greater strength and density than liquid-phase carbons. The small (3 nm and less) pores of gas-phase carbons provide high adsorptive capacity and selectivity for gases and organic vapors. Desirable characteristics of a good gas adsorbent carbon are: high adsorptive capacity per unit volume; high retentive capacity; high preferential adsorption of gases in the presence of moisture; low resistance to gas flow; high strength or breakage resistance; and complete release of adsorbates at increased temperatures and decreased pressures.

The largest single application for gas-phase carbon is in gasoline vapor emission control cannisters on automobiles. Evaporative emissions from both fuel tank and carburetor are adsorbed on the carbon (33,36). An evaporative control carbon should have good hardness, a high vapor working capacity, and a high saturation capacity (37). The working capacity of a carbon for gasoline vapor is determined by the adsorption–desorption temperature differential, the rate at which purge air flows through the carbon cannister, and by the extent to which irreversibly adsorbed, high molecular weight gasoline components accumulate on the carbon (15).

Another application for gas-phase carbons is the purification and separation of gases. Many industrial gas streams are treated with activated carbon to remove impurities or recover valuable constituents. Odors are removed from air in activated carbon air conditioning systems, and gas masks containing activated carbon are used to protect individuals from breathing toxic gases or vapors (38).

The purification of industrial gases is usually performed with gas under pressure in a two-bed system. The carbon is supported in tall vertical towers having one on stream as the other is regenerated, dried, or cooled. Regeneration is accomplished by temperature or pressure change or both, with or without a purge gas.

Activated carbon is used in air conditioning (qv) systems to remove industrial odors and irritants from building inlet air and to remove body, tobacco, and cooking odors, etc from recirculated building air (39). The use of carbon to remove odors allows more economical building operation by permitting higher percentages of recycled air to be used, thus minimizing the need for heating or cooling large amounts of fresh inlet air (40). Manufacturers use carbon to remove objectionable odors from process exhaust gases, and to remove corrosive gases and vapors, eg, sulfur compounds, from intake air to protect electrical switching equipment. Activated carbon also prolongs the storage life of fruits and flowers by adsorbing gases that contribute to odor and deterioration of these products (2).

Since air is circulated by low-pressure fans in air conditioning systems, activated carbon is always used in thin beds from 13–26 mm thick. The carbon is usually supported in cannisters of perforated sheet metal that are either flat, cylindrical, or corrugated. The effective life of the carbon depends on the application but may range from several months to 1.5 yr (41).

Activated carbon is used in nuclear reactor systems to adsorb radioactive gases in carrier or coolant gases, and from the air in reactor emergency exhaust systems (42). Radioactive gas adsorbers are very deep beds of carbon used to adsorb and retain radioactive materials long enough for isotopes to decay to safe levels of radioactivity before they are discharged into the atmosphere. Radioactive iodine and organic iodides are adsorbed by specially impregnated grades of high-activity carbon (43). The radioactive fission products, krypton and xenon, are adsorbed for decay to safe isotopes before they are released to the atmosphere (44–45).

The use of solvent recovery-grade activated carbon for the recovery of volatile organic compounds from process air streams is important to the operating economics of several industries (18,46). A simple activated carbon recovery system consists of an air filter, a blower, two horizontal adsorbers (each containing a bed of carbon 0.3–0.6 m deep), a vapor condenser, and a solvent decanter or continuous still (19).

Some solvent recovery plants are manually operated, but most are automatic, using a time–cycle controller. The newer, larger industrial units allow adsorption cycle times to be controlled according to variable vapor loadings in the process air stream (47).

Specially impregnated grades of activated carbon in cigarette filters adsorb some of the harmful components of tobacco smoke (48). Other applications for activated carbon include use in kitchen range hoods, gas sampling tubes (49), refrigerator deodorizers, and as getters used in pill bottles and vacuum equipment for adsorption of harmful contaminant vapors or to aid in attaining high vacuum.

The catalytic oxidation of compounds by activated carbon has been recognized for many years. It is generally conceded that this action results from the presence of irreversibly adsorbed oxygen on the carbon surface. Many salts, such as ferrous sulfate, sodium arsenite, potassium nitrite, and potassium ferrocyanide, are oxidized in solution by activated carbon; the reaction can be made continuous by passing both solution and air over granular carbon (50).

Hydrogen sulfide in air is oxidized to sulfur when passed over activated carbon. It can be removed from manufactured gas by adding a small amount of air to the gas before bringing it in contact with the carbon (51).

Activated carbon is used as the catalyst in the manufacturing of phosgene (qv) from carbon monoxide and chlorine, and sulfuryl chloride is produced by the reaction of sulfur dioxide with chlorine in the presence of activated carbon.

BIBLIOGRAPHY

"Active Carbon" under "Carbon" in *ECT* 1st ed., Vol. 2, pp. 881–899, by J. W. Hassler, Nuclear Active Carbon Division, West Virginia Pulp and Paper Company, and J. W. Goetz, Carbide and Carbon Chemicals Corporation; "Activated Carbon" under "Carbon" in *ECT* 2nd ed., Vol. 4, pp. 149–158, by E. G. Doying, Union Carbide Corporation, Carbon Products Division.

1. A. J. Juhola, *Carbon* **13**, 437 (Nov. 5, 1975).
2. J. W. Hassler, *Purification with Activated Carbon,* Chemical Publishing Co., Inc., New York, 1974.
3. M. Smisek and S. Cerny, *Active Carbon,* American Elsevier Publishing Company, Inc., New York, 1970.
4. Y. Matsumura, *J. Appl. Chem. Biotechnol.* **25**, 39 (Jan. 1975).
5. R. Prober and co-workers, *AIChE J.* **21**, 1200 (Nov. 1975).
6. N. K. Chaney, *Trans. Am. Electrochem. Soc.* **36**, 91 (1919); U.S. Pat. 1,497,544 (June 10, 1924), N.-K. Chaney.
7. A. B. Ray, *Chem. Metall. Eng.* **30**, 977 (1923).
8. U.S. Pat. 3,876,505 (Apr. 8, 1975), G. R. Stoneburner (to Calgon Corp.).
9. A. C. Fieldner and co-workers, *U.S. Bur. Mines Tech. Paper* **479**, (1930).
10. A. E. Williams, *Min. J.* (London) **227**, 1026 (Dec. 28, 1946).
11. J. Economy and R. Y. Lin, *Appl. Polym. Symp.,* 199 (Nov. 29, 1976); U.S. Pats. 3,769,144 (1973), and 3,831,760 (1974), J. Economy and R. Y. Lin (to Carborundum Co.).
12. *Chem. Eng. News* **52**, 7 (July 22, 1974); W. J. Storck, *Chem. Eng. News* **55**, 10 (Apr. 18, 1977).
13. *Chem. Week* **118**, 44 (Jan. 14, 1976).
14. *Chem. Eng. News,* **55**, 10 (Apr. 18, 1977).
15. U.S. Dept. of Health, Education, and Welfare, *Fed. Reg.* **33**(108), (June 4, 1968).
16. *Chem. Eng. News,* **56**, 10 (Apr. 3, 1978).
17. *Chem. Eng.* (N. Y.) **82**, 113 (Apr. 28, 1975).
18. J. C. Enneking, *Air Pollution Control—Part II, ASHRAE Bulletin CH 73-2,* American Society of Heating, Refrigerating & Air Conditioning Engineers, July, 1973, p. 20.
19. R. R. Manzone and D. W. Oakes, *Pollut. Eng.* **5**, 23 (Oct. 1973).
20. *Annual Book of ASTM Standards, Part 30,* American Society for Testing and Materials, Philadelphia, Pa., 1974.
21. *Japanese Industrial Standard K 1412-1958* (reaffirmed 1967).
22. J. J. Kipling, *Adsorption from Solutions of Non-Electrolytes,* Academic Press, New York, 1965.
23. C. L. Mantell, *Adsorption,* McGraw-Hill Book Company, New York, 1951.
24. *Ind. Res.,* 30 (Jan. 1976).
25. *Chem. Eng.* (N. Y.) **77**, 32 (Sept. 7, 1970).
26. *Pollut. Eng.* **8**, 24 (July 1976).
27. L. D. Friedman and co-workers, *Improving Granular Carbon Treatment,* U.S. Environmental Protection Agency, Grant 17020 GDN, July 1971.
28. *Chem. Week,* 40 (Feb. 16, 1977).
29. G. Rey and co-workers, *Chem. Eng. Prog.* **69**, 45 (Nov. 1973).
30. U.S. Pat. 3,917,821 (Nov. 4, 1975), M. Manes.

31. J. H. Knepshield and co-workers, *Trans. Am. Soc. Artif. Int. Organs* **19,** 590 (1973).
32. T. A. Davis, *Activated Carbon Filters for Artificial Kidney Devices,* Southern Research Institute, Report No. AK-2-72-2208, 1973.
33. J. B. Zadra, *U.S. Bur. Mines Rep. Invest.,* 4672 (1950).
34. U.S. Pat. 3,935,006 (Mar. 19, 1975), D. D. Fischer (to U.S. Department of Interior).
35. D. N. Strazhesko, ed., *Adsorption and Adsorbents,* John Wiley & Sons, Inc., New York, 1973.
36. J. O. Sarto and co-workers, *Society of Automotive Engineers Paper No. 700150,* New York, 1970.
37. R. S. Joyce and co-workers, *Society of Automotive Engineers Paper No. 690086,* New York, 1969.
38. *Am. Ind. Hyg. Assoc. J.* **32,** 404 (June 1971).
39. H. Sleik and A. Turk, *Air Conservation Engineering,* 2nd ed., Connor Engineering Co., Danbury, Ct., 1953.
40. M. Beltran, *Chem. Eng. Prog.* **70,** 57 (May 1974).
41. *ASHRAE Guide and Data Book,* American Society of Heating, Refrigerating & Air Conditioning Engineers, Inc., New York, 1970.
42. R. J. Bender, *Power* **116,** 56 (Sept. 1972).
43. A. G. Evans, *paper presented at 13th AEC Air Cleaning Conference,* 1974, Report DP-MS-74-2 (NSA, 30, 23732).
44. P. J. Geue, *Australian Atomic Energy Commission Research Establishment,* Report AAEC LIB/BIB 401 (1973) (NSA 29,18310).
45. M. N. Myers, *Literature Survey: Sampling, Plate-Out and Cleaning of Gas-Cooled Reactor Effluents,* General Electric Co., Report XDC 61-4-702 (TID 14,469) (NSA 16,6114), 1961.
46. C. R. Wherry, *Activated Carbon, Report 731-2020,* in *Chemical Economics Handbook,* Stanford Research Institute, Menlo Park, Calif., 1969.
47. B. Dundee, *Gravure,* (June 1972).
48. U.S. Pat. 3,355,317 (Mar. 18, 1966), C. H. Keith II and V. Norman (to Liggett and Myers Tobacco Co.).
49. A. Turk and co-workers, eds., *Human Responses to Environmental Odors,* Academic Press, New York, 1974.
50. U.S. Pat. 2,365,729 (Dec. 26, 1944), E. A. Schumacher and G. W. Heise (to Union Carbide Corporation).
51. H. Krill and K. Storp, *Chem. Eng. (N. Y.)* **80,** 84 (July 23, 1973).

General Bibliography

D. G. Hager, *Chem. Eng. Prog.* **72,** 57 (Oct. 1976).
R. A. Hutchins, *Chem. Eng. (N. Y.)* **80,** 133 (Aug. 20, 1973).
W. G. Timpe and co-workers, *Kraft Pulping Treatment and Reuse—State of the Art,* U. S. Environmental Protection Agency, Grant 12040 EJU, Dec. 1973 (EPA-660/2-75-004).
E. W. Long and co-workers, *Activated Carbon Treatment of Unbleached Kraft Effluent for Reuse,* U.S. Environmental Protection Agency, Grant 12040 EJU, Dec. 1973 (EPA-R2-73-164).

R. W. SOFFEL
Union Carbide Corporation

BAKED AND GRAPHITIZED CARBON

Raw Materials

The raw materials used in the production of manufactured carbon and graphite largely determine the ultimate properties and practical applications of the finished products. This dependence can be attributed to the nature of carbonization and graphitization processes.

Throughout the entire process of the thermal conversion of organic materials to carbon and graphite, the natural chemical driving forces cause the growth of larger and larger fused-ring aromatic systems, and ultimately result in the formation of the stable hexagonal carbon network of graphite. Differences in the final materials depend upon the ease and extent of completion of these overall chemical and physical ordering processes.

The first few steps in the carbonization process can be considered as a dehydrogenative polymerization where hydrocarbon molecules lose hydrogen and combine to form larger planar molecular networks. As the process moves toward carbon, solid-state reorganization and recrystallization processes begin to take place. In these processes, a gradual improvement in both in-plane and stacking perfection occurs; the result is the three-dimensionally ordered graphite structure with ABAB stacking (see page 691).

The starting materials in the carbonization process may be partly or entirely aromatic hydrocarbons or heterocyclics derived from coal or petroleum. After some heat treatment, the partially-polymerized products constitute a pitch, which is a complex mixture of many hundreds or even thousands of aromatic hydrocarbons with 3–8 condensed rings and an average molecular weight of 300. Although the individual, pure compounds may melt at fairly high temperatures (>300°C), the pitch mixture acts as a eutectic and softens at a much lower temperature, eg, 50–180°C. Both liquid and solid pitches are isotropic materials, ie, perfectly random in their mutual molecular orientation.

Coke is the next stage in the process of carbonization. Coke consists of aromatic polymers of much higher molecular weights, >3000. In contrast to pitch, coke is a totally infusible solid and is generally anisotropic, ie, the flat molecules within fairly large domains can all have nearly the same orientation.

Coke is well-oriented compared with pitch because an easily oriented liquid crystal or mesophase of aromatic molecules forms during the pitch-to-coke transformation. As isotropic pitch is heated above 400°C, small anisotropic mesophase spheres appear which grow, coalesce, and finally form large anisotropic regions (1). As polymerization continues, these anisotropic regions become very viscous or semisolid and are essentially coke. Such well-oriented coke retains this ordered structure and can be converted to crystalline graphite by high temperature heat treatment. Fine insoluble solids in the pitch are not incorporated within the growing mesophase sphere but tend, instead, to aggregate on the surfaces of the spheres and may modify the coalescence, usually resulting in a less anisotropic coke. Pitches that contain this mesophase material are now generally known as mesophase pitches; the term mesophase is derived from the Greek mesos or intermediate and indicates the pseudo-crystalline nature of this highly oriented anisotropic material.

In order to produce useful carbon and graphite bodies, filler and binder materials are mixed, formed by molding or extrusion, and finally baked, or baked and graphitized, to yield the desired shaped carbon or graphite bodies. More than 30 different raw materials are used in the manufacture of carbon and graphite products. The primary materials, in terms of tonnage consumed, are the petroleum coke or anthracite coal fillers and the coal-tar or petroleum pitch binders and impregnants. Other materials, called additives, are often included to improve processing conditions or to modify certain properties in the finished products.

Filler Materials. *Petroleum Coke.* Petroleum coke is produced in large quantities in the United States as a by-product of the petroleum cracking process used to make gasoline and other petroleum products. It is the largest commercially available source of synthetically produced carbon that can be readily graphitized by heating above 2800°C.

Most of the petroleum coke is produced by the delayed coking process (2). The properties of cokes vary according to the feedstock and to operating variables in the delayed coker unit such as operating pressure, recycle ratio, time, and coker heater outlet temperature.

The cokes to be used in any one product are selected to provide optimum quality. The major quality factors are sulfur content, volatiles, ash, and thermal expansion. High sulfur content is undesirable because of environmental concerns and because it leads to uncontrolled expansion (puffing) during graphitization, resulting in cracking of the product. The volatiles in raw cokes affect the quality of the calcined coke; high levels of volatiles lead to low density and low strength in the calcined product. High ash content is usually undesirable and leads to contamination of the finished product. Also, if these ash contaminants are present in the feedstock from which the coke is made, they impede the formation of desired crystal structure during the coking and lower the strength and density of the final coke product.

High-quality or needle cokes have low thermal expansion, low ash content, high density, and good crystal structure as indicated by x-ray analysis. Such needle cokes are used primarily in the manufacture of large graphite products such as electrodes for electric steel-making furnaces. Cokes of lower quality are known as metallurgical-grade cokes.

Needle cokes are produced from aromatic feedstocks such as decant oils from a refinery catalytic cracking unit or tars made by thermally cracking gas oils. Other feedstocks, such as atmospheric or vacuum residues from a refinery, usually produce sponge cokes. Usually the more aromatic the feedstock, the higher the coke quality will be, since larger, more highly ordered domains of oriented molecules will form during the mesophase stage of the coking reaction.

Sponge cokes are used primarily as fuel and in the manufacture of anodes for the aluminum industry. For every kilogram of aluminum produced, approximately 0.4 kg of petroleum coke is consumed (3).

The properties required in cokes for the aluminum industry are different from those required in needle cokes. Metallic impurities in the coke are more critical and a low ash content is essential since impurities from the coke tend to concentrate in the aluminum as the anode is consumed.

Particularly important impurities are vanadium, nickel, iron, and silicon; other impurities such as sulfur are slightly less critical. Since the carbon consumption per kilogram of aluminum produced affects the economics of the process, bulk density is also an important property (see Aluminum).

Table 1 shows a comparison of properties of typical sponge- and needle-grade cokes.

Natural Graphite. Natural graphite is a crystalline mineral form of graphite occurring in many parts of the world (see Carbon, natural graphite).

Carbon Blacks. Carbon blacks are commonly used as components in mixes to make various types of carbon products (see Carbon, carbon black).

Table 1. Typical Properties of Sponge and Needle Cokes

Property	Aluminum anode grade sponge coke Raw	Aluminum anode grade sponge coke Calcined[a]	Graphite electrode-grade needle coke Raw	Graphite electrode-grade needle coke Calcined[a]
sulfur, wt %	2.5	2.5	1.0	1.0
ash, wt %	0.25	0.50	0.10	0.15
vanadium, ppm	150[b]	200[b]	10	10
nickel, ppm	150[b]	200[b]		20–40
silicon, wt %	0.02	0.02	0.04	0.04
volatile matter, wt %	10–12		8	
resistivity, $\mu\Omega\cdot$m (particles, −35 mesh to +65 mesh Tyler, 0.49 to 0.23 mm)		890		965
real density, g/cm^3		2.06		2.12
bulk density, g/cm^3		0.80		0.88
coefficient of thermal expansion of graphite per °C (30–100°C)		20×10^{-7}		5×10^{-7}

[a] Calcining, discussed later in this section, is a thermal treatment which removes volatiles from the raw materials and shrinks the particles.
[b] Ref. 4.

Anthracite. Anthracite is preferred to other forms of coal in the manufacture of carbon products because of its high carbon-to-hydrogen ratio, its low volatile content, and its more ordered structure. It is commonly added to carbon mixes that are used for fabricating metallurgical carbon products to improve specific properties and reduce cost. Anthracite is used in mix compositions for producing carbon electrodes, structural brick, blocks for cathodes in aluminum manufacture, and in carbon blocks and brick used for blast furnace linings (see Coal).

Pitches. Carbon articles are made by mixing a controlled-size-distribution of coke filler particles with a binder such as coal tar or petroleum pitch. The mix is then formed by molding or extruding and is heated in a packed container to control the shape and set the binder. This brief description of the carbon-making process serves to identify the second most important raw material for making a carbon article, the pitch binder. The pitch binder preserves the shape of the green carbon and also fluidizes the carbon particles, enabling them to flow into an ordered alignment during the forming process. During the subsequent baking steps, the pitch binder is pyrolyzed to form a coke that bridges the filler particles and serves as the permanent binding material. These carbon bridges provide the strength in the finished article and also provide the paths for energy flow through thermal and electric conductance (see Tar and pitch).

A binder used in the manufacture of electrodes and other carbon and graphite products must (1) have high carbon yield, usually 40–60 wt % of the pitch; (2) show good wetting and adhesion properties to bind the coke filler together; (3) exhibit acceptable softening behavior at forming and mixing temperature, usually in the range of 90–180°C; (4) be low in cost and widely available; (5) contain only a minor amount of ash and extraneous matter that could reduce strength and other important physical properties; and (6) produce binder coke that can be graphitized to improve the electrical and thermal properties.

The primary binder material, coal-tar pitch, is produced primarily as a by-product of the destructive distillation of bituminous coal in coke ovens during the production of metallurgical coke. The second largest source of binder is petroleum pitch obtained from the cracking of petroleum in refinery processes. Manufacturers of carbon products require pitches with various softening points, depending upon the products to be made and the processing conditions required in forming and baking. The softening behavior and rheology of the pitch are important factors in establishing the forming conditions of the carbon product (see Rheological measurements). Baking schedules are then developed to improve the carbon yield and to reduce energy input to the furnace.

Petroleum Pitch. Although coal-tar pitch is a main source of binder pitch for the carbon industry, pitches from other precursors are becoming increasingly important. Petroleum pitch has been investigated extensively as a potential source of aluminum cell anode pitch (5). These studies have shown that pitch from petroleum sources, if produced from the proper charging stock, can be the approximate equivalent in quality of coal-tar pitch for manufacture of carbon products. Petroleum pitch is produced by cracking petroleum liquids at 450–500°C, 690–1380 kPa (100–200 psi). The cracking step produces both light and heavy feedstocks. The heavy feedstock is then processed by heat soaking at ca 350°C to form a viscous pitch.

The primary differences (shown in Table 2) between petroleum pitches and coal-tar pitches are in viscosity, benzene insolubles (BI), and quinoline insolubles (QI). In the graphite industry, petroleum pitch is used as an impregnant to increase the density of carbon and graphite products because of its low QI content.

Certain compounds found in some coal-tar and petroleum pitches are carcinogenic. Individuals working with pitches or exposed to fumes or dust should wear protective clothing to avoid skin contact. Respirators should be worn when pitch dust or fume concentrations in the air are above established limits (see Industrial hygiene and toxicology).

Additives. In addition to the primary ingredients in the mixes (the fillers and binders), minor amounts of other materials are added at various steps in the process. Although the amounts of these additives are usually in small percentages, they play an important role in the economics of the process and determine the quality of the final products. Light extrusion oils and lubricants, including light petroleum oils, waxes, and fatty acids and esters, are often added to the mix to improve the extrusion rates and structure of extruded products. Inhibitors are used to reduce the detrimental effects of sulfur in high sulfur cokes. Iron oxide is often added to high sulfur coke to

Table 2. Comparison of Typical Physical Properties of Petroleum and Coal-Tar Pitches

Property	Petroleum	Coal tar
softening point, cube-in-air, °C	120	110
sp gr, 15.5–26.7°C	1.22	1.33
coking value, wt %	51	58
benzene insolubles (BI), wt %	3.6	33
quinoline insolubles (QI), wt %	none	14
ash, wt %	0.16	0.10
sulfur, wt %	1.0	0.8
viscosity in Pa·s[a] at		
160°C	0.8	1.4
177°C	0.3	0.4
199°C	0.1	0.2

[a] To convert Pa·s to poise, multiply by 10.

prevent puffing, the rapid swelling of the coke caused by volatilization of the sulfur at 1600–2400°C. Iron from Fe_2O_3 or other iron compounds prevents this action by forming a more stable iron sulfide, which reduces the gas pressure in the coke particles. Other sulfide-forming compounds such as sodium, nickel, cobalt, and vanadium may also be used. Sodium carbonate is often used in applications requiring a low ash product.

Calcining

Coke and pitch purchased by carbon companies from suppliers must meet rigid specifications. Some materials, such as pitch and natural graphite, may be used as received from suppliers; other materials, such as raw coke and anthracite, require calcining, a thermal treatment to temperatures above 1200°C.

Calcining consists of heating raw filler to remove volatiles and to shrink the filler to produce a strong, dense particle. Raw petroleum coke, eg, has 5–15% volatile matter; when the coke is calcined to 1400°C, it shrinks approximately 10–14%. Less than 0.5% of volatile matter in the form of hydrocarbons remains in raw coke after it is calcined to 1200–1400°C. During calcination the evolving volatiles are primarily methane and hydrogen which burn during the calcining process to provide much of the heat required. The calcining step is particularly important for those materials used in the manufacture of graphite products, such as electrodes, since the high shrinkages occurring in raw coke during the baking cycle of large electrodes would cause the electrode to crack. To prevent partial fusion of the coke during calcining, the volatile content of the green coke is kept below 12%.

Anthracite is calcined at appreciably higher temperatures (1800–2000°C). The higher calcining temperatures for anthracite are necessary to complete most of the shrinkage and to increase the electrical conductivity of the product for use in either Soderberg or prebaked carbon electrodes for aluminum or phosphorus manufacture. Some other forms of carbon used in manufacturing of carbon products, such as carbon black, are also calcined.

The selection of calcining equipment depends upon the temperatures required and the materials to be calcined. Two major types of calcining units are used for cokes: the horizontal rotary drum-type calciner and the vertical rotary hearth-type calciner. Modifications of these have been developed for particular applications and to reduce energy input and product loss.

Rotary-type calciners for calcining raw petroleum coke are similar in design to those used for calcining limestone and cement (6–7).

The conventional rotary calciner is energy-intensive, and 10–15% loss of coke is experienced through oxidation. Newer calciners have been designed to prevent excessive loss of product and to reduce fuel consumption.

Since anthracite must be calcined at higher temperatures than can be reached reasonably in conventional gas-fired kilns, an electrically heated shaft kiln is used to calcine the coal (qv) at temperatures up to 2000°C (8).

BIBLIOGRAPHY

"Baked and Graphitized Products" under "Carbon" in *ECT* 1st ed., Vol. 3, pp. 1–34, by H. W. Abbott, Speer Carbon Company; "Baked and Graphitized Products, Manufacture" under "Carbon" in *ECT* 2nd ed., Vol. 4, pp. 158–202, by L. M. Liggett, Speer Carbon Company.

1. J. D. Brooks and G. H. Tayler, *Chem. Phys. Carbon* **4,** 243 (1968).
2. K. E. Rose, *Hydrocarbon Process,* **50,** 85 (Nov. 7, 1971).
3. C. B. Scott, *Chem. Ind. London,* 1124 (July 1, 1967).
4. C. B. Scott and J. W. Connors, *Light Metals, 1971,* American Institute of Mining, Metallurgical, and Petroleum Engineers, Inc., New York, 1972, p. 277.
5. L. F. King and W. D. Robertson, *Fuel* **47,** 197 (1968).
6. R. F. Wesner, P. T. Luckie, and E. A. Bagdoyan, *Light Metals, 1973,* Vol. 2, American Institute of Mining, Metallurgical, and Petroleum Engineers, Inc., New York, 1973, p. 629.
7. V. D. Allred in ref. 4, p. 313.
8. M. M. Williams, *Light Metals, 1972,* American Institute of Mining, Metallurgical, and Petroleum Engineers, Inc., New York, 1972, p. 163.

General References

E. Wege, *High Temp. High Pressures* **8,** 293 (Nov. 3, 1976).
J. M. Hutcheon in L. C. F. Blackman, ed., *Modern Aspects of Graphite Technology,* Academic Press, New York, 1970, Chapt. 2.
M. L. Deviney and T. M. O'Grady, eds., *Petroleum Derived Carbons,* American Chemical Society, Washington, D.C., 1976.

L. L. WINTER
Union Carbide Corporation

PROCESSING OF BAKED AND GRAPHITIZED CARBON

Raw Material Preparation

Crushing and Sizing. Calcined petroleum coke arrives at the graphite manufacturer's plant in particle sizes ranging typically from dust to 5–8 cm dia. In the first step of artificial graphite production the run-of-kiln coke is crushed, sized, and milled to prepare it for the subsequent processing steps. The degree to which the coke is broken down depends on the grade of graphite to be made. If the product is to be a fine-grained variety for use in aerospace, metallurgical, or nuclear applications, the milling and pulverizing operations are used to produce sizes as small as a few micrometers in diameter. If, on the other hand, the product is to be coarse in character for products like graphite electrodes used in the manufacture of steel, a high yield of particles up to 1.3 cm dia is necessary.

The wide variety of equipment available for the crushing and sizing operations is well-described in the literature (1–2). Roll crushers are commonly used to reduce the incoming coke to particles that are classified in a screening operation. The crushed coke fraction, smaller than the smallest particle needed, is normally fed to a roll or hammer mill for further size reduction to the very fine (flour) portion of the carbon mix. A common flour sizing used in the graphite industry contains particles ranging from 149 μm (100 mesh) to a few micrometers, with about 50% passing through a 74-μm (200-mesh) screen.

For a coarse-grained (particle-containing) graphite, the system depicted in Figure 1 is typical. The run-of-kiln coke is brought in on railroad cars and emptied into pits where the coke is conveyed to an elevator. The elevator feeds a second conveyer which empties the coke into any one of a number of storage silos where the coke is kept dry. The manufacturer usually specifies a maximum moisture content in the incoming coke, at about 0.1–0.2%, to ensure that mix compositions are not altered by fluctuations in moisture content.

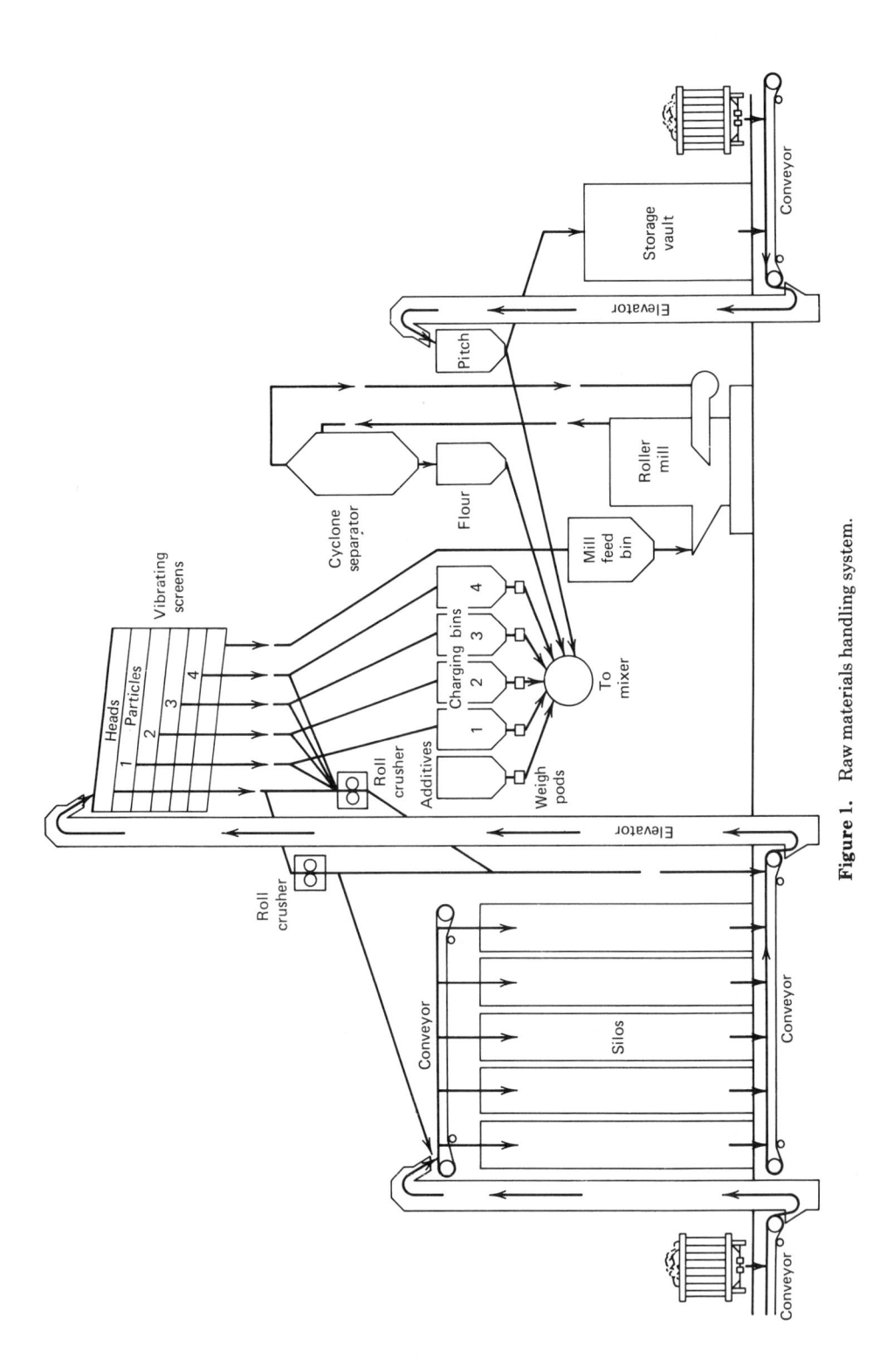

Figure 1. Raw materials handling system.

In the system shown in Figure 1, the oversized coke particles (heads) are diverted to a roll crusher. Most raw materials systems provide the option of further reducing the sizes of particles by passing them through a second crusher directly from the screens and recycling the resulting fractions through the screening system. The undersized coke fractions are transferred to a bin that supplies a mill for production of the flour portion of graphite composition. The mills used in this application may be of impact (hammer) variety or of roller variety. A commonly-used mill consists of a rotating roller operating against a stationary steel ring. The coke is crushed to very fine sizes that are air-classified by a cyclone separator. The sizes larger than those desired in the flour are returned to the mill and the acceptable sizes are fed to a charge bin.

The coal-tar-pitch binder used in graphite manufacture also arrives in railroad cars. If the pitch is shipped in bulk form, the large pieces must be crushed to ca 3 cm and smaller to facilitate uniform melting in the mixer and control of the weighing operation. Many vendors of binder pitches now form their product either by prilling, extruding, or flaking to ensure ease of handling and storage.

The pitch system shown in Figure 1 conveys the incoming pitch through a crusher to an elevator that deposits it into a charging bin. The graphite manufacturer tries to avoid long-term storage of 100°C-softening point pitch because of its tendency to congeal at ambient conditions into masses extremely difficult to break up and handle. Thus, whenever possible, cars of pitch are ordered and used as needed at the carbon plant.

In some plants the pitches are delivered in heated tank cars as liquids.

Proportioning. The size of the largest particle is generally set by application requirements. For example, if a smoothly-machined surface with a minimum of pits is required, as in the case of graphites used in molds, a fine-grained mix containing particles no larger than 0.16 cm with a high flour content is ordinarily used. If high resistance to thermal shock is necessary (eg, in graphite electrodes used in melting and reducing operations in steel plants), particles up to 1.3 cm are used to act as stress absorbers in preventing catastrophic failures in the electrode.

Generally, the guiding principle in designing carbon mixes is the selection of the particle sizes, the flour content, and their relative proportions in such a way that the intergranular void space is minimized. If this condition is met, the volume remaining for binder pitch and the volatile matter generated in baking are also minimized. The volatile evolution is often responsible for structural and property deterioration in the graphite product. In practice, most carbon mixes are developed empirically with the aim of minimizing binder demand and making use of all the coke passed through the first step of the system. From an economic standpoint, accumulation of one size component cannot be tolerated in making mixes for commonly used graphite grades since this procedure will amount to a loss of relatively costly petroleum coke. Typically, a coarse-grained mix may contain a large particle (eg, 6 mm dia), a small particle half this size or smaller, and flour. In this formulation, approximately 25 kg of binder pitch would be used for each 100 kg of coke.

Although binder levels increase as particle size is reduced, and they are greatest in all-flour mixes where surface area is very high, the principle of minimum binder level still applies. The application of particle packing theory to achieve minimum binder level in all-flour mixes is somewhat more complex because of the continuous gradation in sizes encountered (3).

For some carbon and graphite grades, particle packing and minimum pitch level concepts are not used in arriving at a suitable mix design. For relatively small products, eg, where large dimensional changes can be tolerated during the baking and graphitizing operations, high binder levels are often used. Increased pitch content results in greater shrinkage which gives rise to high density and strength in the finished products.

Mixing. Once the raw materials have been crushed, sized, and stored in charging bins and the desired proportions established, the manufacturing process begins with the mixing operation. The purpose of mixing is to blend the coke filler materials and to melt and distribute the pitch binder over the surfaces of the filler grains. The intergranular bond ultimately determines the property levels and structural integrity of the graphite. Thus the more uniform the binder distribution is throughout the filler components, the greater the likelihood for a structurally sound product.

The degree to which mixing uniformity is accomplished depends on factors such as time, temperature, and batch size. However, a primary consideration in achieving mix uniformity is mix design. A number of mechanically agitated, indirectly-heated mixer types are available for this purpose (4–5). Each mixer type operates with a different mixing action and intensity. Ideally, the mixer best suited for a particular mix composition is one that introduces the most work per unit weight of mix without particle breakdown. In practice, only a few mixer types are used in graphite manufacture.

The cylinder mixer is commonly used for coarse-grained mixes. It is equipped with an axial rotating shaft fitted with several radial arms where paddles are attached. The intensity of this mixer is relatively low to avoid particle breakdown, and long mixing times, such as 90 min, are therefore needed to complete the mixing operation. With fine-grained compositions, more intensive mixers may be used with a corresponding reduction in mixing time. Bread or sigma-blade mixers and the high intensity twin-screw mixers of the Werner-Pfleiderer and Banbury variety are examples of the equipment that can be used on fine-grained compositions. For both mixers, temperatures at the time of discharge are 160–170°C.

Following the mixing operation, the hot mix must be cooled to a temperature slightly above the softening point of the binder pitch. Thus the mix achieves the proper rheological consistency for the forming operation and the formed article is able to maintain its shape better as it cools to room temperature. At the end of the cooling cycle, which typically requires 15–30 min, the mix is at 100–110°C and is ready to be charged into an extrusion press or mold.

Forming. One purpose of the forming operation is to compress the mix into a dense mass so that pitch-coated filler particles and flour are in intimate contact. For most applications, a primary goal in the production of graphite is to maximize density; this goal begins by minimizing void volume in the formed (green) product. Another purpose of the forming step is to produce a shape and size as near that of the finished product as possible. This reduces raw material usage and cost of processing graphite that cannot be sold to the customer and must be removed by machining prior to shipment.

The two important methods of forming are extrusion and molding.

Extrusion. The extrusion process is used to form most carbon and graphite products. In essence, the various extrusion presses comprise a removable die attached by means of an adapter to a hollow cylinder, called a mud chamber. This cylinder is charged with mix that is extruded in a number of ways depending on the press design. For one type of press, the cooled mix is introduced into the mud chamber in the form of plugs which are molded in a separate operation. A second type of extruder, called a tilting press, makes use of a movable mud chamber–die assembly to eliminate the need for precompacting the cooled mix. Loading occurs directly from coolers with the assembly in the vertical position; the mixture is extruded with the assembly in the horizontal position. A third type of extrusion press makes use of an auger to force mix through the die. This press is used principally with fine-grained mixes because of its tendency to break down large particles.

The basic steps in the extrusion operation when a tilting press is used are depicted in Figure 2. The cooled mix is usually fed to the press on a conveyor belt where it is discharged into the mud chamber in the vertical position. A ram descends on the filled chamber, tamping the mix to compact the charge. A closing plate located in the pit beneath the press is often used to seal off the die opening thereby preventing the mix from extruding during the application of high tamping pressures. The filling and tamping procedures are repeated until the mud chamber is filled with tamped mix and then rotated back to the horizontal position. The extrusion ram then enters the mud cylinder forcing the mix through the die at 7–15 MPa (69–148 atm). A guillotine-like knife located near the die outlet cuts the extruded stock to the desired length. Round products are rolled into a tank of water where the outer portions are quickly cooled to prevent distortion of the plastic mass. Products having large rectangular cross sections may be transferred from the press to the cooling tank by means of an overhead crane. Water temperatures are regulated to avoid cracking as a result of too rapid cooling. Products with smaller cross sections, such as the 3.2 × 15.2 × 81 cm plates used as anodes in chlorine cells, may be cooled in air on steel tables. Bulk densities of green products range typically from 1.75–1.80 g/cm^3.

The anisotropy, usually observed in graphite products, is established in the forming operation. In extruded products, the anisotropic coke particles orient with

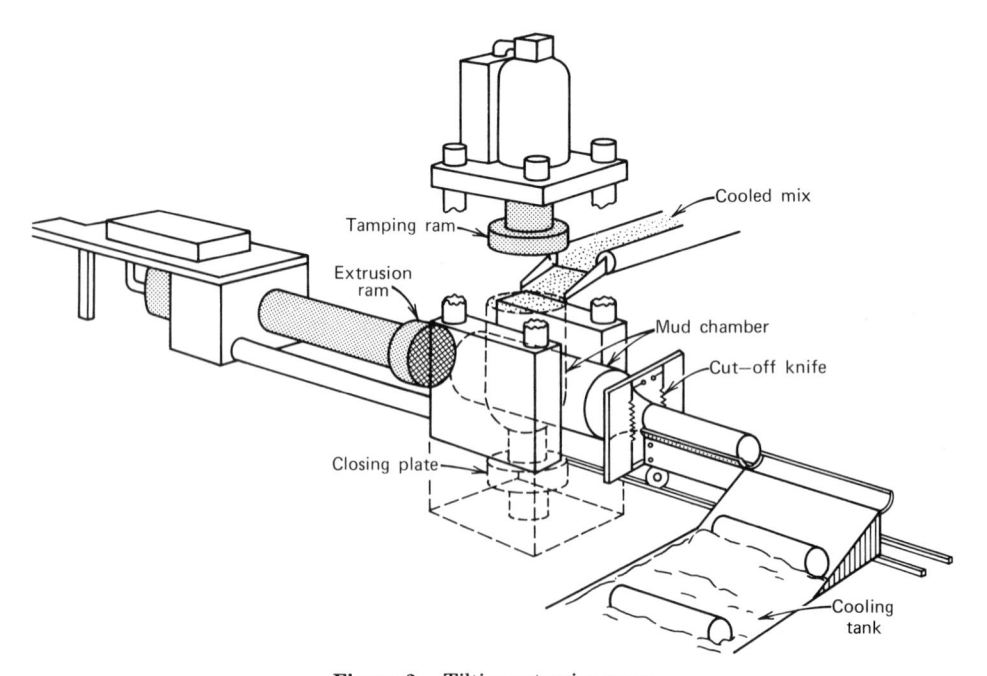

Figure 2. Tilting extrusion press.

their long dimensions parallel to the extrusion direction. The layer planes of the graphite crystals are predominantly parallel to the long dimension of the coke particle. Accordingly, the highly anisotropic properties of the single crystal are translated, to a greater or lesser degree depending on several factors, to the graphite product. The most important of these factors are coke type, particle size, and the ratio of die-to-mud chamber diameters. The more needlelike the coke particle, the greater the difference is between properties with-grain (parallel to the extrusion direction) and cross-grain. The use of smaller particles in the mix design also increases this property difference; the presence of large particles interferes with the alignment process. As the ratio of mud cylinder to die diameter increases, the with-grain to cross-grain ratios of strength and conductivity increase, while the with-grain to cross-grain ratios of resistivity and expansion coefficient decrease. Thus anisotropy is increased for the same coke type and mix design when going from a 0.60 m dia die to a 0.40 m dia die on the same extrusion press. As a result of particle orientation in extruded graphitized products, strength, Young's modulus, and thermal conductivity values are greater, whereas electrical resistivity and coefficient of thermal expansion are smaller in the with-grain direction than in the two cross-grain directions.

Molding. Molding is the older of the two forming methods and is used to form products ranging in size from brushes for motors and generators to billets as large as 1.75 m dia by 1.9 m in length for use in specialty applications.

Several press types are used in molding carbon products. The presses may be single-acting or double-acting, depending on whether one or both platens move to apply pressure to the mix through punched holes in either end of the mold. The use of single-acting presses is reserved for products whose thicknesses are small compared with their cross-sections. As thickness increases, the acting pressure on the mix diminishes with distance from the punch because of frictional losses along the mold wall. Acceptable thicknesses of molded products can be increased by using double-acting presses which apply pressures equally at the top and bottom of the product.

Jar molding is another method used to increase the length of the molded piece and keep nonuniformity within acceptable limits. By this technique the heated mold is vibrated as the hot mix is introduced, thus compacting the mix during the charging operation. Pieces as large as 2.5 m in diameter and 1.8 m in length have been molded in this way; the green densities are comparable with those obtained in extruded materials.

Smaller products, such as brushes and seal rings, are often molded at room temperature from mix that is milled after cooling. When binder levels exceed approximately 30% of the mix, the compacted milled mix has sufficient green strength to facilitate handling in preparation for the baking operation.

In a typical hot-molding operation to form a 1.7 m dia billet 1.3 m long, approximately 7200 kg of mix at 160°C is introduced into a steam-heated mold without cooling. The platens of the press compact the mix at ca 5 MPa (ca 50 atm), holding this pressure for 15–30 min. The cooling step for pieces of this size is the most critical part of the forming operation. Owing to the low thermal conductivity of pitch, 0.13 W/(m·K) (6), and its relatively high expansion coefficient (4.5×10^{-4}/°C at 25–200°C) (7), stresses build rapidly as the outer portions of the piece solidify. If cooling is too rapid, internal cracks are formed which are not removed in subsequent processing steps. As a result, a cooling schedule is established for each product size and is carefully followed by circulating water of various temperatures through the mold for specified time periods.

When the outside of the piece has cooled sufficiently, it is stripped from the mold and the cooling operation continued by direct water spray for several hours. If cooling is stopped too soon, heat from the center of the piece warms the pitch binder to a plastic state, resulting in slumping and distortion. The cooled piece is usually stored indoors prior to baking in order to avoid extreme temperature changes which may result in temperature gradients and damage to the structure. Bulk density of the green billet is usually 1.65–1.70 g/cm^3.

As with extruded products, molded pieces have a preferred grain orientation. The coke particles are aligned with their long dimensions normal to the molding direction. Thus the molded product has two with-grain directions, and one cross-grain direction which coincides with the molding direction. Strength, modulus, and conductivity of molded graphites are higher in both with-grain directions, and expansion coefficient is higher in the cross-grain direction.

Isostatic molding is a forming technique used to orient the coke filler particles randomly, thereby imparting isotropic properties to the finished graphite. One approach to isostatic molding involves placing the mix or blend into a rubber container capable of withstanding relatively high molding temperatures. The container is evacuated, then sealed and placed in an autoclave which is closed and filled with heated oil. The oil is then pressurized to compact the mass which may then be processed in the usual way to obtain isotropic graphite.

Baking. The next stage is the baking operation during which the product is fired to 800–1000°C. One function of this step is to convert the thermoplastic pitch binder to solid coke. Another function of baking is to remove most of the shrinkage in the product associated with pyrolysis of the pitch binder at a slow heating rate. This procedure avoids cracking during subsequent graphitization where very fast firing rates are used. The conversion of pitch to coke is accompanied by marked physical and chemical changes in the binder phase, which if conducted too rapidly, can lead to serious quality deficiencies in the finished product. For this reason, baking is generally regarded as the most critical operation in the production of carbon and graphite.

Several studies discuss the kinetics of pitch pyrolysis and indicate, in detail, the weight loss and volatile evolution as functions of temperature (8–9). During this process weight losses of 30–40% occur, indicating that for every 500 kg of green product containing 20% pitch, 30–40 kg of gas must escape. In terms of gas volume, approximately 150 cm^3 of volatiles at standard conditions must be evolved per g of pitch binder during the baking operation. The product in the green state is virtually impermeable, and the development of a venting porosity early in the bake must be gradual to avoid a grossly porous or cracked structure. The generation of uniform structure during the bake is made more difficult by the poor thermal conductivity of pitch. Long firing times are usually needed to drive the heat into the center of the product which is necessary for pitch pyrolysis and shrinkage. If the heating rates exceed a value which is critical for the size and composition of the product, differential shrinkage leads to splitting scrap. Shrinkage during baking is of the order of 5% and increases with increasing pitch content. Added to these difficulties is the complete loss of mechanical strength experienced by the product in the 200–400°C range where the pitch binder is in a liquid state. To prevent slumping and distortion during this period, the stock must be packed in carefully sized coke or sand which provides the necessary support and is sufficiently permeable to vent the pitch volatiles.

A variety of baking furnaces are in use to provide the flexibility needed to bake a wide range of product sizes and to generate the best possible temperature control. One common baking facility is the pit furnace, so named because it is positioned totally or partially below ground level to facilitate improved insulation. In essence, the pit furnace is a box with ceramic brick walls containing ports or flues through which hot gases are circulated. Traditionally, natural gas has been the fuel used to fire pit furnaces; however, because of natural gas shortages, pit furnaces are being converted to use fuel oil (see Furnaces, fuel-fired).

Another common baking facility is the so-called ring furnace; one form of this is depicted in Figure 3. Two equal rows of pit furnaces are arranged in a rectangular ring. Ports in the furnace walls permit the heated gases from one furnace to pass to the next until the cooled gases are exhausted by a movable fan to a flue leading to a stack. A movable burner, in this case located above one furnace, fires it to a predetermined off-fire temperature. The firing time per furnace is 18–24 h. When the desired temperature has been reached, the burner is moved to the adjacent furnace which has been heated by gases from the most recently fired pit. At the same time, the fan is moved to a furnace that has just been packed. This process continues, with packing,

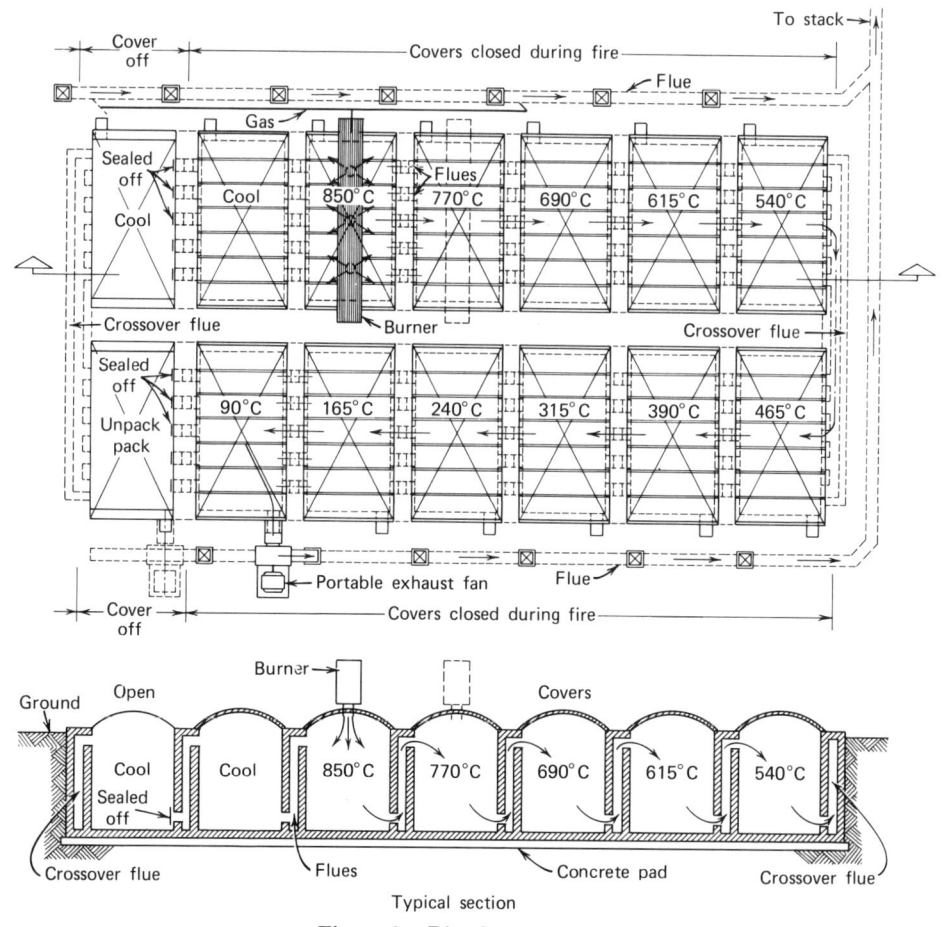

Figure 3. Ring furnace system.

unloading, and cooling stages separating the fan and the burner. Cycle times in this furnace are 3–4 wk. Thermally, the ring furnace is highly efficient but it has the disadvantage that very little control can be exercised over heating rates.

The firing schedules used in the baking operation vary with furnace type, product size, and binder content. A bulk furnace packed with 61 × 81 × 460 cm pieces of specialty graphite may require 6 wk to fire and an additional 3–4 wk to cool. In contrast, very small products, such as seal rings, may be baked in tunnel kilns in a few h. A sagger furnace containing electrodes may require 12–14 d to reach final temperature with an additional 3–5 d to cool. Firing rates early in the baking schedule are reduced to permit pitch volatiles to escape slowly, minimizing damage to the structure. For most carbon products, temperatures must be well below 400°C prior to unpacking to avoid cracking due to thermal shock. The product is scraped or sanded to remove adhering packing materials and is then weighed, measured, and inspected prior to being stored for subsequent processing. Some products that are sold in the baked state are machined at this stage. Baked products include submerged arc furnace electrodes, cathode blocks for the electrolytic production of aluminum, and blast furnace lining blocks.

Impregnation. In some applications the baked product is taken directly to the graphitizing facility for heat treatment to 3000°C. However, for many high performance applications of graphite, the properties of stock processed in this way are inadequate. The method used to improve those properties is impregnation with coal-tar or petroleum pitches. The function of the impregnation step is to deposit additional pitch coke in the open pores of the baked stock, thereby improving properties of the graphite product. Table 1 lists the graphite properties of unimpregnated and impregnated stock 15–30 cm dia and containing a 1.5 mm particle.

Further property improvements result from additional impregnation steps separated by rebaking operations. However, the gains realized diminish quickly, for the quantity of pitch picked up in each succeeding impregnation is approximately half of that in the preceding treatment. Many nuclear and aerospace graphites are multiply pitch-treated to achieve the greatest possible assurance of high performance.

Table 1. Effect of One Pitch Impregnation on Graphite Properties

Property		Unimpregnated	Impregnated
bulk density, g/cm^3		1.6	1.7
	wg[b]	7.4	11.0
Young's modulus[a], GPa	ag[c]	4.4	6.3
	wg	10,000	17,000
flexural strength[a], kPa	ag	7,100	13,000
	wg	5,000	8,100
tensile strength[a], kPa	ag	4,400	7,300
	wg	21,000	34,000
compressive strength[a], kPa	ag	21,000	33,000
	wg	0.40	0.19
permeability, Darcys	ag	0.35	0.16
	wg	1.3	1.5
coefficient of thermal expansion, 10^{-6}/°C	ag	2.7	3.1
	wg	8.8	7.6
specific resistance, $\mu\Omega\cdot$m	ag	13	11

[a] To convert Pa to atm, divide by 101 × 10^3.
[b] With the grain.
[c] Across the grain.

During the baking operation, binder pitch exuding the product surface creates a dense impermeable skin. In addition, the exuding pitch causes packing material to adhere to the baked stock. The skin and the packing material must be removed by sanding, scraping, or machining before the stock can be impregnated on a reasonable time cycle. Unless this operation is properly performed, the impregnant may not reach the center of the product and a so-called dry core will result. When this condition exists, the product usually splits during graphitization as a consequence of the greater concentration of pitch and greater shrinkage in the outer portions of the stock. The likelihood of a dry core increases with the quinoline-insoluble solids content of the impregnant. During the impregnation process, the insolubles form a filter cake of low permeability on the stock surface, reducing the penetrability of the impregnant. Quinoline insolubles significantly greater than 5% reduce the penetration rate and increase the incidence of dry cores.

A schematic diagram of the pitch impregnation process is shown in Figure 4. Before it is placed in an autoclave, the skinned baked stock is preheated to 250–300°C to thoroughly dry it and to facilitate free flow of the molten impregnant into the open pores. The first step in the impregnation process is to evacuate the stock to pressures below 3.5 kPa (26 mm Hg) for a period of 1 h or more depending on the size and permeability of the stock. Unless the stock is adequately evacuated, the remaining air prevents thorough penetration of the impregnant to the center of the product. Heated pitch is then introduced by gravity flow into the autoclave from a holding tank until the charge is completely immersed. The system is then subjected to pressures of 700–1500 kPa (6.9–14.8 atm) for several hours to shorten the time for pitch penetration. When the pressure cycle has been completed, the pitch is blown back to the holding tank by means of compressed air. The autoclave is then opened, and the stock is transferred to a cooler where water and circulating air accelerate the cooling process. After cooling, the stock is weighed to determine the quantity of pitch picked up. If the pickup is below a specified limit, the stock is scrapped. Depending on the density of the baked stock, the pickup is 14–16% on the first impregnation and 7–8% on the second impregnation.

If the stock is to receive a second impregnation, it must be rebaked. In the past, stock containing raw impregnating pitch could be graphitized directly. However, the air polluting effect caused by this practice has made rebaking a necessary preliminary step to graphitization in order to meet the requirements of the EPA.

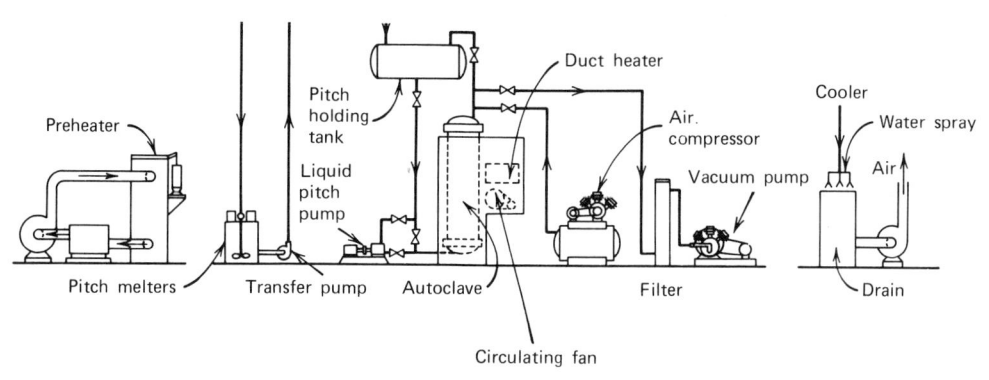

Figure 4. Pitch impregnation system.

Graphitization. Graphitization is an electrical heat treatment of the product to ca 3000°C. The purpose of this step is to cause the carbon atoms in the petroleum coke filler and pitch coke binder to orient into the graphite lattice configuration. This ordering process produces graphite with the intermetallic properties that make it useful in many applications.

Very early in the carbonization of coker feeds and pitch, the carbon atoms are present in distorted layers of condensed benzene ring systems formed by the polymerization of the aromatic hydrocarbons in these materials. The x-ray studies of raw cokes, for example, show that two-dimensional order exists at that early stage of graphite development (10). As the temperature of coke increases, the stack height of the layer planes increases. The layers are skewed about an axis normal to them, however, and it is not until a temperature of ca 2200°C is reached that three-dimensional order is developed. As the graphitizing temperature is increased to 3000°C, the turbostratic (see pp. 690–691) arrangement of the layer planes is effectively eliminated, and the arrangement of the carbon atoms approaches that of the perfect graphite crystal. Depending on the size and orientation of these crystals, the properties of manufactured graphites can be varied controllably to suit a number of critical applications.

The furnace that made the graphite industry possible was invented in 1895 by Acheson (11) and is still in use today with only minor modifications. It is an electrically-fired furnace capable of heating tons of charge to temperatures approaching 3000°C. The basic elements of the Acheson furnace are shown in Figure 5. The furnace bed is made up of refractory tiles supported by concrete piers. The furnace ends are

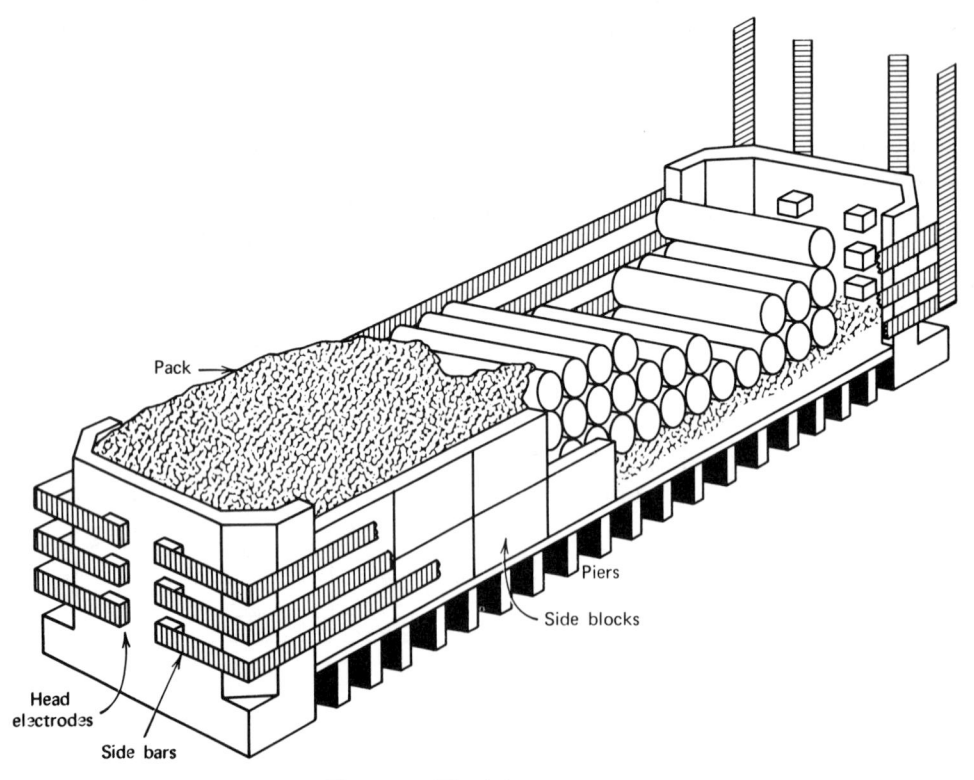

Figure 5. The Acheson furnace.

U-shaped concrete heads through which several graphite electrodes project into the pack. These electrodes, which are water-cooled during operation, are connected by copper bus work to the secondary of a transformer. The product is placed on a layer of metallurgical coke with its long axis transverse to current flow. Although a cylindrical product is shown in Figure 5, any product shape can be graphitized in the Acheson furnace so long as the product pieces are carefully spaced. This feature of the Acheson furnace makes it extremely versatile. The spacing between pieces may vary from less than a centimeter to several centimeters, depending on the shape and size of the product. With the product in place, a coarsely-sized metallurgical coke, called resistor pack, is used to fill the interstices between pieces; most of the heat needed to reach graphitizing temperatures is generated in the resistor material. Once the charge and resistor material are loaded, the furnace is covered with a finer blend of metallurgical coke, sand, and silicon carbide to provide thermal and electrical insulation. Concrete side blocks, usually 0.5–1 m from the charge ends, are used to retain the insulation. The procedure for loading a furnace usually requires one day.

Acheson furnace sizes may vary, depending on the product size and the production rate desired. Typically, the furnace may be 12–15 m by 3–3.5 m. Loads ranging from 35–55 metric tons of product are charged to these furnaces. The transformers used are rated 4000–6000 kW and are capable of delivering up to 60,000 A to the charge. Heating rates are usually 40–60°C/h, the total firing time being approximately three days. At the end of this time, the product temperatures are 2800–3000°C. Total power input varies, depending on the product and load size; for graphite electrodes, total power (energy) inputs average 4.5 kW·h/kg, and total power inputs in excess of 9 kW·h/kg may be used in the thermal purification of nuclear graphites. Following the heating cycle, 8–10 d are needed to cool and unload the furnace. The total cycle time on an Acheson furnace is ca two weeks. The cooling procedure is hastened by the gradual removal of pack with care to leave sufficient cover to prevent oxidation of the product. The insulation and resistor materials are screened to specified sizes and proportions for reuse, and new materials added as necessary. The product is cleaned and inspected prior to being measured and weighed for bulk density and resistivity determinations. If the properties are within specified limits, the product is stored and is ready for machining.

Furnaces other than the Acheson furnace are used commercially, but on a much smaller scale and usually for smaller products. For example, electrographitic brushes are graphitized in tube furnaces, wherein a current-carrying graphite tube is the heating element. These furnaces are particularly useful in the laboratory because of the ease with which they can be loaded and unloaded without the need for handling large quantities of packing material. Inductively-heated furnaces are also used commercially to graphitize a limited number of products, such as some aerospace grades and graphite fibers. These furnaces, also popular in the laboratory, consist basically of a cylindrical graphite shell susceptor positioned inside a water-cooled copper coil. High frequency power supplied to the coil induces current to flow in the susceptor, heating it and causing it to radiate heat to the contained charge (see Furnaces, electric).

More recently, several patents (12–15) have been issued describing a process for graphitization where the carbon charge to be heated is placed in a longitudinal array and covered with insulation to prevent heat losses and oxidation of the charge. An electric current is passed directly through the carbon array, generating within the carbon the heat required to raise the carbon to the graphitization temperatures. These patents describe reduced energy requirements and reductions in process time.

Puffing. In the temperature range of 1500–2000°C, most petroleum cokes undergo an irreversible volume increase known as puffing. This effect has been associated with thermal removal of sulfur from coke and increases with increasing sulfur content. Because of the recent emphasis on the use of low sulfur fuels, many of the sweet crudes that had been used as coker feeds are now being processed as fuels. Desulfurization of the sour crudes available for coking is possible but expensive. The result is an upward trend in the sulfur content of many petroleum cokes, leading to greater criticality in heating rate in the puffing temperature range during graphitization.

Many studies of the puffing phenomenon and of means for reducing or eliminating it have been made (16–18). As a general rule, puffing increases as particle size increases and is greater across the product grain. Depending on particle size and on the product size, heating rates must be adjusted in the puffing range to avoid splitting the product. Fortunately, the use of puffing inhibitors (discussed in the previous section) has eased the problem and has permitted the use of graphitization rates greater than would otherwise be possible.

BIBLIOGRAPHY

"Baked and Graphitized Products" under "Carbon" in *ECT* 1st ed., Vol. 3, pp. 1–34, by H. W. Abbott, Speer Carbon Company; "Baked and Graphitized Products, Manufacture" under "Carbon" in *ECT* 2nd ed., Vol. 4, pp. 158–202, by L. M. Liggett, Speer Carbon Company.

1. F. J. Hiorns, *Br. Chem. Eng.* **15,** 1565 (Dec. 1970).
2. A. Ratcliffe, *Chem. Eng. (N.Y.)* **79,** 62 (July 10, 1972).
3. A. E. Goldman and H. D. Lewis, *U.S. Los Alamos Scientific Laboratory, Report LA 3656,* 1968.
4. W. L. Root and R. A. Nichols, *Chem. Eng. (N.Y.)* **80,** 98 (Mar. 19, 1973).
5. V. W. Uhl and J. B. Gray, *Mixing,* Vol. II, Academic Press, New York, 1967, Chapt. 8.
6. D. McNeil and L. J. Wood, *Industrial Carbon and Graphite; Papers Read at the Conference Held in London, Sept. 24–26, 1957.* Society of Chemical Industry, London, Eng., 1958, p. 162.
7. R. E. Nightingale, *Nuclear Graphite,* Academic Press, New York, 1962, Chapt. 2.
8. M. Born, *Fuel* **53,** 198 (1974).
9. A. S. Fialkov and co-workers, *J. Appl. Chem. USSR* **35,** 2213 (1964).
10. R. E. Franklin, *Acta Cryst.* **4,** 253 (1951).
11. E. G. Acheson, *Pathfinder,* Acheson Industries, Inc., 1965.
12. Ger. Pat. 2,018,764 (Oct. 28, 1971), (to Sigri Elektrographit GmbH).
13. Ger. Pat. 2,316,494 (Oct. 2, 1974), (to Sigri Elektrographit GmbH).
14. Jpn. Kokai 75-86494 (July 11, 1975), (to Toyo Carbon Co. Ltd.).
15. Ger. Pat. 2,623,886 (Dec. 16, 1976), (to Elettrocarbonium SP).
16. M. P. Whittaker and L. I. Grindstaff, *Carbon* **7,** 615 (1969).
17. I. Letizia, *Paper No. CP-69,* presented at the 10th Biennial Conference on Carbon, Lehigh University, Bethlehem, Pa., June 1971.
18. H. F. Volk and M. Janes, *Paper No. 172,* presented at 7th Biennial Conference on Carbon, Cleveland, Ohio, June 1965.

E. L. PIPER
Union Carbide Corporation

PROPERTIES OF MANUFACTURED GRAPHITE

Physical Properties

The graphite crystal, the fundamental building block for manufactured graphite, is one of the most anisotropic bodies known. Properties of graphite single crystals illustrating this anisotropy are shown in Table 1 (1). Anisotropy is the direct result of the layered structure with extremely strong carbon–carbon bonds in the basal plane and weak bonds between planes. The anisotropy of the single crystal is carried over in the properties of commercial graphite, although not nearly to the same degree. By the selection of raw materials and processing conditions, graphites can be manufactured with a very wide range of properties and degree of anisotropy. The range of room temperature properties, attainable for various forms of graphite, is shown in Figures 1 and 2 (1). The range extremities represent special graphites having limited industrial utility, whereas the bulk of all manufactured graphites fall in the bracketed areas marked conventional.

The directional properties of graphite arise in the following way. When the coke aggregate is crushed and sized, the resulting coke particles tend to have one axis longer than the other two. As the plastic mix of particles and binder pitch is formed into the desired shape, the long axis of particles tends to align perpendicularly to the molding force in molded graphites and parallel to the extrusion force in extruded graphites. The particle alignment is preserved during the subsequent processing so that properties of the finished graphite have an axis of symmetry that is parallel to the forming force. Properties in the plane perpendicular to the axis of symmetry are essentially independent of direction. Samples cut parallel to the molding force for molded graphites or perpendicular to the extrusion force for extruded graphites are designated as cross-grain. Samples cut parallel to the molding plane of molded graphites or parallel to the extrusion axis for extruded graphites are designated as with-grain. A number of special test procedures for determining the properties of carbon and graphite have been adopted by ASTM (2).

Manufactured graphite is a composite of coke aggregate (filler particles), binder carbon, and pores. Most graphites have a porosity of 20–30%, although special graphites can be made that have porosity well outside this range. Manufactured graphite is a highly refractory material that has been thermally stabilized to as high as 3000°C. At atmospheric pressure, graphite has no melting point but sublimes at 3850°C, the triple point being approximately 3850°C and 12.2 MPa (120 atm) (3). The strength of graphite increases with temperature to 2200–2500°C; above 2200°C, graphite becomes

Table 1. Room Temperature Properties of Graphite Crystals[a]

Property	Value in basal plane	Value across basal plane
resistivity, Ω·m	40×10^{-4}	ca 6000×10^{-4}
elastic modulus[b], TPa	0.965	0.034
tensile strength (est)[b], TPa	0.096	0.034
thermal conductivity, W/(m·K)	ca 2000	10
thermal expansion, °C^{-1}	-0.5×10^{-6}	27×10^{-6}

[a] Ref. 1.
[b] To convert TPa to psi, multiply by 1.45×10^8.

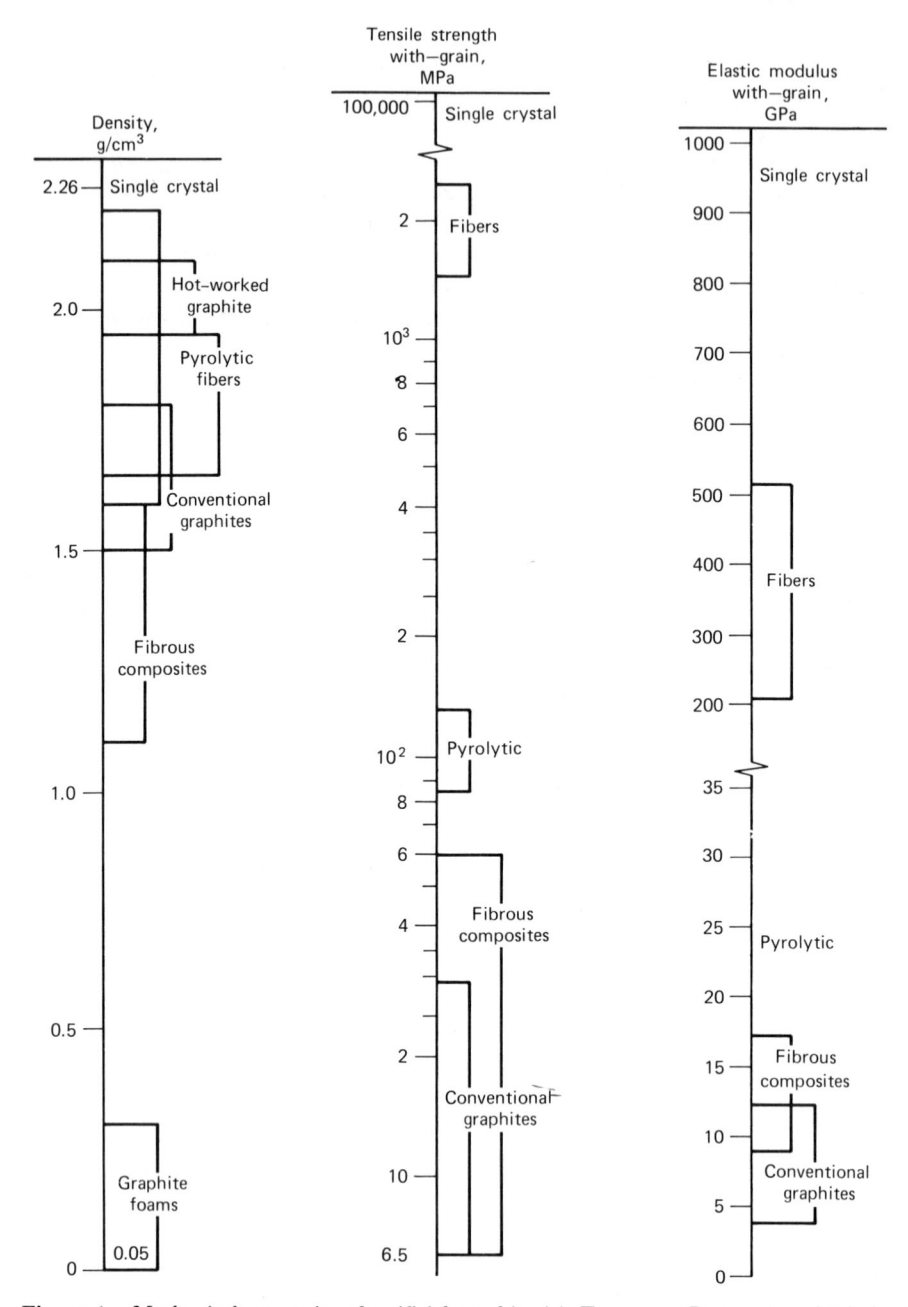

Figure 1. Mechanical properties of artificial graphite (1). To convert Pa to psi, multiply by 145 × 10^{-6}.

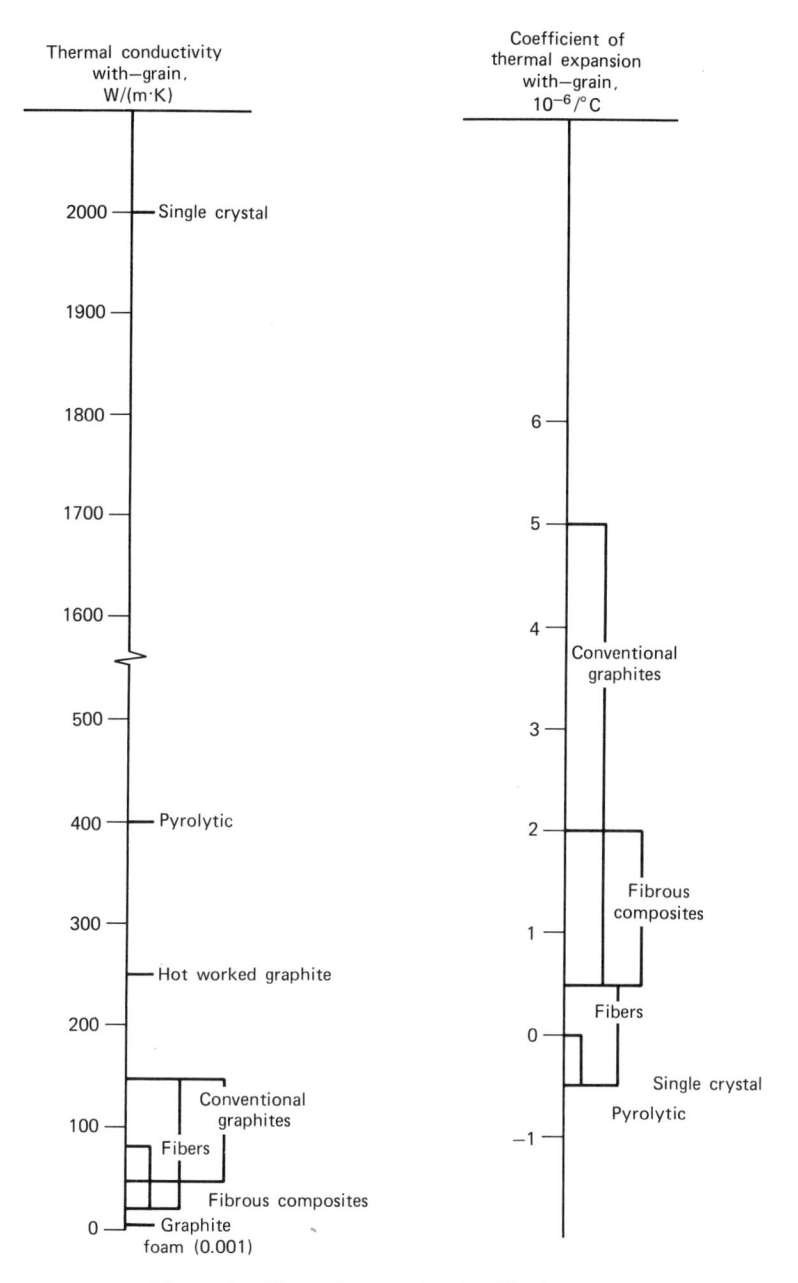

Figure 2. Thermal properties of artificial graphites (1).

plastic and exhibits viscoelastic creep under load (4). Graphite has high resistance to thermal shock, a property that makes it a valuable structural material at higher temperatures than most metals and alloys. For many applications of graphite, one or more of the following characteristics are important: density, elastic modulus, mechanical strength, electrical and thermal conductivity, and thermal expansion.

Electrical Properties. Manufactured graphite is semimetallic in character with the valence and conduction bands overlapping slightly (5–7). Conduction is by means of an approximately equal number of electrons and holes that move along the basal planes. The resistivity of single crystals as measured in the basal plane is approximately 40×10^{-4} Ω·m; this is several orders of magnitude lower than the resistivity across the layer planes (8–10). Thus the electrical conductivity of formed graphite is dominated by the conductivity in the basal plane of the crystallites and is dependent on size, degree of perfection, and orientation of crystallites and on the effective carbon–carbon linkages between crystallites. Manufactured graphite is strongly diamagnetic and exhibits a Hall effect, a Seebeck coefficient, and magnetoresistance. The green carbon body is practically nonconductive; however, heat treatment at 1000°C decreases the resistivity by several orders of magnitude, and thereafter resistivity decreases slowly. After graphitization to over 2500°C, the room temperature electrical resistivity may range from a few hundred to a few tenths Ω·m, depending upon the type of raw materials used. Graphites made from petroleum coke usually have a room temperature resistivity range of 5–15 Ω·cm and a negative temperature coefficient of resistance to about 500°C, above which it is positive. Graphites made from a carbon black base have a resistivity several times higher than those made from petroleum coke, and the temperature coefficient of resistance for the former remains negative to at least 1600°C.

Thermal Conductivity. Compared with other refractories, graphite has an unusually high thermal conductivity near room temperature (11); above room temperature, the conductivity decreases exponentially to approximately 1500°C and more slowly to 3000°C (12). With the grain the thermal conductivity of manufactured graphite is comparable with that of aluminum; against the grain it is comparable to that of brass. However, graphite is similar to a dielectric solid in that the principal mechanism for heat transfer is lattice vibrations. The electronic component of thermal conductivity is less than 1%. Graphite does not obey the Wiedemann-Franz Law; however, at room temperature the ratio of thermal and electrical conductivities is equal to approximately 0.126 when the thermal conductivity is in W/(m·K) and the electrical conductivity is in S(=1/Ω) (13–14). For most graphites, a value of thermal conductivity at room temperature accurate to ±5% can be obtained from the measured value of the electrical conductivity.

Coefficient of Thermal Expansion. The volumetric thermal expansion of manufactured graphite is anomalously low when compared to that of the graphite single crystal. At room temperature, the volume coefficient of thermal expansion of a single crystal is approximately 25×10^{-6}/°C (15–16), whereas those of many manufactured graphites fall in the range of 4–8×10^{-6}/°C. There are exceptions, and some commercially available, very fine-grain, near-isotropic graphites have a volumetric expansion as high as two thirds the value for the single crystal. The low value of volume expansion of most manufactured graphite has been related to the microporosity within the coke particles. The microcracks within the coke particle accommodate the large c-axis expansion of graphite crystallites (17–19) and effectively neutralize it. The coefficient of thermal expansion (CTE) is somewhat sensitive to the filler particle sizing and to the method of processing, but the anisotropy and perfection degree of filler carbon particles largely determine the expansion characteristics of the finished graphite. Except for differences in absolute values, plots of the CTEs of manufactured graphite vs temperature are essentially parallel to each other, showing that the change in CTE with temperature is approximately the same for all graphites at high temperatures. The mean linear coefficient of thermal expansion between room temperature and any final temperature can be obtained by adding the value of CTE for the temperature interval 20–100°C to the appropriate factor which varies from 0 at 100°C to 2.52×10^{-6} at 2500°C (20). This method is valid for stock of any grain orientation.

Mechanical Properties. The hexagonal symmetry of a graphite crystal causes the elastic properties to be transversely isotropic in the layer plane; only five independent constants are necessary to define the complete set. The self-consistent set of elastic constants given in Table 2 has been measured in air at room temperature for highly ordered pyrolytic graphite (21). With the exception of c_{44}, these values are expected to be representative of those for the graphite single crystal. Low values of shear and cleavage strengths between the layer planes compared with very high C–C bond strength in the layer planes, suggest that graphite always fails through a shear or cleavage mechanism. However, the strength of manufactured graphite depends upon the effective network of C–C bonds across any stressed plane in the graphite body. Until these very strong bonds are broken, failure by shear or cleavage cannot take place. Porosity affects the strength of graphite by reducing the internal area over which stress is distributed and by creating local regions of high stress. Because of the complexity of the graphite structure, a simple analytical model of failure has not been derived (22). The stress–strain relation for bulk graphite is concave toward the strain axis. The relaxation of the stress leads to a small residual strain; repeated stressing to larger loads followed by gradual relaxation leads to a set of hysteresis loops contained within the stress–strain envelope (4,23–26). Each successive load causes an increase in the residual strain and results in a decreased modulus for the sample. The residual strain can be removed by annealing the sample to the graphitizing temperature after which its original stress–strain response is restored. In the limit of zero stress, the elastic modulus of graphite is the same in compression and tension, and is equal to the modulus derived from dynamic measurements (27). The modulus of graphite is weakly dependent upon temperature, increasing with temperature to approximately 2000°C and decreasing thereafter. The strain at rupture of most graphites is 0.1–0.2%; however, values of strain at rupture approaching 1.0% have been obtained for specially-processed, fine-grain graphites (28). Graphite exhibits measurable creep under load and at temperatures above 1600°C, but for most applications creep can be neglected below 2200°C. As the temperature is increased above 2500°C, the creep rate increases rapidly and the short-time strength decreases rapidly.

Thermal shock resistance is a primary attribute of graphite and a number of tests have been devised in attempts to establish a quantitative method of measurement (29–30). These tests, which establish very large thermal gradients in small specially shaped samples, continue to give only qualitative data and permit one to establish only the relative order of shock resistance of different graphites. A commonly used thermal shock index is the ratio of the thermal conductivity and strength product to the expansion coefficient and modulus product (31). At high temperatures, values of this index for graphite are higher than for any other refractory material. To show the range of property values of graphite, several properties for a very coarse grain graphite and a very fine grain graphite are given in Table 3 (27,32).

Table 2. Elastic Constants of Graphite [a]

$c_{11} = 1.06 \pm 0.002$	$s_{11} = 0.98 \pm 0.03$
$c_{12} = 0.18 \pm 0.02$	$s_{12} = 0.16 \pm 0.06$
$c_{13} = 0.015 \pm 0.005$	$s_{13} = 0.33 \pm 0.08$
$c_{33} = 0.0365 \pm 0.0010$	$s_{33} = 27.5 \pm 1.0$
$c_{44} = 0.00018 - 0.00035$	$s_{66} = 2.3 \pm 0.2$

[a] Units c_{ij} (stiffness constant) in TPa, s_{ij} (compliance constant) in $(TPa)^{-1}$. To convert TPa to psi, multiply by 1.45×10^8.

Table 3. Properties of Fine and Coarse Grain Graphites[a]

Temperature, °C	Thermal cond. W/(m·K) wg	Thermal cond. W/(m·K) cg	CTE[b], cm/cm ×10⁶/°C wg	CTE[b], cm/cm ×10⁶/°C cg	Specific heat[c] kJ/(kg·K)	Tensile[d] Modulus, GPa wg	Tensile[d] Modulus, GPa cg	Tensile[d] Strength, MPa wg	Tensile[d] Strength, MPa cg	Compression[d] Modulus, GPa wg	Compression[d] Modulus, GPa cg	Compression[d] Strength, MPa wg	Compression[d] Strength, MPa cg
					Fine-grained graphite, 180 μm maximum grain size								
21.1	150	114	2.15	3.10	0.63	11.5	7.9	17.4	15.0	9.7	7.2	26.6	20.1
260	117	93	2.50	3.46	1.30	11.6	8.0	19.3	17.2	10.0	7.4	27.9	21.7
538	91	72	2.82	3.84	1.63	11.7	8.1	21.7	19.7	10.3	7.6	29.3	23.4
816	73	57	3.16	4.12	1.80	11.9	8.3	24.1	22.1	10.6	7.9	30.9	25.2
1093	60	46	3.45	4.45	1.95	12.1	8.6	26.0	24.3	11.4	8.3	32.4	26.9
1371	52	40	3.70	4.69	2.03	12.5	9.0	28.3	26.2	12.4	9.0	35.2	29.3
1649	46	35	3.95	4.91	2.11	13.2	9.6	29.9	27.9	13.4	9.7	38.1	31.6
1927	42	32	4.17	5.16	2.16	13.7	10.5	31.0	29.3	13.4	9.7	32.2	37.2
2204	40	29	4.35	5.39	2.18	11.5	8.4	31.7	30.1	12.1	9.0	37.9	32.6
2482	38	28	4.58	5.71	2.20	8.0	5.9	31.0	29.3	10.0	7.9	32.4	26.6
2760	36	28	4.83	6.04	2.20	5.2	4.3	26.9	24.8	7.9	6.2	26.6	19.1
					Coarse-grained graphite, 6400 μm maximum grain size								
21.1	156	108	0.46	1.03		4.2	2.6	3.75	2.91	3.0	2.6	9.3	12.1
1371	30	22	2.4	3.2		5.8	2.9	5.34	4.54	3.4	2.8	12.0	14.7
1927	24	19	2.7	2.85		6.5	3.7	5.39	4.36	4.3	3.3	14.1	17.4
2427	24	20	3.0	4.2		5.6	3.0	7.32	5.17				

[a] wg = with-grain; cg = cross-grain.
[b] CTE = coefficient of thermal expansion.
[c] To convert J to cal, divide by 4.184.
[d] To convert Pa to psi, multiply by 145 × 10⁻⁶.

Chemical Properties

The impurity (ash) content of all manufactured graphite is low, since most of the impurities originally present in raw materials are volatilized and diffuse from the graphite during graphitization. Ash contents vary from 1.5% for large diameter graphites to less than 10 ppm for purified graphites. Iron, vanadium, calcium, silicon, and sulfur are major impurities in graphite; traces of other elements are also present (33). Through selection of raw materials and processing conditions, the producer can control the impurity content of graphites to be used in critical applications. Because of its porosity and relatively large internal surface area, graphite contains chemically and physically adsorbed gases. Desorption takes place over a wide temperature range, but most of the gas can be removed by heating in a vacuum at approximately 2000°C.

Graphite reacts with oxygen to form CO_2 and CO, with metals to form carbides, with oxides to form metals and CO, and with many substances to form laminar compounds (34–35). Of these reactions, oxidation is the most important to the general use of graphite at high temperatures. Oxidation of graphite depends upon the nature of the carbon, the degree of graphitization, particle size, porosity, and impurities present (36). These conditions may vary widely among graphite grades. Graphite is less reactive at low temperatures than many metals; however, since the oxide is volatile, no protective oxide film is formed. The rate of oxidation is low enough to permit the effective use of graphite in oxidizing atmospheres at very high temperatures when a modest consumption can be tolerated. A formed graphite body alone will not support combustion. The differences in oxidation behavior of various types of graphite are greatest at the lowest temperatures, tending to disappear as the temperature increases. If an oxidation threshold is defined as the temperature at which graphite oxidizes at 1% per day, the threshold for pure graphite lies in the range of 520–560°C. Small amounts of catalyst, such as sodium, potassium, vanadium, or copper, reduce this threshold temperature for graphite by as much as 100°C but greatly increase the oxidation rate in the range of 400–800°C (33). Above 1200°C, the number of oxygen collisions with the graphite surface controls the oxidation reaction. Oxidation of graphite is also produced by steam and carbon dioxide; general purpose graphite has a temperature oxidation threshold of approximately 700°C in steam and 900°C in carbon dioxide. At very low concentrations of water and CO_2, there is also a catalytic effect of impurities on the oxidation behavior of graphite (37).

BIBLIOGRAPHY

"Baked and Graphitized Products" under "Carbon" in *ECT* 1st ed., Vol. 3, pp. 1–34, by H. W. Abbott, Speer Carbon Company; "Baked and Graphitized Products, Manufacture" under "Carbon" in *ECT* 2nd ed., Vol. 4, pp. 158–202, by L. M. Liggett, Speer Carbon Company.

1. E. L. Piper, *Soc. Min. Eng. AIME*, Preprint Number 73-H-14 (1973).
2. *Annual Book of ASTM Standards,* Part 17, American Society for Testing and Materials, Philadelphia, Pa., 1976.
3. N. A. Gokcen and co-workers, *High Temp. Sci.* **8,** 81 (June 1976).
4. E. J. Seldin, *Proceedings of the 5th Conference on Carbon,* Vol. 2, Pergamon Press, New York, 1963, p. 545.
5. B. D. McMichael, E. A. Kmetko, and S. Mrozowski, *J. Opt. Soc. Am.* **44,** 26 (1954).
6. G. A. Saunders in L. C. F. Blackman, ed., *Modern Aspects of Graphite Technology,* Academic Press, New York, 1970, p. 79.

7. J. A. Woollam in M. L. Deviney and T. M. O'Grady, eds., *Petroleum Derived Carbons,* American Chemical Society, Washington, D. C., 1976, p. 378.
8. N. Ganguli and K. S. Krishnan, *Nature (London)* **144,** 667 (1939).
9. A. K. Dutta, *Phys. Rev.* **90,** 187 (1953).
10. D. E. Soule, *Phys. Rev.* **112,** 698 (1958).
11. Y. S. Touloukian and co-eds., *Thermophysical Properties of Matter,* Vol. 2, IFI/Plenum Press, New York, 1970, p. 5.
12. B. T. Kelley, *Chem. Phys. Carbon* **5,** 128 (1969).
13. R. W. Powell and F. H. Schofield, *Proc. Phys. Soc. London* **51,** 170 (1939).
14. T. J. Neubert, private communication quoted by L. M. Currie and co-workers in *Proceedings of the International Conference on the Peaceful Uses of Atomic Energy, Geneva, 1955,* Vol. 8, United Nations, New York, 1956, p. 451.
15. J. B. Nelson and D. P. Riley, *Proc. Phys. Soc. London* **57,** 477 (1945).
16. B. T. Kelley and P. L. Walker, Jr., *Carbon* **8,** 211 (1970).
17. S. Mrozowski, *Proceedings of the 1st and 2nd Conferences on Carbon,* University of Buffalo, Buffalo, N.Y., 1956, p. 31.
18. A. L. Sutton and V. C. Howard, *J. Nucl. Mater.* **7,** 58 (1962).
19. W. C. Morgan, *Carbon* **10,** 73 (1972).
20. *Industrial Graphite Engineering Handbook,* Union Carbide Corporation, Carbon Products Division, New York, 1970, Section 5B.02.03.
21. O. L. Blakslee and co-workers, *J. Appl. Phys.* **41,** 3380 (1970).
22. W. L. Greenstreet, *U.S. Oak Ridge National Laboratory Report ORNL-4327* (Dec. 1968).
23. C. Malmstrom, R. Keen, and L. Green, *J. Appl. Phys.* **22,** 593 (1951).
24. P. P. Arragon and R. Berthier, *Industrial Carbon and Graphite,* Society of Chemical Industry, London, Eng., 1958, p. 565.
25. H. H. W. Losty and J. S. Orchard in *Proceedings of the 5th Conference on Carbon,* Vol. 1, Pergamon Press, New York, 1962, p. 519.
26. G. M. Jenkins, *Br. J. Appl. Phys.* **13,** 30 (1962).
27. E. J. Seldin, *Carbon* **4,** 177 (1966).
28. H. S. Starrett and C. D. Pears, *Southern Research Institute Technical Report, AFML-TR-73-14,* Vol. 1, 1973.
29. J. J. Gangler, *Am. Ceram. Soc. J.* **33,** 367 (1950).
30. E. A. Carden and R. W. Andrae, *Am. Ceram. Soc. J.* **53,** 339 (1970).
31. L. Green, Jr., *J. Appl. Mech.* **18,** 346 (1951).
32. J. K. Legg and S. G. Bapat, *Southern Research Institute Technical Report, AFML-TR-74-161,* 1975.
33. L. M. Currie, V. C. Hamister, and H. G. MacPherson, *Proceedings of the International Conference on the Peaceful Uses of Atomic Energy, Geneva, 1955,* Vol. 8, United Nations, New York, 1956, p. 451.
34. R. E. Nightingale, ed., *Nuclear Graphite,* Academic Press, New York, 1962, p. 142.
35. M. C. Robert, M. Aberline, and J. Mering, *Chem. Phys. Carbon* **10,** 141 (1973).
36. P. L. Walker, Jr., M. Shelef, and K. A. Anderson, *Chem. Phys. Carbon* **4,** 287 (1968).
37. M. R. Everett, D. V. Kinsey, and E. Romberg, *Chem. Phys. Carbon* **3,** 289 (1968).

J. T. MEERS
Union Carbide Corporation

APPLICATIONS OF BAKED AND GRAPHITIZED CARBON

Aerospace and Nuclear Reactor Applications

Graphite is an important material for aerospace and nuclear reactor applications because of a unique combination of thermal, chemical, and mechanical properties that enable its survival under the extremely hostile environments encountered.

Graphite is a lightweight structural material that retains good mechanical strength to extremely high temperatures and is readily machinable and commercially available. It also demonstrates good neutron interaction characteristics and stability under irradiation. The most troublesome problem is oxidation at high temperatures.

Aerospace and nuclear reactor applications of graphite demand both high reliability and reproducibility of properties, and physical integrity of the product. The manufacturing processes require significant additional quality assurance steps that result in high cost.

Aerospace. Graphite has long been employed in rocket nozzles, as wing leading edges, as nose cones, and as structural members for both ballistic and glider types of reentry vehicles (see Ablative materials). Graphite is unique in that it can be used both as a heat sink and as an ablation–sublimation material.

The erosion of graphite in nozzle applications is the result of both chemical and mechanical factors. Changes in temperature, pressure, or fuel-oxidizing ratio markedly affect erosion rates. Graphite properties affecting the erosion resistance include density, porosity, and pore size distribution.

The entrance cap, throat, and exit cone sections in a typical nozzle are frequently made or lined with conventional bulk graphite, especially in small nozzles because a small change in dimension causes a relatively large change in performance. In other designs, the throat may be made of conventional graphite with the entrance cap and exit cone molded of carbon or graphite fibrous materials that serve as reinforcement in conjunction with high-temperature plastic resins. In larger nozzles, all three sections might be made of fibrous, reinforced material owing to the ease of construction as well as the entire assembly being lighter in weight.

Nose cones and wing leading-edge components fabricated of graphite are used on both ballistic and glider types of reentry vehicles. Ballistic missiles are subjected to short-duration and extremely severe friction heating and oxidizing conditions when reentering the atmosphere, whereas glider-type reentry vehicles are exposed to less severe conditions for longer periods. Design technology has overcome any adverse effects of high anisotropy relative to thermal stress.

Nuclear Reactors. Manufactured graphite is the most extensively used material for moderator and reflector materials in thermal reactors. Since its use in the first reactor, CP-1, constructed in 1942 at Stagg Field, University of Chicago, many thousands of metric tons of graphite have been used for this purpose. Recently, great interest has been shown in the use of graphite as a construction material in the High-Temperature Gas-Cooled Reactor (HTGR) system.

Graphite is chosen for use in nuclear reactors because it is the most readily available material with good moderating properties and a low neutron capture cross section. Other features that make its use widespread are its low cost, stability at elevated temperatures in atmospheres free of oxygen and water vapor, and its good heat transfer characteristics, good mechanical and structural properties, and excellent machinability.

Neutron economy in graphite occurs since pure graphite has a neutron capture cross section of only $0.0032 \pm 0.0002 \times 10^{-24}$ cm^2. Taking into account the density of reactor grade graphite (bulk density 1.71 g/cm^3), the bulk neutron absorption coefficient is 0.0003/cm. Thus a slow neutron may travel >32 m in graphite without capture.

The purity of reactor-grade graphite is controlled by raw material selection and subsequent processing and purification. Although high temperature purification is most commonly used, some moderator applications require considerably higher purity levels. This objective is accomplished by halogen purification to remove extremely stable carbides, especially of boron, as volatile halides. The actual purity requirements are determined by the reactor design.

The major effect of high temperature radiation (1) over a long period of time is to produce dimensional changes in the graphite involved. When graphite initially contracts upon exposure to fast neutron doses, the rate of contraction decreases with exposure until it reaches a minimum volume; further exposure causes volume expansion, with the rate of expansion increasing rapidly at neutron doses above 3×10^{22} neutrons/cm^2 (>50 keV) in all bulk graphite tested to date. This behavior is caused by atomic displacements that take place when graphite is exposed to fast neutrons, resulting in anisotropic crystallite growth rates. The crystal expands in the c-axis direction and contracts in the a-axis directions (see page 691). The bulk dimensional change depends upon the geometrical summation of the individual crystallite changes and, hence, is dependent upon the starting materials and the method of fabrication. The extent of radiation damage is also strongly dependent upon the temperature of the graphite during irradiation. The severity of graphite radiation damage at high temperatures was underestimated since the magnitude of this temperature dependence was not recognized until about 1965.

Figure 1 shows the volume change in a conventional nuclear graphite during irradiations at various temperatures of relatively high fluxes. Figure 2 shows the length change in an isotropic nuclear graphite during irradiations at various temperatures at relatively high fluxes. The actual changes in dimensions are, of course, different from grade to grade and depend largely on the degree of anisotropy present in the graphite (1).

Table 1 (2) lists some useful properties of several graphites used for moderators or reflectors in nuclear reactors.

Reactor designers have taken advantage of graphite's properties in applying the material to other than moderator and reflector components, usually in conjunction with some other material.

Combined as an admixture with some forms of boron or other high-neutron-absorbing elements, graphite offers advantages as a neutron shield, control rod, or secondary shutdown material of high temperature stability, without danger of meltdown. In fast reactors, where high-energy neutrons reach the shield region, the presence of carbon atoms slows these neutrons down to energy levels where the probability of capture in the neutron absorber greatly increases. Graphite also serves as a stable matrix for the neutron absorber because it is able to withstand neutron and localized alpha recoil damage, offering protection against gross shield degradation.

Bulk graphites are also used in the HGTR concept to support and surround the active fuel core. These components tend to be large, complex-shaped blocks and have been produced from commercial grades of molded graphites.

In combination with compounds of uranium or thorium, graphite offers advantages as a matrix for fissile or fertile reactor fuel in thermal reactors. In this instance, the graphite serves a dual purpose, as a moderator and as a stable disbursing phase for fuel. Its stability under irradiation and at high temperature aids in minimizing fuel degradation and permits longer useful fuel life. Because of its excellent thermal properties and mechanical integrity, graphite offers an exceptional heat transfer medium for heat removal and also resists thermal shock.

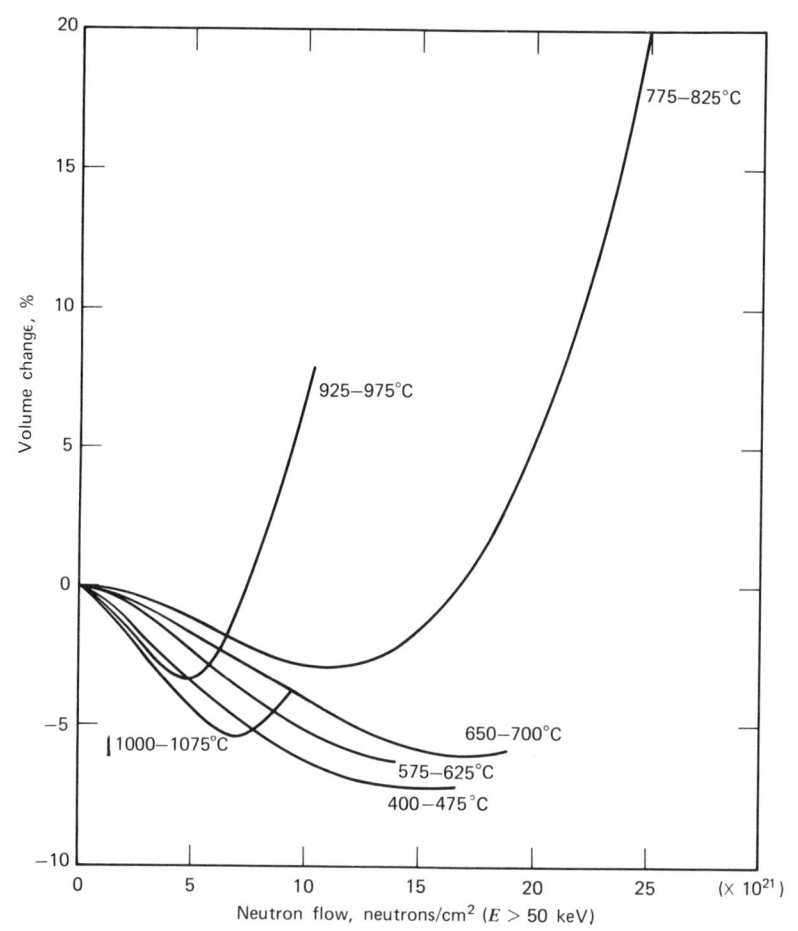

Figure 1. Volume change in anisotropic graphite during GETR (General Electric Test Reactor) irradiations. Courtesy of Oak Ridge National Laboratory, operated by Union Carbide Corporation for the DOE, former Energy Research and Development Administration.

Chemical Applications

The excellent corrosion resistance of carbon and graphite (3) and that of impervious carbon and graphite to acids, alkalies, organics, and inorganic compounds has led to the use of these materials in process equipment where corrosion is a problem. Most of these applications are in chemical process industries but many are in steel, food, petroleum, pharmaceutical, and metal finishing industries.

Other properties, such as the high thermal conductivity of graphite, excellent high temperature stability, and immunity to thermal shock, make these materials useful in applications involving combinations of heat and corrosion, such as heat exchange and high-temperature gas-spray cooling.

Carbon and graphite exhibit varying degrees of porosity, depending on grade, and equipment fabricated of these materials must be operated essentially at atmospheric pressure; otherwise, some degree of leakage must be tolerated. A good example is the carbon brick used to line tanks and vessels handling acids such as hydrofluoric,

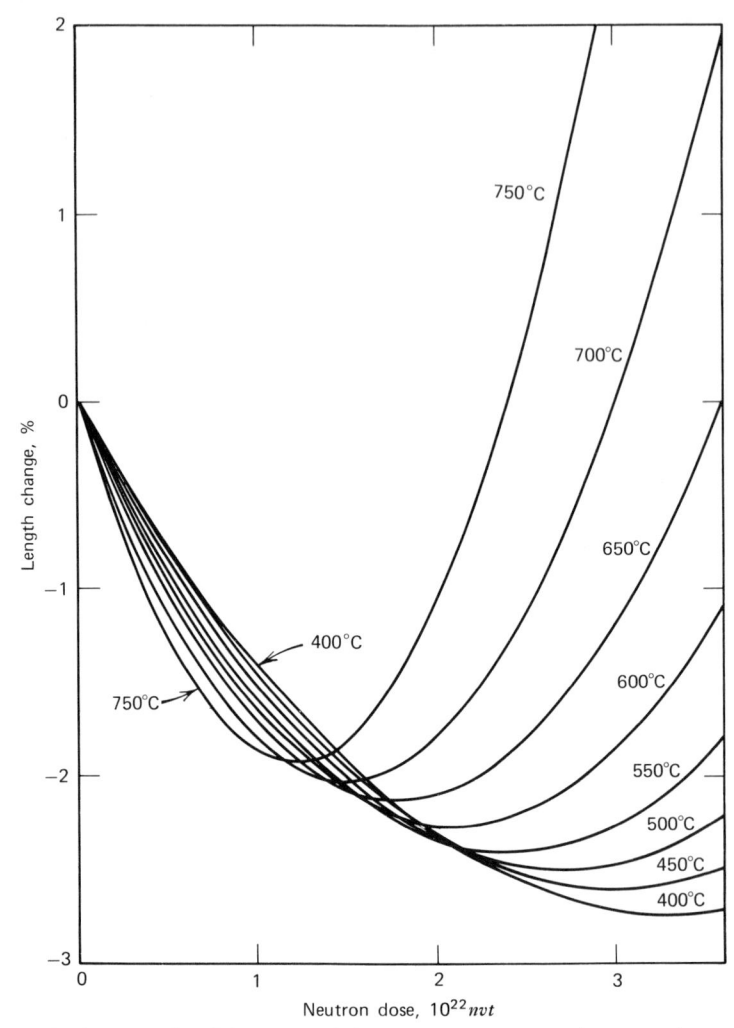

Figure 2. Radiation induced dimensional changes in isotropic graphite at various temperatures. nvt = neutron (density) velocity time. Courtesy of Oak Ridge National Laboratory, operated by Union Carbide Corporation for the DOE, former Energy Research and Development Administration.

nitric–hydrofluoric, phosphoric, sulfuric, and hydrochloric (3). An impervious backing membrane of lead (4), elastomer, or plastic stops the seepage through the brick lining at the outside of the brick. These membrane materials, by themselves, may not withstand the corrosive and temperature conditions in the vessel; however, as a membrane behind the carbon lining, they are protected from adverse temperature and abrasion effects. Carbon linings have provided indefinite life with a minimum of maintenance.

Self-Supporting Structures. Self-supporting structures of carbon and graphite are used in a variety of ways. Water-cooled graphite towers serve as chambers for the burning of phosphorus in air or hydrogen in chlorine. The high thermal conductivity of graphite allows rapid heat transfer to the water film, thus maintaining inside wall temperatures below the graphite threshold oxidation temperature of ca 500°C.

Table 1. Properties of Nuclear Graphites[a]

Property	Anisotropic graphite	Isotropic graphite
density, g/cm^3	1.71	1.82
resistance, $\mu\Omega\cdot$cm	735	1,000
tensile strength, kPa[b]	9,930	16,550
coefficient of thermal expansion (CTE), 10^{-6}/°C		
with grain	2.2	4.8
against grain	3.8	4.6
ansiotropy ratio (CTE ratio)	1.73	0.96
total ash, ppm	740	500
boron content, ppm	0.4	0.3

[a] Ref. 2.
[b] To convert kPa to psi, multiply by 0.145.

Phosphorus combustion chambers for the production of thermal phosphoric acid using cemented graphite block construction have been built 6 m in diameter and 11 m high (5). In other thermal phosphoric acid systems, cement–carbon block structures are used as spray-cooling, hydrating, and absorbing towers.

The immunity of graphite to thermal shock, its high temperature stability, and its corrosion resistance, permit its use for the fabrication of high temperature (800–1650°C), self-supporting reaction vessels such as those used in direct chlorination of metal and alkaline earth oxides. Cemented joints are vulnerable in the units, and it is necessary to machine the graphite components to close tolerances (±0.10 mm) to minimize the cement joint thickness. Graphite is easily machined on commercial metal-machining equipment.

Carbon Raschig-ring tower packing is available in sizes of 10–75 mm dia. Bubble-cap trays up to 3 m dia for hydrochloric–organic stripping towers and packing support structures up to 5.5 m dia for scrubbing towers in pulp and paper mill liquor recovery processes have been installed. Because none of these components requires complex machining or a high degree of imperviousness, carbon rather than graphite is often used in these applications because of its lower cost.

Impervious Graphite. For those applications where fluids under pressure must be retained, impregnated materials are available (3). Imperviousness is attained by blocking the pores of the graphite or carbon material with thermosetting resins such as phenolics, furans, and epoxies. Because the resin pickup is relatively small (usually 12–15 wt %), the physical properties exhibited by the original graphite or carbon material are retained. However, the flexural and compressive strengths are usually doubled. Graphite is also made impervious in a vacuum impregnation process.

Because carbon is difficult to machine, very little impervious carbon equipment is made. However, impervious graphite has been accepted as a standard material of construction by the chemical process industry for the fabrication of process equipment, such as heat exchangers, pumps, valves, towers, pipe, and fittings (6–7).

Many types of impervious graphite shell and tube, cascade, and immersion heat exchangers are in service throughout the world (8). The most common is the shell and tube design where an impervious graphite tube bundle with fixed and floating covers is employed in combination with a steel shell. Whenever parts must be joined, such as the tube to the tube sheet in a shell and tube heat exchanger, very thin resin cement joints are used. These resin cements have the same corrosion-resistant characteristics as the resins used to impregnate the graphite. Because of the high thermal conductivity of graphite, heat exchangers fabricated of impervious graphite have thermal efficiencies equal to metal heat exchangers of equivalent heat transfer area. Heat exchangers up to 1.8 m dia with areas up to 1300 m^2 are commercially available with operating pressures to 690 kPa (100 psi) and temperatures to 170°C (9–11).

Impervious graphite shell and tube heat exchangers are used in boiling, cooling, condensing, heating, interchanging, and cooling–condensing. Large units are used extensively for cooling–condensing wet sulfur dioxide gas in sulfuric acid production plants that burn sludge acid (4). The operation of six units was analyzed to compare actual operation with the calculated design using the Colburn-Hougen analogy; good agreement was reported (12).

These heat exchangers are also used for evaporation of phosphoric acid and rayon spin bath solution; cooling electrolytic copper cell liquor; heating pickle liquor used for descaling sheet steel; boiling, heating, cooling, and absorbing hydrochloric acid and hydrogen chloride; and in many heating and cooling applications involving chlorinated hydrocarbons and sulfuric acid.

Impervious graphite heat exchangers machined from solid blocks are also available (13–14). The solid block construction is less susceptible to damage by mechanical shock, such as steam and water hammer, than are shell and tube exchangers. Block exchangers are limited in size and cost from 50–100% more than shell and tube units on an equivalent area basis.

For in-tank heat exchange, a variety of immersion heat exchangers, such as plate, coil, and steam injection types, are available in impervious graphite. These units are highly efficient, and the heat transfer area required for most services can be easily determined from manufacturers' monographs.

Impervious graphite centrifugal pumps, pipe fittings, and valves were developed because most chemical processes require the movement of liquids.

Impervious graphite pipe and fittings of 25–635 mm ID are used to convey corrosive fluids.

Towers, entrainment separators, thermowells, and rupture disks are fabricated of impervious graphite material. Many equipment items are available from stock. Special equipment can be custom-designed and built, and both standard and special items can be integrated to handle a complete process step. Systems for the absorption of hydrogen chloride in water to produce hydrochloric acid use impervious graphite equipment throughout. Usually, absorption is done in a falling-film absorber (15), a special design adaption of the shell and tube heat exchanger. This approach to absorption of hydrogen chloride (16) was developed and expanded in the United States and is now accepted as the standard.

Stripping hydrogen chloride (13–19) from aqueous hydrochloric acid, and the subsequent production of anhydrous hydrogen chloride, can be efficiently and economically achieved with a series of impervious graphite shell and tube heat exchangers that operate as falling-film reboiler, water and brine-cooled condensers, and bottoms acid cooler. In plants with available chlorine and hydrogen, the production of hydrogen chloride in any form or concentration can be achieved in a system that combines the burning of hydrogen in chlorine in a water-cooled graphite combustion chamber; absorption is carried out in an impervious graphite falling-film absorber, and a train of impervious graphite exchangers is used for stripping and drying (20).

Low Permeability Graphite. Most resin-impregnated impervious graphite materials have a maximum operating temperature limit of 170°C because of resin breakdown above this temperature. Certain special grades with a temperature limitation of 200°C are on the market (21). The chemical industry has developed high temperature processes (370°C and above) where equipment corrosion is a serious problem. Graphite equipment could solve the corrosion problem, but complete fluid containment is usually needed. To meet this need, graphite manufacturers have developed low permeability graphite materials where permeability is reduced by deposition of carbon and graphite in the pores of the base material (21). This material is not limited in its operating temperature, except in oxidizing conditions, and it is used to fabricate high-temperature interchanger ejectors, fused salt cells, fused salt piping systems, and electric resistance heaters.

Several grades of carbon and graphite are commercially available. A controlled combination of high permeability and porosity characterizes these materials. Average pore diameters for typical grades are 0.03–0.12 mm with a total porosity of 48%. Porous graphite is manufactured by graphitizing the amorphous material.

Porous carbon and graphite are used in filtration of hydrogen fluoride streams, caustic solutions, and molten sodium cyanide; in diffusion of chlorine into molten aluminum to produce aluminum chloride; and in aeration of waste sulfite liquors from pulp and paper manufacture and sewage streams.

Electrical Applications

Arc Carbons. A current of electricity produces carbon arc light by leaping across a gap between two carbon electrodes. The typical arc light for visible radiation uses direct current which produces a relatively uniform ball of light across the crater of the positive electrode. The core of the positive electrode contains rare earth compounds to produce white light in the visible portion of the radiation spectrum.

The carbon arc can produce the highest useful brightness of any known artificial light source and provides a color quality matching that of sunlight. The size and shape of the carbon arc light source and the distribution of brilliancy bear an important relation to the optical problems involved in its use. The flat light source at the crater of the positive electrode produces a relatively even brilliancy distribution over a considerable portion of the area. This characteristic permits the light to be collected and directed with a simple reflector or lenses.

The carbon rods used in carbon arc lighting usually consist of two parts, the shell and the core. The shell forms the outer wall and has a central longitudinal hole throughout its length where the core is inserted. The core hole is formed during the extrusion process. The core is usually inserted into the shell after the shell is baked, or alternatively, the shell and core may be extruded simultaneously (22). The core contains, in addition to the conventional carbon filler–binder mix, small percentages of an arc-supporting material, such as potassium, and a flame material such as the metals of the cerium group. For some applications, the carbons are copper-plated to increase electrical conductivity.

Carbon arcs have been used for motion picture projection since the earliest days of the motion picture industry. Owing to the simplicity and reliability of the carbon arc system, thousands of carbon arc lamps are still in use after 20–30 yr. The quantity of light projected by the arc lamps is 10,000–50,000 lm. The amount of light depends on the carbon size and required brightness.

Projector carbon electrodes are manufactured in 7–13.6 mm dia. The diameter of the light ball at the positive crater is proportional to the carbon diameter. Carbon diameters are selected so that the light image at the projector aperture has a side to center brightness of at least 80%.

Peak crater brightness of the d-c high-intensity arc is near 1.8 ncd/m^2. The luminous efficiency is on the order of 25–35 lm/W. Color temperature of the carbon arc light is ca 4000–7000 K, and is dependent on the composition of the core of the positive electrode.

Carbon arc lighting is also used for spotlighting, for searchlights, and in the graphic arts industry for photoengraving, photolithography, and other mechanical reproduction processes. The light sensitized materials, such as dichromates, used for transferring the photographic negative image for subsequent graphic arts operations are affected by radiation in the 0.3–0.5 μm range. Carbon electrodes are specially designed to give a high spectral response in this spectral range.

Light from the carbon arc is used to simulate outdoor exposure for measuring the relative lightfastness, fading, and deterioration of materials such as clothing, paint, plastics, etc. Materials can be tested under precisely controlled conditions which may be repeated often (23–24).

Simulation of the solar radiation in space, by means of the carbon arc, is used for environmental testing of space vehicles, missiles, and their components (25).

Brushes for Motors, Generators, and Slip Rings. Electric motors and generators employ carbon and graphite brushes to maintain electrical contact between an external circuit and the windings of the machine's rotating element. Brushes have several functions, each being essential to the satisfactory operation of the particular machine (26). The brushes must provide electrical contact: they must be good conductors of electricity. The brushes must maintain firm mechanical contact with the rapidly revolving surface of the commutator and function as a bearing material, and they must control the currents resulting from commutation (27). These functions must be fulfilled without destructive sparking, serious commutator wear, or excessive electrical or frictional losses. The brushes must also maintain a smooth commutator surface and a commutator film of suitable characteristics. Only carbon and graphites are capable of meeting all these requirements.

The brushes currently manufactured may be classified into four general types: electrographitic, carbon–graphite, resin-bonded graphite, and metal–graphite. Electrographitic brushes are generally made from calcined blacks and coal-tar pitch. These composites may be supplemented by the addition of graphites and other additives. Mixing and forming follow conventional carbon processing procedures. The final baking temperature approaches 3000°C. Electrographitic brushes have exceptionally good commutating properties and low friction coefficients; they are also efficient in high speed, high current density applications. Carbon–graphite brushes are made from calcined carbon black, calcined coke, graphite, or any combination of the three. They are more abrasive than electrographitic brushes and are generally adapted to lower speed and lower current density applications. Resin-bonded graphite brushes are made from graphite flours and polymeric resins. These brushes have high electrical resistance and are used in applications where very low current densities are required. Metal–graphite brushes can be classified into two general types: those where powdered metal and graphite are mixed and bonded mechanically or chemically and those where carbon–graphite base stock is impregnated with molten metal. Because of the lowered electrical resistance of metal–graphite brushes, they are usually used where high current carrying capacity is required. Table 2 lists typical physical properties of each type of brush.

The performance of brushes in terms of wear and commutation is dependent upon environmental, electrical, and mechanical aspects of each particular application. The environmental aspects which most affect the commutator surface film and consequent performance are temperature and atmosphere. In the absence of water vapor or oxygen, the brushes lose their self-lubricating quality and some impregnant is generally employed to make up for this deficiency. Performance is also dependent on the electrical design of the commutating machine: the location of the brushes with respect to the electrical neutrality of the field poles, the current density required by the application, and the number of commutator bars bridged by the brush face. The mechanical features that affect brush performance include commutator design, brush configuration, holder design, and wear of the moving parts on the commutating machine.

Spectroscopic Electrodes and Powders. Electrodes made of some form of carbon or graphite are employed extensively in optical emission spectrochemical analysis (28–35). The sample to be analyzed (in powder, solid, or liquid form) is excited by means of an electric arc at 5,000–8,000°C or by a spark source at 7,000–20,000°C between two electrodes. At these temperatures, most chemical compounds are dissociated and elements are excited. The chemical elements present are determined qualitatively, semiquantitatively, or quantitatively by measurements of the wavelength and intensity of spectral lines produced by such excitation and dispersed by a suitable optical device.

Graphite and carbon electrodes are used in spectrochemical analysis for several reasons: they are available in high purity grades at reasonable cost; graphite electrodes are easily machinable; they have a uniform structure; they have reasonable electrical and thermal properties; they emit few spectral lines or bands; they do not absorb appreciable moisture; they are chemically inert at room temperature and not wetted even at arc temperatures by most materials; they are porous; they produce an arc with a high excitation potential; they produce a chemically reducing atmosphere in the arc discharge; and they sublime rather than melt.

Graphite is the most universally employed material for spectroscopic electrodes because it is easier than carbon to machine into desired shapes and ship without breakage. Graphite electrodes are available in rods of different diameters and in a variety of preforms, whereas carbon is offered only in rod form. Spectroscopic electrodes are usually supplied in several grades of graphite and one grade of carbon. The graphite grades are of various densities and of different purity levels. The carbon grade is a low crystallinity, high purity material of moderate density. A given spectroscopic grade of graphite or carbon is determined by selection of raw materials and processing conditions.

High purity graphite and carbon powders are also used extensively in spectroscopic techniques as matrix material, either in powder form or as spectroscopic pellets. Graphite carries the sample into the arc thus eliminating excessively long burn times; in addition, fractional distillation of elements from the sample is reduced during exposure. Artificial or synthetic graphite has excellent mixing properties because its almost spherical particles flow more easily and can be transferred and mixed thoroughly with other powders. Natural graphite is specially suited for pelletizing applications because of its flat platelets or needles (depending on whether it is of the

Table 2. Typical Physical Properties of Brushes

Type	Resistivity, $\mu\Omega\cdot m$	Apparent density, g/cm^3	Scleroscope hardness	Flexural strength, MPa[a]	Current capacity, kA/m^2	Contact drop[b]	Coefficient of friction[b]	Abrasiveness[b]
electrographitic	10	1.72	35	14	93	M	M	VL
carbon–graphite	63	1.43	60	34	124	H	L	L
resin-bonded	20	1.80	40	25	62	M	M	M
graphite	1400	1.60	9	6	47	VH	L	VL
metal–graphite	13	1.90	30	31	101	L	M	L
	0.05	6.60	8	103	232	VL	VL	VL
	8	2.50	45	21	116	L	M	M

	VL = very low	L = low	M = medium	H = high	VH = very high
contact drop, V	<0.05	ca 0.2	ca 0.4	ca 0.6	≥0.8
coefficient of friction	≤0.15	≤0.22	≤0.3	≥0.3	

[a] To convert MPa to psi, multiply by 145.
[b]

Madagascar or Sri Lanka (Ceylon) type) which compress along their long slippage plane and cohere to a large contact area.

The areas where optical emission spectrochemical analysis and graphite or carbon spectroscopic products are employed are varied; they include: metallurgy, geology and prospecting, agriculture, wear, metal analysis, the petroleum industry, medicine, forensics, customs work, and astronomy (see Analytical methods).

The temperature of the electrode, the temperature distribution, and the processes occurring in the arc are of concern to the spectroscopist. Vaporization of the electrode material into the arc is a very complex mechanism affected by particle size, particle shape, particle bonding, the degree of crystallization, and other factors. It is possible to control these properties by manufacturing techniques. Selected raw materials are blended under special conditions to prevent contamination. The blend is extruded into rods and subsequently graphitized. From the graphitized rod stock, preformed electrodes are machined to specific dimensions with precision tools and then purified by a halogen gas process that removes the last traces of nearly all contaminants. After the purified electrodes are tested by means of a very sensitive cathode layer spectrographic technique to make sure they meet the purity requirements, they are packed in spectroscopically clean containers. Each step must be carefully executed and controlled to assure a consistent and uniform product, with a total impurity level of less than 6 ppm.

Miscellaneous Electrical Applications. Carbon and graphite can be processed to have the electrical, mechanical, chemical, and thermal properties required for many electrical and electronic applications. Graphite is an ideal electrode material for electrical discharging machining (EDM). Ultrafine grain, high strength grades that are readily machinable to complex shapes are used as electrodes in the EDM process. Fine-grain graphites are used as EDM electrodes in roughing applications where rapid metal removal and electrode cost are of primary concern (see Electrolytic machining methods).

Graphite is employed in a variety of electronic tube applications. The low inherent gas evolution of graphite makes it easily outgassed; it is pure, free from melting and distortion at high temperatures, and conforms closely in radiating characteristics to the theoretical black body (36). Graphite is used for many electronic tube applications (37). Combinations of the electrical and mechanical properties of carbon and graphite make these materials useful in such diverse applications as pantograph contacts and collector shoes, rheostat disks and plates, high-temperature furnace heating elements, telephone and microphone components, and lightning arrestor parts (27).

Electrode Applications

With the exception of carbon use in the manufacture of aluminum, the largest use of carbon and graphite is as electrodes in electric-arc furnaces. In general, the use of graphite electrodes is restricted to open-arc furnaces of the type used in steel production, whereas carbon electrodes are employed in submerged-arc furnaces used in phosphorus, ferroalloy, and calcium carbide production.

Graphite Electrodes. Graphite electrodes are commercially produced in many sizes ranging from 32 mm dia by 610 mm length to 700 mm dia by 2800 mm length. Such electrodes are used in open-arc furnaces for the manufacture of steel (38), iron and steel castings, brass, bronze, copper and its alloys, nickel and its alloys, fused cast refractories, and fused refractory grain. By far the largest use of graphite electrodes is in the manufacture of steel and, as a consequence, the growth of graphite production has been closely related to the growth in electric furnace steel production. A cutaway sketch of an open-arc furnace is shown in Figure 3. Steel is produced by filling the cylindrical shell with ferrous scrap, metallized iron ore, or occasionally, molten pig iron, then melting and refining the metallic charge with the heat derived from the electric arc generated at the tips of the electrodes.

Prior to the mid 1940s, the arc furnace was used almost exclusively for the production of low tonnage, high quality steels such as stainless and alloy steels. Since then its use has been extended to production of the more common high-tonnage steel grades (see Steel). Domestic growth of arc furnace steel production has been dramatic, rising

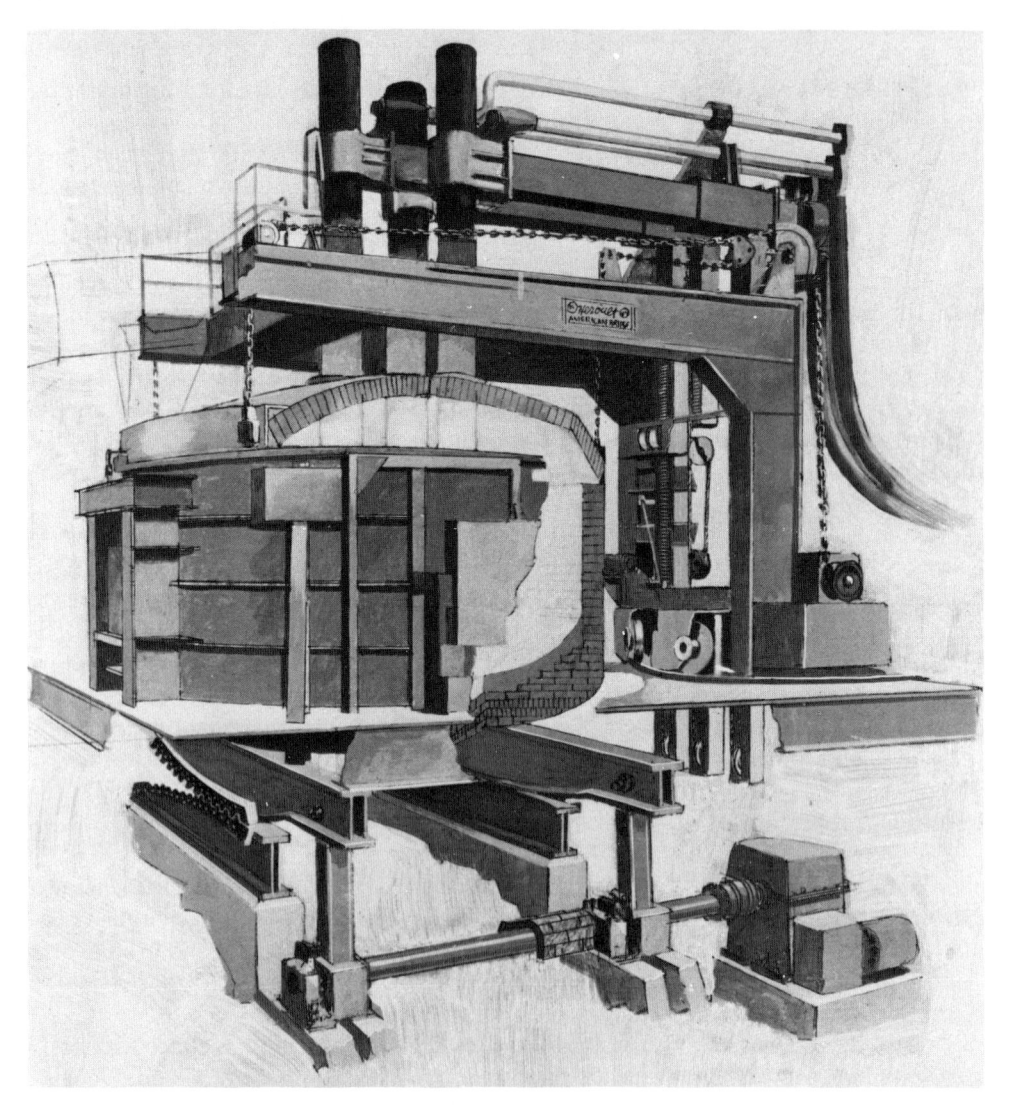

Figure 3. Overall sketch of an electric-arc furnace, cut away to show sections of bottom, sidewall, and roof. Courtesy of American Bridge Division, U.S. Steel Corporation.

from 6% of total domestic steel production in 1950 to 20% in 1975. Over 100 million metric tons of steel were produced in 1975 in electric-arc furnaces in the free world, approximately 20% of total world steel production (39), and these furnaces consumed over 600,000 metric tons of graphite electrodes.

Graphite electrodes are consumed in the melting process. For iron and steel production, the average consumption is ca 5–8 kg/t, depending upon the quality of charge material, the quality of electrodes, and numerous factors related to the operation of the furnace (22). Electrode consumption can be classified into three broad categories: tip consumption, sidewall consumption, and breakage. Roughly half of the observed consumption occurs at the electrode tip where the intensely hot and rapidly moving arc spot produces both vaporization of the graphite and some ejection of small graphite particles. In addition, the electrode tip is eroded by contact with the liquid metal and slag. The rate of incremental tip consumption generally increases when operating currents or power are increased. However, since increased current or power levels generally result in higher productivity, the electrode consumption when expressed in terms of kg/t of metal produced, may exhibit little or no increase. The periphery or sidewall of the hot electrode is slowly consumed by reaction with oxidizing atmospheres both inside and outside the furnace, resulting in a tapering of the electrode toward the arc tip. Sidewall consumption is time dependent and is greatest for low productivity furnaces (40). It is also increased by the use of many fume removal systems and by the use of oxygen in the furnace for assisting melting or refining. Since sidewall consumption may account for 40% or more of total electrode consumption, extensive efforts have been made to reduce this component of consumption through the use of oxidation retardants and electrode coatings. Such efforts have had little success to date, primarily because of the extreme thermal and chemical environment to which the electrode is exposed. A third form of consumption consists primarily of electrode breakage resulting from excessive movement of large masses of scrap during melting or the presence of nonconductors in the charge. Although such breakage usually accounts for less than 10% of net electrode consumption, excessive thermal shock, improper joining practice, and incorrect phase rotation can magnify this form of electrode consumption (41). Although attempts have been made to correlate electrode consumption with relatively small changes in electrode properties, it is apparent that charge quality and furnace operating practice exert a more profound influence on electrode performance. Most notable is the established inverse relationship between furnace productivity and electrode consumption (40).

Graphite electrodes are produced in two broad classifications, regular grade and premium grade. Typical properties of these grades are given in Table 3.

The major differences between the two grades are that the premium grade is made from needle-grade coke and is pitch-impregnated prior to graphitization. The premium grade electrode is used where very high performance is required, such as in the high current operation typical of ultrahigh-powered arc furnaces. The current carrying capacity of an electrode column depends on many characteristics of the furnace operation as well as the characteristics of the electrode and electrode joint. Over the years, significant progress has been achieved in improving the current carrying capacity of electrode columns. For example, the 510 mm dia electrode first introduced in 1938 was designed to carry 26,000 A; by 1961, this same size electrode carried ≤45,000 A and, by 1975, permitted >55,000 A. Such improvements stem primarily from improved raw materials and process technology advancements that are not fully reflected in changes in electrode properties (42).

Table 3. Typical Properties of Regular and Premium Grade Graphite Electrodes[a]

Property	Regular grade	Premium grade
bulk density, g/cm^3	1.58	1.68
resistivity, $\mu\Omega\cdot$m	7.7	6.0
flexural strength, kPa[b], wg	6900	10350
cg	5865	8300
elastic modulus, GPa[b], wg	6.2	7.6
cg	3.5	5.5
coefficient of thermal expansion (CTE), $10^{-6}/°C$		
wg	0.8	0.9
cg	1.7	1.5
thermal conductivity W/(m·K), wg	134	168
cg	67	101
thermal shock parameter[c], wg	186×10^3	254×10^3
cg	66×10^3	102×10^3

[a] wg = with-grain; cg = cross-grain.

[b] To convert Pa to psi, multiply by 1.45×10^{-4}.

[c] Thermal shock parameter = $\dfrac{\text{thermal conductivity} \times \text{strength}}{\text{CTE} \times \text{elastic modulus}}$

In service, graphite electrodes operate at up to 2500 K and are subject to large thermal and mechanical stresses and extreme thermal shock. Graphite is unique in its ability to function in this extreme environment. The relatively low electrical resistance along the length of the electrode minimizes the power loss owing to resistance heating and helps keep the electrode temperature as low as possible. This characteristic is most important in ultrahigh-power furnaces. For such furnaces approximately 30% lower electrode resistivity of premium grade electrodes is usually essential. A high value of the thermal shock parameter is also important (see Table 3); this is enhanced by high strength and high thermal conductivity combined with low elastic modulus and low coefficient of thermal expansion.

The joints between electrodes are an extremely important part of the electrode system, both from the standpoint of resisting the mechanical forces of scrap caves and of carrying high current density without localized overheating (42a). Such joints should possess high strength, especially in flexure, and possess low electrical resistance. Careful assembly and proper torque are vital to good performance (41).

Carbon Electrodes. Carbon electrodes are used primarily in submerged-arc furnaces for the manufacture of ferroalloys, phosphorus, silicon metal, calcium carbide, pig iron, and fused refractory grain. There are two broad types of carbon electrodes. The self-baking Soderberg type consisting of carbon paste is used extensively in the production of calcium carbide and electric-furnace pig iron; and the prebaked electrode finds its major use in production of phosphorus, silicon metal, and several ferroalloys. Prebaked carbon electrodes are produced in sizes ranging from 250 mm dia by 1520 mm length to 1400 mm dia by 2800 mm length. In general, the sizes up to 1140 mm dia are extruded; the larger sizes are molded.

Submerged-arc furnaces (see Fig. 4) differ significantly from open-arc furnaces in both function and operation. The name submerged-arc furnace is derived from the fact that the high temperature reaction zone in such furnaces is always isolated from the walls and roof by cooler charge material, ie, the reaction zone is submerged. As

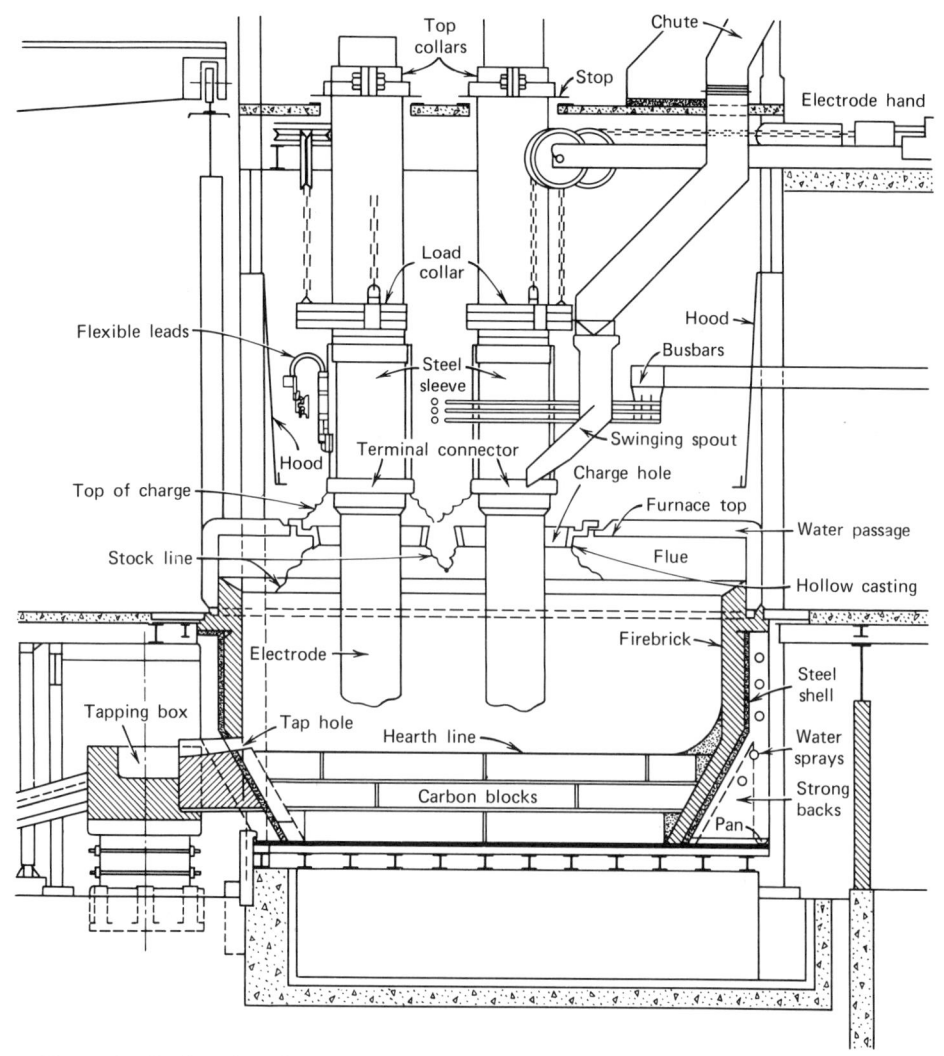

Figure 4. Design of a submerged-arc ferroalloy furnace. Courtesy of Union Carbide Corporation (3).

charge material is consumed in the reaction zone, fresh charge material is added to keep the furnace nearly full. In addition to melting the charge, the furnace also performs the function of a reaction vessel where reduction of oxides is achieved, usually by reaction with carbon. Electrode carbon consumption by reduction reactions varies greatly with the type of product produced and the amount of reducing agent added to the charge. Thus a broad range of specific electrode consumption rates is found in commercial practice, extending from approximately 20 kg/t product for phosphorus or standard ferromanganese to more than 140 kg/t for silicon metal. Unlike open-arc furnaces, the tip of the electrodes in submerged-arc furnaces is often a few meters above the hearth floor so that the volume beneath and around the electrode tip is heated by a combination of resistance heating and arcing (43). Temperatures in this region may reach 2200 K. The volume of this heated reaction zone largely determines furnace productivity. Therefore, to attain the largest possible reaction zone, large diameter electrodes are employed. In many cases, owing to the relatively low electrical currents involved, relatively low cost carbon electrodes may be used rather than graphite which might be prohibitively expensive in the sizes required. However, in certain cases, such as the modern phosphorus furnaces operating at up to 90,000 kW (see Phosphorus), semigraphite electrodes are required. This type of electrode is produced primarily from graphite raw materials bonded with a carbonized pitch binder. In addition to possessing significantly lower electrical resistance, semigraphite electrodes are more easily machined than carbon and have superior thermal shock resistance.

Typical properties of carbon and semigraphite electrodes are shown in Table 4. Strength is an important factor, especially in the larger size electrodes, because of the massive weight of the electrode columns used on many furnaces. Low resistance is an advantage in reducing joule heating losses, although such losses are relatively minor except for very highly powered furnaces. Since relatively large thermal gradients can exist in the submerged electrode, thermal shock or thermal stress resistance is also important in cases of furnaces subjected to transient or intermittent operation and especially where large diameter electrodes are used.

Anode Applications

Graphite has served as the primary material for electrolysis anodes in which chlorine or chlorates are produced at the anode (see Alkali and chlorine products). Recent technological advances, however, have resulted in a dimensionally stable anode (DSA) consisting of precious metal oxides deposited on a titanium substrate that is replacing graphite as the primary anode in mercury cells used in the chlor–alkali industry (40–46).

Although the particles and binder of graphite anodes are subject to oxidative attack under anodic conditions, graphite anodes can provide 270 days of service in diaphragm cell applications. Diaphragm cells account for 75% of the chlor–alkali capacity in the United States (47–48).

Graphite anodes meet the requirements for electrolytic cell applications which include: high degree of insolubility, low initial cost, availability in almost unlimited quantities, few limitations as to size and shape, good electrical conductivity, and high purity to prevent contamination of cell products.

The two basic types of graphite anodes used are plain and impregnated. The purpose of impregnation is to prevent anolyte penetration of the pores in graphite and attendant corrosion inside the anode. For impregnated anodes, base graphite with an initial porosity of 20–30% is given a vacuum-pressure impregnation usually with an oil such as linseed to fill or coat the accessible pores. Proper impregnation provides a 20–50% increase in anode life over unimpregnated graphite.

The electrolysis of fused magnesium chloride for the production of magnesium metal is the second largest use for graphite anodes. The electrolytic process is the current principal method of magnesium production although developments in the metallothermic process and the carbothermic process show potential for economical alternatives (49).

Other applications of graphite anodes include electrolysis of fused chlorides for the production of sodium, lithium, tantalum, and columbium. Graphite anodes are also used for electrolysis of aqueous manganese sulfate and for the anodic deposition of manganese dioxide used as dry battery depolarizer. Carbon anodes are utilized in the electrolytic production of fluorine.

Table 4. Average Properties of Carbon and Semigraphite Electrodes (Measured With the Grain at Room Temperature)

Electrode diameter range, mm	Bulk density, g/cm³	Compressive strength[a], kPa	Flexural strength[a], kPa	Elastic modulus[a], GPa	Specific resistance, μΩ·m	Coefficient of thermal expansion (CTE), 10⁻⁶/°C	Thermal conductivity, W/(m·K)
				Carbon			
425–1125	1.63	11700	4830		50	3.2	16
1250–1375	1.62	11700	3860	3.3	46	3.2	12
				Semigraphite			
875–1125	1.64	12400	5400		24	3.0	34
1250–1375	1.64	12400	4660	3.7	30	3.0	34

[a] To convert Pa to psi, multiply by 1.45×10^{-4}.

Graphite anodes are also used in cathodic protection applications for corrosion prevention of underground and underwater metal structures where low cost, light weight, and excellent electrical conductivity are required. Life of treated graphite ranges from 3–30 yr in cathodic protection applications (50).

Mechanical Applications

Carbon–graphite finds use in many mechanical applications where its natural lubricity, high-temperature mechanical strength, and corrosion resistance give it important advantages over other materials. This lubricity, strength, and corrosion resistance, together with ease of machining, dimensional stability, and high thermal conductivity, make carbon–graphite the material of choice for mechanically supporting loads in sliding or rotating contact. The mechanical applications are: mechanical seals—face, ring, and circumferential types; bearings—carbon cages for roller and ball bearings, carbon sleeve bearings and bushings, carbon thrust bearings or washers, and combination sleeve and thrust bearings; packing rings—steam and water shaft packing rings, and compressor tail-rod packing rings; nonlubricated compressor parts—piston rings, wear rings, segments, scuffer shoes, shaft tail-rod packing rings, pistons, and piston skirts; and miscellaneous applications—flat plate slider parts for support of apparatus and facilitating sliding movement under load, rotor vanes, and metering device parts such as the metering ball and plates.

Carbon–graphite materials employed for mechanical applications are prepared by mixing selected sizes and types of carbon and graphite with binder materials such as pitches and resins. The mixtures are formed into compacts and baked to temperatures of ca 1000–3000°C. Specific raw materials and processing techniques are employed to obtain desired properties for the finished carbon–graphite materials (51).

The successful application of carbon–graphite as a sliding contact is dependent upon the proper use of proprietary additives or impregnants, or both, in the carbon–graphite materials. Carbon–graphite, long considered to be self-lubricating, depends on the presence of adsorbed films of water vapor and/or oxygen for its low friction and low wear properties. This adsorbed boundary layer is soon lost when the operation is conducted at high altitude, high temperature, or in cold, dry air. A substitute boundary layer can be formed by incorporating certain additives or impregnants, or both, such as thermoplastic or thermosetting resins, or metallic sulfides, oxides, or halides. In addition to reducing the friction and wear of the carbon–graphite materials, the additives and impregnants can serve to improve oxidation resistance, provide impermeability to high pressure gases and liquids, and even permit operation under high vacuum conditions (52). Successful operation under high vacuum conditions is a primary requirement of equipment used for exploring outer space.

Carbon–graphite materials will not gall or weld even when rubbed under excessive load and speed. Early carbon materials contained metal fillers to provide strength and high thermal conductivity, but these desirable properties can now be obtained in true carbon–graphite materials that completely eliminate the galling tendency and other disadvantages of metals.

Maintenance of flat faces in rotating and stationary mating seals is important for successful operation. Dimensional stability is necessary in high speed, high-load face seal applications, and carbon–graphite materials with high elastic moduli have been developed to meet the requirements. Distortion of face seals caused by unequal loading or thermal stress leads to high unit loads at contact points, causing high localized friction and additional heat which produces further distortion (53–54).

Carbon–graphite materials are compatible with a wide range of mating materials, such as chrome-plated steel, 440-C stainless, and 300 series stainless (for corrosion resistance) steel, fine-grained cast iron, flame-plated oxides and carbides, ceramics, cermets, and at times, with themselves. However, the importance of rubbing contact has been minimized for certain applications that employ face seals equipped with self-acting lift augmentation (55). For this new generation of seals, pads are machined on the seal face which, during operation, act as a thrust bearing and cause the seal to lift from the counter face and ride on a thin gas film. Ideally, the self-acting seal will experience mechanical wear only during startup and shutdown of the equipment on which it is installed. The advent of this new seal design will enhance the ability of carbon–graphite materials to meet the ever increasing speed, pressure, and temperature requirements for certain applications (56–57).

Metallurgical Applications

Because of their unique combination of physical and chemical properties, manufactured carbons and graphites are widely used in several forms in high temperature processing of metals, ceramics, glass, and fused quartz. A variety of commercial grades is available with properties tailored to best meet the needs of particular applications (58). Industrial carbons and graphites are also available in a broad range of shapes and sizes.

Structural Graphite Shapes. In many metallurgical and other high temperature applications, manufactured graphite is used because it neither melts nor fuses to many common metals or ceramics, exhibits increasing strength with temperature, has high thermal shock resistance, is nonwarping, has low expansion, and possesses high thermal conductivity. However, because of its tendency to oxidize at temperatures above 750 K, prolonged exposure at higher temperatures frequently necessitates use of a nonoxidizing atmosphere. In addition, prolonged contact both with liquid steel and with liquid metals that rapidly form carbides should be avoided.

Some of the more common applications for structural graphite shapes are: (1) hot-pressing molds and dies (59) for beryllium at 1370 K and 6.9 MPa (1000 psi); diamond-impregnated drill bits and saw tooth segments at 1250 K and 13.8 MPa (2000 psi); tungsten and other refractory metals and alloys up to 2370 K and 6.9 MPa (1000 psi); and boron nitride and boron carbide up to 2060 K; (2) molds for metal casting steel railroad car wheels made by the controlled-pressure pouring process (60); steel slabs and billets made by the controlled-pressure pouring process (61); continuous casting of copper and its alloys, aluminum and its alloys, bearing materials, zinc, and gray iron (62–63); centrifugal casting of brasses, bronzes, steels, and refractory metals (64); nickel anodes; welding rods and thermite welding molds; shapes of refractory metals (Ti, Zr, Mo, Nb, W) and carbides; and shapes of gray, ductile and malleable irons (65); (3) foundry accessories including: mold chill plates, core rods, and riser rods; crucible skimmer floats; plunging bells for magnesium additions to ductile iron and desulfurization of blast-furnace hot metal (66–67); stirring rods for nonferrous metals; and railroad brake shoe inserts; (4) injection tubes and nozzles for purifying molten aluminum (68) and other nonferrous metals, desulfurization of blast furnace and foundry iron with calcium carbide or magnesium, and carbon raising of foundry iron with graphite powders; (5) aluminum extrusion components including dies, guides from die openings, run-out table boards, and cooling-rack inserts; (6) rolls for handling metal sheets are used in certain processes because they are self-lubricating and reduce surface marring; (7) immersion thermocouple protection tubes for nonferrous metals; (8) welding electrodes for welding, gouging, and cutting iron and steel, particularly with the aid of an air blast (69); (9) crucibles, either induction or resistance heated, for producing tungsten carbide, beryllium fluoride and beryllium, titanium and zirconium fluoride, semiconductor crystals of germanium and silicon, and for laboratory chemical analysis equipment; (10) ceramic and glass production including: casting molds for fused cast refractories of alumina, magnesia, and chrome–magnesite compositions up to 2650 K (70); mold susceptors for fabricating fused magnesia crucibles; susceptors, electrical resistor elements, fusion crucibles, molds and dies for the production of fused quartz (71); linings for float-glass tanks (72); take-out pads for automatic glass-blowing machines; diablo-wheels for glass tubing production; and linings for hydrofluoric acid tanks for glass etching; (11) boats, trays, and plates for sintering clutch plates, brake disks, and cemented carbides and for the manufacture of semiconductor material and transistors; and (12) furnace jigs for brazing honeycomb panels, automotive ignition points and arms, automotive radiator cores, transistor junction assembly, and glass-to-metal seals.

Electric Heating Elements. Machined graphite shapes are widely used as susceptors and resistor elements to produce temperatures up to 3300 K in applications utilizing nonoxidizing atmospheres. The advantages of graphite in this type of application include its very low vapor pressure (lower than molybdenum), high black body emissivity, high thermal shock resistance, and increasing strength at elevated temperatures with no increase in brittleness. Graphites covering a broad range of electrical resistivity are available and can be easily machined into complex shapes at lower cost than refractory metal elements. Flexible graphite cloth is also used widely as a heating element since its low thermal mass permits rapid heating and cooling cycles. Porous carbon or graphite, and flexible carbon or graphite felts are used for thermal insulation in many high temperature furnaces. Typical applications include molten-iron or steel-holding furnaces, continuous-casting tundishes, liquid-steel degassing units, chemical-reaction chambers, quartz-fusion apparatus, zinc-vaporization chambers, sintering furnaces, vapor deposition units; the felts are also used in the manufacture of semiconductors (qv) (73–74).

Graphite Powder and Particles. Manufactured graphite powders and particles are used extensively in metallurgical applications where the uniformity of physical and chemical characteristics, high purity, and rapid solubility in certain molten metals are important factors (75). The many grades of graphite powders and particles are classified on the basis of fineness and purity. Applications for these materials include facings for foundry molds and steel ingot molds, additives to molten iron to control carbon level and chill characteristics, covering material for molten nonferrous metals and salt baths to prevent oxidation, additives to sintered materials to control carbon level and frictional characteristics of oil-less bearings, and as charge-carbon in steels made in electric arc furnaces.

Refractory Applications

Various forms of carbon and graphite materials have found wide application in the metals industry, particularly in connection with the production of iron and aluminum. Carbon has been used as a refractory material since 1850, although full commercial acceptance and subsequent rapid increase in use has taken place only since 1945 (see Refractories).

Carbon as a Blast Furnace Refractory. The first commercial use of carbon as a refractory for a blast furnace lining took place in France in 1872, followed in 1892 by a carbon block hearth for a blast furnace of the Maryland Steel Company at Sparrows Point, Md. After a period of abated interest, the excellent results obtained with several carbon hearths in Germany and the United States during the late 1930s and early 1940s renewed enthusiasm for this approach. Although initially used only for the hearth floor of blast furnaces, carbon and graphite refractories have been applied successfully to hearth walls, lower and upper boshes, and most recently, to the lower stack of modern high-performance blast furnaces throughout the world (76). More than 360 individual carbon or graphite blast furnace linings have been installed in North America through the end of 1975. Carbon is also used extensively for blast furnace slag troughs and iron runners.

Experience has shown that carbon and graphite refractories have several characteristics contributing to their successful use in blast furnaces. (1) They show no softening or loss of strength at operating temperatures. (2) They are almost immune to attack by either blast furnace slags or molten iron. (3) Their relatively high thermal conductivity, when combined with adequate external cooling, assures solidification of iron and slag far from the furnace exterior, thus promoting long life while maintaining safe lining thicknesses (77). (4) A high level of thermal shock resistance avoids cracking or spalling in service. (5) A positive, low thermal expansion provides both dimensional stability and a tightening of joints in the multiblock linings. However, because of their poor oxidation resistance such refractories must not be exposed to air, carbon dioxide, or water vapor at elevated temperatures (78–79).

The prime requirement for linings in blast furnace hearths or hearth walls is to contain liquid iron and slag safely within the crucible throughout extended periods of continuous operation. This requirement is most readily achieved by providing sufficient cooling of the lining to assure solidification of penetrating iron and slag far from the furnace exterior (80). Where only peripheral cooling is used, a thick carbon bottom is required, the thickness of which is approximately one-quarter of the furnace diameter between the cooling system (77). In North and South America, such hearth bottoms are made from 3 or 4 horizontal layers of long carbon beams, each of which may be up to 6.5 m long and weigh as much as 5 metric tons. Anchoring the ends of these long beams beneath the furnace sidewall prevents flotation of the blocks in the denser molten iron, a phenomenon that has been observed in other parts of the world where long carbon beams are not used. Where short carbon blocks are used in the hearth, whether in a vertical or horizontal orientation, extensive interlocking, keying, and cementing must be employed to prevent loss by flotation. The application of carbon hearth and hearth wall refractories has virtually eliminated the once-prevalent danger of breakouts, a phenomenon that represents a serious threat to both life and property. As of 1976, carbon hearth walls were used in virtually all blast furnaces in North America, and of these, 80 had carbon hearth floors. One such furnace remained in operation after 22 years of service on the original carbon hearth, and another produced over 13 million metric tons of iron on its carbon hearth. In recent years, there has been an increasing tendency to incorporate some form of underhearth cooling in the larger furnaces to enhance hearth life. Medium size furnaces generally use a layer of high conductivity graphite beneath the carbon hearth for improved cooling. Larger furnaces incorporate tightly sealed steel plenums beneath the hearth through which air (or infrequently water) is passed to effect cooling of the hearth (81).

During 1960–1975 the use of carbon and graphite as bosh refractories has tripled. In 1976 ninety North American blast furnaces were operating with carbon boshes, and record performance on a single carbon bosh has exceeded 8 million metric tons of pig iron. As in the hearth and hearth wall, long life is assured by efficient cooling of the carbon bosh lining. 100% External shower cooling is used, and special cements are employed to assure that the high conductivity refractory is in good thermal contact with the cooled bosh shell. Long life of carbon bosh refractories also necessitates a high level of resistance to abrasion and alkali attack (82).

The success of carbon bosh linings has prompted several operators in Europe and Japan to extend the use of this refractory into the lower stack of several large modern blast furnaces where design permits this feature (76).

Refractories in the Aluminum Industry. The Hall-Heroult aluminum cell uses carbon materials for the anode, cathode, and sidewall, since carbon is the only material able to withstand the corrosive action of the molten fluorides used in the process (see Aluminum). The aluminum industry uses more carbon per metric ton of virgin metal than any other industry. Production of 1 t of molten aluminum requires about 500 kg of anode carbon and 7.5–10 kg of cathode carbon.

Because of the very large consumption of anode carbon, aluminum plants usually contain an on-site carbon plant for the manufacture of anodes.

Two types of cathodes are used in aluminum cells: both are produced from blends of calcined anthracite, metallurgical coke, and pitch. Compared with a tamped lining, prebaked carbon block linings provide higher operating strength, higher density, lower porosity, and longer life. Their lower resistance results in a lower voltage drop through the lining that improves the overall electrical efficiency of the cell (83). Prebaked cathodes also exhibit more predictable starting characteristics and possess greatly improved ability to restart successfully after a shutdown. This characteristic helps increase production and service life and usually more than compensates for the higher initial cost of prebaked cathode blocks. The service life of a prebaked cathode is usually 3–4 yr.

Refractories for Cupolas. In many ways, the use of carbon cupola linings has paralleled the application of carbon in the blast furnace. Carbon brick and blocks are used to form the cupola well (84), or crucible, up to the tuyeres. When properly installed and cooled, carbon linings last for many months, or even years of intermittent operation. Their resistance to molten iron and both acid and basic slags provides not only insurance against breakouts but also operational flexibility to produce different iron grades without the necessity of changing refractories. Carbon is also widely used for tap holes, breast blocks, slagging troughs, and dams.

Refractories for Electric Reduction Furnaces. Carbon hearth linings are used in submerged-arc electric reduction furnaces producing phosphorus, calcium carbide, all grades of ferrosilicon, high carbon ferrochromium, ferrovanadium, and ferromolybdenum. They are also used in the production of beryllium oxide and beryllium copper, where temperatures up to 2273 K are required.

Most of the principles pertaining to carbon blast furnace hearths apply as well to hearths for submerged-arc furnaces, although the fact that carbon is an electrical conductor is of importance in the case of electric reduction furnaces. The very long life of carbon linings in this application is attributable to their exceptional resistance to corrosive slags and metals at relatively high temperatures.

BIBLIOGRAPHY

"Baked and Graphitized Products, Uses" under "Carbon" in *ECT* 2nd ed., Vol. 4, pp. 202–243, by W. M. Gaylord, Union Carbide Corporation.

1. P. R. Kasten and co-workers, *U.S. Oak Ridge National Laboratory, ORNL-TM 2136,* Feb. 1969.
2. J. T. Meers and co-workers, *Am. Nucl. Soc. Trans.* **21,** 185 (1975); A. E. Goldman, H. R. Gugerli, and J. T. Meers, *paper presented at NUCLEX 75* Meeting, Basel, Switz., Oct. 6–10, 1975.
3. M. R. Hatfield and C. E. Ford, *Trans. Am. Inst. Chem. Eng.* **42,** 121 (1946).
4. W. M. Gaylord, *Ind. Eng. Chem.* **51,** 1161 (1959).
5. N. J. Johnson, *Ind. Eng. Chem.* **53,** 413 (1961).
6. S. H. Friedman, *Chem. Eng. (N.Y.)* **69**(14), 133 (1962).
7. J. R. Schley, *Chem. Eng. (N.Y.)* **81,** 144 (Feb. 18, 1974); **81,** 102 (Mar. 18, 1974).
8. D. Hills, *Chem. Eng. NY* **81,** 80 (Dec. 23, 1974); **82,** 116 (Jan. 20, 1975).
9. F. L. Rubin, *Chem. Eng. (N.Y.)* **60,** 201 (1953).
10. W. W. Palmquist, *Chem. Eng. Costs Q.* **4,** 111 (1954).
11. C. H. Baumann, *Ind. Eng. Chem.* **54,** 49 (1962).
12. J. F. Revilock, *Chem. Eng. (N.Y.)* **66,** 77 (Nov. 19, 1959).
13. W. M. Gaylord, *Ind. Eng. Chem.* **49,** 1584 (1957).
14. W. S. Norman, A. Hilliard, and C. H. Sawyer, *Materials of Construction in the Chemical Process Industries,* Society of Chemical Industry, London, Eng., 1950, p. 239.
15. J. Coull, C. A. Bishop, and W. M. Gaylord, *Chem. Eng. Prog.* **45,** 525 (1949).
16. W. M. Gaylord and M. A. Miranda, *Chem. Eng. Prog.* **53,** 139 (Mar. 1957).
17. T. F. Meinhold and C. H. Draper, *Chem. Process. Chicago* **23**(8), 92 (1960).
18. C. C. Brumbaugh, A. B. Tillman, and R. C. Sutter, *Ind. Eng. Chem.* **41,** 2165 (1949).
19. C. W. Cannon, *Chem. Ind. (N.Y.)* **65,** 3554 (1949).
20. R. W. Naidel, *Chem. Eng. Prog.* **69,** 53 (Feb. 1973).
21. J. F. Revilock and R. P. Stambaugh, *Chem. Eng. (N.Y.)* **69,** 148 (June 25, 1962).
22. S. Chari and J. Bohra, *Carbon* **10,** 747 (1972).
23. L. I. Nass, *Plast. Tech.* **17,** 91 (Oct. 1971); **18,** 31 (Mar. 1972).
24. R. E. Harrington, *Inst. Environ. Sci. Proc.,* 501 (1968).
25. Institute of Environmental Sciences, *Space Simulation; Proceedings of a Symposium Held in New York City, May 1972,* NASA Special Publication NASA-SP-298, 1972.
26. F. K. Lutz and W. C. Kalb, *Carbon Brushes for Electrical Equipment,* Union Carbide Corporation, New York, 1966.
27. R. Holm, *Electric Contacts; Theory and Application,* 4th ed., Springer Verlag, New York, 1967.
28. E. L. Grove, *Analytical Emission Spectroscopy,* Vol. 1, Marcel Dekker, Inc., New York, 1971.
29. L. H. Ahrens, and S. R. Taylor, *Spectrochemical Analysis,* Addison-Wesley, Reading, Mass., 1961.
30. G. L. Clark, *The Encyclopedia of Spectroscopy,* Reinhold Publishing Corp., New York, 1960.
31. ASTM Committee E-2, *Methods for Emission Spectrochemical Analysis,* 6th ed., American Society for Testing and Materials, Philadelphia, Pa., 1971.
32. ASTM Committee E-2, *Annual Book of ASTM Standards,* Part 42, American Society for Testing and Materials, Philadelphia, Pa.
33. N. H. Nachtrieb, *Principles and Practice of Spectrochemical Analysis,* McGraw-Hill Book Co., New York, 1950.
34. G. R. Harrison, R. C. Lord, and J. R. Loofbourow, *Practical Spectroscopy,* Prentice-Hall, Inc., New York, 1948.
35. E. L. Grove and A. J. Perkins, *Developments in Applied Spectroscopy,* Vol. 9, Plenum Press, New York-London, 1971.
36. G. A. Beitel, *J. Vac. Sci. Technol.* **8,** 647 (Sept.–Oct. 1971).

37. W. A. Kohl, *Handbook of Materials and Techniques for Vacuum Devices,* Reinhold Publishing Corp., New York, 1967, p. 137.
38. C. E. Sims, ed., *Electric Furnace Steelmaking,* John Wiley & Sons, Inc., New York, 1962–1963.
39. *World Steel in Figures,* International Iron and Steel Institute, Brussels, Belg., 1976.
40. W. E. Schwabe, *J. Met.* **24,** 65 (Nov. 1972).
41. *Electric Arc Furnace Digest,* Carbon Products Division, Union Carbide Corporation, 1975.
42. A. Ince, *Ironmaking Steelmaking* **3**(6), 310 (1976).
42a. J. S. Davis and P. Schroth, *AIME Electr. Furn. Steel Proc.* **29,** 145 (1971).
43. V. Paschkis and J. Persson, *Industrial Electric Furnaces and Appliances,* Interscience Publishers Inc., New York, 1960, p. 245.
44. *Chem. Process Chicago* **39**(9), 60 (1976).
45. V. H. Thomas, *J. Electrochem. Soc.* **74,** 618 (1974).
46. S. Puschaver, *Chem. Ind. London,* 236 (Mar. 15, 1975).
47. *Chem. Week* **113,** 32 (Oct. 1973).
48. J. C. Davis, *Chem. Eng. (N.Y.)* **81,** 84 (Feb. 1974).
49. B. S. Gulyanitskii, *Itogi Nauki Tekh.,* (5), 5 (1972).
50. W. W. Palmquist, *Pet. Eng. Los Angeles* **22,** D22 (Jan. 1950).
51. N. J. Fechter and P. S. Petrunich, *Development of Seal Ring Carbon–Graphite Materials,* NASA Contract Reports CR-72799, Jan. 1971; *CR-72986,* Aug. 1971; *CR-120955,* Aug. 1972; and *CR-121092,* Jan. 1973.
52. D. H. Buckley and R. L. Johnson, *Am. Soc. Lubr. Eng. Trans.* **7,** 91 (1964).
53. M. J. Fisher, *Paper D4 presented at International Conference on Fluid Sealing, British Hydromechanics Research Association, Apr. 1961.*
54. C. F. Romine and J. P. Morley, *Mach. Des.* **40,** 173 (Dec. 5, 1968).
55. R. L. Johnson and L. P. Ludwig, *NASA TN-D-5170* (Apr. 1969).
56. L. P. Ludwig, *NASA Contract Report TM-X71588,* 1974.
57. A. Zobens, *Lubr. Eng.* **31,** 16 (Jan. 1975).
58. *Industrial Graphite Engineering Handbook,* Union Carbide Corporation, Carbon Products Div., New York, 1970.
59. R. M. Spriggs in A. M. Alper, ed., *High Temperature Oxides,* Vol. V-3, Academic Press, New York, 1970, p. 183.
60. *J. Met.* **24,** 50 (Nov. 1972).
61. E. A. Carlson, *Iron Steel Eng.* **52,** 25 (Dec. 1975).
62. R. Thomson, *Am. Foundrymen's Soc. Trans.* **79,** 161 (1971).
63. H. A. Krall and B. R. Douglas, *Foundry* **98,** 50 (Nov. 1970).
64. *Foundry* **90,** 63 (Feb. 1962).
65. C. A. Jones and co-workers, *Am. Foundrymen's Soc. Trans.* **79,** 547 (1971).
66. *Foundry* **93,** 132 (Feb. 1965).
67. W. H. Duquette and co-workers, *AIME Open Hearth Proc.* **56,** 79 (1973).
68. *33 Magazine* **13,** 64 (Aug. 1975).
69. L. J. Christensen, *Welding J.* **52,** 782 (Dec. 1973).
70. A. M. Alper and co-workers in A. M. Alper, ed., *High Temperature Oxides,* Vol. V-1, Academic Press, New York, 1970, p. 209.
71. U.S. Pat. 2,852,891 (Sept. 23, 1958), H. J. C. George (to Quartz & Silica, S.A.).
72. U.S. Pat. 3,486,878 (Dec. 30, 1969), R. J. Greenler (to Ford Motor Co.).
73. H. G. Carson, *Ind. Heat.,* (Nov. 1962) and (Jan. 1963).
74. J. G. Campbell, *Second Conference on Industrial Carbon and Graphite,* Society of Chemical Industry, London, Eng., 1966, p. 629.
75. A. T. Lloyd, *Mod. Cast.* **64,** 46 (Dec. 1974).
76. G. Kahlhofer and D. Winzer, *Stahl Eisen* **92**(4), 137 (1972).
77. L. W. Tyler, *Blast-Furnace Refractories,* The Iron and Steel Institute, London, Eng., 1968.
78. F. K. Earp and M. W. Hill, *Industrial Carbon and Graphite,* Society of Chemical Industry, London, Eng., 1958, p. 326.
79. S. Ergun and M. Mentser, *Chem. Phys. Carbon* **1,** 203 (1965).
80. R. D. Westbrook, *Iron Steel Eng.* **30,** 141 (Mar. 1953).
81. S. A. Bell, *J. Met.* **18,** 365 (Mar. 1966).
82. R. J. Hawkins, L. Monte, and J. J. Waters, *Ironmaking Steelmaking* **1,** 151 (Nov. 3, 1974).
83. L. E. Bacon in G. Gerard, ed., *Extractive Metallurgy of Aluminum,* Vol. 2, John Wiley & Sons, Inc., New York, 1969, p. 461.

84. *The Cupola and Its Operation,* American Foundrymen's Society, Des Plaines, Ill., 1965.

General References

Aerospace and Nuclear Reactor Applications

R. M. Bushong, *Aerosp. Eng.* **20,** 40 (Jan. 1963).
C. E. Ford, R. M. Bushong, and R. C. Stroup, *Met. Prog.* **82,** 101 (Dec. 1962).
M. W. Riley, *Mater. Des. Eng.* **56,** 113 (Sept. 1962).
S. Glasstone, *Principles of Nuclear Engineering,* D. Van Nostrand Co., Princeton, N. J., 1955.
R. E. Nightingale, *Nuclear Graphite,* Academic Press, New York, 1962.

Arc Carbons

J. E. Kaufman, ed., *IES Lighting Handbook,* 5th ed., Illuminating Engineering Society, New York, 1972.
G. Kirschstein and co-workers, *Gmelins Handbuch der Anorganischen Chemie, 8th ed., System Number 14, Carbon, Teil B, Lieferung 1,* Verlag Chemie GMBH, Weinheim, Ger., 1967, p. 207.

Chemical Applications

J. R. Schley, *Mater. Prot. Perform.* **9,** 11 (Oct. 1970).
A. Hilliard, *Chem. Ind. London,* 40 (Jan. 10, 1970).
A. R. Ford and E. Greenhalgh in L. C. F. Blackman, ed., *Modern Aspects of Graphite Technology,* Academic Press, London, Eng., 1970, p. 272.

Electrical Applications

F. P. Bowden and D. Tabor, *The Friction and Lubrication of Solids,* Oxford University Press, London, Eng., 1958.
V. Berger and U. Schroeder, *ETZ-A* **8**(4), 91 (1962).
W. T. Clark, A. Conolly, and W. Hirst, *Br. J. Appl. Phys.* **14,** 20 (1963).
H. M. Elsey and C. Lynn, *Electr. Eng. Am. Inst. Electr. Eng.* **68,** 106 (June 1949).
J. K. Lancaster, *Br. J. Appl. Phys.* **13,** 468 (1962).
J. K. Lancaster, *Wear* **6,** 341 (Nov. 5, 1963).
J. W. Midgley and D. G. Teer, *ASME Trans. Ser. D* **85,** 488 (1963).
E. I. Shobert, *Carbon Brushes,* Chemical Publishing Co., Inc., New York, 1965.
W. J. Spry and P. M. Scherer, *Wear* **4,** 137 (Nov. 2, 1961).
K. Binder, *ETZ-B* **86,** 285 (1965).

Electrode Applications

J. R. Bello, *AIME Electr. Furn. Steel Proc.* **29,** 219 (1971).
J. A. Persson, *AIME Electr. Furn. Steel Proc.* **21,** 131 (1973).
W. M. Kelly, *Carbon and Graphite News,* Vol. 5, No. 1, Union Carbide Corp., 1958, p. 1.

Anode Applications

L. E. Vaaler, *Electrochem. Technol.* **5**(5–6), 170 (1967).
J. P. Randin in A. J. Bard, ed., *Encyclopedia of Electrochemistry of the Elements,* Vol. VII, C, V, Marcel Dekker Inc., New York, 1976, Chapt. VII–I.
F. L. Church, *Mod. Met.* **23,** 90 (Aug. 1967).
Eur. Chem. News, (Oct. 1969).
R. R. Irving, *Iron Age* **210,** 64 (Nov. 1972).
V. A. Kolesnikov, *Tsvetnye Met.,* 53 (Nov. 7, 1975).
W. A. Rollwage, *Paper Presented at AIME Annual Meeting,* (Feb. 1967).
V. de Nora, *Chem. Ing. Tech.* **47,** 125 (Feb. 1975).
D. Bergner, *ibid.,* p. 136.

Mechanical Applications

J. W. Abar, *Lubr. Eng.* **201,** 381 (October 1964).
G. P. Allen and D. W. Wisander, *NASA-TN-D-7381* (Sept. 1973).
G. P. Allen and D. W. Wisander, *NASA-TN-D-7871* (Jan. 1975).
P. F. Brown, N. Gordon, and W. J. King, *Lubr. Eng.* **22,** 7 (Jan. 1966).
L. J. Dobek, *NASA Contract Report CR-121177,* (Mar. 1973).
Crane Packing Co., *Packing and Mechanical Seals,* 2nd ed., Morton Grove, Ill., 1966.
J. P. Giltrow, *Composites* **4,** 55 (Mar. 1973).
W. R. Lauzau, B. R. Shelton, and R. A. Waldheger, *Lubr. Eng.* **19,** 201 (May 1963).
G. Oley, *Mech. Eng.* **94,** 18 (Apr. 1972).
R. R. Paxton, *Electrochem. Tech.* **5,** 174 (May–June 1967).
V. P. Povinelli, Jr., *J. Aircr.* **13,** 266 (Apr. 1975).
F. F. Ruhl, A. B. Wendt, and P. N. Dalenberg, *Lubr. Eng.* **23,** 241 (June 1967).
A. G. Spores, *Lubr. Eng.* **31,** 248 (May 1975).
R. D. Taber, J. H. Fuchsluger, and M. L. Rutherber, *Lubr. Eng.* **31,** 565 (Nov. 1975).

R. M. BUSHONG (Aerospace and Nuclear Applications)
R. RUSSELL (Chemical Applications)
B. R. JOYCE (Electrical Applications, Arc Carbons)
P. M. SCHERER (Electrical Applications, Brushes; Anode Applications; and Mechanical Applications)
N. L. BOTTONE (Electrical Applications; Spectroscopic Electrodes)
R. L. REDDY (Electrode Applications; Metallurgical Applications; and Refractory Applications)
Union Carbide Corporation

CARBON FIBERS AND FABRICS

Carbon fibers are filamentary forms (fiber dia 5–15 μm) of carbon (carbon content exceeding 92 wt %) and are characterized by flexibility, electrical conductivity, chemical inertness except to oxidation, refractoriness, and in their high performance varieties, high Young's modulus and high strength. Considerable confusion exists over the terms carbon and graphite fibers. The term graphite fibers should be restricted to materials with the three-dimensional order characteristic of polycrystalline graphite; essentially all commercial fibers are carbon fibers.

Carbon fibers were first made intentionally in 1878 by Edison (1) who pyrolyzed cotton to make incandescent lamp filaments. Interest in carbon fibers remained at a very low level until the late 1950s when commercially useful products, made by carbonizing rayon cloth and felt, were introduced (2). These relatively low strength, low modulus products were followed in 1964 by the development of high modulus, high strength, rayon-based carbon yarns (3) and of intermediate modulus, high strength, polyacrylonitrile (PAN)-based carbon yarns (4). High modulus, intermediate strength fibers made from a mesophase (liquid crystal) pitch precursor were introduced in 1974 as a mat product (5) and in 1975 as continuous yarns (6–7). Carbon fibers can also be made from ordinary (nonmesophase) pitch, but they generally have low strength and low modulus. Other precursor materials, including phenolics, polyacetylene, poly(vinyl alcohol), and polybenzimidazole, have been investigated (8–9). As of early 1977 no evidence was available to indicate that fibers from these latter precursors would become commercially significant.

Most desirable properties of carbon fibers, eg, Young's modulus, electrical and thermal conductivity, and to a lesser extent tensile strength, depend on the degree of preferred orientation. In a high modulus fiber, the carbon layer planes are predominantly parallel to the fiber axis; however, when viewed in cross section, the layers in most carbon fibers are oriented in all directions, although mesophase pitch-based fibers with concentric layers (onionskin structure) and with layers radiating from the fiber axis (radial structure) have been made (7). The high strength, high modulus properties are found only along the fiber axis. Very highly oriented carbon fibers possess a Young's modulus of nearly 900 GPa (130×10^6 psi), very close to that of the graphite single crystal, and tensile strength of over 3.4 GPa (500,000 psi). In commercial production, Young's moduli are usually held below 550 GPa (80×10^6 psi) for three principal reasons: the tensile strength does not always increase proportionally with the modulus, leading to unacceptably low strain-to-failure ratios; the interlamellar shear strength (adhesion between the fiber and the matrix resin) decreases with increasing modulus; and the production costs for very high modulus fibers are high.

Detailed discussions of the carbon fiber structure and the methods of structure determination are reported (10–13). The Young's modulus of carbon fibers depends entirely upon the degree of preferred orientation, as shown in Figure 1. Since carbon fibers of different origin can vary greatly in density, the data in Figure 1 have been normalized to correct for these differences.

The electrical and thermal conductivity also increase with increased orientation (11). The elastic constants of rayon- and PAN-based carbon fibers with Young's moduli ranging from 41–551 GPa ($6–80 \times 10^6$ psi) and of epoxy-matrix composites made from

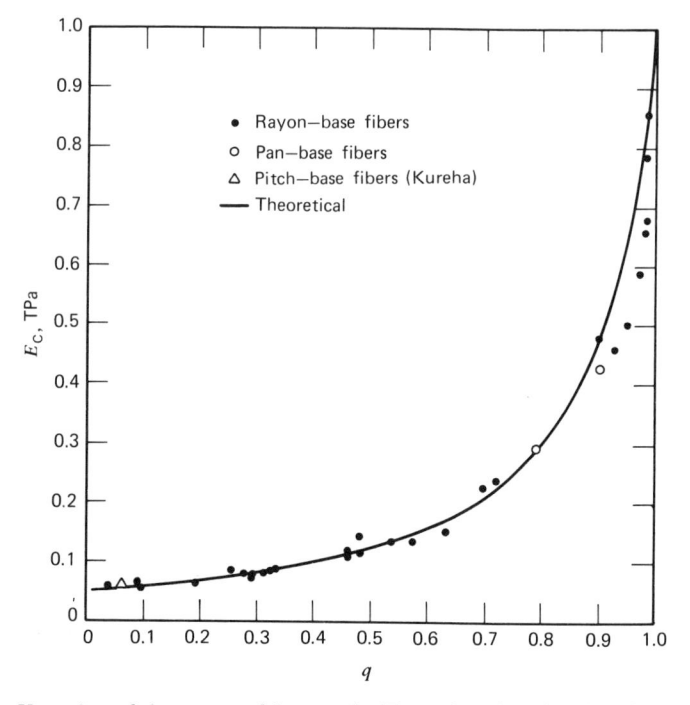

Figure 1. Young's modulus corrected for porosity E_c as a function of preferred orientation q; curve is based on theoretical model (10). To convert TPa to psi, multiply by 145×10^6.

these fibers have been determined (14). The relation between the axial thermal expansion of a variety of commercial carbon fibers to their preferred orientation was measured (15) and the transverse thermal expansion was determined (16).

The tensile strength of highly oriented carbon fibers should, in theory, be approximately 5% of the Young's modulus. Graphite whiskers with tensile strength of over 21 GPa (3×10^6 psi) have been made (17). For commercially produced carbon fibers, the tensile strength is controlled by defects in the fiber structure; these include stress risers in the form of surface irregularities, voids, and particulate inclusions (18–25). The nature and frequency distribution of these limiting defects vary for different carbon fibers, but all fibers show significant increases in single-filament tensile strength (frequently over 100%) when the test-gage length is reduced from 5 cm to 1 mm. The useful tensile strength (translatable into composite properties) of commercial carbon yarn rarely exceeds 3.8 GPa (550,000 psi). In 1977 the standard, low cost, PAN-based carbon fibers had a modulus of 207 GPa (30×10^6 psi) and tensile strengths of 2.4–2.9 GPa (350,000–425,000 psi).

The interlamellar shear strength of carbon fiber composites tends to decrease with increasing fiber modulus. This property is important because a low shear strength results in low transverse composite properties and possibly low compressive strength (26–27). A large number of surface treatments to improve shear strength have been published. Most processes (28) depend on some form of oxidative etching in air, nitric acid, or other oxidizing liquids, anodic oxidation, or the deposition of a low temperature carbon char (29) on the fiber surface. The actual processes used by various producers are closely guarded trade secrets.

There are considerable differences in the process chemistry, production economics, and properties of rayon-, PAN-, and pitch-based carbon fibers. These processes and products are, therefore, discussed separately.

Rayon-Based Carbon Fibers

The production process for rayon-based carbon yarn and cloth involves three distinct steps: preparation and heat treating, carbonization, and optional high temperature heat treatment. Heat treating at 200–350°C to form a char with thermal stability is required for rapid subsequent carbonization. Principally, water evolves during this stage although complex thermal tars also form. These tars can redeposit on the yarn or cloth and render the material brittle upon carbonization. To reduce tar formation and to quicken the process, the heat treatment is frequently carried out in a reactive atmosphere, such as air, or with the aid of chemical impregnants (9). The weight loss during this step is approximately 50–60% and any structure or preferred orientation that may have existed in the starting material is destroyed during the formation of the amorphous char. The yarn or cloth is carbonized next at 1000–2000°C during which additional weight is lost and an incipient carbon layer structure is formed. The material may then be heat treated at temperatures approaching 3000°C. The overall carbon yield is approximately 20–25%. The products have densities of 1.43–1.7 g/cm^3, filament tensile strengths of 345–690 MPa (5–10 $\times 10^4$ psi), and Young's moduli of 21–55 GPa (3–8 $\times 10^6$ psi). They are usually used in the form of cloth for aerospace applications, eg, phenolic impregnated heat shields, and in carbon–carbon composites for missile parts and aircraft brakes. Production statistics are not available; estimated consumption for 1970–1976 was 100,000–250,000 kg/yr. Through 1975 the volume of rayon-based, low modulus yarn and cloth probably exceeded that of all other carbon fibers combined. The price for cloth is $60–120/kg (see Ablative materials).

To produce high modulus, high strength rayon-based fiber, it is necessary to stretch the yarn during the heat treatment process at 2700–3000°C. The high temperature increases the size, perfection, and parallel stacking of the carbon layers, and stretching orients them parallel to the fiber axis. The tensile strength of rayon-based yarn also increases with increasing stretch and Young's modulus, resulting in an almost constant strain-to-failure of approximately 0.5%. The density of the yarn is 1.65 g/cm^3 at a modulus of 345 GPa (50 × 10^6 psi) or 1.82 g/cm^3 at a modulus of 517 GPa (75 × 10^6 psi). There are comprehensive reviews of production processes (10), and the structure and properties of rayon-based yarn (8).

The hot-stretching process is very costly, and the price of high modulus rayon-based yarn is $600–1300/kg. Production volume is relatively low and applications are confined to aerospace uses.

The long range future availability of all rayon-based carbon fiber products depends on the continued production of suitable continuous filament rayon yarn. The rapid decline in the tire cord use of rayon has caused several major rayon manufacturers to discontinue production. Many applications of rayon-based carbon fibers depend on their stability at very high temperatures and the high carbon content of over 99%. In these areas, they are not readily replaceable by the standard PAN-based fibers that have a lower carbon content and change when heated beyond the original process temperature. Mesophase pitch-based fibers can be made with properties very similar to those of rayon-based products and may be expected to replace them in many applications.

PAN-Based Carbon Fibers

Production of PAN-based carbon fibers also involves the three steps of low temperature heat treatment, carbonization, and optional high temperature heat treatment. The principal differences from the rayon-based fiber process are that a well-oriented ladder polymer structure is developed during oxidative heat treatment under tension, and the orientation is essentially maintained through carbonization.

Most producers probably use copolymer PAN fibers containing at least 95% acrylonitrile units. Textile PAN yarns with a high percentage of copolymer content and those containing appreciable quantities of brighteners, such as TiO_2, are not considered suitable precursors. To improve the alignment in the polymer structure and reduce the fiber diameter, the as-spun PAN fibers are frequently stretched by 100–500% at approximately 100°C. Stretching of the polymer also increases the tensile strength and Young's modulus of the resulting carbon fibers (30).

Heat treatment is carried out in an oxidizing atmosphere (usually air) at 190–280°C for 0.5–5 h. Sufficient tension must be applied at this stage to prevent the yarn from shrinking and to develop a high degree of alignment of the ladder polymer chain. Heat treatment is an exothermic process, and precautions against a runaway reaction are required. After heat treatment, the yarn is black, infusible, and can be rapidly carbonized in an inert atmosphere, usually at 1000–1300°C, where the tensile strength reaches a maximum. Hydrogen cyanide is evolved during carbonization, and suitable precautions must be taken for combustion of volatiles. The yield from raw to carbonized yarn is approximately 45–50%. At this stage, the material consists of 92–95%

carbon and most of the remainder is nitrogen. This yarn has a Young's modulus of ca 210 GPa (30 × 10⁶ psi). The modulus can be further increased to 350–520 GPa (50–75 × 10⁶ psi) by heating to above 2500°C, but the tensile strength tends to decrease. In at least one type of PAN-based fiber this decrease in strength is caused by impurities in the raw PAN fibers (24). Fibers spun under clean-room conditions increased in strength as well as modulus upon heating above 2500°C.

The tensile strength of some fibers with 210 GPa (30 × 10⁶ psi) Young's modulus is apparently limited by stress-rising surface irregularities (24), and etching to remove or blunt these flaws can increase the strength (31). PAN-based fibers can also be hot-stretched, with a resulting increase in both tensile strength and Young's modulus. However, because of the high cost, this process is not the usual commercial practice. Detailed reviews of the PAN-based fiber process and fiber characteristics have been published (8–9,32).

Most PAN-based fibers produced in 1976 had a tensile strength of ca 2.4–2.75 GPa (350,000–400,000 psi) and a Young's modulus of 193–241 GPa (28–35 × 10⁶ psi), although fibers with moduli up to 482 GPa (70 × 10⁶ psi) and generally lower strength are available. Fiber densities are usually 1.7–1.8 g/cm³. Product forms include yarns of 1,000, 3,000, 6,000, and 10,000 filaments, tows up to 500,000 filaments, and fabrics of various styles and weave construction.

Although official production statistics are not available, it is estimated that the combined production in the three principal producing countries (United States, Japan, and England) may have risen from 10 metric tons in 1970 to 250 t in 1976. With increased production, the price of the standard 207 GPa (30 × 10⁶ psi) Young's modulus fibers has decreased from approximately $1000/kg before 1970 to $40–80/kg in 1977, although higher modulus fibers and low denier yarns, such as 1000-filament products, are sold at premium prices.

Until about 1972 PAN-based carbon fibers were used almost exclusively in epoxy "prepregs" where structures for the aerospace industry were fabricated. The higher costs had to be justified by weight savings and performance improvements. Long-time reliability experience was meager and aircraft structures were mostly limited to noncritical parts. Fiber price reductions and significant advances in composite fabrication technology had, by 1976, made carbon fiber composites cost competitive with many metal parts in aircraft construction. Confidence in design and reliability has been established; composites are currently used in major flight-critical structures. During the same time period, a second major market for PAN-based carbon fibers developed in the sporting goods industry, first in golf clubs, then in fishing rods, tennis rackets, bows and arrows, skis, and sailboat masts and spars. Applications for textile and computer machinery, automotive and general transportation, and musical instruments are rapidly emerging (see Composite materials).

Mesophase Pitch-Based Fibers

The process for these fibers starts with commercial coal tar or petroleum pitch which is converted through heat treatment into a mesophase or liquid crystal state. Since the mesophase is highly anisotropic, the shear forces acting on the pitch during spinning and drawing result in a highly oriented fiber, with the essentially flat aromatic polymer molecules oriented parallel to the fiber axis. The spun yarn is then thermoset in an oxidizing atmosphere which renders the fibers infusible and amenable to rapid carbonization. The original preferred orientation is maintained during thermosetting, which does not require tensioning, and is further enhanced during carbonization and graphitization. Mesophase pitch-based fibers have a Young's modulus over 690 GPa (100×10^6 psi) when heated (without stretching) to 3000°C. These are the only fibers that can truly be called graphite. The carbon yield from spun to carbonized fiber is 75–85%, depending on the molecular weight of the mesophase pitch and the heat treatment conditions. Fibers with a Young's modulus over 345 GPa (50×10^6 psi) are almost pure (over 99.5%) carbon. Lower modulus fibers may contain small amounts (ca 1%) of sulfur. The fiber density is approximately 2 g/cm^3.

The tensile strength of mesophase pitch-based fibers increases with increasing carbonization temperatures until a modulus of 345 GPa (50×10^6 psi) is reached. It then remains essentially flat to 550 GPa (80×10^6 psi) and rises again with further heat treatment. Single-filament tensile strengths (2 cm gage length) over 3.5 GPa (5×10^5 psi) have been measured. In 1977 the commercial yarn with a 345 GPa (50×10^6 psi) modulus had a tensile strength of approximately 2.0 GPa (3×10^5 psi). The strength-limiting defects consist primarily of voids of 1–2 µm dia and, at this stage of development, of particulate inclusions similar to but more severe than those found in PAN-based fibers. Considerable process details and the microstructure of mesophase pitch-based fibers are described in the literature (7,33).

In 1977 mesophase pitch-based fibers were produced in three forms. One form is a low modulus (138 GPa or 20×10^6 psi), low cost ($17/kg) mat product that is not recommended for structural reinforcement. This material is useful as a veil mat for sheet molding to provide an electrically conductive surface for electrostatic spraying or improved surface appearance or both. Milled mat with a fiber length of ca 1.0 mm is used in injection molding to achieve electrical conductivity, resistance to heat distortion, and improved wear characteristics (for bearing applications).

The second form consists of pitch-based carbon fiber fabrics (12) that have filament moduli of 138–517 GPa (20–75×10^6 psi), depending on process temperature, and present an attractive, low cost alternative to rayon-based cloth.

The third form, a continuous filament yarn with a Young's modulus of 345 GPa (50×10^6 psi) and 2.0 GPa (3×10^5 psi) tensile strength, is by far the lowest priced high modulus carbon yarn. Initial uses are in stiffness-critical applications.

Health and Environmental Considerations

Carbon fibers, like other forms of carbon, are very inert. For this reason they are now used in medical research for body implant studies such as pacemaker leads (see Prosthetic and biomedical devices). The only known medical hazard is a possible skin irritation caused by broken filaments, similar to that caused by glass fibers. Carbon fibers are good electrical conductors. Broken filaments can remain airborne over considerable distances and can cause short circuits in electrical equipment.

BIBLIOGRAPHY

"Baked and Graphitized Products, Uses" under "Carbon" in *ECT* 2nd ed., Vol. 4, pp. 202–243, by W. M. Gaylord, Union Carbide Corporation.

1. U.S. Pat. 223,898 (Jan. 27, 1880), T. A. Edison.
2. U.S. Pat. 3,107,152 (Sept. 12, 1960), C. E. Ford and C. V. Mitchell (to Union Carbide).

3. R. Bacon, A. A. Pallozzi, and S. E. Slosarik, *Technical and Management Conference of the Reinforced Plastics Division, Proceedings of the 21st Annual, Feb. 8–10, 1966, Chicago, Ill.*, Society of the Plastics Industry, Inc., New York, 1966, Sec. 8-E.
4. Brit. Pat. 1,110,791 (Apr. 24, 1968), W. Johnson, L. N. Phillips, and W. Watt (to National Research Corp.).
5. H. F. Volk, *Carbon and Graphite Conference, 4th London International; extended abstracts of paper presented at Imperial College, Sept. 23–27, 1974*, Society of Chemical Industry, London, Eng., 1974, Session IV, Paper 102.
6. H. F. Volk, *Proceedings of the 1975 International Conference on Composite Materials*, Vol. I, Metallurgical Society of AIME, New York, 1976, p. 64.
7. J. B. Barr and co-workers, *Appl. Polym. Symp.* **29**, 161 (1976).
8. P. J. Goodhew, A. J. Clarke, and J. E. Bailey, *Mater. Sci. Eng.* **17**, 3 (1975).
9. H. M. Ezekiel, *U. S. Air Force Materials Laboratory Report AFML-TR-70-100*, Jan. 1971.
10. R. Bacon, *Chem. Phys. Carbon* **9**, 1 (1973).
11. R. Bacon and W. A. Schalamon, *Appl. Polym. Symp.* **9**, 285 (1969).
12. W. Ruland, *Appl. Polym. Symp.* **9**, 293 (1969).
13. B. Harris in M. Langley, ed., *Carbon Fibers in Engineering*, McGraw-Hill Book Co., London, Eng., 1973.
14. R. E. Smith, *J. Appl. Phys.* **43**, 2555 (1972).
15. B. Butler, S. Duliere, and J. Tidmore, *10th Biennial Conference on Carbon*, American Carbon Committee, 1971, Abstract FC-29, p. 45.
16. R. C. Fanning and J. N. Fleck, *10th Biennial Conference on Carbon*, American Carbon Committee, 1971, Abstract FC-30, p. 47.
17. R. Bacon, *J. Appl. Phys.* **31**, 283 (1960).
18. W. N. Reynolds and J. V. Sharp, *Carbon* **12**, 103 (1974).
19. S. G. Burnay and J. V. Sharp, *J. Micros.* **97**, 153 (1973).
20. D. J. Thorne, *Nature (London)* **248**, 754 (1974).
21. W. S. Williams, D. A. Steffens, and R. Bacon, *J. Appl. Phys.* **41**, 4893 (1970).
22. J. W. Johnson, *Appl. Polym. Symp.* **9**, 229 (1969).
23. J. W. Johnson and D. J. Thorne, *Carbon* **7**, 659 (1969).
24. R. Moreton and W. Watt, *Nature (London)* **247**, 360 (1974).
25. W. R. Jones and J. W. Johnson, *Carbon* **9**, 645 (1971).
26. H. M. Hawthorne and E. Teghtsoonian, *J. Mater. Sci.* **10**, 41 (1975).
27. N. L. Hancox, *J. Mater. Sci.* **10**, 234 (1975).
28. D. W. McKee and V. J. Mimeault, *Carbon* **8**, 151 (1973).
29. J. V. Duffy, *U.S. Naval Ordnance Laboratory Report NOLTR-73-153*, Aug. 6, 1973, A.D. 766,782.
30. R. Moreton, *Industrial Carbons and Graphite; 3rd Conference; papers read at the conference held at Imperial College of Science and Technology, London, 14th–17th April, 1970*, Society of Chemical Industry, London, Eng., 1971.
31. J. W. Johnson, *Appl. Polym. Symp.* **9**, 229 (1969).
32. W. Watt, *Proc. R. Soc. London Ser. A* **319**, 5 (1970).
33. U.S. Pat. 4,005,183 (Jan. 25, 1977), L. S. Singer (to Union Carbide).

H. F. VOLK
Union Carbide Corporation

OTHER FORMS OF CARBON AND GRAPHITE

The many forms and applications of graphite already mentioned illustrate the versatility and unique character of this material. Several other forms of graphite with specialized properties have been developed during recent years (1), and although all are characterized by relatively high cost, the specialized properties are sufficiently intriguing that applications have been found resulting in sales of a few millions of dollars in each category.

Carbon and Graphite Foams

The earliest foamed graphite was made from exfoliated small crystals of graphite bound together and compacted to a low density (2–4). This type of foam is structurally weak and will not support loads of even a few newtons per square meter. More recently, carbon and graphite foams have been produced from resinous foams of phenolic or urethane base by careful pyrolysis to preserve the foamed cell structure in the carbonized state. These foams have good structural integrity and a typical foam of 0.25 g/cm^3 apparent density has a compressive strength of 9,300–15,000 kPa (1350–2180 psi) with thermal conductivity of 0.87 W/(m·K) at 1400°C. These properties make the foam attractive as a high-temperature insulating packaging material in the aerospace field and as insulation for high temperature furnaces (see Insulation, thermal). Variations of the resinous-based foams include the syntactic foams where cellular polymers or hollow carbon spheres comprise the major volume of the material bonded and carbonized in a resin matrix.

Pyrolytic Graphite

The carbon commercially produced by the chemical vapor deposition (CVD) process is usually referred to as pyrolytic graphite (5–6). The material is not true graphite in the crystallographic sense, and wide variations in properties occur as a result of deposition methods and conditions; nevertheless, the term pyrolytic graphite is generally accepted and used extensively. Pyrolytic graphite was first produced in the late 1800s for lamp filaments yet received little further attention until the 1950s. Since then pyrolytic graphite has been studied extensively and has been commercially produced in massive shapes by several companies. Commercial applications for pyrolytic graphite, limited by the price of $18 or more per kg, include rocket nozzle parts, nose cones, laboratory ware, and pipe liners for smoking tobacco. Pyrolytic graphite coated on surfaces or infiltrated into porous materials is also used in other applications such as nuclear fuel particles and prosthetic devices (qv).

The greatest quantity of commercial pyrolytic graphite is produced in large, inductively-heated, low pressure furnaces for which natural gas is used as the carbon source. Temperature of deposition is ca 1800–2200°C on a carefully prepared substrate of fine-grained graphite. The properties of pyrolytic graphite are highly anisotropic. The ultimate tensile strength in the *ab* direction is five to ten times greater than that of conventional graphite, and the *c* direction strength is proportionally weaker. The thermal conductivity is even more anisotropic. In the *ab* direction, pyrolytic graphite is one of the very best conductors among elementary materials, whereas in the *c* direction conductivity is quite low. At room temperature the thermal conductivity values are approximately three hundred times as high in the *ab* direction as in the *c* direction. The commercially produced pyrolytic graphite is quite dense, usually 2.0–2.1 g/cm^3, and is quite low in porosity and permeability.

A special form of pyrolytic graphite is produced by annealing under pressure at temperatures above 3000°C. This pressure-annealed pyrolytic graphite exhibits the theoretical density of single crystal graphite; and although the material is polycrystalline, the properties of the material are close to single-crystal properties. The highly reflective, flat faces of pressure-annealed pyrolytic graphite have made the material valuable as an x-ray monochromator (see X-ray techniques).

Glassy Carbon

When carbon is produced from certain nongraphitizable carbonaceous materials, the material resembles a black glass in appearance and brittleness; hence, the terms glassy carbon or vitreous carbon (7–8). Nongraphitizing carbons are obtained from polymers that have some degree of cross-linking, and it is believed that the presence of these cross-linkages inhibits the formation of crystallites during subsequent heat treatments. By pyrolyzing polymers such as cellulosics, phenol–formaldehyde resins, and poly(furfuryl alcohol) under closely controlled conditions, glassy carbon is produced. These carbons are composed of random crystallites, of the order of 5.0 nm across, and are not significantly altered by ordinary graphitization heat treatment to 2700°C.

The properties of glassy carbon are quite different from those of conventional carbons and graphites. The density is low (1.4–1.5 g/cm^3), but the porosity and permeability are also quite low. The material has approximately the same hardness and brittleness as ordinary glass, whereas the strength and modulus are higher than for ordinary graphite. The extreme chemical inertness of glassy carbon, along with its impermeability, make it a useful material for chemical laboratory glassware, crucibles, and other vessels. It has found use as a susceptor for epitaxial growth of silicon crystals and as crucibles for growth of single crystals. Applications have been limited because of high cost owing to technological difficulties in manufacturing, although recently billets of reasonable size and with good properties have become available (9).

Carbon Spheres

Carbon and graphite have been produced for many years in roughly spherical shape in small sizes (1–2 mm dia). Globular carbon of this type was used in telephone receivers in the mid 1930s. Carbon of nearly spherical shape can be produced by mechanically tumbling or rolling a mixture of finely divided carbon or graphite with a resinous binder, screening to select the desired size balls, and carefully carbonizing the selected material. Graphite balls in small sizes have also been made from manufactured graphite by crushing, milling in a chopping-type mill, sizing, and tumbling to abrade away the irregular corners. Usage of these nearly spherical carbon balls has been limited to specialized applications where flowability was at a premium or uniform contact area was important. The telephone resistor already mentioned and shim particles for use in nuclear reactor fuel applications are examples.

Recently, a new type of carbon sphere, usually hollow, has been made from pitch. This technology, developed in Japan (10), produces carbon spheres available in 40–400 μm dia. The spheres are of uniform, very nearly spherical shape and are graphitizable. The particle density is usually 0.20–0.25 g/cm^3, and consequently, attention has been given to application in syntactic carbon foam for high temperature insulation and for lightweight composite structures. The spheres can be activated to produce an adsorptive material of high flow-through capacity. At present, cost has limited the application of these spheres. Although the spheres can be produced for about $1/kg, the cost is greater than for most starting materials for carbon.

BIBLIOGRAPHY

"Baked and Graphitized Products, Uses" under "Carbon" in *ECT* 2nd ed., Vol. 4, pp. 202–243, by W. M. Gaylord, Union Carbide Corporation.

1. R. W. Cahn and B. Harris, *Nature (London)* **221,** 132 (Jan. 11, 1969).
2. R. A. Mercuri, T. R. Wessendorf, and J. M. Criscione, *Am. Chem. Soc. Div. Fuel Chem. Prepr.* **12**(4), 103 (1968).
3. C. R. Thomas, *Mater. Sci. Eng.* **12,** 219 (1973).
4. S. T. Benton and C. R. Schmitt, *Carbon* **10,** 185 (1972).
5. J. C. Bokros, *Chem. Phys. Carbon* **5,** 1 (1969).
6. A. W. Moore, *Chem. Phys. Carbon* **11,** 69 (1973).
7. F. C. Cowtard and J. C. Lewis, *J. Mater. Sci.* **2,** 507 (1967).
8. G. M. Jenkins and K. Kannenmura, *Polymeric Carbons—Carbon Fibre, Glass and Char,* Cambridge University Press, New York, 1976, p. 178.
9. C. Nakayama and co-workers, *Proccedings of the Carbon Society of Japan, Annual Meeting, 1975,* p. 114; C. Nakayama, M. Okawa, and H. Nagashima, *13th Biennial Conference on Carbon, 1977, Extended Abstracts and Program,* American Carbon Society, 1977, p. 424.
10. Y. Amagi, Y. Nishimura, and S. Gomi, *SAMPE 16th National Symposium 1971,* p. 315.

R. M. BUSHONG
Union Carbide Corporation

CARBON BLACK

Carbon black is an important member of the family of industrial carbons. Its various uses depend on chemical composition, pigment properties, state of subdivision, adsorption activity and other colloidal properties. The basic process for manufacturing carbon black has been known since antiquity. The combustion of fuels with insufficient air produces a black smoke containing extremely small carbon black particles which, when separated from the combustion gases, comprise a fluffy powder of intense blackness. The term carbon black refers to a wide range of such products made by partial combustion or thermal decomposition of hydrocarbons in the vapor phase, in contrast to cokes and chars which are formed by the pyrolysis of solids.

This printed page was made with an ink containing a pigment grade of carbon black. Prehistoric cave wall paintings and objects from ancient Egypt were decorated with paints and lacquers containing carbon black. Carbon black was made in China about 3000 BC and records show that it was exported to Japan about 500 AD. The original process consisted of the partial burning of specially purified vegetable oils in small lamps with ceramic covers. The smoke impinged on the covers from which the adhering carbon black was painstakingly removed. Thus carbon black is one of the oldest industrial products. In the United States carbon black has been manufactured for over 100 years. Its use as a pigment continues as an important and growing application, but the major market for the last 50 years has been as a strengthening or reinforcing agent for rubber products, particularly tires.

Relationship of Carbon Black to Other Forms of Industrial Carbons

Carbon exists in two crystalline forms, and numerous so-called amorphous, less-ordered forms. The crystalline forms are diamond and graphite, and the less-ordered forms are mainly cokes and chars. Diamond has a cubic structure in which every carbon atom in the space lattice is bonded by its four valences to adjacent atoms situated in the apexes of a regular tetrahedron around the central atom. The carbon–carbon distance is 0.154 nm, similar to that in aliphatic hydrocarbons. The diamond structure is shown in Figure 1**a**. Diamond has a specific gravity of 3.54 (high for such a low atomic weight element), high refractive index, brilliance and clarity, electrical insulating properties, and a hardness greater than any other known material (see Hardness). The graphite form of carbon has a completely contrasting set of properties: lower density, a grayish-black appearance, softness, slipperiness, and electrical conductivity.

In Figure 1**b** the carbon atoms of the graphite structure are strongly bonded forming large sheets of hexagonal rings in which the electrons are quite mobile. The layer planes are stacked on each other in an ABAB arrangement, ie, layers in which every atom has an atom directly above it separated by one layer. Within a layer plane the carbon atoms are separated by 0.142 nm, comparable to the aromatic carbon separation distance of 0.139 nm in benzene. The mobility of electrons within the layer planes results in a 250-fold higher electrical conductivity in the planar direction than perpendicular to the planes. The bond strengths within the layer planes are over one hundred times stronger than the interlayer van der Waals bonding. The large graphite interplanar distance of 0.335 nm results in a specific gravity of 2.26.

All forms of industrial carbons other than diamond and graphite, including carbon black, can be classified as amorphous carbons characterized by degenerate or imperfect graphitic structures. The parallel layer planes in these carbons are not perfectly oriented with respect to their common perpendicular axis; the angular displacement of one layer with respect to another is random and the layers overlap one another irregularly. This arrangement has been termed turbostratic structure. In this arrangement a separation distance of 0.350–0.365 nm is found for the layer planes. A schematic model of the short-range crystalline order in carbon black is shown in Figure 1**c**. Table 1 lists some of the basic structural and crystallite properties of the various carbons.

Microstructure

X-ray diffraction patterns of carbon black show two or three diffuse rings similar to the more intense rings of natural graphite. Analysis of these patterns by Warren (1) yielded the first model of carbon black microstructure consisting of a random arrangement of crystallites within the particles. A diffuse (002) reflection band provided an estimate of the interplanar spacing and the thickness of the crystallites in the direction perpendicular to the planes. The weak (10) and (11) bands yielded values of the average diameter of the parallel layers. For most furnace and channel-type carbon blacks these values are about 0.35 nm for interplanar spacing (d), 1.2–1.5 nm for average crystallite thickness (L_c), and 1.7 nm for average diameter (L_a). Later studies of carbon black graphitization (2), electron diffraction analysis (3), oxidation studies (4), and high resolution diffracted-beam electron microscopy (5) have led to the conclusion that the microstructure of carbon black consists of a more concentric ar-

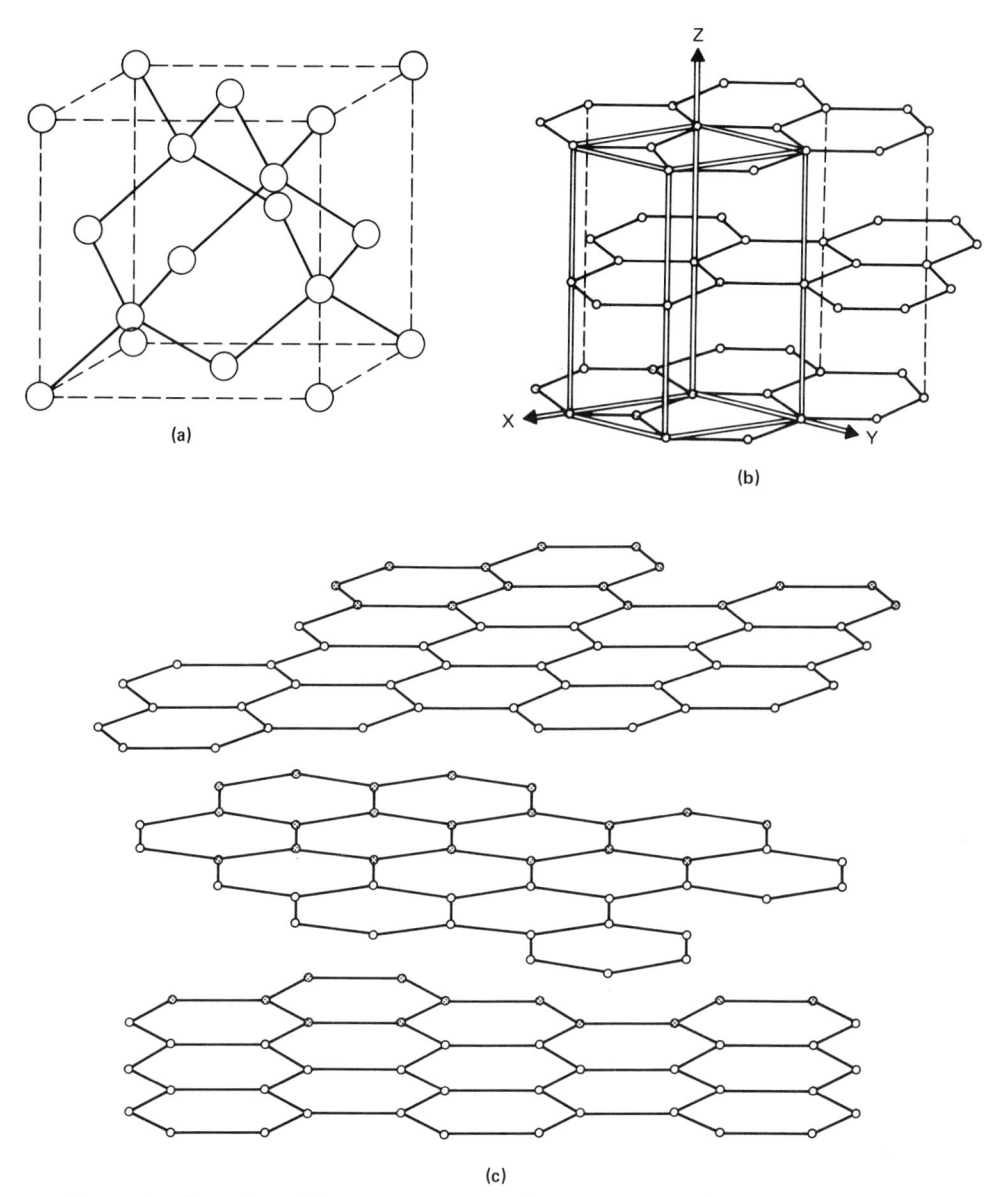

Figure 1. Crystallographic arrangement of carbon atoms in: (**a**) diamond, face-centered cubic structure; (**b**) graphite, parallel layers of hexagons in ordered positions (every third layer is in the same position as the first layer, the second layer corresponds with the fourth, etc); (**c**) carbon black, hexagonal layers are parallel but farther apart and arranged without order—turbostratic arrangement.

rangement of layer planes. Ordinary transmission electron microscopy provides information on the size and shape of carbon black aggregates from their two-dimensional projections as shown in Figure 2 for a rubber-reinforcing grade (N339). High resolution electron microscopy provides information on the microstructure of the aggregates from interference patterns caused by electron diffraction. These patterns are due to a concentric-layer configuration, with an interplanar distance of 0.35 nm, somewhat

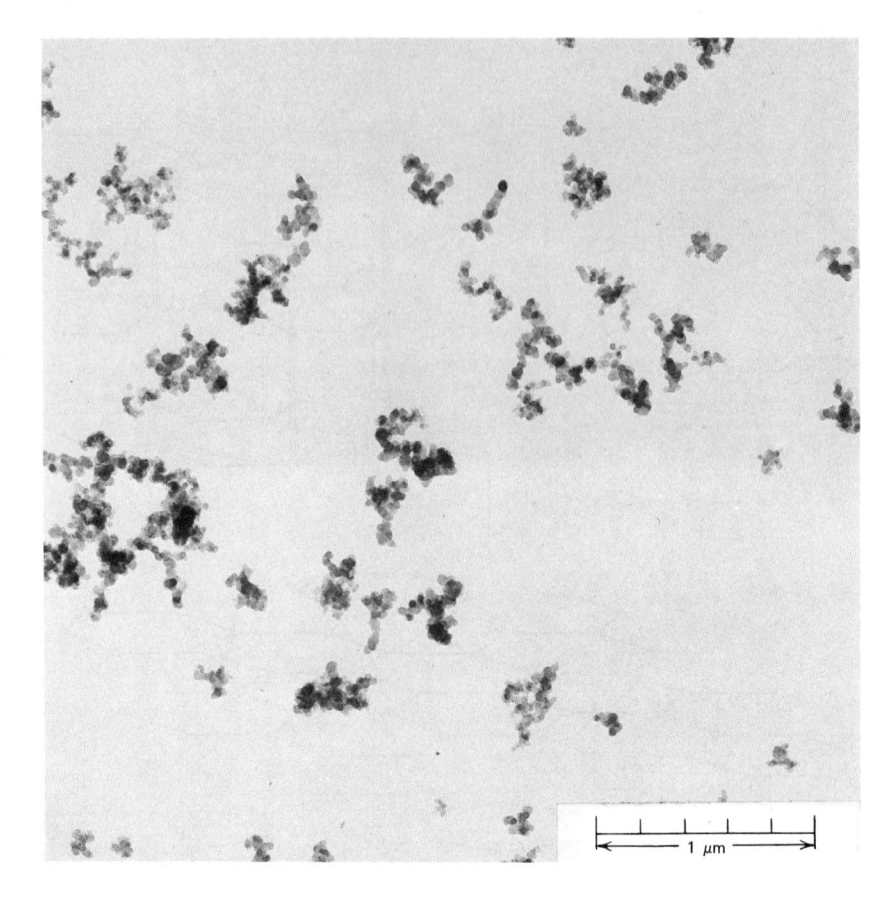

Figure 2. Electron micrograph of furnace black N339 (HAF-HS).

Table 1. Forms of Carbon and Characteristics

Form	Crystal system	Sp gr	C—C distance, nm	Layer distance, nm
diamond	cubical	3.52	0.155	
graphite	hexagonal	2.27	0.142	0.335
carbon black	hexagonal-turbostratic[a]	1.86–2.04	0.142	0.365
cokes (oven and calcined)	hexagonal-turbostratic[a]	1.3–2.1		
chars and activated carbons	hexagonal-turbostratic[a]	1.1–1.3		
fibrous carbon	hexagonal-turbostratic[a]	1.65		
vitreous carbon	hexagonal-turbostratic[a]	1.47		
pyrolytic graphite	hexagonal-turbostratic[a]	1.2–2.2		

[a] Turbostratic crystals have randomly oriented layer planes, see pp. 690–691.

larger than in graphite. Figure 3 shows a high resolution electron micrograph of a portion of a carbon black aggregate. A structural model with concentric-layer orientation proposed by Harling and Heckman (6) is shown in Figure 4. According to this model the interior of the aggregate is less ordered than the surface, has a lower density,

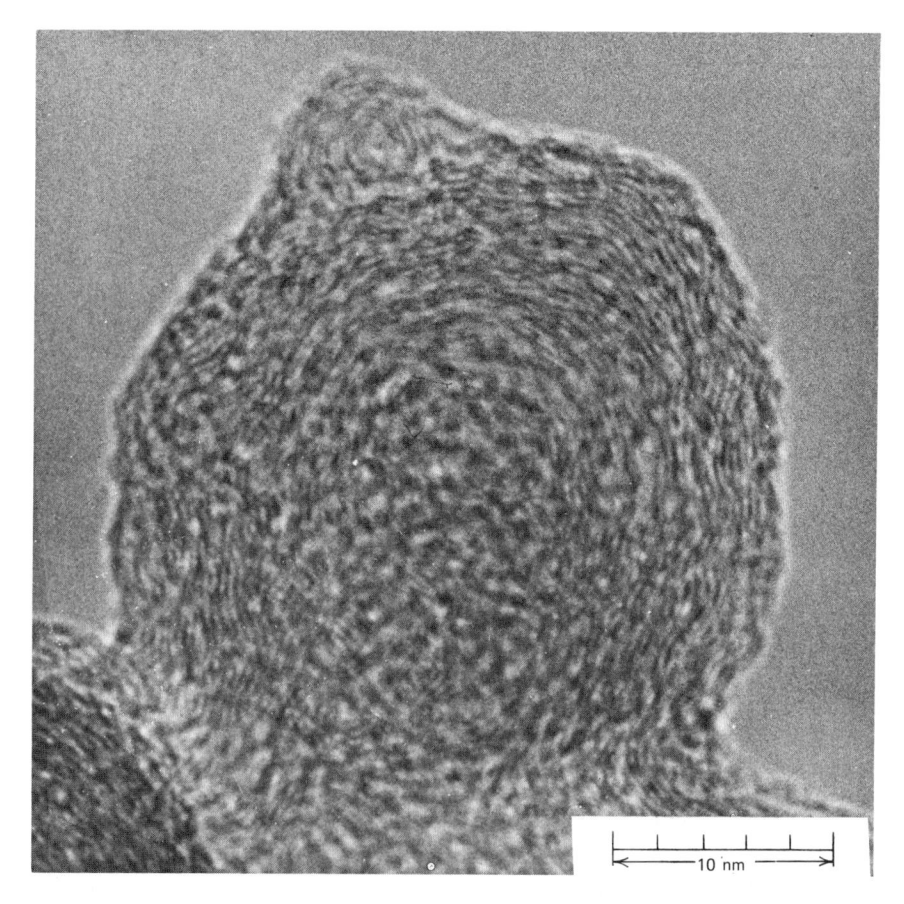

Figure 3. High resolution electron micrograph of carbon black primary aggregate nodule showing concentric orientation of layer planes.

and is chemically more reactive. Graphitization and oxidation studies confirm this view as illustrated by electron micrographs of oxidized and graphitized carbon blacks. Figure 5 shows that during oxidation the more reactive interior carbon is burned away leaving hollow shells of the less reactive surface layers. Figure 6 shows that graphitization causes increased crystallinity. The aggregates take on a capsule-like appearance with hollow centers. These changes are consistent with a concentric-layer structure.

High resolution electron microscopy also reveals that the interiors of some aggregates contain smaller concentric-layer regions which may have been separate entities at some stage of the carbon formation process. This can be seen in Figure 3. These have been called growth centers, and it is assumed that early in the carbon formation process they agglomerated, and continued to grow. The surface layers of these aggregates are distorted and folded in order to conform to aggregate geometry. This structure is clearly evident in the high resolution electron micrograph of graphitized N330 (HAF) in Figure 7.

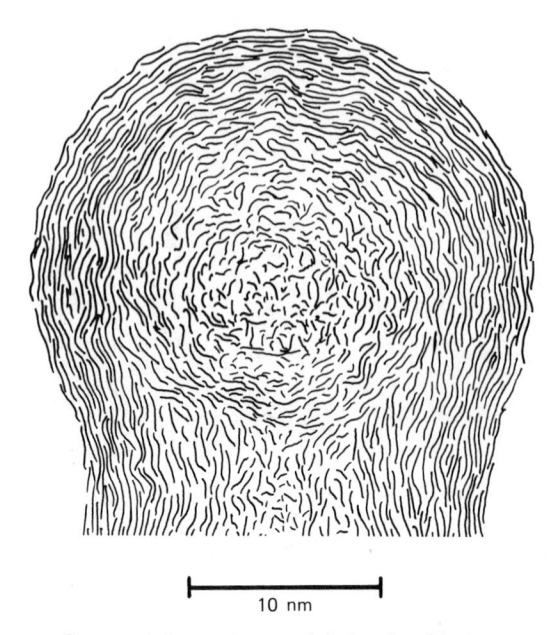

Figure 4. Concentric layer plane model of carbon black microstructure.

Characterization

Carbon blacks differ in particle size or surface area, average aggregate mass, particle and aggregate mass distributions, morphology or structure, and chemical composition. The form of these products, loose or pelleted, is another feature of some special grades. The ultimate colloidal units of cabon black, or the smallest dispersible entities in elastomer, plastic, and fluid systems are called aggregates, fused assemblies of particles. The particle size is related to the surface area of the aggregates. The oil furnace process produces carbon blacks with particle diameters of 10–250 nm and the thermal black process, 120–500 nm.

A primary aggregate is further characterized by its size, the volume of carbon comprising the aggregate, and its morphology. Aggregates have a variety of structural forms. Some are clustered like a bunch of grapes, whereas others are more open, branched, and filamentous.

Figure 8 illustrates the wide range of particle and aggregate sizes of commercial carbon blacks used for rubber reinforcement and for pigment applications.

Surface Area. The three most important properties used to identify and classify carbon blacks are surface area, structure, and tinting strength. Surface areas are measured by both gas- and liquid-phase adsorption techniques and depend on the amount of adsorbate required to produce a monolayer. If the area occupied by a single adsorbate molecule is known, a simple calculation will yield the surface area. The most familiar method is the procedure developed by Brunauer, Emmett, and Teller (BET). All commercial carbon blacks provide well-defined S-shaped isotherms with nitrogen as adsorbate at $-193°C$; the break in the isotherm (B point) corresponding to monolayer coverage V_m is determined and used to calculate the surface area. A useful extension of the BET method, called the "t" (adsorbed layer thickness) method, distinguishes between internal (or porous) surface area and external surface area (8). Most

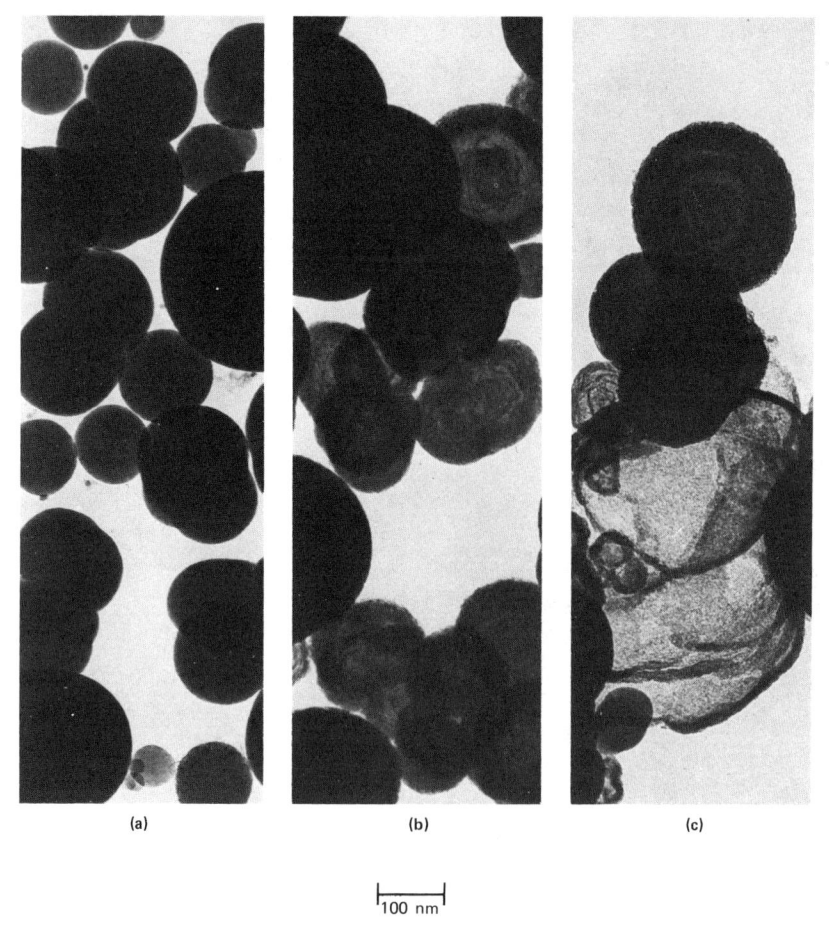

Figure 5. Electron micrographs of medium thermal carbon black particles: (**a**) original; (**b**) air-oxidized to 33% weight loss; (**c**) air-oxidized to 51% weight loss.

rubber-grade carbon blacks are nonporous, so that the BET and "t" methods give identical results. For some porous carbon blacks used for pigments and electrical applications, the "t" method is useful to assess their internal surface areas.

Liquid-phase adsorption methods are widely used for product control and specification purposes. The adsorption of iodine from potassium iodide solution is the standard ASTM method for classification of rubber-grade carbon blacks (9). The method is simple and precise, but its accuracy is affected by the presence of adsorbed hydrocarbons and the degree of surface oxidation. More recently developed surface area methods based on the adsorption of cetyltrimethylammonium bromide (CTAB) or Aerosol OT from aqueous solution are not affected by the same factors as the iodine adsorption test (10). These adsorbate molecules are so large that they cannot penetrate into micropores, and therefore provide an estimate of external surface area. The CTAB surface area method gives values in good agreement with the BET method for nonporous rubber grade blacks.

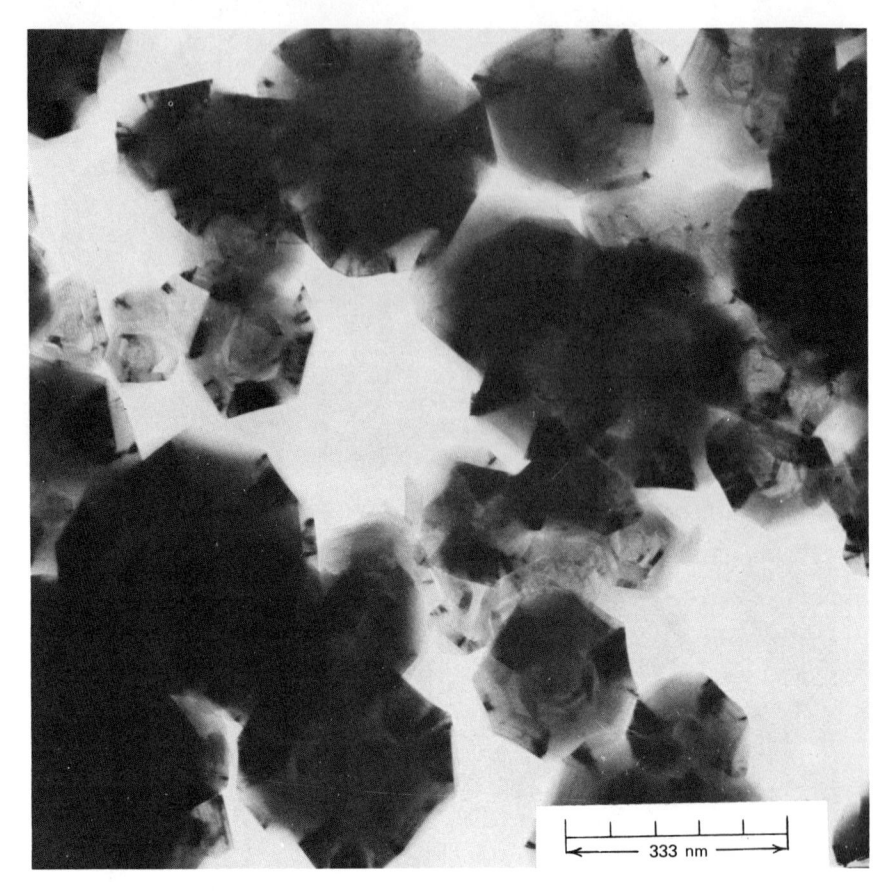

Figure 6. Electron micrograph of graphitized medium thermal carbon black.

Structure. The second most important property of carbon blacks is structure. Structure is determined by aggregate size and shape, the number of particles per aggregate, and their average mass. These characteristics affect aggregate packing and the volume of voids in the bulk material. The measurement of void volume, a characteristic related to structure, is a useful practical method for assessing structure. It is measured by adding linseed oil or dibutyl phthalate (DBP) to carbon black until the consistency of the mixture suddenly changes, at which point almost complete absorption or filling of the voids has occurred. Automated laboratory equipment is used to measure void volume by the DBP absorption (DBPA) method, a standard industry procedure (11). A modification of this procedure uses samples that have been preconditioned by four successive compression and grinding steps to fracture the weaker aggregate structures and more nearly represent the state of carbon black in rubber mixtures (12).

Tinting Strength. Tinting strength is another test method widely used to classify carbon blacks. In this test, a small amount of carbon black is mixed with a larger amount of white pigment, usually zinc oxide or titanium dioxide, in an oil or resinous vehicle to give a gray paste (13) and its diffuse reflectance measured. A carbon black with high tinting strength has a high light absorption coefficient and a low reflectance value. Tinting strength is related to average aggregate solid volume. Centrifugal sed-

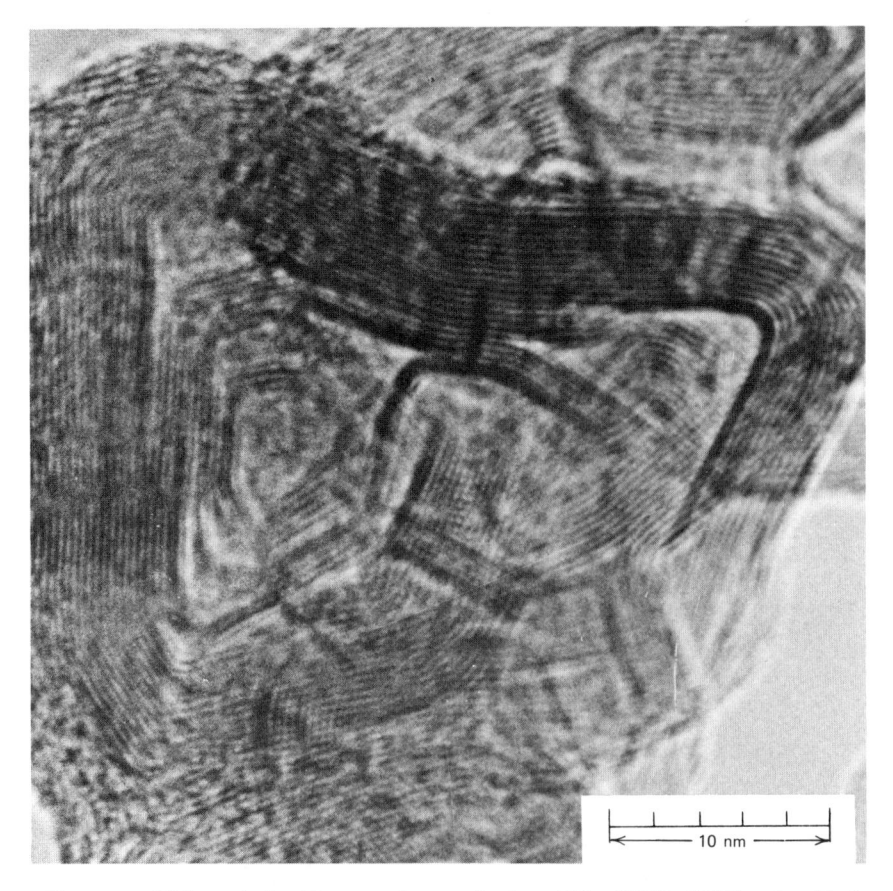

Figure 7. High resolution electron micrograph of graphitized N330 (HAF) carbon black.

imentation of aqueous carbon black suspensions has been used to estimate average aggregate volume and volume distributions (14). Figure 9 shows the correlation of tinting strength with average aggregate Stokes diameter, D_{St}. The Stokes diameter defines a sphere which sediments at the same rate as the actual aggregate.

Properties. Most commercial rubber-grade carbon blacks contain over 97% elemental carbon. A few special pigment grades have carbon contents below 90%. In addition to chemically combined surface oxygen, carbon blacks contain varying amounts of moisture, solvent-extractable hydrocarbons, sulfur, hydrogen, and inorganic salts. Hygroscopicity is increased by surface activity, high surface area, and the presence of salts. Extractable hydrocarbons result from the adsorption of small amounts of incompletely burned hydrocarbons. The combined sulfur content of carbon black has its origin in the sulfur content of the feedstocks. Most of the inorganic salt content comes from the water used for quenching and pelletizing. Table 2 lists the chemical composition of a few typical carbon blacks. Table 3 lists other analytical properties for rubber-grade carbon blacks. Table 4 lists analytical properties of carbon blacks for the ink, paint, plastics, and paper industries.

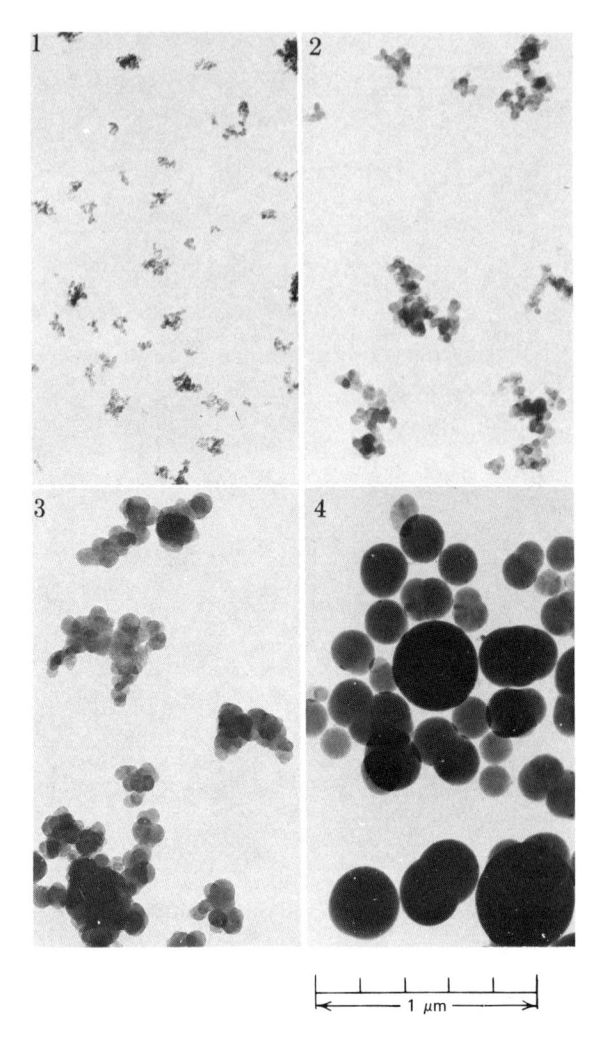

Figure 8. Electron micrographs of a range of carbon blacks: (1) high color pigment grade, 240 m^2/g; (2) rubber reinforcing grade, 90 m^2/g; (3) general purpose rubber grade, 36 m^2/g; (4) medium thermal grade, 7 m^2/g.

Formation Mechanisms

Carbon black is made by partial combustion processes involving flames, and by thermal decomposition processes in the absence of flames. Whether partial combustion or thermal decompsition methods are used, the basic reaction is represented by:

$$C_x H_y \xrightarrow{\Delta} x\, C + \frac{y}{2} H_2$$

With some hydrocarbon gases, such as methane, the reaction is endothermic, whereas with acetylene it is highly exothermic. The decomposition of heavy aromatic feedstocks in the furnace process is slightly endothermic, so that more energy is required than necessary to vaporize and heat the feedstock to reaction temperatures. The actual energy requirement will vary according to the process used, the average temperature of the carbon black formation reaction, the heat losses from the reactors, the degree

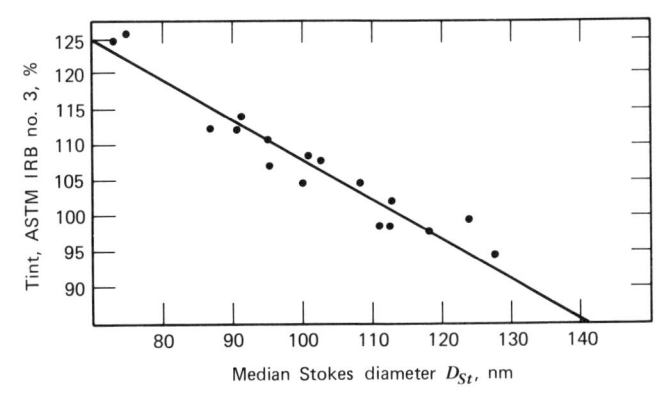

Figure 9. Tinting strength vs Stokes diameter D_{St}.

Table 2. Chemical Composition of Carbon Blacks

Symbol	Carbon, %	Hydrogen, %	Oxygen, %	Sulfur, %	Ash, %
medium thermal (N990) MT	99.4	0.3	0.1	0.0	0.3
semireinforcing furnace (N770) SRF	98.6	0.4	0.2	0.6	0.2
general purpose furnace (N660) GPF	98.6	0.4	0.2	0.6	0.2
fast extruding furnace (N550) FEF	98.4	0.4	0.4	0.7	0.2
high abrasion furnace (N330) HAF	98.0	0.3	0.8	0.6	0.3
superconducting furnace SCF	97.4	0.2	1.2	0.6	0.6
acetylene	99.8	0.1	0.1	0.0	0.0

of preheating of the input streams, the amount of heat recovery from the flue gases, and the extent to which the flue gases are used as auxiliary fuel. Modern furnace carbon black manufacture is quite efficient, reaching levels of 50–70% carbon recovery. The energy required to produce a kilogram of carbon black by the furnace process is in the range of $9-16 \times 10^7$ J/kg ($4-7 \times 10^4$ Btu/lb) depending on the grade of carbon black (see Burner technology).

The mechanism of carbon formation is not well defined. It is likely that no single mechanism can explain carbon black formation from different raw materials and by different processes. The manufacturing process has more of an influence on product quality than the raw material from which it is made. Thus natural gas or liquid aromatic hydrocarbons used in the channel black process produce the same type of carbon black. The decomposition of natural gas and benzene in a thermal process also will produce similar products. Thus a mechanism is called for in which different hydrocarbons in the vapor phase break down into smaller fragments at high temperature and recombine quickly to form particulate carbon. The temperatures and concentrations of reacting species determine the types of carbon black formed. There are several reviews of carbon formation theories (15–18).

Many of the proposed mechanisms of carbon black formation do not account for the microcrystalline morphology of the products. A model reaction mechanism of carbon formation involves sequentially: (1) the raw material hydrocarbon molecules decompose into smaller units or radicals with a loss of hydrogen; (2) these units, which may be monatomic carbon, C_2, C_3, C_6, or C_xH_y radicals, recombine into nuclei or

Table 3. Properties of Rubber-Grade Carbon Blacks

ASTM[a]	Type	Typical I$_2$ adsptn no.[b] D1510, mg/g	Surface area (BET), m^2/g	Particle size, nm[c]	Typical DBPA, mL/100 g D2414	Volatile content, %	Typical pour density, kg/m^3 (lb/ft^3) D1513	Tinting strength[d], % IRB 3 D3265
N110	SAF	145			113		336 (21.0)	128
N121	SAF–HS	120			130		320 (20.0)	
N219	ISAF–LS	118			78		441 (27.5)	124
N220	ISAF	121	115	22	114	1.5	344 (21.5)	114
N231	ISAF–LM	125			91		392 (24.5)	
N234		118			125		320 (20.0)	130
N242	ISAF–HS	123			126		328 (20.5)	119
N293	CF	145			100		376 (23.5)	
N294	SCF	205			106		368 (23.0)	
N326	HAF–LS	82	80	27	71	1.0	465 (29.0)	109
N330	HAF	82	80	27	102	1.0	376 (23.5)	104
N339	HAF–HS	90			120		344 (21.5)	114
N347	HAF–HS	90	90	26	124	1.0	336 (21.0)	104
N351		67			120		344 (21.5)	100
N358	SPF	84			150		288 (18.0)	
N363		66			68		481 (30.0)	
N375	HAF	90			114		344 (21.5)	116
N440	FF	50			60		481 (30.0)	
N472	XCF	270	254	31	178	2.5	256 (16.0)	
N539		42				9		384 (24.0)
N542		44			67		505 (31.5)	
N550	FEF	43	42	42	121	1.0	360 (22.5)	63
N568	FEF–HS	45			132		336 (21.0)	
N650	GPF–HS	36			125		368 (23.0)	
N660	GPF	36			91		424 (26.5)	51
N683	SPF	30			132		336 (21.0)	
N741		20			105		368 (23.0)	
N762		26			62		505 (31.5)	
N765	SRF–HS	31			111		376 (23.5)	54
N774		27			70		497 (31.0)	
N880	FT	13	12	200	35	0.5	673 (42.0)	29
N990	MT	7	8	450	35	0.5	673 (42.0)	17
acetylene black			67	40	260	0.3		

[a] ASTM designations are determined according to *Recommended Practice D2516, Nomenclature for Rubber-Grade Carbon Blacks*.
[b] In general, Method D1510 can be used to estimate the surface area of furnace blacks but not channel, oxidized, and thermal blacks.
[c] Approximate number-average particle size d_n in nm, based on electron microscopy.
[d] Industry reference black, IRB 3, an HAF grade, is taken as 100.

droplets by polymerization and condensation from the vapor phase; (3) the droplets continue to lose hydrogen and grow by further condensation on their surfaces of polymerized species, or the carbon nuclei grow into particles by further condensation of carbon from the vapor phase; (4) carbon particles or droplets flocculate; and (5) further carbon deposition on the carbon particle flocculates or continued loss of hydrogen from the agglomerated droplets result in the formation of primary aggregates. Higher reaction temperatures favor the formation of nuclei, smaller particle size within

Table 4. Properties of Furnace Process Carbon Blacks for Inks, Paints, and Plastics

	Nigro-meter index[a]	Surface area (BET), m²/g	Particle size, nm	Oil (DBP) absorption, mL/100 g Fluffy	Oil (DBP) absorption, mL/100 g Pellets	Tinting strength index	Volatile content, %	Fixed carbon, %	pH	Toluene extract, %
High Color Furnace										
HCF-1	64	560	13	121	105	100	9.5	90.5	3.3	0.08
HCF-2	65	240	14	65	50	115	2.0	98.0	7.0	0.08
HCF-3	69	230	15	70	65	120	2.0	98.0	7.0	0.08
Medium Color Furnace										
MCF-1	74	220	16	120	110	122	1.5	98.5	8.0	0.08
MCF-2	74	210	17	75	68	120	1.5	98.5	8.0	0.08
MCF-3	78	200	18	122	117	118	1.0	99.0	8.0	0.08
Long Flow Furnace	83	138	24	60	55	112	5.0	95.0	3.4	0.10
Medium Flow Furnace	84	96	25	72	70	112	2.5	97.5	4.5	0.10
Conductive Furnace	87	254	30	185	178	82	2.0	98.0	5.0	0.10
Regular Color Furnace										
RCF-1	84	140	19		114	114	1.5	98.5	7.0	0.10
RCF-2	83	112	24	62	60	116	1.0	99.0	7.5	0.10
RCF-3	83	86	25	65		112	1.0	99.0	8.0	0.30
RCF-4	84	94	25	62		110	1.0	99.0	8.5	0.10
RCF-5	87	80	27		70	104	1.0	99.0	7.5	0.10
RCF-6	90	46	36		72	92	1.0	99.0	7.0	0.10
RCF-6A	90	85	27	103	60	100	1.0	99.0	9.0	0.10
RCF-7	93	45	37	95		73	1.0	99.0	9.0	0.10
Low Color Furnace										
LCF-1	94	30	60		64	59	1.0	99.0	8.5	0.10
LCF-2	95	42	41		120	61	1.0	99.0	7.5	0.10
LCF-3	96	35	50		91	49	1.0	99.0	7.5	0.10
LCF-4	99	25	75	71	70	49	1.0	99.0	8.5	0.10

[a] A method for measuring the diffuse reflectance from a black paste with a black tile standard. The low numbers represent the jettest or most intense black grades.

the aggregates, and higher surface area. Aggregate formation is favored by increasing feedstock aromaticity. Particles are formed in close proximity, agglomerate, and form aggregates. The particles within an aggregate are remarkably monodisperse compared to the wide range of sizes apparent for aggregates (Figs. 2 and 8). In practice the various steps described above do not occur in an ideal sequence. For this reason carbon blacks are characterized by broad particle and aggregate size distributions.

Manufacture

Carbon black manufacture has evolved from primitive methods to continuous, high capacity systems using sophisticated designs and control equipment. About 1870 channel black manufacture began in the United States in natural gas producing areas to replace lampblack for pigment use. The discovery of rubber reinforcement by carbon black about 1912 and the growth of the automobile and tire industries transformed carbon black from a small-volume specialty product to a large-volume basic industrial raw material. Today's carbon black production is almost entirely by the oil furnace black process, which was introduced in the United States during World War II. However, the unique relationship of manufacturing process to special performance features of products has prevented the total abandonment of the older processes. The thermal, lamp black, channel black, and acetylene black processes account for less than 10% of the world's production. The easy sea and land transportation of feedstocks for the oil furnace process has made possible worldwide manufacturing facilities convenient to major consumers.

There are currently 30 furnace plant locations in the United States, 25 in Western Europe, 12 in the Eastern bloc countries, and 32 in the rest of the world. Eight United States companies produce carbon black. A recent estimate of their capacities (19) is shown in Table 5. Distribution of world carbon black manufacturing capacities is shown in Table 6.

Oil Furnace Process. The oil furnace process was developed during World War II by the Phillips Petroleum Company. The first grade of black produced was FEF (N500 type), followed by an HAF grade (N300 type) a few years later. The first successful carbon black reactor (Fig. 10) was introduced in the early fifties (20).

The oil furnace process was preceded by the gas furnace process. Gas furnace blacks were manufactured from about 1922 to the 1960s. In this process gas underwent partial combustion in refractory-lined retorts or furnaces, and carbon black was separated from the combustion gases with electrostatic precipitators and cyclone separators. Temperatures in the range of 1200–1500°C were used to produce carbon black grades designated as semireinforcing (SRF), high modulus furnace (HMF), fine furnace (FF), and high abrasion furnace (HAF). Some of these letter designations have been retained for oil furnace blacks. The gas furnace blacks were characterized by low structure levels and low-to-medium reinforcing performance. The carbon recovery of the process depended on the grade of black. The low-surface-area semireinforcing grades had yields of 25–30%, and the yields of the higher-surface-area reinforcing grades were 10–15%. The remarkable success of the oil furnace process was due to its improved yield performance in the range of 50–70%, higher capacities, and its ability to produce a wide range of products.

The use of petroleum oil for the manufacture of carbon black was patented as early as 1922 (21). The reactor contained many of the basic features of present oil

Table 5. United States Carbon Black Capacity, 1976

Manufacturer	Capacity, thousands of metric tons Furnace	Thermal
Ashland Chemical	304	
Arkansas Pass, Tex.	68	
Baldwin, La.	116	
Belpre, Ohio	45	
Mojave, Calif.	27	
Shamrock, Tex.	48	
Cabot Corp.	425	
Big Spring, Tex.	113	
Franklin, La.	91	
Pampa, Tex.	29	
Parkersburg, W. Va.	74	
Ville Platte, La.	118	
Cities Service	343	
Conroe, Tex.	44	
El Dorado, Ark.	37	
Eola, La.	32	
Franklin, La.	71	25
Hickok, Kan.	23	
Mojave, Calif.	24	
Moundsville, W. Va.	71	
Seagraves, Tex.	41	
Continental Oil	194	
Bakersfield, Calif.	35	
Ponca City, Okla.	61	
Sunray, Tex.	43	
Westlake, La.	54	
J. M. Huber Corp.	179	19
Baytown, Tex.	117	
Borger, Tex.	62	
Phillips Petroleum	219	
Borger, Tex.	130	
Orange, Tex.	52	
Toledo, Ohio	36	
Richardson Co.	95	
Addis, La.	45	
Big Spring, Tex.	50	
Thermatomic Carbon		
Sterlington, La.[a]		59
Total	1759	94

[a] Plant closed in 1977.

furnace processes including feedstock preheating and atomization, and water quenching. Another important oil furnace process was patented in 1942 (22). Table 7 compares yields for various processes. Energy requirements for each process are also shown.

Feedstocks. The feedstocks (qv) used by the carbon black industry are viscous, residual aromatic hydrocarbons consisting of branched polynuclear aromatic types mixed with smaller quantities of paraffins and unsaturates. Feedstocks are preferred that are high in aromaticity, free of coke or other gritty materials, and contain low levels of asphaltenes, sulfur, and alkali metals. For a particular plant location, other limi-

Table 6. Distribution of World Carbon Black Capacity [a]

Country	Number of plants	Estimated annual capacity, thousands of metric tons
North America		
Canada	3	164
Mexico	2	88
Total	5	252
South America		
Argentina	1	54
Brazil	3	152
Colombia	2	32
Peru	1	15
Venezuela	1	23
Total	8	276
Europe		
Great Britain	5	300
France	3	197
West Germany	5	269
Holland	2	116
Italy	4	175
Spain	3	91
Sweden	1	33
Yugoslavia	1	20
Total	24	1201
Australia, S. E. Asia		
Australia	3	89
India	5	111
Indonesia	1	3
Japan	9	479
Korea	1	23
Malaysia	1	13
Philippines	1	12
Taiwan	1	15
Total	22	745
Africa	1	60
Middle East		
Iran	1	15
Israel	1	11
Turkey	2	30
Total	4	56
United States	32	1888
remainder of the world [b]		680
Total		5158

[a] Including announced expansions through 1978.
[b] Estimated capacity.

tations usually involve the quantities available on a long-term basis, uniformity, ease of transportation, and cost.

In 1975 over 1.9 GL (5×10^8 gal) of feedstock was used for carbon black production in the United States. Sources of feedstock include petroleum refineries, ethylene plants, and coal coking plants. As a result of the introduction of more efficient petroleum cracking catalysts, decant oil from gasoline production has now become

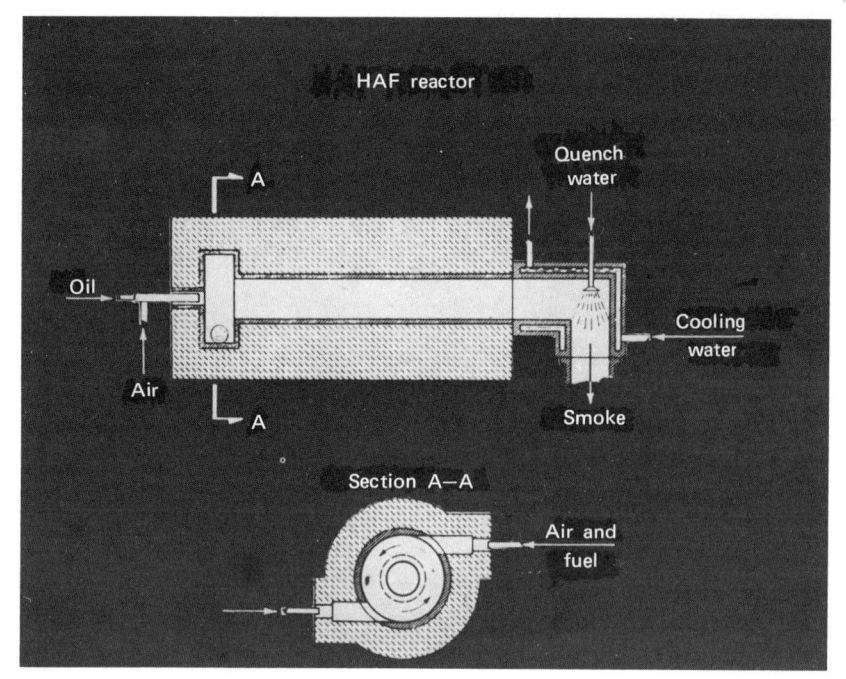

Figure 10. Phillips HAF black reactor (20).

Table 7. **Yields from Different Processes**

Process	Raw material	Commercial yields, g/m³ [a]	Yield % of theoretical carbon content	Energy utilization in manufacture, J/kg [b]
channel	natural gas	8–32	1.6–6.0	1.2–2.3×10^9
gas-furnace	natural gas	144–192	27–36	2.3–3.0×10^8
thermal	natural gas	160–240	30–45	2.0–2.8×10^8
oil-furnace	liquid aromatic hydrocarbons	300–660[c]	23–70	9.3–16×10^7

[a] To convert g/m³ to lb/1000 ft³, divide by 16.
[b] To convert J/kg to Btu/lb, multiply by 0.43×10^{-3}.
[c] In kg/m³ (2.5–5.5 lb/gal).

the most important feedstock in the United States, displacing residual thermal tars, and catalytic cycle stock extracts. Another important raw material is the steam pyrolysis residual tar from the manufacture of ethylene based on naphtha or gas oil. It is expected that ethylene tars will become more important as ethylene capacity increases. In Europe more ethylene process pyrolysis tars are used than in the United States. Coal tars, naphthalene and anthracene oils from coal tars, are also used for carbon black manufacture, but quantities are limited and prices are high.

648   CARBON (CARBON BLACK)

Equipment. Figure 11 is a flow diagram (23) of an oil furnace process. The reactor feeds are preheated feedstock, preheated air, and gas. The carbon-containing combustion products are quenched with water sprays and pass through heat exchangers which preheat the primary combustion air. The product stream is again cooled by a secondary water quench in vertical towers. Carbon black in light fluffy form is separated in bag filters and conveyed to micropulverizers discharging into a surge tank. From the surge tank the carbon black is fed to dry drums for dry pelletization (not shown in Fig. 11) or pin-type wet pelletizers, followed by dryers, to produced pelleted products (see Pelleting).

Different grades of carbon black are made by changing reactor temperatures and residence times as shown in Table 8. The long residence time for thermal black, the largest particle size grade, is obtained with large volume reactors and relatively low gas velocities. The shortest residence times for the highest-surface-area reinforcing grades are obtained using small, high velocity reactors.

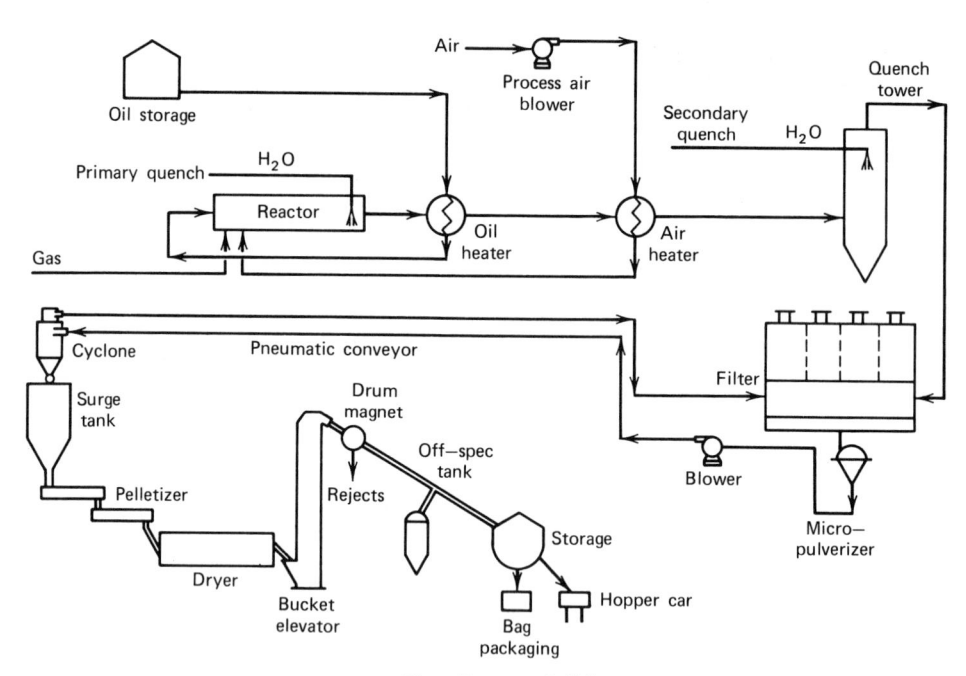

Figure 11.   Flow diagram of oil furnace process.

Table 8.   Time–Temperature Conditions in Carbon Black Reactors

Black grade	Residence time, s	Temperature, °C
thermal (N990)	10	1200–1350
SRF (N700 series)	0.9	1400
HAF (N300 series)	0.031	1550
SAF (N100 series)	0.008	1800

Reactors. The design and construction features of reactors are determined by economics and product specifications. Some reactors can produce over 45 metric tons of carbon black per day. At least two different types of reactors are required to make all the furnace grades for the rubber industry: one for the reinforcing and another for the less reinforcing grades. Additional types are used to produce special pigment grades. During the last decade reinforcing furnace black reactors have undergone substantial design changes to give higher gas velocities and turbulence, more rapid mixing of reactant gases and feedstock, and increased capacity (24–27). There have been fewer advances in the design of the larger reactors used for the less reinforcing, lower-surface-area grades. A notable exception has been the commercial use of huge vertical reactors having a more precise control of turbulence (28–29). Some of the latest reactor designs have refractory-lined metal construction, increased thermal efficiency, more efficient feedstock atomization, and improved quenching.

Carbon black properties, yield, and production rate can be adjusted by controlling the reactor feed variables. Combustion air and primary fuel rate are set to provide 20–50% excess air in the primary combustion flame. The desired temperature determines the feedstock rate. Lowering the feedstock rate produces higher temperatures, lower yields, and higher surface area products. In the mixing zone of the reactor a portion of the feedstock reacts with the excess air to give the required temperature. From 30 to 40% of the feedstock is burned to raise the temperature from 1450°C to 1500–1600°C when producing the most reinforcing types of carbon black. The temperature declines from this level as energy is used in the slightly endothermic hydrocarbon decomposition reactions.

In addition to the control of surface area, it is necessary to control average aggregate size, assessed by the dibutyl phthalate adsorption method (DBPA). Increasing the aromaticity of the feedstock increases the degree of aggregation. One of the most effective methods for reducing carbon black aggregation is by the injection of trace quantities of alkali metal salts into the carbon formation zone (30). This technique is useful for maintaining product DBPA specifications and for the production of low aggregation grades. The effectiveness of alkali metal addition is in the order of ionization potentials. The most commonly used alkali metal is potassium.

Post-reactor processing. The newly-formed carbon black aggregates are protected from attack by the water vapor in the hot reactor stream by rapidly reducing the temperature from 1300–1600°C to about 1000°C with a water spray. The primary water quench represents an unrecoverable energy loss and has the disadvantage of diluting the tail gas with moisture and lowering its heating value. Attempts to use indirect cooling methods to conserve heat energy have not been successful. The product stream is passed through heat exchangers to preheat the primary combustion air. The product is further cooled by a secondary water quench in a vertical tower, lowering its temperature to about 270°C before entering the bag filters. Glass cloth is commonly used in the bag filters. The product stream enters the inside of the bag filter tubes depositing carbon black and the tail gas is redistributed for use as fuel, or is flared or vented to the atmosphere. Periodically a bag filter compartment of tubes is repressurized with tail gas from the outside, releasing the carbon black.

The carbon black is collected in hoppers at the bottom of the bag filter unit, conveyed to grinders or micropulverizers which break up lumps, then conveyed to a cyclone and into the surge tank. The surge tank accumulates and partially compacts the fluffy carbon black and provides a reservoir for maintaining a constant flow to the pelletizers.

Most of the carbon black used by the rubber industry is wet pelletized. Dry pelletization is practiced for some rubber applications and for pigmentation. In the dry process the slightly compacted loose black is conveyed to large rotating steel drums whose rolling motion causes a steady increase in bulk density and the formation of pellets. A fraction of the pelletized product is recirculated and added to the loose black at the drum entry to increase the density of the bed and provide seed material to increase the formation rate.

The wet pellet process uses pin-type agitators enclosed in tubes about 0.5 m dia and about 3 m in length. The pins are arranged as a double helix, or other configuration, around a rotating shaft extending through the center of the tube. Water is added to the loose black through sprays downstream of its entry, resulting in a rapid increase in density while forming wet pellet beads. The water–carbon black ratio is critical; an insufficient amount of water results in a dusty product, and an excess causes a sticky paste which can clog the equipment. The optimum water–black ratio increases with increasing aggregation. For normal aggregation blacks about equal amounts by weight of water and black are used. There are many different types of wet pelletizers. Their water spray pressures, nozzle locations, and rotating shaft speeds influence the characteristics of the pellets.

The wet pellets are conveyed to a dryer which consists of a long, slowly-revolving cylindrical drum heated by natural gas or tail gas burners located under its entire length. A fraction of the heater exhaust gases may be passed over the moving bed of pellets in the drier drum to purge liberated moisture and remove small amounts of carbon black dust formed by pellet breakdown. The wet pellets have an initial moisture content of about 50% on entry to the drier, and after about one hour residence time exit with less than 0.5% moisture. The dried product is screened to remove oversized pellets which are recycled through the pelletization system. The product is conveyed to storage bins for bulk shipment or for bag packing. In the United States as much as 90% of carbon black production is shipped in bulk.

Good pellet quality is necessary for easy loading and unloading of bulk cars, for conveying without appreciable pellet breakdown and dustiness, and for automatic weighing in rubber factories. High bulk density is preferred for economical transportation, cleanliness, and rapid incorporation into rubber mixtures. The maximum bulk density is however limited by the requirement for good dispersion in rubber. Bulk densities of commercial furnace carbon black grades are 256–513 kg/m^3 (16–32 lb/ft^3). Their pellet size distributions are in the range of 125–2000 μm.

New Process Technology. During the early 1970s the industry adopted improved reactor and burner designs which provided more rapid and uniform feedstock atomization, more rapid mixing, and shorter residence times (27). Higher capacities and increases of 6–8% in yields for products rated at equivalent roadwear levels have been reported (31). These new carbon blacks have lower surface areas than the standard grades. Electron microscopy shows fewer large aggregates, a smaller average aggregate size, a narrower aggregate size distribution, and the individual aggregates appear to be more open, branched, and bulky. In addition to these changes in aggregate morphology, the new carbon blacks have higher tinting strengths and give evidence of higher surface activity as shown by an increase in bound rubber formation (14). In rubber, enhanced surface-polymer interaction results in higher tensile strength, modulus, hysteresis, and treadwear. The improvements in pigment and rubber performance are described in recent patents (32) (see Rubber compounding).

Thermal Decomposition Process. The high temperature decomposition of hydrocarbons in the absence of air or flames is the basis for the manufacture of thermal blacks and acetylene black. Thermal black is made by a strongly endothermic reaction requiring a large heat energy input, whereas the acetylene black decomposition reaction is strongly exothermic.

In the thermal process two refractory-lined cylindrical furnaces, or generators, alternate on about a 5-minute cycle between black production and heating, making the overall production continuous from the two generators. The generators are about 4 m in diameter and 10 m high and are nearly filled with an open checkerwork of silica brick. During operation one generator is fired with a stoichiometric ratio of air and fuel, while the other generator, heated in the previous cycle to an average temperature of 1300°C, is fed with natural gas. The product stream contains suspended carbon black, hydrogen, methane, and other hydrocarbons. This stream is cooled with water sprays and the black removed by a bag filter. Carbon recovery is about 30–45% of the total carbon content of the gas used for the heating and production cycles. The gas from the bag filter containing about 90% hydrogen is cooled, dehumidified, compressed, and used as fuel for reheating the generators. It is also used to dilute the natural gas feed for the production of a smaller particle size grade known as Fine Thermal (N880), as well as to fire boilers for plant steam and electricity. The fluffy black is passed through a magnetic separator, screened, and micropulverized. Most of the production is pelletized, packed in paper bags, or loaded directly into bulk hopper cars. Table 9 lists commercial grades of thermal blacks. The N900 series of medium thermal grades are the most widely used. The nonstaining grades N907 and N908 are made at higher average generator temperatures and longer contact times. This virtually eliminates residual hydrocarbon residues which cause staining by diffusion to the surface of many rubber products. A comparison of the analytical properties of medium and fine thermal blacks with carbon blacks made by other processes is shown in Table 10. Thermal blacks have the lowest surface area and lowest aggregation of the commercial carbon blacks. Medium thermal black particles are 400–500 nm and fine thermal blacks 120–150 nm in diameter. The electron micrograph of medium thermal black in Figure 8 shows a predominance of spherical particles in marked contrast to the aggregate nature of furnace black.

A medium thermal carbon black variety is produced in England by a cyclic process similar to the thermal process. This process uses oil for both the heating and production cycles. During the production cycle steam and oil are fed to the hot generator producing a gas containing suspended carbon black. The gas contains about 60% H_2, 25% CH_4, and 15% CO. After water cooling, about 85–90% of the black is removed with cyclones,

Table 9. Thermal Black Grades

ASTM classification	Industry type	Description
N880	FT–FF	fine thermal black, free flowing
N881	FT	fine thermal black
N990	MT–FF	medium thermal black, free flowing
N907	MT–NS–FF	medium thermal, nonstaining, free flowing
N908	MT–NS	medium thermal, nonstaining
N991	MT	medium thermal black

Table 10. Typical Analyses of Carbon Black Grades from Five Different Processes

Property	Type: Symbol: ASTM No.:	Furnace HAF N-330	Thermal MT N-990	Thermal FT N-880	Acetylene Shawinigan	Channel EPC S300	Lampblack Lb
average particle diameter, nm		28	500	180	40	28	65
surface area (BET), m²/g		75	47	13	65	115	22
DBPA, mL/100 g		103	36	33	250	100	130
tinting strength, % SRF		210	35	65	108	180	90
benzene extract, %		0.06	0.3	0.8	0.1	0.00	0.2
pH		7.5	8.5	9.0	4.8	3.8	3.0
volatile material, %		1.0	0.5	0.5	0.3	5	1.5
ash, %		0.4	0.3	0.1	0.0	0.02	0.02
Composition, %							
C		97.9	99.3	99.2	99.7	95.6	98
H		0.4	0.3	0.5	0.1	0.6	0.2
S		0.6	0.01	0.01	0.02	0.20	0.8
O		0.7	0.1	0.3	0.2	3.5	0.8

and the remainder is removed as a wet slurry used in the manufacture of carbon electrodes (33).

Thermal black has had wide use as a low-cost extender for rubber, as well as a functional filler for specially engineered rubber products (see Fillers). Substantial price increases during the last few years, caused by the rise in natural gas prices, have discouraged the extender use. It remains an essential ingredient in many specialty rubber products. The market share of thermal black has been as high as 11%, but in 1976 had declined to less than 5% due to the closing of some production facilities.

Acetylene Black Process. The high carbon content of acetylene (qv) (92%) makes it attractive for conversion to carbon. It decomposes exothermically at high temperatures, a property which was the basis of an explosion process initiated by electrical discharge (34). Acetylene black is made by a continuous decomposition process at 800–1000°C in water-cooled metal retorts lined with a refractory. The process is started by burning acetylene and air to heat the retort to reaction temperature, followed by shutting off the air supply to allow the acetylene to decompose to carbon and hydrogen in the absence of air. The large heat release requires water cooling in order to maintain a constant reaction temperature. The high carbon concentration, high reaction temperature and relatively long residence time produce a unique type of carbon black. After separation from the gas stream it is very fluffy with a bulk density of only 19 kg/m³ (1.2 lb/ft³). Acetylene black is difficult to compact by compression and resists pelletization. Commercial grades are compressed to various bulk densities up to a maximum of 200 kg/m³ (12.5 lb/ft³).

Table 10 lists the properties of acetylene black in comparison with carbon blacks from other processes. It is the purest form of carbon black listed in Table 10 with a carbon content of 99.7%, and a hydrogen content of 0.1%. It has the highest aggregation with a DBPA value of 250 cm³/100 g. X-ray analysis indicates that it is the most crystalline or graphitic of the commercial blacks (35). These features result in a product with low surface activity, low moisture adsorption, high liquid adsorption, and high electrical and thermal conductivities.

A major use for acetylene black is in dry cell batteries because it contributes low electrical resistance and high capacity. In rubber it gives electrically conductive properties to heater pads, heater tapes, antistatic belt drives, conveyor belts, and shoe soles. It is also used in electrically conductive plastics. Some applications of acetylene black in rubber depend on its contribution to improved thermal conductivity, such as rubber curing bags for tire manufacture.

Lampblack Process. Early lampblack processes used large open shallow pans from 0.5 to 2 m in diameter and 16 cm deep for burning various oils in an enclosure with a restricted air supply. The smoke from the burning pans was allowed to pass at low velocities through settling chambers from which it was cleared by motor-driven ploughs. Yields of ca 480–600 kg/m^3 (4–5 lb/gal) of oil were obtained when making rubber grades. Today lampblack substitutes are made by the furnace process. Traditional lampblack manufacture is still practiced, but the quantities produced are small.

Channel Black Process. The channel black process has had a long and successful history, beginning in 1872 and ending in the United States in 1976. Small quantities are still produced in a few scattered plants operating in Germany (roller-process using oil), Eastern Europe, and Japan. Rising natural gas prices, smoke-pollution, low yield, and the rapid development of furnace process grades caused the termination of channel black production in the United States.

The name channel black came from the use of steel channel irons whose flat side was used to collect carbon black deposited from many small flames in contact with its surface. The collecting channels and thousands of flames issuing from ceramic tips were housed in sheet metal buildings, each 35–45 m long, 3–4 m wide, and about 3 m high. The air supply came from the base of these buildings; the waste gases, containing large quantities of undeposited product, were vented to the atmosphere as a black smoke. Carbon black was removed from the channels by scrapers and fell into hoppers beneath the channels. Yields were very low, in the range of 1–5%. The blackest pigment grades had the lowest yields. The product was conveyed from the hot houses to a processing unit where grit, magnetic scale, coke, and other foreign material were removed. From an initial bulk density of 80 kg/m^3 (5 lb/ft^3) it was compacted and pelletized to over 400 kg/m^3 (25 lb/ft^3) for use in rubber. Lower bulk densities were used for pigment applications.

Table 10 lists the properties of easy processing channel black (EPC), a long-time favorite of the rubber industry for tire tread reinforcement. Channel blacks are surface oxidized as a result of their exposure to air at elevated temperatures on the channel irons. Due to surface oxidation, the particles are slightly porous. These features influence performance in most applications. In rubber, the acidity of the oxygen-containing surface groups has a retarding effect on rate of vulcanization. In polyethylene, where carbon black is used as an uv absorber (see Uv absorbers) and to improve weathering resistance, the phenol and hydroquinone surface groups have antioxidant properties (see Antioxidants). In inks, oxygen-containing groups contribute to flow and printing behavior.

Trends in Carbon Black Manufacture (36–37). Since the introduction of the oil furnace process in the 1940s the industry has emphasized the development of new and improved grades to meet customer requirements, and process improvements to increase yields and capacities. Carbon black yields have almost doubled, and reactor capacities have increased about tenfold from 1950 to 1975. The carbon black industry is expected to devote more of its resources in the future to energy conservation, environmental problems, and health and safety aspects of its processes and products. Because of the increasing costs of new production capacity, plant production levels will be maintained closer to rated capacity of existing facilities. The following trends in carbon black manufacture can be expected: increasing use of by-product pyrolysis tars from ethylene manufacture for feedstocks; replacement of natural gas by liquid hydrocarbons for the primary heat source; increased sensible heat recovery from the combustion stream; increased use of tail gas for its heating value; use of higher sulfur content feedstocks; and increasing use of automatic computer control.

Economic Aspects

Carbon black consumption by the rubber industry has grown steadily from the time of the discovery of its outstanding reinforcing properties by S. C. Motte in Great Britain about 1912 (38). The average annual growth rate since 1925 has been about 5.8% in the United States. Figure 12 shows United States consumption from 1887 to 1975 in six distinct growth periods on a semilogarithmic plot.

Outside the United States carbon black consumption has followed a similar pattern with a few notable differences. From 1962–1974, consumption grew at 7.6% for Western Europe and 6.1% for the United States. The United States consumed more carbon black than the rest of the world combined for the period 1915–1965. By 1975 United States consumption accounted for about 42% of world consumption.

The pigment use of carbon black was the only important market until the discovery of rubber reinforcement. The United States market distribution from 1910 to the present is shown in Figure 13 (39). About 75% of the total market depends on motor vehicle applications including tires and automotive products (40). Carbon black prices depend mainly on costs of raw materials, plant equipment, labor, and utilities. Advances in technology and improved process economics had kept prices stable from 1950–1973 but in recent years the cost of feedstock has increased more than fourfold, causing about a twofold increase in the price of the large-volume rubber grades. The bulk price range in the United States for the major rubber grade carbon blacks in 1977 was 28–35¢/kg. Special pigment grades had a very wide price range from 31.4¢/kg for the least expensive low color SRF grades to $4.41/kg for the fluffy high color lacquer and enamel grades. The HAF (N300 series) grade is the most important and now accounts for almost one half the total market. This grade is divided into a number of subgrades with different structure levels. The next important grade is GPF (N660) which was introduced in the late fifties.

Health and Safety Aspects

Carbon is a relatively stable, unreactive element, insoluble in organic or other solvents. Carbon in the form of char has been used in the pharmaceutical industry for many years. Recently, vitreous and pyrolytic carbons have been found to be biocompatible and useful for artificial heart valves, and other medical applications (see Prosthetic and biomedical devices). There is no evidence of carbon black toxicity in humans, despite the fact that it contains trace amounts of some polynuclear aromatic compounds known to be carcinogenic.

During the first years of the carbon black industry, when major production was

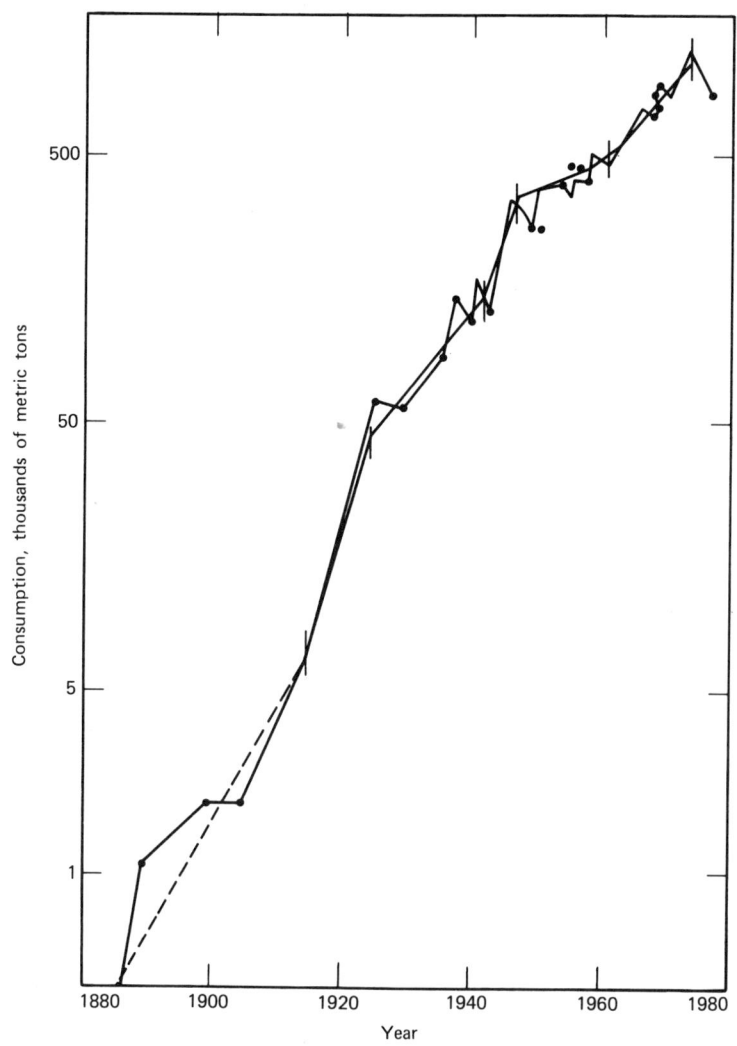

Figure 12. Carbon black consumption in the United States. Growth: 1915–1925, 16%; 1926–1942, 5.2%; 1943–1946, 17%; 1947–1961, 3.3%; 1962–1974, 5.9%.

by the channel black process, workers in the plants were exposed to an atmosphere containing carbon black. Channel black plants in the United States have been replaced by furnace process plants. Because carbon black is fine, light, easily carried by air currents, and jet black, special housekeeping procedures must be used during production and use. In order to reduce dustiness, the carbon black industry supplies its products either in loose compacted forms, or in more highly compacted pelletized forms.

The ingestion, skin contact, subcutaneous injection, and inhalation of channel, furnace, and thermal blacks by various animals causes no significant physiological changes (41–46). Although trace amounts of carcinogens were present, they were at too low a level, or were too inactive in their adsorbed state to cause a health hazard.

656 CARBON (CARBON BLACK)

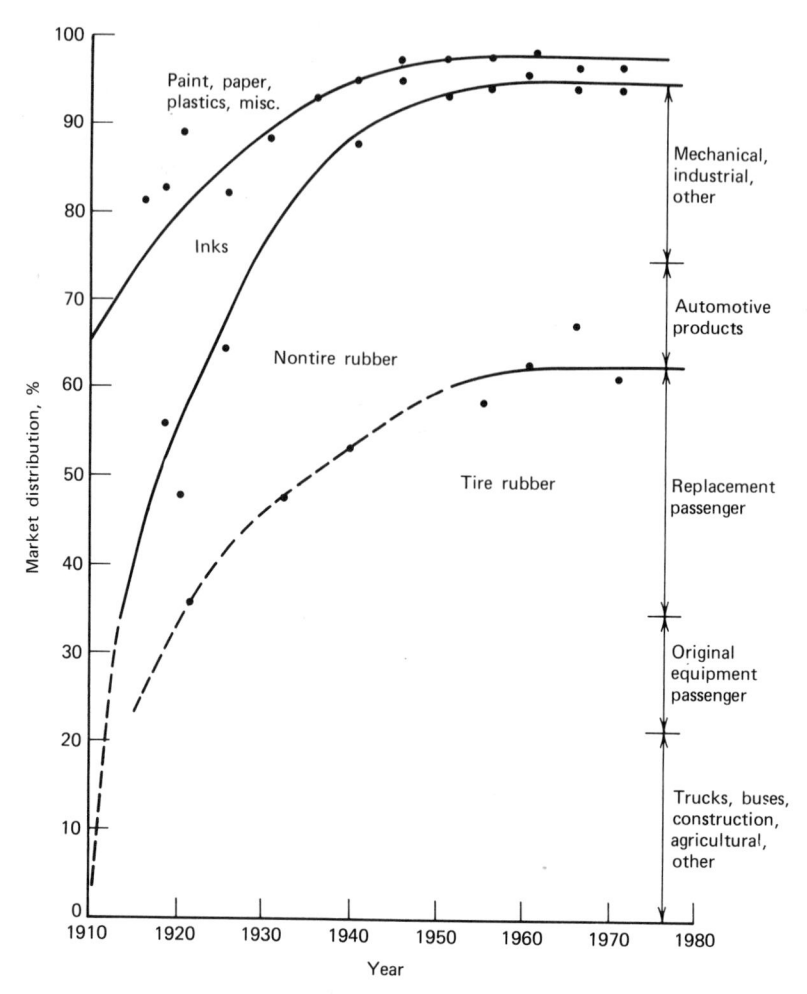

Figure 13. Carbon black market distribution.

Surveys of the health histories of carbon black workers published in 1950 and 1961 (47–48) showed their morbidity rate and observed death rate from cancer to be equal to or lower than those of comparable groups of industrial workers and the general population. Inhalation of high concentrations of carbon black over prolonged periods produced lung changes in animals due primarily to carbon black deposition with minimal or no fibrous tissue proliferation (44).

Carbon black is difficult to ignite, does not undergo spontaneous combustion, and does not produce dust explosions. When ignition does occur from contact with flames, glowing metal, sparks, or lighted cigarettes, it continues slowly with a dull glow. Due to the low conductivity of carbon black, storage fires may go undetected for some time. Fires may be controlled by purging with carbon dioxide. When this is done respirators must be used to avoid breathing carbon monoxide. Even in the absence of a fire, an air line or adequate respirator should be used when entering any confined carbon black storage facilities to avoid breathing possible toxic gases or a low oxygen atmosphere.

The level of toxic substances associated with carbon black is very low. Analytical values for a typical rubber grade semireinforcing carbon black (N774) showed 0.6 ppm phenols, 2.7 ppm lead, less than 0.05 ppm for cyanides, and less than 0.5 ppm for all other heavy metals (49). The metals have their origin in the petroleum feedstocks which have undergone various treatments and catalytic processes. The inorganic content of carbon black also includes such salts as sodium, potassium, calcium, and magnesium sulfates, chlorides, carbonates, and silicates which are contained in the water used for quenching and for wet pelletization.

The combustion of fuels can cause serious problems of atmospheric pollution from the emission of particulate material, sulfur compounds, nitric oxides, hydrocarbons, and other gases. Because of the intense blackness of carbon black it cannot be allowed to escape to the atmosphere even in minute quantities (50). Huge bag filters for complete recovery were put in operation over 30 years ago (see Air pollution control methods).

In the furnace process, tail gas burning to convert CO and H_2S to less noxious CO_2 and SO_2 has been practiced in some plants since 1950. This trend is increasing, stimulated by the necessity to recover the energy content of the tail gases. Not all the sulfur contained in carbon black feedstocks is converted to gases. One third to one half of the sulfur content is incorporated as nonextractable sulfur in the products.

Uses

Rubber. Table 11 lists the major grades used by the rubber industry, the general mechanical properties they contribute to rubber, and some examples of typical uses. A review of carbon black compounding and the types used in rubber products has been published (51) (see Rubber compounding).

Classification. The classification of carbon black grades for rubber was originally based on various performance or property characteristics, including: (*1*) levels of abrasion resistance—high abrasion furnace (HAF), intermediate super abrasion furnace (ISAF), and super abrasion furnace (SAF); (*2*) level of reinforcement—semireinforcing furnace (SRF); (*3*) a vulcanizate property—high modulus furnace (HMF); (*4*) a rubber processing property—fast extrusion furnace (FEF); (*5*) utility—general purpose furnace (GPF) and all-purpose furnace (APF); (*6*) particle size—fine furnace (FF), and large particle size furnace (LPF), fine thermal (FT), and medium thermal (MT); and (*7*) electrically conductive properties (XCF). Within some of these grades there were a variety of subgrades having different aggregate levels, eg, HAF with a high aggregate subgrade (HAF-HS), and a low aggregate subgrade (HAF-LS). The obvious inadequacies of this unwieldy classification procedure led ASTM Committee D-24 on carbon black to establish a letter and number system, shown in Table 3. In the ASTM system the N-series numbers increase as iodine adsorption values or surface areas decrease. The SAF grades have designated numbers from N100 to N199; the ISAF grades, N200 to N299; the HAF grades, N300 to N399; the FF and XCF (with this grade surface areas may be out of order with respect to the ASTM number) grades, N400 to N499; the FEF grades, 500 to N599; the HMF, GPF, and APF grades, N600 to N699; the SRF grades, N700 to N799; fine thermal (FT) has been designated as N880, medium thermal (MT) N990, and nonstaining medium thermal (MT-NS), N907. Aggregate levels or other quality variations within a grade are given arbitrary numbers. Table 3 lists a selected group of carbon black grades, their ASTM number,

Table 11. Applications of Major Rubber Grade Carbon Blacks

ASTM N-Type	Designation	General rubber properties	Typical uses
N990	medium thermal (MT)	low reinforcement, modulus, hardness, hysteresis, tensile strength; high loading capacity and high elongation	wire insulation and jackets, mechanical goods, footwear, belts, hose, packings, gaskets, O-rings, mountings, tire innerliners
N880	fine thermal (FT)	low reinforcement, modulus, hardness, hysteresis, tensile strength; high elongation, tear strength, and flex resistance	mechanical goods, gloves, bladders, tubes, footwear uppers
N700 Series	semireinforcing (SRF)	medium reinforcement, high elongation, high resilience, low compression set	mechanical goods, footwear, inner tubes, floor mats
N660	general purpose (GPF)	medium reinforcement, medium modulus, good flex and fatigue resistance, low heat buildup	standard tire carcass black; tire innerliners and widewalls; sealing rings, cable jackets, hose, soling, and extruded goods; EPDM compounds
N650	general purpose–high structure (GPF–HS)	medium reinforcement, high modulus and hardness, low die swell, smooth extrusion	tire innerliners, carcass, radial belt and sidewall compounds; extruded goods and hose
N550	fast extrusion (FEF)	medium-high reinforcement; high modulus and hardness; low die swell and smooth extrusion	tire innerliners, carcass, and sidewall compounds; innertubes, hose and extruded goods
N326	high abrasion–low structure (HAF–LS)	medium-high reinforcement; low modulus, high elongation, good fatigue resistance, flex resistance, and tear strength	tire belt, carcass, and sidewall compounds
N330	high abrasion (HAF)	medium-high reinforcement; moderate modulus, good processing	tire belt, sidewall, and carcass compounds; retread compounds, mechanical and extruded goods
N339, N347, N375	high abrasion–high structure (HAF–HS)	high reinforcement, modulus, and hardness; excellent processing	standard tire tread blacks
N220	intermediate super abrasion (ISAF)	high reinforcement, tear resistance; good processing	passenger and off-the-road tire treads; special service tires
N110	super abrasion (SAF)	high reinforcement	special tire treads, airplane, off-the-road, racing tires; products for highly abrasive service

their category, and typical analytical properties. Table 12 is a breakdown of the United States carbon black market according to consumption by type in 1975. Almost 80% of the market consisted of three types, HAF, GPF, and FEF.

Figure 14 is a chart of the carbon black product spectrum illustrating performance differences due to surface area, as assessed by iodine absorption, and aggregation or structure, as assessed by DBP absorption.

Properties of Carbon Black–Rubber Compounds. The general effects of different carbon blacks on rubber properties are dominated mainly by surface area and aggregation or structure. High surface area and small particle size impart higher levels of reinforcement as reflected in tensile strength, tear resistance, and resistance to abrasive wear with resulting higher hysteresis and poorer dynamic performance. Higher aggregation gives improved extrusion behavior, higher stock viscosities, improved green strength, and higher modulus values. A summary of the effects of carbon black structure and particle size on rubber processing and vulcanizate properties appears in Table 13.

Table 12. United States Carbon Black Distribution, 1975

Carbon black grade	Market, %
HAF (N300 series)	46.1
GPF (N600 series)	20.3
FEF (N550)	11.5
SRF (N700 series)	8.8
ISAF (N200 series)	8.0
thermal (N800–N900 series)	4.4
SAF (N110)	0.9

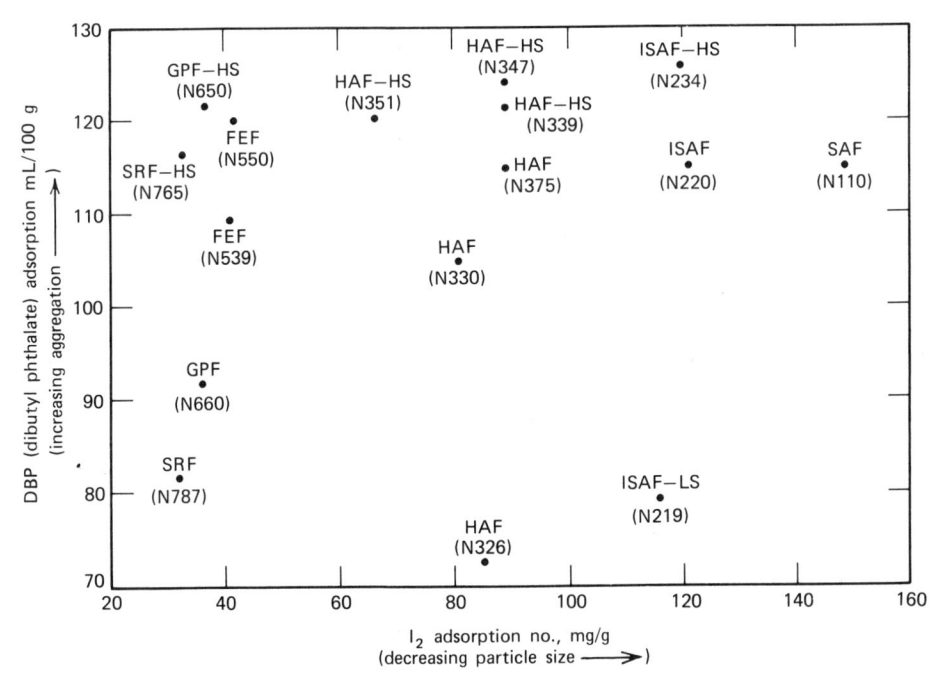

Figure 14. Carbon black grade spectrum.

660   CARBON (CARBON BLACK)

Table 13.   Effect of Carbon Black Colloidal Characteristics on Rubber Processing and Physical Properties

Property	Colloidal characteristics	
	Decreasing particle size	Increasing aggregation
	increasing surface area	increasing aggregate size
	increasing iodine number	increasing DBP absorption
	increasing tinting strength	decreasing bulk density
Rubber processing property		
loading capacity	decreases	decreases
mixing incorporation time	increases	increases
Mooney viscosity	increases	increases
dimension stability (green)	increases	increases
extrusion shrinkage	not significant	decreases
Rubber physical property		
tensile strength	increases	variable
modulus	not significant	increases
elongation	not significant	decreases
hardness	increases	increases
impact resilience	decreases	not significant
abrasion resistance	increases	variable

Rubber property changes produced by carbon black addition depend on loading. Tensile strength and abrasion resistance increase with increased loading to an optimum and then decrease. The optimum is normally in the range of 40–60 phr (parts per hundred of rubber). Increasing the loading level to 35–80 phr produces linear increases in hardness and modulus. The magnitude of these changes increases with structure. These effects are shown in Figure 15 for a group of carbon black grades including HAF (N330) used in tire treads, GPF (N660) and APF (N683) used in tire carcasses, and a medium thermal (MT) (N990) used in mechanical goods. Testing was done in a standard SBR-1500 recipe containing 12 parts of processing oil (52).

Special Carbon Blacks. About 6–7% of the total carbon black production consists of special industrial carbon grades for nonrubber applications (special blacks). Although some of the basic grades produced for the rubber market can be used for special black applications, most of these products are manufactured by methods developed to meet specific use requirements. They sell for a higher average price than the rubber grades. The growth of these markets over the last two decades has been about 4% annually. Of increasing importance in recent years have been applications in plastics where special carbon blacks are required to improve weathering resistance, or to impart antistatic and electrically conductive properties (see Antistatic agents).

About 40% of the special blacks are used in printing inks, 10% in paints and lacquers, 36% in plastics, 2% in paper, and 12% in miscellaneous applications. News inks account for most of the printing ink market. For this use special N300 series (HAF) grades containing 6% mineral oil are made to give rapid dispersion. Medium and high color blacks, the most expensive grades, are used in enamels, lacquers and plastics for their intense jetness. Carbon black grades for these applications are listed in Table 14.

Although the applications for special blacks have not changed greatly from 1965 to 1975, there were major changes in the product line and production technology. During this period the medium and high color grades, which were formerly made by the channel process, were replaced by furnace process grades.

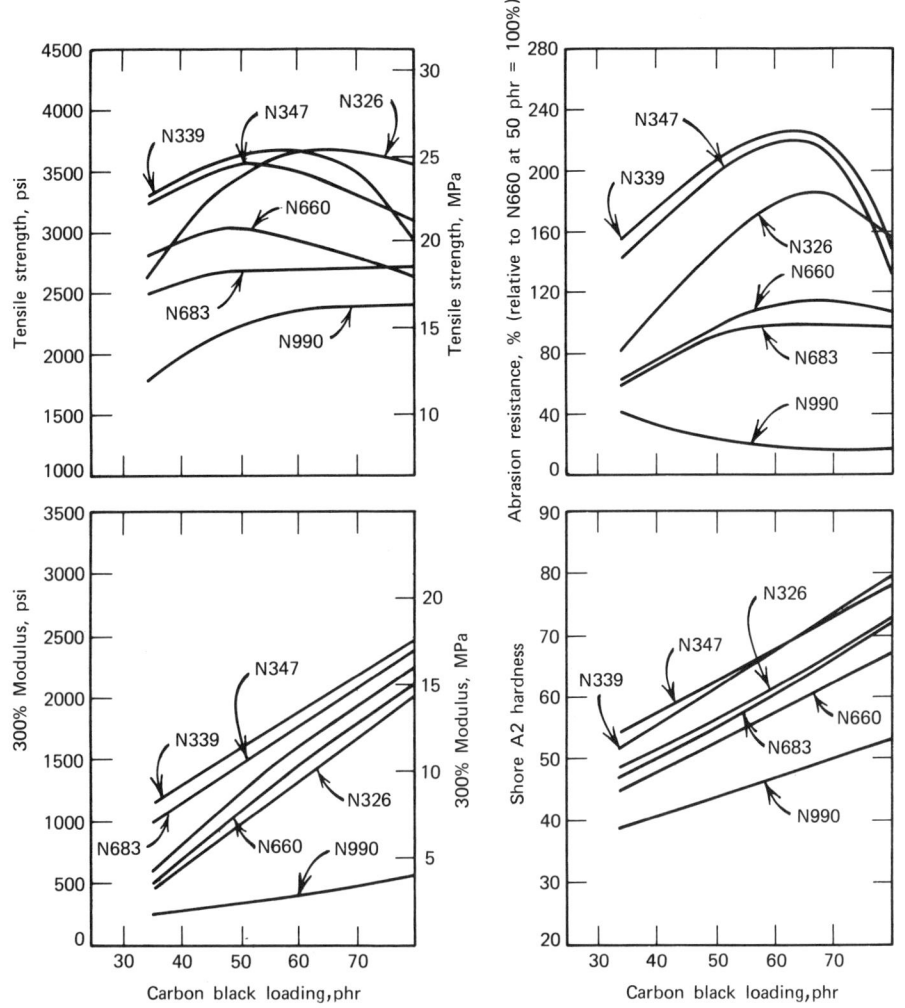

Figure 15. Properties in SBR of N339 (HAF-HS), N347 (HAF-HS), N326 (HAF-LS), N660 (GPF), N683 (APF), N990 (MT). phr = parts per hundred of rubber.

Printing Inks. The printing ink industry uses about 40% of the special blacks production in the United States. The grade and concentration used depend on the type and quality of the ink, and are selected for such factors as degree of jetness, gloss, blueness of tone, viscosity, tack, and rheological properties. Carbon black surface area, structure, degree of surface oxidation, moisture content, and bulk density are the most important factors in performance. Over forty special black grades have been developed having a broad range of properties from 20 m^2/g surface area grades used for inexpensive inks and tinting to oxidized, porous low-aggregation grades of about 500 m^2/g used for high color enamels and lacquers. The color and rheological properties of inks are influenced by both surface area and structure as shown in Table 15. The surface chemistry of carbon blacks affects their behavior in some types of ink. Special flow grades are made by a secondary oxidative aftertreatment process to increase their volatile content, a measure of the chemisorbed surface oxygen complexes. This

CARBON (CARBON BLACK)

Table 14. Types and Applications of Special Carbon Blacks

Type	Surface area, m²/g	DBP absorption, mL/100 g	Volatile content, %	Uses
high color	230–560	50–120	2–10	high jetness for alkyl and acrylic enamels, lacquers, and plastics
medium color	200–220	70–120	1–1.5	medium jetness and good dispersion for paints and plastics; ultraviolet and weathering protection for plastics
medium color, long flow	138	55–60	5	used in lithographic, letterpress, carbon paper, and typewriter ribbon inks; high jetness, excellent flow, low viscosity, high tinting strength, gloss, and good dispersibility
medium color, medium flow	96	70	2.5	used for gloss printing and carbon paper inks; excellent jetness, dispersibility, tinting strength, and gloss in paints
regular color	80–140	60–114	1–1.5	for general pigment applications in inks, paints, plastics, and paper; gives ultraviolet protection in plastics, and high tint, jetness, gloss, and dispersibility in inks and paints
	46	60	1.0	good tinting strength, blue tone, low viscosity; used in gravure and carbon paper inks, paints, and plastics
	45–85	73–100	1.0	main use is in inks; standard and offset news inks
low color	25–42	64–120	1.0	excellent tinting blacks—blue tone; used for inks—gravure, one-time carbon paper inks; also for paints, sealants, plastics, and cements
thermal blacks	7–15	30–35	0.5–1.0	tinting—blue tone; plastics and utility paints
lamp blacks	20–95	100–160	0.4–9.0	paints for tinting—blue tone
conductive blacks	254	180	2.0	conductivity and antistatic applications in rubber and plastics
acetylene	65	250	0.3	conductive and antistatic applications, dry cells, tire curing bags

treatment improves wetting, dispersion, and prevents pigment reagglomeration. Inks made with flow carbon blacks have lower viscosities and a reduced tendency to body-up or thicken during storage. Another factor influencing ease of dispersion is bulk density. Pelleted products (250–500 g/L) are more difficult to disperse than lower density or fluffy products (100–300 g/L).

Different methods of printing require inks with specially designed rheological and drying properties. Table 16 lists the major types of printing inks, typical carbon black grades and concentrations, and the basic ink vehicle composition. Letterpress news inks normally contain from 9–14% carbon black in a petroleum oil. Such inks are relatively fluid, and dry by penetration of the oil into the fibers of the paper. Cost is a paramount factor in news ink production. Large circulation weekly publications printed on coated papers at very high speeds use inks containing solvent, resin, and 10–22% carbon black. Offset or lithographic gloss inks are used in high quality book printing, illustrations, and other material where excellence in detail and transfer are paramount. Such inks must possess maximum tinting strength and covering power. This is achieved by using carbon black concentrations of 15–22%. Gravure inks vary

Table 15. Surface Area and Aggregation Effects on Ink Performance

Ink property	High surface area	Low surface area
masstone color	darker	lighter
tone	browner	bluer
viscosity	higher	lower
tint strength	stronger	weaker
dispersion	more difficult	easier
wetting	more difficult	easier

Ink property	Low aggregation	High aggregation
dispersibility	harder	easier
gloss	higher	lower
wetting	faster	slower
viscosity	lower	higher
color	darker	lighter
undertone	browner	bluer
tint strength	higher	lower
thixotropy	lower	higher

Table 16. Types and Composition of Various Printing Inks

Ink	Grades of carbon black	Carbon black concentration, %	Vehicle composition
letterpress news ink	medium color, medium structure; oil pellets	9–14	mineral oil
letterpress gloss ink (nonheat-set, heat-set)	medium color, low structure, oxidized flow; oil pellets	16–22	nonheat-set or heat-set varnish
lithographic news ink	medium color, medium structure; oil pellets	18–20	mineral oil and varnish
lithographic gloss ink (non-heat-set, heat-set)	medium color, low structure, oxidized flow; oil pellets	15–22	nonheat-set or heat-set varnish
gravure (roto)	low-medium color, low structure	5–15	resins, polymers
flexographic	low to medium structure, oxidized flow grades	12–18	resins, polymers

widely in requirements and formulation because of their diversity of applications. These inks must have good strength, low viscosity, and good gloss. Low-to-medium color, low aggregation carbon blacks are used. Highly fluid flexographic inks designed for printing on nonporous surfaces, such as plastics metal foil, and special coated paper, must dry very rapidly. The most widely used carbon blacks for flexographic inks are the medium-to-long flow, medium color grades (see Inks; Printing processes).

Paints, Lacquers, Enamels, and Industrial Finishes. Carbon black pigments for paints, lacquers, enamels, and industrial finishes fall into three categories, classified as high, medium, and standard color (see also Coatings, industrial; Paints; Pigments). In these applications, jetness or masstone is of primary significance. The coating industry uses a wide range of furnace carbon blacks. High color carbon blacks provide exceptional jetness and gloss for automobile finishes and high grade enamels. Medium color carbon blacks are used in industrial enamels, and the standard grades are used in general utility and industrial paints. Large particle size furnace and thermal grades are used to a limited extent in paints and other coatings. Although they are deficient in jetness, they provide a desirable blue tone. Carbon black is sometimes supplied to the lacquer industry in the form of masterbatch chips. These chips consist of a high loading of carbon black dispersed in a resin.

Plastics. Medium and high color carbon blacks are used for tinting and also to provide jetness in plastics (see Plastics technology). Carbon black is used in polyolefins as a protective agent. Polyolefins without carbon black degrade rapidly on exposure to sunlight. Carbon black is an excellent black body, absorbing both ultraviolet and infrared (53). When dispersed in polyethylene it preferentially absorbs and dissipates the incident radiation (54). Since carbon black also appears to be effective in terminating free radicals it provides protection against thermal degradation (see Heat stabilizers). Clear polyethylene cable jackets become brittle after two years of outdoor exposure, but the same compound containing 2% black showed no change after twenty years' exposure (55) (see Insulation, electric). Medium color furnace blacks are preferred in this application as well as for the weathering protection of plastic pipe.

Polyethylene compounds containing up to 50 wt % of furnace or thermal blacks have been developed. The deterioration of physical properties and embrittlement that would normally accompany such high loadings of black is reduced substantially by cross-linking the compound either by radiation or addition of organic peroxides. This provides greatly increased strength, and makes possible strong, tough compounds for cable coatings or pipe (56–57). Electrically conductive compounds for the cable industry depend on the use of about 30% of electrically conductive grades of carbon black (see Polymers, conductive).

Paper. The paper industry employs carbon black to produce a variety of black papers, including album paper, leatherboard, wrapping and bag papers, opaque backing paper for photographic film, highly conducting and electrosensitive paper, and black tape for wrapping high voltage transmission cables. The use of carbon black is essential when a conducting paper is desired. The loading of black will vary with the application in the range of 2–8 wt % of pulp. To obtain adequate dispersion, the black is employed in the fluffy form and added directly to the beater in a dispersible paper bag to avoid dusting. It may also be added in the form of an aqueous slurry. Aqueous dispersions containing 35% of carbon black are available. Reference 58 is a review of the applications of carbon black in the paper industry (see Paper).

Other Applications. The low thermal conductivity of carbon black makes it an excellent high temperature insulating material. For high temperature insulation up to 3000°C (59), it must be maintained in an inert atmosphere to prevent oxidation. Thermal-grade carbon black has been most widely used.

Carbon black is a source of pure carbon both for ore reduction and carburizing. Carbon brushes and electrodes are fabricated from carbon black.

There are a number of other minor applications of carbon black. For example, it is used as a pigment in the cement industry, in linoleum, leather coatings, polishes, and plastic tile.

BIBLIOGRAPHY

"Carbon Black" under "Carbon" in *ECT* 1st ed., Vol. 3, pp. 34–65, and Suppl. 1, pp. 130–144, by W. R. Smith, Godfrey L. Cabot, Inc.; "Acetylene Black" in *ECT* 1st ed., Vol. 3, pp. 66–69, by B. P. Buckley, Shawinigan Chemicals Ltd.; "Carbon Black" under "Carbon" in *ECT* 2nd ed., Vol. 4, pp. 243–282, by W. R. Smith, Cabot Corporation, D.C. Bean (Acetylene Black), Shawinigan Chemicals Limited.

1. B. E. Warren, *Phys. Rev.* **59,** 693 (1941).
2. E. A. Kmetlso, *Proc. 1st and 2nd Conf. on Carbon,* 21 (1956).
3. V. L. Kasatotshkin and co-workers, *J. Chim. Phys.* **52,** 822 (1964).
4. J. B. Donnet and J. C. Bouland, *Rev. Gen. Caoutch. Plast.* **41,** 407 (1964).
5. W. M. Hess and L. L. Ban, *Norelco Rep.* **13**(4), 102 (1966).
6. D. F. Harling and F. A. Heckman, *International Plastics and Elastomers Conference,* Milan, Italy, Oct. 1968.
7. S. Brunauer, P. H. Emmett, and J. Teller, *J. Am. Chem. Soc.* **60,** 310 (1938).
8. J. H. deBoer and co-workers, *J. Catal.* **4,** 649 (1965).
9. *ASTM D1510-76, Annual Book of ASTM Standards,* American Society for Testing and Materials, Philadelphia, Pa.
10. J. Janzen and G. Kraus, *Rubber Chem. Technol.* **44,** 1287 (1971).
11. Ref. 9, *ASTM D2414-76.*
12. R. E. Dollinger, R. H. Kallinger, and M. L. Studebaker, *Rubber Chem. Technol.* **40,** 1311 (1967).
13. Ref. 9, *ASTM D3265-76.*
14. A. I. Medalia and co-workers, *Rubber Chem. Technol.* **46,** 1239 (1973).
15. H. P. Palmer and C. F. Cullis in P. I. Walker, ed., *Chemistry and Physics of Carbon,* Vol. 1, Marcel Dekker, Inc., New York, 1965, p. 265.
16. J. B. Donnet, "Les Carbones," in *Groupe Francais Etude des Carbones,* Vol. 2, Masson, Paris, Fr., 1962, p. 208.
17. A. Feugier, *Rev. Gen. Therm.* **9,** 105, 1045 (1970).
18. J. Abrahamson, *Nature* **266,** 323 (1977).
19. *Chem. Eng. News,* p. 9 (Apr. 5, 1976).
20. U.S. Pat. 2,564,700 (Aug. 21, 1956), J. C. Krejci (to Phillips Petroleum Co.).
21. U.S. Pat. 1,438,032 (Dec. 5, 1922), W. H. Frost (to Wilckes-Martin-Wilckes Co.).
22. U.S. Pat. 2,292,355 (Aug. 11, 1942), J. W. Ayers (to C. K. Williams and Co.).
23. O. K. Austin, "Commercial Manufacture of Carbon Black," in G. Kraus, ed., *Reinforcement of Elastomers,* Interscience Publishers, New York, 1965.
24. E. M. Dannenberg, *J. Inst. Rubber Ind.* **5,** 190 (1971).
25. Can. Pat. 822,024 (Sept. 2, 1969), G. L. Heller (to Columbian Carbon Co.).
26. U.S. Pat. 3,353,915 (Nov. 21, 1967), B. L. Latham (to Continental Carbon Co.).
27. U.S. Pat. 3,619,140 (Nov. 9, 1971); Re. 28,974 (Sept. 21, 1976), A. C. Morgan and M. E. Jordan (to Cabot Corporation).
28. U.S. Pat. 3,003,855 (Oct. 10, 1961), G. L. Heller and C. I. DeLand (to Columbian Carbon Co.).
29. U.S. Pat. 3,253,890 (May 31, 1966), C. L. DeLand, G. L. DeCuir, and L. E. Wiggins (to Columbian Carbon Co.).
30. U.S. Pats. 3,010,794 (Nov. 28, 1961), G. F. Friauf and B. Thorley (to Cabot Corporation).
31. K. R. Dahmen, "The Carbon Black Furnace Process," *paper presented to the Akron Rubber Group, Technical Symposium, Apr. 15, 1977.*
32. U.S. Pats. 3,725,103 (Apr. 3, 1973); 3,799,788 (Mar. 26, 1974), M. E. Jordan, W. G. Burbine, and F. R. Williams (to Cabot Corporation).
33. H. J. Stern, *Rubber-Natural and Synthetic,* Maclaren & Sons, Ltd., London, Eng., 1954, p. 137.
34. Ger. Pat. 103,862 (June 27, 1899), L. J. E. Hubou.
35. A. E. Austin, *Proc. 3rd Conf. on Carbon,* 389 (1958).
36. E. M. Dannenberg, *Paper No. 42,* Rubber Div., American Chemical Society, San Francisco, Calif., 1976.
37. E. M. Dannenberg, *Plast. Rubber Int.* **3,** 11 (1978).
38. H. J. Stern, *Rubber Age Synth.,* 268 (Dec. 1945–Jan. 1946).
39. *Chemical Economics Handbook,* Stanford Research Institute, Menlo Park, Calif., 1975.
40. H. L. Duncombe, *paper presented to RMA Molded and Extruded Products Division, Hot Springs, Va., Rubber and Plastics News, July 12, 1976.*
41. C. A. Nau, J. Neal, and V. Stembridge, *AMA Arch. Ind. Health* **17,** 21 (1958).

42. C. A. Nau, J. Neal, and V. Stembridge, *AMA Arch. Ind. Health* **18,** 511 (1958).
43. C. A. Nau, J. Neal. and V. Stembridge, *Arch. Environ. Health* **1,** 512 (1960).
44. C. A. Nau and co-workers, *Arch. Environ. Health* **4,** 415 (1962).
45. J. Neal, M. Thornton, and C. A. Nau, *Arch. Environ. Health* **4,** 598 (1962).
46. C. A. Nau, G. T. Taylor, and C. Lawrence, *J. Occup. Med.* **18,** 732 (1976).
47. T. H. Ingalls, *Arch. Ind. Hyg. Occup. Med.* **1,** 662 (1950).
48. T. H. Ingalls and R. Risquez-Iribarren, *Arch. Environ. Health* **2,** 429 (1961).
49. H. J. Collyer, *EPA Conference Paper, Akron, Ohio, Mar. 12, 1975*.
50. E. M. Dannenberg, *Rubber Age* **108**(4), 37 (1976).
51. M. Studebaker, "Compounding with Carbon Black," in G. Kraus, ed., *Reinforcement of Elastomers*, Interscience Publishers, New York, 1965.
52. *Technical Report TG-76-1*, Cabot Corporation, Boston, Mass., 1976.
53. A. J. Wells and W. R. Smith, *J. Phys. Chem.* **45,** 1055 (1941).
54. V. T. Wallder and co-workers, *Ind. Eng. Chem.* **42,** 2320 (1950).
55. W. L. Hawkins, *Rubber Plast. Weekly (London)* **142,** 291 (1962).
56. A. Charlesby, *Atomic Radiation and Polymers,* Pergamon Press, Inc., New York, 1960, Chapt. 13.
57. E. M. Dannenberg, M. E. Jordan, and H. M. Cole, *J. Polym. Sci.* **31,** 127 (1958).
58. I. Drogin, *Pap. Trade J.* **147,** 24 (Apr. 1, 1963).
59. W. D. Schaeffer, W. R. Smith, and M. H. Polley, *Ind. Eng. Chem.* **45,** 1721 (1953).

ELI M. DANNENBERG
Cabot Corporation

DIAMOND, NATURAL

Diamond [7782-40-3], the high-pressure allotrope of carbon, changes to graphite, the high-temperature allotrope of carbon, when heated above 1500°C *in vacuo*. Diamond is metastable and chemically inert at moderate temperatures. Although diamond is the hardest known substance, it is quite brittle and is readily crushed to grit and powder, the form in which approximately 75% of all industrial diamond is used. These properties make diamond of ever-increasing importance in industry and technology as a fixed abrasive in saws, drills, and grinding wheels. In the form of single crystals (stones), it is used in turning and boring tools, wire dies, indenters, tools, to shape and dress conventional abrasive wheels, and in drill crowns.

The high refraction index (2.42) and high optical dispersion give gem diamond its brilliance and fire. Diamond is the most valued of all gems because of these properties and its durability, and also because of its scarcity: the ratio of rock or gravel processed per part of diamond recovered ranges from 10 million to 100 million. Value is dictated by scarcity, color, crystal perfection, and size. The cost may range from approximately $2 to as much as $10,000 per carat (0.2 g).

Diamond, as a natural mineral, ranges widely in purity, soundness, size, color, and shape. Thus, it is necessary to sort diamonds as mined into many categories according to potential uses. Those of near flawless quality having colors of blue, blue-white, water-white, pink, green, or a slight tinge of yellow are of greatest value as gems. Approximately 20% of mined diamonds are classified as gems. Other stones that are sound but poor in color and clarity are used in tools and dies. A good portion of a diamond containing flaws may be cut as a gem. Diamonds unsuitable for use as gem or tool stones are crushed to grit and powder.

Occurrence and Geology

Historically, India and Borneo were the earliest sources of diamond, but their production is now very small. Diamonds were discovered in Brazil in 1670 and have been mined there since 1721.

Diamonds were first found near Kimberley, South Africa, in 1866. Mining was started at Dutoitspan in 1871 and at Wesselton in 1890. The Premier mine was discovered in 1902. Diamonds were found in South West Africa (Namibia) in 1908, near the mouth of the Orange River in 1925, and were discovered in Mwadui, Tanzania, in 1940.

At present, approximately 90% of the world's diamonds are from the continent of Africa and are produced in Angola, Botswana, Zaire, the Central African Republic, Ivory Coast, Ghana, Sierre Leone, South West Africa (Namibia), Tanzania, and the Republic of South Africa. Extensive deposits have been found in Siberia, and the U.S.S.R. has become an important producer. Other significant producers are Brazil, British Guiana, and Venezuela (1).

Diamonds are found in New South Wales, Australia, Colombia, Mexico, and British Columbia. In the United States diamonds have been found in Wisconsin, Indiana, Michigan, California, Georgia, North Carolina, and Arkansas. The most important find is near Murfreesboro, Arkansas, although it has not been commercially significant (2). Diamonds are also found in meteorites (3).

The genesis of diamond is not well established. One theory postulates that diamond was formed in plutonic reservoirs before eruption. The volcanic pipes were probably first filled with kimberlite magma, and while still in a plastic state, the deeper seated, more fluid magma carrying diamonds and other mineral crystals continued to rise up through it and become inhomogeneously mixed with the plastic magma. The resulting rock became the present kimberlite (4–6).

Diamonds are found in ancient volcanic pipes in relatively soft, dark, basic perioditite rock called blue ground or kimberlite. Although these are considered the primary sources, not all diamonds have remained in the pipes. Over geologic ages the pipes weathered and eroded, and their diamonds were deposited in the surrounding areas or transported by rivers and concentrated in alluvial gravels or on marine terraces by the action of the sea.

Forms of Natural Diamonds

In nature, diamonds occur in a wide range of shapes, sizes, and crystal perfection. They occur as near perfect single crystals, twins, dense agglomerates of crystallites, and crystals with foreign inclusions and ingrowths. Many of these forms and shapes have been found particularly useful in industrial applications. Seven of the main diamond forms are listed below.

The most perfectly formed diamond crystal habit is the octahedron, particularly in sizes of $1/4$ carat and smaller. In this habit, angles and faces (111) are well defined.

Commercially, the most frequent habit is called a dodecahedron, but nearly approaches a tetrahexahedron (24 sides). The (110) faces are seldom, if ever, flat. Crystals of this habit vary from symmetrical to elongated and flattened shapes.

The cube is the most infrequent habit for single crystals. Usually, cubes contain ingrowths and spinel twins or have a surface texture resembling cube sugar.

Macles, an industrially useful form, are triangular, flat, pillow-shaped stones, consisting of twin crystals. They are twinned on the (111) plane.

Carbonado (carbons) is a cryptocrystalline material composed of diamond crystallites, graphite, and other impurities. They are extremely strong and difficult to fracture and are used in drill crowns.

Ballas, a near spherical form, consists of a dense growth of randomly oriented crystallites. These crystallites may vary in size from a few micrometers upward. Because of this microstructure, the ballas does not exhibit cleavage and is thus tougher than a single crystal. The ballas finds its best use in the severe applications of dressing large, coarse-grained abrasive wheels (see Abrasives).

Boart usually refers to minutely crystalline gray-to-black pieces of diamond which are most useful when crushed to grit for grinding wheels. The term may also be used to designate all other diamonds of inferior quality.

Properties of Diamond

Structure. Diamond is composed of the single element carbon and has a cubic crystal structure of space group 0_h^7. The lattice constant, a_0, falls between $3.56683 \pm 1 \times 10^{-6}$ nm and $3.56725 \pm 3 \times 10^{-6}$ nm (25°C). The nearest neighbor distance is $1.54450 \pm 0.00005 \times 10^{-1}$ nm (25°C) (7–9). The carbon in diamond is covalently bonded (10).

Classification. Diamonds are classified into four groups based on their optical and electrical properties as types 1a, 1b, 11a, and 11b. Diamond may contain several impurities, some in substantial amounts. Diamond may have mineral inclusions. There have been twenty-two species identified (11). Types 1a and 1b are the most impure with up to 0.2% nitrogen being the major impurity. Other impurities such as silicon, magnesium, aluminum, calcium, iron, and copper are present, generally in amounts below 100 ppm (12–17).

Type 1a diamonds contain 0.1% nitrogen in the form of platelets in the crystal. Most natural diamonds are of this type.

Type 1b diamonds contain up to 0.2% nitrogen in dispersed form. Almost all synthetic diamonds are of this type.

Type 11a diamonds are effectively free of nitrogen. They are very rare and have enhanced optical and thermal properties.

Type 11b diamonds are very pure, are generally blue in color, and extremely rare in nature. They have semiconducting properties thought to be caused by boron atoms in the lattice, which may also be responsible for the blue color (18–22).

Typical Physical Properties. Typical physical properties of natural diamond are listed in Table 1.

Table 1. Physical Properties of Natural Diamond

Property	Value	Reference
density[a], g/cm^3, 25°C	3.51524 ± 0.00005	23
thermal conductivity, W/(cm·K), 20°C[b]		24
Type 1	9	
Type 11a	26	
maximum at 190°C		
Type 1	24	
Type 11a	120	
thermal expansion coefficient[c]		25–26
20°C	0.8 ± 0.1 × 10^{-6}	
−100°C	0.4 ± 0.1 × 10^{-6}	
100–900°C	(1.5 − 4.8) × 10^{-6}	
specific heat (constant volume), J/(mol·K)[d], 20°C	6.184	27–29
refractive index, µm		30
546.1 nm (Hg green)	2.4237	
656.3 nm	2.4099	
226.5 nm	2.7151	
dielectric constant, 27°C, 0–3 kHz	5.58 ± 0.03	31–32
optical transparency		33
Type 11a diamond	225 nm to 2.5 µm > 6 µm	
Type 1 diamond	340 nm to 2.5 µm > 10 µm	
resistivity, 20°C, Ω·cm		34–35
Type 1 and most Type 11a	>10^{16}	
Type 11b	10–10^3	
hardness, Mohs' scale (scratch hardness)[e]	10	36
Knoop scale (indentation hardness)[f,g], kg/mm^2	5,700–10,400	37–41
compressive strength[g,h], kg/mm^2		
average (flawless octahedral diamond)	885	44
maximum	1685	44
modulus of elasticity[g,i], kg/mm^2	118.1 × 10^3	46
Young's modulus[j,k], MPa	1.16 × 10^6	46

[a] Average of 35 diamonds.
[b] For comparison at 20°C, copper has 4 W/(cm·K) conductivity.
[c] For comparison, fused silica at 0–30°C is 0.4 × 10^{-6}.
[d] To convert J to cal, divide by 4.184.
[e] For comparison, WC, W$_2$C, and VC have 9.5 and SiC and Al$_2$O$_3$ have 9 hardness.
[f] Values for SiC (42) and Al$_2$O$_3$ (43) are 2480 and 2100 kg/mm^2, respectively.
[g] To convert kg/mm^2 to psi, multiply by 1422.
[h] Values for SiC and Al$_2$O$_3$ are 57.6 and 300.2 kg/mm^2, respectively (45).
[i] Values for SiC (47) and Al$_2$O$_3$ (48) are 38.7 × 10^3 and 35.9 × 10^3 kg/mm^2, respectively.
[j] To convert MPa to psi, multiply by 145.
[k] Values for SiC (47) and Al$_2$O$_3$ (48) are 0.379 and 0.345 MPa, respectively.

Chemical Properties

Diamond is chemically inert and is not attacked by acids or other chemicals except those that act as oxidizing agents at high temperature. Diamond reacts readily with molten sodium carbonate and nitrate.

Graphitization. Graphitization is the transition of diamond to graphite without the aid of external agents. If diamond is heated in vacuum or an inert atmosphere, graphitization is detected at 1500°C and increases rapidly with increased temperature (49). The octahedral surface graphitizes with an activation energy of 1059 ± 75 kJ/mol (253 ± 18 kcal/mol) and the dodecahedral surface graphitizes more rapidly with an activation energy of 728 ± 50 kJ/mol (174 ± 12 kcal/mol). The activation volume of the graphitization process is about 10 cm^3/mol (49–51). Diamond will form a black surface when heated in air at temperatures as low as 600°C.

Investigation of diamond wear on glass suggests that graphitization of diamond in contact with the glass is the primary mechanism. There may also be some chemical reaction (52). Similar studies on wear of diamonds against metals indicates that wear is related to the melting point of the metal. The ability of metals to form carbides is also involved. Thus, it appears that the mechanism of wear is through graphitization and chemical action at high temperature (53).

Recovery

Pipe mines start as open-cast operations or open-pit mines. As the pits deepen, this method becomes impractical and underground methods are used.

Along the Atlantic Coast of South West Africa (Namibia) and near the mouth of the Orange River, marine terraces, covered with many meters of sand, are mined for diamonds. The most recent operation was the dredging of the shallow ocean floor for diamonds.

Production is also from alluvial deposits in ancient river beds, flood plains, and lake beds now covered with an overburden of soil, sand, gravel, or cemented agglomerates.

Recovery of diamond from blue ground or gravel concentrate is accomplished through progressive crushing, screening, and separation of gangue from diamond and heavy minerals by gravity methods (see Gravity concentration). Jigs and heavy media separators are used. The properties of diamonds that determine the method of final recovery are size, surface conditions, optical, dielectric, and x-ray luminescence.

Grease tables are used where the diamonds are clean or can be cleaned by attrition. A clean diamond surface is not wet by water but will adhere to grease. Some marine and alluvial diamonds have a film of mineral salts that is wettable and must be chemically treated or removed before recovery is practical on grease tables.

A process called skin-flotation is also used where the nonwet diamonds float and the wet gangue material sinks.

For diamonds finer than 0.84 mm (20 mesh), electrostatic separation can be used for separation when hand sorting and other methods are impractical.

A method of separation based upon x-ray luminescence is utilized where a column of single particles of concentrate is fed past an x-ray source. The diamond luminescence is detected by a photomultiplier, and through an associated apparatus, the diamond is deflected by a synchronized air jet from the normal path, and thus separated from other minerals. This method is used extensively for recovery and is superior to an earlier, similar optical method because it is more efficient in detecting dark and discolored industrial diamonds.

Final separation, sorting and classification of all diamond is by hand, utilizing an eye loop of 7 to 10 power.

Diamond Size. The largest gem recovered was the Cullinan, weighing 3106 carats or 621.2 g (5 carats per gram). This was found in the Premier mine in 1905. The average diamond size recovered varies greatly with the mine, and may range from as small as 10 stones per carat, to as large as 1.8 carats for deposits on the Orange River terraces (54–55).

Within one century (1870–1970) of diamond mining in South Africa, 2577 gem diamonds were recovered that were larger than 100 carats and three of which were over 1000 carats (56).

Because of their great value, every effort is made to avoid loss in processing diamond. The largest gem or the most valuable combination of gems is cut from a stone; for use in tools, stones are selected that require the minimum work and have just the required quality. The smallest fragments are converted to powder. Diamonds and powder are reclaimed from worn tools and reused.

Shaping Diamond

Although diamond is the hardest material known, it may be shaped precisely. The process by which the skilled craftsman turns a stone into a brilliant faceted gem or a precisely shaped tool may involve combinations of sawing, cleaving, bruting, grinding, polishing, and drilling. Diamond boart is converted industrially by crushing to carefully sized and shaped grit and powder.

The processes of cleaving, sawing, and grinding of diamonds depend upon the proper direction of force or abrasion. The craftsman can determine the appropriate direction by observing the surface details of a stone. The three prominent planes in diamond are referred to by diamond cutters as: cleavage or 3-point (111); cube or 4-point (100); and dodecahedral or 2-point (110). The abrasion resistance and mechanical properties of the diamond are systematically related to these planes (57).

Diamond exhibits a high degree of anisotropy in abrasion resistance, which is referred to as hardness. Abrasion resistance is systematically related to the crystal structure and corresponds to a similar pattern in the coefficient of friction. This property is frequently described as the hardness vector (see Hardness).

Figures 1 and 2 show diamond crystals referred to the crystallographic axis and indicating the cubic, octahedral, and dodecahedral faces. The arrows of Figures 1 and 2 indicate "hard" and "easy" work directions, respectively. For effective grinding, polishing, and sawing, the abrasion must approximate the easy directions. Slight angular deviations in direction radically change the ability to work the diamond.

Orientation of diamond permits the consistent manufacture of tools designed to best utilize its properties so that cleavage planes may be avoided at critical edges, and a uniform abrasion resistance may be presented to the work around the radius and flanks of shaped tools.

Cleaving is an efficient method of dividing a stone. The operation consists of identifying and marking the cleavage plane (111) which will part the stone into the desired portions. A kerf or notch is abraded into the stone at a point along the mark with a sharp fragment of diamond. A steel blade is inserted into the kerf and given

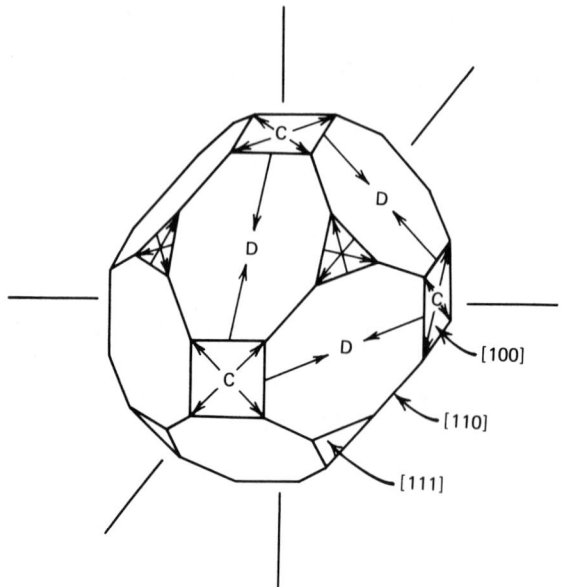

Figure 1. Arrows indicate "hard" directions of three major planes in diamond. Diamond is most resistant to abrasion in these directions.

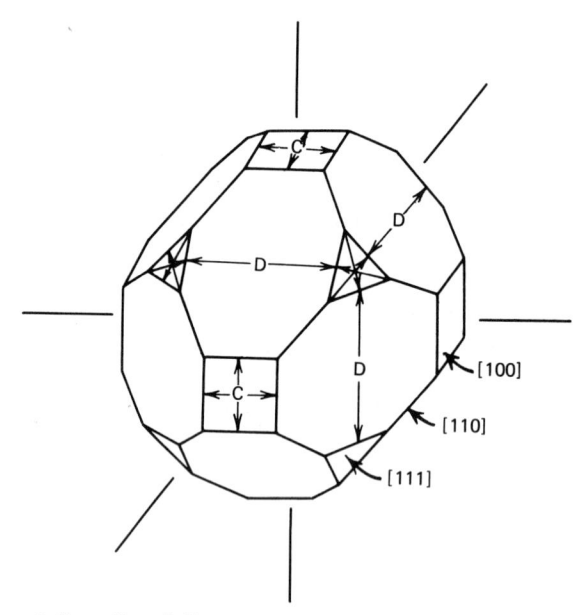

Figure 2. Arrows indicate "easy" direction of three major planes of diamond. Abrasion, sawing, and polishing are more easily accomplished in these directions.

a sharp rap with a mallet. If all steps are correct, the stone parts cleanly as planned.

    Sawing is performed with thin disks of phosphor–bronze alloy of 0.076 to 0.127 mm thickness and from 7.6 to 12.7 cm diameter. These are impregnated with diamond powder. The disk is held between flanges and spins at high speed between pivots. The stone to be sawed is inspected and marked according to the desired position for parting and permissible directions of sawing. The stone is then mounted in the universally adjustable, gravity-operated saw mechanism.

A diamond can be sawed only along a plane parallel to one of the three crystallographic axes. The line of contact or abrasion of the blade must also be tangent to an axis.

Sawing of a clean stone, free of naats or cracks, proceeds quite rapidly. A stone of one carat is usually sawed in an hour or less. Since many stones contain naats, techniques of electrically-assisted abrasive sawing are used.

The same abrasion action is used in grinding and polishing, as in sawing. A diamond may be ground on a diamond-impregnated grinding wheel. Polishing is done on a high-speed cast iron lap or scaife, the surface of which is impregnated with diamond powder. The diamond is held securely in a dop that permits a wide range of angular adjustments so that the facet or radius being ground may be presented in the proper direction to the abrasive action of the wheel or scaife.

Bruting is used to round off corners and to create the girdle on gems and cylindrical, conical, or ball-shaped diamonds. The operation is similar to lathe turning. The turning tool is a piece of diamond, cemented to a hand-held holder which is brought to bear against the spinning diamond. The material removed is in the form of fine chips.

Diamonds are drilled for wire dies. The holes range in size from 0.01 to 3 mm, and are precisely shaped, sized, and polished with diamond powder. In the traditional method of drilling, the die stone is moved with a reciprocating motion against a thin steel needle rotating at high speed. A mixture of olive oil and fine diamond powder is applied to the work area. Diamond die stones are now pierced by laser (58). Traditional methods are used to shape and polish the rough laser-pierced hole (see Lasers).

Diamond powder is manufactured from boart and the lowest quality diamond. This material is progressively crushed, milled, and screened to produce the grit and powder of desired sizes. Every effort is made to produce the desired mesh sizes and shapes without producing excessive fines. Grit and powder down to 44 μm (325 mesh) are usually sized on wire mesh screens. Elutriation and sedimentation (qv) techniques are used to grade subsieve sizes.

World Production

Statistics, for 1971, on world production (in carats) of natural diamond were (59–60):

gem	11,682,000
industrial	37,877,000
total	*49,559,000*
for 1961	34,000,000
for 1950	15,517,344

Diamond Uses

High indexes of refraction, dispersion, rarity, and durability make diamond a highly prized gem.

674 CARBON (DIAMOND, NATURAL)

Hardness and chemical stability make diamond a valuable and strategic industrial material. Its performance as an abrasive in shaping tungsten carbide, ceramics, and other hard materials is unsurpassed. As single crystals, it is used as wire dies and tools for turning, ruling, rock drilling, and abrasive trueing.

The thermal and electrical properties make diamond useful in solid-state applications, such as heat sinks and thermistors.

Every fragment of diamond is utilized and processed, recovered and reprocessed for applications as abrasives and lapping compounds.

BIBLIOGRAPHY

"Diamond" under "Carbon" in *ECT* 1st ed., Vol. 3, pp. 69–80; "Carbon (Diamond, Natural)" under "Carbon" in *ECT* 2nd ed., Vol. 4, pp. 283–294, by H. C. Miller, Super-Cut, Inc.

1. A. A. Linari-Linholm, *Occurrence, Mining and Recovery of Diamonds,* DeBeers Consolidated Mines Ltd., Kenion Press Ltd., Slough Bucks, England, 1973, pp. 2–5.
2. E. H. Kraus, W. F. Hunt, and L. S. Ramsdell, *Mineralogy,* 4th ed., McGraw-Hill Book Co., New York, 1951, pp. 256–263 and 430–431.
3. N. L. Carter and G. C. Kennedy, *J. Geophys. Res.* **69,** 2403 (1964).
4. A. F. Williams, *The Genesis of Diamond,* Ernest Benn Ltd., London, 1932.
5. S. J. Shand, *Eruptive Rocks,* 3rd ed., John Wiley & Sons, Inc., New York, 1947.
6. R. H. Mitchell and J. H. Crocket, *Min. Deposits* **6,** 392 (1971).
7. W. Kaiser and W. L. Bond, *Phys. Rev.* **115,** 857 (1959).
8. K. Lonsdale, *Phil. Trans. R. Soc.* **A240,** 219 (1947).
9. B. J. Skinner, *Am. Mineral.* **42,** 39 (1957).
10. B. Dawson, *Proc. R. Soc. London Ser. A* **298,** 379 (1967).
11. J. W. Harris, *Industrial Diamond Review* **28,** 458 (1968).
12. F. A. Raal, *Am. Mineral.* **42,** 354 (1957).
13. J. F. H. Custers, *Gems Gemol.* **10,** 111 (1957–1958).
14. E. C. Lightowlers in J. Burls, ed., *Science and Technology of Industrial Diamonds,* Vol. 1, Industrial Diamond Information Bureau 1967, p. 27.
15. A. T. Collins and A. W. S. Williams, *Diamond Research,* Industrial Diamond Information Bureau, London, England, 1971, p. 23.
16. J. P. F. Sellschop, *Diamond Research,* Industrial Diamond Information Bureau, London, England, 1975, p. 35.
17. R. M. Chrenko, *Nature Phys. Sci.* **229,** 165 (1971).
18. J. F. H. Custers, *Physica* **18,** 489 (1952).
19. W. Kaiser and W. L. Bond, *Phys. Rev.* **115,** 857 (1959).
20. H. B. Dyer and co-workers, *Phil. Mag.* **11,** 763 (1965).
21. T. Evans and G. Davies, *Diamond Research,* Industrial Diamond Information Bureau, London, England, 1973, p. 2.
22. R. Robertson, J. J. Fox, and A. E. Martin, *Phil. Trans. R. Soc. Ser. A* **232,** 463 (1934).
23. R. Mykolajewycz, J. Kalnajs, and A. Smakula, *J. Appl. Phys.* **35,** 1773 (1964).
24. R. Berman, ed., *Physical Properties of Diamond,* Clarendon Press, 1965.
25. J. Thewlis and A. R. Davey, *Phil. Mag.* **1,** 409 (1956).
26. B. J. Skinner, *Am. Mineral.* **42,** 39 (1957).
27. D. L. Burk and S. A. Friedburg, *Phys. Rev.* **111,** 1275 (1958).
28. J. E. Desnoyers and J. A. Morrison, *Phil. Mag.* **3,** 42 (1958).
29. A. C. Victor, *J. Chem. Phys.* **36,** 1903 (1962).
30. F. Peter, *Z. Phys.* **15,** 358 (1923).
31. S. Whitehead and W. Hacket, *Proc. Phys. Soc.* **51,** 173 (1938).
32. D. F. Gibbs and G. J. Hill, *Phil. Mag.* **9,** 367 (1964).
33. R. Robertson, J. J. Fox, and A. E. Martin, *Phil. Trans. R. Soc.* **A232,** 463 (1934).
34. P. Denham, E. C. Lightowlers, and P. J. Dean, *Phys. Rev.* **161,** (1967).
35. P. J. Kemmey and P. T. Wedepohl, in R. Berman, ed., *Physical Properties of Diamond,* Clarendon Press, 1965.

36. D. Tabor, *Proc. Phys. Soc.* **B67,** 249 (1954).
37. F. Knoop, C. G. Peters, and W. B. Emerson, *J. Res. Nat. Bur. Stand.* **23,** (1939).
38. C. A. Brookes, *Nature* **228,** 660 (1970).
39. C. A. Brookes, *Diamond Research,* Industrial Diamond Information Bureau, London, England, 1971, p. 12.
40. T. N. Loladze, G. V. Bokuchava, and G. E. Davydova, *Ind. Lab.* **33,** 1187 (1967).
41. C. A. Brookes, P. Green, and P. H. Harrison, *Diamond Research,* Industrial Diamond Information Bureau, London, England, 1974, p. 11.
42. E. H. Hull and G. T. Malloy, *J. Eng. Ind. Trans. ASME,* 373 (Nov. 1966).
43. *Conference on Fracture,* Swampscott, Mass., John Wiley & Sons, Inc., New York, 1959, p. 204.
44. R. C. Weast, ed., *Handbook of Chemistry and Physics,* 56th ed., The Chemical Rubber Company, Cleveland, Ohio, 1975.
45. J. N. Plendl and P. J. Gielisse, *Z. Kristall.* **118,** 404 (1963).
46. P. T. B. Shaffer, *Plenum Press Handbooks of High-Temperature Materials,* No. 1 Plenum Press, New York, 1964.
47. H. A. Pearl, U. M. Nowak, and H. G. Deban, *Mechanical Properties of Selected Alloys at Elevated Temperatures,* Bell Aircraft Corporation, U.S. Air Force Contract AF-33 (616)-5760, 1960.
48. *Refractory Ceramics for Aerospace,* American Ceramic Society, Columbus, Ohio, 1964, p. 214/V.
49. G. Davies and T. Evans, *Proc. R. Soc. Ser. A* **328,** 413 (1972).
50. T. Evans and P. F. James, *Proc. R. Soc. Ser. A* **277,** 260 (1964).
51. M. Seal, *Physica Status Solidi* **3,** 658 (1963).
52. F. P. Bowden and H. G. Scott, *Proc. R. Soc. London, Ser. A* **248,** 368 (1958).
53. F. P. Bowden and E. H. Freitag, *Proc. R. Soc. London, Ser. A* **248,** 350 (1958).
54. Ref. 1, p. 1.
55. *International Diamond Annual,* No. 2, Diamond Annual (Pty.) Ltd., Johannesburg, S. Africa, 1972, p. 53.
56. *International Diamond Annual,* Vol. 1, Diamond Annual (Pty.) Ltd., Johannesburg, S. Africa, 1971, p. 74.
57. P. Grodzinski, *Diamond Technology,* 2nd ed., N.A.G. Press Ltd., London, 1953, Chapt. 6.
58. *Production,* (6) 122 (1966).
59. Ref. 1, p. 42.
60. *International Diamond Annual,* Vol. 1, Diamond Annual (Pty.) Ltd., Johannesburg, S. Africa, 1971, p. 13.

General Reference

Y. L. Orlov, *The Mineralogy of the Diamond,* Wiley-Interscience, New York, 1977.

H. C. MILLER
Super-Cut, Inc.

DIAMOND, SYNTHETIC

Interest in the synthesis of diamond [7782-40-3] was first stimulated by Lavoisier's discovery that diamond was simply carbon; it was also observed that diamond, when heated at 1500–2000°C, converted into graphite. In 1880, the British scientist Hannay reported (1) that he made diamond from hydrocarbons, bone oil, and lithium, but no one has been able to repeat this feat (2). About the same time, Moissan believed (3) that he made diamond from hot molten mixtures of iron and carbon, but his experiments could not be repeated (4–5).

The graphite–diamond equilibrium line up to 1200 K was calculated in 1938 (6) by using the observed heat, compressibility, and thermal expansion data of the two components (see lower portion of Fig. 1). Subsequently, estimates of the diamond–graphite equilibrium line were refined and extended (7) and the extrapolation to higher temperatures fits the experimental data (8). It is evident that diamond is not thermodynamically stable below a pressure of about 1.6 GPa (16 kbar) and early investigators were using pressures in their experiments where diamond would have been unstable.

Reproducible Laboratory Diamond Synthesis

In 1955, a team of research workers at General Electric developed the necessary high-pressure equipment and discovered catalytic processes by which ordinary forms of carbon could be changed into diamond.

In the attempt at diamond synthesis (4), much unsuccessful effort was devoted to processes that deposited carbon at low, graphite-stable pressures. Many chemical reactions liberating free carbon were studied at pressures which were then available. New high-pressure apparatus was painstakingly built, tested, analyzed, rebuilt, and sometimes discarded. It was generally believed that diamond would be more likely to form at thermodynamically stable pressures. The refractory nature of carbon indicated that temperatures of about 2000 K and pressures of 5–10 GPa (50–100 kbar) might suffice. Months after operating pressures of ca 7 GPa (70 kbar) had been attained, a reproducible diamond synthesis was finally achieved which is termed the catalyzed diamond synthesis.

Catalyzed Synthesis

In this process, a mixture of carbon (eg, graphite) and catalyst metal is heated high enough to be melted while the system is at a pressure high enough for diamond to be stable. Graphite is then dissolved by the metal and diamond is produced from it. Effective catalysts are Cr, Mn, Fe, Co, Ni, Ru, Rh, Pd, Os, Ir, Pt, and Ta, and their alloys and compounds. If the metal is not molten, graphite is obtained instead of diamond even at pressures high enough to produce diamond. (The exception is tantalum which does not have to be molten to be effective.)

Generally the two requirements that the catalyst be molten and diamond be stable define a pressure–temperature area in which diamond may form with the aid of a particular catalyst system. This diamond-forming region is illustrated in Figure 2 for the nickel–carbon system. The shaded area in which diamond may form is bounded

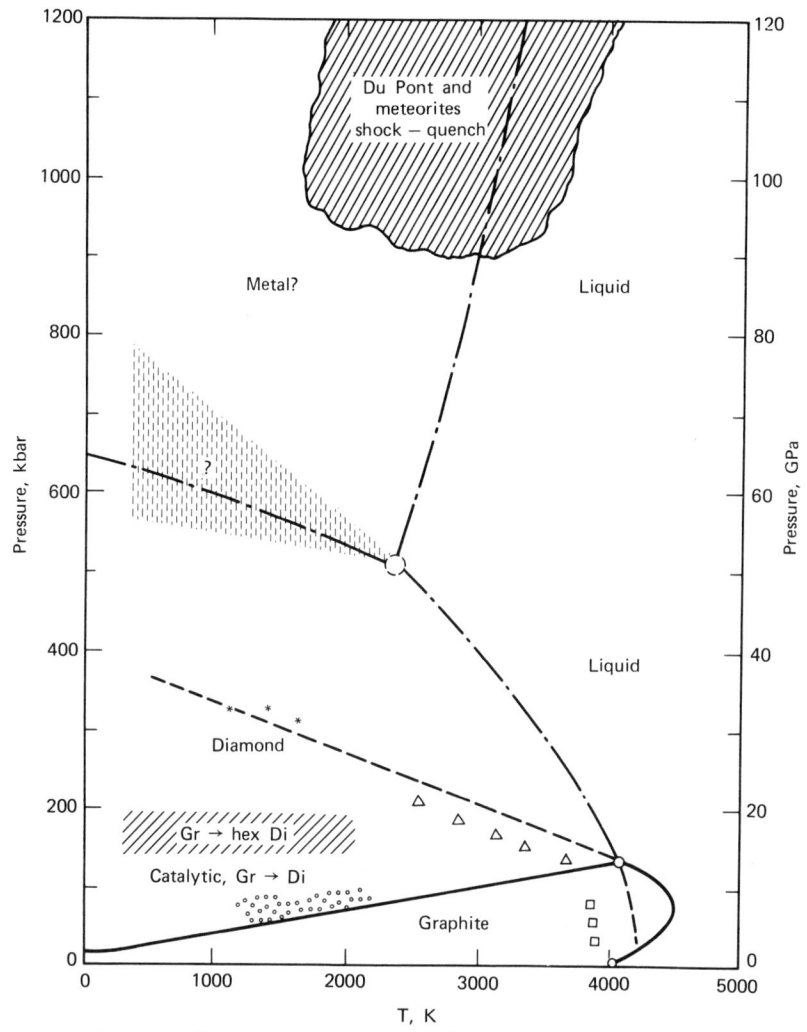

Figure 1. Carbon phase diagram (9). Gr = graphite; Di = diamond.

along the temperature axis by the nickel–carbon eutectic melting line and along the pressure axis by the graphite–diamond equilibrium line. Other catalyst metal systems define similarly shaped regions above the graphite–diamond equilibrium line. The region for platinum, for example, lies at higher pressures and temperatures than that of nickel because of the higher melting temperature of the Pt–C eutectic. Diamonds have been grown at temperatures as low as ca 1500 K and at a pressure of ca 5 GPa (50 kbar), using certain alloys of iron, nickel, and chromium as catalysts; the growth rate then is rather slow and the diamonds are heavily contaminated with metal, carbides, and graphite. Observation of the growth or disappearance of diamond has permitted a closer estimate of the course of the equilibrium line between graphite and diamond as indicated by the dashed line in Figure 2.

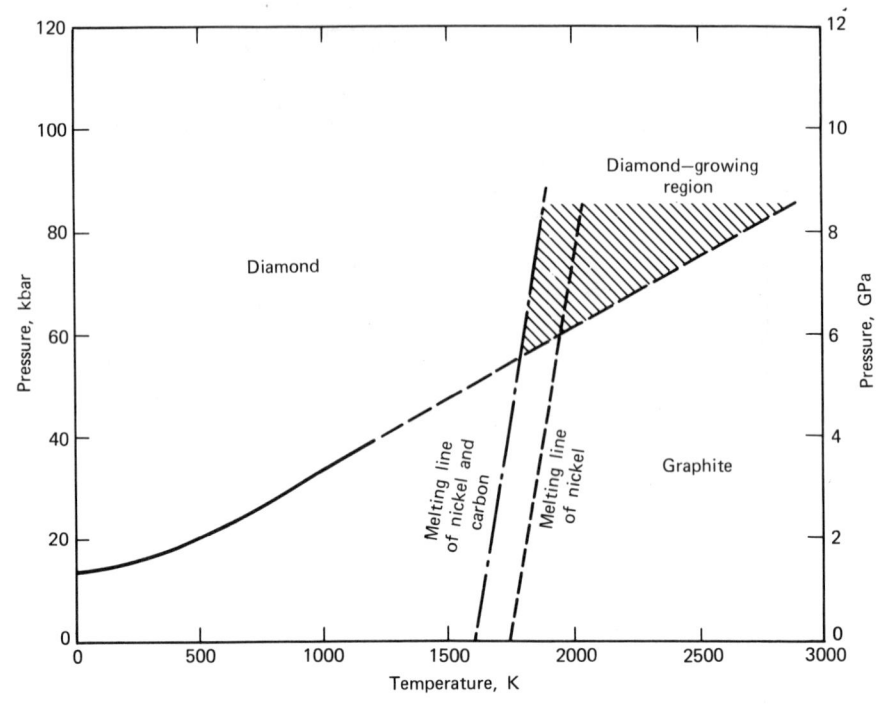

Figure 2. Diamond-forming region for the nickel–carbon system (3).

Apparatus. Many kinds of apparatus have been devised for simultaneously producing the high-pressures and temperatures necessary for diamond synthesis (9). An early, successful design is the belt apparatus (10), shown in cross section in Figure 3.

In the belt apparatus, two opposed, conical punches, made of cemented tungsten carbide and carried in strong steel binding rings, are driven into the ends of a short, tapered chamber which is also made of cemented tungsten carbide supported by strong steel rings. A compressible gasket, constructed in a sandwich-fashion of stone, usually pyrophyllite, and steel cones, seals the annular gap between punch and chamber, distributes stress, provides lateral support for the punch, and permits axial movement of the punches to compress the chamber contents. The reaction zone, usually a cylinder, is buried in pyrophyllite stone in the chamber. The pyrophyllite, a good thermal and electrical insulator, is easily machined and transmits pressure fairly well. The reaction zone is heated electrically with a heavy current.

The reaction zone temperature is measured by introducing thermocouple wires through the compressible gaskets, whereas the pressure is estimated by a calibration technique. A relationship is obtained between the force on the pistons and the pressure in the chamber by loading the reaction zone with certain metals in which abrupt changes of electrical resistance occur at certain pressures, and by noting the piston forces at which these changes occur. Typical pressure changes used are those occurring in bismuth at 2.5 and 7.5 GPa (25 and 75 kbar), in thallium at 3.7 GPa (37 kbar), and in barium at 5.3 GPa (53 kbar); such resistance changes usually coincide with changes of phase or structure in the metals at high pressure.

A belt apparatus is capable of holding pressures of 7 GPa (70 kbar) and temper-

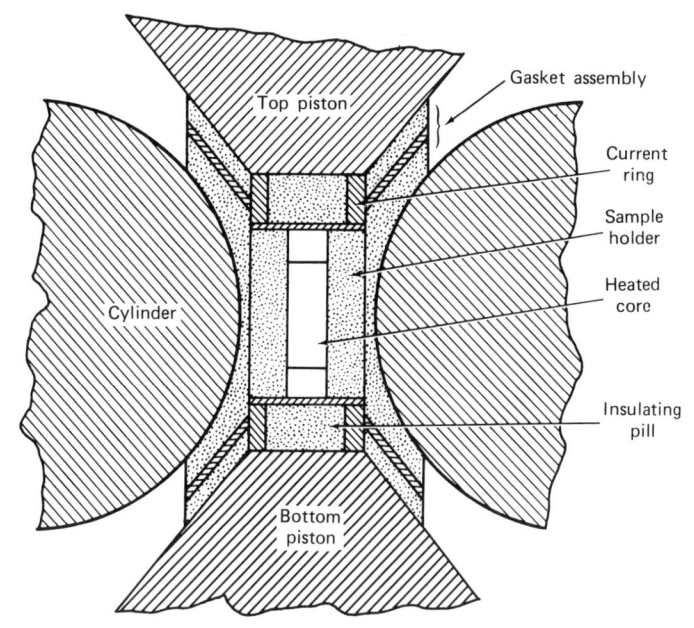

Figure 3. Cross section of belt high-pressure apparatus.

atures of up to 3300 K for periods of hours. The maximum steady-state temperatures are limited by melting of the refractory near the reaction zone (11).

Figure 4 shows an arrangement of carbon and catalyst metal. As the sample is heated at high pressure, the metal next to the graphite usually melts and diamond begins to form there. An exceedingly thin film of molten metal (at most a few thou-

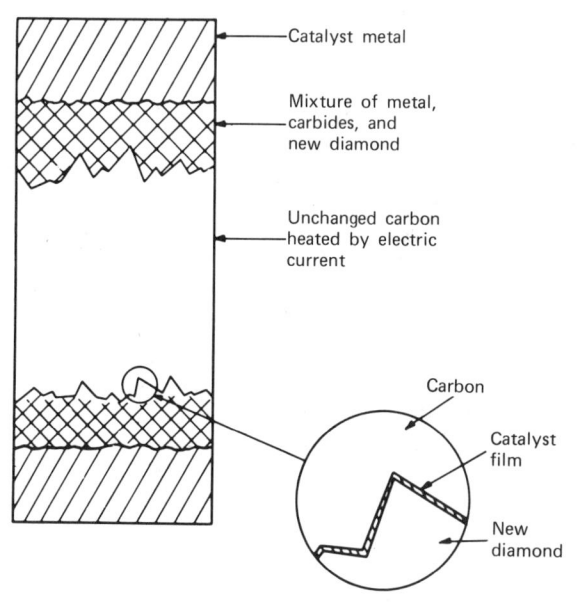

Figure 4. Diamond synthesis cell.

sandths of a cm thick) separates the newly formed diamond from the unchanged graphite. This film advances like a wave through the mass of graphite and transforms it to diamond. The speed at which this film travels depends only slightly upon the temperature but increases very rapidly to at least 1 mm/s as the pressure is increased above that necessary for diamond to be stable. Thus all graphite present in the sample, with the exception of carbon dissolved in the metal, may be transformed to diamond in time intervals of a few minutes to a few hours, depending upon the pressure, temperature, and catalyst system.

A mass of diamond crystals in a metallic matrix remains (see Fig. 5), where most of them are still covered by the catalyst metal film which is dissolved in acids to free the diamonds.

Comprehensive accounts of the methods and phenomena involved in diamond synthesis can be found in references 12 and 13.

Crystal Morphology. Size, shape, color, and impurities are dependent on the conditions of synthesis (12,14). Lower temperatures favor dark-colored, less pure crystals; higher temperatures promote paler, purer crystals. Low pressures (about 5 GPa or 50 kbar) favor the development of the cube faces, whereas higher pressures produce octahedral faces. Nucleation and growth rates increase rapidly as the process pressure is raised above the diamond–graphite equilibrium pressure.

The growing faces of diamond crystals are chemically active and tend to adsorb and incorporate certain impurities, particularly those which may be accommodated in the diamond without greatly straining the host lattice. Common contaminants are graphite, nitrogen, and certain catalyst metals, particularly nickel. The lattice dimensions of nickel and diamond are similar enough so that invisible clusters of nickel,

Figure 5. Mass of freshly formed diamonds.

at least several thousand atoms in diameter, may be included in diamond in an oriented sense. X-ray diffraction studies (12) show that these crystallites are oriented parallel to the diamond host lattice; the composite crystals, usually containing less than 1% of nickel, are ferromagnetic with approximately the same Curie temperature as massive nickel. It appears that even though synthetic diamonds are essentially the same as natural diamonds, there are enough differences between them, mainly in structure and impurity content, to permit an observer to distinguish between them.

Crystal Growth. If diamond seed crystals are placed in the active diamond growing zone of a typical graphite–catalyst metal apparatus, new diamond usually forms on the seed crystals. However, the new growth tends to be uneven in thickness and quality, with gaps or inclusions of foreign material. Such defects probably appear because the main driving force for the nucleation and growth under these conditions is the Gibbs free energy difference between diamond and graphite, which is a function of pressure, temperature, and composition; none of these variables can be sufficiently controlled. However, excellent growth can be obtained if pressure and composition are held relatively constant while the change of composition with temperature is employed as a driving force. In practice, small diamonds are used as the source of carbon in a hotter portion of a molten catalyst metal bath at about 1500°C and 5.5 GPa (55 kbar). Diamond seed crystals are placed in a cooler portion of the bath and the difference in solubility resulting from the difference in temperature causes diamond to recrystallize on the seed crystals in a slow, controlled fashion (15). Growth periods up to a week are used for the larger crystals of about 5 mm or 1 carat (0.2 gram). The process is not commercially feasible, but the control of growth conditions and bath compositions permits the formation of various types of high-quality diamond crystals for property studies (16–17). A few parts-per-million of boron impart a blue color and make the diamonds semiconducting (see Semiconductors). A few dozen parts-per-million of nitrogen give a yellow color. Colorless crystals of outstanding purity, crystal perfection, and thermal conductivity have been made (see Fig. 6).

Direct Graphite-to-Diamond Process

In this process, diamond forms from graphite without a catalyst. The refractory nature of carbon demands a fairly high temperature (2500–3000 K) for sufficient atomic mobility for the transformation, and the high temperature in turn demands a high pressure (above 12 GPa; 120 kbar) for diamond stability. The combination of high temperature and pressure may be achieved statically or dynamically. During the course of experimentation on this process a new form of diamond with a hexagonal (wurtzitic) structure was discovered (18).

Shock Synthesis. When graphite is strongly compressed and heated by the shock produced by an explosive charge, some (up to 10%) diamond may form (19–20). These crystallite diamonds are small (on the order of 1 μm) and appear as a black powder. The peak pressures and temperatures, which are maintained for a few microseconds, are estimated to be about 30 GPa (300 kbar) and 1000 K. It is believed that the diamonds found in certain meteorites were produced by similar shock compression processes which occurred upon impact (5).

Some diamond powder is produced commercially by shock-wave methods. The DuPont process (21), exposes small, well-crystallized graphite lumps in nodular cast iron to the brief, intense pressure generated by a suitable charge of high explosive.

Figure 6. Synthesized high-quality single crystal diamonds.

The graphite lumps are more compressible and reach much higher temperatures than the surrounding iron at the peak pressures, which last for a few microseconds, and part of the graphite turns into diamond. The carbon is cooled rapidly by the iron environment and the new diamond is thereby preserved. After recovery of the mass, the iron is dissolved and the diamond is separated by controlled oxidation of the graphite. The final product is a gray powder with particles ranging in size up to 30 μm.

The annual production of diamond by this process is only a small fraction of total industrial diamond consumption.

Static Pressure Synthesis. Diamond can form directly from graphite at pressures of about 13 GPa (130 kbar) and higher at temperatures of about 3300–4300 K (22). No catalyst is needed. The transformation is carried out in a static high-pressure apparatus in which the sample is heated by the discharge current from a capacitor. Diamond forms in a few milliseconds and is recovered in the form of polycrystalline lumps. From this work the triple point of diamond, graphite, and molten carbon is estimated to lie at about 13 GPa and 4200 K (12,22).

At pressures of 13 GPa many carbonaceous materials decompose when heated and the carbon eventually turns into diamond. The molecular structure of the starting material strongly affects this process. Thus condensed aromatic molecules, such as naphthalene or anthracene, first form graphite even through diamond is the stable form. On the other hand, aliphatic substances such as camphor, paraffin wax, or polyethylene lose hydrogen and condense to diamond via soft, white, solid intermediates with a rudimentary diamond structure (23).

Crystal Structure. A diamond prepared by the direct conversion of well-crystallized graphite, at pressures of about 13 GPa (130 kbar), shows certain unusual reflections in the x-ray-diffraction patterns (18). They could be explained by assuming a hexagonal diamond structure (related to wurtzite) with $a = 0.252$ and $c = 0.412$ nm, space group $P6_3/mmc - D_{6h}^4$ with 4 atoms per unit cell. The calculated density would be 3.51 g/cm^3, the same as for ordinary cubic diamond, and the distances between nearest neighbor carbon atoms would be the same in both hexagonal and cubic diamond, 0.154 nm.

Figure 7 shows the crystal structures of graphite, ordinary cubic diamond, and hexagonal diamond. The layers of carbon atoms lie in flat sheets in graphite, but in diamond the sheets are more wrinkled and lie closer together. Taken separately, the sheets are similar but they may be stacked in various lateral positions and still have bonding between them.

In cubic diamond (zinc-blende structure) the wrinkled sheets lie in the (111) or octahedral face planes of the crystal and are stacked in an ABCABC sequence. In real crystals, this ABCABC sequence continues indefinitely, but deviations do occur. For example, two crystals may grow face-to-face as mirror images; the mirror is called a

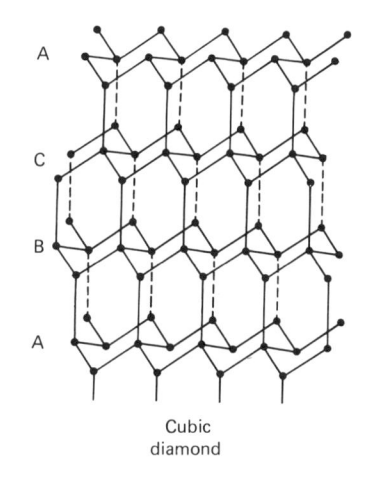

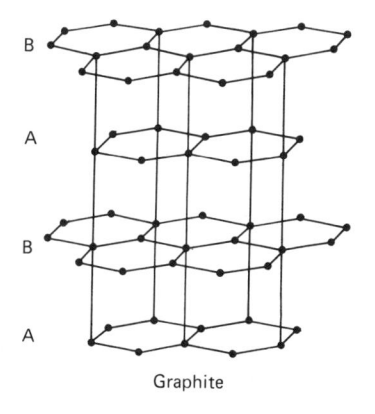

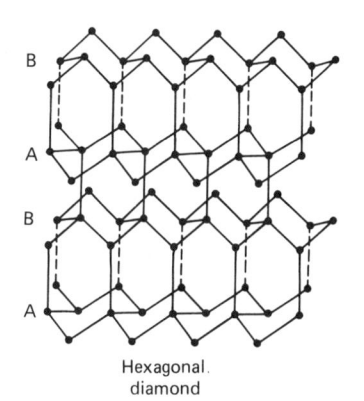

Figure 7. Crystal structures of graphite, ordinary cubic diamond, and hexagonal diamond; A, B, and C are the lateral positions.

twinning plane and the sequence of sheets crossing the mirror runs: ABCABCC-BACBA. Many unusual sequences may exist in real crystals, but they are not easy to study.

In hexagonal diamond (wurtzite structure) the wrinkled sheets are stacked in an ABABAB sequence, as shown in the Figure 7. Looking down on the stack from above, hexagonal holes can be seen formed by the six-membered carbon rings. The crystal has hexagonal symmetry about this axis, hence the name hexagonal diamond.

The sequence of sheets in graphite is also ABAB; however, an examination of the atomic positions shows that they are not simply related to those in either kind of diamond. Thus the simple compression of graphite should not be expected to yield diamond. However, well-crystallized graphite, in which the ABAB sequence extends for at least hundreds of layers, tends to form hexagonal diamond more easily than cubic diamond.

In spite of the close relationship between the two structures, some reshuffling of the atoms is necessary to complete the transformation of graphite into hexagonal diamond crystals of a large enough size to detect. The necessary atomic motions can be provided by heating to about 1500 K, much lower than the 2300–3300 K required to make cubic diamond from ordinary graphite without a catalyst. Even so, various departures from ideality make it difficult to prepare entirely hexagonal diamond material, even when the best graphite is used. The products obtained so far always contain some ordinary diamond as well as some remnant graphite, parts of which are compressed by the nearby diamond regions. Hence many physical properties of hexagonal diamond are not very well known, including its stability relative to cubic diamond.

The same general kinds of effects described here for carbon have also been noted in the direct conversion of hexagonal boron nitride to the denser forms (24) (see Boron compounds, refractory boron compounds).

Hexagonal diamond was found in the Canyon Diablo meteorite and in the shock-made diamond from DuPont, but not in the diamond from Allied Chemical nor in regular abrasive synthetic diamond (25). This new mineral form of natural carbon has been named Lonsdaleite.

Metastable Vapor Phase Deposition

Metastable growth of diamond takes place at low pressures where graphite is thermodynamically stable. The subject has a long history, and work in the United States and the Soviet Union indicates that diamond may form at moderate pressures during decomposition of gases such as methane. In a patented process (26–28), batches of clean diamond powder were alternately exposed to methane at about 1320 K and 10 kPa (0.1 atm) for a few hours followed by hydrogen at 1300 K and 5050 kPa (50 atm). The hydrogen treatment removed graphite. After several such cycles the diamond masses gained in weight by several percent and contained only diamond.

In a similar process for growth of diamond whiskers, a single crystal diamond is heated by radiation during observation under a microscope. The mean growth rate was about 10 μm/h and crystals of 400 by 20 by 20 μm were grown. So far this general method of diamond growth is not used commercially.

The Synthetic Diamond Industry

Soon after the first successful diamond syntheses by the catalyst process, small batches of crystals were prepared in the laboratory. When mounted in abrasive wheels, these first synthetic diamonds performed better than comparable natural diamonds for shaping of ceramics and cemented tungsten carbide. A pilot plant for producing synthetic diamond was established, the efficiency of the operation was increased, production costs declined, and product performance was improved. Today the price of General Electric man-made diamond is competitive with natural diamond prices.

Several kinds of diamonds can be produced, depending upon synthesis conditions, with each kind especially suited for particular uses. For example, the shaping of cemented tungsten carbide is an operation that consumes a large fraction of industrial diamond grit, and friable crystals with many sharp edges and corners, such as shown in Figure 8, mounted in a free-cutting resinoid or vitrified wheel matrix, are the most suitable for this task. The excellent abrasive properties seem to be caused by a better bond between diamonds and wheel matrix, and a greater number of constantly renewed sharp cutting edges per abrasive grain.

When the diamond grit is carried in a sintered, metal matrix in the abrasive tool, tougher, more coherent, block-shaped crystals, as shown in Figure 9, are preferred. The combination of more severe operating conditions and the strong metallic matrix imposes high loads upon each abrasive grain and the grains must be strong enough to resist these high cutting forces without crumbling (see Abrasives).

As of 1977, the bulk of synthetic industrial diamond production consists of the

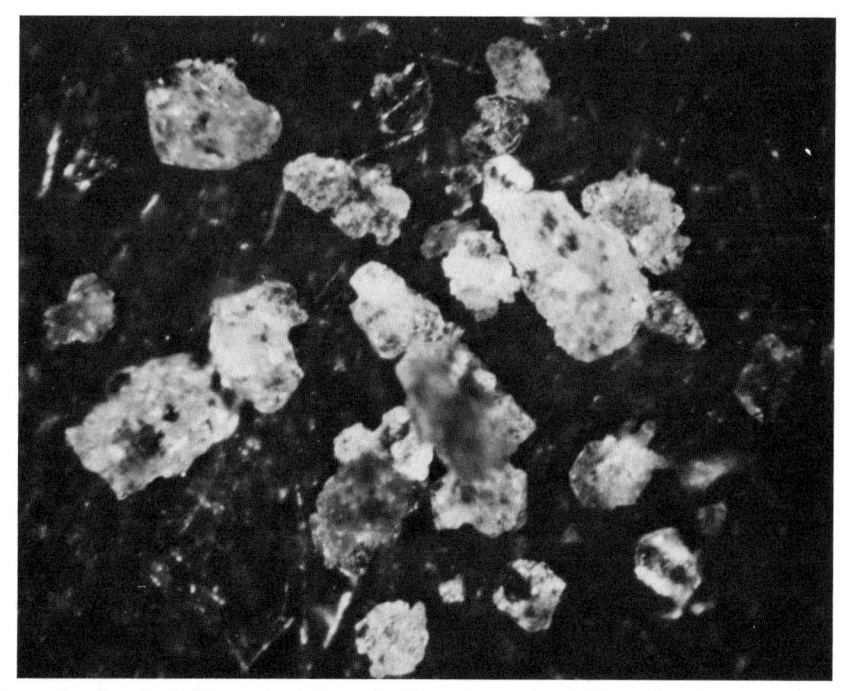

Figure 8. Synthetic diamond grit for resinoid or vitreous bond (free-cutting) abrasive wheels.

Figure 9. Synthetic diamond grit for metal bond abrasive wheels.

smaller crystal sizes up to about 25 mesh (0.7 mm) particle size. This size range has wide utility in industry, and a significant fraction of the world's need for diamond abrasive grit is now met by synthetic production (hundreds of kilograms per year). Because the raw materials are plentiful, synthetic production could, if necessary, supply the world demand for diamond abrasive. Development work continues in order to improve size and utility of the manufactured product and to realize the full potential of diamonds at minimum cost. An appreciable increase in performance has been obtained by coating the diamonds with a thin layer of nickel or copper, before incorporating them into wheels. The thin layer of metal apparently improves adhesion and heat transfer. The General Electric Company is the major producer of synthetic industrial diamonds. DeBeers Consolidated Mines in South Africa (which has a large stake in natural diamonds), and the Soviet Union have recently begun manufacture of synthetic diamonds. DeBeers currently produces synthetic diamond grit in plants in Shannon, Ireland, as well as in South Africa and Sweden. Synthetic, industrial diamond is also made in Japan and in the People's Republic of China.

Semiconducting Diamonds. With the exception of the rare, natural type IIb, diamond is normally a good electric insulator. However, semiconducting diamonds are prepared by adding small amounts of boron, beryllium, or aluminum to the growing mixture, or by diffusing boron into the crystals at high pressures and temperatures. Such diamonds are p-type, with activation energies for conduction usually ranging between 0.1 and 0.35 eV. Addition of boron gives the diamonds a blue color. Some of these crystals have been used as thermistors with a very wide, stable operating temperature range, −200 to 500°C having nominal resistance of 4,000–40,000 ohms (29).

Sintered Diamond Masses. Some natural diamonds known as carbonado or ballas occur as tough, polycrystalline masses. The production of synthetic sintered diamond masses of comparable excellent mechanical properties has only been achieved recently (30). The essential feature is the presence of direct diamond-to-diamond bonding without dependence on any intermediate bonding material between the diamond grains, since no extraneous bonding material can match the stiffness, thermal conductivity, and hardness of diamond. Formidable inherent obstacles are encountered when the sintering of small diamond crystals is attempted (13), and the process must be conducted at high pressure to keep the diamonds from changing into graphite.

Since these masses of polycrystalline diamond possess extensive diamond-to-diamond bonding, they have, in contrast to single-crystal diamond, excellent crack resistance since any crack that begins in one crystal on an easy cracking plane (parallel to an octahedral face), is halted by neighboring crystals that are unfavorably oriented for their propagation.

Natural single diamond crystals and carbonado can now be replaced in many industrial uses by sintered diamond tool blanks. Such tool blanks are available in disks and cores. The disks (or sectors of disks) consist of a thin (0.5–1.5 mm) layer of sintered diamond up to about 13 mm diameter on a cemented tungsten carbide-base block about 3 mm thick. Using diamond abrasive, such blanks can be formed into cutting tools of various shapes. Typical tool blanks are shown in Figure 10. The core blanks have diamond cores up to 6 mm in diameter and 6 mm in length which are encased in a cemented tungsten carbide sleeve up to 13 mm in diameter.

These disks enjoy wide usage as cutting or shaping tools for a variety of hard, abrasive materials such as fiber-reinforced composites, ceramics, rock, silicon-rich aluminum alloys, etc (30). Their toughness, shock resistance, ready availability, and uniform reliable properties have greatly enlarged the use, scope, and durability of diamond cutting tools over that of natural single crystals. They are not suitable for use on ferrous or nickel-base alloys because of the reactivity of diamond with these metals when hot (for such metals, sintered cubic boron nitride tools are useful; see Boron compounds). They show considerable promise as well-drilling bits. The diamond core pieces are commonly pierced and used as wire-drawing dies (31). The uniform polycrystalline character and resistance to bursting of such dies give them improved performance and utility as compared to single crystal diamond dies, and the larger

Figure 10. Sintered polycrystalline diamond cutting tool blanks.

sizes cost less. Some of them have drawn over 160,000 km of copper wire before repolishing was needed.

A few special high pressure pistons with sintered diamond working faces have been made for laboratory experiments. Although the sample volume is very small, pressures of 50 GPa (500 kbar) at temperatures of up to 500°C have been reached with such an apparatus (32).

BIBLIOGRAPHY

"Carbon (Diamond, Synthetic)" under "Carbon" in *ECT* 2nd ed., Vol. 4, pp. 294–303, by R. H. Wentorf, Jr., General Electric Research Laboratory.

1. J. B. Hannay, *Proc. R. Soc.* **30,** 188 (1880); *Nature* **22,** 255 (1880).
2. K. Lonsdale, *Nature* **196,** 104 (1962).
3. H. Moissan, *Compt. Rend.* **118,** 320 (1894); **123,** 206 (1896).
4. H. P. Bovenkerk and co-workers, *Nature* **184,** 1094 (1959).
5. M. E. Lipschutz and E. Anders, *Science* **134,** 2095 (1961).
6. F. D. Rossini and R. S. Jessup, *J. Res. Natl. Bur. Stand.* **21,** 491 (1938).
7. R. Berman and F. Simon, *Z. Electrochem.* **59,** 333 (1955).
8. F. P. Bundy and co-workers, *J. Chem. Phys.* **35,** 383 (1961).
9. R. H. Wentorf, Jr., ed., *Modern Very High Pressure Techniques,* Butterworth & Co., Ltd., London (1962).
10. H. T. Hall, *Rev. Sci. Instrum.* **31,** 125 (1960).
11. G. Davies and T. Evans, *Proc. R. Soc. London A* **328,** 413 (1972).
12. F. P. Bundy, H. M. Strong, and R. H. Wentorf, Jr. in ᴦ. L. Walker and P. A. Thrower, eds, "Methods and Mechanisms of Synthetic Diamond Growth" in *Chemistry and Physics of Carbon,* Vol. 10, M. Dekker, Inc., New York, 1973, pp. 213–263.
13. R. H. Wentorf, Jr., "Diamond Formation at High Pressures" in R. H. Wentorf, Jr., ed., *Advances in High Pressure Research,* Vol. 4, Academic Press, London and New York, 1974, pp. 249–281.
14. H. P. Bovenkerk in F. P. Bundy, W. R. Hibbard, Jr. and H. M. Strong, eds., "Some Observations on the Morphology and Physical Characteristics of Synthetic Diamond" in *Progress in Very High Pressure Research,* John Wiley & Sons, Inc., New York, 1961, pp. 58–69.
15. R. H. Wentorf, Jr., *J. Phys. Chem.* **75,** 1833 (1971).
16. H. M. Strong, *J. Phys. Chem.* **75,** 1837 (1971).
17. H. M. Strong and R. H. Wentorf, Jr., *Naturwissenschaften* **59,** 1 (1972).
18. F. P. Bundy and J. S. Kasper, *J. Chem. Phys.* **46,** 3437 (1967).
19. P. S. DeCarli and J. C. Jamieson, *Science* **133,** 1821 (1961).
20. U.S. Pat. 3,238,019 (Mar. 11, 1966), to P. S. DeCarli (Allied Chemical Corp.).
21. U.S. Pat. 3,401,019 (Sept. 10, 1968) to G. R. Cowan, B. W. Dunnington, and A. H. Holtzman (E. I. du Pont de Nemours & Co., Inc.).
22. F. P. Bundy, *Science* **137,** 1057 (1962); *J. Chem. Phys.* **38,** 631 (1963).
23. R. H. Wentorf, Jr., *J. Phys. Chem.* **69,** 3063 (1965).
24. F. P. Bundy and R. H. Wentorf, Jr., *J. Chem. Phys.* **38,** 1144 (1963).
25. R. E. Hanneman, H. M. Strong, and F. P. Bundy, *Science* **155,** 995 (1967).
26. U.S. Pat. 3,030,187 (Apr. 17, 1962), and 3,030,188 (Apr. 17, 1962), to W. G. Eversole (Union Carbide Co.).
27. U.S. Pat. 3,371,996 (Mar. 5, 1968), to H. J. Hibshman.
28. J. C. Angus, H. A. Will, and W. S. Stanko, *J. Appl. Phys.* **39,** 2915 (1968).
29. R. H. Wentorf, Jr., and H. P. Bovenkerk, *J. Chem. Phys.* **36,** 1968 (1962); "Thermistor Senses Red Heat Temperatures" in *Mat. Eng.,* 85 (Aug. 1968).
30. L. E. Hibbs, Jr. and R. H. Wentorf, Jr., *High Temp.—High Pressures* **6,** 409 (1974).
31. D. G. Flom and co-workers, *Wire Technology,* 19 (Jan. 1975).
32. F. P. Bundy, *Rev. Sci. Instr.* **46,** 1318 (1975).

R. H. WENTORF, JR.
General Electric Co.

NATURAL GRAPHITE

Natural graphite, the mineral form of graphitic carbon, occurs worldwide. It differs from the carbon of coal and of diamond in its predominantly lamellar hexagonal crystal structure. The ore usually contains associated silicate minerals that vary in kind and amount with the source. Except for technical terminology, the name natural graphite is seldom used. It may be simply termed graphite or any of several common names such as: plumbago, black lead, silver lead, carburet of iron, potelot, crayon noir, carbo mineralis, and reissblei. The macrophysical form depends on geological genesis, whereas the properties depend on both the macrophysical form and associated mineral suite. The commercial value depends on specific characteristics such as form, percentage, and kind of mineral suite, and availability. Graphite occurs in widely distributed places as flakes, lumps, and cryptocrystalline masses referred to commercially as amorphous graphite.

Graphite was at one time confused with other minerals of similar appearance, chiefly molybdenite (MoS_2). One common name for graphite is plumbago (leadlike). Until modern times users thought it to contain lead. Conrad Von Gessner in 1565 reported graphite as a separate mineral referring to it as Stimmi Anglicum (1).

Carl Scheele demonstrated in 1779 that graphite oxidized to CO_2, thus proving it to be a form of carbon. Abraham Werner in 1789 named it graphite from the Greek *graphein,* to write. The now depleted Borrowdale graphite mines in Cumberland, England, opened ca 1564 and produced graphite for pencils, called capucines.

One useful classification of graphite depends on the mode of formation which leads to three physically distinct common varieties: flake, lump, and amorphous. The term flake is self-explanatory; flake forms occur disseminated in rock. Lump graphite occurs in fissure-filled veins in pegmatite dikes (also associated with chip and the rarer needle forms). Amorphous graphite occurs in beds that once were coal, but fine-grained, easily ground vein graphite is also classified as amorphous.

All graphite has crystal structure but only certain kinds and sizes of natural graphites are commercially classified as crystalline—a term used for import duty purposes. Throughout this article reference will be made separately to flake, vein (lump), and amorphous forms, all of which are essentially the same crystalline form of carbon. However, fine structured graphites (named cryptocrystalline by Gustav Klar (2)) have been classified as amorphous by the U.S. Government.

Structure

Parallel layers of condensed planar C_6-rings constitute the graphite crystallite. Each carbon atom joins to three neighboring carbon atoms at 120° angles in the plane of the layer. The C–C distance is 0.1414 nm (this bond is 0.1397 nm in benzene); the width of each C_6-ring is 0.2456 nm. Weak van der Waals forces pin the carbons in adjoining layers, thus accounting in part for the marked anisotropic properties of the graphite crystal.

Figure 1 shows a Laue x-ray diffraction pattern of a single natural graphite crystal.

In the completely graphitized crystallite the planar C_6-layers stack in ordered parallel spacing 0.33538 nm apart (d spacing) at room temperatures. The hexagonal form of graphite (Fig. 2) contains the most common stacking order—ABABAB.... A small percentage of graphites exhibit ABCABCABC... stacking order (Fig. 3), resulting in the rhombohedral form. Grinding increases the rhombohedral structure, probably through pressure. Heating above 2000°C transforms the rhombohedral structure to hexagonal, suggesting that the latter is more stable. Impact from explosion can convert rhombohedral graphite to cubic-structured carbon: diamond (see Carbon, diamond, synthetic).

Grinding graphite to particle sizes smaller than ca 0.1 μm reduces the crystallite size to less than 20 nm, at which size two-dimensional ordering replaces the three-dimensional ordering of graphite. The weakened pinning forces permit the planar layers to move further apart and assume progressively random, though parallel, positions with respect to each other. This turbostratic structure (see below) is the characteristic structure of the so-called amorphous carbon that is found in chars. At more than 0.344 nm d spacing, the parallel planar C_6-layers assume a completely random lateral ordering.

When playing cards are bunched into a deck (without evening the sides and ends for redealing), the deck represents turbostratic structure, the structure of amorphous carbon. The parallel cards have no order in the third dimension of the deck. Turbostratic structure requires that each layer (card) in the bunched, uneven deck be sepa-

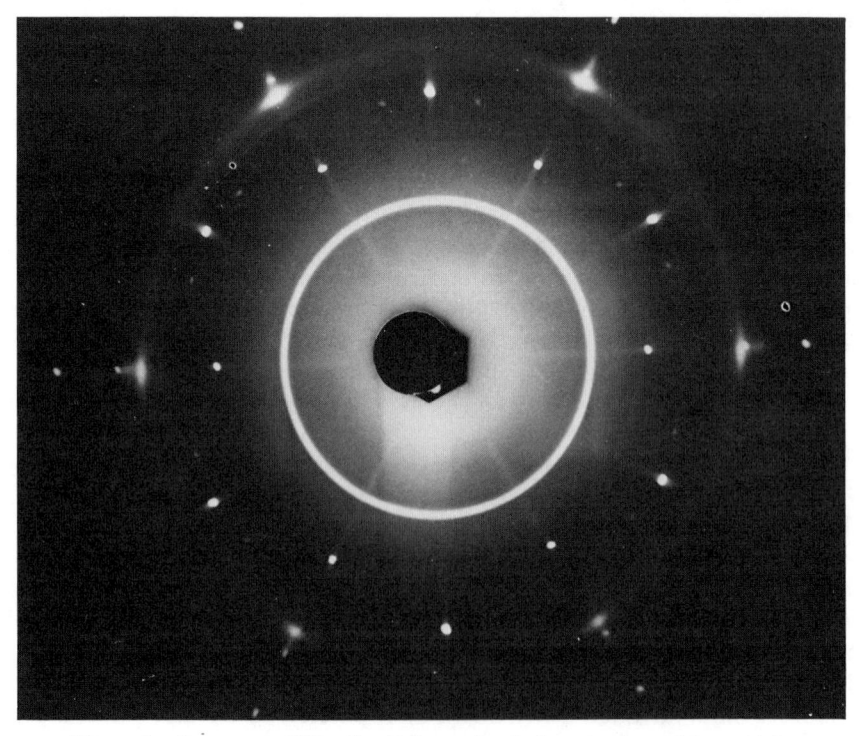

Figure 1. Laue x-ray diffraction pattern of a single natural graphite crystal.

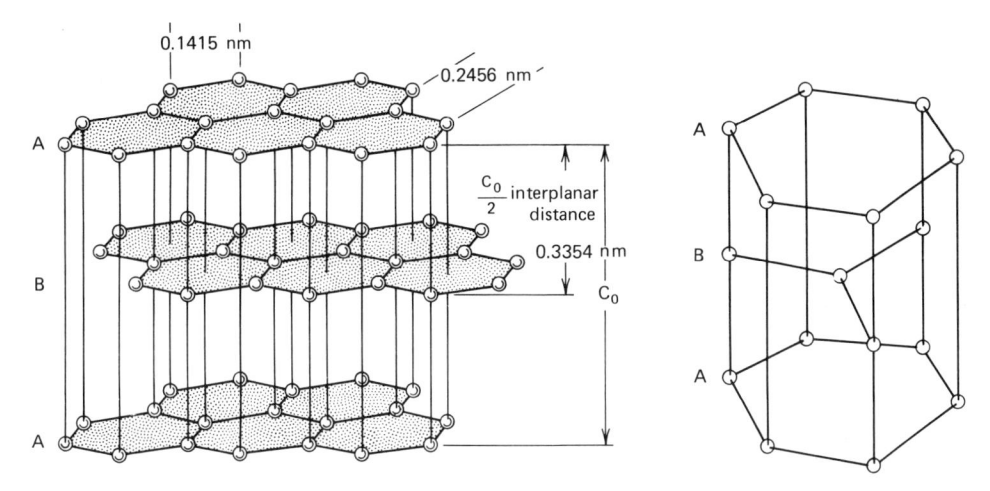

Figure 2. Hexagonal structure of graphite.

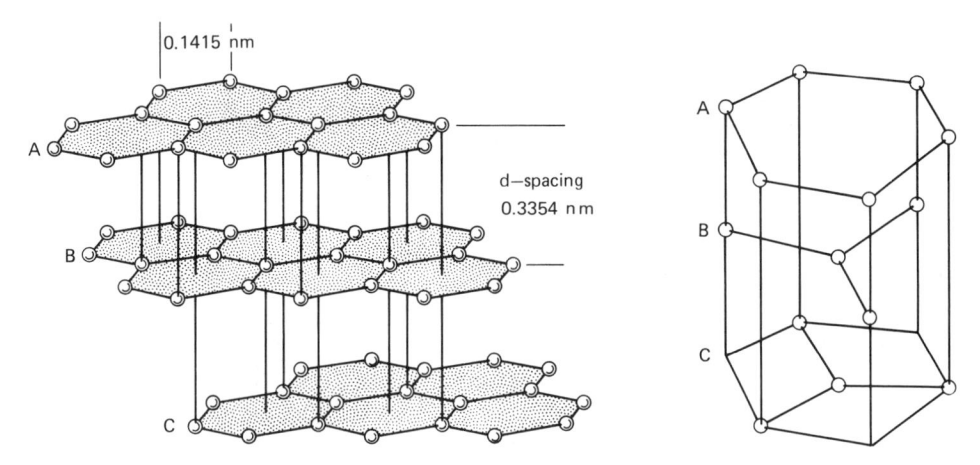

Figure 3. Rhombohedral structure of graphite.

rated further than in the evened deck of hexagonal structure, and at a minimum of 0.344 nm.

Table 1 depicts the effect of different d spacings on graphite structures.

The physical properties of finely ground but highly oriented natural graphites, such as 5 µm Sri Lanka (Ceylon) graphite, differ from those of turbostratic carbons such as chars, carbon blacks, or carbons formed from heavily ground graphites. Fig. 4 is an electron micrograph of 5 µm Sri Lanka graphite; the straight edges and angles of the particles contrast sharply to the rounded shapes of carbon black particles (see Carbon—Carbon black). The micrographs in Figures 5 and 6 show the surface and the edge of Madagascar flake graphite and the differences in density.

Table 1. Effect of Progressive Grinding of Graphite

Sample	Specific surface area, m²/g	Crystallite Thickness, L_c, nm	Crystallite Diameter, L_a, nm	d Spacing, nm
3 μm Sri Lanka (Ceylon) graphite (Dixon 200-10)	11.5	>100	>100	0.3354
Sri Lanka graphite	409	17.2	41.6	0.3356
Sri Lanka graphite	580	0.9		0.378
Sri Lanka graphite	699	0.9		0.380
Monarch 71 carbon black	350	1.5	2.5	0.400

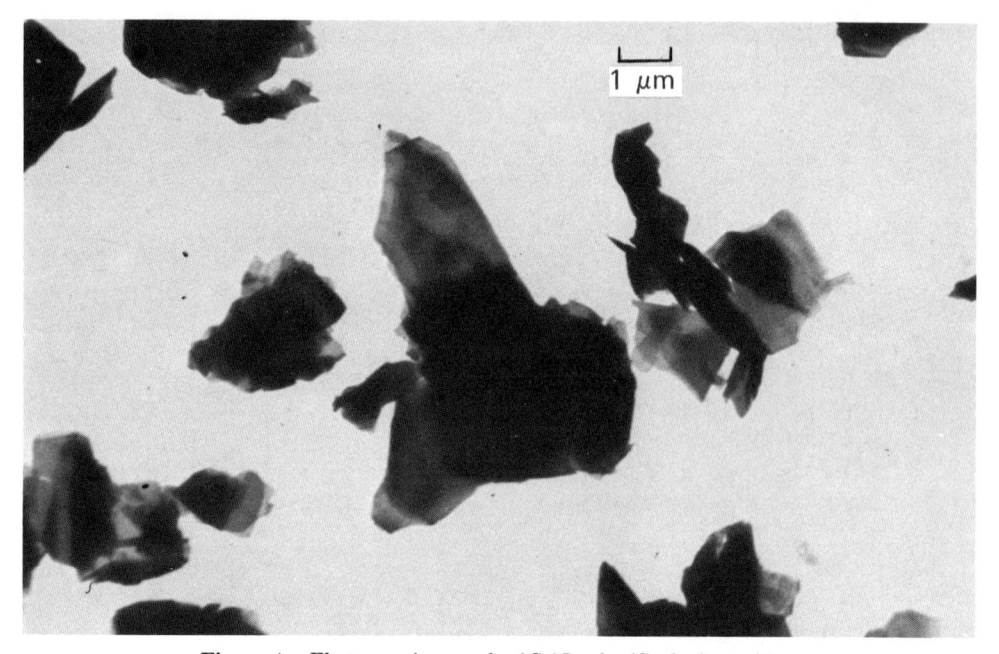

Figure 4. Electron micrograph of Sri Lanka (Ceylon) graphite.

Physical Properties

Solid articles made of natural graphite always require a binder; ie, they are always composites. The influence of the binder, the processing, and the kind of graphite used, together with graphite's strong anisotropism, influence the properties of the composites. In general, the overall quantities are usually greater than those measured along the a axis or along the c axis of the graphite under consideration. Table 2 lists some of the physical properties of natural graphite (see Composite materials).

Graphite's strength increases as the temperature rises. Relief of frozen-in stresses up to ca 2500°C accounts for this unusual property. Plastic deformation occurs above 2500°C.

The coefficient of linear expansion along the a axis changes from slightly negative below 383°C to slightly positive above that temperature; its average along the c axis is 238×10^{-7} between 15 and 800°C.

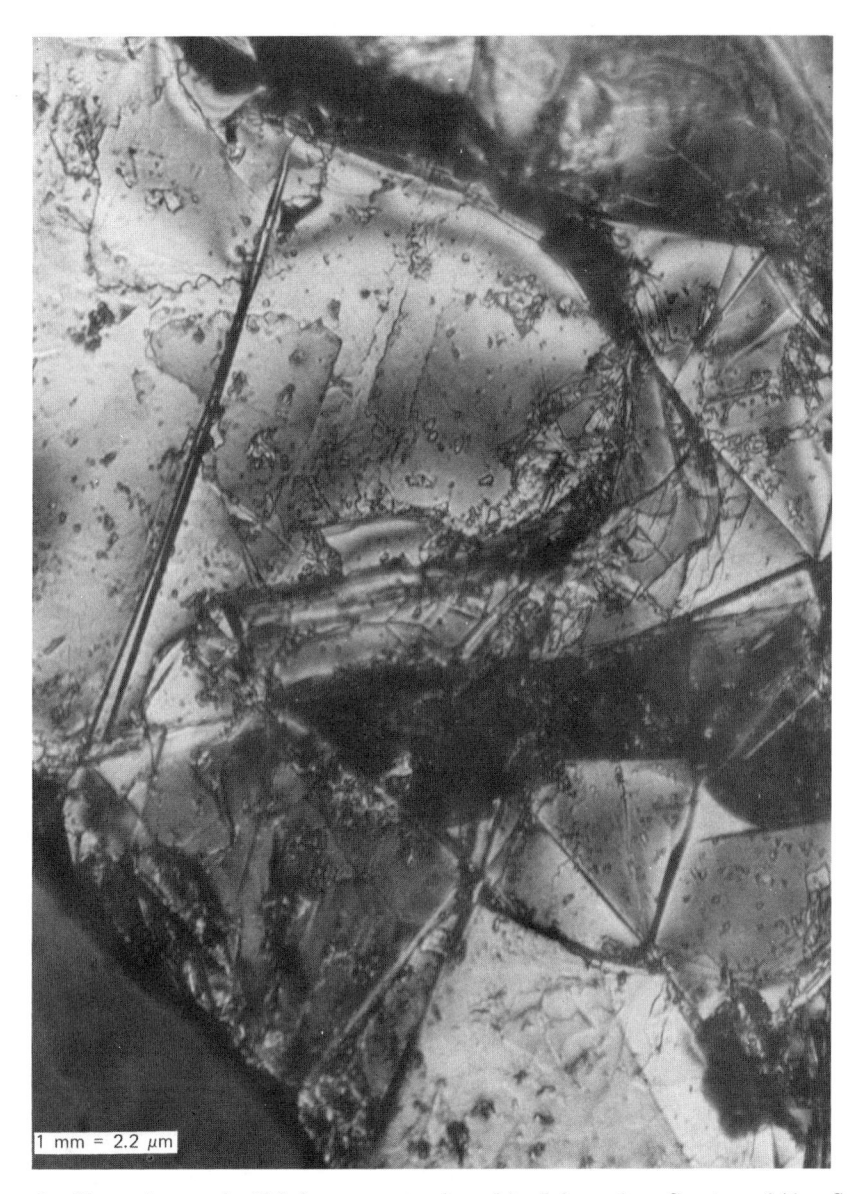

Figure 5. Photomicrograph of Madagascar natural graphite flake surface. Courtesy of Airco Speer.

The thermal conductivity, W/(m·K), along the a axis reaches a maximum of 285 at $-100°$C and falls rapidly with declining temperature. It is 251 at 20°C. Along the c axis it remains ca 837 to very low temperatures.

The specific heat varies markedly with temperature (Fig. 7). The steep rise in C_p above 3500 K probably results from reversibe formation of vacancies or other thermal defects (4).

In thin sections natural graphite is translucent, strongly pleochroic, and uniaxial. It has a negative sign of birefringence and two extinctions per revolution under crossed Nicol prisms. The atomic number of carbon accounts for its low absorption coefficient for x-rays and electrons.

694 CARBON (NATURAL GRAPHITE)

Figure 6. Electronmicrograph of Madagascar graphite flake edge. Courtesy of Airco Speer.

Single graphite crystals exhibit strong temperature-dependent anisotropic properties, both electrical and magnetic. However, this is of academic interest because natural graphite products are not single crystals. Flake graphites enhance anisotropy in bodies where forming processes such as extrusion, pressing, or jiggering align the flakes.

The specific resistance of natural graphite crystals is ca 10^{-4} $\Omega \cdot$cm (room temperature) along the a axis parallel to the network basal plane. The resistance along the c axis (perpendicular to the basal plane) is ca 1 Ω. The c/a axis anisotropy ratio is, therefore, ca 10^4. Screw dislocations within the crystal may short-circuit the current path parallel to the c axis and cause lower anisotropic ratios; separation of planes as in Figure 6 may cause higher anisotropic ratios.

Table 2. Physical Properties of Natural Graphite[a]

density[b], g/mL	
calculated	2.265
experimental, pure Sri Lanka	ca 2.25
compressibility[c], m²/N, Sri Lanka	
at low pressures	4.5×10^{-11}
at high pressures	$<2 \times 10^{-11}$
average	3.1×10^{-11}
shear modulus[c], N/m²	2.3×10^{9}
Young's modulus[c], N/m²	1.13×10^{14}
heat of vaporization[d], kJ/mol	711
sublimation point, K	4000–4015
triple point, K	
graphite–liquid–gas, 10.1 MPa[e]	3900 ± 50
graphite–diamond–liquid, 12–13 GPa[e]	4100–4200
surface energy[d], J/cm²	ca 1.2×10^{-5}

[a] Ref. 3.
[b] The difference between the calculated and experimental values of density is caused by dislocations and imperfections.
[c] To convert N/m² to dyn/cm², multiply by 10.
[d] To convert J to cal, divide by 4.184.
[e] To convert Pa to atm, divide by 1.01×10^{5}.

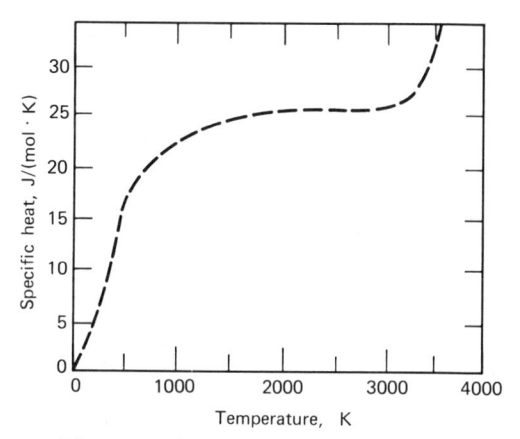

Figure 7. Specific heat of solid graphite.

Graphite is strongly diamagnetic because of its abundance of π electrons. Grinding the crystallite smaller than 20 nm creates a deficiency in π electrons and thus destroys the diamagnetism. The value of the specific magnetic susceptibility for Sri Lanka graphite is ca -6.5×10^{-6} at 20°C, -0.5×10^{-6} along the c axis and -22×10^{-6} along the a axis where it is temperature independent.

Chemical Properties and Graphite Compounds

Graphite burns slowly in air above 450°C, the rate increasing with temperature and exposed area. The particle size and shape govern the ignition temperature. Flake graphites generally resist oxidation better than granular graphites.

Above 800°C graphite reacts with water vapor, carbon monoxide, and carbon dioxide. Chlorine has a negligible effect on graphite, and nitrogen none. Many metals and metal oxides form carbides above 1500°C. These reactions occur with the carbon atom and destroy the graphitic structure (see Carbides). A series of compounds in which the graphite structure is retained, known as graphite compounds, consist of two general kinds: crystal and covalent compounds.

Graphite can be regenerated from the crystal compounds because the graphitic structure has not been too greatly altered. The dark crystal compounds are called intercalation compounds, interstitial compounds, or lamellar compounds because they are formed by reactants that fit in between the planar carbon networks. Each interlayer may be occupied, or every other interlayer, or every third interlayer, etc. Thus the same element, or group, can form a series of distinct compounds. The alkali metals form a variety of such addition compounds: potassium, rubidium, and cesium reactions with graphite are well known. The compounds of sodium and lithium are less well known. When potassium vapor enters graphite interstitially, it forms a series of intercalation compounds such as C_8K [12081-88-8], $C_{24}K$ [12100-36-6], $C_{36}K$ [12103-59-2], and others depending on the sequence of interstitial layers filled.

The C_8K and $C_{24}K$ compounds have an unusual ability to absorb hydrogen, nitrogen, and methane. C_8K catalyzes room-temperature additions of primary and secondary amines to dienes, yielding alkenyl amines. These compounds form by stepwise additions of potassium at increasing vapor pressures. Often several identical treatments are required to complete the reaction stoichiometrically, depending on the particle size of the graphite used. The colors become progressively lighter as the amount of metal constituent increases, from black through blue to bronze.

Compounds of graphite with alkali metals or ammonia are electron donors: the electrical resistance decreases (from that of the original graphite) and the Hall coefficient remains negative. Compounds with the halogens (except fluorine), metal halides, and sulfuric acid (graphite sulfate) are electron acceptors. The electrical resistance decreases but the Hall coefficient changes from negative to positive (5).

Graphite sulfate, long known and early investigated, forms when graphite is warmed in concentrated sulfuric acid containing a small quantity of an oxidizing agent such as concentrated nitric acid. The graphite swells and becomes blue. The compound, approximately $C_{24}^+(HSO_4)^-\cdot 2H_2SO_4$ [12689-13-3], hydrolyzes at once in water and the graphite is recovered. The recovered, washed and dried graphite exfoliates when quickly heated, in the manner of Pharaoh's Serpents (mercuric thiocyanate). Graphite sulfate also forms at the graphite anode while electrolyzing strong sulfuric acid. Turbostratic carbons do not react this way.

Bromine vapor forms C_8Br [12079-58-2] by direct addition to well-oriented graphite. Other halogen and mixed halogen compounds have been prepared.

Some intercalated crystalline compounds find their way into commerce as catalysts for chemical synthesis. Graphite–$FeCl_3$–KCl [56591-80-1] has been used in the synthesis of ammonia (qv). A series of patents were assigned to the Sagami Chemical Research Center (Japan) dealing with graphite complex catalysts for chemical synthesis (6).

Acetic acid (qv) has been synthesized in high yield from methanol and carbon monoxide over a graphite–$RhCl_3$–I_2 compound.

The covalent compounds of graphite differ markedly from the crystal compounds. They are white or lightly colored electrical insulators, have ill–defined formulas and occur in but one form, unlike the series typical of the crystal compounds. In the covalent compounds, the carbon network is deformed and the carbon atoms rearrange tetrahedrally as in diamond. Often they are formed with explosive violence.

Graphite oxide [1399-57-1] (graphitic acid) was prepared and described by Sir Benjamin Brodie in 1859. He heated graphite with a mixture of potassium chlorate and concentrated nitric acid (Brodie's reagent) on a water bath for four days and then repeated the treatment several times after intervening washings and dryings. The process yielded a stable yellow substance that retained the general physical form of the original graphite. It consisted of carbon, hydrogen, and oxygen and it reddened litmus. Since the Brodie method is extremely dangerous, graphite oxide is more safely prepared in the cold by Staudenmaier's method—digesting graphite with a mixture of nitric acid, sulfuric acid, and potassium chlorate. The compound also forms on a graphite anode during electrolysis of a dilute sulfuric acid solution containing an oxidizing compound such as nitric acid. The structure is unknown.

Graphite oxide may explode when heated above 200°C. Below this temperature it converts to a black powder once known as pyrographitic acid. The composition varies with the heat treatment and the end point; according to x-ray diffraction studies it is a form of carbon that reconverts to well-ordered graphite on heating to 1800°C. Before the use of x-rays, chemists used the Brodie reaction to differentiate between graphitic carbons and turbostratic carbons. Turbostratic carbons yield a brown solution of humic acids, whereas further oxidation of graphite oxide produces mellitic acid (1).

$$\text{benzene ring with six } CO_2H \text{ groups}$$

(1)

Fluorine forms covalent compounds with graphite, C_4F [12774-81-1] is prepared by exposing graphite to a mixture of fluorine and hydrogen fluoride at room temperature. Heating graphite fluoride at ca 400°C forms CF [12069-59-9], a gray solid wetted neither by water, alcohol, benzene, nor acetone. These fluorine compounds of graphite explode on heating.

Graphite fluoride continues to be of interest as a high temperature lubricant (6). Careful temperature control at 627 ± 3°C results in the synthesis of poly(carbon monofluoride) [25136-85-0] (6). The compound remains stable in air to ca 600°C and is a superior lubricant under extreme conditions of high temperatures, heavy loads, and oxidizing conditions (see Lubrication). It can be used as an anode for high energy batteries (qv).

Geographic Occurrence

Table 3 summarizes the world's production of natural graphite for 1973–1975 (7). As of 1977, the worked deposits of commercial interest are limited to those of Sri Lanka (Ceylon), Madagascar, Mexico, the Federal Republic of Germany, Austria, the Republic of Korea, Norway, the U.S.S.R., People's Republic of China, and the United States.

Table 3. World Production by Countries (Metric Tons) [a]

Country	1973	1974	1975
Argentina	94	100	41
Austria	17,211	24,655	30,586
Brazil	3,901[b]	4,082[c]	2,722[c]
Burma	183	305	87
People's Republic of China[c]	29,938	29,938	49,896
Federal Republic of Germany	13,525[d]	13,971[d]	16,330[c]
India	19,958[c]	23,061	18,888
Italy	4,161	2,486	1,724[c]
Democratic People's Republic of Korea[c]	77,112	77,112	77,112
Republic of Korea	43,605	45,360	47,232
Malagasy Republic	13,964	17,280	17,774
Mexico	65,393	60,692[e]	60,815
Norway	6,676	9,549	9,979[c]
Romania[c]	5,988	5,988	5,988
Sri Lanka	6,207	9,448[f]	10,429
Republic of South Africa	1,029	726	523
U.S.S.R.[c]	85,277[e]	89,813[e]	89,813
Total	394,222[c]	414,566	439,939
United States	[g]	[g]	[g]

[a] In addition to the countries listed. Czechoslovakia, Japan, Southern Rhodesia, and the Territory of South West Africa (Namibia) are believed to produce graphite, but available information is inadequate for formulation of reliable estimates of output level (8).
[b] Beneficiated product.
[c] Estimate.
[d] Represents marketable production, including some imported graphite.
[e] Revised.
[f] Exports.
[g] Withheld to avoid disclosing individual company confidential data.

The three physically different forms of natural graphite, which provide essentially different commodities, are in Sri Lanka (lump), Madagascar (large flake), the United States (small flake), the Federal Republic of Germany (small flake), Norway (small flake), Mexico (amorphous), and the Republic of Korea (amorphous).

Sources in Asia. Korea. More graphite is mined on the Korean peninsula than any other region in the world. Geologists estimate the reserves of both flake and amorphous but predominantly amorphous graphite on the order of millions of tons. The iron content of Korean graphites is low and the ash is a distinctive white.

Sri Lanka. Sri Lanka, formerly Ceylon, contains the largest known deposits of vein graphite. The extent and manner of the occurrence has inspired much study and although the literature is extensive, little is known about its origin. It occurs in a large area in the southwestern part of the island in metamorphosed Archean sediments known as the Khondalite system. Reserves of minable graphite are believed to be large.

In 1971 the Sri Lanka government nationalized the three operating mines as The State Graphite Corporation. Some commercial disruption occurred but production was gradually resumed.

Sri Lanka graphite is usually completely graphitized and contains both quartz and sulfides. It is favored for lubricants, pencils, and electromotive brushes.

Sources in Europe. *Austria.* Two distinct mineralogical regions, Styria and lower Austria, combine to make Austria one of the world's largest producers of natural graphite. As in all other European deposits, the graphite originated through metamorphosis of carboniferous and bituminous substances. Therefore, the deposits occur in small, lean lenses. With one exception it is mined underground. Styria contains the more important deposits. Graphite from Lower Austria is harder and contains sulfides, carbonates, lime, and pyrrhotite; it is most used in foundries and blast furnaces.

The Federal Republic of Germany. The Passau distict of Bavaria has long produced flake graphite suitable for crucibles such as those used by alchemists in the Middle Ages. In 1250 the inhabitants of Pfaffenreuth were required to pay their tithe with graphite.

Today, Graphitwerk Kropfmuehl A.G. produces flake graphite of high purity suitable for crucibles, pencil leads, and lubricants.

The country rock is part of the "kristallines Grundgebirge", the old gneissic and schistose rocks of the Bohemian basin. The ore contains 20–25% graphite and is beneficiated and processed by flotation, grinding, and sieving.

Norway. A/S Skaland Grafitverk, on Senja Island, is the only operating mine in Norway. The estimated ore reserve is one million tons of 25%-C graphite which occurs as flakes in seams bedded in gneiss. Flotation (qv) raises the carbon content to 80–90%.

Italy. The principal deposits occur near the French border and are amorphous. The main workings lie southwest of San Germano.

Spain. Although Spain produces little graphite, large deposits occur in the province of Jaén and near Toledo.

Sources in Africa. *Madagascar.* The island of Madagascar is an important source of flake graphite of large size and high quality. The widespread reserves are believed to be exceptionally large. The deposits are found with lateritic deposits of iron and bauxite. Graphite, which is resistant to weathering, is found in the weathered residue; thus the graphitic content of 3–10% has been increased by natural leaching.

The flake is large, strong, and flexible and is the best graphite for many refractories. The pyrometric cone equivalent (PCE) of the ash runs from cone 16 to cone 20—higher than the PCE of other flake graphite ashes, indicating that it is more refractory.

Sulfur compounds and carbonates are absent and iron is low; manganese is higher than in most graphites; potassium predominates in the alkali metal content; and volatiles, including water, are about 1.5%.

Other. South West Africa (Namibia) produces a 50%-C graphite in the district of Bethanien. Deposits exist in the Transvaal. A 10% ore is mined in open pits near Kanziku. Two mines operate in the Republic of South Africa, at Cumbu and at Mutali. The ore is high in flake graphite, ca 20%.

Sources in North America. *United States.* In 1977 the Burnet, Texas, mine of the Southwestern Graphite Company was the only one operating. It produces high grade flake. Other mines (many in Alabama) could be reopened. Some 20 other states have deposits that were worked in the past. Most of these deposits are of fine flake.

Canada. Many deposits of commercially important flake occur in Canada's Greenville-type Precambrian limestones and gneisses of Ontario and Quebec. The last operation in Canada, the Black Donald mine, closed in 1954 after 57 years of operation. It produced a high quality flake. Much flake exists in the Canadian shield but economics preclude development.

Mexico. The state of Sonora contains extensive deposits of quality amorphous graphite. There exists, *lit-par-lit*, as many as seven distinct beds of graphite with alternating layers of metamorphosed andalusite-bearing triassic rocks. The mines lie about 400 kilometers south of the U.S. border in the region of Moradillas.

The mineral suite includes micas, clay minerals, tourmaline, and hematite. Pyrite and gypsum sometimes are found. There are considerable dissimilarities in the products of different mines in the mineral suite and the degree of graphitization.

Sources in South America. Itapecerica in Minas Gerais, Brazil, produces some tons of marketable, medium quality, vein graphite, some of which reportedly is refined by flotation to 99% C. The mountains of Pie' de Palo, Argentina, yield graphite.

Identification of Graphite Ores

The ability to identify the source of the graphite depends on what is present in the ore. In practice, samples are nearly always milled to a powder. It is possible, up to a point, for an experienced technologist to distinguish graphite from various commercial sources merely by inspection. As the samples submitted for examination become more and more finely divided in the course of manufacturing and milling operations, the difficulty in making these decisions becomes progressively greater.

Finely divided samples may be identified further by (a) analyses of the graphite ash and (b) identification of the minerals associated with the graphite and comparison with graphites from known sources. Owing to its softness and opaqueness, most of the graphitic carbon must be removed from the sample before analysis by either (a) or (b). There are two general methods for accomplishing this.

In the first method the sample of graphite is ignited gently at 800°C until all the graphitic carbon is consumed, leaving the calcined gangue material. This includes the interlaminated material as well as the larger fragments, some of which may be country rock or lithic fragments as contrasted with minerals indigenous to the graphite deposit. This ash may be separated into light and heavy fractions with a heavy liquid, such as bromoform, or the fines, which constitute the bulk of the ash, may be decanted by repeated flushings with water. The residual material or the minerals separated with a heavy liquid can then be examined under a binocular, petrographic microscope for comparison with known ores.

The disadvantage of this method is that the minerals may be physically or chemically altered during burning. For example, the refractive index of clay minerals is changed; the color, birefringence, and pleochroism of micas is altered; carbonates are destroyed; and the iron sulfides will be oxidized to iron oxides.

To avoid these alterations in the mineral suite, untreated graphite is separated from the accompanying minerals by immersing the sample in a liquid of density intermediate to that of the bulk of the graphitic carbon and the gangue minerals, which then sink together with certain composite grains of graphite attached to a heavy mineral. The specific gravity at which this separation is best effected varies with different graphite samples, but 2.40 is usually satisfactory. About 10 g of graphite is strewn over the surface of about 50 mL of the heavy liquid in a Spaeth sedimentation glass. The slurry is stirred and covered and the whole allowed to stand quietly away from sudden temperature changes until a separation takes place, which may require $>\frac{1}{2}$ h. After separation, the stopcock is turned to the position that retains the heavy mineral concentrate, the light fraction is completly removed from the glass, and the bulk of the heavy liquid is recovered. The concentrate is transferred to a small beaker, washed with benzene, and dried. It is then ready for inspection with the microscope. Often the separated heavy mineral residue grains will be coated with a film of graphite which may interfere with their identification. In such cases the crop may be cautiously washed with soap and water to remove most of the graphite film and then dried. The interlaminated impurities of the graphite flakes, needles, and folia are not included in the sink portion. Since this fraction is usually too finely grained for useful observation, its loss is not serious.

The methods outlined are, in general, preferred among the methods in use for identifying commercial milled graphites by color of ash, grit, color of mark, and feel.

Chemical Analysis

There are no generally accepted methods for the complete analysis of natural graphite. Industrial methods usually emphasize either the carbon content or analysis of the ash. Although the carbon percentage is of considerable importance, it is usually true that the mineral suite is more significant for a specific use, eg, fluxing constituents in graphite must be avoided for refractory uses, and abrasive minerals in graphite must be absent for lubricating uses. Associated minerals are seldom reported other than as "ash."

The simplest analytical procedure is to oxidize a sample in air below the fusion point of the ash. The loss on ignition is reported as graphitic carbon. Refinements are determinations of the presence of amorphous carbon by gravity separation with ethylene bromide or preferably by x-ray diffraction, and carbonates by loss of weight on treating with nitric acid. Corrections for amorphous carbon and carbonates are applied to the ignition data, but loss of volatile materials and oxidation may introduce errors.

Graphite is frequently, although incorrectly, analyzed by the proximate method used for coal in which the volatile material is determined by strongly heating the sample in a covered or luted crucible. Some oxidation of the graphite always occurs so that the value obtained for volatile matter is high and thus the "fixed carbon" is too low. The method lacks both accuracy and precision.

The best indirect, but seldom used, method is to: determine the total moisture separately in a Penfield tube, determine the loss on ignition in air at 825–875°C, and report graphitic carbon as percent loss on ignition (100 −% moisture −% ash). It is desirable to use a platinum dish for ignition loss and it is necessary to spread the graphite in a thin layer. Thermogravimetric analysis (tga) is also used to determine weight losses of free moisture and other volatiles as well as graphite oxidation.

All too often specifications reflect a lack of understanding of what is required of the graphite purchased. Volatile matter, for instance, may be a factor where the graphite is to be heated in use but volatiles are of no significance for mechanical uses at room temperatures. Percent "ash" may be far less important than the material responsible for the ash. An artificial graphite, for instance, whose total ash is 0.5% silicon carbide may be an inferior lubricant to a natural graphite whose total ash is 5.0% fine mica and clay. The results of chemical analysis of graphite residues given in the Table 4 were obtained by following closely the conventional methods for silicate rocks (see Silica).

Table 4. Analysis of Graphite Residues From Principal Sources

Component	Sri Lanka "90%"[a]	Madagascar "86%"	Korean "83%"	Mexican "82%"
SiO_2	57.50	46.26	52.05	50.85
Al_2O_3	6.54	33.16	32.11	29.42
Fe_2O_3	25.07	16.73	4.92	11.74
MgO	1.09	1.46	1.96	0.48
CaO (SrO)	4.63	0.22	1.64	0.73
Na_2O	0.41	0.08	0.68	0.60
K_2O	0.68	0.95	5.05	4.78
CO_2	present	nil	nil	nil
TiO_2	0.41	0.80	1.58	1.27
P_2O_5	0.09	0.17	0.06	0.06
SO_3	0.41	trace		trace
F	0.05		nil	
S (FeS_2)	4.22			
MnO	0.16	0.38	0.04	0.05
BaO	0.06		0.10	0.06

[a] "90%" graphite also contains 0.13% Cu; trace of Pb; 0.014% Zn; 0.016% Ni; 0.011% Co; trace of Zr; trace of As; 0.07% Cr; no Se; no Te.

A method for physically separating turbostratic carbon and graphite involves shaking a sample into suspension in ethylene bromide of sp gr 2.17 and centrifuging. The method is unreliable except where fine carbon and coarse graphite are admixed; it can be an aid in qualitative examination.

The percentage of turbostratic carbon in a graphite sample can be estimated after determining the average d spacing by x-ray diffraction (9) (see Structure).

Three ASTM methods exist for testing graphite; D1553, *Analysis of Graphites Used as Lubricants*; D1367, *Lubricating Qualities of Graphites*; C561, *Test for Ash in Graphites*.

Specifications

The U.S. Government publishes specifications for graphite (10–11). There are no dependable classifications for commercially designated graphite numbers. Domestic flake is classified according to purity with high grade containing 95–96% carbon, and low grade 90–94% carbon. Sri Lanka graphite is classified according to lump, chip, and dust with subclassifications. Air-spun (micronized) graphite is classified according to kind, purity, and average particle size of 2.5, 5, and 10 μm. Amorphous graphite is classified according to locality and carbon content (seldom higher than 85%). The variety of specifications exists because graphite is found worldwide, is mined by many small establishments, and is subject to keen competition among suppliers. Table 5 lists the major U.S. suppliers of natural graphite.

Economic Aspects

For some uses natural graphite is a strategic commodity since no substitutes exist. Imports account for most of the natural graphite used in the U.S. since only one deposit is mined at present. Published prices cover a wide range of specifications (Table 6).

Table 5. Major U.S. Suppliers of Natural Graphite

Company	Location	Grade
Asbury Graphite Co.	Asbury, N.J.	all grades
Cummings–Moore Graphite Co.	Detroit, Mich.	amorphous
The Joseph Dixon Crucible Co.	Jersey City, N.J.	all grades
Southwestern Graphite Co.	Burnet, Texas	flake
Superior Graphite Co.	Chicago, Ill.	all grades
United States Graphite Co.	Saginaw, Mich.	amorphous

Table 6. Prices of Graphite

	1976, $/metric ton[a]	1977, ¢/kg[b]
Flake		
Madagascar	207.23–615.08	
Federal Republic of Germany	257.93–1647.94	
Norway	165.35–272.27	
U.S. no. 1 (90–95% C)		70.5–92.6
Amorphous (80–85% C)		
Republic of Korea (bags)	48.50–55.00	
Mexico (bulk)	40.00	
Crystalline		
U.S. (88–90% C)		39.7–59.5
U.S. (90–92% C)		56.2–60.6
U.S. (95–96% C)		63.9–88.2

[a] Ref. 12.
[b] Ref. 13.

Individual companies vary selling prices by kinds, sizes, and mixtures, depending on the method and extent of physical conditioning. Since the graphite business is highly competitive, both supplier and consumer show a reluctance to discuss negotiated prices.

U.S. import duties on natural graphites, as of 1976, were dropped for so-called most favored nations. For natural graphites from other countries, the duty is 10% *ad valorem*.

Uses

Graphite's many useful properties give rise to a wide variety of products: unctuous, dry lubricant; marks readily, writing and drafting pencils; combination of lubricity and electrical conductivity, motor and generator brushes; excellent weathering properties and inertness, industrial paint pigment; solubility in molten iron, carbon-raiser for steel; poorly wet by most metals and alloys, foundry mold facings; and burns slowly, conducts heat, and retains strength over a large temperature range, refractories such as crucibles, retorts, and stopper heads for steel ladles.

Some additional properties of interest include: hydrophobicity; forms water-in-oil emulsions; carries a negative charge; low photoelectric sensitivity; strongly diamagnetic; infrared absorber.

Table 7 lists the quantities and dollar values of natural graphites used in the U.S. in 1976 by product groups (7,14).

Table 7. Consumption of Natural Graphite in the United States in 1975, by Use[a]

Use	Crystalline Quantity, metric tons	Crystalline Value, 10³ $	Amorphous[b] Quantity, metric tons	Amorphous[b] Value, 10³ $	Total Quantity, metric tons	Total Value, 10³ $
batteries	276	465	351	388	627	853
brake linings	593	415	816	443	1,409	858
carbon products[c]	376	[d]	521	[d]	897	856
crucibles, retorts, stoppers, sleeves, nozzles	2,644	1,120	1,180	596	3,824	1,716
foundries	1,806	[d]	7,002	[d]	8,808	2,174
lubricants[e]	963	595	1,883	664	2,846	1,260
pencils	653	435	143	41	796	476
powdered metals	92	[d]	83	[d]	175	189
refractories	503	84	7,697	1,069	8,200	1,153
rubber	88	50	119	40	207	90
steelmaking	362	81	15,941	3,906	16,303	3,987
other[f]	4,542	1,124	446	282	4,988	1,406
Total	12,898	4,369	36,182	7,430	49,080	15,018

[a] Consumption data incomplete; excludes small consuming firms.
[b] Includes mixtures of natural and manufactured graphite.
[c] Includes bearings and carbon brushes; previously titled "other mechanical products."
[d] Withheld to avoid disclosing individual company confidential data, included with "other."
[e] Includes ammunition, packings, and seed coating.
[f] Includes paints and polishes, antiknock and other compounds, drilling mud, electrical and electronic products, insulation, magnetic tape, small packages, and miscellaneous and proprietary uses.

Steelmaking (carbon raising, hot topping, and ingot wash) and foundry uses (mostly facings) account for about half of the known uses. Natural graphite has its next greatest use (ca 25%) in crucibles and refractories.

Refractories. Natural graphite refractories are either formed and fired ware such as crucibles, or ramming mixes. Graphite imparts high refractoriness, low thermal expansion, excellent heat–shock resistance, high resistance to metal and flux attack, resistance to wetting by molten substances, increased strength and resilience at elevated temperatures, and high thermal and electrical conductivity. The purity of refractory graphites is ca 80–90% graphitic carbon (see Refractories).

Most ramming mixes contain amorphous graphite. The refractory ware group consists chiefly of crucibles for melting metals and alloys, retorts for zinc distillation, stoppers and nozzles for steel pouring ladles; and a miscellaneous group of saggers, slabs, rods, stirrers, and skimmers. Crystalline flake graphite is used almost exclusively because it burns slower than other graphites, suffers less from attrition in manufacturing processes, and imparts a desirable physical structure through orientation of the flake in the forming processes.

Graphite's disadvantage is that it will slowly burn in an oxidizing atmosphere so that some refractories require glazes.

Manufacturers bond their ware with either refractory clays or with mixtures of tars and pitches to form a coke bond.

Crucibles for nonferrous use contain a percentage of silicon carbide; carbon-bonded crucibles run higher in silicon carbide and use finer flake graphite. Graphite for ferrous use contains no silicon carbide because iron attacks it at pouring temperatures (see Carbides).

Clay-bonded ware is processed either by dry pressing or by wet-mud forming, the latter being more common. Either blade mixers or muller mixers are used to minimize the breakdown of the graphite flake. Water is added to provide plasticity in forming. It is customary to age the mix before forming. Usually the mix is deaerated by vacuum mixing. Forming methods follow conventional ceramic methods such as tamping, pressing, jiggering, extruding, and hand forming, or combinations of these methods. The industry uses both air drying and controlled humidity drying. The dried ware is fired either saggered or open depending on the product and the glaze. Glazes may be incorporated in the mix or applied to the kiln-fired body and refired open to develop the glaze. Sometimes glaze is applied before firing the dried form.

Crucibles usually contain 30–50% flake graphite. Stoppers and nozzles, used in bottom-pour ladles to control the flow of molten steel into molds, contain ca 20–30% flake graphite; nozzles may contain as much as 40%. The advantage of graphite stoppers and nozzles to nongraphitic products is their superior resistance to heat shock, their resistance to the erosion of flowing steel, and their property of withstanding deformation without rupture at operating temperatures.

Retorts for the distillation of secondary zinc are either bottle-shaped or straight-sided tubes closed at one end. Some are over 1.5 m in length and weighing upwards of 350 kg; they are the largest of all natural graphite refractory ware.

Foundry facings consume a large tonnage of natural graphite, primarily lump and amorphous. Foundry facings are carbonaceous or mineral powders applied to the surface of sand molds to prevent the molten metal from penetrating into or reacting with the sand. Generally, clay binders are applied dry to green sand molds whereas organic binders are applied wet to dry sand molds.

Lubrication. The slip that readily occurs between layers of graphite planes only partially explains graphite's dry lubricating properties. A suitably adsorbed film such as water must also be present; without it graphite ceases to lubricate and may, although rarely, become abrasive (15). Scrolls (rolled-up layers 1–5 nm dia) may play a part in the lubricity of graphite by acting as rollers between the planar layers.

Graphite lubricants include the dry powder, admixtures with liquid lubricants or greases, volatile liquids compounded with film-forming substances to produce bonded dry films, synthetic resins and powder metal compositions containing graphite for bearings, and finely divided suspensions in liquids (colloidal graphite).

High temperature lubrication, as in some metal forming processes, requires dry graphite. Although the coefficient of friction of graphite is higher than that of petroleum lubricants, it is often added as a safety measure should the carrier lubricant fail (16) (see Lubrication).

Colloidal Graphite. Colloidal graphite refers to a permanent suspension of fine natural or synthetic graphite in a liquid medium. The average particle size is about 1 μm and protective colloids ensure permanency of the suspension. Film-forming binders may also be present. In most aqueous dispersions the particles carry a negative charge. The name semicolloidal is applied to less stable dispersions—those that settle more readily because of larger particle size or because of less effective processing, or both. In stable suspensions, the lower limit of size is controlled by the smallest size that can retain graphitic structure (Fig. 2). Fine particle size is a prerequisite for penetration and for wicking—the ability to be taken up by and passed through a fibrous wick. Colloidal graphite was invented in 1906 by Acheson who coined the word deflocculate for the dispersion (17). Jerome Alexander (18) suggested the term colloidal graphite which became accepted. The diverse uses of colloidal graphite dispersions fall mainly in the classifications of lubrication, electrical uses, and parting compounds, (see Abherents).

Two military specifications cover different types of colloidal graphite: MIL-L-3572, Grade A (2% colloidal graphite in oil), Grade B (10% colloidal graphite in oil), Grade C (10% semicolloidal graphite in oil); and MIL–C26548A aerosol (spray-dry graphite film).

Pencils. The lead pencil (19), so-called because of the "black lead" (natural graphite) on which its marking property depends, is essentially a baked ceramic rod of clay-bonded graphite encased in wood (some plastic-bonded leads are now manufactured). Artificial graphite (see Carbon and artificial graphite) is too harsh for pencils. With the exception of inexpensive pencils, the rod or lead is impregnated with a lubricant to preclude glazing of the point in use. The quality of the lead depends on the quality of the ingredients and the manufacturing process; the degree of hardness depends on the ratio of clay to graphite.

The clays are selected and refined secondary clays; the graphites are of a variety of purities, particle sizes, and kinds. Amorphous graphites are usually used in the cheaper grades of leads. The best leads use a mixture of graphites dictated by carefully controlled testing. A mixture (20) of processed kaolin and bentonite is favored by some pencil lead manufacturers.

Nicholas Jacques Conté invented the process of bonding powdered graphite with clay (21) when English exports to France of capucines, from the famous Borrowdale mine, were stopped because of war. Though the basic concept has changed little since then, the technology has developed enormously because of the growing requirements for better quality and new uses.

A common procedure of lead manufacture is to ball-mill or hammer-mill a water slurry of the clay and graphite, dry the slurry, mix into a stiff dough in an intensive mixer, compact into an extrusion cylinder, and extrude under pressure through a die. The wet strands are dried, packed in saggers, and kiln-fired at temperatures of 800–1100°C. The fired leads are then impregnated with waxes, and fats or fatty acids or both. The waxed leads are surface-cleaned and glued between grooved cedar slats, shaped into pencil sections, lacquered, and imprinted.

Similar formulas are extruded into 0.91 mm and 1.17 mm diameters for mechanical pencil leads, smaller for drawing instruments, and into large diameters of 13 mm (usually hexagonal) for lumber crayons.

The degree of hardness is regulated principally by the clay-to-graphite ratio. Increasing the clay percentage strengthens the lead, thereby increasing its resistance to abrasion with a result that less graphite is deposited on the paper and the mark is less dense. In writing pencils, No. 1 lead contains ca 20% clay, and No. 4 (the hardest) has ca 60% clay. Hardness is a product requirement not a quality factor such as uniformity, smoothness, or strength of the sharpened point.

Indelible leads are mixtures of methyl violet, graphites, and binders (such as gum tragacanth or methyl cellulose), with or without mineral fillers and insoluble soaps. The leads are extruded and dried but not kiln-fired.

Graphites from Mexico, Sri Lanka, Madagascar, Germany, and Texas, individually or combined, comprise most of the graphite used in U.S. pencil leads.

Electrical Uses. Dry cells (see Batteries and electric cells, primary) use graphite to render the nonconductive pyrolusite (MnO_2) conductive through intimate admixture. The degree of graphitization is a factor in that graphites with the same carbon content and from the same locality give different results.

Natural graphite is required in some motor and generator brushes. Its high conductivity, high contact drop, and anisotropy make it particularly useful in brushes for d-c equipment.

Graphite forms the conductive film for some electroplating and electrotyping processes and is applied either dry or wet.

Paint. Some graphites act as reinforcing pigments which aid in the formation of tough, flexible, durable protective coatings (22). The platelike structure of graphite and its "leafing" produce films of low permeability. The paints are used for protecting structural steel and other metal surfaces exposed to unusually rigorous conditions or chemical attack, including water tank interiors. The gray color of graphite is its one drawback, limiting its use to dark-colored paints. Its unctuousness is important for automobile primers to upgrade the sanding properties as well as to improve brushing and flow properties for further coating applications.

Graphite is admixed with other pigments, such as iron oxide, for primers. It may be used singly or combined in intermediate and top coats (see Coatings; Paint).

Powder Metallurgy. In 1976, between one and two billion powder metal pieces were made, including tungsten carbide cutting tools, friction materials and clutch facings (see Brake linings), porous self-lubricating bearings and brushings, ferrous and nonferrous mechanical parts, and electrical components such as metal brushes, contacts, and slip-rings.

Natural graphite is increasingly used in powder metallurgy (qv) primarily for two purposes: as a solid lubricant constituent in bearing products and as a source of carbon in steel products. Graphite is added at the mixing operation in percentages ranging from 0.2 to 25% by weight. It serves to lubricate the die in the compacting operation and to reduce metallic oxides during the sintering operation to form steel, as well as serving as a lubricant during pressing.

Miscellaneous Uses. The following are minor uses of graphite: coating smokeless powder and gunpowder grains to control burning rate and to prevent static sparking from friction between grains; roofing granules; packings; brake linings; gaskets; stove polish, static eliminator; polish for tea leaves and coffee beans; pipe joint compounds; boiler compounds; wire drawing; welding rod coatings; catalysts; oil well drilling muds; lock lubrication; coatings for 8-track tape cartridges; mechanical mounts in cassettes, mercury and silver dry cells; exfoliated flake for gaskets and packings; aircraft disk brakes; catalyst pellet production; O-rings and oil seals; and interior and exterior coatings for cathode ray tubes.

In Europe colloidal graphite is added to lubricating oil for gasoline and diesel engines; and graphite has partially replaced coke in blast furnace production of pig iron.

BIBLIOGRAPHY

"Carbon (Natural Graphite)" in *ECT* 1st ed., Vol. 3, pp. 84–104 by S. B. Seeley and E. Emendorfer, The Joseph Dixon Crucible Co.; Carbon (Natural Graphite)" in *ECT* 2nd ed., Vol. 4, pp. 304–335, by S. B. Seeley, The Joseph Dixon Crucible Co.

1. C. von Gessner, *De Omni Rerum Fossilium Genere,* Tiguri, Zurich, 1565, p. 104.
2. G. Klar, *Die wichtigsten Graphitvorkommen der Welt,* Montan Rundschau, Vienna, 1957.
3. W. N. Reynolds, *Physical Properties of Graphite,* Elsevier Publishing Co. Inc., New York, 1968.
4. J. E. Hove, "Some Physical Properties of Graphite as Affected by High Temperature and Irradiation," *Industrial Carbon Graphite, Papers Conf. London 1957,* 1958, p. 509.
5. G. R. Henning, "Properties of Graphite Compounds," *Proceedings of the Second Conference on Carbon,* University of Buffalo, Buffalo, N.Y., 1956.
6. *Graphite, Minerals Yearbook,* U.S. Bureau of Mines, 1974, p. 8.
7. *Minerals Yearbook,* U.S. Bureau of Mines, 1975.
8. *Chem. Mark. Rep., Chem Mark Abstr.* **68**(6), 281, 613 (1976).
9. G. E. Bacon, "Study of the Structure of Graphite by Diffraction Methods," *Industrial Carbon Graphite, Papers Conf. London 1957* (1958), p. 183.
10. *Graphite, Lubricating Flake,* SS–G–659a; *Graphite Lubricating,* MIL–G–6711; *Graphite, Use in Ammunition,* MIL–G–155a; *Graphite, Dry (For Use in Ammunition)* MIL–G–48771 (PA). Superintendent of Documents, U.S. Government Printing Office, Washington, D.C.
11. *Graphite, Amorphous Lump,* P–21–R; *Graphite, Ceylon Amorphous Lump,* P–21–R3; *Graphite, Crystalline Flake, Crucible Grade* P–21a–R; *Graphite, Crystalline Flake, Lubricating and Packing Grade,* P–22b–R; Superintendent of Documents, U.S. Government Printing Office, Washington, D.C.
12. *Eng. Min. J.* **177**(12), 48 (1976).
13. *Chem. Mark. Rep.* (1977).
14. W. T. Adams, "Graphite," *Minerals Yearbook,* U.S. Bureau of Mines, 1975.
15. R. H. Savage, *J. Appl. Physics* **19**(1), 1 (1948).
16. E. L. Youse, *NLGI (National Lubricating Grease Institute) Spokesman,* **25,** 303 (1962).
17. U.S. Pat. 813,426 (Feb. 5, 1907), E. G. Acheson.
18. J. Alexander, private communication.
19. S. B. Seeley, "Pencils," *Encyclopaedia Britannica,* 15th ed. Encyclopaedia Britannica, Inc., Chicago, Ill.
20. U.S. Pat. 2,986,472 (May 30, 1961), H. H. Murray and H. M. Johnson (to Georgia Kaolin Company).
21. Fr. Pat. 77 (Jan. 30, 1795), N. J. Conte.
22. S. B. Seeley, "Graphite," in T. C. Patton ed., *Pigment Handbook,* Vol. 1, John Wiley & Sons, Inc., New York, 1973, p. 752.

General References

W. T. Adams, "Graphite," *Mineral Facts and Problems,* U.S. Bureau of Mines Bulletin 667, 1975.
"Natural Graphite in 1976," *Mineral Industry Surveys,* U.S. Bureau of Mines 1976.
D. Irving, *Graphite, Mineral Facts and Problems,* U.S. Department of the Interior, Bureau of Mines, Washington, D.C., 1960 ed.,
G. Klar, "Developments in Austria," *Mining Mag. London 95* **88** (Aug. 1956).
G. E. Bacon, "Study of the Structure of Graphite by Diffraction Methods," *Industrial Carbon Graphite, Papers Conf. London* 1957, 1958, pp. 183–185.
A. Grenall and A. Sosin, "Dislocations in Graphite," *Proceedings of the Fourth Carbon Conference,* New York, 1960, pp. 371–402.
P. L. Walker, Jr., and G. Imperial, "Structure of Graphite," *Nature* **180,** 1184, 1957.
A. J. Kennedy, "Graphite as a Structural Material in Conditions of High Thermal Flux," *AGARD M.22,* Sept. 1959.
F. Rusinko and P. L. Walker, "Properties of Molded Ceylon Natural Graphite," *Proceedings Fourth Carbon Conference,* Pergamon Press, Inc., New York, 1960 p. 751–761.
A. R. Ubbelohde and F. A. Lewis, *Graphite and Its Crystal Compounds,* Oxford University Press, London, 1960.
P. L. Walker and S. B. Seeley, "Fine Grinding of Ceylon Natural Graphite," *Proceedings of the Third Biennial Carbon Conference,* Pergamon Press, Inc., New York, 1959 p. 481–494.
P. L. Walker, ed., *Chemistry and Physics of Carbon,* Vols. 1–13, Marcel Dekker, New York, 1965–1977.
Proceedings of the Third Conference (1957) on Carbon, Pergamon Press, Inc., New York, 1959.
Proceedings of the Third Conference (1961) on Carbon, Vol. 1, (1962) and Vol. 2 (1963), Pergamon Press, Inc., New York.
J. G. Hooley and co-workers, articles in *Carbon* (1963–1977).
I. P. Alibert, *Pencil Lead Mines of Asiatic Siberia,* Riverside Press, printed by H. G. Houghton & Co., Cambridge, Mass., 1865.
G. Klar, "Developments in Austria," *Mining Mag. London 95,* 88 (Aug. 1956).
N. Mares, "Sonora Graphite—A Strategic Mineral," *Mexican American Review,* July 1959.
J. E. Reeves, "Graphite," Can. Dept. Mines, Rev. **36,** 1959.
J. J. Schanz, Jr., and A. B. T. Werner, "Summary of the Natural Graphite Industry with Notes on Recent Trends," *Trans. Soc. of Mining Engrs.* 370–380, Dec. 1962.
E. S. Glauch, *Graphite and Its Use in Lubrication,* National Lubricating Grease Institute, Chicago, Ill., 1940.

SHERWOOD B. SEELEY
The Joseph Dixon Crucible Company

CARBONATED BEVERAGES

Manufacture of flavorless, artificially carbonated waters began in Europe at the close of the eighteenth century. Commercial preparation in the United States, chiefly by the druggist or pharmacist, followed within a few years. Establishments for the bottling and sale of carbonated beverages were founded in Philadelphia, New York, New Haven, and other cities. By 1830 flavored soda water was being consumed in the United States.

The U.S. Census of Manufacturers recorded 64 plants bottling mineral waters and pop in 1850, 378 in 1870, and 2763 in 1900. In 1949, 6900 plants produced 860,959,300 cases of beverages to meet the 162 bottles per capita consumption. In 1975, 2258 bottling plants produced 3,916,000,000 cases of 240 mL (8 oz) beverages representing a per capita consumption of 439 bottles, and an estimated wholesale value of 9.4 billion dollars. Cola accounted for 64% of the market, lemon–lime 14%, orange 3%, root beer, ginger ale and grape each 2%, and all other flavors 13%.

A 1975 survey conducted by the National Soft Drink Association indicated that diet soft drinks represented 10% of packaged soft drinks. Cola, the dominant low-calorie flavor, accounted for 51% and lemon–lime 18% of the total diet market (see Sweeteners).

The Standard of Identity refers to soda water as "the class of beverages made by absorbing carbon dioxide in potable water. They either contain no alcohol or only such alcohol, not in excess of 0.5% by weight of the finished beverage, as is contributed by the flavoring ingredient used."

The carbonated soft drink of today is composed of carbonated water, a sweetening agent, acid, flavor, color, and a preservative. These ingredients must be combined in the proper ratio in order to make an appealing and refreshing carbonated beverage. Sweetener, acid, and carbon dioxide ratios vary with the type of flavor. The color used must be approved by the FDA, and must be stable under the conditions to which it will be exposed. The type of flavor, emulsion, alcoholic extract, alcoholic solution, or fruit juice concentrate depends on whether a cola, cloudy citrus beverage, or synthetic fruit is produced. The level of preservative varies according to the quantity of nutrients available (eg, fruit juice), carbon dioxide content, and the pH. Synthetically sweetened carbonated beverages require both a special blend of flavoring ingredients and a unique combination of all other ingredients, except color, in order to produce a pleasing flavor. Once properly blended, the beverage is packaged in a specially designed bottle or can (see Barrier polymers; Packaging materials).

Ingredients

Flavors. Flavors which are alcoholic solutions or extracts, emulsions or fruit juices, are prepared by a specialized industry associated with the beverage industry which also supplies the exact formula to be followed in preparation of the finished syrup.

The acid and color necessary for the finished beverage may be contained in the flavor itself or may be called for in the syrup formula. Flavors are usually supplied in one-, two-, four-, or five-ounce strength (30, 60, 120, or 150 mL, respectively). Strength refers to the amount necessary to produce 3.8 L (1 gal) of bottling syrup. A strength of 1–31 means that 3.8 L of flavor added to 118 L (31 gal) of simple syrup produces 122 L (32 gal) of finished syrup (see Flavors and spices).

Alcoholic Extracts. Alcoholic extracts are made by percolating dry materials with alcoholic solutions or by washing immiscible flavoring oils with water–alcohol mixtures and allowing the oils to separate.

Examples of alcoholic extracts are ginger, grape, and certain lemon–lime flavors. The amount of alcohol introduced into the finished beverage by the use of these extracts is 1/2 vol % or less, depending upon the strength of the extract.

Emulsions. Emulsions (qv) are prepared by mixing the essential oils with a vegetable gum, such as acacia or tragacanth, or other satisfactory emulsifying agent, and homogenizing (see Gums and mucilages). A stable emulsion requires reduction in size of the oil particles to 1–2 µm. Citrus flavors, root beer, and cola are examples of flavor emulsions. Emulsions used in cloudy beverages must be specially treated to stabilize the cloud. For example, the specific gravity of orange oil is 0.84, the specific gravity of the finished beverage, 1.05. This difference causes the orange oil to float to the top where it forms an unsightly white ring. To alleviate this problem, the orange oil is mixed with a suitable weighting oil or substance which is completely miscible with the flavor oils and is approved for this purpose.

The following substances are used for adjusting specific gravities of flavor oil mixtures.

Substances	Specific gravity, 25°/25°
brominated vegetable oil (BVO)	1.23–1.33
glyceryl abietate (ester gum)	1.10
sucrose acetate isobutyrate (SAIB) (not approved in the United States)	1.146
mixture of glycerol tribenzoate and propylene glycerol dibenzoate (polyol benzoates)	1.14–1.20
decaglycerol esters (Caprol PGE)	1.03

In the United States, use of brominated vegetable oil is restricted to 15 ppm in the finished beverage. Ester gum is restricted to 100 ppm.

Particle charge and interfacial tension of the emulsion are also very important in achieving a permanently cloudy beverage.

Alcoholic Solutions. Alcoholic solutions of flavors are made by dissolving flavoring materials in a solution of alcohol and water. The solubility of the particular flavors involved determines the level of alcohol used. Strawberry, cherry, and cream soda are examples of alcoholic solutions.

Fruit Juices. The most common fruit juices (qv) used in the manufacture of carbonated beverages are orange, grapefruit, lemon, grape, and lime. Carbonated beverages containing fruit juice usually contain additional flavor added as an extract, or more commonly, as an emulsion. Some manufacturers use additional ingredients to add desirable taste sensations. Sodium citrate is sometimes used as a buffering agent to give the beverage smoothness and body..

Caffeine Usually 100–135 mg/L (3–4 mg/oz) of caffeine is added to cola drinks. Coffee (qv) contains about 0.55 g/L (16.2 mg/oz) of caffeine. Caffeine is added chiefly for its bitter taste, not for its stimulating properties. The maximum level of caffeine allowed in soft drinks is 0.02% (see Alkaloids).

Extract of Kola Nut. Kola is the dried cotyledon of the *Cola nitida,* a large tree indigenous to western tropical Africa, and largely cultivated in the West Indies and South America. Kola nuts, known in Africa as guru nuts, are reddish to light brown, convex or somewhat globular. The hard and tough cotyledons, 2.5–5 cm in length, yield at least 1% anhydrous caffeine and are used as a caffeinic stimulant where they are grown. In a cola drink, only a negligible amount of caffeine is added from the extract of kola nut.

Other Flavors. Small amounts of ethyl acetate or amyl butyrate may be used to increase the aroma of grape beverages. Capsicum is sometimes used in ginger ales, especially in certain sections of the United States where the public desires a decidedly warm taste. Capsicum is the dried ripe fruit of *Capsicum fructescens,* also referred to as cayenne pepper and African chili. Capsicum extract is either added directly to the syrup, or included in the ginger ale extract.

Acidulants. Acids in carbonated beverages serve several different purposes: (*1*) to impart a sour or tart taste, often imitating the fruit for which the beverage is named; (*2*) to modify the sweetness of the sugar present; (*3*) to act as a preservative in syrups, color solutions, and the finished product; and (*4*) to catalyze the inversion of sucrose in the syrup or beverage.

Citric, phosphoric, malic, and tartaric acids are those most commonly used in carbonated beverages. Fumaric, lactic, and adipic acids are used less often.

Because solutions of citric, tartaric, or phosphoric acids having the same pH taste the same with regard to sourness or tartness, the tartness or acidity in taste of a beverage appears to depend solely upon hydrogen ion concentration. However, the different acids impart other taste sensations which make their use more desirable in some flavors than others. All acids used in beverages must be edible grade or food grade.

Citric Acid. Citric acid, $C_6H_8O_7 \cdot H_2O$, occurs naturally in many fruits and is used in orange, lemon, lime, grapefruit, strawberry, raspberry, cherry, lemon–lime, and sometimes root beer. Citric acid is frequently used as a solution made according to a widely-known formula yielding 50% citric acid solution and consisting of 2.3 kg of USP citric acid made up to 3.8 L (1 gal) of solution. However, some companies use only 1.8 kg of citric acid USP per 3.8 L of solution. With the latter formula, each 30 mL (1 fluid oz) of solution contains 14 g (0.5 oz) of USP citric acid. If anhydrous citric acid is used to replace citric acid USP, only 91.42% as much acid is required or 33.16 kg anhydrous acid for the first formula and 26.54 kg for the second (see Citric acid).

Phosphoric Acid. H_3PO_4 is the cheapest acidulant available, because of its strength as well as cost. The standard commercial grades of phosphoric acid are 75 and 85% solutions. Approximately 0.454 kg of anhydrous citric acid can replace 266 mL (9 oz) of 75% phosphoric acid. Phosphoric acid is used primarily in cola drinks, though many manufacturers also use it in root beer (see Phosphoric acids and phosphates).

Malic Acid. Malic acid, $HOOCCH_2CH(OH)COOH$, has become widely used in fruit-flavored carbonated beverages, particularly the berry flavors. In many instances malic acid is used in combination with citric acid. The level of acid required for substitution varies with the type of flavor, and level of sweetness and carbonation being used (see Hydroxy dicarboxylic acids).

Tartaric Acid. This acid, HOOCCH(OH)CH(OH)COOH, is used in the preparation of grape-flavored beverages. It is used in bottling in the same way that citric acid is used (see Hydroxy dicarboxylic acids).

Fumaric Acid. Fumaric acid, HOOCCH=CHCOOH, is used as a replacement or partial replacement for citric acid in fruit drinks. Replacement of 1.36 kg of citric acid by 0.91 kg of fumaric acid results in an equivalent acid taste. The poor rate of solubility of fumaric acid requires a special method of incorporation into the formula. The maximum level of fumaric acid allowed for food use is 0.3% (as of 1964) (see Maleic anhydride, maleic acid and fumaric acid).

Adipic Acid. Adipic acid (qv), HOOC(CH$_2$)$_4$COOH, can be used in carbonated beverages to impart a mildly tart taste. Its aqueous solutions have the lowest acidity of any of the common food acids. It is sometimes used in combination with other acids in grape flavors.

Sweetening Agents. The sweeteners (qv) used in carbonated beverages fall into two categories: natural, or nutritive, and synthetic, or nonnutritive.

The natural, or nutritive, sweeteners are granulated sucrose, liquid sucrose, liquid invert sugar, high fructose corn syrup, and dextrose. Sucrose, produced from cane or beets, is the primary sugar employed in carbonated beverages (see Sugar; Syrups). It is available in dry or liquid form, generally in a 67% solution. Liquid invert sugar, or sucrose inverted 50% at 75% solids, is also used as a nutritive sweetener in all types of carbonated beverages. Sugar used in beverages ranges from 9 to 15%.

High fructose corn syrup is the purified concentrated aqueous solution of nutritive saccharides obtained from edible corn starch. It is a relatively new nutritive sweetener made available for manufacture of carbonated beverages as well as other food products and complies with Section 26.3, Title 21, *Code of Federal Regulations.* High fructose corn syrup is made with several concentrations of fructose. The original product, available about 1970, contained 42% fructose, 50% dextrose, and 8% other saccharides. Manufacturers now make commercially available products containing 55, 60, and 90% fructose.

The product containing 55% fructose seems the most desirable replacement for sucrose or invert sugar in carbonated beverages. The level of sucrose or invert sugar which can be replaced depends upon the quality of the high fructose corn syrup and the flavor formulation in which it is being used. Achievement of equal sweetness replacement also depends upon the level of fructose present and the type of beverage formulation.

Poor quality sugar has an extremely detrimental effect on the taste, odor, and stability of a carbonated beverage. Extensive work has been done by the National Soft Drink Association in establishing standards for sugar. For detailed specifications on bottler's grade sugar, contact the National Soft Drink Association in Washington, D.C.

The only synthetic or nonnutritive sweetener which is approved by FDA is saccharin. It is normally used to replace sugar in order to produce a low-calorie carbonated beverage. Saccharin is approximately four hundred times as sweet as sucrose and may be used alone or in combination with sucrose.

Water. The ingredient contained in greatest quantity in carbonated beverages is water, constituting 86–92% of the beverage. It is evident, therefore, that the water used must be of highest quality. Municipal water supplies, though potable, frequently contain minerals and vegetation that render the water unsatisfactory for preparation of carbonated beverages. A good bottling water must be clear, colorless, odorless, and free of waterborne organisms. It must have less than 50 ppm alkalinity (expressed as $CaCO_3$), less than 500 ppm total dissolved solids, and less than 0.1 ppm iron or manganese (see Water).

Water that contains suspended matter does not carbonate readily and the beverage made from it rapidly becomes flat. Odors found in water are classified as chemical, from industrial wastes or chemical treatment; fishy, from certain waterborne organisms; septic, from sewage contamination; hydrocarbon, from oil contamination; sulfurous, from hydrogen sulfide; and chlorine, from the presence of free chlorine. These odors alter the flavor of the beverage, often making it extremely disagreeable. Algae and other waterborne organisms produce off-tastes and sediments. An alkalinity exceeding 50 ppm neutralizes sufficient acid in the beverage to cause an insipid taste. Because acid aids in preservation of the beverage, neutralization of the acid also reduces the keeping qualities. Total dissolved solids exceeding 500 ppm give the beverage a brackish taste. Iron or manganese in water may form a sediment in the beverage and contribute to off-taste.

Municipal water is treated in the bottling plant to remove objectionable characteristics. There are a number of possible treatments, and the exact pattern varies according to local requirements. A popular sequence is chlorination, treatment with lime, coagulation, sedimentation, and sand filtration, followed by treatment with activated carbon (see Carbon).

Chlorination. Chlorine (qv) is added for disinfection, and sometimes for oxidation of impurities in the water. Heavy dosages, in the range of 12–15 ppm, are often needed for oxidation of impurities, but a chlorine residual of 0.1–0.3 ppm after a contact period of 10 min is usually satisfactory for disinfection.

Lime. Lime is used to precipitate temporary hardness caused by the bicarbonates of calcium or magnesium.

$$Ca(HCO_3)_2 + Ca(OH)_2 \rightarrow 2\ CaCO_3 + 2\ H_2O$$
$$Mg(HCO_3)_2 + Ca(OH)_2 \rightarrow MgCO_3 + CaCO_3 + 2\ H_2O$$

Since the purpose of this treatment is to reduce alkalinity, and since lime itself is alkaline, it is important to add the lime in exactly the right proportion.

Coagulation. Small quantities of ferrous sulfate or potassium aluminum sulfate are added as coagulants. Natural (or added) alkalinity with these coagulants precipitates hydrated ferric oxide (ferrous oxide is oxidized by dissolved oxygen) or hydrated aluminum oxide.

$$2\ Ca(OH)_2 + 2\ FeSO_4 + 1/2\ O_2 \rightarrow 2\ CaSO_4 + Fe_2O_3.H_2O + H_2O$$
$$3\ Ca(OH)_2 + 2\ KAl(SO_4)_2 \rightarrow K_2SO_4 + 3\ CaSO_4 + Al_2O_3.H_2O + 2\ H_2O$$

Sodium aluminate may also be used as a coagulant; it reacts with calcium or magnesium salts to form calcium or magnesium aluminate.

$$Ca(OH)_2 + 2\ NaAlO_2 \rightarrow 2\ NaOH + Ca(AlO_2)_2$$

These precipitates are gelatinous, and cause coalescence of solid particles followed by flocculation into larger masses of the minute particles that cause color and turbidity. Bacteria and other microorganisms are entrapped in the flocculated masses. The water is allowed to settle, and is then filtered through a sandbed. Turbidity, color, and bacteria and other microorganisms are removed (see Flocculants).

Activated Carbon. Activated carbon is used to remove odors and any remaining color.

Demineralization. Demineralization by means of ion exchangers is sometimes used for removal of alkali, or sometimes with high mineral content for removal of sulfates and chlorides. For the effects of water impurities on carbonated beverages, see Table 1 (see also Ion exchange).

Colors. Visual appeal of a carbonated beverage is largely influenced by its color. Most colors used in carbonated beverages are natural or synthetic (see Colorants for foods, drugs, and cosmetics). The only natural colors present in carbonated beverages are those imparted by the use of natural fruits such as grapes, strawberries, or cherries. Even then, additional synthetic color must be added.

Caramel Color. Caramel color is deemed artificial by the FDA because it is artificially made from natural materials. Caramel color is supplied in varying tinctorial strengths and densities. A foaming caramel is frequently used in root beer, and a nonfoaming caramel in cola drinks. Caramel coloring used in carbonated beverages must be acid proof. A suggested test for acid proof quality follows: acidify a 5% solution of caramel color with 2.5% phosphoric acid or 1% hydrochloric acid and observe clarity after 24 h. If no sediment appears, the caramel is acid proof.

Caramel coloring used in carbonated beverages must have no undesirable taste or odor. Occasionally, constituents used in manufacturing caramel color react with ingredients in the carbonated beverage formula to cause an off-flavor. Therefore, it is necessary to test the caramel color in the formulation in which it will be used in order to determine its acceptability.

Synthetic Colors. The FDA permits eight synthetic colors in food use. Only five of the eight colors (see Table 2) are recommended for carbonated beverages.

Stock solutions of color may be made by dissolving the dry color in water and preserving it with sodium benzoate. Stability of colors in the presence of light, oxidizing agents, reducing agents, metals, chlorine, etc, is important in the manufacture of carbonated beverages (see Table 2).

Ascorbic acid (vitamin C) may be added to beverages as an antioxidant (see Vitamins). Chemically, it acts as a strong reducing agent which is preferentially oxidized,

Table 1. Water Impurities and Their Effect on Beverages

Impurity	Tolerance, max, ppm	Typical effect	Treatment for removal[a]
turbidity	5.0[b]	off-taste, discoloration	C, S, Cl
taste and odor	none	off-taste	C, S, Cl
algae and protozoa	none	off-taste, spoilage, sedimentation	C, S, Cl
bacteria and yeast	none	off-taste, spoilage, sedimentation	C, S, Cl
organic matter	none	off-taste, spoilage, sedimentation	C, S, Cl, O
iron or manganese	0.1	off-taste, discoloration	IX
alkalinity	50	neutralizes beverage acid	lime, hydrogen IX
total solids	500	chlorides cause salty taste, sulfates cause brackish taste	IX

[a] In addition to chemical treatment of any type, all plants should have sand filtration, activated carbon, and composition disk filter units (C = coagulation; S = sedimentation; Cl = chlorination; O = ozone; IX = ion exchange).
[b] Compared with an arbitrary standard based on diatomite.

Table 2. Stability of FD&C Food Color

Color	Light[a]	Acids	Alkalies	Fastness to Oxidizing agents	Fastness to Reducing agents[b]	Ascorbic acid[c]	Reaction to metals	Reaction to tannins
FD&C Red No. 40 (Allura Red)	good	good	blue in hue	good	good	good		
FD&C Yellow No. 5 (Tartrazine)	good	good	fair, darkens	good	poor	poor, appreciable fade after one week	turns dull brown	unaffected
FD&C Yellow No. 6 (Sunset Yellow FCF)	good	good	fair, turns dull red	good	poor	very good, slightly darker	none	unaffected
FD&C Blue No. 1 (Brilliant Blue FCF)	good	good	good	good	fair	good	none	darkens
FD&C Green No. 3 (Fast Green FCF)	good	good	poor, turns violet	fair	poor	good	none	unaffected

[a] Sunlight or ultraviolet.
[b] Any acid and metal reaction producing hydrogen.
[c] Based on 166 ppm of ascorbic acid in water only (166 mg/L or ca 5 mg/oz of solution).

thereby protecting the beverage flavors. The nutritional value of beverages is increased if sufficient ascorbic acid is added. The adult recommended daily allowance (RDA) is 30 mg. The addition of 55 mg of ascorbic acid to a 240 mL (8 oz) bottle of carbonated beverage would supply 100% of the RDA after a normal operating loss of 25 mg due to oxidation. Colors are easily reduced by ascorbic acid in the acidified sugar solution of a carbonated beverage. This fading is retarded by the use of invert sugar or artifical sweeteners.

Preservatives. Many carbonated beverages are satisfactorily preserved by the acid used in the beverage and the carbon dioxide content. Fermentation is inhibited by 3.24 g (fifty grains) of dry citric acid per 3.8 L (1 gal) of 10% sugar solution. Carbonation of a beverage helps prevent mold growth. Disease-producing bacteria will not survive more than a few hours in a typical acidified carbonated beverage.

Carbonated beverages containing fruit juices or little carbonation are preserved with 0.05% sodium benzoate. One tenth of 1% sodium benzoate is sometimes used, although with proper manufacturing procedures this additional amount is not necessary. The sodium benzoate is added during the preparation of the syrup.

Carbon Dioxide. Carbon dioxide (qv) contributes the characteristic pungent taste or bite associated with carbonated beverages. It inhibits the growth of mold and bacteria, and may destroy bacteria, depending on the extent of carbonation used.

Carbonation is measured in terms of volumes. At atmospheric pressure and 15°C a given volume of water absorbs an equal volume of carbon dioxide and is said to contain one volume of carbonation. Solubility is directly proportional to pressure, but decreases as temperature increases.

Carbon dioxide for beverages must be odorless and as pure as possible. Commercial carbon dioxide is usually suitable for carbonated beverages, ie, it is free of impurities which affect taste, odor, and carbonation, such as hydrogen sulfide, mercaptans, and oil. A simple but effective test for the purity of carbon dioxide is to pass 0.45 kg of carbon dioxide through 25 mL of ice-cold water over a one hour period, then taste and smell the water sample.

The carbonator, saturator, or carbo-cooler is the machine used to effect solution of carbon dioxide gas. Gas is furnished to the carbonator under pressure from dry ice converters, drums of liquid carbon dioxide, or a bulk system. Water at 4°C is pumped with the gas into the top of the carbonator in which it flows over baffles under pressure, quickly becomes saturated with the gas, and is piped continuously from the bottom to the filler.

Table 3 lists the ingredients usually contained in carbonated beverages.

Manufacture

Syrup. The complete mixture of all ingredients required to make up carbonated beverages, with the exception of carbonated water, is referred to as the syrup. A solution of sugar in water is referred to as simple syrup; when acid has been added, it is referred to as acidified simple syrup. In a bottling or canning plant, a room equipped with mixing and storage tanks is set aside for syrup mixing. The equipment should be well designed, properly located to facilitate handling of ingredients and cleaning, and constructed of a material such as stainless steel or Monel which will withstand vigorous acidifying and sterilizing agents and the acidic beverage and syrup. In addition to stainless steel, syrup lines may be of Tygon B-44-3 or a sanitary, acid-resistant rubber hose. Metal contamination drastically affects taste.

The following general procedure is recommended for mixing syrup: (1) make

Table 3. Ingredients in Various Carbonated Beverages

Beverage	Flavors	Color	Sugar, %	Acid	Acid, g/L (oz/gal) syrup	CO_2, vol of gas
cola	extract of kola nut, lime oil, lemon oil, orange oil, spice oils, caffeine	caramel	11–12	phosphoric	4.5 (0.6)	3.5
ginger ale	ginger root, oil of ginger, citrus oils	caramel	7–11	citric	7.5 (1)	4–4.5
root beer	oil of wintergreen, vanilla, nutmeg, cloves, anise, sweet birch	caramel	11–13	citric	1.9 (1/4)	3
orange	oil of orange, orange juice	Yellow No. 5	11–13	citric	7.5 (1)	1.5–2.5
		Yellow No. 6	13.5			
grape	methyl anthranilate, ethyl acetate, ethyl butyrate, oil of cognac	Red No. 40	11–13	tartaric or citric	5.6 (3/4)	1–2.5
		Blue No. 1		or citric and malic		
lemon–lime	oil of lemon, oil of lime	none	10–11	citric	7.5 (1)	3.5

simple syrup by dissolving granulated sugar or combining liquid sugar with most of the treated water called for in the formula; (2) filter to remove any sediment contained in the sugar; (3) dissolve preservatives and color in a small amount of hot water, if not already in solution, and mix with simple syrup; (4) add acid, dissolve if solid; (5) add flavor or the flavor, color, and preservative combination provided by the supplier; and (6) add any remaining water required, taking into consideration water displaced by the sugar, water used in rinsing, etc.

Beverage. The final step in the production of carbonated beverages comprises mixing the syrup with water and carbon dioxide in the proper ratio, and packaging in a bottle or can.

Beverages are produced either by the presyrup process or by the premix process. The procedures for canning and bottling are slightly different for each. Figure 1 shows the steps in the canning operation. The operation of the presyrup system and the premix syrup system is the same on a bottling line.

In the presyrup bottling process, a measured amount of beverage syrup is placed in the freshly washed bottle as it passes through the syruper. The bottles then move to the filler where the carbonated water is added. In order to achieve a uniform fill and maintain the proper carbon dioxide content in the beverage, the filler is constructed so that the carbonated water flows slowly and smoothly into each bottle. A constant counterpressure during the filling cycle prevents escape of carbon dioxide. The bottle then moves to the crowner where the crowns are automatically pressed on the bottles, hermetically sealing them, following which the contents of the bottles are mixed by a whirling and end-over-end motion. The finished beverage then passes in front of a light for inspection whereby an electric eye can eject bottles containing foreign matter. The bottles then move on to the case packer and the warehouse.

In the premix process of filling, the proper volumes of syrup and water are automatically measured by a continuous metering system, mixed, cooled, and carbonated simultaneously by being forced through a special type of carbonator, then piped directly to a filler machine from which the bottles move to a crowning section to be capped. The mixing step is omitted and the remaining steps are identical to those in the presyrup process.

Packaging. *Bottles and Bottle Washing.* Seventy-nine percent of carbonated beverages sold are packaged in bottles which range in size from 180 to 1900 mL (6–64 oz). Approximately 51% of packaged beverages sold in 1975 were in a returnable package. This may increase in the future as additional states require returnable containers.

Bulk carbonated beverage sales rose 2% over the 1974 level, constituting 21% of the market. Post-mix beverages represented 73% of bulk sales, premix 27%. Post-mix refers to carbonated soft drinks made at the point of consumption from a blend of syrup, water carbonated at the consumer outlet, and crushed ice. Premix refers to carbonated soft drinks produced in the bottling plant with treated water, maintaining a reasonable degree of carbonation uniformity and controlling syrup–water ratios more accurately than can be done with post-mix. Premix beverages are shipped to point of use in 38 L (10 gal) or smaller stainless steel tanks. These tanks require special refrigeration devices and dispensing valves.

Reusable bottles make bottle washing an important part of production. A carbonated beverage bottle must be of good mechanical strength, attractive in appearance, and sterile for each use.

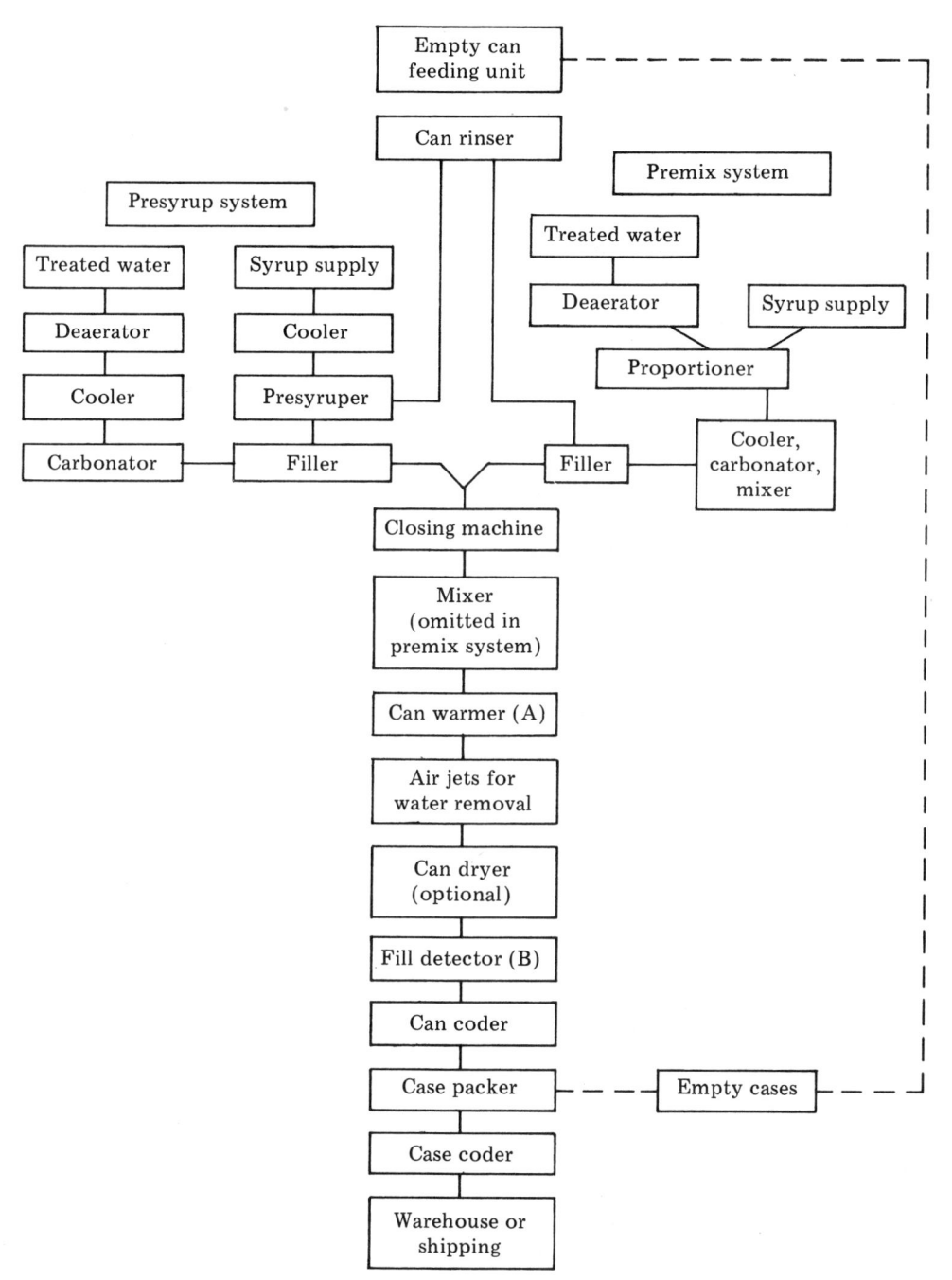

Figure 1. The carbonated beverage canning line. (A) The cans are warmed before drying to prevent condensation. (B) Three types of fill detectors are available—a weighing device, an x-ray detector, and a β-ray detector.

Bottles are cleansed and sterilized with a warm alkaline solution, then rinsed with potable water. The alkaline solution can be composed of caustic soda (sodium hydroxide), sodium carbonate, trisodium phosphate, or sodium metasilicate. Caustic soda is the principal ingredient because it has by far the best germicidal properties. The time and temperature required for bottle sterilization depend almost entirely upon the caustic content. Addition of other alkalies to caustic soda solutions increases somewhat the germicidal strength of the caustic solution; for instance, a 3% alkaline solution containing 1.8% caustic soda has the germicidal effectiveness of a 2.16% caustic soda solution. Milder alkalies are used, however, to improve cleansing properties rather than to increase germicidal efficiency. Sodium carbonate is used to improve the detergent action of the solution. Trisodium phosphate adds good emusification properties and acts as a water softener. Sodium metasilicate prevents the damaging effects of highly alkaline solutions, has good detergent properties, and has special colloidal and dispersive properties attributable to the silicate anion (see Dispersants). Alkali combinations are usually determined by individual plant conditions. The water hardness, labels, label glue, machine, rate of operation, age of bottles, and soil to be removed, determine how the bottles must be washed.

Mechanical strength of the bottles is critically important, especially with the increased use of larger bottles.

The National Soft Drink Association published a *Bottle Specifications Guideline* in 1970 with the expectation that glass container manufacturers would use them in the manufacture and quality control of glass bottles.

The glass industry has a formal system for determining the overall quality of a supply of bottles. Under the system, a supply of bottles is considered to be of satisfactory quality if the percentage of bottles having a particular defect does not exceed certain limits. These defect limits, listed below, are in accordance with the NSDA *Bottle Specifications Guideline* (see Glass).

Bottle defect	*Defective bottles not to exceed, %*
visual defects, class I	0.065
visual defects, class II	0.65
visual defects, class III	1
visual defects, class IV	4
bottle finish	1
bottle height	1
bottle diameter	1
bottle ellipticity	1
bottle weight	0.65
bottle capacity	1
bottle wall thickness	0.65

Cans. During 1975, 640 million cases, containing 24 cans of 355 mL (12 oz) size, of beverage were sold in the United States.

For many years, carbonated beverage cans were made of steel plate of rigidly controlled chemical composition, base weight and temper, that had been coated with a very thin layer of tin. Today, aluminum and tin-face steel as well as tinplate are important to the manufacture of carbonated beverage cans.

The design of the cans has also been changed. There are 7 designs available: three-piece 211 × 413 steel can with soldered side seams; three-piece 211 × 413 tin-free steel can with welded side seam; 209 × 413 three-piece tin-free steel can with welded side seam; 211 × 413 three-piece tin-free steel can with cemented side seam; 209 × 413 three-piece tin-free steel can with cemented side seam; two-piece aluminum can with a 209 diameter end; two-piece drawn and ironed steel can with a 209 diameter. Cans with the 209 diameter end are referred to as necked-in cans. Twenty-one percent of the cans used for carbonated beverages in 1975 were two-piece cans.

Cans and ends are lined with two or more coats of can-lining materials to guard taste, appearance, and wholesomeness of the contents. Conventional solvent-based coatings long used in the container industry are rapidly being replaced by waterborne coatings that comply with new air quality regulations (see Coating materials). The most versatile resins for use with water are the acrylics and polyesters, which have not previously been used on beverage can interiors. However, test results show that cans manufactured with waterborne coatings have good quality, impart no off-flavors to the beverage, and show no significant rise in average iron pick-up.

A can for packaging soft drinks should be sanitary, easily filled and sealed, and corrosion resistant. It must withstand the maximum internal pressure developed by the product during storage and must not impart off-flavors to the product.

Special beverage formulations are not usually required for canned carbonated beverages. However, beverages colored with azo dyes (qv) seem to be more corrosive than beverages which do not contain these dyes. Two of the five FD&C colors used in carbonated beverages, FD&C Yellow No. 6 and FD&C Red. No. 40, are azo dyes.

The shelf life of canned beverages depends on product corrosiveness, air content, metal exposure, and storage temperature. It is estimated that metal containers for carbonated beverages remain acceptable approximately twice as long at 21°C as at 37°C. Excessive air content can result in drastic reductions in shelf life for all types of products. The air in a canned beverage should be less than 2 mL. Incomplete coverage by the protective coating can cause metal exposure at the end radii and body side seams of the can. In addition, microscopic breaks which afford opportunity for corrosion sometimes occur in the coating.

The metal container has both advantages and disadvantages over glass with regard to shelf life. Cans afford complete protection of the product from light and properly canned beverages have only a few milliliters of air per can; bottled beverages contain up to 15 mL. Cans provide a definite advantage for beverages that are sensitive to oxidative deterioration. Metal cans, however, are subject to slight attack by the product, which dissolves trace quantities of metal. This is harmless to the consumer, but does affect the flavor. The level at which dissolved metal becomes detectable depends on the particular product and the sensitivity of the taster.

Stannous chloride, an oxygen scavenger, is added to certain formulations to reduce corrosion. It can be used at levels up to 11 ppm. Stannous chloride is a strong reducing agent and will bleach yellow and red dyes.

Quality Control. Certain quality control tests must be made frequently during production of carbonated beverages. The water should be checked for foreign odors, chlorine, and alkalinity. These frequent tests confirm the efficiency of the water-treatment system.

Density tests are made on the beverage syrup, reported in g/mL or Baumé, and on the finished beverage, reported in Brix (for pure sugar solutions the Brix reading is identical with the percentage of sugar present). Before making the hydrometer test on the finished beverage, it is necessary to remove the carbonation by pouring several times from one vessel to another.

722 CARBONATED BEVERAGES

Carbonation of the finished beverage is also tested by determining its gas pressure and temperature which can be related by charts to the number of volumes of gas present. Gas pressure is determined by inserting a pressure gage through the metal crown. The bottle is shaken to ensure that the liquid and gaseous phases are in equilibrium; otherwise, the pressure reading may be too low.

Colorimetric tests and titrations are made to confirm that the correct amount of color and acid are present in the finished product. Taste tests are made to confirm the flavor. A suggested schedule of control tests for production plants is given in Table 4.

Microbiological tests can be run on the product. Samples should be candled 24 or 48 h following production for detection of nonviable mold, crown dust, glass particles, or other debris.

Bottles and cans should also be tested. Can quality control tests should check the can end, lining, double seam, enamel rating, side seam, weight, visual defects, and diameter. Bottle quality control tests should check annealing, capacity, weight, height, finish, pressure resistance, thermal shock, visual defects, perpendicularity, lubricity, and out-of-roundness.

Spoilage refers to any undesirable change in appearance, color, taste, or odor that takes place in the beverage. Spoilage may be caused by chemical changes in the beverage or the growth of microorganisms. Certain physical effects are brought about as a result of chemical reactions caused by heat or sunlight. Direct sunlight is detrimental to the flavor of almost all beverages, particularly those containing citrus fruit juices. Light causes a flavor change, which in citrus-flavored beverages may be described as oily or terebinthinate. In most other beverages, there is merely a general loss of flavor and character. Some food colors, particularly tartrazine, fade quickly in sunlight. Heat

Table 4. Production Plant Control Tests

Test	Frequency
Baumé of syrup	each batch
syrup pH and titration	each batch
liquid sugar Brix	each load when delivered
Brix of beverage	every 30 min
carbonation of beverage	every 30 min
fill point of package	every 30 min
pH and titration of beverage	
premix lines	each changeover
syrup throw lines	hourly
crown crimp	daily
removal torque, convenience crowns	each run
color	hourly
alkalinity of water	every 2 h
chlorine after purifier	every 2 h
% slurry	daily
% caustic in bottle washer	every 4 h
caustic temperature in washer	every 4 h
caustic carryover[a]	every 4 h
visual inspection	continuous
electronic empty bottle inspector	every 30 min

[a] To confirm that the bottles are thoroughly rinsed, a drop of phenolthalein is added to a bottle ready for filling.

conditions generally encountered are not high enough to produce specific effects, such as a cooked taste. However, heat speeds up chemical reactions, and will therefore hasten loss of freshness.

Chemical changes which take place in beverages are air oxidation, enzyme action, reaction of free chlorine, saponification of esters, hydrolysis, breakdown of the complex molecules in caramel by reactions of acids and minerals, and other chemical reactions of flavoring ingredients.

Free chlorine from the water or incompletely rinsed equipment causes many radical and very undesirable flavor changes and bleaches the color of the beverage. Alkaline waters used in the preparation of a beverage produce a highly insipid beverage with a decided off-odor. This off-odor evidently stems from a reaction with the flavoring ingredients. Odor of beverages gradually changes on standing. This is due in part to the reaction of the acid with the flavoring oils. Caramel has been known to become insoluble after standing a few days. This is attributed to the breakdown of the complex molecules under the action of the acids, and in some cases, mineral salts.

Growth of microorganisms in the finished beverage may cause off-taste, scums, clouds, sediments, etc. Microorganisms important to carbonated beverages are protozoa, algae, molds, bacteria, and yeast. Protozoa, unicellular animals, are always waterborne and enter the beverage from the water supply. Algae are unicellular chlorophyll-bearing plants. Light is required for their continued growth. Sufficient numbers of these plants are frequently introduced through the water supply and cause off-tastes and sediment.

Molds are saprophytic fungi that require oxygen for their growth. A properly carbonated beverage does not normally have enough air to support them. Molds grow quickly on the surface of strong acid solution, on stored syrups, and in improperly cleaned tanks. From these sources, sufficient numbers may be introduced into beverages to produce a musty or moldy odor.

The typical acidified carbonated beverage with an acidity of 1.23 g/L (72 grains/gal) does not support the growth of bacteria as the acid and carbon dioxide present exert a pronounced germicidal effect.

Souring or fermentation (qv) in carbonated beverages is caused by the growth of yeast. Yeasts (qv) are microorganisms which feed on sugar. Yeast growth is considered responsible for 90% of carbonated beverage spoilage. Low-acid or low-carbonated beverages, or beverages with no carbonation are more susceptible to yeast growth, as acids seem to be the main inhibitor. Beverages made from water of high alkalinity sour easily because of neutralization of the acid.

The most insidious contamination, aside from metallic equipment contaminants, because it is not drastic in its action, is contamination with absorbed foreign odors. Off-odors in the atmosphere may be caused by failure to keep the plant clean, or disagreeable, outside industrial odors. Beverage plants must be kept free from offensive odors if their products are to be acceptable.

Oil from the machinery used to manufacture carbon dioxide gas sometimes enters the carbonated beverage. If dry ice converters are not kept drained of oil, the oil passes into the gas line, the carbonator and filler head, and eventually, the finished beverage.

724 CARBONATED BEVERAGES

Energy Conservation. Carbonated beverage plants have developed energy efficiency programs, including processing equipment, building operations, training of personnel and use of selected supplies and raw materials.

The National Soft Drink Association (1101 Sixteenth St., N.W., Washington, D.C. 20036) has published *A Guide to Energy Conservation for the Soft Drink Manufacturer*.

Regulations

Soft drinks are recognized by all responsible authorities as food products. As such, they are subject to those laws and regulations of federal and state governments which generally apply to the production, labeling, and sale of foods.

In 1966 the Commissioner of the FDA established a definition and standard of identity for carbonated soft drinks (soda water) which are either unsweetened or sweetened with a nutritive sweetener. This standard has been revised and published in Title 21 of the *Code of Federal Regulations,* Part 31, as amended through July 31, 1975 (see Regulatory agencies).

BIBLIOGRAPHY

"Carbonated Beverages" in *ECT* 1st ed., Vol. 3, pp. 112–124, by W. T. Miller and J. F. Hale, Nehi Corporation; "Carbonated Beverages" in *ECT* 2nd ed., Vol. 4, pp. 335–352, by Martha B. Jones, Royal Crown Cola Company.

General References

J. J. Riley, *A History of the American Soft Drink Industry,* American Bottlers of Carbonated Beverages, Washington, D.C., 1958, pp. 9, 15, 17.
NSDA/1975 Sales Survey of the Soft Drink Industry, National Soft Drink Association, Washington, D.C., 1976.
P. Becher, "Theoretical Aspects of Emulsification," *American Perfum. Cosmet.* **77**(7), 21 (1962).
D. Melillo, "Stabilization of Cloudy Beverages," *Proceedings of Society of Soft Drink Technologists,* (1976).
B. L. Oser and R. A. Ford, "10 GRAS Substances," *Food Technol.* **31**(1), 65 (1977).
Food Chemicals Codex, 2nd ed., the National Academy of Sciences, National Research Council, Washington, D.C., 1972.
R. L. Hall, "Recent Progress in the Consideration of Flavoring Ingredients Under the Food Additives Amendment—I," *Food Technol.* **14**, 488 (1960).
R. L. Hall, "Name 257 Flavors as GRAS," *Food Process.* **22**(3), 27 (1961).
R. L. Hall and B. L. Oser, "Recent Progress in the Consideration of Flavoring Ingredients Under the Food Additives Amendment—II," *Food Technol.* **15**(12), 20 (1961).
E. Forschbach, "Wanted: A Credo for World Food Laws," *Food Drug Cosmet. Law J.* **18**(2), 93 (1963).
A G. Kitchell, "Control of the Use of Food Additives in the United Kingdom," *Food Drug Cosmet. Law J.* **16**(3), 181 (1961).
The American Soft Drink Journal Blue Book, McFadden Business Publications, Atlanta, Ga., 1962.
M. B. Jacobs, *Manufacture and Analysis of Carbonated Beverages,* Chemical Publishing Co., New York, pp. 89–109, 221–228.
The Colorful World of Hilton Davis, Hilton-Davis Chemical Co. Division, Newark. N.J., Mar. 1976.
A. J. Granata, "Technical Aspects of Carbon Dioxide," *Proceedings of Society of Soft Drink Technologists,* (1969).
Carbon Dioxide for Nehi Products, Technical Services Department, Royal Crown Cola Co., Columbus, Ga.
National Soft Drink Association Bottle Specifications Guideline, the National Soft Drink Association, Washington, D.C., Nov. 1970.

Water for Nehi Products, Technical Services Department, Royal Crown Cola Co., Columbus, Ga., 1953.
L. Landauer and C. E. Scruggs, "New Can Coatings to Meet Air Quality Standards—Status Report," *Proceedings of Society of Soft Drink Technologists,* (1975).
"Technical Aspects of Cans and the Canning of Carbonated Beverages," American Can Company, New York, 1953.
The Canning of Carbonated Beverages, Continental Can Company, New York, 1963.
J. R. Braswell, *Quality Control of Carbonated Beverages,* Royal Crown Cola Co., Columbus, Ga.
Quality Control Procedures, Royal Crown Cola Co., Columbus, Ga., 1970.
J. F. Hale, in M. B. Jacobs, ed., *The Chemistry and Technology of Food and Food Products,* Vol. II, Chapt. XXIII, Interscience Publishers, New York, 1944.
The United States Dispensatory, 24th ed., J. B. Lippincott Company, Philadelphia, Pa., 1947, pp. 610–611.
Use of Ascorbic Acid in Beverages, technical bulletin, Charles Pfizer and Company, Inc., New York.
A Guide to Energy Conservation for the Soft Drink Manufacturer, the National Soft Drink Association, Washington, D.C., 1974.

Martha B. Jones
Royal Crown Cola Corporation

CARBON DIOXIDE

Carbon dioxide [*124-38-9*], CO_2, is a colorless gas with a faintly pungent odor and acid taste. Van Helmont (1577–1644) first recognized carbon dioxide as a distinct gas when he detected its presence as a by-product of both charcoal combustion and fermentation. Today carbon dioxide is a by-product of many commercial processes: synthetic ammonia production, hydrogen production, substitute natural gas production, fermentation, limestone calcination, certain chemical syntheses involving carbon monoxide, and reaction of sulfuric acid with dolomite. Generally present as one of a mixture of gases, carbon dioxide is separated, recovered, and prepared for commercial use as a solid (dry ice), liquid, or gas.

Carbon dioxide is also found in the products of combustion of all carbonaceous fuels, in naturally occurring gases, as a product of animal metabolism, and in small quantities, about 0.03 vol %, in the atmosphere. Its many applications include beverage carbonation, chemical manufacture, fire fighting, food preservation, foundry-mold preparation, greenhouses, mining operations, oil well secondary recovery, rubber tumbling, therapeutical work, and welding. Although it is present in the atmosphere and the metabolic processes of animals and plants, carbon dioxide cannot be recovered economically from these sources.

Physical Properties

Sublimation point, $-78.5°C$ at 101 kPa (1 atm); triple point $-56.5°C$ at 518 kPa (75.1 psia); critical temperature, $31.1°C$; critical pressure, 734 kPa (1071 psia); critical density, 467 g/L; latent heat of vaporization, 348 J/g (149.6 Btu/lb) at the triple point and 235 J/g (101.03 Btu/lb) at $0°C$; gas density, 1.976 g/L at $0°C$ and 101 kPa (1 atm);

liquid density, 914 g/L at 0°C and 0.759 vol/vol at 25°C and 101 kPa pressure of carbon dioxide; viscosity, 0.015 mPa.s (=cP) at 25°C and 101 kPa pressure; heat of formation of carbon dioxide, 393,700 J/mol (373.4 Btu/mol) at 25°C.

An excellent pressure–enthalpy diagram (a large Mollier diagram) over −13 to 500°C and 70–20,000 kPa (10–2900 psi) can be found in ref. 1. The thermodynamic properties of saturated carbon dioxide vapor and liquid from −95°C to the critical point, 31°C, are given in ref. 2. Also given are data for superheated carbon dioxide vapor from −45 to 650°C at pressures from 7 to 7000 kPa (1–1000 psi). A graphical presentation of heat of formation, free energy of formation, heat of vaporization, surface tension, vapor pressure, liquid and vapor heat capacities, densities, viscosities, and thermal conductivities is provided in ref. 3.

Compressibility factors of carbon dioxide from 4 to 200°C and 1400–69,000 kPa (203–10,000 psi) are given in ref. 4.

Chen (5) has reviewed available data on the thermodynamic and transport properties of carbon dioxide and has compiled a table giving specific volume, enthalpy, and entropy values for carbon dioxide at temperatures from −18 to 815°C and at pressures from atmospheric to 27,600 kPa (4000 psia). Diagrams of compressibility factor, specific heat at constant pressure, specific heat at constant volume, specific heat ratio, velocity of sound in carbon dioxide, viscosity, and thermal conductivity have also been prepared (5). The following equation was used (5) to compute viscosity values,

$$10^4 m = \frac{1.554 T^{3/2}}{T + 246/10^{3/T}}$$

where m equals absolute viscosity at zero pressure in pascal seconds (Pa·s) and T equals absolute temperature (Pa·s × 10 = poise). Equations for viscosity at different pressures and thermal conductivity were also provided (5). The vapor pressure function for carbon dioxide in terms of reduced temperature and pressure is as follows:

$$\log P_R = 4.2397 - \frac{4.4229}{T_R} - 5.3795 \log T_R + 0.1832 \frac{P_R}{T_R^2}$$

where P_R equals reduced pressure, which equals P/P_c (P_c, critical pressure, 7.38 MPa or 72.85 atm), and T_R equals reduced temperature which equals T/T_c (T_c, critical temp, 304.2 K) (6). This equation gives accurate vapor pressure values from the triple point to the critical point. A table of reduced density values for carbon dioxide covering the range of reduced pressures from 0.3 to 500 kPa (0.044 to 72.5 psi) and reduced temperatures from 0.712 to 20 K is also supplied (6).

Enthalpy values for carbon dioxide in the critical region, and with temperatures from 150 to 650°C and pressures from 0 to 20 MPa (0 to 200 atm) have been computed (7–8).

Diagrams of isobaric heat capacity (C_p) and thermal conductivity for carbon dioxide covering pressures from 0 to 13,800 kPa (0 to 2000 psi) and 38 to 815°C have been prepared. Viscosities at pressures of 100 to 10,000 kPa (1 to 100 atm) and temperatures from 38 to 815°C have been plotted (9).

Vapor pressure data for solid carbon dioxide are given in Table 1 (10). The sublimation temperature of solid carbon dioxide, −78.5°C at 101 kPa (1 atm), was selected as one of the secondary fixed points for the International Temperature Scale of 1948.

Table 1. Vapor Pressure of Solid Carbon Dioxide[a]

Temperature, °C	Pressure, Pa	Temperature, °C	Pressure, Pa
−188	1.333×10^{-4}	−148	1.333×10^{1}
−182	1.333×10^{-3}	−136	1.333×10^{2}
−175	1.333×10^{-2}	−120	1.333×10^{3}
−167	1.333×10^{-1}	−100	1.333×10^{4}
−159	1.333	−65	1.333×10^{5}

[a] To convert Pa to mm Hg, divide by 133.3.

The solubility of carbon dioxide in water is given in Figure 1 (11). Over the temperature range 0–120°C, the solubilities at pressures below 20 MPa (200 atm) decreased with increasing temperature. From 30 to 70 MPa (300–700 atm) a solubility minimum was observed between 70 and 80°C, with solubilities increasing as temperature increased to 120°C.

Information on the solubility of carbon dioxide in pure water and synthetic sea water over the range −5 to 25°C and 101–4500 kPa pressure (1–44 atm) can be found in refs. 12 and 13.

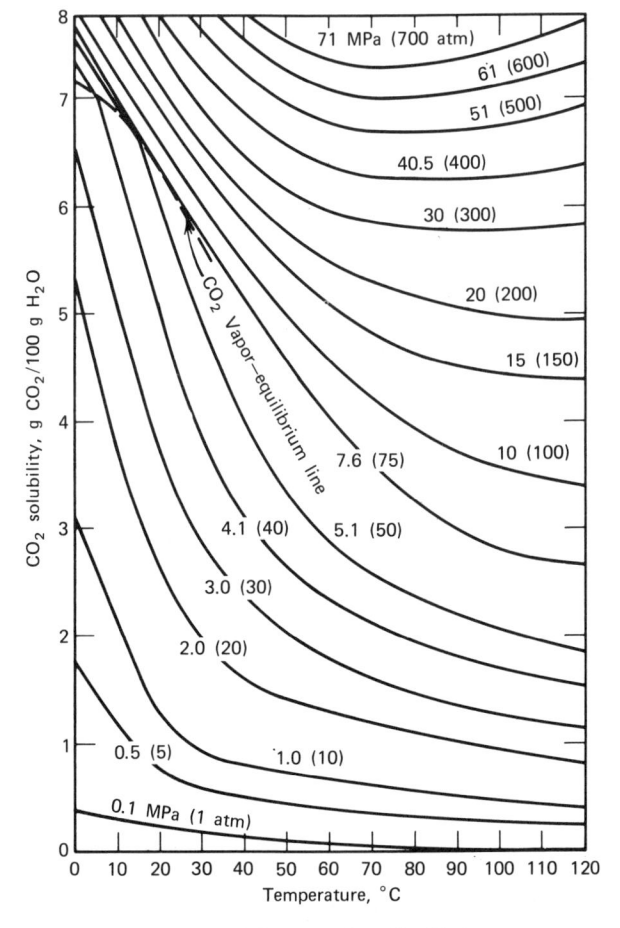

Figure 1. Solubility of carbon dioxide in water.

Price (14) has compiled the following tables of properties of carbon dioxide: enthalpy, entropy, and heat capacity at 0 and 5 MPa (0 and 50 atm, resp) from 0 to 1000°C; pressure–volume product (PV), enthalpy, entropy, and isobaric heat capacity (C_p) from 100 to 1000°C at pressures from 5 to 140 MPa (50–1400 atm).

A more recent compilation includes tables giving temperature and PV as a function of entropies from 0.573 to 0.973 (zero entropy at 0°C, 101 kPa (1 atm) and pressures from 5 to 140 MPa (50 to 1400 atm) (15). Joule–Thomson coefficients, heat capacity differences ($C_p - C_v$), and isochoric heat capacities (C_v) are given for temperatures from 100 to 1000°C at pressures from 5 to 140 MPa.

Chemical Properties

Carbon dioxide, the final oxidation product of carbon, is not very reactive at ordinary temperatures. However, in water solution it forms carbonic acid, H_2CO_3, which forms salts and esters through the typical reactions of a weak acid. The primary ionization constant is 3.5×10^{-7} at 18°C; the secondary is 4.4×10^{-11} at 25°C.

The pH of saturated carbon dioxide solutions varies from 3.7 at 101 kPa (1 atm) to 3.2 at 2370 kPa (23.4 atm).

A solid hydrate [27592-78-5], $CO_2 \cdot 8H_2O$, separates from aqueous solutions of carbon dioxide that are chilled at elevated pressures.

Although carbon dioxide is very stable at ordinary temperatures, when it is heated above about 1700°C, the reaction, proceeds to the right to an appreciable extent (15.8% at 2227°C).

$$2\,CO_2 \rightleftharpoons 2\,CO + O_2$$

This reaction also proceeds to the right to a limited extent in the presence of ultraviolet light and electrical discharges.

Carbon dioxide may be reduced by several means. The most common of these is the reaction with hydrogen.

$$CO_2 + H_2 \rightarrow CO + H_2O$$

This is the reverse of the water gas shift reaction in the production of hydrogen and ammonia (qv). Carbon dioxide may also be reduced catalytically with various hydrocarbons and with carbon itself at elevated temperatures. The latter reaction occurs in almost all cases of combustion of carbonaceous fuels and is generally employed as a method of producing carbon monoxide.

$$CO_2 + C \rightarrow 2\,CO$$

Carbon dioxide reacts with ammonia as the first stage of urea manufacture, to form ammonium carbamate as follows:

$$CO_2 + 2\,NH_3 \rightarrow NH_2COONH_4$$

The ammonium carbamate then loses a molecule of water to produce urea (qv), $CO(NH_2)_2$. Commercially, this is probably the most important reaction of carbon dioxide and it is used worldwide in the production of urea for synthetic fertilizers and plastics (see Carbamic acid).

Radioactive Carbon. In addition to the common stable carbon isotope of mass 12, traces of a radioactive carbon isotope of mass 14 with a half-life estimated at 5568 years are present in the atmosphere and in carbon compounds derived from atmospheric carbon dioxide (see Radioactive elements). Formation of radioactive ^{14}C is thought to be caused by cosmic irradiation of atmospheric nitrogen. The concentration of ^{14}C in atmospheric carbon dioxide is approximately constant throughout the world. The ratio of $^{14}CO_2$ [51-90-1] to $^{12}CO_2$ in the atmosphere, although constant for several hundred years, has decreased in the last sixty years by about 3% because of the influx of carbon dioxide in the atmosphere from the burning of fossil fuels (coal, petroleum, and natural gas). Procedures have been developed for estimating the age of objects containing carbon or carbon compounds by determining the amount of radioactive ^{14}C present in the material as compared with that present in carbon-containing substances of current botanical origin (16). Ages of materials up to 45,000 years have been estimated with the radiocarbon dating technique.

Carbon dioxide containing known amounts of ^{14}C has been used as a tracer in studying botanical and biological problems involving carbon and carbon compounds and in organic chemistry to determine the course of various chemical reactions and rearrangements (see Radiochemical technology). It has also been used in testing gaseous diffusion theory with mixtures of CO_2 and $^{14}CO_2$ at elevated pressures (17).

Environmental Chemistry. Carbon dioxide plays a vital role in the earth's environment. It is a constituent in the atmosphere and, as such, is a necessary ingredient in the life cycle of animals and plants.

In animal metabolism, oxygen from the atmosphere reacts with sugars in the body to produce energy according to the overall formula:

$$C_6H_{12}O_6 + 6\ O_2 \rightarrow 6\ CO_2 + 6\ H_2O + \text{energy}$$

The by-product CO_2 is released to the atmosphere. In metabolism, carbon dioxide from the air is taken into the leaves of the plant. Using energy from light, carbon dioxide reacts with water in the presence of enzymes to produce sugar. This reaction, photosynthesis, is the reverse of the above reaction.

The balance between the animal and plant life cycles as affected by the solubility of carbon dioxide in the earth's water results in the carbon dioxide content in the atmosphere of about 0.03 vol %. As stated above, however, carbon dioxide content of the atmosphere seems to be increasing as increased amounts of fossil fuels are burned. There is some evidence that the rate of release of carbon dioxide to the atmosphere may be greater than the earth's ability to assimilate it. Measurements from the U.S. Weather Bureau show an increase of 1.36% in the CO_2 content of the atmosphere in a five year period and predictions indicate that by the year 2000 the content may have increased by 25% (see Air pollution).

The effects of such an increase, if it occurs, are not known. It could result in a warmer temperature at the earth's surface by allowing the short heat waves from the sun to pass through the atmosphere while blocking larger waves that reflect back from the earth. If the earth's average temperature were to increase by several degrees portions of the polar ice caps could melt causing an increase in the level of the oceans, or air circulation patterns could change, altering rain patterns to make deserts of farmland or vice versa.

On the other hand, it has been demonstrated that the addition of CO_2 to greenhouses increases the growth rate of plants so that an increase in the partial pressure of CO_2 in the air could stimulate plant growth making possible shorter growing seasons and increased consumption of carbon dioxide from the air.

Commercial Production

Sources of carbon dioxide for commercial carbon dioxide recovery plants are (a) flue gases resulting from the combustion of carbonaceous fuels; (b) synthetic ammonia and hydrogen plants in which methane or other hydrocarbons are converted to carbon dioxide and hydrogen ($CH_4 + 2\ H_2O \rightarrow CO_2 + 4\ H_2$); (c) fermentation in which a sugar such as dextrose is converted to ethyl alcohol and carbon dioxide ($C_6H_{12}O_6 \rightarrow 2\ C_2H_5OH + 2\ CO_2$); (d) lime-kiln operation in which carbonates are thermally decomposed ($CaCO_3 \rightarrow CaO + CO_2$); (e) sodium phosphate manufacture ($3\ Na_2CO_3 + 2\ H_3PO_4 \rightarrow 2\ Na_3PO_4 + 3\ CO_2$); and (f) natural carbon dioxide gas wells.

Flue Gases. Figure 2 depicts a typical plant for producing gaseous carbon dioxide from coke, coal, fuel oil, or gas. The fuel is burned under a standard water-tube boiler for the production of 1400–1800 kPa (200–260 psi) steam. At 340°C flue gases containing 10–18% carbon dioxide leave the boiler and pass through two packed towers where they are cooled and cleaned by water. The gases are then passed through a booster blower into the base of the absorption tower. The particular recovery system shown in this figure is the Girbotol amine process (18) although an alkaline carbonate system may also be used. In the tower, carbon dioxide is absorbed selectively by a solution of ethanolamines passing countercurrent to the gas stream (see Alkanolamines). The carbon dioxide-free flue gases pass out of the top of the tower into the atmosphere, the carbon dioxide-bearing solution passes out from the bottom of the tower, through pumps and heat exchangers, into the top of a reactivation tower. Here heat strips the carbon dioxide from the amine solution and the reactivated solution returns through the heat-exchanger equipment to the absorption tower. Carbon dioxide and steam pass through the top of the reactivation tower into a gas cooler in which the steam condenses and returns to the tower as reflux. The carbon dioxide at this point is available as a gas at a pressure of about 200 kPa (2 atm). If liquid or solid carbon dioxide is desired it may be further purified for odor removal before compression.

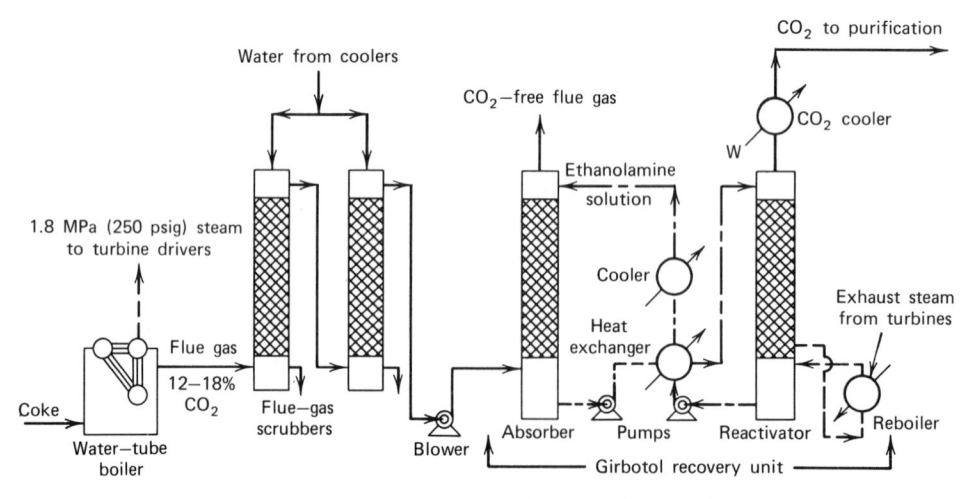

Figure 2. Production of carbon dioxide from coke.

The steam balance in the plant shown in Figure 2 enables all pumps and blowers to be turbine-driven by high-pressure steam from the boiler. The low-pressure exhaust steam is used in the reboiler of the recovery system and the condensate returns to the boiler. Although there is generally some excess power capacity in the high-pressure steam for driving other equipment, eg, compressors in the carbon dioxide liquefaction plant, all the steam produced by the boiler is condensed in the recovery system. This provides a well-balanced plant in which few external utilities are required and combustion conditions can be controlled to maintain efficient operation.

Ammonia and Hydrogen Plants. More carbon dioxide is generated and recovered from ammonia and hydrogen plants than from any other sources. Both plants produce hydrogen and carbon dioxide from the reaction between hydrocarbons and steam. In the case of hydrogen plants, the hydrogen is recovered as a pure gas. For ammonia plants the hydrogen is produced in the presence of air, controlled to give the volume ratio between hydrogen and nitrogen required to synthesize ammonia. In order to produce either product it is necessary to remove the carbon dioxide (19). The annual synthetic ammonia production capacity in the United States was about 15,000,000 metric tons in 1976 (see Ammonia). For each ton of ammonia produced, more than a ton of carbon dioxide is generated. Hence the available carbon dioxide supply from this source is several times as large as the total commercial production of carbon dioxide. A substantial amount of the carbon dioxide recovered from ammonia plants is used for urea production.

Fermentation Industry. Large quantities of carbon dioxide are present in gases given off in the fermentation of organic substances such as molasses, corn, wheat, and potatoes in the production of beer, distilled beverages, and industrial alcohol. These gases may contain impurities such as aldehydes, acids, higher alcohols, glycerol, furfural, glycols, and hydrogen sulfide. Two processes are in general use for removing these contaminants and preparing the carbon dioxide for use. In one process, this is accomplished by the use of activated-carbon adsorbers; the other process is a chemical purification process (see Beer; Beverage spirits, distilled; Ethanol; Fermentation).

The Backus process (20–21) uses active carbon. Carbon dioxide gases from the top of the fermenters are collected in a low-pressure gas holder to even out the flow. Roots-Connersville-type blowers force the gases through Feld scrubbers where they are washed with water to remove the bulk of entrained material, alcohols, aldehydes, etc. The washed gases pass through active-carbon purifiers which adsorb the balance of impurities and then to the compressors. The adsorption process gives off heat, which is removed by water coils imbedded in the carbon. Periodically, the carbon beds must be reactivated to remove accumulated impurities. This is accomplished by passing live steam through the carbon bed and water coils. After steaming, the carbon beds are dried by passing air through them; they are then ready for reuse. In general, two sets of active-carbon purifiers are used, one on stream and the other being reactivated (see Adsorptive separation).

The Reich process (22–23) uses chemical processes to remove impurities. Carbon dioxide from the fermenters is bubbled through a wash box, or catchall, which removes entrained liquids and mash. The gases pass through three packed scrubbing towers. In the first, dilute alcohol solution is passed countercurrent to the gases for de-alcoholizing the gases. The other two towers use water for further removal of alcohol. The alcohol is returned to the alcohol plant for distillation or use in the fermenters. The washed gas passes through a gas holder and blower, which boosts it through the balance of the purification plant. The first stage in this section is a potassium dichromate washer for the oxidation of organic impurities and the removal of hydrogen sulfide. Next, the gas is passed countercurrent to concentrated sulfuric acid for dehydration and dichromate removal. Entrained sulfuric acid is removed by a dry soda ash tower and the residual oxidized material is removed by countercurrent scrubbing with a light oil. The gas is then ready for compression. In some installations the first stage of compression is inserted before the dichromate washer and the purification operations are carried out at 600–800 kPa (87–116 psi).

Lime-Kiln Operation. Gases containing up to 40% carbon dioxide from the lime kiln pass through a cyclone separator which removes the bulk of entrained dust. The gas is then blown through the two scrubbers, which remove the finer dust, is cooled and passes into a absorption tower. Here carbon dioxide may be recovered by the sodium carbonate or Girbotol process.

Sodium Phosphate Manufacturing. Some pure carbon dioxide gas is available as a by-product in plants manufacturing sodium phosphate from sodium carbonate and phosphoric acid. Two carbon dioxide plants were installed prior to 1962 to utilize this by-product gas.

Natural Gas Wells. Natural gas containing high percentages of carbon dioxide has been found in a number of locations: New Mexico, Colorado, Utah, and Washington. Several small plants have been in operation for a number of years producing commercial solid and liquid carbon dioxide from these sources (see Gas, natural).

There are a number of methods of recovering carbon dioxide from industrial or natural gases. The potassium carbonate and ethanolamine processes are most common. In all these processes the carbon dioxide-bearing gases are passed countercurrent to a solution that removes the carbon dioxide by absorption and retains it until it is desorbed by heat in separate equipment.

All of these processes are in commercial use and the most suitable choice for a given application depends on individual conditions. Water could be used as the absorbing medium, but this is uncommon because of the relatively low solubility of carbon dioxide in water at normally encountered pressures. The higher solubility in the alkali carbonate and ethanolamine solutions is due to a chemical combination of the carbon dioxide with the absorbing medium.

Sodium Carbonate Process. This process of recovering pure carbon dioxide from gases containing other diluents, such as nitrogen and carbon monoxide, is based on the reversibility of the following reaction:

$$Na_2CO_3 + H_2O + CO_2 \rightleftharpoons 2\ NaHCO_3$$

This reaction proceeds to the right at low temperatures and takes place in the absorber where the carbon dioxide-bearing gases are passed countercurrent to the carbonate solution. The amount of carbon dioxide absorbed in the solution varies with temperature, pressure, partial pressure of carbon dioxide in the gas, and solution strength. Operating data on this reaction have been obtained by numerous investigators (24).

The reaction proceeds to the left when heat is applied. The reaction takes place in a lye boiler. A heat exchanger preheats the strong lye approaching the boiler and cools the weak lye returning to the absorber. A lye cooler further cools weak lye to permit the reaction to proceed further to the right in the absorber. The carbon dioxide gas and water vapor released from the solution in the boiler pass through a steam condenser where the water condenses and returns to the system. The cool carbon dioxide proceeds to the gas holder and compressors. The absorber is generally a carbon-steel tower filled with coke, Raschig Rings, or steel turnings. The weak solution is distributed evenly over the top of the bed and contacts the gas on the way down. Some plants operate with the tower full of sodium carbonate solution and allow the gas to bubble up through the liquid. Although this may afford a better gas-to-liquid contact, an appreciable amount of power is required to force the gas through the tower.

The lye boiler is usually steam heated but may be direct-fired. Separation efficiency may be increased by adding a tower section with bubble-cap trays.

To permit the bicarbonate content of the solution to build up, many plants are designed to recirculate the lye over the absorber tower with only 20–25% of the solution flowing over this tower passing through the lye boiler. Several absorbers may also be used in series to increase absorption efficiencies.

The sodium carbonate process is used in a number of dry ice plants in the United States, although its operating efficiency is generally not as high as that of processes using other solutions. These plants obtain the carbon dioxide from flue gases as well as lime-kiln gases.

Potassium Carbonate Process. The potassium carbonate process is similar to the sodium carbonate process. However, as potassium bicarbonate is more soluble than the corresponding sodium salt, this process permits a more efficient absorption than the other. The equipment layout is the same and the operation technique is similar.

There are several variations of the potassium carbonate process. The hot potassium carbonate process does not involve cooling of the solution flowing from the boiler to the absorber (25). Absorption takes place at essentially the same temperature as solution regeneration. In its simplest form this process uses an absorption column, a regeneration column, a heated boiler or reboiler, and a circulating pump. This arrangement minimizes energy requirements and capital costs but the higher absorbing temperature does not allow the carbon dioxide removal to be as complete as with lower temperatures. One modification which improves removal, is the use of a split stream flow to the absorber. Part of the solution is cooled and used at the top of the absorber. The balance of the solution, uncooled, is added part way down the absorber. The combined solution from the absorber then flows to the regenerator. In another modification two-stage absorption and regeneration is used. The carbonate solution from the absorber flows to the regeneration column. Part of the solution is withdrawn from the column at some intermediate point and pumped, uncooled, to an intermediate point in the absorber. The remaining solution undergoes a more complete regeneration and is cooled and pumped to the top of the absorber.

Three commercial processes that use these various hot carbonate flow arrangements are the promoted Benfield process, the Catacarb process and the Giammarco–Vetrocoke process (26–29). Each uses an additive described as a promoter, activator or catalyst, which increases the rates of absorption and desorption, improves removal efficiency, and reduces the energy requirement. The processes also use corrosion inhibitors which allow use of carbon-steel equipment. The Benfield and Catacarb processes do not specify additives. Vetrocoke uses boric acid, glycine, or arsenic trioxide, which is the most effective.

These processes have been used in many plants to remove carbon dioxide from ammonia synthesis gas or natural gas. They are most effective if the gas stream being treated is at elevated pressures (1700 kPa (17 atm) or higher). This increases the carbon dioxide partial pressure so that the hot potassium carbonate solution absorbs a substantial amount of carbon dioxide. The stripping tower or regenerator operates at or near atmospheric pressure.

Ethanolamines may also be added to carbonate solutions to improve their performance (30). Vapor pressure-equilibrium data for the K_2CO_3–$KHCO_3$–CO_2–H_2O system are given in reference 31.

Girbotol Amine Process. This process developed by the Girdler Corporation is similar in operation to the alkali carbonate processes. However, it uses aqueous solutions of an ethanolamine (either mono-, di-, or triethanolamine). The operation of the Girbotol process depends on the reversible nature of the following reaction:

$$\underset{\text{monoethanolamine}}{2\ HOC_2H_4NH_2} + H_2O + CO_2 \rightleftharpoons \underset{\text{monoethanolamine carbonate}}{(HOC_2H_4NH_3)_2CO_3}$$

The reaction proceeds in general to the right at low temperatures (27–65°C) and absorbs the carbon dioxide from the gas in the absorber as shown in Figure 2. The amine solution, rich in carbon dioxide, passes from the bottom of the tower through a heat exchanger, where it is preheated by hot, lean solution returning from the reactivator. On passing into the reactivator the solution passes countercurrent to a stream of carbon dioxide and steam, which strips the carbon dioxide out of the solution. By the time the solution reaches the bottom of the tower, where heat is supplied by a steam-heated or direct-fired reboiler, it has been reactivated. This hot solution (100–150°C) passes out of the tower, through the heat exchanger and cooler, and returns to the absorber tower. In the case of flue gases containing oxygen, a small side stream of solution is passed through a redistillation unit (not shown) where the oxidation products are removed and the distilled amine is returned to the process.

The Girbotol amine process is widely used in the United States in the production of carbon dioxide from flue gases. The steam balance described (for Figure 2) is satisfactory as, in general, no steam in excess of that produced in the boiler furnishing the flue gases is required to reactivate the ethanolamine solution.

The amine process or one of the various carbonate processes is used in the majority of CO_2-removal applications. At low pressures the amine process has clear advantages in removal efficiency and installed cost. At higher pressures the increased carbon dioxide partial pressures favor a higher CO_2 content in the absorbing solution (whether carbonate or amine) which permits a lower steam usage per unit of carbon dioxide stripped from the absorbent. Until recently, amine systems have not been able to take full advantage of the benefits of higher absorber pressure because of corrosion problems. The CO_2-rich amine solution is inherently more corrosive than potassium carbonate solutions so that the stronger the amine solution and the greater the carbon dioxide content the worse the corrosion potential. To hold corrosion to a minimum, amine solution designs have been limited to a maximum solution strength of 20 wt% with a maximum CO_2 content of about 0.0374 m^3/L (5 ft^3/gal) of solution. Even then, some plants experience severe corrosion problems requiring replacement of some carbon steel equipment with stainless steel. Carbonate solutions pick up more CO_2 per unit of solution than the amines, resulting in a lower circulation rate and a correspondingly lower heat regeneration requirement.

Overall comparison between amine and carbonate at elevated pressures shows that the amine usually removes carbon dioxide to a lower level at a lower capital cost but requires more maintenance and heat. The impact of the higher heat requirement depends on the individual situation. In many applications, heat used for regeneration is from low temperature process gas suitable only for boiler feed water heating or low pressure steam generation and it may not be useful in the overall plant heat balance.

In recent years Union Carbide has developed Amine Guard which essentially eliminates corrosion in amine systems (32–35). It permits the use of substantially higher amine concentrations and greater carbon dioxide pick-up rates without corrosive attack. This results in an energy requirement comparable to that of the carbonate process and allows the use of smaller equipment for a specific CO_2-removal application thereby reducing the capital cost.

Solubility of carbon dioxide in ethanolamines is affected by temperature, amine solution strength, and carbon dioxide partial pressure. Information on the performance of amines is available in the literature and from amine manufacturers. Values for the solubility of carbon dioxide and hydrogen sulfide mixtures in monoethanolamine and for the solubility of carbon dioxide in diethanolamine are given in references 36 and 37. Solubility of carbon dioxide in monoethanolamine is provided in reference 38. The effects of catalysts have been studied to improve the activity of amines and provide absorption data for carbon dioxide in both mono- and diethanolamine solutions with and without sodium arsenite as a catalyst (39). Absorption kinetics over a range of contact times for carbon dioxide in monoethanolamine have also been investigated (40).

Sulfinol Process. The Sulfinol process was developed during the 1960s to remove carbon dioxide and other acidic gases from gas streams at high partial pressures. It uses a circulating solution with a flow pattern similar to those in the amine and carbonate processes. Regeneration occurs at low pressure, and heat is exchanged between the regenerated solution and the solution from the absorber. The Sulfinol solution is a mixture of sulfolane (tetrahydrothiophene 1,1-dioxide), an alkanolamine and water.

The process is capable of achieving higher solubilities of CO_2 in the solution without the corrosion problems encountered with amine systems before the advent of Amine Guard. The Sulfinol process is used in over 50 plants worldwide, nevertheless, it is used less often than the amine or carbonate processes.

Rectisol Process. This is one of several processes that use a solvent for removing carbon dioxide from gas streams. In the Rectisol process, the solvent is usually methanol and the operating temperature of the absorber is about 0°C. The process uses an absorption column and stripping column with intermediate pumps and heat exchangers. It is necessary to cool the process gas to the absorbing temperature. Split flow of the solvent stream to the absorber is used, with partially stripped solvent from the midsection of the stripper being returned to the midsection of the absorber and fully stripped solvent from the bottom of the stripper being used at the top of the absorber. Stripping is accomplished by pressure release, heat, and contact with an inert stripping gas which is blown through the column.

The processes using physical absorption require a solvent circulation proportional to the quantity of process gas, inversely proportional to the pressure, and nearly independent of the carbon dioxide concentration. Therefore, high pressures could favor the use of these processes. The Rectisol process requires a refrigeration system and more equipment than the other processes.

Purisol Process. This is a solvent process which uses N-methyl-2-pyrrolidinone as the solvent and benefits from high pressure, especially 7000 kPa (69 atm) or higher. All commercial installations are at pressures above 4000 kPa (39 atm).

The processing steps are much the same as those of the Rectisol process; pressure release, heat, and inert stripping gas regenerate the solvent. Absorption, however, is carried out at 38°C and refrigeration is not required.

Fluor Process. This is a solvent process that uses propylene carbonate. Propylene carbonate has a high solubility for CO_2, a low solubility for other light gases, is chemically stable and noncorrosive to carbon steel.

Physical solvent plants represent a higher capital cost than the carbonate or amine process plants but may result in reducing operating costs where high pressures are involved.

Seven plants using the Fluor process for removal of acidic gases were in existence by 1970; 5 natural gas plants, one hydrogen, and one ammonia plant.

Methods of Purification. Although carbon dioxide produced and recovered by the methods outlined above has a high purity, it may contain traces of hydrogen sulfide and sulfur dioxide which cause a slight odor or taste. The fermentation gas recovery processes include a purification stage, but carbon dioxide recovered by other methods must be further purified before it is acceptable for beverage, dry ice, or other uses. The most commonly used purification methods are treatments with (a) potassium permanganate, (b) potassium dichromate, and (c) active carbon.

Potassium Permanganate. Probably the most widely used process for removing traces of hydrogen sulfide from carbon dioxide is to scrub the gas with an aqueous solution saturated with potassium permanganate. Sodium carbonate is added to the solution as buffer. The reaction is as follows:

$$3\,H_2S + 2\,KMnO_4 + 2\,CO_2 \rightarrow 3\,S + 2\,MnO_2 + 2\,KHCO_3 + 2\,H_2O$$

The precipitated manganese dioxide and sulfur are discarded. The solution is used until it becomes spent or so low in potassium permanganate that it is no longer effective and is discarded and replaced. It is customary to place two scrubbers in series, with the liquid flow countercurrent to the gas flow to more efficiently use the permanganate solution. When the solution in the first scrubber is spent, with respect to the gas, the positions of the scrubbers are reversed and the spent scrubber is recharged with fresh solution.

Two types of scrubbers are used. The simpler consists of a vessel half or two-thirds full of solution. The gas feeds into the bottom of the vessel and bubbles up through the solution. The other scrubber is a small packed tower through which the gas stream is passed countercurrent to a recirculating shower of potassium permanganate and soda ash solution. The latter requires a circulating pump and a solution mix chamber, but has the advantage of reducing the pressure drop through the equipment to a minimum. The solution is used until spent and then discarded. Two scrubbers of this type may also be used to improve efficiency.

Potassium Dichromate. This method is similar in application to the potassium permanganate method:

$$K_2Cr_2O_7 + 3\,H_2S + H_2O + 2\,CO_2 \rightarrow 3\,S + 2\,Cr(OH)_3 + 2\,KHCO_3$$

The precipitated chromic hydroxide and sulfur are discarded. This process is used to purify carbon dioxide from fermentation in the Reich process and as a final cleanup after the alkali carbonate or ethanolamine recovery processes (22–23).

Active Carbon. The process of adsorbing impurities from carbon dioxide on active carbon or charcoal has been described in connection with the Backus process of purifying carbon dioxide from fermentation processes. Space velocity and reactivation cycle vary with each application. The use of active carbon need not be limited to the fermentation industries but, where hydrogen sulfide is the only impurity to be removed, the latter two processes are usually employed.

Methods of Liquefaction and Solidification. Carbon dioxide may be liquefied at any temperature between its triple point ($-56.6°C$) and its critical point ($31°C$) by compressing it to the corresponding liquefaction pressure, and removing the heat of condensation.

There are two liquefaction processes. In the first, the carbon dioxide is liquefied near the critical temperature, water is used for cooling. This process requires compression of the carbon dioxide gas to pressures of about 7600 kPa (75 atm). The gas from the final compression stage is cooled to about $32°C$, and then filtered to remove water and entrained lubricating oil. The filtered carbon dioxide gas is then liquefied in a water-cooled condenser.

The second liquefaction process is carried out at temperatures from -12 to $-23°C$, with liquefaction pressures of about 1600–2400 kPa (16–24 atm). The compressed gas is precooled to 4 to $27°C$, water and entrained oil are separated, and the gas is then dehydrated in an activated alumina, bauxite, or silica gel drier, and flows to a refrigerant-cooled condenser (see Drying agents).

Liquid carbon dioxide is stored and transported at ambient temperature in cylinders containing up to 22.7 kg. Larger quantities are stored in refrigerated insulated tanks maintained at $-18°C$ and 2070 kPa (20 atm), and transported in insulated tank trucks and tank rail cars.

Solidification. Liquid carbon dioxide from a cylinder may be converted to "snow" by allowing the liquid to expand to atmospheric pressure. This simple process is used only where very small amounts of solid carbon dioxide are required because less than one-half of the liquid will be recovered as solid.

Solid carbon dioxide is produced in blocks by hydraulic presses. Standard presses produce blocks $25 \times 25 \times 25$ cm, $50 \times 25 \times 25$ cm, or $50 \times 50 \times 25$ cm. A 25-cm cube of dry ice weighs 23 kg, allowing for about 10% sublimation loss during storage and shipment (some 27-kg blocks are also produced). Dry ice is about 1.7 times as dense as water ice, whereas its net refrigerating effect on a weight basis is twice that of water ice. Automation and improved operating cycles have increased dry-ice press capacities so that one $50 \times 50 \times 30$ cm press can produce more than thirty metric tons of dry-ice blocks per day (41).

Liquid carbon dioxide from a supply tank at 700 kPa (7 atm) and $-46°C$ is fed to the press chamber through an automatic feed valve. The pressure in the press is maintained slightly above the triple point (480–550 kPa or 70–80 psi). The quantity fed to the press may be controlled by a timer, or a device which measures the level of liquid in the press chamber. The pressure is reduced and the evolved CO_2 vapor is returned to a recycle system. When the pressure falls below the triple point (518 kPa or 75 psi), the liquid CO_2 solidifies to form carbon dioxide snow. Heat exchange units are used to cool the liquid CO_2 with cold vapors from the press. About 60% of the liquid fed to the press remains as snow when the pressure has dropped close to atmospheric. The hydraulic rams then press the snow to a solid block of dry ice. The block moves along a conveyor and is cut by band saws into four blocks, which are subsequently carried through automatic weighing and packaging machines.

Carbon dioxide is ordinarily dehydrated during the liquefaction cycle to prevent freezeups in the condenser and flow valves in the liquid lines. In some cases brittle or crumbly blocks of dry ice have been formed. This difficulty has been overcome either by varying the residual moisture content of the liquid carbon dioxide, or by injecting minute quantities of colorless mineral oil or diethylene glycol into the liquid carbon dioxide entering the press. If the dry ice is to be used for edible purposes, the additive must meet FDA specifications.

Although liquid carbon dioxide may be stored without loss in tanks and cylinders, dry ice undergoes continuous loss in storage due to sublimation. This loss can be minimized by keeping the dry ice in insulated boxes or bins. Special insulated rail cars and trucks are used for hauling dry-ice blocks. Most plants produce the material at the time it is sold to avoid storage losses and rehandling costs.

Economics

Carbon dioxide production and its total value are presented in Table 2. Throughout the 1950s and 1960s production increased steadily at a rate of 3 to 4% per year. Table 2 shows about an 8% yearly increase and predictions for 1976 through 1980 are for a 10 to 15% annual growth rate (42). The chief reason for this large predicted increase is the rapid increase in the amount of carbon dioxide used for secondary oil recovery (see Uses).

Table 2 indicates peak use of solid carbon dioxide in 1973 but actually dry ice production has declined steadily over the last twenty years. Production in 1955 was ca 520,000 metric tons, about 160% of 1975 production. Much of this decline is caused by the use of liquid carbon dioxide in place of dry ice for refrigeration applications.

Much more carbon dioxide is generated daily than is recovered (43). The decision whether or not to recover by-product carbon dioxide often depends on the distance and cost of transportation between the carbon dioxide producer and consumer. For example, it has become profitable to recover more and more carbon dioxide from CO_2-rich natural gas wells in Texas as the use of carbon dioxide in secondary oil recovery has increased.

Table 2. Carbon Dioxide Production in the United States[a]

Year	Total	Solid	Liquid and gas	Value ($1,000)
1971	1,219	287	932	38,963
1972	1,460	317	1,143	48,375
1973	1,420	337	1,083	44,178
1974	1,636	334	1,302	59,966
1975	1,679	319	1,360	66,633

[a] Quantities in thousands of metric tons.

Toxicity

Although carbon dioxide is a constituent of exhaled air, high concentrations are hazardous. Up to one-half vol% carbon dioxide in air is not considered harmful but carbon dioxide concentrates in low spots because it is one and one-half times as heavy as air. Five vol percent carbon dioxide in air causes a threefold increase in breathing rate and prolonged exposure to concentrations higher than 5% may cause unconsciousness and death.

Ventilation sufficient to prevent accumulation of dangerous percentages of carbon dioxide must be provided where carbon dioxide gas has been released or dry ice has been used for cooling.

Uses

A large portion of the carbon dioxide recovered is used at or near the location where it is generated as an ingredient in a further processing step. In this case, the gaseous form is most often used. Low temperature liquid and solid carbon dioxide are used for refrigeration (qv). Where the producer and the consumer are distant, carbon dioxide may be liquified to reduce transportation cost and revaporized at the point of consumption.

About 40% of the carbon dioxide recovered is used as a raw material in the production of other chemicals, chiefly urea and methanol. Urea (qv) production requires liquid ammonia and gaseous carbon dioxide as feed materials. A urea plant is often located adjacent to an ammonia plant whence it obtains by-product CO_2 as well as ammonia. In the production of methanol (qv), carbon dioxide is reduced to carbon monoxide in the reverse of the water gas shift reaction.

Approximately 35% of carbon dioxide output is used in secondary oil recovery; the technique of adding carbon dioxide to aging or nearly depleted oil wells to increase production (see Petroleum). Most carbon dioxide used for this purpose is obtained from carbon dioxide-rich natural gas wells located near the oil producers. This is the fastest growing use for carbon dioxide and demand may double by 1980 (42). Refrigeration applications account for 10% of recovered carbon dioxide; beverage carbonation 5%, and miscellaneous applications 10%.

Dry Ice. Refrigeration of foodstuffs, especially ice cream, meat products, and frozen foods, is the major use for solid carbon dioxide. Dry ice is especially useful for chilling ice cream products because it can be easily sawed into thin slabs and leaves no liquid residue upon evaporation. Crushed dry ice may be mixed directly with other products without contaminating them and is widely used in the processing of substances that must be kept cold. Dry ice is mixed with molded substances that must be kept cold. For example, dry ice is mixed with molded rubber articles in a tumbling drum to chill them sufficiently so that the thin flash or rind becomes brittle and breaks off. It is also used to chill golf-ball centers before winding.

Dry ice is used to chill aluminum rivets. These harden rapidly at room temperature, but remain soft if kept cold with dry ice. It has found numerous uses in laboratories, hospitals, and airplanes as a convenient and readily available low temperature coolant.

Liquid Carbon Dioxide. The rapid increase in the use of liquid carbon dioxide in recent years is the result of new applications as well as improved facilities for transporting, storing, and handling liquid carbon dioxide. Carbon dioxide manufacturers have developed refrigerated bulk-liquid storage systems which they install and maintain for large consumers. These systems are available in sizes from 2 to 50 tons and have an insulated storage tank which is maintained at 2080 kPa (20.5 atm) and −18°C by a Freon refrigeration unit, with a refrigeration coil in the upper part of the storage tank. An external vaporizer is provided when gaseous carbon dioxide is needed. Liquid-level and pressure relief valves, and a safety rupture disk are provided. The entire assembly is enclosed in a sheet-steel housing mounted on a steel base, and requires nominal power. A twelve-ton unit is supplied with a 1500-W (two-horsepower) refrigeration unit. Vaporizers can be heated by electricity or steam. One kg of steam vaporizes approximately 6 kg of CO_2. The storage tanks are refilled by the CO_2 supplier, either by tank-truck or rail-car delivery.

Ready availability and easy application of bulk liquid carbon dioxide have caused it to replace dry ice in many cases. Liquid CO_2 can be stored without loss and is easily measured or weighed. Liquid carbon dioxide is also used (along with dry ice) for direct injection into chemical reaction systems to control temperature.

Liquid carbon dioxide provides the most readily available method of rapid refrigeration and is used for rapid chilling of loaded trucks and rail cars before shipment. A two to four minute injection of liquid carbon dioxide into a loaded ice cream truck causes the temperature to drop as much as 70°C, flushing the warm air out of the truck and leaving a layer of carbon dioxide snow in the truck which sublimes slowly to provide additional refrigeration. This greatly reduces the load on the truck's mechanical refrigeration system and eliminates the time lag in cooling the truck contents to a safe storage temperature. Test chambers for environmental studies have been effectively cooled at temperatures to −80°C with liquid carbon dioxide, either by direct injection, or by using the liquid to chill circulating refrigerant liquids. The speed with which chilling may be obtained and the low equipment cost required have been the main factors in the selection of liquid carbon dioxide for this service.

Liquid carbon dioxide has been used for many years in the Long-Airdox blasting system for mining coal. A steel cartridge containing liquid carbon dioxide is placed in a hole drilled in the coal seam. A heating mixture in the cartridge is ignited electrically. This vaporizes the carbon dioxide, causing the pressure to increase enough to burst a steel rupture disk and release the carbon dioxide, which shatters the coal. The cartridge is then recovered and reused.

Liquid carbon dioxide is used as a source of power in certain applications. The vapor pressure of liquid carbon dioxide (7290 kPa or 72 atm at 21°C) may be used for operating remote signaling devices, spray painting, and gas-operated firearms. Carbon dioxide in small cylinders is also used for inflating life rafts and jackets.

Fire-extinguishing equipment ranging from hand-type extinguishers to permanent installations in warehouses, chemical plants, ships, and airplanes uses liquid carbon dioxide (see Fire extinguishing agents). In addition to its snuffing action liquid carbon dioxide exerts a pronounced cooling effect helpful in fire extinguishing. It may be used on all types of fires, and leaves no residue but care must be exercised to safeguard against suffocation of personnel.

Carbon dioxide is sometimes added to irrigation water, in the same manner as fertilizer ammonia, in hard water regions. Carbon dioxide is also used with other gases in respiratory problems and in anesthesia.

In addition to chemical synthesis and enhanced oil recovery, gaseous carbon dioxide is used in the carbonated beverage industry. Carbon dioxide gas under pressure is introduced into rubber and plastic mixes, and on pressure release a foamed product is produced. Carbon dioxide and inert gas mixtures rich in carbon dioxide are used to purge and fill industrial equipment to prevent the formation of explosive gas mixtures.

The addition of small amounts of carbon dioxide to the atmosphere in greenhouses greatly improves the growth rate of vegetables and flowers.

Carbon dioxide is widely used in the hardening of sand cores and molds in foundries. Sand is mixed with a sodium silicate binder to form the core or mold after which it is contacted with gaseous carbon dioxide. Carbon dioxide reacts with the sodium silicate to produce sodium carbonate and bicarbonate, plus silicic acid, resulting in hardening of the core or mold without baking.

The use of carbon dioxide gas for shielded arc welding with semiautomatic microwire welding equipment has led to welding speeds up to ten times those obtainable with conventional equipment. No cleaning or wire brushing of the welds is required (44) (see Welding).

Carbon dioxide gas is used to immobilize animals prior to slaughtering them (45). In addition to providing a humane slaughtering technique, this results in better quality meat. The CO_2 increases the animal's blood pressure, thereby increasing blood recovery. The increased accuracy obtainable in the killing operation reduces meat losses due to cut shoulders.

As a weak acid (in aqueous solution) carbon dioxide neutralizes excess caustic in textile manufacturing operations. It does not injure fabrics and is easy to use. Carbon dioxide is also used for neutralizing alkaline waste waters, treating skins in tanning operations, and carbonating treated water to prevent scaling.

Carbon dioxide is used as a chemical reagent in the manufacture of sodium salicylate, basic lead carbonate (white lead), and sodium, potassium, and ammonium carbonates and bicarbonates.

BIBLIOGRAPHY

"Carbon Dioxide" is treated in *ECT* 1st ed. under "Carbon Dioxide", Vol. 3, pp. 125–142, by R. M. Reed and N. C. Updegraff, The Girdler Corporation and "Carbon Dioxide" in *ECT* 2nd ed., Vol. 4, pp 353–369, by R. M. Reed and E. A. Comley, The Girdler Corporation.

1. L. N. Canjar and co-workers, *Hydrocarbon Process.* **45**(1), 139 (1966).
2. *Matheson Gas Data Book*, The Matheson Co., Inc., East Rutherford, N.J., 1961, pp. 81–82.
3. C. L. Yaws, K. Y. Li and C. H. Kuo, *Chem. Eng.* **81**, 115 (1974).
4. B. J. Kendall and B. H. Sage, *Petroleum* (London) **14**, 184 (1951).
5. L. H. Chen, *Thermodynamic and Transport Properties of Gases, Liquids and Solids*, McGraw-Hill Book Co., Inc., New York, 1959, p. 358–369.
6. J. T. Kennedy and G. Thodos, *J. Chem. Eng. Data* **5**(3), 293 (1960).
7. L. B. Koppel and J. M. Smith, *J. Chem. Eng. Data* **5**(3), 437 (1960).
8. P. E. Liley, *J. Chem. Eng. Data* **4**(3), 238 (1959).
9. I. Granet and P. Kass, *Petrol. Refiner* **31**(11), 137 (1952).
10. R. E. Honig and H. O. Hook, *RCA Rev.* **21**, 360 (1960).
11. W. S. Dodds, L. F. Stutzman, and B. J. Sollami, *Chem. Eng. Data Ser.* **1**(1), 92 (1956).
12. P. B. Stewart and P. Munjal, *J. Chem and Eng. Data* **15**, 67 (1970).
13. P. Munjal and P. B. Stewart, *J. Chem and Eng. Data* **16** (2), 170 (1971).
14. D. Price, *Ind. Eng. Chem.* **47**, 1649 (1955).
15. D. Price, *Chem. Eng. Data Ser.* **1**(1), 83 (1956).
16. W. F. Libby, E. C. Anderson, and J. R. Arnold, *Science* **109**, 227 (1949).
17. H. A. O'Hern, Jr., and J. J. Martin, *Ind. Eng. Chem.* **47**, 2081 (1955).
18. U.S. Pat. Re. 18,958 (Sept. 26, 1933), R. R. Bottoms (to The Girdler Corp.) (reissue of 1,783,901).
19. A. V. Slack and G. R. James, eds., *Ammonia Part II*, Marcel Dekker, New York, 1974 (Fertilizer Science and Technology Series, Vol. 2) 1974.
20. U.S. Pat. 1,493,183 (May 6, 1924), A. A. Backus (to U.S. Industrial Alcohol Co.).
21. U.S. Pat. 1,510,373 (Sept. 30, 1924), A. A. Backus (to U.S. Industrial Alcohol Co.).

742 CARBON DIOXIDE

22. U.S. Pat. 1,519,932 (Dec. 16, 1924), G. T. Reich.
23. U.S. Pat. 2,225,131 (Dec. 17, 1940), G. T. Reich.
24. J. H. Perry, ed., *Chemical Engineers Handbook,* 3rd ed., McGraw-Hill Book Co., Inc., 1950, p. 702.
25. U.S. Pat. 2,886,405 (May 12, 1959), H. E. Benson and J. H. Field (to U.S. Gov't.).
26. Ital. Pat. 587,522 (Jan. 16, 1959), G. Giammarco (to S. A. Vetrocoke).
27. U.S. Pat. 2,840,450 (June 24, 1958), G. Giammarco (to S. A. Vetrocoke).
28. Ital. Pat. 518,145 (March 4, 1955), G. Giammarco (to S. A. Vetrocoke).
29. U.S. Pat. 3,086,838 (April 23, 1963) G. Giammarco (to S. A. Vetrocoke).
30. U.S. Pat. 3,144,301 (Aug. 11, 1964) B. J. Mayland (to The Girdler Corp.).
31. J. P. Bocard and B. J. Mayland *Pet. Refiner.* **41**(4), 128 (1962).
32. K. F. Butwell, E. N. Hawkes, and B. F. Mago, *Chem. Eng. Prog.* **69**(2), 57 (1973).
33. *Nitrogen* (96), 33 (1975).
34. *Oil Gas J.* **73**(11), 107 (1975).
35. *Nitrogen* (102), 40 (1976).
36. J. I. Lee, F. D. Otto, and A. E. Mather, *J. Chem and Eng. Data* **20**(2), 161 (1975).
37. J. I. Lee, F. D. Otto, and A. E. Mather, *J. Chem. and Eng. Data* **17**(4), 465 (1972).
38. J. D. Lawson and A. W. Garst, *J. Chem. and Eng. Data* **21**(1), 20 (1976).
39. P. V. Danckwerts and K. M. McNeil, *Trans. Inst. Chem. Engr.* **45**, T32 (1967).
40. E. Sada, H. Kumazawa, and M. A. Butt, *AIChE Journal* **22**(1), 196 (1976).
41. "CO$_2$ Plant: More Capacity, More Finesse," *Chem. Eng.* **62**, 120 (1955).
42. *Chem. Week,* 71 (Aug. 13, 1975).
43. S. Terra, *EPRI J.,* 22 (July/Aug., 1978).
44. J. P. O'Donnel, *Oil Gas J.* **60**(39), 167 (1962).
45. *Animal Immobilization with CO$_2$,* Bull. No. 11, General Dynamics Corporation Chicago, Ill., 1962.

W. Robert Ballou
C&I Girdler Incorporated

CARBON DISULFIDE

Carbon disulfide [75-15-0] (carbon bisulfide), CS$_2$, is a volatile, flammable liquid which is heavier than water. It has many useful properties and has been an important industrial chemical since the nineteenth century. Principal uses are in the manufacture of regenerated cellulose fibers and films, and as a raw material for the manufacture of carbon tetrachloride. It is also used in preparation of various organic sulfur compounds and as a solvent.

Commercial manufacture began about 1880. Until about 1950 the principal method was by the reaction of carbon (wood charcoal) and sulfur. That procedure has been largely supplanted by processes employing the reaction of hydrocarbons and sulfur. Worldwide production of carbon disulfide in 1976 was estimated at over one million metric tons.

Physical Properties

Carbon disulfide is a clear and colorless and has a mild ethereal odor; however, minor impurities impart a disagreeable sulfurous odor. It is slightly soluble in water, and an excellent solvent for many organic compounds. Many properties are known accurately (1–5), including thermodynamic properties (3). Selected physical properties are given in Table 1.

Table 1. Physical Properties of Carbon Disulfide

Property	Value			References
melting point, K	161.11			2
latent heat of fusion, kJ/kg[a]	57.7			2
boiling point at 101 kPa[b], °C	46.25			1
flash point at 101 kPa[b], °C	−30			6
ignition temperature in air, °C				1
10-s lag time	120			
0.5-s lag time	156			
critical temperature, °C	273			1
critical pressure, kPa[b]	7700			1
critical density, kg/m³	378			1
soly H_2O in CS_2 at 10°C, ppm	86			7
at 25°C, ppm	142			7
Liquid at temperature, °C	*0*	*20*	*46.25*	
density, kg/m³	1293	1263	1224	1
specific heat, J/(kg·K)[a]	984	1005	1030	1
latent heat of vaporization, kJ/kg[a]	377	368	355	3,8
surface tension, mN/m(=dyn/cm)	35.3	32.3	28.5	1
thermal conductivity, W/(m·K)		0.161		1
viscosity, mPa·s (=cP)	0.429	0.367	0.305	1
refractive index, n_D	1.6436	1.6276		9
solubility in water, g/kg soln	2.42	2.10	0.48	1
vapor pressure, kPa[b]	16.97	39.66	101.33	1
Gas at temperature, °C[c]	*46.25*	*200*	*400*	
density, kg/m³	2.97	1.96	1.37	3
specific heat, J/(kg·K)[a,d]	611	679	730	4
viscosity, mPa·s (=cP)	0.0111	0.0164	0.0234	5
thermal conductivity, W/(m·K)	0.0073			5,10
Thermochemical data at 298 K[a]				
heat capacity[a], C_p^o, J/(mol·K)		45.48		11
entropy[a], S^o, J/(mol·K)		237.8		11
heat of formation[a], H_f^o, kJ/mol		117.1		11
free energy of formation[a], G_f^o, kJ/mol		66.9		11

[a] To convert J to cal, divide by 4.184.
[b] To convert kPa to atm, divide by 101.3.
[c] At absolute pressure, 101.3 kPa.
[d] $C_p/C_v = 1.21$ at 100°C (1).

A black solid form of carbon disulfide has been reported (12) at high pressures of about 5.5 GPa (5.4 × 10⁴ atm) and about 175°C.

Carbon disulfide is soluble in the usual organic solvents (6,13), and completely miscible with many liquid hydrocarbons, chlorinated hydrocarbons, and alcohols. Sulfur is soluble in carbon disulfide in the range of 17–47 wt % at 0–40°C (14). For saturated solutions of carbon disulfide in liquid sulfur at a CS_2 partial pressure of 101 kPa (1 atm), values from 8.0% CS_2 at 94°C to 2.2% CS_2 at 158°C are reported (15), including a phase diagram for the sulfur–carbon disulfide system. Vapor–liquid equilibrium data for binary mixtures of carbon disulfide and several other compounds at room temperature are available (7), in addition to freezing point data for binary systems of carbon disulfide and 17 different compounds. Spectral transmission data for the ultraviolet (16) and infrared (17) regions have been published.

Chemical Properties

Carbon disulfide is highly flammable (18) and is easily ignited by sparks or hot surfaces. Lower pressure or greater availability of oxygen may cause a considerable lowering of the flash point temperature. The flammability or explosive range of carbon disulfide with air is very wide and depends on conditions. For example, the extreme range is 1.06–50.0 vol % for upward propagation in a glass tube 75 mm in diameter, whereas for downward propagation, the limits are 1.91 and 35.0 vol % (19). The presence of carbon dioxide effectively reduces the upper limit (20). Concentrations of carbon disulfide in air of 4–8% explode with maximum violence, although the thermal energy released in carbon disulfide explosions is low compared to that from other flammable substances. The maximum absolute pressure developed is reported (21) as 730 kPa (7.2 atm).

The equilibrium constant of formation of carbon disulfide increases with temperature, reaching a maximum corresponding to 91% conversion to CS_2 at about 1000 K (11,22–23).

Carbon disulfide forms trithiocarbonates with aqueous alkalies:

$$3\ CS_2 + 6\ KOH \rightarrow K_2CO_3 + 2\ K_2CS_3 + 3\ H_2O \tag{1}$$

Reaction with alcoholic alkali gives dithiocarbonates (xanthates) (see Xanthates):

$$CS_2 + NaOH + C_2H_5OH \rightarrow C_2H_5OC(S)SNa + H_2O \tag{2}$$

With alkali cellulose, the reaction proceeds to form sodium cellulose xanthate:

$$R_{cellulose}ONa + CS_2 \rightarrow R_{cellulose}OC(S)SNa \tag{3}$$

Cellulose xanthate is soluble in caustic soda solution forming viscose solution, a step in the manufacture of viscose rayon.

Carbon disulfide reacts readily with chlorine in the presence of iron catalysts to form carbon tetrachloride:

$$2\ CS_2 + 6\ Cl_2 \rightarrow 2\ CCl_4 + 2\ S_2Cl_2 \tag{4}$$

The reaction can proceed with sulfur monochloride acting as the chlorinating agent:

$$CS_2 + 2\ S_2Cl_2 \rightarrow CCl_4 + 6\ S \tag{5}$$

Carbon tetrabromide is formed by the reaction with bromine. Treatment with chlorine in the absence of iron but with a trace of iodine gives trichloromethanesulfenyl chloride, CCl_3SCl, which is reduced with stannous chloride or tin and hydrochloric acid to thiophosgene (thiocarbonyl chloride), $CSCl_2$ (see Sulfur compounds).

Boiling excess aniline forms thiocarbanilide:

$$CS_2 + 2\ C_6H_5NH_2 \rightarrow SC(NHC_6H_5)_2 + H_2S \tag{6}$$

Primary and secondary amines give substituted ammonium salts of N-substituted dithiocarbamic acids, $RNHC(S)SNH_3R$ and $R_2NC(S)SNH_2R_2$. Ammonium dithiocarbamate is prepared from carbon disulfide and alcoholic ammonia:

$$2\ NH_3 + CS_2 \rightarrow NH_2C(S)SNH_4 \tag{7}$$

Sodium dimethyl dithiocarbamate is obtained with dimethylamine and caustic soda:

$$(CH_3)_2NH + CS_2 + NaOH \rightarrow (CH_3)_2NC(S)SNa + H_2O \tag{8}$$

Other metal salts, such as those of iron and zinc, may be prepared from the sodium salt. Dithiocarbamates form characteristically colored compounds with many heavy-metal ions, a property that may be used in testing for such metals.

Ammonium thiocyanate, NH_4SCN, forms by digesting concentrated ammonia with carbon disulfide. The reaction is promoted by alumina catalysts. Upon heating at about 160°C, thiourea, H_2NCSNH_2, is obtained. With ethylene diamine, ethylenethiourea (2-imidazolidinethione), $\underline{NHCSNHCH_2CH_2}$, is formed in alcoholic solution at 60°C.

Carbon disulfide reacts with certain oxides to produce carbon oxysulfide:

$$CS_2 + MgO \rightarrow COS + MgS \tag{9}$$

Chlorosulfonic acid acts as an oxidizing agent:

$$CS_2 + ClSO_3H \rightarrow COS + HCl + SO_2 + S \tag{10}$$

Urea at 110°C gives carbon oxysulfide and ammonium thiocyanate:

$$CS_2 + (NH_2)_2CO \rightarrow COS + NH_4SCN \tag{11}$$

Carbon disulfide is partially desulfurized by alkyl mercuric hydroxide:

$$CS_2 + RHgOH \rightarrow RHgSH + COS \tag{12}$$

Dithio acids are formed with Grignard reagents:

$$RMgBr + CS_2 \rightarrow RC(S)SMgBr \rightarrow RC(S)SH \tag{13}$$

Carbon disulfide reacts at 40–50°C with aqueous solutions of sodium azide to form sodium azidodithiocarbonate:

$$CS_2 + NaN_3 \rightarrow NaSCSN_3 \tag{14}$$

Dialkyl sulfides are formed by alkanols or dialky ethers and carbon disulfide at elevated temperatures over catalysts such as activated alumina.

Hydrogenation of carbon disulfide yields a variety of compounds, depending upon conditions. For example, at high temperature the following equilibrium reactions occur:

$$CS_2 + 2 H_2 \rightleftharpoons 2 H_2S + C \tag{15}$$

$$CS_2 + 4 H_2 \rightleftharpoons 2 H_2S + CH_4 \tag{16}$$

At 180°C over a reduced nickel catalyst:

$$CS_2 + 2 H_2 \rightarrow \underset{\text{methanedithiol}}{CH_2(SH)_2} \tag{17}$$

At 250°C over a cobalt catalyst:

$$CS_2 + 3 H_2 \rightarrow \underset{\substack{\text{methanethiol}\\\text{(methyl mercaptan)}}}{CH_3SH} + H_2S \tag{18}$$

Other possible reduction products include dimethyl sulfide, thioethers, thioformaldehyde, and thiophene (see Sulfur compounds).

Carbon disulfide reacts almost quantitatively with water vapor at high temperatures to form hydrogen sulfide. The reaction is greatly accelerated by specific catalysts, with conversions as high as 98% over activated alumina at 300–600°C reported

(24). Under milder conditions, carbon oxysulfide and hydrogen sulfide are obtained.

A summary of carbon disulfide reactions and a bibliography are given in reference 14.

Manufacture

The classical process for manufacturing carbon disulfide has been from carbon (charcoal) and sulfur vapor. In the 1950s, hydrocarbon–sulfur processes gradually replaced the charcoal method in the United States, and plants using hydrocarbon processes have also been installed in other countries.

Charcoal–Sulfur Processes. The direct combination of carbon (charcoal) and sulfur vapor at elevated temperatures in either fuel-fired retorts or electric furnaces has been used for nearly 100 years to manufacture carbon disulfide. Sulfur of high quality (low ash) is essential, especially in the retort method, to limit the rate of accumulation of nonvolatile residues that retard heat transfer and obstruct reaction space. Hardwood charcoal with low ash content and of uniform lump size is also preferred. For best yields the charcoal should be prepared under carefully controlled conditions, including charring temperatures of 400–500°C. The charcoal is precalcined at the carbon disulfide plant.

In retort plants the overall endothermic reaction of carbon and sulfur is typically carried out in externally-fired retorts made of cast iron, alloy steel, or refractory brick. The retorts are set in furnaces heated by gas or oil burners to 750–900°C, and operating at slightly above atmospheric pressure. Typical production from one retort is 500–3000 kg/d. About 0.95 kg of sulfur is consumed per kg of carbon disulfide. The theoretical heat required is 1950 kJ/kg (466 kcal/kg) of CS_2, based on carbon and sulfur fed into the reactor at 25°C and carbon disulfide produced at 750°C. There is little opportunity for heat economizing; in practice, external heat must provide many times the theoretical energy input.

Internal electric heating at the reaction zone supplies an efficient energy source and several electric furnace designs for this purpose have been proposed (25–29).

Extensive research efforts have been directed toward other carbon sources such as lignite, peat and brown coal semicokes (30), lignite chars (31–32), pulverized coal (33), and petroleum cokes or pitch cokes (34). Other types of reactors include moving beds of refractory pebbles for superheating sulfur (35), fluidized beds (36–38), and a "whirlpool layer" reactor (39).

Hydrocarbon–Sulfur Processes. By about 1965 hydrocarbons had replaced charcoal as the principal source of carbon for manufacture of carbon disulfide in the United States. The light hydrocarbon gases, especially methane, ethane, and propylene, appear to be favored as they result in fewer by-products. Variations in the system design and product separations (40–42) take advantage of large, continuous processes with low operating cost and high efficiency compared to solid-carbon methods.

Methane and sulfur vapor react as follows:

$$CH_4 + a\, S_x \rightarrow CS_2 + 2\, H_2S \tag{19}$$

$$CH_4 + b\, S_x \rightarrow CS_2 + 2\, H_2 \tag{20}$$

where $ax = 4$, $bx = 2$, and x is between 2 and 8, denoting the presence of sulfur vapor as an equilibrium mixture of S_2, S_6, and S_8.

Equation 19 is thermodynamically very favorable, the calculated equilibrium conversion being greater than 99.9% at 400–700°C (23). Equation 20 is 80% complete at equilibrium and 700°C (43). Since the formation of free hydrogen is suppressed by the presence of excess sulfur, in commercial practice only the kinetics of equation 19 are considered. Catalysts for methane–sulfur reaction include charcoal (44), Cr, W, and Mo compounds (45–46), MgO (47–48), oxides, sulfides, or certain metal salts (49–55), silica gel (56), and alumina gel or bauxite (40,57–65).

The following rate constant expression is derived from reaction rates over a range of space velocities and temperatures (500–700°C), using silica gel catalyst (66).

$$\log k_c = 10.9 - \frac{131.4}{2.303 \, (RT)} \qquad (21)$$

The value of 131.4 kJ (31.4 kcal) per mole is an overall activation energy.

The reaction of diatomic sulfur vapor and methane is exothermic (22,67). However, the total process heat requirement for equation 19 is endothermic, 2950 kJ/kg (705 kcal/kg) of CS_2, based on methane and solid sulfur at 25°C and gaseous products at 600°C. It is advantageous to maximize the dissociation before the reactants reach the catalyst zone by first passing them through a superheater (68). Temperature required to dissociate sulfur may be minimized by the use of several reactors in series (69).

A 1943 patent (59) for converting hydrocarbons to carbon disulfide employs a catalyst, such as silica gel, bauxite, catalytic clays, or alumina, at 450–700°C and a space velocity of 400–3000 h^{-1} (space ratio as total volume of gas at 0°C and 101.3 kPa (1 atm) per volume of space occupied by catalyst). Variations and improvements included the use of olefins such as propylene and 1-butene (70); a process in which the by-product hydrogen sulfide is converted to sulfur by reaction with sulfur dioxide at low temperature, the sulfur being recycled (60); intermittent feeding of sulfur (71); carrying out the reaction in liquid-phase sulfur (61); a complete process including recovery and purification of the product carbon disulfide (65); and a modification claimed to produce carbon disulfide by the reaction of hydrocarbons containing substantial amounts of C_3 and higher hydrocarbons (72).

The Folkins Process. The Folkins process is shown in Figure 1 (40). Hydrocarbon gas, such as methane, is preheated to 480–650°C; the sulfur (10–15% excess) is separately vaporized at the operating pressure employed and then mixed with the methane. The mixture is heated to give an average temperature in the adiabatic catalytic reactor of 580–635°C. Operating pressure is about 250–500 kPa (2.5–5 atm). The products are cooled to approximately 132°C and then charged to the sulfur–gas separator where the major portion of the sulfur contained in the reaction gases is separated and absorbed in a circulating contact stream of liquid sulfur. The gas mixture then passes through a scrubber where remaining sulfur is separated. The products are cooled to about 38°C and passed into the CS_2 absorber. The unabsorbed gas, principally hydrogen sulfide with small amounts of hydrocarbon gas, is passed to a hydrogen sulfide recovery unit wherein the hydrogen sulfide is converted to sulfur (see Sulfur recovery). The rich absorption oil, carrying carbon disulfide and a small amount of hydrogen sulfide, is heated and passed into the stripper, where the carbon disulfide and hydrogen sulfide are removed by fractional distillation. The stripped lean oil is cooled and recirculated. The overhead product is then fed to the stabilizer column, where hydrogen sulfide and remaining hydrocarbon gas are separated overhead and recycled. The

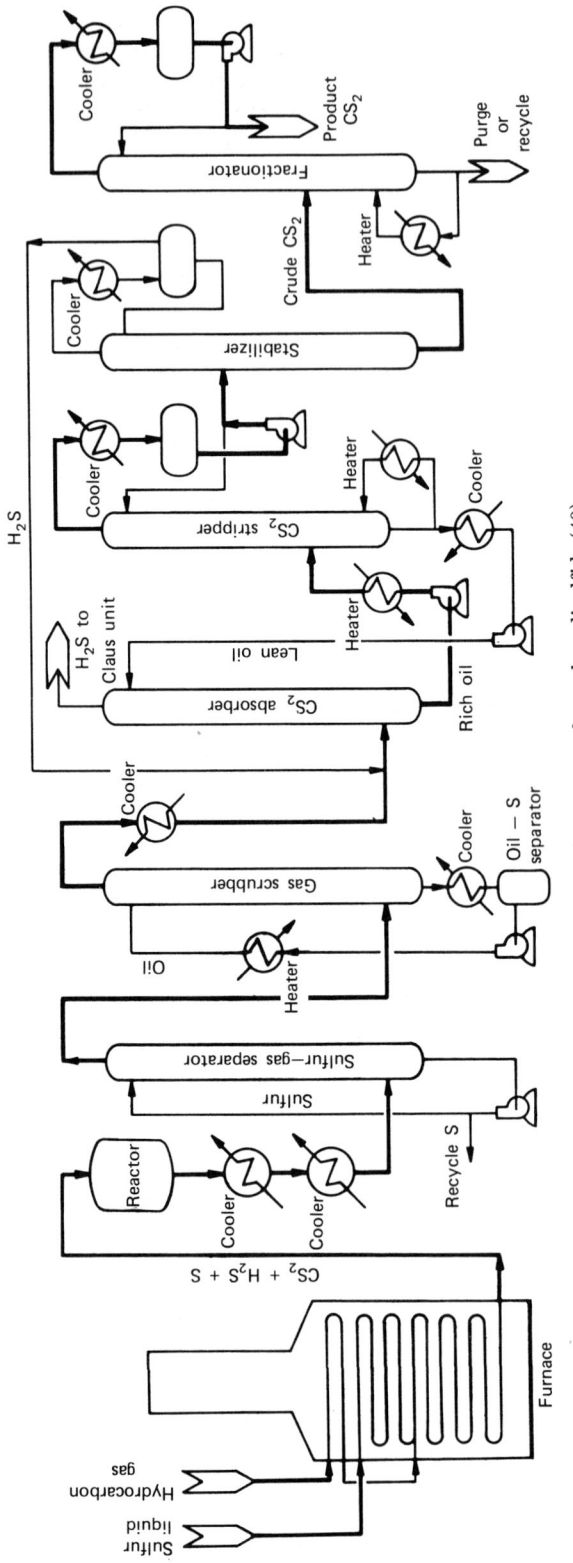

Figure 1. Flow sheet of hydrocarbon process for carbon disulfide (40).

stabilizer underflow, consisting of carbon disulfide and heavy ends, such as absorption oil, is then fed to the fractionator in which pure carbon disulfide is distilled from high boiling fractions.

The preheaters and reactor units are constructed of stainless steel; high-chromium and stabilized nickel-chromium alloys are generally satisfactory (68,72). Natural gas containing about 98% methane prevents undesirable by-product formation from the reaction of higher molecular weight hydrocarbons with sulfur (see Gas, natural). Side reactions occurring in the catalyst bed tend to foul the catalyst and contaminate the unreacted sulfur. This tendency may be alleviated by providing an agglomerating material such as sintered alumina, silica, diatomaceous earth, or similar materials (62). Additional methods for carrying out the catalytic reaction process have also been described (73).

Good yields can also be obtained in the absence of a catalyst (74–75); eg, saturated hydrocarbons at 450–700°C at pressures of 300–2200 kPa (3–22 atm) for 6–72 s can give conversions of 40–100%.

Several French patents (76–78) claim conditions for the noncatalytic reaction of olefins and sulfur to produce carbon disulfide of high purity.

The Folkins process may present certain complications (79). In a simplified version, the reaction is carried out at 1000–2000 kPa (10–20 atm) and 450–750°C. Reaction products flow directly to a stabilizing column in which hydrogen sulfide, remaining hydrocarbon, and any other light-boilers are distilled, and carbon disulfide and excess sulfur and residues are taken as bottoms. The stabilizer bottoms are fed to a finish still where pure carbon disulfide is distilled.

In another variation (80), low pressure is employed, followed by compression of the gaseous reaction mixture before separation and purification.

The process can be improved by mixing a hydrocarbon gas stream with a stream of hot sulfur vapor converging thereon, using a special venturi-type mixing nozzle (81), or mixing sulfur and C_3 and higher hydrocarbons under specified conditions at a pressure of 350–1200 kPa (3.5–12 atm) (82).

Potential Processes. Other hydrocarbons can also be utilized, eg, acetylene and sulfur vapor (83). Carbon disulfide and benzene are obtained from sulfur and hydrocarbons, such as methane, ethane, and ethylene, above 1000°C (84). The reactions of higher molecular weight hydrocarbons with sulfur are complex; intermediates, such as mercaptans, sulfides, and thiophene, tend to be formed together with carbon disulfide. A stoichiometric deficiency of sulfur promotes the formation of unsaturated and intermediate products; eg, thiophene can be produced by the reaction of butane and sulfur (85).

Hydrogen sulfide. The calculated thermodynamic equilibrium for the reaction of hydrogen sulfide and methane (43,86) is 67% at 1100°C and 86.5% at 1288°C.

$$CH_4 + 2 H_2S \rightleftharpoons CS_2 + 4 H_2 \qquad (22)$$

Yields are low because of undesirable side reactions (87).

For reaction of hydrogen sulfide and carbon at 900°C, a 70% conversion to carbon disulfide has been claimed (88–89).

Sulfur dioxide. Methane and sulfur dioxide form carbon disulfide at elevated temperatures in the presence of suitable catalysts, such as lead sulfide on pumice activated with hydrogen chloride (90), giving an 84% yield at 850°C. When anthracite was used instead of methane, a nearly quantitative conversion was obtained at 900–1000°C (91).

Miscellaneous. Carbon oxysulfide as a reactant or intermediate in the preparation of carbon disulfide has been investigated (91–93). Carbon oxides can also serve as starting materials, eg, hydrogen sulfide and carbon monoxide at 600–1125°C (94), or carbon dioxide and hydrogen sulfide over rare earth catalysts at 350–450°C (95).

Handling, Shipping, and Storage

Because of its high volatility, flammability, and toxicity, carbon disulfide must be handled under carefully controlled conditions. In the laboratory, it should be kept away from heat, sparks, and flames, and adequate ventilation should be provided. A self-contained breathing apparatus is necessary if the concentration in air is high. Areas suspected of high concentrations of carbon disulfide vapor should not be entered because of the explosion hazard.

Carbon disulfide is shipped in drums, rail tank cars, tank trucks, and barges. Handling and storage procedures are set forth by the MCA (6). United States DOT regulations require red label (flammable) and specification containers (96). Conditions for transportation on cargo vessels are given in U.S. Coast Guard regulations (97). International Air Transport Association regulations forbid transport of carbon disulfide on either passenger or cargo aircraft (98).

Carbon disulfide may be stored in bulk (14) provided proper design and use procedures of the facilities are observed. The space above the carbon disulfide liquid level in a storage tank must be blanketed with water or inert gas (nitrogen) at all times. Furthermore, the entire tank, if water-blanketed, is preferably submerged completely in a below-grade pit containing water. The tank may be above grade provided it is over a diked area containing water at least equal to the volume of the tank. Storage and handling equipment are generally of conventional carbon steel construction. All parts of a system, including piping, valves, and movable containers, must be earth-grounded and firmly bonded by good electrical conductors to eliminate the possibility of static charge build-up and spark discharge. Carbon disulfide is safely transferred from an enclosed container by displacement with water or nitrogen. General instructions for unloading tank cars are published by the MCA (99).

For fire control, water is most effective when used as fog or spray. Blanketing of a fire with fog or spray cools equipment and prevents reignition. Carbon dioxide and dry chemicals effectively extinguish small fires (6). Foam is ineffective (see Fire-extinguishing agents).

Carbon disulfide is noncorrosive to most commercial structural metals at ordinary temperatures; it is commonly stored and handled in steel equipment. Copper and copper alloys can be corroded by impure carbon disulfide, therefore equipment containing those materials should be used with caution. Above about 250°C, carbon disulfide liquid and vapor become increasingly reactive with iron, steel, and other common metals. For handling and processing at elevated temperatures, high-chromium stainless steel such as types 316, 309, and 310 may be suitable. Glass and ceramic materials are resistant to carbon disulfide at all temperatures.

Economic Aspects

Carbon disulfide manufacture has roughly paralleled that of the viscose rayon industry, its principal user (see Rayon). Table 2 gives production and price data.

Table 2. United States Production and Prices of Carbon Disulfide [a]

Year	Thousands of metric tons	Annual increase, %	$/t
1941		} 6	
↓			
1955	257		120
1959	255		120
↓		} 3.6	
1969	362		107
↓		} constant	
1974	347 av		132
1975	217		150
1976	231		183
1977	229		183

[a] Refs. 100–101.

Carbon tetrachloride accounted for less than 10% of United States carbon disulfide consumption until about 1960. This use increased significantly in the 1960s, with the rapid rise in demand for carbon tetrachloride as an intermediate for fluorocarbon aerosol propellants and refrigerants. In 1974 about 32% of United States carbon disulfide production went to carbon tetrachloride (see Chlorocarbons).

Installed capacities of principal hydrocarbon-based carbon disulfide plants throughout the world are given in Table 3 (102). Commercial plants, typically charcoal, have also been operated in Australia, Austria, Brazil, Bulgaria, Chile, Colombia, Cuba, Czechoslovakia, Egypt, Finland, German Democratic Republic, India, Republic of Korea, Mexico, Norway, Portugal, Rumania, South Africa, Sweden, Turkey, and Yugoslavia.

The commercial future of carbon disulfide is uncertain, in view of the long-term decline in viscose fiber and cellophane demand, in addition to reduced carbon tetrachloride demand caused by restrictions on fluorocarbons as aerosol propellants (see Air pollution; Aerosols; Ozone). Specialty industrial use of carbon disulfide in many other categories should continue.

Specifications and Test Methods

Carbon disulfide obtained from modern processes probably exceeds 99.99% purity. Process control is typically maintained by instrumented continuous fractional distillation. Table 4 summarizes the specifications for technical and reagent grades, with references to analytical methods. For special purposes, certain trace impurities such as benzene, thiophene, or hydrocarbon oil may be specified.

Health and Safety

Carbon disulfide is very toxic and is harmful by inhalation of the vapor, skin absorption of the liquid, or ingestion (6,14,105–107). Carbon disulfide poisoning has disastrous effects upon the nervous system and brain, with symptoms ranging from simple irritability to manic depression. High concentrations of the vapor have an anesthetic effect (6). Other symptoms include headache, nausea, dizziness, restlessness,

Table 3. Principal Manufacturers of Carbon Disulfide by the Hydrocarbon–Sulfur Process in 1977

Producer	Location	Annual capacity, thousands of metric tons
North America		
Stauffer Chemical Company	Delaware City, Del.	159
Stauffer Chemical Company	LeMoyne, Ala.	113
FMC Corp/Allied Chemical Corp.	South Charleston, W. Va.	82
PPG Industries, Inc.	Natrium, W. Va.	30
Penwalt Corporation	Green Bayou, Tex.	16
Cornwall Chemicals	Cornwall, Ontario	23
Thio-Pet Chemicals	Fort Saskatchewan, Alberta	4
Total		427
Western Europe		
Courtaulds, Ltd.	Manchester, Eng.	81
Akzo Chemie G.m.b.H. (Carbosulf)	Cologne, F.R.G.	82
Snia Viscosa S.p.A.	Pavia, Italy	74
Rhône Poulenc	Roches-de-Condrieu, Fr.	120
Rhône Poulenc	Kallo (Antwerp), Belg.	60
Foret SA	Barcelona, Spain	40
Total		457
Other countries		
I. Q. Duperial SAIC	San Lorenzo, Argentina	14
State Authority	Vrasa, Bulg.	20
Nippon Ryutan Kogyo KK	Oita, Jpn.	60
State Authority	Grzybow, Poland	200
State Authority	Sokol, U.S.S.R.	60
State Authority	Volgograd, U.S.S.R.	60
State Authority	Braila, Romania[a]	40
Total		454

[a] Completion scheduled 1980.

euphoria, mucous membrane irritation, and unconsiousness and terminal convulsions (108–109), in addition to fatigue, insomnia, loss of appetite, visual and auditory disturbances, and sensory losses in the extremities (105,108). Carbon disulfide dissolves fat and can extract it from the skin causing dryness and cracking (6). Effects on eyesight have been observed before other symptoms became evident. Studies indicated a gradual and slow increase in the sensitivity of the eyes to light. Alterations in dark adaptation also occurred, in most cases after four years of exposure (110). In industry, repeated brief exposures to high concentrations or prolonged exposures to low concentrations are of much greater importance than single massive exposures (105).

Carbon disulfide concentration in the blood or urine is indicative of exposure and severity (105). A more specific chemical test to evaluate exposure by measuring urine metabolites of CS_2 uses the iodine–azide test developed by Djuric and co-workers (111). Treatment for carbon disulfide poisoning is outlined by Baskin (108) and by Gosselin (112).

Table 5 gives the general effects of carbon disulfide vapor at different concentrations (105). Table 6 summarizes minimum lethal dose and concentration data (113). The recommended limit in workroom air is 20 ppm (approximately 60 mg/m^3) time-weighted average for an 8-h work day and 40-h work week. The short-term exposure limit is 30 ppm in the United States (114), 4 ppm in the Union of Soviet Socialist Re-

Table 4. Carbon Disulfide Specifications

Property	Method	Technical industry (typical)	Technical U.S. Federal[a]	Reagent ACS[b]
specific gravity	pycnometer	1.270–1.272 (15/4°C)	1.262–1.267 (20/20°C)	
residue, %	dry at 60°C	0.002 max	10 mg/100 ml	0.002 max
color	APHA Pt-Co 500 std	<20[c]	special test	10 max
boiling range, °C	ACS distillation		45.5–47.5	1°C incl. 46.3°C
foreign sulfide	$Pb(C_2H_3O_2)_2$, no color	pass		
foreign sulfide			no discolor Cu	Hg drop,[d] no color
sulfite and sulfate	I_2–$BaCl_2$			pass
water, %	Karl Fischer		no turbidity	0.05 max

[a] Ref. 103.
[b] Ref. 104. The USP specifications are the same.
[c] Light transmission vs water = 98% minimum is sometimes used as a specification.
[d] Test is extremely sensitive; material stored for some time, particularly in the presence of light, may fail that test.

Table 5. Effects of Carbon Disulfide[a]

	Concentration in air	
Effect	mg/L	ppm
slight or none	0.5–0.7	160–230
slight symptoms after several hours	1.0–1.2	320–390
symptoms after 30 min	1.5–1.6	420–510
serious symptoms after 30 min	3.6	1150
dangerous to life after 30 min	10.0–12.0	3210–3850
fatal in 30 min	15.0	4815

[a] Ref. 105.

Table 6. Toxic Dose Data for Carbon Disulfide[a]

Species	Route	Dosage
human	oral	14 mg/kg (LDLo)[b]
human	inhalation	4000 ppm/30 min (LCLo)[c]
human	inhalation	50 mg/m³ per 7 yr (TCLo)[d]
rat	intraperitoneal	400 mg/kg (LDLo)[b]
rabbit	subcutaneous	300 mg/kg (LDLo)[b]

[a] Ref. 113.
[b] Lowest published lethal dose.
[c] Lowest published lethal concentration.
[d] Lowest published toxic concentration.

publics, and 10 ppm in Czechoslovakia (115). United States OSHA standards are 20 ppm time-weighted average, 30 ppm ceiling limit, and 100 ppm/30 min peak (113). Carcinogenicity of carbon disulfide has not been reported. Low concentrations of carbon disulfide in drinking water have certain biological effects (116); rabbits receiving water containing 70 mg/L for 6 mo showed changes in phagocyte activity and carbohydrate metabolism. Carbon disulfide was found to be absorbed through the skin from aqueous solutions containing 0.37–1.78 g/L (117).

Uses

The largest single application of carbon disulfide is in the manufacture of viscose rayon, consuming approximately 33% of the United States production in 1974 (see Rayon). Carbon tetrachloride production, an application expected to decline, consumed about 31%. Cellophane film manufacturers used 13%. The remaining 23% was distributed over many uses, including agricultural and pharmaceutical applications with carbon disulfide either as a direct reactant, chemical intermediate, or solvent.

Compounds formed by reaction of carbon disulfide are used extensively in the rubber and polymer industries (see Rubber chemicals). Additional polymer-related uses include solvent and spinning solution applications, and polymerization inhibition of vinyl chloride. Carbon disulfide is employed as a brightening agent in silver and gold electroplating (qv), in electroplating baths for deposition of cobalt, chromium, and zinc, as a gaseous medium in thin film deposition of nickel, as a corrosion inhibitor (see Corrosion), in treatment of metals for wear resistance, rust removal, and extrusion, and in metals removal from waste water. Carbon disulfide is used as sulfiding agent in the synthesis of rare earth sulfides which find application in semiconductors (qv). It is also used as a regenerator for transition metal sulfide catalysts. It is used with amines as development restrainers for Polaroid-type films (see Color photography, instant) and for pretreatment of lithographic silver printing plates for enhanced ink reception (see Printing processes).

Food-related uses include preservation of fresh fruit, in adhesive compositions for food packaging, and as a solvent in the extraction of growth inhibitors. Carbon disulfide is used in the preparation of a large variety of sulfur compounds (qv). Other related applications are as catalyst or catalyst adjuvant or activator in many synthetic reactions.

Solvent properties of carbon disulfide are employed in hydrocarbon conversion, reforming, extraction, dehydration and separation, and in petroleum well cleaning. Other solvent uses include iodine, hydrogen iodide recovery, sulfuric acid reclamation, sulfur extraction, adsorbent reactivation by removal of adsorbate, lubricant separation and extraction, and removal of printing on recycled plastics (see Recycling).

BIBLIOGRAPHY

"Carbon Disulfide" in *ECT* 1st ed., Vol. 3, pp. 142–148, by H. O. Folkins, The Pure Oil Company; "Carbon Disulfide" in *ECT* 2nd ed., Vol. 4, pp. 370–385, by H. O. Folkins, The Pure Oil Company.

1. E. W. Washburn, ed., *International Critical Tables*, McGraw Hill Book Company, New York, 1928, Vol. 3, pp. 23, 213, 231, 248; Vol. 4, p. 447; Vol. 5, pp. 114, 215, 227; Vol. 7, p 213.
2. O. L. I. Brown and G. G. Manov, *J. Am. Chem Soc.* **59,** 500 (1937).
3. L. J. O'Brien and W. J. Alford, *Ind. Eng. Chem.* **43,** 506 (1951).
4. K. A. Kobe and E. G. Long, *Pet. Ref.* **29**(1), 126 (1950).
5. R. W. Gallant, *Hydrocarbon Process.* **49**(4), 132 (1970).
6. *Carbon Disulfide. Chemical Safety Data Sheet SD-12,* Manufacturing Chemists' Association, Washington, D.C., 1967.
7. A. Seidell, *Solubilities of Organic Compounds*, D. Van Nostrand, New York, 1941, Vol. 1, pp. 238, 584; Vol. 2, pp. 10–11.

8. R. C. Weast, ed., *Handbook of Chemistry and Physics,* 55th ed., CRC Press, Cleveland, Ohio, 1974, p. E-32.
9. *Beilstein Handbuch der Organischen Chemie,* Vol. 3, J. Springer, Berlin, Ger., 1921, p. 198.
10. A. Eucken, *Physik Z.* **12,** 1101 (1911); **14,** 324 (1913).
11. D. R. Stull, E. F. Westrum, and G. C. Sinke, *The Chemical Thermodynamic Properties of Organic Compounds,* John Wiley & Sons, Inc., New York, 1969, p. 220.
12. E. G. Butcher, J. A. Weston, and H. Gebbie, *J. Chem. Phys.* **41,** 2554 (1964).
13. A. Seidell and W. F. Linke, *Solubilities of Inorganic and Organic Compounds,* D. Van Nostrand, New York, 1952, p. 101.
14. CS_2, brochure, Stauffer Chemical Company, Westport, Conn., 1975.
15. F. J. Touro and T. W. Wiewiorski, *J. Phys. Chem.* **70,** 3534 (1968).
16. B. J. Zwolinski and co-workers, *Catalog of Selected Ultraviolet Spectral Data, Serial No. 100,* Manufacturing Chemists' Association Research Project, Texas A&M University, College Station, Tex., 1965.
17. B. J. Zwolinski and co-workers, *Catalog of Selected Infrared Spectral Data, Serial No. 321,* Thermodynamics Research Center Data Project, Texas A&M University, College Station, Tex. 1967.
18. J. H. Meidl, *Flammable Hazardous Materials,* Glencoe Press, Beverly Hills, Calif., 1970, pp. 25–40, 173–178.
19. A. G. White, *J. Chem. Soc.* **121,** 1244 (1922).
20. G. Peters and W. Ganter, *Angew. Chem.* **51,** 29 (1938).
21. C. Bondroit, *Rev. Univers. Mines Metall Mec.* **15,** 197 (1939).
22. K. K. Kelley, *U.S. Bur. Mines Bull.* 406, (1937).
23. D. R. Stull, *Ind. Eng. Chem.* **41,** 1968 (1949).
24. R. F. Bacon and E. S. Boe, *Ind. Eng. Chem.* **37,** 469 (1945).
25. H. Rabe, *Chem. Ztg.* **50,** 609 (1926).
26. I. N. Agranovskii, *Khim. Technol. Serougleroda* (2), 8 (1970).
27. G. A. Richter, *Chem. Met. Eng.* **27**(17), 838 (1922).
28. A. A. Pedro and co-workers, *Khim. Volokna* **14**(2), 65 (1972).
29. Can. Pat. 573,356 (Mar. 31, 1959), H. S. Johnson and J. Reid (to Shawinigan Chemicals Limited).
30. R. M. Levit and co-workers, *Khim. Technol. Serougleroda* **1970**(2), 63 (1972).
31. E. A. Sondreal, A. M. Cooley, and R. C. Ellman, *U.S. Bur. Mines Rep. Invest.* 6891, (1967).
32. U.S.S.R. Pat. 139,310 (Sept. 4, 1960), I. Z. Sorokin and co-workers.
33. F. Molyneux, *Ind. Eng. Chem.* **54**(7), 50 (1962).
34. Jpn. Pat. 7,228,319 (July 27, 1972), Y. Harada and M. Miyamoto (to Asahi Chemical Industry Co. Ltd.).
35. Brit. Pat. 642,557 (Sept. 6, 1950), (to The Dow Chemical Co.).
36. U.S. Pat. 2,700,592 (Jan. 25, 1955), T. D. Heath (to Dorr Company).
37. U.S. Pat. 2,443,854 (June 22, 1948), R. P. Ferguson (to Standard Oil Development Company).
38. Brit. Pat. 833,562 (Apr. 27, 1960), H. S. Johnson and J. Reid (to Shawinigan Chemicals Limited).
39. H. Sperling, *Chem. Tech. Berlin* **8,** 405 (1956).
40. U.S. Pat. 2,568,121 (Sept. 18, 1951), H. O. Folkins, C. A. Porter, E. Miller, and H. Hennig (to The Pure Oil Co.).
41. H. W. Haines, Jr., *Ind. Eng. Chem.* **55,** 44 (1963).
42. C. M. Thacker, *Hydrocarbon Process.* **49**(4), 124 (1970); **49**(5), 137 (1970).
43. H. O. Folkins, E. Miller, and H. Hennig, *Ind. Eng. Chem.* **42,** 2202 (1950).
44. Belg. Pat. 630,584 (Aug. 1, 1963), M. Preda.
45. U. Sborgi and E. Giovanni, *Chim. Ind. (Milan)* **31,** 391 (1949); Ital. Pat. 457,263 (May 12, 1950).
46. E. Giovanni, *Ann Chim. Appl.* **39,** 671 (1949).
47. U.S. Pat. 2,616,793 (Nov. 4, 1952), H. O. Folkins and E. Miller (to Food Machinery and Chemical Corp.).
48. U.S. Pat. 2,712,982 (July 12, 1955), K. W. Guebert (to The Dow Chemical Co.).
49. U.S. Pat. 2,536,680 (Aug. 21, 1951), H. O. Folkins and E. Miller (to The Pure Oil Company).
50. U.S. Pat. 2,668,752 (Feb. 9, 1954), H. O. Folkins and E. Miller (to Food Machinery and Chemical Corp.).
51. U.S. Pat. 2,411,236 (Nov. 19, 1946), C. M. Thacker (to The Pure Oil Co.).
52. U.S. Pats. 2,712,984; 2,712,985 (July 12, 1955), K. W. Guebert (to The Dow Chemical Co.).
53. U.S. Pat. 2,187,393 (Jan. 16, 1940), M. DeSimo (to Shell Development Co.).
54. W. J. Thomas and B. John. *Trans. Inst. Chem. Eng.* **45**(3), T119 (1967); *Chem. Eng. (London)* 207, (1967).

55. W. J. Thomas and S. C. Naik, *Trans. Inst. Chem. Eng.* **48**(4–6), 129 (1970).
56. R. C. Forney and J. M. Smith, *Ind. Eng. Chem.* **43**, 1841 (1951).
57. C. M. Thacker and E. Miller, *Ind. Eng. Chem.* **36**, 182 (1944).
58. Y. P. Tret'yakov, *Zh. Chim.* Abstr. No. 6L47 (1962).
59. U.S. Pat. 2,330,934 (Oct. 5, 1943), C. M. Thacker (to The Pure Oil Co.).
60. U.S. Pat. 2,428,727 (Oct. 7, 1947), C. M. Thacker (to The Pure Oil Co.).
61. U.S. Pat. 2,492,719 (Dec. 27, 1949), C. M. Thacker (to The Pure Oil Co.).
62. U.S. Pat. 2,666,690 (Jan. 19, 1954), H. O. Folkins, E. Miller, and H. Hennig (to Food Machinery and Chemical Corp.).
63. U.S. Pat. 2,709,639 (May 31, 1955), H. O. Folkins, E. Miller, and H. Hennig (to Food Machinery and Chemical Corp.).
64. U.S. Pat. 2,565,215 (Aug. 21, 1951), H. O. Folkins and E. Miller (to The Pure Oil Co.).
65. R. A. Fisher and J. M. Smith, *Ind. Eng. Chem.* **42**, 704 (1950).
66. G. W. Nabor and J. M. Smith, *Ind. Eng. Chem.* **45**, 1272 (1953).
67. J. R. West, *Ind. Eng. Chem.* **42**, 713 (1950).
68. U.S. Pat. 2,857,250 (Oct. 21, 1958), R. W. Timmerman, A. G. Draeger, and J. W. Getz (to Food Machinery and Chemical Corp.).
69. U.S. Pat. 2,882,131 (Apr. 14, 1959), J. W. Getz and R. W. Timmerman (to Food Machinery and Chemical Corp.).
70. U.S. Pat. 2,369,377 (Feb. 13, 1945), C. M. Thacker (to The Pure Oil Co.).
71. U.S. Pat. 2,474,067 (June 21, 1949), L. Preisman (to Barium Reduction Corp.).
72. U.S. Pat. 2,661,267 (Dec. 1, 1953), H. O. Folkins, E. Miller, and H. Hennig (to Food Machinery and Chemical Corp.).
73. U.S. Pat. 2,708,154 (May 10, 1955), H. O. Folkins and E. Miller (to Food Machinery and Chemical Corp.).
74. U.S. Pat. 2,882,130 (Apr. 14, 1959), D. J. Porter (to Food Machinery and Chemical Corp.).
75. U.S. Pat. 3,087,788 (Apr. 30, 1963), D. J. Porter (to FMC Corporation).
76. Fr. Pat. 1,482,173 (May 26, 1967), J. Berthoux, J. P. Quillet, and G. Schneider (to Societe Progil).
77. Fr. Pat. 1,493,586 (Sept. 1, 1967), J. Berthoux, J. P. Quillet, and G. Schneider (to Societe Progil).
78. Fr. Pat. 1,573,969 (July 11, 1969), P. Gerin, L. Louat, and J. P. Quillet (to Societe Progil).
79. U.S. Pat. 3,079,233 (Feb. 26, 1963), C. J. Wenzke (to FMC Corporation).
80. U.S. Pat. 3,250,595 (May 10, 1966), D. R. Olsen (to FMC Corporation).
81. U.S. Pat. 3,876,753 (Apr. 8, 1975), J. L. Manganaro, M. Meadow, and S. Berkowitz (to FMC Corporation).
82. U.S. Pat. 3,927,185 (Dec. 16, 1975), M. Meadow and S. Berkowitz (to FMC Corporation).
83. Brit. Pat. 265,994 (Feb. 15, 1926), J. Komlos.
84. U.S. Pat. 1,907,274 (May 2, 1933), T. S. Wheeler and W. Francis (to Imperial Chemical Industries Ltd.).
85. H. E. Rasmussen, R. C. Hansford, and A. N. Sachanen, *Ind. Eng. Chem.* **38**, 376 (1946).
86. H. I. Waterman and C. Van Vlodrop, *J. Soc. Chem. Ind. London* **58**, 109 (1939).
87. U.S. Pat. 2,468,904 (May 3, 1949), C. R. Wagner (to Phillips Petroleum Co.).
88. Brit. Pat. 314,060 (June 23, 1928), H. Oehme (to Chemische Fabrik Kalk Ges.).
89. U.S. Pat. 1,193,210 (Aug. 1, 1916), A. Walter.
90. N. P. Galenko and co-workers, *Gaz. Prom.* **5**(12), 46 (1960).
91. C. W. Siller, *Ind. Eng. Chem.* **40**, 1227 (1948).
92. A. Stock, W. Siecke, and E. Pohlard, *Ber. Dtsch. Chem. Ges. B* **57**, 719 (1924).
93. Ger. Pat. 938,124 (Jan. 26, 1955), H. Welz (to Farbenfabriken Bayer).
94. U.S. Pat. 2,767,059 (Oct. 16, 1956), W. E. Adcock and W. C. Lake (to Stanolind Oil and Gas Co.).
95. Fr. Pat. 1,274,034 (Oct. 20, 1961), (to Hamburger Gaswerke G.m.b.H. and Salzgitter Industriebau G.m.b.H.).
96. *Hazardous Materials Regulations of the Department of Transportation,* U.S. Department of Transportation, Washington, D.C., 1975.
97. *Regulations Governing the Transportation or Storage of Explosive and Combustible Liquids on Board Vessels,* U.S. Coast Guard Department of Transportation, Washington, D.C., 1974.
98. *IATA Restricted Articles Regulations,* 19th ed., International Air Transportation Assoc., Geneva, Switz., 1976.
99. *Unloading Flammable Liquids from Tank Cars, Manual Sheet TC-4,* Manufacturing Chemists' Association, Washington, D.C., 1969.

100. *Synthetic Organic Chemicals, U.S. Production and Sales,* U.S. International Trade Commission, Washington, D.C., annual.
101. *Chemical Pricing Patterns,* 3rd ed., Schnell, New York, 1971 (plus industry sources).
102. *Sulfur* **133,** 23 (Nov.–Dec. 1977).
103. *Carbon Disulfide, Technical Grade, United States Federal Specification, Document No. O-C-131,* Government Printing Office, Washington, D.C., Jan. 29, 1952 and Apr. 14, 1954.
104. *Reagent Chemicals, American Chemical Society Specifications,* 5th ed., American Chemical Society, Washington, D.C., 1974, p. 177.
105. F. A. Patty, *Industrial Hygiene and Toxicology,* 2nd rev. ed., Vol. II, John Wiley & Sons, Inc., New York, 1962, pp. 901–904.
106. R. O. Beauchamp, Jr., *Toxicity and Analysis of Carbon Disulfide. Annotated Bibliography, Report 1974, ORLN-TIRC-74-1,* Toxicological Information Response Center, Oak Ridge, Tenn., 1974.
107. H. Brieger and J. Tessinger, eds., *Toxicology of Carbon Disulfide,* Excerpta Medica Foundation, New York, 1967.
108. A. D. Baskin, ed., *Handling Guide for Potentially Hazardous Materials,* Materials Management and Safety, Chicago, Ill., 1975, Data Forms No. 1778.
109. M. Windholz, ed., *The Merck Index,* 9th ed., Merck, Rahway, N.J., 1976, p. 1819.
110. V. M. Vervel'skaya, *Vestn. Oftal'mol* **80**(2), 59 (1967).
111. D. Djuric, N. Surducki, and I. Berkes, *Br. J. Ind. Med.* **22,** 321 (1965).
112. R. E. Gosselin and co-workers, *Clinical Toxicology of Commercial Products,* 4th ed., Williams and Wilkins, Baltimore, Md., 1976, Sec. III, pp. 83–86.
113. *Registry of Toxic Effects of Chemical Substances,* U.S. Department of Health, Education, and Welfare, Rockville, Md, 1976, p. 303.
114. *Documentation of the Threshold Limit Values for Substances in the Workroom Air,* American Conference of Governmental Industrial Hygienists, Cincinnati, Ohio, 1971 ed., pp. 39–40; 1976 ed., p. 11.
115. A. Hamilton and H. L. Hardy, *Industrial Toxicology,* 3rd ed., Publishing Sciences Group, Acton, Mass., 1974, p. 326.
116. P. B. Vinogradov, *Gig. Sanit.* **31**(1), 13 (1966).
117. T. Dutkiewicz and B. Baranowska, *Int. Congr. Occup. Health, 14th, Madrid, Spain 1963* (2), 715 (1964).

ROBERT W. TIMMERMAN
FMC Corporation

CARBONIC AND CHLOROFORMIC ESTERS

The reaction of phosgene (carbonic dichloride) with alcohols gives two classes of compounds, carbonic esters and chloroformic esters. The carbonic esters (carbonates), ROC(O)OR, are the diesters of carbonic acid. The chloroformic esters (chloroformates, chlorocarbonates), ClC(O)OR, are the esters of hypothetical chloroformic acid, ClCOOH.

The reaction proceeds in stages, first producing a chloroformate, and then a carbonic acid diester. When a different alcohol is used for the second stage, a mixed radical or unsymmetrical carbonate is produced.

$$COCl_2 + ROH \rightarrow ClCOOR + HCl \xrightarrow{R'OH} R'OCOOR + HCl$$

CHLOROFORMIC ESTERS

In the literature, chloroformates are also referred to as chlorocarbonates because of the structural parallel with these acids:

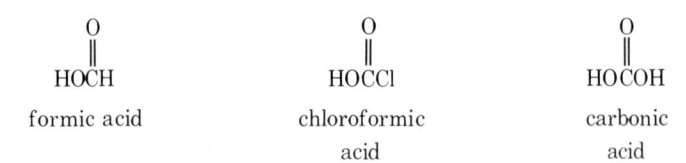

$$\underset{\text{formic acid}}{HOCH=O} \qquad \underset{\text{chloroformic acid}}{HOCCl=O} \qquad \underset{\text{carbonic acid}}{HOCOH=O}$$

Before 1972, chloroformates were indexed in *Chemical Abstracts, Eighth Collective Index,* under formic acid, chloroesters, whereas in the *Ninth Collective Index,* the name carbonochloridic acid, esters, is used. Table 1 lists the common names of the chloroformates, the Chemical Abstracts Service Registry Numbers, and the formulas.

Physical Properties

In general, chloroformates are clear, colorless liquids with low freezing points and high boiling points. They are soluble in most organic solvents, but insoluble in water, although they do hydrolyze slowly. The physical properties of the most widely used chloroformic esters are given in Table 2 (1).

Chemical Properties

Chloroformates are reactive intermediates that combine acid, chloride, and ester functions. They undergo many reactions similar to those of acid chlorides however, the rates are usually slower (2). Reactions of chloroformates and other acid chlorides proceed faster with better yields when alkali hydroxides or tertiary amines are present to react with the HCl as it forms. These bases act as stoichiometric acid acceptors rather than as true catalysts.

Table 1. Chloroformates

Chloroformate	CAS Registry No.	Formula
methyl	[79-22-1]	ClCOOCH$_3$
chloromethyl	[22128-62-7]	ClCOOCH$_2$Cl
dichloromethyl	[22128-63-8]	ClCOOCHCl$_2$
ethyl	[541-41-3]	ClCOOCH$_2$CH$_3$
2-chloroethyl	[627-11-2]	ClCOOCH$_2$CH$_2$Cl
vinyl	[5130-24-5]	ClCOOCH=CH$_2$
isopropyl	[108-23-6]	ClCOOCH(CH$_3$)$_2$
n-propyl	[109-61-5]	ClCOOCH$_2$CH$_2$CH$_3$
allyl	[2937-50-0]	ClCOOCH$_2$CH=CH$_2$
methallyl	[42068-70-2]	ClCOOCH$_2$(CH$_3$)C=CH$_2$
n-butyl	[592-34-7]	ClCOOCH$_2$CH$_2$CH$_2$CH$_3$
sec-butyl	[17462-58-7]	ClCOOCH(CH$_3$)CH$_2$CH$_3$
isobutyl	[543-27-1]	ClCOOCH$_2$CH(CH$_3$)$_2$
n-amyl	[638-41-5]	ClCOOC$_5$H$_{11}$
isoamyl	[628-50-2]	ClCOOCH$_2$CH$_2$CH(CH$_3$)$_2$
neopentyl	[20412-38-8]	ClCOOCH$_2$C(CH$_3$)$_3$
2-ethylhexyl	[24468-13-1]	ClCOOCH$_2$CH(C$_2$H$_5$)(CH$_2$)$_3$CH$_3$
2-octyl	[15586-11-5]	ClCOOCH(CH$_3$)(CH$_2$)$_5$CH$_3$
n-decyl	[55488-51-2]	ClCOOCH$_2$(CH$_2$)$_8$CH$_3$
dodecyl	[24460-74-0]	ClCOOCH$_2$(CH$_2$)$_{10}$CH$_3$
octadecyl	[51637-93-5]	ClCOOCH$_2$(CH$_2$)$_{16}$CH$_3$
phenyl	[1885-14-9]	ClCOOC$_6$H$_5$
benzyl	[501-53-1]	ClCOOCH$_2$C$_6$H$_5$
1-naphthyl	[3759-61-3]	ClCOOC$_{10}$H$_7$
2-naphthyl	[7693-50-7]	ClCOOC$_{10}$H$_7$
ethylene bis	[124-05-0]	ClCOOCH$_2$CH$_2$OOCCl
diethylene glycol bis	[106-75-2]	ClCOOCH$_2$CH$_2$OCH$_2$CH$_2$OOCCl

Stability. The ester moiety determines thermal stability, generally in the following order of decreasing stability: aryl > primary alkyl > secondary alkyl > tertiary alkyl. Moreover, stability is greatly affected by impurities, particularly metallic contaminants. For example, iron, zinc, or aluminum chlorides (but not lead chloride) catalyze decomposition of the lower alkyl chloroformates which are moderately stable in the absence of iron. Free acid in technical-grade chloroformates attacks iron, producing the catalytically active iron salt.

Isopropyl chloroformate is particularly sensitive to iron salts and is also thermally unstable at its boiling point; it must be distilled in glass vessels under vacuum.

Decomposition may also be initiated by tertiary amines such as pyridine or quinoline (3):

$$ClCOOCH_2CH_2R \rightarrow CH_2=CHR + HCl + CO_2$$
$$ClCOOCH_2CH_2R \rightarrow ClCH_2CH_2R + CO_2$$

Reactions with Hydroxylic Compounds. Reaction with water gives the parent hydroxy compound, HCl and CO$_2$, in addition to the symmetrical carbonate formed by the hydroxy compound and the chloroformate.

$$2 \text{ ROCCl} + H_2O \rightarrow \text{ROCOR} + CO_2 + HCl$$
(with C=O groups shown above ROCCl and ROCOR)

Table 2. Physical Properties of Selected Chloroformates

Chloroformate	Mol wt	Sp gr, d_4^{20}	Refractive index, n_D^{20}	Flash point °C TOC[a]	Flash point °C TCC[b]	Viscosity, mPa·s 20°C[c]	bp, °C 2.67 kPa[d]	bp, °C 13.3 kPa[d]	bp, °C 101.3 kPa[d]
methyl	94.50	1.250	1.3864	24.4	17.8				71
ethyl	108.53	1.138	1.3950	27.8	18.3				94
isopropyl	122.55	1.078	1.3974	27.8	23.3	0.65	25.3	47	105
n-propyl	122.55	1.091	1.4045	34.4		0.80	25	57.5	112.4
allyl	120.5	1.1394	1.4223	27.8	31.1	0.71	44	57	
n-butyl	136.58	1.0585	1.4106	52.2	46.0	0.888	36	77.6	
sec-butyl	136.58	1.0493	1.4560	35.6	38.0	0.897	39	69	
isobutyl	136.58	1.0477	1.4079	39.5	34.4	0.88	98	71	
2-ethylhexyl	192.7	0.9914	1.4307		86.0	1.774	122	137	
n-decyl	220.7	0.9732	1.4400	118.3	120.2	3.00	83.5	159	
phenyl	156.57	1.2475	1.5115	80.0	77.0	1.882	103	121	185
benzyl	170.6	1.2166	1.5175	134.0	107.9	2.57	108	123	152
ethylene bis	186.98	1.4704	1.4512	160.0	126.0	4.78		137	
diethylene glycol bis	231.0	1.388	1.4550		182.2	8.76	148	180	

[a] Tag open cup.
[b] Tag closed cup.
[c] mPa·s = cP.
[d] To convert kPa to mm Hg, multiply by 7.50.

Alkali Metal Hydroxides. Addition of less than stoichiometric amounts of metal hydroxides gives the parent hydroxy compound. However, the use of stoichiometric amount of base yields the symmetrical carbonate, particularly in the case of aryl chloroformates (4).

$$2\ \text{ROCCl} + 4\ \text{NaOH} \longrightarrow \text{ROCOR} + \text{Na}_2\text{CO}_3 + 2\ \text{NaCl} + 2\ \text{H}_2\text{O}$$
(where ROCCl and ROCOR have C=O groups)

Aliphatic Alcohols and Thiols. Aliphatic alcohols give carbonates and hydrogen chloride. Frequently, the reaction proceeds at room temperature without a catalyst or hydrogen chloride acceptor. However, faster reactions and better yields are obtained in the presence of alkali metals or their hydroxides, or tertiary amines. Reactions of chloroformates with thiols yield monothiolocarbonates (5–6).

$$\text{ROCCl} + \text{R'OH} \longrightarrow \text{R'OCOR} + \text{HCl}$$

$$\text{ROCCl} + \text{R'SH} \longrightarrow \text{R'SCOR}$$

Phenols are unreactive toward chloroformates at room temperature, whereas at elevated temperatures yield of less than 10% of the expected carbonates are produced. However, under alkaline conditions, high yields are obtained. The addition of agents such as activated carbon, fuller's earth, and alumina, prevents side reactions (7).

$$\text{ArOH} + \text{NaOH} + \text{ROCCl} \longrightarrow \text{ArOCOR} + \text{NaCl} + \text{H}_2\text{O}$$

Heterocyclic Alcohols give carbonates. Furan- and tetrahydrofuran-derived alcohols give carbonates in 75% yield (8). Inorganic bases and tertiary amines increase reactivity and yields.

Reactions With Carboxylic Acids. Esters are the main products in the reaction with carboxylic acids.

$$\text{ROCCl} + \text{R'CO}_2\text{H} \longrightarrow [\text{ROCOCR'}] \longrightarrow \text{ROCR'} + \text{CO}_2 + \text{HCl}$$

The intermediate mixed carboxylic–carbonic anhydrides are very active acylating agents and may be isolated (9–11). Pyrocarbonates (anhydrides of carbonic acids) have been prepared as follows (12):

$$\text{ROCCl} + \text{KOR} + \text{CO}_2 \longrightarrow \text{ROCOCOR} + \text{ROCOR}$$

$$\text{ROCCl} + \text{ROCONa} \longrightarrow \text{ROCOCOR} + \text{NaCl}$$

$$\text{ROCCl} + \text{KOCOR} \longrightarrow \text{ROCOCOR} + \text{KCl}$$

Reactions With Nitrogen Compounds. The reaction with ammonia is the classical method for preparing primary carbamates. Excess ammonia is used to remove the hydrogen chloride as it is evolved (see Carbamic acid).

$$ROOCCl + 2\ NH_3 \rightarrow ROOCNH_2 + NH_4Cl$$

Amines. Primary and secondary aliphatic amines also yield carbamates in the presence of excess amine or inorganic bases. Aromatic primary and secondary amines and heterocyclic amines react similarly although more slowly.

$$ClCOOR + 2\ R'NH_2 \rightarrow R'NHCOOR + R'NH_2 \cdot HCl$$

Tertiary amines give quaternary ammonium compounds (13).

$$ROOCCl + NR_1R_2R_3 \rightarrow [ROOCNR_1R_2R_3]^+Cl^-$$

Amino Alcohols react more rapidly at the amino group than at the hydroxyl group (14). The hydroxycarbamates obtained can be cyclized with base to yield 2-oxazolidones (14). Nonionic detergents may be prepared from polyethylene glycol bis(chloroformates) and long-chain tertiary amino alcohols (14).

Aminophenols are also more reactive at the amino group than at the hydroxyl group. With *o*-aminophenol, benzoxazolones are produced by cyclization of the intermediate carbamate (15).

$$ArOCCl + \underset{NH_2}{\underset{|}{\text{o-aminophenol-OH}}} \rightarrow \underset{NHCOOAr}{\underset{|}{\text{-OH}}} \rightarrow \text{Benzoxazolone} + ArOH$$

Amino Acids. Chloroformates, such as benzyl chloroformate, protect the amine group of amino acids during peptide synthesis. The protective carbamate can be cleaved by hydrogenolysis to free the amine after the carboxyl group has reacted further (16–17).

$$ArOOCCl + H_2NR'CO_2H \rightarrow ArOOCNHR'CO_2H + HCl$$

Acylation. Aryl chloroformates are good acylating agents, reacting with aromatic hydrocarbons under Friedel-Crafts conditions to give the expected aryl esters of the aromatic acid (18) (see Friedel-Crafts reactions).

$$ArOOCCl + Ar'\text{-H} \xrightarrow{AlCl_3} Ar'COOAr + HCl$$

However, alkylation takes place with aliphatic chloroformates under similar conditions (19).

$$ROOCCl + Ar\text{-H} \xrightarrow{AlCl_3} R\text{-}Ar + CO_2 + HCl$$

Miscellaneous Reactions. Epoxy compounds yield chloro-substituted carbonates (20).

$$ROCCl + CH_2CHCH_2Cl \text{ (epoxide)} \rightarrow ROCOCH_2CHCH_2Cl \text{ (with Cl substituent)}$$

Metal peroxides or hydrogen peroxide and base react to give peroxy compounds (21–22).

$$2\ ROCCl + Na_2O_2 \longrightarrow ROC(=O)-OO-C(=O)OR + 2\ NaCl$$

With sodium xanthates, alkyl xanthogen formates are obtained which are useful in ore flotation processes (23).

$$ROCCl + NaSCOR' \longrightarrow ROCSCOR' + NaCl$$

Manufacture

The reaction of phosgene with alcohols or phenols is thoroughly discussed by Matzner and co-workers (1). Alkyl chloroformates are prepared by the reaction of liquid anhydrous alcohols with a molar excess of dry, chlorine-free phosgene at low temperature. Corrosion-resistant reactors, lines, pumps, and valves are required. Materials of construction include glass, porcelain, stainless steel, nickel, lead, or chemically impregnated carbon such as Karbate. Temperatures are kept at 1–10°C for the lower alcohols and may rise to 60°C for the higher alcohols. Hydrogen chloride is evolved as the reaction proceeds and is then absorbed in a tower. Phosgene is usually added to the alcohol. However, when acid acceptors are used, the alcohol is added to an excess of liquid phosgene in order to minimize formation of dialkyl carbonates. Unreacted phosgene is removed from the crude reaction product by vacuum stripping or gas purging. Washing with cold water removes most of the HCl present. A recent patent (24) claims that washing chloroformates with an NaCl solution reduces the iron content. Washing with an aqueous alkali hydroxide solution reduces both HCl and iron content.

Chloroformates of lower primary alcohols are distillable; however, heavy metal contamination should be avoided. The chloroformic esters of higher primary and secondary alcohols and glycols are not ordinarily distillable.

Primary alcohols give yields of chloroformates well above 90%, whereas secondary alcohols give yields of 70–80%.

With an acid acceptor, a solvent such as chloroform, toluene, dioxane, or THF is used to dissolve the chloroformate as it is formed.

Shipping and Storage

Chloroformates are shipped in 0.208 m³ (55-gal), safety-vented or unvented, nonreturnable, polyethylene drums with steel overpack. Isopropyl chloroformate containers are refrigerated, and bulk shipments are made in stainless steel refrigerated tank trucks.

Chloroformates should be stored in a cool, dry atmosphere. They should be transferred from containers to storage tanks or reactors through a closed system using stainless steel and nickel, or glass pumps, lines, and valves.

The U.S. Department of Transportation (Bureau of Hazardous Materials) and the U.S. Coast Guard classify most chloroformates as flammable or combustible liquids, requiring the "Flammable," or "Combustible Liquid," label. However, decyl chloroformate and diethylene glycol bis(chloroformate) can be shipped as Chemicals NOI (Not Otherwise Indicated).

Economic Aspects

Most chloroformate production is used captively and production figures are not available. However, from the available data, the 1976 prices listed in Table 3 can be estimated.

Table 3. Estimated 1978 Prices of Chloroformates, $/kg

Chloroformate	0.208 m^3 (55 gal) drums	Tank trunk or carload
ethyl	1.54	1.34
methyl	1.39	1.19
isopropyl	1.45	1.25
isobutyl	8.80	2.29
ethylhexyl	7.48	1.98
benzyl	5.50	4.40

Specifications and Analysis

Table 4 lists the specifications of commercial chloroformates.

The lower-boiling chloroformates are analyzed by gas–liquid chromatography. Higher molecular weight chloroformates are first hydrolyzed and then analyzed by the Volhard method.

Table 4. Specifications of Commercial Chloroformates

Assay	Value
purity, %	95–98
phosgene, %	<1
iron, ppm	<10
acidity as HCl, %	<1
alcohol or phenol, %	<2

Toxicity

Chloroformates, especially those of low molecular weight, are lachrimators and vesicants and produce effects similar to those of hydrochloric acid or carboxylic acid chlorides. Inhalation of the vapors of lower chloroformates by laboratory animals (25–26) produced a condition similar to pneumonia, which in some cases was fatal. Moreover, chloroformates may hydrolyze in moist air and should be handled with proper ventilation and eye and skin protection.

Table 5 gives the toxicities of some chloroformates (25–26).

Table 5. LD$_{50}$ of Some Chloroformates

Chloroformate	Oral, mg/kg	Dermal, mg/kg	Inhalation, mg/L[a]
methyl	220	>2	<2.84
ethyl	411	>2	<1.62[b]
n-propyl	650	>10.2	319 ppm
isopropyl	177.8	12.1	299 ppm
allyl	178	1470	
phenyl	1581		
benzyl			<200
ethylene bis	1.1 g	2000	5659 ppm
diethylene glycol bis	813	3400	169 ppm

[a] For one hour exposure (mg/L of air).
[b] This value is a highly toxic rating and is equivalent to a Class B poison of the Hazardous Materials Regulations of DOT.

Uses

Chloroformates are versatile intermediates for various products including mixed or symmetrical carbonates (for solvents, plasticizers, or as intermediates for further synthesis), and percarbonates, for use as unsaturated monomer polymerization initiators (see Peroxides). The most widely used are listed below:

Percarbonate	CAS Registry No.
diethyl	[14666-78-5]
diisopropyl	[105-64-6]
di-n-butyl	[16215-49-9]
di-sec-butyl	[19910-65-7]
dicyclohexyl	[1561-49-5]
di-4-tert-butylcyclohexyl	[26523-73-9]
di-n-hexadecyl	[26322-14-5]

Carbamates derived from chloroformates are used to manufacture pharmaceuticals including tranquilizers (27), antihypotensives and local anaesthetics, and pesticides and insecticides (see Carbamic acid).

Ethyl chloroformate is used in the manufacture of ore flotation agents by reaction with various alkyl xanthates (23).

Diethylene glycol bis(chloroformate) [106-75-2] is the starting material for diethylene glycol bis(allyl carbonate) [142-22-3], CR-39 monomer, used in the manufacture of break-resistant optical lenses, which is obtained by the reaction with allyl alcohol (28). Alternatively, it can be made from allyl chloroformate [2937-50-0] and diethylene glycol (29).

Bis(chloroformic esters) condense with diamines to give polyurethanes (29).

Blowing agents for producing foam rubber, polyethylene, and vinyl chloride are made from haloformates, hydrazine, and a base. For example, diisopropyl azodiformate is made from isopropyl chloroformate (30).

CARBONIC ESTERS

As discussed under Chloroformates, chloroformates and alcohols or phenols give carbonic diesters. In addition, the higher diesters can be made from the lower ones by alcoholysis or ester interchange by heating the lower diester with a higher alcohol in the presence of HCl, H_2SO_4, or sodium alcoholate. The lower alcohol formed is removed by fractional distillation. Mixed diesters are prepared by treating a chloroformate with an alcohol or phenol having a different radical.

Carbonates are indexed in *Chemical Abstracts* under carbonic acid, esters. Symmetrical diesters have the prefix di or bis. Unsymmetrical diesters are listed with the two radicals following each other. For example, ethyl phenyl carbonic acid diester is $C_2H_5OCOOC_6H_5$.

Table 6 lists commonly used carbonates, Chemical Abstracts Service Registry Numbers, and formulas.

Properties

The physical properties are given in Table 7. The lower alkyl carbonates are neutral, colorless liquids with a mild odor. Aryl carbonates are normally solids. The alkyl esters, especially the lower ones, hydrolyze very slowly in water. Under alkaline conditions, the rates of hydrolysis are similar to those of the corresponding acetic acid

Table 6. Carbonates

Carbonate	CAS Registry No.	Formula
dimethyl	[616-38-6]	$CH_3OCOOCH_3$
diethyl	[105-58-8]	$C_2H_5OCOOC_2H_5$
divinyl	[7570-02-7]	$CH_2{=}CHOCOOCH{=}CH_2$
di-*n*-propyl	[623-96-1]	$CH_3CH_2CH_2OCOOCH_2CH_2CH_3$
diisopropyl	[6482-34-4]	$(CH_3)_2CHOCOOCH(CH_3)_2$
diallyl	[15022-08-9]	$CH_2{=}CHCH_2OCOOCH_2CH{=}CH_2$
methyl allyl	[35466-83-2]	$CH_3OCOOCH_2CH{=}CH_2$
diisobutyl	[539-92-4]	$(CH_3)_2CHCH_2OCOOCH_2CH(CH_3)_2$
isobutyl propyl	[40882-93-7]	$CH_3CH(CH_3)CH_2OCOOCH_2CH_2CH_3$
di-*tert*-butyl	[34619-03-9]	$(CH_3)_3COCOOC(CH_3)_3$
di-*sec*-butyl	[623-63-2]	$CH_3CH_2CH(CH_3)OCOOCH(CH_3)CH_2CH_3$
di-*n*-butyl	[542-52-9]	$CH_3CH_2CH_2CH_2OCOOCH_2CH_2CH_2CH_3$
hexyl methyl	[39511-75-6]	$CH_3CH_2CH_2CH_2CH_2CH_2OCOOCH_3$
pentyl propyl	[40882-94-8]	$CH_3CH_2CH_2CH_2CH_2OCOOCH_2CH_2CH_3$
didodecyl	[6627-45-8]	$CH_3(CH_2)_{10}CH_2OCOOCH_2(CH_2)_{10}CH_3$
diphenyl	[102-09-0]	$C_6H_5OCOOC_6H_5$
phenyl allyl	[16308-68-2]	$C_6H_5OCOOCH_2CH{=}CH_2$
vinyl ethyl	[7570-06-1]	$CH_2{=}CHOCOOC_2H_5$
ethyl phenyl	[3878-46-4]	$C_2H_5OCOOC_6H_5$
ethylene	[96-49-1]	$\overline{OCOOCH_2CH_2}$
allyl diglycol[a]	[142-22-3]	$O(CH_2CH_2OCOOCH_2CH{:}CH_2)_2$
ditolyl[b]	[41903-18-8]	$CH_3C_6H_4OCOOC_6H_4CH_3$
dibenzyl	[3459-92-5]	$C_6H_5CH_2OCOOCH_2C_6H_5$
di-2-ethylhexyl	[14858-73-2]	$CH_3(CH_2)_3CH(C_2H_5)CH_2OCOOCH_2CH(C_2H_5)$-$(CH_2)_3CH_3$

[a] Diethylene glycol bis(allylcarbonate).
[b] Diethylene glycol bis(tolylcarbonate).

Table 7. Physical Properties of Selected Carbonates

Carbonate	Mol wt	Sp gr, d_4^{20}	Refractive index, n_D	Flash point, °C TOC[a]	Flash point, °C TCC[b]	Viscosity mPa·s[c]	Viscosity temp, °C	bp °C	bp kPa[d]
dimethyl	90.08	1.073	1.3697[e]	21.7	16.7	0.664	20	90.2	101.31
diethyl	118.13	0.975	1.3846[e]	46.1	32.8	0.868	15	23.8	1.33
								69.7	13.33
								126.8	101.31
di-n-propyl	146.18	0.941	1.4022[e]			[f]		165.5–166.6	101.31
diisopropyl	146.18				64			147.0	101.31
diallyl	142.15	0.994	1.4280[e]					97	8.13
								105	13.33
								166	97.31
di-n-butyl	174.14	0.9244	1.4099[g]					207.7	101.31
di-2-ethylhexyl	204.19	0.8974$_{20}^{20}$	1.4352[g]					173	1.33
diphenyl	214.08	1.1215$_{54}^{87}$						302	101.31
diethylene glycol bis(allyl)	274.3	1.143	1.4503	177[h]		9	25	160	0.27
tolyl diglycol	374.4	1.189	1.5229[g]					247–248	0.27
ethylene	88.06	1.3218$_4^{39}$	1.4158[i]					248	101.31

[a] Tag open cup.
[b] Tag closed cup.
[c] mPa·s = cP.
[d] To convert kPa to mm Hg, multiply by 7.5.
[e] Ref. 20.
[f] Brookfield no. 1 spindle; rpm (mPa·s) 10(5), 20(6.5), 50(8.0), 100(12.0).
[g] Ref. 25.
[h] Cleveland open cup.
[i] Ref. 31.

esters. In the presence of a basic catalyst the lower alkyl carbonates may be transesterified, a convenient method for synthesizing higher carbonates. Fiber- and film-forming polycarbonates are produced by transesterifying dialkyl, dicycloalkyl, or diaryl carbonates with alkyl, cycloalkyl, or aryl dihydroxy compounds (32).

Carbonates give carbamic esters with ammonia and amines and further reaction results in the corresponding ureas. With carbamates that are difficult to prepare, especially those made from tertiary alcohols, treatment with ammonia under pressure is required (33). Claisen condensations of alkyl carbonates and esters of aliphatic and aryl-substituted aliphatic acids produce α-carbalkoxy derivatives or derivatives of malonic esters (34).

Manufacture

Carbonates are manufactured by essentially the same method as chloroformates except that more alcohol is required in addition to longer reaction times and higher temperatures. The products are neutralized, washed, and distilled. Corrosion-resistant equipment is important. For secondary alkyl carbonates a tertiary organic base is added as acid acceptor. Diaryl carbonates are prepared from phosgene and two equivalents of the sodium phenolate. Mixed alkyl aryl carbonates can be prepared either by adding the alkyl chloroformate to the sodium phenoxide or by the reaction of aryl chloroformate with alcohol. An improved yield is claimed for the production of dibenzyl carbonate (35) by the reaction of sodium methyl carbonate [6482-39-9], $CH_3OCOONa$, with benzyl chloride in the presence of tertiary amine catalyst.

Alternatively, alkyl carbonates are prepared from saturated aliphatic alcohols with carbon monoxide in the presence of palladium or platinum salts (36) or an organic copper complex catalyst (31). Carbonates are also prepared by the reaction of an alcohol with carbonyl sulfide and amine in an oxidizing atmosphere (37). Chloro-substituted carbonates are made from chloroformates and chlorinated epoxy compounds (20).

The continuous production of high-purity methyl or ethyl carbonate from the alcohol and chloroformates has been patented (38). Chloroformate and alcohol are fed continuously into a Raschig ring-packed column in which a temperature gradient of 72–127°C is maintained between the base and head of the column; HCl is withdrawn at the head and carbonate (99%) is withdrawn at the base.

Dialkyl carbonates can also be made from alkylene carbonates such as ethylene carbonate, and an alcohol in the presence of a basic catalyst (39–40). Polyglycol bis(alkylcarbonates), specifically diethylene glycol esters, are prepared from diethylene glycol bis(chloroformate) with an alcohol in the presence of sodium hydroxide (41). Pyridine is an effective catalyst in the direct synthesis of alkyl carbonates in which phosgene is added to an excess of alcohol at elevated temperatures (42–43). Ethylene carbonate is obtained by reaction of ethylene oxide and CO_2 in the presence of a quaternary ammonium catalyst or by the reaction of the glycol with phosgene (44).

Shipping and Storage

Table 8 lists shipping conditions for the most important carbonates.

Diethylene glycol bis(allyl carbonate) should be stored in a cool place because its viscosity tends to increase with time and it may even polymerize at high temperatures. High humidity should be avoided because of hygroscopicity.

Table 8. Shipping Conditions of Commercial Carbonates

Carbonate	Label	Container, kg Steel drum[a]	Tank truck
diethyl	Flammable Liquid, N.O.S.	181[b]	16,200
dimethyl	Flammable Liquid	200	
dipropyl	Combustible Liquid N.O.S.	193	
diethylene glycol bis(allyl)	Chemicals, N.O.S.	227	

[a] Nonreturnable, 0.208 m^3 (55 gal).
[b] Also available in 0.019 m^3 (5 gal) steel drums, weighing 18 kg.

Economic Aspects

As in the case of the chloroformates, most of the carbonate production is used captively, and production figures are not available. However, from the available data it was possible to estimate the 1976 prices listed in Table 9.

Specifications

Table 10 lists the specifications of the more important commercial carbonates.

Health and Safety

Diethyl, dimethyl, and dipropyl carbonates are fire hazards. They are mildly irritating to skin, eyes, and mucous membranes. Protective clothing, rubber gloves, and goggles should be worn when handling these chemicals. Adequate ventilation should be provided. In case of fire, foam, carbon dioxide, or dry-chemical extinguishing agents should be used, not water. Diethylene glycol bis(allyl carbonate) may be irritating to the skin but it is not classified as a toxic substance, however, it is extremely irritating to the eyes.

Uses

Commercially, the most important carbonate is the diethyl ester. It is used in many organic syntheses, particularly of pharmaceuticals and pharmaceutical intermediates. It is also used as a solvent for many synthetic and natural resins, and in vacuum tube cathode-fixing lacquers. Dimethyl carbonate is used in the synthesis of pharmaceuticals, agricultural chemicals, dyestuffs, and as a specialty solvent. Dipropyl carbonate is also an organic intermediate, a specialty solvent, and is used in pho-

Table 9. Estimated 1978 Prices of Commercial Carbonates, $/kg

Carbonate	0.208 m^3 (55 Gal) drums	Tanktrunk, or carload
diethyl	1.94	1.80
dimethyl	1.58	1.45
diethylene glycol bis(allyl)	3.87	3.63

Table 10. Specifications of Commercial Carbonates

Assay	Diethyl	Dimethyl	Dipropyl	Diethylene glycol bis(allyl)
min, %	98.0	98.0	98.0	94.0
acidity, max %	0.02		0.02	
H_2O, max %	0.10	0.2	0.005	
nonvolatile matter, max %	0.005	0.01		[b]
sp gr, at d_4^{20}	0.973–0.977			1.14–1.16
at d_{20}^{20}		1.070–1.075	0.944–0.948	
ASTM distillation, °C				
range, 90% min	120–128			
none below	120		160	
none above	130		170	
initial bp		87		
dry pt, max		94.0[a]		
density at 20°C, kg/m³	3.69		3.656	4.264–4.400
surface tension at 20°C, mN/m (= dyn/cm)	26.31		29.8	35
viscosity at 25°C, mPa·s (= cP), max				25

[a] Residual methanol, max 0.6%.
[b] Volatility below 150°C at 0.667 kPa (5 mm Hg), 1.0% max.

toengraving as an assist agent for silicon circuitry. Ethylene carbonate is an important solvent for polymers such as polyacrylonitrile and may become an important raw material for synthesizing dialkyl carbonates (39–40). The lower alkyl carbonates and diphenyl carbonate are used in the preparation of polycarbonate resins by transesterification (32). Diethylene glycol bis(allyl carbonate), polymerizes easily because of its two double bonds and is used for colorless, optically clear castings. Polymerization is catalyzed by the use of diisopropyl peroxydicarbonate [105-64-6] (45–47). Such polymers are used in the preparation of safety glasses, light weight prescription lenses, glazing cast sheet, and optical cement (see Allyl monomers and polymers; Polycarbonates).

BIBLIOGRAPHY

"Carbonic Esters and Chloroformic Esters" in *ECT* 1st ed., Vol. 3, pp. 149–154, by Harry L. Fisher, U.S. Industrial Chemicals, Inc.; "Carbonic Esters and Chloroformic Esters" in *ECT* 2nd ed., Vol. 4, pp. 386–393, by William B. Tuemmler, FMC Corporation.

1. M. Matzner, R. P. Kurkjy, and R. J. Cotter, *Chem. Rev.* **64**, 645 (1964).
2. H. K. Hall, Jr., *J. Am. Chem. Soc.* **77**, 5993 (1955).
3. E. S. Lewis, W. C. Herndon, and D. D. Duffey, *J. Am. Chem. Soc.* **83**, 1959 (1961).
4. F. H. Carpenter and D. T. Gish, *J. Am. Chem. Soc.* **74**, 3818 (1952).
5. Ger. Pat. 1,080,546 (Apr. 28, 1960), K. Nagel and F. Korn (to Farbenfabriken Bayer A.-G.).
6. D. D. Reynolds and co-workers, *J. Org. Chem.* **26**, 5119 (1961).
7. Ger. Pat. 857,806 (Dec. 1, 1952), (to I.G. Farbenindustrie A.-G.).
8. J. L. R. Williams and co-workers, *J. Org. Chem.* **24**, 64 (1959).
9. Ger. Pat. 1,133,727 (July 26, 1962) V. Boellert, G. Fritz, and H. Schnell (to Farbenfabriken Bayer A.-G.).
10. D. S. Tarbell and N. A. Leister, *J. Org. Chem.* **23**, 1149 (1958).
11. T. B. Windholz, *J. Org. Chem.* **23**, 2044 (1958).
12. E. F. Degering, G. L. Jenkins, and B. E. Sanders, *J. Am. Pharm. Assoc.* **39**, 824 (1950).

13. T. Hopkins, *J. Chem. Soc.* **117,** 278 (1920).
14. U.S. Pat. 2,649,473 (Aug. 18, 1953), J. A. Chenicek (to Universal Oil Products Company).
15. L. C. Raiford and G. O. Inman, *J. Am. Chem. Soc.* **56,** 1586 (1934).
16. M. Bergmann and L. Zervas, *Berichte* **B65,** 1192 (1932).
17. L. F. Fieser and M. Fieser, *Advanced Organic Chemistry,* Reinhold Publishing Corp., New York, 1961, pp. 1040–1041.
18. W. H. Coppock, *J. Org. Chem.* **22,** 325 (1957).
19. S. Yura and T. Ono, *J. Soc. Chem. Ind. Japan* **48,** 30 (1945).
20. U.S. Pat. 2,518,058 (Aug. 8, 1950), A. Pechukas (to PPG).
21. F. Strain and co-workers, *J. Am. Chem. Soc.* **72,** 1254 (1950).
22. U.S. Pat. 2,370,588 (Feb. 27, 1945), F. Strain (to PPG).
23. U.S. Pats. 2,608,572 and 2,608,573 (Aug. 26, 1952), A. H. Fischer (to Minerec Corp.).
24. U.S. Pat. 3,576,838 (Apr. 27, 1971), J. L. Urness and L. L. Filius (to CPC International Inc.).
25. *Unpublished data,* Chemetron Corporation.
26. F. A. Patty ed., *Industrial Hygiene and Toxicology,* 2nd ed., Vol. 2, *Toxicology,* 1963, Interscience Publishers, Inc., a division of John Wiley & Sons, Inc., New York; G. D. Clayton and F. E. Clayton, eds., 3rd ed., Vol. 1, *General Principles,* 1978.
27. U.S. Pat. 2,937,119 (May 17, 1960), F. M. Berger and B. Ludwig (to Carter Products, Inc.).
28. U.S. Pats. 2,370,565 and 2,370,566 (Feb. 27, 1945), I. E. Muskat and co-workers (to PPG).
29. U.S. Pat. 2,708,617 (May 17, 1955), E. E. Magat and D. R. Strachan (to E. I. du Pont de Nemours & Co., Inc.).
30. U.S. Pat. 3,488,342 (Jan. 6, 1970), C. S. Sheppard and co-workers (to Pennwalt Corporation).
31. U.S. Pat. 3,846,468 (Nov. 5, 1974), E. Perrotti and G. Cipriani (to Snam Progetti S.p.A.).
32. U.S. Pat. 3,022,272 (Feb. 20, 1962), H. Schnell and G. Fritz (to Farbenfabriken Bayer, A.-G.).
33. U.S. Pat. 2,972,564 (Feb. 21, 1961), B. O. Melander and G. Hanshoff (to Aktiebolaget Kabi).
34. U.S. Pat. 2,454,360 (Nov. 23, 1948), V. H. Wallingford and A. H. Homeyer (to Mallinckrodt Chemical Works).
35. U.S. Pat. 2,983,749 (May 9, 1961), J. W. Shepard (to Callery Chemical Company).
36. U.S. Pat. 3,114,762 (Dec. 17, 1963), I. L. Mador and A. U. Blackham (to National Distillers and Chemical Corp.).
37. U.S. Pat. 3,632,624 (Jan. 4, 1972), J. E. Anderson and co-workers (to The Signal Companies, Inc.).
38. Fr. Pat. 2,163,884 (July 27, 1973), (to Société National des Poudres et Explosifs).
39. U.S. Pat. 3,642,858 (Feb. 15, 1972), L. K. Frevel and J. A. Gilpin (to The Dow Chemical Co.).
40. U.S. Pat. 3,803,201 (Apr. 9, 1974), J. A. Gilpin and A. H. Emmons (to The Dow Chemical Co.).
41. U.S. Pat. 2,379,252 (June 26, 1945), I. E. Muskat and F. Strain (to PPG).
42. L. W. Kissinger and co-workers, *J. Org. Chem.* **28,** 2491 (1963).
43. T. N. Hall, *J. Org. Chem.* **33,** 4457 (1968).
44. W. J. Peppel, *Ind. Eng. Chem.* **50,** 767 (1958).
45. F. Strain and co-workers, *J. Am. Chem. Soc.* **72,** 1254 (1950).
46. U.S. Pat. 3,022,281 (Feb. 20, 1962), and Brit. Pat. 851,964 (Oct. 19, 1960), E. S. Smith (to Goodyear Tire and Rubber Company).
47. U.S. Pat. 2,370,588 (Feb. 27, 1945), F. Strain (to PPG).

EDWARD ABRAMS
Chemetron Corporation

CARBONIZATION. See Coal—Coal conversion processes.

CARBON MONOXIDE

Carbon monoxide [630-08-0] (CO), a colorless, odorless, flammable, toxic gas, is produced by steam reforming or partial oxidation of carbonaceous materials. It is used as a fuel, a metallurgical reducing agent, and a feedstock in the manufacture of a variety of chemicals, notably methanol, acetic acid, phosgene, and oxo alcohols. Increased usage of carbon monoxide from coal in chemicals and fuels manufacture is likely if economic coal gasification technology evolves (see Fuels, synthetic).

Carbon monoxide was discovered by Lassonne in 1776 by heating a mixture of charcoal and zinc oxide. It provided a source of heat to the industry and home as a component of town gas and was used as a primary raw material in German synthetic fuel manufacture during World War II; its compounds with transition metals were studied extensively in the rebirth of modern inorganic chemistry in the 1960s. Most recently, carbon monoxide emission from vehicle exhausts has been recognized as a primary source of air pollution (qv).

Physical and Thermodynamic Properties

The physical and thermodynamic properties of carbon monoxide are well documented in a number of excellent summaries (1–8). The thermochemical data cited here are drawn predominantly from references 1–3; physical property data from reference 5. A summary of particularly useful physical constants is presented in Table 1.

Values for the free energy and enthalpy of formation, entropy, and ideal gas heat capacity of carbon monoxide as a function of temperature are listed in Table 2 (1). Thermodynamic properties have been reported from 70–300 K at pressures from 0.1–30 MPa (1–300 atm) (8–9) and from 0.1–120 MPa (1–1200 atm) (10).

Solid carbon monoxide exists in one of two allotropes: a body-centered cubic or a hexagonal structure. The body-centered structure converts into the hexagonal structure at 62 K with a heat of transition of 0.632 kJ/mol (0.151 kcal/mol) (5). The melting point at atmospheric pressure is 68.1 K and rises to 73 K at 20.7 MPa (205 atm) (5).

The vapor pressure of carbon monoxide has been compiled by Stull (11). Liquid phase vapor pressure is represented by equation 1, where P is the pressure in MPa (or atmospheres) and T is the temperature in degrees K (2).

$$\log_{10}P = 4.80780 \text{ (or } 3.81341) - \frac{291.743}{T - 5.15} \quad (1)$$

The compressibility factor, Z, can be calculated (12) for carbon monoxide at pressures up to 8 MPa (80 atm) by (eq. 2):

$$Z = 1 + \frac{BP}{RT} + \frac{(C-B^2)P^2}{(RT)^2} \quad (2)$$

Values for B and C are listed in Table 3. An alternative equation is available (12) to extend the range of calculated compressibilities to pressures of up to 300 MPa (3000 atm). Additional compressibility data over a wide range of conditions have been reported (13–14).

The density of liquid carbon monoxide at various temperatures is listed in Table

Table 1. Physical Properties of Carbon Monoxide

Property	Value
mol wt	28.011
mp	68.09 K
bp	81.65 K
ΔH, fusion (68 K)[a]	0.837 kJ/mol
ΔH, vaporization (81 K)[a]	6.042 kJ/mol
density [273 K, 101.33 kPa (1 atm)]	1.2501 g/L
sp gr, liquid, 79 K[b]	0.814
sp gr, gas, 298 K[c]	0.968
critical temperature	132.9 K
critical pressure	3.496 MPa (34.5 atm)
critical density	0.3010 g/cm^3
triple point	68.1 K/15.39 kPa (−205°C/115.4 torr)
$\Delta G°$ formation (298 K)[a]	−137.16 kJ/mol
$\Delta H°$ formation (298 K)[a]	−110.53 kJ/mol
S° formation (298 K)[a]	0.1975 kJ/(mol·K)
$C_p°$ (298 K)[a]	29.1 J/(mol·K)
$C_v°$ (298 K)[a]	20.8 J/(mol·K)
autoignition temperature	925 K
bond length	0.11282 nm
bond energy[a]	1070 kJ/mol
force constant	1902 N/m (19.02 × 10^5 dyn/cm)
dipole moment	0.374 × 10^{-30} C·m (0.112 debye)
ionization potential	1352 kJ (14.01 eV)
flammability limits in air[d]	
upper limit, %	74.2
lower limit, %	12.5

[a] To convert J to cal, divide by 4.184.
[b] With respect to water at 277 K.
[c] With respect to air at 298 K.
[d] Saturated with water vapor at 290 K.

Table 2. Thermodynamic Data for Carbon Monoxide (Ideal Gas)[a,b]

Temperature, K	$C_p°$, J/(mol·K)	$\Delta H_f°$, kJ/mol	$\Delta G_f°$, kJ/mol	S°, J/(mol·K)
0	0.00	−113.80	−113.80	0.00
100	29.104	−112.45	−120.25	165.74
298	29.142	−110.53	−137.16	197.54
400	29.342	−110.11	−146.34	206.12
500	29.794	−110.02	−155.41	212.72
1000	33.183	−112.01	−200.24	234.42
5000	38.074	−144.41	−524.32	292.74

[a] Ref. 1.
[b] To convert J to cal, divide by 4.184.

4 (5,7). The density of gaseous carbon monoxide (7) can be calculated directly from the equation of state using the compressibility factor at the temperature and pressure of interest.

Values of the viscosity of gaseous carbon monoxide have been reported (5,8,15–16). Vapor viscosity values and their relationship to density have been tabulated over a wide range of conditions (15). Values for liquid carbon monoxide may be obtained from reference 7.

Table 3. Virial Coefficients for Carbon Monoxide [a,b]

Temperature, K	B, (cm³/mol)	C, (cm⁶/mol²)
273.16	−14.19	1781
298.16	−8.28	1674
323.16	−3.40	1591
348.16	0.90	1444
373.16	4.49	1339
398.16	7.52	1264
423.16	10.04	1259

[a] Ref. 12.
[b] R = 8.314 (cm³·MPa)/(mol·K) or 82.06 (cm³·atm)/(mol·deg).

Table 4. Density of Liquid Carbon Monoxide [a]

Temperature, K	Liquid density, g/cm³
68.2	0.847
78.1	0.806
87.2	0.769
94.2	0.734
103.6	0.685
109.1	0.653
121.0	0.566
125.7	0.521
130.0	0.456
131.4	0.422

[a] Refs. 5 and 7.

Extensive listings of thermal conductivity for liquid and gaseous carbon monoxide and relationships of thermal conductivity to pressure appear in the literature (7,17).

The solubility of carbon monoxide in a variety of solvents at 298 K is given in Table 5 (18). Battino (19) has also provided detailed discussion of the solubility of carbon monoxide in water. A compilation of early literature values is given in reference 20.

Carbon monoxide burns readily in air or oxygen. Ignition temperatures are 644–658°C and 637–658°C, respectively. Mixtures of carbon monoxide and air are flammable over a wide range of compositions at atmospheric pressure, Table 6 (21–22). The flammability limits of carbon monoxide–air mixtures change with pressure. As pressure increases, the lower limit rises slowly from 16.3% at 0.1 MPa (1 atm) to 20% at 2 MPa (20 atm), and remains at 20% with pressures to 20 MPa (200 atm), and the upper limit decreases, dropping from 70% at 0.1 MPa to 60% at 2 MPa and to 53% at 10 MPa (100 atm). Carbon monoxide–oxygen mixtures are flammable over a wider range than carbon monoxide–air mixtures. The lower flammability limit remains nearly the same at 16.7%, but the upper flammability limit increases to 93.5% (23). Addition of diluents, particularly carbon dioxide, decreases the range of flammability of carbon monoxide in both air and oxygen (21).

Table 5. Solubility of Carbon Monoxide at 25°C and 101.3 kPa (1 atm) CO Pressure[a]

Solvent	Solubility (mol fraction CO × 10^4)
1-heptane	17.24
cyclohexane	9.91
methylcyclohexane	12.41
benzene	6.68
toluene	8.11
perfluoro-n-heptane	38.75
perfluorobenzene	21.20
carbon tetrachloride	8.76
chlorobenzene	6.47
nitrobenzene	3.72
methanol	3.76
ethanol	4.84
2-methyl-1-propanol	6.52
acetone	7.72
water	0.16

[a] Ref. 18.

Table 6. Flammability Limits of Carbon Monoxide in Dry Air as a Function of Temperature at Atmospheric Pressure[a]

Temperature, K	Lower limit, vol %	Upper limit, vol %
290	16.3	70.0
373	14.8	71.5
473	13.5	73.0
573	12.4	75.0
673	11.4	77.5

[a] Refs. 21–22.

Chemical Properties

The bonding between carbon and oxygen in carbon monoxide is best described by molecular orbital theory. The ten valence electrons from carbon and oxygen fill the lowest energy orbitals of carbon monoxide (^{b}, bonding; *, antibonding) to give the electronic configuration $(\sigma_s^b)^2(\sigma_s^*)^2(\pi_{xy}^b)^4(\pi_z^b)^2$, which predicts one σ- and two π-bonds (24). The π^* and σ^* orbitals of molecular carbon monoxide remain unfilled but available for bonding with transition metal atoms. The bond energy of 1070 kJ/mol (255.7 kcal/mol) is consistent with the triple bond formulation and is the highest observed bond energy for any diatomic molecule. The fundamental absorption in the infrared spectrum of carbon monoxide is located at 2143 cm^{-1}.

The bonding between carbon monoxide and transition metal atoms is particularly important because transition metals, whether deposited on solid supports or present as discrete complexes, are required as catalysts for the reaction between carbon monoxide and most organic molecules. A metal–carbon σ-bond forms by overlapping of metal d-σ orbitals with σ orbitals on carbon. Multiple bond character between the metal and carbon occurs through formation of a metal-to-CO π-bond by overlap of a metal d-π orbital with the empty antibonding orbital of carbon monoxide (Fig. 1).

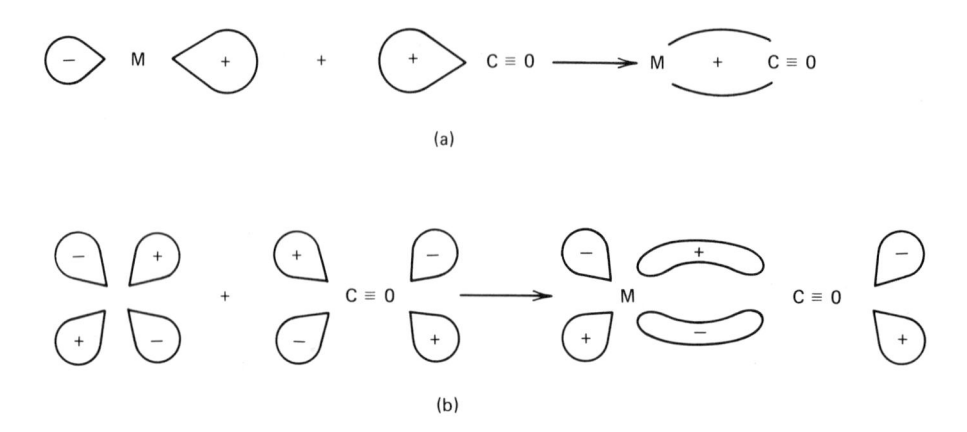

Figure 1. (a) Formation of a σ-bond between a transition metal and carbon monoxide. (b) Metal-to-carbon monoxide π-bond formation.

A weakened carbon–oxygen bond results from the combined σ- and π-bonding, allowing the metal-bonded carbon monoxide to react more readily (see Carbonyls).

For the purpose of this review, reactions of carbon monoxide have been divided into two groups: those used commercially and those used in general organic syntheses. Several bulk industrial compounds produced from carbon monoxide are discussed in individual articles within the *ECT* (see Acetic acid; Acetylene-derived chemicals; Acrylic acid; Isocyanates; Methanol; Oxo process; Phosgene) or in other reviews (25–26). Greater technical detail about the chemistry of carbon monoxide may be found in several other excellent reviews (27–32).

INDUSTRIALLY SIGNIFICANT REACTIONS OF CARBON MONOXIDE

Conversion to Hydrogen (Water Gas Shift Reaction). Carbon monoxide reacts with water over a catalyst to produce hydrogen and carbon dioxide (26). This reaction is used to prepare high purity hydrogen or synthesis gas with a higher hydrogen-to-carbon monoxide ratio than the feed (eq. 3).

$$CO + H_2O \rightarrow H_2 + CO_2 \quad \Delta G_{298\ K} = -28.64 \text{ kJ} (-6.845 \text{ kcal}) \quad (3)$$

The reaction is exothermic, and its equilibrium, unaffected by pressure, favors hydrogen production as reaction temperature is reduced.

The shift reaction is carried out commercially in either a one- or two-stage catalytic process. The first stage, or high temperature shift converter, operates at 315–510°C using a long-lived, reduced iron catalyst. Exit carbon monoxide concentrations of 2–5% are obtained at space velocities of 2000–4000 h^{-1}. Water-to-carbon monoxide ratios of 2–4 are typically used to force the equilibrium toward hydrogen production. In multibed, first-stage reactors, water is sometimes added between beds to remove reaction heat and lower the gas exit temperature to favor higher hydrogen concentrations.

In the second stage, a more active zinc oxide–copper oxide catalyst is used. This higher catalytic activity permits operation at lower exit temperatures than the first-stage reactor, and the resulting product has as low as 0.2% carbon monoxide. For space velocities of 2000–4000 h^{-1}, exit carbon monoxide levels of 0.5% are typical. After bulk carbon dioxide is removed (see Carbon dioxide), residual carbon oxides are hydrogenated to methane using a reduced nickel catalyst.

Oxidation. Carbon monoxide can be oxidized without a catalyst or at a controlled rate with a catalyst (eq. 4) (33). Carbon monoxide oxidation proceeds explosively if the gases are mixed stoichiometrically and then ignited. Surface burning will continue at temperatures above 900°C, but the reaction is slow below 650°C without a catalyst. Hopcalite, a mixture of manganese and copper oxides, catalyzes carbon monoxide oxidation at room temperature; it was used in gas masks during World War I to destroy low levels of carbon monoxide. Catalysts prepared from platinum and palladium are particularly effective for carbon monoxide oxidation at 50°C and at space velocities of 500 to 10,000 h^{-1}. Such catalysts have been used in catalytic converters on automobiles (34) (see Exhaust control).

$$CO + \tfrac{1}{2} O_2 \rightarrow CO_2 \qquad \Delta G_{298\ K} = -257.1 \text{ kJ } (-61.45 \text{ kcal}) \qquad (4)$$

Disproportionation. Carbon monoxide readily disproportionates into elemental carbon and carbon dioxide on a catalyst surface (eq. 5).

$$2\ CO \rightarrow C + CO_2 \qquad \Delta G_{298\ K} = -120.1 \text{ kJ } (-28.70 \text{ kcal}) \qquad (5)$$

This decomposition is thermodynamically favored by decreasing temperature and increasing pressure (35). Decomposition is extremely slow below 400°C in the absence of a catalyst; however, between 400–600°C many surfaces, particularly iron (36), cobalt, and nickel (37), promote the disproportionation reaction.

Phosgene. The reaction between carbon monoxide and chlorine is catalyzed by activated carbon and gives phosgene in nearly quantitative yield (eq. 6) (26).

$$CO + Cl_2 \rightarrow COCl_2 \qquad (6)$$

An equimolar mixture of carbon monoxide and chlorine reacts at 500 K under a slight positive pressure. The reaction is extremely exothermic ($\Delta H_{500\ K} = -109.7$ kJ or -26.22 kcal), and heat removal is the limiting factor in reactor design. Phosgene is often produced on-site for use in the manufacture of toluene diisocyanate (eq. 7).

$$\text{2,4-diaminotoluene} + 2\ COCl_2 \rightarrow \text{bis(carbamoyl chloride)} + 2\ HCl \rightarrow \text{toluene diisocyanate} + 2\ HCl \qquad (7)$$

Methanol. Methanol is manufactured by the reaction between carbon monoxide and hydrogen at 230–400°C and 5–60 MPa (50–600 atm). The reaction is extremely exothermic ($\Delta H_{600\ K} = -100.3$ kJ or -24.0 kcal), and plants must be designed to remove heat efficiently (eq. 8).

$$CO + 2\ H_2 \rightarrow CH_3OH \qquad (8)$$

A low-pressure methanol process was recently developed (38–39) which was carried out at 230–250°C, 5–10 MPa (50–100 atm), space velocities of 20,000–60,000 h^{-1}, and H_2-to-CO ratios of 3. The reaction is catalyzed by a very active zinc–copper–chromite catalyst that allows operation at milder conditions than older catalysts. However, the catalyst is easily poisoned by sulfur or chlorides and very pure synthesis gas must be used.

Acetic Acid. Manufacture of acetic acid by methanol carbonylation has become a leading commercial route to acetic acid (40–41).

$$CH_3OH + CO \rightarrow CH_3COOH \qquad (9)$$

At one time acetic acid was manufactured by methanol carbonylation using an iodide-promoted cobalt catalyst, but vigorous conditions of temperature and pressure (220°C, 48 MPa or 474 atm) were required to achieve yields of up to 60% (40–41). In contrast, an iodide-promoted, homogeneous rhodium catalyst operates at 175–195°C and pressures of 3 MPa (30 atm). These conditions dramatically lower the specifications for pressure vessels. Yields of 99% acetic acid based on methanol are readily attained (see Catalysis).

Hydroformylation. Probably the best known catalytic carbonylation reaction is the hydroformylation, or oxo reaction, for producing aldehydes and alcohols from carbon monoxide, hydrogen, and olefins (eq. 10) (42).

$$R-CH=CH_2 + CO + H_2 \rightarrow RCH_2CH_2CHO \xrightarrow{H_2} RCH_2CH_2CH_2OH \qquad (10)$$

In excess of a million metric tons of oxo products are produced in the United States annually. They are used in the manufacture of plasticizers, solvents, and detergents. The major oxo alcohol product, 2-ethylhexanol, is made from propylene and represents about 75% of the oxo market.

The hydroformylation reaction is carried out in the liquid phase using a metal carbonyl catalyst such as $HCo(CO)_4$ (42), $HCo(CO)_3(PBu_3^n)$ (43), or $HRh(CO)_2[P(C_6H_5)_3]_2$ (44–45). The phosphine-substituted rhodium compound is the catalyst of choice for new commercial plants which can operate at 80–110°C and 0.7–2 MPa (7–20 atm) (45). The major differences among the catalysts are found in their intrinsic activity, their selectivity to straight chain product, their ability to isomerize the olefin feedstock and hydrogenate the product aldehyde to alcohol, and the ease with which they are separated from the reaction medium (42) (see Alcohols, higher aliphatic; Aldehydes).

Acrylic Acid. About one-third of the acrylic acid and ester produced in the United States is manufactured by the Reppe reaction from acetylene, methanol, and carbon monoxide (eq. 11).

$$HC\equiv CH + CH_3OH + CO \rightarrow CH_2=CHCOOCH_3 \qquad (11)$$

The reaction is carried out in the liquid phase at 100–190°C and 3 MPa (30 atm) of carbon monoxide pressure using nickel salt catalyst, or at 40°C and 0.1 MPa (1 atm) using nickel carbonyl as both the catalyst and the source of carbon monoxide. Either acrylic acid or methyl acrylate may be produced directly, depending upon whether water or methanol is used as solvent (27). New technology for acrylic acid production uses direct propylene oxidation rather than acetylene carbonylation because of the high cost of acetylene.

Fischer-Tropsch. Carbon monoxide is catalytically hydrogenated to a mixture of straight-chain aliphatic, olefinic, and oxygenated hydrocarbon molecules in the Fischer-Tropsch reaction (eq. 12) (see Fuels, synthetic).

$$nCO + 2\,nH_2 \rightarrow -(CH_2)_n- + nH_2O \qquad (12)$$

The Fischer-Tropsch process was developed in Germany to manufacture synthetic gasoline during World War II. Carefully promoted iron and cobalt catalysts operating at conditions varying from 190–350°C and 0.7–20 MPa (7–200 atm) were used. Depending on the specific operating conditions, products containing up to 75% liquid hydrocarbon or 55% oxygenated organic molecules were obtained (46–47).

Although the Fischer-Tropsch reaction was the focus of innumerable studies prior to 1960, it received little further attention until the energy crisis of the 1970s spurred new activity in the area. A particularly significant technology coming from this work is a process for converting synthesis gas or methanol into aromatic compounds (48–49). The Fischer-Tropsch synthesis is carried out commercially in South Africa, where an unusual economic environment allows a profitable operation. It is extremely unlikely that Fischer-Tropsch technology will find application in the United States in the foreseeable future (50).

Methanation. Since 1902, when Sabatier discovered that carbon monoxide could be hydrogenated to methane, the methanation reaction has been the subject of intense investigation (eq. 13) (51–52).

$$CO + 3 H_2 \rightarrow CH_4 + H_2O \qquad \Delta G_{500\ K} = -96.5 \text{ kJ } (-23.1 \text{ kcal}) \qquad (13)$$

The methanation reaction is carried out over a catalyst at operating conditions of 230–450°C, 0.1–10 MPa (1–100 atm), and space velocities of 500–25,000 h^{-1}. Although many catalysts are suitable for effecting the conversion of synthesis gas to methane, nickel-based catalysts are used almost exclusively for industrial applications. Methanation is extremely exothermic ($\Delta H_{500\ K} = -214.6$ kJ or -51.3 kcal), and heat must be removed efficiently to minimize loss of catalyst activity from metal sintering or reactor plugging by nickel carbide formation.

The methanation reaction is currently used to remove the last traces (less than 1%) of carbon monoxide and carbon dioxide from hydrogen to prevent poisoning of catalysts employed for subsequent hydrogenation reactions. Major research efforts are underway to perfect processes for conversion of synthesis gas containing large quantities of carbon monoxide (up to 25%) into synthetic natural gas in anticipation of future plants based on coal-supplied synthesis gas.

Nickel Purification. The Mond process for nickel purification is based on the formation of volatile nickel carbonyl, $Ni(CO)_4$, which is stable below 60°C but decomposes rapidly and completely into nickel and carbon monoxide at 180°C (eq. 14). Crude nickel oxide, obtained from roasting nickel sulfide ores, is reduced to impure nickel metal with synthesis gas at 400°C. Under these conditions the reduction is carried out by hydrogen alone. Water is then removed and the exit gas, enriched in carbon monoxide, passes over the impure nickel at 50°C to form nickel carbonyl. The volatile nickel carbonyl leaves the reactor and is decomposed at 180°C, pure nickel is deposited on nickel shot, and the carbon monoxide is recycled for the preparation of additional carbonyl (53).

$$Ni + 4 CO \rightarrow Ni(CO)_4 \qquad (14)$$

GENERAL REACTIONS OF CARBON MONOXIDE

With Hydrogen. In addition to the reactions already discussed, other products may be obtained from synthesis gas depending on the catalyst used. In a liquid-phase high pressure reaction (60 MPa or 600 atm), a rhodium cluster complex catalyzes the direct formation of ethylene glycol, propylene glycol (see Glycols), and glycerol (qv) from synthesis gas (eq. 15) (54). Mixtures of methanol, ethanol (55), acetaldehyde, and acetic acid (56) are formed by using supported rhodium catalysts at 325°C and 17 MPa (168 atm). Rates of reaction for this latter route appear to be too slow for commercial application at this time.

$$2\,CO + 3\,H_2 \rightarrow HOCH_2CH_2OH \qquad (15)$$

With Alcohols, Ethers, and Esters. Carbon monoxide reacts with alcohols, ethers, and esters to give carboxylic acids (qv). The reaction yielding carboxylic acids is general for alkyl (57) and aryl alcohols (58). It is catalyzed by rhodium or cobalt in the presence of iodide and provides the basis for a commercial process to acetic acid (see previous section). Strong base catalyzes the formation of derivatives of formic acid in the reaction between alcohols and carbon monoxide (59). Methyl formate is made at 170–190°C at 1–2 MPa (10–20 atm) pressure, (eq. 16).

$$CH_3OH + CO \xrightarrow{NaOH} HCOOCH_3 \qquad (16)$$

Methanol reacts with carbon monoxide and hydrogen to form ethanol in the homologation reaction. Cobalt carbonyl catalyzes the transformation at 200°C and 30 MPa (300 atm) pressure, and gives yields of less than 75% ethanol. The greatest activity in the homologation reaction is observed for methyl and benzyl alcohols (eq. 17) (60). Reaction between methyl acetate, carbon monoxide, and hydrogen at 135–160°C and up to 10 MPa (100 atm) pressure using a palladium- or rhodium–iodide catalyst leads to the production of ethylidene diacetate (61), (eq. 18). Ethylidene diacetate can be pyrolyzed to vinyl acetate, (eq. 19) (see Vinyl polymers).

$$CH_3OH + CO + 2\,H_2 \rightarrow C_2H_5OH + H_2O \qquad (17)$$

$$2\,CH_3COOCH_3 + 2\,CO + H_2 \rightarrow CH_3CH(OOCCH_3)_2 + CH_3COOH \qquad (18)$$

$$CH_3CH(OOCCH_3)_2 \rightarrow CH_2{=}CHOOCCH_3 + CH_3COOH \qquad (19)$$

With Formaldehyde. The sulfuric acid-catalyzed reaction of formaldehyde with carbon monoxide and water to glycolic acid at 200°C and 70 MPa (700 atm) pressure was the first step in an early process to manufacture ethylene glycol. A recent patent (62) describes the use of liquid hydrogen fluoride as catalyst, enabling the reaction to be carried out at 25°C and 7 MPa (70 atm) (eq. 20).

$$HCHO + CO + H_2O \rightarrow HOCH_2COOH \qquad (20)$$

With Unsaturated Compounds. The reaction of unsaturated organic compounds with carbon monoxide and molecules containing an active hydrogen atom leads to a variety of interesting organic products. The hydroformylation reaction is the most important member of this class of reactions. When the hydroformylation reaction of ethylene takes place in an aqueous medium, diethyl ketone is obtained as the principal product instead of propionaldehyde (63). Ethylene, carbon monoxide, and water also yield propionic acid under mild conditions (175–195°C and 3–7 MPa or 30–70 atm) using cobalt or rhodium catalysts containing bromide or iodide (64–65).

Oxidative Carbonylation. Carbon monoxide is rapidly oxidized to carbon dioxide; however, under proper conditions, carbon monoxide and oxygen react with organic molecules to form carboxylic acids or esters. With olefins, unsaturated carboxylic acids are produced, whereas alcohols yield esters of carbonic or oxalic acid. The formation of acrylic and methacrylic acid (qv) is carried out in the liquid phase at 10 MPa (100 atm) and 110°C using palladium chloride or rhenium chloride catalysts (eq. 21) (66–67).

$$CH_2{=}CH_2 + CO + \tfrac{1}{2}\,O_2 \rightarrow CH_2{=}CHCOOH \qquad (21)$$

Dimethyl carbonate and dimethyl oxalate are both obtained from carbon monoxide, oxygen, and methanol at 90°C and 10 MPa (100 atm) or less. The choice of catalyst is critical; cuprous chloride (68) gives the carbonate (eq. 22); a palladium chloride–copper chloride mixture (69–70) gives the oxalate, (eq. 23). Anhydrous conditions should be maintained by removing product water to minimize the formation of by-product carbon dioxide.

$$2\ CH_3OH + CO + \tfrac{1}{2}\ O_2 \rightarrow (CH_3O)_2CO + H_2O \tag{22}$$

$$2\ CH_3OH + 2\ CO + \tfrac{1}{2}\ O_2 \rightarrow CH_3O-\underset{\underset{O}{\|}}{C}-\underset{\underset{O}{\|}}{C}-OCH_3 + H_2O \tag{23}$$

Isocyanate Synthesis. In the presence of a catalyst, nitroaromatic compounds can be converted into isocyanates, using carbon monoxide as a reducing agent. Conversion of dinitrotoluene toluenediisocyanate into (TDI) with carbon monoxide (eq. 24), could offer significant commercial advantages over the current process using phosgene. The reaction is carried out at 200–250°C and 27–41 MPa (270–400 atm) with a catalyst consisting of either palladium chloride or rhodium chloride complexed with pyridine, isoquinoline, or quinoline and yields are in excess of 80% TDI (71–72).

$$\text{dinitrotoluene} + 6\ CO \rightarrow \text{toluene diisocyanate} + 4\ CO_2 \tag{24}$$

Dimethylformamide. The industrial solvent dimethylformamide is manufactured by the reaction between carbon monoxide and dimethylamine.

$$(CH_3)_2NH + CO \rightarrow (CH_3)_2NCHO \tag{25}$$

The reaction is carried out in the liquid phase using a sodium methoxide catalyst at 60–130°C and 0.5–0.9 MPa (5–9 atm) (73).

Aromatic Aldehydes. Carbon monoxide reacts with aromatic hydrocarbons or aryl halides to yield aromatic aldehydes (see Aldehydes).

$$\text{toluene} + CO \rightarrow \text{p-tolualdehyde} \tag{26}$$

$$\text{chlorobenzene} + CO + H_2 \rightarrow \text{benzaldehyde} + HCl \tag{27}$$

The reaction of equation 26 proceeds with yields of 98% when carried out at 0°C and 0.4 MPa (4 atm) using a boron trifluoride–hydrogen fluoride catalyst (74), whereas conversion of aryl halides to aldehydes in 84% yield requires conditions of 150°C and 7 MPa (70 atm) with a homogeneous palladium catalyst (75).

Metal Carbonyls. Carbon monoxide forms metal carbonyls or metal carbonyl derivatives with most transition metals (76) (see Coordination compounds). Metal carbonyls are used in a variety of industrial applications in addition to their use as catalysts. Methylcyclopentadienylmanganesetricarbonyl (MMT), [CH$_3$C$_5$H$_4$Mn(CO)$_3$], is sold as an antiknock additive but its use in unleaded gasoline (qv was banned by the EPA in 1978; tungsten and molybdenum hexacarbonyls are thermally decomposed to obtain very pure metal films; and numerous carbonyls are used as reagents in organic synthesis.

Polymers. Carbon monoxide forms copolymers with ethylene and suitable vinyl compounds. No large-scale uses for the copolymers or their further reaction products (polyalcohols, polyamines) have been found (77).

Analysis

Many procedures for the analysis of carbon monoxide have been developed based on the reducing properties of the gas (78–79). Qualitative detection of carbon monoxide is achieved by passing the gas through palladium chloride which is either in solution or impregnated on paper. Appearance of black metallic palladium is an indication of the presence of carbon monoxide. Using this technique, concentrations of 100–1000 ppm carbon monoxide in air are easily estimated. Hydrogen, hydrogen sulfide, ethylene, and acetylene also reduce PdCl$_2$ to palladium, interfering with quantitative determination of carbon monoxide.

Several quantitative procedures for concentrations above 0.1 vol % are available. Gas chromatographic analysis (80) is particularly useful because it is fast, accurate, and relatively inexpensive. The standard wet-chemical, analytical method (78) takes advantage of the reaction between iodine pentoxide and carbon monoxide at 150°C.

A variety of instruments are available to analyze carbon monoxide in gas streams at levels from 1 ppm to 90%. One group of analyzers determines the concentration of carbon monoxide by measuring the intensity of its infrared stretching frequency at 2143 cm^{-1}. Another group measures the oxidation of carbon monoxide to carbon dioxide electrochemically. Such instruments are generally lightweight and well suited to applications requiring portable analyzers. Many analyzers are equipped with alarms and serve as work area monitors.

Manufacture and Purification

There are three major commercial processes for the purification of carbon monoxide. Two processes are based upon the absorption of carbon monoxide by salt solutions. The third uses either low temperature condensation or fractionation. All three processes employ similar techniques to remove minor impurities. Particulates are removed in cyclones or by scrubbing. Scrubbing also removes any tars or heavy hydrocarbon fractions. Acid gases are removed by absorption in monoethanolamine or hot potassium carbonate or by other patented removal processes. The prepurified gas stream is then sent to a carbon monoxide recovery section for final purification and by-product recovery.

The selection of a separation process depends upon many factors, not the least of which is the type of gas to be purified (see Separations systems synthesis). Table 7 gives some compositions of typical gases obtained using standard commercial production techniques.

Table 7. Typical Analyses of Carbon Monoxide Sources[a]

Source	Composition vol %, dry basis						
	CO	CO_2	H_2	O_2	N_2	CH_4	Other
blast furnace gas	27.5	11.5	1.0		60.0		
coke oven gas	5.6	1.4	55.4	0.4	4.3	28.4	4.5
water gas	30.0	3.4	31.7	1.2	13.1	12.2	8.4
natural gas, steam reforming	15.5	8.1	75.7		0.2	0.5	
naphtha, steam reforming	6.7	15.8	65.9		2.6	6.3	2.7
partial oxidation, heavy fuel oil	47.0	5.5	4.0		0.3	0.1	0.1
coal gasification	59.4	10.0	29.4		0.6		0.6

[a] Refs. 81–84.

The carbon monoxide concentration of gas streams is a function of many parameters. In general, increased carbon monoxide concentration is found with: (*1*) an increase in the carbon-to-hydrogen ratio in the feed hydrocarbon; (*2*) a decrease in the steam-to-feed-carbon ratio; (*3*) an increase in the synthesis gas exit temperature; and (*4*) avoidance of reequilibration of the gas stream at a temperature lower than the synthesis temperature. Specific improvement in carbon monoxide production by steam reformers is made by recycling by-product carbon dioxide to the process feed inlet of the reformer (85–86). This increases the relative carbon-to-hydrogen ratio of the feed and raises the equilibrium carbon monoxide concentration of the effluent.

Some gas streams contain nitrogen. In these cases it is difficult to achieve high levels of carbon monoxide purity by cryogenic methods because the boiling points of nitrogen and carbon monoxide differ by only 6°C. Therefore, when the nitrogen content of the feed gas exceeds the product purity specification, a salt solution absorption process is preferred for purification. If both high purity carbon monoxide and hydrogen are required, the cryogenic processing route is favored because salt solution processes cannot independently purify the hydrogen stream. As noted earlier, gas composition and system application are primary factors in the selection of a carbon monoxide purification route.

Copper–Liquor Scrubbing. Cuprous ammonium salts of organic acids form complexes with carbon monoxide. Conditions of high pressure and low temperature favor the formation of the complex, whereas low pressure and high temperature tend to release the complexed carbon monoxide from solution. These conditions typify the operation of the absorber-stripper shown in Figure 2. Specific design conditions for the process are given in references 88–90, and an excellent summary of processing considerations is presented in reference 87.

The basic chemistry of the process is represented by equation 28.

$$[Cu(NH_3)_2]^+ + CO + NH_{3(aq)} \rightarrow [Cu(NH_3)_3(CO)]^+ \tag{28}$$

Because the solution is capable of absorbing one mole of carbon monoxide per mole of cuprous ion, it is desirable to maximize the copper content of the solution. The ammonia not only complexes with the cuprous ion to permit absorption, but also increases the copper solubility and thereby permits an even greater carbon monoxide absorption capacity. The ammonia concentration is set by a balance between ammonia vapor pressure and solution acidity. Weak organic acids, (eg, formic, acetic, and carbonic acid) are used because they are relatively noncorrosive and inexpensive. A typical formic acid solution analysis (87) is: Cu^+, 170 g/L; Cu^{2+}, 25 g/L; HCOOH, 110 g/L;

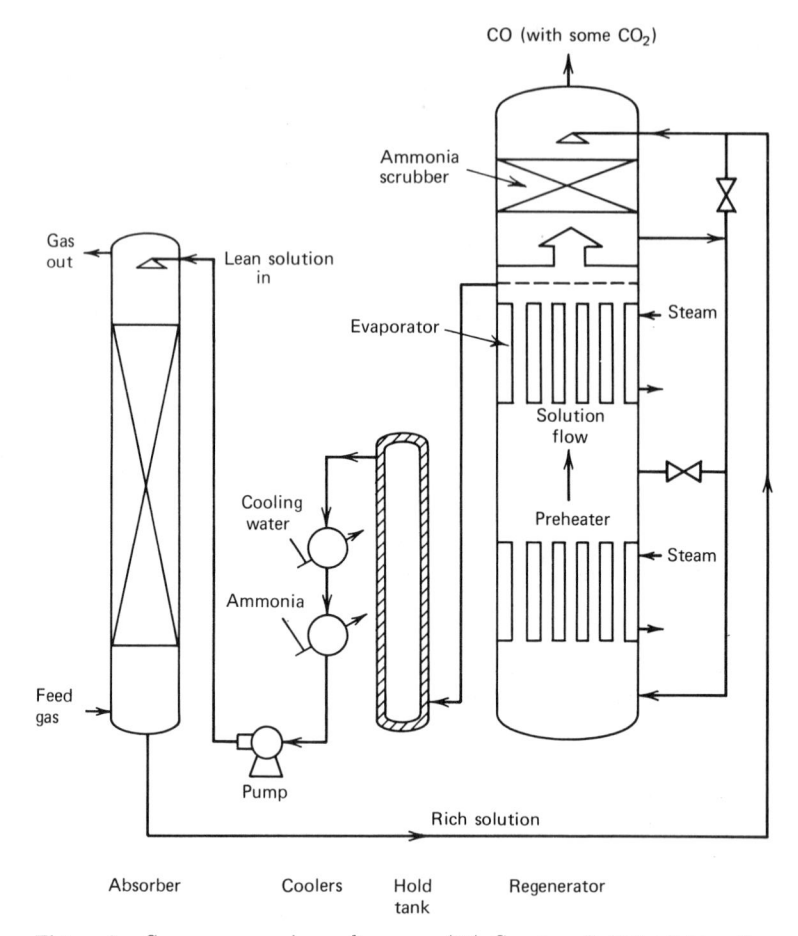

Figure 2. Copper–ammonium salt process (87). Courtesy Gulf Publishing Co.

CO_2, 60 g/L; and NH_3, 140 g/L. This type of solution can absorb between 3 and 20 volumes (STP) of carbon monoxide per volume of solution depending upon process conditions. High carbon monoxide partial pressures and low solution temperatures favor high absorption coefficients.

The preferred ratio of cuprous to cupric ion ranges from 5:1 to 10:1, depending upon system conditions. Too high a level of cuprous ion causes the system to disproportionate and form metallic copper (eq. 29):

$$2\,Cu^+ \rightarrow Cu^{2+} + Cu^0 \tag{29}$$

Cupric ion concentration is kept at an acceptable but low level by direct air oxidation of the solution. Solids formation from sulfides in the feed gas is also possible; therefore, pretreatment for sulfur removal is required.

Carbon dioxide can cause product contamination through ammonium carbonate formation. Ammonium carbonate may also form by oxidation of carbon monoxide by cupric ion (eq. 30):

$$2\,Cu^{2+} + CO + 4\,OH^- \rightarrow 2\,Cu^+ + CO_3^{2-} + 2\,H_2O \tag{30}$$

In both cases, the carbonate ion concentration increases and eventually equilibrates in the system, releasing carbon dioxide in the stripping column and thereby reducing product purity. Hence, a small caustic wash tower is employed to remove any carbon dioxide which is liberated in the stripper.

A small quantity of ammonia vapor is always in equilibrium with the solution. Losses of ammonia from the absorber are negligible owing to the high absorption pressure and the low operating temperature, usually 2.5 MPa (25 atm) and 10–30°C. The greatest potential for ammonia loss comes in the regenerator, where a pressure of 0.1 MPa (1 atm) is typical and stripping temperatures often reach 80°C. The loss is minimized by washing the product gas with the relatively cold saturated solution from the absorber.

Cryogenic Purification. Until recently, the primary commercial technique for purification of carbon monoxide employed either of two basic separation processes: partial condensation or absorption in liquid methane. Both processes operate at cryogenic temperatures and elevated pressures and rely upon the expansion energy in the feed gas to produce most or all of the refrigeration energy required (see Cryogenics). Process selection depends upon product purity specifications for both the carbon monoxide and hydrogen streams. The partial condensation technique is employed to produce carbon monoxide and moderate purity grades of hydrogen (96–98%). The carbon monoxide purity is dictated by the feed impurities, typically methane, which exit with the carbon monoxide. If a higher purity grade of carbon monoxide (99%) is required, an absorption system is used subsequent to the partial condensation unit. A liquid methane scrubbing system is generally applied when high purity standards are set for the hydrogen stream. This technique can produce hydrogen at a purity of 99+% with less than 1 ppm of carbon monoxide (91). High purity levels of carbon monoxide (99+%) can also be obtained.

Flow diagrams for the partial condensation process and the liquid methane scrubbing process are shown in Figures 3 and 4, respectively. As in the case of the salt complexation processes, the cryogenic systems require prepurification of the feed gas. Bulk water, hydrogen sulfide, and carbon dioxide are removed by standard techniques. Final removal of these materials is accomplished by adsorption (qv). After prepurification, the gases are ready for cryogenic processing.

In the partial condensation system, the feed gas is cooled from ambient temperature to approximately 85 K by heat exchange with the exit gases. Cooling condenses the bulk carbon monoxide contained in the feed as well as all of the methane. The uncondensed hydrogen-rich gas stream is removed overhead for further processing. The liquid fraction is then lowered in pressure to remove entrained gases. The low-pressure, carbon monoxide-rich, liquid stream is evaporated, warmed against the incoming feed gas, compressed, and leaves the processing area as the carbon monoxide product. Because there are only two process exit streams, the bulk of the inlet feed contaminants, mainly methane and nitrogen, leave with the carbon monoxide product.

The hydrogen-rich gas from the high-pressure separator is cooled to about 70 K in a second exchanger. This allows additional carbon monoxide and other inlet gases to be condensed and separated from the hydrogen product stream. Hydrogen purities in excess of 95% are usually obtained. The hydrogen is warmed serially against the low temperature and raw-feed gas streams and exits the process at high-pressure. Typically, a portion of this gas stream is expanded to provide the lowest temperature refrigeration for the system. The expanded-hydrogen and low temperature condensate

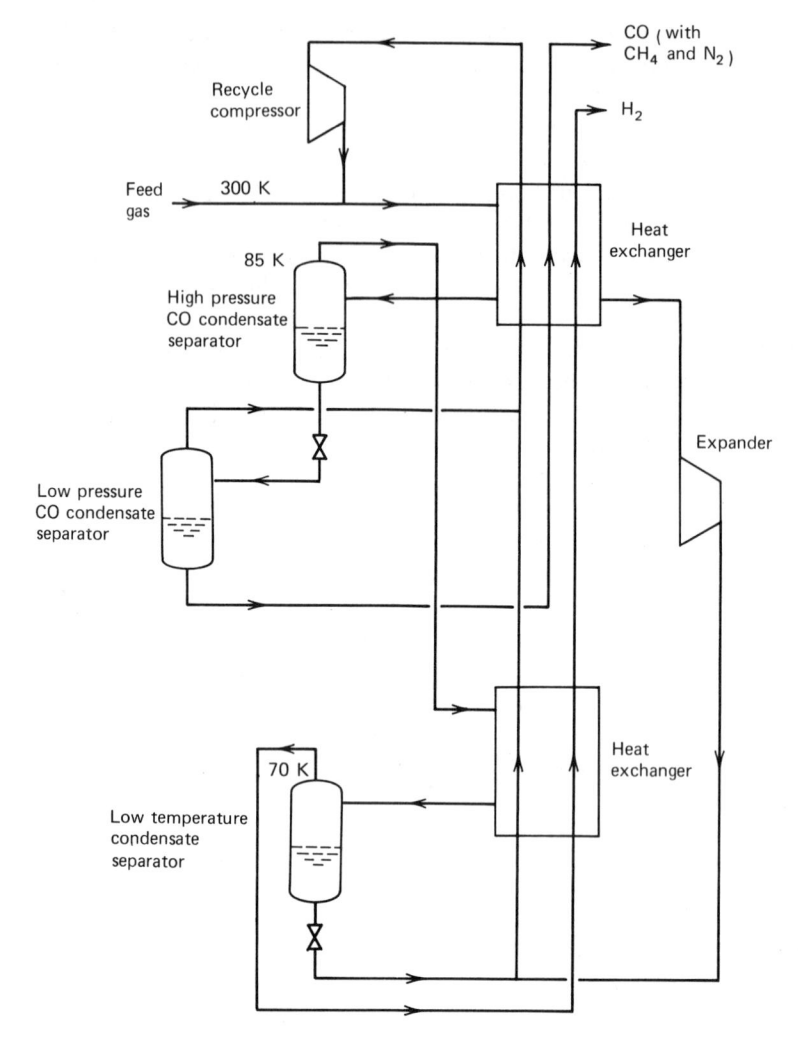

Figure 3. Cryogenic partial condensation process (91). Courtesy Linde A.G.

streams are warmed against the raw hydrogen stream, then combined with the vent gas from the low-pressure CO-condensate separator, and heat-exchanged with the feed gas. The mixture is recompressed and recycled to the feed stream.

The processing techniques employed for the methane wash process are more complex, but produce a higher purity carbon monoxide than those employed by the partial condensation system. In the methane wash process, the inlet gas is cooled to about 90 K in a series of heat exchangers. Bulk carbon monoxide is removed by condensation in a separator, and the hydrogen stream is sent to the methane wash column where it is purified by scrubbing with a liquid methane reflux. The liquor absorbs the carbon monoxide and other gases and carries them to a second column. The purified hydrogen is heat-exchanged against the incoming gas and is available for use at high pressure. A portion of this gas stream is expanded to provide energy for process drives and refrigeration. This hydrogen stream is removed from the system and is available as a low-pressure hydrogen product.

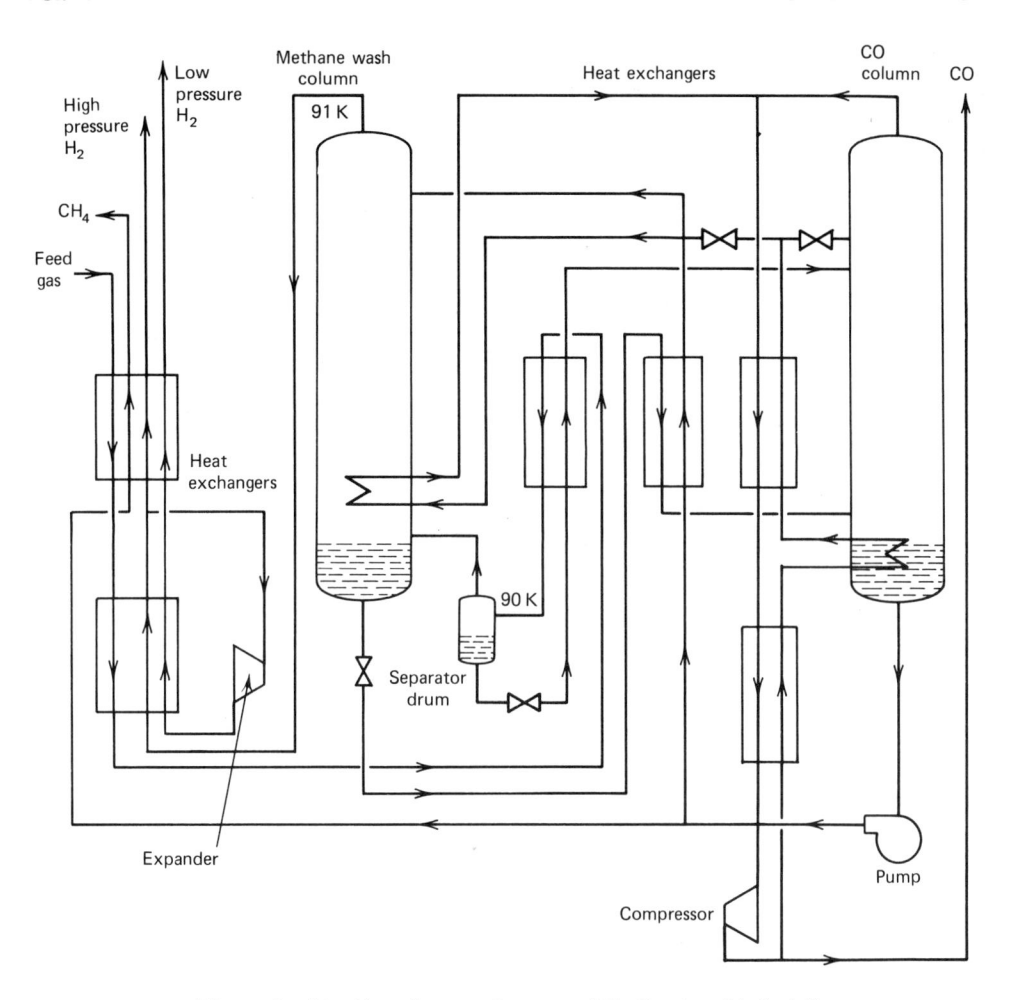

Figure 4. Liquid methane wash process (91). Courtesy Linde A.G.

The carbon monoxide-rich, liquid condensate from the primary separator is expanded and exchanged against the incoming feed and is then sent to a distillation column where the carbon monoxide is purified. The bottoms liquor from the methane wash column is expanded, heat-exchanged, and sent to the bottom section of the distillation column for methane rectification and carbon monoxide recovery. The methane bottom stream is recompressed and recycled to the top of the wash column after subcooling. A sidestream of methane is withdrawn to avoid a buildup of impurities in the system.

The carbon monoxide product is removed from the top of the column and warmed against recycled high pressure product. The warm low pressure stream is compressed, and the bulk of it is recycled to the system for process use as a reboiler medium and as the reflux to the carbon monoxide column; the balance is removed as product. The main impurity in the stream is nitrogen from the feed gas. Carbon monoxide purities of 99.8% are commonly obtained from nitrogen-free feedstocks.

Cosorb Process. The Cosorb process is a recent development in carbon monoxide purification. This process also relies upon the formation of a cuprous complex of carbon monoxide, but uses a nonaqueous organic solvent. The preferred system uses a cuprous tetrachloroaluminate toluene complex in a toluene solvent (92). Many other organometallic complex variants have been proposed (93–95), but have not yet been commercialized.

The Cosorb process is similar in concept to the copper ammonium salt process. Its main advantages over the older salt process are its low corrosion rate, ability to work in carbon dioxide atmospheres, and low energy consumption. The active $CuAlCl_4 \cdot C_6H_5CH_3$ complex is considerably more stable than the cuprous ammonium salt, and solvent toluene losses are much lower than the ammonia losses of the older process (96).

A flow diagram for the system is shown in Figure 5. Feed gas is dried, and ammonia and sulfur compounds are removed to prevent the irreversible buildup of insoluble salts in the system. Water and solids formed by trace ammonia and sulfur compounds are removed in the solvent maintenance section (98). The pretreated carbon monoxide feed gas enters the absorber where it is selectively absorbed by a counter-current flow of solvent to form a carbon monoxide complex with the active copper salt. The carbon monoxide-rich solution flows from the bottom of the absorber to a flash vessel where physically absorbed gas species such as hydrogen, nitrogen, and methane are removed. The solution is then sent to the stripper where the carbon monoxide is released from the complex by heating and pressure reduction to about 0.15 MPa (1.5 atm). The solvent is stripped of residual carbon monoxide, heat-exchanged with the stripper feed, and pumped to the top of the absorber to complete the cycle.

The overhead temperatures of both the absorber and stripper are kept as low as possible to minimize solvent carryover. A temperature of about 38°C is typically used in the high pressure absorber. The overhead temperature in the stripper is set by the

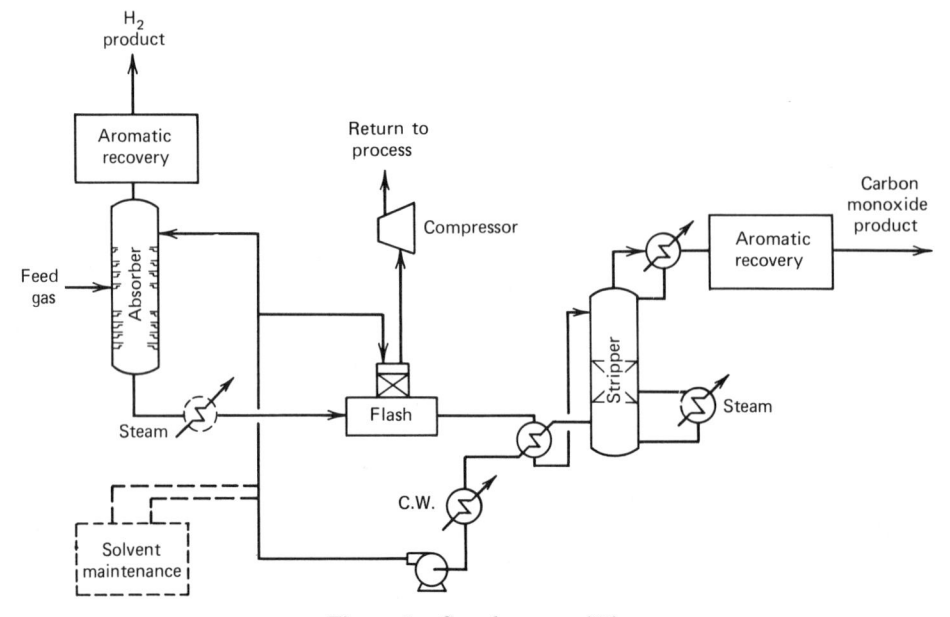

Figure 5. Cosorb process (97).

boiling point of the saturated complex solution and by the operating pressure of the stripper. At a stripping pressure of 0.166 MPa (1.7 atm), a temperature of 105°C is used. The solvent-rich gas from the stripper is cooled to recover as much solvent as possible by condensation prior to the final aromatics-recovery section. Final solvent recovery is accomplished by adsorption on activated carbon (97).

The carbon monoxide purity from the Cosorb process is very high because physically absorbed gases are removed from the solution prior to the low-pressure stripping column. Furthermore, there is no potential for oxidation of absorbed carbon monoxide as in the copper–liquor process. These two factors lead to the production of very high purity carbon monoxide, 99+%. Feed impurities exit with the hydrogen-rich tail gas; therefore, the purity of this coproduct hydrogen stream will depend upon the impurity level in the feed gas.

The high degree of carbon monoxide recovery (99%) is enhanced as absorption pressure is increased. The solution circulation rate is kept low by maximizing the concentration of cuprous complex in solution (eg, 25% wt $CuAlCl_4 \cdot C_6H_5CH_3$ in toluene). Solution temperatures must be controlled to prevent solids formation at these high concentrations. Lack of water in the system coupled with low corrosivity of the solvent system allows the use of high stripper-bottoms temperatures which permit generation of a solution with a very low residual concentration of the carbon monoxide complex. The high carbon monoxide absorption rate of the resulting solution contributes to the excellent monoxide recovery in the system. The low corrosion rate also allows the use of carbon steel equipment throughout the process.

Economic Aspects

Carbon monoxide is produced as a component of synthesis gas or as purified gas by many manufacturers, primarily for on-site process applications. Because very few producers engage in merchant sale of purified gases, published production data are not available. By-product hydrogen price, feedstock prices, and delivery charges as well as location, purity, and volume, affect carbon monoxide pricing. The following prices are illustrative of carbon monoxide derived from typical steam–methane reformer operations for a current Gulf Coast location. Prices range from $0.25 to $0.28/m^3; ($7 to $8/1,000 scf) for volumes of approximately 28,000 m^3/day (1 million scfd) for "over the fence" carbon monoxide, to $0.71/m^3 ($20/1,000 scf) for tube trailer volumes of gas, fob works. Tube trailers generally haul loads of 1,500 to 3,000 m^3 (50,000 to 100,000 scf). Purchases of liquefied carbon monoxide are possible through some vendors; however, volume and availability are restricted. (Gas volumes are based upon equivalent volumes at 0.101 MPa (1 atm) and 21.1°C.)

Carbon monoxide is also available in cylinders at various levels of purity. As in the case of bulk gas sales, price is also a function of delivery location, cylinder size, and purity. Terminology for gas purity has not been standardized by the Compressed Gas Association, thus care must be taken when comparing grades from different suppliers. Table 8 contains the price ranges and purities from supplier catalogue data with prices adjusted to the May 1977 listings (99–101). The price ranges quoted are for the largest available cylinders in each grade provided by the vendor. All prices are quoted fob vendor works. Carbon monoxide must bear a red label during shipment because it is flammable. Additional shipping requirements are contained in the Interstate Commerce Commission regulations.

Table 8. Commercial Price and Purity Data for Cylinder Carbon Monoxide[a]

Grade	Purity, min vol %	Supply pressure, MPa (psig)	Cylinder price, $/cylinder	Gas price[b], $/m³
commercial	98.0–99.0	11.5–13.9 (1650–2000)	43–50	6.22–9.69
C.P.	99.0–99.5	11.5–13.9 (1650–2000)	59–75	8.33–11.90
ultra-high purity	99.8	11.5 (1650)	110–120	21.29–22.20
research	99.97–99.99	11.5 (1650)	280	56.50

[a] Refs. 99–101.
[b] Gas volumes are based upon equivalent volume at 0.101 MPa (1 atm) and 21.1°C.

Health and Safety

Occurrence. Carbon monoxide is a product of incomplete combustion and is not likely to result where a flame burns in an abundant air supply, yet may result when a flame touches a cooler surface than the ignition temperature of the gas. Gas or coal heaters in the home and gas space heaters in industry have been frequent sources of carbon monoxide poisoning when not provided with effective vents. Gas heaters, although properly adjusted when installed, may become hazardous sources of carbon monoxide if maintained improperly. Automobile exhaust gas is perhaps the most familiar source of carbon monoxide exposure. The manufacture and use of synthesis gas, calcium carbide manufacture, distillation of coal or wood, combustion operations, heat treatment of metals, fire fighting, mining, and cigarette smoking represent additional sources of carbon monoxide exposure (102–104).

Toxicity. Carbon monoxide is the most widely spread gaseous hazard to which man is exposed (102). The toxicity of carbon monoxide is a result of its reaction with the hemoglobin of blood. The carboxyhemoglobin (COHb) which is formed displaces oxygen and leads to asphyxiation. Blood levels of COHb as low as 5% impair the function of the brain and reduce visual acuity. Headaches occur when 10–20% of the hemoglobin reacts with carbon monoxide. Increased saturation causes nausea, dizziness, weakness, mental confusion, loss of consciousness, and death. In cases of severe poisoning, the skin and mucous membranes may become cherry red, but the victim usually appears pale (103). Prolonged unconsciousness may result in permanent brain damage. There is little evidence of chronic poisoning from repeated exposure to low concentrations of carbon monoxide. However, because of associated oxygen deprivation, repeated exposure may result in persistent neurologic manifestations such as anorexia, headache, dizziness, and ataxia (104).

The recommended NIOSH limit of 35 ppm is the time-weighted-average exposure to carbon monoxide based on a carboxyhemoglobin level of 5%; this amount of COHb is what an employee engaged in sedentary activity would be expected to approach in eight hours of continuous exposure. The standard does not take into account the smoking habits of a worker; the level of COHB in chronic cigarette smokers has generally been found to be in the 4–5% range prior to carbon monoxide exposure (102). A concentration of 100 ppm is allowable for an exposure of several hours and 400–500 ppm can be inhaled for 1 h without an appreciable effect, whereas 1500–2000 ppm are dangerous and 4000 ppm or more is fatal (103). First aid treatment for carbon monoxide poisoning emphasizes elimination of the gas from the body. Elimination of carbon monoxide occurs solely through the lungs, and although rapid at first, the last traces are difficult to remove. The poisoned patient must be removed to fresh air, kept warm and administered pure oxygen by the best method available. Artificial respiration is necessary whenever breathing is inadequate. Exercise and stimulants, including carbon dioxide, must not be given because they can lead to collapse. A physician must be summoned in all cases of suspected carbon monoxide poisoning (104).

Prevention of carbon monoxide poisoning is best accomplished by providing good ventilation where contamination is a problem. If good ventilation is not possible, a self-contained breathing apparatus, such as a Scott Air-Pak, must be used. The use of gas masks containing an adsorbent is generally not recommended since it is difficult to know when the adsorbent is exhausted.

BIBLIOGRAPHY

"Carbon Monoxide" in *ECT* 1st ed., Vol. 3, pp. 179–191, by D. D. Lee, E. I. du Pont de Nemours & Co., Inc.; "Carbon Monoxide" in *ECT* 2nd ed., Vol. 4, pp. 424–445, by Ralph V. Green, E. I. du Pont de Nemours & Co., Inc.

1. D. R. Stull and H. Prophet, *JANAF Thermochemical Tables,* 2nd ed., NSRDS-NBS 31, U.S. Government Printing Office, Washington, D. C., 1971.
2. *API44-TRC Selected Data on Thermodynamics and Spectroscopy,* Publication 100, 2nd ed., Thermodynamics Research Center, College Station, Texas, 1974.
3. D. R. Stull, E. F. Westrum, and G. C. Sinke, *The Chemical Thermodynamics of Organic Compounds,* John Wiley & Sons, Inc., New York, 1969.
4. F. D. Rossini and co-workers, *Selected Values of Chemical Thermodynamic Properties,* NBS Circular 500, U.S. Government Printing Office, Washington, D. C., 1952.
5. J. Timmermans, *Physico-Chemical Constants of Pure Organic Compounds,* Vol. 1, 1950; Vol. 2, 1965, Elsevier Publishing Co., Inc., New York.
6. K. Raznjevic, *Handbook of Thermodynamic Tables and Charts,* McGraw-Hill Book Co., New York 1976.
7. C. L. Yaws, K. Y. Li, and C. H. Kuo, *Chem. Eng.* **81**(20), 115 (1974).
8. R. H. Perry and C. H. Chilton, eds., *Chemical Engineers Handbook,* 5th ed, McGraw-Hill Book Co., New York, 1973.
9. J. G. Hust and R. B. Stewart, *N.B.S. Tech. Note, 202,* 1963.
10. A. S. Leah in F. Din, ed., *Thermodynamic Functions of Gases,* Vol. 1, Butterworth and Co., Ltd., London, 1956.
11. D. R. Stull, *Ind. Eng. Chem.* **39,** 517 (1947).
12. A. Michels and co-workers, *Physica* **18,** 121 (1952).
13. J. Hilsenrath and co-workers, *Natl. Bur. Stand. U.S.* **564,** (1955).
14. J. M. Smith and H. C. Van Ness, *Chemical Engineering Thermodynamics,* 2nd ed., McGraw-Hill Book Co., New York, 1959, p. 96.
15. A. K. Barua and co-workers, *J. Chem. Phys.* **41,** 374 (1964).
16. G. L. Chierici and A. Paratella, *AIChE J.* **15,** 786 (1969).
17. H. L. Johnston and E. R. Grilly, *J. Chem. Phys.* **14,** 233 (1946).
18. E. Wilhelm and R. Battino, *Chem. Rev.* **73,** 1 (1973).
19. R. Battino and H. L. Clever, *Chem. Rev.* **66,** 395 (1966).
20. W. F. Linke and A. Seidell, *Solubilities of Inorganic and Metal-Organic Compounds,* 4th ed., Vol. 1, American Chemical Society, Washington, D. C., 1958, p. 453.
21. H. F. Coward and G. W. Jones, *U.S. Bur. Mines Bull.* **503,** (1952).
22. C. G. Segeler, ed., *Gas Engineers Handbook,* Industrial Press, New York, 1966, p. 2/73.
23. E. Terres, *J. Gasbeleucht* **63,** 785 (1920).
24. A. K. Holliday, G. Hughes, and S. M. Walker in J. C. Bailar and co-workers, eds., *Comprehensive Inorganic Chemistry,* Vol. 1, Pergamon Press, Oxford, 1973, p. 1225.

25. C. L. Thomas, *Catalytic Processes and Proven Catalysts*, Academic Press, New York, 1970.
26. F. A. Lowenheim and M. K. Moran, *Industrial Chemicals*, 4th ed, Wiley-Interscience, New York, 1975.
27. J. Falbe, *Carbon Monoxide in Organic Synthesis*, Springer-Verlag, New York, 1970.
28. P. N. Rylander, *Organic Synthesis with Noble Metal Catalysts*, Academic Press, New York, 1973.
29. R. F. Heck, *Organotransition Metal Chemistry*, Academic Press, New York, 1974.
30. M. M. T. Khan and A. E. Martell, *Homogeneous Catalysis by Metal Complexes*, Vol. 1, Academic Press, New York, 1974.
31. I. Wender and P. Pino, *Organic Syntheses Via Metal Carbonyls*, Vol. 1, 1968; Vol. 2, 1977, John Wiley & Sons, Inc., New York.
32. G. Henrici-Olive and S. Olive, *Coordination and Catalysis*, Verlag Chemie, New York, 1976, p. 234.
33. P. C. Gravelle and S. J. Teichner, *Adv. Catal.* **20,** 167 (1969).
34. J. E. McEvoy, ed., *Catalysts for the Control of Automotive Pollutants, Advances in Chemistry Series*, Vol. 143, American Chemical Society, Washington, D. C., 1975.
35. C. Gruber in L. Seglin, ed., *Methanation of Synthesis Gas, Advances in Chemistry Series*, Vol. 146, American Chemical Society, Washington, D. C., 1975, p. 31.
36. J. P. Legalland and L. Bonnetain, *C. R. Acad. Sci. Ser. C* **272,** 1919 (1971).
37. G. D. Renshaw, C. Roscoe, and P. L. Walker, *J. Catal.* **22,** 394 (1971).
38. G. C. Humphreys, D. J. Ashman, and N. Harris, *Chem. Econ. Eng. Rev.* **6**(11), 26 (1974).
39. U.S. Pat. 3,923,694 (Dec. 2, 1975), D. Cornthwaite (to Imperial Chemical Industries).
40. J. F. Roth and co-workers, *Chem. Tech.*, 600 (1971).
41. J. Hjortkjaer and V. W. Jensen, *Ind. Eng. Chem. Prod. Res. Dev.* **15,** 46 (1976).
42. F. E. Paulik, *Catal. Rev.* **6,** 49 (1972).
43. U.S. Pats. 3,239,569 and 3,239,570 (Mar. 8, 1966), L. H. Slaugh and R. D. Mullineaux (to Shell Oil Co.).
44. U.S. Pats. 3,527,809 (Sept. 8, 1970), and 3,917,661 (Nov. 4, 1975), R. L. Pruett and J. A. Smith (to Union Carbide).
45. R. Fowler, H. Conner, and R. A. Baehl, *Chem. Tech.*, 772 (1976).
46. H. H. Storch, N. Golumbic, and R. B. Anderson, *The Fischer-Tropsch and Related Syntheses*, John Wiley & Sons, Inc., New York, 1951.
47. H. Pichler, *Adv. Catal.* **4,** 271 (1952).
48. U.S. Pats. 3,894,103–3,894,106 (July 8, 1975), C. D. Chang and A. J. Silvestri (to Mobil Oil Co.).
49. S. L. Meisel and co-workers, *Chem. Tech.*, 86 (1976).
50. G. M. Drissel in A. H. Pelofsky, ed., *Synthetic Fuels Processing*, Marcel Dekker Inc., New York, 1977, p. 319.
51. G. A. Mills and F. W. Steffgen, *Catal. Rev.* **8,** 159 (1973).
52. M. A. Vannice, *Cat. Rev. Sci. Eng.* **14,** 153 (1976).
53. D. Nicholls in ref. 24, Vol. 3, p. 1110.
54. U.S. Pats. 3,833,634 (Sept. 3, 1974), and 3,957,857 (May 18, 1976), R. L. Pruett and W. E. Walker (to Union Carbide).
55. Ger. Offen. 2,503,204 (July 31, 1975), M. M. Bhasin (to Union Carbide).
56. Ger. Offen. 2,503,233 (July 31, 1975), M. M. Bhasin and G. L. O'Connor (to Union Carbide).
57. U.S. Pat. 3,769,329 (Oct. 30, 1973), F. E. Paulik, A. Hershman, W. R. Knox, and J. F. Roth (to Monsanto Co.).
58. U.S. Pat. 3,769,324 (Oct. 30, 1973), F. E. Paulik, A. Hershman, J. F. Roth, and W. R. Knox (to Monsanto Co.).
59. U.S. Pat. 3,928,435 (Dec. 23, 1975), Y. Awane, S. Otsuka, M. Nagata, and F. Tanaka (to Mitsubishi).
60. Ref. 30, Vol. 2, p. 63.
61. Ger. Offen. 2,610,035 (Sept. 23, 1976), N. Rizkalla and C. N. Winnick (to Halcon).
62. U.S. Pat. 3,911,003 (Oct. 7, 1975), S. Suzuki (to Chevron).
63. U.S. Pat. 3,923,904 (Dec. 2, 1975), H. Hara (to Nippon Oil Co.).
64. U.S. Pat. 3,852,346 (Dec. 3, 1974), D. Forster, A. Hershman, and F. E. Paulik (to Monsanto Co.).
65. U.S. Pats. 3,989,747 and 3,989,748 (Nov. 2, 1976), F. E. Paulik, A. Hershman, J. F. Roth, and J. H. Craddock (to Monsanto Co.).
66. U.S. Pats. 3,346,625 (Oct. 10, 1967) and 3,349,119 (Oct. 24, 1967), D. M. Fenton and K. L. Olivier (to Union Oil Co.).
67. U.S. Pat. 3,907,882 (Sept. 23, 1975), W. Gänzler, K. Kabs, and G. Schröder (to Röhm GmbH).

68. U.S. Pats. 3,846,468 (Nov. 5, 1974) and 3,980,690 (Sept. 14, 1976), E. Perrotti and G. Cipriani (to Snam Progetti S.P.A.).
69. Ger. Offen. 2,213,435 (Oct. 11, 1973), W. Gänzler, K. Kabs, and G. Schröder (to Röhm GmbH).
70. U.S. Pats. 3,992,436 (Nov. 16, 1976) and 4,005,128-131 (Jan. 25, 1977), L. R. Zehner (to Atlantic Richfield).
71. U.S. Pat. 3,576,835 (Apr. 27, 1971) E. Smith and W. Schnabel (to Olin).
72. U.S. Pat. 3,832,372 (Aug. 27, 1974), R. D. Hammond, W. M. Clarke, and W. I. Denton (to Olin).
73. U.S. Pat. 2,866,822 (Dec. 30, 1958), H. T. Siefen and W. R. Trutna (to E. I. du Pont de Nemours & Co.).
74. U.S. Pat. 3,948,998 (Apr. 6, 1976), S. Fujiyama, T. Takahashi, S. Kozao, and T. Kasahara (to Mitsubishi).
75. U.S. Pat. 3,960,932 (June 1, 1976), R. F. Heck (to University of Delaware).
76. E. W. Abel and F. G. A. Stone, *Quart. Rev.* **24,** 498 (1970).
77. G. Pieper in H. F. Mark and N. G. Gaylord, eds., *Encyclopedia of Polymer Science and Technology,* Vol. 9, John Wiley & Sons, Inc., New York, 1968, p. 397.
78. H. B. Elkins and L. D. Pagnolto in I. M. Kolthoff, P. J. Elving, and F. S. Stross, eds., *Treatise on Analytical Chemistry,* Part III, Vol. 2, Wiley-Interscience, New York, 1971, p. 61.
79. R. L. Beatty, *U. S. Bur. Mines Bull.* **557,** (1955).
80. R. J. Leibrand, *J. Gas Chromat.* **518,** (1967).
81. H. E. McGannon, *The Making, Shaping and Treating of Steel,* 8th ed., United States Steel Corporation, Pittsburgh, 1964, p. 66.
82. H. H. Lowry, *Chemistry of Coal Utilization,* Vol. II, John Wiley & Sons, Inc., New York, 1945, p. 1743.
83. J. Quibel, *Chem. Process Eng.* **50**(6), 83 (1969).
84. J. F. Farnsworth and co-workers, *AISE Convention Paper,* Philadelphia, Pa., April 22–24, 1974.
85. U.S. Pat. 3,943,236 (Mar. 9, 1976), R. V. Green (to E. I. du Pont de Nemours & Co.).
86. O. J. Quartulli, *Petrol. Petrochem. Intern.* (London) **13** (7), 70 (1973).
87. A. L. Kohl and F. C. Riesenfeld, *Gas Purification,* 2nd ed., Gulf Publishing Company, Houston, 1974.
88. R. Egalon, R. Vanhille, and M. Willemyns, *Ind. Eng. Chem.* **47,** 887 (1955).
89. W. W. Yeandle and G. F. Klein, *Chem. Eng. Prog.* **48,** 349 (1952).
90. N. M. Zhavoronkov and P. M. Reshchikov, *J. Chem. Ind. USSR* **10**(8), 41 (1933).
91. W. Förg, *Linde Reports on Science & Technology,* No. 15, 1970, pp. 20–21.
92. U.S. Pat. 3,651,159 (Mar. 21, 1972), R. B. Long, F. A. Caruso, R. J. DeFeo, and D. G. Walker (to Exxon).
93. U.S. Pat. 3,857,869 (Dec. 31, 1974), R. G. Turnbo (to Tenneco).
94. U.S. Pat. 3,868,398 (Feb. 25, 1975), W. R. Kroll and R. B. Long (to Exxon).
95. U.S. Pat. 3,933,878 (Jan. 20, 1976), W. E. Tyler and M. B. Dines (to Exxon).
96. D. G. Walker, *Chem. Tech.,* 308 (1975).
97. D. J. Haase and D. G. Walker, *Paper, Fourth Joint Chemical Engineering Conference,* Vancouver, B.C., September 1973.
98. U.S. Pat. 3,960,910 (June 1, 1976), J. R. Sudduth and D. A. Keyworth (to Tenneco).
99. Air Products and Chemicals, Inc., *Specialty Gases and Equipment Catalog,* Nov. 1, 1976, pp. 18–19.
100. Matheson Gas Products, *Catalog #30,* Feb. 1975, pp. 28–29.
101. Union Carbide Corporation, *Linde Specialty Gases Catalog,* Dec. 1974, p. 15.
102. *Criteria for a Recommended Standard . . . Occupational Exposure to Carbon Monoxide,* NIOSH-TR-007-72, National Technical Information Service, Springfield, Va., 1972.
103. F. A. Patty in F. A. Patty, ed., *Industrial Hygiene and Toxicology,* Vol. 2, Interscience-Publishers, a division of John Wiley & Sons, Inc., New York, 1963, p. 924.
104. R. E. Gosselin and co-workers, *Clinical Toxicology of Commercial Products,* Williams & Wilkins Co., Baltimore, 1976, p. 86.

CHARLES M. BARTISH
GERALD M. DRISSEL
Air Products and Chemicals, Inc.

CARBON MONOXIDE–HYDROGEN REACTIONS. See Fuels, synthetic.

CARBONYLS

Carbon monoxide (qv), the most important π-acceptor ligand, forms a host of neutral, anionic, and cationic transition metal complexes. There is at least one known type of carbonyl derivative for every transition metal, as well as evidence supporting the existence of the carbonyls of some lanthanides and actinides (1) (see Coordination compounds; Organomettallics).

Carbonyls are useful in the preparation of high purity metals, in catalytic applications, and in the synthesis of organic compounds. Metal carbonyls are employed in the preparation of complexes where the carbon monoxide ligand is replaced by halides, hydrogen, group VB and VIB derivatives, arenes, and many chelating ligands (see Chelating agents). Metal carbonyls and their derivatives are suited for mechanistic studies under favorable experimental conditions. Substitution rates on some metal carbonyls such as those of Cr, Mo, W, and Mn, can be easily measured (2). Detailed mechanistic studies on metal carbonyls have become increasingly important in understanding the factors influencing ligand substitution processes, especially as they apply to catalytic activity. Large clusters are only beginning to be recognized as potential catalysts with favorable heterogenous and homogenous characteristics (see Catalysis).

This survey covers transition metal carbonyls of the type $M_x(CO)_y$, where M is a metal in the zero oxidation state and x and y are integers. The preparation of metal carbonyls is reviewed with emphasis on new techniques utilizing low pressures of carbon monoxide. Emphasis is directed toward the role of metal carbonyls in organic synthesis and in catalytic applications on both a research and industrial scale.

Structure and Bonding of Metal Carbonyls

The literature of the bonding and structure of metal carbonyls is vast. Numerous theoretical approaches have been used in predicting and describing the bonding and structure of metal carbonyls. The Sidgwick concept of effective atomic number, or the inert gas rule, has been particularly useful in predicting the formulas of metal carbonyls. Molecular orbital and ligand field calculations have provided additional insight to the more detailed features of bonding in metal carbonyls.

Bonding. The inert gas rule requires that each metal interact with a sufficient number of CO molecules (each CO supplying a lone pair of electrons) to allow the metal to achieve the electronic structure of the subsequent inert gas in the periodic table. For example, nickel metal with 28 electrons coordinates four CO molecules to form nickel carbonyl containing 36 electrons, the configuration of the inert gas krypton. Nearly every metal forming a carbonyl obeys the inert gas rule. An exception is vanadium which forms a hexacarbonyl in which the number of electrons is 35. This carbonyl has a paramagnetism equivalent to one unpaired electron. Many metals with an odd number of electrons achieve the inert gas configuration by: (*1*) forming a covalent metal–metal bond, (*2*) using CO molecules to act as bridges between two metals with each metal receiving one electron from each CO molecule, or (*3*) using a CO molecule to form a bridge among three metal centers.

The accepted view of bonding in metal carbonyls is one in which charge is donated from the ligand to the metal by a sigma bond and electron density from the metal

d-orbitals is back-donated into the π^* (unoccupied or antibonding) orbitals of the ligand. The electron density in the π^* orbitals of the ligand is dependent to a certain extent on the charge donation from the ligand to the metal, therefore the σ and π bonding is synergistic (Fig. 1).

Bond lengths and infrared spectra support the multiple-bond character of the M—CO bonds (M = metal). Coordination of a CO molecule to a metal center can change the bond order of the C—O bond. According to the description of σ and π bonding given above, increased σ bonding between a metal and CO results in an increased bond order for the C—O bond. Conversely, increased π bonding results in more electron density occupying the π^* orbitals of CO and hence a decrease in the C—O bond order. Changes in the bond order of C—O are reflected in the shifts of the C—O stretching frequencies in the infrared spectrum of a particular metal carbonyl. As the bond order increases, the C—O stretching frequencies shift to higher energies. Compared to the stretching frequency of free CO (2143 cm^{-1}), terminal carbonyl groups in neutral metal complexes have stretching frequencies in the range of 2125–1900 cm^{-1}, showing a reduction in the bond order of CO upon coordination. A qualitative assessment of π back-bonding can be made by changing the electron density on the central metal and noting the change in the C—O stretching frequencies. Assuming the σ bond remains fairly constant (3), any change in the electron density of the metal atom should result in an increase or decrease of electron density flowing through the d-π bonding orbitals. As the charge in the central metal atom is increased, the π^* orbitals of CO must accept more electron density, hence the C—O bond order should decrease. Such a trend is observed in the compounds $Mn(CO)_6^+$, $Cr(CO)_6$, and $V(CO)_6^-$ where the C—O stretching frequencies decrease in the order given as the charge on the central metal increases (4). Bonding in metal carbonyls has been a subject of intense theoretical and experimental interest (5–13).

Structure. The CO molecule coordinates in three ways as shown diagrammatically in Figure 2. Terminal carbonyls are the most common. Bridging carbonyls are common in most polynuclear metal carbonyls. As depicted in Figure 2, metal–metal bonds also play an important role in polynuclear metal carbonyls. The metal atoms in carbonyl complexes show a strong tendency to coordinate all their valence orbitals in forming bonds. These include the nd^5, $(n + 1)s$ and the $(n + 1)p^3$ orbitals. As a result, the inert gas rule is successful in predicting the structure of most metal carbonyls.

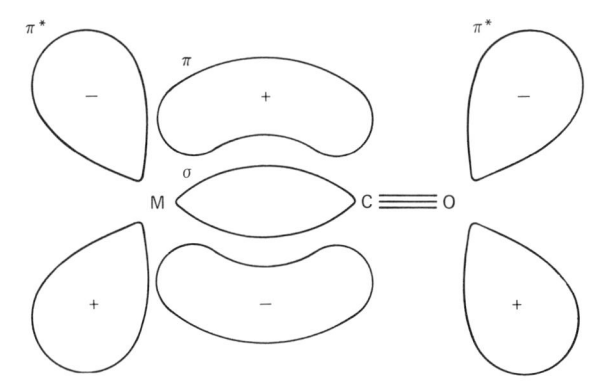

Figure 1. Overlapping of the metal (M) d orbitals with the σ bonding and π^* (antibonding) orbitals of CO.

796   CARBONYLS

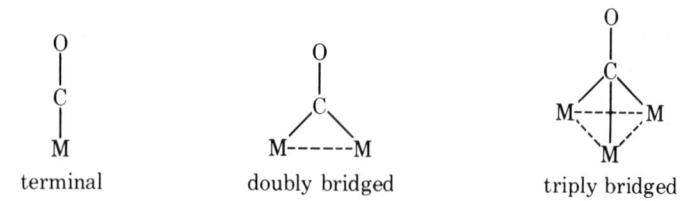

```
    O              O                 O
    ‖              ‖                 ‖
    C              C                 C
    |             ╱ ╲              ╱ | ╲
    M          M------M           M--+--M
                                    ╲|╱
                                     M

 terminal      doubly bridged    triply bridged
```

Figure 2. Bonding modes of CO.

Mononuclear Carbonyls. The lowest coordination number adopted by a metal carbonyl is four. The only representative of this class is nickel carbonyl, the first metal carbonyl isolated (14). The molecule possesses tetrahedral geometry as shown in structure (1). A few transient four-coordinate carbonyls, such as $Fe(CO)_4$, have been detected (15).

Representative pentacarbonyls are restricted to the iron, ruthenium, and osmium group. All three pentacarbonyls possess trigonal bipyramidal structures as shown in structure (2). The pentacarbonyls of ruthenium and osmium are thermally unstable. Osmium pentacarbonyl rapidly polymerizes at room temperature to form polynuclear species. The transient species $Cr(CO)_5$ (16), $Mo(CO)_5$ (17), and $W(CO)_5$ (18) have been investigated.

The neutral complexes of chromium, molybdenum, tungsten, and vanadium are six-coordinate with the CO molecules arranged about the metal in an octahedral configuration as shown in structure (3). Vanadium carbonyl, as previously mentioned, possesses an unpaired electron and would be expected to form a metal–metal bond. Steric hindrance may prevent dimerization. The other hexacarbonyls are diamagnetic.

Polynuclear Carbonyls. Several structures consist of dinuclear metal carbonyls as shown in structures (4)–(6). The metal atoms in $Mn_2(CO)_{10}$, as with technetium and rhenium, are held together by a metal–metal bond and contain ten terminal CO ligands, five coordinated to each atom. The CO ligands of $Mn_2(CO)_{10}$ adopt a staggered configuration as illustrated in structure (4). In $Fe_2(CO)_9$, three CO molecules act as bridging ligands between two iron atoms as shown in structure (5). The structure of $Co_2(CO)_8$ (6) in the solid state is similar to that of $Fe_2(CO)_9$ with one less bridging CO ligand.

Three trinuclear carbonyls are known. Although a linear or a cyclic structure of the metal atoms is possible in a trinuclear complex, the cyclic structure is preferred, as shown in the D_{3h} structure for $Os_3(CO)_{12}$ (7). The molecule possesses a triangular arrangement of osmium atoms with four terminal CO ligands coordinated equally about each osmium atom. The molecule $Ru_3(CO)_{12}$ is also cyclic and is isomorphous with the osmium analogue.

The determination of the structure of $Fe_3(CO)_{12}$ proved to be a difficult problem. An early report on the crystal structure claimed the molecule was a monoclinic prism and established the molecular formula (19). In a later report the structure (8) was shown to be a triangular array of iron atoms with two bridging and ten terminal CO molecules. The presently accepted structure was initially deduced from an x-ray crystal structure of the $Fe_3(CO)_{11}H^-$ analogue (20).

Three tetranuclear carbonyls are known: $Co_4(CO)_{12}$, and the rhodium and iridium analogues. The tetranuclear complex $Co_4(CO)_{12}$ contains a tetrahedral array of cobalt atoms with three bridging and nine terminal CO ligands, as shown in structure (9). The structure in solution was reported to have four bridging CO ligands (21). The rhodium analogue has the same structure as the cobalt complex. The iridium analogue (10) has no bridging ligands, but twelve terminal CO ligands coordinated equally among four iridium atoms.

Heteronuclear Carbonyls and High-Nuclearity Carbonyl Clusters

A few of the heteronuclear metal carbonyls that have been reported are tabulated in Table 1. In most cases the structures of the heteronuclear species are similar to their homonuclear analogues.

Table 1. Representative Heteronuclear Metal Carbonyls

Heteronuclear carbonyl	CAS Registry No.	Refs.
$(CO)_5MnCo(CO)_4$	[35646-82-3]	22–23
$(CO)_5MnRe(CO)_5$	[14693-30-2]	24
$[(CO)_5Mn]_2Fe(CO)_4$	[15668-57-2]	25
$[(CO)_5Mn][(CO)_5Re][(CO)_4Fe]$	[33958-72-4]	26
$(CO)_5ReCo(CO)_4$	[52647-10-6]	27
$[(CO)_5Re]_2Fe(CO)_4$	[16040-31-6]	28
$Ru_2Os(CO)_{12}$	[12389-47-8]	29
$Fe_2Ru(CO)_{12}$	[32311-57-2]	30

Routine syntheses, crystallization, structural identification, and chemical characterization of high-nuclearity clusters can be exceedingly difficult. Usually, several different clusters are formed in any given synthetic procedure, and each compound must be tediously extracted and identified. The problem may be compounded by the instability of a particular molecule. In 1962 Dahl and co-workers characterized the structure of the first high-nuclearity carbide complex formulated as $Fe_5(CO)_{15}C$ (31). This complex was originally prepared in an extremely low yield of 0.5%. This molecule was the first carbide complex isolated and became the forerunner of a whole family of carbide complexes, see structure (11).

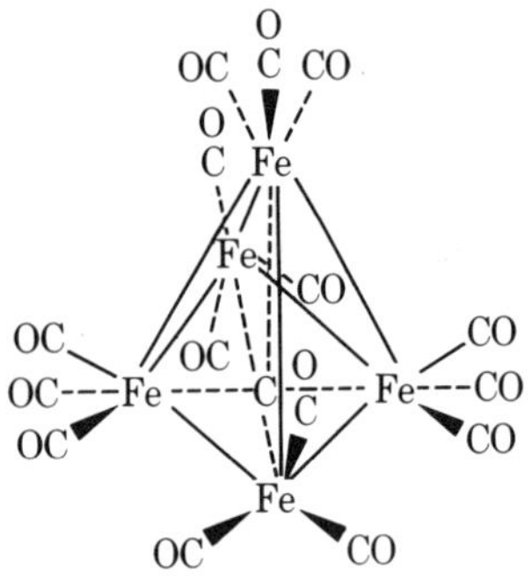

(11)

[11087-47-1]

In 1943 Hieber and Lagally reported that the reaction of anhydrous $RhCl_3$ with CO at 80°C under pressure with a halide acceptor, such as copper, produced a black crystalline product which was formulated as $Rh_4(CO)_{11}$ (32). The correct structure of the complex was determined by Dahl as $Rh_6(CO)_{16}$ (33). This complex has been thoroughly studied and can be considered the forerunner of high-nuclearity chemistry. The rhodium atoms are arranged at the corners of an octahedron with two terminal CO ligands coordinated to each rhodium atom. The molecule has four bridging CO molecules, each coordinated to three rhodium atoms and occupying four of the eight faces of the octahedron in structure (12).

The existence of a central cavity is a characteristic of a cluster containing a regular

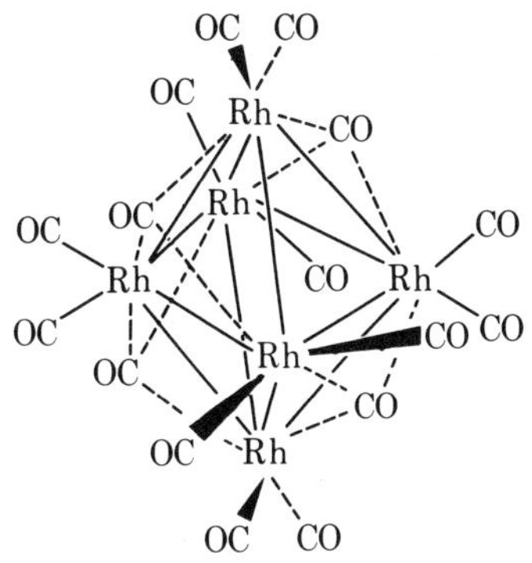

(12)

polyhedron of metal atoms. The cavity is confirmed by the formation of a large number of carbide complexes, as shown in structure (11).

Despite the complexity of transition metal clusters, a basic structural unit is common. A triangular network of metal atoms occurs in most clusters (34). An example of triangulated polyhedra is the molecule $Os_6(CO)_{18}$ (35) shown in structure (13). Nearly half of the reported high-nuclearity clusters contain an octahedron or a distorted octahedron of metal atoms. The large cluster $Rh_{12}(CO)_{30}^{2-}$ contains two octahedral arrays of rhodium atoms connected by a rhodium–rhodium bond and two bridging CO ligands (36). In the complexes $Co_6(CO)_{15}^{2-}$ and $Co_6(CO)_{14}^{4-}$, deformation of the octahedral framework toward a trigonal prism occurs (37–38). Further twisting of the trigonal prism occurs in the dianions of $[Pt_3(CO)_6]_n^{2-}$ ($n = 2, 3, 4, 5$) where the structure assumes a chain of twisted prisms of $Pt_3(CO)_6$ units (39–40).

Only three heptanuclear clusters are known, all containing metal frameworks based on monocapped octahedrons (41). The octanuclear cluster, $Co_8(CO)_{18}C^{2-}$, contains eight cobalt atoms forming a distorted square antiprism (42). One of the largest clusters reported to date is $Rh_{15}(CO)_{28}C_2^-$, in which the fifteen rhodium atoms can be described as a centered tetracapped pentagonal prism (43). Illustrations of the high-nuclearity metal clusters described above are given in a review by Chini (44). References to the x-ray crystal structures for the metal carbonyls discussed in this section are given in Table 2.

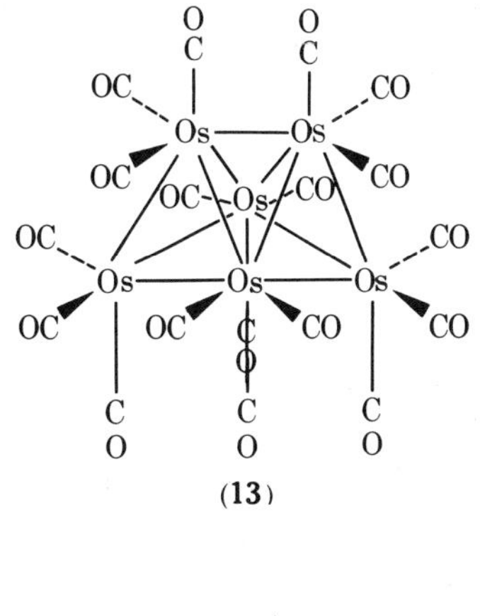

(13)

Table 2. Physical Properties of Metal Carbonyls

Carbonyl formula	CAS Registry No.	Color (solid)	Melting point, °C	Boiling point, °C	Density	M—M distance, nm	Refs.
V(CO)$_6$	[14024-00-1]	blue-green	50 dec	40–50^{15} subl.			45
Cr(CO)$_6$	[13007-92-6]	white	149–155	70–75^{15} subl.	1.77		46
Mo(CO)$_6$	[13939-06-5]	white	150–151 dec		1.96		46
W(CO)$_6$	[14040-11-0]	white	169–170	50 subl.	2.65		46
Mn$_2$(CO)$_{10}$	[10170-69-1]	yellow	151–155	50$^{0.01}$ subl.	1.81	0.293	47
Tc$_2$(CO)$_{10}$	[14837-15-1]	white	159–160	40$^{0.01}$ subl.	2.08	0.3036	48
Re$_2$(CO)$_{10}$	[14285-68-8]	white	177	60$^{0.01}$ subl.	2.87	0.302	47
Fe(CO)$_5$	[13463-40-6]	white	–20	103	1.52		49–51
Fe$_2$(CO)$_9$	[15321-51-4]	yellow	100 dec		2.08	0.2523	52
Fe$_3$(CO)$_{12}$	[12088-65-2]	green–black	140 dec	60$^{0.1}$ subl.	2.00	0.263 (av)	53–54
Ru$_3$(CO)$_{12}$	[15243-33-1]	orange	150 dec		2.75	0.2848	55
Os$_3$(CO)$_{12}$	[15696-40-9]	yellow	224		3.48	0.288	56–57
Co$_2$(CO)$_8$	[10210-68-1]	orange	50–51	45^{10} subl.	1.73	0.2542	58
Co$_4$(CO)$_{12}$	[17786-31-1]	black	60 dec		2.09	0.249	59–60
Rh$_4$(CO)$_{12}$	[19584-30-6]	red	dec		2.52	0.275 (av)	60–61
Rh$_6$(CO)$_{16}$	[28407-51-4]	black	220 dec		2.87	0.2776	33
Ir$_4$(CO)$_{12}$	[11065-24-0]	yellow	210 dec			0.268	62
Ni(CO)$_4$	[13463-39-3]	white	–25	43	1.32		63

Physical Properties

Some physical properties of metal carbonyls are presented in Table 2. Most metal carbonyls are volatile solids which sublime easily. The volatility of metal carbonyls is an important safety consideration. The vapor pressures of many metal carbonyls have been tabulated elsewhere (64).

The thermodynamic properties of simple metal carbonyls have been compiled (65–71). Table 3 lists some of the thermodynamic properties of selected common metal carbonyls.

Preparation

Since the discovery of nickel carbonyl by Mond (14) in 1890, many other metal carbonyls have been prepared. Nickel and iron are the only metals which combine directly with CO to produce metal carbonyls in reasonable yields.

Since transition metals even in a finely-divided state do not readily combine with CO, various metal salts have been used to synthesize metal carbonyls. Metal salts almost always contain the metal in a higher oxidation state than the resulting carbonyl complex. Therefore, most metal carbonyls result from the reduction of the metal in the starting material. Such a process has been referred to as reductive carbonylation. Although detailed mechanistic studies are lacking, the process probably proceeds through stepwise reduction of the metal with simultaneous coordination of CO (72).

To date, neutral metal complexes have been synthesized for nearly every transition metal, and at least one carbonyl derivative is known for every transition metal. All of the neutral complexes that have been reported are listed in Table 4. The following section outlines general methods applied to the preparation of most metal carbonyls. Particular emphasis is placed on low pressure syntheses.

Syntheses from Dry Metals and Salts. As mentioned previously, only metallic nickel and iron react directly with CO at moderate pressures and temperatures to form metal carbonyls. A report has claimed the synthesis of $Co_2(CO)_8$ in 99% yield from cobalt metal and CO at high temperatures and pressures (80–81). The CO has to be absolutely free of oxygen and carbon dioxide or the yield is drastically reduced. Two patents report the formation of carbonyls from molybdenum and tungsten metal (82–83). Ruthenium and osmium do not react with CO even under drastic conditions (84–85).

Table 3. Heats of Combustion and Formation for Some Metal Carbonyls

Carbonyl formula	Heat of combustion[a,b], kJ/mol	Ref.	Heat of formation[a], kJ/mol	Ref.
$Cr(CO)_6$	−1854	66	−1077	66
$Mo(CO)_6$	−2123	66	−982.4	66
$W(CO)_6$	−2250	66	−950.6	66
$Mn_2(CO)_{10}$	−3251	68	−1677	68
$Fe(CO)_5$	−1619	67	−964.0	67
$Ni(CO)_4$	−1181	65	−622.2	71

[a] To convert J to cal, divide by 4.184.
[b] Determined for the combustion to metal oxides with oxygen.

Table 4. Known Metal Carbonyls of the Transition Elements and Their CAS Registry Numbers

IVB	VB	VIB	VIIB	VIII		
Ti(CO)$_6$[a] [61332-66-9]	V(CO)$_6$ [14024-00-1]	Cr(CO)$_6$ [13007-92-6]	Mn$_2$(CO)$_{10}$ [10170-69-1]	Fe(CO)$_5$ [13463-40-6] Fe$_2$(CO)$_9$ [15321-15-4] Fe$_3$(CO)$_{12}$ [12088-65-2]	Co$_2$(CO)$_8$ [10210-68-1] Co$_4$(CO)$_{12}$ [17786-31-1] Co$_6$(CO)$_{16}$[b] [12182-17-1]	Ni(CO)$_4$ [13463-39-3]
	Nb(CO)$_6^-$ [c]	Mo(CO)$_6$ [13939-06-5]	Tc$_2$(CO)$_{10}$ [14837-15-1]	Ru(CO)$_5$ [16406-48-7] Ru$_3$(CO)$_{12}$ [15243-33-1]	Rh$_4$(CO)$_{12}$ [19584-30-6] Rh$_6$(CO)$_{16}$ [28407-51-4]	Pd[d]
	Ta(CO)$_6^-$ [c]	W(CO)$_6$ [14040-11-0]	Re$_2$(CO)$_{10}$ [14285-68-8]	Os(CO)$_5$[e] [16406-49-8] Os$_2$(CO)$_9$[e] [28411-13-4] Os$_3$(CO)$_{12}$ [15696-40-9] Os$_5$(CO)$_{16}$[h] [37190-55-9] Os$_6$(CO)$_{18}$[i] [37216-50-5] Os$_7$(CO)$_{21}$[h] [37190-64-0] Os$_8$(CO)$_{23}$[h] [37190-66-2]	Ir$_4$(CO)$_{12}$ [11065-24-0] Ir$_6$(CO)$_{16}$[f] [56801-74-2]	[Pt$_3$(CO)$_6$]$_n^{2-}$, $n = 1$ or 6[c,g]

[a] Matrix isolated (73).
[b] Ref. 74.
[c] Exists only as anions (75).
[d] All attempts to synthesize this carbonyl have led to reduction to palladium metal.
[e] Decomposes near room temperature (76).
[f] Ref. 77.
[g] Ref. 78.
[h] Ref. 79.
[i] Ref. 35.

The dry method for synthesizing metal carbonyls from salts and oxides has proven very useful in a number of cases. The metal carbonyl is formed in the presence of a suitable reducing agent. In some cases CO itself is the reducing agent. Rhenium (86) and technetium (87–88) carbonyls are conveniently prepared from the oxides. Other suitable reducing agents, such as silver or copper, can be used in the dry state to effect carbonylation.

$$Tc_2O_7 + 17\ CO \rightarrow Tc_2(CO)_{10} + 7\ CO_2$$

$$2\ ReO_3 + 16\ CO \rightarrow Re_2(CO)_{10} + 6\ CO_2$$

$$RuI_3 + 3\ Ag + 5\ CO \rightarrow Ru(CO)_5 + 3\ AgI$$

Syntheses in Solvent Systems. Very few examples of syntheses of metal carbonyls in aqueous solution are reported. An exception is the preparation of Co$_2$(CO)$_8$ from CoSO$_4$ (66% yield) or CoCl$_2$ (56% yield) with CO at 9.6–11 MPa (95–110 atm) in aqueous ammonia at 120°C for 16–18 h (89). Triiron dodecacarbonyl is prepared almost exclusively in aqueous solution. Quantitative yields of Fe$_3$(CO)$_{12}$ have been obtained by oxidizing alkaline solutions of carbonyl ferrates with manganese dioxide (90–92).

Most metal carbonyls are synthesized in nonaqueous media. Reactive metals, such as sodium (75), magnesium (93), zinc (94), and aluminum (95–96), are usually used as reducing agents. Solvents that stabilize low oxidation states of metals and act as electron transfer agents are commonly employed. These include diethyl ether, tetrahydrofuran, and 2-methoxyethyl ether (diglyme).

$$VCl_3 + 4\,Na + 6\,CO \xrightarrow{\text{diglyme}} NaV(CO)_6 + 3\,NaCl$$

$$2\,CrCl_3 + 3\,Mg + 12\,CO \xrightarrow{\text{pyridine}} 2\,Cr(CO)_6 + 3\,MgCl_2$$

$$WCl_6 + 3\,Zn + 6\,CO \xrightarrow{\text{ether}} W(CO)_6 + 3\,ZnCl_2$$

$$WCl_6 + 2\,Al + 6\,CO \xrightarrow{\text{ether}} W(CO)_6 + 2\,AlCl_3$$

Organometallic reagents, such as trialkyl aluminums (72) and sodium benzophenone (97), are quite useful as reducing agents. Alkyl aluminums have been used to synthesize $Cr(CO)_6$, $Mo(CO)_6$, and $W(CO)_6$ in high yields (72). In one case, hydrogen was used effectively as a reducing agent in petroleum ether solvent (98–99).

$$MnX_2 + AlR_3 \rightarrow Mn_2(CO)_{10} + AlX_n$$

$$X = Cl,\,I,\,HCO_2,\,CH_3CO_2 \qquad R = CH_3,\,C_2H_5$$

$$MnCl_2 + Na(\text{benzophenone ketal}) \rightarrow Mn_2(CO)_{10} + 2\,NaCl$$

$$2\,CoCO_3 + H_2 + 8\,CO \xrightarrow{\text{petroleum ether}} Co_2(CO)_8 + 2\,H_2O + 2\,CO_2$$

Condensation of Simple Metal Carbonyls. Some metal carbonyls of lower molecular weight lose CO on heating or uv irradiation leading to the formation of higher molecular weight species. In some cases, as shown below, this method is a useful preparative tool (100–101).

$$2\,Fe(CO)_5 \xrightarrow{h\nu} Fe_2(CO)_9 + CO$$

$$2\,Co_2(CO)_8 \xrightarrow[80-90°C]{\Delta} Co_4(CO)_{12} + 4\,CO$$

Carbonylation by CO Exchange. A few metal carbonyls can be prepared by exchange of CO molecules. The reaction of WCl_6 with $Fe(CO)_5$ in the presence of hydrogen under pressure in diethyl ether results in yields of $W(CO)_6$ as high as 85% (102). The same reaction can be used to synthesize $Mo(CO)_6$ (103).

Synthesis of Heteronuclear and Polynuclear Metal Carbonyls. Heteronuclear metal carbonyls are usually synthesized by either metathesis (24) or condensation (25).

$$NaMn(CO)_5 + Re(CO)_5Cl \rightarrow (CO)_5MnRe(CO)_5 + NaCl$$
$$[14693\text{-}30\text{-}2]$$

$$Mn_2(CO)_{10} + Re_2(CO)_{10} \xrightarrow{h\nu} 2\,(CO)_5MnRe(CO)_5$$

The main synthetic route to high-nuclearity carbonyl clusters is a condensation reaction. The condensation process can be: (*1*) a reaction induced by coordinatively unsaturated species, or (2) a reaction between coordinatively saturated species in different oxidation states. As an example of (1), $Os_3(CO)_{12}$ can be condensed to form a series of higher coordinated species (79).

$$Os_3(CO)_{12} \xrightarrow{h\nu} Os_4(CO)_{13} + Os_5(CO)_{16} + Os_6(CO)_{18} \ldots$$

The interaction of species in different oxidation states can lead to higher-coordinated molecules, as shown below (44,104).

$$Fe_3(CO)_{11}{}^{2-} + Fe(CO)_5 \rightarrow Fe_4(CO)_{13}{}^{2-} + 3\ CO$$
[25767-85-5]

Low Pressure Syntheses. The majority of metal carbonyls are synthesized under high pressures of CO. Early preparations of carbonyls were made under super-pressures of 1 GPa (ca 10,000 atm). Fortunately, refinement of synthetic procedures in the 1960s has eliminated the need for such high pressures and the expensive high pressure equipment required.

Recently, attention has been directed toward finding methods of preparing metal carbonyls under still less drastic conditions. Numerous reports have appeared in the literature concerning low pressure syntheses of metal carbonyls, but the reactions have been restricted primarily to the carbonyls of the transition metals of Groups VIII, IB, and IIB. A procedure for preparing $Mn_2(CO)_{10}$, however, from methylcyclopentadienylmanganese tricarbonyl [12108-13-3] and atmospheric pressures of CO has appeared (105). The carbonyls of ruthenium (106–107), rhodium (108–109), and iridium (110–111) have been synthesized in good yields employing low pressure techniques. In all three cases, very low or even atmospheric pressures of CO effect carbonylation. The examples below represent successful low pressure syntheses.

$$RuCl_3 + CO \rightarrow Ru(CO)_n Cl_m \xrightarrow[CO]{Zn} Ru_3(CO)_{12} + ZnCl_2$$

$$3\ Rh_2(O_2CCH_3)_4 + 22\ CO + 6\ H_2O \rightarrow Rh_6(CO)_{16} + 6\ CO_2 + 12\ CH_3COOH$$

$$Na_3IrCl_6 + CO \rightarrow Ir_4(CO)_{12}$$

$$Rh_2(CO)_4Cl_2 + CO + KOH \xrightarrow{CH_3OH} Rh_4(CO)_{12} + KCl$$

$$Ir(CO)_2Cl(p\text{-toluidine}) + Zn + CO \rightarrow Ir_4(CO)_{12} + ZnCl_2$$

$$[Ru_3(O_2CCH_3)_6(H_2O)_3]O_2CCH_3 \xrightarrow{CO} Ru_3(CO)_{12} + CH_3COOH + CO_2$$

Economic Aspects

The following information represents commercial manufacturers of metal carbonyls. There may be companies not included who produce these compounds on a captive basis.

	Manufacturers
$Ni(CO)_4$	Pressure Chemical Company
$Fe(CO)_5$	GAF Corporation
$Co_2(CO)_8$	Strem Chemicals Inc.
$W(CO)_6$	Strem Chemicals Inc.
$Mo(CO)_6$	and Pressure Chemical Company
$Cr(CO)_6$	

Use of Metal Carbonyls in Organic Synthesis

Coordination of an organic molecule to a metal center can have a pronounced effect on its reactivity, eg, coordination of an alkene to a transition metal can alter its behavior toward nucleophilic or electrophilic attack. An extremely important field of chemistry concerned with stereospecific syntheses has developed around metal complexes that can give products having a high degree of optical purity (see Pharmaceuticals, optically-active).

The number of reactions involving transition metals and organic molecules is seemingly inexhaustible. Only a few of the important organic reactions involving metal carbonyls, either as reagents or catalysts, are described below. Many catalytic reactions involving metal carbonyls have important industrial applications.

Carbonylation of Olefins. The carbonylation of olefins is a process of immense industrial importance. The process includes hydroformylation and hydrosilylation of an olefin. The hydroformylation reaction, or oxo process (qv), leads to the formation of aldehydes from olefins, carbon monoxide, hydrogen, and a transition metal carbonyl. The hydrosilylation reaction involves addition of a silane to an olefin (112–113). One of the most important processes in the carbonylation of olefins uses $Co_2(CO)_8$ or its derivatives with phosphorus ligands as a catalyst. Propionaldehyde (114) and butyraldehyde (qv) (115) are synthesized industrially according to the following equation:

$$RCH=CH_2 + CO + H_2 \xrightarrow[Co_2(CO)_8,\ 100-200°C]{10-31\ MPa\ (1450-4500\ psi)} R(CH_2)_2CHO + RCH(CHO)CH_3$$

The oxo process is not limited to simple olefins. The terminal-to-branched ratio of products can be controlled by ligand addition (116). Butanol is produced from propylene and CO using a similar process (see Butyl alcohols). The catalyst in this case is $Fe(CO)_5$ (117).

$$CH_3CH=CH_2 + 3\ CO + H_2O \xrightarrow{Fe(CO)_5} CH_3CH_2CH_2CH_2OH + 2\ CO_2$$

Tetrarhodium dodecacarbonyl can effect carbonylation of an olefin at atmospheric pressure (118). The rate of hydroformylation of an olefin decreases with increasing alkyl substitution.

$$CH_2=CHCH_3 + Rh_4(CO)_{12} \xrightarrow[toluene]{H_2\ (101\ kPa\ or\ 1\ atm)} (CH_3)_2CHCHO + CH_3CH_2CH_2CHO$$

Straight-chain terminal olefins are more reactive than straight-chain internal olefins, which are more reactive than branched-chain olefins (119).

Carboxylation Reaction. The carboxylation reaction represents the conversion of acetylene (qv) and olefins into carboxylic acids or their derivatives. The industrially important Reppe process is used in the synthesis of β-unsaturated esters from acetylene. Nickel carbonyl is the catalyst of choice (120).

$$HC\equiv CH + CO + ROH \xrightarrow{Ni(CO)_4} CH_2=CHCOOR$$

The process, to which the raw materials are supplied at low pressures, is continuous and gives good yields of acrylates (see Acrylic acid). In the presence of catalytic amounts of $Co_2(CO)_8$, acetylene has been carboxylated in methanol yielding dimethyl succinate as the major product (121).

Allylic compounds can be carboxylated readily with Ni(CO)$_4$, either catalytically or stoichiometrically. In the presence of acetylene, the reaction of allylic compounds with CO, alcohols, and Ni(CO)$_4$ yields dienic carboxylic esters according to the following equation (122):

$$RCH=CHCH_2Cl + HC\equiv CH + CO + R'OH \xrightarrow{Ni(CO)_4} RCH=CHCH_2CH=CHCO_2R' + HCl$$

Olefin Isomerization. Some olefins can be isomerized in the presence of metal carbonyls. The carbonyl can act as a catalyst or a stoichiometric reagent in the reaction (123).

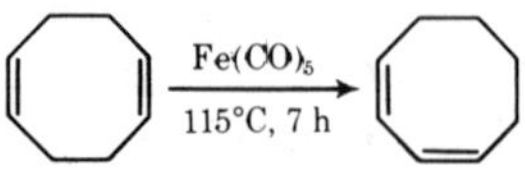

Stabilization of Unstable Intermediates. Transition metals can stabilize normally unstable or transient organic intermediates. Cyclobutadiene has never been isolated as a free molecule, but Pettit isolated and fully characterized an iron tricarbonyl complex of cyclobutadiene by the following reaction (124):

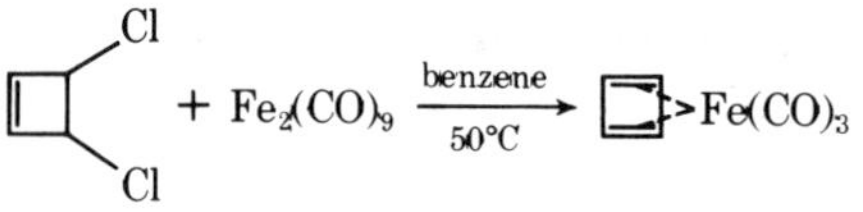

The complex exhibits remarkable stability, and the cyclobutadiene undergoes reactions without destruction of the ring (125). Cyclobutadieneiron tricarbonyl [12078-17-0] can be oxidized to generate cyclobutadiene *in situ* (126).

Coupling and Cyclization Reactions. An important process in organic chemistry is the coupling of alkyl halides to form longer chain hydrocarbons. This process is classically referred to as the Wurtz reaction. Alkyl halides can be coupled using sodium metal in ethyl ether as the solvent, but the yields are normally low and the product impure. One of the most successful reagents for coupling alkyl halides is Ni(CO)$_4$, as shown in the following example (127):

$$2\ CH_3CH=CHCH_2Cl \xrightarrow{Ni(CO)_4} CH_3CH=CH(CH_2)_2CH=CHCH_3$$

The process is versatile and a variety of long chain hydrocarbons can be prepared.

Reactions of acetylene and iron carbonyls can yield benzene derivatives, quinones, cyclopentadienes and a variety of heterocyclic compounds. The cyclization reaction is useful for preparing substituted benzenes. The reaction of *tert*-butylacetylene in the presence of Co$_2$(CO)$_8$ as the catalyst yields 1,2,4-tri-*tert*-butylbenzene (128). The reaction of Fe(CO)$_5$ with diphenylacetylene yields no less than seven different species with a cyclobutadiene derivative being the most important (129–131).

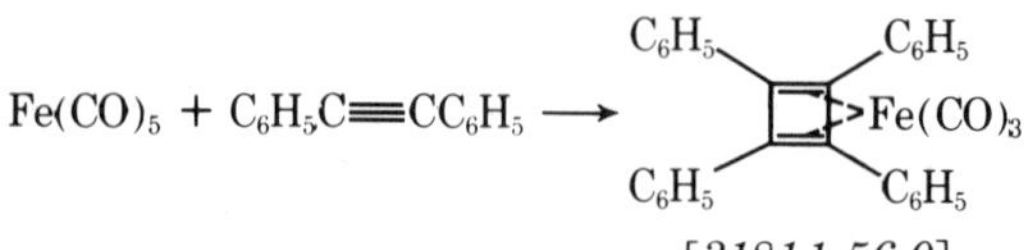

[31811-56-0]

The reaction of $Mn_2(CO)_{10}$ with acetylene yields π-dihydropentalenyl manganese (I) tricarbonyl [67375-48-8] as shown in the following reaction (132):

$$Mn_2(CO)_{10} + HC\equiv CH \longrightarrow \text{[dihydropentalenyl]}-Mn(CO)_3$$

Formation of Functional Groups. Metal carbonyls have been used in a number of cases to synthesize organic molecules containing particular functional groups (133–135). A synthesis of olefins from *gem*-dihalides has been reported, as shown in the following reaction (134):

$$2\,(C_6H_5)_2CCl_2 + Co_2(CO)_8 \rightarrow (C_6H_5)_2C=C(C_6H_5)_2 + 8\,CO + 2\,CoCl_2$$

Certain ketones can be synthesized conveniently from readily available aryl halides such as chloro- or bromobenzene (136–137). Amines react with CO in the presence of metal carbonyls forming *N*-formyl derivatives or substituted ureas (138–139).

$$C_6H_5X + Ni(CO)_4 \xrightarrow{THF} \underset{\text{benzil}}{C_6H_5COCOC_6H_5}$$

Homogenous Hydrogenation Catalysis. Metal carbonyl complexes are generally soluble in organic solvents and can function as homogenous catalysts (see Catalysis). Dicobalt octacarbonyl in the presence of hydrogen and CO will reduce aldehydes to alcohols. In addition to aldehydes, other functional groups, such as imines, nitriles, nitroalkyls, and nitroaryls, can be reduced (140). Aqueous solutions of $Fe(CO)_5$ will reduce nitrobenzenes to anilines (see Amines by reduction), benzil to benzoin and acetylenes to ethylenes (141). Aqueous solutions of $Fe(CO)_5$ will also convert olefins to alcohols at elevated temperatures and CO pressures (142). Fatty esters, such as methyl linoleate (143) and methyl linolenate (144), are reduced to varying degrees by $Fe(CO)_5$ in nonaqueous media at moderate hydrogen pressures.

The use of metal carbonyls in organic synthesis has been thoroughly reviewed (145–155).

Industrial Uses

The main uses of metal carbonyls are in the areas of catalysis and organic synthesis. The Reppe synthesis and Oxo process described above are of enormous industrial importance.

Easily decomposed, volatile metal carbonyls have been used in metal deposition reactions where heating forms the metal and carbon monoxide. Other products such as metal carbides and carbon may also form, depending on the conditions. The commercially important Mond process depends on the thermal decomposition of $Ni(CO)_4$ to form high purity nickel. In a typical vapor deposition process, a purified inert carrier gas is passed over a metal carbonyl containing the metal to be deposited. The carbonyl is volatilized, with or without heat, and carried over a heated substrate. The carbonyl is decomposed and the metal deposited on the substrate. A number of papers have appeared concerning vapor deposition techniques and uses (156–165) (see Film deposition techniques).

An important development in carbonyl chemistry concerns the use of metal carbonyls as antiknock compounds in gasoline (qv). The Ethyl Corporation has marketed methylcyclopentadienylmanganese tricarbonyl (MMT) for years as an antiknock compound. Unfortunately, owing to its higher cost, it has not completely replaced tetraethyllead (166) but it was used extensively in unleaded gasoline until it was banned by the EPA in 1978.

There is much current research to find alternative routes to synthetic fuels from less expensive and more abundant raw materials (see Fuels, synthetic). In the United States, there has been particular interest directed toward coal (qv) which is abundant. An important method for converting coal into a more useful source of energy in certain applications is the water gas reaction (the high temperature reaction of carbon (coal) and water to give carbon monoxide and hydrogen). Water gas is an important industrial fuel and source of hydrogen (see Hydrogen energy). There has been considerable interest directed toward converting water gas into simple organic compounds. The reaction of carbon monoxide with hydrogen to produce ethylene glycol in the presence of $Rh_6(CO)_{16}$ or other metal carbonyl derivatives as catalysts has been reported (167). At present, ethylene glycol used as an antifreeze (qv) is produced from ethylene, which is itself a product of petroleum. The catalytic reaction described above provides a means of synthesizing ethylene glycol from coal.

Other valuable catalytic systems are being investigated, such as the ruthenium dodecacarbonyl-catalyzed water gas shift reaction shown in the following equation (168):

$$CO + H_2O \xrightarrow[CH_3CH_2OCH_2CH_2OH,\ KOH,\ 100°C]{Ru_3(CO)_{12}} CO_2 + H_2$$

The water gas shift reaction (169) may also be involved in the carbonylation of acetylene with water as the hydrogen source to produce hydroquinone (see Acetylene-derived chemicals).

$$2\ C_2H_2 + 3\ CO + H_2O \xrightarrow[THF,\ 190°C,\ 15.2\ MPa\ (150\ atm)]{Ru_3(CO)_{12}} C_6H_4(OH)_2 + CO_2$$

An interesting development in the use of metal carbonyl catalysts is the production of hydrocarbons from carbon monoxide and hydrogen. The reaction of carbon monoxide and hydrogen in a molten solution of sodium chloride and aluminum chloride with $Ir_4(CO)_{12}$ as a catalyst yields a mixture of hydrocarbons, ethane is the major product (170).

Toxicology and Safety

Exposure to metal carbonyls can present a serious health threat. Nickel carbonyl is considered to be one of the most poisonous inorganic compounds. However, the toxicological information available on metal carbonyls is restricted to the more common, commercially important compounds such as $Ni(CO)_4$ and $Fe(CO)_5$. Other metal carbonyls are considered potentially dangerous, especially in the gaseous state, by analogy to nickel and iron carbonyls. Table 5 lists data concerning toxicological studies on a few common metal carbonyls (171).

The toxic symptoms from inhalation of nickel carbonyl are believed to be caused by both nickel metal and carbon monoxide. In many acute cases the symptoms are headache, dizziness, nausea, vomiting, fever, and difficulty in breathing. If exposure is continued, unconsciousness follows with subsequent damage to vital organs and death. Iron pentacarbonyl produces symptoms similar to nickel carbonyl but is considered less toxic than nickel carbonyl.

Table 5. Metal Carbonyls by Inhalation, 30 Minute LC$_{50}$ Values for Rats[a]

Metal carbonyl	Carbonyl, mg/m^3	Metal, mg/m^3	Toxicity relative to Ni(CO)$_4$	Ref.
Ni(CO)$_4$	240	85	1	172
HCo(CO)$_4$	560	165	0.52	172
Fe(CO)$_5$	910	260	0.33	173
	2190	625 (mice)	0.14	173
Co$_2$(CO)$_8$	1400[b]	480[b]	0.17	174

[a] Ref. 171.
[b] A concentration that for 60 min resulted in no toxic signs.

When heated to about 60°C, nickel carbonyl explodes. For both iron and nickel carbonyl, suitable fire extinguishers are water, foam, carbon dioxide, or dry chemical. Large amounts of iron pentacarbonyl also have been reported to ignite spontaneously (175). Solutions of molybdenum carbonyl have been reported to be capable of spontaneous detonation (176). Recent sources of the toxicity of industrial chemicals including metal carbonyls are references 177–179.

BIBLIOGRAPHY

"Carbonyls" in *ECT* 1st ed., Vol. 3, pp. 201–205, by Carl Mueller, General Aniline & Film Corporation; "Carbonyls" in *ECT* 2nd ed., Vol. 4, pp. 489–510, by Jack C. Hileman, El Camino College.

1. R. K. Sheline and H. Mahnke, *New Synth. Methods* **3**, 203 (1975).
2. G. R. Dobson, *Acc. Chem. Res.* **9**, 300 (1976).
3. K. G. Caulton and R. F. Fenske, *Inorg. Chem.* **7**, 1273 (1968).
4. W. Hieber and T. Z. Kruck, *Z. Naturforsch.* **16B**, 709 (1961).
5. M. H. D. Stiddard and L. M. Haines, *Adv. Inorg. Chem. Radiochem.* **12**, 121 (1969).
6. W. A. G. Graham, *Inorg. Chem.* **11**, 315 (1968).
7. R. P. Stewart and P. M. Treichel, *Inorg. Chem.* **7**, 1942 (1968).
8. R. J. Angelici and M. D. Malone, *Inorg. Chem.* **6**, 1731 (1967).
9. D. J. Darensbourg and T. L. Brown, *Inorg. Chem.* **7**, 959 (1968).
10. A. F. Schreiner and T. L. Brown, *J. Am. Chem. Soc.* **90**, 3366 (1968).
11. E. W. Abel and co-workers, *J. Chem. Soc. Chem. Commun.*, 900 (1968).
12. D. C. Carroll and S. P. McGlynn, *Inorg. Chem.* **7**, 1285 (1968).
13. D. A. Brown and R. M. Rawlinson, *J. Chem. Soc. A*, 1530 (1969).
14. L. Mond, C. Langer, and F. Quincke, *J. Chem. Soc.* **57**, 749 (1890).
15. I. W. Stolz, G. R. Dobson, and R. K. Sheline, *J. Am. Chem. Soc.* **84**, 3589 (1962).
16. P. J. Hay, *J. Am. Chem. Soc.* **100**, 2411 (1978).
17. M. Paliakoff, *J. Chem. Soc. Faraday Trans. 2* **73**, 569 (1977).
18. M. A. Graham, A. J. Rest, and J. J. Turner, *J. Organomet. Chem.* **24**, C54 (1970); G. A. Ozin and A. Vander Voit, *Prog. Inorg. Chem.* **19**, 105 (1975); R. N. Peritz and J. J. Turner, *Inorg. Chem.* **14**, 202 (1975).
19. R. Brill, *Z. Kristallogr.* **77**, 36 (1931).
20. L. F. Dahl and J. F. Blount, *Inorg. Chem.* **4**, 1373 (1965).
21. D. L. Smith, *J. Chem. Phys.* **42**, 1460 (1965).
22. K. K. Joshi and P. L. Paulson, *Z. Naturforsch.* **17B**, 565 (1962).
23. T. Kruck and M. Höfler, *Chem. Ber.* **97**, 2289 (1964).
24. N. Flitcroft, D. Huggins, and H. D. Kaesz, *Inorg. Chem.* **3**, 1123 (1964).
25. E. H. Schubert and R. K. Sheline, *Z. Naturforsch.* **20B**, 1306 (1965).
26. G. O. Evans and R. K. Sheline, *J. Inorg. Nucl. Chem.* **30**, 2862 (1968).
27. T. Kruck and M. Höfler, *Angew. Chem.* **76**, 786 (1964).
28. G. O. Evans, J. P. Hargaden, and R. K. Sheline, *J. Chem. Soc. Chem. Commun.*, 186 (1967).

29. B. F. G. Johnson and co-workers, *J. Chem. Soc. Chem. Commun.*, 861 (1968).
30. B. W. Yawney and F. G. A. Stone, *J. Chem. Soc. A*, 502 (1969).
31. E. H. Braye and co-workers, *J. Am. Chem. Soc.* **84**, 4633 (1962).
32. W. Hieber and H. Lagally, *Z. Anorg. Allg. Chem.* **251**, 96 (1943).
33. E. R. Corey, L. F. Dahl, and W. Beck, *J. Am. Chem. Soc.* **85**, 1202 (1963).
34. R. B. King, *J. Am. Chem. Soc.* **94**, 95 (1972).
35. R. Mason, K. M. Thomas, and D. M. P. Mingos, *J. Am. Chem. Soc.* **95**, 3802 (1973).
36. V. G. Albano, P. L. Bellon, and G. F. Ciani, *J. Chem. Soc. Chem. Commun.*, 1024 (1969).
37. V. G. Albano, P. Chini, and V. Scatturin, *J. Organomet. Chem.* **15**, 423 (1968).
38. V. G. Albano and co-workers, *J. Organomet. Chem.* **16**, 461 (1969).
39. J. C. Calabrese and co-workers, *J. Am. Chem. Soc.* **96**, 2614 (1974).
40. G. Longoni and co-workers, *J. Am. Chem. Soc.* **97**, 5034 (1975).
41. V. G. Albano and co-workers, *J. Organomet. Chem.* **88**, 381 (1975).
42. V. G. Albano and co-workers, *J. Chem. Soc. Chem. Commun.*, 859 (1975).
43. *Ibid.*, 299 (1974).
44. P. Chini, G. Longoni, and V. G. Albano, *Adv. Organomet. Chem.* **14**, 285 (1976).
45. G. Natta and co-workers, *Atti Accad. Naz. Lincei Cl. Sci. Fis. Mat. Nat. Rend.* (8), **27**, 107 (1959).
46. W. Rudorff and U. Hofman, *Z. Phys. Chem. B.* **28**, 351 (1933).
47. L. F. Dahl, E. Ishishi, and R. E. Rundle, *J. Chem. Phys.* **26**, 1750 (1957).
48. M. F. Bailey and L. F. Dahl, *Inorg. Chem.* **4**, 1140 (1965).
49. A. W. Hanson, *Acta. Crystallogr.* **15**, 930 (1962).
50. J. Donohue and A. Caron, *Acta. Crystallogr.* **17**, 663 (1964).
51. H. M. Powell and R. V. G. Evans, *J. Chem. Soc.*, 286 (1939).
52. F. A. Cotton and J. M. Troup, *J. Chem. Soc. Dalton Trans.*, 800 (1974).
53. C. H. Wei and L. F. Dahl, *J. Am. Chem. Soc.* **91**, 1351 (1969).
54. F. A. Cotton and J. M. Troup, *J. Am. Chem. Soc.* **96**, 4155 (1974).
55. M. R. Churchill, F. J. Hollander, and J. P. Hutchinson, *Inorg. Chem.* **16**, 2655 (1977).
56. E. R. Corey and L. F. Dahl, *Inorg. Chem.* **1**, 521 (1962).
57. M. R. Churchill and B. G. DeBow, *Inorg. Chem.* **16**, 878 (1977).
58. G. G. Sumner, H. P. Klug, and L. E. Alexander, *Acta. Cryst.* **17**, 732 (1964).
59. C. H. Wei and L. F. Dahl, *J. Am. Chem. Soc.* **88**, 1821 (1966).
60. C. H. Wei, *Inorg. Chem.* **8**, 2384 (1969).
61. C. H. Wei, G. R. Wilkes, and L. F. Dahl, *J. Am. Chem. Soc.* **89**, 4792 (1967).
62. G. R. Wilkes, *Diss. Abst.* **XXVI**, 5029 (1966).
63. J. Ladell, B. Post, and I. Fankuchen, *Acta. Crystallogr.* **5**, 795 (1952).
64. F. Calderazzo, R. Ercoli, and G. Natta in I. Wender and P. Pino, eds., *Organic Synthesis via Metal Carbonyls*, Wiley-Interscience, New York, 1968, Chapt. 1.
65. A. K. Fischer, F. A. Cotton, and G. Wilkinson, *J. Am. Chem. Soc.* **79**, 2044 (1957).
66. *Ibid.*, **78**, 5168 (1956).
67. *Ibid.*, **81**, 800 (1959).
68. W. D. Good, D. M. Fairbrother, and G. Waddington, *J. Phys. Chem.* **62**, 835 (1958).
69. *Natl. Bur. Stand. U.S. Circ.* **500**, (1961).
70. H. A. Skinner, *Adv. Organomet. Chem.* **2**, 49 (1964).
71. R. H. T. Bleyerveld, Th. Hohle, and K. Vrieze, *J. Organomet. Chem.* **94**, 281 (1975); J. A. Conner, *J. Organomet. Chem.* **94**, 195 (1975); K. Wade, *Inorg. Nucl. Chem. Lett.* **14**, 71 (1978).
72. H. E. Podall, J. H. Dunn, and H. Shapiro, *J. Am. Chem. Soc.* **82**, 1325 (1960).
73. R. Busby, W. Klotzbücher, and G. A. Ozin, *Inorg. Chem.* **16**, 822 (1977).
74. P. Chini, *Inorg. Chem.* **8**, 1206 (1969).
75. R. P. M. Werner and H. E. Podall, *Chem. Ind. London*, 144 (1961).
76. J. R. Moss and W. A. G. Graham, *J. Chem. Soc. Chem. Commun.*, 835 (1970).
77. L. Malatesta, G. Caglio, and M. Angoletta, *J. Chem. Soc. Chem. Commun.*, 532 (1970).
78. C. Longoni and P. Chini, *J. Am. Chem. Soc.* **98**, 7225 (1976).
79. C. R. Eady, B. F. G. Johnson, and J. Lewis, *J. Organomet. Chem.* **37**, C39 (1972).
80. Brit. Pat. 298,714 (July 28, 1927), (to I. G. Farbenind, A.G.).
81. Brit. Pat. 307,112 (Dec. 3, 1927), (to I. G. Farbenind, A.G.).
82. Ger. Pat. 531,402 (June 21, 1930), M. Naumann.
83. Ger. Pat. 547,025 (Jan. 13, 1931), M. Naumann.
84. W. Manchot and W. J. Manchot, *Z. Anorg. Allg. Chem.* **226**, 385 (1936).
85. W. Hieber and H. Stallman, *Z. Electrochem.* **49**, 288 (1943).

86. W. Hieber and H. Fuchs, *Z. Anorg. Allg. Chem.* **248,** 256 (1941).
87. J. C. Hileman, D. K. Huggins, and H. D. Kaesz, *J. Am. Chem. Soc.* **83,** 2953 (1961).
88. W. Hieber and C. Huget, *Angew. Chem.* **73,** 579 (1961).
89. W. Reppe, *Ann. Chem.* **582,** 116 (1953).
90. W. Hieber and G. Brendel, *Z. Anorg. Allg. Chem.* **289,** 324 (1957).
91. *Inorg. Synth.* **VIII,** 181 (1966).
92. *Inorg. Synth.* **VII,** 193 (1963).
93. G. Natta and co-workers, *J. Am. Chem. Soc.* **79,** 3611 (1957).
94. K. N. Anisinov and A. N. Nesmeyanov, *Dokl. Akad. Nauk SSR* **26,** 57 (1940).
95. U.S. Pat. 2,544,194 (May 22, 1951), D. T. Hurd (to General Electric Co.).
96. U.S. Pat. 2,557,744 (June 19, 1951), D. T. Hurd (to General Electric Co.).
97. R. D. Clossen, L. R. Buzbee, and G. G. Eche, *J. Am. Chem. Soc.* **80,** 6167 (1958).
98. I. Wender, H. Greenfield, and M. Orchin, *J. Am. Chem. Soc.* **73,** 2656 (1951).
99. I. Wender, H. W. Sternberg, S. Metlin, and M. Orchin, *Inorg. Synth.* **5,** 190 (1957).
100. E. Speyer and H. Wolf, *Chem. Ber.* **69,** 1424 (1927).
101. L. Mond, H. Hirtz and M. D. Cowap, *J. Chem. Soc.,* 798 (1910).
102. A. N. Nesmeyanov and co-workers, *Zh. Neorg. Khim.* **4,** 249 (1959).
103. *Ibid.,* p. 503.
104. W. Hieber and F. H. Schubert, *Z. Anorg. Allg. Chem.* **338,** 32 (1965).
105. R. B. King, J. C. Stokes, and T. F. Korenowski, *J. Organomet. Chem.* **11,** 641 (1968).
106. *Inorg. Synth.* **XVI,** 47 (1976).
107. *Inorg. Synth.* **XVI,** 45 (1976).
108. *Inorg. Synth.* **XVI,** 49 (1976).
109. G. F. Stuntz and J. R. Shapley, *Inorg. Nucl. Chem. Lett.* **12,** 49 (1976).
110. *Inorg. Synth.* **XIII,** 95 (1972).
111. P. Chini and S. Martinengo, *Inorg. Chim. Acta.* **3,** 299 (1969).
112. M. F. Lippert, *J. Organomet. Chem.* **136,** 73 (1977).
113. G. K. I. Magomedov and co-workers, *J. Organomet. Chem.* **149,** 29 (1978).
114. H. Adkins and G. Krsek, *J. Am. Chem. Soc.* **71,** 3051 (1949).
115. S. Brervis, *J. Chem. Soc.,* 5014 (1964).
116. H. Adkins and G. Krsek, *J. Am. Chem. Soc.* **70,** 383 (1948).
117. W. Reppe and H. Vetter, *Justus Liebigs Ann. Chem.* **582,** 133 (1953).
118. P. Chini, S. Martinengo, and G. Garlaschelli, *J. Chem. Soc. Chem. Commun.,* 709 (1972).
119. I. Wender and co-workers, *J. Am. Chem. Soc.* **78,** 5401 (1956).
120. W. Reppe, *Justus Liebigs Ann. Chem.* **582,** 1 (1953).
121. P. Pino, *Gazz. Chim. Ital.* **81,** 625 (1951).
122. Y. Sakakibara and T. Nakamura, *J. Soc. Org. Synth. Chem. (Jpn.)* **23,** 757 (1965).
123. J. E. Arnett and R. Pettit, *J. Am. Chem. Soc.* **83,** 2954 (1961).
124. R. Pettit and J. Henery, *Org. Synth.* **50,** 21 (1970).
125. J. D. Fitzpatrick and co-workers, *J. Am. Chem. Soc.* **87,** 3254 (1965).
126. L. Watts and R. Pettit, *Adv. Chem. Ser.* **(62),** 549 (1966).
127. I. D. Webb and G. T. Borcherdt, *J. Am. Chem. Soc.* **73,** 2654 (1951).
128. W. Hubel and C. Hoogzand, *Chem. Ber.* **93,** 103 (1960).
129. W. Hubel and co-workers, *J. Inorg. Nucl. Chem.* **9,** 204 (1959).
130. A. Nakamura, *Mem. Inst. Sci. Ind. Res. Osaka Univ.* **19,** 81 (1962).
131. A. Nakamura and N. Hogihara, *Nippon Kagaku Zasshi* **84,** 339 (1963).
132. T. H. Coffield, K. G. Ihrman and W. Burns, *J. Am. Chem. Soc.* **82,** 4209 (1960).
133. C. E. Coffey, *J. Am. Chem. Soc.* **83,** 1623 (1961).
134. D. Seyferth and M. D. Millar, *J. Organomet. Chem.* **38,** 373 (1972).
135. H. Alper and D. Des Roches, *J. Org. Chem.* **41,** 806 (1976).
136. N. L. Bould, *Tetrahedron Lett.,* 1841 (1963).
137. I. Rhee, M. Ryang, and S. Tsutsumi, *J. Organomet. Chem.* **9,** 361 (1967).
138. H. W. Sternberg and I. Wender, *Int. Conf. Coord. Chem.,* 35 (1959).
139. B. D. Dombek and R. L. Angelici, *J. Organomet. Chem.* **134,** 203 (1977).
140. A. Misonou and I. Ogata, *J. Soc. Org. Synth. Chem. (Jpn.)* **22,** 975 (1964).
141. H. W. Sternberg and co-workers, *J. Am. Chem. Soc.* **78,** 3621 (1956).
142. H. W. Sternberg, W. R. Markby, and I. Wender, *J. Am. Chem. Soc.* **79,** 6116 (1957).
143. E. Frankel and co-workers, *J. Org. Chem.* **29,** 3292 (1964).
144. E. Frankel, E. Emken, and U. Davison, *J. Org. Chem.* **30,** 2739 (1965).

145. A. J. Chalk and J. F. Harrod, *Adv. Organomet. Chem.* **6,** 119 (1968).
146. H. Alper, ed., *Transition Metal Organometallics in Organic Synthesis,* Academic Press, Inc., New York, 1976.
147. M. Ryang, *Organometallic Chem. Rev. A.* **5,** 67 (1970).
148. M. Ryang and S. Tsutsumi, *Synthesis* (2), 55 (1971).
149. C. W. Bird, *Transition Metal Intermediates in Organic Synthesis,* Logos Press, London, Eng., 1968.
150. H. W. Sternberg and I. Wender, *Proc. Intern. Conf. Coord. Chem. Chem. Soc. Spec. Publ.* (13), 35 (1959).
151. C. W. Bird, *Chem. Revs.* **62,** 283 (1962).
152. I. Wender, H. W. Sternberg, and M. Orchin in P. H. Emmett, ed., *The Oxo Reaction in Catalysis,* Vol. 5, Reinhold, New York, 1957.
153. G. N. Schrauzer, ed., *Transition Metals In Homogenous Catalysis,* Marcel Dekker Inc., New York, 1971.
154. H. Alper, *J. Organomet. Chem. Libr.* **1,** 305 (1976).
155. I. Wender and P. Pino, eds., *Organic Syntheses Via Metal Carbonyls,* Vol. 2, Wiley-Interscience, New York, 1977.
156. W. A. G. Graham and A. R. Gatti, "Organometallic Compounds in the Preparation of High Purity Metals" in M. S. Brooks and J. K. Kennedy, eds., *Ultra Purification of Semiconductor Materials,* Electronic Research Directorate, U.S. Air Force, New York, 1962.
157. *Carbonyl Iron Powders,* General Aniline and Film Corporation, New York, 1962.
158. H. E. Podall and M. M. Mitchell, Jr., *Ann. N.Y. Acad. Sci.* **125,** 218 (1965).
159. B. B. Owen and R. T. Wibber, *Ann. Inst. Mining Metall. Eng. Inst. Metals Div. Metals Technol.* **15,** 2306 (1948).
160. H. Lux and G. Illmann, *Chem. Ber.* **92,** 2364 (1959).
161. U.S. Pat. 2,685,532 (Aug. 3, 1954), P. Parolyk (to Commonwealth Engineering Co. of Ohio).
162. U.S. Pat. 2,785,082 (Mar. 12, 1957), P. J. Clough and P. Godley (to National Research Corp.).
163. U.S. Pat. 2,815,302 (Dec. 3, 1957), B. Ostrofsky and J. W. Ballard (to Commonwealth Engineering Co. of Ohio).
164. U.S. Pat. 2,793,140 (Mar. 21, 1957), B. Ostrofsky and J. W. Ballard (to Commonwealth Engineering Co. of Ohio).
165. U.S. Pat. 2,913,413 (Nov. 17, 1959), J. E. Brown (to Ethyl Corp.).
166. R. M. Whitcomb, *Non-Lead Antiknock Agents for Motor Fuels, Chemical Technology Review No. 49,* Noyes Data Corp., Park Ridge, N. J., 1975.
167. Ger. Pat. 2,426,411 (Jan. 9, 1975), W. E. Walker and R. L. Pruett (to Union Carbide Corp.); U.S. Pats. 3,878,214; 3,878,290; 3,878,292 (Apr. 15, 1975), W. E. Walker, E. S. Brown, and R. L. Pruett (to Union Carbide Corp.).
168. R. M. Laine, R. G. Rinker, and P. C. Ford, *J. Am. Chem. Soc.* **99,** 252 (1977); H. Kang and co-workers, *J. Am. Chem. Soc.* **99,** 8323 (1977).
169. P. Pino and co-workers, *Chem. Ind.,* 1732 (1968).
170. G. C. Demitras and E. L. Muetterties, *J. Am. Chem. Soc.* **99,** 2796 (1977).
171. F. A. Patty, ed., *Industrial Hygiene and Toxicology,* 3rd ed., Vol. 2, Wiley-Interscience, New York, 1977, p. 1108.
172. E. D. Palmes and co-workers, *Am. Ind. Hyg. Assoc. J.* **20,** 453 (1953).
173. F. W. Sunderman, B. West, and J. F. Kincaid, *AMA Arch. Ind. Health* **19,** 11 (1959).
174. H. M. Armit, *J. Hyg.* **9,** 249 (1909).
175. R. K. Sheline and K. S. Pitzer, *J. Am. Chem. Soc.* **72,** 1107 (1950).
176. B. B. Owen, J. English, Jr., H. G. Cassidy, and C. V. Dundon, *Inorg. Syn.* **3,** 156 (1950).
177. H. E. Christensen, ed., *Toxic Substance List,* U.S. Dept. of Health, Education, and Welfare, Washington, D.C., 1973.
178. N. I. Sax, *Dangerous Properties of Industrial Chemicals,* Reinhold, New York, 1975.
179. A. Hamilton and L. Hardy, *Industrial Toxicology,* Publishing Sciences Group, Inc., Acton, Mass., 1974.

General References

Reviews

F. A. Cotton and G. Wilkinson, *Advanced Inorganic Chemistry*, Wiley-Interscience, New York, 1972, pp. 682–709.
J. A. Mattern and S. J. Gill, "Metal Carbonyls and Nitrosyls" in J. C. Bailar, Jr., ed., *Chemistry of Coordination Compounds*, Reinhold Publishing Corp., New York, 1956, p. 509.
N. A. Belozerskii, *Metal Carbonyls*, Gosudarst. Nauch. Tekh. Izdatel. Lit. Chernog. Chernoy i Tsvetnoi Met., Moscow, U.S.S.R., 1958.
J. Chatt, P. L. Paulson, and L. M. Venanzi, "Metal Carbonyls and Related Compounds" in H. Zeiss, ed., *Organometallic Chemistry, ACS Monograph 147,* Reinhold Publishing Corp., New York, 1960.
H. J. Eméleus and J. S. Anderson, "Coordination Compounds IV, Metal Carbonyls and Other Pi Bonded Complexes" in *Modern Aspects of Inorganic Chemistry*, 3rd ed., D. Van Nostrand Co., Inc., Princeton, N.J., 1960, p. 252.
N. V. Sidgwick, *The Chemical Elements and Their Compounds,* Vol. 1, Oxford University Press, London, Eng., 1960.
H. E. Podall, *J. Chem. Educ.* **38,** 187 (1961).
W. Hieber, *Angew. Chem.* **73,** 364 (1961).
H. D. Kaesz, *J. Chem. Educ.* **40,** 159 (1963).
E. W. Abel, *Quart. Rev. Chem. Soc.* **17,** 133 (1963).
J. C. Hileman, *Prep. Inorg. React.* **1,** 77 (1964).
M. R. Churchill and R. Mason, *Adv. Organomet. Chem.* **5,** 93 (1967).
P. Chini, *Inorg. Chim. Act. Rev.* **2,** 31 (1968).
R. B. King, *Organometallic Synthesis,* Academic Press, Inc., New York, 1968.
M. I. Bruce and F. G. A. Stone, *Angew. Chem. Intern. Ed. Eng.* **7,** 427 (1968).
E. W. Abel and F. G. A. Stone, *Quart. Rev. Chem. Soc.* **23,** 325 (1969).
E. W. Abel and F. G. A. Stone, *Quart. Rev. Chem. Soc.* **24,** 498 (1970).
M. I. Bruce, *J. Organomet. Chem.* **44,** 209 (1972).
R. D. Johnston, *Inorganic Chemistry*, Series 1, Vol. 6, Butterworths, London, Eng., University Park Press, Baltimore, Md., 1972, p. 1.
S. C. Tripathi, S. C. Srivastava, R. P. Mani, and A. K. Shrimal, *Inorg. Chim. Acta* **15,** 249 (1975).
S. C. Tripathi, S. C. Srivastava, R. P. Mani, and A. K. Shrimal, *Inorg. Chim. Acta.* **17,** 257 (1976).
D. McIntosh and G. A. Ozin, *Inorg. Chem.* **16,** 51 (1977).
"Specialist Periodical Reports" in *Inorganic Chemistry of Transition Elements*, Vols. 1–4, The Chemical Society, London, Eng., 1969 to date.

Metal carbonyl clusters

M. C. Baird, *Prog. Inorg. Chem.* **9,** 1 (1968).
R. B. Penfold, *Perspect. Struct. Chem.* **2,** 71 (1968).
B. P. Biryukov and Y. I. Struchkov, *Russ. Chem. Rev.* **39,** 789 (1970).
P. Chini, *Pure Appl. Chem.* **23,** 489 (1970).
R. D. Johnston, *Adv. Inorg. Chem. Radiochem.* **13,** 471 (1971).
R. B. King, *Prog. Inorg. Chem.* **15,** 288 (1972).
H. D. Kaesz, *Chem. Br.* **9,** 344 (1973).
M. H. B. Stiddard, *Rev. Chim. Miner.,* 801 (1966).
E. W. Abel and S. Tyfield, *Adv. Organomet. Chem.* **8,** 117 (1970).
S. Martinengo and co-workers, *J. Organomet. Chem.* **59,** 379 (1973).
C. U. Pittman, Jr., and R. C. Ryan, *Chem. Tech.* **8,** 170 (1978).

Metal carbonyl hydrides

M. L. H. Green, *Angew. Chem.* **72,** 719 (1960).
M. L. H. Green and D. J. Jones, *Adv. Inorg. Chem. Radiochem.* **7,** 115 (1965).
A. Ginsberg, *Prog. Tran. Met. Chem.* **1,** 111 (1965).
H. D. Kaesz and R. B. Saillant, *Chem. Rev.* **72,** 231 (1972).

814 CARBONYLS

Derivatives of metal carbonyls

R. B. King, *Adv. Organomet. Chem.* **2**, 157 (1964).
R. Pettit and G. F. Emerson, *Adv. Organomet. Chem.* **1**, 1 (1964).
T. A. Manuel, *Adv. Organomet. Chem.* **3**, 181 (1965).
R. G. Hayter, *Prep. Inorg. React.* **2**, 211 (1965).
R. L. Pruett, *Prep. Inorg. React.* **2**, 187 (1965).
G. R. Dobson, I. W. Stolz, and R. K. Sheline, *Adv. Inorg. Chem. Radiochem.* **8**, 1 (1966).
E. O. Fischer and H. Werner, *Metal π-Complexes*, Elsevier, New York, 1966.
R. F. Heck, *Adv. Organomet. Chem.* **4**, 243 (1966).
E. W. Abel and D. C. Crosse, *Organomet. Chem. Rev.* **2**, 443 (1967).
D. A. Brown, *Inorg. Chim. Acta. Rev.* **1**, 35 (1967).
M. W. Anker, R. Colton, and I. B. Tomkins, *Rev. Pure App. Chem.* **18**, 23 (1968).
M. I. Bruce and F. G. A. Stone, *Prep. Inorg. React.* **4**, 177 (1968).
W. Hubel in I. Wender and P. Pino, eds., *Organic Synthesis via Metal Carbonyls*, Vol. 1, Wiley-Interscience, New York, 1968, Chapt. 2.
F. G. A. Stone in G. A. V. Elsworth, A. G. Maddock, and A. G. Shayse eds., *New Pathways in Inorganic Chemistry*, Cambridge University Press, Cambridge, Mass., 1968, p. 283.
R. S. Dickson and P. J. Fraser, *Adv. Organomet. Chem.* **12**, 323 (1974).
Dietmar Seyferth, *Adv. Organomet. Chem.* **14**, 97 (1976).
F. A. Cotton, *Prog. Inorg. Chem.* **21**, 1 (1976).

FRANK S. WAGNER
Strem Chemicals Inc.

CARBOXYLIC ACIDS

Survey, 814
Manufacture, 835
Analysis and standards, 845
Economic aspects, 853
Fatty acids from tall oil, 859
Branched-chain acids, 861
Trialkylacetic acids, 863

SURVEY

Carboxylic acids from the smallest, formic, to the 22-carbon fatty acids, eg, erucic, are economically important; several million metric tons are produced annually. The shorter-chain aliphatic acids are colorless liquids. Each has a characteristic odor ranging from sharp and penetrating (formic and acetic acids) or vinegary (dilute acetic) to the odors of rancid butter (butyric acid) and goat fat (the 6–10-carbon acids). At room temperature, the *cis*-unsaturated acids through C_{18} are liquids and the saturated aliphatic acids from decanoic through the higher acids and *trans*-unsaturated acids are solids. The latter are higher melting because of their greater symmetrical structure (eg, elaidic acid).

Most higher acids were not known until the beginning of the 19th century, although vegetable oils and animal fats were commonly used in ancient times. Since the outstanding work of Chevreul (1786–1889) in establishing the nature of the 18-carbon, naturally occurring fatty acids, hundreds of long-chain fatty acids have been isolated from natural sources and characterized.

Both odd- and even-numbered alkanoic acids of molecular formula $C_nH_{2n}O_2$ up through hexanoic acid occur naturally. Only the even-numbered higher acids, most often the C_{18} acids, occur naturally (Table 1). Formic (qv), acetic (qv), propionic, and butyric acids are manufactured in large quantities from petrochemical feedstocks. The higher fatty acids are derived from animal fats, vegetable oils, or fish oils. Some higher saturated fatty acids with significant industrial applications are pelargonic, lauric, myristic, palmitic, and stearic acids.

In the alkenoic series of molecular formula $C_nH_{2n-2}O_2$, acrylic (qv), methacrylic (qv), undecylenic, and oleic acids have important applications (Table 2). Acrylic and methacrylic acids have a petrochemical origin, and undecylenic and oleic acids have natural origins.

The polyunsaturated aliphatic monocarboxylic acids having industrial significance include sorbic, linoleic, linolenic, eleostearic, and various polyunsaturated fish acids (Table 3). Of these, only sorbic acid (qv) is made synthetically. The other acids, except those from tall oil, occur naturally as glycerides and are used in this form.

The shorter-chain alkynoic (acetylenic) acids are common in laboratory organic syntheses, and several long-chain acids occur naturally (Table 4).

Many substituted fatty acids, particularly methacrylic, 2-ethylhexanoic, and ricinoleic acids, are commercially significant. Several substituted fatty acids exist naturally (Table 5). Fatty acids with a methyl group in the penultimate position are called iso acids, and those with a methyl group in the antepenultimate position are called anteiso acids (1) (see Carboxylic acids, branched chain acids).

Some naturally occurring fatty acids have alicyclic substituents: the cyclopentenyl-containing chaulmoogra acids, notable for their use in treating leprosy (see Chemotherapeutics, antimycotic), and the cyclopropenyl or sterculic acids (Table 6).

The prostaglandins (qv) constitute another class of fatty acids with alicyclic structures. These are of great biological importance and are formed by *in vivo* oxidation of 20-carbon polyunsaturated fatty acids, particularly arachidonic acid. The several prostaglandins, eg, PGE_1, have different degrees of unsaturation and oxidation when compared to the parent compound, prostanoic acid.

prostanoic acid
[2,51,51-18-9]

PGE_1
[745-65-3]

Aromatic carboxylic acids are produced annually in amounts of several million metric tons. Several aromatic acids occur naturally (eg, benzoic (qv), salicylic (qv), cinnamic (qv), and gallic acids), but those used in commerce are produced synthetically. These acids are generally crystalline solids with relatively high melting points, attributable to the rigid, planar, aromatic nucleus (see also Phthalic acids).

Table 1. Physical Properties of the Straight-Chain Alkanoic Acids

Value of n in $C_nH_{2n}O_2$	Systematic name (Trivial name)[a]	CAS Registry Number	Mol wt	mp, °C	bp, °C	Density, d_4^{20}	Refractive index, n_D^{20}	Heat of formation, $\Delta H°_{298}$, kJ/mol[b]	Specific heat, J/g[b]	Heat of fusion, kJ/mol[b]	Surface tension, mN/m[c] at 70°C	Viscosity, mPa·s[d] (°C)	Flash point, °C[e]	Heat of combustion, ΔH_R^{25}, kJ/mol[b] (liq)
1	methanoic (formic)	[64-18-16]	46.03	8.4	100.5	1.220	1.3714							
2	ethanoic (acetic)	[64-19-7]	60.05	16.6	118.1	1.049	1.3718							
3	propanoic (propionic)	[79-09-4]	74.08	−22	141.1	0.992	1.3874	−511.2 (l)	2.34 (l)		20.7	1.099 (20)	54	−1,536
4	butanoic (butyric)	[107-92-6]	88.11	−7.9	163.5	0.959	1.39906	−534.1 (l)	2.16 (l)		21.8	1.538 (20)	66	−2,194
5	pentanoic (valeric)	[109-52-4]	102.13	−34.5	187.0	0.942	1.4086	−559.2 (l)				2.30 (20)	96 (OC)	−2,837.8
6	hexanoic ([caproic])	[142-62-1]	116.16	−3.4	205.8	0.929	1.4170	−584.0 (l)	2.23 (l)	15.1	23.4	3.23 (20)	102 (OC)	−3,492.4
7	heptanoic ([enanthic])	[111-14-8]	130.19	−10.5	223.0	0.922	1.4230	−609.1 (l)		15.0		4.33 (20)	132	−4,146.9
8	octanoic ([caprylic])	[124-07-2]	144.21	16.7	239.7	0.910	1.4280	−635.2 (l)	2.62 (s)	21.4	23.7	5.74 (20)	132 (OC)	−4,799.9
9	nonanoic (pelargonic)	[112-05-0]	158.24	12.5	255.6	0.907	1.4322	−658.5 (l)	2.91 (s)	20.3		8.08 (20)		−5,456.1

No.	Name	CAS	MW	mp	bp	density	n							
10	decanoic ([capric])	[334-48-5]	172.27	31.6	270.0	0.8953^{30}	1.4169^{70}	−685.2 (l)		28.0		4.30 (50)		−6,108.7
11	undecanoic ([undecylic])	[112-37-8]	186.30	29.3	284.0	0.9905^{25}	1.4202^{70}	−736.7 (s)		25.1	25.0	7.30 (50)		−6,762.3
12	dodecanoic (lauric)	[143-07-7]	200.32	44.2	298.9	0.883	1.4230^{70}	−775.6 (s)	1.80 (s)	36.6	26.6	7.30 (50)		−7,413.7
13	tridecanoic [tridecylic]	[638-53-9]	214.35	41.5	312.4	0.8458^{80}	1.4252^{70}							
14	tetradecanoic (myristic)	[544-63-8]	228.38	53.9	326.2	0.858^{60}	1.4273^{70}	−835.9 (s)	1.60 (s)	44.8	27.4	5.83 (70)		−8,721.4
15	pentadecanoic ([pentadecylic])	[1002-84-2]	242.40	52.3	339.1	0.8423^{80}	1.4297^{70}							
16	hexadecanoic (palmitic)	[57-10-3]	256.43	63.1	351.5	0.8534^{62}	1.4309^{70}	−892.9 (s)	1.80 (s)	54.4	28.2	7.80 (70)		−10,030.6
17	heptadecanoic (margaric)	[506-12-7]	270.46	61.3	363.8	0.853^{60}	1.4324^{70}							
18	octadecanoic (stearic)	[57-11-4]	284.48	69.6	376.1	$0.8476^{69.3}$	1.4337^{70}	−949.4 (s)	1.67 (s)	63.2	28.9	9.87 (70)		−11,342.4
19	nonadecanoic ([nonadecylic])	[646-30-0]	298.51	68.6	$299_{100}{}^f$	0.8771^{24}	1.4512^{25}	−1013.3 (s)		71.0			196	−12,646.2
20	eicosanoic (arachidic)	[506-30-9]	312.54	75.3	$(203–205)_1{}^f$	0.8240^{100}	1.4250^{100}	−1063.7 (s)		78.7				−13,976
22	docosanoic (behenic)	[112-85-6]	340.59	79.9	$306_{20}{}^f$	0.8221^{100}	1.4270^{100}							
24	tetracosanoic (lignoceric)	[557-59-5]	368.65	84.2		0.8207^{100}	1.4287^{100}							

Table 1 (continued)

Value of n in $C_nH_{2n}O_2$	Systematic name (Trivial name)[a]	CAS Registry Number	Mol wt	mp, °C	bp, °C	Density, d_4^{20}	Refractive index, n_D^{20}	Heat of formation, ΔH°_{298}, kJ/mol[b]	Specific heat, J/g[b]	Heat of fusion, kJ/mol[b]	Surface tension, mN/m[c], at 70°C	Viscosity, mPa·s[d] (°C)	Flash point, °C[e]	Heat of combustion, ΔH_R^{25}, kJ/mol[b] (liq)
26	hexacosanoic (cerotic)	[504-46-7]	396.70	87.7		0.8198^{100}	1.4301^{100}							
28	octacosanoic (montanic)	[506-48-9]	424.75											
30	triacontanoic (melissic)	[506-50-3]	452.81											
33	tritriacontanoic (psyllic)	[38232-03-0]	494.89											
35	pentatriaconta- noic (ceroplas- tic)	[38232-05-2]	522.94											

[a] Brackets signify a trivial name no longer in use.
[b] To convert J to cal, divide by 4.184.
[c] mN/m = dyn/cm.
[d] mPa·s = cP.
[e] OC = open cup.
[f] To convert kPa to mm Hg, multiply by 7.5.

Table 2. Physical Properties of the Straight Chain Alkenoic Acids

Value of n in $C_m H_{2n-2} O_2$	Systematic name (Trivial name)	CAS Registry Number	Mol wt	mp, °C	bp, °C	Density, d_4^{20}	Refractive index n_D^{20}
3	propenoic (acrylic)	[79-10-7]	72.06	12.3	141.9	1.0621^{16}	1.4224
4	trans-2-butenoic (crotonic)	[107-93-7]	86.09	72	189	1.018	$1.4228^{79.7}$
4	cis-2-butenoic (isocrotonic)	[503-64-0]	86.09	14	171.9	1.0312^{15}	1.4457
4	3-butenoic (vinylacetic)	[625-38-7]	86.09	−39	163	1.013_{15}^{15}	1.4257^{15}
5	2-pentenoic (β-ethylacrylic)	[626-98-2]	100.12				
5	3-pentenoic (β-pentenoic)	[5204-64-8]	100.12				
5	4-pentenoic (allylacetic)	[591-80-0]	100.12	−22.5	188–189	0.9809	1.4281
6	2-hexenoic (isohydroascorbic)	[1191-04-4]	114.14				
6	3-hexenoic (hydrosorbic)	[4219-24-3]	114.14	12	208	0.9640^{23}	1.4935
7	trans-2-heptenoic	[10352-88-2]	128.17				
8	2-octenoic	[1470-50-4]	142.20				
9	2-nonenoic	[3760-11-0]	156.23				
10	4-decenoic (obtusilic)	trans [57602-94-5] cis [505-90-8]	170.25				
10	9-decenoic (caproleic)	[14436-32-9]	170.25				
11	10-undecenoic (undecylenic)	[112-38-9]	184.28	24.5	275	0.9075_4^{25}	1.4464
12	3-dodecenoic (linderic)	trans [4998-71-4]	198.31				
13	tridecenoic	[28555-21-7]	212.33				
14	9-tetradecenoic (myristoleic)	cis [544-64-9]	226.36				
15	pentadecenoic	[26444-04-2]	240.39				
16	cis-9-hexadecenoic (cis-9-palmitoleic)	[373-49-9]	254.41				
16	trans-9-hexadecenoic (trans-9-palmitoleic)	[10030-73-6]	254.41				
17	9-heptadecenoic	[10136-52-4]	268.44				
18	cis-6-octadecenoic (petroselinic)	[593-39-5]	282.47	30		0.8681^{40}	1.4533^{40}
18	trans-6-octadecenoic (petroselaidic)	[593-40-8]	282.47	54			
18	cis-9-octadecenoic (oleic)	[112-80-1]	282.47	13.6	$234_{15}{}^a$	0.8905	1.4582^3
18	trans-9-octadecenoic (elaidic)	[112-79-8]	282.47	43.7	$234_{15}{}^a$	0.8568^{70}	1.4405^{70}
18	cis-11-octadecenoic	[506-17-2]	282.47	14.5			
18	trans-11-octadecenoic (vaccenic)	[693-72-1]	282.47	44		0.8563^{70}	1.4406^{70}

Table 2 (continued)

Value of n in $C_mH_{2n-2}O_2$	Systematic name (Trivial name)	CAS Registry Number	Mol wt	mp, °C	bp, °C	Density, d_4^{20}	Refractive index, n_D^{20}
20	cis-9-eicosenoic (godoleic)	[506-31-0]	310.52				
22	cis-11-docosenoic (cetoleic)	[506-36-5]	338.58				
22	cis-13-docosenoic (erucic)	[112-86-7]	338.58	34.7	$281_{30}{}^a$	0.85321^{70}	1.44438^{70}
22	trans-13-docosenoic (brassidic)	[506-33-2]	338.58	61.9	$265_{15}{}^a$	0.85002^{70}	1.44349^{70}
24	cis-15-tetracosenoic (selacholeic)	[506-37-6]	366.63				
26	cis-17-hexacosenoic (ximenic)	[544-84-3]	394.68				
30	cis-21-triacontenoic (lumequeic)	[67329-09-3]	450.79				

^{a} To convert kPa to mm Hg, multiply by 7.5.

Nomenclature

Acyclic monocarboxylic acids are named substitutively by dropping the *e* from the parent hydrocarbon name and adding *oic acid*. Cyclic carboxylic acids are named substitutively by adding the suffix *carboxylic acid* to the name of the cyclic hydrocarbon. Positions of substituents, eg, amino-, chloro-, hydroxy-, methyl-, oxo-, in acyclic alkanoic acids are indicated by number (not Greek letter) locants, with the numbering starting from the carbon of the carboxyl group. The trivial names caproic, caprylic, and capric have been abandoned and should no longer be used for the respective hexanoic, octanoic, and decanoic acids. Substituted derivatives should be named systematically and not on the basis of the trivial name for the parent compound, eg, 12-hydroxyoctadecanoic, *not* 12-hydroxystearic acid. Acid halides of acyclic acids are named by dropping the ending *ic acid* and adding the suffix *oyl* or *yl* to either the hydrocarbon name or the acid trivial name, eg, ace*tyl*, hexano*yl*, and stearo*yl* chlorides from the respective names ace*tic acid*, hexa*ne*, and stea*ric acid*. Esters are named by replacing the ending *ic acid* with the suffix *ate*. The alcohol portion of the ester is named by replacing the *ane* ending of the parent hydrocarbon name with the suffix *yl*.

The alkyl radical name of an ester is separated from the carboxylate name, eg, methyl formate for $HCOOCH_3$. Amides are named by changing the ending *oic acid* to *amide* for either systematic or trivial names (eg, hexan*amide*, acet*amide*). Mono-, di-, and triunsaturated fatty acids are named with Arabic numeral locants and with the suffix *-enoic, -adienoic,* and *-atrienoic* acid in place of the *ane* ending of the saturated hydrocarbon name, eg, 10,12,14-octadecatrienoic acid.

Shorthand notations have been developed to avoid repetitive systematic names of unsaturated fatty acids. For example, linolenic or *cis*-9-, *cis*-12-, *cis*-15-octatrienoic acid can be represented by 18:3(9c,12c,15c). The Greek letter Δ has been used to in-

Table 3. Some Polyunsaturated Fatty Acids

Total number of carbon atoms	Systematic name (Trivial name)	CAS Registry Number	Mol wt	mp, °C	bp, °C	Refractive index, n_D^{20}
Dienoic acids ($C_nH_{2n-4}O_2$)						
5	2,4-pentadienoic (β-vinylacrylic)	[626-99-3]	98.10			
6	2,4-hexadienoic[a] (sorbic)	[22500-92-1]	112.13	134.5		
10	2,4-decadienoic	trans [3036-33-2]	168.24			
12	2,4-dodecadienoic	trans [24738-48-5]	196.29			
18	cis-9,cis-12-octadecadienoic (linoleic)	[60-33-3]	280.45	−5	$202_{1.4}$[b]	1.4699
18	trans-9,trans-12-octadecadienoic (linolelaidic)	[506-21-8] [26764-24-9]	280.45 336.56	28–29		
22	9,13-docosadienoic					
Trienoic acids ($C_nH_{2n-6}O_2$)						
16	6,10,14-hexadeca trienoic (hiragonic)	[4444-12-6]	250.38			
18	cis-9,cis-12,cis-15-octadecatrienoic (linolenic)	[463-40-1]	278.44	−10 to −11.3	$157_{0.001}$[b]	1.4800
18	cis-9,trans-11,trans-13-octadecatrienoic (α-eleostearic)	[13296-76-9]	278.44	48–49	235_{12}[b]	1.5112
18	trans-9,trans-11,trans-13-octadecatrienoic (β-eleostearic)	[544-73-0]	278.44	71.5		1.5002
18	cis-9,cis-11,trans-13-octadecatrienoic (punicic)	[544-72-9]	278.44			
18	trans-9,trans-12,trans-15-octadecatrienoic (linolenelaidic)	[28290-79-1]	278.44			
Tetraenoic acids ($C_nH_{2n-8}O_2$)						
18	4,8,12,15-octadecatetraenoic (moroctic)	[67329-10-6]	276.42			
18	9,11,13,15-octadecatetraenoic (α-parinaric)	[593-38-4]	276.42			
18	9,11,13,15-octadecatetraenoic (β-parinaric)	[18841-21-9]	276.42			
20	5,8,11,14-eicosatetraenoic (arachidonic)	[27400-91-5]	304.47			
Pentaenoic acids ($C_nH_{2n-10}O_2$)						
22	4,8,12,15,19-docosapentaenoic (clupanodonic)	[2548-85-8]	330.51			

[a] $\Delta H_{298}^\circ = -393.5$ kJ/mol $\left(= -\dfrac{393.5}{4.184}\text{ kcal/mol}\right)$; flash pt (OC) = 127 °C.

[b] To convert kPa to mm Hg, multiply by 7.5.

Table 4. The Acetylenic Fatty Acids

Total number of carbon atoms	Systematic name (Trivial names)[a]	CAS Registry Number	Mol wt
3	propynoic (propiolic propargylic)	[471-25-0]	70.05
4	2-butynoic (tetrolic)	[590-93-2]	84.07
5	4-pentynoic	[6089-09-4]	98.10
6	5-hexynoic	[53293-00-8]	112.13
7	6-heptynoic	[30964-00-2]	126.16
8	7-octynoic	[10297-09-3]	140.18
9	8-nonynoic	[30964-01-3]	154.21
10	9-decynoic	[1642-49-5]	168.24
11	10-undecynoic ([dehydro-10-undecylenic])	[2777-65-3]	182.26
18	6-octadecynoic (tariric)	[544-74-1]	280.45
18	9-octadecynoic (stearolic)	[506-24-1]	280.45
18	17-octadecene-9, 11-diynoic (isanic, erythrogenic)	[506-25-2]	274.40
18	trans-11-octadecene-9-ynoic (ximenynic)	[557-58-4]	278.44
22	13-docosynoic (behenolic)	[506-35-4]	336.56

[a] Brackets indicate a trivial name no longer in use.

dicate presence and position of double bonds, eg, a $\Delta^{9,12,15}$ fatty acid, but it should never be used in a systematic name.

Physical Properties

Melting points, boiling points, densities, and refractive indexes for carboxylic acids vary widely depending upon molecular weight, structure, and the presence of unsaturation or other functional group (Tables 1, 2, and 5). In addition, some useful constants for alkanoic acids are listed in Table 1.

Some constants for selected unsaturated and substituted acids are given in Table 7.

Equations for the specific heats for solid and liquid states of palmitic acid are, respectively (J/g) (2):

$$Cp = 1.604 + 0.00544t \quad (-73 \text{ to } 40°C)$$
$$Cp = 1.936 + 0.00734t \quad (63 \text{ to } 92°C)$$

For stearic acid, the equations are (3):

$$Cp = 1.7886 + 0.00754t \quad (-120 \text{ to } 65°C)$$
$$Cp = 1.7861 + 0.00754t \quad (70 \text{ to } 78°C)$$

Table 5. Some Substituted Acids

Total number of carbon atoms	Systematic name (Trivial name)	CAS Registry Number	Mol wt	mp, °C	bp, °C	Density, d_4^{20}	Refractive index, n_D^{20}
4	2-methylpropenoic (methacrylic)	[79-41-4]	86.09	16	163	1.0153	1.4314
4	2-methylpropanoic (isobutyric)	[79-31-2]	88.10	−47	154.4	0.9504	1.3930
5	2-methyl-cis-2-butenoic (angelic)	[565-63-9]	100.12	45	185	0.9539[76]	1.4434[47]
5	2-methyl-trans-2-butenoic (tiglic)	[80-59-1]	100.12	65.5	198.5	0.9641[76]	1.4329[76]
5	3-methyl-2-butenoic (β,β-dimethyl acrylic)	[541-47-9]	100.12				
5	2-methylbutanoic	[116-53-0]	102.13				
5	3-methylbutanoic (isovaleric)	[503-74-2]	102.13	−37.6	176	0.93319[17.6]	1.40178[22.4]
5	2,2-dimethylpropanoic (pivalic)	[75-98-9]	102.13	35.5	163	0.905[50]	
8	2-ethylhexanoic	[149-57-5]	144.21		220	0.9031[25]	1.4255[28]
14	3,11-dihydroxy-tetradecanoic (ipurolic)	[36138-54-2]	260.37				
16	2,15,16-trihydroxy-hexadecanoic (ustilic)	[557-44-8]	304.43				
16	9,10,16-trihydroxy-hexadecanoic (aleuritic)	[533-87-9]	304.43				
16	16-hydroxy-7-hexadecenoic (ambrettolic)	[506-14-9]	270.41				

Table 5 (continued)

Total number of carbon atoms	Systematic name (Trivial name)	CAS Registry Number	Mol wt	mp, °C	bp, °C	Density, d_4^{20}	Refractive index, n_D^{20}
18	12-hydroxy-cis-9-octadecenoic (ricinoleic)	[141-22-0]	298.47	5.0, 7.7, 16.0	$226_{1.3}$[a]	0.9496^{15}	1.4145^{15}
18	12-hydroxy-trans-9-octadecenoic (ricinelaidic)	[540-12-5]	298.47	52–53	$240_{1.3}$[a]		
18	4-oxo-9,11,13-octadecatrienoic (licanic)	[17699-20-6]	292.42				
18	9,10-dihydroxy-octadecanoic	[120-87-6]	316.48	90			
18	12-hydroxy-octadecanoic (dl)	[106-14-9]	300.48	79			
18	12-oxooctadecanoic	[925-44-0]	298.47	81.5			
18	18-hydroxy-9,11,13-octadecatrienoic (kamlolenic)	[4444-93-3]	294.43	77–78			
18	12,13-epoxy-9-octadecenoic (vernolic)	[31263-20-4]	296.45				
18	8-hydroxy-trans-11-octadecene-9-ynoic (ximenynolic)	[557-58-4]	294.43				
18	8-hydroxy-17-octadecene-9,11-diynoic (isanolic)	[64144-78-1]	290.40				

[a] To convert kPa to mm Hg, multiply by 7.5.

Table 6. Some Fatty Acids with Alicyclic Substituents

Total number of carbon atoms	Common name	CAS Registry Number	n		Mol wt
6	aleprolic	[2348-89-2]	0		112.13
10	aleprestic	[2348-90-5]	4		168.24
12	aleprylic	[24874-21-3]	6	$(CH_2)_n$	196.29
14	alepric	[2519-24-6]	8		224.34
16	hydnocarpic	[459-67-6]	10	CO_2H	252.40
18	chaulmoogric	[502-30-7]	12		280.45
18	malvalic (halphenic)	[503-05-9]	6	$(CH_2)_nCO_2H$	280.45
19	sterculic	[738-87-4]	7	$(CH_2)_7CH_3$	294.48
19	lactobacillic	[503-06-0]		$(CH_2)_8CO_2H$ / $(CH_2)_5CH_3$	296.49

Table 7. Some Constants for Selected Unsaturated and Substituted Acids

Acid	Heat of formation, $\Delta H°_{298}$, kJ/mol[a]	Flash point, °C[b]	Heat of combustion, ΔH_R^{25}, kJ/mol[a] (liq)
A. Unsaturated acids			
acrylic	−384.3 (l)	49 (OC)	
crotonic	−431.6 (s)	88 (OC)	
2-pentenoic	−447.5 (s)		
3-pentenoic	−435.3 (s)		
4-pentenoic	−431.6 (l)		
undecenoic	−572.2	146 (OC)	−6614
elaidic	−786.1 (s)		−11154
oleic	−802.9 (l)	189	−11228
brassidic	−938.9		−23775
erucic	−854.9 (s)		−13797
B. Substituted acids			
methacrylic	−435.0 (l)		
isobutyric	−534.1 (l)		−2166
pivalic	−565.1 (s)		−2834

[a] To convert J to cal, divide by 4.184.
[b] OC = open cup.

The viscosity of a mixture of fatty acids depends upon the average chain length, $\bar{n}$, and can be calculated from the equations (4):

$$70°C, \log \eta = -0.602802 + 0.134844\bar{n} - 0.00259(\bar{n})^2$$
$$90°C, \log \eta = -0.510490 + 0.101571\bar{n} - 0.001628(\bar{n})^2$$

where η = viscosity in mPa·s (= cP).

Heats of combustion for liquid alkanoic acids at 25°C are given by the equation (5):

$$-\Delta H_R^{25} = 654.4n - 430.4 \text{ J/mol} \ (n \geq 5)$$

Crystallographic properties of solid alkanoic acids significantly affect many of their other properties (6–8). For example, heat of crystallization, melting point, and solubility depend upon whether the acid is even- or odd-numbered, and vary alternately in a homologous series. Other physical properties such as boiling point, density (liquid), and refractive index depend upon molecular rather than crystal structure and change in a regular manner according to molecular weight.

The long-chain alkanoic acids and their derivatives are polymorphic, with the unit cell containing dimers formed by hydrogen bonding between carboxyl groups.

$$R-C\begin{matrix}O---H-O\\ \\O-H---O\end{matrix}C-R$$

In crystallizing fatty acids, solvent polarity does not influence crystal form so much as temperature and concentration (9). Infrared (9,12) and wide-line nmr spectra (13) as well as x-ray methods (10–11) can be used to detect the various crystalline forms.

Alkenoic acids also have polymorphic crystalline forms. For example, both oleic and elaidic acids are dimorphic with melting points of 13.6 and 16.3°C for oleic, and 43.7 and 44.8°C for elaidic acid (14).

The higher fatty acids undergo decarboxylation and other undesirable reactions when heated at their boiling points at atmospheric pressure. Hence, they are distilled at reduced pressure (15–16). Methyl esters boil at lower temperatures than acids at the same pressure as the result of the absence of hydrogen bonding (17).

A procedure for calculation of the vapor pressures of fatty acids at various temperatures has been described (18).

Formic, acetic, propionic, and butyric acids are miscible with water at room temperature. Solubility in water decreases rapidly for the higher alkanoic acids as the chain length increases (Table 8) (19). The solubility in water (at pH 2–3) for unionized acids is given by the following relationship:

$$\log S = -0.6n + 2.32$$

Table 8. Solubilities of Alkanoic Acids in Water and Organic Solvents

Number of carbon atoms in RCOOH	Solubility at 20°C, g/100 g solvent				
	Water	Acetone	Benzene	Cyclohexane	n-Hexane
4	∞				
5	3.7				
6	0.968				
7	0.244				
8	0.068				
9	0.026	∞	∞	∞	∞
10	0.015	407	398	342	290
11	0.0093	706	663	525	
12	0.0055	60.5	93.6	68	47.7
13	0.0033	78.6	117	100	
14	0.0020	15.9	29.2	21.5	11.9
16	0.00072	5.38	7.3	6.5	3.1
18	0.00029	1.54	2.46	2.4	0.5

where S = solubility in mol/L and n = number of carbon atoms (20).

The hydrophilic nature of the carboxyl group balanced against the hydrophobic nature of the hydrocarbon chain allows long-chain fatty acids to form monomolecular films at aqueous liquid–gas, liquid–liquid, or liquid–solid interfaces (18).

The solubility of water in fatty acids (0.92% for stearic acid at 68.7°C) is greater than the solubility of the acid in water (0.0003% for stearic acid at 20°C), and this solubility tends to increase with increasing temperature (21).

Solubilities of aliphatic acids in organic solvents demonstrate another example of the alternating effect of odd- vs even-numbered acids (Table 8).

An important chemical characteristic of unsaturated acids is the iodine value (IV) which indicates the average degree of unsaturation. It is equal to the number of grams of iodine absorbed under standard conditions by 100 grams of the unsaturated acid.

Unsaturation in a fatty acid increases its solubility in organic solvents, and the differences in solubilities between saturated and unsaturated acids can be used to separate these acids (Table 9).

Formic acid is the most acidic straight-chain alkanoic acid. Alkanoic acids containing more than 9 carbon atoms have too low solubility in water to permit accurate measurement of dissociation (Table 10). The acidity of 2-chloroalkanoic acids is much greater than that of formic acid, and trichloroacetic acid is comparable to the mineral acids in acid strength.

Chemical Properties

The alkanoic acids, with exception of formic acid, undergo typical reactions of the carboxyl group. Formic acid has reducing properties, and does not form an acid chloride or an anhydride. The hydrocarbon chain of alkanoic acids undergoes the usual

Table 9. Solubilities of Fatty Acids in Organic Solvents at Various Temperatures

Fatty acid	Temperature, °C	Solubility, g/100 g solvent		
		Acetone	Toluene	n-Heptane
16:0	10	1.60	1.41	0.30
	0	0.66	0.36	0.08
	−10	0.27	0.086	0.02
	−20	0.10	0.018	0.005
18:0	10	0.54	0.390	0.080
	0	0.11	0.080	0.018
	−10	0.023	0.015	0.004
	−20	0.005	0.003	
18:1 (9c)	−20	5.2		2.25
	−30	1.68	3.12	0.66
	−40	0.53	0.96	0.19
	−50	0.17	0.28	0.05
18:1 (9t)	−10		0.86	0.19
	−20	0.26	0.20	0.06
	−30	0.092	0.056	0.019
18:2 (9c,12c)	−50	4.10		0.98
	−60	1.20		0.20
	−70	0.35		0.042

Table 10. Dissociation Constant[a] for Straight Chain and Chlorinated Alkanoic Acids at 25°C

Acid	Substituent None	2-Cl
formic	21.0	
acetic	1.81	1.4×10^{-3}
propionic	1.32	160.0
butyric	1.50	140.0
pentanoic	1.56	
hexanoic	1.40	

[a] $K \times 10^5$ except where noted.

reactions of hydrocarbons except that the carboxyl group exerts considerable influence on the site and ease of reaction. The alkenoic acids in which the double bond is not conjugated with the carboxyl group show typical reactions of internal olefins. All three types of reactions are industrially important.

Reactions of the carboxyl group include salt and acid chloride formation, esterification, pyrolysis, reduction, and amide, nitrile, and amine formation.

Salt formation occurs when the carboxylic acid reacts with an alkaline substance (22).

$$RCOOH + MOH \rightarrow RCOO^- + M^+ + H_2O$$
$$(M = Li, Na, K, NH_4, R_4N, etc)$$

The alkaline substance can also be an oxide, hydroxide, or carbonate of a metal of higher valence such as Ca, Mg, Zn, or Al. The saponification of fats and oils with caustic soda or potash gives water-soluble soaps. Water-insoluble, metallic soaps are prepared by fusion, precipitation, or direct solution of metal (see Driers and metallic soaps). Fusion gives fine, dense, but slightly off-color metallic soaps useful as driers. The precipitation method forms fluffy, finely divided soaps of excellent color and is most important for producing metal stearates. Lithium 12-hydroxyoctadecanoate is an important constituent of many greases.

Acid chlorides are prepared with reagents such as PCl_3, $SOCl_2$, $(COCl)_2$, and $COCl_2$ (23); preparation with thionyl chloride follows the reaction:

$$RCOOH + SOCl_2 \rightarrow RCOCl + SO_2 + HCl$$

Fatty acid chlorides are very reactive and can be used instead of conventional methods to facilitate production of amides and esters. Imidazoles are effective recyclable catalysts for the reaction with phosgene (qv) (24).

Esterification is one of the most important reactions of fatty acids (25). Several types of esters are produced including those resulting from reaction with monohydric alcohols, polyhydric alcohols, ethylene or propylene oxide, and acetylene or vinyl acetate. The principal monohydric alcohols used are methyl, ethyl, propyl, isopropyl, butyl, and isobutyl alcohols (26).

Sulfuric acid or hydrogen chloride may be used to catalyze esterification, and weight ratios of alcohol to fatty acid of 2–4 corresponding to molar ratios of 10–20 may be used to drive the equilibrium reaction to completion. Stoichiometric quantities of acid and alcohol can be used with hexyl and higher alcohols because the water of reaction is removed by azeotropic distillation with toluene or xylene. With long-chain fatty alcohols, water may be removed by azeotropic distillation, sparging with an inert gas, or subjecting the reaction to reduced pressure. Esters are also prepared by alcoholyses of animal fats or vegetable oils in the presence of alkaline or acidic catalysts. Esters of monohydric alcohols are used for plasticizers and in cosmetics.

Esterification with polyhydric alcohols (qv) such as ethylene-, propylene-, diethylene-, and polyethylene glycols (see Glycols), glycerol(qv), pentaerythritol, and certain carbohydrates is a more complex reaction because of immiscibility problems. A good reaction requires temperatures of 230–235°C and vigorous agitation in contrast to the milder conditions used for the simple alcohols. Temperatures higher than 235°C cause polyols to condense to ethers and to decompose. Nearly stoichiometric quantities of glycol and fatty acid are needed to make either monoesters or diesters. Product water is removed by reduced pressure, azeotropic distillation, or sparging. Monoesters are usually formed by reaction of ethylene or propylene oxide with the fatty acid:

$$RCOOH + n\ CH_2\underset{O}{-\!\!-\!\!-}CH_2 \longrightarrow RCOO\!-\!(CH_2CH_2O)_n\!-\!H$$

The product is a mixture of various polyoxyethylene chain lengths (27–29). Glycol diesters are used as vinyl plasticizers; the monoesters as surface-active agents and viscosity modifiers for alkyd resins.

Glycerol esterifications are still more complex (30–33). Even with excess glycerol, a mixture of mono-, di-, and triglycerides is formed because of the limited solubility of glycerol in the reaction product.

Compositions of the reaction mixture can be calculated on a statistical basis assuming equivalence of the three hydroxyl groups and no isomer formation (31). Glycerides are important as surface-active agents; triolein is used to some extent as a plasticizer (see Surfactants; Plasticizers).

Pentaerythritol with its four primary hydroxyl groups is used for the preparation of tetraesters and presents little difficulty except for its high melting point (263°C, when pure). Pentaerythritol tetraesters are used in synthetic drying oils and alkyds. Esters derived from trimethylolalkanes and dipentaerythritol are also used in alkyds. Esterification may take place *in situ* during preparation of the alkyd (see Alkyd resins).

Sorbitol is the most important higher polyol used in direct esterification of fatty acids. Esters of sorbitans and sorbitans modified with ethylene oxide are extensively used as surface-active agents. Interesterification of fatty acid methyl esters with sucrose yields biodegradable detergents; and with starch, thermoplastic polymers (34).

Vinyl esters are prepared by the reaction of a fatty acid with either acetylene in direct condensation or vinyl acetate by acidolysis.

Reduction of fatty acids to alcohols is obtained by catalytic hydrogenation over a copper chromite catalyst at high temperatures (325°C) and pressures (24 MPa or 3500 psig) (35):

$$RCOOH + 4\ H_2 \rightarrow RCH_2OH + H_2O$$

The yield of fatty alcohol is ca 90%. Fatty alcohols may also be prepared by high-pressure catalytic hydrogenolysis of either a glyceride or a methyl ester. A copper chromite catalyst is used at 270–300°C and 34.6 MPa (5000 psig) of hydrogen pressure. If a glyceride is used, the yield of glycerol is relatively low because of hydrogenolysis to propylene glycol and isopropyl alcohol. The saturated fatty alcohols thus produced are used primarily in the production of detergents.

Reduction of glycerides or other esters with sodium and a secondary alcohol such as cyclohexanol or 4-methyl-2-pentanol was at one time a second commercial method for producing fatty alcohols:

$$RCOOR' + 4\,Na + 2\,R''OH \rightarrow RCH_2ONa + R'ONa + 2\,R''ONa$$

$$RCH_2ONa \xrightarrow{H_2O} RCH_2OH + NaOH$$

Though more costly than hydrogenolysis, this method gives high yields of glycerol and unsaturated fatty alcohols if the original fatty ester is unsaturated.

Selective hydrogenation of the carboxyl or ester group in preference to the olefinic unsaturation also produces unsaturated alcohols. Copper–cadmium and zinc–chromium oxides seem to provide most selectivity (36–40). Copper chromite catalysts are not selective. Reduction of red oil-grade oleic acid has been accomplished in 60–70% yield and with high selectivity with Cr–Zn–Cd, Cr–Zn–Cd–Al, or Zn–Cd–Al oxides (41). The reduction may be a homogeneously catalyzed reaction as the result of the formation of copper or cadmium soaps (42).

Pyrolysis of either saturated or unsaturated fatty acids leads to mixtures of hydrocarbons, olefins, and cyclic compounds (43). Pyrolysis of fatty acid salts or vegetable oils has been used to make hydrocarbon fractions suitable as petroleum substitutes. Various products are obtained by pyrolysis of fatty acid salts depending upon the metal salt or catalyst used. Calcium salts generally form symmetrical ketones; aldehydes are formed in the presence of calcium formate, and hydrocarbons in the presence of excess calcium hydroxide. Magnesium and lead salts produce ketones in improved yield over the calcium salts. Zinc oxide promotes hydrocarbon rather than ketone formation. Vapor-phase pyrolysis of fatty acids or esters over thorium or cesium oxides produces ketones in high yields. Homogeneous decarbonylation of fatty acid chlorides with $PdCl_2$ catalyst occurs readily at 185–200°C to make olefins in high yield (44).

Pyrolysis is used to produce undecenoic acid from ricinoleic acid:

$$CH_3(CH_2)_5CHOHCH_2CH=CH(CH_2)_7COONa \xrightarrow{\Delta} CH_3(CH_2)_5CHO + CH_2=CH(CH_2)_8COONa$$

Undecenoic acid is the starting point for making 11-aminoundecanoic acid and nylon-11 (see Castor oil).

Reactions of ammonia and amines with carboxylic acids result in the formation of a variety of products (45–46). Ammonium salts ($RCOONH_4$) are prepared with or without solvent, by reaction with anhydrous ammonia. Ammonium salts readily decompose to acid ammonium salts ($RCOONH_4 \cdot RCOOH$), particularly at temperatures higher than 50°C. Amides are formed at 150–200°C by reaction of the acid with ammonia at reduced pressures:

$$RCOOH + NH_3 \rightarrow RCONH_2 + H_2O\uparrow$$

Amides are also formed by the reaction of an acid chloride with ammonia or an amine:

$$RCOCl + R'NH_2 \rightarrow RCONHR' + HCl$$

Ammonolysis or aminolysis of an ester can be used to make the respective amide or *N*-substituted amide:

$$\text{RCOOR}' + \text{R}''\text{NH}_2 \rightarrow \text{RCONHR}'' + \text{R}'\text{OH}$$

Ammonium acetate and sodium methoxide are effective catalysts for the ammonolysis of soybean oil (47). Polyfunctional amines and amino alcohols such as ethylenediamine, ethanolamine, and diethanolamine react to give useful intermediates. Ethylenediamine can form either a monoamide or a diamide depending upon the mole ratio of reactants. With an equimolar ratio of reactants and a temperature of ≥250°C, a cyclization reaction occurs to give imidazolines with ethylenediamine (48):

$$\underset{\text{O}}{\overset{\|}{\text{R}\text{C}}}\text{NHCH}_2\text{CH}_2\text{NH}_2 \xrightarrow{\Delta} \text{R}\!-\!\!\left\langle\begin{array}{c}\text{N} \\ \text{N} \\ \text{H}\end{array}\right] + \text{H}_2\text{O}$$

Ethanolamine produces oxazolines.

Fatty amines are made by dehydration of amides to nitriles at 280–330°C, followed by hydrogenation of the nitrile over nickel or cobalt catalysts:

$$\text{RCONH}_2 \xrightarrow{\Delta} \text{RCN} + \text{H}_2\text{O}$$

$$\text{RCN} + 2\,\text{H}_2 \xrightarrow[\text{NH}_3]{\text{Ni},\text{H}_2} \text{RCH}_2\text{NH}_2$$

The presence of ammonia during hydrogenation suppresses formation of secondary amines and inhibits hydrogenation of double bonds in unsaturated nitriles. Fatty amines are used as corrosion inhibitors, flotation agents, quaternary salts for sanitizing agents and textile fabric softeners, and surface-active agents.

Acyl aminimides have recently proved useful as surface-active and antimicrobial agents and as an intermediate for isocyanate preparation (49):

$$\underset{}{\overset{\text{O}}{\underset{\|}{\text{RC}}}}\!\!\diagdown_{\text{N}^-\!\!-\!\!\text{N}^+(\text{CH}_3)_3}$$

Reactions of the hydrocarbon chain in alkanoic acids include α-sulfonation and halogenation (50–53). The α-sulfonated fatty ester salts have excellent lime-dispersing properties and are valuable surface-active agents.

Reactions of the double bonds include isomerization and conjugation, cyclization, various addition reactions including hydrogenation, pyrolytic and oxidative cleavage, metathesis, and various polymerization reactions (50).

Geometrical isomerization of *cis*- to *trans*-alkenoic acids occurs by photosensitization (54) or with heat treatment in the presence of catalysts such as SO_2, I_2, or HNO_2.

Positional isomerization occurs most often during partial hydrogenation of unsaturated fatty acids; it also occurs in strongly basic or acidic solution and by catalysis with metal hydrides or organometallic carbonyl complexes. Concentrated sulfuric or, better, 70% perchloric acid treatment of oleic acid at 85°C produces γ-stearolactone from a series of double bond isomerizations, hydration, and dehydration steps (55).

Conjugation as well as geometric and positional isomerization occur when an alkadienoic acid such as linoleic acid is treated with strong base at an elevated temperature. Cyclic fatty acids result from cyclization of linolenic acid in strong base at about 250°C (56). Conjugated fatty acids undergo the Diels-Alder reaction with many dienophiles including ethylene, propylene, acrylic acid, and maleic anhydride.

Addition to the double bonds occurs readily with hydrogen halides, hypohalous, sulfuric, or formic acids (57):

$$—CH=CH— + HX \rightarrow —CH_2CHX—$$

where X = F, Cl, Br, I, OCl, OBr, OSO_3H, HCOO, etc

Addition of a weaker acid such as acetic acid takes place if the reaction is catalyzed by a sulfonic acid ion-exchange resin (58). Addition of halogens, mixed halogens (eg, ICl—the Wijs reagent) or halogen-like compounds (eg, NOCl), and thiocyanogen [$(SCN)_2$] occurs easily and is the basis of several analytical methods for determining total unsaturation (59–60). Addition of hydrogen occurs only in the presence of an active catalyst such as nickel at moderately elevated temperatures and pressures (61). Other reagents that add to the double bond include carbon monoxide and hydrogen (eg, the oxo reaction) (62–63), carbon monoxide and water (eg, hydrocarboxylation) (63), various carbon free-radical compounds (64), dialkylphosphonates (65), formaldehyde (66), and mercuric acetate (67). Dihydroxylation of a double bond may be brought about by peroxy acids or treatment with alkaline permanganate (68). Addition of oxygen is carried out with peroxy acids in the presence of strong acid catalysts to make epoxidized compounds (69).

Cleavage of an alkenoic acid can be carried out with permanganate, a permanganate–periodate mixture, periodate or with nitric acid, dichromate, ozone, or, if the unsaturation is first converted to a dihydroxy compound, lead tetraacetate (68,70). Oxidative ozonolysis is a process for the manufacture of azelaic and pelargonic acids (71):

$$CH_3(CH_2)_7CH=CH(CH_2)_7COOH \xrightarrow{O_3} CH_3(CH_2)_7COOH + HOOC(CH_2)_7COOH$$
oleic acid pelargonic acid azelaic acid [123-999]

Alkali fusion of oleic acid at about 350°C in the Varrentrapp reaction causes double bond isomerization to a conjugated system with the carboxylate group followed by oxidative cleavage to form palmitic acid (72). In contrast, alkali fusion of ricinoleic acid is the commercial route to sebacic acid (73):

$$CH_3(CH_2)_5CHOHCH_2CH=CH(CH_2)_7COOH \xrightarrow{NaOH} CH_3(CH_2)_5CHOHCH_3 + HOOC(CH_2)_8COOH$$
sebacic acid [111-20-6]

Metathesis of oleic acid to produce a C_{18} straight-chain dibasic acid can be carried out at 70°C with a $WCl_6 \cdot Sn(CH_3)_4$ catalyst (74) or with rhenium heptoxide promoted by $Sn(CH_3)_4$ (75).

Polymerization takes a variety of forms including dimerization or trimerization to polybasic acids. The internal double bond in oleic acid and other unsaturated fatty acids can participate, but only to a limited extent in copolymerization with ethylene (76). Conjugated linoleate esters form co-oligomers with styrene by cationic catalysis (77). The resulting dibasic acid or ester can be condensed with ethyleneamine oligomers to make reactive polyamides useful for producing frothed epoxy compositions and as a catalyst for cross-linking rigid urethane foams (78). Conjugated linoleate esters are readily copolymerized with styrene and acrylonitrile by free-radical catalysis (79). Methyl eleostearate, however, inhibits copolymerization with either styrene or acrylonitrile. Polymerization of unsaturated fatty acids in drying oils through autoxidative mechanisms is the basis for a significant use of these materials in the paint industry (80).

BIBLIOGRAPHY

"Acids, Carboxylic" in *ECT* 1st ed., Vol. 1, pp. 139–151, by E. F. Landau, Celanese Corporation of America; "Acids, Carboxylic" in *ECT* 2nd ed., Vol. 1, pp. 224–239.

1. S. Abrahamsson, S. Ställberg-Stenhagen, and E. Stenhagen, in R. T. Holman, ed., *Progress in the Chemistry of Fats and Other Lipids*, Vol. 7, Pt. 1, Pergamon Press, Oxford, Eng., 1964, pp. 1–164.
2. T. L. Ward and W. S. Singleton, *J. Phys. Chem.* **56,** 696 (1952).
3. W. S. Singleton, T. L. Ward, and F. G. Dollear, *J. Am. Oil Chem. Soc.* **27,** 143 (1950).
4. F. Fernandez-Martin and F. Montes, *J. Am. Oil Chem. Soc.* **53,** 130 (1976).
5. N. Adriaanse, H. Dekker, and J. Coops, *Rec. Trav. Chim. Pays-Bas* **84,** 393 (1965).
6. R. T. O'Connor, in K. S. Markley, ed., *Fatty Acids*, Pt. 1, 2nd ed., Interscience Publishers, Inc., New York, 1960, pp. 285–378.
7. E. S. Lutton, in K. S. Markley, ed., *Fatty Acids*, Pt 4, 2nd ed., Interscience Publishers, Inc., New York, 1967, pp. 2583–2641.
8. K. Larsson, *J. Am. Oil Chem. Soc.* **43,** 559 (1966).
9. A. V. Bailey and co-workers, *J. Am. Oil Chem. Soc.* **49,** 419 (1972).
10. A. V. Bailey and co-workers, *J. Am. Oil Chem. Soc.* **52,** 196 (1975).
11. D. Chapman, *The Structure of Lipids*, John Wiley & Sons, Inc., New York, 1965.
12. D. Mitcham, A. V. Bailey, and V. W. Tripp, *J. Am. Oil Chem. Soc.* **50,** 446 (1973).
13. A. V. Bailey and R. A. Pittman, *J. Am. Oil Chem. Soc.* **48,** 775 (1971).
14. J. A. Harris, *J. Am. Oil Chem. Soc.* **44,** 737 (1967).
15. W. S. Singleton, in K. S. Markley, ed., *Fatty Acids*, Pt. 1, 2nd ed., Interscience Publishers, Inc., New York, 1960, pp. 499–607.
16. E. Jantzen and W. Erdmann, *Fette Seifen* **54,** 197 (1952).
17. H. Stage, *Fette Seifen* **55,** 217 (1953).
18. E. L. Lederer, *Seifensieder–Ztg.* **57,** 67 (1930).
19. W. S. Singleton, in K. S. Markley, ed., *Fatty Acids*, Pt. 1, 2nd ed., Interscience Publishers, Inc., New York, 1960, pp. 609–678.
20. G. H. Bell, *Chem. Phys. Lipids* **10,** 1 (1973).
21. C. W. Hoerr, W. O. Pool, and A. W. Ralston, *Oil Soap* **19,** 126 (1942).
22. J. Levy, in E. S. Pattison, ed., *Fatty Acids and Their Industrial Applications*, Marcel Dekker, Inc., New York, 1968, pp. 209–220.
23. N. O. V. Sonntag, in K. S. Markley, ed., *Fatty Acids*, Pt. 2, 2nd ed., Interscience Publishers, Inc., New York, 1961, pp. 1127–1151.
24. C. F. Hauser and L. F. Theiling, *J. Org. Chem.* **39,** 1134 (1974).
25. K. S. Markley, in K. S. Markley, ed., *Fatty Acids*, Pt. 2, 2nd ed., Interscience Publishers, Inc., New York, 1961, pp. 757–984.
26. V. Sreeramulu and P. B. Rao, *Ind. Eng. Chem., Prod. Res. Dev.* **12,** 483 (1973).
27. H. Grossmann, *Tenside Deterg.* **12,** 16 (1975).
28. M. Bareš and co-workers, *Tenside Deterg.* **12,** 155 (1975).
29. M. Vares and co-workers, *Tenside Deterg.* **12,** 162 (1975).
30. R. O. Feuge, *J. Am. Oil Chem. Soc.* **39,** 521 (1962).
31. R. O. Feuge and A. E. Bailey, *Oil Soap* **23,** 259 (1946).
32. R. O. Feuge, E. A. Kraemer, and A. E. Bailey, *Oil Soap* **22,** 202 (1945).
33. A. T. Gros and R. O. Feuge, *J. Am. Oil Chem. Soc.* **41,** 727 (1964).
34. M. L. Rooney, *Polymer* **17,** 555 (1976).
35. K. S. Markley, in K. S. Markley ed., *Fatty Acids*, Pt. 2, 2nd ed., Interscience Publishers, Inc., New York, 1961, pp. 1187–1305.
36. H. Bertsch, H. Reinheckel, and E. Konig, *Fette, Seifen, Anstrichm.* **69,** 387 (1967).
37. *Ibid.*, p. 731.
38. H. Bertsch, H. Reinheckel, and K. Haage, *Fette, Seifen, Anstrichm.* **71,** 357 (1969).
39. *Ibid.*, p. 785.

40. *Ibid.*, p. 851.
41. R. S. Klonowski and co-workers, *J. Am. Oil Chem. Soc.* **47,** 326 (1970).
42. B. Stouthamer and J. C. Vlugter, *J. Am. Oil Chem. Soc.* **42,** 646 (1965).
43. N. O. V. Sonntag, in K. S. Markley, ed., *Fatty Acids,* Pt. 2, Interscience Publishers, New York, 1961, pp. 985–1072.
44. T. A. Foglia, I. Schmeltz, and P. A. Barr, *Tetrahedron* **30,** 11 (1974).
45. S. H. Shapiro, in E. S. Pattison, ed., *Fatty Acids and Their Industrial Applications,* Marcel Dekker, Inc., New York, 1968, pp. 77–154.
46. N. O. V. Sonntag, in K. S. Markley, ed., *Fatty Acids,* Pt. 3, Interscience Publishers, New York, 1964, pp. 1551–1715.
47. W. L. Kohlhase, E. H. Pryde, and J. C. Cowan, *J. Am. Oil Chem. Soc.* **48,** 265 (1971).
48. R. N. Butler, C. B. O'Regan, and P. Moynihan, *J. Chem. Soc. Perkin Trans* 386 (1976).
49. W. J. McKillip, ed., *Advances in Urethane Science and Technology,* Vol. 3, Technomic Publishing Co., Inc., Westport, Conn., 1974, pp. 81–107.
50. H. J. Harwood, *Chem. Rev.* **62,** 99 (1962).
51. W. Stein and H. Baumann, *J. Am. Oil Chem. Soc.* **52,** 323 (1975).
52. N. O. V. Sonntag, in K. S. Markley, ed., *Fatty Acids,* Pt. 2, Interscience Publishers, Inc., New York, 1961, pp. 1073–1185.
53. M. Chals and R. Perron, *J. Am. Oil Chem. Soc.* **48,** 595 (1971).
54. K. S. Markley, in K. S. Markley, ed., *Fatty Acids,* Pt. 1, Interscience Publishers, Inc., New York, 1960, pp. 251–283.
55. J. S. Showell, D. Swern, and W. R. Noble, *J. Org. Chem.* **33,** 2697 (1968).
56. J. P. Friedrich and R. E. Beal, *J. Am. Oil Chem. Soc.* **39,** 528 (1962).
57. N. O. V. Sonntag, in K. S. Markley, ed., *Fatty Acids,* Pt. 2, Interscience Publishers, Inc., New York, 1961, pp. 1073–1185.
58. L. T. Black and R. E. Beal, *J. Am. Oil Chem. Soc.* **44,** 310 (1967).
59. H. A. Boekenoogen, *Analysis and Characterization of Oils, Fats, and Fat Products,* Interscience Publishers, Inc., New York, 1964.
60. W. E. Link, ed., *Official and Tentative Methods of the American Oil Chemists' Society,* American Oil Chemists' Society, 508 South Sixth Street, Champaign, Ill. 61820, 1976.
61. E. N. Frankel and H. J. Dutton, in F. D. Gunstone, ed., *Topics in Lipid Chemistry,* Vol. 1, Logos Press, London, Eng., 1970, pp. 161–276.
62. E. H. Pryde, E. N. Frankel, and J. C. Cowan, *J. Am. Oil Chem. Soc.* **49,** 451 (1972).
63. E. N. Frankel and E. H. Pryde, *J. Am. Oil Chem. Soc.* **54,** 873A (1977).
64. E. Roe, D. A. Konen, and D. Swern, *J. Am. Oil Chem. Soc.* **42,** 457 (1965).
65. R. Sasin and co-workers, *J. Am. Chem. Soc.* **81,** 6275 (1959).
66. E. P. DiBella, *J. Am. Oil Chem. Soc.* **42,** 199 (1965).
67. M. Naudet and E. Ucciani, in F. D. Gunstone, ed., *Topics in Lipid Chemistry,* Vol. 2, Logos Press, London, Eng., 1971, pp. 99–158.
68. D. Swern, in K. S. Markley, ed., *Fatty Acids,* Pt. 2, Interscience Publishers, New York, 1961, pp. 1307–1385.
69. D. Swern, in D. Swern, ed., *Organic Peroxides,* Vol. 2, Wiley-Interscience, New York, 1971, pp. 355–533.
70. L. A. Goldblatt, in K. S. Markley, ed., *Fatty Acids,* Pt. 5, Interscience Publishers, New York, 1968, pp. 3657–3684.
71. E. H. Pryde and J. C. Cowan, in F. D. Gunstone, ed., *Topics in Lipid Chemistry,* Vol. 2, Logos Press Ltd., London, Eng., 1971, pp. 1–98.
72. M. F. Ansell, A. N. Radziwell, and B. C. L. Weedon, *J. Chem. Soc. C,* 1851 (1971).
73. E. H. Pryde and J. C. Cowan, in J. K. Stille and T. W. Campbell, eds., *Condensation Monomers,* Wiley-Interscience, New York, 1972, pp. 74–89.
74. E. Verkuijlen and C. Boelhouwer, *Fette, Seifen, Anstrichm.* **78,** 444 (1976).
75. E. Verkuijlen, *J. Chem. Soc. Chem. Commun.,* 198 (1977).
76. C. J. Vetter, Jr., *Aust. Chem. Eng.* **3** (1968).
77. J. Baltes, *Fette, Seifen, Anstrichm.* **66,** 942 (1964).
78. W. E. Richardson and C. H. Smith, *J. Am. Oil Chem. Soc.* **51,** 499 (1974).
79. F. R. Mayo and C. W. Gould, *J. Am. Oil Chem. Soc.* **41,** 25 (1964).
80. H. Wexler, *Chem. Rev.* **64,** 591 (1964).

EVERETT H. PRYDE
United States Department of Agriculture

MANUFACTURE

This section deals mainly with those acids commonly known as fatty acids that have 6 to 24 carbon atoms, different degrees of unsaturation, and are found in substantial amounts in natural fats and oils.

Carboxylic acids containing fewer than six carbon atoms are not classified as fatty acids. These acids, notably formic, acetic, and propionic acids are produced synthetically in large amounts from petrochemicals; methods of their manufacture are described elsewhere (see Acetic acid; Formic acid; Oxo process).

The C_6–C_{24} acids, or fatty acids, are obtained from animal tallows and greases, vegetable, coconut, palm, and marine oils (Table 1). They are also produced synthetically from petroleum sources. For many years fatty acids have been produced by oxidation of hydrocarbons. A list of actual and potential methods for synthesizing fatty acids follows.

Catalytic Oxidation of Paraffinic Hydrocarbons. Air-oxidation processes were developed during World War II in Germany and more recently in the Union of Soviet Socialist Republics. These countries operate 6–8 plants with an annual production of ca 106 metric tons. The acids are used in the manufacture of soap, allowing edible uses of natural fats and oils (see Hydrocarbon oxidation).

Catalytic air oxidation produces mixtures of odd and even carbon atoms with many impurities; recovery of pure acids is almost impossible.

Oxidation of Olefins. Many manufacturing processes are based on the reaction of carbon monoxide, water, and olefins (see Oxo process).

Oxidation or Carboxylation of Ethylene-Growth Compounds. Compounds obtained from ethylene by the Ziegler process contain an even number of carbon atoms. They are more easily converted to alcohols and acids than naturally occurring compounds (see Alcohols; Ziegler-Natta catalysts). The cost of producing the acids from the Ziegler process prevents them from competing with natural fatty acid. Furthermore, the cost of ethylene has increased more rapidly than the costs of natural oils and fats.

Oxidation of Natural Fats. Dibasic and straight chain monobasic acids are produced by oxidation of unsaturated fatty acids with ozone or other oxidizing agents, eg, oleic acid oxidation produces azelaic, $HO_2C(CH_2)_7CO_2H$, and pelargonic acids, $CH_3(CH_2)_7CO_2H$.

Alkali Fusion of Alcohols. Treatment of fatty alcohols with alkalies at 300°C and ca 5.6 MPa (800 psig) produces the corresponding fatty acid plus hydrogen.

Telomerization. Telomerization of ethylene and butadiene with formic and acetic acids, methyl acrylate, or methyl chloride produces acids of an odd number of carbon atoms. Few of the above processes, with the exception of the paraffin oxidation plants operated in the Union of Soviet Socialist Republics, have furnished synthetic fatty acids in commercial amounts. Although several large petroleum companies have distributed pilot plant quantities of such acids in the C_{12}–C_{14} range over the past decade, these acids are no longer available.

The synthetic processes mentioned above are reviewed in detail (1).

Saponification of Natural Fats

The first step in the manufacture of fatty acids from natural fats and oils is the saponification or hydrolysis of the triglyceride shown by the following reaction:

Table 1. Average Fatty Acid Composition and Constants of Fats and Oils

No. of carbons alkanoic acids	Coconut	Palm kernel	Babassu	Murumuru	Palm	Tallow-beef	Lard	Rape seed	Mustard seed	Olive	Peanut	Sesame	Corn
6	0.2	trace	0.1	1.1									
8	8.0	2.7	6.5	1.6									
10	7.0	7.0	2.7	42.6									
12	48.0	46.9	45.8	36.8	1.0	3.0	2.0	1.5	2.0	trace	7.0	8.5	7.5
14	17.5	14.1	19.9	4.5	42.5	29.0	23.5	0.5	trace	9.0	5.0	4.5	3.5
16	8.8	8.8	6.9	2.2	3.5	18.5	11.4		0.5	2.3	4.0	0.6	0.5
18	2.0	1.3	trace		4.0				1.5	0.2			
20								0.5			3.0		
22					trace			1.0					0.2
24													
unsaturated acids													
14:1 (9c)													
16:1 (9c)	6.0	18.5	18.1	10.8	43.0	46.5	51.5	23.9	25.0	82.5	60.0	47.4	46.3
18:1 (9c)	2.5	0.7		0.4	9.5	3.0	11.6	19.8	19.5	6.0	21.0	39.0	42.0
18:2 (9c, 12c)								1.8					
18:3 (9c, 12c, 15c)													
18:3 (9t, 11c, 13t)													
18:3 (9, 11, 13), 4-oxo-													
18:1 (9c), R-12-hydroxy-													
22:1 (13c)								51.0	51.5				
20, unsaturated													
22, unsaturated													

No. of carbons alkanoic acids	Cottonseed	Soybean	Sunflower	Walnut	Linseed	Perilla	Castor	Tung	Oiticica	Whale	Menhaden	Sardine	Herring
6													
8													
10													
12	0.5	6.5	3.5	trace	0.2					8.0	7.0	5.0	7.0
14	21.0	4.5	3.0	4.6	5.6	7.5	2.0	4.0	6.5	11.0	16.0	14.0	8.0
16	2.0	0.7	0.6	1.0	3.5	trace		1.5	5.5	2.5	1.0	3.0	trace
18	trace			trace	0.6								
20													
22													
24													
unsaturated acids													
14:1 (9c)		trace	0.4	0.1						1.5	trace	trace	trace
16:1 (9c)	33.0	33.5	34.0	17.8	21.0	8.0	8.6	15.0	5.0	17.0	17.0	12.0	18.0
18:1 (9c)	43.5	52.5	58.5	73.3	24.0	38.0	3.5			34.0	27.0	10.0	9.0
18:2 (9c, 12c)		2.3		3.3	45.0	46.5				9.0	trace	15.0	13.0
18:3 (9c, 12c, 15c)										trace			
18:3 (9t, 11c, 13t)								79.5	5.0				
18:3 (9, 11, 13), 4-oxo-									78.0				
18:1 (9c), R-12-hydroxy-							85.9						
22:1 (13c)													
20, unsaturated										5.0	20.0	22.0	20.0
22, unsaturated										12.0	12.0	19.0	25.0

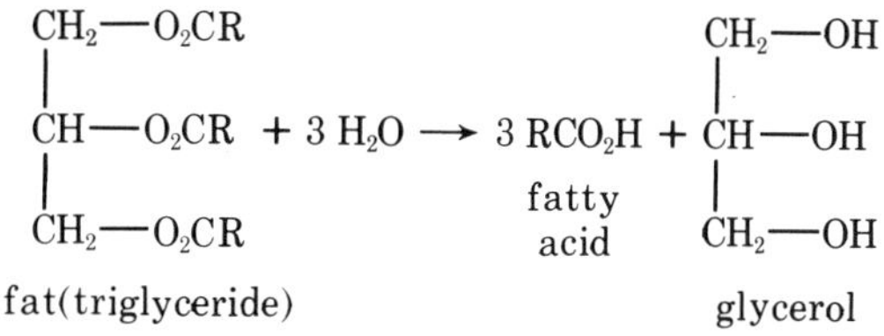

$$\begin{array}{c}\text{CH}_2\text{—O}_2\text{CR}\\|\\\text{CH—O}_2\text{CR}\\|\\\text{CH}_2\text{—O}_2\text{CR}\end{array} + 3\ \text{H}_2\text{O} \rightarrow 3\ \text{RCO}_2\text{H} + \begin{array}{c}\text{CH}_2\text{—OH}\\|\\\text{CH—OH}\\|\\\text{CH}_2\text{—OH}\end{array}$$

fat(triglyceride)   fatty acid   glycerol

where R has various amounts of unsaturation. Many low-grade fats and oils require prior treatment with H_2SO_4 or H_3PO_4 for removal of impurities (2–3) (see Fats and fatty oils).

Saponification of a fat is stepwise, ie, removal of one acid group forms a diglyceride, another acid removal leaves a monoglyceride, and finally free glycerol is produced as shown above (4–5).

In the past, hydrolyses were often nonreversible processes with alkalies or alkaline earths. The resulting soap was washed with brine solutions or water, then neutralized with acid to recover the fatty acids. Another method involved treatment with concentrated sulfuric acid followed by hydrolysis at 100°C. These expensive procedures were replaced by methods that reduced the amounts of labor, energy, and required chemicals.

Less base is necessary with autoclaving under reduced pressure. The dibasic CaO and ZnO are preferred since they yield water-insoluble soaps that can be separated from the aqueous glycerol. In order to prevent an equilibrium that prevents complete hydrolysis, the glycerol is removed as it is formed (6–7).

Hydrolysis can occur in acidic medium with less than the stoichiometric quantity of acid. Shortly after 1900, E. Twitchell (8) developed a method of hydrolysis at atmospheric pressure using a sulfonated mixture of oleic and naphthenic acids (the Twitchell reagent). Petroleum sulfonates are also acceptable catalysts. These catalysts form water-in-oil emulsions; thus the water is carried to the neutral oil and hydrolysis occurs. The usual method employs a mixture of ca 60% neutral oil, 40% water, 1% Twitchell reagent, and 0.5% sulfuric acid. After boiling with open steam for about 24 h the sweet water is removed, 25% fresh water containing 0.5% sulfuric acid is added, and the boiling is continued.

The Twitchell reagent requires large acid-resistant tanks and emits acidic, odorous fumes.

Batch pressure-splitting with or without catalysts is another common method of hydrolysis. The solubility of water in oil increases with temperature; at sufficiently high temperatures, catalysts are not required to carry the water into the oil phase. At 240°C fat splits to ca 92% in 2–4 h in a batch autoclave. The reaction slows down and stops unless the sweet water is replaced with fresh water.

In the 1930s investigators at Procter and Gamble (7) and Colgate-Palmolive-Peet (9) independently developed methods for removing glycerol by countercurrent washing in a tall pressure vessel (Fig. 1). The tower contains disengaging zones to collect the fatty acid at the top and the sweet water at the bottom. The reaction center in the middle contains a continuous fat phase. As shown in Figure 2, fatty acids are 10–15% water soluble at 245–260°C. The sweet water is replaced by water containing less glycerol. The water added to the top of the tower falls through the reaction zone. In

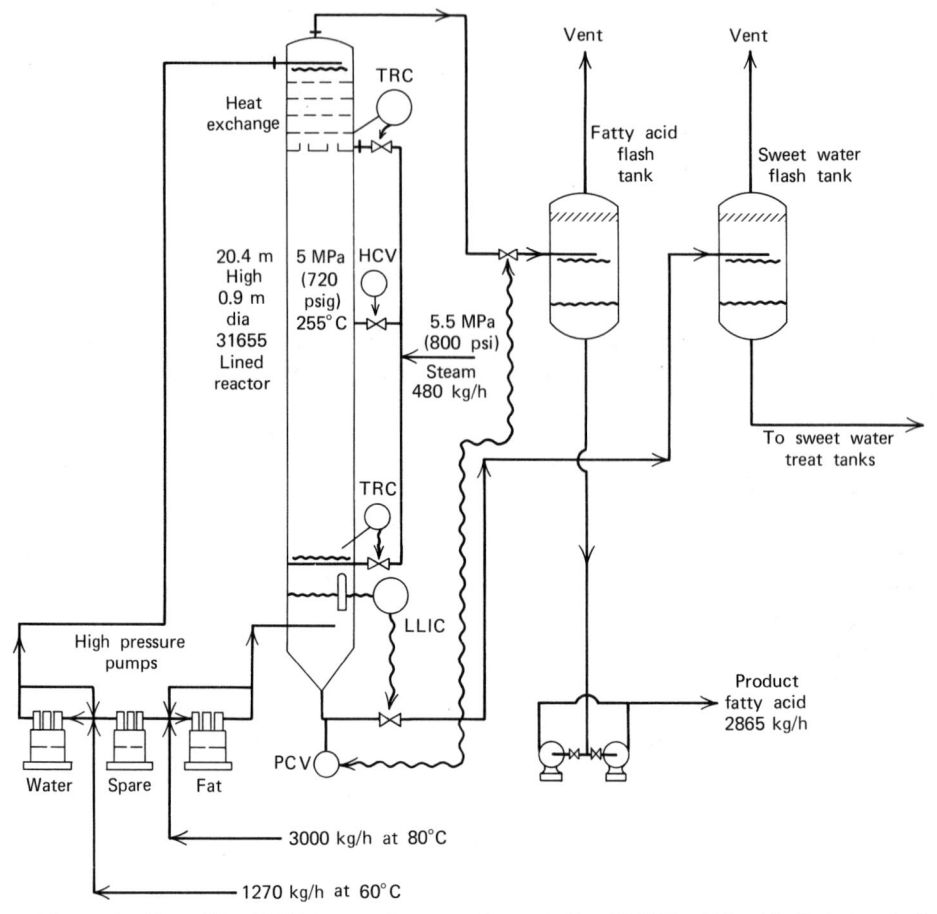

Figure 1. Fat splitter. TRC, temperature recorder controller; LLIC liquid level indicator controller; PCV, pressure control valve; and HCV, heat control valve.

order to maintain the water in the liquid phase, pressure of about 4.9 MPa (700 psig) is supplied by oil and water injection pumps. The temperature of the fat is raised to the reaction temperature by contact with live steam. The reaction requires little, if any, additional heat when the fat and water are at reaction temperatures. Most towers are built with an additional steam injection site at midpoint of the reaction zone to compensate for any heat loss, a well insulated tower does not require additional steam (10–11). A sweet water having 16–20% glycerol is produced from a tower 18–24 m high.

The cost of the fatty acid produced in a countercurrent tower is usually less than the cost of the original fat because the glycerol credit pays for the labor, overhead, and energy required.

The temperatures employed are high for highly unsaturated fats and some loss of unsaturation occurs. In addition, proteinaceous material retained in the original fat breaks down. Some of the breakdown products are fat soluble and remain with the fatty acid, others are removed with the sweet water (see Fig. 3).

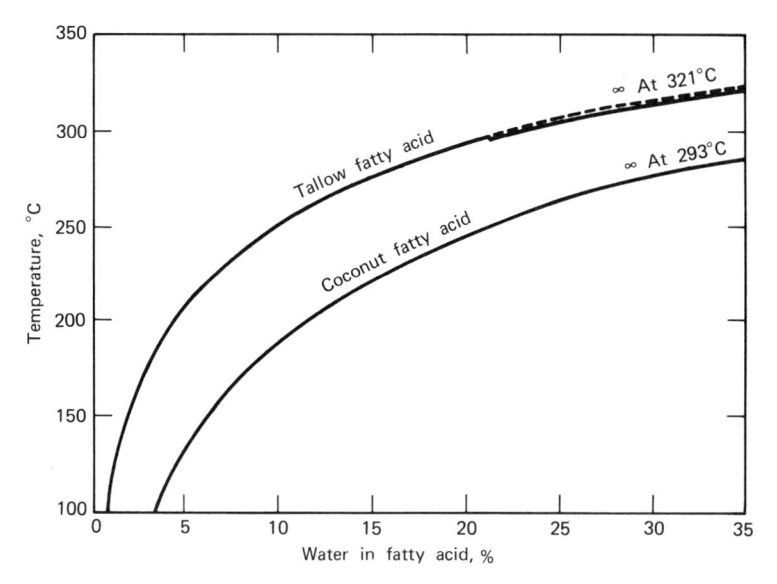

Figure 2. Water solubility of fatty acids (12).

The fatty acid is usually distilled before use. For production of high grade toilet soap it is flash distilled (see Soap). The removal of 1–2% of low boiling impurities improves the odor and heat stability.

The fatty acids obtained from the splitter are C_6–C_{24} and have varying degrees of unsaturation. The methods of separation depend on the physical properties of the acids, eg, the relationship of vapor pressures and length of carbon chain are shown in Figure 4. Acids of different chain lengths can be separated by fractional distillation. The degree of unsaturation does not greatly change the boiling points for acids having the same chain length. Hence they cannot be economically separated by distillation. The degree of unsaturation alters melting points (see page 821).

Glycerol Recovery

Sweet water from a continuous countercurrent fat splitter contains ca 12–20% glycerol, emusified fat, slight amounts of soluble acids and proteinaceous material, and a trace of inorganic salts. Its pH is 4.5–5. The sweet water may be concentrated to some degree with acid-resistant evaporators but it is usually allowed to settle so that the insoluble fatty materials may be skimmed. It is filtered after treatment with lime which precipitates the dissolved fatty acids. The excess lime is removed by soda ash (Na_2CO_3) at pH 8. A second filtration removes $CaCO_3$. This treatment can be carried out in a single step but extreme care is necessary since excess lime causes scale formation in the evaporator tubes.

Crude glycerol from saponification (88% glycerol) varies in color from pale yellow to brown. It is distilled or deionized to make the commercial Chemical Pure Grade or Dynamite Grade (see Glycerol).

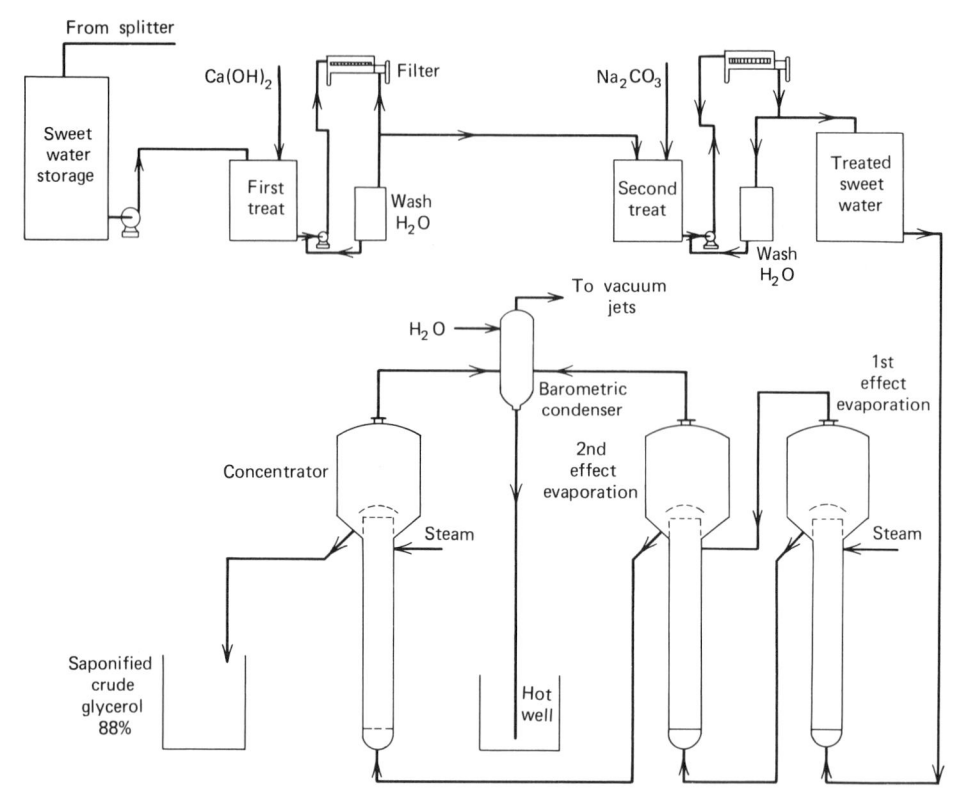

Figure 3. Treatment and concentration of sweet water for saponification of crude glycerol.

Crystallization

Separation of oleic, stearic, and monoleic acids is usually done by crystallization (qv), either with or without solvents. Cotton seed, soya bean, corn oil, and other liquid edible oils are winterized by lengthy treatments at low temperatures to allow the more saturated triglycerides to crystallize and drop out of solution; the resulting slurries are then filtered.

Liquid acids are removed from a mass of crystals by cold pressing. The solid fraction is hot pressed to remove the remaining liquid acids. Pressing was a standard method of manufacture of stearic and oleic acids for many years; hence the names single-, double-, and triple-pressed stearic acids.

Polar solvents such as acetone or methanol allow saturated acids, eg, stearic or palmitic acid, to crystallize almost quantitatively while the unsaturated acids, eg, oleic or linoleic acid, remain dissolved in the solvent. Separation can be accomplished by filtration. Acetone and crude or distilled fatty acid are mixed in a ratio of ca 3–4 L solvent to 1 L fatty acid and chilled in a double pipe chiller. Internal scrapers, rotating at low rpm remove the crystals from the chilled surface. The mix is cooled at −10 to −15°C and separated by a vacuum rotary filter. The filter is sprayed with cold acetone to remove free oleic acid. The solvents are removed by flash evaporation and steam stripping.

Separation without solvents was developed by Henkel in the FRG (13). This

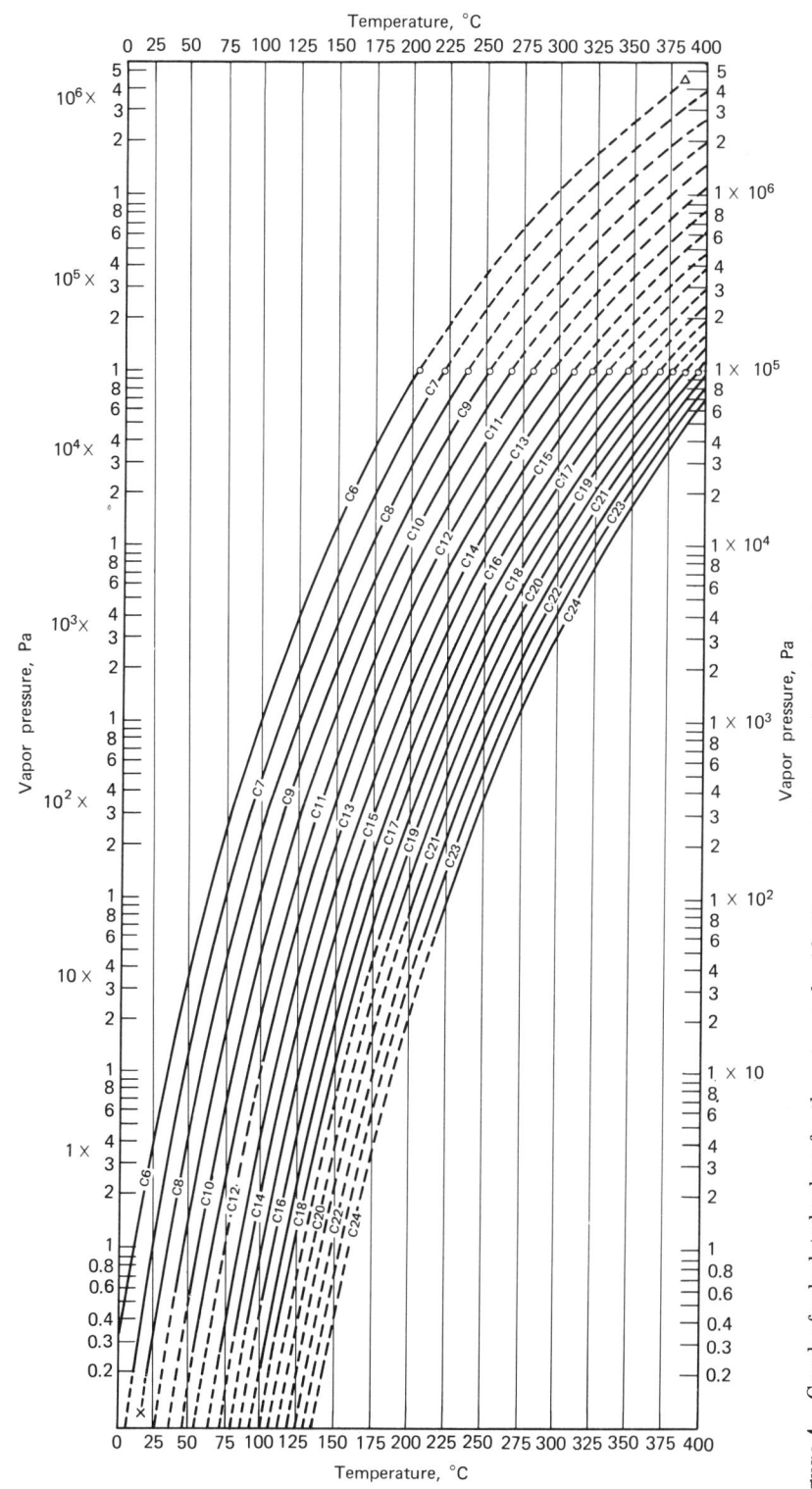

Figure 4. Graph of calculated values for the saturated acids. To convert Pa to mmHg, divide by 133. - - - = estimated values, x = melting point, o = normal boiling point, and △ = critical point.

process involves chilling fat or fatty acid in water containing a detergent, eg, sodium decyl sulfate. The crystals, coated with a film of detergent, remain with the water phase when the mixture is centrifugally separated. The oil phase is free of crystals and moisture. This process is not as complete as the solvent process because more liquid remains with the crystals. In the separation of tallow acids, the stearic acid fraction may have an iodine value of 13–15, whereas solvent separation produces a cake fraction of 4–7 IV. However, because today's market demands a fully saturated grade of stearic acid (<1.0 IV), and the product from either method must be hydrogenated, the Henkel process is gaining in popularity.

Hydrogenation

The solid fractions from crystallization must be hydrogenated to obtain saturated acids with an IV of less than one. Hydrogenation is not difficult and is usually accomplished by the batch method (see Fig. 5). Batch autoclaves vary in size; vessels of 20 t are used. Although it is possible to hydrogenate at pressures of 790 kPa (100 psig), the preferred operating range is 1.48–3.20 MPa (200–450 psig). Nickel is the preferred hydrogenation catalyst; others are Co, Pt, Pd, Cr, and Zn.

At 150°C the autoclave reaction is exothermic. For every kg of C_{18} acid the reduction of 1 IV releases 7.1 J (1.7 cal) which raises the temperature 1.58°C. Additional catalyst or more heat, or both, are applied if the catalyst becomes inactivated before the desired IV is reached. The charge is then cooled and filtered to remove the catalyst.

Several continuous processes using stationary beds or powder catalysts added to the feed have been reported; an example is shown in Figure 6 (14).

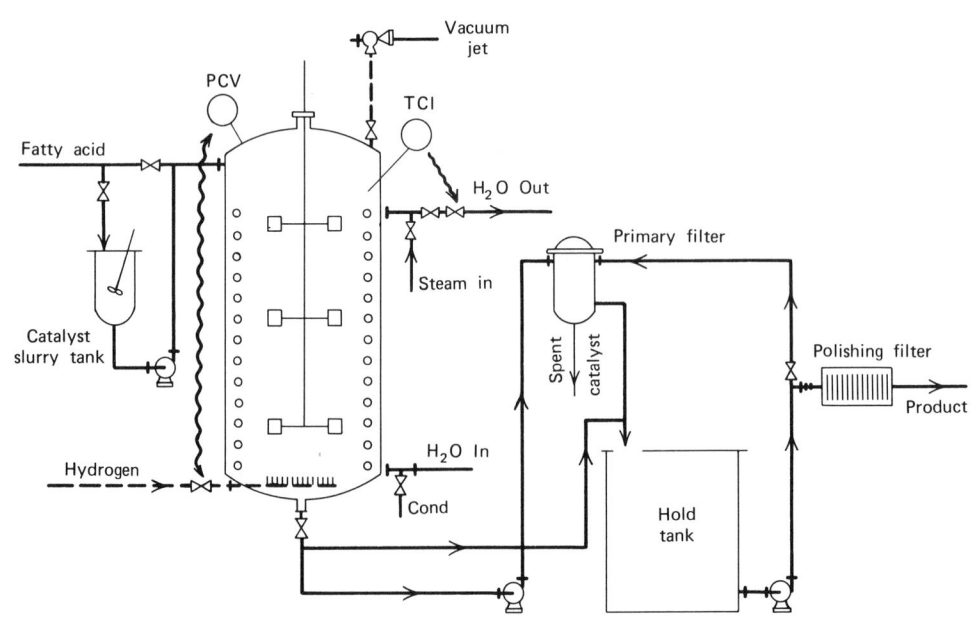

Figure 5. Batch hydrogenator dead end. PCV, pressure control valve; TCI, temperature control indicator.

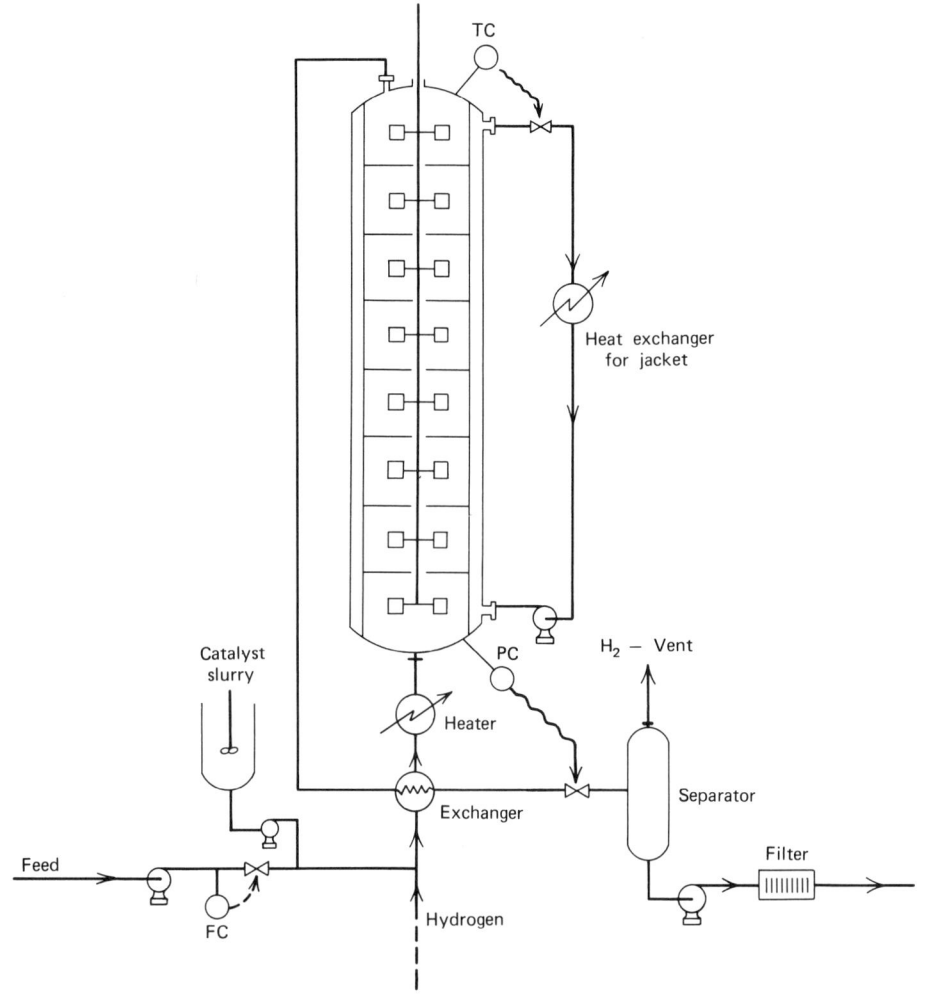

Figure 6. Continuous hydrogenerator. FC, flow controller; PC, pressure controller; TC, temperature controller.

Distillation

For distillation (qv) the preferred method is film evaporation which can be achieved by a falling-film evaporator or steam injection into the heating tubes (15–16).

Present pollution control laws have caused criticism of the use of large amounts of water in contact condensers (injected steam). The use of steam is unnecessary with the falling-film heater which maintains a thin film over the heating surface. Mechanical means of maintaining a thin film over the heating surface, such as white film evaporators and spinning disks, are used increasingly in the distillation of heat-sensitive substances.

Fractional distillation of fatty acids (17) presents problems because the devices needed to obtain staged vapor–liquid equilibrium zones cause a drop in pressure.

Packings of various types, eg, pall rings, mesh expanded metal, etc, can reduce the pressure drop; however, if packed columns are subject to slight changes in operational conditions, equilibrium conditions are drastically changed, and sharpness of separation is immediately sacrificed (18). Packed columns are used only when necessary for thermal stability, ie, when bottom temperatures of 260°C are exceeded.

The quality of a distilled acid is enhanced by efficient degassing, drying, or deodorization prior to vaporization. The vaporization temperature and the retention time under heat must be minimized to reduce thermal decomposition. The dissolved oxygen in the raw stock reacts forming compounds that lower the quality and contribute to bad odor and heat instability if the stock is not thoroughly degassed at a temperature lower than its boiling point. Unwanted impurities must be concentrated, separated, and removed.

Steam cannot be used when temperatures in excess of 300°C are required; eg, for separation of tall oil containing large percentages of rosin acids pressures of more than 14 MPa (2000 psig) are required (19). Dowtherm can be used for heating liquid or vapor organic materials; it is an excellent means of heating all types of fatty acid stills. Leakage of Dowtherm into the still causes odor problems because its boiling point is in the range of coco fatty acids and it is difficult to remove from the product (20).

Fatty acids are corrosive at high temperatures. Types 316L and 317 stainless steels are in general use; however, it is important to specify these steels with at least 2.5% min molybdenum content. High nickel and chrome alloys such as Hastelloy and Carpenter 20 are excellent for pumps.

To comply with environmental standards, fatty acid producers have had to minimize the formation of odorous substances by conducting distillations under conditions that reduce thermal breakdown and oxidation of fatty acid, prevent air leakage into hot acids, and use effective separating devices to remove odorous compounds.

BIBLIOGRAPHY

"Manufacture from Fats" under "Fatty Acids (Survey)" in *ECT* 1st ed., Vol. 6, pp. 231–236, by H. J. Harwood and E. F. Binkerd, Armour and Company; "Manufacture" under "Fatty Acids" in *ECT* 2nd ed., Vol. 8, pp. 825–830, by W. C. Ault, U.S. Department of Agriculture.

1. N. E. Bednarcyk and W. L. Erikson, *Fatty Acid Synthesis and Applications, Chemical Technology Series No. 9,* Noyes Corporation, Park Ridge, N. J., 1973; *Synthetic Fatty Acids—Neo Acids, Oxo Process, Paraffin Oxidation,* Fatty Acid Producers Council SDA, New York, Jan. 1966; N. O. V. Sonntag and K. T. Zilch in E. S. Pattison, ed., *Fatty Acids and Their Industrial Applications,* Marcel Dekker, Inc., New York, 1968, Chapt. 17, pp. 353–367; M. Fefer and A. J. Rutkowski, *J. Am. Oil Chem. Soc.* **45,** 5 (Jan. 1968); K. T. Zilch, *J. Am. Oil Chem. Soc.* **45,** 11 (Jan. 1968); N. O. V. Sonntag, *J. Am. Oil Chem. Soc.* **45,** 14 (Jan. 1968); J. J. Langford, *Source Materials for Synthetic Fatty Alcohols and Acids,* Paper—SDA Convention, Jan. 22, 1969, pp. 4–8; K. T. Zilch, *Comm. Dev. J.* **1**(1), 15 (Feb. 1969); *Soap Perfum. Cosmet.* **42**(8), 565 (Aug. 1969); O. W. Boyd, *Rend. Yearb.,* 22 (1970).
2. *Baileys Industrial Oil and Fat Products,* The Interstate Printers & Publishers, Inc., Danville, Ill., 1964, pp. 719–727.
3. A. J. C. Anderson, *Refining of Oil and Fats,* Pergamon Press, New York, 1962, p. 32.
4. L. Lascaray, *J. Am. Oil Chem. Soc.* **29,** 362 (1952).
5. A. Sturzenegger and H. Sturm, *Ind. Eng. Chem.* **43,** 510 (1951).
6. V. Mills and H. K. McClain, *J. Am. Oil Chem. Soc.* **29,** 318 (1949).
7. U.S. Pat. 2,156,863 (May 2, 1939), V. Mills (to Procter & Gamble).
8. U.S. Pat. 601,603 (Mar. 29, 1898), E. Twitchell.
9. U.S. Pat. 2,139,589 (Dec. 6, 1938), M. Ittner (to Colgate-Palmolive-Peet).
10. K. C. D. Hickman and N. D. Embree, *Ind. Eng. Chem.* **40,** 135 (1948).
11. P. C. Nygren and G. K. S. Connelly, *Chem. Eng. Prog.* **67**(3), 49 (1971).
12. V. Mills and H. K. McClain, *Ind. Eng. Chem.* **41,** 1982 (1949).

13. U.S. Pat. 2,800,493 (July 23, 1957), W. Stein and co-workers. (to Henkel and Cie, G.m.b.H.).
14. Lurgi Gesellschaft für Wärme und Chemotechnik MBH, *Continuous and Discontinuous Hydrogenation of Fatty Acids and Neutral Oils,* Frankfurt am Main, FRG, 1970.
15. H. Stage, *German and French Fat Symposium—Strasbourg, Fr., 14th ACV Research Report,* 1974.
16. H. Stage, *Chem. Age India* **25,** 8, 581 (1974).
17. U.S. Pats. 2,054,096 (Sept. 15, 1936), 2,224,984 (Dec. 17, 1940), 2,322,056 (June 15, 1943), and 2,674,570 (Apr. 6, 1954), R. H. Potts and co-workers (to Armour and Co.).
18. P. Reich and co-workers, *AICHE 63rd Annual Meeting, Chicago, Ill., Nov. 29–Dec. 3, 1970.*
19. R. H. Potts, "Distillation of Tall Oil," *paper presented at AOCS Meeting, New Orleans, La., May 1957.*
20. R. H. Potts, *J. Am. Chem. Soc.* **33,** 545 (1956).

General References

F. D. Gunstone, *An Introduction to the Chemistry and Biochemistry of Fatty Acids and Their Glycerides,* Halsted Press, London, *Eng.*, 1975.
K. S. Markley, ed., *Fatty Acids: Their Chemistry and Physical Properties,* 2nd ed., Interscience Publishers, Inc., a division of John Wiley & Sons, Inc., New York, 1960–1968.
E. S. Pattison, ed., *Fatty Acids and Their Industrial Applications,* Marcel Dekker, Inc., New York, 1968.

RALPH H. POTTS
Armak Company

ANALYSIS AND STANDARDS OF FATTY ACIDS

Composition and Chemical Properties. The fatty acids most commonly encountered as solids or liquids in natural fats have common names related to their natural occurrence. A list of commonly occurring fatty acids is given on page 848.

Most fatty acids arise from processing selected fats and oils such as coconut, soybean, cottonseed, corn, and other vegetable oils, and lard and beef tallow. Such fatty acids are originally present as triesters of glycerol (triglycerides) with other mixed lipid constituents such as sterols (see Vegetable oils; Fats and fatty oils).

Standardized methods are available in the Official and Tentative Methods of The American Oil Chemists' Society (AOCS) to determine the following chemical and physical properties of fats, oils, and fatty acids: titer, acid and iodine values, color, stability, saponification value, unsaponifiables, and fatty acid composition (1).

Titer: (AOCS tentative method Cc12-59). This test measures the solidification point of fatty acids in °C as specified in the method.

Acid value: (AOCS official method Da14–48). The acid value is the number of milligrams of potassium hydroxide necessary to neutralize fatty or rosin acids in 1 g of sample.

Iodine value: (AOCS official method Da15–48). The iodine value, a measure of the unsaturation of fatty acids, is expressed as the number of centigrams of iodine absorbed per gram of sample (% iodine absorbed) or the number of grams of iodine absorbed per 100 g of sample.

Color. The color of an oil reflects its quality; the highest quality oil approaches a water-white color. Color can be measured spectrophotometrically as in AOCS Cc 13c-50 or Tb 2a64.

Moisture: (AOCS method tb 1a-64). This method measures the moisture and any other material volatile under the conditions of the test.

Heat stability. The measurement of color development as a function of heating the fat.

Saponification value: (AOCS tLla-64T). The saponification value is a measure of the alkali reactive groups in fatty acids and oils and is expressed as the number of milligrams of potassium hydroxide that reacts with one gram of sample.

Unsaponifiables: (AOCS Tk 1a-64T). Unsaponifiables are materials found in fatty acids and oils, eg, aliphatic alcohols (C_{12} and higher), sterols, pigments, and hydrocarbons that cannot react with caustic alkalies but are soluble in ordinary solvents.

Fatty acid composition. The fatty acid composition of fatty acid mixtures with 8–24 carbon atoms can be determined for both saturated and unsaturated fatty acids by glc (AOCS tentative method Ce1-62).

Fatty acid composition, although not normally a specification of fatty acids, is now routinely employed to describe the properties of fatty acids. The specifications outlined above were necessary since the use of most fatty acids arose prior to the accepted use of gas chromatography. Knowledge of acid composition has been available since the early forties through the use of high-vacuum fractional distillation for the separation of saturated and monoenoic fatty acids (2), and ultraviolet spectrophotometry for the more unsaturated acids (3). Methods for determinations of physical and chemical properties of fatty acids are also described by ASTM in the *Standard Methods for Analysis of Fats and Oils* (4) and by the Association of Official Analytical Chemists (5).

The melting points of fatty acid mixtures vary according to the chain lengths, crystalline structures, and proportions of saturated fatty acids in the mixture. Crystalline structures are termed polymorphic since they are easily altered by heat treatment (6). Fatty acid mixtures with the same titer may have varying iodine values. For instance, pure oleic acid has the same iodine value (ca 90) as a 1:1 mixture of stearic (IV = 0) and linoleic acids (IV = 180).

Both acid and saponification values are measures of average molecular weight. However, mixtures of fatty acids can differ widely in composition yet yield very similar values. These values are also influenced by the unsaponifiable content and cannot be relied upon for quantitative information concerning composition.

Gas–liquid chromatography (glc), is now routinely used for quantitative analysis of fatty acid mixtures. Information concerning the exact percentage composition of any saturated fatty acid may be obtained with modern liquid phases such as the cyano silicone Sp 2340 in under 30 minutes (7). Fatty acid separations are accomplished both on the basis of molecular weight and polarity using a highly selective stationary phase, such as that mentioned above, as a partitioning agent. A separation for the methyl esters of a vegetable oil fatty acid mixture is shown in Figure 1. Such analyses can be readily obtained for most fatty materials (1).

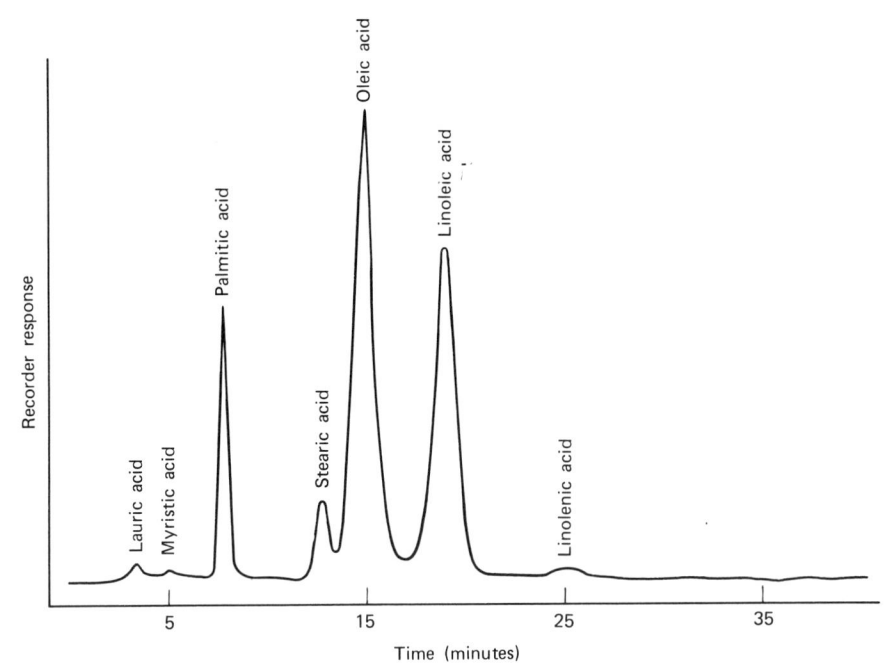

Figure 1. Gas–liquid chromatogram of the methyl esters of partially hydrogenated soybean oil.

Other qualitative and semiquantitative methods for the analysis of fatty acids are: paper chromatography, thin layer chromatography (tlc), and high pressure liquid chromatography (hplc).

Both paper chromatography and tlc have been used for fatty acid separations, usually by the reversed-phase technique (8). These techniques require less sophisticated equipment than either glc or hplc, but they are not useful for quantitative analyses (9). Both qualitative and quantitative separations of fatty acids can be accomplished with hplc. For instance, Schofield (10) has reported the separation of fatty acids on an octadecylsilane-modified silica support, and Water's Associates have developed a column specifically for fatty acid analysis (11).

The compositions of fatty acids reflect their sources. Tables 1 and 2 compare the composition of fatty acids with the common fats and oils from which they are derived. The only fatty acids available in high purity are saturated, eg, lauric, myristic, palmitic, and stearic acids.

There are many markets for unfractionated fatty acids, each of which has the same fatty acid composition as the parent fat or oil. These fatty acids, termed soapstocks, are obtained as by-products from alkali treatment during refining of an oil. Other sources are less expensive oils such as coconut, and lard and beef tallow. The availability of fatty acids as by-products of refining has diminished with increased use of solvent refining; and demand for fatty acids as fat sources for the animal feed industry has increased. Other sources of fatty acids, such as natural oils (Table 2) and tall oil acids, have become more popular. Tall oil acids are similar in composition to soybean oil fatty acids. The specifications and compositions of some common fatty acid mixtures are given in Tables 1 and 3.

The unsaturated fatty acid portions of commercial fatty acids are complex mix-

Table 1. Approximate Percent Chemical Composition of Commercial Fatty Acids

Source of fatty acid	Saturated fatty acids %						Unsaturated fatty acids %			
	Octanoic	Decanoic	Lauric	Myristic	Palmitic	Stearic	Palmitoleic	Oleic	Linoleic	Linolenic
animal				3	29	18	3	40	4	1
coco, (stripped)		1	55	22	11	3		6	2	
coco	6	7	50	19	9	2		1	6	
corn				1	25	4		26	50	2
cottonseed				1	7	4		26	43	1
soya				1	16	4		28	54	5
linseed				1	7	2		23	17	51
hydrogenated tallow			2	9	21	60		5		

Table 2. Mean Fatty Acid Composition of Common Fats and Oils, %

Fatty acid	Coconut	Palm	Palm kernel	Lard	Cottonseed	Corn	Soybean	Linseed	Rapeseed	Beef tallow
caproic	0.5		0.5							
caprylic	8		4							
capric	7		5							
lauric	48		50	1						0.1
myristic	17	2	15		0.5				0.5	3
palmitic	9	42	7	28	21	8	8	0.5	2	29
stearic	2	4	2	13	2	3.5	4	21	0.9	20
palmitoleic	0.2		0.5	3					0.1	2
oleic	6	43	15	46	29	46	28	22	18	42
erucic										
linoleic	2	9	1	6	45	42	54	17	22	2
linolenic				0.7	2		5	51		0.5
erucic									51	

Table 3. Approximate Percent Composition of Stearic Acids

Grades of stearic acid	Lauric	Myristic	Pentadecanoic	Fatty Acid, % Palmitic	Margaric	Stearic	Arachidic	Oleic
commercially pure stearic				7	2	90	1	
special cp stearic				7	2	90	1	
single-pressed stearic		2	1	52	2	38		5
double-pressed stearic		2	1	52	2	39		4
triple-pressed stearic		2	1	52	2	43		
hydrogenated tallow		2	1	29	1	66	1	
rubber grade stearic	2	9	1	21	1	60	1	5
stearic–palmitic		4	1	21	1	72	1	

Table 4. Specifications of Commercial Stearic Acids

Grade of stearic acid	Titer, °C min	Titer, °C max	Iodine value min	Iodine value max	Acid value min	Acid value max	Color, Lovibond, 13.3 cm cell max	Moisture, %	Heat stability, Lovibond	Unsaponifiable, %
commercially pure stearic	65.5	68.0		1.0	195	200	1.0 R–5 Y	0.2	3.5 R–25 Y	0.5
special cp stearic	65.5			0.5	195	200	0.5 R–1.5 Y	0.2	1.5 R–5.0 Y	0.5
single-pressed stearic	53.3	54.2	5.0	10.0	207	210	2.0 R–15 Y	0.5		0.8
double-pressed stearic	54.0	54.6	4.5	7.0	208	211	0.5 R–2 Y	0.5	3.0 R–20 Y	0.5
triple-pressed stearic	55.0	56.0		0.5	208	211	0.5 R–2 Y	0.5	1.0 R–7 Y	0.5
hydrogenated tallow	57.0	61.0		1.0	201	206	1.0 R–5 Y	0.5	2.5 R–15 Y	0.5
rubber grade stearic	55.0	62.0		9.0	195	208	8.0 R–40 Y	0.5		2.0
stearic–palmitic	60.0	64.0		1.0	198	205	1.0 R–5 Y	0.5	3.5 R–25 Y	0.5

Table 5. Approximate Compositions of Commercial Unsaturated Fatty Acids

Fatty acid (commercial mixture)	Saturated fatty acids, %							Unsaturated fatty acids, %			
	Myristic	Pentadecanoic	Palmitic	Margaric	Stearic	Myristoleic	Palmitoleic	Oleic	Linoleic	Linolenic
low polyunsaturated oleic	3	1	3	1		1	7	79	4	1
white oleic	3	1	3	1.5		1.5	6	77	6	1
5° titer, red oil	3	1	3	1.5		1.5	6	76	6	1
8–11° titer red oil	3	1	4	1.5	2	1.5	6	73	6	1
distilled animal	3	1	29	0.5	18	0.5	3	40	4	1
linoleic	1		4				1	33	60	1

Table 6. Approximate Specifications of Commercial Unsaturated Acids

Fatty acid (commercial mixture)	Titer, °C		Iodine value		Acid value		Color Lovibond max	Moisture, %	Unsaponifiables, %
	min	max	min	max	min	max			
low polyunsaturated oleic		7.0	84.0		200	204	1.0 R–8 Y	0.4	1.0
white oleic		5.0		95.0	200	204	1.3 R–9 Y	0.4	1.0
5° titer red oil		5.0		95.0	199	204	1.0 R–7 Y	0.4	1.0
8–11° titer red oil	8.0	11.0		95.0	199	204	1.0 R–7 Y	0.4	1.5
distilled animal	40.0	44.0		60.0	201	206	1.5 R–10 Y	0.5	1.5
linoleic		5	140	145	195	200	3 Gardner	0.5	1.0

Table 7. Approximate Compositions of Other Commercial Saturated Fatty Acids

Commercial acid	Fatty acid, %										
	Caproic	Caprylic	Capric	Lauric	Myristic	Pentadecanoic	Palmitic	Margaric	Stearic	Oleic	Linoleic
commercially pure caprylic (cp)	5	92	3								
98% min cp caprylic	0.5	99	0.5								
cp lauric, 99%		1	97	2							
cp myristic				1	98						
stripped coco			1	55	22		11		3	6	2
distilled coco		5	6	52	19		9	1	2	6	1
cp palmitic					1	1	92	1	5		
cp palmitic, 97%							98		2		
cp palmitic, 80%					1	1	81	2	15		
palmitic eutectic					2	1	66	1	30		

Table 8. Approximate Specifications of Other Commercial Saturated Fatty Acids

Commercial acid	Titer, °C		Iodine value		Acid value		Color, Lovibond, max	Moisture, %, max	Heat stability, max	Unsaponifiables
	min	max	min	max	min	max				
caprylic cp	8	12		0.7	387	392	1.0 R-5 Y	0.2	4.0 R-40 Y	0.2
98% min cp caprylic	15			0.5	385	390	0.5 R-3 Y	0.2	2.5 R-15 Y	0.2
cp capric	29	32		0.5	323	329	0.8 R-3 Y	0.2	2.5 R-10 Y	0.2
cp lauric, 99%	41.5	44.0		0.5	278	282	0.5 R-3 Y	0.2	1.5 R-10 Y	0.2
cp myristic	52.0			0.5	244	249	0.5 R-2 Y	0.2	1.0 R-5 Y	0.2
stripped coco	27.5	29.5	8.0	13.0	252	258	1.0 R-7 Y	0.3	1.5 R-8 Y	0.5
distilled coco	23.0	26.0		10.0	265	275	1.0 R-7 Y	0.3		0.5
cp palmitic	59.0	61.0		0.5	216	220	0.5 R-2 Y	0.2		0.3
cp palmitic, 97%	61.6			0.5	216	220	0.5 R-2 Y	0.2	1.0 R-6 Y	0.3
cp palmitic, 80%	56.0	58.0		1.0	214	218	0.6 R-3.5 Y	0.2	2.0 R-12 Y	0.4
palmitic–eutectic	53.0	55.0		0.5	211	213	0.5 R-2 Y	0.2	1.5 R-7 Y	0.4

tures of positional or geometric isomers. These arise during processing operations such as hydrogenation and deodorization. Other changes to produce new isomers occur during storage.

Today fatty acids are produced by a combination of crystallization and fractional distillation. Most saturated stearic and oleic fatty acid types come from animal fats such as inedible tallow. They have retained the names reminiscent of earlier production methods. For example, the eight or more available grades of stearic acid (Table 4) are referred to as either single-, double-, or triple-pressed stearic acid. Since the composition of most vegetable oils is primarily of the C_{18} variety, stearic acid can be readily prepared from them by hydrogenation, crystallization, and distillation. Several types of oleic acid, low poly(unsaturated) oleic, crystallized white oleic, crystallized red oil, and distilled animal acids, are also available. These acids vary primarily in the amount of linoleic and oleic acid they contain. The compositions and specifications for these acids are shown in Tables 5 and 6.

Other common saturated fatty acids are octanoic, decanoic, lauric, myristic, and palmitic acids which are usually obtained from fractional distillation of coconut oil. Specification and composition data for these acids are shown in Tables 7 and 8.

Certain longer chain length fatty acids such as the C_{20} and C_{22} are available from hydrogenated fish oils and rapeseed oil, sources that contain relatively large amounts of these chain lengths in unsaturated forms.

Resistance to Deterioration. Fatty acid deterioration is related to the total unsaturation of the acid mixture. The extent and relative speed of deterioration decreases in the series of acids: linoleic, oleic, saturated acids (12).

Such deterioration arises as a result of free radical reactions resulting in polymerization and chain cleavage (12). Rancid odor and flavor, objectionable in food and cosmetics, are caused by short chain (C_1–C_{10}) aldehydes and ketones.

The AOCS method td3a-64 measures the color stability of fatty acids after heating under the specified conditions of the test; the specification used is the maximum allowable color of the sample after the test (1). A standardized method for determining fat stability, the active oxygen method (AOCS CD 12–57, ref. 1), measures the time required for a sample to attain a predetermined peroxide value under the specific conditions of the test within hours. The length of time is assumed to be an index of resistance to rancidity, although the exact relationships between peroxide value, actual rancidity, and oxidative stability have not been established.

BIBLIOGRAPHY

"Separation and Analysis" under "Fatty Acids (Survey)" in *ECT* 1st ed., Vol. 6, pp. 228–231, by T. H. Hopper, U.S. Department of Agriculture; "Analysis and Standards" under "Fatty Acids" in *ECT* 2nd ed., Vol. 8, pp. 830–839, by W. C. Ault, U.S. Department of Agriculture.

1. *Official and Tentative Methods of the American Oil Chemists' Society,* 3rd ed., American Oil Chemists' Society, Champaign, Ill.
2. K. S. Markley, *Fatty Acids,* 2nd ed., Part 3, Interscience Publishers, a division of John Wiley & Sons, Inc., New York, 1964, p. 1983.
3. *Ibid.,* p. 379.
4. *Standard Methods for the Analysis of Fats and Oils,* Fat Commission, IUPAC, Paris, Fr., 1954.
5. W. Horowitz, ed., *Official Methods of Analysis of the Assoc. of Official Analytical Chemists,* 12th ed., AOAC, Washington, D.C., 1975.
6. F. D. Gunstone, *An Introduction to the Chemistry of Fatty Acids and Their Glycerides,* 2nd ed., Halsted Press, John Wiley & Sons, Inc., New York, 1975, p. 69.

7. D. Ottenstein, D. A. Barthley, and W. R. Supina, *J. Chromatogr.* **119,** 401 (1976).
8. E. Lederer and M. Lederer, *Chromatography,* 2nd ed., Elsevier Scientific Publishing Company, Amsterdam, The Netherlands, 1957.
9. *Ibid.,* p. 177.
10. C. R. Scnolfield, *J. Am. Oil Chem. Soc.* **52,** 36 (1975).
11. *Separation of Fatty Acids,* Waters Associates Bulletin, Milford, Mass.
12. W. O. Lundberg, ed., *Autoxidation and Antioxidants,* Vols. 1 and 2, Interscience Publishers, New York, 1961 and 1962.

General References

D. Swern, ed., *Bailey's Industrial Oil and Fat Products,* 3rd ed., Interscience Publishers, a division of John Wiley & Sons, Inc., New York, 1964.
E. W. Eckey, *Vegetable Fats and Oils,* Reinhold Publishing Corporation, New York, 1954.
T. P. Hilditch and P. N. Williams, *The Chemical Constitution of Natural Fats,* 4th ed., John Wiley & Sons, Inc., New York, 1964.

EDWARD G. PERKINS
Jniversity of Illinois

ECONOMIC ASPECTS

Several aliphatic carboxylic acids are produced on a large scale (1976 production in thousand metric tons): formic (23) (qv), acetic (997) (qv), propionic (33), butyric (1.2, reported sales in 1970), acrylic (98) (qv), and 2-ethylhexanoic (5.4, as various metal salts). The 1976 production of higher fatty acids, including both saturated (see Table 1) and unsaturated (see Table 2), totals about 550 thousand metric tons (see Table 3). Prices for the major carboxylic acids cover a wide range ($0.35–2.82/kg) (see Table 4).

Propionic acid is made by the liquid phase oxidation of propionaldehyde, which in turn is made by application of the oxo synthesis (qv) to ethylene. Propionic acid can also be made by oxidation of propane or by hydrocarboxylation of ethylene with CO and H_2 in the presence of a rhodium (1) or iridium (2) catalyst. Markets for propionic acid are distributed as follows: grain preservation, ca 28%; in the form of the calcium or sodium salt as an antifungal agent in bread and other foods, ca 28%; cellulose propionate plastics, ca 20%; herbicides such as 2,2-dichloropropionic acid [75-99-0], 2-(2,4-dichlorophenoxy)propionic acid [120-36-5], and 3',4'-dichloropropionanilide, ca 17%; and miscellaneous uses, ca 7%. Grain preservation may become a more prominent market for propionic acid as such or in the form of methanediol dipropionate (3–4).

Butyric acid is made by air oxidation of butyraldehyde which is obtained by application of the oxo synthesis to propylene. Annual consumption of butyric acid has reached 20,000 t, however markets have gradually declined as cellulose acetate butyrate plastics have been replaced with other types of thermoplastic molding compounds.

Table 1. Saturated Fatty Acid Production and Disposition[a] **(1000 Metric Tons)**

Fatty acid	1965	1975	1976
Production			
stearic, 40–50% stearic content	30.4	42.8	56.0
hydrogenated animal and vegetable oils			
60°C max titer and min IV 5	40.5	39.1	46.8
57°C min titer and max IV under 5	38.5	45.1	57.5
minimum stearic content of 70%		12.3	13.9
high palmitic, over 60% palmitic, IV max 12	3.5	3.8	3.8
hydrogenated fish and marine mammal	4.9	2.3	3.3
lauric-type acids, IV min 5, saponification value min 245	10.8	26.7	31.4
fractionated fatty acids			
C_{10} or lower, including decanoic		6.4	8.3
	8.1		
lauric and/or myristic content of 55% or more		6.8	7.3
Total	136.7	185.3	228.3
Disposition			
domestic shipments	128.2	142.0	185.6
captive consumption	12.0	70.3	59.5
export shipments	1.3	3.6	3.1
Total	141.5	215.9	248.2

[a] Source is Fatty Acid Producer's Council.

Table 2. Unsaturated Fatty Acid Production and Disposition[a] **(1000 Metric Tons)**

Fatty acid	1965	1975	1976
Production			
oleic, red oil	51.4	53.6	70.1
animal fatty acids, other than oleic, IV 36 to 80	18.8	51.4	61.6
vegetable or marine, IV max of 115	4.4	0.2	4.3
unsaturated fatty acid, IV 116 to 130	7.3	8.8	7.9
unsaturated fatty acid, IV over 130	5.2	5.5	9.7
tall oil fatty acids	135.3	133.0	169.6
Total	222.4	252.5	323.2
Disposition			
domestic shipments	172.5	83.2	101.0
captive consumption	31.4	44.0	61.0
export shipments	24.8	2.0	3.6
Total	228.7	129.2	165.6

[a] Source is Fatty Acid Producer's Council.

Table 3. Total Production and Disposition of All Fatty Acids[a] **(1000 Metric Tons)**

	1965	1975	1976
production	359.1	437.8	551.5
disposition			
domestic shipments	300.7	225.2	286.6
captive consumption	43.4	114.3	120.5
export shipments	26.1	5.6	9.8
Total disposition	370.2	345.1	416.9

[a] Source is Fatty Acid Producer's Council.

Table 4. Prices and Major Producers of Commercial Carboxylic Acids

Acid[a]	Price[b], $/kg 1976	1977	Major producers
acetic	0.353	0.375	Borden, Celanese, Tennessee Eastman, Monsanto, Union Carbide
acrylic	0.772–1.10	0.772–1.10	Celanese, Rohm & Haas, Union Carbide
benzoic	0.683–1.10	0.728	Kalama, Velsicol
butyric	0.705–1.10	1.10	Celanese, Tennessee Eastman, Texas Eastman
castor oil acids, dehydrated	1.43–1.10	1.85	NL Industries
coconut oil acids[c]			Armak, Ashland, Emery, Humko, Lever,
distilled	1.15	1.21	Procter & Gamble, Welch Holme &
double distilled	0.771	1.26	Clark
corn oil acids[d]	0.661	0.882	Ashland, Emery, Glyco, Humko
cottonseed oil acids[e]	0.617	0.705	Acme Hardesty, Armak, Ashland, Emery, Humko, Welch Holme & Clark
formic	0.353	0.353	Celanese, Union Carbide
isobutyric	0.948	1.65	Tennessee Eastman
lauric (dodecanoic)	0.728	1.41	Armak, Ashland, Emery, Humko, Procter & Gamble
linseed oil acids[f]			Ashland, Procter & Gamble
distilled	1.08	1.06	
water-white	1.12	1.04	
methacrylic	1.04	1.10	DuPont, Rohm & Haas
myristic (tetradecanoic)	0.661	1.61	Armak, Ashland, Emery, Humko
oleic			Armak, Ashland, Darling, Emery,
single-distilled (red)	0.617	0.860	Glyco, Hercules, Humko, Van Waters & Rogers, Welch Holme & Clark
double-distilled (white)	0.705	0.948	
palm oil acids[g]			Armak, Ashland, Emery, Michel,
single-distilled	0.772	0.772	Procter & Gamble
double-distilled	0.683	0.683	
palmitic, 90% (hexadecanoic)	0.640	0.992	Armak, Ashland, Emery, Glidden-Durkee, Glyco, Humko, Procter & Gamble
pelargonic (nonanoic)	1.08	1.08	Emery, Givaudin
phthalic anhydride	0.573	0.593	Allied, BASF Chevron, Exxon, Hooker, Koppers, Monsanto, Stepan, Union Carbide, U.S. Steel
propionic	0.441–1.10	0.441–1.10	Celanese, Tennessee Eastman, Monsanto, Union Carbide
ricinoleic	1.50–1.10	1.65	ICN Pharmaceuticals
salicylic	1.70	1.70	Dow, Monsanto, Sterling Drug, Tenneco
soybean oil acids[h]			Ashland, Emery, Glidden-Durkee, Glyco,
single-distilled	0.926	0.948	Humko, Petrochemicals, Procter &
double-distilled	0.728	0.970	Gamble, Welch Holme & Clark
stearic			Armak, Ashland, Darling, Emery, Glyco,
single-pressed	0.530	0.904	Humko, Petrochemicals, Procter &
double-pressed	0.551	0.926	Gamble, Van Waters & Rogers, Welch
triple-pressed	0.573	0.948	Holme & Clark
tall oil acids[i]			Arizona, Ashland, Emery, Glidden-Durkee,
2% or more rosin	0.485	0.485	Hercules, Union Camp, Welch Holme &
less than 2%	0.529	0.529	Clark, Westvaco
tallow fatty acids[j]			Armak, Ashland, Darling, Emery, Glyco,
technical	0.485	0.639	Humko, Petrochemicals, Procter &
hydrogenated	0.507	0.794	Gamble

Table 4. (continued)

Acid[a]	Price[b], $/kg 1976	1977	Major producers
terephthalic, dimethyl ester			Amoco, DuPont, Eastman, Hercules, Hoechst, Mobil
undecylenic (10-undecenoic)	2.82	2.82	NL Industries

[a] In alphabetical order as listed in the *Chemical Marketing Reporter*. The systematic name is given in parentheses as necessary to identify the acid.
[b] Price for tank car quantities when available in such amounts.
[c] Coconut oil has the average composition: 12:0, 48.2%; 14:0, 16.6%; 16:0, 8.0%; 18:0, 3.8%; 18:1, 5.0%; and 18:2, 2.5%.
[d] Corn oil has the average composition: 16:0, 11.5%; 18:0, 2.2%; 18:1, 26.6%; and 18:2, 58.7%.
[e] Cottonseed oil has the average composition: 16:0, 25.0%; 18:0, 2.8%; 18:1, 17.1%; and 18:2, 52.7%.
[f] Linseed oil has the average composition: 16:0, 5.5%; 18:0, 3.5%; 18:1, 19.1%; 18:2, 15.3%; and 18:3, 56.6%.
[g] Palm oil has the average composition: 16:0, 46.8%; 18:0, 3.8%; 18:1, 37.6%; and 18:2, 10.0%.
[h] Soybean oil has the average composition: 16.0, 10.5%; 18:0, 3.2%; 18:1, 22.3%; 18:2, 54.5%; and 18:3, 8.3%.
[i] Tall oil fatty acids have the average composition: 18:0, 3%; 18:1, 48%; 18:2, 37%; and conjugated 18:2, 5%.
[j] Tallow fatty acids have the average composition: 14:0, 3.0%; 16:0, 26.2%; 16:1, 2.6%; 18:0, 22.4%; 18:1, 43.1%; and 18:2, 1.4%.

Table 5. Production of Selected Fatty Acid Derivatives as Surface-Active Agents[a] (1000 Metric Tons)

Fatty acid derivatives	1965	1974	1975	1975 Unit value, $/kg
Anionic surface-active agents				
carboxylic acids and salts	407.5	380.4	333.2	0.838
sulfated acids, amines, and esters				0.926
sulfated alcohols				1.76
sulfated ethoxylated alcohols		106.1	126.5	0.661
sulfated fats and oils	13.4	14.3	14.2	0.661
Cationic surface-active agents				
acyclic amine oxides		18.5	17.6	2.65
polyamine condensates with carboxylic acids		11.2	11.4	1.34
amines and polyamines	22.0	51.3	29.9	1.41
acyclic quaternary ammonium salts	8.1	31.1	30.0	1.12
Nonionic surface-active agents				
carboxylic acid amides		40.4	37.3	1.12
carboxylic acid esters	66.6	118.9	100.9	1.32
nonbenzenoid ethers	86.2	274.7	255.3	0.728

[a] Source is Synthetic Organic Chemicals, United States Production and Sales, TC 206, US ITC 776 and 804, United States International Trade Commission.

Isobutyric acid is made from isobutyraldehyde, a major product in the synthesis of butyraldehyde. Certain butyrate and isobutyrate esters have pleasant, fruity odors and are useful as flavoring agents.

Butyraldehyde is also used to make 2-ethylhexanoic acid which has important applications, in metal salt form, for catalyzing oxidation and drying of paint films (see

Table 6. Production of Fatty Acid Derivatives as Plasticizers[a] (Metric Tons)

Fatty acid derivative	1965	1974	1975	1975 Unit value, $/kg
adipic acid esters, total	21,664	29,110	21,986	1.03
di(2-ethylhexyl) adipate	6,670	18,435	13,745	0.970
diisodecyl adipate	4,346	1,248	895	1.06
diisopropyl adipate			186	1.52
n-octyl n-decyl adipate	4,565		3,406	1.01
all other adipate esters	6,082	9,427	3,754	1.26
complex linear polyesters and polymeric plasticizers	18,264	28,614	17,404	1.48
epoxidized esters, total	34,430	69,818	44,251	1.10
epoxidized soybean oil	22,446	57,582	35,208	1.08
all other epoxidized esters	11,984	12,237	9,042	1.26
isopropyl myristate	668	2,267	1,125	1.19
isopropyl palmitate	457	3,377	766	1.26
oleic acid esters, total	4,296	5,728	3,951	0.970
butyl oleate	1,409	1,387	580	1.04
glyceryl trioleate	1,209		1,807	0.926
methyl oleate	1,193	1,298	1,221	0.860
propyl oleates	393	408	142	0.926
all other oleates	91	2,636	201	1.94
sebacic acid esters	4,719	3,477	2,546	
stearic acid esters, total	3,491	6,878	5,353	0.904
n-butyl stearate	1,745	3,524	2,704	0.750
all other stearates	1,745	3,354	2,649	1.06

[a] Source is Synthetic Organic Chemicals, United States Production and Sales, TC 206, US ITC 776 and 804, United States International Trade Commission.

Driers). In 1975 these included the following (production in metric tons): calcium (878), cobalt (1396), lead (802), zinc (269), zirconium (934), and other (1115) salts at prices ranging from $0.99/kg of lead salt to $2.62/kg of cobalt salt.

Sources for the higher fatty acids include coconut oil for lauric acid, inedible tallow and grease for stearic and oleic acid, tall oil for oleic and linoleic mixtures (see Fatty acids from tall oil), various vegetable oils (qv) for mixtures of unsaturated fatty acids, and castor oil (qv) for ricinoleic acid. Many of these sources are by-products of other industries: tallow and grease from the meat industry (see Meat products); tall oil (qv) from the kraft paper industry, unsaturated fatty acids from the soapstocks resulting from alkali refining of edible oils, and fish oil and acids from preparation of fish meal. Vegetable oil soapstocks are frequently available at a lower cost than the original oil or the fatty acids made from the oil: corn oil soapstock, $0.33/kg; cottonseed oil soapstock, $0.71/kg; soybean oil soapstock, $0.33–0.35/kg (see Soybeans and other seed proteins).

A major use for the higher fatty acids and their derivatives is as surface-active agents, synthetic detergents, and soaps (see Table 5). The nonbenzenoid surfactants of the total market, derived in a large part but not entirely from fatty acids, has increased from 34% in 1964 through 57% in 1965, and 72% in 1966–1972 to 82% in 1975. This dramatic turn-around in market pattern reflects environmental concerns and the superior degradability of straight-chain carbon compounds. Straight-chain compounds are available not only from fatty acids and their derivatives but also from the mixture of linear alcohols obtained by telomerization of ethylene (see Soap; Surfactants and detersive systems).

858   CARBOXYLIC ACIDS (ECONOMIC ASPECTS)

Table 7. Production of Selected Fatty Acid Derivatives[a]

Material	Production, metric tons 1965	1974	1975	1975 Unit value, $/kg
alcohols, C_{12} and higher, unmixed	167,158[b]	70,782	100,576	0.617
alcohols, mixtures, total		300,716	261,168	0.639
distearyl 3,3'-thiodipropionate		928	727	2.58
dodecyl mercaptan	5,693	10,462	7,698	1.19
erucamide	205	1,504	1,106	4.01
lauroyl chloride	4,321	1,035	624	
stearic acid salts[c], total	17,031	37,576	26,393	1.26
aluminum stearates, total	2,481	1,931	1,175	1.48
barium stearate		318	173	1.48
calcium stearate	5,749	20,692	15,132	1.08
magnesium stearate	1,075	2,718	1,784	1.46
zinc stearate	5,452	10,122	6,924	1.46
all other	2,274	1,794	1,206	1.59
all other salts of organic acids		76,841	63,650	1.28
sulfurized lard oil for lubricants	1,162	1,569	4,364	0.595

[a] Source is Synthetic Organic Chemicals. United States Production and Sales, TC 206, USITC 776 and 804, United States International Trade Commission.
[b] C_{10} alcohols and higher.
[c] Excluding sodium and potassium salts.

    Another major use for higher fatty acids is in alkyd resins (qv). Although vegetable oils are the main products used (195,000 t consumed in 1975, down from 420,000 t consumed in 1966), about 68,000 t of fatty acids are also consumed in making coatings. At least half of the fatty acids used in alkyds were tall oil fatty acids.
    Plasticizers (qv) constitute another large market for fatty acids and their derivatives (see Table 6). Epoxidized esters of tall oil fatty acids and particularly epoxidized soybean oil are valuable plasticizer/stabilizers for poly(vinyl chloride).
    Production for selected fatty acid derivatives is given in Table 7. Dodecyl mercaptan is an important polymerization regulator in the manufacture of synthetic rubber (see Elastomers, synthetic). Erucamide is an important antiblock agent for polyethylene film. About 300 t of fatty esters are used in flavor and perfume materials (see Flavors and spices; Perfumes).
    Several aromatic carboxylic acids are produced on a large scale (1976 production in thousand metric tons): benzoic (29) (qv), phthalic anhydride (319) (see Phthalic acids), salicylic (14) (qv), and terephthalic as the dimethyl ester (2,093) (see Phthalic acids). Phthalic anhydride markets are distributed as follows: plasticizers for poly(vinyl chloride), ca 50%; unsaturated polyester resins, ca 25%; and alkyd resins, ca 20%. Dimethyl terephthalate markets are confined almost exclusively to polyester textile fibers (88%) and film (7%) (see Polyester fibers; Polyesters).
    Carboxylic acids will continue to be a significant part of the chemical industry and the advanced industrial economy. However, as petroleum resources decline, more emphasis will be placed on the acids based on renewable resources. The extent of such a shift will depend on the marketplace, ie, on the relative prices for starting materials derived from fats and oils vs petrochemicals (5–6).

BIBLIOGRAPHY

"Economic Aspects" under "Fatty Acids" in *ECT* 2nd ed., Vol. 8, pp. 839–845 by E. H. Pryde, U. S. Department of Agriculture.

1. U.S. Pat. 3,989,747 (Nov. 2, 1976), J. H. Craddock, J. F. Roth, A. Hershman, and F. E. Paulik (to Monsanto Company).
2. U.S. Pat. 3,989,748 (Nov. 2, 1976), F. E. Paulik, A. Hershman, J. F. Roth, and J. H. Craddock (to Monsanto Company).
3. *Chem. Mark. Rep.,* (Mar. 8, 1976).
4. U.S. Pat. 4,012,526 (Mar. 15, 1977), D. L. Kensler, Jr., G. K. Kohn, and D. G. Walgenbach (to Chevron Research Company).
5. E. H. Pryde, L. E. Gast, E. N. Frankel, and K. D. Carlson, *Polym. Plast. Technol. Eng.* **7,** 1 (1976).
6. E. H. Pryde in D. S. Siegler, ed., *Crop Resources,* Academic Press, New York, 1977, pp. 25–45.

EVERETT H. PRYDE
United States Department of Agriculture

FATTY ACIDS FROM TALL OIL

Tall oil fatty acids (TOFA) primarily consist of oleic and linoleic acids and are obtained by the distillation of crude tall oil. Crude tall oil, a by-product of the kraft pulping process, is a mixture of fatty acids, rosin acids and unsaponifiables (1). These components are separated from one another by a series of distillations (2). Several grades of TOFA are available depending on rosin, unsaponifiable content, color, and color stability. Typical compositions of tall oil fatty acid products are shown in Table 1 (see Tall oil).

At present, tall oil fatty acids are produced by eight companies using twelve fractionating plants in the United States, one in Canada, thirteen in Europe, three in Japan, and at least one in Russia. Worldwide crude tall oil fractionating capacity in 1974 was estimated at slightly over 1.4 million metric tons and the fractionating

Table 1. Typical Fatty Acid Composition of Tall Oil Products

	CAS Registry No.	Crude, %	Crude fatty acid, %	<2% rosin in fatty acid, %	Distilled tall oil, %
Fatty acids (normalized to 100%)					
$C_{16}H_{32}O_2$	[57-10-3]	6.3	1.6	0.4	
$C_{17}H_{34}O_2$	[506-12-7]	1.5	0.7	0.7	
$C_{18}H_{36}O_2$	[57-11-4]	1.5	2.2	2.3	1.4
$C_{18}H_{34}O_2$	[112-80-1]	39.8	42.3	46.4	22.9
$C_{18}H_{32}O_2$	[60-33-3]	34.0	34.8	36.3	22.0
$C_{18}H_{32}O_2$ (isomers)	[26764-25-0]	10.2	12.7	10.3	24.6
$C_{19}H_{38}O_2$	[646-30-0]	1.2	1.1	1.1	1.1
$C_{20}H_{36}O_2$	[25448-01-5]	4.6	4.7	2.4	28.0
Rosin acids		40	7	1	30
Unsaps		8	2.5	1.5	2

Table 2. Production[a] and Prices[b] of TOFA

Year	Crude tall oil, 10^3 t/yr	TOFA, 10^3 t/yr	Approx average price of TOFA, ¢/kg
1965	533	135	17–20
1969	674	168	17–22
1973	719	187	20–30
1974	659	165	33–82
1975	557	133	46–75
1976	673	169	39–57

[a] Ref. 3.
[b] Ref. 4.

Table 3. TOFA Utilization, %

Use	1967	1975
intermediate chemicals	29.5	46.2
protective coatings	29.4	25.1
soaps and detergents	12.3	11.3
flotation	10.9	4.8
hard floor coverings	1.3	1.0
other uses	16.6	11.6
Total	100.0	100.0

capacity of the United States at just under 900,000 t. Domestic production of TOFA during the last twelve years is shown in Table 2. Production peaked in 1972 and showed a general decline in the years that followed; in 1975 production of TOFA was at a twelve-year low. Increased use of recycled paper, which produces no crude tall oil, has caused this decline in production of TOFA. The prices for TOFA remained fairly constant until 1974.

Tall oil fatty acids have many applications (Table 3). The most widely growing area of application is intermediate chemicals, which include dimer acids (qv) (5) and epoxidized TOFA esters (6). Other areas of significant use are protective coatings (7), soaps and detergents (8), and ore flotation (qv) (9).

BIBLIOGRAPHY

"Fatty Acids from Tall Oil" under "Fatty Acids" in *ECT* 2nd ed., Vol. 8, pp. 845–847, by W. C. Ault, U.S. Department of Agriculture.

1. L. G. Zachary, H. W. Bajok, and G. J. Eveline, eds., *Tall Oil and Its Uses*, Pulp Chemicals Association, New York, 1965.
2. U.S. Pat. 3,216,909 (Nov. 9, 1965), D. F. Bress (to Foster Wheeler Corp.).
3. Pulp Chemicals Association, 60 East 42 St., New York, N.Y. 10017.
4. Chemical Division, Sales Department, Union Camp Corporation, 1600 Valley Road, Wayne, N.J. 07470.
5. E. C. Leonard, ed., *The Dimer Acids*, Humko Sheffield Chemical Co., Memphis, Tenn., 1975.
6. D. B. S. Min and S. S. Chang, *J. Am. Oil Chem. Soc.* **49,** 675 (1972).

7. K. S. Ennor, *J. Oil Colour Chem. Assoc.* **51**, 485 (1968).
8. F. D. Snell and L. Reech, *J. Am. Oil Chem. Soc.* **27**, 289 (1950).
9. G. E. Agar, *J. Am. Oil Chem. Soc.* **44**, 396A (1967).

ROBERT W. JOHNSON, JR.
Union Camp Corporation

BRANCHED-CHAIN ACIDS

Branched-chain acids contain at least one branching alkyl group attached to the carbon chain which causes the acid to have different physical, and in some cases different chemical, properties than their corresponding straight-chain isomers. For example, stearic acid has a melting point of about 69°C, whereas isostearic acid has a melting point of about 5°C. Some properties of commercial branched-chain acids are shown in Table 1.

Manufacturing procedures for most branched-chain acids are well known. For example, oxo process acids are manufactured from branched-chain olefins using

Table 1. Properties and Prices of Branched-Chain Acids

Branched-chain acid (trivial name)	CAS Registry No.	Mol wt	bp, °C 101.3 kPa[a]	mp, °C	Approx price (1977), $/kg, bulk, fob	Major producers in the United States
2-methylpropanoic (isobutyric)	[79-31-2]	88	155	−46.1	0.871	American Hoechst Eastman
2-methylbutanoic (isopentanoic)	[116-53-0]	102	180.3	−48	0.772	Union Carbide
3-methylbutanoic isovaleric	[503-74-2]	102	175–176.5	ca −30	3.20	American Hoechst
2,2-dimethylpropanoic (neopentanoic)	[75-98-9]	102	163–165	34.4	1.37	Exxon
(isooctanoic)	[25103-52-0]	144	235–241	<−80	1.34	American Hoechst
2-ethylhexanoic	[149-57-5]	144	224–230	<−70	0.805	Eastman Union Carbide
(isononanoic)	[26896-18-4]	158	232–246	ca −70	0.992	
2,2-dimethyloctanoic (neodecanoic)	[129662-90-6]	172	147–150/20	<−40	0.948	Exxon
(isopalmitic)	[4669-02-7]	256		ca 10	2.65	Hexagon
(isostearic)	[2724-58-5]	284	192–204/5	ca 7	1.52	Emery Union Camp

[a] To convert kPa to mm·Hg, multiply by 7.5.

CARBOXYLIC ACIDS (BRANCHED-CHAIN)

carbonylation followed by oxidation (1) (see Oxo process). Neo-acids are prepared from selected olefins using carbon monoxide and an acid catalyst (2) (see Trialkylacetic acids). 2-Ethylhexanoic acid is manufactured by an aldol condensation of butyraldehyde followed by an oxidation of the resulting aldehyde (3). Isopalmitic acid is probably made by an aldol condensation of octanal. Isostearic acid is produced from the monomeric acids obtained in the dimerization of unsaturated C_{18} fatty acids (4).

Oxo process acids

$$CH_3CHCH=CH_2 + CO \xrightarrow{H_2} CH_3CHCH_2CH_2CHO \xrightarrow{[O]} CH_3CHCH_2CH_2CO_2H$$
$$\quad\;\; | \qquad\qquad\qquad\qquad\;\; | \qquad\qquad\qquad\qquad\;\; |$$
$$\quad CH_3 \qquad\qquad\qquad\qquad CH_3 \qquad\qquad\qquad\qquad CH_3$$
$$\text{(major product)}$$

Neo-acids

$$CH_3C=CH_2 \xrightarrow{H^+} CH_3\overset{+}{C}CH_3 \xrightarrow{CO} CH_3\overset{+}{C}—CO \xrightarrow{H_2O} CH_3CCO_2H$$
$$\quad\;\; | \qquad\qquad\quad | \qquad\qquad\quad\;\; | \qquad\qquad\qquad\;\; |$$
$$\quad CH_3 \qquad\qquad CH_3 \qquad\qquad\; CH_3 \qquad\qquad\qquad CH_3$$

with CH_3 groups on the top of the central carbons.

2-Ethylhexanoic via aldol condensation

$$CH_3CH_2CH_2CHO \xrightarrow{OH^-} CH_3CH_2CH_2CH—CHCHO \xrightarrow{-H_2O}$$
$$\qquad\qquad\qquad\qquad\qquad\qquad\quad\;\; | \qquad\;\; |$$
$$\qquad\qquad\qquad\qquad\qquad\qquad\; OH \quad CH_2CH_3$$

$$CH_3CH_2CH_2CH=CCHO \xrightarrow[\text{catalyst}]{H_2} CH_3CH_2CH_2CH_2CHCHO \xrightarrow{[O]} CH_3CH_2CH_2CH_2CHCO_2H$$
$$\qquad\qquad\qquad\;\; | \qquad\qquad\qquad\qquad\qquad\qquad\quad | \qquad\qquad\qquad\qquad\qquad\qquad\; |$$
$$\qquad\qquad\qquad CH_2CH_3 \qquad\qquad\qquad\qquad\qquad\quad CH_2CH_3 \qquad\qquad\qquad\qquad\qquad CH_2CH_3$$

Isostearic acid from monomer acids

$$\text{monomeric acids} \xrightarrow[\text{catalyst separation}]{H_2} \xrightarrow{\text{solvent}} \{\text{stearic acid + isostearic acid}\}$$

Branched-chain acids have a wide variety of industrial uses as paint driers (5), vinyl stabilizers (6), and cosmetic products (7). Cobalt and manganese salts of 2-ethylhexanoic acid and neodecanoic acid are used as driers for paint, varnishes, and enamels; lithium, magnesium, calcium, and aluminum salts of 2-ethylhexanoic acid are used in the formulation of greases and lubricants. Derivatives of isostearic acid have been used as pour point depressants in 2-cycle engine oils, as textile lubricants, and in cosmetic formulations.

The hazards of handling branched-chain acids are similar to those encountered with other aliphatic acids of the same molecular weight. Eye and skin contact as well as inhalation of vapors of the shorter chain acids should be avoided.

BIBLIOGRAPHY

"Branched-Chain Saturated Acids" under "Fatty Acids (Branched-Chain)" in *ECT* 1st ed., Vol. 6, pp. 259–262, by M. D. Reiner (in part), and J. A. Field (in part), Union Carbide and Carbon Corporation; "Branched-Chain Acids" under "Fatty Acids" in *ECT* 2nd ed., Vol. 8, pp. 849–850, by W. C. Ault, U.S. Department of Agriculture.

1. *Oxo-Synthesis Products,* Farbwerke Hoechst A.G., Frankfurt, F.R.G., 1971.
2. M. Fefer and A. J. Rutkowski, *J. Am. Oil Chem. Soc.* **45,** 5 (1968).
3. U.S. Pat. 2,779,808 (Jan. 29, 1957), A. C. Whitaker (to Gulf Research and Development Company).
4. U.S. Pat. 2,812,342 (Nov. 5, 1957), R. M. Peters (to Emery Industries).
5. A. Fisher, *J. Am. Oil Chem. Soc.* **43,** 469 (1966).
6. M. Fefer and G. Rubin, *Mod. Plast.*, (Apr. 1970).
7. *Ashland Chemicals Formulary for the Cosmetics Industry, Bulletin 1155,* Ashland Chemical Co., Columbus, Ohio.

ROBERT W. JOHNSON, JR.
Union Camp Corporation

TRIALKYLACETIC ACIDS

The trialkylacetic acids comprise a range of synthetically produced acids characterized by the structure:

$$\text{R}-\underset{\underset{\text{R}''}{|}}{\overset{\overset{\text{R}'}{|}}{\text{C}}}-\text{CO}_2\text{H}$$

No member of the series has been reported as naturally occurring. Commercial and research quantities of various members of the series have been manufactured in the United States and Europe since the early 1960s. These acids have been marketed as Versatic (Shell) and as neo-acids (Exxon). Some of the acids have been produced from mixed isomer feedstocks, resulting in products that are not single compounds. Such products include Versatic 911 and 1519. Since 1975, manufacture has been restricted to two of the more commercially useful products. These are (*1*) the first member of the series, trimethylacetic acid, or 2,2-dimethylpropanoic acid (pivalic or neopentanoic acid, or Versatic 5), and (*2*) the C_{10} acid, neodecanoic acid (Versatic 10).

Three of the largest individual outlets for trialkylacetic acids are for the manufacture of the vinyl, glycidyl, as well as peroxy esters.

PIVALIC ACID

Physical Properties

Pivalic acid [75-98-9] is a solid at room temperature. Essentially dry pivalic acid can be obtained by drying over molecular sieves immediately before use. The commercially available product is usually more than 99% pure.

$$(CH_3)_3CCO_2H$$
pivalic acid

The physical properties of a typical grade of pivalic acid are given in Table 1.

Chemical Properties

The reactions of pivalic acid are characterized by steric hindrance at the carboxyl group caused by the alpha substituents. Although reactions proceed less readily than with straight chain acids, the derivatives prepared are correspondingly more resistant to hydrolysis and oxidation.

Chlorination. Chlorination of pivalic acid produces mono- or dichloropivalic acids. Monochloropivalic acid [13511-38-1] may be prepared in high yield by the reaction of pivalic acid with sulfuryl chloride catalyzed by benzoyl peroxide (1).

Monochloropivalic acid can also be prepared by photochlorination at reflux temperature (2). Monochlorination of pivalic acid followed by amination yields aminopivalic acid [19036-43-2]. Dichlorination yields a mixture of products.

Esterification. Pivalic acid can be esterified by conventional techniques such as sulfuric acid catalysis, although higher than normal levels are needed. Methyl pivalate can be prepared by the reaction of methanol with pivalic acid in the presence of *p*-toluene sulfonic acid. Vinyl pivalate is prepared in high yields by treating pivalic acid with acetylene in the presence of a zinc pivalate catalyst and a Lewis acid (eg, $ZnCl_2$) as a cocatalyst (3). Direct esterification of pivalic acid with ethylene glycol produces both glycol monopivalate and glycol dipivalate. The preparation and properties of a range of diesters based on trialkylacetic acids and glycols are described by Lederle (4).

An important ester produced from pivalic acid is *tert*-butyl peroxypivalate, a compound used in polymer manufacture.

Table 1. Typical Physical Properties of Commercially Available Pivalic Acid

Property	Value
mp, °C	33.5
bp, °C	163–164
acid value, mg KOH/g	544
(theoretical for $C_5H_{10}O_2$ = 549)	
color, Pt–Co (Hazen) of molten material	50
sp gr, 40/20°C	0.905
water, wt %	0.12
flash point, closed cup, °C	64
refractive index, n_D^{50}	1.388

Reduction. Pivalic acid can be reduced to 2,2-dimethylpropanol by treatment with LiAlH$_4$ in ether (5).

Oxidation. 2,2,5,5-Tetramethyladipic acid may be produced via an oxidative coupling reaction of pivalic acid by generating hydroxyl free radicals from H$_2$O$_2$ and an oxidizable metal ion (eg, Fe^{2+}) in the presence of hydrogen and a noble metal hydrogenation catalyst at a temperature of 40–42°C (6).

Preparation of Sulfur-Containing Derivatives. The cyclic anhydride of sulfopivalic acid is obtained by treating pivalic acid with chlorine and sulfur dioxide with uv radiation at 20°C, scavenging with nitrogen and heating the solution for 3 h at 70–80°C (7). A large number of sulfur-containing derivatives of pivalic acid have been prepared using pivalolactone as the starting material.

Other Reactions. Alkaline hydrolysis of monochloropivalic acid with excess caustic soda solution produces, under reflux, hydroxypivalic acid [4835-90-9]; with excess ethyl alcohol, the ethyl ester is obtained. Lactones can be produced from hydroxypivalic acid or directly from the monochloropivalic acid. Abstraction of sodium chloride from the sodium salt produces the lactone.

$$\underset{\underset{H_2C-Cl}{|}}{\overset{\overset{CH_3}{|}}{CH_3-C-COOH}} \longrightarrow \underset{\underset{H_2C-Cl}{|}}{\overset{\overset{CH_3}{|}}{H_3C-C-COONa}} \longrightarrow \underset{\underset{H_2C-O}{|}}{\overset{\overset{CH_3}{|}}{H_3C-C-C=O}}$$

Pivaloyl chloride is prepared by reaction of pivalic acid with phosphorus trichloride or pentachloride, thionyl chloride, or phosgene. Pivalic anhydride [1538-75-6], another useful intermediate, can be made by the reaction of pivalic acid with acetic anhydride.

Polymerization. Polymerization of the vinyl ester results in a polymer having greater hydrolytic and solvent stability as well as a higher softening point than poly(vinyl acetate). Copolymers of vinyl pivalate and trichloroethylene, vinyl acetate, methyl methacrylate, and vinyl chloride have also been reported (8).

Manufacture

Tertiary carboxylic acids were first manufactured on an industrial scale in the early 1960s (9–11). The chemistry underlying the industrial processes was based upon the carboxylation of olefins with carbon monoxide and water in the presence of strong acid catalysts (12). Pivalic acid can be manufactured on an industrial scale from isobutylene or diisobutylene with carbon monoxide and water at a temperature below 100°C and elevated CO pressure (see Oxo process).

The reaction is thought to proceed via a carbenium and acylcarbenium ion mechanism:

Tertiary carbenium ions are formed preferentially to secondary or primary ions. Consequently almost no 3-methylbutyric acid is formed in this reaction. The reaction mechanism is preceded by a depolymerization step when diisobutylene is used as the feedstock. Side reactions may occur depending on the reaction conditions applied (temperature, CO pressure, catalyst composition).

Economic Aspects and Shipment

Annual worldwide production of pivalic acid is estimated at a few thousand metric tons; prices for the drummed product averaged $2000/t in 1978.

Pivalic acid is shipped in heated tank cars, heated tank trucks, and drums.

Health and Safety Factors

Pivalic acid is only slightly toxic by ingestion (oral LD_{50} in rats is 0.9–1.9 g/kg) or skin absorption (dermal LD_{50} in rats is 1.9–3.6 g/kg), and because of its low volatility is relatively harmless when inhaled at normal ambient temperature (ie, around 20°C). The principal handling hazard is from skin and eye irritation. At elevated temperatures the vapors will be irritant. Any such contact should therefore be avoided.

If strong irritation occurs following eye contact, the eyes should be immediately flushed with water, and medical advice obtained. Contaminated skin should be washed with soap and water. Pivalic acid will burn although it is not in the flammable range; fire should be curbed with carbon dioxide or dry sand.

Uses

Polymers and Resins. tert-Butyl peroxypivalate is suitable as a free radical initiator for the polymerization of vinyl chloride at about 50°C (13–21). The peroxypivalate can also be used in chlorination of poly(vinyl chloride) (22) and curing of unsaturated polyester resins (23) (see Vinyl polymers; Polyesters).

Polymers can be grafted with other materials in the presence of tert-butyl peroxypivalate. Some properties of poly(vinyl pivalate) and of the copolymer of vinyl pivalate with vinyl acetate (both made in bulk polymerizations) are given in Table 2. Properties of poly(vinyl acetate), polystyrene, and poly(methyl methacrylate) are listed for comparison.

Pharmaceuticals. Pivalic esters have advantages in slow release drugs because they are hydrolyzed slowly (see Pharmaceuticals, controlled release).

A concentrated solution of prednisolone pivalate may replace hydrocortisone to relieve pain of rheumatoid arthritis. It is claimed to be more effective, more soluble, and to have minimal side effects (24).

Benzoyl pivalate has been used in the treatment of pulmonary tuberculosis to increase capillary stability (25). Pivalic acid has exhibited anticonvulsive properties when injected into mice and rats (26). Application of dexamethasone pivalate has effected a clear improvement in the conventional clinical parameters of rheumatoid arthritis (27). Psoriasis has been treated with a paste containing flumetasone pivalate as the active ingredient (28).

Pivaloyl chloride is employed in the production of the antibiotic pivampicillin (see Antibiotics, β-lactams).

Table 2. Some Properties of Poly(vinyl Pivalate) and of a Copolymer of Vinyl Pivalate With Vinyl Acetate

Property	Poly(vinyl pivalate)	50/50 Copolymer of poly(vinyl pivalate) and poly(vinyl acetate)	50/50 Copolymer of poly(vinyl pivalate) and poly(vinyl maleate)	Poly-(vinyl acetate)	Poly-styrene	Poly-(methyl methacrylate)
molding time, min	3	3	3	3	3	
molding temp, °C	160	160	160	120	180	
Vicat softening point, °C						
at 0.1 mm penetration	73	54	60	35	95	112
at 0.5 mm penetration	76	57	66	40	99	116
at 1.0 mm penetration	79	58	69	42	100	120
Barcol hardness, 23°C	12	10	5	5	28	48
tensile strength, kg/cm²	295	383		136	345	
elongation at rupture, %	2	5		415	2	

Cosmetic Preparations. Polymers containing allyl pivalate are used as hair sprays and wave sets (29). The hydrobromide and hydrochloride salts of pivalic esters of scopolamines and atropine can be used in antiperspirants (30–31) (see Cosmetics).

Fuels, Lubricants, and Transmission Fluids. Automotive brake fluids have been formulated from polyethylene glycol monopivalates (32–33). Pivalic acid and its lithium salts can be used in mixtures to increase the octane number of motor fuel compositions (34–35) (see Gasoline). *tert*-Butyl pivalate increases the octane number of a gasoline, either leaded or unleaded, when 0.5–0.75% volume is added to the mixture (36).

Agricultural Applications. Nematodes are controlled by contacting an infested environment with a copper or mercury salt of pivalic acid on vermiculite (37) (see Poisons, economic). Pivalic reputedly has phytotoxic properties and can be used as a broad-base, post-emergent herbicide (38). Pivalic acid also attracts the olive fly (39); and its octyl and nonyl esters are highly effective attractants to yellow jacket wasps (40) (see Herbicides; Insect control technology).

Metal Extractants. The solvent extraction of copper(II) with pivalic acid has been investigated (41). Methyl and isopropyl pivalate are used for selectively separating by liquid–liquid extraction (qv) one or more metals from an aqueous medium containing a mixture of metal salts, eg, iron(III) from copper(II) chlorides (42).

Miscellaneous Applications. Some pivalates, eg, tin pivalate, have antifungal properties and, accordingly, have been used as wood preservatives. Ammonium pivalate is effective in preventing the reversible isomerization between 3,4-dichloro-1-butene and 1,4-dichloro-2-butene (43). The oxidation of toluene by cobalt perchlorate is promoted by pivalic acid which is oxidized at the same time (44). The oxidation of cumene (qv) into cumene hydroperoxide by molecular oxygen is greatly accelerated by the presence of alkali metal salts of various acids including pivalic acid (45). Aqueous dispersions of polymers containing vinyl pivalate can be used to impregnate cement mixtures in order to improve waterproofness, alkali-resistance, weathering resistance, and mechanical strength (46).

CARBOXYLIC ACIDS (TRIALKYLACETIC)

Barium pivalate is employed as a coating of phosphors to improve lumen output and maintenance in fluorescent lamps (47). Cigarette filters that are effective in removing volatile acidic compounds such as hydrogen sulfide and hydrogen cyanide from tobacco smoke may contain up to 50% zinc pivalate (48).

C_{10} TRIALKYLACETIC ACIDS

Physical Properties

The C_{10} trialkylacetic acids are mobile liquids at room temperature. Their physical properties are given in Table 3.

Chemical Properties

The C_{10} trialkylacetic acids are a mixture of a large number (at least 27) of highly branched isomers of monocarboxylic acids, at least 98% of which have the tertiary structure represented on p. 863.

Analyses have shown that at least one of the three R groups is methyl. Like pivalic acid, reactions proceed less readily with the C_{10} trialkylacetic acids than with straight chain acids, but the derivatives formed are more stable.

Manufacture

The C_{10} acids are manufactured from C_9 branched-chain olefin feedstocks with the same process and catalyst systems as those described for pivalic acid. Because of the carbenium ion mechanism mentioned, these synthetic acids are composed of a mixture of highly branched (almost all tertiary) isomers of C_{10} acids. Side reactions

Table 3. Typical Physical Properties of Commercially Available C_{10} Trialkylacetic Acids[a]

Property	Versatic 10 [52627-73-3]	Neodecanoic [26896-20-8]	Neodecanoic technical grade
mp, °C	<−40	<−40	<−40
acid value, mg KOH/g	320	320	295–320
theoretical for $C_{10}H_{20}O_2$ = 326			
unsaponifiable matter, %	0.4		
color, Pt–Co (Hazen)	70	75	2[b]
density at 20°C, g/cm³	0.910		
sp gr, 20/20°C		0.913	0.913
kinematic viscosity, mm²/s (= cSt)			
20°C	45	35.7	
50°C	10		
60°C		7	
water, wt %	0.03	0.1	1.0
flash point, closed cup, °C	129		
flash point, open cup, °C		152	88
refractive index, n_D^{20}		1.4385	
ionization constant, $K_a \times 10^{-6}$			
25°C		4.2	

[a] Refs. 49–50.
[b] Gardner scale.

that may occur are disproportionation of olefins resulting in the formation of higher or lower acids, and dimerization or polymerization.

Economic Aspects and Shipment

The C_{10} acids are produced in larger volume than pivalic acid. Annual worldwide production reaches several tens of thousands of metric tons. Prices in 1978 were in the $600–1000/t range.

The C_{10} acids are shipped in bulk sea vessel, tank cars, tank trucks, and drums.

Toxicology

The C_{10} trialkylacetic acids have toxicities similar to pivalic acid: oral LD_{50} in rats 2.1 g/kg, dermal LD_{50} in rats 3.6 g/kg. The same health and safety factors apply.

Uses

Surface-Coating Industries. The vinyl and glycidyl esters of C_{10} trialkylacetic acids have wide application in the surface-coating industries. The metal salts of C_{10} acids are very effective and widely used paint driers (qv). As such, they have to some extent replaced the corresponding metal naphthenates because of their lighter color, low odor, and shorter drying times (51).

Vinyl Stabilizers. The C_{10} acids are also widely employed as metal carriers in vinyl stabilizer production. Most vinyl stabilizers are packages containing organic salts of barium, cadmium, and zinc. Significant advantages have been reported in the use of the C_{10} trialkylacetic acids over 2-ethylhexanoic acid as the carrier for cadmium and zinc salts (52) (see Heat stabilizers).

Metal Solvent Extraction. Together with other carboxylic acids, the C_{10} acids are a cation exchange reagent for metal solvent extraction, and extract metal ions in the order of their basicity. Thus the extraction of metal ions with carboxylic acids is in the same order as the metal hydroxides precipitated on neutralization. For the C_{10} trialkylacetic acids this is $Fe(III) < Al(III) < Cu(II), Zn(II) < Be(II) < Ni(II), Co(II) < Mn(II) < Ca(II), Mg(II)$.

Glycidyl and Vinyl Esters

Physical properties of the commercially available glycidyl and vinyl esters are given in Table 4.

The most important reaction of the glycidyl ester is with carboxylic acids at 100–150°C, giving diacyl-substituted glycerides, in the production of alkyds (see Alkyd resins) and acrylic resins. That of the vinyl ester is the persulfate-initiated emulsion copolymerization with vinyl acetate in paint latex manufacture.

The glycidyl ester, Cardura E10, is manufactured by the reaction of Versatic 10 with epichlorohydrin under alkaline conditions followed by purification. The product is shipped in bulk or mild steel drums; it must be protected from contact with atmospheric moisture in storage.

870 CARBOXYLIC ACIDS (TRIALKYLACETIC)

Table 4. Typical Physical Properties of Cardura E10 and VeoVa 10

Property	Cardura E10	VeoVa 10
initial bp at 101 kPa (1 atm), °C	249	210
freezing point, °C	−60	−20
density at 25°C, g/cm^3	0.959	0.8745
viscosity at 25°C, mPa·s (= cP)	7.13	2.51
vapor pressure at 100°C, kPa (mmHg)	0.33(2.5)	2.82(21.2)
150°C, kPa (mmHg)	2.29(17.2)	18.67(140)
flash point, closed cup, °C	126	75
solubility in water at 20°C, %	0 01	<0.1

The vinyl ester, VeoVa 10, is the product of the reaction of Versatic 10 with acetylene; it is supplied, stabilized with 5 ppm of the monoethyl ether of hydroquinone, in bulk or lacquer-lined steel drums. Stainless steel or low copper-grade aluminum tanks are recommended for bulk storage.

Cardura E10 (Shell) is produced at Pernis, The Netherlands, at an annual rate of several thousand metric tons; prices in 1978 were $1.58–1.81/kg.

VeoVa 10 (Shell) is also produced at Pernis and a second plant was brought on stream at Moerdijk, The Netherlands, in 1976. Annual production is several tens of thousands of metric tons and prices in 1978 were $1.13–1.36/kg.

BIBLIOGRAPHY

"Trialkylacetic" under "Fatty Acids" in *ECT* 2nd ed., Vol. 8, pp. 851–856, by E. J. Wickson, Enjay Laboratories.

1. M. S. Kharasch and H. C. Brown, *J. Am. Chem. Soc.* **62,** 925 (1940).
2. A. Bruylants and co-workers, *Bull. Soc. Chim. Belg.* **61,** 366 (1952).
3. U.S. Pat. 3,455,998 (July 15, 1969), H. J. Arpe (to Shell Oil Co.).
4. H. F. Lederle, *J. Chem. Eng. Data* **15**(1), 193 (1970).
5. R. F. Nystrom and W. G. Brown, *J. Am. Chem. Soc.* **69,** 2548 (1947).
6. U.S. Pat. 3,076,846 (Feb. 5, 1963), W. R. McLellan (to E. I. du Pont de Nemours & Co., Inc.).
7. Brit. Pat. 1,012,325 (Dec. 8, 1965), (to Schering A.G.).
8. U.S. Pat. 2,381,338 (Aug. 7, 1945), W. R. Cornthwaite and N. D. Scott (to E. I. du Pont de Nemours & Co., Inc.).
9. *Chem. Trade J. Chem. Eng.* **153,** 890 (1963).
10. Neth. Pat. 100,296 (Jan. 15, 1962), M. J. Waale and J. M. Vos (to Shell Internationale Research Mij N.V.).
11. E. J. Wickson and R. R. Moore, *Hydrocarbon Process.* **43**(11), 185 (1964).
12. K. E. Moeller, *Brennst. Chem.* **45**(5), 129 (1964); **45**(7), 209 (1964).
13. *Brochure No. 1/74 E.0266*, Noury en van der Lande N.V., Deventer, Holland.
14. U.S. Pat. 3,558,578 (Jan. 26, 1971), P. Kraft and co-workers (to Stauffer Chemical Co.).
15. Ger. Offen. 1,915,386 (Oct. 1, 1970), E. Gulbins and co-workers (to Badische Anilin- und Soda-Fabrik A.G.).
16. Ger. Offen. 1,940,475 (Feb. 18, 1971), J. Bauer and co-workers (to Wacker-Chemie A.G.).
17. Ger. Offen. 2,037,043 (Feb. 10, 1972), O. Schott and co-workers (to Badische Anilin- und Soda-Fabrik A.G.).
18. Ger. Offen. 2,039,010 (Mar. 11, 1971), F. A. Cox and D. R. Glenn (to Goodyear Tire and Rubber Co.).
19. Ger. Offen. 2,062,130 (Feb. 10, 1972), L. Feiler and S. F. Gelman (to Stauffer Chemical Co.).
20. Ger. Offen. 2,125,015 (Dec. 2, 1971), C. Lambling and J. Boissel (to Produits Chimiques Pechiney, St. Gobain).

21. U.S. Pat. 3,420,807 (Jan. 7, 1969), J. B. Harrison and co-workers (to Wallace & Tiernan Inc.).
22. Fr. Pat. 1,511,458 (Jan. 26, 1968), R. Rettoce and G. Gatta (to Montecatini Edison S.p.A.).
23. H. Schwarzer and H. Twittenhoff, *Kunststoffe* **59**(12), 981 (1969).
24. N. Cardoe, *Proc. R. Soc. Med.* **52**, 1109 (1959).
25. G. Rossini, G. Piazza, and G. Nessig, *Minerva Med.* **49**, 1620 (1958).
26. G. Carraz, *Agressologie* **8**(1), 13 (1967).
27. J. L. Kalliomaki, *Curr. Ther. Res., Clin. Exp.* **9**(7), 327 (1967).
28. Ger. Offen. 1,817,288 (Aug. 7, 1969), E. G. Weirich (to CIBA Ltd.).
29. Ger. Offen. 2,513,808 (Oct. 9, 1975), C. Papantoniou and J. C. Grognet (to Oreal SA).
30. Brit. Pat. 940,279 (Oct. 30, 1963), (to Procter & Gamble Ltd.).
31. F. S. Kilmer Macmillan and co-workers, *J. Invest. Dermatol.* **43**, 363 (1964).
32. Brit. Pat. 1,319,695 (June 6, 1973), (to Maruzen Oil Co. Ltd.).
33. U.S. Pat. 3,799,878 (Mar. 26, 1974), Y. Kanatsu and co-workers (to Maruzen Oil Co. Ltd.).
34. Fr. Pat. 2,058,502 (July 2, 1971), A. Duval and co-workers (to L'Enterprise de Recherches et d'Activites Petrolières (ELF)).
35. U.S. Pat. 2,935,973 (May 10, 1960), C. A. Sendy and J. H. Werntz (to E. I. du Pont de Nemours & Co., Inc.).
36. Fr. Pat. 2,050,841 (May 7, 1971), A. Duval and co-workers (to L'Enterprise de Recherches et d'Activites Petrolières (ELF)).
37. U.S. Pat. 3,558,782 (Jan. 26, 1971), A. J. Rutkowski (to Esso Research & Engineering Co.).
38. Brit. Pat. 1,243,987 (Aug. 25, 1971), A. J. Rutkowski (to Esso Research & Engineering Co.).
39. G. N. Stavrakis and R. H. Wright, *Can. Entomol.* **106**(3), 333 (1974).
40. T. P. McGovern and co-workers, *J. Econ. Entomol.* **63**(5), 1534 (1970).
41. W. J. Haffenden and G. J. Lawson, *J. Inorg. Nucl. Chem.* **29**(4), 1133 (1967).
42. Neth. Appl. 68, 15,902 (May 11, 1970), (to Shell Internationale Research Mij N.V.).
43. Jpn. Pat. 68, 09,729 (Apr. 22, 1968), R. Kabayishi and co-workers (to Electro Chemical Industrial Co., Ltd.).
44. T. A. Cooper and co-workers, *J. Chem. Soc., B, Phys. Org.* (9), 793 (1966).
45. Fr. Addn. 76,585 (Feb. 22, 1962), (to Société des Usines Chimiques Rhône-Poulenc).
46. Jpn. Pat. 74, 98,420 (Sept. 18, 1974), M. Katata and co-workers (to Sumitomo Chemical Co., Ltd.).
47. U.S. Pat. 2,951,767 (Sept. 6, 1960), N. C. Beese (to Westinghouse Electric Corp.).
48. U.S. Pat. 3,403,690 (Oct. 1, 1968), H. G. Horswell and T. W. C. Tolman (to Brown and Williamson Tobacco Corp.).
49. Shell International Chemical Co., Ltd., *Data Sheet 2D 041, Versatic 10,* Aug. 1977.
50. Exxon Chemical Co., *Neo-Acids,* CHE-75-1679.
51. M. Fefer and A. Lauer, *J. Am. Oil Chem. Soc.* **45**, 479 (1968).
52. M. Fefer and G. Rubin, *Mod. Plast.* **47**(4), 178 (1970).

J. W. Parker
M. P. Ingham
R. J. Turner
J. H. Woode
Shell International Chemical Co., Ltd.

CARBOXYLIC ACIDS, POLYBASIC ACIDS. See Dimer acids.

CARBURIZING. See Metal surface treatment.

CARCINOGENS. See Industrial hygiene and toxicology.

CARDBOARD. See Paper.

CARDIOVASCULAR AGENTS

Antihypertensive agents, 872
Cardiac glycosides, 894
Antiarrhythmic agents, 902
Antiatherosclerotic agents, 909
Antianginal agents, 915
Peripheral vasodilators, 920

Cardiovascular agents are drugs that affect the circulatory system of the body. This complex network consists of the heart and a system of elastic blood vessels (the aorta, arteries, capillaries, and veins) that carry the blood from the heart to and from the various organs (1). Cardiovascular agents elicit their responses by actions on the heart and blood vessels, but may in some instances principally affect the autonomic and central nervous systems or the kidneys. Their mechanisms of action are discussed in the following sections under their principal uses.

ANTIHYPERTENSIVE AGENTS

Hypertension (diastolic blood pressure > 12.7 kPa [95 mm Hg]) is a disease known to affect 10–15% of the adult population of the United States and is thought to have been identified only in 50–75% of afflicted persons (2). Therapy by drug treatment, surgery, or both, is aimed at reducing arterial pressure as close to normal as possible (diastolic blood pressure < 11.3 kPa [85 mm Hg]) in order to prevent or postpone cardiovascular complications, ameliorate congestive heart failure, reverse retinopathy, and prolong life. The Framingham study (3–4), among others, showed that elevated blood pressure and the incidence of myocardial infarction and stroke are directly related. Furthermore, the Veterans Administration Cooperative Study (5–6) has demonstrated that drug therapy is effective in reducing morbidity and mortality.

Over 90% of hypertension cases are of unknown etiology (essential hypertension). The known causes of the other 5–10% of hypertension cases are renal arterial disease, coarctation of the aorta, pheochromocytoma and primary aldosteronism; this is most likely among children and adults who become hypertensive after the age of fifty. Some of these conditions are relieved or corrected by surgery. The etiology of essential or primary hypertension, the most likely cause of elevated blood pressure in a 40 year old, is as yet unidentified and generally requires lifelong drug therapy. The principal physiological abnormality in essential hypertension is increased total peripheral resistance, yet no defect of the effector organ has been found.

Physiology and Biochemistry of Blood Pressure Regulation. The level of arterial blood pressure is nearly constant and maintained despite the variation in requirements for blood flow through vascular beds such as kidneys, skeletal muscle and gastrointestinal tract, exercise, and postural changes. The maintenance of mean arterial pressure is a result of homeostatic mechanisms that control cardiac output and peripheral resistance. The heart's pumping action must overcome the resistance of peripheral arterioles (vasomotor tone) to the blood flow into the arterial circuit and to the capillaries. Blood pressure is directly proportional to the product of the flow per unit time and the resistance to flow. Blood pressure is also controlled by cardiac output which is dependent on blood volume. Various metabolic and hormonal factors determine the vasomotor tone: the sympathetic nervous system, the adrenal catecholamines, and angiotensin II are known vasoconstricting factors; the parasympathetic nervous system, the kallikrein–kinin system (7a) and the prostaglandins (qv) (7b–7c) are known vasodilating factors.

Sympathetic and Parasympathetic Nervous Systems. *The Peripheral Nervous System.* The arterioles are normally controlled by the sympathetic nervous system. The short, preganglionic fibers from the thoracolumbar outflow of the spinal column synapse in sympathetic ganglia and release acetylcholine (1) at the cholinergic synapse (8). The postganglionic fibers carry the nerve impulses to adrenergic nerve terminals in the vascular smooth muscle of the arterioles which release norepinephrine. The receptors on smooth muscle responding to the action of catecholamines, are termed α and β receptors (9). Activation of α-adrenergic receptors in the blood vessels by catecholamines causes constriction; activation of β-adrenergic receptors causes relaxation. The most potent excitatory neurotransmitter, norepinephrine (7) (10), is synthesized from tyrosine (2) in the sympathetic neurons by the sequence shown in Figure 1. Hydroxylation of tyrosine by tyrosine hydroxylase (11) to levodopa (3) is followed by decarboxylation by dopa decarboxylase (12) to dopamine (5). Dopamine, after being taken up in storage granules, is oxidized by dopamine β-oxidase to norepinephrine (13). Norepinephrine, released from the storage granules, enters the synaptic cleft where it mainly binds to α-receptors and causes contraction of vascular smooth muscle (see Neuroregulators). The effect of this is to increase peripheral resistance and blood pressure. The concentration of norepinephrine available for this effect is governed by the rates of synthesis from tyrosine, metabolism by monoamine oxidase (MAO) to 3,4-dihydroxymandelic acid (9), metabolism by catechol-O-methyl transferase (COMT) to normetanephrine (8), and transfer in and out of storage granules (14). It is thought that the release of norepinephrine by nerve impulses is inhibited by a feedback loop involving α-adrenergic receptors (15). The loop is interrupted when α-receptor antagonists block norepinephrine and other α-agonists from the receptor sites, thereby increasing the release of norepinephrine. Epinephrine (10) (qv) is formed from norepinephrine by phenylethanolamine N-methyltransferase catalysis in the adrenal medulla (16) (see Epinephrine). The alternative biosynthetic pathway, tyrosine → tyramine (4) → octopamine (6) → norepinephrine, is of lesser importance.

$$(CH_3)_3\overset{+}{N}CH_2CH_2OCOCH_3 \ OH^-$$
(1) acetylcholine

β-Adrenergic receptors involved in the cardiovascular system are known to exist in arteries, arterioles, and the heart, and it has been well-established that β-adrenergic blocking agents lower blood pressure (17), but the mechanism is still undecided. Proposed mechanisms are: effects on central β-adrenoceptors; resetting of baroreceptors; inhibition of uptake of norepinephrine; reductions in cardiac output, plasma volume, and venous return; and inhibition of renin release. One of the effects that may contribute to antihypertensive activity is interference with adrenergic neuronal

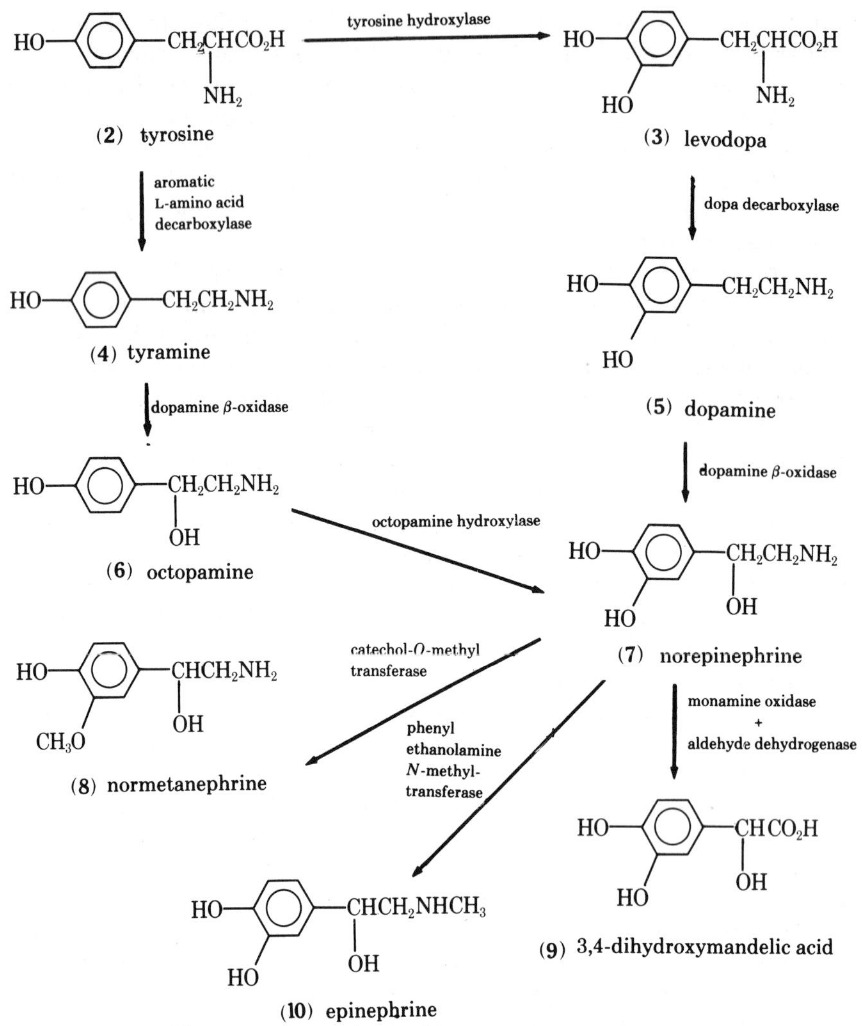

Figure 1. Catecholamine biosynthesis and metabolism.

function. It has been suggested (18) that two different presynaptic mechanisms are involved in norepinephrine release. Prejunctional β-receptors, activated by low concentrations of norepinephrine in the synaptic cleft, release increased amounts of norepinephrine per nerve stimulus. Norepinephrine release by prejunctional α-receptors is inhibited by high concentrations of norepinephrine by a negative feedback mechanism. If synaptic neurons of blood vessels also have prejunctional β-receptors, their blockade could inhibit release of norepinephrine and, consequently, reduce vasomotor tone.

The Central Nervous System (CNS). Adrenergic mechanisms operating in sympathetic nervous system control of arterial blood pressure are also found in the CNS (19–20). In addition to norepinephrine (7) (21), other neurotransmitters involved in central control are epinephrine (10) (22–23), dopamine (5) (20), and serotonin (11) (24–25). Histamine (12) (26) may also play a role (see Neuroregulators). Arterial blood pressure is controlled reflexively by central adrenergic nerves actuated by the arterial

(11) serotonin
(5-hydroxytryptamine)

(12) histamine

baroreceptors located mainly in the carotid sinus and aortic arch. When blood pressure is increased the baroreceptor reflexes cause a decreased sympathetic tone and increased vagal tone. However, the continued elevation of blood pressure in hypertensive states is not sensed by the baroreceptors. They appear to be reset and, acting as if pressure were normal, fail to respond (27).

Hypothalamic regulation of blood pressure is thought to be one of the functions of adrenergic neurons that originate in the locus coerulus (28) and terminate in the posterior hypothalamus (29). Stimulation of central α-adrenergic receptors is considered to be one of the mechanisms whereby antihypertensive drugs elicit their effects (30–32), and these α-receptors are located in the pontomedullary area (nucleus tractus solitarius) (33) and in the posterior hypothalamic area (34–35). As in the sympathetic nervous system (15), control of blood pressure has been shown to occur by a feedback mechanism present in the hypothalamus that involves presynaptic α-receptor-mediated release of norepinephrine (34).

Where it is well-accepted that α-adrenergic receptors are involved in the central regulation of blood pressure, it is, however, not known with certainty that β-adrenergic receptors are involved or if they even exist in the CNS. Evidence has been provided for the central action of β-blockers (36) that are known to be effective in lowering blood pressure. They are thought to cause a reduction in sympathetic nervous tone (37–38). β-Receptors have been postulated to exist in the CNS (39–40); blockade of β-like receptors in the CNS causes reduction in blood pressure (41). It has been suggested that β-receptors in the medullary region may control heart rate (42).

The False Neurotransmitter Hypothesis. The antihypertensive agent, methyldopa (13), a potent inhibitor of dopa decarboxylase *in vitro* (43) and *in vivo* (44), depletes norepinephrine from its storage sites; this was first assumed to be the mechanism by which it lowered blood pressure. However, the prolonged lowering of blood pressure produced by methyldopa seems to be independent of the transient inhibition of dopa decarboxylase (45). In addition, other powerful inhibitors of dopa decarboxylase fail to lower blood pressure in man (46). One concept that has been advanced to explain some effects of methyldopa is the false neurotransmitter hypothesis (47). Methyldopa is thought to enter the sympathetic neurons where it undergoes metabolism (46–48) according to the scheme shown in Figure 2: methyldopa is decarboxylated by dopa decarboxylase to α-methyldopamine (14) which is hydroxylated by dopamine β-oxidase to α-methylnorepinephrine (15). α-Methylnorepinephrine, the false neurotransmitter, displaces norepinephrine, is stored in the storage granules, and released into the synaptic cleft and binds to α-receptors where it is less effective as a neurotransmitter than norepinephrine (49). However, this fails to explain many effects of methyldopa. There is strong evidence that the main action of methyldopa is the stimulation of α-receptors in the vasomotor center in the central nervous system, thus inhibiting sympathetic nerve transmission.

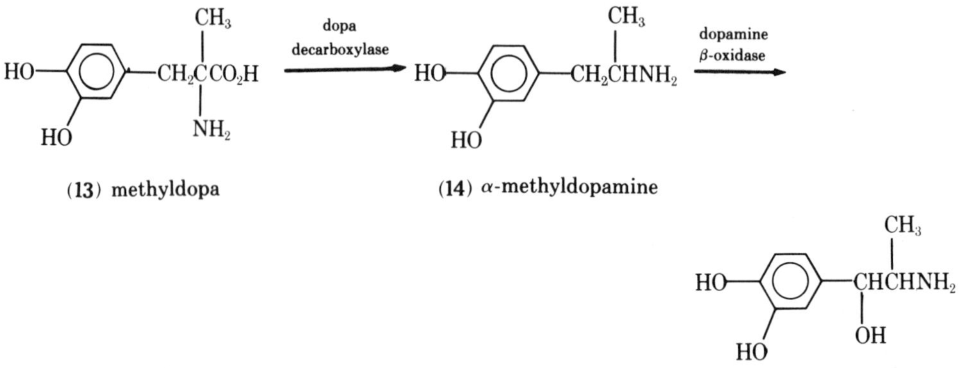

876 CARDIOVASCULAR AGENTS

Figure 2. Metabolism of methyldopa.

The Renin–Angiotensin System. Renin, a circulating proteolytic enzyme, has been recognized since its discovery (50) as having the capability of elevating blood pressure. The renin–angiotensin system (51) is also recognized as a major factor in the homeostasis of sodium, potassium, and water (52) (see Polypeptides). Its role in the maintainance of blood pressure and involvement in the pathogenesis of hypertension is complex and not fully understood.

Renin, an acid protease (53) of ca 40,000 mol wt, is produced and released into the blood by the granules of the juxtaglomerular apparatus of the kidney. The major cause of this is the response of baroreceptors located in the afferent arteriole to a fall in blood pressure. It is the first autoregulatory step to restore blood pressure. The release of renin and the events that follow are involved in the control of extracellular fluid volume and the regulation of excretion of electrolytes by the kidney.

Renin (Figure 3) acts on a plasma α_2-globulin, angiotensinogen (16) of ca 60,000 mol wt, which is synthesized in the liver, to split off a decapeptide, angiotensin I (AI) (17), having very little biological activity (54–55). AI is split very rapidly by converting enzyme that is mainly present in the lung to an octapeptide, angiotensin II (AII) (18), which on a weight basis is the most potent pressor substance known (as little as 2 ng/(kg·min) causes a detectable increase in blood pressure) (52). Both AI and AII are rapidly metabolized in the circulation by peptidases with the formation of the biologically active heptapeptide, angiotensin III (AIII, [des-Asp¹]angiotensin II) (19), and inactive smaller fragments. AII elevates blood pressure mainly by constricting smooth muscle (arteries and arterioles) and by stimulating the release of aldosterone from the adrenal cortex (56). Aldosterone causes renal sodium retention and, hence, elevates blood pressure. AIII specifically stimulates release of aldosterone (57), and may act at the same site as AII to mediate the renin–angiotensin system (58). In addition to response to baroreceptor detection of a fall in blood pressure, renin release is stimulated by adrenal catecholamines via β_1 receptors in the region of the juxtaglomerular cells (59–60). Renin secretion is apparently inhibited by circulating AII and AIII by a feedback mechanism. Hyperkalemia also suppresses renin release.

Antihypertensive Drug Therapy. Rational management of hypertension by drugs requires knowledge of the extent of vascular disease and the hypertensive mechanisms obtained in the patient, such as cardiac output, vascular resistance, aortic impedance, blood volume, renin levels, neural activity, and adrenal steroid levels. It is necessary

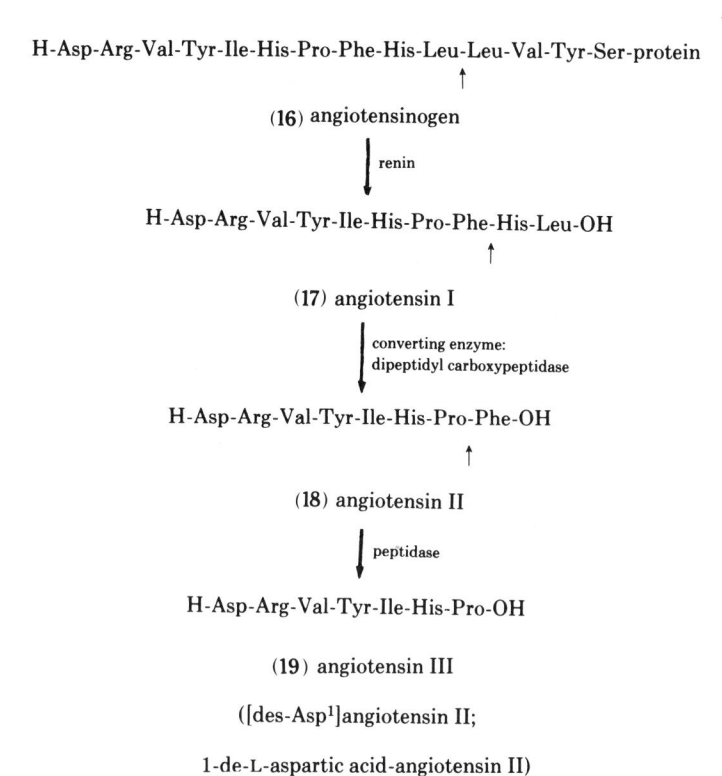

```
H-Asp-Arg-Val-Tyr-Ile-His-Pro-Phe-His-Leu-Leu-Val-Tyr-Ser-protein
                                        ↑
              (16) angiotensinogen
                         │ renin
                         ↓
H-Asp-Arg-Val-Tyr-Ile-His-Pro-Phe-His-Leu-OH
                                ↑
              (17) angiotensin I
                         │ converting enzyme:
                         │ dipeptidyl carboxypeptidase
                         ↓
H-Asp-Arg-Val-Tyr-Ile-His-Pro-Phe-OH
                           ↑
              (18) angiotensin II
                         │ peptidase
                         ↓
H-Asp-Arg-Val-Tyr-Ile-His-Pro-OH

              (19) angiotensin III

              ([des-Asp¹]angiotensin II;

              1-de-L-aspartic acid-angiotensin II)
```

Figure 3. Renin and angiotensins. Arrows indicate affected peptide linkages.

to identify patients with renal parenchymal disease, renal vascular disease, hyperaldosteronism and pheochromocytoma, in order to direct drug treatment and surgery if required. With increasing knowledge of the physiological and biochemical mechanisms involved in various types of hypertension and the pharmacology of various drugs, rational therapy (61) is now becoming possible. On the other hand, because it is very difficult and often impossible to determine the precise mechanisms operating in each patient, empirical therapy is usually employed.

In the stepped-care program (62), a widely used approach effective in most patients when diastolic pressure is between 13.7–18.7 kPa (103–140 mm Hg), an oral diuretic (usually a thiazide type) (Tables 1 and 2) is administered in the first step. In more than 30% of patients, blood pressure is normalized to ≤18.7/12 kPa (140/90 mm Hg). If this does not occur within two months, addition of the direct-acting vasodilator hydralazine, or a sympathetic depressant, propranolol, methyldopa (13) or reserpine, is taken as the second step. In malignant hypertension (diastolic pressure >17.3 kPa [130 mm Hg]) this combination of two drugs is the start of therapy. If this two-drug combination therapy is not successful, a third drug is added: if hydralazine was used in the second step, methyldopa is added in the third; if methyldopa or reserpine was used in the second step, hydralazine is added in the third. Individualized therapy is advised in the event that unsatisfactory results are obtained with the triple combination. Neuronal blocking agents, such as guanethidine, are the last drugs to be added to the regimen.

If enough information is known to allow characterization of the nature of the

Table 1. Thiazide Diuretics

2H-benzo-1,2,4-thiadiazine 1,1-dioxides:

Drug	R	Oral dosage
(20) benzthiazide	CH₂SCH₂—C₆H₅	50–75 mg twice daily
(21) chlorothiazide	H	500–750 mg twice daily

3,4-dihydro-1,2,4-thiadiazine 1,1-dioxides:

Drug	R	R′	R″	Oral dosage
(22) bendroflumethiazide	CH₂—C₆H₅	H	CF₃	10–15 mg once daily
(23) cyclothiazide	CH₂-norbornenyl	H	Cl	2–4 mg once daily
(24) hydrochlorothiazide	H	H	Cl	50–75 mg twice daily
(25) hydroflumethiazide	H	H	CF₃	50–75 mg twice daily
(26) methyclothiazide	CH₂Cl	CH₃	Cl	10–15 mg once daily
(27) polythiazide	CH₂SCH₂CF₃	CH₃	Cl	4–8 mg once daily
(28) trichlormethiazide	CHCl₂	H	Cl	4–12 mg once daily

hypertensive disease the patient is suffering from, a rational approach to therapy can be used. For this purpose antihypertensive drugs can be classified into three categories, (1) diuretics, (2) drugs that depress the sympathetic nervous system, and (3) vasodilators. They may act centrally or peripherally, and by one or more of the mechanisms previously described.

Diuretics. *Primary Aldosteronism.* Diuretics (qv) are indicated for the treatment of primary aldosteronism. After the initial administration of spironolactone to correct hypokalemia, a thiazide or thiazide-type diuretic is added to lower blood pressure by producing a negative salt and water balance.

Hypervolemic Essential Hypertension. Essential hypertension that is salt- and water-dependent, or low renin hypertension, responds to diuretic therapy.

Renal Parenchymal Disease. Diuretics are used to treat the hypertension accompanying this disease, often salt- and water-dependent and sometimes renin-dependent.

Sympathetic Nervous System Depressants. *Orthostatic Hypertension.* This neurally mediated hypertension responds to treatment with clonidine, guanethidine, methyldopa (17), or reserpine. The β-blockers, such as propranolol, are not useful since α-adrenergic mechanisms are not blocked.

Table 2. Thiazide-Like Diuretics

Drug	Structure	Oral dosage
(29) chlorthalidone	[structure]	50–100 mg once daily
(30) metolazone	[structure]	2.5–10 mg once daily
(31) quinethazone	[structure]	50–75 mg twice daily

Pheochromocytoma. Prior to surgical removal of the tumor or in the event that the tumor is inoperable (metastasized-malignant), hypertension is controlled with α- and β-adrenergic blocking agents such as phentolamine and propranolol.

Inhibition of Renin Release. Vasodilators and diuretics often cause increased release of renin which reduces their antihypertensive effect. Hyperreninemia is also associated with malignant hypertension and renal arterial stenosis. Addition of a sympatholytic, especially propranolol, to the antihypertensive drug regimen inhibits renin release.

Reflex Tachycardia and Cardiac Output. These adverse reactions produced by vasodilator drugs are countered by β-adrenergic blocking agents.

Vasodilators. *High Vascular Resistance Hypertension.* Vasodilators, such as hydralazine, minoxidil, and prazosin, reduce peripheral resistance by direct relaxation of arterial and arteriolar smooth muscle and are given orally in combination with a diuretic, or with a diuretic and a sympatholytic.

Hypertensive Emergencies. Intravenous administration of the vasodilators, diazoxide or sodium nitroprusside, is very effective in lowering elevated blood pressure to normal.

Inhibitors of Angiotensin II and of Its Formation. Angiotensin II-dependent hypertension should be alleviated by agents that block converting enzyme or that are specific antagonists of angiotensin II.

Antihypertensive Drugs. **Diuretics.** Three classes of diuretics are used in the treatment of hypertension, (1) the thiazides and related nonthiazides, (2) the loop diuretics, and (3) the potassium-sparing diuretics.

Thiazide and Related Nonthiazide Diuretics. The so-called thiazide diuretics (Table 1) are 7-sulfamoyl-2H-benzo-1,2,4-thiadiazine 1,1-dioxides (20) and (21), and their 3,4-dihydro derivatives (22–28). The thiazide-like diuretics (Table 2) include the (1-hydroxy-3-oxo-1-isoindolinyl)benzenesulfonamide (29), and the 1,2,3,4-tetrahydro-4-oxo-6-quinazolinesulfonamides (30) and (31).

The thiazides are usually the first type of drug prescribed for the treatment of mild or moderate hypertension. Initially, the antihypertensive effect is a result of a reduction in plasma volume with consequent decrease in cardiac output, probably through an initial negative sodium balance (63). Urinary excretion of sodium and water is increased by inhibition of reabsorption of sodium in the distal tubules and in the cortical segment of the ascending limb of Henle's loop. Although plasma volume and cardiac output return to normal on continued treatment, the lowering in blood pressure persists (64). The mechanism of the antihypertensive action of diuretics is not known, but may be caused by altered sodium metabolism in the vascular wall (65). They are known to dilate arterioles of smooth muscle (65) and to alter the water and electrolyte content of the walls of arterioles (64).

The thiazide diuretics, although possessing antihypertensive properties of their own, find greatest use in combination with other antihypertensive drugs, such as hydralazine, methyldopa, and reserpine, by potentiating their action and reducing the dose requirement. This has the beneficial effect of reducing the incidence and severity of adverse side effects. They are well tolerated and effective in lowering blood pressure in both supine and standing positions. Long term antihypertensive effects may result from reduction in vascular reactivity to sympathetic stimulation, thereby preventing compensation for reduction in plasma volume. When given with an antihypertensive drug of the vasodilator type, they counteract secondary fluid retention and may diminish development of drug resistance.

The main side effects of thiazide diuretics are hypokalemia, increase in concentrations of blood uric acid, serum calcium or blood urea, hyperglycemia, weakness, dizziness, fatigue, leg cramps, and gastrointestinal disturbances. Plasma renin activity is often increased and stays elevated (66).

The thiazide and related nonthiazide diuretics vary in potency on a weight basis and in duration of action, but at equieffective doses there are no differences in toxicity. Since they have very flat dose response curves, fixed combinations with other antihypertensive drugs are widely used for therapy (see Table 3).

Table 3. Available Fixed Combinations of Antihypertensive Drugs with Diuretics

Antihypertensive	mg	Diuretic	mg
clonidine·HCl	0.1 or 0.2	chlorthalidone	15
deserpidine	0.125	hydrochlorothiazide	25 or 50
deserpidine	0.25	hydrochlorothiazide	25
deserpidine	0.25 or 0.5	methyclothiazide	5
guanethidine·½H$_2$SO$_4$	10	hydrochlorothiazide	25
hydralazine·HCl	25	hydrochlorothiazide	15
methyldopa	250	chlorothiazide	150 or 250
methyldopa	250	hydrochlorothiazide	15 or 25
Rauwolfia serpentina	50	bendroflumethiazide	4
reserpine	0.125	benzthiazide	50
reserpine	0.125	chlorothiazide	250 or 500
reserpine	0.25	chlorthalidone	50
reserpine	0.1 or 0.125	hydrochlorothiazide	25 or 50
reserpine	0.25	polythiazide	2
reserpine	0.125	quinethazone	50
reserpine	0.1	trichlormethiazide	2 or 4
spironolactone	25	hydrochlorothiazide	25
syrosingopine	0.5 or 1	hydrochlorothiazide	25
triamterene	50	hydrochlorothiazide	25

Loop Diuretics. The loop diuretics, ethacrynic acid (32) and furosemide (33) (see Table 4), are more potent and act more rapidly than the thiazides. They inhibit the reabsorption of sodium by acting primarily on the medullary portion of the ascending limb of the Henle loop where they block the active transport of sodium. They also inhibit reabsorption of sodium in the proximal and distal tubules (67). They are used in hypertensive patients who have become refractory to the thiazide diuretics and for patients with severely impaired renal function. Both drugs can cause hyperglycemia, hyperuricemia, and transient deafness. Furosemide is usually preferred to ethacrynic acid because it is less ototoxic, has a broader dose-response curve, and produces fewer side effects. Ethacrynic acid is less likely to cause hyperglycemia than furosemide.

Table 4. Loop Diuretics

Drug	Structure	Oral dose
(32) ethacrynic acid	[structure: C$_2$H$_5$, CH$_2$=C(CO)-phenyl(Cl,Cl)-OCH$_2$CO$_2$H]	25–50 mg twice daily
(33) furosemide	[structure: H$_2$NSO$_2$-phenyl-NH-CH$_2$-furan, CO$_2$H]	10–150 mg four times daily

Bumetanide (34) and indapamide (35) are new diuretics that inhibit sodium and chloride transport in the ascending loop of Henle. At 1–3 mg orally daily, bumetanide causes a significant diuresis and increased excretion of sodium, chloride, and also potassium. In congestive heart failure, bumetanide and furosemide at equipotent doses have a similar effect on electrolytes in serum and urine and on water excretion (68). In patients with secondary hyperaldosteronism, bumetanide at 1 mg bid (twice daily) and furosemide at 40 mg bid have comparable effects on urine volume, and sodium, chloride, and potassium content of urine (69). Indapamide has been shown to be a long-acting antihypertensive drug that acts by decreasing vascular resistance (70). Indapamide is saluretic without being kaliuretic and does not elevate serum calcium, uric acid, or glucose (71).

(34) bumetanide

(35) indapamide

Potassium-Sparing Diuretics. Spironolactone (36) and triamterene (37) (Table 5) are potassium-sparing diuretics which interfere with reabsorption of sodium at the distal exchange sites. Hence sodium excretion is promoted while potassium is conserved. Spironolactone, an antagonist of the corticosteroid aldosterone, has only weak antihypertensive activity whereas triamterene, which interferes directly with electrolyte transport in the distal tubules, has no significant antihypertensive activity. Both drugs are used mainly in conjunction with the thiazide diuretics in order to reduce potassium excretion and minimize alkalosis (72).

Side effects of these drugs are occasional gastrointestinal disturbances, muscular

Table 5. Potassium-Sparing Diuretics

Drug	Structure	Oral dose
(36) spironolactone		25–100 mg four times daily
(37) triamterene		25–75 mg four times daily

weakness, and skin rashes. Spironolactone sometimes produces gynecomastia in men and menstrual disturbances in women. Triamterene may increase blood urea nitrogen and serum uric acid.

Spironolactone and triamterene are available in fixed combinations with hydrochlorothiazide.

Drugs that Depress the Sympathetic Nervous System. Depression of the sympathetic nervous system, either centrally or peripherally, is the predominant blood pressure lowering mechanism of the majority of antihypertensive drugs. Inhibition of any steps involved in the formation and disposition of norepinehrine (Figure 1) has potential for antihypertensive action.

Inhibition of adrenergic nerve transmission by preventing impulses from the sympathetic nerve terminals from reaching the effector site (arterioles) is produced by so-called adrenergic neuronal blocking agents (73). This has the effect of decreasing sympathetic tone (vasodilatation) with consequent decreased peripheral resistance, increased peripheral flow, and reduction in blood pressure. In contrast with the sympathetic nervous system, where norepinephrine is released at the sympathetic postganglionic fibers and into the synaptic cleft, acetylcholine is released at the parasympathetic postganglionic end organs. Blocking of parasympathetic and sympathetic impulses at the first level of the ganglia (ganglionic blockade) results in the desired reduction in blood pressure (74) by interference with the adrenergic system, but at the same time produces undesirable side effects because of interference with parasympathetic (cholinergic) control of many visceral functions: urinary retention, constipation, dry mouth, blurred vision, and impotence. Orthostatic hypotension and weakness are adverse side effects that may occur from sympathetic blockade.

Drugs that lower blood pressure mainly by sympatholytic action are: methyldopa; *Rauwolfia serpentina* and reserpine and related alkaloids; propranolol and other β-adrenergic blocking agents; clonidine; guanethidine; α-adrenergic blocking agents; ganglionic blocking agents; and guanabenz.

Methyldopa. Methyldopa (13), 3-hydroxy-α-methyl-L-tyrosine, was initially synthesized as the DL form (75) as an inhibitor of dopa decarboxylase, and was shown to be a potent inhibitor (43–44). The antihypertensive activity and biochemical effects were shown to reside in the L isomer (76). The L isomer (13) was synthesized (77) from 3-methoxy-4-hydroxyphenylacetone (38) by initial conversion to the α-aminopropionitrile (39) which was resolved by conversion to the d-10-camphorsulfonate dioxanate (40). The L-α-aminonitrile was liberated from its salt with ammonia and hydrolyzed with acid to give (13).

[Reaction scheme:]

HO—⟨C₆H₃⟩(CH₃O)—CH₂COCH₃ →[KCN, NH₄Cl, NH₄OH] HO—⟨C₆H₃⟩(CH₃O)—CH₂C(NH₂)(CH₃)CN →[d-10-camphorsulfonic acid]
(38) → (39)

d-10-camphorsulfonate dioxanate solvate (40) →[(1) NH₃, (2) 45% HCl] HO—⟨C₆H₃⟩(HO)—CH₂C(NH₂)(CH₃)CO₂H
(13) methyldopa
(3-hydroxy-α-methyl-L-tyrosine)

Methyldopa, which is mainly used for treatment of moderate hypertension (78), is given orally at 0.5–2 g/d. Lower doses are effective when given with thiazide diuretics to counteract sodium and fluid retention.

Side effects are sedation (which usually disappears on continued treatment), mental depression, dryness of mouth, nausea, vomiting, diarrhea, and impotence. Orthostatic hypotension is rarely seen. Methyldopate hydrochloride (41) may be given intravenously, usually 250–500 mg every 6 h for hypertensive crisis.

HO—⟨C₆H₃⟩(HO)—CH₂C(NH₂)(CH₃)CO₂C₂H₅·HCl

(41) methyldopate hydrochloride

Fixed combinations of methyldopa are available with chlorothiazide or hydrochlorothiazide.

Rauwolfia Serpentina, and Reserpine and Related Alkaloids. *Rauwolfia serpentina* (L.) Benth, one of more than 130 *Rauwolfia* species of the *Apocynaceae* family (79), is a small climbing shrub indigenous to India. Its medicinal properties have been known in folk medicine for centuries (80), but its use as an antihypertensive agent was first reported in 1918 (81), and additionally for treatment of psychosis in 1931 (82). A systematic investigation of the components of the roots of *Rauwolfia serpentina* was reported soon thereafter (83–84), and further investigations led to the isolation (85) from the resinous fraction of a weakly basic indole alkaloid, reserpine, having hypotensive and sedative properties (86). It was shown to be a derivative of 3-epialloyohimban, thus differing from yohimban in the stereochemistry of the C/D/E ring fusions: 3β-H and 20α-H (*cis* D/E ring fusion) (87–88). In the preferred conformation (42) the 16, 17, and 18-substituents are equatorial; this offsets the unfavorable steric effect of ring C (see Alkaloids).

Although the powdered root of *Rauwolfia serpentina* is still used for the treatment of hypertension, the pure alkaloid reserpine and its relatives account for the major applications of this class of drug (see Table 6). Deserpidine (44), from *Rauwolfia canescens,* differs from reserpine in that it lacks the 11-methoxy substituent. Rescinnamine (45), isolated from *Rauwolfia serpentina,* has the 18-(3,4,5-trimethoxycinnamoyl) substituent instead of the 18-(3,4,5-trimethoxybenzoyl) substituent of reserpine. Syrosingopine (46) is obtained by methanolysis of reserpine to give methyl

884   CARDIOVASCULAR AGENTS

(42) preferred conformation of reserpine

Table 6. *Rauwolfia Serpentina* and Its Antihypertensive Alkaloids and Derivatives.

Name	R	R′	Oral dosage
(43) reserpine	OCH$_3$	—CO—C$_6$H$_2$(OCH$_3$)$_3$ (3,4,5-trimethoxybenzoyl)	0.1–0.25 mg
(44) deserpidine	H	—CO—C$_6$H$_2$(OCH$_3$)$_3$	0.25 mg
(45) rescinnamine	OCH$_3$	—COCH=CH—C$_6$H$_2$(OCH$_3$)$_3$	0.25–2 mg
(46) syrosingopine	OCH$_3$	—CO—C$_6$H$_2$(OCH$_3$)$_2$(OCO$_2$C$_2$H$_5$)	0.75–3 mg
Rauwolfia serpentina	(mixture)		50–300 mg

(47) methyl reserpate

(48)

(49)

reserpate (47), having a free hydroxyl at C-18, followed by esterification with 3,5-dimethoxy-4-ethoxycarbonyloxybenzoyl chloride in pyridine (89). The brilliant total synthesis of reserpine by R. B. Woodward and co-workers (90), was carried out via bromolactone (49). The synthesis of deserpidine via the same intermediate was later carried out by Czech chemists (91) (see Alkaloids).

Reserpine and related alkaloids are orally effective, sympathetic depressants that are used for the treatment of mild to moderate hypertension. They are seldom used alone for their antihypertensive effect is weak. They are most useful in combination with oral thiazide diuretics. They act by preventing the incorporation of norepinephrine into storage granules, thus allowing its degradation by catechol-O-methyl transferase and monoamine oxidase (Fig. 2). Thus, release of norepinephrine at sympathetic neuroeffector junctions and in the CNS are prevented. Since α-adrenergic receptors in vascular smooth muscle are not activated, there is a decrease in vascular resistance with a lowering of blood pressure. Depletion of serotonin in the brain also occurs. At high concentrations there may be some direct effects on arteriolar smooth muscle.

Side effects are sedation, nasal congestion, edema, bradycardia, diarrhea, gastric hyperacidity, and parkinsonian-like rigidity. The danger of increased incidence of breast cancer has been pointed out but has not been supported in many statistical studies. Occasionally, severe depression occurs.

The following fixed combinations with thiazide and thiazide-like diuretics are available: reserpine + benzthiazide, or chlorothiazide, or chlorthalidone, or hydrochlorothiazide, or polythiazide, or quinethazone, or trichlormethiazide; deserpidine with hydrochlorothiazide or methyclothiazide; syrosingopine with hydrochlorothiazide; *Rauwolfia serpentina* with bendroflumethiazide.

Propranolol and Other β-Adrenergic Blocking Agents. Classification of adrenergic receptors by Ahlquist in 1948 (92) as α and β provided a rational understanding of the various direct actions of catecholamines, and an explanation for blockade by existing drugs of α-adrenergic receptors. β-Adrenergic receptors, located in the heart, arteries, arterioles of skeletal muscle, and the bronchi, have been classified as β_1 (heart) and β_2 (vascular and tracheal smooth muscle) (93). Stimulation of β receptors in the heart causes an increase in heart rate and contractility, increase in conduction velocity, and shortening of the refractory period; stimulation of β-receptors in arterioles causes vasodilatation. β-Adrenergic receptor blockers are known to lower blood pressure, and propranolol is considered a safe and effective antihypertensive drug (94). Other β-blockers (see Table 7) are also effective (95).

A selective blockade of $β_1$ receptors decreases heart rate and myocardial contractility, with attendant lowering of cardiac output, without causing bronchoconstriction. β-Adrenergic blocking agents vary in potency, selectivity for $β_1$ and $β_2$ receptors, β-receptor stimulant (intrinsic sympathomimetic) activity, and direct action on cell membranes (membrane stabilizing, local anesthetic, quinidine-like).

The discovery of the β-adrenergic blocking action of dichloroisoproterenol (106) (61), which reverses the agonist activity of the sympathomimetic amine, isoproterenol (62) was followed by the development of the β-blocker, pronethalol (63) (107). The latter, a phenethanolamine, which was reported to cause tumors in animals, was superceded by many new β-blockers, most of which have the aryloxypropanolamine structure (Table 7). These are synthesized by the following route: the phenol (64) is condensed with epichlorohydrin (65) in the presence of a base and the resulting epoxide (66) reacts with an amine, usually isopropylamine, to give the aryloxypropanolamine (67).

A number of β-blockers have been prepared in optically active forms, either by synthesis from chiral intermediates or by resolution of the racemic mixture that results by most synthetic routes. These include dichloroisoproterenol (108), pindolol (109), practolol (110), propranolol (103,108), pronethalol (108), and timolol (111). β-Blocker and antihypertensive activity were found to be properties of levorotatory (S) isomers (see Pharmaceuticals, optically active).

The mechanism whereby propranolol and other β-blockers lower blood pressure is not fully understood, but β-blockade seems to be an important property, whereas the presence or absence of membrane stabilizing or intrinsic sympathomimetic properties is of no consequence. Whereas *dl*-propranolol has both antihypertensive and β-blocker properties, *d*-propranolol (103,108) has neither (95,112). Decrease in cardiac output, inhibition of renin release, and inhibition of sympathetic activity may be important factors, as well as central inhibition of neuronal activity. High doses of propranolol may involve an effect on the CNS, lessening sympathetic outflow (113–114).

β-Adrenergic blocking agents are best used with a thiazide diuretic, and the efficiency of blood pressure lowering is further enhanced by the addition of a vasodilator such as hydralazine. This latter combination is unusually effective in treatment of moderate to severe hypertension. The oral dose of propranolol for chronic hypertension is 60–480 mg daily.

Although usually well tolerated, side effects of propranolol consist of bradycardia, lethargy, weakness, and gastrointestinal disturbances. When bronchial asthma or congestive heart failure are present, propranolol is contraindicated. A cardioselective β-blocker, however, can be used on asthmatic patients. Practolol has been shown to cause oculomucocutaneous syndrome and peritonitis, and is no longer in general use (115).

Clonidine. The initial clinical observation (116) that the α-adrenergic stimulant tetrahydrozoline (68) is orally effective in lowering blood pressure in hypertensive patients led to the development of related imidazolines, including clonidine (69) (117–118). Clonidine is synthesized in high yield by reaction of 2,6-dichloroaniline (70) with *N*-acetylimidazolidinone (71) and phosphorus oxychloride, followed by ethanolysis of the intermediate imidazoline (72) (119). Clonidine has been shown to exist as (69b) with the double bond in conjugation with the phenyl ring (120).

Table 7. Antihypertensive β-Adrenergic Blocking Agents

	Structure	Reference to synthesis
Cardioselective drug		
(50) acebutolol	n-C₃H₇CONH–C₆H₃(COCH₃)–OCH₂CH(OH)CH₂NHCH(CH₃)₂	96
(51) atenolol	H₂NCOCH₂–C₆H₄–OCH₂CH(OH)CH₂NHCH(CH₃)₂	97
(52) metoprolol	CH₃OCH₂CH₂–C₆H₄–OCH₂CH(OH)CH₂NHCH(CH₃)₂	98
(53) practolol	CH₃CONH–C₆H₄–OCH₂CH(OH)CH₂NHCH(CH₃)₂	99
(54) tolamolol	2-Cl-C₆H₄–OCH₂CH(OH)CH₂NHCH₂CH₂O–C₆H₄–CONH₂	100
Noncardioselective drug		
(55) alprenolol	2-(CH₂CH=CH₂)-C₆H₄–OCH₂CH(OH)CH₂NHCH(CH₃)₂	101
(56) oxprenolol	2-(OCH₂CH=CH₂)-C₆H₄–OCH₂CH(OH)CH₂NHCH(CH₃)₂	101
(57) pindolol	(4-indolyl)–OCH₂CH(OH)CH₂NHCH(CH₃)₂	102
(58) propranolol	(1-naphthyl)–OCH₂CH(OH)CH₂NHCH(CH₃)₂	103–104
(59) sotalol	CH₃SO₂NH–C₆H₄–CH(OH)CH₂NHCH(CH₃)₂	105
(60) timolol	(4-morpholinyl)-(1,2,5-thiadiazol-3-yl)–OCH₂CH(OH)CH₂NHC(CH₃)₃	101

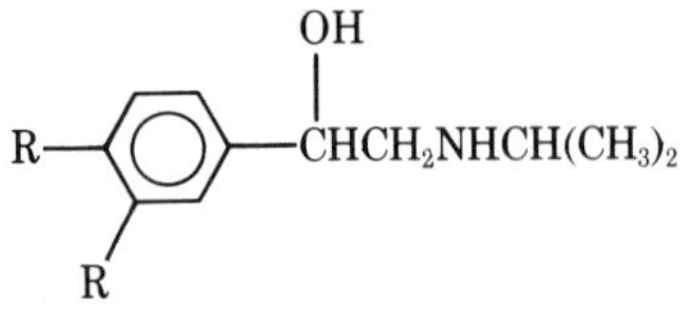

888 CARDIOVASCULAR AGENTS

(61) dichlorisoproterenol R = Cl
(62) isoproterenol R = OH

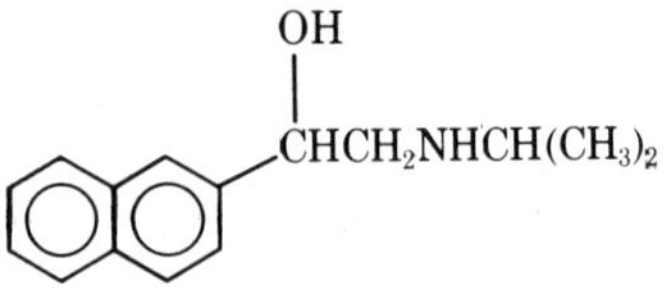

(63) pronethalol

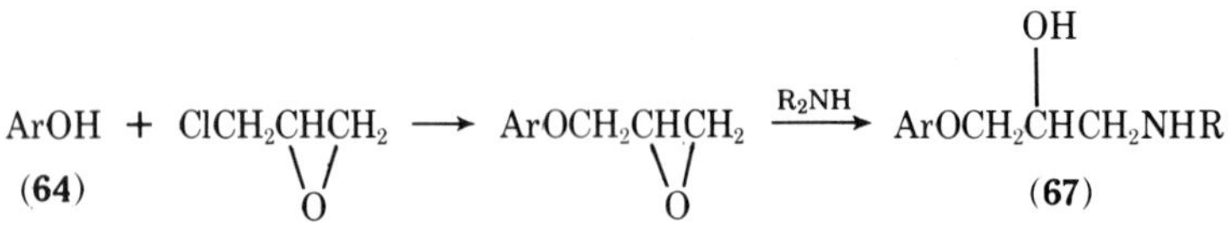

$$\text{ArOH} + \text{ClCH}_2\text{CHCH}_2\underset{\underset{O}{\diagdown\diagup}}{} \longrightarrow \text{ArOCH}_2\text{CHCH}_2\underset{\underset{O}{\diagdown\diagup}}{} \xrightarrow{R_2NH} \text{ArOCH}_2\overset{OH}{\underset{|}{C}}\text{HCH}_2\text{NHR}$$

(64)   (65)   (66)   (67)

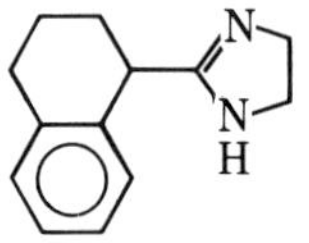

(68) tetrahydrozoline

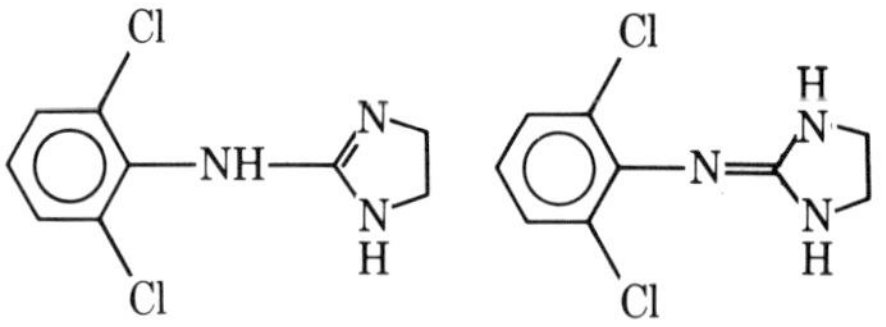

(69a)   (69b)
(69) clonidine

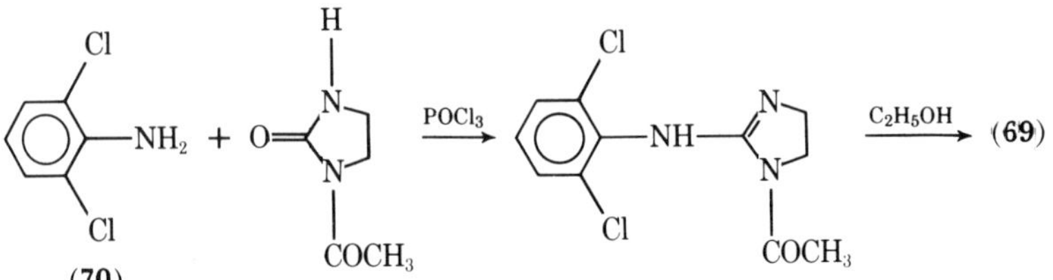

(70) + (71) $\xrightarrow{POCl_3}$ (72) $\xrightarrow{C_2H_5OH}$ (69)

    Clonidine hydrochloride, used for treating moderate to severe hypertension, lowers blood pressure by a central action. It is thought to bind to α-receptors in the vasomotor center of the medulla, thus inhibiting sympathetic outflow from the brain. It also produces some peripheral α-adrenergic blockade. Plasma renin activity is lowered in some patients with essential hypertension, but how this relates to its antihypertensive activity is not known (33,121). The daily oral dose is usually 0.2–0.8 mg. Fixed combinations with hydrochlorothiazide or chlorthalidone are available.
    Major side effects are drowsiness, dryness of mouth, and constipation; rarely dizziness and orthostatic hypotension.
    Guanethidine. This potent antihypertensive drug, which is of value in treating patients with severe malignant hypertension, is classified as an adrenergic neuronal blocking agent. It probably acts by gaining access to adrenergic nerve terminals by the same mechanism as norepinephrine (122). There it displaces norepinephrine from storage granules (123) and is released instead of norepinephrine by nerve stimulation. There is a resultant reduction in cardiac output and peripheral resistance, thereby

lowering blood pressure. The initial oral dose is 12.5 mg/d, and may be increased as needed up to 300 mg/d. The drug has a long half-life, thus it may take from 7–14 d to observe the maximum effect. It should be given with an oral diuretic to counteract sodium and water retention. A fixed combination with hydrochlorothiazide is available.

Side effects are orthostatic and exercise hypotension, bradycardia, diarrhea, weakness, and retrograde ejaculation or impotence.

Guanethidine is synthesized (124) by alkylation of octahydroazocine (73) with chloroacetonitrile, followed by catalytic reduction of the resulting N-cyanomethyl derivative (74) to give diamine (75). The diamine reacts with S-methylisothiourea to give guanethidine (76).

α-Adrenergic Blocking Agents. α-Adrenergic receptors, primarily located in the arterioles of the skin and mucosa, abdominal viscera, kidney, and salivary glands, cause contraction of smooth muscle on response to the action of catecholamines. These receptors are selectively blocked by α-adrenergic blocking agents, thereby lowering peripheral resistance with consequent lowering of blood pressure (125). Although not useful in treating essential hypertension, they are effective in blocking the effects of an excess of circulating catecholamines resulting from pheochromocytoma. The noncompetitive α-blocker phenoxybenzamine (77) is preferred to the competitive blocker phentolamine (78) hydrochloride and mesylate because of its longer duration of action. Both drugs are given in combination with a β-blocker which counteracts excess reflex cardiac stimulation. The oral dosage of phenoxybenzamine hydrochloride is 10–30 mg three times daily; phentolamine hydrochloride oral dosage is 50 mg 4–6 times daily. Phentolamine mesylate, 5 mg iv (intravenous), is given for presurgical treatment of pheochromocytoma.

Side effects most commonly experienced are reflex tachycardia and orthostatic hypotension. Oral dosing may cause nasal congestion, nausea, vomiting, and diarrhea.

890 CARDIOVASCULAR AGENTS

Ganglionic Blocking Agents. These sympatholytic agents inhibit synaptic transmission at autonomic ganglia, but produce severe side effects by affecting many visceral functions under parasympathetic control. Their use is limited to intravenous administration for hypertensive crisis.

Trimethaphan camsylate (**79**), a decahydro-2-oxo-[1′,2′:1,2]thieno[3,4]imidazol-5-ium salt with (+)-β-camphorsulfonic acid (126), has displaced the earlier drugs, pentolinium tartrate (**80**) and hexamethonium chloride (**81**). Intravenous trimethaphan camsylate is used to produce controlled hypotension during neuro- and some cardiovascular surgery.

(**79**) trimethaphan camsylate

(**80**) pentolinium tartrate

$(CH_3)_3\overset{+}{N}(CH_2)_6\overset{+}{N}(CH_3)_3 \ \ 2\ Cl^-$
(**81**) hexamethonium chloride

Side effects are orthostatic hypotension, urinary retention, anorexia, nausea, dryness of mouth, and mydriasis.

Guanabenz. Guanabenz stimulates the central α-adrenergic receptors and, additionally, has some adrenergic neuron blocking action, thus causing a decrease in sympathetic flow (127–128). Guanabenz is effective in mild to moderately severe hypertension at doses of 4–16 mg twice daily (129–130). Side effects are dry mouth, dizziness, and emesis; no tachycardia, postural hypotension, evidence of sodium retention, or ECG abnormalities are seen.

Guanabenz (**84**) is synthesized in the presence of acetic acid, followed by condensation of 2,6-dichlorobenzaldehyde (**82**) with aminoguanidinium bicarbonate (**83**) by treatment with alkali (131).

(**82**)   (**83**)   (**84**) guanabenz

Vasodilators. Lowering of peripheral resistance with resultant lowering of blood pressure, can occur by direct action of drugs on vascular smooth muscle or by central α-adrenergic stimulation. Since direct-acting vasodilators do not affect sympathetic reflexes, little or no postural or exercise-induced hypotension result. They are generally given with another antihypertensive drug since, alone, they are not very effective in lowering blood pressure at doses that are safe. The reflex tachycardia that results from the action of antihypertensive vasodilators requires concomitant administration of a β-adrenergic blocking agent.

Hydralazine. Hydralazine, which has been in use for about 25 yr for the treatment of hypertension, is an orally effective vasodilator that reduces peripheral resistance by a direct relaxation of arterial smooth muscle (132–133). Cardiac output and heart rate are increased and renal, coronary, and splanchnic flows are increased (134) because reflex controls are not blocked. Also, sodium and water are retained. In order to control tachycardia, hydralazine is given with a β-blocker or reserpine. Combined therapy with a thiazide diuretic controls sodium and water retention. The daily dose of 200–300 mg (400 mg max) is usually given in four divided doses, but because of its long duration of action the total daily dose can be given in two divided doses with the same efficacy (135).

Side effects that are usually seen during early stages of treatment are headache, gastrointestinal disturbances, flushing, and rash. On prolonged treatment a reversible lupus erythematosus-like syndrome requires discontinuance of the drug.

The fixed combinations available are those with hydrochlorothiazide, reserpine, or hydrochlorothiazide + reserpine.

Hydralazine is synthesized (136) from phthalaldehydic acid (85) by reaction with hydrazine to give 1(2H)-phthalazinone (86) which is chlorinated to give 1-chlorophthalazine (87). This is converted to its hydrazine, hydralazine (88).

Minoxidil. Minoxidil is a potent, direct-acting vasodilator that is useful in the treatment of severe hypertension resulting from increased vascular resistance (137–138), and is effective in combination with propranolol and hydrochlorothiazide for hypertension that does not respond adequately to treatment with β-adrenergic blocking agents and thiazide diuretics (139).

Side effects are tachycardia, elevations of renin levels, sodium and water retention, and increase in facial hair growth (hypertrichosis).

Minoxidil is synthesized (140) by a sequence of reactions wherein ethyl cyanoacetate is condensed with guanidine under base catalysis to give the diaminopyrimidinol (89) which is chlorinated, and the resultant chloropyrimidine (90) is converted to the N-oxide (91) by reaction with m-chloroperbenzoic acid (MCPBA). Reaction of (91) with piperidine gives minoxidil (92).

Diazoxide. Like other direct-acting vasodilators, diazoxide relaxes vascular smooth muscle. It is particularly useful for hypertensive crises (141), but must be administered rapidly since it apparently binds to serum proteins. Coadministration with a β-blocker is required to prevent tachycardia, increased cardiac output, and elevation of renin levels. The iv dose of diazoxide is 250–300 mg or 5 mg/kg of body weight.

Major side effects are sodium and water retention, controlled by diuretic administration, usually furosemide given iv 30–60 min prior to injection of diazoxide. Other side effects are hyperglycemia and hyperreninemia.

Diazoxide (93), a nondiuretic 2H-1,2,4-benzothiadiazine 1,1-dioxide that lacks

892 CARDIOVASCULAR AGENTS

(89) → (90) → (91) → (92) minoxidil

the sulfonamido substituent of the diuretics, is synthesized (142) by the following sequence of reactions:

(93) diazoxide

Sodium Nitroprusside. Sodium nitroprusside (94) is a rapid acting, intravenous peripheral vasodilator that acts directly on vascular smooth muscle (143). It is used for severe malignant hypertension and, when given intravenously, consistently lowers blood pressure within one to two min. It is the drug of choice in hypertensive crisis The dosage of the dihydrate is 60 μg/min iv.

$$Na_2Fe(CN)_5NO$$
(94) sodium nitroprusside
[sodium nitrosylpentacyanoferrate (III)]

Side effects are nausea, vomiting, palpitations, headache, sweating, and restlessness.

Prazosin. Prazosin is an orally effective antihypertensive agent for treating mild to moderate hypertension when used alone, and for severe hypertension when used concomitantly with a thiazide diuretic and a β-blocker (144). Some studies suggest that the mode of action involves direct relaxation of vascular smooth muscle (145) and, in addition, selective antagonism of post-synaptic α-adrenergic receptors (146). Other studies indicate an antisympathetic mechanism (147). The daily oral dose of prazosin is 1–15 mg.

Side effects of postural hypotension, weakness, and headache during the first weeks of treatment mainly disappear on continued treatment.

Prazosin is synthesized (148) by reaction of the 2,4-dichloro-6,7-dimethoxyquinazoline (95) with ammonia at room temperature, which selectively displaces the more reactive 4-chlorine to give 4-amino-2-chloro-6,7-dimethoxyquinazoline (96). Reaction of (96) with 2-furoyl, piperazine at elevated temperatures gives prazosin (97).

Antagonists of Angiotensin II and Its Formation. Saralasin (sar[1]-ala[8]-angiotensin II) (98), a competitive antagonist of angiotensin II, and teprotide (99) (a modified bradykinin B potentiator and angiotensin I-converting enzyme inhibitor) (149), have been found useful for the diagnosis (150) and treatment of hypertension resulting from high renin levels (151–153). Administration of sar[1]-ala[8]-angiotensin II produces a fall in blood pressure in sodium-depleted patients with renal artery stenosis (154). Application of these angiotensin II antagonists is limited since they are active only by the iv route.

H-Sar-Arg-Val-Tyr-Val-His-Pro-Ala-OH
(98) saralasin
(sar[1]-ala[8]-angiotensin II; P-113, sarenin)
H-5-oxo-Pro-Trp-Pro-Arg-Pro-Gln-Ile-Pro-Pro-OH
(99) teprotide
(SQ 20881; 2-L-tryptophan-3-de-L-leucine-4-de-L-proline-8-L-glutamine–bradykinin potentiator B)

SQ 14,225 (D-2-methyl-3-mercaptopropanoyl-L-proline) (100) (155) is a newly developed orally active inhibitor of angiotensin-converting enzyme that has been shown to lower blood pressure in hypertensive animals. In healthy men, single oral doses up to 20 mg had no effect on resting blood pressure and heart rate, and no adverse effects were found. Blockage of the response to iv-administered angiotensin I was found to be dose-related. The pressor response to angiotensin II was not changed. Plasma renin activity was increased after administration of SQ 14,225 at doses of 5–20 mg, indicating blockade of angiotensin II feedback inhibition of renin secretion (156).

(100) SQ 14,225
(D-2-methyl-3-mercaptopropanoyl-L-proline; captopril)

CARDIAC GLYCOSIDES

Cardiac glycosides, used in modern medicine as cardiotonics, are found in plants of the *Digitalis, Strophanthus,* and *Scilla* (Squill) species, and in the skin of certain toads. Plant extracts containing these substances have been used by natives in various parts of the world as arrow poisons and in the Middle Ages as poisons in trial by ordeal. The ancient Egyptians and Romans employed the sea onion or squill as a heart tonic and as a rat poison, and the Chinese have for centuries used the dried skin of the common toad as a drug. Introduction of Digitalis (purple foxglove) in Western medicine as a cardiac drug was described in 1785 by Withering (157).

The cardiac glycosides are made up of a steroid aglycone (genin) in combination with 1–4 mol sugar. Some of the sugars are acetylated. The genins, which are generally convulsive poisons, have some cardiotonic activity of their own, but the saccharide moieties contribute to their potency and transport. The glycosides are extraordinarily powerful on a weight basis in their ability to contract heart muscle (the average daily oral therapeutic dose of digitoxin in man is 0.05–0.2 mg), but they also have a very low margin of safety.

The beneficial effects that the cardiac glycosides demonstrate in congestive heart failure are the result of increased cardiac efficiency with attendant increase in blood flow to other organs, especially the kidneys. Diuresis is increased with reduction in edema; venous pressure, blood volume, and heart size are decreased. These effects result from increased myocardial contractile force (positive inotropic effect) in the failed heart (158). The cardiac glycosides also slow the ventricular rate in atrial fibrillation or flutter (159). Digitalis terminates and prevents the recurrence of supraventricular tachycardia (160). It is useful for the treatment of sinus tachycardia and supraventricular and ventricular premature beats (161) when primarily caused by congestive heart failure.

Mode of Action. The mechanism of the positive inotropic action of digitalis is thought to result from its strong inhibition of the Na–K pump at the cellular membrane (162) by binding specifically to Na–K ATPase (the enzyme that mediates the pumping of sodium ions out of the heart muscle cell and potassium ions into the cell) (163–164). The energy for this process is the Na–K ATPase-catalyzed splitting of ATP (adenosine triphosphate) in the presence of calcium ions. Calcium ions needed for this process enter the cell (in conjunction with entry of sodium ions) when the cell membrane depolarizes and, in addition, bound calcium within the cell is released so that the contractile mechanism is activated. The reverse occurs on relaxation of the muscle. Inhibition of Na–K ATPase consequently interferes with the Na–K pump and results in enhanced influx of calcium, which may be coupled with accumulation of sodium ions or decrease of potassium ions in the heart muscle cell (165). Many of the toxic effects of digitalis are explainable on the basis of excessive loss of potassium ion.

In ventricular myocardial conducting (Purkinje) fibers, digitalis increases the rate of spontaneous depolarization during phase 4 (diastole) of the transmembrane potential, thereby increasing impulse formation (automaticity) (166). The automaticity of the sino-atrial (S-A) node is not directly affected by digitalis (167), and the atrioventricular (A-V) node only minimally or not at all (168). The drug reduces conduction velocity by decreasing maximum diastolic voltage (resting potential) in A-V node and Purkinje fibers, which has the effect of decreasing the rate of rise of the action potential (phase 0, systole), which is related to the resting potential or critical firing potential (maximum diastolic voltage) (169).

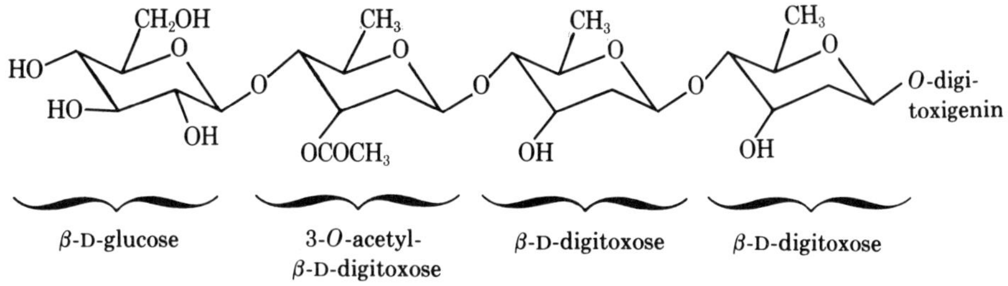

[Structure of lanatoside A shown: β-D-glucose — 3-O-acetyl-β-D-digitoxose — β-D-digitoxose — β-D-digitoxose — O-digitoxigenin]

β-D-glucose   3-O-acetyl-   β-D-digitoxose   β-D-digitoxose
                β-D-digitoxose

(101) lanatoside A
(digilanide A)

base ↓ (− CH₃CO₂H)

β-D-glucose-(β-D-digitoxose)₃-digitoxigenin

(102) deacetyllanatoside A
(deacetyldigilanide A;
purpurea glycoside A)

digilanidase ↓ (− D-glucose)

(β-D-digitoxose)₃-digitoxigenin

(103) digitoxin

HCl ↓ (− 3 D-digitoxose)

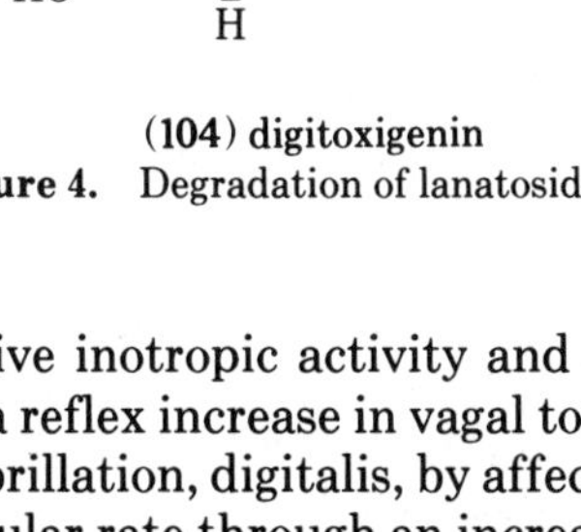

(104) digitoxigenin

Figure 4. Degradation of lanatoside A.

In addition to its positive inotropic activity and effects on automaticity and conduction, digitalis causes a reflex increase in vagal tone with a decrease in sympathetic tone (170). In atrial fibrillation, digitalis, by affecting vagal and extravagal influences, reduces the ventricular rate through an increase in the effective refractory period of the A-V transmission system and through an increase in atrial frequency (171).

Chemistry. Medicinally important cardiac glycosides (see Figs. 4–9) that occur in or are isolated from leaves of *Digitalis lanata* Ehrh. and *Digitalis purpurea* L. (family *Scrophulariacea*) are the primary glycosides, lanatosides A (**101**), B, and C, and deacetylanatosides A (**102**), and B. Seeds of *Strophanthus kombé* Oliv. (family

896 CARDIOVASCULAR AGENTS

β-D-glucose-(3-O-acetyl-β-D-digitoxose)-(β-D-digitoxose)$_2$-O-gitoxigenin

(105) lanatoside B
(digilanide B)

| base ($-$ CH$_3$CO$_2$H)
↓

β-D-glucose-(β-D-digitoxose)$_3$-O-gitoxigenin

(106) deacetyllanatoside B
(deacetyldigilanide B;
purpurea glycoside B)

digilanidase | ($-$ D-glucose)
↓

(β-D-digitoxose)$_3$-O-gitoxigenin

(107) gitoxin

HCl | ($-$ 3 D-digitoxose)
↓

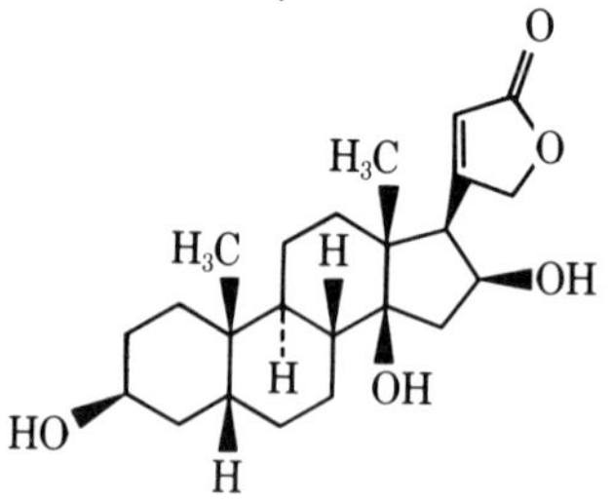

(108) gitoxigenin

Figure 5. Degradation of lanatoside B.

Apocynaceae) yield K-strophanthoside (113), K-strophanthin-β (114), and cymarin (115) (Fig. 7). Ouabain (118) (Fig. 8) is obtained from seeds of *Strophanthus gratus* (Wall. & Hock.) Baill. (family *Apocynacaea*), and scillaren A (120) (Fig. 9) from *Urginea (Scilla) maritima* (L.) Baker (family *Liliceae*). These glycosides are degraded to their component parts in stepwise fashion by appropriate sequences of acidic, enzymatic, and basic hydrolysis (172–173).

The genins of the cardiac glycosides all have the cyclopentanoperhydrophenanthrene backbone of the steroids and bile acids. They are substituted with a 3β-hydroxyl group (the point of attachment of the sugar moiety) in common with many steroids, but uniquely have a 14β-hydroxyl group with consequent C/D-*cis* ring fusion. Each has a 17β-butenolide or 17β-pentadienolide substituent and there may be, additionally, hydroxyl, carbonyl, epoxide, and unsaturation functions in the molecule. The genins of *Digitalis* and *Strophanthus* sp. are derivatives of card-20(22)-enolide and, thus, have the A/B-*cis* and B/C-*trans* ring fusions of the α-androstanes; the genins of *Scilla* sp. are derivatives of bufa-4,20,22-trienolide, wherein the B/C ring fusion is the usual *trans* of androstanes (see Steroids).

Hydrolytic cleavage (enzymatic or acid, or both) releases the sugars from the β--glucose-(3-O-acetyl-β-D-digitoxose)-(β-D-digitoxose)$_2$-O-digoxigenin

(109) lanatoside C
(digilanide C)

base ↓ (− CH$_3$CO$_2$H)

β-D-glucose-(β-D-digitoxose)$_3$-O-digoxigenin

(110) deacetyllanatoside C
(deacetyldigilanide C)

digilanidase ↓ (− D-glucose)

(β-D-digitoxose)$_3$-O-digoxigenin

(111) digoxin

HCl ↓ (− 3 D-digitoxose)

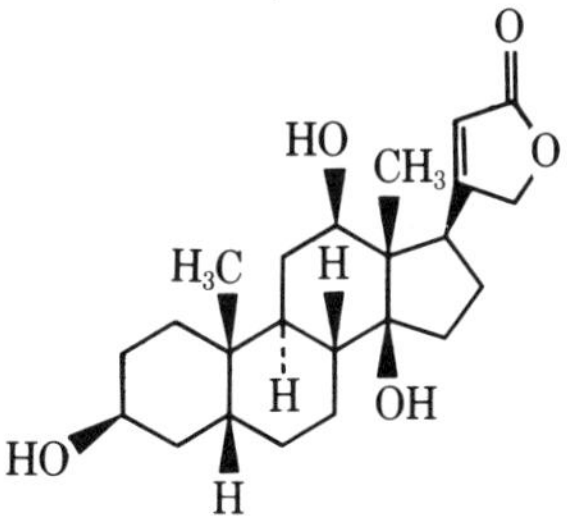

(112) digoxigenin

Figure 6. Degradation of lanatoside C.

aglycone, and this can be carried out in a stepwise fashion to remove one sugar at a time, or else the polysaccharide can be removed in some instances as a unit. The sugars obtained from the medically important glycosides are D-glucose (124), L-rhamnose (125), D-digitoxose (126) and D-cymarose (127), and, except in the case of K-strophanthoside (113) where the glucose units are linked 1-6, have the 1-4 linkage between the sugars. The D-glycosides belong to the β-series and the L-glycosides to the α-series (174).

Hydrolysis of lanatosides A (101), B (105), and C (109) with calcium hydroxide liberates 1 mol of acetic acid from each, thereby producing deacetyllanatosides A (102), B (106), and C (110), respectively (Figs. 4, 5, and 6). When the latter are subjected to the action of the enzyme digilanidase, 1 mol D-glucose is liberated from each, with the formation of the medicinally important secondary glycosides, digitoxin (103), gitoxin (107), and digoxin (111), respectively. Acid hydrolysis of each of these glycosides then liberates 3 mol D-digitoxose and produces the genins, digitoxigenin (104), gitoxigenin (108), and digoxigenin (112), respectively.

The primary glycosides of *Strophanthus kombé* are K-strophanthoside (113) and K-strophanthin-β (114) (Fig. 7). Hydrolysis of K-strophanthoside with β-glycosidase gives K-strophanthin-β (114) and one mole of D-glucose. A second mole of D-

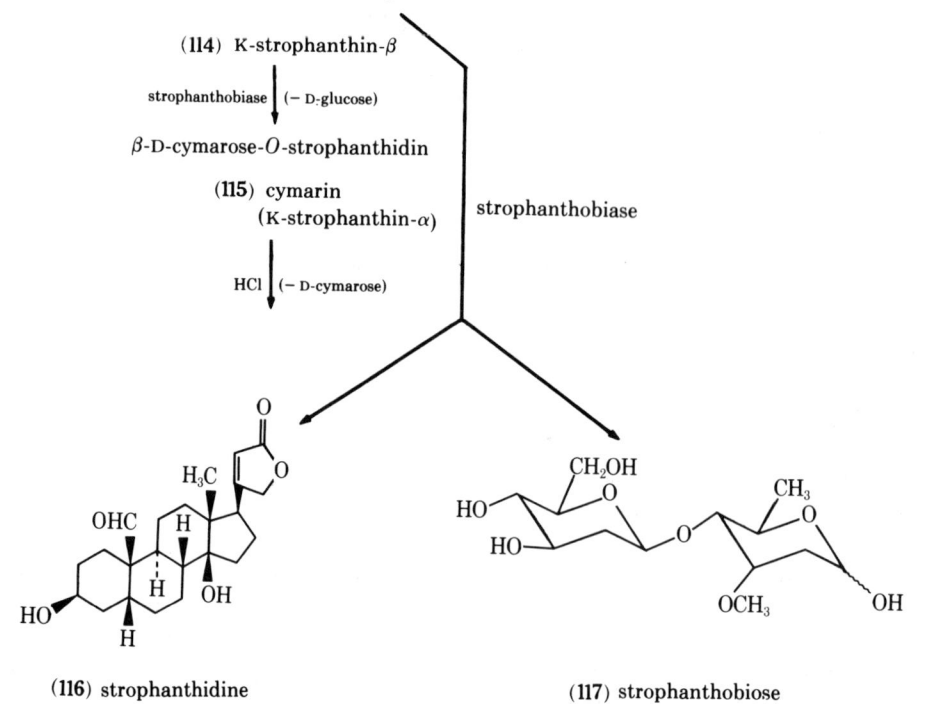

Figure 7. Degradation of K-strophanthoside.

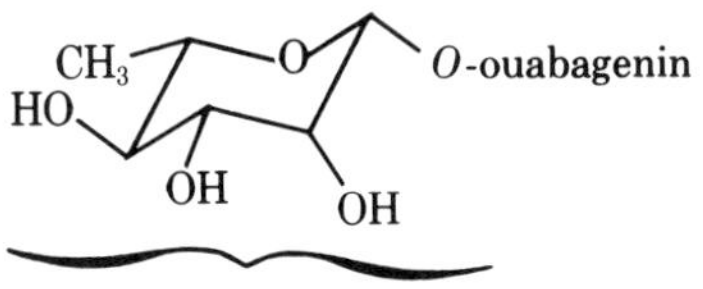

α-L-rhamnose

(**118**) ouabain
(G-strophanthin)

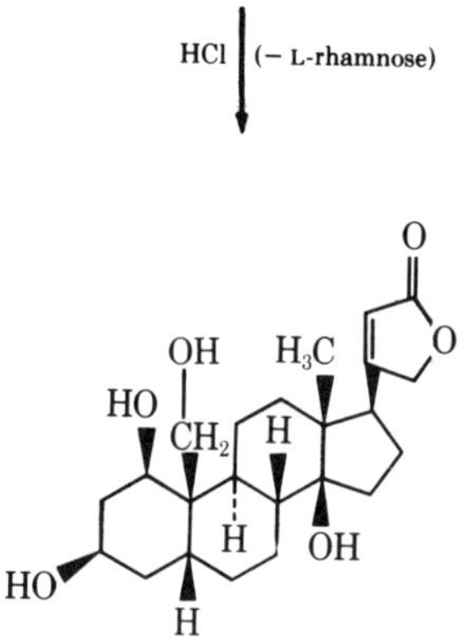

(**119**) ouabagenin
(G-strophanthidin)

Figure 8. Degradation of ouabain.

glucose is liberated by the subsequent action of strophanthobiase on (**114**) to give cymarin (**115**). Finally, this glycoside is split by acid to give strophanthidin (**116**) and one mole of D-cymarose. Alternatively, K-strophanthin-β is hydrolyzed by strophanthobiase to give strophanthidin and the disaccharide strophanthobiose (**117**). Ouabagenin (**119**) and L-rhamnose are the acid hydrolysis products from ouabain (**118**), the glycoside that occurs in *Strophanthus gratus* (Fig. 8).

Scillaren A (**120**), the primary glycoside of *Urginea maritima,* yields proscillaridin A (**121**) and one mole of D-glucose when subjected to the action of scillarenase or strophanthobiase. Further enzymatic hydrolysis liberates one mole of L-rhamnose to produce the genin, scillarenin (**122**). Alternatively, hydrolysis of scillaren A with scillabiase gives scillarenin and the disaccharide, scillabiose (**123**) (Fig. 9).

Therapeutic Indications. ***Congestive Heart Failure.*** The principal use of the cardiac glycosides is for the treatment of congestive heart failure. These drugs increase cardiac efficiency and enhance myocardial contractility (positive inotropic effect), resulting in increased cardiac output, lowered venous pressure, decrease in heart size, and counteraction of reflex tachycardia. Diuresis is promoted which reduces blood volume and edema.

Arrhythmias. Digitalis is the drug of choice for atrial flutter or fibrillation since it slows A-V conduction and produces some A-V block. Normal sinus rhythm may result. Paroxysmal supraventricular tachycardia may respond to digitalis, especially

900 CARDIOVASCULAR AGENTS

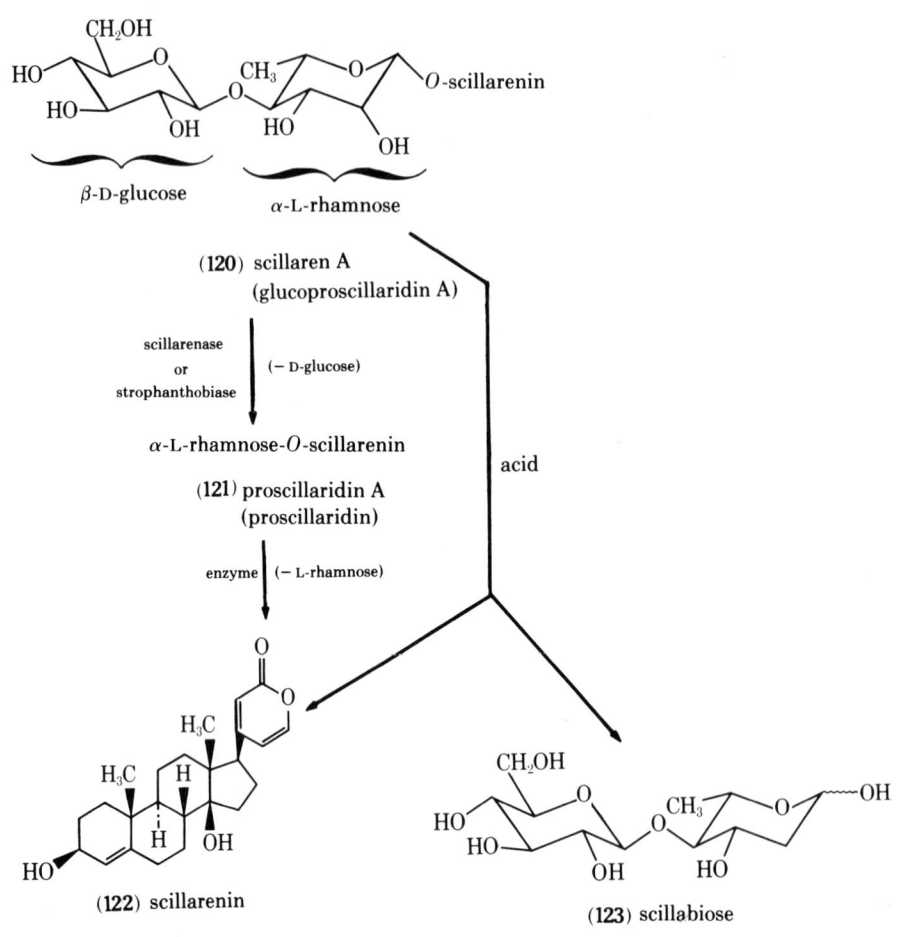

Figure 9. Degradation of scillaren A.

(120) scillaren A (glucoproscillaridin A)
(121) proscillaridin A (proscillaridin)
(122) scillarenin
(123) scillabiose
(124) D-glucose
(125) L-rhamnose
(126) D-digitoxose
(127) D-cymarose

when resistant to lesser measures. Treatment of sinus tachycardia resulting from heart failure may be beneficial.

Preparations for Clinical Use. *Digitalis Powdered Leaf and Gitalin.* The dried leaf of *Digitalis purpurea* L. (family *Scrophulariaceae*) contains 0.2–0.4% of digitoxin as its principal cardiac glycoside component. It is listed in the USP, 18th Revision, where the pigeon assay for potency is described (100 mg is equivalent to not less than 1 USP Digitalis Unit; one USP Digitalis Unit represents the potency of 100 mg of USP Digitalis Reference Standard). Owing to the variation in glycoside content of the crude drug, with attendant variation in physiological response, and the unreliability of the pigeon assay, digitalis has been dropped from USP XIX (official from July 1, 1975). This drug has mainly been supplanted by the pure glycosides.

Gitalin is a water soluble glycoside mixture from *Digitalis purpurea* that consists of digitoxin, gitoxin, and 16-formylgitoxin.

Pure Single Glycosides. The most used, pure, individual cardiac glycosides are acetyldigitoxin, deacetyllanatoside C, digitoxin, digoxin, lanatoside C, and ouabain. All have the same therapeutic applications, although their pharmacokinetic characteristics differ because of differences in gastrointestinal absorption, lipophilicity, serum protein binding, metabolism, and excretion. Dosages for therapy and pharmacokinetic values are listed in Table 8.

Adverse Reactions. Since all cardiac glycosides and digitalis preparations are toxic, and since the ratio of therapeutic dose to toxic dose is small, the incidence of serious adverse reactions is very high (20% in hospitalized patients on digitalis) (175). Early stages of digitalis toxicity are disturbances of cardiac rhythm (ventricular ectopic beats, bradycardia) and anorexia. Increasing intoxication is evidenced by ventricular beats, headache, weakness, nausea, and vomiting. Severe toxic reactions are junctional or ventricular tachycardia, paroxysmal atrial tachycardia with varying degrees of A-V block, A-V dissociation, diarrhea, confusion, personality changes, and ocular disturbances, including blurred vision and photophobia. Ventricular fibrillation and high-degree conduction blocks are symptoms of very severe toxicity.

Severe toxicity caused by digitalis is treated by withholding digitalis and diuretics, and treating cardiac failure if present. Potassium salts, either orally or iv, are administered, with the exception that they are not given iv in the presence of A-V block or renal failure. Digitalis-induced arrhythmias are treated with phenytoin, lidocaine, and propranolol (58).

Table 8. Dosages and Average Pharmacokinetic Values for Cardiac Glycosides and Digitalis Preparations

Drug	Average digitalizing dose (adult) Oral, mg	Average digitalizing dose (adult) Intravenous, mg	Daily oral maintenance dose, mg	Oral absorption, %	Onset of action, h	Duration of action, d
acetyldigitoxin	1.6–2.2[a]		0.1–0.2	75	4–6	14–21
deacetyllanatoside C		2[a]		40	1–2	3–6
digitoxin	1.2–1.4[a]	1.2–1.4[a]	0.1	90–100	3–8	14–21
digoxin	1–1.5	0.5–1	0.125–0.5	50–90	1–2	3–6
lanatoside C	6[a]	1.6	1.6	10–40		2–6
ouabain		0.25–0.5		0	½–1½	2–4
digitalis powdered leaf	1000–1500		50–200	20–40	6–8	18–21
gitalin	6[a]		0.25–1.25	50–100	4–6	8–14

[a] Divided doses.

ANTIARRHYTHMIC AGENTS

The cardiac cycle consists of contraction (systole) and relaxation (diastole), including a brief period of inactivity before the next cycle. In the normal heart the beat originates at the automatic, or pacemaker, cells of the sinoatrial (S-A) node. The transmembrane potential steadily drifts from the resting potential to the threshold potential during diastole (phase 4 depolarization), at which time the fiber rapidly depolarizes (phase 0 of the action potential). The potential rises rapidly to ca 20 mV. The generated impulse excites the adjacent atrial muscle that contracts as it is stimulated. The atrioventricular (A-V) node, in turn, is stimulated and the impulse is rapidly transmitted to all parts of the ventricles through nodal tissue, causing the entire ventricle to contract nearly all at once. Repolarization of the pacemaker cells then occurs with a drop in potential, at first rapidly, and then more slowly, through phases 1, 2, and 3 of repolarization until the resting potential (phase 4) of ca −80 mV is reached.

During the depolarization phases 1, 2, and 3 (the refractory period), the cell is nonresponsive or less responsive to propagation of an impulse. The events during depolarization consist sequentially of the absolutely refractory period when the membrane cannot be excited, the effective refractory period (ERP) when the membrane can be excited but cannot propagate an impulse, the relatively refractory period when an impulse can be propagated at lowered velocity, and finally the supranormal period. Effectively, the ERP is the minimal time interval between the propagation of impulses (176).

Origin of Arrhythmias. Cardiac arrhythmias can arise from disturbances of impulse formation, disturbances of conduction, or both (177).

Arrhythmia Induced from Ectopic Impulses. Automaticity, the ability of specialized cardiac fibers to spontaneously generate an impulse, is characteristic of those found in the S-A node, atrial fibers connecting the S-A node to the A-V node, the bundles of His, the bundle branches, and the Purkinje fibers. S-A nodal automaticity is much faster than in other conducting fibers, thus its firing rhythm is dominant. Automaticity is not normally exhibited by fibers of ordinary atrial and ventricular muscle, but the rate of spontaneous firing can increase in these fibers when pathological conditions cause formation of ectopic foci. Ectopic sites can initiate firing if the rate of phase 4 depolarization is increased, resting membrane potential is reduced, or threshold excitation is reduced. Premature beats or extrasystoles result when the rate of firing from ectopic sites in the atria, ventricles, or the A-V tissue, exceeds that of the S-A node, and then may initiate sustained tachyarrhythmias. These can be abolished by drugs that depress automaticity or the rate of diastolic depolarization.

Arrhythmias From Abnormal Conduction. Conduction velocity of an impulse depends primarily on (a) the maximum rate of depolarization during phase 0 of the action potential, which is determined by the transmembrane potential at the time of initiation of the upstroke, and (b) the amplitude of the action potential. The transmembrane potential depends on intracellular and extracellular gradients and cell membrane permeability for sodium, potassium, calcium, and chloride (178–180). Rapid entry of sodium across the cell membrane, which occurs during diastole (phase 4) when the threshold potential is reached, results in a sudden loss of negative potential which produces the upstroke (phase 0) of the action potential.

Most clinical arrhythmias result from disturbances of conduction. The decreasing capability of conducting fiber to propagate an impulse results from decreasing amplitude and a decrease in the maximum rate of membrane depolarization. When this decremental conduction in one direction is impaired, as from reduced membrane potential in an ischemic area (most commonly in the junction of Purkinje fibers with ventricular contractile cells), unidirectional block occurs, with the continuation of retrograde conduction. Normally, both terminal branches of Purkinje fibers conduct impulses to ventricular muscle, causing ventricular depolarization. When there is a combination of unidirectional block and failure of propagation of action potential, circus movement of polarization occurs, with reentry. The impulse in the injured fiber fails to propagate past the ischemic zone, but the same impulse is conducted in reverse through the normal branch and reexcites the ventricular muscle shortly after the initial excitation. The coupled ventricular depolarizations may be the cause of bigeminal rhythms.

Actions of Antiarrhythmic Drugs. Antiarrhythmic drugs, by affecting the electrophysiology of heart muscle, act mainly by (a) diminishing automaticity by reducing the rate of phase 4 diastolic depolarization and, hence, depressing repetitive firing of impulses, and (b) altering conduction velocity and refractory period, thereby inhibiting disorders of reentry and reciprocal excitation.

The effects of the most-used antiarrhythmic drugs on electrophysiologic properties of antiarrhythmic drugs are shown in Table 9 (properties of digitalis are discussed on p. 895). Antiarrhythmic drugs affect conduction velocity or membrane responsiveness in different ways, both qualitatively and quantitatively. Type 1 drugs decrease automaticity and delay conduction velocity (189), but otherwise differ: quinidine, procainamide hydrochloride, bretylium tosylate, and propranolol (58) or other β-adrenergic blocking agents. Type 2 drugs decrease automaticity, but have little effect on or in low concentrations can increase, conduction velocity: lidocaine and phenytoin.

Antiarrhythmic Drugs. *Quinidine.* Quinidine (128), one of the alkaloids (qv) of *Cinchona* sp. (family *Rubiaceae*), is the dextrorotatory stereoisomer of quinine, differing only in configuration at C-9. Quinidine occurs in cinchona barks to the extent of 0.25–3%, but is mainly obtained by partial racemization of quinine (129). Thus, quinine is refluxed with an alkali metal alkoxide and quinidine is isolated from the

Table 9. Electrophysiologic Properties of Antiarrhythmic Drugs

Drug	Automaticity of ecotopic pacemakers[a]	Effective refractory period (ERP)[a]	Conduction velocity (A-V node and Purkinje fibers)[a]	Reference no.
type 1				
quinidine	↓	↑	↓	181
procainamide	↓	↑	↓	182
propranolol	↓	↓	↓	183–186
bretylium tosylate	→↑	↑	→	187
type 2				
lidocaine	↓	↑	→↑	182, 188
phenytoin	↓	↑	→↑	189

[a] ↑ = Increased, ↓ = decreased, and → = no change.

(128) quinidine   **(129)** quinine   **(130)**

resulting mixture as the D-tartaric acid salt (190). Quinidine has been totally synthesized, utilizing the key intermediate, N-benzoylmeroquinene methyl ester (**130**) (191).

Therapeutic effects: quinidine prevents premature atrial and ventricular contractions, restores normal sinus rhythm in atrial flutter and paroxysmal atrial fibrillation (after normal rate has been established with digitalis), terminates ventricular tachycardia when not associated with complete heart block, maintains normal sinus rhythm after d-c cardioversion of atrial fibrillations or flutters, and prevents postcountershock arrhythmias.

Dosages: quinidine gluconate, 200–400 mg im (intramuscular), 25 mg/min iv, < 800 mg total; quinidine sulfate, 200–400 mg orally every 6 h.

Adverse effects: direct irritant action causes diarrhea, nausea, and vomiting; other manifestations of cinchonism are headache, vertigo, palpitations, etc.

Procainamide. The local anesthetic, procaine hydrochloride (**131**), was shown (192) to elevate the threshold to electrical stimulation when applied to the myocardium of animals. Later it was applied topically to the exposed myocardium during surgery to reduce premature ventricular and atrial contractions. However, central nervous system toxicity and its short duration of action led to development of procainamide hydrochloride (**132**) (this is more stable to esterases) and its introduction as an antiarrhythmic drug (193).

Therapeutic effects: same as for quinidine, but used when adverse reactions of quinidine preclude patient use; preferred over quinidine for intravenous therapy of ventricular premature extrasystoles or ventricular tachycardia because of more rapid action (lidocaine is drug of choice for emergency management).

Dosages: procainamide hydrochloride, 250–500 mg im, 100 mg iv every 5 min, <1000 mg total; 250–500 mg orally every 6 h.

Adverse reactions: similar to those of quinidine; anorexia, nausea, and vomiting, but less than with quinidine; reversible lupus erythematosis syndrome.

Propranolol and Other β-Adrenergic Blocking Agents. The effects of β-adrenergic stimulation on cardiac electrophysiology are reflected in increased automaticity, increased effective refractory period, and increased conduction velocity. These result in sinus tachycardia, ectopic focal stimulation, and reentrant ectopic activity. β-Adrenergic blocking drugs reverse these effects (effects of propranolol are shown in Table 9) by competitively blocking sympathetic stimulation and by direct action on cell membranes (membrane stabilizing effect) (194–195). In addition to propranolol

$H_2N-\langle\bigcirc\rangle-CO_2CH_2CH_2N(C_2H_5)_2\cdot HCl$ $H_2N-\langle\bigcirc\rangle-CONHCH_2CH_2N(C_2H_5)_2\cdot HCl$

(**131**) procaine hydrochloride   (**132**) procainamide hydrochloride

(58), the following β-blockers have been shown in the clinic to reverse arrhythmias resulting from sympathetic stimulation: acebutolol (50), atenolol (51), metoprolol (52), practolol (53), tolamolol (54), alprenolol (55), oxprenolol (56), pindolol (57), sotalol (59), and timolol (60) (see Table 7 for structures).

Therapeutic effects: terminates catecholamine-induced arrhythmias, slows sinus tachycardia, converts paroxysmal atrial or A-V nodal tachycardias to normal sinus rhythm, converts to sinus rhythm atrial fibrillation or flutter resulting from acute myocardial infarction, reverses digitalis-induced arrhythmias.

Dosages: propranolol hydrochloride, 0.5–3 mg iv; 10–40 mg orally 3–4 times daily.

Adverse reactions: bradycardia, congestive heart failure, cardiac arrest (in patients with A-V block), bronchospasm.

Bretylium. Bretylium tosylate (133), an adrenergic neuronal blocking agent that is no longer used for treatment of hypertension because of rapid tolerance buildup, is used for refractory ventricular arrhythmias. It is an o-bromobenzyl quaternary ammonium salt and is synthesized as follows (196):

$$\text{N,N-dimethyl-o-bromobenzylamine} + \text{ethyl } p\text{-tosylate} \longrightarrow \text{(133) bretylium tosylate}$$

Therapeutic effects: terminates ventricular tachycardia and fibrillation; terminates refractory ventricular fibrillation.

Dosages: bretylium tosylate, 3–5 mg/kg im or iv, 2–5 mg/kg every 8 or 12 h for maintainance; 300–600 mg orally every 8 to 12 h.

Adverse Reactions: hypotension, nausea, vomiting, diarrhea.

Lidocaine. Lidocaine, a widely used local anesthetic, is a reverse amide of procainamide. It is synthesized from 2,6-xylidine (134) by chloroacetylation to give the chloroacetanilide (135) which reacts with diethylamine to give lidocaine (136) (197).

Therapeutic effects: drug of choice for immediate control of ventricular premature extrasystoles and ventricular tachycardia; particularly useful for ventricular arrhythmias during cardiac surgery or after acute myocardial infarction; reverses digitalis-induced arrhythmias; only useful for short-term treatment.

(134) → ClCH₂COCl → (135) → (C₂H₅)₂NH → (136) lidocaine

Dosages: lidocaine hydrochloride, 1–2 mg/kg of body weight iv, not exceeding 50–100 mg in a single dose; for immediate effect, iv, bolus.

Adverse reactions: action on CNS is drowsiness, paresthesias, muscle twitch, convulsions, respiratory depression, coma; depression of myocardial contractions from large doses.

Phenytoin (*Diphenylhydantoin*). Phenytoin is an anticonvulsant drug used as an antiarrhythmic agent. The synthesis from benzophenone (**137**) involves addition of cyanide ion and ammonia (198) to give an intermediate amidine (**138**), and finally phenytoin (**139**).

Therapeutic effects: reverses digitalis-induced arrhythmias (particularly ventricular arrhythmias), prevents postcountershock arrhythmias in digitalized patients.

Dosages: phenyltoin, 100 mg orally every 6 h; phenytoin sodium, 100 mg iv every 5 min, <1000 mg total; 1 g orally 1st day, 300–600 mg 2nd day, 100 mg 3 times daily for maintenance.

Adverse reactions: fatigue, dizziness, nausea, vomiting.

Digitalis. *Therapeutic effects:* drug of choice for atrial flutter or fibrillation.
Dosages: see Table 8.

Atropine Sulfate Hydrate (140). This blocker of the parasympathetic neurotransmitter, acetylcholine, increases heart rate, thus it is used to treat reversible bradyarrhythmias resulting from acute myocardial infarction. The iv dose is 0.4–2 mg every 1–2 h. Side effects are dryness of the mouth, cycloplegia, and mydriasis. Atropine is extracted from *Atropa belladonna* L., *Datura stramonium* L., and other *Solonaceae,* and results from racemization of the natural alkaloid *l*-hyoscyamine (199).

Edrophonium Chloride (141). This rapid acting cholinesterase inhibitor is used to terminate supraventricular tachycardia that does not respond to vagal maneuvers (200). The iv dose is 5–10 mg. Adverse effects are salivation, sweating, myosis, and gastrointestinal disturbances.

(**137**)   (**138**)   (**139**) phenytoin

(**140**) atropine sulfate hydrate

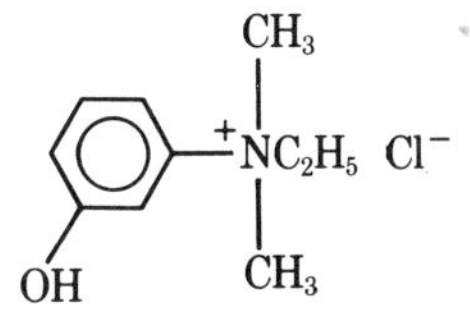

$$\underset{\underset{\text{OH}}{}}{\bigcirc}-\overset{+}{\text{N}}\underset{\underset{\text{CH}_3}{|}}{\overset{\overset{\text{CH}_3}{|}}{\text{C}_2\text{H}_5}}\quad \text{Cl}^-$$

(141) edrophonium chloride

***Phenylephrine Hydrochloride* (142).** Phenylephrine hydrochloride, a potent constrictor of arteries and arterioles, is used to control supraventricular tachycardia in patients who do not respond to carotid massage. The iv dose is 0.25–0.5 mg. The major adverse effect is excessive hypertension.

Phenylephrine is synthesized in racemic form from *m*-hydroxybenzaldehyde by a route that involves the Curtius rearrangement of a β-hydroxy hydrazide to give an isocyanate that spontaneously cyclizes to an oxazolidinone. Methylation and acid hydrolysis gives *dl*-phenylephrine (201).

Disopyramide. Disopyramide (143) was synthesized in an attempt to develop a quinidine-like antiarrhythmic agent. It was found to be less toxic than quinidine and more effective in reversing arrhythmias following myocardial infarction. Disopyramide phosphate increases the electrical threshold, prolongs the effective refractory period, and slows the conduction velocity in isolated rabbit atria. Paroxysmal ventricular tachycardia is converted to sinus rhythm with bolus iv administration of 2 mg/kg, and multiple premature ventricular contractions are suppressed (202). Reversion of ventricular tachycardia to sinus rhythm occurs on iv administration, and recurrent attacks are prevented by 800 mg/d orally (203). Atrial fibrillation and ventricular extrasystoles respond to treatment with disopyramide phosphate. The drug is well tolerated, with side effects limited to vagolytic action, and may be suitable for prolonged use.

Verapamil. Verapamil, a coronary vasodilator and clinically effective antianginal agent (204), is synthesized by reaction of the phenylacetonitrile (144) with sodium amide, followed by condensation of intermediate (145) with the γ-chloropropylamine (146) to give *dl*-verapamil (147) (205). Verapamil reduces the ventricular rate in patients with atrial fibrillation (206) and prolongs the PR interval in patients with sinus rhythm. Its action on the A-V node in slowing impulse transmission is thought to result from its specific calcium-antagonistic action in heart muscle (207). It reverts paroxysmal supraventricular tachycardia (208) and is very effective in reciprocating

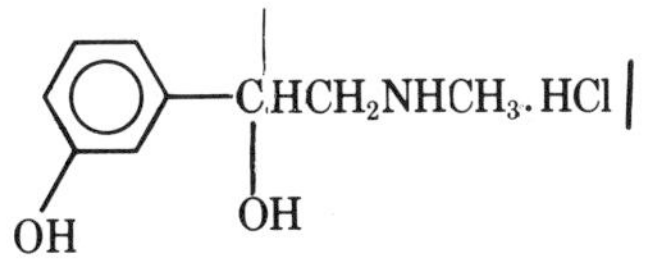

(142) phenylephrine hydrochloride

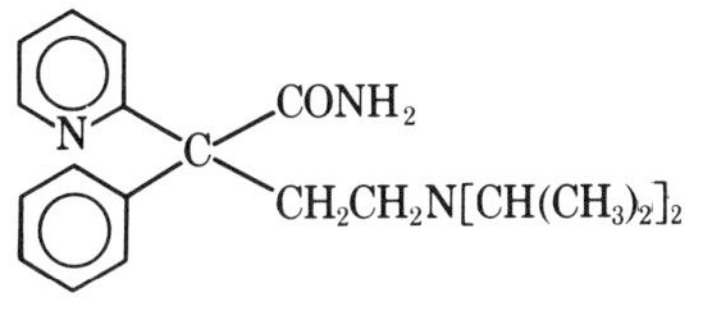

(143) disopyramide

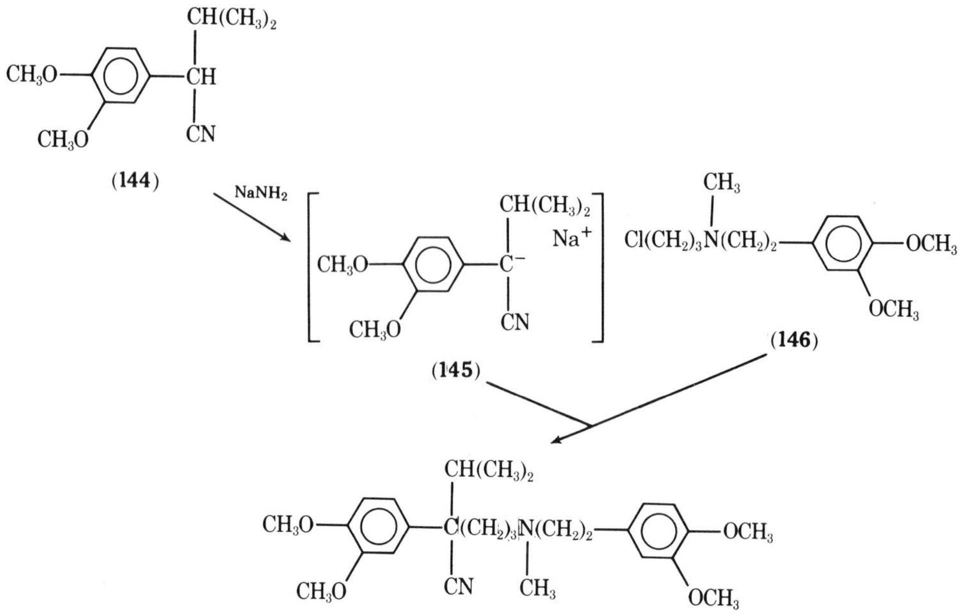

908   CARDIOVASCULAR AGENTS

(144)

(145)

(146)

(147) verapamil

tachycardias. The ventricular rate is reduced in atrial fibrillation and flutter, but immediate reversion to sinus rhythm does not occur. The drug is given iv at a rate of 1 mg/min to a maximum dose of 10 mg. Side effects most commonly seen are bradycardia, hypotension, heart block, and cardiac arrest (209).

Canrenoate Potassium. The aldosterone antagonist, canrenoate potassium (148), at a mean dose of 525 mg iv suppresses frequent ventricular premature beats and ventricular bigeminal and trigeminal rhythms presumably resulting from digitalis toxicity (210). However, studies on the effects of canrenoate potassium on Na–K ATPase and cardiac functions indicate no antagonism of digitalis induced arrhythmias (211).

Mexiletine Hydrochloride. Mexiletine hydrochloride (149), like procainamide, is effective in preventing recurrence of ventricular arrhythmias resulting from myocardial infarction, once these arrhythmias have been suppressed by iv lidocaine

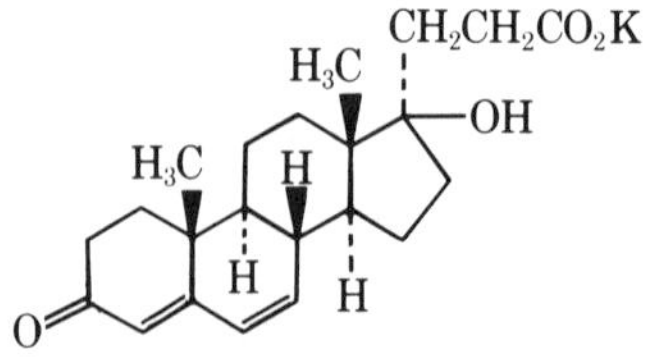

(148) canrenoate potassium

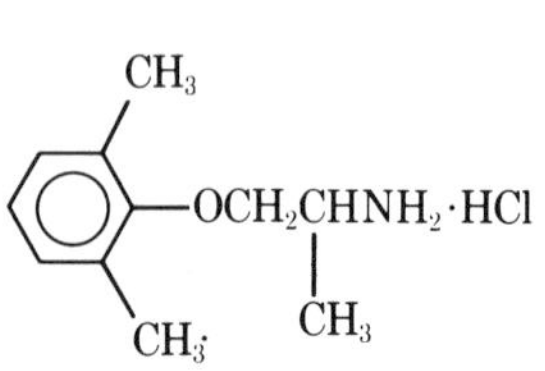

(149) mexiletine hydrochloride

(212–213). Unlike procainamide, it has a long half-life and few side effects (214). When given orally at 150–350 mg every 8 h, it is effective in controlling ventricular arrhythmias over a long time period and is well tolerated (215). Side effects that are dose-related and usually occur during the early treatment period, are gastrointestinal disturbances, including nausea, vomiting, unpleasant taste, and hiccups, CNS effects, including drowsiness, confusion, dizziness, and blurred vision, and cardiovascular effects, including hypotension, sinus bradycardia, and atrial fibrillation.

ANTIATHEROSCLEROTIC AGENTS

The most common cause of death in the United States is a result of the complications stemming from atherosclerosis. These are coronary heart disease (over a million heart attacks and 600,000 deaths annually in the 1970s), cerebral vascular disease, and peripheral vascular diseases, and they account for 50–60% of all deaths.

Formation of Atherosclerotic Lesions. A common occurrence of the aging process is the development from early childhood of atheromatous plaques that are incorporated in the walls of the coronary, aortal, cerebral, vertebral, and renal arteries, and the principal arteries to the legs. These lesions begin as gelatinous elevations or fatty streaks and progress to pearly-white fibrous plaques or atheromatous plaques containing a central core of lipid (mainly cholesterol) overlaid with a fibrous cap of connective tissue (mostly collagen and elastic fibers) (216–217). Several theories have been offered to explain the origin of these lesions: thrombogenic, inflammatory, lipid and insudation (217–218). They also include initiation by vascular injury of diverse origin (bacterial, viral, hemodynamic, and anoxic) and by platelet aggregation and release of factors by platelets that alter vascular permeability, etc (219). The injury induces platelet adhesiveness and aggregation and thrombus formation and, if the thrombus is not removed, it is invaded by fibromuscular connective tissue derived from collagen-synthesizing modified smooth muscle cells (220). Incorporation into the vessel wall occurs by being covered by a layer of endothelial cells (221). Smooth muscle cell proliferation, influenced by low density lipoproteins (LDL), a platelet factor, and hormones (222–223), enlarges the atherosclerotic plaque. The plaque becomes a site for further thrombus formation and deposition of platelets, followed by incorporation into the arterial walls. Mature plaques are calcified. Arteries are occluded in the mural thrombi that form and circulation to downstream tissues is reduced. A critical situation exists when this process involves coronary arteries (217). Myocardial ischemia results when the lumen of a coronary artery is 75% obstructed by the atherosclerotic plaque, and the transmural infarct (extending from the inner to the outer surface of the myocardial wall and involving at least $\frac{1}{2}$ of the wall thickness) that may be produced, is associated with the major complications that are seen. Irreversible cell changes resulting in cell death are the consequences of complete ischemia for more than 60 min (224).

Relationship Between Atherosclerosis and Lipid Levels. Studies in animals show that lesions similar to atherosclerosis in man are produced when plasma cholesterol levels are raised. In the rhesus monkey, for example, this is induced by dietary fat (butter) and cholesterol (225). Elevated serum cholesterol (226) and LDL (227–228) levels in man are documented as being associated with coronary heart disease, as are elevated triglyceride levels (228). Increased LDL serum concentration is characteristic of familial hypercholesterolemia and diabetes, where there is a marked increase in the incidence of coronary heart disease. The Framingham study showed persons with serum cholesterol levels greater than 300 mg % have four times as great an incidence of coronary heart disease as those with less than 200 mg % (229). Lowering of plasma cholesterol levels, as was experienced in certain European countries during World War II, probably as a result of diet change, may have been associated with a decrease in incidence of coronary heart disease (230). The above provides a rationale for lowering of serum lipid levels, either by diet or by drugs, for reducing the incidence of atherosclerosis.

Evidence is now accumulating that human atherosclerosis can be reversed to a large extent and, more importantly, can be prevented almost completely (231). These conclusions are based on autopsy observations, angiography, or measurements of coronary and peripheral circulations before and after therapy with diet, or partial ileal bypass (232–233). A recently reported study on a limited number of patients with types II and IV hyperlipoproteinemia, who were treated with clofibrate, clofibrate + neomycin, or tibric acid, showed regression of atherosclerotic lesions that were significantly correlated with lowering of serum cholesterol (234). Plaque size and luminal diameter were quantitated by successive arteriograms.

Plasma Lipoproteins. The major blood lipids, cholesterol, cholesterol esters, triglycerides, and phospholipids, are solubilized and transported through the plasma as lipoproteins (235–237). These are lipid–protein complexes varying in molecular weight from 186,000 for high density lipoprotein-2 (HDL-2) to 10^9–10^{10} for chylomicrons (see Table 10). Lipoproteins vary in the ratio of lipid to protein and, therefore, differ in density and can be separated by ultracentrifugation. They can also be separated by electrophoresis. These physical properties provide a means of classification: chylomicrons (least dense), very low density (VLDL), low density (LDL), and high density (HDL). They are also found to vary in composition: chylomicrons and VLDL are rich in triglycerides; most of the plasma cholesterol is contained in LDL. Since the ratio of lipoproteins is characteristic of the pathological primary hyperlipoproteinemias, it is useful to classify the latter based on lipoprotein patterns (235,238) (Table 11).

Disorders involving lipids are the result of deranged lipoprotein metabolism (235) where there is either an overproduction or faulty catabolism of one or more lipoproteins. The hyperlipoproteinemia that results may be caused (239) by (a) abnormally high levels of chylomicrons entering from the intestine carrying increased amounts of exogenous triglycerides or cholesterol, (b) elevated VLDL synthesized and released from the liver and small intestine, carrying increased amounts of endogenous triglycerides or cholesterol, (c) faulty catabolism because of defects in clearing enzyme tissue level catabolism or bile acid biosynthesis, and (d) abnormalities in the properties of the lipoproteins, including plasma solubility. Serum lipoprotein elevation may also arise from secondary causes, such as diabetes mellitus, myxedema, renal disease, liver disease, alcoholism, stress, and excessive dietary intake.

Antiatherosclerotic Drugs. Treatment of hyperlipoproteinemia is almost always initiated with a suitable diet since this often alleviates the condition without the use of potentially toxic drugs. Drugs, if needed, are used in combination with diet for an additive effect. Drug therapy is aimed at correcting specific lipoprotein elevation either by affecting production of the lipoprotein or effecting its removal.

Table 10. Properties and Composition of Plasma Lipoproteins

Lipoprotein family	Mol wt	Density	Paper electrophoretic mobility	Composition, wt %				
				Free cholesterol	Cholesterol esters	Triglycerides	Phospholipids	Protein
chylomicrons	10^9–10^{10}	<0.95	origin	3.1	6.0	81.3	7.1	2.5
very low density, VLDL	5×10^6	<0.95–1.006	pre-β	6.0	16.2	51.8	17.9	7.1
low density, LDL	2.3×10^6	1.006–1.063	β	7.5	39.4	9.3	23.1	20.7
high density, HDL	HDL-2: 186,000 HDL-3: 340,000	1.063–1.21	α	2.0	17.4	8.1	26.1	46.4

CARDIOVASCULAR AGENTS

Table 11. Types of Primary Hyperlipoproteinemias

Type	Elevated lipoprotein family
I	chylomicrons
IIa	LDL, β-lipoproteins
IIb	LDL and VLDL
III	IDL, intermediate density β-lipoproteins
IV	VLDL, pre-β-lipoproteins
V	VLDL and chylomicrons

Drugs That Decrease Lipoprotein Production. Clofibrate. Among a group of aryloxyisobutyric acid esters that reduced total plasma lipid and cholesterol levels in rats, clofibrate was chosen as the one with best therapeutic index (240). It is synthesized by condensation of *p*-chlorophenol with acetone and chloroform and the resultant acid is esterified to give clofibrate (**150**) (241).

Clofibrate is most useful in the treatment of type III and some cases of types IV and V hyperlipoproteinemias, since its main action is to reduce VLDL both by inhibiting its synthesis and by increasing its clearance (242). Levels of VLDL are lowered within 2–5 d of starting therapy, and intermediate density β-lipoprotein (IDL) levels also fall; the effect on LDL is variable. During the Coronary Drug Project (243) the mean plasma cholesterol concentration was reduced only 6% in men given 1.8 g/d of clofibrate, and the plasma triglyceride level was reduced 22%, and this preferential lowering of triglyceride level over cholesterol level has also been observed in other studies. The usual total daily dose is 2 g.

Clofibrate is well tolerated, but occasional side effects are nausea, diarrhea, drowsiness, weakness, and giddiness. Elevations in creatinine phosphate and glutamic oxalacetate are infrequently seen, with attendant severe muscle cramps, stiffness, weakness, and muscle tenderness. In several studies (243–244) lowering was seen in serum alkaline phosphatase, RBC (red blood cell count), WBC (white blood cell count), and hematocrit, and elevations in SGOT (serum glutamic oxaloacetic transaminase), SGPT (serum glutamic pyruvic transaminase), BUN (blood urea nitrogen), and prothombin time. A study of 8,341 men with previous heart attack who received clofibrate for six years shows a 54% excess incidence of gallbladder disease over placebo controls (245).

Nicotinic Acid. The vitamin, nicotinic acid, is an effective lipid-lowering agent at high doses for hyperlipoproteinemias characterized by elevated VLDL or its products, IDL and LDL.

Nicotinic acid (**152**) is synthesized either by oxidation of β-picoline (**151**) (246) or of nicotine (**153**) (247) (see Vitamins).

$$Cl-\underset{}{\bigcirc}-OH + (CH_3)_2CO + CHCl_3 \longrightarrow Cl-\underset{}{\bigcirc}-O\underset{CH_3}{\overset{CH_3}{\underset{|}{C}}}CO_2H \longrightarrow$$

$$Cl-\underset{}{\bigcirc}-O\underset{CH_3}{\overset{CH_3}{\underset{|}{C}}}CO_2C_2H_5$$

(**150**) clofibrate

(151) → [O] → (152) nicotinic acid ← Conc. HNO$_3$ ← (153)

Its mechanism of action involves inhibition of the synthesis of VLDL and subsequently IDL and LDL. Lowering of plasma triglyceride levels are seen within 4–6 h of administration of nicotinic acid. In large doses given chronically, both triglyceride and cholesterol levels are lowered, the latter after several days because LDL has a longer half-life that VLDL (248). Cholesterol lowering was 15–30% (249) on hypercholesterolemics but lowering of triglycerides is variable, although more than 60% was seen (250). Oral dosing is initiated with 100 mg three times daily, and then increasing to 3–9 g/d. The most common side effect is intense, cutaneous flushing and itching; however, this no longer occurs after several weeks of dosing.

Patients with previous history of myocardial infarction were given nicotinic acid during the Coronary Drug Project and there was significantly decreased incidence of recurrent myocardial infarction, but there was no effect on overall mortality. The incidence of arrhythmias and gastrointestinal disturbances were increased.

Drugs That Increase Lipoprotein Catabolism. *Cholestyramine Resin.* The bile sequestering quaternary ammonium salt anion-exchange resin, cholestyramine (154) is a high mol wt (av >10^6) copolymer of styrene, which bears a quaternary ammonium moiety, with 2% of divinylbenzene. It is used to treat primary type II hyperlipoproteinemia, particularly IIa where LDL alone is elevated. The resin acts by binding bile acids in the small intestine, thereby decreasing their reabsorption into the enterohepatic circulation. The resin is unchanged in passing through the gastrointestinal tract. Although there is an increase in cholesterol catabolism to bile acids in the liver in response to the interruption in bile acid reabsorption, this does not explain the lowering of plasma cholesterol levels since there is often a concomitant increase in *de novo* synthesis of cholesterol to compensate for the increased cholesterol catabolism (251). Although both plasma cholesterol and LDL cholesterol are reduced (20.6% and 27.3%, respectively) (252) by treatment with cholestyramine, it mainly decreases LDL (253). Cholestyramine may elevate cholesterol levels in hypercholesterolemia resulting from increased VLDL or IDL, (250,254) and so it is contraindicated in types III, IV, and V. The usual adult dose is 12–24 g daily in divided doses.

The most common side effect is constipation; other gastrointestinal effects are nausea, vomiting, cramps, and abdominal distention.

D-Thyroxine Sodium. The sodium salt of the dextrorotatory enantiomer of thyroxine, D-thyroxine sodium (155), is used for types II and III hypolipoproteinemia since it lowers LDL (255), primarily by accelerating LDL catabolism. Serum cholesterol levels are reduced 15–25% with doses of 4–8 mg/d.

Side effects are those of hypermetabolism and, most seriously, cardiotoxicity.

(154) cholestyramine resin

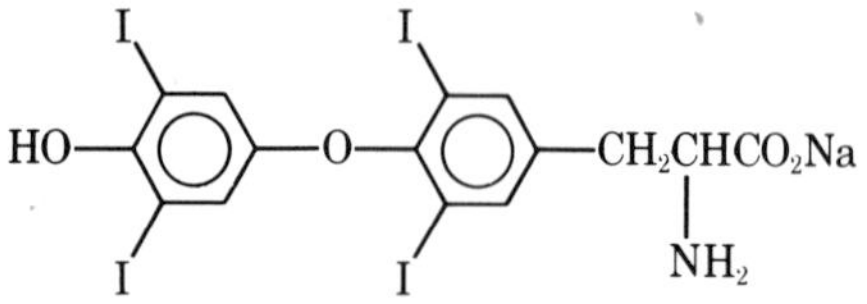

(155) D-thyroxine sodium

Colestipol. The high molecular weight anion-exchange resin, colestipol, a co-polymer of *N*-(2-aminoethyl)-1,2-ethanediamine with (chloromethyl)oxirane, is a bile-sequestering agent like cholestyramine in lowering plasma levels of cholesterol. Reductions of 7–18% are seen with a dose of 5 g given three times daily (256). There is little if any effect on triglyceride levels (257).

Side effects, constipation, and lack of palatability are mild.

Neomycin. The antibiotic neomycin, isolated from *Streptomyces fradiae* (258), is a complex from which neomycins A, B (156), and C have been isolated. Given orally at 0.5–2 g daily, 15–25% reductions in cholesterol levels are seen with no effect on triglycerides (259–260). It appears to have a mechanism similar to that of cholestyramine, of increasing the excretion of bile acids and clearance of LDL. It is thought to form insoluble complexes with bile acids in the intestine.

Side effects are diarrhea, and serious oto- and nephrotoxicity in a few cases.

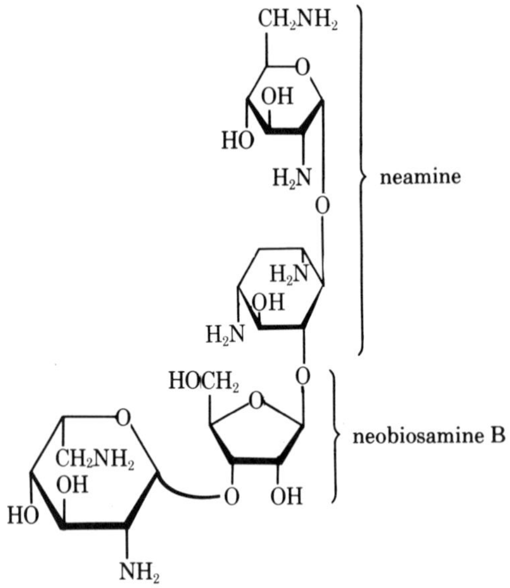

(156) neomycin B

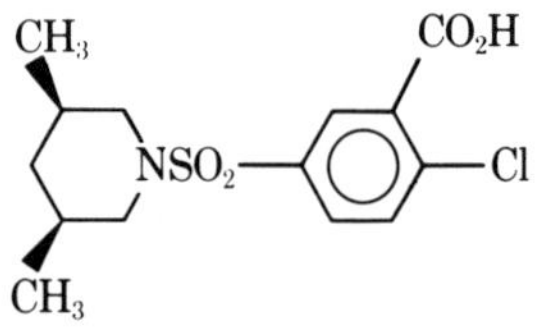

(157) tibric acid

Miscellaneous Drugs. *Tibric Acid.* Tibric acid (**157**) is a sulfonamidobenzoic acid that lowers serum triglycerides and cholesterol in type IV hyperlipoproteinemia. Its main effect is in lowering triglycerides (261). Type IV patients, who failed to respond to diet therapy, were aided by 500 mg bid of tibric acid; both serum triglycerides and cholesterol were lowered, with a greater effect on the former. There was no rebound on terminating therapy. The drug is well tolerated.

PAS-C. p-Aminosalicylic acid (**158**) is effective in type II hyperlipoproteinemia, causing lowering of plasma triglycerides and cholesterol when given at 6–8 g daily (262). PAS-C, a highly purified preparation with ascorbic acid (**159**), is effective at 8–9 g/d in type IIa and IIb (263). Gastrointestinal side effects are reported.

(**158**) *p*-aminosalicylic acid

(**159**) ascorbic acid

ANTIANGINAL AGENTS

The clinical syndrome of angina pectoris, typically a pressing, tight, or burning sensation in the precordial region, sometimes with pain radiating to the left arm or to the neck or back, was first described in 1768 by W. Heberden (264). The pain is characteristically brought on by consumption of heavy meals, exercise, anxiety, or other stress, and is quickly relieved by rest. The pain results when the myocardial oxygen supply is insufficient to meet the metabolic demand and produces myocardial hypoxia. This condition can be alleviated by either increasing the supply of oxygen or reducing the demand.

Oxygen supply to the heart depends mainly on coronary blood flow and oxygen content. Blood passing into the myocardium is essentially fully saturated with oxygen, and oxygen is nearly maximally extracted from it in passing through the myocardium. Increase in oxygen supply is possible in the normal heart by means of increased coronary blood flow which is controlled by mean aortic blood pressure, heart rate, and stroke volume. Flow is increased as a result of dilation of coronary arteries by an autoregulatory process which responds to lowered oxygen content or increased oxygen demand. Conditions that constrict the artery or limit its dilation, as from a fixed proximal stenosis, prevent an increase in coronary blood flow required to meet the increased oxygen demand produced by exercise, etc, and results in anginal pain. In any event, during hypoxia, the coronary arteries are fully dilated.

Reduction of myocardial oxygen demand is a feasible means of alleviating myocardial hypoxia with attendant anginal pain. Therapy consists of correction of obesity, cessation of smoking, avoidance of strenuous exercise and tension, and institution of a carefully controlled regimen of conditioning exercise. Medical treatment consists of reduction of hypertension, heart rate, myocardial contractility, and heart size. Reduction of serum lipoproteins is carried out by diet control and by drug therapy in order to attempt to prevent the further development of coronary atherosclerosis or to aid in its regression (231).

Antianginal Drugs. Drugs that have been clearly shown to be of value in alleviating angina pectoris are the sublingually administered nitrates and the orally administered β-adrenergic blocking agents. The latter are used for protracted prophylaxis of angina pectoris. A number of nitrates are claimed to be effective when given orally and to be long-acting, but these claims are not well established. A number of additional drugs have shown activity in limited clinical trials (see Alcohols, polyhydric).

Nitrates. *Sublingual Nitrates.* Of the sublingual nitrates (**160–165**) (Table 12), nitroglycerin (**160**) is the choice drug (265), since it dramatically and rapidly relieves the pain of an attack of angina pectoris.

Nitroglycerin and the other nitrates (Table 12) are thought to relieve anginal pain by reducing myocardial oxygen requirement through reduction in ventricular volume and myocardial tension and by lowering arterial blood pressure. These effects result from reduction in venous tone and from lowering peripheral vascular resistance. Also, they may both increase and redistribute myocardial blood flow to ischemic areas.

Side effects seen during beginning treatment are flushing, dizziness, and headache. Occasionally hypotension and reflex tachycardia from vasodilatation occur; also, orthostatic hypotension. Tolerance may occur on prolonged use.

Combination therapy with β-adrenergic blocking agents may give additive effects since they have different mechanisms of action.

Nitroglycerin is alternatively administered by the oral route and by topical ointment. Oral administration requires much higher dosage than sublingual because of rapid metabolism in the liver which also occurs with the other orally-administered nitrates (Table 12). Topically administered (usually to the skin of the chest) nitroglycerin (266) is long acting, but tolerance may develop. Long exposure to nitroglycerin by munitions workers is known to produce nitrate dependence, with anginal attacks and, infrequently, myocardial infarction and sudden death on withdrawal.

Long-Acting Nitrates. Isosorbide dinitrate (**161**) and erythrityl tetranitrate (**162**) are administered sublingually and are effective within 2–5 min in relieving angina pectoris; however, although their duration of action is longer than nitroglycerin, their onset of action is less rapid.

Nitrates, when administered orally [erythrityl tetranitrate, mannitol hexanitrate (**163**), pentaerythritol tetranitrate (**164**), trolnitrate phosphate (**165**) (Table 12)], are claimed to have a long duration of action and, hence, provide long-term prophylaxis. Large oral doses are required to produce vasodilatation, and the possibility of cross tolerance, especially to nitroglycerin, is known to occur.

β-Adrenergic Blocking Agents. Because β-adrenergic blocking agents reduce heart rate and contractility resulting from sympathetic stimulation, particularly from exercise (267), myocardial oxygen consumption is reduced and the pain of angina pectoris is delayed or prevented (268–269). In exercise-tolerance tests, such as the treadmill exercise test (270), administration of propranolol or other β-blockers (Table 7) delays the onset of anginal pain and S-T segment (the interval of the electrocardiogram between the end of the QRS complex and the beginning of the T wave; usually isoelectric) depression of the electrocardiogram, symptoms of hypoxia. Long term therapy reduces the frequency of anginal attacks, nitroglycerin consumption, and may have the effect of reducing the incidence of myocardial infarction.

Combination therapy with sublingual nitrates results in additive effects of the two types of drugs because of their different modes of action. The β-blockers reduce the reflex tachycardia induced by the nitrates that are used by the patient as required

Table 12. Antianginal Nitrates

Structure no.	Drug	Structure, R = ONO$_2$	Dosage, mg	Route	Frequency
(160)	nitroglycerin	CH$_2$R–HCR–CH$_2$R	0.15–0.6 2.5 or 6.5	sublingual oral topical	every 5 min every 12 h
(161)	isosorbide dinitrate	(bicyclic isosorbide structure)	2.5–5 5–30	sublingual oral	4 times daily
(162)	erythrityl tetranitrate	CH$_2$R–HCR–HCR–CH$_2$R	5–15 10–30	sublingual oral	3 times daily
(163)	mannitol hexanitrate	CH$_2$R–RCH–RCH–HCR–HCR–CH$_2$R	30–60	oral	every 4–6 h
(164)	pentaerythritol tetranitrate	C(CH$_2$R)$_4$	40–160	oral	divided doses
(165)	trolnitrate phosphate	$\left[\overset{+}{\text{HN}}(\text{CH}_2\text{CH}_2\text{R})_3\right]_2 \cdot \text{HPO}_4^{2-}$	10	oral	every 6–12 h

918   CARDIOVASCULAR AGENTS

to prevent or relieve anginal symptoms. The nitrates also counter β-blocker-induced decrease in coronary blood flow, and increase in heart size and prolongation of ventricular systole.

The oral dose of propranolol for the treatment of angina is 10 mg 3–4 times daily on initiation of therapy and maintainance with 40–100 mg 4 times daily. Side effects are described under Antihypertensive Agents, p. 872, and Antiarrhythmic Agents, p. 902.

Miscellaneous Agents. *Amyl Nitrite.* Amyl nitrite (**166**) is usually classified with the rapid-acting nitrates and was the first of this group of compounds to be used for the treatment of angina pectoris. It is a liquid with a high vapor pressure, administered by inhalation, and acts rapidly to alleviate anginal pain. Because of its unpleasant penetrating odor, high price, and side effects (reflex tachycardia, orthostatic effects, headache), it is not used very often today.

Perhexiline Maleate. A (dicyclohexyl)ethylpiperidine that is synthesized by catalytic reduction of (dicyclohexyl)ethylpyridine (**167**) is perhexiline (**168**) (271). It is a dilator of femoral and coronary arteries and lowers heart rate, and increases myocardial blood flow and oxygen consumption in animals (272). In addition, it does not depress the myocardium or cause arrhythmias (273). In man, perhexiline maleate is moderately diuretic, and increases exercise tolerance in patients with angina pectoris who show electrocardiographic evidence (S-T segment depression) of myocardial ischemia (274). In a large study of patients with angina pectoris who were given 200–400 mg daily, there was a 70–75% mean reduction in angina attacks after one month of treatment and complete suppression of angina in 24% of patients with parallel decrease in consumption of nitroglycerin. These results were similar to β-adrenergic blocker therapy (275). The mild side effects are mainly dizziness and nausea. A recently reported adverse reaction is proximal myopathy (276).

Lidoflazine. Synthesis by acylation of piperazine derivative (**169**) with chloroacet-2,6-xylidide (**170**) gives lidoflazine (**171**), an acyl derivative of 2,6-xylidine

(**167**)   (**168**) perhexiline

(F—⌬—)$_2$CH(CH$_2$)$_3$N⌒NH + ClCH$_2$CONH—⌬

(**169**)   (**170**)

(F—⌬—)$_2$CH(CH$_2$)$_3$N⌒NCH$_2$CONH—⌬

(**171**) lidoflazine

similar to the antiarrhythmic and local anesthetic lidocaine (136) (277). In addition to some antiarrhythmic properties, lidoflazine is a long acting coronary vasodilator with a long delayed (several weeks) onset of action (278). Although its mechanism of action is not known, it is thought that it does not preferentially dilate coronary arteries, and result in coronary steal which could enhance myocardial hypoxia (279). In postinfarction patients given 60 mg three times daily, maximum work load is increased in bicycle ergometric tests with significant lowering of heart rate and systolic blood pressure (280). With maintainance doses of 360–450 mg/d, lidoflazine subjectively decreases the frequency and severity of anginal attacks with lowered intake of nitroglycerin. Objective improvement is shown by increase in exercise tolerance (278).

Side effects are headache, gastrointestinal disturbances, flushing, cardiac depression, and conduction disturbances.

Nifedipine. Nifedipine (174) is one of the few pharmaceutically safe nitroaromatic compounds. It is synthesized in one step by the Hantsch dihydropyridine synthesis in high yield by refluxing together a solution of o-nitrobenzaldehyde (172), methyl acetoacetate (173) and ammonia in methanol (281). On a weight basis it is one of the most potent coronary dilators in animals (282). Its negative inotropic effect, thought to be caused by interference with calcium transport, decreases cardiac work and oxygen demand (282). Clinical efficacy has been demonstrated at 30 mg/d in patients with angina pectoris by subjective and objective measurements: anginal pain, exercise tolerance, and amount of nitroglycerin used (283). Decrease in S-T segment depression is also seen (284).

Side effects are headache and flushing and, at high doses, cardiac depression.

Verapamil. Verapamil (147), an antiarrhythmic agent, is a short-acting coronary vasodilator (285) and hypotensive agent in animals, but unlike other nonnitrate vasodilators it ultimately produces a reduction in heart rate and contracility of long duration (286), thus making it of value for the treatment of cardiac arrhythmias and in coronary artery disease (287). The primary mechanism for its cardiodepressant (negative inotropic) effect and, the consequent lowering of oxygen consumption, is thought to be inhibition of the movement across the myocardial cell membrane of calcium and other ions needed for ATPase action (288). Racemic verapamil suppresses the slow component of the action potential by blocking the electrical activity of sinus node cells, apparently by inhibiting cyclic 3′,5′-AMP (289). (−)-(S)-Verapamil has been shown to inhibit the slow calcium-channel whereas (+)-(R)-verapamil interferes with the fast sodium-channel.

The (+) and (−)-enantiomers of verapamil were synthesized from (+) and (−)-acids (175) and (176), respectively, and (+)-verapamil was shown to have the R configuration and (−)-verapamil the S configuration (290).

Verapamil is effective in ischemic heart disease as shown by treadmill exercise evaluation (287). At 120 mg three times daily, verapamil is as effective as propranolol at 100 mg three times daily. Side effects are not reported (291).

$$\underset{(175)\ S\text{-}(+)}{\overset{\underset{|}{\overset{CH_3\ \ CH_3}{\diagdown\diagup}}}{\underset{HO_2C\ \ CH_3}{\overset{C}{\diagup\ \ \diagdown CH_2CO_2H}}}} \qquad \underset{(176)\ R\text{-}(-)}{\overset{\underset{|}{\overset{CH_3\ \ CH_3}{\diagdown\diagup}}}{\underset{HO_2C\ \ CH_2CO_2H}{\overset{C}{\diagup\ \ \diagdown CH_3}}}}$$

PERIPHERAL VASODILATORS

The peripheral circulation is impaired under certain conditions, and vasodilator drugs are prescribed in an attempt to restore circulation to ischemic areas. However, most drugs are nonselective in their action and produce generalized vasodilatation which can prove deleterious by shunting blood away from the affected area to normal areas in the extremities. The most common vascular disorders affect circulation to the skin, skeletal muscle, and cerebrum, but drug treatment has only been well established in cutaneous disorders.

Mechanism of Action. α-Adrenergic receptors are present in coronary, skin, skeletal, and cerebral arterioles and cause constriction in response to sympathetic nerve stimulation. β-Adrenergic receptors, present in coronary and skeletal arterioles, cause dilatation in response to sympathetic stimulation. Vasodilating drugs that act by α-adrenergic blockade are effective in cutaneous disorders and are used to treat *Raynaud's disease,* a vasospastic disorder with little or no organic involvement and characterized by marked reduction in blood flow to the skin. On the other hand, in secondary Raynaud's phenomenon, wherein vascular impairment results from arterial and collagen diseases, effective vasodilator therapy may be possible, provided narrowing of arterioles has not occurred to a significant extent.

In early atherosclerosis, where circulation through large arteries of skeletal muscle is impaired, the small resistance vessels are fully dilated by autoregulatory processes responding to exercise and ischemia. Intermittent claudication, exercise-pain resulting from inadequate blood flow to meet the increased metabolic requirement of skeletal muscle, is rarely relieved by vasodilator drugs. Although pronounced vasodilatation in skeletal muscle can be produced by drugs that stimulate β-adrenergic receptors, areas not already dilated by ischemia are mainly affected (292).

Vasodilator drugs have not been shown to be of value in treating advanced arteriosclerosis obliterans since no consistent increase in blood flow in the ischemic skin lesions has been demonstrated. In fact, by shunting blood away from the diseased area, vasodilators may be contraindicated (293).

Certain cerebrovascular disorders are reported to respond to vasodilator drug therapy. Impaired memory and intellectual function, depression, and psychosis, are symptoms of cerebral atherosclerosis or senile brain disease (see Memory-enhancing agents). Psychosis and behavioral disturbances frequently respond to the phenothiazine tranquilizers, and tricyclic antidepressants are useful for treating depression. Although most drugs that have been tried in an attempt to improve intellectual function and memory are ineffective, the dihydrogenated ergot alkaloids and some direct-acting vasodilators are still in use.

Vasodilator Drugs. *Sympathetic Inhibitors.* The agents that act on sympathetic nerve terminals are reserpine (**43**) and phenoxybenzamine (**77**) hydrochloride.

Direct-Acting Vasodilators. The direct-acting vasodilators are nicotinic acid,

Table 13. Alphabetical List of Cardiovascular Agents Referred to in the Text

Compound	Structure number	CAS Registry No.	Compound	Structure number	CAS Registry No.
acebutolol	(50)	[37517-30-9]	deacetyllanatoside C	(110)	[17598-65-1]
acetyldigitoxin		[25395-32-8]	desperpidine	(44)	[131-01-1]
alprenolol	(55)	[13655-52-2]	diazoxide	(93)	[364-98-7]
amyl nitrite	(166)	[110-46-3]	dichloroisoproterenol	(61)	[59-61-0]
angiotensin I	(17)	[9041-90-1]	digitoxin	(103)	[71-63-6]
angiotensin II	(18)	[11128-99-7]	digoxin	(111)	[20830-75-5]
angiotensin III	(19)	[12687-51-3]	dioxyline	(183)	[147-27-3]
atenolol	(51)	[29122-68-7]	disopyramide	(143)	[3737-09-5]
atropine sulfate hydrate	(140)	[55-48-1]	dopamine	(5)	[51-61-6]
bendroflumethiazide	(22)	[73-48-3]	edrophonium chloride	(141)	[116-38-1]
benzthiazide	(20)	[91-33-8]	epinephrine	(10)	[51-43-4]
bretylium tosylate	(133)	[61-75-6]	erithrityl tetranitrate	(162)	[7297-25-8]
bumetanide	(34)	[28395-03-1]	ethacrynic acid	(32)	[58-54-8]
canrenoate potassium	(148)	[2181-04-6]	ethaverine	(185)	[486-47-5]
chlorothiazide	(21)	[58-94-6]	formylgitoxin		[3261-53-8]
chlorthalidone	(29)	[73-36-1]	furosemide	(33)	[54-31-9]
cholestyramine resin	(154)	[11041-12-6]	gitoxin	(107)	[4562-36-1]
clofibrate	(150)	[637-07-0]	guanabenz	(84)	[23256-50-0]
clonidine	(69)	[4205-90-7]	guanethidine	(76)	[55-65-2]
colestipol		[26658-42-4]	guanethidine sulfate (2:1)		[60-02-6]
cyclandelate	(178)	[456-59-7]	hexamethonium chloride	(81)	[60-25-3]
cyclothiazide	(23)	[2259-96-3]	hydergine		[8067-24-2]
cymarin	(115)	[508-77-0]	hydralazine	(88)	[86-54-4]
deacetyllanatoside A	(102)	[19855-40-4]	hydrochlorothiazide	(24)	[58-93-5]
deacetyllanatoside B	(106)	[19855-39-1]			

Table 13. (continued)

Compound	Structure number	CAS Registry No.	Compound	Structure number	CAS Registry No.
hydroflumethiazide	(25)	[135-09-1]	neomycin B	(156)	[119-04-0]
indapamide	(35)	[26807-65-8]	neomycin C		[66-86-4]
isoproterenol	(62)	[7683-59-2]	nicotinic acid	(152)	[59-67-6]
isosorbide dinitrate	(161)	[87-33-2]	nicotinyl alcohol	(181)	[100-55-0]
isoxsuprine	(179)	[395-28-8]	nifedipine	(174)	[21829-25-4]
lanatoside A	(101)	[17575-20-1]	nitroglycerin	(160)	[55-63-0]
lanatoside B	(105)	[17575-21-2]	norepinephrine	(7)	[51-41-2]
lanatoside C	(109)	[17575-22-3]	nylidrin	(180)	[447-41-6]
levodopa	(3)	[59-92-7]	ouabain	(118)	[630-60-4]
lidocaine	(136)	[137-58-6]	oxprenolol	(56)	[6452-71-7]
lidocaine hydrochloride		[73-78-9]	papaverine	(182)	[58-74-2]
lidoflazine	(171)	[3416-26-0]	papaverine hydrochloride		[61-25-6]
mannitol hexanitrate	(163)	[15825-70-4]	PAS-C { p-aminosalicyclic acid	(158)	[65-49-6]
methyclothiazide	(26)	[135-07-9]	{ ascorbic acid	(159)	[50-81-7]
methyldopa	(13)	[555-30-6]	pentaerythritol tetranitrate	(169)	[78-11-5]
methyldopate hydrochloride	(41)	[2508-79-4]	pentolinium tartrate	(80)	[52-62-0]
methyl reserpate	(47)	[2901-66-8]	perhexiline	(168)	[6621-47-2]
metolazone	(30)	[17560-51-9]	phenoxybenzamide	(77)	[59-96-1]
metroprolol, (±)	(52)	[54163-88-1]	phentolamine	(78)	[50-60-2]
mexiletine hydrochloride	(149)	[5370-01-4]	phentolamine hydrochloride		[73-05-2]
minoxidil	(92)	[38304-91-5]	phentolamine mesylate		[65-28-1]
naftidrofuryl	(184)	[3200-06-4]	phenylephrine hydrochloride	(142)	[61-76-7]
naftidrofuryl maleate		[31329-57-4]	phenytoin	(139)	[57-41-0]
neomycin	(184)	[1404-04-2]	phenytoin sodium		[630-93-3]
neomycin A		[3947-65-7]	pindolol, (±)	(57)	[21870-06-4]

polythiazide	(27)	[346-18-9]	sodium nitroprusside	(94)	[14402-89-2]
practolol	(53)	[6673-35-4]	sotalol	(59)	[3930-20-9]
prazosin	(97)	[19216-56-9]	spironolactone	(36)	[52-01-7]
procainamide hydrochloride	(132)	[614-39-1]	SQ 14,225	(100)	[62571-86-2]
procaine hydrochloride	(131)	[51-05-8]	k-strophanthin-β	(114)	[560-53-2]
pronethalol	(63)	[54-80-8]	k-strophanthoside	(113)	[11005-63-3]
propranolol	(58)	[525-66-6]	suloctidil	(186)	[54063-56-8]
propranolol hydrochloride		[318-98-9]	syrosingopine	(46)	[84-36-6]
proscillaridin A	(121)	[466-06-8]	teprotide	(99)	[35115-60-7]
quinethazone	(31)	[73-49-4]	D-thyroxine sodium	(155)	[137-53-1]
quinidine	(128)	[56-54-2]	tibric acid	(157)	[37087-94-8]
quinidine gluconate		[7054-25-3]	timolol, (S)	(60)	[26839-75-8]
quinidine sulfate		[6591-63-5]	timolol maleate, (S)		[26921-17-5]
quinine	(129)	[130-95-0]	tolamolol	(54)	[38103-61-6]
renin		[9015-94-5]	tolazoline	(177)	[59-98-3]
rescinnamine	(45)	[24815-24-5]	triamterene	(37)	[396-01-0]
reserpine	(43)	[50-55-5]	trichlormethiazide	(28)	[133-67-5]
saralasin	(98)	[34273-10-4]	trimethaphan camsylate	(79)	[68-91-7]
scillaren A	(120)	[11003-70-6]	trolnitrate phosphate	(165)	[588-42-1]
serotonin	(11)	[50-67-9]	verapamil	(147)	[52-53-9]

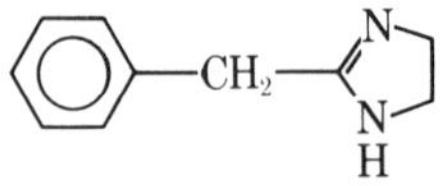

924 CARDIOVASCULAR AGENTS

(177) tolazoline

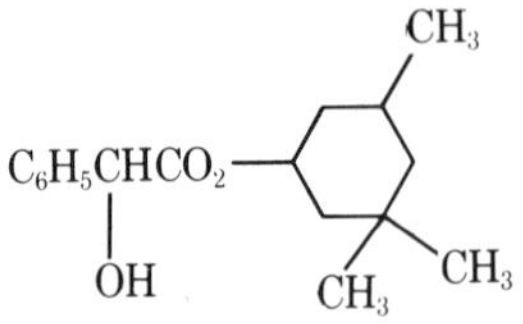

(178) cyclandelate

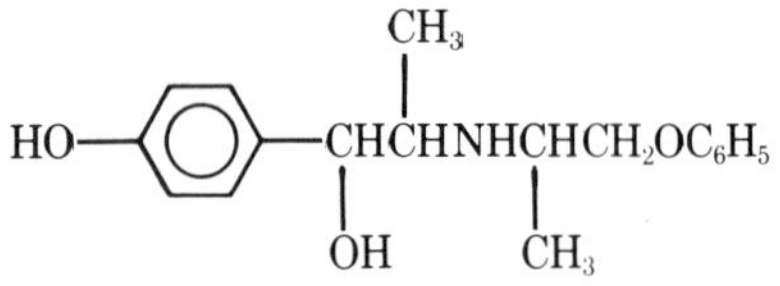

(179) isoxsuprine

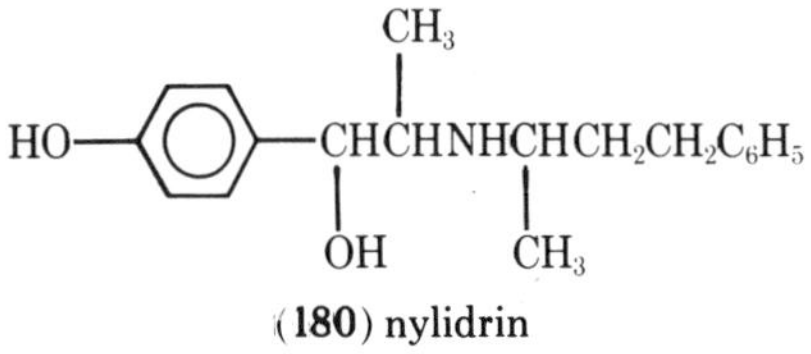

(180) nylidrin

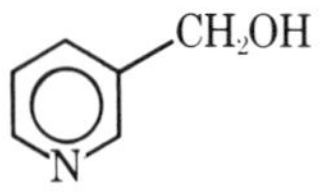

(181) nicotinyl alcohol

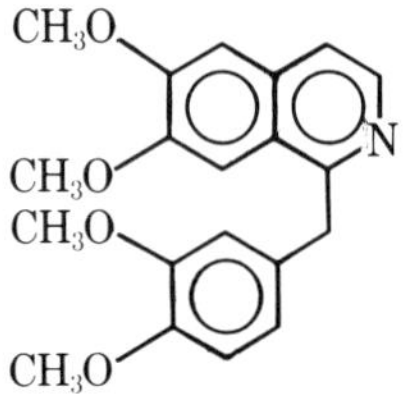

(182) papaverine

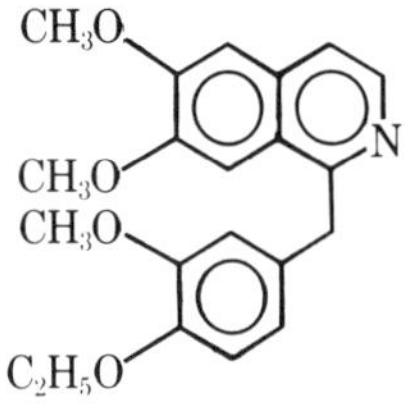

(183) dioxyline

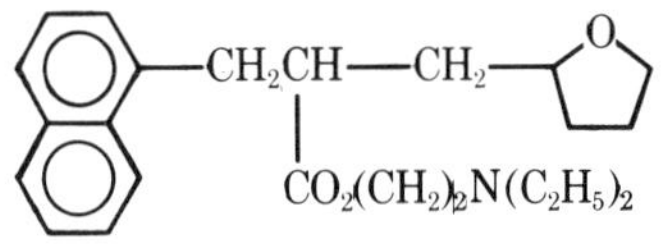

(184) naftidrofuryl

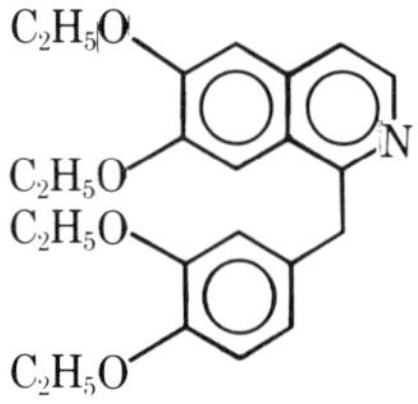

(185) ethaverine

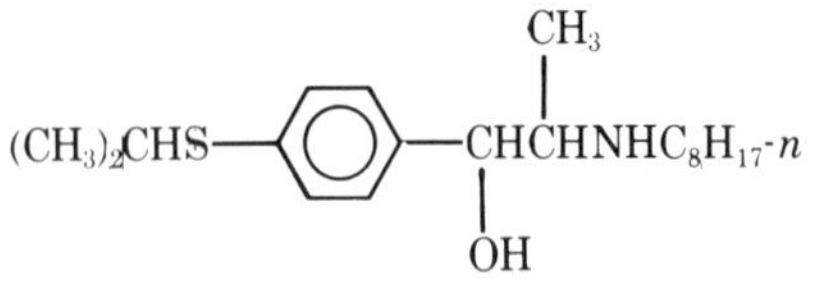

(186) suloctidil

compounds (177)–(186), as well as hydergine, a mixture of hydrogenated ergot alkaloids.

Table 13 gives an alphabetical listing of cardiovascular agents, with structure numbers, and Chemical Abstract Registry numbers mentioned in this article.

BIBLIOGRAPHY

"Cardiovascular Agents" in *ECT* 1st ed., Vol. 3, pp. 211–224, by Walter Modell, Cornell University Medical College; "Cardiovascular Agents" in *ECT* 2nd ed., Vol. 4, pp. 510–524, by Walter Modell, Cornell University Medical College.

1. R. M. Berne and M. N. Levy, *Cardiovascular Physiology*, 2nd ed., C. V. Mosby Co., St. Louis, Mo., 1922; J. P. Henry and J. P. Meehan, *The Circulation. An Integrative Physiologic Study*, Year Book Publishers, Inc., Chicago, Ill., 1971.
2. S. Hatano, I. Shigematsu, and T. Strasser, *Hypertension and Stroke Control in the Community*, World Health Organization, Geneva, 1976; *National Conference on High Blood Pressure Education*, DHEW Publication No. (NIH) 73-486, Washington, D.C., 1973.
3. W. B. Kannel and co-workers, *J. Am. Med. Assoc.* **214,** 301 (1970).
4. W. B. Kannel, M. J. Schwartz, and P. M. McNamara, *Dis. Chest* **56,** 43 (1969).
5. U.S. Veterans Administration Cooperative Study Group on Antihypertensive Agents, *J. Am. Med. Assoc.* **202,** 1028 (1967).
6. U.S. Veterans Administration Cooperative Study Group on Antihypertensive Agents, *J. Am. Med. Assoc.* **213,** 1143 (1970).
7. (a) D. L. Wilhelm, *Ann. Rev. Med.* **22,** 63 (1971); (b) H. B. Barner and co-workers, *Am. Heart J.* **85,** 584 (1973); (c) B. F. Robinson and co-workers, *Clin. Sci.* **44,** 367 (1973).

8. H. H. Dale, *J. Pharm. Exp. Ther.* **6,** 147 (1914).
9. R. P. Ahlquist, *Am. J. Physiol.* **153,** 586 (1948).
10. U.S. von Euler, *Noradrenaline: Chemistry, Physiology, Pharmacology and Clinical Aspects,* Charles C. Thomas, Springfield, Il., 1956.
11. T. Nagatsu, M. Levitt, and S. Udenfriend, *J. Biol. Chem.* **239,** 2910 (1964).
12. P. Holtz, R. Heise, and K. Lüdtke, *Arch. Exp. Pathol. Pharmakol.* **191,** 87 (1938).
13. L. T. Potter and J. Axelrod, *J. Pharmacol. Exp. Ther.* **142,** 299 (1963).
14. C. B. Ferry, *Ann. Rev. Pharmacol.* **7,** 185 (1967).
15. K. Starke in E. Usdin and S. Synder, eds., *Frontiers in Catecholamine Research,* Pergamon Press, New York, 1973, p. 561.
16. J. Axelrod, *Pharmacol. Rev.* **18,** 95 (1966).
17. F. O. Simpson, *Drugs* **7,** 85 (1974).
18. E. Adler-Graschinsky and S. Z. Langer, *Brit. J. Pharmacol.* **53,** 43 (1975).
19. G. Haeusler in E. Usdin and S. Synder, eds., *Frontiers in Catecholamine Research,* Pergamon Press, Oxford, 1973.
20. R. J. Neumayr, B. D. Hare, and D. N. Franz, *Life Sci.* **14,** 793 (1974).
21. A. Scriabine, B. V. Clineschmidt, and C. S. Sweet, *Ann. Rev. Pharm. Toxicol.* **16,** 113 (1976).
22. T. Hokfelt, K. Fuxe, and M. Goldstein, *Brain Res.* **62,** 461 (1973).
23. J. M. Saavedra and co-workers, *Nature* **248,** 695 (1974).
24. J. P. Chalmers, *Cir. Res.* **36,** 469 (1975).
25. J. P. Chalmers and L. M. H. Wing, *Clin. Exp. Pharmacol. Physiol. Suppl. 2* **2,** 195 (1975); G. Lambert, E. Friedman, and S. Gershon, *Life Sci.* **17,** 915 (1975); L. Finch, *Clin. Exp. Pharmacol. Physiol.* **2,** 503 (1975).
26. L. Finch and P. Hiicks, *Brit. J. Pharmacol* **55,** 274P (1975).
27. J. W. McCubbin, J. H. Green, and I. H. Page, *Circ. Res.* **4,** 205 (1956).
28. N. E. Andén and co-workers, *Acta Physiol. Scand.* **67,** 313 (1966).
29. H. Przuntek and A. Philippu, *Nauyn-Schmiedeberg's Arch. Pharmacol.* **276,** 119 (1973).
30. W. Kobinger and A. Walland, *Eur. J. Pharmacol.* **16,** 120 (1971).
31. H. Schmitt, H. Schmitt, and S. Fenard, *Eur. J. Pharmacol.* **14,** 98 (1971).
32. P. A. van Zwieten, *J. Pharm. Pharmacol.* **25,** 89 (1973).
33. P. A. van Zwieten, *Prog. Pharmacol.* **1,** 1 (1975).
34. A. Philippu, W. Roensberg, and H. Przuntek, *Nauyn-Schmiedeberg's Arch. Pharmacol.* **278,** 373 (1973).
35. A. Dahlström and K. Fuxe, *Acta Physiol. Scand. Suppl. 232* **62,** 1 (1964); *Acta Physiol. Scand. Suppl. 247* **64,** 1 (1965).
36. G. J. Kelliher and J. P. Buckley, *J. Pharm. Sci.* **59,** 1276 (1970).
37. D. S. Davis and J. L. Reid, eds., *Central Actions of Drugs in Blood Pressure Regulation,* Pitman Medical, 1975.
38. P. J. Lewis and G. Haeusler, *Nature* **256,** 440 (1975).
39. N. N. Share and K. I. Melville, *Int. J. Neuropharmacol.* **4,** 149 (1965).
40. R. K. Srivastava and co-workers, *Eur. J. Pharmacol.* **21,** 222 (1973).
41. M. D. Day and A. G. Roach, *Nature* **242,** 30 (1973).
42. A. Ito and S. M. Shanberg, *J. Pharm. Exp. Ther.* **189,** 392 (1974).
43. T. L. Sourkes, *Arch. Biochem. Biophys.* **51,** 444 (1954).
44. J. A. Oates and co-workers, *Science* **131,** 1890 (1960).
45. S. M. Hess and co-workers, *J. Pharmacol.* **134,** 129 (1961).
46. L. Gillespie, Jr., and co-workers, *Circulation* **25,** 281 (1962).
47. M. D. Day and M. J. Rand, *J. Pharm. Pharmacol.* **15,** 221 (1963).
48. H. Weisbach, W. Lovenberg, and S. Udenfriend, *Biochem. Biophys. Res. Commun.* **3,** 225 (1960).
49. I. J. Kopin, *Ann. Rev. Pharm.* **8,** 377 (1968).
50. R. Tigerstedt and P. G. Bergman, *Skand. Arch. Physiol.* **8,** 223 (1898).
51. H. Gavras, J. A. Oliver, and P. J. Cannon, *Ann. Rev. Med.* **27,** 485 (1976).
52. J. H. Laragh and J. E. Sealy, *Handbook of Physiology,* American Physiological Society, Washington, D.C., 1973, Sect. 8, p. 831.
53. L. T. Skeggs and co-workers, *Fed. Proc.* **36,** 1755 (1977).
54. I. H. Page and O. M. Helmer, *J. Exp. Med.* **71,** 485 (1940).
55. E. Braun-Menéndez, J. C. Fasciolo, and L. F. Leloir, *J. Physiol. London* **98,** 283 (1940).
56. F. Gross, *Acta Endocr. Copenhagen, Suppl.* **124,** 41 (1968).
57. T. L. Goodfriend and M. J. Peach, *Circ. Res. Suppl. 1* **36, 37,** 38 (1975).
58. R. H. Freeman and co-workers, *Fed. Proc.* **36,** 1766 (1977).

59. F. R. Bühler and co-workers, *Am. J. Cardiol.* **36,** 653 (1975).
60. R. Davies and co-workers, *Clin. Sci. Mol. Med.* **49,** 14P (1975).
61. H. P. Dustan, *Mod. Concepts Cardiovas. Dis.* **45,** 97 (1976).
62. *Executive Summary of the Task Force Reports to the Hypertension Information and Education Advisory Committee,* National High Blood Pressure Education Program, National Institutes of Health, Bethesda, Md., Sept. 1, 1973, pp. 1–20.
63. H. P. Dustan, R. C. Tarazi, and E. L. Bravo, *Arch. Intern. Med.* **133,** 1007 (1974).
64. J. Conway and H. Palermo, *Arch. Intern. Med.* **111,** 203 (1963).
65. L. Tobin, *Ann. Rev. Pharmacol.* **7,** 399 (1967).
66. R. C. Tarazi, H. P. Dustan, and E. D. Frohlich, *Circulation* **41,** 709 (1970).
67. M. Goldbert in J. Orloff and R. W. Berliner, eds., *Renal Physiology, Handbook of Physiology,* American Physiology Society, Washington, D.C., 1973, Sect. 8, p. 1003.
68. C. Kourouklis, O. Christensen, and D. Augoustakis, *Curr. Med. Res. Opinion* **4,** 422 (1976).
69. R. F. Maronde and M. Quinn, *Clin. Pharm. Ther. Abstr.* **21,** 110 (1977).
70. J. Kyncel and co-workers, *Arzneim. Forsch. Drug-Res.* **25,** 1491 (1975).
71. G. Onesti and co-workers, *Clin. Pharm. Ther. Abstr.* **21,** 113 (1977).
72. K. B. Hansen and A. D. Bender, *Clin. Pharm. Ther.* **8,** 392 (1976).
73. A. L. A. Boura and A. F. Green, *Ann. Rev. Pharmacol.* **5,** 183 (1965).
74. W. D. M. Paton, in J. H. Moyer, ed., *Hypertension, The First Hahnemann Symposium on Hypertensive Vascular Disease,* W. B. Saunders Co., Philadelphia, Pa., 1959, pp. 365–375.
75. G. A. Stein, H. A. Brouner, and K. Pfister, III, *J. Am. Chem. Soc.* **77,** 700 (1955).
76. A. Sjoerdsma and S. Udenfriend, *Biochem. Pharmacol.* **8,** 164 (1961).
77. D. F. Reinhold and co-workers, *J. Org. Chem.* **33,** 1209 (1968).
78. V. S. Aoki and W. R. Wilson, *Am. Heart J.* **79,** 798 (1970).
79. R. E. Woodson, Jr. and co-workers, *Rauwolfia,* Little, Brown, Boston, 1957.
80. H. J. Bein, *Pharmacol. Rev.* **8,** 435 (1956).
81. R. K. Kirtikar and B. D. Basu, *Indian Medicinal Plants,* Part II, Sudhindra Nath Basu, Allahabad, India, 1918, p. 777.
82. G. Sen and K. C. Bose, *Indian Med. World* **2,** 194 (1931).
83. S. Siddiqui and R. H. Siddiqui, *J. Indian Chem. Soc.* **8,** 667 (1931).
84. L. van Itallie and A. J. Steenhauer, *Arch. Pharm.* **270,** 313 (1932).
85. J. M. Mueller, E. Schlittler, and H. J. Bein, *Experientia* **8,** 338 (1952).
86. H. J. Bein, *Experientia* **9,** 107 (1953).
87. P. A. Diassi and co-workers, *J. Am. Chem. Soc.* **77,** 4687 (1955).
88. E. E. van Tamelen and P. D. Hance, *J. Am. Chem. Soc.* **77,** 4692 (1952).
89. R. A. Lucas and co-workers, *J. Am. Chem. Soc.* **77,** 1071 (1955).
90. R. B. Woodward and co-workers, *J. Am. Chem. Soc.* **78,** 2023 (1956); *Tetrahedron* **2,** 1 (1958).
91. L. Blaha and co-workers, *Coll. Czech. Chem. Commun.* **25,** 237 (1960).
92. H. Kewitz and co-workers, *Eur. J. Clin. Pharmacol.* **11,** 79 (1977).
93. B. Levy and B. E. Wilkenfeld, *Fed. Proc.* **29,** 1362 (1970).
94. F. J. Zacharias and co-workers, *Am. Heart J.* **83,** 755 (1972).
95. B. N. C. Prichard in P. R. Saxena and R. P. Forsyth, eds., *Beta-Adrenergic Blocking Agents,* American Elsevier Publishing Co., Inc., New York, 1976, p. 213.
96. S. Afr. Pat. 68,08,345 (June 12, 1967), K. R. H. Wooldridge and B. Basil (to May and Baker Ltd.).
97. U.S. Pat. 3,663,607 (May 16, 1972), A. M. Barrett and co-workers (to ICI Ltd.).
98. Ger. Pat. 2,106,209 (Aug. 26, 1971), A. E. Brandstrom and co-workers (to Aktiebolaget Hassle).
99. U.S. Pat. 3,408,387 (Oct. 29, 1968), R. Howe and L. H. Smith (to ICI Ltd.).
100. J. Angstein and co-workers, *J. Med. Chem.* **16,** 1245 (1973).
101. H. L. Yale and co-workers, *J. Am. Chem. Soc.* **72,** 3710 (1950).
102. F. Seemann and co-workers, *Helv. Chim. Acta* **54,** 2411 (1971).
103. R. Howe and R. G. Shanks, *Nature* **210,** 1336 (1966).
104. A. F. Crowther and L. H. Smith, *J. Med. Chem.* **11,** 1009 (1968).
105. R. E. Uloth and co-workers, *J. Med. Chem.* **9,** 88 (1966).
106. C. E. Powell and L. H. Slater, *J. Pharm. Exp. Therap.* **122,** 480 (1958).
107. Brit. Pat. 909,357 (Oct. 31, 1962), J. S. Stephenson (to ICI Ltd.).
108. R. Howe and B. S. Rao, *J. Med. Chem.* **11,** 1118 (1968).
109. Ger. Pat. 1,905,881 (Sept. 25, 1969), F. Troxler and A. Hofmann (to Sandoz Ltd.).
110. J. C. Danilewicz and J. E. G. Kemp, *J. Med. Chem.* **16,** 168 (1973).
111. B. K. Wasson and co-workers, *J. Med. Chem.* **15,** 651 (1972).

112. H. J. Waal-Manning, *Proc. Univ. Otago Med. Sch.* **48**, 80 (1970).
113. B. E. Karlberg and co-workers, *Brit. Med. J.* **1**, 251 (1976).
114. J. W. Hollifield and co-workers, *N. Engl. J. Med.* **295**, 68 (1976).
115. Editorial, *Brit. Med. J.* **1**, 529 (1977).
116. F. A. Finnerty, J. H. Buchholtz, and R. L. Guillandeu, *Proc. Soc. Exp. Biol. Med.* **94**, 376 (1957).
117. Belg. Pat. 623,305 (Apr. 5, 1963) (to C. H. Boehringer Sohn).
118. D. E. Hutcheon and co-workers, *Arch. Int. Pharmacodyn. Ther.* **147**, 146 (1964).
119. U.S. Pat. 3,931,216 (Jan. 6, 1976), R. Franzmair (to Boehringer Ingelheim GmbH).
120. B. Rouet, G. Leclerc, and C. G. Wermuth, *Chim. Ther.* **5**, 545 (1973).
121. W. Kobinger and L. Pichler, *Eur. J. Pharmacol.* **27**, 151 (1974).
122. J. R. Mitchell and J. Oates, *J. Pharm. Exp. Ther.* **172**, 100 (1970).
123. C. I. Twist, *Adv. Drug. Res.* **4**, 133 (1967).
124. R. A. Maxwell, R. P. Mull, and A. J. Plummer, *Experientia* **15**, 267 (1959).
125. M. Nickerson and N. K. Hollenberg in W. S. Root and F. G. Hofmann, eds., *Physiological Pharmacology, Vol. 4, The Nervous System—Part D: Autonomic Nervous System Drugs,* Academic Press, New York, 1967, p. 243.
126. L. O. Randall, W. C. Peterson, and G. Lehmann, *J. Pharm. Exp. Ther.* **97**, 48 (1949).
127. R. S. Shah, B. R. Walker, and R. H. Helfant, *Clin. Res.* **23**, 571A (1975).
128. T. Baum and A. T. Shropshire, *Eur. J. Pharmacol.* **37**, 31 (1976).
129. F. G. McMahon and co-workers, *Clin. Pharm. Ther.* **21**, 272 (1977).
130. J. A. F. deSilva, J. P. de O'Capecia, and I. Cohen, *Pharmatherapeutica* **1**, 1 (1976).
131. U.S. Pat. 3,816,531 (June 11, 1974), W. F. Bruce and T. Baum (to American Home Products).
132. J. Druey and J. Tripod in E. Schlittler, ed., *Antihypertensive Agents,* Academic Press, New York, 1967, p. 223.
133. G. G. Rowe and co-workers, *J. Clin. Invest.* **34**, 115 (1955).
134. E. D. Fries, *N. Engl. J. Med.* **266**, 607 (1962).
135. K. O'Malley and co-workers, *Clin. Pharm. Ther.* **18**, 581 (1975).
136. J. Druey and B. H. Ringler, *Helv. Chim. Acta* **34**, 195 (1951).
137. C. A. Chidsey, *J. Clin. Sci. Mol. Med.* **45**, 171s (1973).
138. D. W. Du Charme and co-workers, *J. Pharm. Exp. Ther.* **184**, 662 (1973).
139. R. K. Bryan and co-workers, *Am. J. Cardiol.* **39**, 796 (1977).
140. U.S. Pat. 3,461,461 (Aug. 12, 1969), W. C. Anthony and J. J. Ursprung (to Upjohn Co.).
141. F. Finnerty, *Am. Heart J.* **75**, 559 (1968).
142. A. A. Rubin and co-workers, *Science* **133**, 2067 (1961).
143. I. H. Page and co-workers, *Circulation* **11**, 188 (1955).
144. A. Schirger, G. Sheldon, and G. Sheps, *J. Am. Med. Assoc.* **237**, 989 (1977).
145. A. Scriabine and co-workers, *Experientia* **24**, 1150 (1968).
146. D. Cambridge, M. J. Davey, and R. Massingham, *Brit. J. Pharm.* **59**, 514P (1977).
147. H. F. Oates and co-workers, *Arch. Int. Pharmacodyn.* **224**, 239 (1976).
148. U.S. Pat. 3,663,706 (May 16, 1972), H.-J. E. Hess (to Pfizer).
149. M. A. Ondetti and co-workers, *Biochemistry* **10**, 4033 (1971).
150. D. H. P. Streeten and co-workers, *N. Engl. J. Med.* **292**, 657 (1975).
151. H. R. Brunner and co-workers, *Lancet.* **2**, 1045 (1973).
152. F. M. Bumpus and co-workers, *Circ. Res. Suppl.* **32**, 150 (1973).
153. H. Gavras and co-workers, *N. Engl. J. Med.* **291**, 817 (1974).
154. T. Philipp, H. Zschiedrich, and A. Distler, *Kidney Int. Abstr.* **11**, 218 (1977).
155. M. A. Ondetti, B. Rubin, and D. W. Cushman, *Science* **196**, 441 (1977).
156. R. K. Ferguson and co-workers, *Lancet.* **1**, 775 (1977).
157. W. Withering, *An Account of the Foxglove and Some of Its Medicinal Uses: With Practical Remarks on Dropsy and Other Diseases,* C. G. J. and J. Robinson, London, 1785. Reprinted in *Med. Class.* **2**, 305 (1937).
158. M. I. Ferrer, R. J. Conroy, and R. M. Harvey, *Circulation* **21**, 372 (1960).
159. G. K. Moe and J. A. Abildskov, *Circ. Res.* **14**, 447 (1964).
160. O. S. Narula, *Circulation* **47**, 872 (1973).
161. B. Lown, J. V. Temte, and W. J. Arter, *Circulation* **47**, 1364 (1973).
162. I. M. Glynn, *Pharmacol. Rev.* **16**, 381 (1964).
163. H. Matsui and A. Schwartz, *Biochim. Biophys. Acta* **151**, 655 (1968).
164. A. Askari, ed., *Ann. N. Y Acad. Sci.* **242**, 1 (1974).
165. G. A. Langer, *Ann. Rev. Med.* **28**, 13 (1977).

166. M. R. Rosen and co-workers, *Circulation* **47**, 681 (1973).
167. R. E. Ten Eick and B. F. Hoffman, *Circ. Res.* **25**, 365 (1969).
168. B. J. Sherlag and co-workers, *Am. Heart J.* **81**, 227 (1971).
169. Y. Watanabe and L. S. Dreifus, *Am. J. Physiol.* **211**, 1461 (1966).
170. S. Bellet, *Clinical Disorders of the Heart Beat*, Lea and Febiger, Philadelphia, 1971, p. 1066.
171. G. K. Moe and A. E. Farah in L. S. Goodman and A. Gilman, eds., *The Pharmacological Basis of Therapeutics*, 5th ed., Macmillan Publishing Co., Inc., New York, 1975, p. 653.
172. L. F. Fieser and M. Fieser, *Steroids*, Reinhold Publishing Corp., New York, 1959.
173. P. G. Marshall in S. Coffey, ed., *Rodd's Chemistry of Carbon Compounds*, 2nd ed., Elsevier Scientific Publishing Co., Amsterdam, The Netherlands, 1970, p. 360.
174. W. Klyne, *Proc. Biochem. Soc.* **47**, xli (1950).
175. *AMA Drug Evaluations*, 3rd ed., Publishing Sciences Group, Inc., Littleton, Mass., 1977.
176. G. K. Moe and J. A. Abildskov in ref. 171, p. 683.
177. B. F. Hoffman and J. T. Bigger in J. R. DiPalma, ed., *Drill's Pharmacology in Medicine*, 4th ed., McGraw-Hill Book Co., New York, 1971, p. 824.
178. B. F. Rusy, *Med. Clin. N. Am.* **58**, 987 (1974).
179. G. W. Beeler, Jr. and H. Reuter, *J. Physiol. Lond.* **207**, 191 (1970).
180. A. Fleckenstein, *Ann. Rev. Pharmacol. Toxicol.* **17**, 149 (1977).
181. H. L. Conn, Jr. and R. J. Luchi, *Am. J. Med.* **37**, 685 (1964).
182. J. T. Bigger, Jr. and R. H. Heissenbuttel, *Prog. Cardiovas. Dis.* **11**, 515 (1969).
183. D. H. Singer and R. C. Ten Eick, *Prog. Cardiovasc. Dis.* **11**, 488 (1969).
184. L. D. Davis and J. V. Temte, *Circ. Res.* **2**, 661 (1968).
185. D. Gibson and E. Sowton, *Prog. Cardiovasc. Dis.* **12**, 16 (1969).
186. A. G. Wallace and co-workers, *Circ. Res.* **18**, 140 (1966).
187. J. T. Bigger, Jr. and C. C. Jaffee, *Am. J. Cardiol.* **27**, 82 (1971).
188. J. T. Bigger, Jr. and W. J. Mandel, *J. Clin. Invest.* **49**, 63 (1970).
189. M. R. Rosen and B. F. Hoffman, *Circ. Res.* **32**, 1 (1973).
190. W. E. Doering, G. Cortes, and L. H. Knox, *J. Am. Chem. Soc.* **69**, 1700 (1947).
191. M. Uskoković, J. Gutzwiller, and T. Henderson, *J. Am. Chem. Soc.* **92**, 203 (1970).
192. F. R. Mautz, *J. Thoracic Surg.* **5**, 612 (1936).
193. L. C. Mark and co-workers, *J. Pharm. Exp. Ther.* **102**, 5 (1951).
194. M. R. Rosen and B. F. Hoffman, *Circ. Res.* **32**, 1 (1973).
195. J. P. P. Stock, *Am. J. Cardiol.* **18**, 444 (1966).
196. U.S. Pat. 3,038,004 (June 5, 1962), F. C. Copp and D. Stephenson (to Wellcome Found.).
197. U.S. Pat. 2,441,498 (May 11, 1948), N. F. Löfgren and B. J. Lundqvist (to Aktiebolaget Astra).
198. U.S. Pat. 2,409,754 (Oct. 22, 1946), H. R. Henze (to Parke, Davis and Co.).
199. W. Schneider, *Arch. Pharm.* **284**, 306 (1951).
200. U.S. Pat. 2,647,924 (Aug. 4, 1953), J. A. Aeschlimann and A. Stempel (to Hoffmann-La Roche).
201. E. D. Bergmann and M. Sulzbacher, *J. Org. Chem.* **16**, 84 (1951).
202. D. A. Deano and co-workers, *Clin. Pharm. Ther.* **19**, 106 (1976).
203. J. Hulting and G. Rosenhamer, *Acta Med. Scand.* **200**, 209 (1976).
204. G. Sandler, G. A. Clayton, and S. G. Thornicroft, *Brit. Med. J.* **3**, 224 (1968).
205. U.S. Pat. 3,261,859 (July 19, 1966), F. Dengel (to Knoll).
206. O. Storstein and K. H. Landmark, *Acta Med. Scand.* **198**, 483 (1975).
207. D. P. Zipes in L. S. Dreifus and W. Likoff, eds., *Cardiac Arrhythmias 25th Hahnemann Symposium*, Grune and Stratton, New York, 1973, p. 55.
208. J. Vohra, D. Hunt, and G. Sloman, *Med. J. Australia* **2**, 417 (1975).
209. J. K. Vohra, *Drugs* **13**, 219 (1977).
210. B. K. Yeh, B. N. Chiang, and P. Sung, *Am. Heart J.* **92**, 308 (1976).
211. S. I. Baskin and co-workers, *Proc. Soc. Exp. Biol. Med.* **143**, 495 (1973).
212. R. G. Talbot and co-workers, *Lancet ii*, 399 (1973).
213. N. P. S. Campbell and co-workers, *Lancet ii*, 404 (1973).
214. R. W. F. Campbell and co-workers, *Lancet i*, 1257 (1975).
215. R. G. Talbot, D. G. Julian, and L. F. Prescott, *Am. Heart J.* **91**, 58 (1976).
216. E. B. Smith, P. H. Evans, and M. D. Downham, *J. Atheroscl. Res.* **7**, 171 (1967).
217. W. Dobbs and H. J. Povalski in M. Antonaccio, ed., *Cardiovascular Pharmacology*, Raven Press, New York, 1977, p. 461.
218. M. D. Haust and R. H. More in R. W. Wissler and J. C. Geer, eds., *The Pathogenesis of Atherosclerosis*, Williams and Wilkins, Baltimore, Md., 1972, p. 1.

219. J. F. Mustard and M. A. Packham, *Thromb. Diath. Haemorrh.* **33**, 444 (1975).
220. M. D. Haust, R. H. More, and H. Z. Movat, *Am. J. Pathol.* **37**, 377 (1960).
221. A. B. Chandler in R. J. Jones, ed., *Thrombosis and the Development of Atherosclerotic Lesions,* Springer-Verlag, New York, 1970, p. 88.
222. R. Ross and J. A. Glomset, *Science* **180**, 1332 (1973).
223. R. Ross and co-workers, *Proc. Nat. Acad. Sci. USA* **71**, 1207 (1974).
224. N. Savranoglu, R. J. Boucek, and G. G. Casten, *Am. Heart J.* **58**, 726 (1959).
225. C. B. Taylor and co-workers, *Arch. Path.* **74**, 16 (1962).
226. M. M. Gertler, S. M. Garn, and J. Lerman, *Circulation* **2**, 205 (1950).
227. J. W. Gofman and co-workers, *Circulation* **2**, 161 (1950).
228. M. J. Albrink and E. B. Mann, *Arch. Intern. Med.* **103**, 4 (1959).
229. W. B. Kannel, W. P. Castelli, and P. M. McNamara, *J. Occup. Med.* **9**, 611 (1967).
230. H. Malmros, *Acta. Med. Scand. Suppl. 246* **138**, 137 (1950).
231. R. W. Wissler and D. Vesselinovitch, *Mod. Concepts Cardiovasc. Dis.* **46**, 27 (1977).
232. *Nutr. Rev.* **35**, 104 (1977).
233. H. Buchwald, R. B. Moore, and R. L. Varco in D. Kritchevsky, R. Paoletti, and W. L. Holmes, eds., *Lipids, Lipoproteins and Drugs,* Plenum Press, New York and London, 1975, p. 221.
234. R. Barndt, Jr. and co-workers, *Ann. Int. Med.* **86**, 139 (1977).
235. D. S. Fredrickson and R. I. Levy in J. B. Stanbury, J. B. Wyngaarden, and D. S. Fredrickson, eds., *The Metabolic Basis of Inherited Disease,* 3rd ed., McGraw-Hill Book Co., New York, 1972, p. 545.
236. D. S. Fredrickson, R. I. Levy, and R. S. Lees, *N. Engl. J. Med.* **276**, 34 (1967).
237. W. L. Holmes, R. Paoletti, and D. Kritchevsky, *Adv. Exp. Med. Biol.* **26**, 1 (1972).
238. WHO Memorandum, *Circulation* **45**, 501 (1972).
239. R. I. Levy and T. Langer, *Mod. Treat.* **6**, 1313 (1969).
240. J. M. Thorp and W. S. Waring, *Nature (London)* 948 (1962).
241. Brit. Pat. 860,303 (Feb. 1, 1961), W. G. M. Jones, J. M. Thorp, and W. S. Waring, (to ICI Ltd.).
242. B. M. Wolfe and co-workers, *J. Clin. Invest.* **52**, 2146 (1973).
243. Coronary Drug Project, *J. Am. Med. Assoc.* **231**, 360 (1975).
244. P. Bielmann and co-workers, *Int. J. Clin. Pharmacol.* **15**, 166 (1977).
245. Coronary Drug Project Research Group, *N. Engl. J. Med.* **296**, 1185 (1977).
246. A. Ladenburg, *Ann.* **301**, 117 (1898).
247. S. M. McElvain, *Org. Synth.* **4**, 49 (1925).
248. T. Langer and R. I. Levy, *Clin. Res. Abstr.* **18**, 458 (1970).
249. D. Kritchevsky in K. F. Gey, L. A. Carlson, and H. Bern, eds., *Metabolic Effects of Nicotinic Acid and Its Derivatives,* Hans Huber Verlag, Gmbh, Stuttgart and Vienna, 1971.
250. W. B. Parsons in H. R. Casdorph, ed., *Treatment of Hyperlipidemic States,* C. C. Thomas Publishers, Springfield, Ill., 1971, p. 333.
251. T. Langer, R. I. Levy, and D. Fredickson, *Circulation* **40**, III-14 (1969).
252. R. I. Levy and co-workers, *Ann. Int. Med.* **79**, 51 (1973).
253. L. Oro and co-workers, *Postgrad. Med. J. Suppl. 8* **51**, 76 (1975).
254. R. I. Levy, *Ann. Rev. Pharmacol. Toxicol.* **17**, 499 (1977).
255. E. H. Strisower, *Fed. Proc.* **21**, 96 (1962).
256. J. R. Ryan, A. K. Jain, and F. G. McMahon, *Clin. Pharm. Ther.* **21**, 116 (1977).
257. J. R. Ryan and A. Jain, *J. Clin. Pharmacol.* **12**, 268 (1972).
258. S. A. Waksman and H. A. Lechevalier, *Science* **109**, 305 (1949).
259. G. A. Leveille and co-workers, *Am. J. Clin. Nutr.* **12**, 421 (1963).
260. R. W. B. Schade and co-workers, *Acta Med. Scand.* **199**, 175 (1976).
261. P. Bielmann and co-workers, *Clin. Pharm. Ther.* **17**, 606 (1975).
262. P. J. Barter and co-workers, *Ann. Intern. Med.* **81**, 619 (1974).
263. P. T. Kuo and co-workers, *Circulation* **53**, 338 (1976).
264. W. Heberden, *Med. Trans. R. Coll. Phys.* **2**, 59 (1768).
265. O. Miller and K. Rørvik, *Brit. Heart J.* **20**, 302 (1958).
266. M. E. Davidov and W. J. Mroczek, *Angiology* **27**, 205 (1976).
267. S. E. Epstein, B. F. Robinson, and R. L. Kahler, *J. Clin. Invest.* **44**, 1745 (1965).
268. S. Wolfson and R. Gorlin, *Circulation* **40**, 501 (1969).
269. B. N. C. Prichard, *Drugs* **7**, 55 (1974).
270. P. L. McHenry and C. Fisch, *Mod. Concepts Cardiovasc. Dis.* **46**, 21 (1977).
271. U.S. Pat. 3,038,905 (June 12, 1962), F. A. Palopoli and W. L. Kuhn (to Richardson-Merrell, Inc.).

272. W. J. Hudak, R. E. Lewis, and W. L. Kuhn, *J. Pharm. Exp. Ther.* **173,** 371 (1970).
273. Z. Vera and co-workers, *Clin. Pharm. Ther.* **18,** 623 (1975).
274. L. Alcocor, J. Aspe, and E. Arce-Gomez, *Curr. Ther. Res.* **15,** 349 (1973).
275. J. D. F. Lockhart and H. C. Masheter, *Brit. Med. J.* **1,** 1530 (1976).
276. I. W. Tomlinson and F. D. Rosenthal, *Brit. Med. J.* **1,** 1319 (1977).
277. Neth. Pat. 6,507,312 (Dec. 10, 1965), (to Janssen).
278. V. Bernstein and D. I. Peretz, *Curr. Ther. Res.* **14,** 483 (1972).
279. R. J. Marshall and J. R. Parratt, *Clin. Exp. Pharmacol. Physiol.* **1,** 99 (1974).
280. L. J. Meilink-Hoedemaker, J. Pool, and M. M. Muste-Heijns, *Acta. Med. Scand.* **199,** 17 (1976).
281. S. Afr. Pat. 68,01,482 (Aug. 7, 1968), J. Bossert and W. Vater (to Farbenfabriken Bayer).
282. W. Vater and co-workers, *Arzneim. Forsch. Drug. Res.* **22,** 1 (1972).
283. E. Kimura, G. Mabuchi, and H. Hikuchi, *Arzneim. Forsch. Drug Res.* **22,** 365 (1972).
284. A. Loos and M. Kattenbach, *Arzneim. Forsch. Drug. Res.* **22,** 358 (1972).
285. W. G. Naylor and D. Krikler, *Postgrad. Med. J.* **50,** 441 (1974).
286. B. G. Benfey, K. Greeff, and E. Heeg, *Brit. J. Pharmacol.* **30,** 23 (1967).
287. V. Balasubramanian and co-workers, *Postgrad. Med. J.* **52,** 143 (1976).
288. A. M. Watanabe and H. R. Besch, *J. Pharm. Expt. Ther.* **191,** 241 (1974).
289. W. Tuganowski and co-workers, *Experientia* **33,** 642 (1977).
290. H. Ramuz, *Helv. Chim. Acta* **58,** 2050 (1975).
291. B. Livesley and co-workers, *Brit. Med. J.* **1,** 375 (1973).
292. M. J. Allwood, *Clin. Sci.* **22,** 279 (1962).
293. W. Tuganowski and co-workers, *Experientia* **33,** 642 (1977).

Leon Goldman
Lederle Laboratories
American Cyanamid Company

CARNALLITE, KCl.MgCl$_2$.6H$_2$O. See Potassium compounds.

CARNAUBA WAX. See Wax.

CARNOTITE, K$_2$O.2UO$_3$.V$_2$O$_5$.3H$_2$O. See Uranium and uranium compounds.

CAROTENOIDS. See Vitamins—Vitamin A.

CASEHARDENING. See Metal surface treatments.

CASEIN. See Milk products.

KIRK-OTHMER

ENCYCLOPEDIA OF CHEMICAL TECHNOLOGY

THIRD EDITION

VOLUME 4

BLOOD, COAGULANTS AND ANTICOAGULANTS
TO
CARDIOVASCULAR AGENTS

A WILEY-INTERSCIENCE PUBLICATION
John Wiley & Sons
NEW YORK • CHICHESTER • BRISBANE • TORONTO

Copyright © 1978 by John Wiley & Sons, Inc.

All rights reserved. Published simultaneously in Canada.

Reproduction or translation of any part of this work
beyond that permitted by Sections 107 or 108 of the
1976 United States Copyright Act without the permission
of the copyright owner is unlawful. Requests for
permission or further information should be addressed to
the Permissions Department, John Wiley & Sons, Inc.

Library of Congress Cataloging in Publication Data:

Main entry under title:
    Encyclopedia of chemical technology.

    At head of title: Kirk-Othmer.
    "A Wiley-Interscience publication."
    Includes bibliographies.
    1. Chemistry, Technical—Dictionaries. I. Kirk,
Raymond Eller, 1890–1957. II. Othmer, Donald Frederick,
1904–   III. Grayson, Martin. IV. Eckroth, David.
V. Title: Kirk-Othmer encyclopedia of chemical technology.

TP9.E685   1978   660'.03   77-15820
ISBN 0-471-02040-0

Printed in the United States of America